The Archaeology of Roman Southern Pannonia

The state of research and selected problems in the Croatian part of the Roman province of Pannonia

Edited by

Branka Migotti

BAR International Series 2393
2012

Published in 2016 by
BAR Publishing, Oxford

BAR International Series 2393

The Archaeology of Roman Southern Pannonia

ISBN 978 1 4073 0985 9

Translated by Valér Bedő, Tomislav Bilić, Danijel Dzino, Branka Migotti, Sanjin Mihalić, Miroslav Nađ, Mirko Sardelić and Vlasta Vyroubal

Proofread by Mirta Jambrović and Branka Migotti

BAR Publishing is the trading name of British Archaeological Reports (Oxford) Ltd. British Archaeological Reports was first incorporated in 1974 to publish the BAR Series, International and British. In 1992 Hadrian Books Ltd became part of the BAR group. This volume was originally published by Archaeopress in conjunction with British Archaeological Reports (Oxford) Ltd / Hadrian Books Ltd, the Series principal publisher, in 2012. This present volume is published by BAR Publishing, 2016.

Printed in England

BAR
PUBLISHING

BAR titles are available from:

BAR Publishing
122 Banbury Rd, Oxford, OX2 7BP, UK
EMAIL info@barpublishing.com
PHONE +44 (0)1865 310431
FAX +44 (0)1865 316916
www.barpublishing.com

In memory of Ivan Knezović (1977-2011)

Table of Contents

FOREWORD

Quis enim dubitat quin multis iam saeculis, ex quo uires illius ad Romanum nomen accesserint, Italia quidem sit gentium domina gloriae uetustate, sed Pannonia uirtute? (Mam. Paneg. Max. X (II), 2.2)

Who can doubt that, although Italy is the queen of all peoples on the basis of its old glory, on the basis of valour Pannonia is supreme?

This book emerges as the result of the work on the project *Romanisation and Christianisation of the Croatian portion of Roman southern Pannonia*, conducted by B. Migotti in the period 2007-2011, and financed by the Ministry of Science, Education and Sport of the Republic of Croatia. Some of the authors in this volume were researchers in the mentioned project (Alka Domić Kunić, Tino Leleković, Zsolt Mráv, Ivan Radman-Livaja, and Zoran Wiewegh), while the remainder were asked to participate on account of their research interests in the history or archaeology of southern Pannonia (Tomislav Bilić, Marko Dizdar, Danijel Dzino, Hana Ivezić, Ivan Knezović, Tanja Lolić, Miroslav Nađ, Anita Rapan Papeša, Ante Rendić-Miočević, and Mirko Sardelić). This book's title is reminiscent of that edited by A. Lengyel and G. T. B. Radan in 1980 (The Archaeology of Roman Pannonia), which established an overall frame and a structure of Roman civilisation in the province of Pannonia. A comparison, however, is out of place, as the *ARP* by far surpasses the present volume in both the thematic variety and scope, as well as in a systematic approach and its textbook profile. Therefore, the ARP could have only served as a stimulus to this volume, and not its model.

This book's aims correspond to those set by the abovementioned project, which, for practical reasons, was focused on the Croatian portion of southern Pannonia, that is, the territory of northern Croatia. No doubt, it is more acceptable and more to the point to study Roman provincial civilisation within the individual provinces' boundaries; some commentators have even pointed to the absurdity of doing otherwise. On the other hand, the approach as taken here has some advantages. First of all, local scholars have easier access to the material and the literature, especially if the latter is in the so-called minor languages. Also, there should be no obstacle to include such (territorially limited) studies into broader syntheses. The scholarly background and goals of the research that has generated this volume were above all rooted in the need to present the archaeology of southern Pannonia more realistically, that is, as better founded in the archaeological evidence. In other words, the idea was to prevent its „loosing" or even partial distortion of facts in discussions of Pannonia as a whole; this used to happen quite often, as the assumptions based on north-Pannonian evidence were automatically applied to its southern portion. The reasons for this are twofold. Firstly, it is a lower rate of research into the Croatian portion of Pannonia as compared to its northern (Hungarian) part. Such position of north Croatian Roman archaeology was partly caused by the fact that it was somewhat overshadowed by the splendid architectural evidence of the Roman period on the Croatian Adriatic coast (Istria and Dalmatia). Another reason for the inadequate information on north Croatian Roman archaeology is a dearth of works and discussions by Croatian authors in foreign languages. Consequently, the aim of this volume is to improve on this picture by addressing topics that were available for discussion, with no pretension to a systematic approach in either a choice of subjects or the way they are presented. Such coverage is just not realistic at this stage of research, given also that it was partly conditioned by contributors available. Consequently, some subjects (funerary stones, economy, commerce and traffic, religion, typology of basic categories of small finds, particularly pottery, etc.) are noticeably missing to the obvious disadvantage of the volume. Otherwise, it was not left at the authors' liberty to choose their titles; as a matter of fact, the contributors were chosen as experts in their particular area of interests. In doing so, the editor was much limited by the lack of available experts (or their engagement in other commitments) for individual segments of the Roman archaeology of northern Croatia. Furthermore, the majority of the discussions were affected by a still limited body of evidence and in some cases by the level of expertise of their authors, resulting in uneven quality of the chapters. Admittedly, the research in the archaeology of the Croatian southern Pannonia needs to be continued along the lines of specialised pursues of individual subjects.

In the last two decades, which correspond with the existence of the Republic of Croatia as an independent state, the archaeological research has made a considerable progress in both the quantity of work and the quality of methods applied. However, the majority of the excavations were rescue or, at, best, protective,[1] while those systematic were comparatively few. Furthermore, such ample work, particularly huge excavations in advance of the building of high roads and other infrastructure, could not have been accompanied by a prompt study of small finds. In other words, all we still have at the moment is mostly preliminary reports, while full publications are practically non-existent. Therefore, in spite of the quantity of the material evidence amassed recently, its usability in the discussions here was limited, and so was their contribution towards a full understanding of the history and archaeology of the area of study. As transpires from the title, this book's main concern is with the archaeological material in the narrow sense of the word, epigraphy included. This is partly based on the fact that the archaeological evidence for southern Pannonia strongly predominates over the written sources, with the latter forming a rounded corpus with not much hope for new finds, while the former is bound to rise in scope from day to day. Another reason for giving archaeology a complete precedence over history in this volume comes from recent popularity of historical narratives that tend to lean on the archaeological evidence, but still use it fragmentarily and more as an illustration than an equal substance. Given, however, that the archaeologists tend to do the same with historical sources, not least in this volume, it was decided to include as a separate theme a compilation of the written sources on southern Pannonia (Domić Kunić, Sardelić), free of archaeological insertions. On the other hand, a classic survey of the political history of the area in question has not been included as a discrete chapter, as it can be found in various other publications in foreign languages (the *ARP,* the *CAH*), while this volume remains a material- based reading. Overall, direct comments on the material were encouraged over theoretical discussions; the latter were pursued in only one paper (Dzino and Domić Kunić) and, to a certain extent, in the editor's Introduction. As already mentioned, the essays were generally limited to the material of the Croatian portion of southern Pannonia. This means that Poetovio and Sirmium, two most important southern-Pannonian towns apart from Siscia, were not discussed, but were still mentioned in various chapters. As a matter of fact, the contributors were at their liberty to go beyond the given territorial framework whenever they felt the need to do so in order to be able to present more clearly and accurately the themes pertaining directly to the area under study. The territorial definition *the Croatian portion of (southern Pannonia)* has been omitted from the chapters' titles, given that it is implicit in the subtitle of the volume.

Although the participation of all the contributors is appreciated, my thanks go especially to those authors who had no obligations as researchers in the abovementioned project. For the photos and drawings used in the paper of Rapan Papeša I owe sincere thanks to Ante Rendić-Miočević, ex-director of the AMZ, further to Mladen Radić (MSO) and Dragan Radojević (GMSo), and for the photo of a Scordiscan coin for the contribution by M. Dizdar I am grateful to Jacqueline Balen, directress of the AMZ. Other acknowledgements can be found in individual chapters. Finally, I would like to thank Rajka Makjanić for her generous editorial help and patience during the final stage of the preparation of this volume.

Branka Migotti

[1] On the meaning and ambiguous uses of these terms see various authors, in: Katalin Bozóki-Ernyey (ed.), *European Preventive Archaeology,* Papers of the EPAC Meeting (Vilnius 2004), Budapest 2007.

List of Contributors

Tomislav Bilić, PhD
Arheološki muzej u Zagrebu
Trg Nikole Šubića Zrinskog 19
HR-10000 Zagreb
tbilic@amz.hr

Marko Dizdar, PhD
Institut za arheologiju
Ljudevita Gaja 32
HR-10000 Zagreb
marko.dizdar@iarh.hr

Alka Domić Kunić, PhD
Hrvatska akademija znanosti i umjetnosti
Odsjek za arheologiju
Ante Kovačića 5
adomic@hazu.hr

Danijel Dzino, PhD
Australian Research Council Australian Postdoctoral Fellow
Department of Ancient History
Macquarie University, Sydney
danijel.dzino@mq.edu.au

Hana Ivezić
Josipa Gigla 5
HR-10000 Zagreb
hana.ivezic@gmail.com

Tino Leleković, PhD
Hrvatska akademija znanosti i umjetnosti
Odsjek za arheologiju
Ante Kovačića 5
HR-10000 Zagreb
lelekovic@gmail.com

Tatjana Lolić
Ministarstvo kulture Republike Hrvatske
Konzervatorski odjel u Zagrebu
Mesnička 49
HR-10000 Zagreb
tanja_lolic@yahoo.com

Branka Migotti, PhD
Hrvatska akademija znanosti i umjetnosti
Odsjek za arheologiju
Ante Kovačića 5
HR-10000 Zagreb
branka.migotti@zg.t-com.hr

Zsolt Mráv, MA
Magyar Nemzeti Múzeum
Múzeum krt. 14-16
H-1088 Budapest
zsoltmrav@gmail.com

Miroslav Nađ
Arheološki muzej u Zagrebu
Trg Nikole Šubića Zrinskog 19
HR-10000 Zagreb
miroslavnad@gmail.com

Ivan Radman-Livaja, PhD
Arheološki muzej u Zagrebu
Trg Nikole Šubića Zrinskog 19
HR-10000 Zagreb
iradman@amz.hr

Anita Rapan Papeša
Gradski muzej Vinkovci
Trg bana Josipa Šokčevića 16
HR-32 100 Vinkovci
anita@muzejvk.hr

Ante Rendić-Miočević
Poljička poljana 29 A
HR-10000 Zagreb
rendicmiocevic@xnet.hr

Mirko Sardelić
Hrvatska akademija znanosti i umjetnosti
Odsjek za povijesne znanosti
Strossmayerov trg 2
HR-10000 Zagreb
sardelicmirko@yahoo.com

Zoran Wiewegh, MA
Ministarstvo kulture Republike Hrvatske
Uprava za zaštitu kulturne baštine
Runjaninova 2
HR-10000 Zagreb
zoran.wiewegh@min-kulture.hr

Abbreviations

Museums

AMO	Arheološki muzej Osijek / The Archaeological Museum Osijek, Croatia
AMZ	Arheološki muzej u Zagrebu / The Archaeological Museum in Zagreb, Croatia
GMB	Gradski muzej Bjelovar / The Town Museum Bjelovar, Croatia
GMSi	Gradski muzej Sisak / The Town Museum Sisak, Croatia
GMSo	Gradski muzej Sombor / The Town Museum Sombor, Serbia
GMVa	Gradski muzej Varaždin / The Town Museum Varaždin, Croatia
GMVi	Gradski muzej Vinkovci / The Town Museum Vinkovci, Croatia
GMVu	Gradski muzej Vukovar / The Town Museum Vukovar, Croatia
KHM	Kunsthistorisches Museum, Wien, Austria
MBP	Muzej Brodskog Posavlja / The Museum of the Brodsko Posavlje, Slavonski Brod, Croatia
MGK	Muzej grada Koprivnice / The Town Museum Koprivnica, Croatia
MNM	Magyar Nemzeti Múzeum / The Hungarian Nationa Museum, Budapest, Hungary
MSO	Muzej Slavonije Osijek / The Museum of Slavonia Osijek, Croatia
MP	Muzej Prigorja / The Museum of the Prigorje, Sesvete, Croatia
MV	Muzej Vojvodine / The Museum of Vojvodina, Novi Sad, Serbia
NMB	Narodni muzej Beograd / The National Museum Belgrade, Serbia
NMLj	Narodni muzej Ljubljana / The National Museum Ljubljana, Slovenia
NMP	Národní muzeum Praha / The National Museum Prague, Czech Republic
NMZ	Narodni muzej Zagreb / The National Museum Zagreb, Croatia
ZMS	Zemaljski muzej Sarajevo / The Land Museum Sarajevo, Bosnia and Herzegovina

Literary sources

Agrip. *Comm.*	M. Vipsanius Agrippa, *Commentarii geographici* (*Geographic Commentaries*); *M. Vipsani Agrippae fragmenta ad Chorographiam spectantia*, in: A. Riese, *Geographi Latini Minores*, Hildesheim, Zürich and New York, 1995.
Ambrosiast. *Quaest.*	Ambrosiaster, *Quaestiones Veteris et Novi Testamenti* (*Questions on the Old and New Testaments*); Ambrosiaster, *Quaestiones Veteris et Novi Testamenti*, ed. A. Souter, Corpus Scriptorum Ecclesiasticorum Latinorum 50, Vienna, 1908.
Ambr. *Epist.*	Ambrose, *Epistulae* (*Letters*); Ambrosius, *Epistulae*, Corpus Scriptorum Ecclesiasticorum Latinorum 82.1-2, ed. O. Faller and M. Zelzer, Vienna, 1968-1990; Saint Ambrose *Letters*, translated by sister Mary Melchior Beyenka, OP, New York, 1954.
Amm. Marc.	Ammianus Marcellinus, *Res gestae* (*The Deeds*); Ammianus Marcellinus, *Rerum Gestarum Libri*, 3 Vol., English translation by John C. Rolfe, The Loeb Classical Library, Cambridge, Mass. and London, 1950-1952.
Anon. Raven.	Anonymus Ravennatis, *Cosmographia* (*Cosmography*); *Ravennatis Anonymi Cosmographia et Guidonis Geographica*, eds. M. Pinder and G. Parthey, Berlin, 1860; *La Cosmographie du Ravennate*, ed. Y. Janvier, Collection Latomus 235, Bruxelles, 1997.
Anon. Vales.	Anonymus Valesii, *Excerpta Valesiana* (*The Excerpts of Valesius*); Ammianus Marcellinus, *History*, vol. 3, English translation by John C. Rolfe, The Loeb Classical Library, Cambridge, Mass. and London, 1952.
App.	Appian of Alexandria, Ῥωμαική ἱστορία (*Roman History*), chapter Ἰλλυρικά (*Illyrian History*); Appian, *Roman History*, translated by Horace White, The Loeb Classical Library, Cambridge, Mass. and London, 1979.
App. *B. C.*	Appian of Alexandria, Ῥωμαική ἱστορία (*Roman History*), chapter Ῥωμαικῶν ἐμφυλί ων (*The Civil Wars*); Appian, *Roman History. The Civil Wars*, translated by Horace White, The Loeb Classical Library, Cambridge, Mass. and London, 1913.

Arr.	Flavius Arrianus (Arrian), Ἀνάβασις Ἀλεξάνδρου (*Anabasis of Alexander*); Flavius Arrianus, *The Campaigns of Alexander*, translated by Aubrey de Sélincourt, Penguin Books, London, 1976.
Ath.	Athenaeus, Δειπνοσοφισταί (*The Gastronomers*); *Athenaei Naucratitae deipnosophistarum libri xv*, Vol. II., ed. Georg Kaibel, Leipzig, 1887 [Stuttgart, 1965].
Aug. *RG*	C. Iulius Caesar Octavianus Augustus (Octavian Augustus), *Index rerum gestarum* (*The Deeds of the Divine Augustus*); *Res gestae divi Augusti*, text, translation, and commentary by Alison E. Cooley, Cambridge, 2009; Octavianus Augustus, *Res gestae divi Augusti* / Πράξεις Σεβαστοῦ θεοῦ / *Djela božanskog Augusta*, translated by Robert Matijašić, Latina & Graeca, Zagreb, 2007.
Aur. Vict.	Aurelius Victor, *Liber de Caesaribus* (*Book of the Emperors*); Sexti Aurelii Victoris *Liber de Caesaribus*, praecedunt *Origo gentis Romanae* et *Liber de viris illustribus urbis Romae*, subsequitur *Epitome de Caesaribus*, Leipzig, 1961; Aurelius Victor, *De caesaribus*, translated by H. W. Bird, Liverpool University Press, 1994.
Auson *Cent. nupt.*	D. Magnus Ausonius, *Cento nuptialis* (*A nuptial cento*); *Opere di Decimo Magno Ausonio*, a cura di Agostino Pastorino, Torino, 1971; *Ausonius, Decimus Magnus*, in two volumes, translated by Hugh G. Evelyn-White, Heinemann Ltd., London, 1919-1921.
Auson. *Epigr.*	D. Magnus Ausonius, *Epigrammata* (*Epigrams*); N. M. Kay, *Ausonius: Epigrams: Text with Introduction and Commentary*, Duckworth Publishers, London, 2001.
Cassiod. *Var.*	Flavius Magnus Aurelius Cassiodorus, *Variae (epistolae)* (*Various Letters*); Cassiodorus: *Variae*, translated by S. J. B. Barnish, Liverpool University Press, Liverpool, 1992.
Claud.	Claudius Claudianus, *De consulatu Stilichonis* (*On the Consulship of Stilicho*); *Claudian*, with an English translation by Maurice Platnauer, in two volumes, The Loeb Classical Library, Cambridge, Mass., 1922.
Dio	Dio Cassius Cocceianus, Ῥωμαική ἱστορία (*Roman History*); Cassius Dio Cocceianus, *Historiae Romanae quae supersunt* I-V, ed. U. P. Boissevain, Berlin, 1895-1931; Dio Cassius Cocceianus, *Roman History*, translation by Earnest Cary, The Loeb Classical Library, Cambridge, Mass., 1914-1927.
Enn. *Paneg. Theod.*	Magnus Felix Ennodius, *Panegyricus ad Theodericum* (*Panegyric to Theoderic*); Ennodius, Magnus Felix, *Panegyricus dictus clementissimo regi Theoderico*, Magni Felicis Ennodii opera, ed. F. Vogel, MGH, Auctores antiquissimi 7, Berlin 1885, 204-213; *Der Theoderich-Panegyricus des Ennodius*, ed. Christian Rohr, Hannover, 1995.
Epit. de Caes.	*Epitome de Caesaribus* (*A Booklet about the Emperors*); Sexti Aurelii Victoris *Liber de Caesaribus*, praecedunt *Origo gentis Romanae* et *Liber de viris illustribus urbis Romae*, subsequitur *Epitome de Caesaribus*, Leipzig, 1961; *Epitome de cesaribus*, translated by Thomas M. Banchich, Canisius College Translated Texts, Number 1, 2009, 2nd edition.
Eutr. *Brev.*	Eutropius, *Breviarium ab Urbe condita* (*Abridgement of Roman History*); Eutropii *Breviarum ab Urbe condita*, recognovit Carolus Santini, Teubner, Stuttgart and Leipzig, 1979; Eutropius, *Breviarum ab Urbe condita*, translated by H. W. Bird, Liverpool, 1993.
Exp. tot. mund.	*Expositio totius mundi et gentium* (*Review of the Whole World and Its Nations*); *Expositio totius mundi et gentium*, in: A. Riese, *Geographi Latini Minores*, Hildesheim, Zürich and New York, 1995; *The Expositio totius mundi et gentium, its geography and its language*, MA thesis by Jesse E. Woodman, The Ohio State University, 1964.
Fest. *Brev.*	Sextus Rufus (Rufius Festus), *Breviarium rerum gestarum populi Romani* (*Short Review of the History of the Roman People*); Sextus Rufus, *Breviarium rerum gestarum populi Romani*, ed. C. Wagener, Leipzig-Prag, 1886; Sextus Rufus, *The Breviarium of Festus*, ed. John William Eadie, London, 1967; *Breviarium of the Accomplishments of the Roman People*, translated by Thomas M. Banchich and Jennifer A. Meka, Canisius College Translated Texts, no. 2, Buffalo, NY, 2001.
Flor. *Epit.*	L. (or P.) Annaeus Florus, *Epitome bellorum omnium annorum DCC* (*Epitome of Roman History*); Lucius Annaeus Florus, *Epitome of Roman history*; translated by E. S. Forster, The Loeb Classical Library, Putnam, New York, 1929; Lucius Annaeus Florus, *Epitome ex Tito Livio / Epitome bellorum omnium annorum DCC*, ed. H. Malcovati, Roma, 1938; *Flori Epitomae de Tito Livio* / Lucije Anej Flor: *Dvije knjige izvadaka iz Tita Livija*, translated by Josip Miklić, Latina et Graeca, Zagreb, 2005.
FGrHist	*Fragmente der griechischen Historiker*. Vol. II: *Zeitgeschichte A: Universalgeschichte und Hellenika*. Text., ed. Felix Jacoby, Berlin, 1926, Leiden, 1996.

Fron. *Str.*	Sex. Iulius Frontinus, *Strategemata* (*Stratagems*); Iuli Frontini *Strategemata*, recensuit Robert I. Ireland, Leipzig, 1990; Sextus Julius Frontinus, *The Stratagems. Aqueducts of Rome*, translated by Charles E. Bennett, The Loeb Classical Library, Cambridge, Mass., 1980; Sextus Iulius Frontinus, *Strategemata*, ed. D. Savić, Demetra, Zagreb, 2004.
Front. *Princ. Hist.*	M. Cornelius Fronto, *Principia Historiae* (*The Principles of History*); Fronto, *Correspondence*, in 2 vols., translated by C. R. Haines, The Loeb Classical Library, Cambridge, Mass., 1919-1920.
Galen	Aelius (or Claudius) Galen of Pergamon, Περὶ φυσικῶν δυνάμεων (*On the Natural Faculties*) or Πρὸς Πίσωνα περὶ τῆς θηριακῆς (*On Viper's Antivenom – Dedicated to Piso*); *Galen on the Natural Faculties*, translated by A. J. Brock, The Loeb Classical Library, London, 1952.
Hadr. *Carm.*	P. Aelius Hadrianus (Hadrian), *Carmina* (*Poems*); http://www.forumromanum.org/literature/hadriane.html
Hdn. *Hist.*	Herodian of Syria, Τῆς μετὰ Μάρκον βασιλείας ἱστορία (*History of the Roman Empire since the Death of Marcus Aurelius*); Herodianus, *Ab excessu divi Marci libri octo*, ed. K. Stavanhagen, Leipzig, 1922; Herodian *Historia*, in two volumes, English translation by C. R. Whittaker, London, 1969.
Hier. *Chron.*	Hieronymus, *Chronicon* (*The Chronicle*); Hieronymus, *Chronicon, Die Chronik des Hieronymus – Hieronymi Chronicon*, ed. Rudolf Helm, Berlin 1956; Malcolm Drew Donalson, *A translation of Jerome's Chronicon with historical commentary*, Mellen University Press, 1996.
Hier. *Ep.*	Hieronymus, *Epistulae* (*Letters*); *The letters of St. Jerome*, translated by C. C. Mierow with notes by T. C. Lawler, Newman Press, Westminster, Md, 1963.
Hier. *Vir. Ill.*	Hieronymus, *De Viris Illustribus* (*On Illustrious Men*); *De viris inlustribus liber. Accedit Gennadii catalogus virorum inlustrium*, ex recensione Guilelmi Herdingii, Teubner, Leipzig, 1879; Saint Jerome, *On illustrious men*, translated by Thomas P. Halton, Catholic University of America Press, Washington, D. C., 1999.
Hyg. Grom.	Hyginus Gromaticus, *Constitutio limitum* (*On Setting up Centuriation*); Hyginus Gromaticus, *Constitutio limitum*, in: *Corpus agrimensorum Romanorum*, recensuit Carolus Thulin, vol. I, fasc. I: *Opuscula agrimensorum veterum*, Leipzig, 1913.
Isid. *Etym.*	Isidorus Hispalensis, *Etymologiae seu Origines* (*The Etymologies* or *The Origins*); Isidori Hispalensis Episcopi *Etymologiarvm sive originvm libri XX* / recognovit brevique adnotatione critica instrvxit W. M. Lindsay, Oxford, 1911; *The Etimologies of Isidore of Seville*, translated by Stephen A. Barney et al., Cambridge, 2006.
Itin. Ant.	*Itinerarium Antonini Augusti* (*The Antonine Itinerary*); *Itinerarium Antonini Augusti et Hierosolymitanum*, ex libris manu scriptis ediderunt G. Parthey et M. Pinder, Berlin, 1848.
Itin. Burd.	*Itinerarium Burdigalense* (*The Bordeaux Itinerary*); Itineraria Romana I: *Itineraria Antonini Augusti et Burdigalense*, Leipzig, 1929.
Jord. *Get.*	Jordanes, *Getica* (*The Goths History*); *Jordanis Romana et Getica*, recensuit Th. Mommsen, Berlin, 1882; Jordanes, *Getica*, *The Gothic history of Jordanes*, English version with an introduction and a commentary by Charles Christopher Mierow, Merchantville, NJ, 2006.
Jord. Rom.	Jordanes, *De summa temporum vel origine actibusque gentis Romanorum* (*On Totality of Time, or On Origin and Deeds of the Romans*); *Jordanis Romana et Getica*, recensuit Th. Mommsen, Berlin, 1882; English translation by Brian T. Regan at http://www.harbornet.com/folks/theedrich/Goths/Romana.htm
Julian *Mis.*	Flavius Claudius Iulianus Apostata, Μισοπώγων (*Beard-Hater*); Julien: *Misopogon*, translated by Christian Lacombrade, Les Belles Lettres, Paris, 2002.
Just. *Epit.*	M. Iunianus Iustinius, *Epitome Historiarum Philippicarum* (*Epitome of the Philippic History*); *Epitome of the Philippic History of Pompeius Trogus*, translated by the Rev. John Selby Watson, London, 1853.
Liv.	Titus Livius, *Ab Urbe condita* (*History of Rome*); Titus Livy, *History of Rome*, The Loeb Classical Library, Cambridge, Mass. and London, 1997.
Liv. *Per.*	Titus Livius, *Periochae* (*Summary*); Titus Livy, *Summaries, Fragments, General Index*, translated by A. C. Schlesinger, The Loeb Classical Library, Cambridge, Mass. and London, 1987.
Luc.	M. Annaeus Lucanus, *Bellum civile sive Pharsalia* (*Civil war*); Lucan, *The Civil War*

	(*Pharsalia*), translated by James D. Duff, The Loeb Classical Library, Cambridge, Mass. and London 1962.
Mam. *Paneg. Max.*	Claudius Mamertinus, *Panegyricus Maximiano Augusto dictus* (*Panegyric to Maximian Augustus*); Mamertinus, *Panegyricus Maximiano Augusto*, In Praise of Later Roman Emperors, *The Panegyrici Latini*, translated by R. A. B. Mynors, C. E. V. Nixon and Barbara Saylor Rodgers, University of California Press, 1994.
Mam. *Paneg. gen.*	Claudius Mamertinus, *Panegyricus genethliacus Maximiano Augusto dictus* (*The Birthday Panegyric to Maximian Augustus*); Claudius Mamertinus, *Panegyricus genethliacus Maximino Augusto dictus*, in: *Panegyrici Latini XII*, ed. R. A. B. Mynors, Oxford, 1964.
Martinus *In Basil.*	Martinus Dumiensis, *Versus B. Martini Dumiensis In Basilica* (*Poem of the Blessed Martin of Braga, In Basilica*); *Bibliotheca Latina Mediae et Infimae Aetatis*, eds. Johann Albert Fabricius, Johann Domenico Mansi, Christian Schöttgen, Firenze, 1858-1868.
Nic. Dam.	Nicolaus of Damascus, Περὶ τοῦ βίου Καίσαρος τοῦ Σεβαστοῦ καὶ τῆς αὐτοῦ ἀγωγῆς (*On Life and Education of the Emperor Augustus*); Nicolaus Damascenus, *Fragmenta*, ed. K. Müller, FHG 3, Didot, Paris, 1841-1870 (348-464); F. Jacoby, *Die Fragmente der griechischen Historiker*, frg. 90, Berlin, 1923-1959; http://www.attalus.org/translate/nicolaus1.html
Not. Dign. Occ.	*Notitia Dignitatum* (*The Register of Dignitaries*); *Notitia dignitatum omnium tam civilium quam militarium in partibus Occidentis*, in: *Notitia dignitatum. Accedunt Notitia urbis Constantinopolitanae et Laterculi provinciarum*, ed. Otto Seeck, Berlin, 1876 [Frankfurt, 1962].
Obs. *De prod.*	Iulius Obsequens, *Liber de prodigiis* (*Book of Prodigies*); Iulius Obsequens, *A Book of Prodigies after the 505th Year of Rome*, in: Titus Livy, *Summaries, Fragments, General Index*, translated by A. C. Schlesinger, The Loeb Classical Library, Cambridge, Mass. and London, 1987.
Oros. *Hist.*	Paulus Orosius, *Historiae adversus paganos* (*The History against Pagans*); *Pauli Orosii Historiarum adversum paganos*, libri VII, recensuit Carolus Zangemeister, Vienna, 1882; Orosius, *Seven books of history against the pagans*, translated by A. T. Fear, Liverpool University Press, 2010.
Ov. *Cons.*	P. Ovidius Naso, *Consolatio ad Liviam* or *Epicedium Drusi* (*Consolation to Livia*, or *Eulogy for Drusus*); Ovid, *The Art of Love and Other Poems*, translated by J. H. Mozley, The Loeb Classical Library, London and New York, 1929.
Ov. *Ep.*	P. Ovidius Naso, *Epistulae ex Ponto* (*Letters from the Black Sea*); Ovid, *Tristia. Ex Ponto*, English translation by Arthur Leslie Wheeler, The Loeb Classical Library, Oxford and London, 1924.
Ov. *Tr.*	P. Ovidius Naso, *Tristia* (*Sorrows*); Ovid, *Tristia. Ex Ponto*, English translation by Arthur Leslie Wheeler, The Loeb Classical Library, Oxford and London, 1924.
Pacat.	Pacatus, *Panegyricus Latini Pacati Drepani dictus Theodosio* (*Panegyric to Theodosius by Latinius Pacatus Drepanius*); *In Praise of Later Roman Emperors, The Panegyrici Latini*, translated by R. A. B. Mynors, C. E. V. Nixon and Barbara Saylor Rodgers, University of California Press, 1994.
Pan. Lat.	*Panegyrici Latini* (*Latin Panegyrics*); *In Praise of Later Roman Emperors, The Panegyrici Latini*, translated by R. A. B. Mynors, C. E. V. Nixon and Barbara Saylor Rodgers, University of California Press, 1994.
Paul *HL*	Paulus Diaconus, *Historia Langobardorum* (*The History of the Lombards*); Paulus Diaconus, *Historia Langobardorum*, ed. Georg Waitz, Monumenta Germaniae Historica Scriptores rerum Langobardicarum et Italicarum saec. VI.-IX., Hannover, 1878; Paul the Deacon, *History of the Langobards*, translated by William Dudley Foulke, University of Pennsilvania, 1907; Pavao Đakon, *Povijest Langobarda*, translated by R. Šćerbe and H. Šugar, Zagreb, 2010.
Plin. *NH*	C. Plinius Secundus Maior, *Naturalis historia* (*Natural History*); Pliny the Elder, *Natural History*, translated by H. Rackham, The Loeb Classical Library, Cambridge, Mass. and London, 1961.
Plin. *Tra.*	C. Plinius Caecilius Secundus Minor, *Panegyricus Traiani* (*Panegyric to Traian*); Pliny the Younger, *Letters and Panegyricus*, English translation by Betty Radice, The Loeb Classical Library, Cambridge, Mass. and London, 1969.
Polyb.	Polybius of Megalopolis, Ἱστορίαι (*The Histories*); Polybius, *The Histories*, translated by W. R. Paton, The Loeb Classical Library, Cambridge, Mass. and London, 2010.

Pomp. Mela	Pomponius Mela, *Chorographia* (*Chorography*); Pomponius Mela, *De Chorographia*, ed. C. Frick, Leipzig, 1880; Pomponius Mela, *De chorographia libri tres*, ed. Piergiorgio Parroni, Storia e Letteratura, vol. 160, Rome, 1984.
Pomp. Trog.	Pompeius Trogus, *Historiae Philippicae* (*The Philippic History*); see Just. *Epit.*
Posid.	Poseidonius, *fragmenta* (fragments); *Posidonius*, Vol. I: *The Fragments*, eds. Ludwig Edelstein and Ian G. Kidd, CCTC 13, Cambridge, 1972.
Procop.	Procopius, Ὑπέρ τῶν πολέμων λόγοι (*History of the Wars*); Procopius, *History of the Wars*, with an English translation by H. B. Dewing, in seven volumes, The Loeb Classical Library, London, 1914-1940.
Prud. *Peristeph.*	Prudentius, *Hymnus in honorem Quirini martyris, episcopi ecclesiae Siscianae* (*A Hymn in Honour of Quirinus, Martyr and Bishop of the Church of Siscia*); *Peristephanon* VII, Prudentius, translated by H. J. Thomson, in 2 Vols, The Loeb Classical Library, London and Cambridge, Mass., 1949-1953.
Ptol.	Claudius Ptolemaeus, Γεωγραφικὴ ὑφήγησις (*Geography*); Claudius Ptolemaeus, *Geographia*, ed. C. Müller, Paris, 1883.
SHA	*Scriptores Historiae Augustae* (*The Historia Augusta*); *Scriptores Historiae Augustae*, English translation by David Magie, in 3 Vols, The Loeb Classical Library, London and New York, 1922-1932; *Scriptores Historiae Augustae*, translated by Daniel Nečas Hraste, Antibarbarus, Zagreb, 1994.
Solin.	Solinus, *Collectanea rerum memorabilium*, cogn. Th. Mommsen, Berlin, 1864.
Sozom.	Sozomenos, *Historia Ecclesiastica* (*The Ecclesiastical History*); Sozomenos, *Historia Ecclesiastica*, *Kirchengeschichte*, übersetzt und eingeleitet von Günter Christian Hansen, Brepols Publishers, 2004; *The Ecclesiastical History of Sozomen*, also *The Ecclesiastical History* of Philostorgius, translated by Edward Walford, London, 1855.
Stat. *Silv.*	P. Papinius Statius, *Silvae* (*Woods*); Statius, *Silvae*, translated by D. R. Shackleton Bailey, The Loeb Classical Library, Cambridge, Mass. and London, 2003.
Strab.	Strabo of Amasia, Γεωγραφικά (*Geography*); Strabo, *The Geography*, translated by H. L. Jones, The Loeb Classical Library, Cambridge, Mass. and London, 1969.
Suda	*Suidae lexicon: A–Ω*; *Index*, ed. Ada Adler, Lexicographi Graeci 5, München and Leipzig, 2001.
Suet.	C. Suetonius Tranquillus, *De vita Caesarum* (*Twelve Caesars*); Caius Suetonius Tranquillus, *De vita caesarum*, ed. M. Ihm, Lipsiae, 1908; Suetonius, *The Lives of the Caesars*, vols. 1-2, translated by J. C. Rolfe, New York, 1928.
Sulp. Sev. *Chron.*	Sulpicius Severus, *Chronica* (*Chronicle*); *Sulpicii Severi libri qui supersunt*, recensuit et commentario critico instruxit Carolus Halm, apud C. Geroldi filium, Wien, 1866; *The works of Sulpitius Severus*, translated, with preface, and notes, by Alexander Roberts, Parker, Oxford, 1894.
Tab. Peut.	*Tabula Peutingeriana*. Codex Vindobonensis 324 I-II, ed. E. Weber, Graz, 1976; *Tabula Peutingeriana: le antiche vie del mondo*, a cura di Francesco Prontera, L. S. Olschki, Firenze, 2003.
Tac. *Ann.*	Cornelius Tacitus, *Annales sive Ab excessu divi Augusti* (*The Annals*); P. Cornelius Tacitus, *Annales sive Ab excessu divi Augusti*, translated by John Jackson, The Loeb Classical Library, Cambridge, Mass. and London, 1969.
Tac. Germ.	Cornelius Tacitus, *De origine et situ Germanorum* (*On the Origin and Situation of the Germans*); P. Cornelius Tacitus: *Agricola. Germania. Dialogus*, translated by M. Hutton and W. Peterson, The Loeb Classical Library, Cambridge, Mass. and London, 1914 (revised edition).
Tac. *Hist.*	Cornelius Tacitus, *Historiae* (*The Histories*); Tacitus, *Histories*, translated by Clifford H. Moore, The Loeb Classical Library, Cambridge, Mass. and London, 1925; *Historiae Taciti* / Tacit: *Historije*, translated by Josip Miklić, Latina et Graeca, Zagreb, 1987.
Tib. *Paneg.*	Albius Tibullus, *Panegyricum Messallae* (*Panegyric to Messalla*); Albius Tibullus, *Panegyricus Messallae* (*Corpus Tibullianum*), in: *Catullus. Tibullus. Pervigilium Veneris*, translated by J. P. Postgate, The Loeb Classical Library, London and Cambridge, Mass., 1968.
Vell.	C. Velleius Paterculus, *Historia Romana* (*Roman History*); C. Velleius Paterculus, *Historia Romana*, translated by Frederick W. Shipley, The Loeb Classical Library, Cambridge, Mass. and London, 1967; C. Vellei Paterculi *Historiae Romanae* / Gaj Velej Paterkul: *Rimska povijest*, translated by Josip Miklić, Latina et Graeca, Zagreb, 2006.

Vitruv. Vitruvius Pollio. *De architectura* (*On Architecture*); Vitruvius: *On Architecture*, translated by Frank Granger, The Loeb Classical Library, Cambridge, Mass. and London, 1931.

Zon. Ioannes Zonaras, Ἐπιτομὴ ἱστοριῶν (*Extracts of History*); Ioannis Zonarae *Epitome historiarum*, cum Caroli Ducangii suisque annotationibus edidit Ludovicus Dindorfius, Teubner, Leipzig, 1868-1875; *The History of Zonaras: From Alexander Severus to the Death of Theodosius the Great*, translated by Thomas Bancich and Eugene N. Lane, Routledge, London and New York, 2009.

Zos. Zosimus, Ἱστορία νέα (*The New History*); Zosime, *Histoire nouvelle*, texte établi et traduit par François Paschoud, Paris, 1971; Zosimus, *New History*, translated by Ronald T. Ridley, Sydney, 1982.

Bibliography

ACIAC	*Acta Congressus Internationalis Archaeologiae Cristianae*
AAntHung	*Acta Antiqua Academiae Scientiarum Hungaricae*, Budapest
AArchHung	*Acta Archaeologica Academiae Scientiarum Hungaricae*, Budapest
AÉ	*L'Année Épigraphique*, Paris
AI	*Archaeologia Iugoslavica*, Beograd
AIA	*Annales Instituti Archaeologici*, Zagreb
AIJ	Viktor Hoffiller and Balduin Saria, *Antike Inschriften aus Jugoslavien,* Heft I, *Noricum und Pannonia Superior*, Zagreb 1938
AJA	*American Journal of Archaeology*, Boston
AK	*Archäologisches Korrespondenzblatt*, Mainz am Rhein
AN	*Aquileia Nostra,* Aquileia
AnOs	*Anali Zavoda za umjetnički i znanstveni rad u Osijeku,* Osijek
ANRW	*Aufstieg und Niedergang der römischen Welt*, Berlin-New York
Antaeus	*Antaeus*, Budapest
AP	*Arheološki pregled*, Beograd, Ljubljana
AR	*Alba Regia*, Székesfehérvár
ArchAdr	*Archaeologia Adriatica,* Zadar
ArchÉrt	*Archaeologia Értesitő*, Budapest
ARP	A. Lengyel and G. T. B. Radan (eds.), *The Archaeology of Roman Pannonia*, Budapest -Lexington 1980
ARR	*Arheološki radovi i rasprave,* Zagreb
Athenaeum	*Athenaeum*, London
AV	*Arheološki vestnik*, Ljubljana
BAR	*British Archaeological Reports*, Cambridge
BJb	*Bonner Jahrbücher*, Bonn
BK	*Balácai Közlemények*, Veszprém
BudRég	*Budapest Régiségei*, Budapest
CAH	*Cambridge Ancient History*, Cambridge
CBI	*Corpus der griechischen und lateinischen Beneficiarier-Inschriften des Römischen Reiches*, E. Schallmayer, K. Eibl, J. Ott, G. Preuss, and E. Wittkopf (eds.), Die römische Weihebezirk von Osterburken I, Forschungen und Berichte zur Vor- und Frühgeschichte in Baden-Württemberg, Band 40, Stuttgart, 1990
Chiron	*Chiron*, München
CIL	*Corpus Inscriptionum Latinarum*, Berlin
CommArchHung	*Communicationes Archaeologicae Hungariae*, Budapest
DissPann	*Dissertationes Pannonicae*, Budapest
DjANUBiH	*Djela Akademije nauka i umjetnosti Bosne i Hercegovine*, Sarajevo
DjCBI	*Djela Centra za balkanološka ispitivanja*, Sarajevo
DMBg	*Dissertationes et Monographiae*, Beograd
DMZg	*Dissertationes et Monographiae*, Zagreb
FA	*Folia Archaeologica*, Budapest
FMRÖ	*Die Fundmünzen der römischen Zeit in Österreich*
FPA	Péter Kovács (ed.), *Fontes Pannoniae Antiquae in aetate Severorum* IV, Budapest, 2007
GGMS	*Godišnjak Gradskog muzeja Sisak*, Sisak
GodANUBiH	*Godišnjak Akademije nauka i umjetnosti Bosne i Hercegovine,* Sarajevo

GodCBI	*Godišnjak Centra za balkanološka ispitivanja*, Sarajevo
GSM	*Glasnik slavonskih muzeja*, Vukovar
GZMS	*Glasnik Zemaljskog muzeja*, Sarajevo
GZPSKV	*Građa za proučavanje spomenika kulture Vojvodine,* Novi Sad
HAG	*Hrvatski arheološki godišnjak*, Zagreb
HAnt	*Histria Antiqua,* Pula
Historia	*Historia,* Stuttgart, Wiesbaden
ILJug	Ana Šašel and Jaro Šašel, *Inscriptiones Latinae quae in Iugoslavia inter annos MCMXL et MCMLX repertae et editae* sunt, *Situla* 5, 1963; *iidem*, *Inscriptiones Latinae quae in Iugoslavia inter annos MCMLX et MCMLX X repertae et editae sunt*, *Situla* 19, 1978; *iidem*, *Inscriptiones Latinae quae in Iugoslavia inter annos MCMLXII et MCMXL repertae et editae sunt*, *Situla* 25, 1986
ILS	*Inscriptiones Latinae selectae,* Berlin 1982-1916
ILSl	Milan Lovenjak (ed.), *Inscriptiones Latinae Sloveniae 1. Neviodunum*, Ljubljana, 1998
InscrIt	*Inscriptiones Italiae*, Roma, 1931-
IzdHAD	*Izdanja Hrvatskog arheološkog društva*
Lihnid	*Lihnid*, Ohrid
JRA	*Journal of Roman Archaeology*, Portsmouth RI
JRGZM	*Jahrbuch des Römisch-Germanischen Zentralmuseums Meinz*, Meinz am Rhein
JRS	*Journal of Roman Studies,* London
KiM	*Katalogi in monografije*, Ljubljana
Klio	*Klio*, Berlin
KZ	*Križevački zbornik*, Križevci
MAI	*Mitteilungen des Archäologischen Instituts der Ungarischen Akademie der Wissenschaften*, Budapest
MatSADJ	*Materijali Saveza arheoloških društava Jugoslavije*
MV	*Muzejski vjesnik*, Koprivnica, Varaždin, Čakovec, Bjelovar, Kutina
NC	*Numismatic Chronicle,* London
Numizmatičar	*Numizmatičar*, Beograd
Numizmatika	*Numizmatika*, Zagreb
NV	*Numizmatičke vijesti*, Zagreb
NVe	*Numizmatični vestnik*, Ljubljana
NZ	*Numismatische Zeitschrift*, Wien
ObHAD	*Obavijesti Hrvatskog arheološkog društva*, Zagreb
Obol	*Obol*, Zagreb
OpArch	*Opuscula archaeologica*, Zagreb
OZ	*Osječki zbornik*, Osijek
Pannonia I, II	M. Šašel Kos and P. Scherrer (eds.), *The Autonomous Towns of Noricum and Pannonia.* I: *Situla* 41, 2003; II: *Situla* 42, 2004
PIAZ	*Prilozi Instituta za arheologiju u Zagrebu*, Zagreb
PIR[2]	*Prosopographia Imperii Romani*, Berlin
PJZ	Alojz Benac (ed.), *Praistorija jugoslavenskih zemalja* 1-5, Sarajevo
PZ	*Podravski zbornik*, Koprivnica
RE	*Realencyclopädie der Classischen Altertumswissenschaft*, Stuttgart
RFFZd	*Radovi Filozofskog fakulteta u Zadru*, Zadar
RINMS	Marjeta Šašel Kos, *The Roman Inscriptions in the National Museum of Slovenia*, *Situla* 35,1997
RIU	*Römische Inschriften Ungarns*, Budapest
RJAZU (*RHAZU*)	*Rad Jugoslavenske (Hrvatske) akademije znanosti i umjetnosti,* Zagreb
RMD	Margaret Roxan and Paul Holder, *Roman Military Diplomas*, Vols. I-IV, London, 1978-2003
RVM	*Rad Vojvođanskih muzeja*, Novi Sad
SchrBalk	*Schriften der Balkankomission,* Wien
SEP	*Studia Epigraphica Pannonica*, Budapest
SHP	*Starohrvatska prosvjeta*, Zagreb, Split
SSlav	*Scrinia Slavonica*, Slavonski Brod
Starinar	*Starinar*, Beograd
TRAC	*Proceedings of the Annual Theoretical Roman Archaeology Conference*
Tyche	*Tyche,* Wien

VAHD (*VAPD*)	*Vjesnik za arheologiju i historiju* (*povijest*) *dalmatinsku,* Split
VAMZ	*Vjesnik Arheološkog muzeja u Zagrebu*, Zagreb
VHAD	*Vjesnik hrvatskog arheološkog društva*, Zagreb
VMKH	*Vijesti muzealaca i konzervatora Hrvatske,* Zagreb
ZČ	*Zgodovinski časopis*, Ljubljana
ZM	*Zalai Múzeum*, Zalaegerszeg
ZN	*Zeitschrift für Numismatik*, Berlin
ZPE	*Zeitschrift für Papyrologie und Epigraphik*, Bonn
ŽAnt	*Živa antika*, Skopje

Introduction and Commentaries

Branka Migotti

1. Introduction

This paper is structured as an *introduction*, although it might well have followed the chapters as an *afterword*. It has been conceived as a combination of a discussion of the basic facts (borders, the state of research, chronological frame on the area of study, and a comment on various issues considered in individual chapters, especially in their controversial aspects. Therefore, it is not the individual papers that are commented as such, but the chosen archaeological themes of south-Pannonian archaeology, as seen against the background of the province as a whole; more often than not the issues are discussed or are touched upon in more than one paper. Urban architecture and rural settlements are not commented as an individual topic, but are addressed within chronological sections, mostly *The Severan period* and *The 4th century*. The same is true of economy, trade and commerce, with hints at various aspects of these scattered among the majority of texts. A broad historical narrative in its strict chronological frame is missing as a separate contribution, as the basic facts of south-Pannonian Roman history, written in English, can be found elsewhere within the history of the whole of the province.[1] In this volume, more or les detailed historical narrative is dispersed between various chapters according to their specific subjects, mostly in Domić Kunić, Dizdar, Dzino and Domić Kunić, Radman-Livaja (on the army), Rapan Papeša, and Sardelić. Otherwise, historical and archaeological surveys written by Croatian authors in foreign languages are mostly not very extensive and are, more often than not, concerned with the whole territory of present-day Croatia, including therefore part of the Roman province of Dalmatia, as well as Istria, the latter as part of Italy.[2] More recent foreign commentators include facts on southern Pannonia in the wider narrative of the history of the whole of Illyricum, or, rather, the Danube and Balkan provinces. The advantage of such studies is their striving towards a more extensive use of the archaeological evidence, and of course, the *big picture* achieved. On the other hand, there is also a serious disadvantage to such works, in spite of their authors' scholarly excellence: the lack of the substantial insight into the archaeology of individual areas and too much inevitable generalisation.[3] *Generalisation* in this context means above all the use of the archaeological evidence for northern Pannonia as relevant for the whole of the province, or even for its southern portion specifically. A telling example concerns the state of research of the cemeteries. While W. Schmidt would posit that Pannonia is one of the leading provinces in this regard,[4] the truth is that it only concerns the Hungarian part of the province, while the same is hardly true of northern Croatia, in spite of recent improvements – see Leleković, in this volume, section 1. Generalisations or misconceptions were mostly due to the lack of data, but sometimes they seem to have resulted from the lack of a more careful insight.[5] While Roman provincial narratives in various European countries have preferably emerged as parts of their national histories, Hungarian scholars have until recently produced studies and books on Pannonia as a whole, covering also its Austrian, Slovenian, Croatian, and Serbian portions.[6] While such approach seems to be the only right one theoretically, in practice it meets a major obstacle: the lack of the archaeological evidence and/or its inaccessibility for the territories outside Hungary.[7] Possibly for this very reason in a recent study of Hungarian archaeology, its Roman portion has been limited to the territory of Hungary, with towns from other modern countries, Croatia included, mentioned in passing and in exclusively historical contexts.[8]

2. The borders of the Pannonian provinces in their territorial and chronological frame

The Roman province of Pannonia encompassed parts of the territories of modern countries of Hungary, Austria, Slovenia, Croatia, Bosnia and Herzegovina, and Serbia. The

---

1 Mócsy 1962; Mócsy 1974; Barkóczi 1980. It should be noted that the data on the Croatian portion of Pannonia in these and similar books mostly, and not very extensively at that, concern the towns of Siscia, Mursa, and Cibalae.

2 Matijašić, Čače, and Rendić-Miočević 1998; Kuntić-Makvić 1999; Zaninović 1999.

3 Poulter 1996; Wilkes 2005; Caldwell 2007. Cf. also A European Research Group on the History and Archaeology of the Balkans, coordinated by J. Luc Lamboley (Isaacs 2007, 204). A synthesis as the final aim of a research was stressed by A. Mócsy (1974, xx) and mentioned also by J. J. Wilkes (2005, 125).

4 Schmidt 2001, 251. Curiously, P. Zsidi (2003, 251) still claims that the study of cemeteries in Hungary has been neglected as compared to settlement research.

5 Cf. for instance in Poulter 1996, where a conspicuous dearth of information on the Croatian portion of Pannonia is matched by an assertion (quotation missing) that in the 4th century Siscia enlarged its perimeter (p. 106), while on the map of the Danube provinces (p. 102) there is no Mursa. A specific case is the study of the so-called 3rd-century crisis in the western province (Witschel 2004), with Pannonia altogether omitted.

6 Mócsy 1962; Mócsy 1974; Fitz 1977; *ARP*.

7 The risk of such omissions is lesser in studies that are not so dependant on the archaeological evidence (epigraphy excluded), or those that concentrate on narrower subject matters, such as, for instance, Fitz 1993-1995 and Kovács 2004.

8 Various authors in: Visy 2003c, 204-280, specifically Nagy (2003).

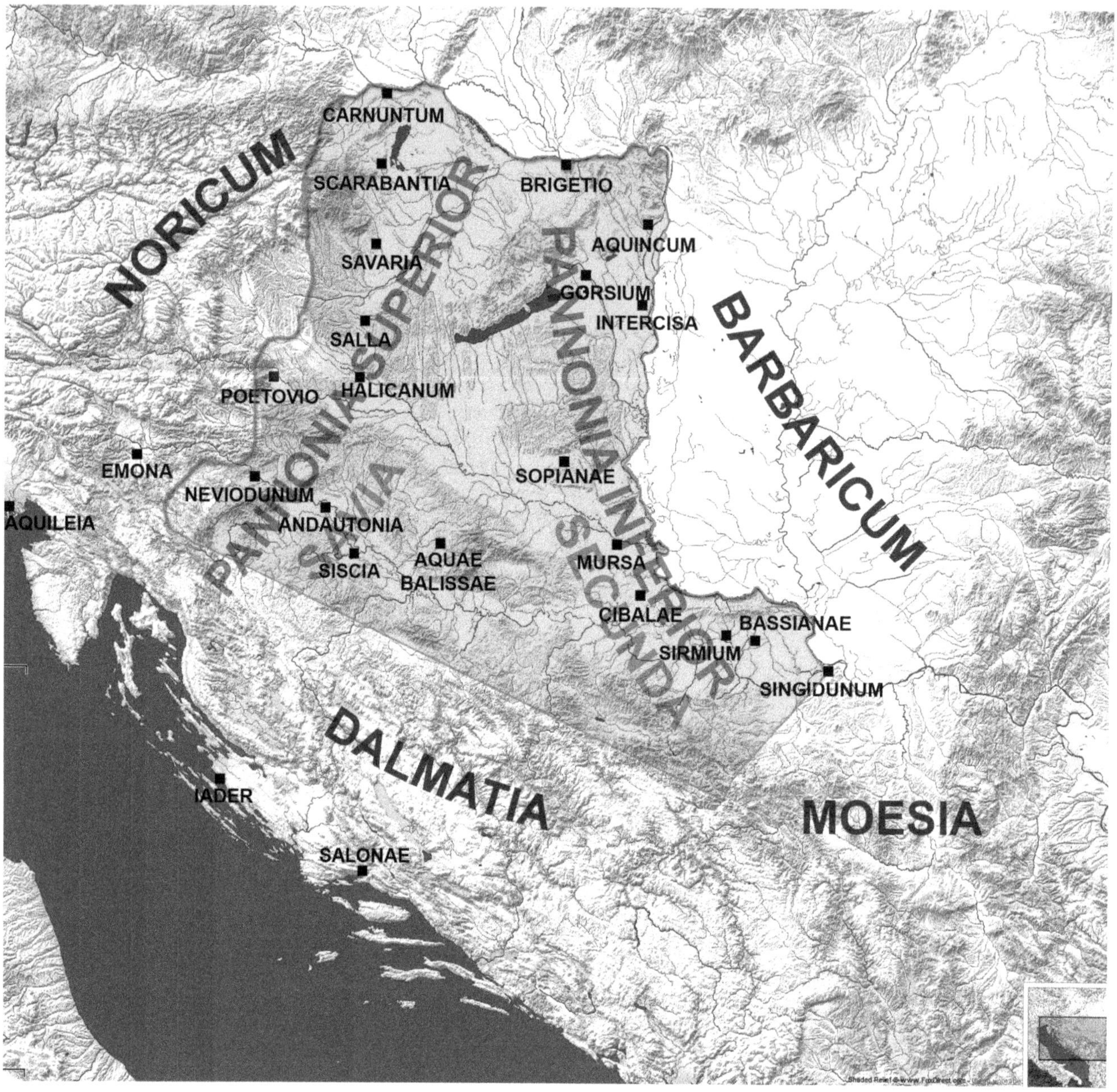

Fig. 1. Map of the province of Pannonia (after Google Maps, modified by T. Leleković).

majority of its northern portion is in Hungarian territory (fig. 1), while the largest part of southern Pannonia between the rivers Sava, Drava and Dunav is in modern Croatia; the remainder is divided between Slovenia, Bosnia and Herzegovina, and Serbia. Major Roman towns in the Croatian portion of Pannonia were the following: *coloniae Siscia* (Sisak), *Mursa* (Osijek) and *Cibalae* (Vinkovci), *municipium Andautonia* (Ščitarjevo) and *Municipium Iasorum / Aquae Balissae* (Daruvar), and *civitas Iovia* (Ludbreg). The most important south-Pannonian towns outside Croatia were the *colonia Poetovio* and the *municipium Neviodunum* in Slovenian territory and the *colonia Sirmium* and the *municipium Bassiana* in Serbian territory (fig. 2).

The prevailing opinion is that the province of Pannonia was created by the division of Illyricum between the years 8 and 10 AD, while at the beginning of the 2nd century the emperor Trajan split the territory into two provinces: Pannonia Superior and Pannonia Inferior.[9] Although the background to this division was political and administrative, it has also been suggested that the inter-provincial border was adjusted to the existing division between the native populations.[10] This suggestion is logical and seems to be exemplified also in southern Pannonia, where the inter-

[9] Mócsy 1962, 583-586; Barkóczi 1980, 89-94; Fitz 1993, 32-37; Zaninović 1999, 69; Gruen 1996,178, 188-189; Wilkes 1996, 554. According to another opinion, Pannonia was established as a civilian province only in the time of the emperor Vespasian: Šašel Kos 2005, 242; Šašel Kos 2010, 220.

[10] On the reasons for the division see Mócsy 1974, 92-94, and on its ethnic implications see Barkóczi 1980, 94 and Dizdar, in this volume, section 2. The border between the two provinces followed roughly a northwest line, nearly touching the east shores of Lake Balaton (*Lacus Pelso*), which remained in Pannonia Superior (Mócsy 1974, 93, fig. 16; Soproni 1980b, 59, fig. 1; Fitz 1993, 372; *Pannonia* II, map on p. 9).

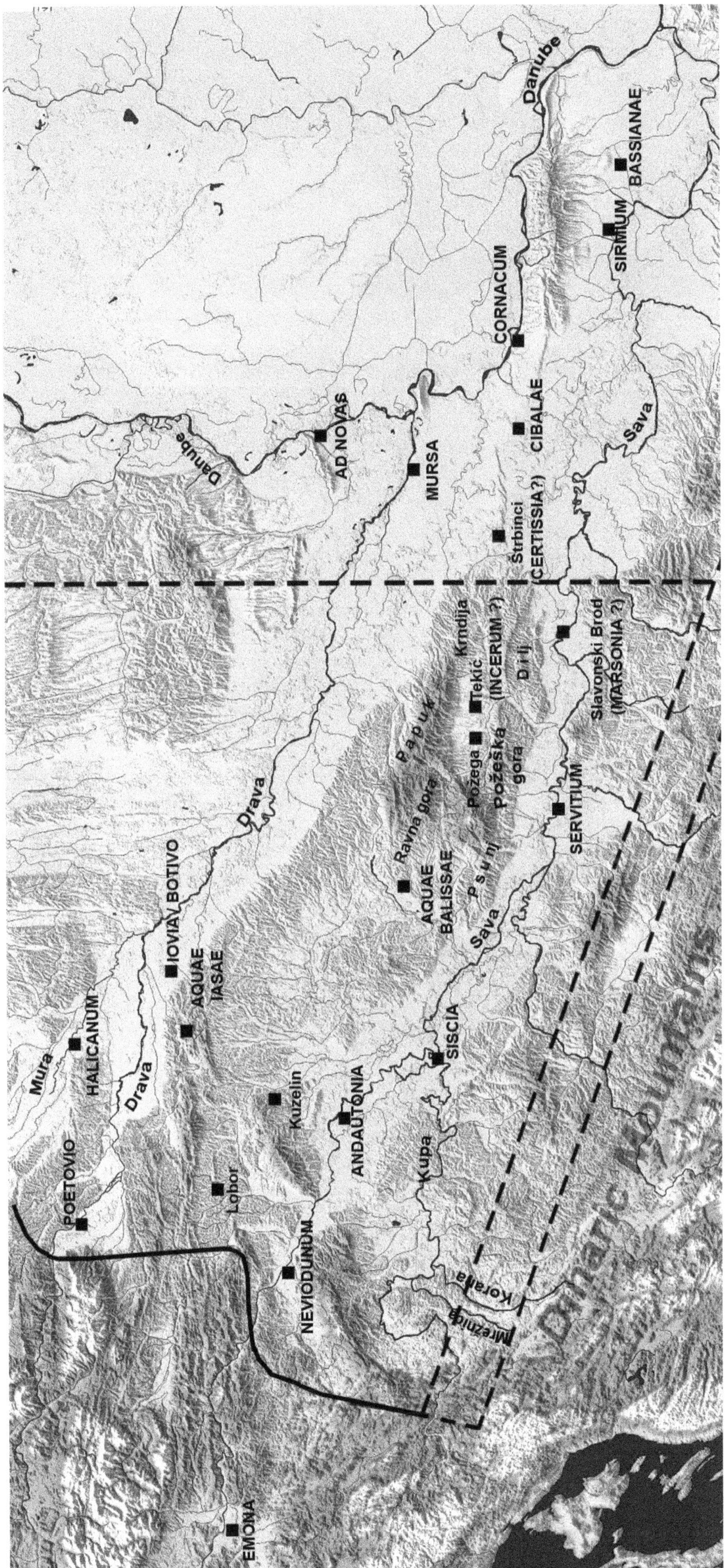

Fig. 2. Map of the Croatian part of the province of Pannonia (after Google Maps, modified by T. Leleković).

provincial border probably followed the division between the Breuci and Oseriates. Nevertheless, this issue remains unresolved for several reasons. Firstly, the areas occupied by different populations can be only very roughly outlined, as no such detailed information is accurately adduced in ancient written sources. Secondly, the nature of ethnic communities is inherently unstable and shifting both social and geographic frames. Finally, the absolute observance of pre-Roman tribal territories would contradict a well-proved Roman policy of splitting the territorial integrity of the native communities at an early stage of the administration of newly conquered lands, if the need for such method arose.[11] A formal division between the northern and southern portions of Pannona did not exist prior to the time of the emperor Diocletian, when, between the years 294 and 296, it was formed into four provinces: *Pannoniae Prima*, *Valeria*, *Savia*, and *Secunda*.[12] As the initial north-south line of division remained, the border between Pannoniae Savia and Secunda followed its lower stretch, probably leaving in Pannonia Savia the Požega Valley and the mountains on its eastern fringe, save for the easternmost slopes of Krndija and Dilj.[13] However, it remains unclear whether from there the border took a western course towards the Sava, or whether it proceeded along the straight north-south line; in any case, the border reached the river at an undetermined point between Bosanska Gradiška (Servitium) and Slavonski Brod (Marsonia).[14] Although the border between northern Pannoniae (Prima and Valeria) and those southern (Savia and Secunda) is not recorded in the written sources, it is customarily put on the River Drava.[15] However, the westernmost stretch of the northern border of the Croatian part of Pannonia is on the River Mura, a Drava's tributary. Given this, a minor portion of the area discussed here stretches into northern Pannonia, bordering on the west with Slovenia. This border, on the other hand, does not coincide with the division between the territories of the Roman towns of Poetovio and Neviodunum (Slovenia) on the one hand, and those of Siscia, Andautonia, and Iovia (Croatia) on the other.[16] These issues are too complex to be discussed here at any lengths; even though, the eastern border of the Poetovian territory still needs to be addressed. This question is equally controversial, but should be tackled on account of its repercussions for an intriguing issue of the territorial and administrative appurtenance of the important and well-researched Roman town of Aquae Iasae. There is no doubt that the Poetovian *ager* stretched into Croatian territory, but just how far to the east remains unknown. If the hypothesised location of *municipium Halicanum* in present-day Sveti Martin na Muri is correct, its territory must have been formed on account of the Poetovian *ager*.[17] As far as Aquae Iasae is concerned, the prevailing opinion is that in terms of administration this spa settlement belonged to Poetovio, on the basis of several inscriptions that the *res publica Poetovionensis* set up there in the 2nd century.[18] However, such evidence is not conclusive for the establishment of the administrative appurtenance of this kind, as instances of the inhabitants of one town setting up inscriptions in another pertaining to a different *ager*, are well known.[19] In view of that, the administrative appurtenance of Aquae Iasae to Poetovio during the early and middle Imperial periods remains inconclusive. Even it this was so, the circumstances changed by the end of the 3rd century, when in the emperor Diocletian's administrative reforms Poetovio was allocated to the province of Noricum.[20] The *Itinerarium Burdigalense sive Hierosolymitanum* (333-334) fixes the border between Noricum and Pannonia Savia on the bridge across the Drava, some 15 km east of Poetovio. This, at the same time, was most probably the eastern limit of the Poetovian *ager*.[21] At the time when Poetovio became a Norican town, Aquae Iasae remained in Pannonia, as transpires from the inscription set up in the latter settlement by the governor of Pannonia Prima on behalf of the emperor Constantine I.[22] This should lead to the conclusion that Aquae Iasae most probably remained outside the Poetovian *ager*. Such conclusion is based on the fact that in defining provincial boundaries the Romans most usually observed the earlier division between local administrative communities.[23] There is yet another argument in favour of the suggested solution: the foundation of Iovia as a new municipality in the surroundings of Aquae Iasae. Iovia was in the Jerusalem itinerary (333-334) recorded as a *civitas*, denoting its municipal status, acquired probably at an unknown time in the course of the 3rd century. In the *Itinerarium Antonini* (3rd century) Iovia is not yet recorded as a *civitas*, suggesting that this form of civil rank was bestowed on it in the later 3rd or the early 4th centuries. On the other hand, the Antonine itinerary does not record the municipal character of the settlements in a consistent manner, leaving a possibility that Iovia became a *civitas* at any point in the 3rd century.[24] Although its *ager* remains

---

11 Fitz 1980, 141; Salway 2000, 7; Gros 2008, 65. See also Dizdar in this volume (n. 82).

12 Mócsy 1962, 588; Barkóczi 1980, 109; Fitz 1994, 1175-1183; Kuntić-Makvić 1999, 82.

13 The mountains surrounding the Požega Valley (Dilj, Krndija, Papuk, Požeška gora, Psunj, and Ravna gora) are in this volume referred to as the Slavonian Mountains, although it is not their official common name.

14 See fig. 1 in Rapan Papeša, in this volume. In some authors this border is pushed towards west, leaving the largest portion of the Požega Valley in Pannonia Secunda. However, the former solution is more realistic: Pavan 1955, 432; Mócsy 1974, fig. 59; Soproni 1980b, 58; Hajnóczy *et al.* 1999, 7; Schejbal 2004, 100; *Pannonia* II, 9; Bratož 2008, 408-409; Ciglenečki 2008, 440, fig. 2; Matijašić 2009, 222.

15 Some authors hypothesised the border between Pannoniae Prima and Savia as stretching south of the Drava, which seems less likely. Cf. Tóth 1994, 261, Abb. 1; Gračanin 2011, 53, n. 26.

16 Cf. Horvat *et al.* 2003, 154, Fig. 1, 159-160; Lovenjak 2003, 93.

17 Cf. *AIJ*, 193-194; Póczy 1980, 244 and n. 9; Kos and Šašel Kos 2000, 292; Horvat *et al.* 2003, 154, fig. 1; Redő 2003, 194, fig. 1.

18 Until recently only two such inscriptions had been known: *AIJ*, 122, 206, no. 461; Gorenc and Vikić 1975, 42, fig. 7. See also Horvat *et al.* 2003, 154, fig. 1, 160. However, in the excavations in 2011 new evidence came to light (Dora Kušan Špalj, pers. com).

19 Jeličić-Radonić 2007, 57; Buora 2008b, 158;Tatscheva 2005, 215. An example of the mid-Italian town of Castrum Novum that allowed another town to use its baths, raising a *vectigal balnearum* for that service (*CIL* IX 5144 – see also Starac 2000, 155), gives an insight into possible reasons for a special relationship between the *res publica Poetovionensis* and Aquae Iasae.

20 Horvat *et al.* 2003,157.

21 Bratož 2004, 333.

22 *AIJ*, 210, no. 469.

23 Salway 2000, 20.

24 It has been assumed that the Antonine itinerary was composed in the early 3rd century, to be updated later in the same century: Soproni 1980a, 213; Nagy 2004, 77. On the civic status of Iovia see Mócsy 1974, 222 and Migotti 2002, 55.

unknown, it should have been created on account of the Poetovian territory, and its western limit probably followed the border between Noricum and Pannonia. Its northern limit was on the Drava, and its territory bordered on the north-west with that of Sal(l)a (Zalaegerszeg, Hungary) and on the north-east with Halicanum (Sveti Martin na Muri?).[25] It is possible that the *ager* of Iovia was much smaller in comparison with earlier municipalities (*coloniae* and *municipia*), given that the rise in number of towns, accompanied by reduction of their perimeters, is a typical feature of late antique urbanism.[26] On balance, in Later Antiquity Aquae Iasae was almost certainly situated in Iovian territory.

The western portion of Pannonia south of the Sava is in Croatian territory, while the middle and eastern stretches are in northern Bosnia, the latter bordering on the Roman province of Dalmatia. The line of the border is, however, a matter of debate; it is usually described vaguely as (*at a small distance*) *south of the Sava*. From the accompanying maps it transpires that some authors put this border some 30 km south of the Sava, while others see it as lying further south, nearer the Adriatic hinterland.[27] The latter assumption seems to be realistic, at least as concerns the western (Croatian) part of the border. Recent researches into the stone quarrying and stone cutting of the area have revealed that these activities were directed towards Siscia, and that both stone blocks and (semi)-finished products were transported via the River Kupa and its tributaries, especially the Mrežnica and Korana.[28] In addition to that, a recent insight into the pre-Roman material culture of the Colapiani (Illyrian-Pannonian ethnic community) suggests the stretching of Pannonia further south, towards the upper course of the Kupa.[29] Indirectly, these data suggest that also in Bosnia and Herzegovina the border was nearer to the upper than to the lower courses of the Sava tributaries. [30] The question is whether the so-hypothesised border can be connected with the ethnographical and geographical data of the Roman historiographers, given that they relate of the Pannonians in the immediately pre-conquest period as occupying the lands from the Danube and Drava all the way to the Dinaric Mountains.[31] The eastern limit of the area under study is the state border shared by Croatia and Serbia. In its northern stretch the Danube was the dividing line between the Roman Empire and the *Barbaricum*, while in the south the territories of Cibalae and Sirmium met a little west of the present-day state border, roughly on a north-south line from Sotin (Cornacum) towards the Sava.[32]

As has already been noted, modern state or regional borders make a less convenient frame for the study of Roman provincial history and archaeology. Nevertheless, in the example of northern Croatia the state and regional territory roughly coincides with the portion of Pannonia showing some specific characteristics, which singled it out from the rest of the province. A favourite geo-strategic position, a particular ethnic substratum and early Romanisation – all these earned southern Pannonia the status of a kind of *Pannonia proprie dicta*.[33]

3. The state of knowledge of south-Pannonian archaeology

This section does not aim to offer a comprehensive survey of the state of research and a full bibliography. Instead, a selection of studies on Pannonia as a whole in foreign languages will be briefly reviewed, with comments highlighting the way southern Pannonia is treated in this literature. While a centuries-long tradition of writing in foreign languages has earned Hungarian archaeology an unquestionable advantage in European scholarship, the same does not apply to northern Croatia.[34]

The 18th- and 19th-century surveys of Roman urbanism, architecture, inscriptions and coinage, written in Latin, figure as the earliest literature on south-Pannonian Roman archaeology, with two authors standing out. The first one is Luigi Ferdinando Marsigli, the Italian scholar and military architect, who in his six-volume work *Danubius Pannonico-mysicus observationibus geographicis, astronomicis, hydrographicis, historicis, physicis perlustrarus et in sex tomos digestus* (Hagae-Amstelodami 1726) produced measurements, drawings and descriptions of Roman towns on the Danube *limes* and in other areas of Pannonia; among them are Mursa, Cibalae, and Siscia (volume 2). Despite the frequent mention of Marsigli's data in the Croatian historical and archaeological literature since the 18th century, its potential has, until quite recently, never been fully exploited in terms of testing through archaeological evidence.[35] With the new approach to Marsigli's data, especially in the case of Siscia, it is now possible to witness this author's accuracy, as had already been noted in Hungarian literature. The same is true of Matija Petar Katančić, the Croatian historian and archaeologist; of his several archaeological studies on southern Pannonia, two should be singled out: *Dissertatio de columna miliaria ad Eszekum reperta* (1782) and *Specimen philologiae et geographiae Pannoniorum in quo de origine lingua Croatorum simul de Sisciae, Andautonii, Nevioduni, Poetovionis urbium in Pannonia olim celebrium et his interiectarum via militari mansionum*

25 See n. 17.
26 Brown 1996, 47; Borhy 2003, 417-419; Zanini 2003, 214.
27 Wilkes 1969, a map attached; Soproni 1980b, 58-59, fig. 1; Bojanovski 1988, 35 and a map attached; Chevalier 1995, 15-16; Bratož 2004, 333; Domić Kunić 2006, 65; Sanader 2007, 70, Abb. 61; Dzino and Domić Kunić, in this volume, section 4.
28 Perkić 2002, 118; Migotti 2005, 371.
29 Čučković 2008, 80.
30 Ivan Knezović (pers. com.) believed that in the given circumstances the provincial border was set at the point(s) where the rivers become navigable, which roughly coincides with the above picture of the Pannonian-Dalmatian border. We planned on resuming this discussion in the autumn of 2011, but to our utter dismay, Ivan never returned from his summer holidays that year.
31 See Dzino and Domić Kunić in this volume (ns. 53 to 58).
32 Iskra-Janošić 2004, 177; Mirković 2004, 148.
33 Cf. Mócsy 1974, 63; *ARP*, 24-25; Alföldy 1994, 27; Wilkes 2005, 137-138; Domić Kunić 2006, 64-65; Domić Kunić, in this volume, n. 2.
34 *ARP*, 19-20.
35 Pinterović 1978, 158-159; *ARP*, 19, 30; Iskra-Janošić 2004, 178; Vukelić 2009, 94. See also Lolić and Wiewegh in this volume, 6.2.1.

situ disseritur ... Istri adcolarum geographia vetus e monumentis epigraphicis, marmoribus, numis, tabellis eruta et commentariis illustrata a patre Math. Petro Katancsich ..., Pars I (1826).[36] Early Christianity of northern Croatia was in the past given even less attention than the other Roman-period subject matters. After the pioneering work of the Italian scholars Daniele Farlati, Jacopo Coleti, and Filip Riceputi (*Illyricum sacrum*, Venetiis 1751-1819), north-Croatian early Christianity used to appear as an indiscriminate and archaeologically unsubstantiated part of the history of Pannonian Christianity as a whole, until recently written nearly exclusively by Hungarian authors.[37]

In both sub-modern books on Pannonia (Mócsy 1974 and the *ARP*), the Croatian portion of the province was not given much space, particularly so in terms of the archaeological evidence. A. Mócsy specifically remarked on the low rate of research of the Sava-Drava area.[38] In spite of the extensive archaeological research in Hungary after World War II, the editors of the *ARP* still complained that the book's frame remained historical-analytical rather than archaeological. If this was true for the book as a whole, it was even truer of the data pertaining to north-Croatian section, given a low rate of archaeological research at that time.[39] Shortcomings notwithstanding, both books realised their authors' wishes: they remained in use and have still not been replaced by more recent titles of the same profile. Due to a vast scope of themes and an easy-to-survey structure, the *ARP* still figures as a textbook of a kind. On the other hand, Mócsy's in-debt and perceptive analyses have once and for all defined and partly resolved several issues of Pannonian Roman archaeology. Nevertheless, the archaeology-based Roman history of Pannonia needs updating.[40] Admittedly, the debt of Croatian archaeology is in this respect heavier than that of the other countries participating in the heritage of Roman Pannonia. As much as the responsibility for the lack of information on the archaeological evidence from northern Croatia is prevalently its own, it is not entirely so. On the occasion of some conferences or exhibitions occurring after the disintegration of Yugoslavia, Croatian archaeologists were not invited to participate. They were, for example, absent from the conference *La Pannonia e l'Impero Romano,* organised by the Hungarian Academy and the Austrian Cultural Centre in Rome in 1994.[41] Another near-similar example is the exhibition *Religions and cults in Pannonia*, organised by the Hungarian National Cultural Fund and the City of Székesféhervár in 1996, which, fortunately, still ended by including the Croatian material.[42] In the volume accompanying the exhibition, the paper by A. Rendić-Miočević and M. Šegvić covered the cults in the Croatian part of Pannonia, and it was chosen not to tackle this theme in the present volume.[43] Two major exhibitions on the emperor Constantine the Great, one held in Rimini in 2005 and another in Trier in 2007, were organised without including material from Croatia. Such omissions should not be expected for two reasons. Firstly, in both exhibitions the number of participating countries was huge, including those of the Pannonian milieu. Secondly, the early Christian evidence from northern Croatia is accessible to the wider archaeological community.[44] Any exhibition as the two just mentioned should at least include a building inscription from the spa settlement of Aquae Iasae, dated 321-324/6.[45] Besides making clear that at the time Aquae Iasae belonged to the province of Pannonia, it reveals the fact that it was exactly the emperor Constantine who ordered a rebuilding of the settlement, previously destroyed in a fire. Furthermore, the inscription is composed in a way to adumbrate a specific religious atmosphere, resulting from Constantine's wavering between Christianity and the cult of Sol Invictus. As a matter of fact, even in the monographs on Pannonia, Aquae Iasae is mentioned very seldom and cursorily, despite the fact that archaeological excavations had been going on there regularly since the 1950s.[46]

Let us return to the literature on southern Pannonia, produced by Croatian authors. A book on the entire Croatian scientific achievements and cultural heritage, evaluated in a European context, includes three titles on Roman antiquity.[47] Although all three were written to meet high academic standards, they do not surpass the scope of fairly limited overviews, comprising the whole territory of Croatia. This means that Pannonian themes could not be given either enough space or a necessary clarity and coherence. This is even truer of the book on classical antiquity in Croatia. Although it, too, was written by well-esteemed experts in their respective fields of Greek and Roman archaeology in Croatia, Pannonian themes once again remained in the shadow of the splendid Mediterranean monumental heritage, while the book was anyway intended rather for the wider audience than the narrower academic community.[48] Croatian archaeologists contributed to the conference *The Autonomous Towns of Noricum and* Pannonia, held in 1999 in Brdo near Kranj (Slovenia). In two edited volumes (Marjeta Šašel Kos and Peter Scherrer) concerning Pannonia, the following Roman towns in Croatia were covered: Andautonia, Siscia, Aquae

36 Kuntić-Makvić and Šegvić 1992.
37 Migotti 2010, 22-26.
38 Mócsy 1974, 169.
39 *ARP*, 18, 29.
40 The already mentioned title (Visy 2003) is worthwhile, but given its vast scope, the chapters covering Roman archaeology still make an overview and not an in-depth study.
41 Hajnóczi 1994. Absent were also scholars from Slovenia and the former Yugoslavia.
42 Fitz 1998. The Croatian participation had not been envisaged initially, but was effected through the intervention of Ante Rendić-Miočević, the then director of the AMZ (pers. com.). See Rendić-Miočević and Šegvić 1998.
43 Not many important finds or papers on this subject appeared in the meantime. However, several unpublished and one published MA / PhD theses submitted to the Faculty of Philosophy in Zagreb dealt with various aspects of Roman religions and cults in southern Pannonia: military religions and the god Silvanus (Perinić 2003; Perinić Muratović 2008), and Cybele and Attys (Nikoloska 2010).
44 Migotti 1994; Migotti 1997.
45 *CIL* III, 4121; *AIJ*, 210, no. 469; Demo 1994, 110, no. 139. Symptomatically, this inscription was not mentioned in a discussion on the emperor Constantine's attitude towards the financing of public buildings in towns of the Empire (Lewin 2001, 28-30).
46 Mócsy 1974, 144; *ARP*, 165, 167, 264. On the excavation in Varaždinske Toplice see Gorenc and Vikić 1979.
47 Kuntić-Makvić 1999; Suić 1999; Zaninović 1999.
48 Sanader 2007.

Balissae, Iovia, Mursa, and Cibalae.[49] The two volumes is a worthwhile and still indispensable reading, but it does not amount to a comprehensive synthesis of Pannonian Roman archaeology as a whole, but of the province's individual autonomous towns (*coloniae, municipia* and one *civitas*). Besides, the towns in Croatia were discussed on the basis of a fairly scanty evidence, given that the boom of archaeological excavations, relating to the new legislative frameworks in development and the cultural heritage protection started just about the turn of the millennium, ushering in a new era of archaeological research in Croatia.

It would be idealistic to assume that the results of the excavations in the last decade would significantly mark this book. Those were mostly rescue and protective excavations, carried out on many sites by various state and private institutions, including quite a number of archaeologists; the vast amount of the evidence gathered is yet to be systematised, studied and prepared for publication.[50] At this stage it was possible to use only preliminary reports, while some authors were able to use the results of their own excavations (Dizdar, Knezović, Leleković, Rapan Papeša, Lolić and Wiewegh). Finally, it should be strongly pointed out that the evidence from those excavations is most often patchy by its inherent nature, and should hardly be used for drawing final conclusions, even when (or if) it is once fully published; for this we need systematic archaeological research. On balance, this book does not aspire to a synthesis; it is rather an edited collection of papers aimed at a better understanding of south Pannonian Roman archaeology. It certainly suffers from many shortcomings and in some aspects remains incomplete or even not to high academic standards. On the other hand, it is hoped that it does not suffer from incorrect spelling or misplacement of sites, which abound in the literature produced outside Croatia and covering southern Pannonia „in passing".

4. Chosen issues of south Pannonian archaeology

4.1. Ethnic and cultural identities

In the last couple of decades or so the notion of identitie(s) has become one of the most discussed and also the sorest issues of social sciences and cultural studies, archaeology included. The key question is whether it is possible to recognise ethnic communities on the basis of fragmentary material evidence, given a socially constructed and fluent nature of ethnic and other identities. Since the answer should be negative, ethnic identity is preferably viewed as a component of a wider conception of cultural identity, while the assumptions on the role of material evidence still waver. It is necessary to comment on it briefly, as several papers either touch on this issue or have it as the main subject, taking, however, different approaches: Dzino and Domić Kunić, Dizdar, Rapan Papeša, and Radman-Livaja (Anthroponymy). With one exception, the starting point of all papers is that the identification of ethnicity through archaeological evidence is a workable concept, at least theoretically. One paper (Dzino and Domić Kunić) takes a somewhat mitigated post-modern stance, contrary to the former. While accepting the post-modern theoretical approach in relation to the concept of ethnic identity, the authors do not reject the use of the archaeological evidence in rounding the picture of identities, whatever the latter mean. Given that in the paper in question the theoretical issues and the analyses of the historical sources absolutely prevail over the discussion of the archaeological evidence, the definition of the paper given by the authors themselves (*an outline of the different identity-narratives rather than a full and thorough exploration of all available sources*) seems to be accurate.[51] As such, it makes a reading of great academic subtlety and a very useful theoretical frame to be further completed with the material evidence. Secondly, it offers a good possibility to compare theory-based and material-based discussions on the same topic.

In addressing the issue of identity, the prevailing (or, perhaps, just more eagerly advertised) approach today, especially in American and British archaeology, is that post-modern.[52] However, it sometimes suffers from over-generalisations, as it does not seem justified to consider the issue of ethnicity indiscriminately in all periods and in all phases of the Roman Empire and the adjacent areas. An example worth quoting is that of extremely ethnically unstable *Germania magna* in the 1st-2nd centuries, with peoples constantly moving, forming new communities and changing names, and therefore leaving a corresponding imprint on the contemporary written sources: *Given the state of the sources, the most recent archaeological studies have lost all concern to attribute the elements uncovered by the excavations ... to clearly identified individual peoples, preferring to describe the first and second centuries CE in terms of a variety of civilisations and cultures.*[53] Despite this, quite a large number of archaeologists, especially those (continental) European, remain true to the possibility of associating the archaeological evidence (plus that anthropological and historical) with ethnicity; arguably, this is a deliberate choice and not the lack of insight into archaeological theory. This attitude should be acceptable, as long as it shuns a *deterministic automatism*, as Maurizio Buora defined it conveniently.[54] To put it simplistically: not everyone with a Latin name was Roman, nor does a pair of Gothic fibulae in a grave necessary denote a Gothic lady. With such frame of mind as a starting point, it seems possible to generally differentiate between the autochthonous Roman provincial finds from those Barbarian, at least at a general level. It should be much

---

49 *Pannonia* I, II.

50 Preliminary reports have been published regularly in the yearly journal of the Ministry of Culture of the Republic of Croatia: *HAG* 1-6 (2004-2009), 2005-2010.

51 The paper could not fully benefit from the potential of the inscribed stone grave monuments, given a partial state of knowledge of this material and the fact that it has been currently studied by B. Migotti (see also Migotti, in press).

52 On the vast body of literature on this topic, especially Anglo-Saxon, see Dzino and Domić Kunić in this volume.

53 Schuster 2008, 116.

54 Buora 2008a, 80; Buora 2008b, 129.

harder, although not *a priori* impossible, to systematise some of the Barbarian finds according to a more precise ethnic affiliation. Throughout, it is necessary to bear in mind the intermingling of ethnical cultures and new forms springing from them, as well as imitations, trading and deliberate borrowing of *the others*' artefacts; unresolved or even irresolvable issues should not altogether distract from research in this direction. Finally, ethnic and cultural identities should rather be viewed as complementary than opposed, and a research into this phenomenon remains a challenge to archaeological scholarship. The above example of *Magna Germania* is not fully comparable to the circumstances of the area between the Drava and Sava. The picture of the ethnic groups, named *tribes* in the written sources, at the initial stage of the Roman conquest was far from simple. Nevertheless, a bigger picture emerges, featuring the powerful and strongly „Celticised" Taurisci and Scordisci at the far western and eastern ends of the territory, respectively, and a host of smaller Pannonian communities in between (the Iasi, Andizetes, Colapiani, Breuci) that were also touched by the La Tène culture, although less substantially. This picture corresponds with the starting point of research, formulated several decades ago on the basis of the written sources, and still accepted today in its broad frame.[55] However, the research now is entirely directed towards a verification of the written sources through the archaeological record. It should be borne in mind that the research into the archaeological evidence for the initial phase of the Roman occupation of northern Croatia has only just begun. Even so, it managed to reveal the potential for a better understanding of the relationship between the Romans and the Pannonian peoples in the context of the latter's higher or lower rate of „Celtisation" (see Dizdar, in this volume). One of the sore points in this understanding is the low rate of archaeological research into Segestica/Siscia, given that it was the centre of an originally Pannonian ethnic area. Quite a new insight has been gained recently into the subject of the pre-Roman and Roman anthroponymy of the area, based on the research of a large collection of inscribed lead tags from Siscia (see Radman-Livaja, in this volume, notes 40-41). Nevertheless, it has turned out that the onomastic material is, expectedly, still not diagnostic enough (or just not researched enough) of the differences between various components of pre-Roman name systems in southern Pannonia, that is, Celtic, Illyrian, and Pannonian. As far as the archaeological evidence is concerned, the Scordiscan material from the eastern part of the area under study has so far been better researched. In view of that, a Celtic settlement on the site of Jakopovec-Blizna in north-west Croatia, in the hypothesised Tauriscan territory, should receive more attention than has been the case so far.[56] As far as the period from the later 4th to the end of the 6th centuries is concerned, various Barbarian peoples were either passing or were temporarily or permanently settled in the area between the Drava and Sava; to the majority of them at least a broad ethnic-cultural *facies* can be attributed on the basis of the material evidence.[57] Even more important and equally intriguing in terms of cultural anthropology is the fact that in the period from the 4th to the 6th centuries a mixture (a kind of a *koinē*) of late Roman and Barbarian material cultures developed.[58] Arguably, its individual components point to ethnic origins of ultimate cultural affiliations of its bearers.

The term *Romanisation*, which is closely related to the discussion of ethnic identities, has been more and more abandoned in modern theoretical archaeology; it has either been used with quotation marks, or replaced by terms such as *acculturation* or the like. Nevertheless, the word *Romanisation* has been used in this book in all papers but that of Dzino and Domić Kunić, who explain the controversy of the term by quoting a vast body of relevant literature (note 4). They particularly mark the inappropriateness of the terms *assimilation* (as used by A. Mócsy and J. J. Wilkes) and *surface Romanisation* (as employed by D. Rendić-Miočević). In the face of this, I would like to remind of an „old-fashioned" claim of Lengyel and Radan, that the social process which went hand in hand with the step-by-step military conquest of Pannonia should be named *Romanisation*, and, further, that Pannonia never adapted itself completely to the Roman way of life.[59] It seems that modern archaeological theory, again prevalently American and British, anachronistically finds the expressions connected with war and conquest, as well as with ensuing imperial and colonial politics, as politically incorrect and therefore unsuitable in scholarly discourse.[60] Dzino and Domić Kunić are right in their claim that „Romanisation" was much more complicated than this word or its substitutes of the similar meaning can express. But if the older literature is read carefully and without preconceptions, it turns out that, for example, D. Rendić-Miočević was fully aware of the complex nature of this process. The expressions such as *surface Romanisation*, *denationalisation* or various other "negative" connotations of Romanisation, were this author's response to the neglected state of research of the interaction of the two cultures, Roman and native, and the critique of the assertions that in the process the Illyrians lost all their ethnic and cultural attributes. His final conclusion was that Romanisation was a two-way process of taking and giving, resulting in new cosmopolitan forms of material and spiritual culture.[61] Does this not sound very much like that the opponents of the Romanisation-theory would claim? If a process ended as acculturation and a willing or even enthusiastic acceptance of Roman cultural templates, it still does not mean that it did not start as a military conquest and an imperialistic and colonialist imposition of a foreign way of life. On balance, *Romanisation* is as good a term as any modern one for the process in question, as long

55 Lengyel and Radan 1980, 22-23; Šašel Kos 2005, 378. See also in this volume: Dizdar, section 2, and Radman-Livaja, Anthroponymy.
56 Bekić 2006, 114-118; Dizdar, in this volume, n. 10.

57 As transpires from the essay of Rapan Papeša in this volume. See also Tomičić 2000.
58 Fischer, Precht, and Tejral 1999; Tejral 1999, 235; Cracco Ruggini 2008.
59 Lengyel and Radan 1980, 24, 26.
60 Cf. Reece 2000, 270.
61 Rendić-Miočević 1989, 422-437.

as its various facets are implied.[62] Needless to say, the term remains „in vogue" in the same circles (though not exclusively) that accept the concept of ethnic cultures as reflected in archaeological material.

4.2. Written sources versus archaeology

To discuss the controversial and endlessly debated issue of a priority of the written sources over the archaeological material, or vice versa, would be out of place here, as anywhere else for that matter. Although this volume is above all based on the archaeological evidence, the majority of the authors often refer to the written sources. Within such framework, examples of a disparity between the two sources of information are evident, and are especially revealing within the topic of (over-exaggerated) destructions in wars and sieges as posited by written records and more often than not challenged by archaeology.[63] In the present context, a good example is Siscia, the town that according to Zosimus should have been reduced to ruin in an attack of the emperor Magnentius in the mid 4th century (Sardelić, in this volume, section 4.). Differing approaches are also at work in the example of the acceptance or otherwise of the locations of itinerary settlements. While historians will not question common identifications such as, for instance, Slavonski Brod with Marsonia (Domić Kunić, note 44 and section 5.3.), some of the archaeologists point to the lack of material evidence so far for such identification (Lolić and Wiewegh, section 2; Leleković and Rendić, notes 116 and 117; Leleković, section 2.2.). While it is true that the exact location of Marsonia still remains unknown, the fact that two military diplomas[64] come from Slavonski Brod or its nearest vicinity only confirms the fact known from the itineraries, namely that Marsonia *was* located somewhere in the area. Another controversy concerning the provincial appurtenance relates to Emona. In a historical contribution (Domić Kunić, sections 4.2. and 6.5.) this town is considered to be Pannonian, as some literary sources would have it, although the epigraphic evidence testifies to the contrary.[65] As noted in the paper by M. Sardelić (section 1.), archaeological findings and epigraphic materials are essential for any proper use of written sources. This claim sounds right theoretically, and has been an omnipresent maxim of historical research in recent decades. However, the written sources are too scanty and the archaeological material too vast and under-systematised to always integrate satisfactorily, even if it were not for the difference in approach. On balance, the written source remains important for the archaeologist, but neither as taken at face value nor necessarily as an obligation to be accommodated "by force" into the archaeological record.

Given the customary generalisations concerning the province of Pannonia as a whole versus its southern portion, it was necessary to gather all written sources exclusively on southern Pannonia in one place, consciously omitting those lacking a narrower geographic specification. This does not mean that the ancient writings that fail to disclose the narrower area of their concern within Pannonia do not contain information also on its southern portion. Such potential data were to a certain extent willingly sacrificed here in order to get a clearer and more consistent and precise picture of southern Pannonia, as reflected in ancient historiography and as opposed to the northern portion of the province. Admittedly, it was not easy or even possible for the authors to stick to such conception in every single detail. Nevertheless, the idea was in both papers (Domić Kunić, Sardelić) carried out at least in broad lines, with both authors evidently feeling strongly inclined to include more of the unspecific data, especially if the circumstances suggested their probable southern-Pannonian affiliation. For example, knowing the south-Pannonian landscape, and having in mind that St. Ambrose's qualification of Pannonia as the breadbasket of the Empire most probably refers to southern Pannonia (Sardelić, in this volume, note 7), how can one escape thinking that the *terra dives in omnibus* from the *Expositio totius mundi et gentium* also concerns the same area? This is because the anonymous author's Pannonian coordinates given in this quotation are Sirmium, Noricum and the Danube as a border towards the Sarmatians.[66]

The written sources were in both chapters structured in thematic wholes, hoping that these would further help to clarify and better define the Roman-side picture of the studied area.[67] So compiled written material was meant to be made accessible in one place to make it easier for the archaeologists to correlate with the material evidence. Nevertheless, one specific branch of documentary material – Roman itineraries – were not tackled in any extension in the chapters on the written sources, although they are mentioned in passing both in these and in some other chapters. This is because the broad stretches of the main roads, as given in the itineraries, are known even if not proved archaeologically, while the reconstruction of local communications remains tentative and altogether in the realm of archaeology.[68] The two papers on the written sources differ drastically in extensiveness, leaving a misleading impression that the early period is much better covered than Late Antiquity. This, however, is rather due

62 At the level of historical research the major theoretical controversy of today's social sciences and humanities was very conveniently expressed by H. Gračanin (2011, 10), as namely a criticism of the post-modernistic deconstructions of concepts with no alternatives offered.

63 Cf. Dizdar, in this volume, section 3. A discrepancy between the written sources and archaeological evidence has by now become proverbial: Alston 2010, 345; Alessandro Marano 2011, 187; Mirković 2011, 91; Stoyanov 2011.

64 Miškiv 1998, 4.

65 Šašel Kos, 2003; Šašel Kos 2005, 90, n. 23.

66 Drexhage 1983, 343. The obviously incorrect statement of Noricum as a Pannonian town is of no relevance in the present argumentation, all three geographical points being consistent with southern Pannonia.

67 The late Roman period in northern Croatia, based on the written sources and backed by selected archaeological evidence, was discussed as a separate subject in many articles and, finally, in a book by H. Gračanin (2011). The early Imperial period, either as a separate subject or within the wider area of Illyricum (Pannonia and Dalmatia), was discussed in several articles by A. Domić Kunić (see literature in the paper by the same author in this volume).

68 This theme is quite well covered in recent publications in Croatian, although many controversies still remain - cf. Milotić 2010; Gračanin 2011, 29-48. See also in this volume: Leleković and Rendić-Miočević (section 6.), and Lolić and Wiewegh (section 2).

to disparate scholarly personalities of the two authors and different approaches taken; while A. Domić Kunić exercised wordiness and an inclination towards using the sources more freely in terms of the given limitations, M. Sardelić was expressly over-concise. Such approaches also reflect the scope of knowledge and the experience in the subject matters covered by the authors in question. The proportional shortness of the paper by M. Sardelić further results from two facts: firstly, that the history of early Christian Church as the most important segment of south-Pannonian late antique history has already been covered by published materials in foreign languages, and is consequently given only in a concise form here, and secondly, that for the reasons already stated itineraries have not been tackled in any extent. On balance, the two phases of the Roman period in southern Pannonia are fairly equally covered by the written sources, as transpires from the number of quotations given, or, rather, from the number of authors represented, some of those from the earlier period producing more quotations to an individual author – see Sardelić in this volume, section 7. Symptomatically, in the process southern Pannonia transformed from a hostile wilderness of harsh climate, uncultured if brave brutes, and deceitful negotiators (Domić Kunić, sections 2.2. and 4.3.), into a fertile land of trustworthy soldiers and, occasionally, poeticised individuals (Sardelić, sections 5. and 6.).

4.3. The Military

The subject of the military as a component of an archaeological synthesis of any Roman provincial society goes without saying. In the example of Pannonia this is even truer, given the importance that the Danube and Balkan provinces gained in Late Antiquity exactly on account of their armies and the military, who were even named *Illyriciani*.[69] In spite of that, the state of research into the military architecture in the Croatian part of Pannonia is next to nil. However, owing to Mirjana Sanader, a project has been started of gathering relevant data on the *limes* for the inclusion on the UNESCO World Heritage List and of a research of the *limes* architecture and its logistics.[70] In the chapters by Leleković and Rendić-Miočević (section 2.1.) and Knezović (section 2.1.), another so-called *limes* was mentioned, that on the Drava River, but it was neither part of the official defence system of the Roman Empire, nor it features obligatory *limes* constructions in terms of archaeology.[71] Knezović put forward coin finds as a possible corroboration of the presence of army formations in Andautonia, having probably in mind that a military camp was supposed there on the basis of an early presence of Italian civilians.[72] Although the majority of the written reports on the occupation of the Croatian part of Pannonia are related to Siscia, the location of the military architecture in this town remains unresolved. The same is true of other towns that should have hosted military formations, or should even have developed from such architecture, such as Andautonia, Mursa, and Cibalae (see Radman-Livaja in this volume, The Roman army, section 4.). On the other hand, there are graves in some of those towns and the settlements on the Danube *limes* that point to the military (see Leleković in this volume, sections 2.2., 3.1.2., 3.4., 4.4.). On balance, military equipment and items of arms and weapons from northern Croatia are so far much better researched than the architecture.[73] On the other hand, the potential of this evidence still cannot be fully exploited, as it consists mostly of chance finds. Let us illustrate this by the example of small finds from Siscia. On the basis of brooches, R. Koščević claimed that the military component of the town probably prevailed over that civilian, while I. Radman-Livaja established that the military equipment was represented by only some six per cent of the collection of Siscian small finds kept in the AMZ, which possibly contradicts the former opinion.[74] An understanding of this issue is further complicated by the fact that apart from the early period, Roman troops were certainly present in Siscia in Late Antiquity, most probably from the time of Gallienus (Radman-Livaja in this volume: The Roman army, note 221). On the other hand, if the military material stems from an unknown archaeological context, it does not necessarily testify the presence of military installations; it can equally result from the production for the needs of the army.[75] Just this one example is a sufficient indicator of the potential of the material evidence of the military presence in the region for the research into its social history. One line of further research should be directed towards a combined study of small finds and epigraphy, aimed at a better understanding of the interrelationship between the military and the civil society from the early Principate to Late Antiquity.[76]

4.4. The Coinage

Republican and early Imperial coins, either as hoards or as individual finds, were in two papers brought as an archaeological support of the epigraphic or historical data. In this way I. Knezović claimed the existence of an early military camp in Andautonia (section 2.5.), while in early coin finds M. Dizdar (section 2.) saw a corroboration of the battles fought during the occupation of the eastern part of the territory between the Sava and Drava. However, neither of these claims is well founded, for two reasons. Firstly, the Republican and early Imperial coins were in use as late as the 3rd century, and secondly, such coins are customarily found in layers that postdate the time of their minting.[77] Despite the facts that in Later Antiquity the Siscian mint was one of the most important in the Empire, and that the Numismatic Collection of the AMZ keeps more than 40,000

[69] Mócsy 1974, 183-212; Alföldy 1994, 30; Caldwell 2007, 26-36. See also Dzino and Domić Kunić in this volume (ns. 129-137).
[70] Sanader 2003. On the low rate of research of the Croatian portion of the *limes* see Wilkes 2005, 131-132.
[71] Cf. Šašel Kos 2005, 468.
[72] Fitz 2003, 50.

[73] Radman-Livaja 2004; Radman-Livaja 2010.
[74] Koščević1980, 37; Radman-Livaja 2004, 119.
[75] Cf. Jilek and Breeze 2007, 211.
[76] Overall, it is considered that the integration of soldiers into the provincial societies advanced with the passage of time (Gardner 2007, 19, 29; Alston: 2010, 345; Scheidel 2011; Dzino and Domić Kunić in this volume, n. 129).
[77] Mirnik 1981, 52-58; Schachinger 2006, 84, 91; Gardner 2006, 76. See also Bilić in this volume, ns. 195-197, 224, and 293-296.

coins, mostly from northern Croatia and Syrmia, so far no substantial survey of the coin circulation in southern Pannonia has been produced. It is therefore no wonder that in such survey, related to the province as a whole, the evidence from northern Croatia was hardly considered at all.[78] Hopefully, the two numismatic papers (Bilić, Nađ) will make a beginning of a systematic study and publishing of coin circulation, but also of coinage as an indicator of political and economic processes in Roman-period northern Croatia. The importance of these two papers does not result only from the first systematisation of the vast material, but also from the inclusion of some so-far unpublished finds, which improve on the results of other authors' work. For instance, Bilić (ns. 45 and 46) brought 2 Scordiscan coins from Siscia, correcting in this way Dizdar's assumption (note 50) that the westernmost point of such finds was Donja Dolina, which is actually some 80 km to the east of Sisak. Unfortunately, the survey of the money circulation from 193 AD onwards (Nađ, in this volume) has been incomplete, that is, based on chosen sites and not on all available material. The lack of the information on Cibalae and the Požega Valley is a particularly sore point in the present context.[79] Fortunately, the complete evidence, at least that from the AMZ, has currently been revised and studied, and will hopefully be published in the foreseeable future. Given the lack of the paper on the economy of the province, and given that the numismatic chapters did not tackle this issue either, it should at least be noted that Pannonia's "passive commercial balance" was, especially in the 3rd and 4th centuries, compensated by the money pumped into the province through soldiers' payments and state officials' salaries, as well as state building projects.[80] While this is true for the whole of Pannonia, its southern portion was privileged in terms of the circulation of commodities and people on account of two main traffic arteries, those along the Sava and Drava, passing through the area and connecting it to the Danube basin.

4.5. The Marcomannic wars

On the basis of coin finds, Knezović (in this volume, 2.6.3.) suggests that a layer of burning in Varaždinske Toplice possibly indicates a destruction of Aquae Iasae in the Marcomannic wars, the same being tentatively suggested for Ludbreg/Iovia (2.6.1.). This is possible, although the fire could have been started any other way and caused equally by a human or natural factor. Arguably, Pannonia suffered devastations, as transpires from both the historical sources and archaeological evidence. However, generalisations such as a quotation from Mócsy that *the losses suffered by Pannonia were heavy and various* should be avoided,[81] unless accompanied by specific evidence. Nevertheless, it was exactly A. Mócsy who cautioned against a trap set to Pannonian archaeology by the Marcomannic wars: an urge to ascribe to those wars each and every layer of destruction from the end of the 2nd century. In the next decades a number of authors followed in Mócsy's steps, looking for other potential causes of destruction, from chance fires to climatic factors.[82] In the light of proverbial exaggerations in descriptions of all kinds of catastrophes in the historical sources,[83] and because of the abovementioned tendency towards generalisations in the archaeological literature, it is important to note P. Zsidi's research of the architecture of Aquincum. She managed to demonstrate two things: firstly, that not all the town was destroyed, but only some its sections, and secondly, that some building works during the Severi were not rebuilding of destroyed architecture, but renovations and adaptations of standing structures.[84] This example clearly shows that a great caution is necessary in establishing exactly which areas and settlements of the province were indeed affected. Here certainly belong the whole stretch of the Danube *limes* and part of the province of Noricum; the latter was exposed during the Barbarian incursion towards north Italy.[85] On the basis of coin finds, P. Kos concluded that the southeastern (Slovenian) part of Noricum did not suffer from Barbarian incursions.[86] In the light of that, the destruction of Aquae Iasae in the Marcomannic wars seems less likely. On the other hand, Mursa and Cibalae, situated in the immediate hinterland of the *limes*, seem to have been more exposed to attacks by the Iazyges and Sarmatians.[87] Burnt layers from the end of the 2nd century in Mursa have customarily been connected with the Marcomannic wars.[88] Significantly,

[78] Mócsy 1974, 208; Bíró-Sey 1980, 344-348; Dukat and Mirnik 2004, 17. Only the coinage of the Germanic rulers was published extensively (Demo 1981 and Demo 1994 – quoted in Rapan Papeša, in this volume). Attention has also been given to coin circulations as reflected in hoards (Mirnik 1981), as well as to Byzantine coinage (Mirnik 1996; Göricke-Lukić 1998).

[79] On the Požega Valley see n. 150. An abundance of late Roman coins from Cibalae is mentioned in Radman-Livaja in this volume (The Roman Army, n. 229).

[80] Fitz 1980, 331.

[81] Mócsy 1974, 194.

[82] The following quotations do not concern only the Macomannic wars, but also various disturbances during the 3rd century: Mócsy 1974, 194; Genser 1994, 270; Zsidi 2002, 133; Kandler *et al.* 2004, 18; Wilkes 2005, 170; Bratož 2007a, 339; Knight 2007, 14-15; Hekster 2008, 32; Scheidel 2008. Extensively on the Marcomannic wars see Stuppner and Tejral 1994. Significantly, contradictions between the written sources and archaeological evidence are particularly revealing in reports on war destructions: Stoyanov 2011.

[83] Cf. Müller 2003, 289.

[84] Zsidi 2002. A similar situation was noticed at one site in Sisak (Škrgulja 2009, 287), while such possibility also transpires from Lolić and Wiewegh's observation (in this volume, section 9) that some of the Severan architecture in Siscia was built upon the ruins of buildings from the Flavian period.

[85] Mócsy 1974, 183-202; Haider 1994; Lőrinz 1994; Pollak 1994; Scherrer 1994; Wilkes 2005, 139.

[86] P. Kos, quoted in T. Bilić (in this volume, n. 302). The majority of authors see coin hoards as indicators of pending wars (Mócsy 1974, 205; Bíró-Sey 1980, 341; Fitz 1982, 61; Caldwell 2007, 239, *passim*; Varbanov 2007; Hekster 2008, 33). Some commentators, however, seek other reasons behind the hoards (Mócsy 1974, 206; Witschel 2004, 258; Reece 2006, 76-77; Schachinger 2006, 62). J. J. Wilkes (2005, 175) cautions that site-finds are even more important in the reconstruction of historical processes and events than are hoards. Nevertheless, if we bear in mind that during the troubled times of the later 2nd century even far-away Salona on the Dalmatian coast strengthened its walls (Haider 1994, 27), the idea of hoards indicating the pending Marcomannic incursions in the whole of Pannonia seems acceptable.

[87] See Bilić (in this volume, n. 303, and section 5.).

[88] Pinterović 1978, 59-60; Sunko, Katavić, and Jerončić 2008, 25; Vodička Miholjek 2008, 33-34. The same layer was noted by T. Leleković (2009, 47). On the basis of burnt *terra sigillata* D. Gabler (1994, 357) counted Mursa among the towns that suffered destruction.

it has been supposed that Siscia got town walls only during the Severi, while Mursa's southern town walls, the only so far archaeologically attested, seem to predate the Marcomannic wars (see Lolić and Wiewegh in this volume, 6.2.1.). Some destruction has been postulated for Cibalae, although with no specific archaeological evidence adduced.[89] While Cibalae probably did not suffer heavier damages, the destructions in Mursa should not be conceived of as catastrophic either; having in mind the above example of Aquincum, only its sections were probably pulled down. This should be a good case in point against ascribing each 2nd-century layer of destruction to the Marcomannic wars. Future research into this issue in northern Croatia should therefore rely on the data on architecture, sourced from recent rescue excavations, given that the evidence from the coin hoards is inconclusive.[90] At a purely theoretical level, suspicion is allowed as to the possibility of such quick recovery of Pannonia in the period immediately following the wars, during the Severi, had the province been so systematically destroyed.

4.6. The Severan period

The only non-commissioned paper and the only one dealing specifically with the Severan period is that of Z. Mráv, whose author was at liberty to choose the subject matter, as long as it touched on southern Pannonia. As a mater of fact, the four Severan inscriptions discussed in this paper stem from both northern and southern Pannonia, adducing some lesser-known but important data in the light of new interpretations. Firstly, it is the fact that the scope of the Severan imperial building in Pannonia was comparable to that in the Severi's north-African homeland. Generally, Pannonian Roman scholarship acknowledges the fact that the rule of the Severan House was beneficial for Pannonia, referring to this period as *the great age*.[91] Nevertheless, until recently some misconceptions persisted in the literature on various subjects. For example, the notion of the Marcomannic wars as bringing the "end of the world" used to hinder a correct understanding of the Norico-Pannonian sculptural art of the period. The theory that the extinction of the Antonine House brought about a significant reduction and a final deterioration of sculptural art has only recently been abandoned in favour of a definition of the Norico-Pannonian artistic and cultural circle of the Antonine-Severan period as an integral phenomenon.[92] By and large, the idea of a prosperous Severan era in Pannonia is a common knowledge among the scholars from the area, but less so in the wider circle of commentators. In the recent book on the Severan culture,[93] in the chapter dedicated to the urbanism and architecture of the western provinces (pp. 290-326), the provinces of Pannonia and Moesia were given 1-odd page (pp. 322-323). In Introduction, S. Swain postulated that the subject matter of urbanism and architecture in the northern and western provinces was *unrewarding for historians*, commenting the paper on Severan urbanism by A. Wilson in the same volume, in which Pannonia and Moesia together were given a total of one page.[94] If Swain's claim is true of historians, it is not of the archaeologists of the province of Pannonia. It is commonly held that during the Severans the towns on the Danube *limes* flourished most, due to their previous destruction in the Marcomannic wars. In terms of archaeology, this is evidenced by an increased rate of all kinds of public and private constructions (residential, infrastructural, religious, funerary, etc.), accompanied by a rise in the number of inscriptions.[95] However, the abovementioned example of the Severan building in Aquincum for reasons other than previous destructions brings attention to the need of such research of inner-Pannonian towns, those in the Drava-Sava area included, as two of them (Mursa and Cibalae) actually belonged to the *limes* area in the wider sense. The Severan imperial inscriptions from Cibalae further support the idea of different reasons for building, as the town was probably not destroyed in the Marcomannic wars. Otherwise, the two inscriptions from Cibalae, as interpreted by Mráv, bring to mind a doubt (unjustified?) on the origin of the town's colonial rank: despite the title *Aurelia* that speaks against Septimius Severus as the founder of the *colonia Cibalae*, the latter idea has still been suggested.[96]

A. Mócsy ingeniously explained a specific relationship between the number of inscriptions and prosperity of a region or a province, postulating that it was not a mere number of inscriptions that testified well-being, as the *epigraphic culture* was circumstantial. He exemplified this by the Severan epigraphy in Pannonia, claiming that the prosperity of the province transpired not through the total number of inscriptions, but through an increased number of votive and building inscriptions on account of funerary.[97] Regrettably, no such research on the epigraphic material has been done in northern Croatia. Nevertheless, the archaeological evidence seems to corroborate some devastation in Mursa during the Marcomannic wars, while its flourishing during the Severans has been suggested by recent excavations, which revealed luxurious residential constructions.[98] A town that should have benefited from the Severan rule is Siscia, given its doubled colonial

[89] Iskra-Janošić 2004, 170-178. No burnt layer from the period has been noticed in Vinkovci, either in earlier or in recent rescue excavations (Hrvoje Vulić, pers. com.).

[90] As was already mentioned in n. 85, the hoards from the western part of the area of study are preferably not taken as proof for the incursions, contrary to the eastern part. However, only four hoards, dated to the periods of the reign of Antoninus Pius and Marcus Aurelius, stem from Mursa and its wider area (Bilić in this volume, section 3).

[91] Mócsy 1974; Barkóczi 1980, 103; Fitz 1982; Alföldy 1994, 29. See also Wilkes 2005, 125, 140; Caldwell 2007, 25; Egri 2007, 104.

[92] Kremer 2001, 17-18; Newby 2007, 204.

[93] Swain, Harrison, and Elsner 2007.

[94] Swain, Harrison, and Elsner 2007, 15-16; Wilson 2007, 322-323.

[95] Barkóczi 1973, 108; Mócsy 1974, 217-24, particularly 236; Poulter 1996, 112. On the advantages of the border position in the 3rd century see Corbier 2008, 424.

[96] The idea was put forward by S. Dimitrijević (1979, 233), but was recently mentioned as possible by M. Nagy (2003, 223).

[97] Mócsy 1974, 230-236.

[98] Leleković 2009, 47-49. According to T. Leleković (2011, 59, 202), Mursa definitely prospered in the Severan period, although the evidence is not clear enough on the point whether this was accompanied by a territorial

title (*Flavia* and *Septimia*). When Mócsy postulated that Siscia was among rare towns that prospered in the time of the Severi, he was not able to support his claim by the archaeological evidence.[99] Recently, archaeological excavations have revealed architecture from the early 3rd century in several places, including the town walls.[100] Severan constructions in Cibalae are well illustrated by imperial inscriptions (see Mráv in this volume). Earlier, the architectural development and even an expansion of the town territory have been postulated for the period in question, although this suggestion was not accompanied by adequate material evidence.[101] Once systematised, the results of rescue excavations in the last two decades should provide more substantial material proofs for the Severan Cibalae. As far as the town walls of the major north-Croatian Roman towns are concerned, some data is available only for Siscia, Mursa, and Cibalae. While the perimeter of Siscia, as known from earlier publications, seems to be accurate in broad lines, those of Mursa and Cibalae have recently been challenged, especially the former.[102] The perimeter of Cibalae, as known from the earlier literature, has been accepted in this volume as well (Knezović, sections 2.3. and 4; Lolić and Wiewegh, section 6.2.1.), but this opinion either is not unanimously shared.[103] Similar controversial points relating to architecture can be found in various authors in this volume. For example, while Knezović (section 2.5.) speaks about sections of the Andautonian fortifications, Lolić and Wiewegh (section 6.2.1.) claim their complete absence in this town. The same authors disagree on the main building materials: while Lolić and Wiewegh (section 10.) hold that brick prevailed in south-Pannonian building, Knezović (section 4.) asserts that it is only true of the eastern part of the area of study, while stone was more commonplace in Andautonia, Iovia, and Aquae Iasae, with Siscia mostly featuring a combination of brick and stone.

Let us return for a moment to the paper on the Severan imperial inscriptions. Apart from supporting the otherwise well-known data by new evidence, and revealing some so-far unrecognised facts about Aquincum, Cibalae, and Brigetio, this paper seriously challenges several archaeological stereotypes. By dating three out of four inscriptions to the period before the year 202, it has shown that not all Severan imperial buildings should be connected with the emperor Septimius' journey through Pannonia in 202.[104] Also, it transpires from Mráv's paper that the imperial benevolence was not the privilege of border towns, but was equally directed to the whole of the province. Cibalae is one example of this, as it was not the border town in the narrow meaning of the word. Another example is Siscia, although so far it is mostly its title *Septimia* that speaks in favour of a special imperial attitude towards this town. Finally, quite a few inscriptions from Aquae Balissae should be a sign of gratitude for some beneficence done, as their number by far surpasses the towns' obligations in honouring members of the imperial family. One such inscription was set up in Rome by eight praetorians (three from Cibalae and five from Mursa) in honour of Alexander Severus and his mother.[105] On balance, such inscriptions further suggest that the imperial building was not only spurred by previous destructions, but, among others, by a gratitude for loyalties in winning the throne, as was rightly brought to the fore in Mráv's paper.

Imperial milestones should also be considered as the evidence for the imperial participation in the province's well being. Three or four such finds stem from the area of study, one from Siscia and two (or three) from Mursa.[106] Apart from the architecture, stone grave monuments and inscriptions, simple in-ground depositions are an important demographic indicator. It should be kept in mind that *a lack of building need not necessarily be equated with «stagnation», and that it may rather be symptomatic of the very distinctive rhythms of late-antique urbanism.*[107] This is very true of the under-researched towns in the area discussed here. Recent research into the cemeteries of Siscia and Mursa has shown the lack of a major Severan layer in both towns, although our knowledge on Mursa in this respect has been much improved recently (Leleković, in this volume, section 3.2.). However, this data cannot be finally substantiated (or invalidated) by the evidence collected previously, as quite a lot of it is chance finds, or they stem from old excavations with graves not precisely dated.[108] On the other hand, quite a few among the funerary stones from Siscia, mostly chance finds, date from the Severan period.[109] The fact that at the same time the number of contemporary earth-pit graves is scanty, or rather, hardly distinguishable from 4th-century burials, only reveals a precarious nature of the scrappy archaeological record at this stage of research

expansion in comparison with the Antonine times, or just in terms of the scope of building and population growth. Further substantiation of the flourish of Mursa in the given period is expected when more of the results from the major archaeological excavations in Osijek in the last decade are fully published. On the restoration of the presumed *proseucha* in Mursa in the Severan period see Mócsy 1974, 228.

99 Mócsy 1974, 217. An expansion of the town in the Severan period was postulated by M. Hoti (1992,145), with Lolić and Wiewegh (in this volume, section 9) discussing the town's flourishing along the same lines, but also claiming that it took place within the urban frame set in the previous period.

100 Lolić and Burmaz 2006, 176; Lolić and Wiewegh in this volume, 6.2.1.

101 Iskra-Janošić 2004, 180. Additional proofs in this direction should be gained by the results of rescue excavations in the last two decades, once they are systematised.

102 For all three towns see Lolić and Wiewegh, in this volume, 6.2.1. For the eastern walls of Mursa see Göricke-Lukić 2011, 187 and Leleković 2011, 63-77.

103 According to Hrvoje Vulić of the GMVi (pers. com.), it seems that neither the eastern stretch of the perimeter nor the nature of the defensive structures (ditches?, earthen walls?, streams?) can be taken as finally proved; the same goes for the presumed double stretch of the Cibalitan fortifications.

104 It used to be an "archaeological fashion" to interpret various constructions with emperors' visits (Gardner 2007, 248).

105 *CIL* VI 2883.

106 Siscia: *CIL* III 15199 (=*AIJ* 608); Mursa: *CIL* III 6469 and 3830 – see also Pinterović 1978, 71.

107 Keay 2004, 743. The above quotation is relative to a site in Spain, but it is actually universal. Significantly, 4th-century Sopianae has yielded a very limited evidence for the architecture, while cemeteries abound (Gábor *et al.* 2004, 287).

108 Göricke-Lukić 2011, 7, *passim*; Leleković, in this volume, section 3.2.

109 Migotti 2007a, nos. II.1.1., II.2.1,, II.3.2.-4., II.5.1., Migotti, in press, no. 30.

of south-Pannonian towns. At least as far as Siscia is concerned, the imbalance between the architecture and cemeteries seems to be haphazard. According to Leleković (sections 3.2., 3.3., and 3.4.), the lack of Severan-period burials is true of the whole of northern-Croatian territory, apart from Mursa, with some increase in the overall number of both inhumations and incinerations coming recently from the rural areas and the Danube *limes*.

4.7. "The 3rd century crisis"

As the divided opinions on the so-called 3rd-century crisis demonstrate, this phenomenon can hardly be defined individually in various provinces, let alone in the Empire as a whole. The period in question (235-284) is sometimes called *the period of the military emperors*, and is unanimously considered as a decline in relation to the Severan era. However, the questions how hard was the crisis and even whether it was a crisis at all remain open to debate, although the recent tendency is towards a more optimistic view.[110] In any case, it transpires that the only sensible approach to this phenomenon is through the archaeological evidence from individual regions or provinces.[111] The same applies to the use of the 2nd- and 3rd-century plague in the context of demography. The commentators agree that although the plague hit seriously, neither its scale nor its "itineraries" are sufficiently known.[112] It is hardly likely that Pannonia was not affected (the emperor Claudius Gothicus died of plague in Sirmium), but the written sources fail to mention this province among the areas that suffered most, like Italy and Rome, Gaul or Germany.[113] Had Pannonia as a whole been so hard hit by the so-called Antonine plague, the Severan period could have hardly amounted to a *great age* (see above). Opinions on the *3rd-century crisis* in Pannonia specifically are as discrepant as are those on the Empire as a whole.[114] Irrespective of the assessment of the nature of the *crisis*, the authors customarily considered Pannonia as a whole in this respect. Needless to say, the data from the Croatian territory was almost entirely missing, with the exception of the Siscian mint, started by the emperor Gallienus. In the present context a crucial issue is a possibility that the life conditions in the second part of the 3rd century were different on the *limes* from those in the interior of the province. War disturbances in south Pannonia notwithstanding, the prosperity of its outlying towns (Poetovio, Neviodunum, Sirmium) indirectly proves that this period did not mean a huge regression in relation to the Severan era.[115] The following features should be adduced as more direct proofs of the 2nd half of the 3rd century as anything but a catastrophic crisis: the opening of the mint in Siscia and the lack of archaeological proofs for its destruction in the given period; an increase in the number of graves in Mursa, as well as the existence of funerary stones and sarcophagi in the whole territory; honorary inscriptions to emperors, set up by Adautonia, Aquae Balissae, and Cibalae.[116] According to P. Kovács, the majority of the Pannonian milestones from 260 until the time of Constantine the Great stem from the *limes*.[117] The fact that as many as five milestones come from northern Croatia (three from the territory of Andautonia and two each from those of Siscia and Mursa) renders the above claim somewhat questionable.[118] The architecture (or the lack of it) of this period has remained a moot point. However, an example from Mursa possibly demonstrates a misleading nature of the patchy archaeological record: at one place outside the town walls a sumptuous residential architecture gave place to a large cemetery in the 2nd half of the 3rd century.[119] Should this be interpreted as a demographic rise or fall? If, on the other hand, the example of Siscia is considered, once again we are facing the discrepancy between the (small) number of graves excavated and quite a few funerary stones, mostly chance finds.[120] On balance, it is not possible to postulate firmly that in the 2nd half of the 3rd century southern Pannonia was in a better position to prosper than was the northern portion of the province. For instance, the flourishing of the villa settlements in the 2nd half of the 3rd century was postulated for the whole of Pannonia.[121] Even though, it is possible that the circumstances in the Croatian portion of Pannonia Inferior were less favourable than in Pannonia Superior, for two reasons: a lesser exposure of the latter to Gothic incursions and the position of Siscia, which must have benefited from the previous generosity of the Severans and the mint established in the town by Gallienus.

4.8. Late Antiquity

4.8.1. The 4th century

Arguably, the Danube provinces, Pannonia included, gained in importance in the 3rd and 4th centuries. In the light of this, the historian and archaeologist is faced with an unrewarding task of alleviating the dichotomy between the notions, both attached to the period in question, of the political-military crisis and decline on the one hand and importance and

110 Mócsy 1974, 183-212; Witschel 2004, 254; Wilkes 2005, 160; Knight 2007, 8-21; S. Swain, in: Swain, Harrison, and Elsner 2007, 16; Hekster 2008; Vingo 2010, 261.

111 Corbier 2008, 437.

112 Bowmann 2008, 324; Corbier 2008, 398.

113 Drinkwater 2008, 50 (on Claudius Gothicus' death); Frier 2000, 815 (on the spread of the Antonine plague). Occasional recurrences of plague in the 3rd century are mostly related to Rome and the eastern provinces: Campbell 2008, 111; Drinkwater 2008, 40, *passim*.

114 Mócsy 1974, 208, 263-265; Barkóczi 1973, 108-110; Barkóczi 1980, 109; Fitz 1982, 53-62. P. Kovács (2004, 185) defined the period 260-582 as marked by the decline of inscriptions, which at that time are to be found only along the *limes* and in some major interior towns.

115 On Poetovio and Neviodunum see Bratož 2007a, 339; on Sirmium see Mócsy 1974, 202, *passim*.

116 On the Siscian mint see Mócsy 1974, 208, 343 and Nađ (in this volume, section 2.), on the cemeteries of Mursa: Leleković 2011, on stelae and sarcophagi: Migotti 2007a and Migotti (in press), on the inscriptions in honour of emperors: *AIJ*, nos. 477 and 478 (Andautonia); *AIJ* no. 586 (Aquae Balissae) and Iskra-Janošić 2001, 63 (Cibalae).

117 Kovács 2004, 190.

118 Andautonia: *AIJ*, no. 602; Siscia: *AIJ*, nos. 604 and 605 (= *CIL* III 15203 and 15201); Mursa: *CIL* III 6465 – see also Pinterović 1978, 71. Two milestones from the territory of Andautonia (one each of Philip the Arab and Decius) were recently found at the outskirts of Zagreb (unpublished).

119 Leleković 2009, 49; Leleković, in this volume, section 3.2.

120 Migotti 2007, nos. II.2.2., II.3.1., II.3.5; Migotti, in press, nos. 31, 33, 34. See also Leleković in this volume, section 3.2.

121 Gábor 2005, 419.

prosperity on the other.[122] However, the late antique period is not discussed here as a separate subject. Such chapter would probably make more sense if combined with the discussion of early Christianity, which has been omitted because of its adequate coverage by publications in English elsewhere.[123] Therefore, topics of Late Antiquity (apart from early Christianity) are interspersed among various chapters, but will be commented in a separate section in this paper. First of all, some theoretical facts should be brought to mind. Late Antiquity as a historical and civilisational turning point is in historical and archaeological scholarship connected with the Middle Ages, be it in terms of continuities or discontinuities, rather than with the Roman Imperial period.[124] However, another approach should also be taken into account: *The excessive specialization by which scholars have tended to concentrate either on the Early or on the Late Empire has led them to exaggerate the contrast between the two periods at the expense of the continuities, and, in consequence, to stress unduly the links between Late Antiquity and the Middle ages.*[125] This quotation makes a good introduction to the discussion of the 4th century as a crucial period of Late Antiquity in Pannonia, as it turns to be firmly rooted in the 3rd century. Regrettably, the material of the second half of the 3rd century remains the least explored in terms of its differentiation from the 4th-century finds (see section 4.7.) Political and military calamities notwithstanding, the 4th century was also the time of prosperity for some segments of Pannonian society. The commentators generally agree that prosperous were the towns along the *limes* and some of those in the interior, that is, western and southern Pannonia. A certain disagreement concerns the question of which exactly, apart from Sirmium, were those possibly thriving towns, with Savaria, Scarbantia, Neviodunum, Gorsium/Herculia, and Sopianae mentioned as candidates for this role. If Siscia is noted at all in this context, it is rarely and hesitatingly, and with practically no support in the archaeological record; yet the town was a provincial capital and no obvious reasons existed for its decline at that time.[126] Dzino and Domić Kunić (in this volume, note 128) point to Siscia and Mursa among flourishing towns in the 4th century. The circumstances changed later in the century, when even some towns in the interior suffered destructions in war clashes.[127] The same should be true of the towns in northern Croatia (Andautonia, Siscia, Cibalae, and Mursa), situated along the main traffic route, though perhaps less so of Aquae Balissae, given its more protected position in the interior. A newly found treasure in Vinkovci, containing precious silver gold-plated items of tableware, dated preliminary to the end of the 4th century, possibly indicates some such war threat.[128] The numismatic evidence suggests that in the later 4th century the emperor Valentinian I's reconstruction of the buildings along the *limes* brought recovery to the border towns, as well as to some in the interior.[129] The picture obtained from the epigraphic evidence is equally evasive, which is only expected, as it cannot in itself be more than just one indicator of prosperity or the lack of it. For the entire period 260-582, a decline of the *epigraphic culture* was claimed, within which only the border towns and few of those in the interior yielded more inscriptions.[130] If grouped in narrower chronological phases, the epigraphic evidence would probably be able to give more detailed information on the dynamics of life in the province in the above-mentioned period. Nevertheless, it is significant in this context that from the time of Constantine I the majority of the milestones, which had previously abounded along the *limes*, now became concentrated on the two imperial roads in southern Pannonia, which followed the courses of the Rivers Sava and Drava.[131]

Some data on south-Pannonian towns as recorded in the written sources should be viewed against the archaeological evidence, despite the latter's meagreness. In this way we are informed that during the pending war with the emperor Theodosius I in 388, the usurper Magnus Maximus wanted to restore the once powerful Siscian walls, and that in the same war Theodosius won the granaries of Siscia, but lost the mint to Magnus Maximus.[132] Rescue archaeological excavations in Sisak so far demonstrated time and again that at several places the architecture was built, destroyed, rebuilt and adapted, but they never proved a consistent destruction level belonging to one clearly defined period. Theoretically, some of the destruction spotted could be tentatively connected with the historical data from the later 4th century, but even more significant is the fact that in the 4th century the Siscian walls were indeed repaired, and some of the destroyed residential architecture was adapted for further use. The late Roman phase of the *horreum* abutting on the southern town walls can also be tentatively connected with the above-mentioned written data; significantly, the 4th-century *horreum* was restructured to receive a new, or a more complex (?) function.[133] About Mursa we learn that

[122] Fitz 1982, 64; Alföldy 1994, 38-39; Caldwell 2007, iii, *passim*; Migotti 2008, 85, n. 186. Curiously, in Roman Britain the 4th century has also been perceived as both "a golden age" and a time of "decline and fall": Gardner 2007, 20.

[123] Jarak 1994; Migotti 1994; Migotti 1997; Buzov 2010. As far as my knowledge goes, there is only one new early Christian find since the 1990s: a fragmentary stone slab with a Chi-Rho, recently recovered at the site of Kamenica near Vinkovci (pers.com. Hrvoje Vulić). On Kamenica as a potentially important Christian site see Migotti 1997, 22.

[124] Cameron 2003; Gardner 2007, 26; Heinrich-Tamáska 2011.

[125] Lepelley 1996, 66; for the contrary opinion in the same volume see Poulter (1996, 101), who claims that the 4th-century Danube town was much different from that of the 1st to 3rd centuries. On the other hand, A. Gardner (2007, 244) stresses that many significant elements of 4th-century practices were prefigured in the earlier period.

[126] Mócsy 1974, 267, 297-321, especially 310-311; Barkóczi 1980, 114; Poulter 1996, 100-120; Šašel Kos 1996, 161-163; Witschel 2006, 224, n. 7; Caldwell 2007.

[127] Šašel Kos 1996, 162.

[128] This quite sensational find occurred in protective archaeological excavations in March of 2012.

[129] C. H. Caldwell (2007, 175) does not specify those towns. However, given the emperor's Cibalitan descent, constructions in Pannonia Secunda should be supposed. On Valentinan's role in the reconstruction of the *limes* and the flourish of the towns see also Mócsy 1974, 291-293, and Šašel Kos 1996, 154-165.

[130] Kovács 2004, 185.

[131] Kovács 2004, 190.

[132] Caldwell 2007, 211-212 (walls), 219, 223 (granaries).

[133] On the walls and *horreum* see Lolić and Wiewegh, in this volume (6.2.5). The term *horreum* has been used throughout, to compensate for the fact that we cannot say with certainty whether the building was a granary or a storehouse for various commodities. On the reconstruction and adaptation of the residential architecture see Škrgulja 2009.

the brush and trees covered the area previously cleared for building.[134] Admittedly, this clearing occurred in the aftermath of some of the destructions Mursa had sustained from the later 2nd to the end of the 4th centuries; recent excavations in Osijek indeed revealed some indications for the circumstances as described in the written source.[135] On the other hand, archaeology has not yet managed to assert several very important data from the 5th-century historian Zosimus: on the town walls of Mursa and two buildings (a *basilica martyrum* and a stadium or amphitheatre) situated outside these walls.[136] Although the destruction of Aquae Iasae in the war between Theodosius and Magnus Maximus has not been recorded in the sources, it seems to have been substantiated the archaeological evidence.[137] However, the scope of 4th-century destructions in the whole area under research has not yet been established by archaeological excavations; the evidence is too mutilated and fragmentary, and, above all, not yet fully studied. On the other hand, even preliminary reports testify to another aspect of the life in 4th-century southern Pannonia: neither towns nor the country experienced a serious decline in the intensity of settlement. Sites from the 1st and 2nd centuries with no traces of further existence are rare, with the exception of cemeteries, while the reverse (later settlements with scanty or even no remains of earlier phases) is commonplace.[138] This is even more significant in the light of the fact that late antique layers, if not topped by medieval ones, which is mostly the case in our territory, are the first to be destroyed by human and natural factors.[139] This, in its turn, accounts for the fact that the most numerous among late Roman evidences are graves and small finds; the latter were most often recovered from sunken features, such as refuse pits and the like. The architecture is mostly found either in its lowest level or in foundations, or even in foundation "negatives", while late Roman adaptations cannot always be clearly discerned. This picture is, however, not so straightforward or uniform, and has its various facets, especially as concerns urban archaeology with its multiple stratigraphic layers reaching to the present day. While, for instance, late antique architecture sustained most damage in Mursa, Severan-period and 4th-century buildings are the best preserved in Siscia and Iovia (in this volume: Knezović, sections 2.2. and 2.6.1; Lolić and Wiewegh, section 9.), let alone Aquae Iasae. The last-named town is a revealing example of what late Roman urban architectural *facies* might have been, as the town was not overlaid by later architecture, and its 4th-century layer is the best preserved (in this volume: Knezović, section 2.6.3; Lolić and Wiewegh, sections 6.2.3. and 6.2.4.). As has been already mentioned, Lolić and Wiewegh, as well as Knezović, suggest that the best preserved layers in Siscia are those Severan and late Roman. Notwithstanding this, several sites inside Sisak, excavated in the last decade, produced only "negatives" of late Roman buildings, which were destroyed by later development and especially "brick hunting"; only those of late antique structures survive (in some measure) that were not overlaid by later architecture. On balance, a dearth of late antique burials in Siscia, Iovia, and Aquae Iasae cannot be but accidental.[140] Although in recent excavations layers of the 4th-century destruction were isolated in this town and in some others (Andautonia, Siscia, Mursa, Cibalae), this fact is not in itself sufficient to judge the accuracy of the statement that both Pannoniae Savia and Secunda were paralysed by the war between Theodosius and Magnus Maximus.[141] Firstly, the cause(s) of the destructions cannot be identified with certainty. Secondly, and even more important, destruction should not be perceived as total (with, perhaps, the exception of Aquae Iasae) and was everywhere followed by reconstruction.[142] Consequently, it is more realistic to speak about a decline in the quality of life than its overall intensity. This statement should not necessarily be in contradiction with the assertion that 4th-century architectural remains are scanty (Lolić and Wiewegh in this volume: section 10.), since the authors claim that the 4th-century evidence is still present throughout the territory in question. Finally, the importance of southern Pannonia in the given period is testified by a conspicuous number of imperial inscriptions (both building and road) from the beginning to the end of the 4th century, found in several towns or in their territories, mostly Mursa and Siscia.[143]

Although the research of Roman cemeteries in northern Croatia has been much improved recently, it still lags behind that in Hungary.[144] Curiously, the proportion between

[134] Caldwell 2007, 154. On the destruction of Mursa in the later 4th century see Bratož 2011, 109.

[135] At one place outside the town the abandoned and decayed Severan residential architecture was in the 2nd half of the 3rd century overlaid by a cemetery (Leleković 2009, 49; Leleković, in this volume, section 3.2.).

[136] Bulat 1989. See also in this volume: Lolić and Wiewegh, section 6.2.6.; Leleković, section 4.2., and Sardelić, section 5. For semantic ambiguities in the description of this building, which led to its various naming as a *stadium* or *amphitheatre* in the archaeological literature, see the quotation of Zos. 2.49-50 in Sardelić, in this volume.

[137] Knezović, in this volume (2.6.3.); Mócsy 1974, 345. It has been suggested that the inhabitants had left Aquae Iasae in advance of the war, as the destruction layer failed to yield small finds. A similar situation has been recorded, although not explained, in Brigetio (L. Borhy, in: Visy 2003c, 230).

[138] On early cemeteries see Leleković, in this volume, section 2. Even if those cemeteries belonged to the settlements extinguished or abandoned in the 3rd century, this cannot be considered as a definite proof of a decrease in the intensity of life in Late Antiquity. Overall, the picture of the Roman-period settlement of northern Croatia has not been fully studied and all reconstructions remain tentative.

[139] Steinklauber 2002, 85; Ottományi 2995, 365; Filipović and Katavić 2006, 10; Scherrer 2011, 112. A dearth of a more modest urban and rural residential architecture in 3rd- and 4th-century northern Pannonia was by A. Mócsy (1974, 317) ascribed to dry-wall buildings without foundations; see also Gabler 2003, 242-243 and Müller 2003, 290. Such architecture has not been recovered on northern-Croatian sites.

[140] No cemetery of Aquae Iasae has so far been discovered, while late Roman burials in Iovia and Siscia are scanty: Leleković, in this volume, section 4.2.

[141] Barkóczi 1980, 117-118.

[142] Kušan Špalj and Nemeth-Ehrlich 2008, 201; Sunko, Katančić, and Jerončić 2008, 25; Škrgulja 2009, 288; Vulić, Rapan Papeša, and Krznarić Škrivanko 2009. A comparison with some sites in Roman Britain is useful, as excavations there have disclosed examples of certain parts of settlements going into disuse, with others being rebuilt at the same time (Gardner 2007, 170, *passim*).

[143] Siscia: *AIJ* nos. 564, 606, 607, and 608 (= *CIL* III 10851, 15202, 15200, and 15199); Mursa *CIL* III 10648 – see also *ILJug*, no. 1072 and Pinterović 1978, 89, pl. 19: 1, 90, pl. 19: 2); Aquae Balissae or Iovia: *AIJ*, no. 582 (= *CIL* III 13392).

[144] Until nearly a decade ago, the total number of researched graves in Hungary already surpassed 10,000 (Zsidi 2003, 251), while the number of burials as listed in Leleković in this volume, section 1.) is 2,265. This

inhumations and incinerations is evidently discrepant between northern and southern Pannonia, as the number of 3rd- and 4th- century graves in the Hungarian portion of Pannonia, as noted in 1996, was about twenty times bigger than those from the 1st and the 2nd centuries, while in Croatia this ratio is about 1.5 times in favour of inhumations.[145] In view of all we know about Pannonian Late Antiquity at this stage, the ratio relating to north-Croatian finds should be accidental, but the real proportions, and the reasons behind them, are yet to be researched.

Most commentators postulate that 4th- century *villae rusticae* were the most conspicuous settlement feature of the period in Pannonia, and that they prospered on account of the towns and the municipal life in general. Nevertheless, a huge literature on this subject, especially where villas are addressed within broader issues, abounds in generalisations, evasiveness and even contradictions about the fate (prosperity or decline) of villas in the 4th century.[146] Needless to say, the evidence from Croatia is either entirely missing from these discussions, or the villas of the area are mentioned only in passing and mostly on the basis of the written sources.[147] Zosimus reports on *villae* between Siscia and Sirmium at unspecified locations in the Sava valley, while Ammianus Marcellinus mentions one such building in the surroundings of Cibalae, owned by Gratianus, father of the emperors Valentinianus and Valens.[148] Although the majority of those villas should be expected in Pannonia Secunda on account of the vicinity of Sirmium, the fact that an administrator of the *res private* existed only in Pannonia Savia suggests not one but a cluster of such buildings in the wider surroundings of Siscia.[149] In any case, it is not the commentators who should be blamed for the lack of mention of northern-Croatian villas in the literature, but the state of research. On the other hand, some of this architecture is well-known, although probably mostly in Croatia (T. Leleković and A. Rendić-Miočević, in this volume). As this section deals with the 4th century, it should be pointed out that quite a few villas that seem to have bene built in the 3rd or 4th centuries, or at least continued life in Late Antiquity on the same level as before, are situated in the central part of the area under research, that is, in the Požega Valley (Brusnik, Striježevica, Imrijevci, Vetovačke livade, Velika, Tekić, etc). Among other things, this might have to do with the position of the area away from the two main roads along the Sava and Drava, which were often used by the armies in the 4th century.[150] It is hardly a mere chance that in this area (probably nearby the village of Tekić) an imperial roadside *mansio* grew, which is recorded in the Antonine itinerary.[151] Coin finds from the Požega Valley, mostly 3rd- and 4th-century hoards, which strongly prevail over earlier finds, were tentatively put in the context of pending war clashes.[152] While this should be true eventually, such evidence in itself testifies to the prosperity of the area throughout the 4th century. On balance, it is even realistic to hypothesise the existence of an early Christian bishopric in the area, although no mention of it exists in historical sources. Finally, it should be noted that 4th-century destructions such as those in the towns, have either not been spotted in rural settlements or have not been reported in the literature.

An interesting observation has been made by T. Leleković in relation to the villas of the Požega Valley and the area around Daruvar: the lack of *vici*, which, on the other hand, have been recognised and researched in recent rescue and protective excavations in other parts of northern Croatia (Leleković and Rendić-Miočević, in this volume, section 6.). Such circumstances can broadly be compared with the observations of M. Suić, made long ago on the settlement pattern in the coastal region of the province of Dalmatia. Suić suggested that the lack of urban settlements with municipal status in the fertile area in the hinterland of Zadar (Roman Iader) was conditioned by the fact that the land had been taken by huge villa estates.[153] Just possibly similar establishments, though far more modest,[154] could have prevented the development of major *vici* in central-northern Croatia. This does not mean that small villages of the autochthonous inhabitants did not exist besides *villae*; they can be conceived of as inconspicuous wooden or wattle and daub constructions. Speaking of wooden architecture, it should be noted that in northern Croatia such features have been very rarely established prior to recent archaeological excavations, and that mostly in towns (see in this volume: Knezović, section 2.1.-2. 6; Lolić and Wiewegh, section 6.1.). Only quite recently has wooden architecture been detected on rural sites and, accordingly, the *vicus* type of settlement recognised (Leleković, in: Leleković and Rendić-Miočević in this volume, section 3.). Another important information relating to *vici* (from the same quotation of Leleković) concerns economy, as commercial horse breeding has been hypothesised in two such settlements. Should this presumption prove right, it would fit in well with the fact that Pannonia was known for quality-horse breeding from pre-Roman times and during the Roman Empire.[155] Back to *villae*: it has been presumed that both in Dalmatia and in the Danube provinces villa

is a considerable lag even if the difference in the total areas of these two territories are taken into consideration, and especially if we bear in mind that in the more than a decade the number of burials in Hungary must have increased considerably. As far as Croatia is concerned, the number given is approximate due to the scrappy documentation on early finds and the lack of published data on recent excavations.

145 Gesztelyi 1996, 55.

146 Mócsy 1974, 299-302, 308; Thomas 1964; Thomas 1980; Mulvin 2002 (pp. 39 and 42 for contradictory opinions on the destiny of villas in Late Antiquity); Gábor 2005, 419; Wilkes 2005, 165, n. 102; Caldwell 2007, 138, *passim*: Egri 2007, 105.

147 Symptomatically, of 150-odd villa sites from Pannonia mentioned in Thomas (1964), only six are from northern Croatia. See in this volume: Leleković and Rendić-Miočević, section 1, and Dzino and Domić Kunić, n. 138.

148 Caldwell 2007, 149-151.

149 Mócsy 1974, 306.

150 Migotti 2009a, 133. On a potential bishopric see Migotti 1997, 98.

151 *Incero* (*sed mansio Augusti in praetorio est*). Cf. Soproni 1980a, 214. On the *praetorium* as a rare form of a roadside settlement see Wilson 2006, 212.

152 Mirnik 1979, 83.

153 Suić 1976, 57-59.

154 As observed by M. Egri (2007, 108), geographical and climatic conditions prevented development of huge villa estates in the Danube provinces.

155 *Exp. tot. mun.* 57; Bökönyi 1983, 338.

estates were mostly owned by colonists from Italy and the western provinces, and veterans.[156] So far, we only have one epigraphic testimony to such ownership from the Požega area – a stele of a legionary of local origin from Brusnik.[157]

4.8.2. The 5th and 6th centuries

In the Croatian archaeology of late Roman southern Pannonia, the 4th century has customarily been considered as the end of Roman provincial civilisation.[158] The 5th and 6th centuries were both in the university education and museum practice labelled as the early medieval or the so-called migration period (see Rapan Papeša in this volume). While it is true that the beginning of the Middle Ages is not universal in terms of chronology, the safest criterion for defining the introduction of medieval civilisation should be the arrival of the population(s) that remained permanently in a given territory. In some parts of the Pannonian provinces Roman rule, or at least the claim of Roman emperors to the territories lost to Barbarians, persisted in one or another form until the arrival of the Avars in the second half of the 6th century.[159] Although the first stage of Avar rule in southern Pannonia developed under a political surveillance and a cultural influence of the Roman Empire, the turn of the 6th and 7th centuries is here taken as the end of Roman civilisation in Pannonia. This was done so because together with the Avars came the Slavs, who were to remain in this area permanently.[160] Therefore, the period considered in this volume stretched from the 2nd century BC to the end of the 6th century AD. A syntagm often used to denote the stretch of time from the 3rd to the 6th centuries is *migration period*.[161] This definition has not been used here because of its slightly misleading overtone: it disregards the Roman existence in Pannonia till the end of the 6th century, substantiated by both the historical record and the archaeological evidence.

The evidence for a continuous life of the Romanised native population in the context of the Barbarian presence in 5th-and 6th-century Pannonia is still a sore question of the archaeology of Roman Pannonia.[162] This issue was summarised by A. Mócsy in several points: 1. a hypothesis that the Romans were concentrated in the towns; 2. the difficulty in differentiating between 4th- and 5th-century finds, due to the dearth of coins from the 5th century; 3. secure 5th-century finds as exclusively Barbarian; 4. Barbarian finds as absent from the towns before the 6th century; 5. the lack of Roman finds of the whole period in question as explainable by a gradual emigration of Roman population.[163] Since the knowledge of the 5th- and 6th-century period has much progressed in the past decades and the evidence for the Romans somewhat increased, the above-mentioned scheme figures as less straightforward and dependable now.[164]

The Croatian archaeology has started to accept, although not unanimously, the fact that the end of the 4th century was not the end of Roman life in southern Pannonia. Nevertheless, commentators are faced with the same problem of chronology as encountered by Hungarian archaeologists: the lack of 5th-century coins.[165] This is reflected in preliminary excavation reports, in which the sites producing late Roman finds are dated from the 1st to the end of the 4th centuries. In those less frequent examples of the chronology stretching into the 5th century, it is not always clearly stated whether this is the result of the specific chronology of small finds and/or coins, or whether it comes from the common belief that the life went on in the province.[166] In the light of this, the importance of the study of pottery clearly transpires; such work has been going on in northern Croatia for quite some time now, but has not yet produced systematic results. Recently, a received opinion that later 4th- to 6th-century graves within town perimeters, especially individual Barbarian interments, denote a complete abandonment of urban life, has been reconsidered; clearly, such situations rather reflect a reorganisation of town structures and ways of life.[167] Adaptations and restructuring leading to gradual transformations, rather than a complete decay and abandonment, have been claimed for 5th-century layers in other parts of the Roman Empire, on the presumption of low-quality building and the use of perishable materials. Given that some continuity has been detected in the surrounding areas, it would be odd to presume that southern Pannonia presents a completely different picture as an answer to similar circumstances in the 5th and 6th-centuries.[168] As a matter of fact, any presumption of neatly divided periods and a sudden disappearance of a

[156] Suić 1976, 57; Mulvin 2002, 56; Mulvin 2002, 56; Egri 2007, 104; Busuladžić 2011, 113-115.
[157] *AIJ*, no. 590.
[158] Migotti 1997, 105-106; see also Radman-Livaja in this volume: The Roman Army, section 6).
[159] Barkóczi 1980, 120; Salamon and Sós 1980, 397, 406; Jarnut 2000; Schwarcz 2000; Mirković 2011, 93-94.
[160] Migotti 1997, 4-5; Cameron, Ward-Perkins and Whitby 2000, xvii-xix; Vida 2003, 283; Daim 2008; Quast 2008, 277; Vingo 2010, 275-276.
[161] Visy 2003c, 281. A better approach seems to be applied in Salamon and Sós (1980, 397), with the role of the Empire taken into consideration.
[162] The terms *barbaricum* and *barbarian* prevail in the literature, especially recent; occasional quotation marks with the latter word suggest its ambiguous meaning. Here the words *Barbaricum* and *Barbarian* have been used throughout, with the intention to stress their basic meaning in the realm of political-ethnical geography and not cultural-anthropological characterisation. The same attitude was taken in the catalogue of the recent major international exhibition on the Barbarians (Aillagon 2008, 32, *passim*). See also Kuntić-Makvić 1999, 81-85; various authors, in: Visy 2003c, 207 -289; Kovács 2004, 185; Gardner 2007, 26.
[163] Mócsy 1974, 343-345, 353, 357. On the issue of 4th-and 5th-century chronology see Barkóczi 1980, 119 and Müller 2003, 288.
[164] Salamon and Sós 1980; various authors, in Visy 2003c, 263-301; Quast 2008; Vingo 2010.
[165] While coins would make a safe proof for the existence of a site at a given period, it should be noted that there were no 5th-century coins on some Norican sites whose existence at that time is proved by the written sources (Scherrer 2011, 106).
[166] Various authors in the *HAG*: 2 (2005), 38, 85; 3 (2006), 199, 292; 4 (2007), 25; 5 (2008), 295.
[167] Migotti 1997, 22; Kušan Špalj and Nemeth-Ehrlich 2008, 201. Restructuring rather than a complete decay and decline has been claimed for much of the late Roman architecture in Britain (Gardner 2007, 101, 252-253). Cf. also Mirković 2011, 91-93 on Sirmium and Brogiolo 2011, 201 on north-Italian towns.
[168] On northern Pannonia cf. Borhy 2011, on Sirmium: Mirković 2011, 91-92, on Venetia: Alessandro Marano 2011, on Noricum: Ubl 1982, 85-86 and Stuppner 2011, 129, on Dacia Ripensis: Petković 2011, on Britannia: Gardner 2007, 101.

culture, even if spurred by sudden turbulences or political events, would be out of line with the reality of historical processes.[169] This means that the period in question in northern Croatia has just not been investigated properly so far, or that it has not been given enough attention in recent protective excavations. Or perhaps it has, but we are still waiting for more thorough analyses and publications of these finds.[170]

In the last several decades the study of the 5th- and 6th-century period in northern Croatia has been impaired in two ways: firstly, by not being given sufficient scholarly attention in itself, and secondly, by its inappropriate separation from Late Antiquity.[171] The last to tackle this period with committed engagement was Zdenko Vinski (1913-1996), who long ago pointed to the need for a search into the differences between the material cultures of the native Romans and Barbarian settlers.[172] Regrettably, Vinski's work has never been continued on such high academic level, nor has our understanding of this subject matter become much deeper since. The relationship between the Roman and Barbarian archaeological evidence has recently been addressed in the context of the early Christianity of southern Pannonia, with the intention to stress continuity of life in the province in the 6th century.[173] The essay by A. Rapan Papeša in this volume is prevalently a compilation of the more or less known facts and already published evidence, but conveniently put together to produce a picture of the period as a whole, which had not been done before. Apart from that, some new finds were considered as well, which are currently being studied.[174] The author has entitled her paper *Early mediaeval Barbarian elements in late antique southern Pannonia*, although she also tackled the evidence of the Romans. Once the basic approach to the issue of identities accepted (see above, section 4.1.), some of the finds from Rapan Papeša's paper need to be commented. As pointed out in Rapan Papeša's section *Roman provincial culture (late 4th-end of 6th c.)*, a differentiation between the Barbarian and Roman material evidence is problematic due to a high rate of interconnectedness of these two civilisations, particularly in terms of Barbarian integration.[175] No wonder therefore that some artefacts, presented the chapter in question as Roman, for example the brooch with bent foot (figs. 18-20) or torques (fig. 5), would by other commentators be considered as Germanic.[176] Contrarily, some finds that were in the Croatian literature customarily taken as Barbarian, are rightly proclaimed as Roman provincial, such as a bird-head staff (fig. 43)[177] and cross-shaped plate brooch (fig. 44).[178] On the other hand, a silver spoon (fig. 72) is put among the Germanic evidence, which has been a received opinion in Croatian scholarship; moreover, the spoon used to be taken for an Arian liturgical device.[179] The eponymous findspot (the famous treasure of Desana, Italy) for this type of spoon is a typical example of a cultural-ethnical blend, yielding a combination of objects of Romano-provincial culture and those of the Ostrogothic origin. The spoons of the type Desana from this treasure were estimated as belonging to the Roman cultural tradition, while their cultic usage is highly questionable; notwithstanding the engraved cross, the spoon from Sisak could have been item of a household tableware rather than a Christian (Arian) cult device.[180] Therefore, this artefact can in no way be a testimony to Arian Christianity in Siscia, as was often proposed, or an Ostrogothic artefact in the first place. The idea was otherwise tempting, given that during the Ostrogothic rule in Pannonia Savia (504-535) Siscia was both the seat of the Gothic governor (*comes*) and a bishopric, with a certain number of Germanic inhabitants of probable Arian religious affiliation.[181]

Apart from the material ascribed to the Romans in Rapan Papeša's contribution, some additional finds should be considered as Romano-provincial, such as crossbow brooches, given that their use was especially frequent at the end of the 4th century and the beginning of the 5th, and that they were worn by both the Romans and Barbarians. Although quite a lot of such finds stem from northern Croatia, they have never been studied as a whole and their total number is unknown.[182] Also, no find that would explicitly point to the use of such brooch by a Barbarian is known in the region of study, although the appearance of crossbow brooches at Kuzelin (Rapan Papeša, in this volume, notes 69-72) is possibly indicative in this regard. Further, two cemeteries from the 2nd half

---

169 Cf. Gardner 2007, 252; Cool 2009, 712; Heinrich-Tamáska 2009; Stuppner 2011, 129.
170 Symptomatically, Croatian archaeologists were not present at the conference held in 2010 in Keszthely (see Heinrich-Tamáska 2011).
171 This artificial disciplinary divide between Late Antiquity (4th century) and early Middle Ages (5th and 6th centuries) further slowed down the study of early Christianity of southern Pannonia: Migotti 1997, 6, 104. On the ambiguous classification of 4th- and 5th-century finds see Kiss 1999, 218.
172 Vinski 1968, 149.
173 Migotti 1997, 106.
174 See Rapan Papeša in this volume, ns. 117, 132 and 146. A certain number of finds, mostly bought in recent decades for the AMO, was not available for publication here as it is currently under study by Mladen Radić of the MSO.
175 This is best reflected in common (mixed) cemeteries: Barkóczi 1980, 118; Salamon and Sós 1980, 399; Halsall 2002; Müller 2003, 291; Cracco Ruggini 2008; Stickler 2011, 499-502.
176 Brooch: Tejral 1988, 236, Abb 7: 19 and 8: 10, 11; Vaday 1989, 86-89; Buora 2008a, 82; Filipović 2010, 22. Torques: Tejral 1988, 237; Vaday 1989, 48-49; Riffaud-Longuespé 2008, 322-323. Ambiguity of the Romano-Barbarian culture is well exemplified through the torques, as it was both a classic Barbarian piece of jewellery and an item of Roman military decorations (Vasić 2001, 177-178; D. Quast, in: Aillagon 2008, 224-225.
177 It is true that this motif was used also among the Barbarians, but its Roman origin should not be doubted, as witnessed by imperial staffs dating from the early Imperial period: Cadario 2011, 210, fig. Fig. 1; Tortorella 2011, 310, fig. 10.
178 Cross-shaped brooches used to be ascribed to the Barbarians on the account of dotted-circle motifs, regardless of the fact that the cross was a Roman-Christian symbol, while the decoration in question is most probably a common pre-Christian symbolical heritage, with possible connotations on the Sol worship: Migotti 1997, 103.
179 A contrary opinion (Migotti 1997, 76) was mostly ignored in the literature.
180 Aimone 2008, 378. See also the literature in Migotti 1997, 76.
181 Šašel 1992, 740-745 = 1978, 251-253; Jarak 1994, 34; Migotti 1997, 102; Schwarcz 2000, 68-69.
182 On the crossbow brooches from northern Croatia see Migotti 1994, 203 and Migotti 2008, and on Barbarians as wearers of such objects see Buora 2008a and Buora 2008b, 156-157.

of the 4th century and possibly the first half of the 5th, both in Pannonia Secunda, should be considered as part of the Romano-Barbarian mixed culture: Zmajevac in the immediate hinterland of the *limes* and Štrbinci in its wider area.[183] Some of the finds from Zmajevac were specifically ascribed to the Goths: a silver brooch with bent foot and silver pennanular appliqués.[184] If we consider a conception of Barbarian affiliation in the wider sense, some more finds could be added to the group: an iron distaff and a bone whorl from grave 67, and possibly also a two-row comb from grave 19.[185] None of the evidence from Štrbinci can be counted among Barbarian finds, although some of them belong to the provincial cultural heritage with Barbarian (specifically Sarmatian or perhaps even local pre-Roman) "overtones", for example parts of belt sets and jewellery made of iron, a necklace made of bronze links, a bronze bracelet with conical endings, conical glasses, a fragment of chain mail, a boar's eye-tooth and various forms of glass beads (opaque multi-coloured, gold-in-glass, and imitations of flattened square-sectioned cornelian beads), and those made of amber, also of various forms.[186]

Some traditional misconceptions of archaeology add to the difficulties in the understanding of Romano-Barbarian cultural and religious relationships. One of them is the attitude towards grave goods, specifically vessels as containers for foods and drinks. The majority of commentators would claim the latter habit as non-Christian. If, however, Roman Christians in the 4th century cherished such habit, and the same is true of the Barbarians in the 6th century, how founded indeed is the claim that the lack of vessels in 5th-century graves should by itself designate Christians?[187] Jaroslav Šašel pointed to an untenable preconception that the organised town life in Pannonia was extinguished after the disintegration of the Hun Empire.[188] In spite of the fact that this opinion is above all related to Pannonia Savia and its capital Siscia, it did not find much response in the Croatian archaeological literature. As a matter of fact, the lack of the mention of 5th- and 6th-century Siscia is common to all literature on Pannonia and/or Illyricum of the period, quoted in this paper.[189] The fact that the Siscian mint resumed work not only in 408-423, but also in 489-526, has been underplayed.[190] As far as Pannonia Secunda is concerned, the towns such as Mursa and Cibalae are customarily considered as completely disintegrated or even non-existent, given the lack of the archaeological evidence for the contrary.[191] On the basis of the 5th- and 6th-century finds, A. Rapan Papeša (in this volume) claims this to be wrong.[192] It is true that her theory is somewhat invalidated by the random nature of mostly non-contextual finds, and the fact that even individual finds are taken as proof for the existence of Barbarian groups in the towns or in separate Barbarian settlements. Likewise, the use of the Roman coinage is not a sufficient proof that the Barbarians accepted the entire economic structure of the Roman Empire.[193] All we know about the Barbarians in the area of study comes from a rather modest amount of written sources and similarly limited archaeological evidence: chance finds of coinage and other small artefacts, with only a somewhat larger number of systematically excavated Gepidic graves and some settlement structures.[194] Contrary to northern Pannonia, next to nothing is known of the settlement structures of other Barbarian peoples, and the lack of finds from the country is a particularly sore point. Recently, attempts were made at identifying Germanic pottery of the 5th to 7th centuries among late Roman and early medieval (Slavic) finds in northwestern Croatia.[195] In any case, the results of the so-far investigation of the 5th and 6th centuries as presented here have shown the potential for further research. At this stage, enough is known to contradict the historical sources' and archaeologists' claims about a total devastation of Pannonia by the end of the 4th century and to demonstrate the continuity of life in southern Pannonia till the later 6th century. Apart from the Gepidic finds from Pannonia Secunda, the archaeological evidence mostly pertains to Siscia and its wider area. Nevertheless, the majority of the towns and settlement known from before are mentioned (some with distorted or changed names) in the Ravenna Cosmography. Notwithstanding all controversy about this source, including its date, there are no doubts that its author leaned on the information from the 6th century.[196] The mention of south-Pannonian towns in the Ravenna Cosmography can therefore be considered as a further proof of the continuity of life in the area after the end of the 4th century. It transpires from recent research that

183 See Leleković, in this volume, section 4.2. On the Romano-Barbarian culture see n. 58.

184 Filipović 2010, 22, grave 60.

185 On the distaff and the whorl as possibly Barbarian grave goods see: Vágó and Bóna 1976, 203; Böhme 1996, 92-93; Müller 2010, 231. A. Rapan Papeša (in this volume, n. 125) claims that the comb as a grave good is a Gepidic custom, which is in line with the presumption that the Romans would not put a comb into the grave (Quast 2008, 278). On the other hand, two-row combs are preferably connected with the Roman population, while unilateral (bell-shaped and triangular) pieces are usually ascribed to the Barbarians (Biró 1994, 35-36; Petković 2008, 353, 362). Even though, two-row combs are customarily considered as typical Germanic finds (Straub 2011, 329, Abb. 2: 6-4; Ódor 2011, 348 and 351, fig. 4.). On balance, the ethnic-cultural appurtenance of quite a number of late antique small finds (jewellery and dress accessories) remains controversial.

186 Migotti and Perinić 2001, 150-155, pl. 10: 4, 6, pl. 13: 4, 5, pl. 21: 3, Migotti 2004, 194-197, pl. 14: 2, pl. 26: 1, 2; Migotti 2007b; Migotti 2009a, 156-162, pl. 39: 1, pl. 40: 4, 6.

187 Migotti 1997, 37. See also Müller 2003, 292; Tóth 2003, 296.

188 Šašel 1992, 761 = 1980, 13.

189 K. Póczi (1980, 254) mentions the existence of the Siscian see just in passing, while both A. Mócsy (1974, 352) and P. De Vingo (2010, 276) even claim Sirmium to be the only Christian see in southern Pannonia, which continued till the end of the Roman period.

190 Bíró-Sey 1980, 345; Dukat and Mirnik 2004, 113.

191 Pinterović 1978, 100-101; Iskra-Janošić 2004, 170-171; Jarak 1998, 784. Otherwise, gradual depopulation and decay was postulated for all Pannonian towns Mócsy (1974, 353-354; Barkóczi 1980, 116-120; Müller 2003, 289-290).

192 Unlike Hungarian archaeologists, who have been committed to the research of the period in question since long, this issue has never become the subject of a more engaged discussion by Croatian scholars. See n. 171.

193 In the context of Barbarian coinage, one Byzantine coin was brought (n. 144) from Pannonia Savia. It should be noted that Pannonia Secunda has yielded 60-odd coins of the emperor Justinian: Göricke-Lukić 1998.

194 All that Rapan Papeša (in this volume, section 4.) mentions in regard to settlements is the fact that Barbarian peoples respected the pre-existing settlement structures. Elsewhere (Gračanin and Rapan Papeša 2011, 20) we learn that some Gepidic settlement-structures have been researched in Cibalae.

195 Bekić 2006, 123-124. Germanic pottery from Vinkovci (Cibalae) has recently been more systematically studied by Danijela Roksandić of the Faculty of Philosophy in Zagreb.

196 Cf. Tóth 1989, 207; Gračanin 2011, 233-249.

more fortified settlements of the type of Kuzelin and Lobor should be expected in the hilly and mountainous areas of northern Croatia.[197] Also, first attempts were made at the isolation of the pottery of the autochthonous Romans of some 6th- century lowland sites.[198] If more of such evidence were available, it could possibly be used in the research into identity issues. Dzino and Domić Kunić (in this volume) claim that in Late Antiquity the notion of Pannonian *patria* did remain, while at the same time Roman and Pannonian identities became more or less meaningless.[199] Still, the possibility should be taken into account that Romano-Pannonian self-consciousness remained meaningful as the opposite to Barbarian, but now in a religious context of Christian orthodoxy versus Barbarian Arianism or both Roman and Barbarian paganism.[200] This, however, will for a long time (or permanently?) remain a moot question. How, for example, should we explain an instance of a Pannonian identity which was still at the end of the 8th century deeply rooted in the conception of a homeland, irrespective of religion?[201]

Acknowledgements

For precious information on various subjects or the literature my sincere thanks are due to Zoran Gregl, Dora Kušan Špalj, Marjeta Šašel Kos, and Hrvoje Vulić. I am also grateful to Tomislav Bilić for his very useful comments on numismatic issues, and to Alka Domić Kunić and Mirko Sardelić for their invaluable advice concerning the written sources in the chapters throughout the volume.

Bibliography

Aillagon 2008 Jean-Jacques Aillagon (ed.), *Rome and the Barbarians. The Birth of a New World*, Exhibition Catalogue, Milano.

Alföldy 1994 Géza Alföldy, La Pannonia e l'Impero romano, in: Hajnóczy 1994, 25-40.

Alessandro Marano 2011 Yuri Alessandro Marano, The towns of the central and eastern *Venetia* in the Ostrogothic period, in : Heinrich-Tamáska 2011, 173-194.

Alston 2010 Richard Alston, Review (A. Lee, War in Late Antiquity: A Social History, Malden, MA), *JRS* 100, 345-346.

Barkóczi 1973 László Barkóczi, Beiträge zur Steinbearbeitung in Pannonien am Ende des 3. und zu Beginn des 4. Jahrhunderts, *FA* 24, 67-112.

Barkóczi 1980 László Barkóczi, History of Pannonia, in: *ARP*, 85-124.

Bekić 2006 Luka Bekić, *Zaštitna arheologija u okolici Varaždina. Arheološka istraživanja na autocesti Zagreb – Goričan / Rescue Archaeology in the Varaždin Environs. Archaeological Research on the Zagreb Goričan Highway and its Access Roads*, Zagreb.

Bíró 1994 Mária T. Bíró, *The Bone Objects of the Roman Collection, Catalogi Musei Nationalis Hungarici, Series Archaeologica* II, Budapest.

Bíró-Sey 1980 Katalin Bíró-Sey, Currency, in : *ARP*, 337-348.Bojanovski 1988 Ivo Bojanovski, *Bosna i Hercegovina u antičko doba*, Sarajevo.

Borhy 2003 László Borhy, Überlegungen zur Änderung des Begriffes *civitas* in der Spätantike, in: Á. Szabo and E. Tóth (eds.), *Pannonica provincialia et archaeologia. Studia sollemnia Eugenio Fitz octogenario dedicata,* Budapest, 415-424.

Borhy 2003 László Borhy, Die letzten Jahrzente der Erforschung des spätrömisch-pannonischen Limes seit Sándor Sopronis „Die letzten Jahrzente ..." – ein Überblick, in: Heinrich-Tamáska 2011, 29-46.

Bowman 2008 Alan K. Bowman, Egypt from Septimius Severus to the death of Constantine, in: Bowman, Garnsey, and Cameron 2008, 313-326.

Bowman, Champlin, and Lintott 1996 Alan K. Bowman, Edward Champlin, and Andrew Lintott (eds.), *The Augustan Empire, 43 B.C.- A.D. 69, CAH* 10.

Bowman, Garnsey, and Cameron 2008 A. Bowman, P. Garnsey, and A. Cameron (eds.), *The Crisis of Empire, A.D. 193-337, CAH* 12.

Böhme 1996 Horst Wolfgang Böhme, Söldner und Siedler im spätantiken Nordgallien, in: *Die Franken – Wegbereiters Europas, 5. bis 8. Jahrhundert n. Chr*, Exhibition Catalogue, Mannheim-Berlin-Mainz, 91-101.

Bökönyi 1983 Sándor Bökönyi, Trade of domestic animals between Pannonia and Italy, *Savaria* 16 (1982), Szombothely, 335- 339.

Bratož 2000 Rajko Bratož (ed.), *Slowenien und die Nachbarländer zwischen Antike und karolingischer Epoche. Anfänge der slowenischen Ethnogenese, Situla* 39/1.

Bratož 2004 Rajko Bratož, Le persecuzioni dei cristiani delle provincie danubiane e balcaniche sotto Diocleziano, *Quaderni giuliani di storia 25*, Trieste, 261-342.

Bratož 2007a Rajko Bratož, *Rimska zgodovna. Del 1 - Od začetkov do nastopa cesarja Dioklecijana, Zbirka zgodovinskega časopisa* 33, Ljubljana.

Bratož 2007b Rajko Bratož, Izseljavanje prebivalstva iz Zahodnega Ilirika v 5. in 6. stoletju, in: T. Šenk (ed.), *Vilfanov spominski zbornik*, Ljubljana, 247-284.

Bratož 2008 Rajko Bratož, La diffusione del cristianesimo tra la *Venetia et Histria* e l'Illirico», in: *Cromazio*, 406-415.

Bratož 2011 Rajko Bratož, La Chiesa aquileiese e l'Illirico occidentale al tempo di Cromazio, in: P. F. Beatrice and A. Peršič (eds.), *Chromatius of Aquileia and His Age, Instrumenta patristica et mediaevalia* 57, Turnhout, 103-143.

Brogiolo 2011 Gian Pietro Brogiolo, From ancient to early

[197] Northwestern Croatia: Šimek 2007; the Slavonian Mountains: Migotti 2009b, 132.

[198] Bekić 2006, 213

[199] According to G. H. Caldwell (2007, 221) this phenomenon transpires also from 4th-century panegyrists.

[200] For the religious component in creating late Roman ethnic identities see Liebeschuetz 2011, 490.

[201] Paulus Diaconus reports that two centuries after settling in north Italy, Pannonians retained their ethnic identity and a sense of community in the face of the surrounding Romans (Bratož 2007b, 272).

mediaeval town: reflections starting from the case of Verona, in: Heinrich-Tamáska 2011, 195-210.
Brown 1996 Peter Brown, *The Making of Late Antiquity*, Cambridge, Mass.-London.
Bulat 1989 Mirko Bulat, Novi podaci za baziliku mučenika u Mursi, *Lihnid* 7, 195-202.
Buora 2008a Maurizio Buora, L'origine dei Goti secondo Jordanes e la ricerca archeologica, in: Piussi 2008, 80-85.
Buora 2008b Maurizio Buora, Militari in Aquileia e nell'arco alpino orientale, in: Piussi 2008, 154-161.
Busuladžić 2011 Adnan Busuladžić, *Rimske vile u Bosni i Hercegovini / Roman Villas in Bosnia and Herzegovina*, Sarajevo.
Buzov 2010 Marija Buzov, The Topography and the Archaeological Material of the Early Christian Period in Continental Croatia, *Classica et Christiana* 5/2, Iaşi, 299-334.
Cadario 2011 Matteo Cadario, Il linguaggio dei corpi nel ritratto romano, in: La Rocca, Presicce, and Lo Monaco 2011, 209-221.
Caldwell 2007 Craig H. Caldwell, *Contesting Late Roman Illyricum: Invasions and Transformations in the Danubian-Balkan Province*, Unpublished PhD thesis, Princeton University.
Cameron 2003 Averil Cameron, Ideologies and Agendas in Late Antique Studies, in: Lavan and Bowden 2003, 3-21.
Cameron, Ward-Perkins,and Whitby 2000 Averil Cameron, Bryan Ward-Perkins, and Michael Whitby, *Preface*, in: A. Cameron, B. Ward-Perkins, and M. Whitby (eds.), *Late Antiquity: Empire and Successors, A.D. 425-600*, *CAH* 14, xii-xx.
Cambi and Marin 1998 Nenad Cambi and Emilio Marin (eds.), *L'Époque de Justinien et les problèmes de VI*[e] *et VII*[e] *siècles*, *ACIAC* 13/1-3, Città del Vaticano-Split.
Campbell 2008 Brian Campbell, The army, in: Bowman, Garnsey, and Cameron 2008, 110-130.
Chevalier 1996 Pascale Chevalier, *Ecclesiae Dalmatiae*, *Salona* II, Rome-Split (1995).
Ciglenečki 2008 Slavko Ciglenečki, Monumenti del primo cristianesimo nella Savia e nella provincia Valeria Media, in: *Cromazio*, 440-447.
Cool 2009 Hilary E. M. Cool, A new (theoretical) look at Late Roman Britain, Review, *JRA* 22/2, 710-714.
Corbier 2008 Mireille Corbier, Coinage, society and economy, in: A. Bowman, P. Garnsey, and A. Cameron 2008, 393-439.
Cracco Ruggini 2008 Lellia Cracco Ruggini, Rome and the Barbarians in Late Antiquity, in: Aillagon 2008, 204-215.
Čučković 2008 Lazo Čučković, Sjeverna granica Japoda, in: *Arheološka istraživanja u Lici*, *IzdHAD* 23, Zagreb, 79-82.
Daim 2008 Falco Daim, The Avars, in: Aillagon 2008, 413-417.
Demandt and Engemann 2007 Alexander Demandt and Josef Engemann (eds.), *Konstantin der Grosse – Imperator Caesar Flavius Constantinus*, Exhibition Catalogue, Trier.
Demo 1994 Željko Demo (ed.), Od Nepobjedivog sunca do Sunca pravde. Rano kršćanstvo u kontinentalnoj Hrvatskoj / *From the Invincible Sun to the Sun* of Justice. Early Christianity in Continental Croatia, Exhibition Catalogue, Zagreb.
De Vingo 2010 Paolo de Vingo, Shifting populations in late antiquity. Germanic populations, nomads and the transformation of the Pannonian *limes*, *AArchHung* 61/1, 261-282.
Dimitrijević 1979 Stojan Dimitrijević, Archäologische Topographie und Auswahl archäologischer Funde vom Vinkovcer Boden, in: *Corolla memoriae Iosepho Brunšmid dicata*, *IzdHAD* 4, Vinkovci, 201-275.
Domić Kunić 2006 Alka Domić Kunić, Bellum pannonicum (12.-11. pr. Kr.). Posljednja faza osvajanja južne Panonije, *VAMZ* 39, 59-160.
Donati and Gentili 2005 Angela Donati and Giovanni Gentili (eds.), *Costantino il Grande. La civiltà antica al bivio tra Occidente e Oriente*, Exhibition Catalogue, Milano.
Drexhage 1983 Hans-Joachim Drexhage, Die «Expositio totius mundi et gentium». Eie Handelsgfeographie aus dem 4. Jh. n. Chr., *Savaria* 16 (1982), Szombathely, 341-346.
Dukat and Mirnik 2004 Ivan Mirnik and Zdenka Dukat, *Arheološki muzej u Zagrebu. Numizmatička zbirka – vodič*, Zagreb.
Egri 2007 Mariana Egri, The role of local elites in economic exploitation of the Danube provinces, in: Vagalinski 2007, 103-112.
Erdkamp 2011 Paul Erdkamp (ed.), *A Companion to the Roman Army*, Chichester.
Filipović 2010 Slavica Filipović, *Kasnoantička nekropola u Zmajevcu – Mocsolás / A Late Antique Necropolis in Zmajevac – Mocsolás*, Exhibition Catalogue, Osijek.
Filipović and Katavić 2006 Slavica Filipović and Vedran Katavić, *OS- V- PF. Rezultati arheoloških istraživanja vojarna – Poljoprivredni fakultet*, Exhibition Catalogue, Osijek.
Fischer, Precht, and Tejral 1999 Thomas Fischer, Gundolf Precht, and Jaroslav Tejral (eds.), *Germanen beiderseits des spätantiken Limes*, Köln-Brno.
Fitz 1977 Jenő Fitz , Die Eroberung Pannoniens, *ANRW* 2/6, 543-556.
Fitz 1980 Economic life, in: *ARP*, 323-335.Fitz 1982 Jenő Fitz, *The Great Age of Pannonia (A.D. 193-284)*, Budapest.
Fitz 1993-1995 Jenő Fitz, *Die Verwaltung Pannoniens in der Römerzeit* I, II, 1993; III, 1994; V, 1995 (Budapest).
Fitz 1998 Jenő Fitz (ed.), *Religions and Cults in Pannonia*, Székesféhervár.
Fitz 2003 Jenő Fitz, Die Städte Pannoniens, in: *Pannonia* I, 47-52.
Frier 2000 Bruce W. Frier, Demography, in: A. K. Bowman, P. Garnsey, and D. Rathbone, *The High Empire, A. D. 70-192*, *CAH* 11, 788-816.
Gabler 1994 Dénes Gabler, Über die Aussagekraft der

Terra Sigillata-Funde bezüglich der Zerstörungen in den Provinzen, in: Stuppner and Tejral 1994, 355-369.

Gabler 2003 Dénes Gabler, Rural settlements in Pannonia, in: Visy 2003, 241-243.

Gábor *et al.* 2004 Olivér Gábor, Gábor Kárpáti, István Lengvári, and Csaba Pozsárkó, Sopianae, in: *Pannonia* II, 269-294.

Gábor 2005 Olivér Gábor, Suburbanum Sopianarum and Roman Villas around Sopianae, *BK* 9, 413-427.

Gardner 2007 Andrew Gardner, *An Archaeology of Identity. Soldiers and Society in Late Roman Britain*, Walnut Creek, CA.

Genser 1995 Kurt Genser, Zur Topographie und Chronologie der ländlichen Besiedlung in der Provinz Noricum, *BK* 3 (1994), 270-277.

Gesztelyi 1996 Tamás Gesztely, Pannonische Siegelringe: Kontinuität und Diskontinuität, in: G. Bauchhenß (ed.), *Akten des 3. internationalen Kolloquiums über Probleme des provinzialrömischen Kunstschaffens*, Köln-Bonn, 127-139.

Gorenc and Vikić 1975 Marcel Gorenc and Branka Vikić, Das fünfundzwanzigjährige Jubiläum der Untersuchungen der antiken Lokalität Aquae Iasae (Varaždinske Toplice), *AI* 16, 32-50.

Göricke-Lukić 1998 Hermine Göricke-Lukić, Justinijanov novac iz Slavonije i Baranje, in: Cambi and Marin 1998/2, 1145-1155.

Göricke-Lukić 2011 Hermine Göricke-Lukić, *Nekropole rimskodobne Murse*, Osijek.

Gračanin 2011 Hrvoje Gračanin, *Južna Panonija u kasnoj antici i ranom srednjovjekovlju (od konca 4. stoljeća do konca 11. stoljeća)*, Zagreb.

Gračanin and Rapan Papeša 2011 Hrvoje Gračanin and Anita Rapan Papeša, Postrimski grad u južnoj Panoniji: primjer Cibala, *SSlav* 11, 7-30.

Gross 2008 Pierre Gross, City and Territorial Organisation, in: Aillagon 2008, 63-67.

Gruen 1996 Erich S. Gruen, The expansion of the empire under Augustus, in: A. K. Bowman, E. Champlin, and A. Lintott 1996, 147-197.

Haider 1994 Peter W. Haider, Historische Aspekte der römerzeitlichen Paßstraßen über die Hohen Tauern, in: Stuppner und Tejral 1994, 23-32.

Hajnóczy 1994 Gábor Hajnóczy (ed.), *La Pannonia e l'Impero Romano*, Roma.

Hajnóczy *et al.* 1999 Gyula Hajnóczy, Tamás Mezős, Mihály Nagy, and Zsolt Visy (eds.), *Itinerarium Hungaricum* I, *Pannonia Hungarica Antiqua*, Budapest.

Halsall 2002 Guy Halsall, The origins of the *Reihengräberzivilisation*: forty years on, in: J. Drinkwater and H. Elton (eds.), *Fifth-century Gaul: a crisis of identity?*, Cambridge, 196-207.

Heinrich-Tamáska 2009 Orsolya Heinrich-Tamáska, Bemerkungen zur Transformation spätantiker Strukturen in Pannonien am Beispiel von Keszthely-Fenékpuszta, *Acta Archaeologica Carpathica* 42-43 (2007-2008), Kraków, 199-229.

Heinrich-Tamáska 2011 Orsolya Heinrich-Tamáska (ed.), *Keszthely-Fenékpuszta im Kontext spätantiker Kontinuitätsforschung zwischen Noricum und Moesia, Castellum Pannonicum Pelsonense* 2, Budapest-Leipzig-Keszthely-Rahden/Westf.

Hekster 2008 Oliver Hekster, *Rome and its Empire AD 193-284*, Edinburgh.

Horvat *et al.* 2003 Jana Horvat, Milan Lovenjak, Andreja Dolenc Vičič, Marija Lubšina-Tušek, Marijana Tomanič-Jevremov, and Zorka Šubic, Poetovio. Development and Topography, in: *Pannonia* I, 153-189.

Isaacs 2007 Ann Katherine Isaacs (ed.), *Immigration and emigration in Historical Perspective*, Pisa.

Iskra-Janošić 2001 Ivana Iskra-Janošić, *Urbanizacija Cibala i razvoj keramičarskih središta*, Zagreb-Vinkovci.

Iskra-Janošić 2004 Ivana Iskra-Janošić, Colonia Aurelia Cibalae. Entwicklung der Stadt, in: *Pannonia* II, 169-195.

Jarak Mirja Jarak, The History of Early Christian Communities in Continental Croatia, in : Demo 1994, 17-39.

Jarak 1998 Mirja Jarak, «Pannonian» Benedictine Possessions in the 6th Century, in: Cambi and Marin 1998/2, 779-786.

Jarnut 2000 Jörg Jarnut, Die Langobarden zwischen Pannonien und Italien, in: Bratož 2000, 73-79.

Jeličić-Radonić 2007 Jasna Jeličić-Radonić, Natpis carice Faustine iz zvonika splitske katedrale, *VAPD* 100, 49-61.

Jilek and Breeze 2007 Sonja Jilek and David Breeze, The Detritus of Life: the Contribution of Small Finds to Understanding Smaller Military Installations, in: R. Hingly and S. Willis (eds.), *Roman Finds: Context and Theory,* Proceedings of a conference held at the University of Durham, Oxford, 199-213.

Kandler *et al.* 2004 Manfred Kandler, mit Beiträgen von Franz Humer und Heinrich Zabehlicky, Carnuntum, in: *Pannonia* II, 11-66.

Keay 2004 Simon J. Keay, The late-antique complex of Centecelles (Tarragona), Review, *JAR* 17/2, 741-743.

Kiss 1999 Atilla Kiss, Glasfunde aus Gräbern des frühmittelalterlichen Karpatenbeckens (400-1000), *Antaeus* 24 (1997-1999), 217- 241.

Kos and Šašel Kos 2000 Peter Kos and Marjeta Šašel Kos, Map 20 Pannonia- Dalmatia, in: *The Barrington Atlas of the Greek and Roman World*, vol. 1 (ed. R. Talbert), Princeton-Oxford, 286-309.

Koščević 1980 Remza Koščević, *Antičke fibule s područja Siska*, Zagreb.

Kovács 2004 Péter Kovács, The late Roman Epigraphy in Pannonia (260-582), in: G. Németh and I. Piso (eds.), *Epigraphica II. Mensa rotunda epigraphiae Dacicae Pannonicaeqve, Hungarian Polis Studies* 11, Debrecen, 185-195.

Kremer 2001 Gabrielle Kremer, *Antike Grabbauten in Noricum*, Wien.

Kuntić-Makvić 1999 Bruna Kuntić-Makvić, Greek and Roman Antiquity, in: Supičić 1999, 73-91.

Kuntić-Makvić and Šegvić 1992 Bruna Kuntić-Makvić and

Marina Šegvić, Katančićev opis Siscije, *OpArch* 16, 165-181.

Kušan Špalj and Nemeth-Ehrlich 2008 Dora Kušan Špalj and Dorica Nemeth Ehrlich, Lokalitet: Ščitarjevo – dvorište osnovne škole, *HAG* 4 (2007), 198-202.

La Rocca, Parisi Presicce, and Lo Monaco 2011 Eugenio La Rocca, Parisi Presicce, and Annalisa Lo Monaco (eds.), *Ritratti. Le tante facce del potere*, Exhibition Catalogue, Roma.

Lavan and Bowden 2003 Luke Lawan and William Bowden (eds.), *Theory and Practice in Late Antique Archaeology*, Leiden-Boston.

Leleković 2009 Tino Leleković, Lokalitet: Osijek – Trg bana J. Jelačića, *HAG* 5 (2008), 45-51.

Leleković 2011 Tino Leleković, Antičke nekropole Siscije i Murse, Unpublished PhD thesis, Zagreb.

Lengyel and Radan 1980 Alfonz Lengyel and George T. Radan, *Introduction*, in: *ARP*, 17-31.

Lepelley 1996 Claude Lepelley, The survival and fall of the classical city in Late Roman Africa, in: Rich 1996, 50-76.

Lewin 2001 Ariel Lewin, Urban public building from Constantine to Julian: the epigraphic evidence, in: L. Lavan (ed.), *Recent Research in Late Antique Urbanism*, Portsmouth RI, 27-37.

Liebeschuetz 2011 Wolf Liebeschuetz, Wardlords and Landlords, in: P. Erdkamp 201, 479-494.

Lolić and Burmaz 2006 Tatjana Lolić and Josip Burmaz, Lokalitet: Sisak – Sv. Kvirin, *HAG* 2 (2005),175-177.

Lovenjak 2003 Milan Lovenjak, Municipium Flavium Latobicorum Neviodunum, in: *Pannonia* I, 93-105.

Lőrinz 1994 Barnabás Lőrinz, Die Truppenbewegungen am pannonischen Limes während der Markomannenkriege, in: Stuppner and Tejral 1994, 51-57.

Matijašić 2009 Robert Matijašić, *Povijest hrvatskih zemalja u antici do cara* Dioklecijana, Zagreb.

Matijašić, Čače, and Rendić-Miočević 1998 Robert Matijašić, Slobodan Čače, and Ante Rendić- Miočević, Istria, Dalmazia e Pannonia nell'antichità, in: *Trasparenze imperiali. Vetri romani dalla Croazia*, Milano-Roma (1997), 19-34.

Migotti 1994 Branka Migotti, The Archaeological Material of the Early Christian Period in Continental Croatia, in: Demo 1994, 41-67.

Migotti 1997 *Evidence for Christianity in Roman Southern Pannonia (Northern Croatia). A catalogue of finds and sites, BAR Intern. Ser.* 684.

Migotti 1998 Branka Migotti, Od kulta Sola do kršćanstva u Varaždinskim Toplicama (*Aquae Iasae*), *RFFZd* 37(24), 51-88.

Migotti 2002 Early Christianity in Aquae Iasae (Varaždinske Toplice) and Iovia (Ludbreg) in Pannonia Savia, *ZM* 11, 51-66.

Migotti 2004 Branka Migotti, Kasnoantička nekropola na Štrbincima kod Đakova – iskopavanja u 2001, *ARR* 14, 141-246.

Migotti 2005 Branka Migotti, The Ash-chest of Marcus Aurelius Glabrio from Siscia reconsidered, in: *Illyrica antiqua. Ob honorem Duje Rendić- Miočević*, Zagreb, 367-384.

Migotti 2007a Branka Migotti, Rimski sarkofazi Siscije, *GGMS* 7, 7-30.

Migotti 2007b Branka Migotti, Fragments of a Chain Mail as Amulets in the Late Roman Cemetery at Štrbinci near Đakovo, *PIAZ* 24, 203-212.

Migotti 2008 Branka Migotti, Lukovičaste fibule s portretima na području Rimskog carstva / *The Crossbow Brooches with Portraits in the Roman Empire*, Zagreb.

Migotti 2009a Branka Migotti, Kasnoantičko groblje na Štrbincima kod Đakova iskopavanja u 2004. i 2005., *ARR* 16, 107-224.

Migotti 2009b Branka Migotti, Kasna antika i rano kršćanstvo, in: V. Kusin and B. Šulc (eds.), *Slavonija, Baranja i Srijem – vrela europske Civilizacije*, vol. 1, Zagreb, 129-135.

Migotti 2010 Branka Migotti, Die Geschichte der frühchristliche Archäologie in Nordkroatien, *Römische Quartalschrift* 105/1-2, 22-31.

Migotti (in press) Branka Migotti, The Stelae of Northern Croatia, in: *Funerary Sculpture of the Western Illyricum and Neighbouring Regions of the Roman Empire,* Papers of the International Conference (Split, 2009).

Migotti and Perinić 2001 Branka Migotti and Ljubica Perinić, Nekropola na Štrbincima u svjetlu kasnoantičkog horiznta Panonije, *ARR* 13, 103-204.

Milotić 2010 Ivan Milotić, *Rimska cestovna baština na tlu Hrvatske*, Zagreb.

Mirković 2004 Miroslava Mirković, Sirmium, in: *Pannonia* II, 145-156.

Mirković 2011 Miroslava Mirković, Kontiunität und Diskontinuität bei der Entwicklung der Stadt *Sirmium*, in: Heinrich-Tamáska 2011, 87-96.

Mirnik 1979 Ivan Mirnik, Novčane ostave Požeške kotline, *Vjesnik Muzeja Požeške kotline* 2-3, Slavonska Požega, 83-91.

Mirnik 1981 Ivan Mirnik, *Coin Hoards in Yugoslavia, BAR Intern. Ser.* 95.

Mirnik 1996 Ivan Mirnik, Novac istočnog rimskog Carstva u numizmatičkoj zbirci Arheološkog muzeja u Zagrebu, *VAMZ* 28-29 (1995- 1996), 159-228.

Miškiv 1998 Jesenka Miškiv, *Rimska vojnička diploma iz Slavonskog Broda*, Slavonski Brod.

Mócsy 1962 András Mócsy, *s. v.* Pannonia, in: *RE Suppl.* 9, 516-776.

Mócsy 1974 András Mócsy, *Pannonia and Uper Moesia*, London-Boston.

Mulvin 2002 Lynda Mulvin, *Late Roman Villas in the Danube-Balkan Region, BAR Intern. Ser.* 1064.

Müller 2003 Robert Müller, The Romanized Population in the 5th and 6th centuries, in: Visy 2003c, 289-291.

Müller 2010 Robert Müller, *Die Gräberfelder vor der Südmauer der Befestigung von Keszthely-Fenékpuszta, Castellum Pannonicum Pelsonense* 1, Budapest-Leipzig-Keszthely- Rahden.

Nagy 2003 Mihály Nagy, The Early Development of Pannonian Towns, in: Visy 2003c, 221-223.

Nagy 2004 Mihály Nagy, Mogetiana, in: *Pannonia* II, 75-83.
Newby 2007 Zahra Newby, Art at the crossroads? Themes and styles in Severan Art, in: Swain, Harrison, and Elsner 2007, 201-249.
Nikoloska 2010 Aleksandra Nikoloska, *Aspects of the Cult of Cybele and Attis on the Monuments from the Republic of Croatia, BAR Intern. Ser.* 2086.
Ódor 2011 János Gábor Ódor, The 5th-century cemetery and settlement at Mözs (Tolna county, Hungary) – some issues concerning the «East-Germanic» period in Transdanubia, in: Heinrich-Tamáska 2011, 347-359.
Ottományi 2005 Katalin Ottományi, Römerzeitliche Vicus in Budaörs, *BK* 9, 361-374.
Pavan 1955 Massimiliano Pavan, *La provincia Romana della Pannonia Superior, Atti della Academia Nazionale dei Lincei, Memorie, Classe di Scienze morali, storiche e filologiche,* ser. 8, vol. 6, fasc. 5, Roma.
Perinić 2003 Ljubica Perinić, *Vojnički kultovi na području Hrvatske (rimskih provincija Dalmaci i Panonije),* Unpublished MA thesis, Zagreb.
Perinić Muratović 2008 Ljubica Perinić Muratović, *Podrijetlo i narav kutla boga Silvana u rimskim provincijama Dalmaciji i Panoniji*, Unpublished PhD thesis, Zagreb.
Perkić 2002 Domagoj Perkić, *Grad mrtvih u Bubijevoj* jami kod Barilovića, *HAnt* 8, 103-130.
Petković 2008 Sofija Petković, Unilateral Antler Combs from *Romuliana, Starinar* 56 (2006), 353-365.
Petković 2011 Sofija Petković, Late Roman *Romuliana* and Mediaeval Gamzigrad from the end of the 4th to 11th centuries AD, in:
Heinrich-Tamáska 2011, 267-284. Pinterović 1978 Danica Pinterović, *Mursa i njeno područje u antičko doba*, Osijek.
Piussi 2008 Sandro Piussi (ed.), *Cromazio di Aquileia 388-408*, Exhibition Catalogue, Milano.
Póczy 1980 Klára Póczy, Pannonian cities, in: *ARP*, 239-274.
Pollak 1994 Marianne Pollak, Auswirkungen der Markomannenkriege beiderseits des westnorischen Limes, in: Stuppner and Tejral 1994, 431-437.
Poulter 1996 Andrew Poulter, The use and abuse of urbanism in the Danubian prvinces during the Later Roman Empire, in: Rich 1996, 99-135.
Quast 2008 Dieter Quast, Pannonia in the Fifth Century, in: Aillagon 2008, 276-279.
Radman-Livaja 2004 Ivan Radman-Livaja, *Militaria Sisciensia. Nalazi rimske vojne opreme iz Siska u fundusu Arheološkog muzeja u Zagrebu / Militaria Sisciensia. Finds of the Roman Military Equipment in Sisak in the Holdings of the Archaeological Museum in Zagreb*, Zagreb.
Radman-Livaja 2010 Ivan Radman-Livaja (ed.), *Nalazi rimske vojne opreme u Hrvatskoj / Finds of the Roman Military Equipment in Croatia*, Zagreb.
Redő 2003 Ferenc Redő, Municipium Aelium Salla, in: *Pannonia* I, 191-235.
Reece 2000 Richard Reece, Afterword, in: J. Pearce, M. Millet, and M. Struck (eds.), *Burial, Society and Context in the Roman World*, Oxford, 270-272.
Reece 2006 Richard Reece, *The Coinage of Roman Britain*, Stroud.
Rendić-Miočević and Šegvić 1998 Ante Rendić-Miočević and Marina Šegvić, *Religions and cults in south Pannonian regions*, in: Fitz 1998, 7-22.
Rendić-Miočević 1989 Duje Rendić-Miočević, *Iliri i antički svijet*, Split.
Rich 1996 John Rich (ed.), *The City in Late Antiquity*, London.
Riffaud-Longuespé 2008 Philippe Riffaud-Longuespé, The Treasure of Pouan (France), in: Aillagon 2008, 322-323.
Salamon and Sós 1980 Ágnes Salamon and Ágnes Cs. Sós, Pannonia – Fifth to ninth centuries, in : *ARP*, 397-425.
Salway 2000 Peter Salway, Roman Oxfordshire, *Oxoniensia* 64 (1999), 1-22.
Sanader 2003 Mirjana Sanader, The Ripa Pannonica in Croatia, in: Visy 2003, 135-142.
Sanader 2007 Mirjana Sanader (ed.), *Kroatien in der Antike*, Mainz.
Sanader 2010 Mirjana Sanader, O problemima topografije hrvatskog dijela Dunavskog limesa na temelju novijih arheoloških istraživanja / On the problem of topography of the Croatian part of the Danube limes based on recent aechaeological excavations, in: Radman-Livaja 2010, 221-233.
Schachinger 2006 Ursula Schachinger, *Der antike Münzumlauf in der Steiermark*, *Denkschriften der Österr. Akad. Der Wissensch.* 341, Wien.
Scheidel 2008 Walter Scheidel, Review (J. Haas, Die Umweltkreise des 3. Jahrhunderts n. Chr. Im Nordenwesten des Imperium Romanum ..., Stuttgart 2006), *JRS* 98, 268-269.
Scheidel 2011 Walter Scheidel, Marriage, Families, and Survival: Demographic Aspects, in: Erdkamp 2011, 417-434.
Schejbal 2004 Berislav Schejbal, Municipium Iasorum (Aquae Balissae), in: *Pannonia* II, 99-129
Scherrer 1994 Peter Scherrer, Der grosse Markomanneneinfall des Jahres 170 und seine Folgen im Lichte der neuen Ausgrabungen in Aelium Cetium/ St. Polten, in: Stuppner and Tejral 1994, 447-455.
Scherrer 2011 Peter Scherrer (mit einem Beitrag von Berhnard Schrettle), *Noricum* in der Spätantike – Zu den Forschungen des vergangenen Jahrezhnts, in: Heinrich-Tamáska 2011, 103-127.
Schmidt 2001 Wolfgang Schmidt, Spätantike Gräberfelder in den Nordprovinzen des römischen Reiches und das Aufkommen christlichen Bestattungsbrauchtums. Tricciana (Ságvár) in den Provinz Valeria, *Saalburg Jahrbuch* 50 (2000), Saalburg, 213-441.
Schuster 2008 Jan Schuster, The Germans between the Rhine and the Vistula (First and Second Centuries), in: Aillagon 2008, 116-118.
Schwarcz 2000 Andreas Schwarcz, Der Nordadria- und Westbalkanraum im 6. Jahrhundert zwischen Goten und Byzantinern, in: Bratož 2000, 59-70.
Soproni 1980a Sándor Soproni, Roads, in: *ARP*, 207-217.

Soproni 1980b Sándor Soproni, Geography of Pannonia, in: *ARP*, 57-63.

Starac 2000 Alka Starac, *Rimsko vladanje u Histriji I Liburniji II, Monografije i katalozi* 10/2, Pula.

Steinklauber 2002 Ulla Steinklauber, Am Westrand Pannoniens, wo sind die Christen der Steiermark?, *ZM* 11, 85-91.

Stickler 2011 Timo Stickler, The *Foederati*, in: Erdkamp 2011, 495-514.

Stoyanov 2011 Yuri Stoyanov, Archaeology Versus Written Sources, in: *Terra antiqua balcanica et mediterranea. Miscellanea in honour of Alexander Minchev, Acta Musei Varnaensis* 8/1, 351-358.

Straub 2011 Péter Straub, Angaben zum hunnenzeitlichen ostgermanischen Fundhorizont in Südtransdanubien – ausgehend von der Nekropole in Keszthely-Fenékpuszta, in: Heinrich-Tamáska 2011, 325-345.

Stuppner 2011 Alois Stuppner, Zur Kontinuität in der Spätantike am norisch- pannonischen limes in Niederösterreich, in: Heinrich-Tamáska 2011, 129-155.

Stuppner and Tejral 1994 A. Stuppner and J. Tejral (eds.), *Markomannenkriege – Ursachen und Wirkungen*, Brno.

Suić 1976 Mate Suić, *Antički grad na istočnom Jadranu*, Zagreb.

Suić 1999 Mate Suić, Continuity of urban civilisation in the Croatian lands, in: Supičić 1999, 93-115.

Sunko, Katavić, and Jerončić 2008 Ana Sunko, Vedran Katavić, and Tomislav Jerončić, Lokalitet: Osijek – Huttlerova (Silos), *HAG* 4 (2007), 20-26.

Supičić 1999 Ivan Supičić (ed.), *Croatia in the Early Middle Ages. A cultural Survey*, London-Zagreb.

Swain 2007 Simon Swain, Introduction, in: Swain, Harrison and Elsner 2007, 1-26.

Swain, Harrison, and Elsner 2007 Simon Swain, Stephen Harrison, and Jaś Elsner (eds.), *Severan Culture*, Cambridge.

Šašel 1992 Jaroslav Šašel, *Opera selecta* (eds. R. Bratož and M. Šašel Kos), *Situla* 30.

Šašel Kos 1996 Marjeta Šašel Kos, The Defensive Policy of Valentinian I in Pannonia – A Reminiscence of Marcus Aurelius?, in: R. Bratož (ed.), *Westillyricum und Nordostitalien in der spätrömischen Zeit*, *Situla* 34, 145-173.

Šašel Kos 2003 Marjeta Šašel Kos, Emona was in Italy, not in Pannonia, in: *Pannonia* I, 11-19.

Šašel Kos 2005 Marjeta Šašel Kos, *Appian and Illyricum*, *Situla* 43.

Šašel Kos 2010 Marjeta Šašel Kos, The early urbanization of Noricum and Pannonia, in: L. Zerbini (ed.), *Roma e le province del Danubio*, Soveria Manelli, 209-230.

Šimek 2007 Marina Šimek, Lokalitet: Lonja – Gradišće, *HAG* 3 (2006), 141-143.

Škrgulja 2009 Rosana Škrgulja, Lokalitet: Sisak – Rimska 25, *HAG* 5 (2008), 286-288.

Tatscheva 2005 Margarita Tatscheva, Die Munizipalisierung in den Provinzen Moesia Superior und Moesia Inferior (Mitte des 2. – Mitte des 3. Jhs.), in: M. Mirković (ed.), *Römische Städte und Festungen an der Donau*, Beograd, 211-217.

Tejral 1988 Jaroslav Tejral, Zur Chronologie der frühen Völkerwanderungszeit im mittleren Donauraum, *Archaeologia Austriaca* 72, Wien, 223-304.

Tejral 1999 Jaroslav Tejral, Die spätantiken militärischen Eliten beiderseits der norisch-pannonischen Grenze aus der sicht der Grabfunde, in: Fischer, Precht, and Tejral 1999, 217-292.

Thomas 1964 Edit B. Thomas, *Römische Villen in Pannonien. Beiträge zur pannonischen Siedlungsgeschichte*, Budapest.

Thomas 1980 Edit B. Thomas, Villa settlements, in: *ARP*, 275-321.

Tomičić 2000 Željko Tomičić, Der Untergang der Antike und deren Nachlebensformen in Südpannonien (Nordkroatien), in: Bratož 2000, 255-297.

Tomičić and Dizdar 2007 Željko Tomičić and Marko Dizdar, Lokalitet: Ilok – dvor knezova iločkih, *HAG* 3 (2006), 38-41.

Tortorella 2011 Stefano Tortorella, divi e dive del passato: l'apoteosi imperiale, in: La Rocca, Presicce, and Lo Monaco 2011, 303-313.

Tóth 2003 Ágnes B. Tóth, The Gepids, in: Visy 2003, 294-298.

Tóth 1994 Endre Tóth, Provincia Valeria Media, *AArchHung* 41, 197- 226.

Tóth 1994 Endre Tóth, Das Christentum in Pannonien bis zum 7. Jahrhundert nach den archäologischen Zeugnissen, in: E. Bosh and H. Wolf (eds.), *Das Christentum im bairischen Raum von den Anfängen bis ins 11. Jahrhundert*, Köln-Weimar-Wien, 241-272.

Ubl 1982 Hannsjörg Ubl, Die archäologische Erforschung der Severinsorte und das Ende der Römerzeit im Donau-Alpen-Raum, in: *Severin zwischen Römerzeit und Völkerwanderung*, Catalogue Exhibition, Linz, 71-97.

Vagalinski 2007 Lyudmil F. Vagalinski (ed.), *The Lower Danube in Antiquity (VI C BC – VI C AD)*, Sofia.

Vágó and Bóna 1976 Eszter Vágó and István Bóna, *Die Gräberfelder von Intercisa. Die Spätrömische Südfriedhof*, Budapest.

Varbanov 2007 Varbin Varbanov, Barbarian invasions in the provinces of *Moesia Inferior* and *Moesia Superior* between 138-192 (according to the numismatic data), in: Vagalinski 2007, 153-170.

Vaday 1989 Andrea H. Vaday, *Die sarmatischen Denkmäler des Komitat Szolnok*, *Antaeus* 17-18 (1988-1989).

Vaday 2008 Andrea H. Vaday. The Sarmatians, in: Aillagon 2008, 134-136.

Vasić 2001 Miloje Vasić, Osvrt na nalaz IV veka iz Starčeva, *Zbornik Narodnog muzeja* 17/1, Beograd, 175-201.

Vida 2003 Tivadar Vida, Cross-roads of Peoples and Cultures, in: Visy 2003c, 283-284.

Vinski 1968 Krstoliki nakit epohe seobe naroda u Jugoslaviji, *VAMZ* 3, 103-168.

Visy 2003a Zsolt Visy, *The Ripa Pannonica in Hungary*, Budapest.

Visy 2003b Zsolt Visy (ed.), *The Roman Army in Pannonia. An Archaeological Guide of The Ripa Pannonica*, Budapest.

Visy 2003c Zsolt Visy (ed.), *Hungarian Archaeology at the Turn of the Millenium*, Budpest.

Vodička Miholjek 2008 Kristina Vodička Miholjek, Osijek – Vukovarska 49 (Mursa, Pristanište i Vijenac I. Meštrovića), *HAG* 4 (2007), 32-34.

Vukelić 2009 Vlatka Vukelić, Počeci sustavnih arheoloških istraživanja u Sisku – druga polovina 19. stoljeća, *GGMS* 9, 93-126.

Vulić, Rapan Papeša, and Krznarić Škrivanko 2009 Hrvoje Vulić, Anita Rapan Papeša, and Maja Krznarić Škrivanko, Lokalitet: Vinkovci – Korzo, *HAG* 5 (2008), 95-99.

Wilkes 1969 John J. Wilkes, *Dalmatia*, London.

Wilkes 1996 John J. Wilkes, The Danubian and Balkan provinces, in: Bowman, Champlin, and Lintott 1996, 545-585.

Wilkes 2005 J. J. Wilkes, The Roman Danube: An Archaeological Survey, *JRS* 95, 124-225.

Wilson 2007 Andrew Wilson, Urban development in the Severan empire, in: Swain, Harrison, and Elsner 2007, 290-326.

Wilson 2006 Roger J. A. Wilson, What's New in Roman Baden-Würtenberg?, *JRS* 96, 198-212.

Witschel 2004 Christian Witschel, Re-evaluating the Roman West in the 3rd c. A. D., *JRA* 17/1, 251-281.

Zanini 2003 Enrico Zanini, The Urban Ideal and Urban Planning in Byzantine New Cities of the Sixth Century A.D., in: Lavan and Bowden 2003, 196-223.

Zaninović 1999 Marin Zaninović, The Illyrians on Croatian soil in Antiquity, in: Supičić 1999, 55-71.

Zsidi 2002 Paula Zsidi, Transformation of the town structure in the civil town of Aquincum during the Severi (A. D. 193-235), *AArchHung* 53, 131-149.

Zsidi 2003 Paula Zsidi, Burials, in: Visy 2003c, 251-261.

Literary Sources Before the Marcomannic Wars*

Alka Domić Kunić

1. Introduction

The passing of time has not favoured the preservation of the original written sources from the Graeco-Roman world. Only a few sources survived to the present day, and even these are not preserved as complete works, but rather as fragments and quotations. In this diminished canon very few texts mention Pannonia, its inhabitants, geography, and history. As far as we can tell Graeco-Hellenistic writers had little knowledge of the continental part of Europe, while the Roman historical sources (as well as the Greek ones from the Roman period) mention Pannonia only as one of the conquered territories, the war narrative being yet another chance to glorify Rome. Even so, reports on Pannonia were not limited to the subject of war; some of them touch on Pannonian natural and political geography, economy, ethnography, as well as cultural anthropology. However, these data are sporadic and tainted with prejudice of the Mediterranean world, and as such should be taken with caution.

This paper deals with certain authors from the Greek- and Latin-speaking circle, whose common interest was the Pannonian territory. Their information on Pannonia are gathered and systematised according to specific topics, with information relating to southern Pannonia of the early Imperial period taken into account. The purpose of this paper is not to critically discuss the quality and accuracy of the given information, as this was done in the paper by Dzino and Domić Kunić in this volume, but to bring the facts as related in the sources. The term *Pannonia* discussed in this paper refers mostly to southern Pannonia, with the north border on the Drava River, with the stress on the territory of northern Croatia (see note 23). The purpose of occasional stepping out of the given geographical frame is to gain additional insight of specific elements of Pannonian geography or history. (fig. 1; for the distribution of ethnic communities in southern Pannonia see in this volume: Dizdar, figt. 1; Dzino and Domić Kunić, fig. 1; Radman-Livaja, Anthroponymy, fig. 1).

2. Geography and natural resources

2.1. Space

There were two basic ways in which classical cartographers placed Pannonia in space. Some authors chose the geographical approach, primarily taking into consideration the natural borders - rivers, mountain ranges, etc. The others put Pannonia into political borders determined by Roman administration, which often coincided with the natural borders. Florus places the people of Pannonia between the Sava and Drava, the two big Pannonian rivers (Flor. *Epit.* 2.24). The same information is repeated by two later authors, Jordanes (Jord. *Rom.* 216; 243) and Isidore of Seville (Isid. *Etym.* 14.4,16). This information originates from Titus Livy, a contemporary of the first Roman emperor Octavian Augustus (31 BC - AD 14), who was already in his time considered as an authority on classical historiography.[1] The Drava in the north and the Sava in the south determine the territory which was considered to be originally Pannonian (perhaps even in the ethnic sense) as early as the late Republic period and was the first Roman conquered territory north of the Alps. From there Pannonia spread to the east - to the part of the Danube course were it is joined by the Drava and the Sava Rivers, and to the north - to the great Danube bend in Hungary. To paraphrase Pliny the Elder and Pomponius Mela,[2] *Pannonia proprie dicta*, or the "real" Pannonia, covered that space between the Drava and the Sava for which Pliny (Plin. *NH* 3.25,147) states to be 120 000 steps wide (approximately 180 km).[3] Even though other geographical and historical sources do not specifically mention Pannonian territory, they do imply of such ethnic and geographical determination of Pannonia - Velleius Paterculus claims that in 6 AD, the year of the great Pannonian revolt (also known as the *Bellum Batonianum*, the war of two Batos) "all Pannonia" rose to arms (Vell. 2.110,2). From the following events (apart from Paterculus, Dio Cassius also wrote on the subject) it is obvious that the revolt spread to the territory south of the Drava, without including the territory between the Drava and the Danube. The Drava River as the north border of Pannonian territory is mentioned in Pliny, who describes the territory north of the Drava in one chapter (Plin. *NH* 3.24,146), and Pannonia

* The chronologies and titles of the sources that have been used are included in the list of full quotations following the text.

[1] The title of Florus' book is *Epitome ex Tito Livio* (*Summary by Titus Livy*), with a possible variant *Epitome bellorum omnium annorum DCC* (*Summary of all wars in 700 years*), which implies that Florus used the information provided by Livy.

[2] Both authors use the term *Illyrii proprie dicti* ("Illyrians in the true sense of the word") when referring to an ethnic community that lived in Illyricum, from where the name spread to the territory which Rome called Illyricum - *dein sunt quos proprie Illyrios vocant* (Pomp. Mela, 2.56), *proprie dicti Illyrii* (Plin. *NH* 3.22,144); Suić 1976. However, the parallel with the Pannonians and Pannonia has not been confirmed in the written sources.

[3] In reality, the distance between the Sava and the Drava does not exceed 100 kilometres at any point.

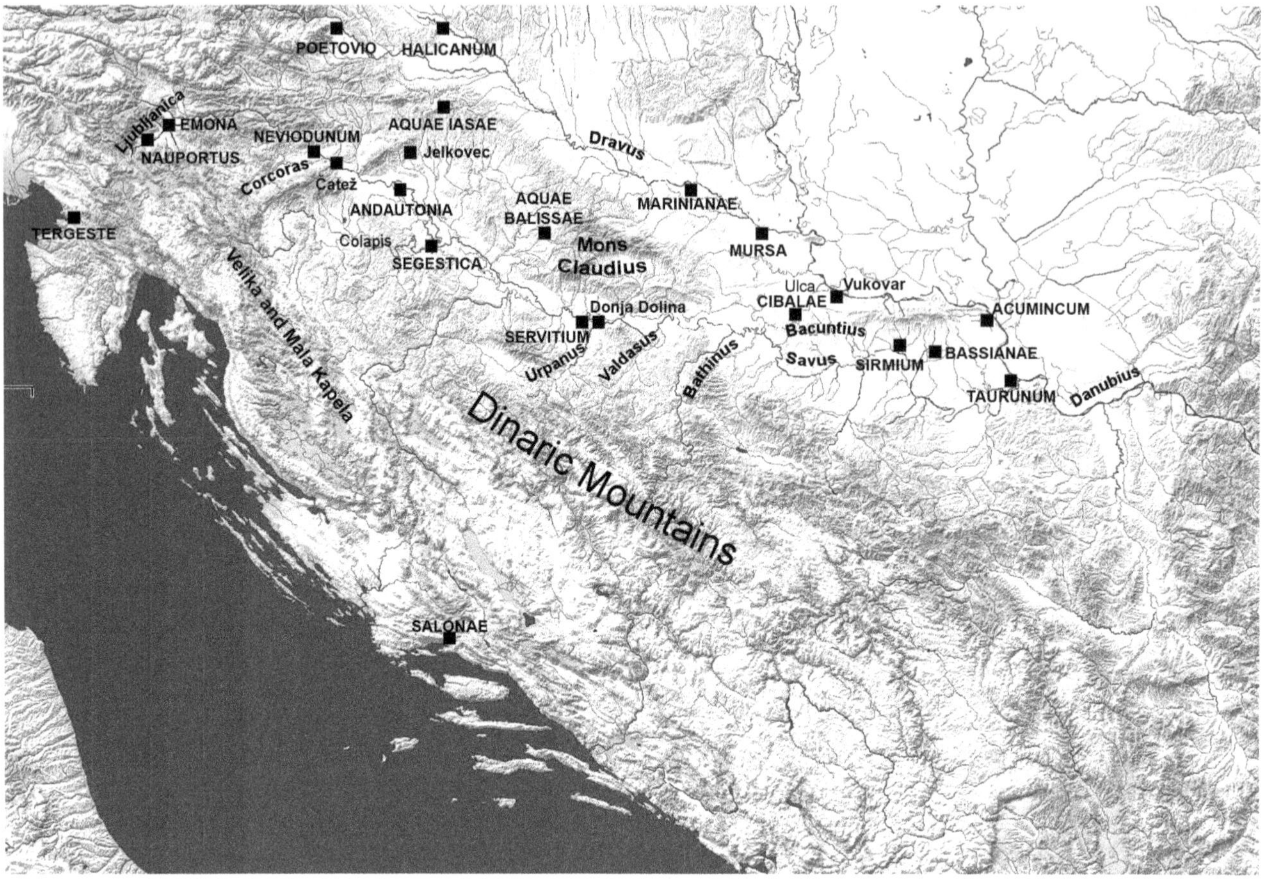

Fig. 1. Map of the geography of southern Pannonia based on the written sources of the 1st and 2nd centuries AD (after Google Maps, modified by T. Leleković).

in another (Plin. *NH* 3.25,147-148). The region north of the Drava, Transdanubia, will not be part of the Roman state until the reign of the emperor Claudius (AD 41-54). The territory was annexed peacefully since it had already depended on Rome for a considerable time, both politically and economically.[4] Since Claudius' time the region between the Drava, the Sava, and the Danube represents the southern part, while Transdanubia represents the northern part of the Roman province of Pannonia.

The geographical approach to setting the borders of Pannonia was adopted by Strabo and Appian - the former being the contemporary of Octavian Augustus and his military activities in southern Pannonia around 30 BC, and the latter, even though a whole century younger, depends on Augustus as his main source (App. 3.14).[5] A unique approach to the subject has Pliny's *Natural History* (the only entirely preserved classical encyclopaedia), which among else contains information hinting at the borders of Pannonian territory. Pliny combines the geographical principle with information accepted from official documents. When he lists the communities that live along the Drava and the Sava (Plin. *NH* 3.25,147), seeing Pannonia as divided between two Celtic blocs - the Taurisci west of *Mons Claudius* and the Scordisci south of these mountains (Plin. *NH* 3.25,148), it is possible that he is using some illustrative source, perhaps the Marcus Vipsanius Agrippa's world map (even though, and one can not be certain - the source might as well be of Hellenistic origin).[6] Pliny accepted the alphabet listing of various communities in Pannonia, which already included the communities north of the Drava (Plin. *NH* 3.25, 148), sourced from the official documents that were drawn up primarily for the purposes of collecting tax and for recruitment. It is important to take into consideration the list of communities according to *conventus iuridici* (Plin. *NH* 3.21,142-143) in the province of Dalmatia because it contains several communities that were perceived by Rome as Pannonian in the ethnic sense. Strabo's and Appian's southernmost area of Pannonia, bordered with Dalmatia, stretches from the Iapodians in the west to the Scordisci and Dardanians in the east, extending all the way to the territory of the Delmatae and Ardiaei (Strab. 7.5,3; 7.5,10; App. 3.14). Pannonia with its north border on the Danube and expanding deep into modern-day Bosnia in the south, towards the Roman territory known as Dalmatia (originally Illyricum), should be viewed as a *provincia* in the literal sense of the word

[4] The annexation of Transdanubia: Tóth 1976, 200; Fitz 1977, 543; 1990; Mócsy 1979.
[5] Appian's sources: Šašel Kos 2005, 38-41, 393-397.

[6] Agrippa's world map has been the subject of many studies: Detlefsen 1906; Hanslik 1961; Dilke 1985, 44-50. On Pannonia and Dalmatia: Domić Kunić 2003, 173, 180-181; 2004, 130-131. Possible Hellenistic sources are Posidonius or Polybius.

– as the territory that is yet to be conquered or controlled, rather than as an already established administrative unit of the Roman Empire. Strabo's information is contemporary to such understanding of Pannonia, while Appian also sources from the same period. Pannonia is separated from Dalmatia by a mountain range - a fact agreed upon by both Strabo and Pliny,[7] the difference being that Strabo has the Bosnian mountains and perhaps the Dinaric massif in mind (Strab. 7.5,3), while Pliny thinks of another mountain range to which Velika and Mala Kapela in the north Adriatic hinterland are attached (Plin. *NH* 3.25,147). Claudius Ptolemy also places Pannonia to the north of Dalmatia ("Histria and Illyria"), viewing the whole region as Roman administrative units (provinces) (Ptol. 2.14,1; 2.15,1).

2.2. Geographical and climatic characteristics

Classical authors emphasise three main geographical features of Pannonia - mountains, forests, and swamps. The only way to communicate through this unapproachable terrain (as seen by the Mediterranean world) was along the river courses. The two largest south Pannonian rivers, the Drava and the Sava, flow in the west-east direction; with its south confluents the latter connected Pannonia with the Adriatic coast, and, by implication, the Mediterranean. There are three main rivers in Pannonia: the Danube, the Drava, and the Sava; they surround the area which was considered to be the core of Pannonian territory in the late Republic period - in this respect the Danube is only marginal, since only small part of its course connects the Drava and Sava in the easternmost part of the discussed territory. As already stated, the geographical determination of Pannonia as the region between the Drava and Sava originates from Livy - the earliest information being found in Florus' *Summary* (Flor. *Epit.* 2.24), and the same being repeated five hundred years later in the works of Jordanes (Jord. *Rom.* 216; 243) and Isidore (Isid. *Etym.* 14.4,6). Pliny managed to summarise the main characteristics of the Drava and Sava, referring to the Drava course as "a rather violent stream", and to the Sava as being "more gentle" (Plin. *NH* 3.25,147),[8] while Florus, Jordanes, and Isidore characterised both rivers simply as fast. From the Roman point of view, the main advantage of these rivers was the fact that they were navigable and therefore suitable for commerce; accordingly, these two rivers, along with the Danube, were the transport backbone of Europe, connecting its western and eastern parts.

The Greek understanding of Pannonian hydrography was rather deficient and faulty, and was the subject of criticism already in the Roman period.[9] The Roman Republic adopted the Greek geographical experience along with its prejudices, and was using distorted information until it gained its own knowledge on the region in question. Octavian Augustus dramatically changed the Roman perception of Pannonian territory. During his reign the trading activities directed towards the Pannonian territory were established and reinforced, which was soon followed by concrete steps to connect politically this area to Italy. It is only from then on that one can talk of Rome's clear view not only on hydrography, but also on other geographical characteristics of Pannonia. Augustus' generals have brought first-hand information on geographical features of Pannonian territory, the distribution and strength of ethnic communities, natural resources, and everything else that was important to the Empire regarding a newly conquered territory. A reflection of this information has been preserved by Strabo and Appian. Both authors focus on Segestica, perhaps the most important trading hub in the western part of the region as well as the main strategic goal for Octavian Augustus during his Pannonian expedition in the 30s BC. While Appian gives only basic information on Segestica as a Pannonian settlement located on the River Sava (App. 4.22), Strabo puts weight on the settlement's location on the tributary of several navigable rivers (Strab. 7.5,2; cf. 4.6,10; 5.1,8) and emphasises the importance of Pannonian river communications (the Drava, Krka, Ljubljanica, Kupa, Sava, and Danube) as the unavoidable segment of the European communication network. In the attempt to give his readers a clear picture of the Pannonian hydrography, but not knowing the region first-hand, Strabo found himself entangled in the contradictions and inaccuracies, combining the unverified information with correct facts.[10] According to him, the important trading centre Nauportus (today's Vrhnika in Slovenia) is located in the vicinity of the River Corcoras (the Krka) which flows into the Sava (Strab. 4.6,10; 7.5,2). In reality, Vrhnika lies on the River Ljubljanica, 35 km as the crow flies from the Krka spring with which it has no connection. In the Greek perception Ljubljanica was indeed considered to be one the Sava's river branches; in these terms Nauportus is indeed "near" the Krka. The Krka does flow into the Sava, but the Sava is not the Drava's confluent, just as the Drava does not flow into the Noarus in the vicinity of Segestica (Strab. 7.5,12).[11] From the context of Strabo's narrative one can conclude that the Noarus, otherwise completely unknown river (mentioned only here) is actually the Sava: the Noarus flows past Segestica (Strab. 7.5.2; 7.5.12), is joined by the Colapis

[7] It is interesting to note that Pliny the Elder did not seem to know Strabo's work: Dilke 1985, 62; Clarke 1997; Domić Kunić 2004, 148-149, 159, 165. However, it seems that both authors (Pliny and Strabo) used the same sources, at least as far the region southwest of the Alps is concerned - most probably Polybius and Posidonius. About Strabo: Dzino 2006; 2008.

[8] The hydronym *Drava* is derived from the Indo-European root **dra-*, **dreu-* ("run", "hurry"), and the closest meaning is "a fast current", while the root **seu-* means "rain", "liquid", therefore *Sava* means simply "a fast stream" (Mayer 1959, 42, 101-102; Pokorny 1959, 204-206, 912-913). In such context Pliny's etimology is perfectly correct.

[9] The Greek knowledge on the region around the Danube and the reasons of its distortion over time: Domić Kunić 2003, 22-29; 2005; 2006, 67-69; 2009, 221-222. Example of the criticism in the Roman period: Plin. *NH* 3.18,127-128.

[10] Strabo relied on the outdated Hellenistic sources, Eratosthenes and Timasthenes (3rd c. BC) who "were to a considerable extent ignorant of Italy, the Adriatic sea, the Pontus, and the regions beyond them on the north" (Strab. 2.1,41); Pannonia can be found in "the regions beyond them on the north", Domić Kunić 2009, 221-222.

[11] The mention of the Drava might be due to a transcriber's mistake.

(the Kupa) (Strab. 7.5,2),[12] and flows into the Danube "near the country of the Scordisci" who controlled the lower reaches of the Drava and the Sava (Strab. 7.5,2; 7.5,12). The final point of the river traffic and Italic trade in Pannonia was the Danube, which is clearly accentuated by Strabo (Strab. 4.6,10; 7.5,2). The Danube, with its middle reaches determining the northern Pannonian border, passes through southern Pannonia in a small segment of its course. With the Drava and Sava as its confluents, the Danube actually outlines the southern-Pannonian territory. This easternmost part of southern Pannonia was dependant on the political alliance with the Scordisci, to which the sources ascribe mostly Celtic ethnic characteristics (Strab. 7.5,12). Strabo quite well knows the entire Danube course, from its spring to the river mouth, which can be seen in many places in his narrative in the 7th book, largely dedicated to the regions around the Rhine and the Danube. Appian, on the other hand, only states that the Sava flows into the Danube (Pliny adds that the Sava mouth is located near Taurunum, today's Zemun, Plin. *NH* 3.25,148) and that the Danube is called Ister in its lower reaches (App. 4.22). Unlike the Drava, Sava and Danube, other Pannonian rivers have not drawn such attention of classical geographers. Apart from Strabo's rather confused description of the river courses, which most likely includes a mention of the Ljubljanica (which flows past Nauportus), it is only Pliny who also mentions other Pannonian rivers and the location of their confluents. He states that the Colapis (the Kupa) flows into the Sava near Siscia, and the Bacuntius (the Bosut) near Sirmium. Apart from these two rivers, he also mentions the Valdasus (the Ukrina?) and the Urpanus (the Vrbas). It is a rather selective review of Pannonian hydrography, which Pliny justifies with the claim that it is only the mentioned rivers that are "noteworthy" (Plin. *NH* 3.25,148). His *Naturalis historia* contains one very valuable detail - information on two Sava's river islands. The first one is Segestica, actually a piece of land surrounded by the Kupa and the Sava courses. The second is the otherwise unknown Metubarbis, which Pliny states to be "the largest known island formed by the river" (Plin. *NH* 3.25,148). Metubarbis should be looked for in the lower reaches of the Sava, somewhere in eastern Slavonia or Syrmia, an area that was in the Roman period covered by swamps, so elevated parts of the land along the Sava and Drava courses could indeed have appeared as river islands.[13]

South Pannonia was also characterised as being profusely forested. Pliny the Elder refers to Pannonia as "acorn-producing" (Plin. *NH* 3.25,147), implying that this region was dominated by oak forests. Appian's reference to Pannonia as a wooded land (App. 4.22) added, an image of vast thick forests starts to emerge. Forestation of Pannonia is mentioned by other sources, using this characteristic as an argument to justify the stereotypical image of its inhabitants: Florus is impressed with "gloomy forests" in the land of the Scordisci (southern Pannonia) and compares this frightening landscape with the crudeness of its inhabitants (Flor. *Epit.* 1.39). One hundred years later Herodian testifies that the "limitless dense forests" extended to southern Pannonia, that is, the western part of the territory under study (Hdn. *Hist.* 8.1,1). Aurelius Victor mentions "vast forests" in today's Transdanubia (northern Pannonia) (Aur. Vict. 40.9-10).[14] It is therefore not hard to imagine the Pannonia of two thousand years ago: it was a vast region of centennial forests (oak in the lowlands, beech on the hills), swamps and rivers.[15] Thick deciduous forests are often accompanied by swampy land, which in south Pannonia was further fed by frequent overflowing of the rivers. One can conclude that there was hardly any space without swamps, but the classical sources mention only two larger swampy areas - one in the western and one in the eastern part of the region.[16] The swamp Lugeum which Strabo places between Tergeste and Nauportus (Strab. 7.5,2) can in some of its parts be identified with the Ljubljana swamp area (Ljubljansko barje), between the Sava and the Ljubljanica south of Ljubljana (Emona). It was probably in this area that the rebellion of the Pannonian legions took place in 14 AD. One of the reasons for the discontent among the soldiers was poor living conditions during the military service in the swampy and rugged terrain (Tac. *Ann.* 1.17; cf. Dio 57.4). The other swampy area mentioned in the sources extended over the space between Vukovar and Vinkovci (the Volcaean marshes), and was formed by the frequent overflowing of the River Vuka; it was there that a notorious battle during the Pannonian revolt took place in AD 7. It was only because of the iron Roman discipline that an almost catastrophic defeat of the Roman legions was turned into a victory (Vell. 2.112; Dio 55.32). Throughout the Roman period, the settlements in this part of Pannonia (Mursa, Cibalae, Sirmium) were surrounded with swamps, which were used as means of defence as well as isolation.[17]

---

[12] Strabo is wrong when he says that the Colapis (the Kupa) flows into the Noarus (the Sava) "immediately below Nauportus" (Strab. 7.5,2). Possibly, this is his or the transcriber's mistake, as in that place Segestica should be mentioned instead of Nauportus. Since the distance from Vrhnika to Sisak is 170 km as the crow flies, any other mistake but a *lapsus calami* is hard to imagine. About Noarus see Šašel Kos 2002; Domić Kunić 2003, 30-31, 81-83.

[13] The suggested etymology of the word *Metubarbis* (Gr. μετά = „between", Gr. βόρβος, Slav. *bara* = „mud", Mayer 1935a, 7) contributes to the presumption that the area in question is indeed some kind of natural elevation in a swamp. Metubarbis should be located in the territory of Spačva (east of Županja) or somewhere near the Drina River mouth (Domić Kunić 2003, 164, n. 15).

[14] Florus informs us that the general Caius Scribonius Curio reached the Danube but did not dare to go any further because he was afraid of impenetrable forests; Aurelius Victor wrote about the emperor Galerius having the vast forests cut down and Lake *Pelso* drained, while Herodian follows the direction of Maximinus' march through Pannonia towards Aquileia.

[15] The rest of the once thick forests in Krško polje (surroundings of the Neviodunum / Drnovo near Krško) is today's swampy area of Krakovski gozd; Moslavina (western Slavonia) is rich in names that imply the existence of ancient forests, a number of charred oak material has been found in the area around the Sava, and on the site of the lake-dwelling of Donja Dolina on the Sava animal bone and remains of acorn had been found (Domić Kunić 2003, 163, n. 12 and 13).

[16] About the swamps in southern Pannonia and ameliorative works in the Roman period: Domić Kunić 2003, 162-165; 2006, 61-62.

[17] In the names with the root *murs-*, *mars-* (*Mursa, Mursella, Marsonia*) the Indo-European root **merg-* "to rot" (Mayer 1935b, 80-81) is recognised. According to Mayer (1935b, 86) the root **pan-* (from which *Pannonia* is derived) could mean "swamp" (Mayer 1935b, 86). If correct, such interpretation points to a typical feature of the Pannonian landscape and at the same time explains the origin of the name *Pannonia*.

Another characteristic of Pannonian relief is its elevated areas. This applies to the area between the Drava, Sava, and Danube which is in all its length interspersed by hills, rather than to the northern, mostly flat part of the province (Transdanubia). Strabo mentions Pannonia as "mountainous country" (Strab. 7.5,10), which mostly refers to today's Bosnia where he places the largest Pannonian ethnic communities (Strab. 7.5,3),[18] but also to the region of southern Pannonia. According to Strabo, the same relief characteristics can be found "in the whole country situated above (the seaboard)" (Strab. 7.5,10). Pannonian "mountain ranges" are mentioned by somewhat later Statius (Stat. *Silv.* 5.2,132-137), identifying Pannonia with the mountain area. That this is not just a poetic image, but a widely accepted stereotype, is confirmed by two poets of Augustus' time. Tibullus states that Pannonians are scattered upon the "icy Alps" (Tib. *Paneg.* 3.7,108-109),[19] and the Pannonian people are similarly viewed by Ovid who calls them "the inhabitants of mountain ridges" (Ov. *Cons.*, 389-390).[20] Florus compares south Pannonia, the territory of the Scordisci, to their inhabitants: "their haunts among the woods and mountains harmonised well with their fierce temper" (Flor. *Epit.* 1.39,2). Although without direct confirmation in the written sources, the mountains around Požega are a physical barrier that contributed to the ethnic division. The only mention of this is in Pliny's *Natural History*: Mons Claudius (which is identified as the Požega or Slavonian Mountains) is a borderline between the two larger (political) units: the Taurisci in the west, and the Scordisci in the east (Plin. *NH* 3.25,148). It is as if Pliny runs his finger over the map from west to east, coming across the Mons Claudius and notices that the Taurisci are located "behind" and Scordisci "in front of" it, regarding the direction in which he examines the map.[21]

Impenetrable mountains, thick and frightening centennial forests, and vast swamps were only the external manifestations of the crude Pannonian climate - as was perceived by the Greek and Roman geographers and historians, more accustomed to the mild and fertile Mediterranean region. While Galen scientifically and objectively places the Pannonians (hidden under the name of "Illyrians" in his writings) among the people that live in the cold and wet climate (Galen *De temp.*, 1.618,10), Strabo and Dio Cassius subjectively depreciate the continental climate and its main characteristic - long and cold winters with frequent snows (Strab. 7.5,10; Dio 49.36,2-4). Regretfully, Velleius Paterculus did not leave us a promised monograph on Pannonia and '"the situation of their country and its rivers" (Vel. 2.69,2-3). Other classical sources keep referring to the coldness (Tib. *Paneg.* 3.7,108-109) and rivers frozen in winter (Plin. *Tra.* 12.4; Hdn. *Hist.* 6.7,6).[22] It is a fact that the Roman mind worked in stereotypes, overlooking certain details that would have provided a clearer and more accurate view on specific territories and their inhabitants. In an almost identical manner Caesar perceived certain communities in Gallia, and Tacitus a population to which the Romans mostly referred to as the Germans. What bothered most geographers and historians was the lack of Mediterranean crops - vines and olives, without which life in the Mediterranean was unthinkable (Strab. 7.5,10; Dio 49.36,2; cf 50.28,4).[23] Therefore, Dio Cassius concludes that Pannonians lead "the most miserable existence" and "are not well off as regards either soil or climate", startled by the fact that they consume barley and millet, which in the Roman world were only used for feeding horses, or as a disciplinary measure in the army.[24] On balance, it transpires that in the eyes of the Mediterranean world Pannonia was a region of unfavourable geographic and climate conditions. A crude continental climate created the traits that were to the classical writers both abhorrent and admirable - the courage and belligerence of the people which inhabit the region. These divided feelings of admiration and abhorrence were perhaps best expressed by Dio Cassius: "For all that they are considered the bravest of all men of whom we have knowledge; for they possess nothing that makes an honourable life worth while" (Dio 49.36,3).

2.3. Natural resources

Even though the Mediterranean world thought of Pannonia as unsuitable for living, the region was much milder than described by the written sources. Pliny the Elder referred to Pannonia as "acorn-producing" (Plin. *NH* 3.25,147), which suggests rich resources of oak forests in its southern stretches. The image of Pannonia as a region suitable for agriculture and forestry comes to the surface in the narratives of the late antique authors. By the end of the 1st century AD Pannonia was a representative example of a province abounding in a variety of lands and a model for the land division and evaluation in other parts of the Empire. Hyginus lists types of land in Pannonia (Hyg. Grom. 28) and presents the main resources with which the province contributed to the entire Empire. However, Pannonian economy was not based only on fields, forests and pastures - Pliny the Elder mentions two advantageous Pannonian species, *saliunca* (probably valerian spikenard, *valeriana Celtica* Linnaei, Plin. *NH* 21.20,43), and *asarum* (wild ginger, Plin. *NH* 27.47). As far as fauna is concerned, Pliny mentions European wisent (Plin. *NH* 8.16,40), which

---

18 Among the communities that Strabo calls Pannonian are the Daesitiates who lived in the mountains between the upper flows of the Vrbas and the Drina; cf. Vell. 2.115,4.

19 The word *Alpes* means the whole mountain range from the Alps to the Adriatic sea: Mala and Velika Kapela, Velebit and Dinara, even though Rome was familiar with the names of the individual mountains.

20 Ovid alludes to the military success of Octavian Augustus BC 35-33 when the region of southern Pannonia came under Roman rule. Accordingly, the mountains mentioned by Ovid should be looked for in the wider area of the upper reaches of the Drava and Sava.

21 Domić Kunić 2004, 130, 154-155. On the term *Slavonian Mountains* see Migotti in this volume, note 13, fig. 2.

22 Pliny the Younger and Herodian discuss the Danube.

23 When discussing Pannonia we must bear in mind that in most cases it is difficult (if not impossible) to establish with certainty which information relate specifically to southern Pannonia; some of them certainly relate to the province as the whole.

24 During Octavian's Delmataean stage of the war in 33 BC, a part of the Roman army war punished with meals of millet instead of wheat (Dio 49.38,4).

was hunted primarily for its fur and horns (cf. Paul *HL* 2.28), and the "Illyrian pig with strong hooves" (Plin. *NH* 11.106,255) - a citation that can very easily be referring to Pannonia (and not Dalmatia), since it was by the same author characterised as "acorn-producing", which in turn implies the richness of oak forests, as wild boars' natural habitat. Sources mention two game animals - bear ("Pannonian bear", Luc. 6.220-223) and wild boar ("Pannonian boars", Hadr. *Carm.* 4).

3. Ethnic distribution

3.1. Illyrians and/or Pannonians?

There is no mention of the Pannonians in the pre-Roman (Hellenistic) geography and ethnography. It was Appian who puts the Pannonians in the context of a mythical lineage (App. 1.2), placing their eponymous ancestor on the genealogical tree along with the ancestors of the other communities in Illyria (later to become Roman Illyricum). The family tree starts with the mythical progenitors Polyphemus and Galatea; Pannonius can be found as the descendant of Illyrius through Autarieus. Despite today's understanding of family trees as being unreliable, genealogical charts were a common apparatus which historiography used to decipher ancestry and kinship from distant past, and to give credit to contemporary aspirations for a specific territory.[25] The term *Pannonians* covered several identity categories, depending on the period in which it was used.[26] Distinctly ethnically coloured in the beginning, the term transformed to a purely administrative category of the inhabitants of the province of Pannonia. Calling the inhabitants of the region *Paeonians* instead of *Pannonians*, Greek erudites brought some confusion to the classical historiography. Appian uses this alternative name, emphasizing that he Roman variant (*Pannonii*) is different, while Polybius and Strabo only mention Παννόνιοι.[27] The term *Pannonius* denoting "the inhabitant of the province of Pannonia" is introduced as early as the first half of the 1st century AD (concomitantly with the forming of an individual province of Pannonia from Roman Illyricum), but both terms (*Illyricum* and *Pannonia*) are used evenly, though quite arbitrarily, as can be seen in the narrative of Velleius Paterculus, the contemporary of the events.

3.2. Ethnic diversity

Speaking of ethnic composition of Pannonia (like any other province), one should bare in mind that the information preserved in literary sources has to be interpreted according to the Roman perception - ethnonyms that we know from the Roman sources are not the names of ethnic, but rather territorial (political) communities formed after the region fell under Roman rule. We cannot be certain if the territorial communities (*civitates peregrinae*) spatially coincided with the territory of the ethnic community with which they shared the name, nor can we know which ethnonyms were only adopted and reproduced in the written form, and which "ethnonyms" are a Roman construction. As far as we can tell from the available archaeological and historical data, the Pannonian region was inhabited by several ethnic communities.[28] Generally speaking, the ethnic composition of Pannonia was zoned - the northern part (Transdanubia) being primarily Celtic, the middle part (southern Pannonia) consisting of both Celtic and indigenous components, while the southernmost area (south of the Sava) was deprived of such Celtic influence and was, conditionally termed, "ethnically purest", even though influence from all sides (the Delmatae and Illyrians from the south and the Celts from the north) can be recognised there. According to Pliny, the Taurisci bordered with the Scordisci at Mons Claudius (Plin. *NH* 3.25,148), forming two Celtic "blocks" (political alliances actually) on the edges of the indigenous population. This population was by the time of the late Republic dominated by the militarily and economically more developed Celts, and was more or less Celtisized.[29] The name *Taurisci* denoted several Pannonian ethnic communities - according to Pliny the area around the Drava was inhabited by the Serreti, Serapilli and Iasi, while the Varciani, Colapiani and Oseriates settled around the Sava. The Eastern Celtic bloc, the Scordisci, who, according to Strabo, were ethnically inhomogeneous (being a mixture of the Celtic and indigenous population), dominated the communities east of the Slavonian mountains (Mons Claudius) - the Andizetes along the Drava, and the Breuci and Amantini along the Sava. All of these ethnonyms emerged during the Roman conquest of the region between the Drava, Sava and Danube (35-33 BC and 12-9 BC), as Rome got a better insight into the region. Overall, the subject of Pannonian ethnicities is yet to be better understood and fully explored. It would be particularly interesting to establish whether there is some kind of pattern or regularity in the historical sources regarding the use of the names of the ethnic communities as opposed to the administrative names, as inhabitants of the province of Pannonia, and to find out whether the ethnic names we know today are indigenous ethnonyms or descriptive names invented by the Greeks or the Romans.[30]

Following a geographical source (Marcus Vipsanius Agrippa's *orbis pictus*, or perhaps *formula provinciae*?), Pliny the Elder is the only author who mentions the communities in the Drava valley called the Serreti and the Serapilli (Plin. *NH* 3.25,147). Bearing in mind the order of listing the settlements along the Drava and Sava (from west to east), the Serreti and the Serapilli should be located somewhere in the upper reaches of the Drava River - the former near Poetovio and the latter perhaps near

[25] And vice versa - the genealogy gave certain credibility to all members of the family tree, especially if "proved" any relation to Rome: Woolf 2011, 38-44. On Appian's genealogical tree see Šašel Kos 2005, 119-126.
[26] See Dzino and Domić Kunić, in this volume, sections 3 and 5.
[27] Šašel Kos 2005, 375-383.

[28] For more details see Dzino and Domić Kunić, in this volume.
[29] It would be better to say: La Tèneised, because we cannot be certain of the Celtic ethnic influence in the creation of the identity of Pannonian communities – cf. Dzino and Domić Kunić, in this volume.
[30] What is meant here is mostly ethnonyms such as the Colapiani ("inhabitants in the River Kolapis valley").

Halicanum (Sv. Martin na Muri).[31] Down the river was the territory of the Iasi, which Claudius Ptolemy places in the eastern part of Upper Pannonia (Ptol. 2.14,2). According to Ptolemy the border between Upper and Lower Pannonia, established in Trajan's time, cut through the Croatian part of southern Pannonia in the north-south direction all the way to Servitium (Bosanska Gradiška) on the Sava. Apart from Pliny and Ptolemy, the Iasi are not mentioned in any other literary source, but are quite well documented in the epigraphy. Owing to inscriptions dating to the first half of the 3rd century[32] the eastern and the western border of their *civitas* can be established: Aquae Iasae (Varaždinske Toplice) and Aquae Balissae (Daruvar), respectively. A military diploma dating to AD 85 testifies to recruitments of the Iasi in Roman auxiliary troops.[33] The upper reaches of the Drava near Mursa were held by the Andizetes, the eastern neighbours of the Iasi. Pliny places them in the easternmost part of the region around the Drava (Plin. *NH* 3.25,147), to Strabo they are just another Pannonian community (Strab. 7.5,3), while Ptolemy locates them in Upper Pannonia (Ptol. 2.15,2), which suggests that the emperor Trajan used an already existing delineation between the Iasi and Andizetes as a borderline between Upper and Lower Pannonia. In the *Sebasteion*, Augustus' shrine in the city of Aphrodisias on the west coast of Asia Minor (today's Geyre in Turkey), the Andizetes are mentioned in a special context. Along with at least one more community from Dalmatia (and supposedly some other from Pannonia as well), and many others from the rest of the Empire, the Andizetes depicted as a female personification were used to promote Augustus' successful foreign policy.[34]

According to Pliny, the Colapiani were the western neighbours of the Breuci in the area around the Sava (Plin. *NH* 3.25,147); Ptolemy places them in the southwestern part of Upper Pannonia (2.14,2). It is interesting that they are not mentioned by any other author from the first half of the 1st century AD, that is, by Strabo, Velleius Paterculus, or Appian (who sourced from Octavian Augustus). On the other hand, at that time they are already mentioned in epigraphy as members of Roman auxiliary troops - two members of the *ala Pannoniorum*, drafted during the Tiberius' or Claudius' reign and a member (somewhat younger) of *cohors I Ulpia Pannoniorum*; the latter's tombstone is dated to the reign of Trajan or Hadrian, or perhaps of Antoninus Pius.[35] The name *Colapiani* is derived from the hydronym C*olapis* and denotes "the inhabitants along the Colapis river (the Kupa)", but was also used to determine members of ethnic, or, rather, territorial community, as is evident from its mention in the official context of funerary stones. The Colapiani most probably belong to the Strabo's (Strab. 4.6,10, 7.5,2) and Appian's (App. 4.17, 4.22-24) Segestani[36] who were, as already established, identified as true ("*proprie dicti*") Pannonians. Pliny's seemingly incorrect assertion that the Colapiani ("inhabitants along the Kupa") hold part of the Sava course leads us to the fact that they actually hold the river mouth where the Kupa flows into the Sava (it is there that their main centre Segestica is located). West of the Colapiani, one should look for the Varciani - an indigenous community which adopted La Tène elements, as mostly transpires from their onomastics.[37] Ptolemy placed the Varciani in Upper Pannonia "towards north" (Ptol. 2.14,2), and Pliny included them in the alphabet list of the Pannonian communities (Plin. *NH* 3.22,148).[38] Just like the Colapiani, they are not mentioned in the writings of Strabo, Velleius Paterculus, or Appian. However, in Strabo they might be hidden under the name *Taurisci* (Strab. 4.6,10; 7.5,2), given that they are mentioned in a votive inscription as being situated between the *Sisciani* (in the east) and *Latobici* (in the west).[39] Three Varciani are documented as members of Roman auxiliary troops - two served during the reign of Claudius and Nero, respectively, and the third during the second half of the 1st century AD.[40] The Varciani gave their name to at least two equestrian cohorts (*cohortes Varcianorum equitatae*), and at least one infantry cohort that they shared with the Latobici (*cohortes Latobicum et Varcianorum*). Like other units that bore ethnic names, members of specific eponymous communities were recruited first, followed by the members of other communities.[41]

Pliny and Ptolemy mention one other Pannonian *civitas* - the Oseriates. They can be found in Pliny's alphabet list (Plin. *NH* 3.22,148) and therefore their location is impossible to determine, while Ptolemy places them in Upper Pannonia, "beneath" the Iasi (Ptol. 2.14,2). The mention of the Oseriates in the list of Dolabella's roads through Dalmatia is questionable - debatable is the reconstruction of the part of the inscription which bears the names of two communities on the either sides of the Bathinus (the Bosna) River in northern Dalmatia, at its border with Pannonia.[42] If the reconstruction is correct, the Oseriates could be settled along the right (south) bank of the Sava, all the way to the

[31] Soproni 1979, 95; Vidović 1990, 32.

[32] *CIL* III 4121 (*Aquae Iasae* / Varaždinske Toplice), *AIJ* 586 (*Aquae Balissae* / Daruvar), *AIJ* 587 (*Aquae Balissae* / Daruvar). On the Iasi: Schejbal 2003; 2004.

[33] *CIL* XVI 31 (Beleg near Kaposvár): Fronto, son of Scenus, was recruited around AD 60.

[34] Smith 1988.

[35] From the 1st c. AD: *CIL* III 4372 (*RIU* 255) (*Arrabona* / Győr): *Bato Buli f(ilius)*, and *CIL* III 4376 (*Arrabona* / Győr): *[...]crali f(ilius)*; from the 2nd c. AD: *CIL* III 11227 (*Carnuntum* / Deutsch-Altenburg): *[... L]iccai [f(ilius)]*.

[36] Cf. Dio 49.36.

[37] Cf. Vercaius (*CIL* III 9796 *Sinodium* / Balina glavica, Dalmatia), Iantumarus Andedunis (*CIL* XVI 4, unknown place in Pannonia).

[38] Pliny must have deliberately left out from his alphabet list those communities he had already mentioned when discussing the geography around the Sava and Drava, the ones he called "the principal peoples" (Plin. *NH* 3.25,147); likewise he does not mention any of the communities from the alphabet list when discussing geography (Plin. *NH* 3.25,148).

[39] *AE* 1909, 235 (*AE* 1938, 13) (*Samaria*, Iudaea), the inscription dated to Hadrian's period (AD 119-138) or the time of the second Partian war (between AD 197 and 201). The Sisciani (the residents of Siscia, in the place of former Segestica) can be found at the confluence of the Rivers Kupa and Sava, and the Latobici in the Slovenian part of the Sava valley); the Varciani were in between, around the Sava near Zagreb.

[40] *CIL* III 9796 (*Sinodium* / Balina glavica, Dalmatia); *CIL* XVI 4 (*Vindobona* / Vienna); *ILJug* 116 (*AE* 1950, 110) (*Bigeste* / Humac, Dalmatia).

[41] For *cohortes Varcianorum* cf. Bogaers (*sine anno*).

[42] *CIL* III 3201 (10159) + 3198b (10156b) (*Salona*). The contested part of the inscription: *HBIS[...]RIBUS*, usually reconstructed as: *Breucos Oseriatibus*.

Bosna mouth, which was held by the Breuci. The centre of the Oseriates is believed to be the lake dwelling at Donja Dolina on the Sava, half way between Nova Gradiška and the mouth of the Vrbas.[43] The territory of the Breuci is outlined on the basis of archaeological, epigraphic and literary sources. The Breuci occupied the Brodsko Posavlje and south-eastern Slavonia, which made them one of the largest communities in southern Pannonia.[44] The context in which the Breuci are mentioned in the literary sources testifies to their inner strength shown in the clashes with the Roman legions on at least two occasions.[45] Strabo counts them among the Pannonian communities (Strab. 7.5,3), Pliny places them among the "principal peoples" in the eastern neighbourhood of the Colapiani (Plin. *NH* 3.25,147), and Ptolemy locates them in Lower Pannonia (Ptol. 2.15,2). According to Suetonius, the Breuci were the main target of Tiberius' Pannonian expedition (Suet. *Tib.* 9), and from the reports of Dio Cassius it is evident that they played one of the main roles in the Pannonian rebellion (Dio 55.29; 55.34). Eight cohorts which originally recruited Breuci during the Iulio-Claudian reign are documented epigraphically, and individual Breuci have been evidenced as serving in other auxiliary troops - all were recruited during the 30s and 40s of the 1st century, in the reign of Tiberius and Claudius.[46]

The identification of the Cornacates which can be found in Pliny's alphabet list (Plin. *NH* 3.25,148) is still questionable. The Cornacates should be comprehended as a separate territorial (ethnic?) entity, even though there is a possibility that they are "the residents of Cornacum" (today's Sotin near Vukovar) in the region ascribed to the Breuci[47] - the same as *Segestani* might have been the "residents of Segestica" but belonged to the Colapiani in the ethnic sense (App. 4.17, 4.22-24; cf. Strab. 4.6,10, 4.5,2), and the *Sirmienses*, "the residents of Sirmium", were ethnically the Amantini (Plin. *NH* 3.25,148). It seems that Rome still perceived the Cornacates, Segestani, and Sirmienses as separate territorial (ethnic?) units, as has also been attested epigraphically.[48]

The south-easternmost part of Pannonia (today's Syrmia) was held by the Amantini, a strong and rather influential community if one is to believe the (extremely scanty) literary sources. Pliny mentions them on two occasions - in the alphabet list of the Pannonian communities, and as a territorial community (*civitas peregrina*) with its centre in Sirmium (Plin. *NH* 3.25,148); Ptolemy places them in the north-west part of Lower Pannonia (Ptol. 2.15,2). A boy belonging to the community of the Amantini is mentioned in the unfortunately lost cenotaph from Putinci, east of Sremska Mitrovica (*Sirmium*) (*CIL* III 3224). He was one of the hostages taken to Italy as a part of a peace pledge after a war defeat - most probably after Tiberius' Pannonian war of 12-9 BC, or the Pannonian rebellion of AD 6-9.[49]

The Rome of Augustus' time used the term *Pannonians* for some populations south of the Sava as well - according to such perception the border with the non-Pannonian population in the eastern Adriatic inland was the Dinara mountain range. Strabo designates the communities of the Maezaei, Ditiones and Daesitiates as Pannonians (Strab. 7.5,3), led by their participation in the great Pannonian rebellion.[50] Suetonius mentions only the Daesitiates, giving them the Pannonian identity (Suet. *Tib.* 20) - in this case probably only in the administrative sense. As early as the Iulio-Claudian time, the Pannonian communities south of the Sava, as well as those in the region between the Drava, Sava, and Danube, were recruited into the auxiliary troops and navy; as far as we know, the Maezaei, Ditones, and Daesitiates were treated differently, in that they did not form "ethnic" *cohortes* and *alae*, but were joined to other troops.[51]

Pannonia was strongly influenced by the Celts.[52] According to tradition, the waves of Celtic culture in the 4th or 3rd centuries BC overflowed the valleys of the Danube, Drava, and Sava, thus causing disturbance in Pannonia and whole Illyricum (Pomp. Trog. in Just. *Epit.* 24.4; 32.2; cf. Strab. 7.5,6; 7.5,11). After initial conflicts the Pannonian Celts accepted coexistence with the indigenous population in southern Pannonia, leaving an indelible cultural (and also ethnical?) imprint on it. The communities that inhabited the valleys of the Sava and Drava expressed their mixed Celtic-Pannonian identity through various aspects of material and

[43] Marić 1964; Čović 1987; Bojanovski 1988, 335-338; Schejbal 2003; Potrebica 2003; Domić Kunić 2006, 79-80.

[44] It is believed that the easternmost and westernmost edges of the Breucan territory have been confirmed epigraphically: Slavonski Brod (*Marsonia*) and Vukovar respectively, on the basis of four diplomas from the reigns of Nero and Vespasian: Miškiv 1998 (*Marsonia* / Slavonski Brod); *CIL* XVI 17 (Grabarje near Slavonski Brod); Dušanić 1999 (*RMD* 202) (Vukovar); Dušanić 1978 (Negoslavci near Vukovar). The Breuci might have held Cibalae in the northeast, where they bordered with the Andizetes: Domić Kunić 2006, 78.

[45] These two occasions are Tiberius' conquest of Breucan territory (12-9 BC), and the Pannonian rebellion (AD 6-9).

[46] The Breuci in *cohortes Breucorum*: Dušanić 1978 (*RMD* 79) (Negoslavci near Vukovar); *CIL* XIII 12062 (*Ara Ubiorum* / *Colonia Agripinensis* / Cologne, Upper Germany); Breuci in other auxiliary troops: Dušanić 1999 (*RMD* 202) (Vukovar); *CIL* XIII 4377 (*RIU* 256) (*Arrabona* / Györ); *CIL* XIII 7510 (*Bingium* / Bingerbrück, Upper Germany). Possibly a veteran, identified as a Pannonian, could be a Breucus, which is suggested by the findspot of his diploma - Grabarje near Slavonski Brod (*Marsonia*) (*CIL* XVI 17). If the term *Pannonius* is here indeed devoid of administrative meaning ("the inhabitant of the province of Pannonia"), it would possibly mean that Rome considered the Breuci, as well as the Colapiani, as the original Pannonians.

[47] Zaninović 2003, 445-446; cf. Domić Kunić 2006, 79.

[48] *Civitas Cornacatium*: *CIL* V 6985 + 6986 (*Taurinium* / Torino); *Cornacatus*: *CIL* XVI 2 (191) (Bela Crkva near *Sirmium*).

[49] Sirmium was possibly taken from the Amantini during Tiberius' war: Domić Kunić 2006, 108-110.

[50] This is probably the source of Strabo's determination of the Pirustae as a Pannonian community. The border of Strabo's Pannonian territory with Illyrian territory (Strab. 7.5,3) coincides with the borderline set by Appian, stretching from the Iapodes (around the Una River and in the Lika) to the Dardani (*Naissus* / Niš, Serbia) (App. 3.14; 4.22) and corresponding with the natural border between Pannonia and Dalmatia, which follows the border between the Adriatic and the Posavina (the Sava valley) watershed: Domić Kunić 2006, 81.

[51] The Maezaei in *cohortes Delmatarum*: *CIL* VIII 9377 and 9384 (*Caesarea* / Cherchell, Mauretania, early Principate); The Daesitiate in *legio Adiutrix*, formed by the members of the Praetorian fleet: *CIL* XVI 11 (*Herculaneum*, AD 70); the Daesitiate in *cohors Belgarum*: *CIL* III 9739 (*Tilurium* / Gardun, Dalmatia, the age of Antonini); a Ditio in *cohors Delmatarum*: *CIL* XIII 7508 (*Bingium* / Bingerbrück, Upper Germany, at the latest around AD 74); the Ditio in the Praetorian fleet: *CIL* V 541 (*Tergeste* / Trieste).

[52] Dzino 2008.

spiritual culture. Written sources testify to two Celtic blocs in Pannonia - the Taurisci and the Scordisci, bordering on Mons Claudius (Plin. *NH* 3.25,148); this cultural division is supported by recent archaeological investigations.[53] Written sources identify the western bloc, the Taurisci, with the Norici and other communities near the Alps (Strab. 4.6,9; Plin. *NH* 3.20,133); the eastern bloc, the Scordisci, are seen as a mixture of Celtic newcomers and indigenous population which Strabo calls the Illyrians or Thracians (Strab. 7.5,2). It seems that the Celtic newcomers imposed some sort of domination (military or political) on the local population, which is reflected in the disappearance from the historical record of the names of the indigenous ethnic communities, united under the common (political) names of the Taurisci and Scordisci. Under the name *Taurisci,* one should also look for the communities west of the Slavonian Mountains (the Varciani, and perhaps the Iasi and Oseriates), and the term *Scordisci* should also denote the communities east of this natural barrier (the Andizetes and Amantini, and perhaps even the Breuci). While the Taurisci entered historiography as entrepreneurs interested in trading and mining (Polyb. in Strab. 4.6,12), the Scordisci, depicted as robbers of Greek temples, were for nearly a century and a half a threat to the Roman positions in the Balkans, endangering the borders of the province Macedonia. The wars with the Scordisci are mentioned in fragmentarily preserved writings of Polybius and Livy, and Suetonius testifies to their final outcome: in the Pannonian war of 12-9 BC the Scordisci were already involved on the Roman side, fighting the Breuci. Strabo is indispensable for determining the territory of the Scordisci in the time of the wars with Rome (Strab. 7.5,2; 7.5,10; 7.5,12). A tombstone of a *praefectus Scordiscorum*, found in Stari Slankamen (*Acumincum*) in the easternmost part of southern Pannonia, testifies to the fate of the political (military) alliance of the Scordisci after being defeated by the Roman legions.[54] After the Scordiscan defeat, other names of the indigenous communities come to the surface - the Breuci around the Sava, the Andizetes around the Drava, and the Amantini in Syrmia.

4. Social and cultural anthropology

4.1. Pre-Roman society

Appian claims that the Pannonian society is organised on the tribal principle, without a clearly defined central power (App. 4.22). Communities are gathered as the need arises, in the times of greater danger or crisis, which is clearly confirmed by the reports on the wars that Octavian Augustus waged in Pannonia. In these circumstances the leaders of the communities now united for the purpose of defence are chosen in order to coordinate the movements of the "armies" of individual communities (cf. App. 4.22). Some help in filling the gaps in our knowledge of Pannonian social structure is furnished by archaeology, and one should not neglect epigraphy, which along with archaeological finds and literary sources gives us invaluable information. The few known *principes*, the leaders of the individual communities in Pannonia (as well as Dalmatia), testify to the survival of the indigenous class society and its adjustment to the new conditions directed by Roman administration.

4.2. Settlements

The same Appian's paragraph (App. 4.22) gives us important information about the settlements in Pannonia in the time of the conflict with the Roman legions at the end of the 1st century BC. There were no towns in Pannonia yet at that time - Pannonia knows only of villages, constructed most probably of wood, a material highly available in this intensely wooded area. Dio Cassius testifies to the rural character of Pannonia, mentioning the "villages in the plain" (the Sava or the Kupa) (Dio 49.37,1). Appian's claim that the Pannonians live "throughout the countryside" (App. 4.22), or Tibullus' that they are scattered "on the icy Alps" (Tib. *Paneg.* 3.7,106-112) is a common phrase used for the population which is unfamiliar with the Mediterranean urbanisation – Greek and Roman historiography used the same terms to characterise Gallic and Germanic communities, emphasising strongly the difference between the "primitive" Barbaric societies in relation to the civilised Mediterranean world.

Owing to the results of archaeological investigations, the credibility of the historical sources is confirmed, while the contours of the settlement geography of pre-Roman Pannonia are becoming clearer. The region between the Drava, Sava, and Danube was a region scattered with settlements located on hills or elevated fluvial terraces, protected from seasonal floods. Such places were the only ones suitable for living in the landscape filled with forests and swamps. It seems that the western part of the area of study held no hill-top settlements similar to those in neighbouring Dalmatia, which the written sources call "strongholds in the mountains" (Vell. 2.115,4), or "mountain fortresses" (Dio 55.30,6). The general term τεῖχος, "stronghold" (Dio 55.34,5), where Bato the Daesitiatian captured Bato the Breucan, does not help in determining the character of that settlement - perhaps it was not a settlement at all, but a military fortification. Nevertheless, one town is mentioned in southwestern Pannonia - Segestica, the largest centre in this part of the region. Appian calls it πόλις ἐχυρά ("a fortified town", App. 4.22) and describes its defence system (rivers, ditches, and fortified walls - App. 4.22-24). To this agrees Dio's report stating that Segestica (which he also calls a town) is fortified with strong walls, palisades, and ditches (Dio 49.37). There were several trading and traffic centres in southern Pannonia (Poetovio and Mursa on the Drava, Emona, Nauportus, and Sirmium on the Sava), but the sources mention them only in the context of the Roman period, when these pre-Roman settlements were already transformed into urban centres by Roman standards. Archaeological excavations revealed several settlements

[53] Dizdar 2003; 2006.
[54] Mócsy 1957, 488.

(*oppida*)[55] in the part of the southeastern Pannonia ascribed to the Scordisci. The system of *oppida* along the Danube was transformed by the Romans during the 1st century AD into the borderline fortifications (*castella*). Strabo mentions two Scordiscan "towns" (πόλεις), but does not give their location (Strab. 7.5,12).

4.3. Psychophysical traits of the Pannonians

The Pannonians were not exempt from the stereotypical image of the Barbarian - most characteristics ascribed by the Romans and Greeks were valid for any other population which the Mediterranean world thought to be barbaric. The Pannonians are seen as hearted and belligerent by Appian (App. 4.24), Velleius Paterculus (Vell. 2.115,4), Statius (Stat. *Silv.* 1.4,78), Dio Cassius (Dio 50.28,4), and Herodian (Hdn. *Hist.* 2.9,11). Testimonies of the first three authors are most probably directed to the population of southern Pannonia,[56] while Dio Cassius and Herodian refer to entire Pannonia. Mentioning "warlike temper" and "wonderful knowledge of fighting" of the Daesitiates during the Pannonian rebellion, Velleius seems to express both abhorrence and admiration; a positive attitude will become more prominent in late antique authors, who perceive the Pannonians as the inhabitants of the Roman Empire rather than the Barbarians. The roots of this belligerence and courage, that in the Roman mind bordered with madness, was to be found in the environment (cool climate, thick forests and impenetrable swamps), as was the usual practice in the Greek and Roman scholarly circles. In the case of the Pannonians, such point of view is represented by Dio Cassius and Claudius Mamertinus, and by Florus in the case of the Scordisci (Flor. *Epit.* 1.39,3).[57] In the eyes of the Roman world the Pannonians were considered to be deceitful. They were characterised as such by Tibullus, a contemporary of the Octavian's military campaign in 35-33 BC (Tib. *Paneg.* 3.7,109). Judging by the reports on the conquest of Segestica (App. 4.23; Dio 49.37,2), the deceitfulness of the Pannonians can be connected to the fact that after the fall of the town the residents of Segestica did not abide to the initial agreement with Octavian and jeopardised the Roman troops settled there. Compared to the Scordisci, the Pannonians, that is, La Tène population in the southeastern Pannonia, did well in this characterisation. All of the historical sources agree on the fact that the Scordisci were an extremely cruel and intimidating people both by their physical strength and the horrible customs that are discussed centuries after.[58]

[55] About the fortified settlements in eastern Slavonia and Syrmia see Majnarić-Pandžić 1984; Popović 2003.

[56] Appian's information originates from Octavian Augustus and is dated to the time of the conquest of southern Pannonia. Velleius is a contemporary of the introduction of Roman rule in the area, and Statius writes about the time when northern Pannonia (north of the Drava) was annexed and therefore uses mostly outdated information.

[57] About the belief that the landscape affects psychophysical traits of the population: Woolf 2011, 34-35, 44-48.

[58] For example, Pomp. Trog. in Just. *Epit.* 24.4; Flor. *Epit.* 1.39,2-4; Oros. *Hist.* 4.23,18; Arr. 1.4,6-8.

5. The Roman conquest

5.1. Earlier unsuccessful wars (2nd century BC)

The purpose of this paper is not to discuss whether Rome had a plan for the conquest of the region around the Danube (to which southern Pannonia belongs), or if it acted according to the immediate needs of the foreign policy. The foundation of Aquileia in 181 BC might be a direct indicator that Rome's expansionist plans were in a measure long-term. Aquileia was supposed to protect southeast border of Italy, which was in danger by raids from Noricum, Iapodia, and Pannonia (cf. Strab. 5.1,8; App. 4.18; Dio 49.34,1-2; 54.20,1-3; 54.22,1-2). Aquileia was also a base for the Roman military and merchant expansion towards the Danube. Strabo informs us that the Roman presence in the areas around the Drava and the Danube was initially for pure economical reasons (Strab. 4.6,10; 4.6,12; 5.1,8; 7.5,2). However, in the mid 2nd century BC Rome undertook the first steps to gain complete control over southern Pannonia. It is possible that the first conflict with the Pannonians, traditionally dated to 156 BC, was either due to the need to protect Italy from yet another attack from the direction of the Alps, or it was a concrete military action in order to gain control over the trading routes leading through southern Pannonia towards the Danube. This first war was reconstructed from few information provided by Polybius (Polyb. frg. 64 = *Suda*, frg. 122) and Appian (App. 4.22; 3.14); all that is known is that at the beginning of the war the Pannonians conquered a certain stronghold and that the action finished as a disaster for Rome, since (the consul?) Cornelius was defeated. The goal was to overtake Segestica, as can bee seen in Appian's narrative. Appian reports on the next attempt to conquer Segestica, dated to 119 BC. His rather confusing report on that war (App. 2.10; cf. 4.22) is the result of his lack of knowledge of the early history of Pannonia, which he admits himself (App. 3.14). It seems that also this war was not favourable for Rome, because Appian decidedly claims that Segestica was only conquered by Octavian Augustus. We can only guess what valuable information on these early conflicts is lost in Livy's (and Polybius') books, missing exactly in the parts covering the period in question.

Around the time of the first known conflict with the Pannonians, the wars with the Delmatae and Scordisci began. This event is only briefly mentioned in the summary of Livy's *History* created by Iulius Obsequens (*liber* 16). It is possible that the Scordisci were involved in this war only as Delmataean mercenaries that fought the Roman legions in Delmataean territory[59] - another half a century will pass before the battlefield is moved to Slavonia and Syrmia, the core territory of the Scordisci. A series of wars with the Scordisci followed in the 2nd and 1st centuries BC, and Pannonia enters the scene only with Octavian's military expedition in 35-33 BC. Like in Pannonia, the start was disgraceful for the Roman legions, as Livy reports of the Roman defeat in 140s BC (Liv. *Per.* 54). With time the

[59] Domić Kunić 2006, 87, n. 12.

conflicts became more frequent, and by the end of the 2nd century the wars were waged almost every season, with each side taking its turn in winning.[60] Until the beginning of the 1st century BC all conflicts took place in the area of the Vardar's lower flow, and the purpose was to withstand the attacks on the northern border of the province of Macedonia. The victory of consul Lucius Cornelius Scipio Asiagenus in the 80s BC (App. 1.5) was an overturn in the Roman strategy with the Scordisci - the conquest of the Scordiscan core territory, eastern Slavonia and Syrmia, began. Florus informs us that Caius Scribonius Curio was the first general to lead his legions as far as the Danube bank, or the very centre of the territory of the Scordisci (Flor. *Brev*. 1.39,6); Augustus' short comment on expanding the borders of Illyricum to the Danube might relate to this event (Aug. *RG* 30). It is not certain in which conflict the Scordisci were finally defeated - it probably happened between 16 BC, when the last known conflict with the Scordisci on the northern Macedonian border occurred (Dio 54.20,3), and 12 BC, when the Scordisci participated as allies in Tiberius' war against the Breuci (Dio 54.31,3; Suet. *Tib*. 9). With the defeat of the Scordisci, eastern Pannonia around the Danube was conquered. Rome already controlled the upper reaches of the Danube, opening thereby the way for conquering the whole of southern Pannonia. The remaining part of the Danube course, the great Danube bend, both politically and economically secured the basis for the conquest of the land between the Drava and Sava. Rome will acquire Transdanubia with diplomacy and not by sword,[61] but before that it was necessary to gain control over southern Pannonia.

5.2. Octavian's Illyrian military campaign (35-33 BC)

A well planned and successful military campaign to the west part of the region between the Drava, Sava, and Danube took place 84 years after the last attempt of conquering Segestica. Segestica was Octavian's main goal - conquering this important political and economical centre gave Rome control over entire southwest Pannonia.[62] The course of Octavian's military campaign is well known owing to Appian (App. 4.16 - 5.28), who used Octavian's notes (reports to the Senate and memoirs, which military leaders used to write glorifying their campaigns[63]), and Dio Cassius (Dio 49.35-38), who complements Appian's report. The whole of Illyricum was a part of Roman strategic planning - from the coast and its hinterland to the areas around the Kupa, to the Sava and the Drava. After the introductory part of Appian's narrative[64] and reports on the warfare with the Iapodes, a detailed description of the siege and conquest of Segestica follows (App. 4.22-24). The description contains the testimony of the advances of the Roman legions along the Kupa - a territory inhabited by the Colapiani, to which belongs Segestica, Octavian Augustus' main strategic goal. Even though the sources do not mention other Octavian's goals in Pannonia ("other Paiones" that Octavian defeated were the ones that came to help Segestica under siege, App. 4.23), it is highly unlikely that he would miss out on the chance to gain control over the entire western part of southern Pannonia, when once conquered the area around the Sava; possible battles in the territory around the Drava have not been mentioned in the written sources. Dio Cassius follows Appian's narrative, adding an occasional detail but not disrupting the unity of the two reports.[65] Specific details, drawn out of context, are provided by Strabo, Pliny the Elder, Suetonius, Florus, Rufius Festus, and the poet Tibullus, but they do not contribute significantly to Appian's and Dio's narratives.[66] It is a true shame that the lost parts of Livy's *History* are exactly the ones dealing with this period - it is evident from the preserved summaries that Livy reported on the whole course of the war (Liv. *Per.* 131-132). It is also a pity that Strabo, a contemporary of the events, did not venture into a more detailed description - the nature of his work (geography and ethnography) obviously left no space for that.[67] Augustus successfully accomplished his goals - pacification of the Delmatae (App. 5.25-28; cf. 4.17; Dio 49.43,8), penalty measures on some coastal and island communities (App. 4.16), conquest of the Iapodian territory (App. 4.16; 4.18-21; Dio 49.34,2; 49.35), and the conquest of the western part of southern Pannonia (App. 4.22-24; cf. 4.17; Dio 49.36-38). The control over this expanded Roman territory was gained, and the borders of Illyricum were moved further north, to the Drava. Sources testify of the triumph assigned to Octavian Augustus by the Senate for this acquisition (Dio 49.38; 51.21,5; Liv. *Per*. 133). Yet, the job in Pannonia was far from finished. Florus implies of a need for new wars, justified by the arrogance of all northern Barbarians (Flor. *Epit*. 2.21). Among the Barbarians he lists the "Illyrians, Pannonians, and Delmatae". Octavian's military campaign itself had a very interesting justification: a strategic need to turn Segestica (presumably) into a base for the anticipated war against the Dacians (App. 4.22-23; cf. Strab. 4.6,10; 7.5,2) on the one hand, and the need to train legions for a forthcoming conflict with Marc Antony (Dio 49.36) on the other.

---

60 Liv. *Per*. 56 (between 136 and 134 BC); Eutr. *Brev*. 4.24 (118 or 114 BC); Liv. *Per*. 63 (between 114 and 111 BC); Eutr. *Brev*. 4.27 (109 BC).

61 Transdanubia became part of the Empire during Claudius' reign (AD 41-54); Mócsy 1974, 48-49; 1979; Fitz 1977; 1990, 337-338; Tóth 1977.

62 Concisely about Octavian's campaign see Domić Kunić 2006, 91-100 (with previous literature); Šašel Kos 2005, 393-471; Dzino 2010, 99-116.

63 The first detailed description of Pannonia probably originates from Augustus' notes, related by Appian (App. 3.14; 4.22).

64 In the introductory part of the report on Octavian's expedition Appian brings the classification of military actions, based on the intensity and the expedience in gaining the goal, respectively (App. 4.16-17). With the Segestani and dubious δαισιοι(τε), which might be the Daesitiates - that is, with the only two Pannonian communities mentioned in that list, Octavian had most trouble (App. 4.17).

65 The Pannonian phase of the expedition (siege and conquest of Segestica) is described in Dio 49.37-38; cf. 50.24,3-5; 50.28,3-4 (Octavian's speech to the army before the battle with Marc Antony in 31 BC); 53.7,1 (Octavian's speech in the Senate in 27 BC).

66 Strabo mentions the conquest of Iapodia (Strab. 7.5,4) and of Delmataean settlements (Strab. 7.5,5); Pliny - Octavian being wounded, probably during the Iapodian phase of the expedition (Plin. *NH* 7.1,48; cf. App. 4.20; Dio 49.35); Suetonius - defeat of the Delmatae (Suet. *Aug*. 20); Florus - the conquest of Iapodia (Flor. *Epit*. 2.23); Rufius Festus - the conquest of Iapodia and Pannonia (Fest. *Brev*. 7); Tibullus alludes to the Pannonian faithlessness, already mentioned in the paper.

67 Strabo was a young man during Octavian's campaign (35-33 BC), middle-aged during Tiberius' war (12-9 BC), and an old man at the time of the Pannonian rebellion (AD 6-9).

5.3. Tiberius' Pannonian war (12-9 BC)

Literary sources report on new disturbances in Pannonia and areas around the Alps from 16 BC onwards. Rome successfully responds to the Pannonian and Norican raids in Histria in the years 16 and 15 BC (Strab. 4.6,9; Dio 54.20,2-3).[68] The next season (14 BC) disturbances continue and the Pannonians were defeated again (Dio 54.24,3; cf. Flor. *Epit.* 2.24; Jord. *Rom.* 243). The context of the mentioned narratives gives the only logical solution as to which Pannonians this refers to - they are the same Pannonians Octavian Augustus defeated some twenty years before. It is impossible to identify them more closely (Varciani? Iasi? Colapiani?); it was possibly a combined action of several communities, but unfortunately, we will never be able to know that. The only thing that is certain is that this initial "disturbances", as the sources call them,[69] turned into a true military campaign under the leadership of Tiberius, Augustus' *legatus*. Tiberius' war in Pannonia is poorly documented, which is explained by the general negative approach of the historical sources towards Tiberius as an emperor.[70] The silence in historiography has unjustly reduced this important war to a series of disputes with the rebels. Nevertheless, it was a true military campaign in every sense, and it gained Rome control over the remaining parts of southern Pannonia, as well as Bosnia.

The most complete report on Tiberius' war in Pannonia has been left by Dio Cassius (Dio 54.24,3; 54.28,1-2; 54.31; 54.34,3; 54.36,2-3; 55.2,4). Velleius Paterculus is very concise, but with a reason - as a participant of the war he promised to write a whole monograph on it, but nothing is known about it (Vell. 2.39,3; 2.96,2-3). Other sources produce some details, which to a certain measure help to complete the picture of the war actions (Suet. *Aug.* 20 and 21, *Tib.* 9 and 14; Aug. *RG* 30; Fron. *Str.* 2.1,5; Liv. *Per.* 141; Flor. *Epit.* 2.24; Fest. *Brev.* 7; Eutr. *Brev.* 7.9). Dio Cassius describes the starting operations, led by Marcus Vipsanius Agrippa, Augustus' most capable and most trusted commander-in-chief (Dio 54.28,1-2), to proceed with the further development of the war (Dio 54.31), in which he mentions the "Pannonians", whose main adversaries were the Scordisci - it transpires from Suetonius' narrative that these *Pannonians* are actually the Breuci (Suet. *Tib.* 9). According to Dio Cassius, the war was at the same time going on in Dalmatia; in spite of anti-Tiberian mood of his sources (and possibly himself as well), this short report cannot hide Tiberius' big war successes (Dio 54.34,3-4), which, however, were not adequately remunerated - he was denied a true triumph and had to be satisfied either with an ovation (Vell. 2.96,3) or with triumph decorations (Dio 54.31; 54.34,3-4; 55.2,4). The remaining communities in southern Pannonia came under Roman rule on this occasion - apart from the Breuci mentioned in the sources, these were the Oseriates with their centre in Donja Dolina, the Andizetes with their centre in Mursa, and the Amantini whose centre was in Sirmium. These communities are not mentioned in the reports on the Pannonian war, nor is there any mention of the communities south of the Sava (the Ditiones, Maezaei, and Daesitiates), even though by all likelihood they came under Roman rule on this occasion as well.[71] Dio Cassius specifically states that most of the men fit to fight were sold into slavery after the war (Dio 54.31), which could mean that Rome did not yet undergo (mass) recruitment of the Pannonian defeated population. This is confirmed by epigraphic evidence; the earliest known case of recruitment among the Pannonians (*Dases Dasmeni f. Cornacatus*)[72] is dated to the time of Tiberius' reign, which began 25 years later. As already stated before, Tiberius, supposedly, was not permitted to celebrate a true triumph - anti-Tiberian propaganda interpreted this as a sign of Augustus' averseness towards his (best) general (Dio 54.31); instead, he was awarded with a triumph of a lower rank for his accomplishments in the Pannonian war (Vell. 2.96; Dio 55.2,4; Suet. *Tib.* 9).[73]

For the years that followed, historiography reports only of peaceful periods. The report on Dacians crossing the frozen Danube and robbing Pannonia (Dio 54.36,2-3) might refer not to the newest Roman acquisition in southern Pannonia, but to Transdanubia, which the Roman foreign policy perceived as part of Pannonia, being already subdued to Rome through the system of friendships and alliances.

5.4. Pannonian rebellion (Bato's rebellion) of AD 6-9

Tiberius' war in Pannonia was the last Roman military campaign in Illyricum, which by this time expanded from the eastern Adriatic coast to the Drava. A period of the reinforcement of Roman control and organisation of the infrastructure in the province ensued. It is perfectly logical to presume that tax collecting among the defeated communities started immediately because one of the reasons for the rebellion (which happened some 15 years later) was the discontent created by high taxes and recruitment for the purposes of war against the Marcomanni (Dio 55.29; 56.16). Velleius Paterculus sees the reasons for the rebellion differently: according to his interpretation, the rebellion was a wantonness of the Barbarians who cannot get accustomed to peace and prosperity (Vell. 2.110). The great Pannonian, or Bato's rebellion (6-9 AD) expanded to entire southern Pannonia, later to extend to Dalmatia, as directly attested by

[68] At the same time the Scordisci threaten Macedonian northern border again (Dio 54.20,1-3).
[69] All these events, occurring from 16 BC on, Dio Cassius persistently calls "disturbances". If it is the same Pannonians Octavian defeated in 30s BC, Rome could have indeed referred to their war actions as to disturbances (the actions took place in territory under Roman rule). Tiberius used the disturbances as an excuse to conquer the still unconquered eastern part of southern Pannonia.
[70] Domić Kunić 2006, 102-113.

[71] Domić Kunić 2006:106 for the Oseriates, 107-108 for the Andizetes, 108-110 for the Amantini, 110-113 for the Ditiones, Maezaei, and Daesitiates. The Daesitiates might have been defeated during Octavian's war of 35-33 BC, according to a dubious citation in Appian: δαισιοιτε (App. 4.17). One might conclude that the Daesitiatian involvement in the war was as Delmataean allies and that they were defeated in the territory of the Delmatae, rather than in their own territory.
[72] *CIL* XVI 2 Bela Crkva near Sremska Mitrovica (*Sirmium*). This military diploma is dated before AD 54, which implies that Dases was recruited some time before AD 29.
[73] Domić Kunić 2006, 115.

Velleius Paterculus who took part in fighting the rebellion (Vell. 2.110),[74] and Strabo, a contemporary of the events (Strab. 7.5,3). Strabo's list of Pannonian "tribes" is actually a list of the communities that took part in the rebellion – it is the only way to explain the inclusion of the Pirustae, who cannot be ascribed the Pannonian ethnic identity.

Because it caused a lot of fear and suspense for the Romans, Bato's rebellion is well documented in Roman historiography. The rebellion turned into a long-term war for the rescue and preservation of the newly acquired territory in southern Pannonia. The extent of the crisis and panic can be seen in Suetonius (Suet. *Tib.* 16), Pliny the Elder (Plin. *NH* 7.45,149) and Velleius Paterculus (Vell. 2.110; 2.111; 2.113).[75] Velleius' report is rather inadequate, which is a great shame considering the amount of first-hand information that he must have had. His *Roman History* is structured as a brief survey of general history, with no room for details. A promised separate report on Bato's rebellion (Vell. 2.114,4) was either never written, or was lost during the course of time. The course of the war can be reconstructed fairly accurately combining the narratives of Velleius Paterculus (Vell. 2.110-115; 2.117; 2.121) and Dio Cassius (Dio 55.29-30; 55.32-34; 56.16-17), and specific details brought by Suetonius (Suet. *Tib.* 16; 20), Rufius Festus (Fest. *Brev.* 7), and Jordanes (Jord. *Rom.* 216). Only three most important episodes will be mentioned: the battle of Volcaean marshes, which nearly ended as a great Roman catastrophe (Dio 55.32; Vell. 2.112), the capitulation of Pannonia (Dio 55.34), and the fall of Delmataean hill-fort Andetrium, which marked the end of the war (Dio 56.12,3-14,7; cf. Vell. 2.115). The end of "the most serious of all foreign wars since those with Carthage", as described by Suetonius (Suet. *Tib.* 16), was followed by great victory festivities. The honours earned by certain *legati* are mentioned in the inscription from Tivoli (Antique *Tibur*) northeast of Rome (*CIL* XIV 3613), as well as by several historians (Vell. 2.115-116; Dio 56.17; Suet. *Tib.* 17). The triumph itself is noted in the *Fasti Praenestini*,[76] and the course of the ceremony is described by Ovid (Ov. *Ep.* 2.1,19-118; 2.2,75-92), Velleius Paterculus (Vell. 2.121,2-3) and Suetonius (Suet. *Tib.* 20).

The final pacification of Pannonia and Dalmatia marked the beginning of a real control over the Illyrian territory between the Drava and the Adriatic Sea. Historiography is full of short statements of the conditions required for introducing Roman administration to the region, to mention but a few: Vell. 2.39,3; Suet. *Aug.* 21; Eutr. *Brev.* 7.9; Fest. *Brev.* 3.4; Nic. Dam. *FGrH* frg. 125(1); Ov. *Tr.* 2.1,225-230; Oros. *Hist.* 6.21,12-27.

6. Roman administration

6.1. Organisation of the province - the division of Illyricum

Octavian's Illyrian campaign 35-33 BC and Tiberius' Pannonian war 12-9 BC expanded the territory of Illyricum, moving its northern border to the Drava.[77] The area between the Sava and Drava made the northern stretch of the province, with the Posavina as the border with Dalmatia on the south.[78] The official name of the province was *Illyricum*,[79] although alternative names *Dalmatia* and *Pannonia* were concomitantly used. Appian states that Illyricum is a κοινή (community) within which he recognises several ethnic entities - he lists the Paiones, the Raeti, the Norici, the European Mysians (App. 1.6). The size of the territory of the province of Illyricum[80] and political and administrative problems resulting from such vastness resulted in the division of Illyricum into two provinces.[81] It is not certain when exactly this happened - either during Bato's rebellion, or shortly after the rebellion. The main argument for this lies in the report of Velleius Paterculus who calls Marcus Valerius Messalla Messallinus *praepositus Illyrico* (Vell. 2.112,2) in the initial stage of the rebellion in AD 6; three years later, during the final operations (AD 9), Caius Vibius Postumus is referred to as the *praepostus Delmatiae* (Vell. 2.116,2). Therefore, Illyricum must have been divided into Dalmatia and Pannonia some time between AD 6 and 9. Even though this theory is broadly accepted, it has no serious arguments: Illyricum and Dalmatia do not necessarily mean two different entities in the writings of Velleius Paterculus - historiography is full of similar synonyms. It is considered

[74] Velleius was involved in the Pannonan phase of the war, 6-8 AD (Vell. 2.111); his brother Magius Celer Velleianus, who participated in its final phase, provided the report on the Dalmatian phase, 8-9 AD (Vell. 2.115).

[75] Rome was in crisis during and immediately after the war, because the absence of Germanic legions, engaged in the suppression of the rebellion, prompted the Cheruscan leader Arminius to a revolt which ended with the annihilation of all three legions in Lower Germany, which were stationed on the Rhine under the command of Publius Quinctilius Varus. Bato's rebellion and the massacre of Varus' legions were the reasons Octavian Augustus abandoned the plan to move the northern border from the Rhine to the Elbe.

[76] *Ti. Caesar curru triumphavit ex Ilurico* (*Fasti Praenestini, Calendarium*, dated to October 23, AD 12). Other citations in the *Fasti* of the towns Antium and Praeneste might be referring to that triumph, but also to the triumph after the Pannonian war BC 12-9: *ob res in Ilyrico bene gestas* (*Fasti Antiates, Calendarium, Inscr. It.* IV, 1, p. 122); *Ti. Caesar ex Pan[nonia et Delmatia triumph]avit* (*Fasti Praenestini, Inscr. It.* XIII, 2, p. 114); *Ti. Aug. Inlyrico vicit* (*Fasti Antiates, Inscr. It.* IV, 1, p. 122).

[77] Transdanubia, the region between the Drava and the great Danube bend, was not under Roman rule at that time but was connected to Rome through diplomacy, and political and economical dependency. Annexation will take place about 50 years later, during Claudius' reign.

[78] A milestone (*CIL* 3201 + 3198b) from Salona implies of that border, which some time later became the official border between the two provinces, and the building of Dalmatian roads in the time of the governor Publius Cornelius Dolabella (14-20 AD). Five roads which started in Salona radiated through the Dalmatian inland towards the northern border with Pannonia - for one of them it is stated that the direction is *ad fines provinciae Illyrici* (*CIL* III 3198a + 3200).

[79] The official name *Illyricum* is found in the *Fasti* of several towns in Italy, along with the triumph(s) approved to Tiberius. Variants of the name are: *Inlyricum* (*Inscr. It.* IV, 1, p. 123), *Ilyricum* (*Inscr. It.* IV, 1, p. 122), *Iluricum* (*Fasti Praenestini, Calendarium*). It seems that the official terminology changed rather slowly, because even during Claudius' and Nero's reign Pannonia was, in an official context, called Illyricum - cf. diplomas from Bela Crkva (*CIL* XVI 2), Vukovar (Dušanić 1999 = *RMD* 202) and *Vindobona* / Vienna (*CIL* XVI 4); cf. also Dio 12.19 (Zon. 8.19).

[80] According to the general Roman perception, Illyricum extended to the Danube from the very beginning. Till Claudius' reign Roman rule expanded to the Drava, while Transdanubia was only a zone of interest. With the annexation of Transdanubia the Roman rule extended to the Danube.

[81] Šašel Kos 2010; Dzino 2010, 156-167.

that during the division of Illyricum the two new provinces had the official names Upper Illyricum (southern province) and Lower Illyricum (northern province), but there is no evidence of that. The theory is based on an (today lost) inscription to Publius Cornelius Dolabella, first governor of the southern province (*CIL* III 1741 *Epidaurus* / Cavtat). The inscription was badly damaged in the place where the name of the province is mentioned, and this part has been reconstructed as *civitates superioris provinciae Hillyrici*, but that is pure guessing.[82] According to such reconstruction, the southern province was at least for some time officially called *Illyricum Superius* (Upper Illyricum), implying that the northern province was called *Illyricum Inferius* (Lower Illyricum), although the latter remains hypothetical. At the same time, the southern province was called *Dalmatia / Delmatia* and the northern province *Pannonia*.[83] The use of the alternative name in historiography was confirmed even prior to the division of Illyricum - the terms *Dalmatia* and *Pannonia* are used at that time (unofficially?) to denote the territory south and north of the Sava, respectively. Velleius' *Illyricum* and *Delmatia* are synonyms for the southern half of the still unified province.[84] It is unlikely that Rome would undergo a demanding task of the administrative organisation of the restless territory during the rebellion, while the main aim was to preserve Illyricum in the fist place. It is more probable that the division took place after the suppression of the uprising, when all conditions to organise the pacified territory were acquired. The historical sources reporting on the rebellion confirm that Pannonia was still part of Illyricum during the revolt. Suetonius mentions Illyricum (Suet. *Tib.* 17), and Dio Cassius explicitly calls Mesallinus the governor of Dalmatia *and* Pannonia (Dio 55.29,1) - a (still) undivided province. It is not certain when the two provinces were officially given the names *Dalmatia / Delmatia* and *Pannonia* - the name *Illyricum* (for Pannonia) is still officially used in the military diploma dating to AD 60 (*CIL* XVI 4), while the first official evidence for the name *Pannonia* dates from AD 68/69 (*CIL* X 6225 *Fundi*).[85]

As already seen, the names *Illyricum / Illyrii*, *Pannonia / Pannonii* and *Dalmatia / Delmatae* are still used for a long time in Roman historiography in a rather promiscuous manner. In some cases the situation is pretty clear - the administrative meaning of the terms *Pannonians* ("inhabitants of the province Pannonia") and Dalmatians ("inhabitants of the province Dalmatia") can be found in Velleius Paterculus (Vell. 2.115), who calls the Pirustae and the Daesitates "*Delmatae*", and in Tacitus (Tac. *Hist.* 3.12; 3.50), whose *Delmatae / Dalmatae* and *Pannonii* are widely attested epigraphically in the service of the praetorian fleets; under the name Dalmatians ("inhabitants of the province Dalmatia") Dio Cassius lists the Daesitates (Dio 55.29; 55.30,4; 55.31,1) and Maezaei ("a Dalmatian tribe", Dio 55.32,4; cf. 56.11,1 - the Maezaeian settlement called Splonum is located "among other places in Dalmatia"), while the Breuci are "a Pannonian tribe" (Dio 55.29,3) and therefore inhabitants of the province Pannonia. The same can be applied to Dio's Dalmatians and Pannonians (Dio 55.33,1; 55.34; 56.12,1) as well as to Dalmatia and Pannonia (Dio 55.34; 56.12,1). Since the provinces Dalmatia and Pannonia are well established in the 3rd century AD, it is considered that Dio Cassius was using contemporary terminology.[86] It is not certain whether Velleius' Pannonia and Dalmatia (Vell. 2.110,2), as well as "the tribes of Pannonians and the races of Dalmatians" (Vell. 2.69,2) should be regarded as specific geographical areas (something like Strabo's and Appian's Pannonia and Dalmatia) or as Roman administrative units. The dilemma is evident when Velleius mentions the "provinces" Raetia, Vindelicia, Noricum, Pannonia, and the Scordisci (Vell. 2.39,3). I am more inclined to understand these names in terms of geographical regions, or in the best case as Roman *provinciae* in the original sense (territory assigned to a military commander that is yet to be conquered or pacified). If there were no mention of the Scordisci, it would be perfectly logical to regard the remainder as already organised administrative units (provinces).

The province of Pannonia met the fate of Illyricum at the beginning of 2nd century AD, when Trajan (AD 98-117) divided Pannonia into two separate provinces: Upper (western) Pannonia and Lower (eastern) Pannonia, with the border following the great Danube bend and the eastern Balaton margin, crossing the Drava at Donji Miholjac (*Marinianae*) and extending to Nova Gradiška (*Servitium*) over the east side of the Slavonian Mountains. Pannonia as divided is mentioned for the first time by the geographer Claudius Ptolemy (Ptol. 2.14; 2.15), and the authors of the *Imperial history* (as well as epigraphy) recorded the name of the first known governor of Lower Pannonia - the future emperor Hadrian (*SHA*, *Hadr.* 3.9).

[82] Novak 1965.

[83] As a rule, Pannonia was not alternately called *Illyricum* - that name was reserved for Dalmatia, an old Roman occupancy from where the name *Illyricum* spred to the north. Illyricum in the meaning of *Pannonia* is found, for instance, in Tacitus (Tac. *Ann.* 1.46; 1.53) and Suetonius (Suet. *Tib.* 25), as well as in late antique writings. Besides, the name *Illyricum* denoted the whole area along the south bank of the mid and lower Danube reaches (cf. Tac. *Hist.* 1.76; 2.86, and several later sources - e. g. *SHA*, *M. Ant.* 14.6; 22.1).

[84] E. g. Vell. 2.110,1 - in this context Pannonia and Dalmatia are possibly colloquial names for the two parts of the province Illyricum (at that time - AD 6 - Illyricum was definitely still undivided), rather than for the two provinces - bearing in mind that Velleius wrote his *Roman History* 20 years after the Pannonian rebellion, when the division of Illyricum and the existence of the two provinces are certain. For a similar statement see Vell. 2.112,2. The first suggested option is confirmed by the mention of the *praepositus* (Vell. 2.112,2) and legions in Illyricum (Vell. 2.125,1), which again might suggest that the province was still undivided. Other parts of his narrative, where he mentions a Dalmatian governor (Vell. 2.116,1) and the triumph over the Pannonians and Delmatae (but not the Illyrians or Illyricum) are closer to the second option. In any case, the problem is open for discussion.

[85] Lucius Tampius Flavianus, in whose *cursus honorum* among other things is the service *of legatus Augusti pro praetore Pannoniae*, is mentioned in that inscription: Fitz 1993, 150. Next official confirmation of the name *Pannonia* is found in a military diploma from AD 71 (*CIL* XVI 14, unknown site in Pannonia).

[86] However, one should bear in mind another possibility: that these are not administrative but political units, and that the names *Pannonians* and *Dalmatians* refer to ethnic units which were part of military / political alliances in which the Pannonii and Delmatae respectively were dominant; these ethnic names possibly passed on to the communities subordinated to them.

6.2. Provincial administration - principes, praefecti, praepositi and legati Augusti

Like in many other segments of the organisation of a conquered territory, Rome exhibited a high level of tolerance and political wisdom by including the local elite in the administration. In that way, the possibility of discontent or rebellion among the population was brought to a minimum, and the local elites were an ideal bridge between the indigenous population and the imposed force. Those persons were authorised by the Romans to govern their communities if they showed the will to cooperate with Rome; the trial period, in which Rome wanted to test their adaptability to the system of Roman administration, lasted for at least two generations.[87] In the meantime, in recently conquered and not completely organised Pannonia, military administration led by a *praepositus* was introduced. Dio Cassius testifies to discontent caused by the actions of some of the *praepositi* in Illyricum, which led to the rebellion of wide proportions - high taxes and recruitment for the war in Germania were the immediate reasons for the rebellion (Dio 55.29), and the defeated leader Bato the Daesitiatian drew attention to the abuse of power of a Roman governor (Dio 56.16,3).[88] *Praefecti*, or higher officers (most often centurions) were assigned to assist the *praepositus*. Their task was to install order and preserve peace in particular parts of the province. While examples from Dalmatia are plenty, the only two mentioned probable *praefecti* from Pannonia are the one of the Colapian community, and the other of two large communities of the Maezaei and Daesitiates.[89] After all required conditions for the local *principes* to take control were fulfilled, military administration was replaced by a regular administrative apparatus, in which once again, the local elite found their place at the head of their respective ethnic community, or individual territorial units within it (*civitas peregrina*), or individual urban centres. The earliest examples are dated to the time of the Flavians: *principes* of the communities of the Iasi, Breuci, and Andizetes, and *princeps et praefectus Scordiscorum.*[90] Most of these *principes* had only just acquired Roman citizenship - most probably during the reign of Vespasian, whose endeavours in the sphere of urbanisation and general Romanisation of Pannonia are very well known. The institution of *princeps* was not completely abolished during the advanced Roman period; moreover, most of the testimonies are dated to 2nd century AD; examples in Pannonia are the *princeps Daesitiatium*, a person given citizenship by the Flavians, and a dubious *princeps civitatis Breucorum*, an Amantinus holding Iulian citizenship.[91] According to available historical sources, the military administration was in power all throughout the Iulio-Claudian reign, and was not replaced with regular administration with a governor (*legatus Augusti*) in charge till the rule of the Flavians. The chronology of governorship in Pannonia during the 1st century AD is rather incomplete because the literary and epigraphic sources are quite poor in this respect.[92] Let us just mention Domitian's governor Lucius Funisulanus Vettonianus whose name is inscribed in the honorary inscription originating from the Andautonian territory (*CIL* III 4013, Petrovina Turopoljska).

6.3. Roads and river routes

One of the main pre-conditions for introducing administration in the newly conquered territories was the road infrastructure. Main-road routes through Dalmatia and Pannonia were laid out by Augustus, and finished during Tiberius' and Claudius' reign. Their main purpose was to connect military strongholds, improve on the supplying of the military and enable the movement of the troops through the recently conquered territory - originally, those roads were *viae militares*. Only later were the same roads used as traffic routes.

The building of roads through the province of Dalmatia is documented in two inscriptions from Salona, where five roads are mentioned - four started in Salona as their *caput viarum* and connected the southernmost (coastal) part of the province with its northern border with the province of Pannonia.[93] These roads terminated in the territory which was ethnically speaking Pannonian - the first in the land of Daesitiates, the second at the river Bosna which was the border between the territories held by the Breuci and Oseriates,[94] the third in the land of the Ditiones, while the fourth stretched all the way to the border of the "province Illyricum" (Dalmatia). At the same time as the roads in Dalmatia were built, connections between military strongholds in Pannonia were reinforced. Immediately after the suppression of Bato's rebellion, the Romans started building roads along the courses of the Drava and Sava in order to connect the legionary camps in Emona, Poetovio, and Siscia and to secure communication throughout the area, all the way to Mursa and Sirmium at the eastern end of the province. Two main Pannonian roads, which followed the Sava and the Drava courses, respectively, were initially built for military purposes, that is, for the distribution and transportation of troops. Both are noted in itineraries, with Aquileia as their origin, which testifies to their importance in the later period, when they were used to

87 What is meant here is the time between the pacification of Pannonia (after the suppression of the Pannonian rebellion of AD 9) and the first epigraphic evidences for the local *principes* (end of 1st c.); Wilkes 1969, 104, 266-267, 287-288.

88 Cf. Dio 57.10,5.

89 Probable *praefectus Colaphianorum* (*sic!*): *CIL* III 14387f *Heliopolis* / Baalbek, Syria, between AD 45 and 63 (Bojanovski 1988, 61, 334); *praefectus civitatis Maezeiorum et civitatis Daesitiatium*: *CIL* IX 2564 *Bovianum Undecimanorum*, Samnium, probably AD 75 (Rendić-Miočević 1962, 329-330; Wilkes 1969, 289).

90 *Principes* of the Iasi, Breuci, and Andizetes are mentioned as witnesses in the diploma issued to a veteran of the Ravenna fleet, from the community of the Eravisci (*RMD* 205, unknown findspot); *princeps (et) praefectus Scordiscorum*: *ILJug* 280 *Acumincum* / Novi Slankamen (Mócsy 1957, 488).

91 *Princeps Daesitiatium*: *ILJug* 1582 Breza near Sarajevo (Čremošnik and Sergejevski 1930); *princeps Breucorum*: Bulat 1989, no. 7 (a damaged inscription from Mursa, whose restoration is questionable).

92 Fitz 1993.

93 *CIL* III 3198a + 10156 + 3200; *CIL* III 3201 + 3198b + 10159 + 10156b. Publius Cornelius Dolabella was the governor of the province of Dalmatia (AD 14-20) in the time when the roads were built.

94 This part is dubious because the inscription was badly damaged exactly in the place where the two communities were mentioned.

connect the Empire to the East. The road along the Drava River passes through several settlements, whose importance will in the 1st and 2nd centuries AD be proved by their status - Poetovio in the westernmost part will become a colony during Trajan's reign, and Mursa in the easternmost during the reign of Hadrian. These two important strategic points were connected by series of stations along the Drava River, whose names are mentioned in the itineraries, while their locations are suggested by sporadic archaeological finds. The building of the road along the Sava, started in Augustus' time, was continued by Tiberius (and finished by the time of the Flavians), at the same time when Publius Cornelius Dolabella's roads in Dalmatia were built. As early as the 2nd and 1st centuries BC the most frequent communication between Italy and the region around the Sava was the Emona - Siscia section of the road, which was probably finished immediately after Augustus' campaign of 35-33 BC, was the most travelled route between Italy and the Sava region (Strab. 7,5,2). The stations along this route are documented in the itineraries, and confirmed by archaeological finds. Building and maintenance of roads were the task of the provincial army. The Dalmatian roads were built by the *vexillarii* of the VIIth and XIth legions, which is evident from the already mentioned inscription from Salona (*CIL* III 3198a + 10156 + 3200). The road building in Pannonia is mentioned in Tacitus' report on the rebellion of the Pannonian legions in AD 14 (Tac. *Ann.* 1.20).

River routes through Pannonia were as important as the roads - Strabo mentions a combination of river and land routes in the region between the Drava and Sava that were dominant at least till the end of 1st century BC (Strab. 4.6,10); in another place he mentions Pannonian rivers in terms of the traffic connection between Italy and the Danube area (Strab. 7.5,2). Strabo explicitly claims that Segestica was the centre of the river-transport network through Pannonia; the importance of Segestica as a traffic centre is seen in Octavian's campaign (App. 4.22). After the consolidation of Roman power in Pannonia, a river fleet that was part of the Pannonian fleet, and which included also Moesian ships, patrolled along the Sava.

6.4. Romanisation and the award of citizenship

Romanisation is a process that closely followed the organisation of life in a territory that had become part of the Roman Empire. It was not necessarily a process planned or imposed by the Romans, but rather a natural development, an expression of coexistence of the indigenous population and the representatives of Roman rule. Romanisation first touched the local elites, who had it in their interest to preserve dominance in their communities, a task that in new conditions could have been carried out only by adjusting to the Roman rule and show willingness to cooperate; this was the only way to keep legitimacy of power in the new circumstances.

Since Pannonia was involved in extensive trade with Italy as early as the late Republic period, the Latin language was already present in the region before the Romans included it in their imperialist plans. Velleius Paterculus claims that Latin language was widely spread in Pannonia as early as the beginning of 1st century AD: the language was known to "all the Pannonians" (Vell. 2.110,5) - a very interesting information considering the fact that 40 years had passed since the west part of southern Pannonia fell under Roman rule, and only 15 years since its eastern half was conquered. Velleius is probably exaggerating when he says "all Pannonians" - the Latin language was most probably the *lingua franca* for particular classes or groups in Pannonian society, such as the members of the local elites, auxiliaries originating from Pannonia, and local tradesmen.[95] Pannonia was already Romanised to a certain degree even before Bato's rebellion, which leads us to the conclusion that the process of Romanisation in southern Pannonia closely followed the expansion of Roman rule. The expansion started with Augustus' conquest of the western part of southern Panonnia, and finished with Tiberius' Pannonian war, when the eastern part of the region of study finally fell under Roman rule. According to inscriptions dedicated to the members of a higher social class incorporated in the administration, the first wave of the awarding of Roman citizenship in southern Pannonia happened in the time of the Flavians. One example is Titus Flavius Proculus - *princeps praefectus Scordiscorum* (*ILJug* 280 *Acumincum* / Stari Slankamen) (second half of the 2nd century AD).

6.5. From territorial districts to municipalities

Hardly anything is known about the internal organisation of ethnic communities before the establishment of Roman power - every attempt at reconstruction is based on historical (mainly epigraphic) sources, which are rather limited and anachronistic. One such example is the cenotaph of a 10- year-old Amantinian, a member of the *gens Undius* and *centuria secunda* (*CIL* III 3224 Putnici near *Sirmium* / Sremska Mitrovica). The cenotaph is dated to the beginning of 1st century AD and gives us some insight as to how the local communities were organised. Social structure of the indigenous population was subject to various changes dictated by Roman administration. From the 1st century AD onwards, former social units based on kinship were replaced by territorial units called *civitates peregrinae*. It is not certain whether these units were formed according to the pre-Roman tribal division, or whether Rome set new borders; one should therefore be careful when identifying an indigenous population with a Roman administrative district. Pliny's *Natural History* is a unique source of information for the Pannonian territorial units (Plin. *NH* 3.25,148). The alphabet list[96] does not allow their geographical locations with any certainty; however, it does tell us that Pannonia had an already established administration during Pliny's time (the Flavians). All the

---

95 Džino 2009a, 37.

96 A number of communities were left out from the alphabet list – Pliny felt no need to mention them again, since he lists them in the section where he discusses geography (Plin. *NH* 3.25,147).

listed Pannonian *civitates peregrinae* enter the category of *civitates stipendiariae* - communities without Roman civil rights, and obliged to pay taxes.[97] Most of them can be found in Ptolemy, who outlines their boundaries (Ptol. 2.14,2 for Upper Pannonia; Ptol. 2,15,2 for Lower Pannonia; Ptol. 2.16,5 for the communities in the province of Dalmatia[98]). Pliny mentions only one Pannonian territorial community by name - *civitas Sirmiensium et Amantinorum* (Plin. *NH* 3.25,148), while some others can be found in epigraphy. Three such communities are recorded in inscriptions: *civitas Cornacatium* and *civitas Breucorum* (it is not certain whether the former was originally a separate community, or whether it was subsequently singled out from the latter), and *civitas Colapianorum*.[99] Inscriptions where *principes* or *praefecti* of certain communities are mentioned are also related to peregrine territorial districts, but without the use of the word *civitates*, such as the Iasi, Breuci, Andizetes, and Scordisci.[100] Roman administration mostly complied with the already set borders between the peregrine communities. However, this rule was not unanimously followed, as is exemplified by the Scordisci, a large mixed population inhabiting the vast territory that expanded from the Sava and Danube confluence deep towards south along the course of the Morava. After the division of Illyricum, the Scordiscan territory became part of as many as three neighbouring provinces: Pannonia, Moesia, and Dalmatia. During the 1st century AD the abovementioned eponymous *civitas Scordiscorum* was attested in Pannonia, while the Celegeri are mentioned in Moesia, and the Dindari in Dalmatia - the general presumption is that both these communities were singled out from the Scordiscan original community.[101] The name Scordisci denoted a political alliance that included several ethnic communities, named after the people that were dominant both politically and militarily. After the dissolution of the alliance, the names of other communities, so far hidden behind the name of the Scordisci, came to the surface, and Rome probably only accepted the existing situation.

Pliny mentions two colonies in Pannonia - Emona and Siscia (Plin. *NH* 3.25,147). Unfortunately, Augustus' or Tiberius' colony *Iulia Emona*, an important traffic and trade centre since the prehistoric period, steps out of the geographical frame of this paper. The same goes for Nauportus (Vrhnika) and Neviodunum (Drnovo near Krško) - both were very important stations on the route that connected Italy and the Danube region, as was Poetovio (Ptuj) on the Drava. Siscia gained the rank of a colony during the Flavian period (most probably during Vespasian's reign).[102] In this way, the extreme strategic and traffic importance of the nearby native settlement Segestica was confirmed; Rome was fully aware of the importance of Segestica already in the late Republic period (cf. Strab. 4.6,10; 7.5,2). Strabo and Pliny differentiate the settlement of Segestica from the stronghold of Siscia (Strab. 7.5,2; Plin. *NH* 3.25,148),[103] and Appian and Dio Cassius mention the impressive defence system of Segestica (App. 4.22; Dio 49.37,2-4). All of this reveals the preconditions this most important centre in southern Pannonia fulfilled in order to gain the status of a town in the Roman sense of the word. Siscia as a *colonia Flavia* is mentioned also in inscriptions, and epigraphy testifies to an intensive urbanisation during Trajan's reign, as well as to a number of municipalities formed almost a century later, during Septimius Severus.[104]

During the Flavian reign the province of Pannonia went through its first phase of urbanisation and municipalisation. Vespasian and his heirs were fully aware of the strategic and trading importance of the River Sava valley, therefore assigning the rank of a colony to two crucial stations on the Sava (Siscia nad Sirmium), and the status of a *municipium* to two settlements in the northwestern Sava stretches: Neviodunum (Drnovo near Krško) and Andautonia (Ščitarjevo near Velika Gorica). All of these towns lie on the itinerary road Emona - Siscia - Sirmium, the most important and the oldest Roman traffic artery through Pannonia founded immediately after the conquest of south Pannonia (second half of 1st century BC).[105] The Flavian *municipia* Neviodunum and Andautonia are testified epigraphically - Andautonia is not mentioned till the turn of 2nd and 3rd centuries AD (*CIL* II 4008). The borders of *muncipium Andautonia* are suggested by two milestones - the northern by the milestone from Jelkovec near Sesvete (*AIJ* 602) and the western by the milestone found in Čatež (*ILJug* 1176).[106] Pliny overlooked both Pannonian *municipia* (as well as the *colonia Sirmium*) by mentioning their names in the alphabet list of the tax-burdened *civitates peregrinae* (*Latovici,*

[97] About characterisation of particular types of settlements and communities see Detlefsen 1908, 100.

[98] Elsewhere, Pliny mentions Pannonian territorial districts within *conventus iuridici* in the province of Dalmatia, wich included peregrine communities there. The Ditiones and Maezaei belonged to the *conventus* centred in Salona (Plin. *NH* 3.22,142), and the Daesitiates to the *conventus* centred in Narona (Plin. *NH* 3.22,143).

[99] Cornacates: *CIL* V 6985 + 6986 (*Taurinum* / Torino). About the possible affiliation of the Cornacates with the Breuci see Bojanovski 1988, 337 and 339; on the Breuci see Bulat 1989, 7 (*Mursa*) on the basis of a badly damaged inscription, rendering the mention of the Breuci very dubious. On the Colapiani see *CIL* III 14387f (*Heliopolis* / Baalbek, Syria), again a dubious inscription where the Colapiani might be recognised (Bojanovski 1988, 61 and 334).

[100] The Iasi, Breuci, and Andizetes: *RMD* 205 (unknown site) - *principes* of the territorial districts are mentioned as witnesses in the diploma issued in AD 71; the Scordisci: *ILJug* 280 (*Acumincum* / Stari Slankamen) - *princeps praefectus Scordiscorum*.

[101] Wilkes 1969, 172 and 288.

[102] The status of a colony was at the same time assigned to Sirmium, the most important centre in southeastern Pannonia. Pliny calls Sirmium an *oppidum*, the centre of the *civitas Sirmiensium et Amantinorum* (Plin. *NH* 3.25,148); it appears to be an oversight, since to Pliny the information that Sirmium had been given the status of a colony was new, that is, dating from his own time.

[103] This fortification played a crucial role during the Pannonian (Bato's) rebellion as the main stronghold of Roman legions (Vell. 2.113).

[104] *Colonia Flavia Siscia*: *CIL* III 4471, *CIL* III 11029, *CIL* VI 2644, *CIL* VI 2689; organization in Trajan's time: *CIL* III 6476, *CIL* III 10856; *colonia Septimia Siscia Augusta*: *CIL* III 4193. On Segestica / Siscia see Durman 1992; Hoti 1992; Zaninović 1996.

[105] This road is mentioned by Strabo (Strab. 7.5,2).

[106] Here are some examples: *CIL* III 3925 (Velike Malence): *municipium Latobicorum*; *CIL* III 3913 (Velike Malence): *municipium Flavium Neviodunum*; *CIL* III 4008 (*AIJ* 474) (Petrovina Turopoljska): *municipium Andautoniensium*; *CIL* III 4010 (*AIJ* 477) (Stenjevec): *respublica Andautoniensium*; *CIL* III 4011 (*AIJ* 478) (Ščitarjevo): *respublica Andautoniensium*; *AIJ* 602 (Jelkovec near Sesvete): *a Siscia XXX*; *ILJug* 1176 (Čatež): *ab A[ndautonia]*. About Andautonia: Zaninović 1996; Margetić 1999.

Varciani, Plin. *NH* 3,25,148. The next phase of urbanisation and municipalisation in Pannonia occured during Hadrian's reign (AD 117-138). The rank of *municipium* was awarded to the centres of some *civitates peregrinae*: Aquae Balissae / Daruvar (*civitas Iasorum*), Cibalae / Vinkovci (*civitas Cornacatium*) and Bassianae / Petrovci (part of the *civitas Scordiscorum*). The status of a colony was awarded to Mursa (Osijek); there is no evidence in any literary sources (Ptolemy calls it Μουρσία, Ptol. 2.15,4), but there is plenty of epigraphic evidence for the colonial status of Mursa. On the other hand, Cibalae (Ptolemy's Κιβαλίς, Ptol. 2.16,7) is only mentioned sporadically, at the end of Antiquity.[107] According to Roman administrative system, the territory of the town was divided into districts (*pagi*), which were further divided into villages (*vici*).[108] A few *pagi* and *vici* within the territory of certain Pannonian *municipia* are attested epigraphically, for example: *vicus Cocconetes* within the *pagus Iovista* in the territory of Aquae Balissae (*CIL* VI 3297 *Roma*) and *vicus Budalia* and *vicus [...] dianus* within the *pagus Martius* in the territory of Sirmium (*CIL* VI 32713 *Roma*; *CIL* V 829 *Aquileia*).

7. Economy in Pannonia – "a land rich in all fruits"

Strabo testifies to a lively trade between Italy and the Danube region as early as the late Republic period (2nd and 1st centuries BC). The trade with Pannonia started to flourish with the establishment of the colony of Aquileia in 181 BC (Strab. 5.1,8; 4.6,10). Italian goods reached Pannonia through combined caravan and river routes across the southeastern Alps. Nauportus was an important centre where goods were loaded from carriages to ships and then sent to their final destination along the Pannonian rivers (Strab. 4.6,10: 7.5,2; cf. 5.8,1). Strabo points to Segestica as an important trading centre and road junction that lies on a river traffic route, as well as on one of the oldest and the most important roads that connected Italy and the Danube region - the later itinerary road Aquileia - Emona - Siscia - Sirmium (Strab. 7.5,2). According to Strabo Italy exchanged livestock and slaves for Mediterranean products (Strab. 5.1,8). The same items are listed in the late antique *Exposito totius mundi et gentium* (*Exp. tot. mund.* 57), but the trade in Pannonia was not dependant only on these items – goods also "from other countries" reached Segestica (Strab. 7.5,2). It is an unrewarding task to guess at what "other countries" Strabo had in mind - the answer can be provided only by archaeological research. Pliny implies that the Baltic was Europe's (and "mostly Pannonian") main supplier of amber (Plin. *NH* 37.11,43 and 45). Pannonia was an agricultural country despite the fact that most of its territory was covered by swamps and centennial forests. One could not grow olives or quality grapevine, or other crops that the Mediterranean world was used to, but the situation was far from one depicted by Strabo (Strab. 7.5,10) and Dio Cassius (Dio 49.36,2-3). By the end of the 1st and the beginning of the 2nd centuries AD Pannonia was already a model province for the land division and evaluation - according to Hyginus Gromaticus (Hyg. Grom. 28) Pannonia had a variety of lands: fields, pastures, forests, and meadows.

Strabo mentions the value of Pannonian livestock (Strab. 5.1,8), pointing to yet another branch of economy - cattle breeding. In view of this, one might understand the answer of Bato the Daesitiatian to Tiberius' inquiry about the reasons for the rebellion (Dio 56.16); a comparison of a Roman governor with a shepherd and people with a herd is suitable for a cattle-breeding milieu.

The image of Pannonia depicted by classical sources is the result of a subjective and stereotypical point of view of the world of the "Others", those who inhabit the world beyond the Mediterranean *oecumene*. This subjective image changed quite a lot with time and Pannonian indisputable qualities - its geopolitical importance, as well as the richness of its natural and human resources - were recognised and confirmed in Late Antiquity.

Sources[109]

App. 1.6 (2nd c. AD)

Ῥωμαῖοι δὲ καὶ τούσδε καὶ Παίονας ἐπ' αὐτοῖς καὶ Ῥαιτοὺς καὶ Νωρικοὺς καὶ Μυσοὺς τοὺς ἐν Εὐρώπῃ, καὶ ὅσα ἄλλα ὅμορα τούτοις ἐν δεξιᾷ τοῦ Ἴστρου καταπλέοντι ᾤκηται, διαιροῦσι μὲν ὁμοίως τοῖς Ἕλλησιν ἀπὸ Ἑλλήνων, καὶ καλοῦσι τοῖς ἰδίοις ἑκάστους ὀνόμασι, κοινῇ δὲ πάντας Ἰλλυρίδα ἡγοῦνται, ὅθεν μὲν ἀρξάμενοι τῆσδε τῆς δόξης, οὐκ ἔσχον εὑρεῖν, χρώμενοι δ' αὐτῇ καὶ νῦν, ὅπου καὶ τὸ τέλος τῶνδε τῶν ἐθνῶν, ἀπὸ ἀνίσχοντος Ἴστρου μέχρι τῆς Ποντικῆς θαλάσσης, ὑφ' ἓν ἐκμισθοῦσι καὶ Ἰλλυρικὸν τέλος προσαγορεύουσιν. ὅπως δὲ αὐτοὺς ὑπηγάγοντο Ῥωμαῖοι, ὡμολόγησα μὲν καὶ περὶ Κρήτης λέγων οὐχ εὑρεῖν τὰς ἀκριβεῖς τῶν πολέμων ἀρχάς τε καὶ προφάσεις, καὶ ἐς τοῦτο τοὺς δυναμένους τι πλέον εἰπεῖν παρεκάλουν· ὅσα δ' αὐτὸς ἔγνων, ἀναγράψω.

The Romans differentiate these peoples and in addition also the Paeones, the Raeti, the Norici, the European Mysians and other neighbouring tribes who live on the right bank of the Ister, in a similar manner as the Greeks differentiate the Greeks; each people is called by its own name, and together they are all considered to be living in Illyria. When and how such a concept was adopted I could not discover, although it is still current, and thus also the tax which is collected from these nations extending from the source of the Ister to the Pontic Sea is leased as one tax and is called Illyrian. I have already stated in the book on Crete that I could not discover how the Romans conquered them and what exactly were

[107] For the *municipium Iasorum* see *ILJug* 1132 (*Aquae Balssae* / Daruvar) - Hadrian's period, *CIL* VI 3297 (*Roma*); *respublica Iasorum*: *AIJ* 586, *AIJ* 587 – 3rd c. AD. For the *municipium Cibalensis* see *CIL* III 3267 (*Cibalae* / Vinkovci). For Cibalae in literary sources (as *oppidum Pannoniae*) see: Ammianus Marcellinus (4th c. AD); Zosimus (5th-6th c. AD). Examples for *Aelia Mursa*: *CIL* VI 3214 (*Roma*), *CIL* VI 3235 (*Roma*). For Hadrian's *municipia* see Mócsy 1974, 119, 143-145.

[108] Laffi 1973, 40.

[109] Literary sources are given alphabetically, for easier handling.

the causes and pretexts for the wars, and I urged those who could say more on the subject to do so. I shall write down as much as I myself have learned.

App. 2.10 (2nd c. AD)

(...) ἐοίκασι δὲ καὶ Σεγεστανοὶ Λευκίῳ Κόττᾳ καὶ Μετέλλῳ, ἀμφότεροι δ' οὐ πολὺ ὕστερον ἀποστῆναι.

(...) and it also seems that Lucius Cotta and Metellus defeated the Segestani, although not long afterwards both peoples [Iapodes and Segestani] revolted.

App. 3.14 (2nd c. AD)

Οἱ δὲ Παίονές εἰσιν ἔθνος μέγα παρὰ τὸν Ἴστρον, ἐπίμηκες ἐξ Ἰαπόδων ἐπὶ Δαρδάνους, Παίονες μὲν ὑπὸ τῶν Ἑλλήνων λεγόμενοι καὶ ῥωμαϊστὶ Παννόνιοι, συναριθμούμενοι δὲ ὑπὸ Ῥωμαίων τῇ Ἰλλυρίδι, ὡς προεῖπον. Διὸ καὶ περὶ τῶνδέ μοι δοκεῖ νῦν κατὰ τὰ Ἰλλυρικὰ εἰπεῖν. (...) Ἐπεὶ δ' ἐπὶ τοὺς Παίονας ἐστράτευσε Κορνήλιος, κακῶς ἀπαλλάξας μέγα δέος Παιόνων Ἰταλοῖς ἅπασιν ἐνεποίησε καὶ ἐς πολὺ τοῖς ἔπειτα ὑπάτοις ὄκνον, ἐπὶ Παίονας ἐλαύνειν. Τὰ μὲν δὴ πάλαι τοσαῦτα περὶ Ἰλλυριῶν καὶ Παιόνων ἔσχον εὑρεῖν· ἐν δὲ τοῖς ὑπομνήμασι τοῦ δευτέρου Καίσαρος, τοῦ κληθέντος καὶ Σεβαστοῦ, παλαιότερον μὲν οὐδὲν οὐδ' ἐν τοῖσδε περὶ Παιόνων εὗρον (...)

The Paeones are a large nation along the Ister who extend from the Iapodes to the Dardanians. The Greeks call them Paeones, the Romans Pannonians, and they include them, as I have noted previously, among the Illyrians; thus it seems proper that I should also speak of them in my survey of Illyrian history. (...) Afterwards, when Cornelius campaigned against the Paeones and narrowly avoided a total defeat, the entire Italian population was overcome by great fear of the Paeones, and for a long time later consuls hesitated to undertake military expeditions against them. This is as much as I succeeded in learning about the Illyrians and the Paeones in the early period. I also did not find anything earlier about the Paeones in the Memoirs of the second Caesar, surnamed Augustus.

App. 3.15 (2nd c. AD)

Ἰλλυριῶν δέ μοι φαίνεται χωρὶς τῶν εἰρημένων ἐθνῶν καὶ ἕτερα Ῥωμαίων προϋπακοῦσαι. Καὶ ὅπως μέν, οὐκ ἔγνων (οὐ γὰρ ἀλλοτρίας πράξεις ὁ Σεβαστός, ἀλλὰ τὰς ἑαυτοῦ συνέγραφεν), ὡς δ' ἀποστάντας ἐς τοὺς φόρους ἐπανήγαγεν καὶ ἑτέρους, ὡς ἀρχῆθεν ἔτι ὄντας αὐτονόμους, εἷλε καὶ πάντας ἐκρατύνατο, ὅσοι τὰς κορυφὰς οἰκοῦσι τῶν Ἄλπεων, βάρβαρα καὶ μάχιμα ἔθνη καὶ κλοπεύοντα τὴν Ἰταλίαν ὡς γείτονα.

In addition to those mentioned, I think that other Illyrian peoples had come under Roman authority even before this. I do not know in what manner, since Augustus did not describe the actions of others, but rather his own: how he forced the peoples who had resisted into again paying tribute, and how he defeated others who were still independent and conquered all the barbarian and bellicose tribes living among the peaks of the Alps who had plundered neighbouring Italy.

App. 4.17 (2nd c. AD)

Μάλιστα δ' ἠνώχλησαν αὐτὸν Σαλασσοί τε καὶ Ἰάποδες οἱ πέραν Ἄλπεων καὶ Σεγεστανοὶ καὶ Δαλμάται καὶ Δαίσιοί τε καὶ Παίονες, ὄντες ἑκὰς τοῖς Σαλασσοῖς (...)

The greatest difficulties were caused to him by the Salassi, Iapodes on the other side of the Alps, and the Segestani, and further the Dalmatae, the Daesii and the Paeones, who live far away from the Salassi.

App. 4.22 (2nd c. AD)

Ἐς δὲ τὴν Σεγεστικὴν γῆν οἱ Ῥωμαῖοι, δὶς πρότερον ἐμβαλόντες, οὔτε ὅμηρον οὔτε ἄλλο τι εἰλήφεσαν· ὅθεν ἦσαν ἐπὶ φρονήματος οἱ Σεγεστανοί. Ὁ δὲ Καῖσαρ αὐτοῖς ἐπῄει διὰ τῆς Παιόνων γῆς, οὔπω Ῥωμαίοις οὐδὲ τῆσδε ὑπηκόου γενομένης. Ὑλώδης δέ ἐστιν ἡ Παιόνων καὶ ἐπιμήκης ἐξ Ἰαπόδων ἐπὶ Δαρδάνους. Καὶ οὐ πόλεις ᾤκουν οἱ Παίονες οἵδε, ἀλλ' ἀγροὺς ἢ κώμας κατὰ συγγένειαν· οὐδ' εἰς βουλευτήρια κοινὰ συνῄεσαν, οὐδ' ἄρχοντες αὐτοῖς ἦσαν ἐπὶ πᾶσιν. Οἱ δ' ἐν ἡλικίᾳ μάχης ἐς δέκα μυριάδας συνετέλουν· ἀλλ' οὐδ' οὗτοι συνῄεσαν ἀθρόοι δι' ἀναρχίαν. Προσιόντος δ' αὐτοῖς τοῦ Καίσαρος ἐς τὰς ὕλας ὑποφυγόντες τοὺς ἀποσκιδναμένους τῶν στρατιωτῶν ἀνῄρουν. Ὁ δὲ Καῖσαρ, ἕως μὲν ἤλπιζεν αὐτοὺς ἀφίξεσθαι πρὸς αὐτόν, οὔτε τὰς κώμας οὔτε τοὺς ἀγροὺς ἐλυμαίνετο, οὐκ ἀπαντώντων δὲ πάντα ἐνεπίμπρη καὶ ἔκειρεν ἐπὶ ἡμέρας ὀκτώ, ἐς ὃ διῆλθεν ἐς τὴν Σεγεστανῶν, καὶ τήνδε Παιόνων οὖσαν, ἐπὶ τοῦ Σάου ποταμοῦ, ἐν ᾧ καὶ πόλις ἔστιν ἐχυρά, τῷ τε ποταμῷ καὶ τάφρῳ μεγίστῃ διειλημμένη, διὸ καὶ μάλιστα αὐτῆς ἔχρῃζεν ὁ Καῖσαρ, ὡς ταμιείῳ χρησόμενος ἐς τὸν Δακῶν καὶ Βαστερνῶν πόλεμον, οἳ πέραν εἰσὶ τοῦ Ἴστρου, λεγομένου μὲν ἐνταῦθα Δανουβίου, γιγνομένου δὲ μετ' ὀλίγον Ἴστρου. Ἐμβάλλει δ' ὁ Σάος ἐς τὸν Ἴστρον· καὶ αἱ νῆες ἐν τῷ Σάῳ Καίσαρι ἐγίγνοντο, αἳ ἐς τὸν Δανούβιον αὐτῷ τὴν ἀγορὰν διοίσειν ἔμελλον.

Although the Romans had previously twice invaded the country of Segestica, they took neither hostages nor anything else, due to which the Segestani became very arrogant. Caesar advanced against them through the country of the Paeones, which was also not under Roman authority at that time. The country of the Paeones is wooded and extends from the Iapodes to the Dardanians. These Paeones did not live in cities, but rather according to clans throughout the countryside and in villages. They did not gather for consultation, nor did they have collective leaders. The total of men capable of figthting was 100,000, but because of the absence of a common government they were not united into a single unit. When Caesar approached them, they took refuge in the forests and killed those soldiers who did not keep in step. As long as Caesar hoped that they would surrender of their own free will, he did not order the destruction of either their villages or their fields. As they did not come to negotiate, he ordered everything to be burnt and destroyed, which continued for eight days until he entered the country of the Segestani. This is also a part of the Paeonian region along the Savus river; a fortified town also lies along the river, which is protected as much by the river as by an extremely large ditch. Caesar particularly wished to take possession of the town for this reason, as he intended to utilize it as a supply base for a war against the Dacians and the Bastarnae, who live on the other side of the Ister. The river there is called the Danube, and somewhat lower it becomes the Ister. The Savus flows into it; Caesar had ships built on this river to transport provisions to the Danube.

App. 4.23 (2nd c. AD)

Διὰ μὲν δὴ ταῦτα τῆς Σεγέστης ἔχρῃζεν ὁ Καῖσαρ· προσιόντι δ' αὐτῷ οἱ Σεγεστανοὶ προσέπεμψαν, πυνθανόμενοι, τίνος χρῄζει. Ὁ δὲ φρουρὰν ἐσαγαγεῖν ἔφη καὶ ὁμήρους ἑκατὸν λαβεῖν, ἵν' ἀσφαλῶς ταμιείῳ τῇ πόλει χρῷτο ἐπὶ Δᾶκας. Ἤιτει δὲ καὶ σῖτον, ὅσον δύναιντο φέρειν· ταῦθ' οἱ μὲν πρωτεύοντες ἠξίουν δοῦναι· ὁ δὲ δῆμος ἐξαγριαίνων τὰ μὲν ὅμηρα διδόμενα περιεῖδεν, ὅτι ἴσως οὐ παρὰ σφῶν, ἀλλὰ τῶν πρωτευόντων παῖδες ἦσαν, προσιούσης δὲ τῆς φρουρᾶς τὴν ὄψιν οὐκ ἐνεγκόντες ὁρμῇ μανιώδει τὰς πύλας αὖθις ἀπέκλειον καὶ αὐτοὺς τοῖς τείχεσιν ἐπέστησαν. Ὁ οὖν Καῖσαρ τόν τε ποταμὸν ἐγεφύρου καὶ χάρακας καὶ τάφρους πάντοθεν ἐποιεῖτο, ἀποτειχίσας δ' αὐτοὺς δύο χώματα ἔχου. Οἷς ἐπέδραμον μὲν οἱ Σεγεστανοὶ πολλάκις, οὐ δυνηθέντες δ' ἑλεῖν λαμπάδας καὶ πῦρ πολὺ ἄνωθεν ἐπέβαλλον. Προσιούσης δ' αὐτοῖς Παιόνων ἑτέρων βοηθείας ὁ Καῖσαρ ὑπαντήσας ἐνήδρευσεν αὐτήν· καὶ οἳ μὲν ἀνῃρέθησαν, οἳ δ' ἔφυγον, καὶ οὐδεὶς ἔτι Παιόνων ἐβοήθει.

For these reasons Caesar wanted to seize Segesta. As he was approaching the town, the Segestani sent messengers to ask what he wanted. He replied that he wished to quarter a military garrison and take one hundred hostages, so that he could safely use the town as a supply base in the war against the Dacians. He also requested that they bring him as much food as they could provide. The local leaders were prepared to comply with him in this matter, but the population was infuriated; they endured the delivery of the hostages, probably because they were not their children but rather those of the leaders. However, they could not bear the sight of the garrison when it approached, and in an enraged assault they again shut the gates and placed themselves on the walls. Caesar then built a bridge across the river and constructed palisades and ditches on all sides. When he had thus confined them, he had two ramparts erected. The Segestani attacked them repeatedly, but since they could not occupy them, they threw torches and great quantities of fire onto them from above. As help was approaching from other Paeones, Caesar intercepted them in an ambush; some of them he killed, others fled and none of the Paeones sent aid after this.

App. 4.24 (2nd c. AD)

Οἱ Σεγεστανοὶ δέ, πᾶσαν πολιορκίαν ὑποστάντες, ἡμέρᾳ τριακοστῇ κατὰ κράτος ἐλήφθησαν καὶ τότε πρῶτον ἤρξαντο ἱκετεύειν. Καὶ αὐτοὺς ὁ Καῖσαρ ἐπαίνῳ τε τῆς ἀρετῆς καὶ ἐλέῳ τῆς ἱκεσίας οὔτε ἔκτεινεν οὔτε ἀνέστησεν, ἀλλὰ χρήμασιν ἐζημίωσε καὶ τῆς πόλεως μέρος διατειχίσας ἐσήγαγεν ἐς αὐτὸ φρουρὰν πέντε καὶ εἴκοσι σπειρῶν. Καὶ ὃ μὲν τάδε ἐργασάμενος ἐς Ῥώμην ἀνέζευξεν, ὡς ἦρος ἐπανήξων ἐς τὴν Ἰλλυρίδα. Φήμης δ' ἐπιδραμούσης, ὅτι τὴν φρουρὰν οἱ Σεγεστανοὶ διέφθειραν, ἐξέθορε χειμῶνος, καὶ τὸ μὲν τέλος τῆς φήμης ψευδὲς εὗρε, τὴν δὲ αἰτίαν ἀληθῆ. Γεγόνεσαν γὰρ ἐν κινδύνῳ, τῶν Σεγεστανῶν αὐτοὺς ἄφνω περιστάντων, καὶ πολλοὺς τὸ αἰφνίδιον ἀπωλωλέκει, τῆς δ' ἐπιούσης προελθόντες ἐκράτουν τῶν Σεγεστανῶν. Ὁ οὖν Καῖσαρ ἐπὶ Δαλμάτας μετῄει, γένος ἕτερον Ἰλλυριῶν, Ταυλαντίοις ὅμορον.

The Segestani endured the entire period of siege and on the thirtieth day they were subjugated with force; only then did they begin to beg for mercy. Caesar expressed recognition of their bravery and, taking pity on those who had implored, he neither had them killed nor exiled, but rather only imposed a monetary fine; he had a part of the town walled off and stationed a military detachment of twenty-five cohorts in it. Having completed that, he left for Rome, to return to Illyria again in the spring. As a rumour was circulating that the Segestani had killed the garrison, he set off on the journey as early as mid-winter. He ascertained that the end of the rumour was fictitious, but that grounds for worry had actually existed. The soldiers had found themselves in danger, as the Segestani had suddenly surrounded them and many had lost their lives in the unexpected chaos; the following day they attacked and defeated the Segestani. Caesar consequently set off against the Dalmatae, also one of the Illyrian peoples who have a border with the Taulantii.

Aug. RG (1st c. BC / 1st c. AD)

Παννονίων ἔθνη, οἷς πρὸ ἐμοῦ ἡγεμόνος στράτευμα Ῥωμαίων οὐκ ἤνγισεν, ἡσσηθέντα ὑπὸ Τιβερίου Νέρωνος, ὃς τότε μου ἦν πρόγονος καὶ πρεσβευτής, ἡγεμονίαι δήμου Ῥωμαίων ὑπέταξα τά τε Ἰλλυρικοῦ ὅρια μέχρι Ἴστρου ποταμοῦ προήγαγον.

Pannoniorum gentes, quas ante me principem populi Romani exercitus nunquam adit, devictas per Ti. Neronem, qui tum erat privignus et legatus meus, imperio populi Romani subieci protulique fines Illyrici ad ripam fluminis Danui.

The tribes of the Pannonians, to which no army of the Roman people had ever penetrated before my principate, having been subdued by Tiberius Nero who was then my stepson and my legate, I brought under the sovereignty of the Roman people, and I pushed forward the frontier of Illyricum as far as the bank of the river Danube.

Dio 12.19 (in Zon. 8.19) (Dio: 2nd/3rd c. AD; Zonaras: 12th c. AD)

Τὸ δ' Ἰλλυρικὸν ὄνομα πάλαι μὲν ἐν ἄλλοις ἐπεκέκλητο, ὕστερον δὲ ἐς τὴν ἄνω μεταβέβηκεν ἤπειρον καὶ ὑπὲρ τὴν Μακεεδονίαν τήν τε Θρᾴκην τὴν ἐντὸς τοῦ Αἵμου καὶ τὴν πρὸς τῇ Ῥοδόπῃ, καὶ ἔστιν ἐν τῷ μέσῳ τούτων τῶν ὀρῶν καὶ τῶν Ἄλπεων τοῦ τε Αἴνου ποταμοῦ καὶ τοῦ Ἴστρου μέχρι τοῦ Εὐξείου πόντου, καί πῃ καὶ ἐπέκεινα τοῦ Ἴστρου νέμεται.

The name Illyricum was anciently applied to different regions, but later it was transferred to the interior of the mainland and to the region above Macedonia and the part of Thrace lying this side of Haemus and next to Rhodope. It lies between these mountains and the Alps, also between the river Aenus and the Ister, extending as far as the Euxine Sea; indeed, at some points it extends even beyond the Ister.

Dio 49.36 (2nd/3rd c. AD)

Ἐπεὶ δ' οὖν οὗτοί τε ἀπωλώλεσαν καὶ οἱ ἄλλοι κατεστράφατο μηδὲν ἀξιόλογον πράξαντες, ἐπὶ Παννονίους ἐπεστράτευσεν, ἔγκλημα μὲν οὐδὲν αὐτοῖς ἐπιφέρων (οὐδὲ γὰρ οὐδ' ἠδίκητό τι ὑπ' αὐτῶν), ἵνα δὲ δὴ τοὺς στρατιώτας ἀσκῇ τε ἅμα καὶ ἐκ τῶν ἀλλοτρίων τρέφῃ, πᾶν τὸ τῷ κρείττονι τοῖς ὅπλοις ἀρέσκον δίκαιον ἐς τοὺς ἀσθενεστέρους ποιούμενος. Οἱ δὲ δὴ Παννόνιοι νέμονται μὲν πρὸς τῇ Δελματίᾳ, παρ' αὐτὸν τὸν Ἴστρον, ἀπὸ Νωρικοῦ μέχρι τῆς Μυσίας τῆς ἐν τῇ Εὐρώπῃ, κακοβιώτατοι δὲ ἀνθρώπων ὄντες (οὔτε γὰρ γῆς οὔτε ἀέρων εὖ ἥκουσιν· οὐκ ἔλαιον, οὐκ οἶνον, πλὴν ἐλαχίστου καὶ τούτου κακίστου, γεωργοῦσιν, ἅτε ἐν χειμῶνι πικροτάτῳ τὸ πλεῖστον διαιτώμενοι, ἀλλὰ τάς τε κριθὰς καὶ τοὺς κέγχρους καὶ ἐσθίουσιν ὁμοίως καὶ πίνουσιν) ἀνδρειότατοι δ' οὖν διὰ πάντων ὧν ἴσμεν νομίζονται· θυμικώτατοι γὰρ καὶ φονικώτατοι, οἷα μηδὲν ἄξιον τοῦ καλῶς ζῆν ἔχοντες, εἰσί. Ταῦτα δὲ οὐκ ἀκούσας οὐδ' ἀναγνοὺς μόνον, ἀλλὰ καὶ ἔργῳ μαθὼν ὥστε καὶ ἄρξας αὐτῶν, οἶδα· μετὰ γάρ τοι τὴν ἐν τῇ Ἀφρικῇ ἡγεμονίαν τῇ τε Δελματίᾳ, ἧς ποτε καὶ ὁ πατήρ μου χρόνον τινὰ ἦρξε, καὶ τῇ Παννονίᾳ τῇ ἄνω καλουμένῃ προσετάχθην, ὅθεν ἀκριβῶς πάντα τὰ κατ' αὐτοὺς εἰδὼς γράφω. Ὀνομάζονται δὲ οὕτως ὅτι τοὺς χιτῶνας τοὺς χειριδωτοὺς ἐξ ἱματίων τινῶν ἐς πάννους ἐπιχωρίως πως καὶ κατατέμνοντες καὶ προσαγορεύοντες συρράπτουσι. καὶ οἱ μὲν εἴτ' οὖν διὰ τοῦτο εἴτε καὶ δι' ἄλλο τι οὕτως ὠνομάδαται· τῶν δὲ δὴ Ἑλλήνων τινὲς τἀληθὲς ἀγνοήσαντες Παίονάς σφας προσεῖπον, ἀρχαίου μέν που τοῦ προσρήματος τούτου ὄντος, οὐ μέντοι καὶ ἐκεῖ, ἀλλ' ἔν τε τῇ Ῥοδόπῃ καὶ πρὸς αὐτῇ τῇ Μακεδονίᾳ τῇ νῦν μέχρι τῆς θαλάσσης. Ὑφ' οὗπερ καὶ ἐγὼ ἐκείνους μὲν Παίονας τούτους δὲ Παννονίους, ὥσπερ που καὶ αὐτοὶ ἑαυτοὺς καὶ Ῥωμαῖοί σφας καλοῦσι, προσαγορεύσω.

When these, then, had perished and the rest had been subdued without performing any exploit of note, Caesar made a campaign against the Pannonians. He had no complaint to bring against them, not having been wronged by them in any way, but he wanted both to give his soldiers practice and to support them at the expense of an alien people, for he regarded every demonstration against a weaker party as just, when it pleased the man who was their superior in arms. The Pannonians dwell near Dalmatia along the very bank of the Ister from Noricum to Moesia and lead the most miserable existence of all mankind. For they are not well off as regards either soil or climate; they cultivate no olives and produce no wine except to a very slight extent and a wretched quality at that, since the winter is very rigorous and occupies the greater part of their year, but drink as well as eat both barley and millet. For all that they are considered the bravest of all men of whom we have knowledge; for they are very high-spirited and bloodthirsty, as men who possess nothing that makes an honourable life worth while. This I know not from hearsay or reading only, but I have learned it from actual experience as once their governor, for after my command in Africa and in Dalmatia (the latter position my father also held for a time) I was appointed to what is known as Upper Pannonia, and hence it is with exact knowledge of all conditions among them that I write. Their name is derived from the fact that their sleeved tunics are made by stitching together pieces of old clothes which they cut up into strips in a way peculiar to themselves and call *panni*. This is their name, whether the reason be what I have stated or some other; but certain of the Greeks in ignorance of the truth have called them Paeones, an appellation which, though no doubt old, does not, however, apply to that country, but rather to Rhodope, close to the present Macedonia, as far as the sea. Therefore I also shall call the people of the latter district Paeones, but the others Pannonians, just as both they themselves and the Romans do.

Dio 49.37 (2nd/3rd c. AD)

Ἐπὶ οὖν τούτους ὁ Καῖσαρ τότε στρατεύσας τὸ μὲν πρῶτον οὔτε τι ἐδῄου οὔτε τι ἥρπαζε, καίπερ καὶ τὰς κώμας αὐτῶν τὰς ἐν τοῖς πεδίοις ἐκλιπόντων· ἤλπιζε γὰρ ἐθελοντὰς αὐτοὺς ὑπάξεσθαι· ὡς δὲ καὶ προχωροῦντα αὐτὸν πρὸς Σισκίαν ἐλύπησαν, ὠργίσθη καὶ τήν τε χώραν σφῶν ἔκαιε καὶ πάνθ' ὅσα ἐδύνατο λεῖαν ἐποιεῖτο. Πλησιάσαντος δ' αὐτοῦ τῇ πόλει οἱ ἐπιχώριοι παραυτίκα μὲν ὑπὸ τῶν δυνατῶν ἀναπεισθέντες ὡμολόγησάν τε αὐτῷ καὶ ὁμήρους ἔδοσαν, μετὰ δὲ τοῦτο τάς τε πύλας ἀπέκλεισαν καὶ ἐς πολιορκίαν κατέστησαν. Εἶχον μὲν γὰρ καὶ τείχη ἰσχυρά, τὸ δ' ὅλον ποταμοῖς δύο ναυσιπόροις ἐθάρσουν. Ὁ γὰρ Κόλοψ ὀνομαζόμενος παρ' αὐτὸν τὸν περίβολον παραρρέων ἐς τὸν Σάουον ὀλίγον ἀπέχοντα αὐτοῦ ἐμβάλλει καὶ νῦν πᾶσαν τὴν πόλιν ἐγκεκύκλωται, Τιβερίου τάφρῳ τινὶ μεγάλῃ ἐς τοῦτο αὐτὸν καταστήσαντος, δι' ἧς ἐς τὸ ἀρχαῖον αὖθις ῥεῖθρον ἐπανέρχεται. Τότε δὲ τῇ μὲν τοῦ Κόλοπος παρ' αὐτὰ τὰ τείχη παρεξιόντος, τῇ δὲ τοῦ Σάουου ὀλίγον ἄπωθεν παραρρέοντος διάκενόν τι κατελέλειπτο, ὃ καὶ σταυρώμασι καὶ ταφρεύμασιν ὠχύρωτο. Ὁ οὖν Καῖσαρ πλοῖα παρὰ τῶν ταύτῃ συμμάχων ποιηθέντα λαβών, καὶ διά τε τοῦ Ἴστρου ἐς τὸν Σάουον καὶ δι' ἐκείνου ἐς τὸν Κόλοπα αὐτὰ ἀγαγών, προσέβαλέ σφισι τῷ πεζῷ ἅμα καὶ ταῖς ναυσί, καί τινας καὶ ναυμαχίας ἐν αὐτῷ ἐποιήσατο. Καὶ γὰρ οἱ βάρβαροι μονόξυλα πλοῖα ἀντικατασκευάσαντες διεκινδύνευον, καὶ ἔν τε τῷ ποταμῷ ἄλλους τε συχνοὺς καὶ τὸν Μηνᾶν τὸν τοῦ Σέξτου ἐξελεύθερον ἀπέκτειναν, καὶ ἐν τῇ γῇ ἰσχυρῶς αὐτὸν ἠμύνοντο, μέχρις οὗ τῶν συμμάχων τινὰς ἐνηδρεῦσθαί τε καὶ ἐφθάρθαι ἐπύθοντο· τότε γὰρ ἀθυμήσαντες ἐνέδοσαν. Καὶ οὕτως ἁλόντων ἐκείνων καὶ τὸ ἄλλο Παννονικὸν ὁμολογίᾳ προσηγάγετο.

It was against this people, then, that Caesar at that time conducted a campaign. At first he did not devastate or plunder at all, although they abandoned their villages in the plain; for he hoped to make them his subjects of their own free will. But when they harassed him as he advanced to Siscia, he became angry, burned their country, and took all the booty he could. When he drew near the city, the natives for the moment listened to their leaders and made terms with him and gave hostages, but afterwards they shut their gates and underwent a siege. For while they possessed strong walls also, yet they placed their whole confidence in two navigable rivers. The one named the Colops flows past the very circuit of the wall and empties into the Savus not far distant; it has now encircled the entire city, for Tiberius gave it this shape by constructing a great canal through which it comes back to its original channel. But at that time between the Colops on the one hand, which flowed past the very walls, and the Savus on the other, which flowed at a little distance, a gap had been left which had been fortified with palisades and ditches. Caesar secured boats made by the allies in that vicinity, and after towing them through the Ister into the Savus, and through that stream into the Colops, he assailed the enemy with his infantry and ships together, and had some naval battles on the river. For the barbarians prepared in turn some boats made of single logs, with which they risked a conflict; and thus on the river they killed Menas, the freedman of Sextus, besides many others, while on the land they vigorously repulsed the invader, until they ascertained that some of their allies had been ambushed and destroyed. Then they lost heart and yielded; and when they had been captured in this manner, the remainder of Pannonia was induced to capitulate.

Dio 49.38,1 and 3 (2nd/3rd c. AD)

Καὶ μετὰ ταῦτα ἐκεῖ μὲν Φούφιον Γέμινον σὺν δυνάμει τινὶ κατέλιπεν, αὐτὸς δὲ ἐς τὴν Ῥώμην ἀνεκομίσθη, καὶ τὰ μὲν ἐπινίκια ψηφισθέντα οἱ ἀνεβάλετο (...) Καὶ Παννονίους μὲν ὁ Γέμινος, καίτοι τῆς Σισκίας ἐκπεσών, ὅμως μάχαις ἀνεκτήσατο (...)

After this he left Fufius Geminus there with a small force and himself returned to Rome. The triumph which had been voted to him he deferred (...) Geminus, although expelled from Siscia, nevertheless recovered Pannonia by a few battles (...)

Dio 50.24,3 and 5 (2nd/3rd c. AD)

Τὸ γάρ τοι Ῥωμαίους τε ὄντας καὶ τῆς πλείστης καὶ ἀρίστης οἰκουμένης ἄρχοντας καταφρονεῖσθαι καὶ καταπατεῖσθαι πρὸς γυναικὸς Αἰγυπτίας (...) ἀνάξιον δὲ καὶ ἡμῶν αὐτῶν τῶν τοὺς Γαλάτας κατεστραμμένων, τῶν τοὺς Παννονίους κεχειρωμένων, τῶν μέχρι τοῦ Ἴστρου προκεχωρηκότων, τὸν Ῥῆνον διαβεβηκότων, ἐς Βρεττανίαν πεπεραιωμένων.

"For that we who are Romans and lords of the greatest and best portion of the world should be despised and trodden under foot by an Egyptian woman (...) is unworthy of ourselves, who have subjugted the Gauls, subdued the Pannonians, advanced as far as the Ister, crossed the Rhine, and passed over the sea into Britain."

Dio 50.28,3-4 (2nd/3rd c. AD)

Ἦ πρὸς μὲν Ταυρίσκους καὶ Ἰάπυδας καὶ Δελμάτας καὶ Παννονίους προθυμότατα αὐτοὶ ὑμεῖς οἱ νῦν παρόντες ὑπὲρ ὀλίγων τινῶν τειχῶν καὶ γῆς ἐρήμου πολλάκις ἐμαχέσασθε, καὶ πάντας αὐτοὺς καίτοι πολεμικωτάτους ὁμολογουμένως ὄντας ἐχειρώσασθε.

"Against the Taurisci and Iapydes and Dalmatians and Pannonians you yourselves who are now present battled most zealously, often to win a few walls and a barren land; and you subdued all these people, though they are admittedly most warlike."

Dio 51.21,5 (2nd/3rd c. AD)

Ἑώρτασε δὲ τῇ μὲν πρώτῃ ἡμέρᾳ τά τε τῶν Παννονίων καὶ τὰ τῶν Δελματῶν, τῆς τε Ἰαπυδίας καὶ τῶν προσχώρων σφίσι, Κελτῶν τε καὶ Γαλατῶν τινων.

As for the triumph, Caesar celebrated on the first day his victories over the Pannonians and Dalmatians, the Iapydes and their neighbours, and some Germans and Gauls.

Dio 53.7,1 (2nd/3rd c. AD)

Τί γαρ ἄν τις καὶ παρεξετάσειεν αὐτῷ: τὴν Γαλατίας ἅλωσιν ἢ τὴν Παννονίας δούλωσιν ἢ τὴν Μυσίας χείρωσιν ἢ τὴν Αἰγύπτου καταστροφήν:

„What achievement, indeed, could one compare with these acts of ours? The conquest of Gaul, the enslavement of Pannonia, the subjugation of Moesia, the overthrow of Egypt?"

Dio 54.20,2-3 (2nd/3rd c. AD)

(...) καὶ οἱ Παννόνιοι τήν τε Ἰστρίαν μετὰ Νωρίκων κατέδραμον, καὶ αὐτοί τε πρός τε τοῦ Σιλίου καὶ τῶν ὑποστρατήγων αὐτοῦ κακωθέντες αὖθις ὡμολόγησαν, καὶ τοῖς Νωρίκοις αἴτιοι τῆς αὐτῆς δουλείας ἐγένοντο. (...) καὶ ἡ Μακεδονία ὑπό τε τῶν Δενθελητῶν καὶ ὑπὸ τῶν Σκορδίσκων ἐπορθήθη.

The Pannonians in company with the Norici overran Istria; but the former, upon being discomfited by Silius and his lieutenants, both came to terms again themselves and caused the Norici to be subjected to the same slavery. (...) Macedonia was ravaged by the Dentheleti and the Scordisci.

Dio 54.24,3 (2nd/3rd c. AD)

Τότε δὲ οἵ τε Παννόνιοι νεωτερίσαντες αὖθις ἐχειρώθησαν (...)

At this time the Pannonians revolted again and were subdued

Dio 54.28,1-2 (2nd/3rd c. AD)

Κἀν τούτῳ τὸν Ἀγρίππαν ἐκ τῆς Συρίας ἐλθόντα τῇ τε δημαρχικῇ ἐξουσίᾳ αὖθις ἐς ἄλλα ἔτη πέντε ἐμεγάλυνε καὶ ἐς τὴν Παννονίαν πολεμησείουσαν ἐξέπεμψε, μεῖζον αὐτῷ τῶν ἑκασταχόθι ἔξω τῆς Ἰταλίας ἀρχόντων ἰσχῦσαι ἐπιτρέψας. Καὶ ὃς τὴν μὲν στρατείαν καίτοι τοῦ χειμῶνος, ἐν ᾧ Μᾶρκος τε Οὐαλέριος καὶ Πούπλιος Σουλπίκιος ὑπάτευον, ἐνεστηκότος ἐποιήσατο, ἐκπλαγέντων δὲ τῶν Παννονίων πρὸς τὴν ἔφοδον αὐτοῦ καὶ μηδὲν ἔτι νεωτερισάντων ἐπανῆλθε, καὶ ἐν Καμπανίᾳ γενόμενος ἐνόσησε.

Meanwhile he increased the power of Agrippa, who had returned from Syria, by giving him the tribunician power again for another five years, and he sent him out to Pannonia, which was eager for war, entrusting him with greater authority than the officials outside Italy ordinarily possessed. And Agrippa set out on the campaign in spite of the fact that the winter had already begun (this was the year in which Marcus Valerius and Publius Sulpicius were the consuls); but when the Pannonians became terrified at his approach and gave up their plans for rebellion, he returned, and upon reaching Campania, fell ill.

Dio 54.31 (2nd/3rd c. AD)

Ὡς δ' οὖν ὁ Ἀγρίππας, ὅνπερ που δι' ἀρετὴν ἀλλ' οὐ δι' ἀνάγκην τινὰ ἠγάπα, ἐτεθνήκει, καὶ συνεργοῦ πρὸς τὰ πράγματα πολὺ τῶν ἄλλων καὶ τῇ τιμῇ καὶ τῇ δυνάμει προφέροντος, ὥστε καὶ ἐν καιρῷ καὶ ἄνευ φθόνου καὶ ἐπιβουλῆς πάντα διάγεσθαι, ἐδεῖτο, τὸν Τιβέριον καὶ ἄκων προσείλετο· (...) τήν τε Ἰουλίαν οἱ ἠγγύησε καὶ ἐπὶ τοὺς Παννονίους αὐτὸν ἐξέπεμψε· τέως μὲν γὰρ τὸν Ἀγρίππαν φοβηθέντες ἡσύχασαν, τότε δὲ τελευτήσαντος αὐτοῦ ἐπανέστησαν. Καί σφας ὁ Τιβέριος, πολλὰ μὲν τῆς χώρας πορθήσας πολλὰ δὲ καὶ τοὺς ἀνθρώπους κακώσας, ἐχειρώσατο, τοῖς Σκορδίσκοις, ὁμόροις τε αὐτῶν καὶ ὁμοσκεύοις οὖσι, συμμάχοις ὅτι μάλιστα χρησάμενος. Καὶ τά τε ὅπλα σφῶν ἀφείλετο, καὶ τῆς ἡλικίας τὸ πλεῖον ἐπ' ἐξαγωγῇ ἀπέδοτο. Καὶ αὐτῷ διὰ ταῦτα ἡ μὲν βουλὴ τά γε ἐπινίκια ἐψηφίσατο, ὁ δ' Αὔγουστος ταῦτα μὲν οὐκ ἐπέτρεψεν ἑορτάσαι, τὰς δὲ τιμὰς τὰς ἐπινικίους ἀντέδωκε.

When now Agrippa, whom he loved because of his excellence and not because of any kinship, was dead, Augustus felt the need of an assistant in the public business, one who would far surpass all the others in both rank and influence, so that he might transact all business promptly and without being the object of envy and intrigue. Therefore he reluctantly chose Tiberius (...) and having bethroted Julia to him, he sent him out against the Pannonians. This people had for a time been quiet through fear of Agrippa, but now after his death they had revolted. Tiberius subdued them after ravaging much of their country and doing much injury to the inhabitants, making as much use as possible of his allies the Scordisci, who were neighbours of the Pannonians and were similarly equipped. He took away the enemy's arms and sold most of the men of military age into slavery, to be deported from the country. For these achievements the senate voted him a triumph, but Augustus did not permit him to celebrate it, though he granted him the triumphal honours instead.

Dio 54.34,3-4 (2nd/3rd c. AD)

Ὁ τε Τιβέριος τούς τε Δελμάτας νεοχμώσαντας καὶ τοὺς Παννονίους μετὰ τοῦτο πρός τε τὴν ἐκείνου καὶ πρὸς τὴν τοῦ πλείονος στρατοῦ ἀπουσίαν νεωτερίσαντας ἐχειρώσατο, πολεμῶν τε ἅμα ἀμφοτέροις, καὶ τοτὲ μὲν τῇ τοτὲ δὲ τῇ μεθιστάμενος, ὥστε καὶ τῶν ἄθλων τῶν αὐτῶν τῷ Δρούσῳ τυχεῖν. Κἀκ τούτου καὶ ἡ Δελματία τῇ τοῦ Αὐγούστου φρουρᾷ, ὡς καὶ ὅπλων τινῶν ἀεὶ καὶ δι' ἑαυτὴν καὶ διὰ τὴν τῶν Παννονίων γειτονίαν δεομένη, παρεδόθη.

Tiberius subdued the Dalmatians, who began a rebellion, and later the Pannonians, who likewise revolted, taking advantage of the absence of himself and the larger part of his army. He made war upon both of them at once, shifting now to one front and now to the other. As a result of his success he gained the same prizes as Drusus. After this Dalmatia was given over into the keeping of Augustus, because of the feeling that it would always require armed forces both on its own account and because of the neighbouring Pannonians.

Dio 54.36,2-3 (2nd/3rd c. AD)

Ἐψηφίσθη μὲν οὖν τὸν Ἰανὸν τὸν γέμινον ὡς καὶ πεπαυμένων τῶν πολέμων (ἀνέῳκτο γάρ) κλεισθῆναι, οὐ μέντοι καὶ ἐκλείσθη· οἵ τε γὰρ Δακοὶ τὸν Ἴστρον πεπηγότα διαβάντες λείαν ἐκ τῆς Παννονίας ἀπετέμοντο, καὶ οἱ Δελμάται πρὸς τὰς ἐσπράξεις τῶν χρημάτων ἐπανέστησαν. Καὶ τούτους μὲν ὁ Τιβέριος ἐκ τῆς Γαλατίας, ἐς ἣν μετὰ τοῦ Αὐγούστου ἐσεληλύθει, καταπεμφθεὶς ἀνεκτήσατο (...)

It was voted that the temple of Janus Geminus, which had been opened, should be closed, on the ground that the wars had ceased. It was not closed, however, for the Dacians, crossing the Ister on the ice, carried off booty from Pannonia, and the Dalmatians rebelled against the exactions of tribute. Against these people Tiberius was sent from Gaul, whither he had gone in company with Augustus; and he reduced them again to submission.

Dio 55.2,4 (2nd/3rd c. AD)

Ὁ δὲ δὴ Τιβέριος τῶν τε Δελματῶν καὶ τῶν Παννονίων ὑποκινησάντων τι αὖθις ζῶντος ἔτι αὐτοῦ κρατήσας, τά τε ἐπὶ τοῦ κέλητος ἐπινίκια ἔπεμψε, καὶ τοῦ δήμου τοὺς μὲν ἐν τῷ Καπιτωλίῳ τοὺς δ' ἄλλοθι πολλαχόθι ἐδείπνισε.

Tiberius, while Drusus was yet alive, had overcome the Dalmatians and Pannonians, who had once more begun a rebellion, and he had celebrated the equestrian triumph, and had feasted the people, some on the Capitol and the rest in many other places.

Dio 55.28,7 (2nd/3rd c. AD)

Αἴτια δὲ τοῦ καίπερ παρασπονδήσασί σφισι δι' ὀλίγου αὖθις τὴν εἰρήνην δοθῆναι τά τε τῶν Δελματῶν καὶ τὰ τῶν Παννονίων, μειζόνως τε ταραχθέντα καὶ ὀξείας ἐπιστροφῆς δεηθέντα, ἐγένετο.

The reason that peace was granted them [Germans] a second time, in spite of their having broken their truce so soon, was that the Dalmatians and Pannonians were in a state of great disturbance and required sharp attention.

Dio 55.29,1-3 (2nd/3rd c. AD)

Ταῖς γὰρ ἐσφοραῖς τῶν χρημάτων οἱ Δελμάται βαρυνόμενοι τὸν μὲν ἔμπροσθε χρόνον καὶ ἄκοντες ἡσύχαζον· ὡς δ' ὅ τε Τιβέριος ἐπὶ τοὺς Κελτοὺς τὸ δεύτερον ἐστράτευσε, καὶ Οὐαλέριος Μεσσαλῖνος ὁ τότε καὶ τῆς Δελματίας καὶ τῆς Παννονίας ἄρχων αὐτός τε σὺν ἐκείνῳ ἐστάλη καὶ τὸ πολὺ τοῦ στρατοῦ συνεξήγαγε, καί τινα καὶ σφεῖς δύναμιν πέμψαι κελευσθέντες συνῆλθόν τε ἐπὶ τούτῳ καὶ τὴν ἡλικίαν σφῶν ἀνθοῦσαν εἶδον, οὐκέτι διεμέλλησαν, ἀλλ' ἐνάγοντος αὐτοὺς ὅτι μάλιστα Βάτωνός τινος Δησιδιάτου τὸ μὲν πρῶτον ὀλίγοι τινὲς ἐνεωτέρισαν καὶ τοὺς Ῥωμαίους ἐπελθόντας σφίσιν ἔσφηλαν, ἔπειτα δὲ ἐκ τούτου καὶ οἱ ἄλλοι προσαπέστησαν. καὶ μετὰ τοῦτο καὶ Βρεῦκοι Παννονικὸν ἔθνος, Βάτωνα καὶ αὐτοὶ ἕτερον προστησάμενοι, ἐπί τε τὸ Σίρμιον καὶ ἐπὶ τοὺς ἐν αὐτῷ Ῥωμαίους ὥρμησαν. Καὶ ἐκεῖνο μὲν οὐκ ἐξεῖλον (αἰσθόμενος γὰρ τῆς ἐπαναστάσεως αὐτῶν Καικίνας Σεουῆρος ὁ τῆς πλησιοχώρου Μυσίας ἄρχων ἐπῆλθέ τε αὐτοῖς διὰ ταχέων περὶ τὸν Δράουον ποταμὸν οὖσι καὶ συμβαλὼν ἐνίκησεν), ἀναμαχέσεσθαι δέ πῃ διὰ βραχέος, ἐπειδὴ καὶ τῶν Ῥωμαίων συχνοὶ ἐπεπτώκεσαν, ἐλπίσαντες πρὸς παράκλησιν συμμάχων ἐτράποντο.

The Dalmatians, chafing under the levies of tribute, had hitherto kept quiet, though unwillingly. But when Tiberius made his second campaign against the Germans, and Valerius Messallinus, the governor of Dalmatia and Pannonia at the time, was sent out with him, taking most of his army along, the Dalmatians, too, were ordered to send a contingent; and on coming together for this purpose and beholding the strength of their warriors, they no longer delayed, but, under the vehement urging of one Bato, a Desidiatian, at first a few revolted and defeated the Romans who came against them, and then the rest also rebelled in consequence of this success. Next the Breucians, a Pannonian tribe, put another Bato at their head and marched against Sirmium and the Romans in that town. They did not capture the place, however, for Caecina Severus, the governor of the neighbouring province of Moesia, marched rapidly against them, when he heard of their uprising, and joining battle with them near the river Dravus, vanquished them; but hoping in some way to renew the struggle soon, since many of the Romans also had fallen, they turned their attention to summoning their allies and were getting together as many as they could.

Dio 55.30 (2nd/3rd c. AD)

Πυθόμενος οὖν ταῦθ' ὁ Τιβέριος, καὶ φοβηθεὶς μὴ καὶ ἐς τὴν Ἰταλίαν ἐσβάλωσιν, ἔκ τε τῆς Κελτικῆς ἀνέστρεψε, καὶ τὸν Μεσσαλῖνον προπέμψας αὐτὸς τῷ πλείονι τοῦ στρατοῦ ἐφείπετο. Αἰσθόμενος δὲ τῆς προσόδου αὐτῶν ὁ Βάτων ἀπήντησε τῷ Μεσσαλίνῳ, καίπερ μηδέπω καλῶς ἔχων, καὶ ἐπικρατέστερος αὐτοῦ ἐν παρατάξει γενόμενος ἔπειτ' ἐξ ἐνέδρας ἐνικήθη. Κἀκ τούτου πρός τε τὸν Βάτωνα τὸν Βρεῦκον ἦλθε, καὶ κοινωσάμενος αὐτῷ τὸν πόλεμον ὄρος τι Ἀλμὰν κατέλαβε· κἀνταῦθα πρὸς μὲν τοῦ Ῥυμητάλκου τοῦ Θρᾳκός, προπεμφθέντος ἐπ' αὐτοὺς ὑπὸ τοῦ Σεουήρου, βραχείᾳ τινὶ μάχῃ ἡττήθησαν, πρὸς δ' αὐτὸν ἐκεῖνον ἰσχυρῶς ἀντέσχον. Καὶ μετὰ ταῦτα τοῦ τε Σεουήρου ἐς τὴν Μυσίαν διά τε τοὺς Δακοὺς καὶ διὰ τοὺς Σαυρομάτας πορθοῦντας αὐτὴν ἀπάραντος, καὶ τοῦ Τιβερίου τοῦ τε Μεσσαλίνου ἐν Σισκίᾳ ἐγχρονισάντων, τήν τε συμμαχίδα σφῶν ἐπέδραμον καὶ συχνοὺς προσαπέστησαν. Καὶ ἐς μὲν χεῖρας, καίπερ τοῦ Τιβερίου πλησιάσαντός σφισιν, οὐκ ἦλθον αὐτῷ, ἄλλοσε δὲ καὶ ἄλλοσε μεθιστάμενοι πολλὰ ἐπόρθησαν· τῆς τε γὰρ χώρας ἐμπείρως ἔχοντες καὶ κούφως ἐσκευασμένοι, ῥᾳδίως ὅπῃ ποτὲ ἐβούλοντο ἐχώρουν. Καὶ ἐπειδή γε ὁ χειμὼν ἐνέστη, πολὺ πλείω ἐκακούργησαν· καὶ γὰρ καὶ ἐς τὴν Μακεδονίαν αὖθις ἐνέβαλον. Καὶ τούτους μὲν ὅ τε Ῥυμητάλκης καὶ ὁ ἀδελφὸς αὐτοῦ Ῥασκύπορις μάχῃ κατέλαβον· οἱ δὲ δὴ ἄλλοι τῇ μὲν χώρᾳ σφῶν πορθουμένῃ μετὰ τοῦτο ἐπί τε Καικιλίου Μετέλλου καὶ ἐπὶ Λικινίου Σιλανοῦ ὑπάτων οὐκ ἐπήμυναν, ἐς δὲ τὰ ἐρυμνὰ ἀναφυγόντες ἐκεῖθεν ὅπῃ παρείκοι καταδρομὰς ἐποιοῦντο.

Now when Tiberius learned of this, fearing that they might invade Italy, he returned from Germany, sending Messallinus ahead and following himself with most of his army. But Bato learned of their approach, and although not yet well, went to meet Messallinus; and though he proved stronger than Messallinus in open conflict, he was afterward defeated by an ambuscade. Thereupon he went to Bato, the Breucian, and making common cause with him in the war, occupied a mountain named Alma. Here they were defeated by Rhoemetalces, the Thracian, who had been sent ahead against them by Severus, but resisted Severus himself vigorously. Later, when Severus withdrew to Moesia, because the Dacians and Sarmatians were ravaging it, and Tiberius and Messallinus were tarrying in Siscia, the Dalmatians overran the territory of their allies and caused many more to revolt. And although Tiberius approached them, they would engage in no pitched battle with him, but kept moving from one place to another, causing great devastation; for, owing to their knowledge of the country and the lightness of their equipment, they could easily proceedd wherever they pleased. And when winter set in they did much greater damage, for they even invaded Macedonia again. As for these forces, now, Rhoemetalces and his brother Rhascyporis checked them by a battle; and as for the others, they did not come to the defence of their country when it was later ravaged (in the consulship of Caecilius Metellus and Licinius Silanus), but took refuge in the mountain fortresses, from which they made raiding expeditions whenever the chance offered.

Dio 55.32,3-4 (2nd/3rd c. AD)

(...) τοῦ δὲ δὴ Γερμανικοῦ ἐς τὴν Παννονίαν ἐλθόντος καὶ στρατευμάτων πολλαχόθεν ἐκεῖσε συνιόντων, τηρήσαντες οἱ Βάτωνες τὸν Σεουῆρον ἐκ τῆς Μυσίας προσιόντα ἐπέπεσον αὐτῷ ἀπροσδόκητοι, στρατοπεδευομένῳ πρὸς τοῖς Οὐολκαίοις ἕλεσι, καὶ τοὺς μὲν ἔξω τοῦ ταφρεύματος ἐφόβησαν καὶ κατήραξαν ἐς αὐτό, δεξαμένων δέ σφας τῶν ἔνδον ἡττήθησαν. Καὶ μετὰ τοῦθ' οἱ Ῥωμαῖοι νεμηθέντες, ὅπως πολλαχῇ ἅμα τῆς χώρας κατατρέχωσιν, οἱ μὲν ἄλλοι οὐδὲν ἄξιον λόγου τότε γε ἔδρασαν, ὁ δὲ δὴ Γερμανικὸς Μαζαίους Δελματικὸν ἔθνος μάχῃ νικήσας ἐκάκωσεν.

After Germanicus reached Pannonia and armies were assembling there from many sides, the two Batos waited until Severus approached from Moesia and then fell upon him unexpectedly, while he was encamped near the Volcaean marshes. They frightened the pickets outside the ramparts and drove them back inside, but when the men in the camp stood their ground, the attackers were defeated. After this the Romans were divided into detachments, in order that they might overrun many parts of the country at once; most of these detachments did nothing worthy of note, at least not at that time, but Germanicus conquered in battle and harassed the Mazaei, a Dalmatian tribe.

Dio 55.33,1-2 (2nd/3rd c. AD)

Μάρκου δὲ δὴ Φουρίου μετὰ Σέξτου Νωνίου ὑπατεύσαντος ἐπεθύμησαν μὲν καὶ οἱ Δελμάται καὶ οἱ Παννόνιοι συμβῆναι διὰ τὸ τὸ μὲν πρῶτον λιμῷ, εἶτα καὶ νόσῳ ἀπ' αὐτοῦ, πόαις τέ τισιν ἀλλοκότοις καὶ ῥίζαις χρώμενοι, πονηθῆναι, οὐκ ἐπεκηρυκεύσαντο δὲ κωλυσάντων τῶν μηδεμίαν παρὰ τῶν Ῥωμαίων ἐλπίδα σωτηρίας ἐχόντων, ἀλλὰ καὶ ὣς ἀντεῖχον. Καὶ Σκενόβαρδός τέ τις προσποιησάμενος μεταστήσεσθαι, καὶ πέμψας κατ' αὐτὸ τοῦτο πρὸς Μάνιον Ἔννιον φρούραρχον Σισκίας ὡς ἕτοιμος μὲν ὢν αὐτομολῆσαι, δεδιὼς δὲ μὴ προπάθῃ <...>

In the consulship of Marcus Furius and Sextus Nonius, the Dalmatians and Pannonians desired to make terms, because they were afflicted first by famine and then by disease that followed it, since they were using for food roots and strange herbs. They did not, however, make any overtures, being hindered by those who had no hope of being spared by the Romans, but even in their distress still resisted. And one, Scenobardus, who had pretended he was going to change sides and with reference to this very matter had sent to Manius Ennius, the commander of the garrison in Siscia, as if he were ready to desert, became afraid that he might suffer harm beforehand <...>

Dio 55.34,3-7 (2nd/3rd c. AD)

Πρὸς μέντοι τὰς τῶν πολέμων διαχειρίσεις οὕτως ἔρρωτο ὥσθ', ἵν' ἐγγύθεν καὶ ἐπὶ τοῖς Δελμάταις καὶ ἐπὶ τοῖς Παννονίοις πᾶν ὅ τι χρὴ συμβουλεύειν ἔχῃ, πρὸς Ἀρίμινον ἐξώρμησε. (...) ἐν δὲ τούτῳ ὁ Βάτων ὁ Βρεῦκος, ὁ τόν τε Πίννην προδοὺς καὶ μισθὸν τούτου τὴν ἀρχὴν τῶν Βρεύκων λαβών, ἑάλω τε ὑπὸ τοῦ ἑτέρου Βάτωνος καὶ διεφθάρη· ἐπειδὴ γὰρ ὑποπτεύσας τι ἐς τὸ ὑπήκοον ὁμήρους καθ' ἕκαστον τῶν φρουρίων περιιὼν ᾔτει, μαθὼν τοῦτ' ἐκεῖνος ἐνήδρευσέ που αὐτόν, καὶ μάχῃ κρατήσας κατέκλεισεν ἐς τεῖχος, καὶ μετὰ τοῦτ' ἐκδοθέντα ὑπὸ τῶν ἔνδον λαβὼν παρήγαγέ τε ἐς τὸ στράτευμα, καὶ καταψηφισθέντα ἀποθανεῖν ἐν χερσὶν ἐποίησε. Γενομένου δὲ τούτου συχνοὶ τῶν Παννονίων ἐπανέστησαν, καὶ αὐτοῖς ὁ Σιλουανὸς ἐπιστρατεύσας τούς τε Βρεύκους ἐνίκησε καὶ τῶν ἄλλων τινὰς ἀμαχεὶ προσεποιήσατο. Ἰδὼν οὖν ταῦτα ὁ Βάτων τῆς μὲν Παννονίας οὐδεμίαν ἔτ' ἐλπίδα ἔσχε, τὰς δὲ ἐς τὴν Δελματίαν ἐξ αὐτῆς ἐσόδους φρουραῖς διαλαβὼν ἐκείνην ἐπόρθει. καὶ οὕτω καὶ οἱ λοιποὶ τῶν Παννονίων, ἄλλως τε καὶ τῆς χώρας σφῶν ὑπὸ τοῦ Σιλουανοῦ κακουμένης, ὡμολόγησαν, πλὴν καθ' ὅσον ληστικά τινα οἷα ἐκ ταραχῆς τοσαύτης ἐπὶ πλεῖον κακουργοῦντα διεγένετο, ὅπερ που καὶ ἀεὶ ὡς εἰπεῖν παρά τε τοῖς ἄλλοις καὶ παρ' ἐκείνοις μάλιστα συμβαίνει.

Yet he [Augustus] was so vigorous when it came to directing campaigns against the enemy that he proceeded to Ariminum in order that he might be near at hand to give all necessary advice in regard to both the Dalmatians and the Pannonians. (...) Meanwhile, Bato, the Breucian, who had betrayed Pinnes and had received the right to rule over the Breucians as his reward, was captured by the other Bato and put to death. The Breucian, it seems, had been somewhat suspicious of his subject tribes and had gone round to each of the garrisons to demand hostages; and the other, learning of this, lay in wait for him somewhere or other, defeated him in battle, and shut him up in a stronghold. Later, when the Breucian was delivered over by those inside, he took him and brought him before the army, and then, when he had been condemned, put him to death on the spot. After this many of the Pannonians rose in revolt, and Silanus made a campaign against them, conquered the Breucians, and won over some of the others without a battle. Bato, on seeing this, gave up all hope of Pannonia, but occupied the passes leading to Dalmatia with garrisons and ravaged that country. Then at last the remainder of the Pannonians also came to terms, chiefly for the reason that their country was being harried by Silvanus. However, certain bands of brigands continued their forays for a long time, as was natural after so great a disturbance; indeed, this nearly always happens, not only among other peoples, but especially in the case of these tribes.

Dio 56.16 (2nd/3rd c. AD)

Κἀν τούτῳ καὶ ὁ Βάτων, Σκευᾶν τὸν υἱὸν πρὸς Τιβέριον πέμψας, παραδώσειν οἱ καὶ ἑαυτὸν καὶ τοὺς σὺν αὑτῷ πάντας ὑπέσχετο, ἂν τῆς ἀδείας τύχῃ. καὶ μετὰ τοῦτο πίστιν λαβὼν νυκτός τε ἐς τὸ στρατόπεδον αὐτοῦ ἐσῆλθε, καὶ τῇ ὑστεραίᾳ ἐπὶ βήματος αὐτῷ καθημένῳ προςαχθεὶς ὑπὲρ μὲν ἑαυτοῦ οὐδὲν ἐδεήθη, ἀλλὰ καὶ τὴν κεφαλὴν προέτεινεν ὥστ' ἀποκοπῆναι, ὑπὲρ δὲ τῶν ἄλλων πολλὰ ἀπελογήσατο. Καὶ τέλος ἐρωτηθεὶς ὑπὸ τοῦ Τιβερίου "τί ὑμῖν ἔδοξε καὶ ἀποστῆναι καὶ ἐπὶ τοσοῦτον ἡμῖν χρόνον ἀντιπολεμῆσαι;" ἔφη ὅτι "ὑμεῖς τούτων αἴτιοί ἐστε· ἐπὶ γὰρ τὰς ἀγέλας ὑμῶν φύλακας οὐ κύνας οὐδὲ νομέας ἀλλὰ λύκους πέμπετε." Ὁ μὲν οὖν πόλεμος τοῦτο τὸ τέλος ἔσχε, πολλῶν μὲν καὶ ἀνδρῶν, πλείστων δὲ δὴ καὶ χρημάτων ἀπολομένων· πάμπολλά τε γὰρ ἐς αὐτὸν στρατόπεδα ἐτράφη καὶ λεία ἐλαχίστη ἑάλω.

In the meantime Bato sent his son Sceuas to Tiberius, promising to surrender both himself and all his followers if he obtained pardon. And when he later received a pledge, he came by night to Tiberius' camp and on the following day was led before him as he sat on a tribunal. Bato asked nothing for himself, even holding his head forward to await the stroke, but in behalf of the others he made a long defence. Finally, upon being asked by Tiberius why his people had taken it into their heads to revolt and to war against the Romans so long, he replied: "You Romans are to blame for this; for you send as guardians of your flocks, not dogs or shepherds, but wolves". In this way the war was ended after the loss of many men and immense treasure; for ever so many legions were maintained for this campaign and but very little booty was taken.

Dio 56.17 (2nd/3rd c. AD)

Ἀνήγγειλε δὲ καὶ τότε τὴν νίκην ὁ Γερμανικός, καὶ ἐπ' αὐτῇ τῷ μὲν Αὐγούστῳ καὶ τῷ Τιβερίῳ τό τε <τὸ> τοῦ αὐτοκράτορος ὄνομα προςθέσθαι καὶ τὸ τὰ ἐπινίκια πέμψαι, ἄλλαι τέ τινες τιμαὶ καὶ ἁψῖδες ἐν τῇ Παννονίᾳ τροπαιοφόροι δύο ἐδόθησαν (ταῦτα γὰρ ἀπὸ πολλῶν τῶν ψηφισθέντων σφίσιν ὁ Αὔγουστος ἐδέξατο), τῷ δὲ δὴ Γερμανικῷ αἵ τε νικητήριοι τιμαί, ὅπερ που καὶ τοῖς ἄλλοις στρατιάρχοις ὑπῆρξε, καὶ αἱ στρατηγικαί, τό τε τὴν γνώμην πρώτῳ μετὰ τοὺς ὑπατευκότας ἀποφαίνεσθαι, καὶ τὸ τὴν ὑπατείαν θᾶσσον παρὰ τὸ νενομισμένον λαβεῖν. καὶ τῷ Δρούσῳ δὲ τῷ τοῦ Τιβερίου υἱεῖ, καίτοι μὴ μετασχόντι τοῦ πολέμου, καὶ ἐς τὸ συνέδριον συμφοιτᾶν πρὶν βουλεῦσαι, καὶ ἐπειδὰν ταμιεύσῃ γνώμην πρὸ τῶν ἐστρατηγηκότων ποιεῖσθαι, ἐψηφίσθη.

On this occasion, also, Germanicus announced the victory; and because of it Augustus and Tiberius were permitted to add the title of *imperator* to their other titles and to celebrate a triumph, and they received other honours, besides, notably two triumphal arches in Pannonia; for these were the only distinctions of the many voted to them that Augustus would accept. Germanicus received the *ornamenta triumphalia*, a distinction which fell likewise to the other commanders, and also the rank of a praetor, as well as the privilege of giving his vote immediately after the ex-consuls and of holding the consulship earlier than custom allowed. To Drusus, also, the son of Tiberius, even though he had taken no part in the war, was granted the privilege of attending the sittings of the senate before becoming a member of that body and of voting ahead of the ex-praetors as soon as he should become quaestor.

Dio 57.4,1-2 (2nd/3rd c. AD)

Ἐθορύβησαν μὲν γὰρ καὶ οἱ ἐν τῇ Παννονίᾳ στρατιῶται, ἐπειδὴ τάχιστα τῆς τοῦ Αὐγούστου μεταλλαγῆς ᾔσθοντο· καὶ συνελθόντες ἐς ἓν τεῖχος, καὶ ἐκεῖνο κρατυνάμενοι, πολλὰ καὶ στασιαστικὰ ἔπραξαν. Τά τε γὰρ ἄλλα καὶ τὸν ἄρχοντά σφων Ἰούνιον Βλαῖσον ἀποκτεῖναι ἐπεχείρησαν, τούς τε δούλους αὐτοῦ συλλαβόντες ἐβασάνισαν. Τό τε σύμπαν οὔθ' ὑπὲρ ἑκκαίδεκα ἔτη στρατεύεσθαι ἤθελον, καὶ δραχμὴν ἡμερησίαν φέρειν τά τε ἆθλα εὐθὺς αὐτοῦ ἐν τῷ στρατοπέδῳ λαμβάνειν ἠξίουν, ἀπειλοῦντες, ἂν μὴ τύχωσιν αὐτῶν, τό τε ἔθνος ἀποστήσειν καὶ ἐπὶ τὴν Ῥώμην ἐλάσειν.

For the troops in Pannonia had mutinied as soon as they learned of the death of Augustus, and coming together into one camp and strengthening it, they committed many rebellious acts. Among other things they attempted to kill their commander, Junius Blaesus, and arrested and tortured his slaves. Their demands were, in brief, that their term of service should be limited to sixteen years, that they should be paid a denarius per day, and that they should receive their prizes then and there in the camp; and they threatened, in case they did not obtain these demands, to cause the province to revolt and then to march upon Rome.

Dio 57.10,5 (2nd/3rd c. AD)

Αἰμιλίῳ γοῦν Ῥήκτῳ χρήματά ποτε αὐτῷ πλείω παρὰ τὸ τεταγμένον ἐκ τῆς Αἰγύπτου ἧς ἦρχε πέμψαντι ἀντεπέστειλεν ὅτι κείρεσθαί μου τὰ πρόβατα, ἀλλ' οὐκ ἀποξύρεσθαι βούλομαι.

In fact, when Aemilius Rectus once sent him [Tiberius] from Egypt, which he was governing, more money than was stipulated, he sent back to him the message: "I want my sheep shorn, not shaven".

Dio 59.18,4 (2nd/3rd c. AD)

Καλουίσιος δὲ δὴ Σαβῖνος ἔν τε τοῖς πρώτοις τῆς βουλῆς ὢν καὶ τότε ἐκ τῆς ἐν τῇ Παννονίᾳ ἀρχῆς ἀφιγμένος, ἥ τε γυνὴ αὐτοῦ Κορνηλία γραφέντες (...) οὐχ ὑπέμειναν τὴν κρίσιν, ἀλλ' ἑαυτοὺς προανάλωσαν.

Thus, Calvisius Sabinus, one of the foremost men in the senate, who had just returned from governing Pannonia, was indicted together with his wife Cornelia. (...) These two did not stand trial but despatched themselves before the time fixed.

Enn. Paneg. 7.206 (5th/6th c. AD)

Ulca est tutela Gepidarum, quae vice aggerum munit audaces et in jugorum montem latus provinciae quibusdam muris complectitur, nullo ariete frustrandis.

The Ulca River is the defence to the Gepids, which surrounds the brave like a moat, and on the slopes of the hills encircles the flanks of the province; it cannot be broken by siege engines of any kind.

Eutr. Brev. 4.24 (4th c. AD)

Ab urbe condita anno sexcentesimo tricesimo quinto C. Cato consul Scordiscis intulit bellum ignominioseque pugnavit.

In the six hundred and thirty-fifth year from the building of the city, the consul Caius Cato made war upon the Scordisci, and fought with them to his dishonour.

Eutr. Brev. 4.27 (4th c. AD)

A Minucio Rufo in Macedonia Scordisci et Triballi (victi sunt).

The Scordisci and Triballi (were defeated) in Macedonia by Minutius Rufus.

Eutr. Brev. 7.9 (4th c. AD)

(...) Romano adiecit imperio Aegyptum, Cantabriam, Dalmatiam saepe ante victam, sed penitus tunc subactam, Pannoniam, Aquitaniam, Illyricum, Raetiam, Vindelicos et Salassos in Alpibus, omnes Ponti maritimas civitates, in his nobilissimas Bosphorum et Panticapaeum. (...) Hoc tamen bellum [Germanicum] per Drusum, privignum suum, administravit, sicut per Tiberium, privignum alterum, Pannonicum (...).

(Octavian Augustus) added to the Roman empire Egypt, Cantabria, Dalmatia, often before conquered but only then entirely subdued, Pannonia, Aquitania, Illyricum, Rhaetia, the Vindelici and Salassi on the Alps, and all the maritime cities of Pontus, among which the two most noble were Bosporus and Panticapaeon. (...) This [German] war however he carried on by the agency of his step-son Drusus, as he had conducted the Pannonian war by that of his other step-son Tiberius (...).

Exp. tot. mund. 57 (4th c. AD)

Deinde Pannonia regio, terra dives in omnibus, fructibus quoque et iumentis et negotiis, ex parte et mancipiis. Et semper habitatio imperatorum est. Habet autem et civitates maximas, Sirmium quoque Noricum: unde et vestis Noricus exire dicitur. Haec Pannonia regio.

Then there is Pannonia, a land rich in all things - crops, cattle, slaves, and a little trade. The dwelling place of the emperors are always here. Its greatest cities are Sirmium and Noricum; the Noric overcoat is reported to come from the latter. This is the area of Pannonia.

Fest. Brev. 3.4 (4th c. AD)

Accesserunt Romano orbi Alpes Maritimae, Alpes Cottiae, Raetiae, Norica, Pannoniae, Moesiae, et omnis ora Danubii in provincias est redacta.

(...) the Maritime Alps, Cottian, Raetian, and Norican Alps, the Pannonias, and the Moesias accrued to the Roman world, and the entire bank of the Danube was reduced to provinces.

Fest. Brev. 7 (4th c. AD)

Sub Iulio Octaviano Caesare Augusto per Alpes Iulias iter factum est. Alpinis omnibus victis, Noricorum provinciae accesserunt. Bathone Pannoniarum rege subacto in dicionem nostram Pannoniae venerunt. Amantinis inter Savum et Dravum prostratis regio Saviensis ac secundorum loca Pannoniarum obtenta sunt.

Under Julius Octavian Caesar Augustus, a road was made through the Julian Alps; when all the Alpini had been conquered, the provinces of the Norici were added. After Batho, king of the Pannonians, had been subdued, the Pannonias came under our sway. After the Amantini between the Save and Drave had been laid low, the area adjoining the Save and environs of Pannonia Secunda were obtained.

Flor. Epit. 1.39,2-4 (2nd c. AD)

Nihil interim per id omne tempus residuum crudelitatis fuit in captivos saevientibus: litare dis sanguine humano, bibere in ossibus capitum, cuiusque modi ludibriis foedare mortem tam igni quam fumo, partus quoque gravidarum mulierum extorquere tormentis. Saevissimi omnium Thracum Scordisci fuere, sed calliditas quoque ad robur accesserat: silvarum et montium situs cum ingenio consentiebant.

Throughout the period of their advance they left no cruelty untried, as they vented their fury on their prisoners; they sacrificed to the gods with human blood; they drank out of human skulls; by every kind of insult inflicted by burning and fumigation they made death more foul; they even forced infants from their mothers' wombs by torture. The cruellest of all the Thracians were the Scordisci, and to their strength was added cunning as well; their haunts among the woods and mountains harmonized well with their fierce temper.

Flor. Epit. 2.24 (2nd c. AD)

Pannonii duobus acribus fluviis, Dravo Savoque vallantur. Populati proximos intra ripas se recipiebant. In hos domandos Vinnium misit. Caesi sunt in utrisque fluminibus. Arma victorum non ex more belli cremata, sed capta sunt et in profluentem data, ut Caesaris nomen eis qui resistebant sic nuntiaretur.

The Pannonians are protected by two swiftly-flowing rivers, the Drave and the Save; after ravaging the territory of their neighbours, they used to withdraw behind the banks of these streams. Caesar sent Vinnius to subdue them, and they were defeated on both rivers. The arms of the conquered enemy were not burnt, as was the usual custom in war, but broken to pieces and hurled into the current, that the fame of Caesar might thus be announced to those who were still resisting.

Fron. Str. 2.1,15 (1st/2nd c. AD)

Ti. Nero adversus Pannonios, cum barbari feroces in aciem oriente statim die processissent, continuit suos passusque est hostem nebula et imbribus, qui forte illo die crebri erant, verberari. Ac deinde, ubi fessum stando et pluvia non solum sed et lassitudine deficere animadvertit, signo dato adortus superavit.

In the campaign against the Pannonians, when the barbarians in warlike mood had formed for battle at the very break of day, Tiberius Nero held back hand his own troops, and allowed the enemy to be hampered by the fog and be drenched with the showers, which happened to be frequent that day. Then, when he noticed that they were weary with standing, and faint not only from exposure but also from exhaustion, he gave the signal, attacked and defeated them.

Galen I, p. 618, 1.10 (2nd c. AD)

Ὅσοι δ' ἔμπαλιν τούτοις ὑγρὰν καὶ ψυχρὰν χώγαν ἐποικοῦσιν, Ἰλλύριοί τε καὶ Γερμανοὶ καὶ Δαλμάται καὶ Σαυρομάται καὶ σύμπαν τὸ Σκυθικὸν εὐαυξεῖς μετρίως καὶ λεπτὰς καὶ εὐθείας καὶ πυρράς:

On the other hand, the people who inhabit humid and cold land - the Illyrians, Germans, Dalmatians, Sauromatae and all Scythians - they are moderately tall, slender, of noble stature and red-haired.

Hdn. Hist. 8.1,5-6 (2nd/3rd c. AD)

Ἅπερ ὄρη ὑπερμήκη ὥσπερ τεῖχος Ἰταλίας ἡ φύσις ἤγειρεν, ὑπερνεφῆ μὲν τὸ ὕψος, ἐπιμηκέστατα δέ, ὡς πᾶσαν Ἰταλίαν διειληφότα καθήκειν ἐν μὲν τοῖς δεξιοῖς Ἰταλίας μέρεσιν ἐς τὸ Τυρρηναῖον πέλαγος, ἐν δὲ τοῖς λαιοῖς ἐς τὸν Ἰόνιον κόλπον. σκέπεται δὲ ὕλαις δασείαις καὶ πυκναῖς, τὰς δὲ διεξόδους ἐστένωται ἢ διὰ κρημνοὺς ἐς βάθος μέγιστον ἀπερρωγότας ἢ διὰ πετρῶν τραχύτητα·

Nature has provided this long mountain range as a kind of fortification for Italy, whose peaks reach above the clouds, and so extensive that it completely cuts off Italy, running from the Tyrrhenian Sea in the West to the Ionian Gulf in the East. The mountains are covered with dense, thick forests, and have passes hemmed in by overhanging cliffs that end abruptly in precipitous drops, or by rocky outcrops.

Hyg. Grom. 28 (1st/2nd c. AD)

Multi huius modi agrum more colonico decimanis et kardinibus diviserunt, hoc est per centurias, sicut in Pannonia: mihi [autem] videtur huius soli mensura alia ratione agenda. (...) Certa [enim] pretia agris constituta sunt, ut in Pannonia arvi primi, arvi secundi, prati, silvae glandiferae, silvae vulgaris, pascuae.

Many people have thus parcelled the field out, according to the colonial practice, into the decumani and cardines, namely into the centuriae, as in Pannonia. Yet, it seems to me that the land of this kind ought to be measured in a different way. (...) Certain categories of fields have been established, as in Pannonia: plough lands of the first class, plough lands of the second class, meadows, acorn-producing forests, common forests, pastures.

Isid. Etym. 14.4,16 (6th/7th c. AD)

Pannonia ab Alpibus Apenninis est nuncupata, quibus ab Italia secernitur, regio viro fortis et solo laeta duobus satis acribus fluviis, Dravo Savoque, vallata.

Pannonia is named after the Apennine Alps, by which it is

separated from Italy. It is a region strong in men that enjoys good soil and it is surrounded by two rivers that are very rough, the Drava and the Sava.

Jord. Rom. 216 (6th c. AD)

Illyriam autem Gentione suo rege Macedonibus auxiliantibus vicit Romanorum Lucius praetor et in provinciam redegit. Dardanos Mysosque Curion primum proconsul edomuit primusque omnium Romanorum Danubium amnem usque profectus cuncta eius loca vastavit. Pannonum quoque regem in certamine superans idem Lucius redegit in provinciam utrasque Pannonias. Amantinos autem, qui inter Saum Draumque flumina insident, rege eorum interempto ipsa vice Romanam fecit provinciam.

Illyricum, under its king Gentius, who helped the Macedonians, was conquered by Lucius, praetor of the Romans, and reduced to a province. For the first time Curio the proconsul subjugated the Dardani and the Mysi; and being the first of all Romans to advance all the way to the Danube river, he laid waste to its whole region. Likewise overcoming the king of the Pannonians in battle, the same Lucius turned both Pannonias into a province. Moreover, after their king had been killed, he made the Amantini, who live between the Sava and Drava rivers, into a province in their turn.

Jord. Rom. 243 (6th c. AD)

Pannonii vero duobus acribus fluviis Drao Savoque vallantur. Contra quos Vini<ci>um misit, qui eos plus velociter vicit, quam eorum flumina cursu rapido currunt.

The Pannonians are protected by two swiftly-flowing rivers, the Drave and the Save. Against them he sent Vini<ci>us who defeated them in a victory more swift than the flow of their rivers.

Liv. Per. 54 (1st c. BC / 1st c. AD)

In Scordiscis cladis accepta.

A disaster befell the Romans among the Scordisci.

Liv. Per. 56 (1st c. BC / 1st c. AD)

M. Cosconius praetor in Thracia cum Scordiscis prospere pugnavit.

Praetor Marcus Cosconius fought successfully against the Scordisci in Thrace.

Liv. Per. 63 (1st c. BC / 1st c. AD)

C. Porcius consul in Thracia male adversus Scordiscos pugnavit. (...) Livius Drusus consul adversus Scordiscos, gentem a Gallis oriundam, in Thracia feliciter pugnavit.

Consul Gaius Porcius lost a battle against the Scordisci in Thrace. (...) Consul Livius Drusus fought successfully in Thrace against the Scordisci, a people of Gallic descent.

Liv. Per. 131 (1st c. BC / 1st c. AD)

Caesar seditionem veteranorum cum magna pernicie motam inhibuit, Iapydas et Dalmatas et Pannonios subegit.

Caesar checked an uprising among the veterans which caused great damage, and subdued the Iapydae, Dalmatians, and Pannonians.

Liv. Per. 133 (1st c. BC / 1st c. AD)

Caesar (...) in urbem reversus tres triumphos egit, unum ex Illyrico, alterum ex Actiaca victoria, tertium de Cleopatra.

Caesar returned to Rome to celebrate three triumphs, one for the campaign in Illyricum, a second for his victory at Actium, and the third over Cleopatra.

Liv. Per. 141 (1st c. BC / 1st c. AD)

Dalmatas et Pannonios Nero, frater Drusi, subegit.

Nero, Drusus' brother, subdued the Dalmatians and Pannonians.

Luc. 6.220-223 (1st c. AD)

Pannonis haud aliter post ictum saevior ursa, / cum iaculum parva Ligys ammentavit habena, / se rotat in volnus telumque irata receptum / impetit et secum fugientem circumit hastam.

Even so, when the Libyan has sped his javelin straight by means of a little thong, the Pannonian bear, infuriated by the wound, whirls round towards the injured part; in her rage she attacks the weapon that has struck her, and pursues in a circle the spear that flies along with her.

Mam. Paneg. Max. 2.2,29-34 (3rd/4th c. AD)

Unde igitur ordiar? Commemorabo nimirum patriae tuae in rempublicam merita? Quis enim dubitat quin multis iam saeculis, ex quo vires illius ad Romanum nomen accesserint, Italia quidem sit gentium domina gloriae vetustate, sed Pannonia virtute?

Where, then, I shall begin? Shall I recall, indeed, the services of your native land to the State? For who doubts that for many centuries now, ever since its strength was added to the Roman name, while Italy indeed may have been the mistress of nations by virtue of the antiquity of her glory, Pannonia has been in valor?

Mam. Paneg. gen. 3.3,21-28 (3rd/4th c. AD)

Non enim in otiosa aliqua deliciisque corrupta parte terrarum nati institutique estis, sed in his provinciis quas ad infatigabilem consuetudinem laboris atque patientiae fracto licet oppositus hosti, sed armis tamen semper instructus miles exercet, in quibus omnis vita militia est, quarum etiam feminae ceterarum gentium viris fortiores sunt.

For you were not born and raised in some quiet part of the world, a land enfeebled by luxury, but in those provinces whose border, exposed to the enemy (although a beaten one) and always arrayed in arms, has taught them the tireless habit of toil and patience, in provinces where all of life in military service, whose women even are braver than the men of other lands.

Nic. Dam. FGrH, frg. 125.1 (1st c. BC)

(...) ἔτι ἔνδηλος γίνεσθαι τῇ φιλανθωπίᾳ, ἔπεισεν ἑαυτοῦ ἀκροᾶσθαι, ὧν δὴ πρότερον οὐδὲ ὀνόματα ἠπίσταντο οἱ ἄνθρωποι, οὐδέ τινοσ ὑπήκοοι ἐγένοντο διὰ μνήμης, ἡμερωσάμενος, ὁπόσοι ἐντὸς Ῥήνου ποταμοῦ κατοικοῦσιν, ὑπέρ τε τὸν Ἰόνιον πόντον καὶ τὰ Ἰλλυριῶν γένη· Παννονίους αὐτοὺς καὶ Δάκας καλοῦσιν.

By making himself known through kindness he persuaded them to obey him. The names of some of them he had never heard before, nor had they been subject within the memory of anyone, but he subdued them: all those that live as far as the Rhine and beyond the Ionian Sea and the Illyrian peoples. These are called Pannonians and Dacians.

Oros. Hist. 6.19,3 (4th/5th c. AD)

Caesar Illyricum Pannoniam partemque Italiae bellis subegit ac domuit.

Caesar subdued Illyricum, Pannonia, and part of Italy by force, bringing them under his control.

Oros. Hist. 6.21,12-27 (4th/5th c. AD)

Norici Illyrii Pannonii Dalmatae Moesi Thraces et Daci Sarmatae plurimique et maximi Germaniae populi per diversos duces vel superati vel repressi vel etiam obiectu maximorum fluminum, Rheni Danuviique, seclusi sunt. (...) Pannonios novo motu intumescentes Tiberius privignus Caesaris cruentissima caede delevit. Idemque continuo Germanos bello arripuit, e quibus quadraginta milia captivorum victor abduxit. Quod revera bellum maximum et formidulosissimum quindecim legionibus per triennium gestum est, nec fere ullum maius bellum, sicut Suetonius adtestatur, post Punicum fuit.

The Norici, Illyrii, Pannonii, Dalmatae, Moesi, Thraces, and the Daci Sarmatae, the largest and strongest peoples of Germany, were either overcome or subdued by different generals and shut in by the mightiest of rivers, the Rhine and the Danube. (...) Tiberius, the stepson of Caesar, who had most cruelly slaughtered and destroyed the Pannonians when they had risen in a new revolt, immediately engaged the Germans in war. As a conqueror he carried off forty thousand of them as captives. This truly was a great and most formidable war waged by fifteen legions over a period of fifteen years; nor had there been, according to the testimony of Suetonius, another conflict equally great since the Punic wars.

Ov. *Cons.* 389-390 (1st c. BC / 1st c. AD)

Armeniusque fugax et tandem Dalmata supplex / summaque dispersi per iuga Pannonii.

(...) and the Armenian, ready to flee, and the Dalmatian, at last suppliant, / and the Pannonians scattered over the mountain summits.

Ov. Tr. 2.1,225-230 (1st c. BC / 1st c. AD)

Nunc tibi Pannonia est, nunc Illyris ora domanda, / Raetica nunc praebet Thraciaque arma metum (...)

Now Pannonia, now the Illyrian coast's to be subdued, / now Raetia and the war in Thrace concerns you (...)

Ov. Ep. 2.2,75-92 (1st c. BC / 1st c. AD)

Adde triumphatos modo Paeonas, adde quietis / subdita montanae brachia Dalmatiae. / Nec dedignata est abiectis Illyris armis / Caesareum famulo vertice ferre pedem. / Ipse super currum placido spectabilis ore / tempora Phoebea virgine nexa tulit, / quem pia vobiscum proles comitavit euntem, / digna parente suo nominibusque datis, / fratribus adsimilis quos proxima templa tenentis / Divus ab excelsa Iulius aede videt. / His Messalinus quibus omnia cedere debent / primum laetitiae non negat esse locum. / Quicquid ab his superest venit in certamen amoris, / hac hominum nulli parte secundus erit. / Hanc colet ante diem qua, quae decreta merenti, / venit honoratis laurea digna comis. / Felices quibus o licuit spectare triumphos / et ducis ore deos aequiperante frui!

Then there's the triumph over Paeonia, there are / raised arms in mountainous Dalmatia lowered in peace: / Illyria, a servant now, throwing down her weapons, / and not refusing to set her head beneath Caesar's foot. / Tiberius himself appeared in his chariot, calm-faced, / his forehead wreathed with Apollo's Daphnian laurel. / His loyal sons worthy of their father and the names granted / them, followed him, attended by you two brothers, / like the Twins of the neighbouring temple / whom divine Julius views from his lofty shrine. / Messalinus does not deny the supreme place / of joy, to those before whom all must yield, / whatever's left concerns battles of affection, / in these he'll take second place to no man. / He'll celebrate this day above all others, on which / the laurel of merit's rightly set on an honoured brow.

O happy are those allowed to see the triumph, / and savour the general's godlike countenance!

Ov. Ep. 2.1,1-2, 19-48 (1st c. BC / 1st c. AD)

Huc quoque Caesarei pervenit fama triumphi, / languida quo fessi vix venit aura Noti. (...)

Gratia, Fama, tibi per quam spectata triumphi / incluso mediis est mihi pompa Getis! / Indice te didici nuper visenda coisse / innumeras gentes ad ducis ora sui, / quaeque capit vastis inmensum moenibus orbem, / hospitiis Romam vix habuisse locum. / Tu mihi narrasti, cum multis lucibus ante / fuderit adsiduas nubilus auster aquas, / numine caelesti solem fulsisse serenum / cum populi vultu conveniente die / atque ita victorem cum magnae vocis honore / bellica laudatis dona dedisse viris / claraque sumpturum pictas insignia vestes / tura prius sanctis inposuisse focis / Iustitiamque sui caste placasse parentis, / illo quae templum pectore semper habet, / quaque ierit, felix adiectum plausibus omen / saxaque roratis erubuisse rosis; / protinus argento versos imitantia muros / barbara cum pictis oppida lata viris / fluminaque et montes et in altis proelia silvis / armaque cum telis in strue mixta sua / deque tropaeorum quod sol incenderet auro / aurea Romani tecta fuisse fori / totque tulisse duces captivos addita collis / vincula paene hostis quot satis esse fuit. / Maxima pars horum vitam veniamque tulerunt, / in quibus et belli summa caputque Bato. / Cur ego posse negem minui mihi numinis iram, / cum videam mitis hostibus esse deos?

The news of Caesar's triumph has reached this place / as well, where the south wind's breath barely comes. (...)

Thanks to you, Fame, though I'm imprisoned / among the Getae, I've seen the glorious triumph. / Your report told me how countless peoples / recently gathered to gaze on the leader's face: / and Rome whose vast walls compass the wide world, / scarcely had room to hold her many guests. / You told me how the sun shone brightly, / by heaven's power, and the day matched / the faces of the crowd, though, for days before, / a cloudy southerly poured down endless rain, / and how the victor made warlike gifts to the heroes, / honouring them in a mighty voice, how, wearing embroidered robes, glorious insignia, / he first scattered incense on the sacred fires / purely to placate his father's Justice, / which always occupies a temple in his heart, / how he heard happy omens of applause as he went / and the stones were red with dew-wet roses: / silver fascimiles of conquered walls were carried / before him, barbarian towns with defeated men, / rivers, mountains and battles in thick forests, / mingled piles of shields and spears, / the buildings of the Roman forum gilded / by the gold of trophies, glittering in the sun, / and so many captive chieftains, chained by the neck, / they were almost enough to form an enemy host. / Most of them were granted life and pardon, / among them Bato, high chieftain of the war. / Why should I fail to believe the divine anger could lessen / towards me, when I see the gods merciful to an enemy?

Paul HL 2.8 (8th c. AD)

Nec mirum, cum usque huc Pannonia pertingat, quae horum animantium ferax est. Denique retulit mihi quidam veracissimus senex, tale se corium in hoc monte occisi bisontis vidisse, in quo quindecim, ut aiebat, homines, unus iuxta alium potuissent cubare.

They say wild oxen graze upon it, and no wonder, since at this point it touches Pannonia, which is productive of these animals. In fine, a certain very truthful old man related to me that he had seen the hide of a wild ox killed on this mountain of such size that in it fifteen men, as he said, could lie one against the other.

Plin. NH 3.18,127-128 (1st c. AD)

Formio amnis, ab Ravenna CLXXXIX, anticus auctae Italiae terminus, nunc vero Histriae; quam cognominatam a flumine Histro in Hadriam effluente e Danuvio amne eodemque Histro exadversum Padi fauces, contrario eorum percussu mari interiecto dulcescente, plerique dixere falso, et Nepos etiam Padi accola. Nullus enim ex Danuvio amnis in mare Hadriaticum effunditur. Deceptos credo quoniam Argo navis flumine in mare Hadriaticum descendit non procul Tergeste; nec iam constat quo flumine. Umeris travectam Alpes diligentiores tradunt, subisse autem Histro, dein Savo, dein Nauporto, cui nomen ex ea causa est, inter Aemonam Alpesque exorienti.

It has been stated by many authors, even including Nepos, who lived on the banks of the Po, that Formio takes its name from the stream called Hister flowing out of the river Danube (which also has the name of Hister) into the Adriatic, opposite the mouths of the Po, and that their currents, colliding from contrary directions, turn the intervening sea into a pool of fresh water; but these statements are erroneous, for no river flows out of the Danube into the Adriatic. I believe that they have been misled by the fact that the ship Argo came down a river into the Adriatic not far from Tergeste, but it has not hitherto been decided what river this was. More careful writers say that the Argo was portaged on men's shoulders across the Alps, but that she had come up the Hister and then the Save and then the Nauportus, a stream rising between Emona and the Alps, that has got its name from this occurrence.

Plin. NH 3.25,147-148 (1st c. AD)

Inde glandifera Pannoniae, qua mitescentia Alpium iuga per medium Illyricum a septentrione ad meridiem versa molli in dextra ac laeva devexitate considunt. Quae pars ad mare Hadriaticum spectat appellatur Delmatia et Illyricum supra dictum; ad septentriones Pannonia vergit: finitur inde Danuvio. In ea coloniae Aemona, Siscia. Amnes clari et navigabiles in Danuvium defluunt Draus e Noricis violentior, Saus ex Alpibus Carnicis placidior, centum viginti milia intervallo, Draus per Serretes, Serapillos, Iasos, Andizetes, Saus per Colapianos Breucosque.

Populorum haec capita; praeterea Arviates, Azali, Amantini, Belgites, Catari, Cornacates, Eravisci, Hercuniates, Latovici, Oseriates, Varciani, mons Claudius, cuius in fronte Scordisci, in tergo Taurisci. Insula in Savo Metubarbis, amnicarum maxima. Praeterea amnes memorandi Colapis in Saum influens iuxta Sisciam gemino alveo insulam ibi efficit quae Segestica appellatur, alter amnis Bacuntius in Saum Sirmio oppido influit, ubi civitas Sirmiensium et Amantinorum. Inde quadraginta quinque milia Taurunum, ubi Danuvio miscetur Saus; supra influunt Valdasus, Urpanus, et ipsi non ignobiles.

Then come the acorn-producing lands of Pannonia, where the chain of the Alps gradually becomes less formidable, and slopes to the right and left hand with gentle contours as it traverses the middle of Illyricum from north to south. The part looking towards the Adriatic is called Delmatia and Illyricum mentioned above, while the part stretching northward is Pannonia, terminating in that direction at the Danube. In it are the colonies of Emona and Siscia. Famous navigable rivers flowing into the Danube are the Drave from Noricum, a rather violent stream, and the Save from the Carnian Alps which is more gentle, there being a space of 120 miles between them; the Drave flows through the Serretes, Serapilli, Iasi and Andizetes; the Save through the Colapiani and Breuci. These are the principal peoples; and there are besides the Arviates, Azali, Amantini, Belgites, Catari, Cornacates, Eravisci, Hercuniates, Latovici, Oseriates and Varciani, and Mount Claudius, in front of which are the Scordisci and behind it the Taurisci. In the Save is the island Metubarbis, the largest known island formed by a river. Other noteworthy rivers are the Colapis, which flows into the Save near Siscia, where its channel divides and forms the island called Segestica, and another river the Bacuntius, flowing into the Save at the town of Sirmium, the capital of the Sirmienses and Amantini. From Sirmium it is 45 miles to Taurunum, where the Save joins the Danube; tributaries flowing into the Danube higher up are the Valdasus and the Urpanus, themselves also not inconsiderable streams.

Plin. NH 11.106,255 (1st c. AD)

Sues in Illyrico quibusdam locis solidas habent ungulas.

The swine in some places in Illyricum have solid hoofs.

Plin. NH 37.11,43 (1st c. AD)

Adfertur [sucinum] a Germanis in Pannoniam maxime provinciam, et inde Veneti primum, quos Enetos Graeci vocaverunt, famam rei fecere proximique Pannoniae et agentes circa mare Hadriaticum.

Amber is imported by the Germans into Pannonia, more particularly; from whence the Veneti, by the Greeks called Eneti, first brought it into general notice, a people in the vicinity of Pannonia, and dwelling on the shores of the Adriatic Sea.

Polyb. frg 64 (= Suda, frg. 122) (Polybius: 2nd c. BC; Suda: 10th c. AD)

[Ὁρμητήριον] Πολύβιος˙ τὸ δὲ φρούριον οἱ Παννόνιοι κατ᾽ ἀρχὰς τοῦ πολέμου λαβόντες ὁρμητήριον ἐπεποίηντο καὶ εἰς ὑποδοχὴν τῶν λαφύρων ἐξῃρήκεσαν.

Stronghold, Polybius. The Pannonians have captured this fort at the beginining of the war, made it their stronghold and chose it for storing the plunder.

Ptol. 2.14 (2nd c. AD)

Κατέχουσι δὲ τὴν ἐπαρχίαν ἐν μὲν τοῖς πρὸς ἄρκτους μέρεσιν <Ἄζαλοι> μὲν δυσμικώτεροι, <Κύτνοι> δ᾽ ἀνατολικώτεροι, ἐν δὲ τοῖς μεσημβρινοῖς <Λατόβικοι> μὲν ὑπὸ τὸ Νωρικὸν, <Οὐαρκιανοὶ> δὲ τὰ πρὸς ἀνατολὰς, ἐν δὲ τοῖς μεταξὺ Βόϊοι μὲν πρὸς δυσμὰς καὶ ὑπ᾽ αὐτοὺς <Κολαιτιανοὶ, Ἰάσσιοι> δὲ πρὸς ἀνατολὰς καὶ ὑπ᾽ αὐτοὺς <Ὀσεριάτες>.

That province [Upper Pannonia] is held by the Azali in the outmost north-west, the Kytnoi in the north-east, the Latobici in the south beneath the Noricum, the Varciani towards the north, and in-between the Boi towards the west and below them the Kolanianoi (*sic*!), the Iasioi towards the east and the Oseriates below them.

Ptol. 2.15 (2nd c. AD)

Κατέχουσι δὲ ταύτην τὴν ἐπαρχίαν ἐν μὲν τοῖς δυσμικοῖς μέρεσιν Ἀμαντινοὶ ἀρκτικώτεροι, ὑφ᾽ οὓς <Ἑρκουνιάτες>, εἶτα <Ἀνδίζητες>, εἶτα <Βρεῦκοι>· ἐν δὲ τοῖς ἀνατολικοῖς ἀρκτικώτατοι μὲν <Ἀραυίσκοι>, μεσημβρινώτεροι δὲ <Σκορδίσκοι>.

The whole province [Lower Pannonia] is held by the Amantini in the north-west, the Hercuniates below them, then the Andizetes and the Breuci. The Eravisci are in the north-east, the Scordisci in the south.

SHA Hadr. 3.9 (4th c. AD)

Legatus postea praetorius in Pannoniam inferiorem missus Sarmatas compressit (...)

Next he was sent as praetorian legate to Lower Pannonia, where he held the Sarmatians in check (...)

SHA M. Ant. 14.6 (4th c. AD)

Denique transcensis Alpibus longius processerunt composueruntque omnia, quae ad munimen Italiae atque Illyrici pertinebant.

Finally, they crossed the Alps, and pressing further on, completed all measures necessary for the defence of Italy and Illyricum.

SHA M. Ant. 22.1 (4th c. AD)

Gentes omnes ab Illyrici limite usque in Galliam conspiraverant, ut Marcomanni Varistae Hermunduri et Quadi Suebi Sarmatae Lacringes et Buri hi aliique cum Victualis Osi Bessi Cobotes Roxolani Bastarnae Alani Peucini Costoboci.

Then, from the borders of Illyricum even into Gaul, all the nations banded together against us – the Marcomanni, Varistae, Hermunduri and Quadi, the Suebians, Sarmatians, Lacringes and Buri, these and certain others together with the Victuali, namely, Osi, Bessi, Cobotes, Roxolani, Bastarnae, Alani, Peucini, and finally, the Costoboci.

Stat. Silv. 1.4,76-80 (1st c. AD)

Pannoniusque ferox arcuque horrenda fugaci / Armenia (...)

(...) and the bold Pannonian and Armenia's dire retreating bowmen (...)

Stat. Silv. 5.2,136-137 (1st c. AD)

An iuga Pannoniae mutatoresque domorum / Sauromatas quaties?

Will you make Pannonian mountains tremble, and the Sauromatae that shift their dwelling?

Strab. 4.6,10 (1st c. BC / 1st c. AD)

Μεθ' οὓς ἡ Σεγεστικὴ πόλις ἐν πεδίῳ, παρ' ἣν ὁ Σάος παραρρεῖ ποταμὸς ἐκδιδοὺς εἰς τὸν Ἴστρον. Κεῖται δὲ ἡ πόλις εὐφυῶς πρὸς τὸν κατὰ τῶν Δακῶν πόλεμον· ἡ δ' Ὄκρα τὸ ταπεινότατον μέρος τῶν Ἄλπεών ἐστι καθ' ὃ συνάπτουσι τοῖς Κάρνοις, καὶ δι' οὗ τὰ ἐκ τῆς Ἀκυληίας φορτία κομίζουσιν ἁρμαμάξαις εἰς τὸν καλούμενον Ναύπορτον, σταδίων ὁδὸν οὐ πολὺ πλειόνων ἢ τετρακοσίων· ἐκεῖθεν δὲ τοῖς ποταμοῖς κατάγεται μέχρι τοῦ Ἴστρου καὶ τῶν ταύτῃ χωρίων. Παραρρεῖ γὰρ δὴ τὸν Ναύπορτον [Κορκόρας] ποταμὸς ἐκ τῆς Ἰλλυρίδος φερόμενος πλωτός, ἐκβάλλει δ' εἰς τὸν Σάον, ὥστ' εὐμαρῶς εἰς τὴν Σεγεστικὴν κατάγεται καὶ τοὺς Παννονίους καὶ Ταυρίσκους. Συμβάλλει δ' εἰς τὸν Σάον κατὰ τὴν πόλιν καὶ ὁ Κόλαπις· ἀμφότεροι δ' εἰσὶ πλωτοί, ῥέουσι δ' ἀπὸ τῶν Ἄλπεων.

After the Iapodes comes Segestica, a city in the plain, past which flows the river Savus, which empties into the Ister. The situation of the city is naturally well-suited for making war against the Daci. The Ocra is the lowest part of the Alps in that region in which the Alps join the country of the Carni, and through which the merchandise from Aquileia is conveyed in wagons to what is called Nauportus (over a road of not much more than four hundred stadia); from here, however, it is carried down by the rivers as far as the Ister and the districts in that part of the country; for there is, in fact, a river which flows past Nauportus; it runs out of Illyria, is navigable, and empties into the Savus, so that the merchandise is easily carried down to Segestica and the country of the Pannonii and Taurisci. And the Colapis too joins the Savus near the city; both are navigable and flow from the Alps.

Strab. 5.1,8 (1st c. BC / 1st c. AD)

Ἀκυληία δ' ἥπερ μάλιστα τῷ μυχῷ πλησιάζει, κτίσμα μέν ἐστι Ῥωμαίων ἐπιτειχισθὲν τοῖς ὑπερκειμένοις βαρβάροις (...) Ἀνεῖται δ' ἐμπόριον [τοῖς τε Ἐνετοῖς καὶ] τοῖς περὶ τὸν Ἴστρον τῶν Ἰλλυριῶν ἔθνεσι· κομίζουσι δ' οὗτοι μὲν τὰ ἐκ θαλάττης, καὶ οἶνον ἐπὶ ξυλίνων πίθων ἁρμαμάξαις ἀναθέντες καὶ ἔλαιον, ἐκεῖνοι δ' ἀνδράποδα καὶ βοσκήματα καὶ δέρματα.

Aquileia, which is nearest of all to the recess of the (Adriatic) Gulf, was founded by the Romans as a fortress against the barbarians who were situated above it (...) Aquileia has been given over as an emporium for those tribes of the Illyrians that live near the Ister; the latter load on wagons and carry inland the products of the sea, and wine stored in wooden jars, and also olive oil, whereas the former get in exchange slaves, cattle, and hides.

Strab. 7.5,2 (1st c. BC / 1st c. AD)

Τὸ δὲ λοιπὸν ἔχουσι Παννόνιοι μέχρι Σεγεστικῆς καὶ Ἴστρου πρὸς ἄρκτον καὶ ἕω· πρὸς δὲ τἆλλα μέρη ἐπὶ πλέον διατείνουσιν. Ἡ δὲ Σεγεστικὴ πόλις ἐστὶ Παννονίων ἐν συμβολῇ ποταμῶν πλειόνων, ἁπάντων πλωτῶν, εὐφυὲς ὁρμητήριον τῷ πρὸς Δακοὺς πολέμῳ· ὑποπέπτωκε γὰρ ταῖς Ἄλπεσιν, αἳ διατείνουσι μέχρι τῶν Ἰαπόδων, Κελτικοῦ τε ἅμα καὶ Ἰλλυρικοῦ ἔθνους· ἐντεῦθεν δὲ καὶ ποταμοὶ ῥέουσι πολλοὶ καταφέροντες εἰς αὐτὴν τόν τε ἄλλον καὶ τὸν ἐκ τῆς Ἰταλίας φόρτον. Εἰς γὰρ Ναύπορτον ἐξ Ἀκυληίας ὑπερτιθεῖσι τὴν Ὄκραν εἰσὶ στάδιοι τριακόσιοι πεντήκοντα, εἰς ἣν αἱ ἁρμάμαξαι κατάγονται, τῶν Ταυρίσκων οὖσαν κατοικίαν· ἔνιοι δὲ πεντακοσίους φασίν. Ἡ δ' Ὄκρα ταπεινότατον μέρος τῶν Ἄλπεών ἐστι τῶν διατεινουσῶν ἀπὸ τῆς Ῥαιτικῆς μέχρι Ἰαπόδων· ἐντεῦθεν δ' ἐξαίρεται τὰ ὄρη πάλιν ἐν τοῖς Ἰάποσι καὶ καλεῖται Ἄλβια. Ὁμοίως δὲ καὶ ἐκ Τεργέστε κώμης Καρνικῆς ὑπέρθεσίς ἐστι διὰ τῆς Ὄκρας εἰς ἕλος Λούγεον καλούμενον. Πλησίον δὲ τοῦ Ναυπόρτου ποταμός ἐστι Κορκόρας ὁ δεχόμενος τὰ φορτία· οὗτος μὲν οὖν εἰς τὸν Σάβον ἐμβάλλει, ἐκεῖνος δ' εἰς τὸν Δράβον, ὁ δὲ εἰς τὸν Νόαρον κατὰ τὴν Σεγεστικήν. Ἐντεῦθεν δ' ἤδη ὁ Νόαρος * πλήθει προσλαβὼν τὸν διὰ τῶν Ἰαπόδων ῥέοντα ἐκ τοῦ Ἀλβίου ὄρους Κόλαπιν συμβάλλει τῷ Δανουίῳ κατὰ τοὺς Σκορδίσκους. Ὁ δὲ πλοῦς τὰ πολλὰ τοῖς ποταμοῖς ἐπὶ τὰς ἄρκτους ἐστίν· ὁδὸς δ' ἀπὸ Τεργέστε ἐπὶ τὸν Δανούιον σταδίων ὅσον χιλίων καὶ διακοσίων. Ἐγγὺς δὲ τῆς Σεγεστικῆς ἐστι καὶ ἡ Σισκία φρούριον καὶ Σίρμιον ἐν ὁδῷ κείμεναι τῇ εἰς Ἰταλίαν.

The remainder of the country in question is held by the Pannonii as far as Segestica and the Ister, on the north and east, although their territory extends still farther in the other directions. The city Segestica, belonging to the Pannonians, is at the confluence of several rivers, all of them navigable, and is naturally fitted to be a base of operations for making war against the Dacians; for it lies beneath that part of the Alps which extends as far as the country of the Iapodes, a tribe which is at the same time both Celtic and Illyrian. And thence, too, flow rivers which bring down into Segestica much merchandise both from other countries and from Italy. For if one passes over Mount Ocra from Aquileia to Nauportus, a settlement of the Taurisci, whither the wagons are brought, the distance is three hundred and fifty stadia, though some say five hundred. Now the Ocra is the lowest part of that portion of the Alps which extends from the country of the Rhaeti to that of the Iapodes. Then the mountains rise again, in the country of the Iapodes, and are called Albian. In like manner, also, there is a pass which leads over Ocra from Tergeste, a Carnic village, to a marsh called Lugeum. Near Nauportus there is a river, the Corcoras, which receives the cargoes. Now this river empties into the Savus, and the Savus into the Dravus, and the Dravus into the Noarus near Segestica. Immediately below Nauportus the Noarus is further increased in volume by the Colapis, which flows from the Albian Mountain through the country of the Iapodes and meets the Danuvius near the country of the Scordisci. The voyage on these rivers is, for the most part, towards the north. The road from Tergeste to the Danuvius is about one thousand two hundred stadia. Near Segestica, and on the road to Italy, are situated both Siscia, a fort, and Sirmium.

Strab. 7.5,3 (1st c. BC / 1st c. AD)

Ἔθνη δ' ἐστὶ τῶν Παννονίων Βρεῦκοι καὶ Ἀνδιζήτιοι καὶ Διτίωνες καὶ Πειροῦσται καὶ Μαζαῖοι καὶ Δαισιτιᾶται, ὧν Βάτων ἡγεμών, καὶ ἄλλα ἀσημότερα μικρά, ἃ διατείνει μέχρι Δαλματίας σχεδὸν δέ τι καὶ Ἀρδιαίων ἰόντι πρὸς νότον· ἅπασα δ' ἡ ἀπὸ τοῦ μυχοῦ τοῦ Ἀδρίου παρήκουσα ὀρεινὴ μέχρι τοῦ Ῥιζονικοῦ κόλπου καὶ τῆς Ἀρδιαίων γῆς ... μεταξὺ πίπτουσα τῆς τε θαλάττης καὶ τῶν Παννονίων ἐθνῶν.

The tribes of the Pannonii are: the Breuci, the Andizetii, the Ditiones, the Peirustae, the Mazaei, and the Daesitiatae, whose leader is Bato, and also other small tribes of less significance which extend as far as Dalmatia and, as one goes south, almost as far as the land of the Ardiaei. The whole of the mountainous country that stretches alongside Pannonia from the recess of the Adriatic as far as the Rhizonic Gulf and the land of the Ardiaei is Illyrian, falling as it does between the sea and the Pannonian tribes.

Strab. 7.5,10 (1st c. BC / 1st c. AD)

Ἡ δ' ὑπερκειμένη ταύτης πᾶσα ὀρεινὴ καὶ ψυχρὰ καὶ νιφόβολός ἐστιν, ἡ δὲ προσάρκτιος καὶ μᾶλλον, ὥστε καὶ τῶν ἀμπέλων σπάνιν εἶναι καὶ ἐν ταῖς ὑψώσεσι καὶ ἐν τοῖς ἐπιπεδωτέροις. ὀροπέδια δ' ἐστὶ ταῦτα ἃ κατέχουσιν οἱ Παννόνιοι, πρὸς νότον μὲν μέχρι Δαλματέων καὶ Ἀρδιαίων διατείνοντα, πρὸς ἄρκτον δὲ ἐπὶ τὸν Ἴστρον τελευτῶντα, πρὸς ἕω δὲ Σκορδίσκοις συνάπτοντα ... τῇ δὲ παρὰ τὰ ὄρη τῶν Μακεδόνων καὶ Θρᾳκῶν.

But the whole of the country situated above this [the coast] is mountainous, cold, and subject to snows, especially the northerly part, so that there is a scarcity of the vine, not only on the heights but also on the levels. These latter are the mountain-plains occupied by the Pannonians; on the south they extend as far as the country of the Dalmatians and the Ardiaei, on the north they end at the Ister, while on the east they border on the country of the Scordisci, that is, on the country that extends along the mountains of the Macedonians and the Thracians.

Strab. 7.5,11 (1st c. BC / 1st c. AD)

(...) ἐπῆρξαν καὶ τῶν ἄλλων Θρᾳκῶν τε καὶ Ἰλλυριῶν· κατελύθησαν δ' ὑπὸ Σκορδίσκων πρότερον, ὕστερον δ' ὑπὸ Ῥωμαίων, οἳ καὶ τοὺς Σκορδίσκους αὐτοὺς κατεπολέμησαν πολὺν χρόνον ἰσχύσαντας.

(The Autariatae) held sway also over the rest of the Thracians and the Illyrians; but they were overpowered, at first by the Scordisci, and later on by the Romans, who also subdued the Scordisci themselves, after these had been in power for a long time.

Strab. 7.5,12 (1st c. BC / 1st c. AD)

Ὤικησαν δ' οὗτοι παρὰ τὸν Ἴστρον διῃρημένοι δίχα, οἱ μὲν μεγάλοι Σκορδίσκοι καλούμενοι οἱ δὲ μικροί· οἱ μὲν μεταξὺ δυεῖν ποταμῶν ἐμβαλλόντων εἰς τὸν Ἴστρον, τοῦ τε Νοάρου τοῦ παρὰ τὴν Σεγεστικὴν ῥέοντος καὶ τοῦ Μάργου (τινὲς δὲ Βάργον φασίν), οἱ δὲ μικροὶ τούτου πέραν, συνάπτοντες Τριβαλλοῖς καὶ Μυσοῖς. Εἶχον δὲ καὶ τῶν νήσων τινὰς οἱ Σκορδίσκοι· ἐπὶ τοσοῦτον δ' ηὐξήθησαν ὥστε καὶ μέχρι τῶν Ἰλλυρικῶν καὶ τῶν Παιονικῶν καὶ Θρᾳκίων προῆλθον ὀρῶν· κατέσχον οὖν καὶ τὰς νήσους τὰς ἐν τῷ Ἴστρῳ τὰς πλείους, ἦσαν δὲ καὶ πόλεις αὐτοῖς Ἑόρτα καὶ Καπέδουνον.

The Scordisci lived along the Ister and were divided into two tribes called the Great Scordisci and the Little Scordisci. The former lived between two rivers that empty into the Ister - the Noarus, which flows past Segestica, and the Margus (by some called the Bargus), whereas the Little Scordisci lived on the far side of this river, and their territory bordered on that of the Triballi and the Mysi. The Scordisci also held some of the islands; and they increased to such an extent that they advanced as far as the Illyrian, Paeonian, and Thracian mountains; accordingly, they also took possession of most of the islands in the Ister. And they also had two cities – Heorta and Capedunum.

Suet. Aug. 20 (1st/2nd c. AD)

Reliqua per legatos administravit, ut tamen quibusdam Pannonicis atque Germanicis aut interveniret aut non longe abesset, Ravennam vel Mediolanium vel Aquileiam usque ab urbe progrediens.

His other wars he carried on through his generals, although he was either present at some of those in Pannonia and Germany, or was not far from the front, since he went from the city as far as Ravenna, Mediolanum, or Aquileia.

Suet. Aug. 21 (1st/2nd c. AD)

Domuit autem partim ductu partim auspiciis suis Cantabriam, Aquitaniam, Pannoniam, Delmatiam cum Illyrico omni, item Raetiam et Vindelicos ac Salassos, gentes Inalpinas.

In part as leader, and in part with armies serving under his auspices, he subdued Cantabria, Aquitania, Pannonia, Dalmatia, and all Illyricum, as well as Raetia and the Vindelici and Salassi, which are Alpine tribes.

Suet. Tib. 9 (1st/2nd c. AD)

Exin Raeticum Vindelicumque bellum, inde Pannonicum, inde Germanicum gessit. Raetico atque Vindelico gentis Alpinas, Pannonico Breucos et Dalmatas subegit (...) Quas ob res et ovans et curru urbem ingressus est, prius, ut quidam putant, triumphalibus ornamentis honoratus, novo nec antea cuiquam tributo genere honoris.

Next he carried on war with the Raeti and Vindelici, then in Pannonia, and finally in Germany. In the first of these wars he subdued the Alpine tribes, in the second the Breuci and Dalmatians (...) Because of these exploits he entered the city both in an ovation and riding in a chariot, having previously, as some think, been honoured with the triumphal regalia, a new kind of distinction never before conferred upon anyone.

Suet. Tib. 16 (1st/2nd c. AD)

Sed nuntiata Illyrici defectione transiit ad curam novi belli, quod gravissimum omnium externorum bellorum post Punica, per quindecim legiones paremque auxiliorum copiam triennio gessit in magnis omnium rerum difficultatibus summaque frugum inopia. Et quamquam saepius revocaretur, tamen perseveravit, metuens ne vicinus et praevalens hostis instaret ultro cedentibus. Ac perseverantiae grande pretium tulit, toto Illyrico, quod inter Italiam regnumque Noricum et Thraciam et Macedoniam interque Danuvium flumen et sinum maris Hadriatici patet, perdomito et in dicionem redacto.

But when the revolt of Illyricum was reported, he was transferred to the charge of a new war, the most serious of all foreign wars since those with Carthage, which he carried on for three years with fifteen legions and a corresponding force of auxiliaries, amid great difficulties of every kind and the utmost scarcity of supplies. But though he was often recalled, he none the less kept on, for fear that the enemy, who were close at hand and very strong, might assume the offensive if the Romans gave ground. He reaped an ample reward for his perseverance, for he completely subdued and reduced to submission the whole of Illyricum, which is bounded by Italy and the kingdom of Noricum, by Thrace and Macedonia, by the Danube, and by the Adriatic sea.

Suet. Tib. 17 (1st/2nd c. AD)

Nam sub id fere tempus Quintilius Varus cum tribus legionibus in Germania periit, nemine dubitante quin victores Germani iuncturi se Pannoniis fuerint, nisi debellatum prius Illyricum esset. Quas ob res triumphus ei decretus est multi[que] et magni honores. Censuerunt etiam quidam ut Pannonicus, alii ut Invictus, nonnulli ut Pius cognominaretur. Sed de cognomine intercessit Augustus, eo contentum repromittens, quod se defuncto suscepturus esset. Triumphum ipse distulit maesta civitate clade Variana; nihilo minus urbem praetextatus et laurea coronatus intravit positumque in Saeptis tribunal senatu astante conscendit ac medius inter duos consules cum Augusto simul sedit; unde populo consalutato circum templa deductus est.

For it was at just about that time that Quintilius Varus perished with three legions in Germany, and no one doubted that the victorious Germans would have united with the Pannonians, had not Illyricum been subdued first. Consequently a triumph was voted him and many high honours. Some also recommended that he be given the surname of Pannonicus, others of Invictus, others of Pius. Augustus however vetoed the surname, reiterating the promise that Tiberius would be satisfied with one which he would receive at his father's death. Tiberius himself put off the triumph, because the country was in mourning for the disaster to Varus; but he entered the city clad in the purple-bordered toga and crowned with laurel, and mounting a tribunal which had been set up in the Saepta, while the senate stood alongside, he took his seat beside Augustus between the two consuls. Having greeted the people from this position, he was escorted to the various temples.

Suet. Tib. 20 (1st/2nd c. AD)

A Germania in urbem post biennium regressus triumphum, quem distulerat, egit prosequentibus etiam legatis, quibus triumphalia ornamenta impetrarat. Ac prius quam in Capitolium flecteret, descendit e curru seque praesidenti patri ad genua summisit. Batonem Pannonium ducem ingentibus donatum praemiis Ravennam transtulit, gratiam referens, quod se quondam cum exercitu iniquitate loci circumclusum passus esset evadere. Prandium dehinc populo mille mensis et congiarium trecenos nummos viritim dedit. Dedicavit et Concordiae aedem, item Pollucis et Castoris suo fratrisque nomine de manubiis.

After two years he returned to the city from Germany and celebrated the triumph which he had postponed, accompanied also by his generals, for whom he had obtained the triumphal regalia. And before turning to enter the Capitol, he dismounted from his chariot and fell at the knees of his father, who was presiding over the ceremonies. He sent Bato, the leader of the Pannonians, to Ravenna, after presenting him with rich gifts; thus showing his gratitude to him for allowing him to escape when he was trapped with his army in a dangerous place. Then he gave a banquet to the people at a thousand tables, and a largess of three hundred sesterces to every man. With the proceeds of his spoils he restored and dedicated the temple of Concord, as well as that of Pollux and Castor, in his own name and that of his brother.

Tac. Ann. 1.17 (1st/2nd c. AD)

Ac si quis tot casus vita superaverit, trahi adhuc diversas in terras ubi per nomen agrorum uligines paludum vel inculta montium accipiant (...)

If a soldier survives so many risks, he is still dragged into remote regions where, under the name of lands, he receives soaking swamps or mountainous wastes (...)

Tac. Ann. 1.20 (1st/2nd c. AD)

Interea manipuli ante coeptam seditionem Nauportum missi ob itinera et pontes et alios usus, postquam turbatum in castris accepere, vexilla convellunt direptisque proximis vicis ipsoque Nauporto, quod municipii instar erat (...)

Meanwhile the companies which previous to the mutiny had been sent to Nauportus to make roads and bridges and for other purposes, when they heard of the tumult in the camp, tore up the standards, and having plundered the neighbouring villages and Nauportus itself, which was like a town (...)

Tac. Hist. 2.86 (1st/2nd c. AD)

Tampius Flavianus Pannoniam, Pompeius Silvanus Dalmatiam tenebant, divites senes (...)

Tampius Flavianus was the governor of Pannonia, Pompeius Silvanus of Dalmatia. They were both rich and advanced in years.

Tac. Hist. 3.12 (1st/2nd c. AD)

Lucilius Bassus classi Ravennatis praefectus ambiguos militum animos, quod magna pars Dalmatae Pannoniique erant, quae provinciae Vespasiano tenebantur, partibus eius adgregaverat.

Lucilius Bassus, prefect of the Ravenna fleet, finding that the troops wavered in purpose, from the fact that many were natives of Dalmatia and Pannonia, provinces held for Vespasian, had attached them to the Flavianist party.

Tac. Hist. 3.50 (1st/2nd c. AD)

(...) sex milia Dalmatarum, recens dilectus, comitabantur; ducebat Pompeius Silvanus consularis (...) Ad has copias e classicis Ravennatibus legionariam militiam poscentibus optumus quisque adsciti: classem Dalmatae supplevere.

A recent levy of 6000 Dalmatians was attached to the legion. They were under the command of Pompeius Silvanus, a man of consular rank (...) To these forces were added the elite of the marines of the Ravenna fleet, who demanded permission to serve in the legions. The crews were made up with Dalmatians.

Tib. Paneg. 3.7,106-112 (1st c. BC)

At non per dubias errant mea carmina laudes: / nam bellis experta cano. Testis mihi victae / fortis Iapydiae miles, testis quoque fallax / Pannonius, gelidas passim disiectus in Alpes (...)

But my verses do not stray among unproved distinctions. / I sing what wars have shown. I have a witness / in the gallant soldiery of vanquished Iapydia; a witness also in the cunning / Pannonian, scattered far and wide over the icy Alps (...)

Pomp. Trog. in: Just. Epit. 24.4 (Pompeius Trogus: 1st/2nd c. AD, Justin: 3rd c. AD)

Ex his portio in Italia consedit quae et urbem Romanam captam incendit et portio Illyricos sinus ducibus avibus - nam augurandi studio Galli praeter ceteros callent - per strages barbarorum penetravit et in Pannonia consedit (...) Ibi domitis Pannoniis per multos annos cum finitimis varia bella gesserunt.

Of these adventurers part settled in Italy, and took and burnt the city of Rome; and part penetrated into the remotest parts of Illyricum under the direction of a flight of birds (for the Gauls are skilled in augury beyond other nations) making their way amidst great slaughter of the barbarous tribes, and fixed their abode in Pannonia. (...) After having subdued the Pannonians, they carried on various wars with their neighbours for many years.

Pomp. Trog. in: Just. Epit. 32.3 (Pompeius Trogus: 1st/2nd c. AD, Justin: 3rd c. AD)

Ex his manus quaedam in confluente Danuvii et Savi consedit Scordiscosque se appellari voluit. (...) Ex gente Tectosagorum non mediocris populus praedae dulcedine Illyricum repetivit spoliatisque Histris in Pannonia consedit.

Of these, a certain number settled at the conflux of the Danube and Save, and took the name of Scordisci. (...) Of these Tectosagi, no small number, attracted by the charms of plunder, repaired to Illyricum, and, after spoiling the Istrians, settled in Pannonia.

Vell. 2.39,3 (1st c. AD)

Raetiam autem et Vindelicos ac Noricos Pannoniamque et Scordiscos novas imperio nostro subiunxit provincias. Ut has armis (...)

(Tiberius) also added to our empire as new provinces Raetia, Vindelicia, Noricum, Pannonia, and the Scordisci. These he conquered by arms.

Vell. 2.96,2-3 (1st c. AD)

Subinde bellum Pannonicum, quod inchoatum ab Agrippa, Marco Vinicio, avo tuo consule, magnum atroxque et perquam vicinum imminebat Italiae, per Neronem gestum est. Gentes Pannoniorum Delmatarumque nationes situmque regionum ac fluminum numerumque et modum virium excelsissimasque et multiplices eo bello victorias tanti imperatoris alio loco explicabimus: hoc opus servet formam suam. Huius victoriae compos Nero ovans triumphavit.

Shortly after, the Pannonian war, which had been begun by Agrippa in the consulate of your grandfather, Marcus Vinicius, was conducted by Nero, a war which was important and formidable enough, and on account of its proximity a menace to Italy. In another place I shall describe the tribes of the Pannonians and the races of Dalmatians, the situation of their country and its rivers, the number and extent of their forces, and the many glorious victories won in the course of this war by this great commander; my present work must keep to its design. After achieving this victory Nero celebrated an ovation.

Vell. 2.110 (1st c. AD)

Praeparaverat iam hiberna Caesar ad Danubium (...) cum universa Pannonia, insolens longae pacis bonis, adulta viribus, Delmatia omnibusque tractus eius gentibus in societatem adductis consilii, arma corripuit. Tum necessaria gloriosis praeposita neque tutum visum abdito in interiora exercitu vacuam tam vicino hosti relinquere Italiam. Gentium nationumque, quae rebellaverant, omnis numerus amplius octingentis milibus explebat; ducenta fere peditum colligebantur armis habilia, equitum novem. Cuius immensae multitudinis, parentis acerrimis ac peritissimis ducibus, pars petere Italiam decreverat iunctam sibi Nauporti ac Tergestis confinio, pars in Macedoniam se effuderat, pars suis sedibus praesidium esse destinaverat. Maxima duobus Batonibus ac Pinneti ducibus auctoritas erat. In omnibus autem Pannoniis non disciplinae tantummodo, sed linguae quoque notitia Romanae, plerisque etiam litterarum usus et familiaris animorum erat exercitatio. Itaque hercules nulla umquam natio tam mature consilio belli bellum iunxit ac decreta patravit. Oppressi cives Romani, trucidati negotiatores, magnus vexillariorum numerus ad internecionem ea in regione, quae plurimum ab imperatore aberat, caesus, occupata armis Macedonia, omnia et in omnibus locis igni ferroque vastata. Quin etiam tantus huius belli metus fuit, ut stabilem illum et firmatum tantorum bellorum experientia Caesaris Augusti animum quateret atque terreret.

Caesar had already arranged his winter quarters on the Danube (...) when all Pannonia, grown arrogant through the blessings of a long peace and now at the maturity of her power, suddenly took up arms, bringing Dalmatia and all the races of that region into her alliance. Thereupon glory was sacrificed to necessity; and it did not seem to Tiberius a safe course to keep his army buried in the interior of the country and thus leave Italy unprotected from an enemy so near at hand. The full number of the races and tribes which had rebelled reached a total of more than eight hundred thousand. About two hundred thousand infantry trained to arms, and nine thousand cavalry were being assembled. Of this immense number, which acted under the orders of energetic and capable generals, one portion had decided to make Italy its goal, which was connected with them by the line of Nauportus and Tergeste, a second had already poured into Macedonia, while a third had set itself the task of protecting their own territories. The chief authority rested with the two Batones and Pinnes as generals. Now all the Pannonians possessed not only a knowledge of Roman discipline but also of the Roman tongue, many also had some measure of literary culture, and the exercise of the intellect was not uncommon among them. And so it came to pass, by Hercules, that no nation ever displayed such swiftness in following up with war its own plans for war, and in putting its resolves into execution. Roman citizens were overpowered, traders were massacred, a considerable detachment of veterans, stationed in the region which was most remote from the commander, was exterminated to a man, Macedonia was seized by armed forces, everywhere was wholesale devastation by fire and sword. Moreover, such a panic did this war inspire that even the courage of Caesar Augustus, rendered steady and firm by experience in so many wars, was shaken with fear.

Vell. 2.111 (1st c. AD)

Habiti itaque dilectus, revocati undique et omnes veterani, viri feminaeque ex censu libertinum coactae dare militem. Audita in senatu vox principis, decimo die, ni caveretur, posse hostem in urbis Romae venire conspectum. Senatorum equitumque Romanorum exactae ad id bellum operae, pollicitati. Omnia haec frustra praeparassemus, nisi qui illa regeret fuisset. Itaque ut praesidium ultimum res publica ab Augusto ducem in bellum poposcit Tiberium. Habuit in hoc quoque bello mediocritas nostra speciosi ministerii locum. Finita equestri militia designatus quaestor necdum senator aequatus senatoribus, etiam designatis tribunis plebei, partem exercitus ab urbe traditi ab Augusto perduxi ad filium eius. In quaestura deinde remissa sorte provinciae legatus eiusdem ad eundem missus sum. Quas nos primo anno acies hostium vidimus! Quantis prudentia ducis opportunitatibus furentes eorum vires universas elusimus, fudimus partibus! Quanto cum temperamento simul civilitatis res auctoritate imperatoria agi vidimus! Qua prudentia hiberna disposita sunt! Quanto opere inclusus custodiis exercitus nostri, ne qua posset erumpere inopsque copiarum et intra se furens viribus hostis elanguesceret!

Accordingly levies were held, from every quarter all the veterans were recalled to the standards, men and women were compelled, in proportion to their income, to furnish freedmen as soldiers. Men heard Augustus say in the senate, that, unless precautions were taken, the enemy might appear in sight of Rome within ten days. The services of senators and knights were demanded for this war, and promised. All these our preparations would have been vain had we not had the man to take command. And so, as a final measure of protection, the state demanded from Augustus that Tiberius should conduct the war. In this war also my modest abilities had an opportunity for glorious service. I was now, at the end of my service in the cavalry, quaestor designate, and though not yet a senator I was placed upon a parity with senators and even tribunes elect, and led from the city to Tiberius a portion of the army which was entrusted to me by Augustus. Then in my quaestorship, giving up my right to have a province allotted me, I was sent to Tiberius as legatus Augusti. What armies of the enemy did we see drawn up for battle in that first year! What opportunities did we avail ourselves of through the foresight of the general to evade their united forces and rout them in separate divisions! With what moderation and kindness did we see all the business of warfare conducted, though under the authority of a military commander! With what judgement did he place our winter camps! How carefully was the enemy so blockaded by the outposts of our army that he could nowhere break through, and that, through lack of supplies and by disaffection within his own ranks, he might gradually be weakened in strength!

Vell. 2.112 (1st c. AD)

Felix eventu, forte conatu prima aestate belli Messalini opus mandandum est memoriae. Qui vir (...) praepositus Illyrico subita rebellione cum semiplena legione vicesima circumdatus hostili exercitu amplius viginti milia fudit fugavitque et ob id ornamentis triumphalibus honoratus est. Ita placebat barbaris numerus suus, ita fiducia virium, ut ubicumque Caesar esset, nihil in se reponerent. Pars exercitus eorum, proposita ipsi duci et ad arbitrium utilitatemque nostram macerata perductaque ad exitiabilem famem, neque instantem sustinere neque cum facientibus copiam pugnandi derigentibusque aciem ausa congredi occupato monte Claudio munitione se defendit. At ea pars, quae obviam se effuderat exercitui, quem A. Caecina et Silvanus Plautius consulares ex transmarinis adducebant provinciis, circumfusa quinque legionibus nostris auxiliaribusque et equitatui regio (quippe magnam Thracum manum iunctus praedictis ducibus Rhoemetalces, Thraciae rex, in adiutorium eius belli secum trahebat) paene exitiabilem omnibus cladem intuit: fusa regiorum equestris acies, fugatae alae, conversae cohortes sunt, apud signa quoque legionum trepidatum. Sed Romani virtus militis plus eo tempore vindicavit gloriae quam ducibus reliquit, qui multum a more imperatoris sui discrepantes ante in hostem inciderunt, quam per exploratores, ubi hostis esset, cognoscerent. Iam igitur in dubiis rebus semet ipsae legiones adhortatae, iugulatis ab hoste quibusdam tribunis militum, interempto praefecto castrorum praefectisque cohortium, non incruentis centurionibus, e quibus etiam primi ordinis cecidere, invasere hostes nec sustinuisse contenti perrupta eorum acie ex insperato victoriam vindicaverunt.

An exploit of Messalinus in the first summer of the war, fortunate in its issue as it was bold in undertaking, must here be recorded for posterity. This man (...) was in command in Illyricum, and, at the sudden outbreak of the rebellion, finding himself surrounded by the army of the enemy and supported by only the twentieth legion, and that at but half its normal strength, he routed and put to flight more than twenty thousand, and for this was honoured with the ornaments of a triumph. The barbarians were so little satisfied with their numbers and had so little confidence in their own strength that they had no faith in themselves where Caesar was. The part of their army which faced the commander himself, worn down according as it suited our pleasure or advantage, and reduced to the verge of destruction by famine, not daring to withstand him when he took the offensive, nor to meet our men when they gave them an opportunity for fighting and drew up their line of battle, occupied the Claudian mountain and defended itself behind fortifications. But the division of their forces which had swarmed out to meet the army which the consulars Aulus Caecina and Silvanus Plautius were bringing up from the provinces across the sea, surrounded five of our legions, together with the troops of our allies and the cavalry of the king (for Rhoemetalces, king of Thrace, in conjunction with the aforesaid generals was bringing with him a large body of Thracians as reinforcements for the war), and inflicted a disaster that came near being fatal to all. The horsemen of the king were routed, the cavalry of the allies put to flight, the cohorts turned their backs to the enemy, and the panic extended even to the standards of the legion. But in this crisis the valour of the Roman soldier claimed for itself a greater share of glory than it left to the generals, who departing far from the policy of their commander, had allowed themselves to come into contact with the enemy before they had learned through their scouts where the enemy was. At this critical moment, when some tribunes of the soldiers had been slain by the enemy, the prefect of the camp and several prefects of cohorts had been cut off, a number of centurions had been wounded, and even some of the centurions of the first rank had fallen, the legions, shouting encouragement to each other, fell upon the enemy, and not content with sustaining their onslaught, broke through their line and wrested a victory from a desperate plight.

Vell. 2.113 (1st c. AD)

Iunctis exercitibus, quique sub Caesare fuerant quique ad eum venerant, contractisque in una castra decem legionibus, septuaginta amplius cohortibus, decem alis et pluribus quam ecem veteranorum milibus, ad hoc magno voluntariorum numero frequentique equite regio, tanto denique exercitu, quantus nullo umquam loco post bella fuerat civilia, omnes eo ipso laeti erant maximamque fiduciam victoriae in numero reponebant. At imperator, optimus eorum quae agebat iudex et utilia speciosis praeferens quodque semper eum facientem vidi in omnibus bellis quae probanda essent, non quae utique probarentur sequens, paucis diebus exercitum, qui venerat, ad refovendas ex itinere eius vires moratus, cum eum maiorem, quam ut temperari posset, neque habilem gubernaculo cerneret, dimittere statuit; prosecutusque longo et perquam laborioso itinere, cuius difficultas narrari vic potest, ut neque universos quisquam auderet adgredi et partem digredientium, suorum quisque metu finium, universi temptare non possent, remisit eo, unde venerant, et ipse asperrimae hiemis initio regressus Sisciam legatos, inter quos ipsi fuimus, partitis praefecit hibernis.

When the two armies were united, that is to say the troops which had served under Caesar and those which had come to reinforce him, and there were now gathered together in one camp ten legions, more than seventy cohorts, fourteen troops of cavalry and more than ten thousand veterans, and in addition a large number of volunteers and the numerous cavalry of the king – in a word a greater army than had ever been assembled in one place since the civil wars – all were finding satisfaction in this fact and reposed their greatest hope of victory in their numbers. But the general, who was the best judge of the course he pursued, preferring efficiency to show, and, as we have so often seen him doing in all his wars, following the course which deserved approval rather than that which was currently approval, after keeping the army which had newly arrived for only a few days in order to allow it to recover from the march, decided to send it away, since he saw that it was too large to be managed and was not well adapted to effective control. And so he sent it back whence it came, escorting it with his own army on a long exceedingly laborious march, whose difficulty can hardly be described. His purpose in this was, on the one hand, that no one might dare to attack his united forces, and, on the other, to prevent the united forces of the enemy from falling upon the departing division, through the apprehension of each nation for its own territory. Then returning himself to Siscia, at the beginning of a very hard winter, he placed his lieutenants, of whom I was one, in charge of the divisions of his winter quarters.

Vell. 2.114 (1st c. AD)

Hiems emolumentum patrati belli contulit, sed insequenti aestate omnis Pannonia reliquiis totius belli in Delmatia manentibus pacem petiit. Ferocem illam tot milium iuventutem, paulo ante servitutem minatam Italiae, conferentem arma, quibus usa erat, apud flumen nomine Bathinum prosternentemque se universam genibus imperatoris, Batonemque et Pinnetem excelsissimos duces, captum alterum, alterum a se deditum iustis voluminibus ordine narrabimus, ut spero. Autumno victor in hiberna reducitur exercitus, cuius omnibus copiis a Caesare M. Lepidus praefectus est (...)

The winter brought the reward of our efforts in the termination of the war, though it was not until the following summer that all Pannonia sought peace, the remnants of the war as a whole being confined to Dalmatia. In my complete work I hope to describe in detail how those fierce warriors, many thousand in number, who had but a short time before threatened Italy with slavery, now brought the arms they had used in rebellion and laid them down, at a river called the Bathinus, prostrating themselves one and all before the knees of the commander; and how of their two supreme commanders, Bato and Pinnes, the one was made a prisoner and the other gave himself up. In the autumn the victorious army was led back to winter quarters. Caesar gave the chief command of all the forces to Marcus Lepidus (...)

Vell. 2.115 (1st c. AD)

Illa aestas maximi belli consummavit effectus: quippe Perustae et Desidiates Delmatae, situ locorum ac montium, ingeniorum ferocia, mira etiam pugnandi scientia et praecipue angustiis saltuum paene inexpugnabiles, non iam ductu, sed manibus atque armis ipsius Caesaris tum demum pacati sunt, cum paene funditus eversi forent.

This campaign brought the momentous war to a successful conclusion; for the Perustae and Desiadates, Dalmatian tribes, who were almost unconquerable on account of the position of their strongholds in the mountains, their warlike temper, their wonderful knowledge of fighting, and, above all, the narrow passes in which they lived, were then at last pacified, not now under the mere generalship, but by the armed prowess of Caesar himself, and then only when they were almost entirely exterminated.

Vell. 2.117,1 (1st c. AD)

Tantum quod ultimam imposuerat Pannonico ac Delmatico bello Caesar manum, cum intra quinque consummati tanti operis dies funestae ex Germania epistulae caesi Vari trucidatarumque legionum trium totidemque alarum et sex cohortium <allatae sunt>, velut in hoc saltem tantummodo indulgente nobis fortuna, ne occupato duce <...> et causa <et> persona moram exigit.

Scarcely had Caesar put the finishing touch upon the Pannonian and Dalmatian war, when, within five days of the completion of this task, dispatches from Germany brought the baleful news of the death of Varus, and of the slaughter of three legions, of as many divisions of cavalry, and of six cohorts — as though fortune were granting us this indulgence at least, that such a disaster should not be brought upon us when our commander was occupied by other wars.

Vell. 2.121,2-3 (1st c. AD)

In Urbem reversus, iam pridem debitum, sed continuatione bellorum dilatum ex Pannoniis Delmatisque egit triumphum. Cuius magnificentiam quis miretur in Caesare? Fortunae vero quis non miretur indulgentiam? Quippe omnis eminentissimos hostium duces non occisos fama narravit, sed vinctos triumphus ostendit; quem mihi fratrique meo inter praecipuos praecipuisque donis adornatos viros comitari contigit.

On his return to the city he celebrated the triumph over the Pannonians and Dalmatians, long since due him, but postponed by reason of a succession of wars. Who can be surprised at its magnificence, since it was the triumph of Caesar. Yet who can fail to wonder at the kindness of fortune to him? For the most eminent leaders of the enemy were not slain in battle, that report should tell thereof, but were taken captive, so that in his triumph he exhibited them in chains. It was my lot and that of my brother to participate in this triumph among the men of distinguished rank and those who were decorated with distinguished honours.

Bibliography

Bogaers (*sine anno*) Julianus Egidius Bogaers, *Auxiliaria*, http://repository.ubn.ru.nl:8080/bitstream/2066/26507/1/26507___.PDF

Bojanovski 1988 Ivo Bojanovski, *Bosna i Hercegovina u antičko doba, DjANUBiH* 66.

Bulat 1989 Mirko Bulat, Novi rimski natpisi iz Osijeka, *OZ* 20, 31-51.

Clarke 1997 Katherine Clarke, In Search of the Author of Strabo's Geography, *JRS* 87, 92-110.

Čović 1987 Borivoj Čović, Grupa Donja Dolina - Sanski Most, in: A. Benac (ed.), *PJZ* 5, 232-286.

Čremošnik and Sergejevski 1930 Gregor Čremošnik and Dimitrije Sergejevski, Gotisches und römisches aus Breza bei Sarajevo, *Novitates* 9, Sarajevo, 8-16.

Detlefsen 1906 Detlef Detlefsen, *Ursprung, Einrichtung und Bedeutung der Erdkarte Agrippas, Quellen und Forschungen zur alten Geschichte und Geographie* 13, Berlin.

Detlefsen 1908 Detlef Detlefsen, *Die Geographie Afrikas bei Plinius und Mela und ihre Quellen*, Quellen und Forschungen zur alten Geschichte und Geographie 14, Berlin.

Dilke 1985 Oswald Ashton Wentworth Dilke, Greek and Roman Maps, in: H. H. Scullard (ed.), *Aspects of Greek and Roman Life*, London.

Dizdar 2003 Marko Dizdar, Prilog poznavanju kasnog latena u istočnoj Slavoniji, *OpArch* 27, 337-349.

Dizdar 2006 Marko Dizdar, Nalaz staklenih narukvica latenske kulture u Podravini, *PIAZ* 23, 67-128.

Domić Kunić 2003 Alka Domić Kunić, Plinijeva geografija i etnografija Ilirika (s osobitim obzirom na panonski dio iliričkog prostora), unpublished PhD thesis, University of Zagreb.

Domić Kunić 2004 Alka Domić Kunić, Literarni izvori za iliričke provincije (Dalmaciju i osobito Panoniju) u *Naturalis historia* Plinija Starijeg, *VAMZ* 37, 119-171.

Domić Kunić 2005 Alka Domić Kunić, The Danube Region in Legend - Prehistorical Trade across Pannonia, in: *Illyrica antiqua. Ob honorem Duje Rendić.Miočević* 219-228.

Domić Kunić 2006 Alka Domić Kunić, *Bellum Pannonicum* (12.-11. pr. Kr.). Posljednja faza osvajanja južne Panonije, *VAMZ* 39, 59-164.

Domić Kunić 2009 Alka Domić Kunić, Jugoistočna Panonija u antičkim literarnim vrelima, in: V. Kusin and B. Šulc (eds.), *Slavonija, Baranja i Srijem. Vrela europske civilizacije*, Zagreb, 221-231.

Durman 1992 Aleksandar Durman, O geostrateškom položaju Siscije, *OpArch* 16, 117-131.

Dušanić 1978 Slobodan Dušanić, A Military Diploma of A. D. 65, *Germania* 56, Mainz, 461-475.

Dušanić 1999 Slobodan Dušanić, An Early Diploma Militare, *Starinar* 49 (1998), 51-62.

Dzino 2005 Danijel Dzino, Sabaiarius: Beer, wine and Ammianus Marcellinus, in: W. Mayer, and S. Trzcionka (eds.), *Feast, fast or famine: Food and drink in Byzantium*, *Byzantina Australiensia* 15, Brisbane, 57-68.

Dzino 2006 Danijel Dzino, Welcome to the Mediterranean semi-periphery: The place of Illyricum in book 7 of Strabo, *ŽAnt* 56, 113-128.

Dzino 2008 Danijel Dzino, Strabo 7.5 and imaginary Illyricum, *Athenaeum* 96/1, 173-192.

Džino 2008 Danijel Džino, The Celts in Illyricum - whoever they may be: The hybridization and construction of identities in Southeastern Europe in the fourth and third centuries BC, *OpArch* 31 (2007), 49-68.

Džino 2009a Danijel Džino, The *Bellum Batonianum* in contemporary historiographical narratives. In a search for the post-modern Bato the Daesitiate, *ARR* 16, 29-45.

Džino 2009b Danijel Džino, „Dezidijati": Identitetski konstrukt između antičkih i suvremenih percepcija, *GodCBI* 38 (36), 75-95.

Dzino 2010 Danijel Dzino, *Illyricum in Roman Politics 229 BC - AD 68*, Cambridge, 2010.

Fitz 1977 Jenő Fitz, Die Eroberung Pannoniens, *ANRW* 2/6, Berlin-New York, 543-556.

Fitz 1990 Jenő Fitz, La direttrice Emona - Aquincum, in: *La Venetia nell'area padano-danubiana. Le vie di communicazione*, (Convegno internazionale, Venezia, 6-10 aprile 1988), Padova, 337-347.

Fitz 1993 Jenő Fitz, *Die Verwaltung Pannoniens in der Römerzeit*, Vol. I, Budapest.

Hanslik 1961 Rudolf Hanslik, *Marcus Vipsanius Agrippa*, *RE* 9, A1, *coll.* 1226-1275.

Hoti 1992 Marina Hoti, Sisak u antičkim izvorima, *OpArch* 16, 133-163.

Laffi 1973 Umberto Laffi, Sull'organizzazione amministrativa dell'Italia dopo la guerra sociale, *Vestigia* 17 (*Akten des VI. Internationalen Kongresses*

für Griechische und Lateinische Epigraphik, München 1972), München, 37-53.
Lolić 2003 Tatjana Lolić, *Colonia Flavia Siscia*, in: *Pannonia* I, 131-152.
Lörincz 2001 Barnabás Lörincz, *Die römischen Hilfstruppen in Pannonien während der Prinzipatszeit. Teil I: Die Inschriften*, Wien.
Lovenjak 2003 Milan Lovenjak, *Municipium Flavium Latobicorum Neviodunum*, in: *Pannonia* I, 93-105.
Majnarić-Pandžić 1984 Nives Majnarić-Pandžić, Prilog problematici kasnolatenskih utvrđenih naselja u Slavoniji, *OpArch* 9, 23-34.
Margetić 1999 Lujo Margetić, O municipiju Andautonia, *RHAZU* 37, 5-24.
Marić 1964 Zdravko Marić, Donja Dolina, *GZMS* 19, 5-128.
Mayer 1935a Anton Mayer, Ime Mursa, *VHAD*, 16/1-2, 5-10.
Mayer 1935b Anton Mayer, Iasi, *VHAD*, 16/1-2, 69-82.
Mayer 1959 Anton Mayer (ed.), *Die Sprache der alten Illyrier, Band II. Etymologisches Wörterbuch des Illyrischen Grammatik der illyrischen Sprache*, Wien.
Mirković 2004 Miroslava Mirković, *Sirmium*, in: *Pannonia* II, 145-156.
Miškiv 1998 Jesenka Miškiv, Rimska vojnička diploma iz Slavonskog Broda, *VAMZ* 30-31 (1997-1998), 83-101.
Mócsy 1957 András Mócsy, Zur Geschichte der peregrinen Gemeinden in Pannonien, *Historia* 6/4, 488-498.
Mócsy 1974 András Mócsy, *Pannonia and Upper Moesia. A History of the Middle Danube Provinces of the Roman Empire*, London-Boston.
Mócsy 1979 András Mócsy, Illyricum északi határa Claudius elött, *ArchÉrt* 106/2, 177-186.
Nemeth-Ehrlich and Kušan Špalj 2003 Dorica Nemeth-Ehrlich and Dora Kušan Špalj, *Municipium Andautonia*, in: *Pannonia* I, 107-129.
Novak 1965 Grga Novak, *Quaestiones Epidauritanae*, *RJAZU* 339, 97-108.
Pinterović 1978 Danica Pinterović, *Mursa i njeno područje u antičko doba*, Osijek.
Pokorny 1959 Julius Pokorny, *Indogermanisches etymologisches Wörterbuch*, I. Band, Bern-München.
Popović 2003 Petar Popović, Gradina na Bosutu kod Vašice, *OpArch* 27, 311-320.
Potrebica 2003 Hrvoje Potrebica, Požeška kotlina i Donja Dolina u komunikacijskoj mreži starijeg željeznog doba, *OpArch* 27, 217-242.
Rendić-Miočević 1962 Duje Rendić-Miočević, Princeps municipii Riditarum. Uz novi epigrafski nalaz u Danilu, *ARR* 2, 315-334.
Schejbal 2003 Berislav Schejbal, Nova razmatranja o *Aquae Balissae* i narodu Jaza, *OpArch* 27, 393-416.
Schejbal 2004 Berislav Schejbal, *Municipium Iasorum* (*Aquae Balissae*), in: *Pannonia* II, 99-129.
Smith 1988 R. R. R. Smith, *Simulacra gentium*: The *ethne* from the Sebasteion at Aphrodisias, *JRS* 78, 50-77.
Soproni 1979 Sándor Soproni, *Municipium Halicanum*, *FA* 30, 91-98.
Suić 1976 Mate Suić, Illyrii proprie dicti, *GodCBI* 13 (11), 179-196.
Šašel Kos 1994 Marjeta Šašel Kos, Savus and Adsalluta, *AV* 45, 99-122.
Šašel Kos 2002 Marjeta Šašel Kos, The Noarus River in Strabo's Geography, *Tyche. Beiträge zur alten Geschichte, Papyrologie und Epigraphik* 17, Wien, 145-153.
Šašel Kos 2005 Marjeta Šašel Kos, *Appian and Illyricum*, *Situla* 43.
Šašel Kos 2010 Marjeta Šašel Kos, Pannonia or Lower Illyricum?, *Tyche* 25, 123-130.
Tóth 1976 Endre Tóth, Pannónia provincia kialakulásához, *ArchÉrt* 103/9, 197-202.
Tóth 1977 Endre Tóth, „... *protulique fines Illyrici ad ripam fluminis Danuvii*", *AV* 28, 278-287.
Vidović 1990 Josip Vidović, Osvrt na jedan brončani nalaz antike i antičku urbanizaciju područja Međimurja, *MV* 13, Bjelovar, 30-32.
Wilkes 1969 John J. Wilkes, *Dalmatia. History of the Provinces of the Roman Empire*, London.
Woolf 2011 Greg Woolf, *Tales of the Barbarians. Ethnography and Empire in the Roman West*, Chichester.
Zaninović 1996 Marin Zaninović, Andautonia i Siscia u odrazu flavijevske politike, in: *Arheološka istraživanja u Zagrebu i zagrebačkoj regiji. Arheologija i obnova, IzdHAD* 17, Zagreb, 59-64.
Zaninović 2003 Marin Zaninović, Breuci od Sirmija do Marsonije, *OpArch* 27, 443-449.

Literary Sources from the End of the 2nd to the End of the 6th Century

Mirko Sardelić

1. Introduction

This article is a continuation of the overview of the literary sources related to southern Pannonia, the first part of which is dealt with by Alka Domić Kunić in this volume. As it is stated in the introduction to that first part, the aim of the work is merely to present a collection of all sources relevant to the area in a documentary fashion and put it into the historical context. The in-depth analysis of the problems of their origin, content and reception would require a separate volume.[1] Although the focus is set on the part of southern Pannonia which is within the borders of the present-day Republic of Croatia, to completely exclude sources mentioning the provincial capital, Sirmium (present-day Sremska Mitrovica, Serbia) or other parts of both Pannonias, would be an insensitive intervention into the natural historical unit. However, respecting the frame of the task, the inclusion of such sources was strictly limited either to the events that affected the whole province, or to some features and activities most likely to have been common in the whole region or among all the inhabitants.

The structure of the work is as follows: the historical overview is divided into thematic sections supported by sources; the contents in brackets in the main text are the sources included in the bilingual list. In rare cases, some sources are also quoted in footnotes; these either do not belong to the group of sources dealing with southern Pannonia or they simply serve as argumentation or clarification of the related text. Some segments of history lack primary literary materials; these have been reconstructed using the results of other disciplines. I mention these sources only briefly, so that the reader may have insight into the most important questions concerning the province in the four centuries dealt with here, with reference to the bibliography on the topic. The scarcity of the primary sources is such that a wide interdisciplinary approach is required – archaeological findings and epigraphic materials, just to start with – to make any proper use of them. However, in the present paper they are listed and treated as a separate source of material, as much as is possible.[2]

2. Geography, roads, natural resources

The Pannonian plain is surrounded by the Carpathian Mountains to the north and the north-east, the Alps to the west and the Dinarides in the south. In the time of the Roman conquest, at the end of the 1st century BC, the part which is within the borders of the Republic of Croatia was included in the province of Illyricum. During later divisions of the Empire, it was the southern part of both the provinces of Pannonia Superior and Pannonia Inferior, and later occupied approximately the area of the province Pannonia Savia and the western part of Pannonia Secunda. It is surrounded by the Drava River in the north, the Sava River in the south and the Danube River, into which both rivers flow, in the east. Jordanes wrote on the position of the province in relation to other surrounding provinces of the Empire, and on its geographical features, i.e. that it is stretched in a long plain (Jord. *Get.* 264). In the early Middle Ages (early 7th century) Isidore of Seville puts Pannonia between the Rivers Drava and Sava and gives a more exhaustive list of neighbouring provinces than Jordanes (Isid. *Etym.* 14.4,16). The Sava River is also mentioned as a boundary between Pannonia and Moesia (Dio 51.27,3).

Fertile land was made arable by draining many swamps; of the ones that remained the sources single out a particular one, *palus Hiulca* in the Vuka valley (Dio 55.32,3; *Epit. de Caes.* 41.5). Zosimos reports that the major part of the road toward Cibalae leads through deep marshes (Zos. 2.18), and also gives the marshes' width to be five stadia. Ennodius in the 5th century mentions the River Ulca (Vuka, most likely referring to its marshes) as a natural protection used by the Gepids (Enn. *Paneg. Theod.* 7.206). The swamp that Jordanes (Jord. *Get.* 30 and 35) refers to as *stagnum Morsianum* or *lacus Morsianus* should not be identified with the Vuka marshes as some do,[3] but should be looked for elsewhere.[4] The Romans cut down a part of the temperate rain forest of Slavonian (pedunculate) oak (*Quercus robur L.*), which stretched between the Rivers

[1] Some of the ancient authors were, in more or less detailed way, commented on in works such as Šašel Kos 1986 (bilingual, Slovenian and English) and Gračanin 2005a. In the latter there are very useful bibliographical references to the studies on Cassius Dio, Herodian, Zosimos, Aurelius Victor, *Epitomae de caesaribus, Historia Augusta* and Eutropius. More on Jordanes and his *Getica* see in Heather 1991, 7-67.

[2] I would like to thank Alka Domić Kunić, Hrvoje Gračanin and Branka Migotti for their valuable commentaries on the text. I also owe gratitude to Rachel S. Anderson and Vedran Nikolić for reading it patiently.

[3] Soproni 1980, 61.

[4] L. Margetić (2005, 115-117) argues that it should be a body of water somewhere around the mouth of the River Danube. The same argumentation can be found in Šišić 1925, 187, referring to some older authors; see also Mayer 1935, 5-6; Curta 2001, 42; identification with the one between the Drava and Danube Rivers would locate the southeastern border of Slavic settlement right there in the time of Jordanes' account (mid–6th century). Cf. Gračanin 2008, 20 and Gračanin 2011, 25, 127-128.

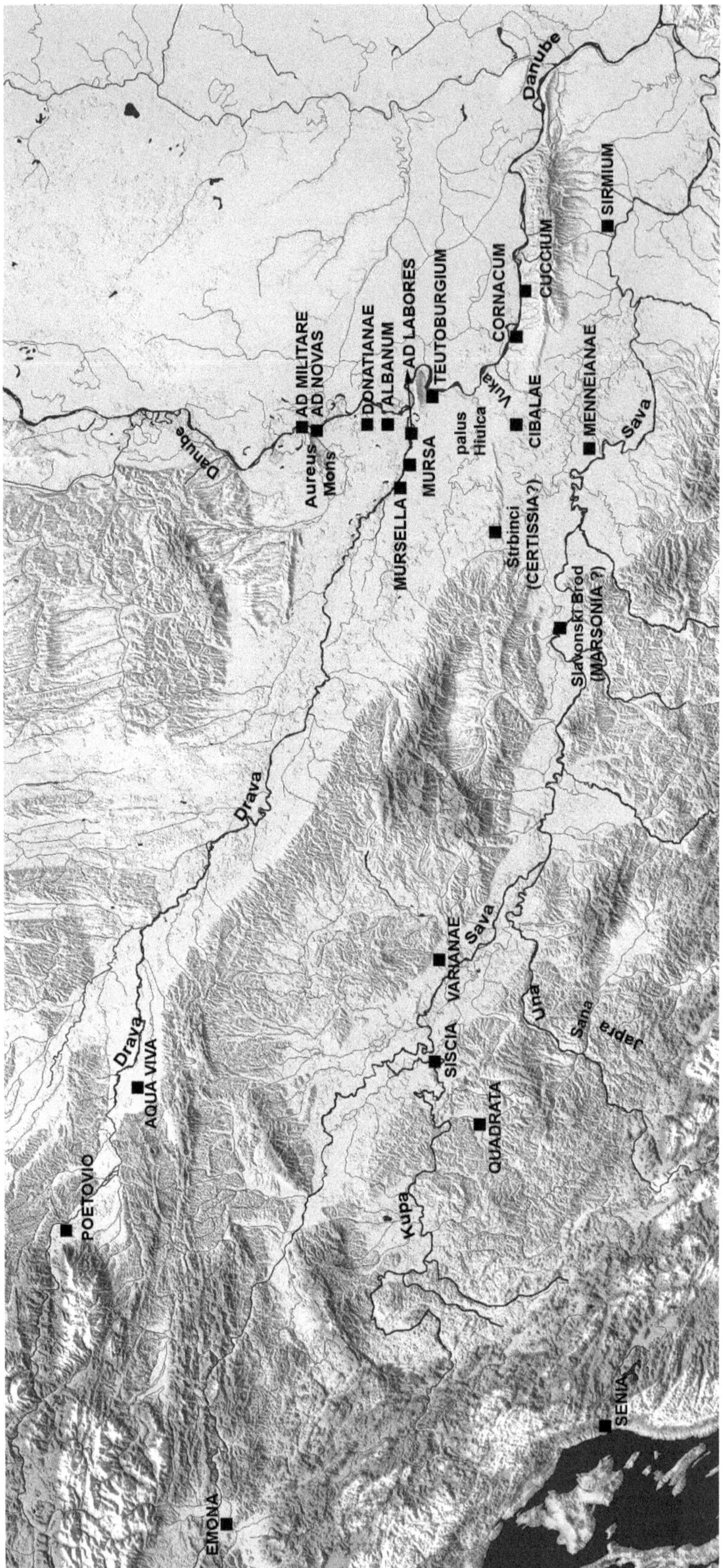

Fig. 1. Map of the geography of southern Pannonia based on the written sources of the 3rd to 6th centuries AD (after Google Maps, modified by T. Leleković).

Sava and Drava. (Claud. 199-200).[5] The sources also report that some parts of the land were covered with vineyards (*Epit. de Caes.* 37.3; Eutr. *Brev.* 9.17; *SHA* Probus, 18.8; Claud. 200).

As an early 3rd century governor of Upper Pannonia, Dio Cassius gives a first hand report on the low-quality wine, and harsh Pannonian climate with long winters (Dio 49.36,2). It should be taken into account that although the capital Carnuntum was at the far north of the province, the report on the weather is most likely applicable to the entire Upper Pannonia.[6] In the same paragraph Dio also describes the diet of Pannonians, who use barley and millet both for eating and drinking. Ammianus Marcellinus also reports on a beverage made of barley or other grains, consumed by the poor of Illyricum (Amm. Marc. 26.8,2). At the turn of the 5th century Claudius Claudianus writes more on wine, this time specifically mentioning the Pannonians living along the Sava River, i.e. those in the very south of the province who cut down forests, cultivate lands and plant vineyards along the banks of the Danube (Claud. 2.190-201). In the time of St. Ambrose of Milan (339-397), Pannonia was renowned for its fertility and the grain production (Ambr. *Epist.* 18.21). Given the fact that St. Ambrose was very much involved in ecclesiastic matters of the Sirmium archbishopric and travelled there on several occasions, his account most likely refers to southern Pannonia.[7] There is also a record dating from approximately the same time period, the mid 4th century, which praises Pannonia for its opulence (*Exp. tot. mund.* 57). In two occurrences, authors report on the multitude of cattle: horses (*Exp. tot. mund.* 57) and oxen (Paul *HL* 2,8).[8] Fertility of the land was emphasised by Isidore in the above-mentioned excerpt (Isid. *Etym.* 14.4,16). *Tabula Peutingeriana* contains the toponym Ad Labores Pont(is) Ulcae (*Tab. Peut.* VI,2), 12 miles east of the Mursa maior settlement, which testifies to the work done to regulate the flow of the River Vuka. Unfortunately, there are no preserved records in literary sources of swamp drainage in the Croatian portion of Pannonia, as there are for the Sirmium area;[9] however there ought to have been some as well.[10]

The above-mentioned map, *Tabula Peutingeriana* (4th century), is one of the three major sources from Late Antiquity that enable a reconstruction of the provincial road network. The other two are the itineraries, namely Antonine's (*Itinerarium Antonini Augusti*, 3rd century) and *Itinerarium Burdigalense*, dating from 333/334. There were two major roads that transversed southern Pannonia. The first one led from Aquileia via Poetovio further along the River Drava – this road was a branch of the Amber road, which connected the Mediterranean with the Baltic Sea – while the other road connected the two main provincial centres, leading from Siscia to Sirmium along the River Sava (*Tab. Peut.* VI). The former (*Itin. Burd.* 561,7-563,7), having passed Poetovio, crossed the River Drava, coming into Aqua Viva (most likely Petrijanec near Varaždin), proceeding along the right (south) bank of the River Drava all the way to Mursa and Cibalae, two other important crossroads.[11] In addition to these main ones, there is also a record of an alternative west-east route between the rivers, from Siscia via Certissia to Cibalae and Sirmium (*Itin. Ant.* 260,2-261,3). In Varianae it branched, the northern road going via Aquae Balissae (*Itin. Ant.* 265,5-10), and the southern one going via Menneianae; ultimately both roads rejoined at Incerum. There were also two other routes perpendicular to these; in the west there was a road from Aqua Viva via Andautonia to Siscia (*Itin. Ant.* 265,11-266,3), and in the east from Marsonia to Cibalae via Certissia (*Tab. Peut.* VI).[12] From Siscia southwards there was a road via Quadrata to Senia on the Adriatic coast (*Itin. Ant.* 274,4-7). There was also another important military road along the *limes*, which connected the Danube forts (*Itin. Ant.* 243,2-5).

There is an early medieval geographic work worth mentioning here, because of the interest and controversies it arose among the commentators, given its importantance for the discussion on the continuity of life in Pannonia after the changes that came along with the Barbarians during the 5th and the 6th centuries. It is the *Ravenna Cosmography* (*Ravennatis Anonymi Cosmographia*), a list of place-names of the known world compiled by a cleric from Ravenna in the period that has been variously dated to between the 7th and the 8th centuries. The problem of the authorship and the compiler's sources is just one in the series of controversies related to this important piece of ancient historical geography.[13]

3. The ethnographic overview

Pannonia is likely to have been one of the most complex areas of Europe in regard to ethnic relations. It is the area where one witnesses the dynamic interaction of the Mediterranean and continental world, the Western and Eastern Roman Empire, the sedentary peoples of Europe and the nomadic peoples of Asia. Of the latter, all of those who in antiquity and in the Middle Ages invaded Pannonia managed to establish at least a short-lived dominance over

5 These two most important natural features of southern Pannonia enjoy special protection: the remains of the oak temperate rain forest (called Prašnik) as a nature reserve, and the marshland habitat as the park of nature Kopački rit. On logging in the forests north from the River Drava see Aur. Vict. 40.9.

6 See the reports on weather on the Danube (in Valeria) by Amm. Marc. 19.11,4 and Hdn. Hist. 1,6. On severe frosts in the northern Danube area see Amm. Marc. 30.5,14. According to the present-day climatological tables, Vienna (closest to Carnuntum) and Zagreb (just north of Siscia) differ in precipitation, but have almost the same average temperatures during the year, as they share the same continental climate.

7 Jarak 1994, 36 (English text 176).

8 The wild ox mentioned by Paul is most likely the European bison (*Bison bonasus*), given that the author writes of having heard that the hide of the animal is of such size that fifteen men could lie in it.

9 Aur. Vict. 37.4, *SHA* Probus 21.2.

10 See also Gračanin 2011, 27-28.

11 More on the roads through Pannonia in Graf 1936, 42-128; Soproni 1980b, 207-217; Škrivanić 1977; Bojanovski 1984; in detail, specifically on southern Pannonia: Gračanin 2010. See also Gračanin 2011, 29-48.

12 See also Andrić 2002, 121-123.

13 The list of Pannonian places is in the book IV (Europe), ch. 19 – see: Anon. Raven. 213-221. On the Cosmography see: Funaioli 1914; Gračanin 2011, 29-30, *passim*.

the area, starting with the Huns (4th and 5th centuries), Avars (6th century), Hungarians (9th century) to the Mongols in the 13th century.

When the Romans conquered Pannonia, a large number of Illyrian and Celtic peoples were dwelling in the area.[14] Settlements attracted a substantial number of Italic traders and veterans, and Romanisation took place, especially in the urban areas. However, one of the consequences of the Marcomannic Wars was a change of demographic structure. Pannonia Inferior received a great number of immigrants in the form of Thracian, African and Syrian troops which reinforced the weakened legions, although there were other immigrants, mostly from the East.[15] That was only the beginning of the even greater demographic changes that started taking place with the retreat from Dacia in 271 when Pannonia became even more exposed as the border province that defended the Empire.

From the late 3rd century onwards, Pannonia was ravaged by, and then settled, by various Germanic peoples and waves of nomadic attackers from the Black Sea steppe. The Goths dwelt to the north and the east of the province and enjoyed *foederati* status with the Empire in the 3rd and the 4th centuries. The importance of their units in defence against other peoples was acknowledged by Jordanes.[16] During the reign of the emperor Gallienus (260-268) the Black Sea Sarmatians and Germanic Quadi pillaged across Pannonia (Eutr. *Brev*. 9.8; Jord. *Rom*. 287), and even though the Sarmatians settled in the area later on several occasions, they did not have significant impact on the province.

In the summer of 374, the Quadi and other peoples crossed the Danube and invaded eastern Pannonia, laying waste, killing, and enslaving people (Amm. Marc. 29.6,6; Hier. *Chron*. AD 375). Soon after, in 380, Alateus' and Safrax's Ostrogothic-Alanic-Hunnic group pillaged and plundered northern and southern Pannonia (Jord. *Get*. 140). After that, all three peoples inhabited the *limes* area, most likely north of the River Drava, as imperial *foederati* and repopulated the very towns they plundered with their soldiers (Pacat. 32.3-4).[17] Numerous groups of the Barbarians settled in as *foederati*, although in difficult times they tended to join the new waves of attackers; therefore, they were quite unreliable as defenders of the border area.[18]

The Vandals and Alans inhabited »both Pannonias« until the beginning of the 5th century, when, fearing the return of the Goths, they relocated to Gallia (Jord. *Get*. 161). The Huns, after having ruled for fifty years, left Pannonia in 427, but only for a short while (Jord. *Get*. 166). In 433 the Empire renounced most of Pannonia in favour of the so–called Great Huns who, until 441, subdued the towns in Pannonia Secunda that they did not hold until that point. There are no remains that would indicate more intensive settlement of the Huns in southern Pannonia after they subdued the towns.[19] After the fall of the Hunnic Alliance in 453, Pannonia Secunda was taken over by the Ostrogoths, with the permission of the Eastern Roman Empire (Jord. *Get*. 264). The Gepids took advantage of discontinuity between two periods of the Ostrogothic rule (473-504) and took over the eastern part of Pannonia Secunda (of their presence around the River Vuka writes Enn. *Paneg. Theod*. 28). On the other hand, Pannonia Savia was settled by the Svevi, which led to the common misinterpretation of the name of the province Savia as Svevia (Jord. *Get*. 273).[20] Jordanes tells of Hunimund, chief of the Svevi, who attacked neighbouring Dalmatia (Jord. *Get*. 273). Eventually, the Ostrogoths under the leadership of Theoderic (471-526), with the support of the Empire, took control of the major part of Pannonia Secunda and Pannonia Savia. Theoderic relocated most of the Gepids from Pannonia Secunda to Ostrogothic Gallia (523-526).[21] After his death, the Ostrogothic position weakened and they retreated from Pannonia after the war with the Eastern Roman Empire (535-537). The Gepids seized that opportunity to occupy the towns of Sirmium and Cibalae.[22]

A relatively short episode in Pannonian history was marked by the Lombard presence. During the reign of king Vacho (510-540) the Lombards invaded from the north into the provinces of Pannonia Prima and Pannonia Valeria, and took over the Svevi lands.[23] After the death of Theoderic in 527, the Empire recognised the Lombard sovereignty over the lands north of the River Drava and made the inhabitants *foederati*. In 551, somewhere between Cibalae and Sirmium, there was a decisive battle in which the Lombards were victorious over the Gepid forces and they took possession of Pannonia to the west of the Gepid-ruled Sirmium, having already obtained the permission of the Empire (Procop. *BG* 3,33, Paul *HL* 2,1).[24] For some time there was a truce between the Lombards and Gepids; then the Lombards overcame their enemies, but they could enjoy their victory only for a short while (Paul *HL* 2,7).[25]

In 567/568, a significant demographic change occurred in Pannonia; after the final collapse of the Gepid state in the Carpathian valley, some of them joined the Lombards during their move towards Italy, others fled to the Eastern

[14] See section 3.2. (Ethnic diversity) in the article by Alka Domić Kunić in this volume, and Fitz 1980a, 141-142.
[15] Fitz 1980a, 151; on the early Syrian settlers see Mócsy 1974, 227-230, on the settlers from the East see Fitz 1982, 32.
[16] Jord. *Get*. 111-112. General overview of the Gothic-Roman relations see in Heather 1991.
[17] Jord. *Get*. 141-142; Gračanin 2006, 85; Gračanin 2011, 218.
[18] Gračanin 2006, 86.

[19] Gračanin 2011, 222.
[20] Cf. Gračanin 2011, 223 – the author indicates that this misinterpretation can also be read in Cassiodorus, *Variae* (4.49, 5.14,1 and 5.15) and Procopius' *Bellum Gothicum* (1.15,25-26 and 1.16, 8-9); see also Gračanin 2006, 88-89.
[21] Gračanin 2007, 21-22.
[22] On the Gepid occupation of Sirmium and their settlement along the both banks of the River Danube see Procopius, *Bellum Vandalicum*, 1.2,6. On other settlements in Pannonia see Gračanin 2007, 30-34; on Cibalae, for which there are no literary sources, only the archaeological evidence, see Gračanin 2007, 31, n. 114.
[23] Gračanin 2007, 35.
[24] Christie 1992, 330.
[25] For more details on the Lombards dwelling in Southern Pannonia see Gračanin 2007, 35-45.

Roman Empire and still others become assimilated under the Avars. The Svevi from Savia also join the Lombards in their migration westwards. The Avars and Slavs expanded throughout the provinces of Pannonia and Dalmatia and became the dominant ethnic element, which slowly assimilated the existing Romanised Barbaric population.[26]

4. Political history: external and internal wars

The Marcomannic wars ended in the year 180;[27] these conflicts demonstrated the vulnerability of the northeast border and were merely an introduction to the dramatic invasions of different peoples into the Empire in the following centuries. Since that time almost half of the Roman armies were stationed alongside the borders on the Danube and Rhine (14 out of 33 legions); this had even more impact on the internal affairs of the Empire than on its defence.[28] Such a concentration of armies in Pannonia, with the path to Italy wide open, was too much of a temptation for many aspirants to the throne to resist. The result was that Pannonia became not only the site where the new emperors emerged, a battlefield of different aspirants to the throne, but also the first line of defence against known and unknown Barbarians from the East.

In 193 the governor of the Upper Pannonia, L. Septimius Severus, assumed the imperial power with the help of the Pannonian army (Hdn. *Hist.* II, 9, 11).[29] During the time of his reign (until 211) there are no traces of any significant military activity. Not only did he manage to recover the provinces from the consequences of the Marcomannic wars, but also the records show that the military camps and the towns along the *limes* grew in size and importance. Having proved their allegiance to the emperor, the towns of Aquincum, Brigetio, Carnuntum and Siscia were elevated in status.[30]

However, the relatively peaceful period did not last long: the Quadi and Marcomanni pillaged Pannonia Superior, and the Sarmatians depopulated the upper course of the Danube in 227/228, as was recorded on the Siscia inscriptions (*CIL* III 10843, 15181).[31] In the following decades, the Barbarians invaded more frequently and the Roman army, in addition to guarding the borders, started acting as a decisive factor in internal struggles for power: Maximin the Thrax (235-238) was the first in a line of military emperors from Illyricum (Hdn. *Hist.* VII, 8, 11). In the following 50 years there were more than 20 'legitimate' emperors, while just during the reign of Gallienus more than ten usurpers tried to seize the throne.[32] After Gallienus' death, the emperors from Pannonia marked the period that followed (Aur. Vict. 39.26; Mam. *Paneg. Max.* 10 (2), 2.2).

The simultaneous attack of the Goths on Dacia 269-270 and the invasion of the Vandals, Svevi and Sarmatians of Pannonia led to a turning point in the history of the Empire. The emperor Aurelius managed to drive out the Barbarians from Pannonia, but he also completely abandoned Dacia in 271, depriving the Empire, among others, of its rich gold deposits, and exposing Pannonia directly to Barbarian invasions which already had started during the reign of Gallienus (Eutr. *Brev.* 9.8).[33]

The army reorganized by Diocletian (283-305) kept the Germanic peoples and the Sarmatians out of Pannonia. In 290s, the Emperor repelled one of the Sarmatian attacks appointing Galerius to end the Sarmatian wars and to put into practice his reforms of the Empire.[34] Diocletian's reforms were of economic, financial and military nature, and included the modernisation of the *limes* forts.[35] The reforms regarding Pannonia, in terms of administrative division, were as follows: Sirmium became the seat of the caesar Galerius, and the provinces of Noricum, Dalmatia and Pannonia merged into the diocese of Illyricum (*dioecesis Pannoniarum* or *Illyricum occidentale*). Furthermore, the provinces were divided into two units: Pannonia Superior north of the River Drava was called Prima and had the capital Savaria; the part of the province south of the Drava was called Savia, with the capital in Siscia. The northern part of Pannonia Inferior, with its capital in Sopianae, was named Valeria, and its southern part, between the Rivers Drava and Sava and along the Danube, was named Pannonia Secunda, with the capital Sirmium (*Not. Dign. Occ.* I,40-42).[36] Although the division took place in the beginning of the 4th century, one can still find the old names of the provinces in some literary sources of later date, because it took a while for the new names to become common, which happened towards the end of the century.

In the turbulent period just after the Diocletian's abdication, the emperor in the West, Constantine I, and the Augustus in the East, Licinius, had a decisive clash in southern Pannonia. In 314 Constantine appropriated the Siscia mint from Licinius; in the same year (or in 316) the battle near Cibalae occurred: Constantine led the Gallic armies to victory over Licinius' armies from Illyricum. Licinius retreated to Sirmium, and then all the way back to Thrace; he ceded the Danube provinces to Constantine (Eutr. *Brev.*

[26] On the expansion of the Slavic peoples see Curta 2001 and Gračanin 2008.
[27] For more on the Marcomannic wars (166-180) see Birley 1975 and Kerr 1995. On the military history of Pannonia in the period that followed see Fitz 1962.
[28] Mócsy 1974, 194.
[29] In *SHA* Sever IV, 2 there is a mention of Severus being proconsul in »both Pannonias«, but this is most likely not accurate, in regard both to the function and the time sequence (i.e. in relation to the office of proconsul in Sicily, which he held prior to, not after the Pannonian office). At this place it is worth mentioning that *SHA* is generally considered as a rather unreliable compilation of sources – see more in Barnes 1978 and Syme 1983. For comments on the accounts of Herodian and the one in *SHA* see Šašel Kos 1986, 362-365.
[30] There are no literary sources on this, just the epigraphic ones, like: *CIL* III 4193. See also Barkóczi 1980, 102; Hoti, 1992, 145.
[31] Hoti 1992, 146.
[32] Birley 1975, 257-258.
[33] Barkóczi 1980, 105.
[34] Barkóczi 1980, 109, 112.
[35] For more on the reinforcements in the mid-4th century see Christie 2008, 554.
[36] The division can be reconstructed from Laterculus Veronensis (297 AD or the beginning of the 4th century) and *Notitia Dignitatum* (early 5th century). See also Andrić 2002, 124-125 and Gračanin 2011, 49-50.

10.5),[37] holding power in the East until 324, when he was finally forced to surrender his purple robe.

Southern Pannonia again became the stage of an internal clash in the mid–4th century; Zosimos gives an account of the struggle between Constantius II (337-361) and the usurper Magnentius.[38] Vetranius interfered into the conflict by proclaiming himself the emperor, backed by troops in Mursa (Zos. 2.43). Constantius II made his camp in Cibalae, expecting the battle to take place there, while Magnentius made a surprise attack on Siscia, reducing it to ruin and plundering it hugely (Zos. 2.48; 2.49). Although it seemed that Magnentius gained the upper hand, in the decisive battle near Mursa in the autumn of 351 Constantius was victorious and Magnentius fled to Italy, only to be killed in Aquileia. Eutropius ends his report on these events with a lamentation over the casualties and the exhausted resources of the Empire, all of which could have been used for new victories or keeping the peace (Eutr. *Brev.* 10.12; also Oros. *Hist.* 7.29,12). More details on the battle and the casualties on both sides may be found in Zonaras (Zon. XIII, 8), who used some older reports.[39]

In the second half of the 4th century, the Huns incited the movement of the Germanic and some of the nomadic peoples towards the borders of Pannonia.[40] During the time of the emperor Constantius II (late 350s) the Sarmatians and Quadi invaded both Pannonias and Upper Moesia (Amm. Marc. 17.12,1). At the time of Valentinian's ascension to the throne (AD 364), both Pannonias were raided by the Quadi (Amm. Marc. 26.4,5).[41] They devastated and sacked southern Pannonia again in 374, laying siege to Sirmium, as a result of their rage over the killing of their king Gabinius (Amm. Marc. 29.6,6). In the battle of Hadrianopolis in 378, the Goths defeated the emperor Valens, moving into Illyricum and Pannonia in the following year.[42] From that period (between 366 and 384) dates the quotation, accredited to Ambrosiaster, on the irreparable damage done to Pannonia (Ambrosiast. *Quaest.* 115,49). The damage was ever-increasing: in 385 the Alans and Huns rebelled against their imperial commanders and allied themselves with the Sarmatians, which led to pillaging all over Savia and Secunda.[43] At the same time, the Empire was engulfed in a civil war: in 383 the usurper Magnus Maximus ousted the emperor Gratian and had him executed. In 388 the emperor Theodosius marched from Thessalonike and led the Gothic, Hunnic and Alanic cavalry, gathered in Pannonia, to defeat Maximus along the River Sava near Siscia (Pacat. 34, 1); shortly after, he defeated him in a battle near Poetovio as well (Ambr. *Epist.* 40,23).[44]

After Alatheus and Safrax attacked Pannonia and defeated the duke of Illyricum, Vitalian, in 380, the emperor Gratian allowed the Ostrogoths, Alans, and Huns to settle in Pannonia as imperial *foederati* in the same year (Zos. 4.34).[45] The complex relations between the Empire and the Visigothic *foederati* of Moesia and Dacia devolved into discontent, which led to an open rebellion of the Visigoths. In 401 the rebel leader, king Alaric, started a campaign towards Aquileia, passing through Sirmium (Jord. *Get.* 147) and the area between the Sava and Drava Rivers, where he got support from the Huns and Visigoths who inhabited those parts.[46] After Stilichon defeated his armies in northern Italy in 402, Alaric's Visigoths took the same route back, spending some time in »Barbaric parts near Dalmatia and Pannonia«.[47] After various internal struggles within the Empire, Alaric made his way back to Italy from Epirus six years later, eventually laying siege to Rome and sacking the town. His brother-in-law Ataulf joined the campaign, leading a great number of the Ostrogoths from Pannonia Superior to Italy, thus ending the first period of Gothic presence in Pannonia.[48]

In 427 the members of the Hunic alliance were driven out from Pannonia (Jord. *Get.* 166),[49] only to be replaced by the Great Huns as early as 433 when Aetius ceded to them parts of the province by a new *foederati* pact.[50] The Hunnic presence also affected the boundaries between the two parts of the Empire.[51] In the campaign of 440/1 the Huns conquered Viminatium and Sirmium. In 451, Attila invaded Gallia and sacked Aquileia the following year. Attila died in 453, leaving his sons unable to keep his extensive domain in one piece. As early as 454 the Germanic peoples rebelled against the Huns and divided the territories among themselves: the Gepids took the land between the Tisza and Danube, and the Ostrogoths took over Pannonia, with the permission of the Empire (Jord., *Get.* 264).[52] In the 460s the Huns disappeared as a force in Pannonia: some were driven out by the Goths (Jord. *Get.* 273), and others joined some other ethnic groups.[53]

By the end of the 5th century the Ostrogoths became discontent with their position in Pannonia (Jord. *Get.* 283); some of them went to Italy and some tried to achieve recognition as *foederati* by force. The king of the Ostrogoths, Theoderic, defeated the Gepids between

37 Barkóczi 1980, 111; Caldwell 2007, 61-66.
38 See also Gračanin 2010, 43-44.
39 See Šašel 1971, and some more recent studies on the course of the battle and its consequences: Gračanin 2003; Caldwell 2007, 142-160. For a more detailed overview of the events in the period from AD 376 to 382 see Heather 2005, chapter 4 (The War on the Danube), 145-190.
40 Thompson 1996, 26-35; Andrić 2002, 130.
41 This passage of Amm. Marc. is commented on in Tomlin 1979.
42 Gračanin 2006, 84.
43 Hoti 1992, 152.
44 Williams and Friell 1995, 63; Lotter 2003, 83-84; Gračanin 2005b, 16-17.

45 Klemenc 1953; Andrić 2002, 130; Gračanin 2006, 85; Mócsi 1974, 341-342 places the *foederati* in Savia, in the Drava River valley.
46 Cf. Gračanin 2011, 62.
47 Sozom. 9.4.4; Gračanin 2011, 63.
48 Gračanin 2006, 90.
49 Gračanin 2005b, 25. For analysis of the sources and more details on those who drove them out and how they did so, see Maenchen-Helfen 1973, 77-80.
50 Salamon and Sós 1980, 397-400; Maenchen-Helfen 1973, 87-90; Andrić 2002, 131.
51 Gračanin 2005b, 29.
52 Gračanin 2006, 91. For more on the Ostrogoths in Pannonia in the period 456/457-473 see Wolfram 1988, 258-268.
53 Maenchen-Helfen 1973, 168. For a more detailed view of the end of Hunnic domination and its impact on the Empire see Heather 2005, chapter 8 (The Fall of the Hunnic Empire), 351-384.

Mursa and Cibalae in 489 (Enn. *Paneg. Theod.* 28).[54] In their negotiations with the Empire, the Ostrogoths gained recognition of their possession of a part of Pannonia Secunda and of Pannonia Savia, the latter merged with Dalmatia under the administrative rule of a Gothic *comes*. After the death of king Theoderic in 526, the Eastern Empire tried to re-establish its influence in southern Pannonia. In a large-scale military effort against the Goths in Italy, Dalmatia and Pannonia (535-537) the emperor Justinian succeeded in completely forcing out the Goths from the area between the Rivers Sava and Drava.

5. Towns, emperors, military camps

Many literary sources dealing with Pannonia Secunda refer to its capital Sirmium, but as it is not the part of this overview, this work will focus on the other two important towns, Mursa and Cibalae. As far as Pannonia Savia is concerned, most sources refer to its most important centre, Siscia.

Two features were significant for the topography of Cibalae (present-day Vinkovci): the town was located on top of a hill,[55] and the road that led to the town passed through marshlands, which were some five stades wide (*Epit. de Caes.* 41.5; Zos. 2.18). The status of colony was given to the town at the latest in the beginning of the 3rd century, during the reign of either Septimius Severus or Caracalla.[56] On two occasions it was the site of decisive battles between aspirants to the throne: in 324 Constantine defeated Licinius there (*Epit. de Caes.* 41,5; Oros. *Hist.* 7.28,18-19; Zos. 2.18), and the town was the camp of Constantius II just prior to his battle with the usurper Magnentius near Mursa (Eutr. *Brev.* 10.5). Cibalae was particularly notable for its tradition of spawning imperial guards and also for the fact that it was the birthplace to two emperors, the brothers Valentinian (364-375) and Valens (375-378) (*Epit. de Caes.* 45.1; Zos. 3.36).[57] When, in his absence, Valentinian was proclaimed emperor by the troops in Bithynia, there were two Pannonians there, Aequitius and Leo, who put much effort into and succeeded in ensuring that the army, »always ready for a change, does not elect someone else, present«.[58] During the reign of the two Pannonian emperors, there was a sudden increase in the administrative positions held by the natives of Illyricum, loyal to the emperors.[59] Their father Gratian, also born in Cibalae, well-known for his strength being equal to that of five men (Amm. Marc. 30.7,2; *Epit. de Caes.* 45.1), was a military man who rose to the title of *protector domesticus*.[60]

In the literary sources from the 3rd and the 4th centuries, Mursa[61] is mentioned only regarding the two events, both times as the battlefield for two civil wars. In the first one in 260, the emperor Gallienus defeated Ingenius, a contestant for the throne backed by the Moesian legions (Eutr. *Brev.* 9.8; Aur. Vict. 33.1-2). In March of 350 Vetranius (Zos. 2.43), commander of the Illyrian infantry (*magister peditum*), was proclaimed emperor at Mursa, only to withdraw from the position by handing over his troops to Constantius II who had just came back from Persia. In a great battle outside the town in 351, which turned into the bloodbath lamented by Eutropius, Constantius defeated the usurper Magnentius (Eutr. *Brev.* 10.12; Zos. 2.49-50).[62] The sources make mention of two edifices that stood outside the Mursa town walls. The first one is a *stadium* (in some scholarly articles sometimes regarded to as amphitheatre), which had already lost its function and was covered by the bush in the time of the 351 Mursa battle, when Magnentius used the location to conceal four companies of his men to make an ambush on Constantius (Zos. 2.49-50). The other one is the so-called *basilica martyrum* (Sulp. Sev. *Chron.* II,38). Neither of the two buildings has been yet archaeologically attested.[63] In the period of the late Empire, Mursa was also mentioned as the seat of the prefect of the Danubian fleet (*Not. Dign. Occ.* XXXII,52).

The town of Siscia[64] was located in the very southwestern corner of the Pannonian plain, with its importance rising as the Barbarians got closer to the borders of Italy. Siscia, named *colonia Flavia* in the first century, was elevated to the status of *colonia Septimia Siscia Augusta* after it supported Septimius Severus in his claim to imperial power.[65] Zosimos (Zos. 2.45) gives an account of spacious fields around the town. Its exceptional position is mainly due to the proximity of both the Kupa and Sava Rivers, the latter having the most convenient crossing just outside the town,[66] and the proximity of the rich iron deposits in the notable mining area of the Sava-Japra valley.[67] The latter fact was one of the key reasons for the establishment of the Siscia mint in 260 (or 262, but without a doubt during the reign of Gallienus).[68] Its production started with two *officinae*, but as the demand increased in the late 3rd and the 4th centuries there were six *officinae* operating.[69] By the end of the 4th century, six mints supplied the Western Empire with coins, the Siscia mint among them, as attested by the mention of the officer in charge, *Procurator monetae Siscianae* (*Not.*

54 On the Ostrogoths under Theoderic's reign and a more detailed overview of Pannonian history at the time, see Gračanin 2006, 104-113. On the Gepids in southern Pannonia, till their banishment, see Gračanin 2007, 8-18.

55 For an argument against such location see Brunšmid 1902, 126. For more on the topography of the town see Rapan Papeša 2011.

56 Iskra-Janošić 2001, 59, 62-63; Lenski 2002, 37 states that, according to inscriptions, Cibalae became a colony in the 2nd century and that the area of the town was 56 hectares.

57 Cf. Lenski 2002, 36-37.

58 Amm. Marc. 26,1,6.

59 Lenski 2002, 57-58.

60 Lenski 2002, 46-47.

61 Colonia Aelia Mursa (*CIL* III 3560). For more details on the town and its inscriptions see Brunšmid 1900 and Pinterović 1978.

62 The battle of Mursa was mentioned also in Amm. Mar. 15.5,33; see also Caldwell 2007, 142-160.

63 More on both edifices see: Pinterović 1980; Bulat 1989.

64 Colonia Septimia Siscia Augusta (*CIL* III 4193). On the history of the town and its inscriptions see Zaninović 1981; Hoti 1992; Koščević and Makjanić 1995.

65 *CIL* 4193; Hoti 1992, 145.

66 See Durman 1992.

67 See Škegro 1999.

68 Mócsy 1974, 208; Hoti 1992, 147-148; Lenski 2002, 42. See also Bíró-Sey 1980, 343.

69 During the time of the emperor Claudius (268-270) the third and the fourth were added (Nenadić 1986, 93), and during Aurelianus (270-275) the fifth and the sixth (Hoti 1992, 148).

Dign. Occ. XI,39). That was the reason Siscia was attacked and sacked by Magnentius during the struggle for imperial power in 351 (Zos. 2.49). Another imperial battle took place near the town in 388 (Pacat. 34,1), when Emperor Theodosius defeated the usurper Magnus Maximus (383-388).[70] In the 5th century the province of Savia was settled first by the Huns, and then, in the 2nd part of the century, by the Ostrogothic *foederati*. Cassiodorus confirms that the Ostrogothic king Theodosius ruled over »Siscia or Savia« in the beginning of the 6th century.[71] At that time Siscia was the seat of the Ostrogothic *comes*, up until 537, the last record of Ostrogothic activity in southern Pannonia.[72]

Military camps of the later Imperial period, along with their garrisons, were listed in the *Notitia dignitatum*; some of them were also noted on the *Tabula Peutingeriana* and on the Antonine's itinerary, along the *limes* road. The list is as follows:[73] Ad Militare (*Not. Dign. Occ.* XXXIII,45), present-day Batina (Skela), in the very north-eastern corner of Croatia, on the Danube. Ad Novas (*Not. Dign. Occ.* XXXII,28) is present-day Zmajevac, Aureus Mons (*Not. Dign. Occ.* XXXII,45) might be the name for the present-day Banovo Brdo, or a camp that still has not been located. Albanum (*Not. Dign. Occ.* XXXII,29), present-day Lug, was the seat of the Dalmatian cavalry. Donatianae (*Tab. Peut.* VI,1) was situated on the hill near the village of Vardarac, Mursella (*Tab. Peut.* VI,1) has been located on Kraljevo Brdo, by the village of Petrijevci. Ad Labores (*Tab. Peut.* VI,2) was situated by the present-day village Nemetin, and it is still uncertain if it was a fort or just a guarded bridge across the River Drava. *Teutoburgium castellum* (*Not. Dign. Occ.* XXXII,23; 30; 47; *Tab. Peut.* VI, 2), present-day Dalj, was a fort on the bank of the Danube which had slowly been destroying as the bank eroded. The location of Cornacum (*Not. Dign. Occ.* XXXII,22; 31; *Tab. Peut.* VI,2) has been established by present-day Sotin and Cuccium (*Not. Dign. Occ.* XXXII,25; 32; *Tab. Peut.* VI,3) near present-day Ilok.

6. Cultural anthropology, cults and religion

The geographer Gaius Julius Solinus (3rd century) in his »Miracles of the world« describes Pannonia as a land of strong men, rich in fertile soil (Solin. 21,2). Dio Cassius writes that the Pannonians are believed to be the bravest among men, but that they are also bloodthirsty, lacking any values that would comprise a honourable life. He ends his description with an etymology of the Pannonian name, deriving it not from an Illyrian word, as he should have done, but from the Latin word *panni* for the sleeved tunics the Pannonians make from the strips of old clothes (Dio 49.36,2-5). According to Aurelius Victor, even all of the Illyrian emperors, and there were many, were void of any *humanitas* (Aur. Vict. 39.26). Ausonius, however, praises Cibalae-born Valentinian as *eruditus*, a well-educated man (Auson. *Cent. nupt.* XXIV,1,11). It should be noted at this place that, although Valentinian was indeed considered one of the last great emperors, Ausonius was the tutor of his son Gratian, and was held in highest possible honour by both emperors. Nevertheless, the Emperor must have been educated to a certain extent when he proposed a poetry competition to the Bordeaux poet (although Zos. 3.36 claims the opposite). Herodian mentions the Pannonians twice. The first reference is related to the ascension of Septimius Severus to the imperial throne, saying that the Pannonians supported his claim not understanding that they were being deceived (Hdn. *Hist.* II,9,11). According to Herodian, they are men of strong physique, able and experienced warriors, but witless and slow minded. However, he does continue his account by acknowledging in them the quality of being very loyal – the emperor Maximinus had complete trust in his Pannonian troops, holding them worthy to lead the march on Italy (Hdn. *Hist.* VII, 8, 11).

Pannonia is a region of many interconnected cults. Apart from the worship of the gods of the Roman pantheon, the cult of Silvanus, an indigenous and highly popular deity in all of Pannonia, was so well established that the number of inscriptions devoted to him was almost on a par with those of Jupiter.[74] Moreover, the river deities Danube, Drava and Sava were especially revered. During the 3rd century the salvation religions (Isis, Cybele, Attis) started to spread their influence, especially in the colonies of Siscia, Mursa and Cibalae. During the same period the cult of Mithras was also gaining ground in Pannonia. Of all of these and other forms of religious practice, there is unfortunately no evidence preserved in the literary sources, so the researchers have to rely on the ample archaeological findings and epigraphic material.[75]

The scholars suppose, based on the complete lack of sources for the earlier period and the assumed similarity of development in the greater area, that the Christianisation of the provinces Pannonia Savia and Secunda started taking place relatively late, after the mid-3rd century.[76] By the end of the 4th century the Christian population in southern Pannonia was organized in four bishoprics: in the west, in Pannonia Savia, they were in Siscia and presumably Iovia, and in the east, in Pannonia Secunda, the bishoprics Mursa and Cibalae.[77] The ecclesiastic centre of both provinces was Sirmium which was elevated to the rank of an archbishopric in the 4th century.[78] Under its jurisdiction were also parts

[70] Klemenc 1953; see also Lotter 2003, 83-84 and Gračanin 2011, 60-61.
[71] Cass. *Var.* 4. 49.
[72] Procop. 1.16,8.9; Lotter 2003, 126; Gračanin 2011, 102.
[73] Based on the *Notitia Dignitatum* and *Tabula Peutingeriana*. This is just a list, on the camps see more in Visy 2003, 135-142 and Sanader 2003. For more on the army in Pannonia see Kovács 2004.

[74] Mócsy 1974, 250-252; Rendić-Miočević and Šegvić 1998, 9.
[75] For a concise overview on the religions and cults in southern Pannonia, see Rendić-Miočević and Šegvić 1998. See also Mócsy 1974, 250-259; Selem 1980; Fitz 1980b 161-175; Thomas 1980, 177-193; Fitz 1982, 35-41.
[76] Jarak 1984, 19-20 (158-159). This, however, might be contested if some new evidence appears. One is particularly tempted to move the Christianisation to the earlier date because of the mention of a martyr from Cibalae, bishop Eusebius, who was most probably martyred as early as the mid-3rd century.
[77] Jarak 1984, 18 (157); Bratož 1986.
[78] There are no extant literary sources that explicitly confirm the elevation to archbishopric: it has been reconstructed from the titles of the Sirmium prelates.

of Pannonia north of the River Drava; however there is no preserved written evidence, although the archaeological findings confirm the existence of bishoprics. After the town was taken by the Huns in 441, from the mid-5th century the importance of the Sirmium archbishopric declined and the influence of the archbishopric of Aquileia, which took over the jurisdiction over the western parts of the province, was on the rise. The archbishopric of Salona also gained importance, notably at the beginning of the 6th century when it took jurisdiction over Savia.[79]

The first reliable evidence for the Pannonian Christian communities date back to the time of the persecutions. Some light on the earliest Pannonian martyrs is shed by the Martyrology of St. Jerome (*Martyrologium Hieronymianum*), a very respectable source, as it was compiled relatively close to the events, in the 5th or the 6th century.[80] The analysis of the Martyrology shows that a fairly large number of martyrs was from the Pannonian provinces (many of them being from Sirmium), especially relatively to the number of the Dalmatian or Norican martyrs. The origin of the names of the martyrs attests to the rather variegated ethnic composition of the Pannonian Christian communities.[81] One of the most notable ones, St. Quirinus, bishop of Siscia, was martyred not long after the time of Diocletian's persecutions, in 308, under the emperor Galerius, as mentioned by Prudentius in a powerful poem consisted of 18 cinquains (Prud. *Peristeph.* VII,1-10).[82]

The other important source for research on the early ecclesiastic history of Pannonia is the list of bishops of Sirmium, which is complete for the 4th century.[83] For other bishoprics unfortunately there are no preserved lists, and there are only sporadic accounts of single bishops from the earliest times. Fortunately, there are extant records for some of the successors of martyr Quirinus in the see of the Siscia bishopric, as the bishops participated in the councils of Serdica (347, Marcus), Aquileia (381, Constantius) and Salona (530, Joannes; 533, Constantinus). Further evidence of the ecclesiastical activity in southern Pannonia lies in the fact that there were three Pannonian bishops participating in the Serdica council: Euterius of Sirmium, Marcus of Siscia, and Valent of Mursa.[84] There were also three bishops at the following council in Aquileia.

One of the most important issues within the Pannonian church in the 4th century was, without a doubt, Arianism, supported by the emperor Valens and, later on, the empress Justina, wife of Valentinian I. One of the influential Arian bishops of the East was Valens of Mursa, present at all important councils, starting with the First council of Tyre (335), over at least three decades. St. Ambrose of Milan was very zealously engaged in fighting Arianism in Pannonia, as attested in his *Vita*, by Paulinus.[85] Following the decrees of the councils of Sirmium (378) and Aquileia (381), the Arian fraction was suppressed, but did not completely disappear from Pannonia. It even revived somewhat at the beginning of the 6th century, during the period of Ostrogoth and Gepid rule.[86] However, after the mid-6th century not only Arianism, but also the Christian church as a whole suffered a significant decline in Pannonia, due to a new wave of invaders from the north-east.[87]

7. Concluding remarks

The concluding remarks will also take into account the chapter written by Alka Domić Kunić, because these two chapters form a natural unit. Furthermore, in some segments they intertwine because the historiographers of later periods wrote about the events that had happened several centuries prior to their time, often basing their histories on works of authors from different periods.

Titus Livius was the first one to define Pannonia as the region between the Sava and the Drava Rivers, a definition that remained in use well into the Early Middle Ages when it appears as such in the works of Jordanes and Isidore of Seville. Strabo, Appian and Pliny the Elder wrote about the general geographical features of Pannonia as early as the 1st century, when the province was created. The historiographers of Late Antiquity, except for Jordanes, were not concerned with geography any more; moreover, they were not prone to giving any precise indications. Most of them often use the term 'both Pannonias' (Amm. Marc. 17.12,1; 26.4.5; *SHA* Severus IV,2; Jord. *Get.* 161), as if referring to a commonplace rather than a clear indication of the authors' familiarity with the province and its geography. In other words, there is no convincing impression that the users are absolutely sure that something really took place in both provinces. Therefore, one should be careful in any attempt to establish the exact location of the events, especially in regard to the fact that at some point Pannonia was a part of the 'older' province of Illyricum, and from the late 4th century a constituent part of the prefecture of the same name, with the capital in Sirmium (Jord. *Rom.* 218; Fest. *Brev.* 8).[88] This fact linked the terms Illyricum and Pannonia so closely together that it is not rare to find instances of the same author using both terms to designate

[79] Bratož 1983, 270.

[80] On the Martytology and the Pannonian martyrs see Jarak 1984, 21-32 (160-172).

[81] Jarak 1984, 24 (162).

[82] Jarak 1984, 30-31 (170-171). See also Buzov 2003. For more on Prudentius' work see Roberts 1993.

[83] Mócsy 1974, 329; Bratož 1983, 268.

[84] Bishop Valens was mentioned on several occasions by Sozom. (3,12; 3,13; 4,6; 4,15).

[85] Paul. *Vita*, 11.

[86] Gračanin 2011, 116, 263.

[87] On the beginnings and development of the Christian communities in southern Pannonia, Pannonian martyrs, as well as bishops at the councils, up until the 6th century decline – in short, a complete concise overview of the Christianity in southern Pannonia in Late Antiquity – see (bilingual) Jarak 1994, 17-39 (Croatian), 155-179 (English). See also Mócsi 1974, 325-336; Fitz 1980b 168-169; Thomas 1980, 193-204; Bratož 1986; Hoti 1992, 149-150, 152-153.

[88] The name *Illyricum* had an interesting history on the territory it used to denote long after the fall of the Roman Empire. It was used in ecclesiastical hierarchy – in the 18th century Daniele Farlati compiled the *Illyricum sacrum*, a history of the churches in the area which once was the Roman province of Illyricum. In the beginning of the 19th century the Illyrian provinces were the constituent part of the Napoleon's French Empire, and the name was also used a few decades later in the formative phase of the Croatian nation.

separate administrative units in one chapter, and only one term as a synonym for the other in the next.[89]

Dio Cassius (Dio 49.37) was very precise in locating Sisak in regard to the Rivers Sava and Kupa, but that should not be surprising given the fact that he was *legatus* both in Dalmatia and Upper Pannonia (both offices held in the period 223-228). He also wrote (Dio 49.36,2-4) about the rather hostile climate, but since he points out the lack of typical Mediterranean plants, namely olives and grape vines, it becomes obvious that this piece of information was related to the very north of the province, since several other authors mention vineyards between the Danube and the Sava (*Epit. de Caes.* 37.3; Eutr. *Brev.* 9.17; *SHA* Probus, 18.8; Claud. 206). Not only was the province of Savia rarely mentioned in the sources, but when such mentions occurred on several occasions by Jordanes (Jord. *Get.* 273; Jord. *Rom.* 218), Procopius and Cassiodorus, it was often misnamed *Svavia* because of the Germanic inhabitants, the Svevi.[90] There is another inconsistency in nomenclature related to (the inhabitants of) Pannonia: while Polybius and Strabo wrote *Pannonioi*, Herodian and some other later Greek authors wrote *Paiones*, the name also used for the inhabitants of the land west of Thrace. This common mistake had already been addressed by Appian (App. 3.14) and Cassius Dio (Dio 49.36).[91]

More than half of all of the writings of early authors concerning southern Pannonia deal with military history. Cassius Dio provided many *in extenso* records, while Titus Livius, Velleus Paterculus and Appian, who lived closer to the time of events, also provided some useful information. That is by no means unusual, given the reputation of Illyricum which in the 1st century BC was considered a deadly military 'training ground' where soldiers sharpened their skills and commanders gained experience and valuable political points – it is the region where Octavian Augustus himself was wounded, and which was conquered and tamed only after fierce resistance. The campaign (12-9 BC) of Tiberius, another emperor who fought in Pannonia, was recorded in detail in the work of Dio Cassius whose reports on Pannonian warfare make up a quarter of all segments (27 of 108) dealt with by Alka Domić Kunić. The peace was established in the first decade of the 1st century AD after the rebellion of the two Batons was crushed, the records of which were preserved in the reports by Dio Cassius and Velleius Paterculus. On the other hand, in the second part of the overview, the largest group of the sources, second only to those that tell of Barbarian activity, describes the battles of the civil wars that on three occasions (AD 314 or 316, 351 and 388) took place in or in the vicinity of the Pannonian towns of Siscia, Mursa and Cibalae. Moreover, in all three wars the decisive battles in which the emperors defeated the usurpers took place just outside of two of these towns.

As far as the Barbarian presence in southern Pannonia is concerned – and to this part of history almost one third of all listed segments from Late Antiquity is dedicated – Jordanes' accounts deal with the Ostrogoth, Hunnic, Vandal and Alan activity in the region during the 4th and the beginning of the 5th century. Actually, the information Jordanes gave is almost exclusively concerned with the ethnic groups which alternately inhabited or possessed the *de facto* rule in Pannonia. That makes sense considering his work is in fact the abridged version of the (lost) history of the Goths by Cassiodorus. Without his contribution, the knowledge of the history of Barbarian raids and settlements would be very limited because he is the author of the majority of excerpts concerned with this part of history. Some pieces of information on the ravaging of both Pannonias in the 3rd century is given by Ammianus Marcellinus, while Paul the Deacon completes the ethnic picture for the 6th century with his accounts of the short period of the Lombard presence in Pannonia. One just needs to be cautious with the very doubtful mention of the duration of their dominance which is probably due to an error of the copier.[92] The accounts of other authors are either scarce or give a rather vague image of the difficulties that were the result of the Barbarian raids.

Most of the reports were given by the historiographers and the geographers – alongside others such as Galen, who was a physician, or Saint Ambrose, who was a bishop – but in both periods there were several poets who mentioned Pannonia and its inhabitants in their poems. In the earlier period, among 30 listed authors, there were four poets: Tibullus, Ovidius, Lucan and Statius. All their mentions are scanty, stereotyped images, but the audience must have been familiar with the references. In Late Antiquity, five out of 29 authors were poets: Mammertinus and Pacatus mention Pannonia in the panegyrics to the emperors, Claudianus refers to four of its most characteristic features: the Danube, the Sava, its forests and vineyards. The lawyer, provincial governor and Christian poet Prudentius wrote in honour of 14 martyrs, and one of his powerful hymns tells of the martyrdom and the miracles of St. Quirinus of Siscia. In the introduction to a centon, a poem composed from verse taken from other poets, Ausonius wrote that it had been composed upon the behest of Emperor Valentinianus, who wanted to challenge the poet in composition of this poetic form. An emperor in contest with the famous poet is a somewhat idealised image, but it opposes, at least to a certain extent, the stereotype that all the emperors from Illyricum were capable soldiers and firm rulers, but lacking in education.

To summarise, it should again be noted that in this two-part overview and list of sources dealing with the history of southern Pannonia, it was not possible to address all the studies on individual authors, let alone discuss the wide range of problems related to each of them.[93] When making use of these sources – as it should be done with any primary material, especially one this scarce – scholars should take

[89] For such an example in Herodian's work see Gračanin 2005a, 288-289.
[90] Cf. Gračanin 2006, 100.
[91] For more on this see Domić Kunić in this volume, section 3.1.

[92] See Gračanin 2007, 36.
[93] Some studies are mentioned in note 1.

into consideration two facts. Firstly, it is the information about the author, his position and motivation for writing, i.e. the context of the origin of the work. Secondly, most authors of these historiography works drew upon numerous sources which they incorporated into their writings. Hopefully a more comprehensive study, one with a subtle analysis of the authors and their writings related to the history of southern Pannonia, will be produced, supported by numerous references from the rather opulent recent bibliography.[94]

Sources

Ambrosiast. *Quaest.* 115,49 (4th century AD)

(…) quid dicemus de Pannonia, quae sic erasa est, ut remedium habere non possit?

(…) what will we say about Pannonia, which was so devastated that there can be no relief?

Ambr. *Epist.* 18.21 (4th century AD)

Frumentum Pannoniae, quod non severant, vendiderunt et secunda Raetia fertilitatis suae novit invidiam.

They sold the grain of Pannonia which they did not sow, and Rhaetia Secunda incurred hostility owing to her fertility (…)

Ambr. *Epist.* 40.23 (4th century AD)

Ille [Maximus] igitur statim a Francis, a Saxorum gente, in Sicilia, Sisciae, Petavione: ubique denique terrarum victus est.

He [Maximus] was soon conquered by the Franks, by the Saxon nation, in Sicily, at Siscia, at Pettau; in fact, everywhere.

Amm. Marc. 17.12,1 (4th century AD)

Augusto inter haec quiescenti per hiemem apud Sirmium, indicabant nuntii graves et crebri, permixtos Sarmatas et Quados, vicinitate et similitudine morum armaturaeque concordes, Pannonias Moesiarumque alteram cuneis incursare dispersis.

As Augustus [Constantius] meanwhile was taking his winter rest at Sirmium, frequent serious reports showed that the Sarmatians and the Quadi, who were in agreement because they were neighbours and had like customs and armour, had united and were raiding the Pannonias and Second Moesia in detached bands.

Amm. Marc. 26.4,5 (4th century AD)

Hoc tempore velut per universum orbem Romanum, bellicum canentibus bucinis, excitae gentes saevissimae, limites sibi proximos persultabant. Gallias Raetiasque simul Alamanni populabantur; Sarmatae Pannonias et Quadi; Picti, Saxonesque et Scotti, et Attacotti Brittannos aerumnis vexavere continuis; Austoriani Mauricaeque aliae gentes, Africam solito acrius incursabant; Thracias et Pannonias diripiebant praedatorii globi Gothorum.

At this time, as if the war trumpets were sounding throughout the whole Roman world, the most savage peoples roused themselves and poured across the nearest frontiers. At the same time the Alamanni were devastating Gaul and Raetia, the Sarmatae and Quadi Pannonia, while the Picts, Saxons, Scots and Attacotti were harassing the Britons with constant disasters. The Austoriani and other Moorish tribes raided Africa more fiercely than ever and predatory bands of Goths were plundering Thrace and Pannonia.

Amm. Marc. 26.8,2 (4th century AD)

Est autem sabaia ex ordeo vel frumento, in liquorem conversis, paupertinus in Illyrico potus.

Now Sabai is a drink of the poorer people in Illyricum, a liquor made from barley or some other grain.

Amm. Marc. 29.6,6 (4th century AD)

Cuius rei tam atrocis disseminatus rumor ilico per diversa et Quados et gentes circumsitas efferavit, regisque flentes interitum, in unum coactas misere vastatorias manus, quae Danubium transgressae, cum nihil exspectaretur hostile, occupatam circa messem agrestem, adortae sunt plebem, maioreque parte truncata, quicquid superfuit, domum cum multitudine varii pecoris abduxerunt.

The report of so atrocious a deed at once spread abroad on all sides and roused the Quadi and the tribes around them to madness. Weeping for the death of the king, they mustered and sent out devastating bands, which crossed the Danube while no hostility was anticipated, and fell upon the country people, who were busy with their harvest; most of them they killed, the survivors they led home as prisoners, along with a quantity of all kinds of domestic animals.

Amm. Marc. 30.7,2 (4th century AD)

Natus apud Cibalas, Pannoniae oppidum, Gratianus maior ignobili stirpe, cognominatus est a pueritia prima Funarius, ea re quod nondum adultus, venalem circumferens funem, quinque militibus eum rapere studio magno conatis, nequaquam cessit (...)

His [Valentinian's] father, the elder Gratianus, was born at Cibalae, a city of Pannonia, of a humble family, and from his early boyhood was surnamed Funarius, because when he was not yet grown up and was carrying around a rope

94 There is very useful book in Croatian that should be taken into account, which deals in great detail with southern Pannonia in Late Antiquity, with large number of references to the sources and the bibliography on the topic: Gračanin 2011.

for sale, and five soldiers tried with all their might to tear it from him, he gave way not an inch (...)

Aur. Vict. 33.1-2 (4th century AD)

Sub idem tempus Licinius Gallienus cum a Gallia Germanos strenue arceret, in Illyricum properans descendit. Ibi Ingebum, quem curantem Pannonios comperta Valeriani clade imperandi cupido incesserat, Mursiae devicit moxque Regalianum, qui receptis militibus, quos Mursina labes reliquos fecerat, bellum duplicaverat.

At the same time, although he was strenuously attempting to drive the Germans out of Gaul, Licinius Gallienus hurriedly descended on Illyricum. There at Mursa he defeated Ingebus, the governor of Pannonia, who had conceived a desire to be emperor after learning of Valerian's disaster, and subsequently Regalianus, who had renewed the war after rallying the soldiers who had survived the disaster at Mursa.

Aur. Vict. 39.26 (4th century AD)

His sane omnibus Illyricum patria fuit: qui, quamquam humanitatis parum, ruris tamen ac militiae miseriis imbuti satis optimi reipublicae fuere.

Illyricum was actually the native land of all of them: so although they were deficient in culture, they had nevertheless been sufficiently schooled by the hardships of the countryside and of military service to be the best men for the state.

Auson. *Cent. nupt.* (4th century AD)

Sed quid facerem? Iussum erat: quodque est potentissimum imperandi genus, rogabat, qui iubere poterat, sanctus imperator Valentinianus, vir meo iudicio eruditus. Nuptias quondam eiusmodi ludo descripserat, aptis equidem versibus et compositione festiva. Experiri deinde volens, quantum nostra contentione praecelleret, simile nos de eodem concinnare praecepit.

But what was I to do? It was written by command, and at the request (which is the most pressing kind of order!) of one who was able to command – the sainted Emperor Valentinian, a man, in my opinion, of deep learning. He had once described a wedding in a *jeu d'esprit* of this kind, wherein the verses were to the point and their connections amusing. Then, wishing to show by means of a competition with me the great superiority of his production, he bade me compile a similar poem on the same subject.

Claud. 2 190-201 (4-5th century AD)

Hinc Poeni cumulant laudes, quod rura tyranno
libera possideant; hinc obsidione solutus
Pannonius potorque Savi, quod clausa tot annis
oppida laxatis ausus iam pandere portis
rursum cote novat nigras rubigine falces
exesosque situ cogit splendere ligones
agnoscitque casas et collibus oscula notis
figit et impresso glaebis non credit aratro,
exsectis, inculta dabant quas saecula, silvis
restituit terras et opacum vitibus Histrum
conserit et patrium vectigal solvere gaudet,
inmunis qui clade fuit.

Here Carthaginians crown thy praise, because they possess their lands delivered from the tyrant's rule; there the Pannonian, freed from the blockade, and he who drinks the Save, grateful because he now dare throw open the gates of cities closed for so many years. Such sharpen once more upon the whetstone their sickles dark with rust and cause their mattocks, foul with want of use, to shine as of old. Each sees again his well-remembered cottage, kisses his native hills, and can scarce believe real the furrows cut by his heavy plough. He hews down the forests and renders again fit for cultivation fields which generations had let run wild. Once more he covers the banks of the Danube with vineyards and rejoices to pay the taxes his forefathers paid, for it was bloodshed that brought immunity.

Dio 49.36,2-6 (3rd century AD)

Οἱ δὲ δὴ Παννόνιοι νέμονται μὲν πρὸς τῇ Δελματίᾳ, παρ' αὐτὸν τὸν Ἴστρον, ἀπὸ Νωρικοῦ μέχρι τῆς Μυσίας τῆς ἐν τῇ Εὐρώπῃ, κακοβιώτατοι δὲ ἀνθρώπων ὄντες (οὔτε γὰρ γῆς οὔτε ἀέρων εὖ ἥκουσιν· οὐκ ἔλαιον, οὐκ οἶνον, πλὴν ἐλαχίστου καὶ τούτου κακίστου, γεωργοῦσιν, ἅτε ἐν χειμῶνι πικροτάτῳ τὸ πλεῖστον διαιτώμενοι, ἀλλὰ τάς τε κριθὰς καὶ τοὺς κέγχρους καὶ ἐσθίουσιν ὁμοίως καὶ πίνουσιν) ἀνδρειότατοι δ' οὖν διὰ πάντων ὧν ἴσμεν νομίζονται· θυμικώτατοι γὰρ καὶ φονικώτατοι, οἷα μηδὲν ἄξιον τοῦ καλῶς ζῆν ἔχοντες, εἰσί. ταῦτα δὲ οὐκ ἀκούσας οὐδ' ἀναγνοὺς μόνον, ἀλλὰ καὶ ἔργῳ μαθὼν ὥστε καὶ ἄρξας αὐτῶν, οἶδα· μετὰ γάρ τοι τὴν ἐν τῇ Ἀφρικῇ ἡγεμονίαν τῇ τε Δελματίᾳ, ἧς ποτε καὶ ὁ πατήρ μου χρόνον τινὰ ἦρξε, καὶ τῇ Παννονίᾳ τῇ ἄνω καλουμένῃ προσετάχθην, ὅθεν ἀκριβῶς πάντα τὰ κατ' αὐτοὺς εἰδὼς γράφω. Ὀνομάζονται δὲ οὕτως ὅτι τοὺς χιτῶνας τοὺς χειριδωτοὺς ἐξ ἱματίων τινῶν ἐς πάννους ἐπιχωρίως πως καὶ κατατέμνοντες καὶ προσαγορεύοντες συρράπτουσι. Καὶ οἱ μὲν εἴτ' οὖν διὰ τοῦτο εἴτε καὶ δι' ἄλλο τι οὕτως ὠνομάδαται· τῶν δὲ δὴ Ἑλλήνων τινὲς τἀληθὲς ἀγνοήσαντες Παίονάς σφας προσεῖπον, ἀρχαίου μέν που τοῦ προσρήματος τούτου ὄντος, οὐ μέντοι καὶ ἐκεῖ, ἀλλ' ἔν τε τῇ Ῥοδόπῃ καὶ πρὸς αὐτῇ τῇ Μακεδονίᾳ τῇ νῦν μέχρι τῆς θαλάσσης. ὑφ' οὗπερ καὶ ἐγὼ ἐκείνους μὲν Παίονας τούτους δὲ Παννονίους, ὥσπερ που καὶ αὐτοὶ ἑαυτοὺς καὶ Ῥωμαῖοί σφας καλοῦσι, προσαγορεύσω.

The Pannonians dwell near Dalmatia along the very bank of the Ister from Noricum to Moesia and lead the most miserable existence of all mankind. For they are not well off as regards either soil or climate; they cultivate no olives and produce no wine except to a very slight extent and a wretched quality at that, since the winter is very rigorous and occupies the greater part of their year, but drink as well as eat both barley and millet. For all that they are considered the bravest of all men of whom we have knowledge; for they are very high-spirited and bloodthirsty, as men who possess nothing that makes an honourable life worth while. This I know not from hearsay or reading only, but I have learned it from actual experience as once their governor, for after my command in Africa and in Dalmatia (the latter position my father also held for a time) I was appointed to what is known as Upper Pannonia, and hence it is with exact knowledge of all conditions among them that I write. Their name is derived from the fact that their sleeved tunics are made by stitching together pieces of old clothes which they cut up into strips in a way peculiar to themselves and call panni. This is their name, whether the reason be what I have stated or some other; but certain of the Greeks in ignorance of the truth have called them Paeones, an appellation which, though no doubt old, does not, however, apply to that country, but rather to Rhodope, close to the present Macedonia, as far as the sea. Therefore I also shall call the people of the latter district Paeones, but the others Pannonians, just as both they themselves and the Romans do.

Dio 51.27,3 (3rd century AD)

(…) καὶ μετὰ ταῦτ᾽ ἐς τὸ τῆς Μυσίας ὄνομα πάνθ᾽ ὅσα ὁ Σάουος ἐς τὸν Ἴστρον ἐμβάλλων, ὑπέρ τε τῆς Δελματίας καὶ ὑπὲρ τῆς Μακεδονίας τῆς τε Θρᾴκης, ἀπὸ τῆς Παννονίας ἀφορίζει, συγκεχώρηκεν.

(…) there have been included under the name of Moesia all the tribes living above Dalmatia, Macedonia, and Thrace, and separated from Pannonia by the Savus, a tributary of the Ister.

Enn. *Paneg. Theod.* VII (28) (5/6th century AD)

Vlca fluvius est tutela Gepidarum, quae vice aggerum munit audaces et in iugorum more latus provinciae quibusdam muris amplectitur nullo ariete frustrandis.

The Ulca River is the defence of the Gepids, which surrounds the brave like a moat, and on the slopes of the hills encircles the flanks of the province; it cannot be broken by siege engines of any kind.

Epit. de Caes. 37.3 (4th century AD)

Vineas Gallos et Pannonios habere permisit. Opere militari Almam montem apud Sirmium et Aureum apud Moesiam superiorem vineis conseruit.

He allowed the Gauls and Pannonians to have vineyards. By the labor of the soldiery, he planted vineyards on Mount Alma near Sirmium and on Mount Areum in Moesia Superior.

Epit. de Caes. 41.5 (4th century AD)

Verum enimvero ut imperia difficile concordiam custodiunt, discidium inter Licinium Constantinumque exoritur; primumque apud Cibalas iuxta paludem Hiulcam nomine Constantino nocte castra Licinii irrumpente Licinius fugam petiit Byzantiumque fuga volucri pervenit.

But, indeed, as imperia preserve concord with difficulty, a rift arose between Licinius and Constantine; and first, near Cibalae, beside a lake named Hiulca, when Constantine burst into Licinus' camps by night, Licinius sought escape and, by a swift flight, reached Byzantium.

Epit. de Caes. 45.1 (4th century AD)

Valentinianus imperavit annos duodecim minus diebus centum. Huius pater Gratianus, mediocri stirpe ortus apud Cibalas, Funarius appellatus est, eo quod venalicium funem portanti quinque milites nequirent extorquere.

Valentinian ruled ten years less ten days. His father, Gratianus, sprung from modest stock near Cibalis, was called Funarius ["Trace-horse"] because five soldiers were unable to wrest a slave-market rope from him while he was carrying it.

Eutr. *Brev.* 9.8 (4th century AD)

Gallienus, cum adulescens factus esset Augustus, imperium primum feliciter, mox commode, ad ultimum perniciose gessit. Nam iuvenis in Gallia et Illyrico multa strenue fecit, occiso apud Mursam Ingenuo, qui purpuram sumpserat, et Trebelliano. (...) Pannonia a Sarmatis Quadisque populata est (...).

Gallienus, although he had been made Augustus as quite a young man, at first governed the empire successfully, afterwards appropriately, but at the end disastrously. For, as a young man he achieved a great deal with vigour in Gaul and Illyricum. At Mursa he killed Ingenuus, who had seized the purple, and Trebellianus. (…) Pannonia was ravaged by the Sarmatians and Quadi (…).

Eutr. *Brev.* 9.17 (4th century AD)

Vineas Gallos et Pannonios habere permisit, opere militari Almam montem apud Sirmium et Aureum apud Moesiam superiorem vineis conseruit et provincialibus colendos dedit.

He allowed the Gauls and Pannonians to have vineyards and with military labour he planted Mount Alma at Sirmium and Mount Aureus in Upper Moesia with vines and gave them to the provincials for cultivation.

Eutr. *Brev.* 10.5 (4th century AD)

Ac primo [Constantinus] eum [Licinium] in Pannonia

secunda ingenti apparatu bellum apud Cibalas instruentem repentinus oppressit omnique Dardania, Moesia, Macedonia potitus numerosas provincias occupavit.

First he [Constantinus] overwhelmed him suddenly in Pannonia Secunda, while Licinius was making vast preparations for war at Cibalae, gained possesion of the whole of Dardania, Moesia and Macedonia and occupied numerous other provinces.

Eutr. *Brev.* 10.12 (4th century AD)

Non multo post Magnentius apud Mursam profligatus acie est ac paene captus. Ingentes Romani imperii vires ea dimicatione consumptae sunt, ad quaelibet bella externa idoneae, quae multum triumphorum possent securitatisque conferre.

Not long afterwards Magnentius was overtwhelmed in a battle at Mursa, and almost captured. Vast forces of the Roman empire were destroyed in that conflict, forces which were sufficient for any foreign wars and which might have provided many a triumph and much security.

Exp. tot. mun. 57 (4th century AD)

Deinde Pannonia regio, terra dives in omnibus, fructibus quoque et jumentis et negotiis, ex parte et mancipiis. Et semper habitatio imperatorum est. Habet autem civitates maximas, Sirmium quoque et Noricus, unde et vestis norica exire dicitur. Haec Pannonia regio. Quae adjacet trans flumen Danubium, gens barbarorum Sarmatum est.

Then there is Pannonia, a land rich in all things – crops, cattle, slaves and a little trade. The dwelling place of the emperors are always here. Its greatest cities are Syrmium and Noricum; the Noric overcoat is reported to come ftom the latter. This is the area of Pannonia; across the Danube from it lies the Sarmatian tribe of barbarians.

Fest. *Brev.* 8 (4th century AD)

Prouincias habet Illyricus XVIII: Noricorum duas, Pannoniarum duas, Valeriam, Sauiam, Dalmatiam, Moesiam <superiorem, Dardaniam>, Daciarum duas; et in dioecesi Macedonica prouinciae sunt septem: Macedonia, Thessalia, Achaia, Epiri duae, Praeualis, Creta.

Illyricus has 18 provinces: two of Noricas, two of Pannonias, Valeria, Savia, Dalmatia, Moesia Superior, Dardania, two of Dacias, and in the Macedonic diocese are seven provinces: Macedonia, Thessaly, Achaea, two Epiruses, Praevalis, and Crete.

Hdn. *Hist.* II,9,11 (2nd-3rd centuries AD)

ὥσπερ δὲ τὰ σώματα οἱ ἐκεῖσε ἄνθρωποι γενναιότατοί τε καὶ μεγάλοι εἰσὶ καὶ πρὸς μάχας ἐπιτήδειοι καὶ φονικώτατοι, οὕτω καὶ τὰς διανοίας παχεῖς καὶ μὴ ῥᾳδίως συνεῖναι δυνάμενοι, εἴ τι μετὰ πανουργίας ἢ δόλου λέγοιτο ἢ πράττοιτο. Πιστεύσαντες γοῦν τῷ Σεβήρῳ προσποιουμένῳ χαλεπαίνειν καὶ θέλειν ἐπεξελθεῖν τῷ Περτίνακος φόνῳ ἐπέδοσαν αὑτούς, ὡς αὐτοκράτορά τε ἀποδεῖξαι καὶ τὴν ἀρχὴν ἐγχειρίσαι.

The inhabitants of the district of Pannonia are tall men of fine physique, natural and fierce fighters, but intellectually dull and slow-witted when it comes to crafty words of subtle actions. Accepting Severus' protests of indignation and desire for the revenge of Pertinax's murder, they put themselves in his charge by declaring him emperor and entrusting him with the supreme power.

Hdn. *Hist.* VII,8,11 (2nd-3rd centuries AD)

Αἰφνιδίου γὰρ γενομένης τῆς ἐπ' Ἰταλίαν ὁδοῦ, οὐκ ἐκ προνοίας, ὥσπερ εἰώθει, ἀλλ' ἐξ αὐτοσχεδίου καὶ ἐπειγούσης ὑπηρεσίας τὰ χρειώδη τῷ στρατῷ ἠθροίζετο ἔδοξεν οὖν αὐτῷ προπέμψαι τὰς τῶν Παιόνων φάλαγγας· ἐκείνοις γὰρ μάλιστα ἐπίστευεν, οἵπερ αὐτὸν καὶ πρῶτοι ἀνεῖπον βασιλέα ἐθελονταί τε ὑπὲρ αὐτοῦ κινδυνεύσειν ὑπισχνοῦντο. Ἐκέλευσεν οὖν αὐτοῖς φθάσαι τὴν λοιπὴν δύναμιν καὶ τὰ ἐν Ἰταλίᾳ χωρία προκαταλαβεῖν.

The march was somewhat slow because of all the baggage carts and supplies choking all the roads. The reason for this was that his [Maximinus'] expedition into Italy was a sudden decision without the usual advance planning; supplies for the army were being collected en route as aid was being rushed to him. So Maximinus decided to send ahead the Pannonian legions, in whom he had special confidence (they were the ones that first recognized him as emperor) and who had volunteered to take on a dangerous mission for him.

Hier. *Chron.* (329F) 2390 XI. ad AD 375 (4th-5th centuries AD)

Quia superiore anno Sarmatae Pannonias uastauerant, idem consules permansere.

Because in the previous year the Sarmatians had devastated the Pannonian provinces, the same Consuls remained in office.

Isid. *Etym.* 14.4,16-17 (6th-7th centuries AD)

Pannonia ab Alpibus Appenninis est nuncupata, quibus ab Italia secernitur, regio viro fortis et solo laeta, duobus satis acribus fluviis, Dravo Savoque, vallata. Coniungitur autem cum Norico et Raetia; habentes ab oriente Moesiam, ab Euro Istriam; ab Africo vero Alpes Appenninos habent, ab occasu Galliam Belgicam, a septentrione Danubii fontem, vel limitem qui Germaniam Galliamque secernit. Istriam Ister amnis vocavit, qui eius terram influit. Ipse est Danubius. Habet autem Istria a septentrione Pannoniam.

Pannonia is named after the Apennine (Appenninus) Alps, by which it is separated from Italy. It is a region strong in men that enjoys good soil and it is surrounded by two rivers that are very rough, the Drava (Dravus) and the Sava (Savus). It is adjacent to Noricum and Raetia; east of it lies Moesia, southeast Istria, southwest the Apennine Alps, in the West Belgian Gaul, in the north the source of the Danube and the border that divides Germania and Gaul. The River Ister, which flows through this land, gives Istria its name. It is also known as the Danube. Pannonia lies to the north of Istria.

Itin. Ant. 131,1-4 (3rd century AD)

Mursa civitas mpm XXVI; Cibalas civitas mpm XXIII; Ulmos vicus mpm XXII; Sirmi civitas mpm XXVI

Itin. Ant. 232,4-6 (3rd century AD)

Ulmos mpm XXVI; Cibalis mpm XXIII; Mursa mpm XXII

Itin. Ant. 243,2-5 (3rd century AD)

Cucci mpm XVI; Cornaco mpm XVI; Teotiburgio mpm XVI; Mursa mpm XVI

Itin. Ant. 260,1-261,3 (3rd century AD)

Quadrata mpm XXVIII; Siscia mpm XXVIIII; Varianis mpm XXIII; Menneianis mpm XXVI; Incero mpm XXVIII; (sed mansio Augusti in praetorio est); Picentino mpm XXV; Leucono mpm XXVI; Cirtisa XII; Cibalis mpm XXII; Ulmos mpm XXII; Sirmi mpm XXVI

Itin. Ant. 265,5-10 (3rd century AD)

Item a Siscia Mursa mpm CXXXIIII sic; Varianis mpm XXIIII; Aquis Balissis mpm XXXI; Incero mpm XXV; Stravianis mpm XXIIII; Mursa mpm XXX

Itin. Ant. 265,11-266,3 (3rd century AD)

A Poetovione Siscia mpm C sic; Aqua viva XVIIII; Pyrri mpm XXX; Dautonia mpm XXIIII; Siscia mpm XXVII

Itin. Ant. 267,1-3 (3rd century AD)

Ulmo mpm XXVI; Cibalis mpm XXIIII; Mursa mpm XXII

Itin. Ant. 268,3-5 (3rd century AD)

Ulmo mpm X; Cibalis mpm XXII; Cirtisia mpm XXIV

Itin. Ant. 274,4-7 (3rd century AD)

Romula mpm X; Quadrata mpm XIV; Ad fines XIV; Siscia XXI

Itin. Burd. 561,7-563,7 (4th century AD)

Mutatio Ramista mil VIIII; mansio Aqua viva mil VIIII; Mutatio Popolis mil X; Civitas Iovia mil VIIII; Mutatio Sunista mil VIIII; (562,1) Mutatio Peritur mil XII; Mansio Lentolis mil XII; Mutatio Cardono mil X; Mutatio Cocconis mil XII; Mansio Serota mil X; Mutatio Bolenta mil X; Mansio Maurianis mil VIIII; Intras Pannoniam superiorem; Mutatio Serena mil VIII; Mansio Vereis mil X; Mutatio Iovalia mil VIII, Mutatio Mersella mil VIII; Civitas Mursa mil X; (563,1) Mutatio Leutuoano mil XII; Civitas Cibalis mil XII; Mutatio Celena mil XI; Mansio Ulmo mil XI; Mutatio Spaneta mil X; Mutatio Vedulia mil VIII; Civitas Sirmium mil VIII

Jord. *Get.* 140 (6th century AD)

Sed Theodosio Principe paene tunc usque ad desperationem aegrotanti datur iterum Gothis audacia divisoque exercitu Fritigernus ad Thessaliam praedandam, Epiros et Achaiam digressus est, Alatheus vero et Safrac cum residuis copiis Pannoniam petierunt.

But as the Emperor Theodosius fell so sick at this time that his life was almost despaired of, the Goths were again inspired with courage. Dividing the Gothic army, Fritigern set out to plunder Thessaly, Epirus and Achaia, while Alatheus and Safrac with the rest of the troops made for Pannonia.

Jord. *Get.* 147 (6th century AD)

Mox ergo antefatus Alaricus creatus est rex, cum suis deliberans suasit eos suo labore quærere regna quam alienis per otium subiacere, et, sumpto exercitu per Pannonias – Stilicone et Aureliano consulibus – et per Sirmium dextroque latere quasi viris vacuam intravit Italiam nulloque penitus obsistente ad pontem applicavit Candidiani, qui tertio miliario ab urbe aberat regia Ravennate.

Now when this Alaric was made king, he took counsel with his men and persuaded them to seek a kingdom by their own exertions rather than serve others in idleness. In the consulship of Stilicho and Aurelian he raised an army and entered Italy, which seemed to be bare of defenders, and came through Pannonia and Sirmium along the right side. Without meeting any resistance, he reached the bridge of the river Candidianus at the third milestone from the royal city of Ravenna.

Jord. *Get.* 161 (6th century AD)

Nam Vandali vel Alani, quos superius diximus permissu principum Romanorum utramque Pannoniam resedere, nec ibi sibi metu Gothorum arbitrantes tutum fore, si reverterentur, ad Gallias transierunt.

Now the Vandals and the Alani, as we have said before, had been dwelling in both Pannonias by permission of the Roman Emperors. Yet fearing they would not be safe even here if the Goths should return, they crossed over into Gaul.

Jord. *Get.* 166 (6th century AD)

Nam duodecimo anno regni Valiae, quando et Hunni post pene quinquaginta annorum invasa Pannoniam a Romanis et Gothis expulsi sunt, (…)

Now in the twelfth year of Valia's reign the Huns were driven out of Pannonia by the Romans and Goths, almost fifty years after they had taken possession of it.

Jord. *Get.* 264 (6th century AD)

Gothi vero cernentes Gepidas Hunnorum sedes sibi defendere Hunnorumque populum suis antiquis sedibus occupare, maluerunt a Romano regno terras petere quam cum discrimine suo invadere alienas, accipientesque Pannoniam, quae in longo porrecta planitiae habet ab oriente Moesiam superiorem, a meridie Dalmatiam, ab occasu Noricum, a septentrione Danubium, ornata patria civitatibus plurimus, quarum prima Syrmis, extrema Vindomina.

Now when the Goths saw the Gepidae defending for themselves the territory of the Huns, and the people of the Huns dwelling again in their ancient abodes, they preferred to ask for lands from the Roman Empire, rather than invade the lands of others with danger to themselves. So they received Pannonia, which stretches in a long plain, being bounded on the East by Upper Moesia, on the south by Dalmatia, on the west by Noricum and on the north by the Danube. This land is adorned with many cities, the first of which is Sirmium and the last Vindobona.

Jord. *Get.* 273 (6th century AD)

Quod conperto Gothi ibi, ubi erant, expeditionemque solventes, quam contra Sadagis collegerant, in Hunnos convertunt et sic eos suis a finibus ingloriosos pepulerunt, ut iam ex illo tempore qui remanserunt Hunni et usque actenus Gothorum arma formident. Quiescente vero tandem Hunnorum gente a Gothis Hunumundus Suavorum dux dum ad depraedandas Dalmatias transit, armenta Gothorum in campis errantia depraedavit, quia Dalmatia Suaviae vicina erat nec a Pannonios fines multum distabat, praesertim ubi tunc Gothi residebant.

Then the Goths at once abandoned the expedition they had planned against the Sadagis, turned upon the Huns and drove them so ingloriously from their own land that those who remained have been in dread of the arms of the Goths from that time even down to the present day. When the tribe of the Huns was at last subdued by the Goths, Hunimund, chief of the Suavi, who was crossing over to plunder Dalmatia, carried off some cattle of the Goths which were straying over the plains; for Dalmatia was near Suavia and not far distant from the territory of Pannonia, especially that part where the Goths were then staying.

Jord. *Rom.* 218 (6th century AD)

Illyricus autem cuncta per partes quidem et membra devicta ad unum tamen corpus apta est, quae habet intra se provincias XVIII et sunt Norici duo, Pannonias duas, Valeria, Suavia, Dalmatia, Moesia superior, Dardania, Dacias duas, Macedonia, Thessalia, Achaia, Epiros duos, Praevales, Creta, simul XVIII.

But the whole of Illyria, conquered indeed by parts and piecemeal, was nonetheless combined into a single body which has within itself 18 provinces: and there are two Noricums, two Pannonias, Valeria, Swevia, Dalmatia, Moesia superior, Dardania, two Dacias, Macedonia, Thessalia, Achaja, two Epiruses, Praevales, Creta: altogether 18.

Jord. *Rom.* 287 (6th century AD)

(...) Graeciam Gothi vastaverunt, Quadi et Sarmatae Pannonias invaserunt, (...)

(…) the Goths devastated Greece, the Quadi and Sarmatæ invaded the Pannonias (…)

Jord. *Rom.* 307 (6th century AD)

Valentinianus et Valens regnaverunt annis XIII mensibus V. Nam Pannones erant, Cibalenses, utrimque germani.

Valentinian and Valens reigned for 13 years and 5 months. For they were Pannonians, born at Cibalae, brothers on both sides.

Mam. *Paneg. Max.* X (II), 2.2 (4th century AD)

Vnde igitur ordiar? Commemorabo nimirum patriae tuae in rem publicam merita? Quis enim dubitat quin multis iam saeculis, ex quo uires illius ad Romanum nomen accesserint, Italia quidem sit gentium domina gloriae uetustate, sed Pannonia uirtute?

Where, then, shall I begin? Shall I recall, indeed, the services of your native land to the State? For who doubts that for many centuries now, ever since its strenght was added to the Roman name, while Italy indeed may have been the mistress of nations by virtue of the antiquity of her glory, Pannonia has been in valor?

Not. Dign. Occ. (5th century AD?)

I,41 (Dux) Pannoniae secundae

I,51 (Consular) Pannoniae

I,82-83 (Corrector) per Pannoniam unum: Saviae

II,29 (Praefectum) Pannoniae secundae

II,30 (Praefectum) Saviae

V,6 (Insignia magistri peditum) Pannoniciani

V,136 (Dux limitum) Pannoniae secundae

V,149 (Legio palatina) Pannoniciani seniores

XI,10 Rationalis summarum Pannoniae secundae, Dalmatiae et Saviae

XI,24 Praepositus thesaurorum Siscianorum, Saviae

XI,39 Procurator monetae Siscianae

XII,21 Procurator rei priuatae per Saviam

XXXII,1 Dux Pannoniae

XXXII,21-22 Sub dispositione uiri spectabili ducis prouinciae secundae ripariensis et (siue) Saviae: Cuneus equitum scutariorum, Cornacii

XXXII,23 Cuneus equitum Dalmatarum, Teutiborgio

XXXII,25 Cuneus equitum promotorum, Cuccis

XXXII,28 Equites Dalmatae, Novas

XXXII,29 Equites Dalmatae, Albano

XXXII,30 Equites promoti, Teutibarcio

XXXII,31 Equites Dalmatae, Cornaco

XXXII,32 Equites sagittarii, Cuccis

XXXII,43 Auxilia ascarii, Tauruno siue Marsonia

XXXII,45 Praefectus legionis sextae Herculeae cohortis quintae partis superioris, Aureo monte

XXXII,47 Praefectus legionis sextae Herculeae, Teutiborgio

XXXII,52 Praefectus classis Histricae, Mursae

XXXII,55 Praefectus classis Aegetensium siue secundae Pannonicae,... nunc Sisciae

XXXII,57 Tribunus cohortis tertiae Alpinorum, Sisciae

XXXIII,45 Equites Flavianenses, Ad Militare

Oros. *Hist.* 7.28,18-19 (4th-5th centuries AD)

Licinius, repentina rabie suscitatus, omnes Christianos e palatio suo iussit expelli. Mox bellum inter ipsum Licinium et Constantinum efferbuit. Sed Constantinus Licinium, sororis suae uirum, in Pannonia primum uicit, deinde apud Cibalas oppressit uniuersaque Graecia potitus Licinium crebris bellis terra marique adsurgentem et repressum tandem ad deditionem coegit.

Licinius, overtaken by a sudden madness, ordered that all Christians be expelled from his court, and soon war broke out between him and Constantine. Constantine first defeated Licinius, who was his sister's husband, in Pannonia, then vanquished him at Cibalae and, after gaining control of all of Greece, finally forced Licinius, who had launched many unsuccessful attacks on him by land and sea, to surrender.

Oros. *Hist.* 7.29,12-13 (4th-5th centuries AD)

Sequitur bellum illud horribile inter Constantium Magnentiumque apud Mursam urbem gestum, in quo multa Romanorum uirium profligatio etiam in posterum nocuit. Magnentius tamen uictus aufugit ac non multo post apud Lugdunum propria se manu interfecit.

There then followed a terrible battle between Constantius and Magnentius which was fought by the city of Mursa. The squandering of Roman might there weakened even future generations. Magnentius, though defeated, escaped; however, he killed himself soon afterwards at Lyons.

Pacat. 32.3-4 (4th century AD)

Postremo populis barbarorum ultroneam tibi operam ferre uouentibus commilitum munus indulges, ut et limiti manus suspecta decederet et militi auxiliator accederet; qua tua benignitate pellectae omnes Scythicae nations tantis examinibus confluebant, ut quem remiseras tuis, barbaris uidereris imperasse dilectum. O res digna memoratu! Ibat sub ducibus uexillisque Romanis hostis aliquando Romanus, et signa contra quae steterat sequebatur, urbesque Pannoniae quas inimica dudum populatione uacuauerat miles implebat. Gothus ille et Chunus et Halanus respondebat ad nomen et alternabat excubias et notari infrequens uerebatur.

Finally you granted the privileged status of fellow soldiers to the barbarian peoples who promised to give you voluntary service, both to remove the frontier troops of dubious loyality, and to add reinforements to your army. Attracted by your kindness, all the Scythian nations flocked to you in such great numbers that you seemed to have imposed a levy upon barbarians from which you exempted your subjects. O event worthy of memory! There marched under Roman leaders and banners the onetime enemies of Rome, and they followed standards which they had once opposed, and filled with soldiers the cities of Pannonia which they had not long ago emptied by hostile plundering. The Goth, the Hun and the Alan responded to their names, and stood watch in their turn, and were afraid of being marked down as absent without leave.

Pacat. 34,1 (4th century AD)

Testis es Siscia, testis pulcherrimi, Saue, conflictus (...)

You are witness, Siscia, and you are witness, Save, of that finest of conflicts (...)

Paul *HL* 1.22 (8th century AD)

Waltari ergo cum per septem annos regnum tenuisset, ab hac luce subtractus est. Post quem nonus Audoin regnum adeptus est. Qui non multo post tempore Langobardos in Pannoniam adduxit.

Waltari, therefore, when he had held the sovereignty for seven years, departed from this life, and after him Audoin was the ninth who attained the kingly power, and he, not long afterwards, led the Langobards into Pannonia.

Paul *HL* 2.1 (8th century AD)

Omnique tempore quo Langobardi Pannoniam possederunt, Romanae rei publicae adversus aemulos adiutores fuerunt.

During all the time the Langobards held Pannonia, they were the allies of the Roman state against its rivals.

Paul *HL* 2.7 (8th century AD)

Tunc Alboin sedes proprias, hoc est Pannoniam, amicis suis Hunnis contribuit, eo scilicet ordine, ut, si quo tempore Langobardis necesse esset reverti, sua rursus arva repeterent. Igitur Langobardi, relicta Pannonia, cum uxoribus et natis omnique supellectili Italiam properant possessuri. Habitaverunt autem in Pannonia annis quadraginta duobus.

Then Alboin bestowed his own abode, that is, Pannonia, to his friends the Huns on this condition: that if at any time it should be necessary for the Langobards to return they should take back their own fields. Then the Langobards, having left Pannonia, hastened to take possession of Italy with their wives and children and all their goods. They dwelt in Pannonia forty-two years.

Paul *HL* 2.8 (8th century AD)

Ferunt, in hoc monte bisontes feras enutriri. Nec mirum, cum usque huc Pannonia pertingat, quae horum animantium ferax est. Denique retulit mihi quidam veracissimus senex, tale se corium in hoc monte occisi bisontis vidisse, in quo quindecim, ut aiebat, homines, unus iuxta alium potuissent cubare.

They say wild oxen graze upon it, and no wonder, since at this point it touches Pannonia, which is productive of these animals. In fine, a certain very truthful old man related to me that he had seen the hide of a wild ox killed on this mountain of such size that in it fifteen men, as he said, could lie one against the other.

Procop. *BG* 3.33,10-11 (6th century AD)

Λαγγοβάρδας δὲ βασιλεὺς Ἰουστινιανὸς ἐδωρήσατο Νωρικῷ τε πόλει καὶ τοῖς ἐπὶ Παννονίας ὀχυρώμασί τε καὶ ἄλλοις χωρίοις πολλοῖς καὶ χρήμασι μεγάλοις ἄγαν. Διά τοι τοῦτο ἐξ ἠθῶν τῶν πατρίων Λαγγοβάρδαι ἐξαναστάντες ἐνθένδε ποταμοῦ Ἴστρου ἱδρύσαντο, Γηπαίδων οὐ πολλῷ ἄποθεν. Καὶ αὐτοὶ μὲν Δαλμάτας τε καὶ Ἰλλυριοὺς μέχρι τῶν Ἐπιδάμνου ὁρίων ληϊσάμενοι ἠνδραπόδισαν, (...)

Now the Emperor Justinian had bestowed upon the Lombards the city of Noricum and the strongholds of Pannonia, as well as many other cities and a very great amount of money. It was because of this that the Lombards departed from their ancestral homes and settled on the south side of the Ister River, not far from the Gepaedes. They then, in their turn, plundered the population of Dalmatia and Illyricum as far as the boundaries of Epidamnus, taking captives (...)

Prudent. *Peristeph.* VII,1-10 (4th-5th centuries AD)

Insignem meriti virum
Quirinum, placitum Deo,
urbis moenia Sisciae
concessum sibi martyrem
complexu patrio fovent.

hic sub Galerio duce,
qui tunc Illyricos sinus
urgebat ditionibus,
fertur catholicam fidem
inlustrasse per exitum. (...)

The walls of Siscia's city in fatherly embrace
cherish Quirinus, a man of illustrious merit and
pleasing to God, a martyr granted to them.

He, under the rule of Galerius, whose dominion was then
heavy on Illyricum's winding shore, by his death
shed lustre, as they tell, on the Catholic faith. (...)

SHA Probus XVIII, 8 (4th century AD?)

Gallis omnibus et Hispanis ac Britannis hinc permisit, ut vites haberent vinumque conficerent. Ipse Almam montem in Illyrico circa Sirmium militari manu fossum lecta vite conseruit.

Hence he granted permission to all the Gauls and the Spaniards and Britons to cultivate vineyards and make wines, and he himself planted chosen vines on Mount Alma near Sirmium in Illyricum, after having had the ground dug up by the hands of the soldiers.

Solinus 21.2 (3rd century AD)

(...) Pannonia viro fortis et solo laeta (...)

(...) Pannonia strong in men and prosperous in land (...)

Sulp. Sev. *Chron.* II, 38 (4th-5th century AD)

Nam eo tempore, quo apud Mursam contra Magnentium armis certatum, Constantius descendere in conspectum pugnae non ausus in basilica martyrum extra oppidum sita, Valente tum eius loci episcopo in solatium assumpto, diversatus est.

For at that time, when a battle was fought at Mursa against Magnentius, Constantius had not the courage to go down to witness for himself the conflict, but took up his abode in a church of the martyrs which stood outside the town, Valens who was then the bishop of the place being with him to keep up his courage.

Zon. XIII,8 (12th century)

Ἕωθεν δὲ ὁ Κωνστάντιος ἐπί τινος ἀναβὰς λόφου, καὶ τὴν παρακειμένην πεδιάδα, ἀλλὰ μὴν καὶ τὸν παραρρέοντα ποταμὸν νεκρῶν ἰδὼν γέμοντα, εἰς φανερὰ κατήνεκτο δάκρυα, οὐ μᾶλλον διὰ τὴν νίκην ἡδόμενος ὅσον διὰ τὸν τῶν πεσόντων δακνόμενος ὄλεθρον. Λέγονται γὰρ ἐκ μὲν τῶν ἐκείνου περὶ τριάκοντα πεσεῖν χιλιάδας, ἁπασῶν ἀριθμουμένων εἰς ὀγδοήκοντα, ἐκ δὲ τῶν Μαγνεντίου τριάκοντα καὶ ἓξ οὐσῶν χιλιάδων διαφθαρῆναι τὰς εἴκοσι πρὸς ταῖς τέσσαρσιν. αὐτίκα τοίνυν τοὺς μὲν ἀνῃρημένους τῶν πεσόντων ταφῆς ἀξιωθῆναι πάντας ἐκέλευσε, μὴ διακρινομένων τῶν οἰκείων ἢ τῶν πολεμίων, τοὺς δ' ἔτι ἐμπνέοντας ἐπιμελείας τυχεῖν καὶ ἰατρικῆς θεραπείας.

At daybreak, Constantius went up on a certain mound and surveyed the nearby plain and the river which flowed past it, both filled with bodies. He was clearly moved to tears, not so much pleased by the victory as vexed because of the destruction of those who had fallen. Out of total of 80,000, 30,000 of his own men were said to have fallen; 24,000 of Magnentius' men out of 36,000 perished. He ordered that all the dead among the fallen be buried and that no distinction be made between his own men or those of the enemy. Those who were still breathing were given help and received medical care.

Zos. 2.18 (5th-6th centuries AD)

Ἀλλὰ Λικίννιος μὲν εἰς Κίβαλιν συνήθροιζε τὸν στρατόν· πόλις δὲ αὕτη Παιονίας ἐστίν, ἐπὶ λόφου κειμένη· στενὴ δὲ ὁδὸς ἡ ἐπὶ τὴν πόλιν ἀνάγει, ἧς τὸ πολὺ μέρος ἐπέχει λίμνη βαθεῖα, σταδίων πέντε τὸ εὖρος ἔχουσα, τὸ δὲ λειπόμενον ὄρος ἐστίν, ἐν ᾧ καὶ ὁ λόφος ἐφ' οὗπερ ἡ πόλις· ἐντεῦθεν πεδίον ἀναπεπταμένον ἐκδέχεται πολύ τι καὶ ἐς ἄποψιν ἄπειρον· ἐν τούτῳ Λικίννιος ἐστρατοπεδεύετο, τὴν φάλαγγα τὴν οἰκείαν εἰς μῆκος ἐκτείνων ὑπὸ τὸν λόφον, ὡς ἂν μὴ ἀσθενῆ δόξαιεν εἶναι τὰ κέρατα.

Licinius established his headquarters at Cibalis, a hillside city in Pannonia. The road leading to it is narrow, with a deep marsh (some five stades wide) lying alongside a great part of it and the rest mountainous, including the hill on which the city stands. Below it is a spacious plain, boundless to behold. Here Licinius encamped, spreading his battle line right up to the hill, to protect his wings.

Zos. 2.43 (5th-6th centuries AD)

Μαγνεντίου τοίνυν τὴν ἀρχὴν ἔχοντος καὶ τῶν ὑπὲρ τὰς Ἄλπεις ἐθνῶν καὶ τῆς Ἰταλίας αὐτῆς κυριεύοντος, τὴν στρατηγίαν τῶν ἐν Παιονίαις στρατοπέδων Βετρανίων ἔχειν ταχθείς, ὅτι Μαγνέντιος εἰς βασιλείαν ἤχθη πυθόμενος, εἰς τὴν αὐτὴν ἦλθεν ἐπιθυμίαν καὶ ψήφῳ τῶν αὐτόθι ταγμάτων βασιλεὺς ἀνεδείχθη, καὶ ἐν τῇ Μούρσῃ Παιονίας οὔσῃ πόλει διέτριβε·

Magnentius thus gained power and was master of the provinces beyond the Alps and even of Italy. Vetranio, Magister Militum in Pannonia, hearing that Magnentius had been promoted to the emperorship, was overcome with the same ambition. He was proclaimed emperor by vote of the legions with him and stayed at Mursa, a city in Pannonia.

Zos. 2.45 (5th-6th centuries AD)

Τῶν δὲ στρατοπέδων ἐν Παιονίαις συνδραμόντων καὶ περὶ Μοῦρσαν πόλιν ἀλλήλων ἐγγὺς γενομένων, ἐν τοῖς περὶ τὰ Ἄδρανα τέμπεσι καθίσας ὁ Μαγνέντιος λόχον ἐπιπέμπει τινὰ τοῖς Κωνσταντίου στρατηγοῖς ἄγγελον, τὴν πορείαν ἀναβαλλόμενον, ὅπως ἂν Σισκίαν οἱ ἐναντίοι καταλάβωσιν· ἐν ταύτῃ γὰρ αὐτόν, ὡς ἀναπεπταμένα πεδία ἐχούσῃ, μάχην συνάψειν.

The armies met in Pannonia and came close to each other near the city of Mursa. Magnentius set up an ambush in the valley near Adrans and sent a messenger to Constantius' generals to delay their progress, so that the enemy might reach Siscia; for he intended to join battle here where the country was open.

Zos. 2.48 (5th-6th centuries AD)

Τούτων ἀκηκοὼς τῶν λόγων Κωνστάντιος ἀπέστησε μὲν τῆς διώξεως τοὺς οἰκείους, ἐνεδίδου δὲ Μαγνεντίῳ τὴν ἑαυτοῦ στρατιὰν εἰς τὰ μεταξὺ Νωρικοῦ καὶ Παιονίας καὶ Μυσίας καὶ Δακίας πεδία μεταστῆσαι, τῶν μὲν δυσχωριῶν ἀπαλλαγῆναι βουλόμενος, ὡς δὲ τῇ ἵππῳ πλεονεκτῶν ἐν ἱππασίμοις αὐτὸν καταπολεμῆσαι χωρίοις· ἐκβάσης <δὲ> τῆς σκέψεως εἰς ἔργον αὐτῷ, τόπον ἐπιτήδειον εἰς τὸ πληρῶσαι τὸ σπουδαζόμενον ἐνόμισεν εἶναι τὴν Κίβαλιν, ἐν ᾗ καὶ Κωνσταντῖνος μάχην Λικιννίῳ συνάψας ἐνίκησε κατὰ κράτος. Θέσιν γὰρ ἐχούσης τῆς πόλεως οἵαν ἀφηγούμενος τὰ τότε πραχθέντα διεξῆλθον, μέρος μὲν τοῦ στρατεύματος εἶχεν ἐν ταύτῃ, χάρακα δὲ μεταξὺ τοῦ λόφου τοῦ τὴν πόλιν ἔχοντος καὶ τοῦ μέχρι τοῦ Σάου ποταμοῦ πεδίου κατασκευάσας, πᾶν μὲν ὅσον οὐ διέζωστο τῷ ποταμῷ τάφρῳ βαθείᾳ καὶ χάρακι πυκνῷ περιέβαλεν, ὅσον δὲ ὁ ποταμὸς περιεῖχε ναυσὶ συμπλακείσαις ἀλλήλαις ἐγεφύρωσε, διαλύων τὴν γέφυραν, ὁπότε δοκοίη, καὶ πάλιν ῥᾷστα συνάπτων.

When Constantius heard what they had said, he held his own men back and allowed Magnentius to move his army to the plains running through Noricum, Pannonia, Moesia and Dacia. He thought that Cibalis, where Constantine battled and conquered Licinius, was a suitable place for his purpose. Keeping part of his army in the city, he constructed a wall between the hill on which it stood and the plain down to the river Savus, and enclosed everything not encircled by the river by a deep ditch and a solid palisade. And the surrounding river was bridged by boats which he could break when he wanted to and easily join again.

Zos. 2.49-50 (5th-6th centuries AD)

(...) Μαγνέντιος δὲ προαγαγὼν τὸν στρατόν, Σισκίαν ἑλὼν ἐξ ἐφόδου κατέσκαψεν, ἐπελθὼν δὲ πάντα τὰ περὶ τὸν Σάον χωρία καὶ λείαν πλείστην ἀπαγαγὼν ἐπὶ τὸ Σίρμιον ἤλαυνεν, ἀμαχητὶ καὶ ταύτην αἱρήσειν οἰόμενος. Διαμαρτὼν δὲ τῆς ἐγχειρήσεως (ἀπεκρούσθη γὰρ ὑπό τε τοῦ πλήθους τῶν οἰκητόρων καὶ τῶν ἐπιτεταγμένων τὴν πόλιν φυλάττειν στρατιωτῶν) ἐπὶ τὴν Μοῦρσαν ἐχώρει ἅμα τῇ δυνάμει πάσῃ· τῶν δὲ κατὰ τὴν πόλιν ἀποκλεισάντων αὐτῷ τὰς πύλας καὶ εἰς τὰς ἐπάλξεις ἀναβάντων ἠπόρει περὶ τὸ πρακτέον ὁ Μαγνέντιος, οὔτε μηχανὰς ἔχων οὔτε ἄλλως πλησιάσαι τῷ τείχει δυνάμενος. Κατηκοντίζετο γὰρ βέλεσί τε καὶ λίθοις ὑπὸ τῶν ἐπὶ τῶν ἐπάλξεων μαχομένων· ἐπεὶ δὲ καὶ Κωνσταντίῳ τὰ περὶ τῆς πολιορκίας ἠγγέλθη, μετὰ πάσης ἐξαναστὰς τῆς δυνάμεως παρεγένετο βοηθήσων τῇ πόλει κινδυνευούσῃ, παραμείψας [δὲ] Κίβαλιν καὶ γῆν ἅπασαν ὅσην ὁ Δρᾶος παραρρεῖ ποταμός. Μούρσῃ τε πλησιάσας ὁ Μαγνέντιος πῦρ προσῆγε ταῖς πύλαις, ὡς ἂν τὰ ξύλα τοῦ καλύπτοντος σιδήρου δαπανηθέντος, εἴξαντα τῷ πυρὶ τὴν εἰς <τὴν> πόλιν εἴσοδον ἀναπετάσαιεν τῷ στρατῷ· τοῦτο μὲν οὖν αὐτῷ πρὸς ἔργον οὐκ ἤχθη, τῶν ἐπὶ τοῦ τείχους ὕδατι πολλῷ κατασβεννύντων τὸ πῦρ· τὸν δὲ Κωνστάντιον προσαγαγόντα τῇ Μούρσῃ πυθόμενος ἕτερόν τι μηχανᾶται τοιόνδε. Στάδιον ἦν τι πρὸ τῆς πόλεως τῶν ἀφορισθέντων ἄνωθεν τῇ τῶν ὁπλομάχων ἁμίλλῃ, πανταχόθεν ὕλαις ἀμφηρεφές· τούτῳ Κελτῶν φάλαγγας τέσσαρας ἐναπέκρυψεν, ἐγκελευσάμενος, ἐπειδὰν ἐφισταμένου Κωνσταντίου μέλλοι [τοῖς] πρὸ τῆς πόλεως συνάπτεσθαι μάχῃ, τοῖς ἐναντίοις ἐπελθεῖν ἀδοκήτοις, ὥστε αὐτοὺς ἐν μέσῳ ληφθέντας πανωλεθρίᾳ διαφθαρῆναι· (...) Τούτῳ τῷ τρόπῳ τῆς Μαγνεντίου διαπεσούσης ἐπιβουλῆς καὶ τῆς ἐνέδρας εἰς τοὐναντίον αὐτῷ περιστάσης, εἰς ταὐτὸν ἀμφότερα γενόμενα τὰ στρατεύματα συνέπιπτεν ἀλλήλοις ἐν τῷ πρὸ τῆς Μούρσης πεδίῳ· μάχης τε γενομένης οἵα σχεδὸν οὔπω πρότερον ἐν [τούτῳ] τῷ πολέμῳ φαίνεται γεγονυῖα, πίπτουσι μὲν ἐξ ἑκατέρου πλεῖστοι.

Magnentius then led out his army and taking Siscia at the first attack, razed it to the ground. When he had overrun the whole Savus valley and carried off considerable booty, but foiled in his attempt by the inhabitants and the garrison, he returned to Mursa with his whole army. Its inhabitants, however, shut the gates against him and mounted the battlements, which placed Magnentius in a dilemma, since he had no siege engines or any other means of approaching the wall, and was under fire from its defenders with arrows and stones. When Constantius heard news of the siege, he left Cibalis and the Dravus river valley, and set out with his whole force to come to the assistance of the threatened city. Meanwhile, Magnentius closed in on Mursa and set fire to the gates, thinking that if the iron-bound wood yielded to the fire, he would open up the way into the city for his army, but this did not eventuate because the men on the wall quenched the fire with plenty of water. So when he heard of Constantius' approach, he devised the following scheme: there was a stadium outside the city, one of those originally designed for gladiatorial contests, which was all overgrown with woods. Here he concealed four companies of Gauls and ordered them, when Constantius came up with the intention of engaging outside the city, to attack him unawares, so as to surround and annihilate his men.(...) Thus Magnentius' scheme fell through and it was he who was defeated in the ambush. Then the armies met and joined battle in the plain outside Mursa. This conflict was the fiercest so far in the course of this war and great numbers fell on both sides.

Zos. 3.36 (5th-6th centuries AD)

(...) ἤνεγκαν δὲ τὴν ψῆφον ἐπὶ Οὐαλεντινιανόν, ὃς ἐκ μὲν Κιβάλεως ὥρμητο (πόλις δὲ αὕτη Παιονική), πολέμων δὲ μετασχὼν οὐκ ὀλίγων παιδεύσεως οὐδεμιᾶς μετεσχήκει.

So they voted for Valentinian, a native of Cibalis in Pannonia, who was experienced soldier but quite uneducated.

Bibliography

Andrić 2002 Stanko Andrić, Savsko-dravsko međurječje od kasne antike do ranog srednjeg vijeka – Južna Panonija u doba velike seobe naroda, *SSlav* 2, 117-167.

Barkóczi 1980 László Barkóczi, History of Pannonia, in: *ARP*, 85-124.

Barnes 1978 Timothy David Barnes, *The Sources of the* Historia Augusta, *Collection Latomus*, vol. 155, Bruxelles.

Birley 1975 Anthony R. Birley, The Third Century Crisis in the Roman Empire, *Bulletin of the John Rylands Library* 58, Manchester, 253-281.

Bíró-Sey 1980 Katalin Bíró-Sey, Currency, in: *ARP*, 337-348.
Bojanovski 1984 Ivo Bojanovski, Prilozi za topografiju rimskih i predrimskih komunikacija i naselja u rimskoj provinciji Dalmaciji (IV): Rimska cesta Siscia-Sirmium (Tab. Peut.) i njena topografija, *GodCBI* 22 (20), 145-265.
Bratož 1983 Rajko Bratož, Crkvenopolitični in kulturnozgodovinski odnosi med Sirmijem in Akvilejo, *ZČ* 37, 259-272.
Bratož 1986 Rajko Bratož, Razvoj organizacije zgodnjekrščanske cerkve na ozemlju Jugoslavije od 3. do 6. stoletja. Kratek oris, *ZČ* 40, 363- 395.
Brunšmid 1900 Josip Brunšmid, Colonia Aelia Mursa, *VHAD* 4 (1899-1900), 21-42.
Brunšmid 1902 Josip Brunšmid, Colonia Aurelia Cibalae, *VHAD* 6, 117-166.
Bulat 1989 Mirko Bulat, Novi podaci za baziliku mučenika u Mursi, *Lihnid* 7, 195-202.
Buzov 2003 Marija Buzov, Kvirin – sisački biskup i mučenik, *Riječi, časopis za književnost, kulturu i znanost Matice hrvatske Sisak*, 1-2, Sisak, 1-27.
Caldwell 2007 Craig H. Caldwell, *Contesting Late Roman Illyricum: Invasions and Transformations in the Danubian-Balkan Provinces*, Unpublished PhD thesis, Princeton University.
Christie 1992 Neil Christie, The survival of Roman settlement along the Middle Danube: Pannonia from the fourth to the tenth century AD, *Oxford Journal of Archaeology* 11, Oxford, 317-339.
Christie 2008 Neil Christie, 'From the Danube to the Po: Late Roman Defence in Pannonia and Italy', in: A. Poulter (ed.), *The Transition to Late Antiquity, on the Danube and Beyond* (*Proceedings of the British Academy* 141) Oxford-New York, 547-577.
Curta 2001 Florin Curta, *The Making of the Slavs*, Cambridge.
Durman 1992 Aleksandar Durman, O geostrateškom položaju Siscije, *OpArch* 16 117-131.
Errington 2006 Robert Malcolm Errington, *Roman imperial policy from Julian to Theodosius*, University of North Carolina Press.
Fitz 1962 Jenő Fitz, A military history of Pannonia from the Marcomann wars to the death of Alexander Severus (180-235), *AArchHung*, 14, 25-112.
Fitz 1980a Jenő Fitz, Population, in: *ARP*, 141-159.
Fitz 1980b Jenő Fitz, The Way of Life, in: *ARP*, 161-175.
Fitz 1982 Jenő Fitz, *The Great Age of Pannonia* (AD 193-284), Budapest.
Fitz 1998 Jenő Fitz (ed.), *Religions and Cults in Pannonia*, Szekesfehervar.
Funaioli 1914 Gino Funaioli, *s. v.* Ravenna, *RE* IA/1, 305-310.
Gračanin 2003 Hrvoje Gračanin, Bitka kod Murse 351. i njezin odjek, *SSlav* 3, 9-29.
Gračanin 2005a Hrvoje Gračanin, Illyricum of the 2nd and 3rd centuries AD in the works of Latin and Greek historians, in: *Illyrica Antiqua: ob honorem Duje Rendić-Miočević*, Zagreb, 287-298.
Gračanin 2005b Hrvoje Gračanin, Huni i Južna Panonija, *SSlav* 5, 9-47.
Gračanin 2006 Hrvoje Gračanin, Goti i Južna Panonija, *SSlav* 6, 83-126.
Gračanin 2007 Hrvoje Gračanin, Gepidi, Heruli, Langobardi i Južna Panonija, *SSlav* 7, 7-64.
Gračanin 2008 Hrvoje Gračanin, Slaveni u ranosrednjovjekovnoj Južnoj Panoniji, *SSlav* 8, 13-54.
Gračanin 2010 Hrvoje Gračanin, Rimske prometnice i komunikacije u kasnoantičkoj južnoj Panoniji, *SSlav* 10, 9-69.
Gračanin 2011 Hrvoje Gračanin, *Južna Panonija u kasnoj antici i ranom srednjovjekovlju (od konca 4. do konca 11. stoljeća)*, Zagreb.
Graf 1936 András Graf, *Übersicht der antiken Geographie von Pannonien, DisPann* I/5.
Heather 1991 Peter J. Heather, *Goths and Romans 332-489*, Oxford.
Heather 2005 Peter J. Heather, *The Fall of the Roman Empire*, London.
Hoti 1992 Marina Hoti, Sisak u antičkim izvorima, *OpArch* 16, 133-163.
Iskra-Janošić 2001 Ivana Iskra-Janošić, *Urbanizacija Cibala i razvoj keramičarskih središta*, Zagreb-Vinkovci.
Jarak 1994 Mirja Jarak, The history of early Christian communities in continental Croatia (17-39 Croatian, 155-186 English), in: Ž. Demo (ed.), *Od nepobjedivog sunca do sunca pravde / From the Invincible Sun to the Sun of Justice*, Exhibition Catalogue, Zagreb.
Kerr 1995 William G. Kerr, *A Chronological Study of the Marcomannic Wars of Marcus Aurelius*, Princeton.
Klemenc 1953 Josip Klemenc, Teodozijev pohod proti Maximusu iz Siscije do Petovija, *ZČ* 6-7 (Kosov zbornik) (1952-1953), 78-88.
Koščević and Makjanić 1995 Remza Koščević and Rajka Makjanić, *Siscia, Pannonia Superior*, *BAR Intern. Ser.* 621.
Kovács 2004 Péter Kovács, The late Roman army in Pannonia, *AantHung* 44(1), 115-122.
Lenski 2002 Noel Lenski, *Failure of Empire: Valens and the Roman State in the Fourth Century A.D.*, University of California Press, Berkeley.
Lotter 2003 Friedrich Lotter (with R. Bratož and H. Castritius), *Völkerverschiebungen im Ostalpen-Mitteldonau-Raum zwischen Antike und Mittelalter (375- 600)*, Berlin-New York.
Maenchen-Helfen 1973 Otto J. Maenchen-Helfen, *The World of the Huns; studies in their history and culture*, Berkeley.
Margetić 2005 Lujo Margetić, Etnogeneza Slavena, *RHAZU* 492, 89-143.
Mayer 1935 Antun Mayer, Ime Mursa, *VHAD* 16, 5-10.
Migotti 1994 Branka Migotti, The archaeological material of the early Christian period in continental Croatia (Croatian 41-70, English 187-209), in: Ž. Demo (ed.), *Od nepobjedivog sunca do sunca pravde / From the Invincible Sun to the Sun of Justice*, Exhibition Catalogue, Zagreb.

Mócsy 1974 András Mócsy, *Pannonia and Upper Moesia*, London.

Pinterović 1978 Danica Pinterović, Mursa *i njeno područje u antičko doba*, Osijek.

Pinterović 1980 Danica Pinterović, "Basilica Martyrum" u Mursi, in: *Gunjačin zbornik* (eds. I. Erceg *et al.*), 59-66.

Rapan Papeša 2011 Anita Rapan Papeša, Topography of Cibalae in Late Antiquity, *OpArch* 35 189-224.

Rendić-Miočević and Šegvić 1998 Ante Rendić-Miočević and Marina Šegvić, Religions and cults in south Pannonian regions, in: Fitz 1998, 7-13.

Roberts 1993 Michael Roberts, *Poetry and the Cult of the Martyrs. The* Liber Peristephanon *of Prudentius*, Ann Arbour.

Salamon and Sós 1980 Ágnes Salamon and Ágnes Cs. Sós, Pannonia – fifth to ninth centuries, in: *ARP*, 397-425.

Sanader 2003 Mirjana Sanader, Rimske legije i njihovi logori u hrvatskom dijelu panonskog limesa, *OpArch* 27, 463-468.

Selem 1980 Petar Selem, *Les religions orientales dans la Pannonie Romaine: partie en Yougoslavie, Etudes préliminaires aux religions orientales dans l'Empire Romain* 85, Leiden.

Soproni 1980a Sándor Soproni, Geography of Pannonia, in: *ARP*, 57-63.

Soproni 1980b Sándor Soproni, Roads, in: *ARP*, 207-217.

Syme 1983 Ronald Syme, *Historia Augusta Papers*, Oxford.

Šašel 1971 Jaroslav Šašel, The Struggle between Magnentius and Constantius II for Italy and Illyricum, *ŽAnt* 21, 205-216.

Šašel Kos 1986 Marjeta Šašel Kos, *Zgodovinska podoba prostora med Akvilejo, Jadranom in Sirmijem pri Kasiju Dionu in Herodijanu*, Ljubljana.

Šišić 1925 Ferdo Šišić, *Povijest Hrvata u vrijeme narodnih vladara*, Zagreb.

Škegro 1999 Ante Škegro, Eksploatacija željeza u Bosni u rimsko doba, *Prilozi Instituta za istoriju u Sarajevu* 26 (28), Sarajevo, 17-45.

Škrivanić 1977 G. Škrivanić, Roman Roads and Settlements in the Balkans, in: F. W. Carter (ed.), *An historical geography of the Balkans*, London-New York-San Francisco, 115-145.

Thomas 1980 Edit B. Thomas, Religion, in: *ARP*, 177-206.

Thompson 1996 Edgar Arthur Thompson, *The Huns*, revised and with an afterword by Peter Heather, Oxford.

Tomlin 1979 R. S. O. Tomlin, Ammianus Marcellinus 26.4.5-6, *The ClassicalQuarterly* (New Series), 29/02, 470-478.

Várady 1969 László Várady, *Das Letzte Jahrhundert Pannoniens (376-476)*, Amsterdam.

Visy 2003 Zsolt Visy (ed.), *The Roman Army in Pannonia*, Budapest.

Williams and Friell 1995 Steven Williams and Gerard Friell, *Theodosius, The Empire at Bay*, New Haven.

Wolfram 1988 Herwig Wolfram, *History of the Goths*, translated by Thomas J. Dunlop, Berkeley.

Zaninović 1981 Marin Zaninović, Siscia u svojim natpisima, in: *Arheološka istraživanja u Zagrebu i njegovoj okolici, IzdHAD* 6, Zagreb, 1981, 201-208.

PANNONIANS: IDENTITY-PERCEPTIONS FROM THE LATE IRON AGE TO LATER ANTIQUITY

Danijel Dzino

Alka Domić Kunić

1. Introduction[1]

This paper will discuss ancient Pannonian identity-narratives and their transformations until Late Antiquity. As far as we know, Pannonian identity first appears in the written sources as an outsider's depiction of the indigenous communities living in what will become Roman southern Pannonia and northern Dalmatia. After the Roman conquest, the narratives of Pannonianess become more complex and develop into what we can today see as a set of different outside labels, and internal self-perceptions relating to the roman province(s) of Pannonia, their regions, and individual communities. The focal point will be, in tune with this whole volume, Pannonian narratives from the southern parts of the province. It is impossible to treat Pannonian identities here in full detail – such an encompassing study would need a whole monograph, rather than just a single chapter. What we offer here is more an outline of the different identity-narratives rather than a full and thorough exploration of all available sources.

This study is long overdue. The research of identities, in the Danubian provinces of Rome, has been rather slow in incorporating new methodological frameworks used in the research of identity of comparatively similar societies of the Roman West.[2] Pannonian identities are excellent examples for the research of identity-transformations in pre-modern societies. Such studies show identity rather nicely as a contextual and fluctuating social construct, existing within several different historical contexts and being claimed by different groups, either to label the outsider, or to accept and reclaim outside labeling and incorporating them within their own sense of belonging. The task of exploring Pannonian identity-narratives also touches upon some important debates arising in modern scholarship dealing with the ancient history of the Mediterranean and continental Europe. The first issue concerns the limitation of our knowledge about group-identities and social models in which late Iron-Age communities in the Roman West operated.[3] The other problem is that of so-called 'Romanization'. In particular, what were the identity-strategies chosen by those Iron-Age communities after they were included in the Roman political framework – the assessment of short-term and long-term impacts of the change on the construction of their identities.[4]

The issue of identity has for quite some time attracted increasing attention of scholarship in archaeology, classical studies and ancient history. It is not surprising that major attention is focused on social and cultural group identities, and in particular on ethnicity, because our understanding of the past is in many ways shaped by our perception of the groups mentioned in the sources as distinct ethnicities. This rise of interest comes as the result of the more substantial absorption of the developments in theoretical frameworks of social anthropology from the 1960s and 1970s, defining ethnicity as a contextual, fluent and changeable social phenomenon, constructed through the perception of differences with other groups. Also, scholarship dealing with the ancient world benefited significantly from the incorporation of three essential 'post-' theoretical frameworks (post-modern, post-structuralist and post-colonial) dealing with relations between the control of discourse and power, interaction of different cultures as an active two-sided process, and research of individuals and groups with multiple identities.[5]

Some generalities and basic literature should be noted regarding the construction of social and cultural identities, in order to present the theoretical framework in which this paper will operate. Of first and foremost importance is to (re)state that most social identities, such as ethnicity, represent subjective feelings of commonality or belonging with other people, which is socially constructed as a product of interaction with a different group – the 'Other'.[6] Social identity is usually self-constructed internally as a group

[1] This paper is part of the Discovery Project on identity-transformation in Illyricum, sponsored by the *Australian Research Council*.

[2] It is worth mentioning the encompassing survey of the scholarship on the archaeology and history of the Danubian provinces in Wilkes 2005. More advanced work has been done recently on Dacian identity, see Oltean 2007; 2009. The situation is better regarding Norican and Norican-Pannonian identities e.g. Scherrer 2004; Pochmarski 2004; Kremer 2004 and also Hales 2010.

[3] See for example Thurston 2009, or Hill 1995; 2006; Bevan 1999; Haselgrove and Moore 2007.

[4] On Romanization, from its rejection to the basic defence of its framework see e.g. Mattingly 1997b; 2004; Millett 1990; Woolf 1992; 1998; Keay and Terrenato 2001; Alföldy 2005; Hingley 1996; 2005; Cecconi 2006; Pitts 2007, etc.

[5] See the concise and ecompassing recent overview in Hodos 2010.

[6] Barth 1969; Yelvington 1991, see also Barth 1989.

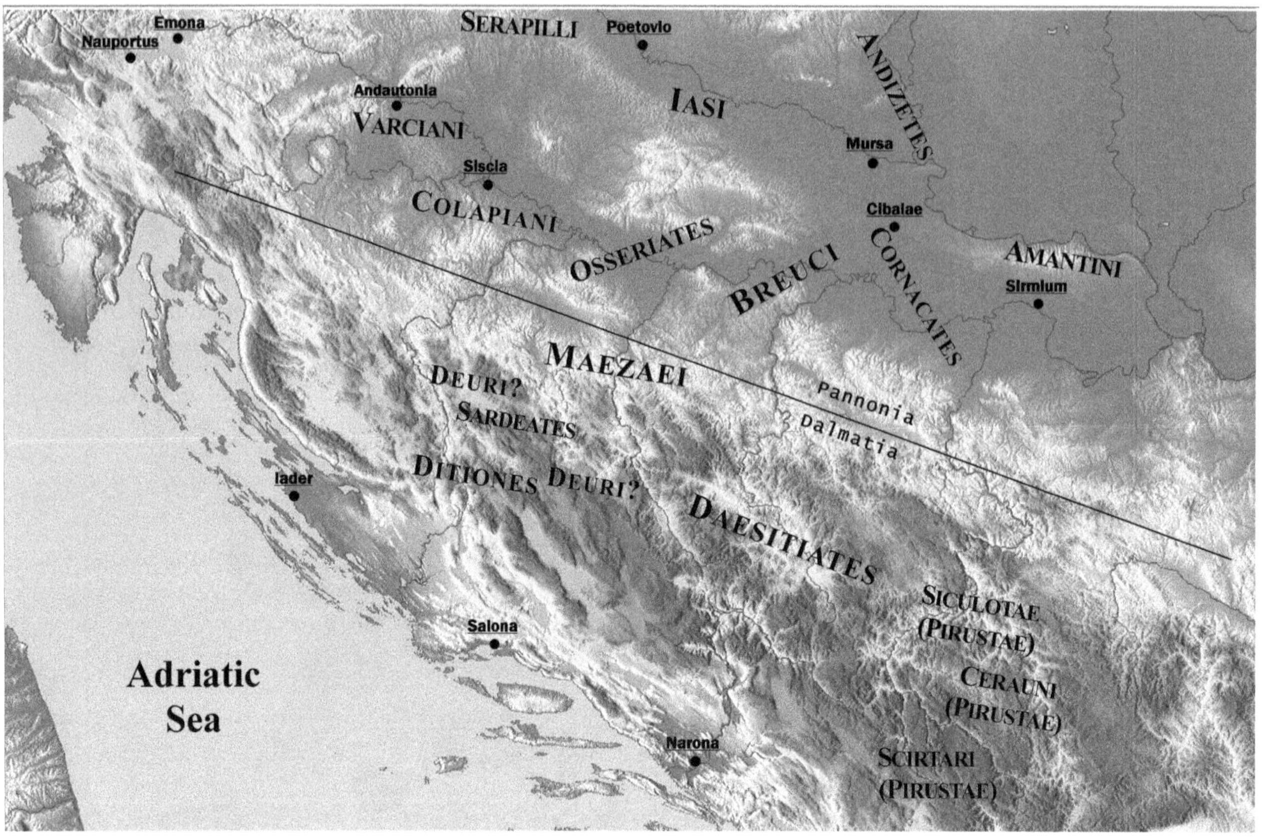

Fig. 1. The peregrine civitates composed from the groups perceived as the Pannonii in pre-conquest period (after Google Maps, modified by D. Dzino).

identity (ethnic, cultural, sub-cultural, regional), but it also might be imposed by the outsiders as a label, especially if the outsiders are in a position of power-domination.[7] An important constructive element of identity is the cultural *habitus*, basically a simplified keyword for sets of shared cultural practices and a subliminal disposition towards certain perceptions and social practices that are implanted into an individual's sense of self.[8] *Habitus* does not represent social or cultural identity *per se*, however its articulation, especially in the context of power-struggles with other communities, can lead to the construction of a distinct sense of common social, regional, political or ethnic identity. Similarities in *habitus* do not always result with a common identity. On the other hand, differences in *habitus* are never an obstacle to the establishment of the shared sense of identity.[9] The notion of culture as a reified, homogenizing and static feature that people are simply born into has in recent decades been replaced with a view that culture is a fluid, perpetually creative process that recasts existing cultural symbols and creates new meanings in different chronological periods.[10] Finally, it should be said that social and cultural identities are often perceived as a binary (either-or) feature by the outsiders; a taxonomic category which can be distinguished from other similar categories, while in truth they represent multiple belongings (both and-and),[11] thus crossing the boundaries imposed by the outside observer.

Problems associated with the exploration of identity of communities from the past are numerous, especially regarding the development of clear and proper terminology and the application of the correct methodology, as each identity-construction develops in unique circumstances. The most significant difficulty is distinguishing different kinds of social and cultural identity in scholarly interpretations of texts and material culture.[12] The same identity-label can represent many different identities depending on the historical and situational contexts for which they are used: for example, the label 'the Romans' in antiquity, can represent civic, ethnic, cultural, sub-cultural, religious or social identity when used in different contexts and claimed by different groups.

Textual evidence necessitates the development of complex keys for reading the ancient texts within their literary, cultural and historical contexts. The most significant problem with earlier scholarship is that it treated written

[7] Jenkins 1997, 53-56.
[8] Bourdieu 1977; Bentley 1987.
[9] Yelvington 1991, 168. For the research of ancient identities, *habitus* and social conditions are covered in Jones 1996, 68-70; 1997, 120.
[10] It is not necessary to 'liquidate culture' as a concept (Bayart 1996 esp. 166; Apparadui 1996, 12), but rather see it as a fluid and inventive process, Clifford 1988, 15; Bauman 1999.
[11] Kearney 1995, 556-559.
[12] See the valiant attempts of Lucy 2005 in sorting out social identities and ethnicity.

sources from antiquity as reports of facts - fellow scholars - while today we can see that those texts are embedded in different narratives, genres and cultural contexts and should be read in certain ways.[13] In the case of communities that did not have control of written discourse, such as the Pannonians, written sources must be taken with the utmost caution. They report not only perceptions of outside observers, but also project discourse on power-domination and exclusion in the ancient Mediterranean world, which constructed the cultural inferiority of indigenous populations inside the genre of 'Barbarian ethnography'. This 'barbarity' represented a symbolic mirror for reflecting the 'civilization' of the writers who belonged to the elite circle of what we call Graeco-Roman, or the ancient Mediterranean cultural circle.[14]

From the Roman period we are somewhat luckier, as there is more direct evidence from the indigenous people of Pannonia. Epigraphy reflects individual identities slightly more accurately than the elite-dominated written sources, especially in funeral and mortuary contexts such as tombstone inscriptions which we have in abundant numbers. However, it is also important to be aware that there were different strategies which individuals could assume in showing their identity on tombstones. The tombstones reflect not only the identity of the deceased in one particular moment, but also the ways in which the community reclaimed that individual as its member in a mortuary context.[15] Also, similar to the recent research of Batavian, Dacian and Dalmatian identity, Pannonian identity in the epigraphic record is most abundantly revealed in diasporic contexts, outside of the province. The reason for this is understandable – there was no need for presenting identities within the province, but communicating it towards the 'Other', which occurs in the diasporic context.[16]

Using material culture as the primary tool for the reconstruction of ethnicity and identity of a given group, which is most frequently the case with the identity of the Pannonians, is a very difficult task. Earlier scholarship usually conceptualised material culture as secondary to the written sources, trying to adjust material evidence to the framework determined by those written sources. Material evidence has also been seen as a distinct historical entity in scholarship, so scholars deliberately ethnicised certain types of objects or burial practices in the so-called culture-history framework, and they automatically became signifiers of ancient ethnic identities.[17] Nevertheless, it is certainly true that material culture on occasion embodies 'emblemic styles', which are channels used to communicate stylistic messages by non-verbal means about a conscious affiliation and a shared communal or social identity of an individual.[18] Certain artifacts have had clear meanings in the past which were indeed used to articulate and contest social identities, actively communicating and defining a wide range of identities, including ethnicity.[19] However, these meanings were often embedded in social contexts about which we know too little to read with more certainty. Thus, more recent archaeological theory is, at very least, suspicious about the possibilities that interactive social phenomena, such as ancient ethnicity, can be determined only through the analysis of material culture, and is still looking for proper ways in which to reconcile material records with identity narratives.

2. Earlier views of the indigenous population in southern Pannonia and the Dalmatian hinterland

There is a limited amount of information at our disposal about the prehistoric and proto-historic indigenous population from the region between the Rivers Sava and Drava. They were illiterate societies before the Roman arrival, and no member of Pannonian elite in Roman times attempted to reconstruct historical memories of their ancestors from pre-Roman times, as for example did the members of Gallo-Roman aristocracy such as Pompeius Trogus. Written sources dating from the pre-conquest period are almost non-existent and are only available through later writers. For this reason material culture remains the most reliable evidence for the pre-Roman past of these communities.

Earlier scholarship viewed the indigenous population not too differently from the late Republican and early Imperial writers: a loose but distinctive ethno-cultural group named *Pannonii*[20] which never really fitted into the existing scholarly taxonomy of other ancient pseudo-ethnic categories. They are defined either through exclusion, as non-Celtic communities of Pannonia, or as an 'impure' and messy ethno-cultural mixture which could not be classified any other way – *"... a fringe races of Illyrians, Thracians and Celts"*.[21] More recent historiography sees the *Pannonii* as an ethno-cultural 'ethnic complex', like the other prehistoric groups from the wider region, such as the Thracians, Illyrians, the North-Adriatic complex (Liburni, Histri, Iapodes) and Celts.[22] This attempt to scholarly classify indigenous populations from the pre-Roman period into distinctive ethno-cultural groups has used a

[13] Marincola 1997; Shuttleworth-Kraus 1999; Potter 1999; Mellor 1999 etc.

[14] Wells 1999 is perhaps the most popular work on the topic, earlier Dauge 1981; Hartog 1980. See also the new research on Roman ethnography as a literary category: Murphy 2004; Dench 2005, 93-151 and the overview in Syed 2005, or for example Clarke 1999 on Hellenistic geography (Strabo, Posidonius and Polybius) as a genre.

[15] Morris 1992, 156-173; Hope 1997; 2001; 2003 etc.

[16] Van Driel-Murray 2002, Roymans 2004; Derks 2009 (Batavian); Oltean 2009 (Dacian); Dzino 2010b (Dalmatian). Pannonian auxiliarics and sailors are discussed in Domić Kunić 1988; 1996a; 1996b; 1996c.

[17] Criticisms of culture-history approach are numerous: see Shennan 1989; Jones 1996; 1997; Graves-Brown *et al.* 1996; Brather 2004 etc.

[18] For the notion of 'emblemic style', see the works of Wiessner 1983; 1989; 1990.

[19] See in different historical contexts Curta 2007a; Antonaccio 2001; 2010.

[20] They are usually refered to in the scholarship as the *Pannonii*. This should be distinguished from the term Pannonians which depicts the Pannonian identity from Roman times – we will follow this convention in the paper, and italicise the term.

[21] Mócsy 1974, 4-5., cf. recently Colombo 2010, 202: " *né puri Celti, né meri Iliri, ma piuttosto Celto-Iliri*".

[22] Benac 1987b, followed by Olujić 2004; 2007, 11-24 and Matijašić 2009, 30-50.

methodological 'tripod', combining the evidence provided by ancient written sources, indigenous anthroponymy (the research of indigenous personal names and evidence for frequency of certain names in certain regions from Roman period inscriptions), and the geographical extension of late Iron Age archaeological cultures in the region.[23]

The evidence from the written sources always represented the most essential part of this framework, despite the fact that the overwhelming majority of written evidence post-dates the Roman conquest. The placement of the indigenous groups called the *Pannonii* in a defined geographic space by ancient authors, namely Strabo and Appian, was usually taken at face value. It was considered an objective report of their self-designation, but today we cannot see it as much more than an outsider perception of the: *"... internal network of relationships based on ethnicity and kinship."*[24] Anthroponymic research of indigenous personal names from the Roman period and written sources defined the group called Delmato-Pannonian, underlining similarities in names appearing in southern Pannonia and central and south-western regions of Dalmatia.[25] However, archaeology lacked the evidence to see the Pannonii from the historical sources and anthroponymic research as a distinct archeological culture. They have been seen as the descendants of the late Bronze/early Iron Age Urnfielders, but the appearance of La Tène north of the Sava in c. fourth century BC, interrupted these earlier traditions and resulted in visible differences in material culture between the communities from the Sava valley and those south of the Sava, which belong to the Central Bosnian archaeological culture.[26]

The foundations of this particular framework rely on several, mostly outdated scholarly assumptions, and it is here neither the time nor the place to criticize all of them more comprehensively.[27] The most questionable is certainly the assumption that ethnicity is the most significant group-designation throughout history as it was in nineteenth and twentieth century European nation-states. The other questionable assumption is the view that ethnicity and/or cultural identity are predetermined and regulated sets of values and symbols, which can be objectively recognized by the outside observers, or through material evidence. Collective identities are communicative, discursive constructions grounded in the commonality of self- and world-descriptions.[28] Without really knowing patterns of identity-discourses of self-identification that might give the pre-Roman Pannonii a collective identity throughout history (if those discourses ever existed), the scholarship attempted to invent new 'objective', taxonomic identity-categories such as 'people-making community', 'ethnic complex', etc. This reflected modernistic discourse on ethnicity in these communities, rather than the identity-narratives of the actual population in antiquity.

The Greeks and Romans perceived the 'Barbarian' *ethne* by different, sometimes contradicting criteria. Occasionally they depicted linguistic groups, regional or political communities, or were simply being anachronistic – their depiction in the sources did not necessarily reflect a self-expressed ethnicity, as we might understand it today.[29] Thus, the assumption that the sources accurately depict the ethnography of the *Pannonii* is difficult to maintain. Furthermore, personal indigenous names from Roman-era inscriptions and tombstones should also be taken with caution. They do not necessarily reflect shared traditions of the indigenous population, but rather, the identity-discourse of certain indigenous sub-cultural and regional groups formed after the Roman conquest, such as provincial elites or the military and their families who are most frequently represented in the inscriptions.[30] Research into some other 'ethnic complexes' from antiquity, such as the linguistic groups we might conveniently call the 'Celts' or 'Germans', reveals that Graeco-Roman perceptions of cultural or ethnic similarity was not necessarily shared amongst the communities labeled in a way we could today call ethnic.[31]

3. Material culture and the earliest concepts of proto-historic Pannonian communities (fig. 1)

The material record from the Iron Age shows a world of constantly changing communities in what was to become Roman southern Pannonia and northern/central Dalmatia. The Early Iron Age cultural groups appearing in the region in a wider context of the Hallstatt world, such as Central-Danubian groups (Dalj, Bosut, Srijem) as well as the Martijanec-Kaptol, Budinjak, Donja Dolina-Sanski Most complex, Colapiani and Central Bosnian culture experienced significant changes of their cultural identities in the mid-later Iron Age. The Martijanec-Kaptol and Budinjak groups disappeared due to insufficiently explained reasons in the sixth century BC, while the disappearance of the Central-Danubian groups is related to the expansion of La Tène cultural matrices into the Pannonian plains.[32] The influence of the La Tène in the Late Iron Age was such that

[23] On the Iron Age groups in the western part of the Balkan peninsula in English see: Wilkes 1992, 40-206; Šašel Kos 2005, 219-244; Dzino 2010a, 36-43, in German see Lippert 2004 and Pavic 2010.

[24] Šašel Kos 2005, 376-378, quote from 378.

[25] Katičić 1963; 1965, 69-73 for the Central Dalmatian or Pannonian-Dalmatian anthroponymic group, see the general overview of anthroponymic research in Wilkes 1992, 74-87; Šašel Kos 2005, 228-231. For the analysis of the written sources, see below.

[26] Marić 1964a (links with the Urnfielders), see also Wilkes 1992, 202-207. Milin 2003 unconvincingly attempted to reconcile material with the anthroponymic and paleolinguistic evidence. See below for the analysis of the material evidence.

[27] See Dzino 2011

[28] Straub 2002, see also above.

[29] Wells 2001, 105-118; see also the example of ethnicising political alliances in Hammond 2000, and the projection of identities into the past e.g. Isayev 2010.

[30] See Pitts 2007, 700.

[31] The assessment of the sources for ancient Germans as accurate ethnographic works have been tarnished by Goffart 1988; 2002 (Late Antiquity) and Lund 1988; 1991 (Tacitus), so that the validity of the term 'Germans', except as linguistic group, has been questioned by Goffart 2006. A similar re-examination of the sources for the 'Celts' can be found in Chapman 1992, 30-52; James 1999a; Wells 1999, 99 ff.; Collis 2003, etc.

[32] Vinski-Gasparini 1987; Čović 1976, 171-237; 1987, 481-530; Vasić 1987; Potrebica 2003 as well as the contributions in Balen-Letunić 2004, 35-210.

the communities between the Sava and Drava Rivers used to be seen in the archaeological record as middle ground between the so-called 'Eastern' and 'Western' 'Celtic' groups, which developed at the confluence of the Sava and Danube (the Scordisci) and the southeastern Alps (the Taurisci). The appearance of La Tène was in the earlier scholarship attributed to the massive migration of the ethnically distinct 'Celts' in the fourth century BC, as the series of conjunctures from scattered written sources might suggest, but the scholarship is starting to rethink these assumptions, accepting that we are dealing with small-scale movements of highly mobile groups, rather than massive migrations.[33] The change of cultural matrices do not appear only in the material record, but also in the change to burial customs and the ways elites chose to define themselves, developing stronger emphasis on warrior attributes. However, these La Tène influences impacted various sub-regions in different ways, and it is today becoming clear that the region between the Sava and Drava shows the strong and peculiar ways that these matrices were re-negotiated in local communities.[34] The settlements in Segestica, which we know from written sources, and Donja Dolina, unknown from written sources, appear as the most important western Pannonian centres, taking into account that the importance of Segestica increased and Donja Dolina diminished after the appearance of La Tène.[35]

It is very likely that the Sava valley and the plains between the Sava and Drava were politically directly or indirectly dominated by the political alliances we know from the sources as the Scordisci and Taurisci, formed sometime in the early third century BC. This is suggested by the differences in the material culture of eastern and western parts of this region, and also by Pliny's passage mentioning the *Mons Claudius* (the Slavonian Mountains) anachronistically as a boundary between the Scordisci and Taurisci, probably describing an earlier line of demarcation between their zones of influence.[36] The extent of the political dominance of the Scordisci and Taurisci might be affected by their military defeats against the Romans in the second and first centuries BC, but it seems that the rise of the Dacian kingdom under Burebista crushed the political dominance of these two alliances in the forties BC.[37] These defeats probably shifted the centres of political power and enabled changes in the political and identity structures of the wider region.

The earlier scholarship downplayed the appearance of the La Tène finds south of the Sava, suggesting that there was no evidence for Celtic settlement in the hinterland, which could be linked to the La Tène finds.[38] However, existing material evidence shows the strong impact of La Tène south of the Sava, especially through frequent occurrence of the La Tène or La Tène-like brooches.[39] The appearance of the brooches might suggest change in the ways social status and clan-affiliations were advertised, taking into account the visibility of the brooches on clothes. There is an indication that burial customs change from inhumation to cremation before the Roman arrival – but the poor state of research should caution us in making more conclusive generalisations about the burials before the fourth century BC.[40] A recently published cemetery at Kamenjača near Breza, in the valley of the River Bosna, has revealed examples of active negotiations between Greek-influenced symbols of elite identity such as the Illyrian helmets, and a strong presence of northern status identifiers such as La Tène brooches and cremation burials.[41] Exchanges between the Sava valley and the southern Central-Dalmatian (Gorica) culture in Dalmatian Zagora and western Herzegovina[42] might suggest the formation of links between the clans, in particular after the Delmatian alliance was formed, which relied on the supply of metal from the hinterland, whether through political control, alliance or family and clientship networks.

There is only one written source which undoubtedly refers to pre-conquest Pannonians from a contemporary perspective, and that is the fragment of Polybius, which mentions the Pannonian war, preserved in the tenth century Byzantine lexicon known as Suda.[43] This fragment was related by earlier scholarship to the first documented Roman interactions with this region, mostly linked with the unsuccessful expedition of the otherwise unidentified Roman magistrate 'Cornelius' against the group called the Segestani (Segesticani), which inhabited Segestica – the locality of Pogorelec in modern-day Sisak. This expedition is sometimes dated to the 150s BC, but there is nothing more to clarify this more, except Polybius' fragment and Appian's mentioning of Cornelius. Only one more Roman encounter with the region is noted before the first century

[33] Todorović 1980; Majnarić-Pandžić 2009, Blečić Kavur and Kavur 2010, cf. Dzino 2007. See also re-examination of similar perceptions for 'Celtic flooding' of Thrace in Emilov 2005; 2007; 2010.

[34] Majnarić-Pandžić 1996a; 2009; Dizdar 2003; 2004; Dizdar and Potrebica 2002; 2005.

[35] Donja Dolina: Marić 1964b; Wilkes 1992, 51-54; Segestica: Buzov 1993, 48-52; Lolić 2003, 135-138; Radman-Livaja 2007, 159-170. For the La Tène settlements in this region see Todorović 1971 or Majnarić-Pandžić 1984; 1996b.

[36] Plin. *NH* 3.148, see Domić Kunić 2003; 2006, 74-77; Dizdar and Potrebica 2005 on the *Mons Claudius*. On the Taurisci see: Guštin 1996; Božič 1999; Graßl 2000, and the Scordisci: Božič 1981; Tasić 1992; Popović 1993. On the term *Slavonian Mountains* see Migotti in this volume, note 13, fig. 2.

[37] Šašel Kos 2005, 326-29, 490-492 (Roman wars with the *Taurisci* and *Scordisci*). Burebista defeated the alliance of the Boii and Taurisci and "*the Celts mingled with the Thracians and Illyrians*" – most probably the Scordisci (Strab. 7.3.11, 7.5.2), see Dobesch 1994; Lica 2000, 65-78; Graßl 2001; Šašel Kos 2005, 152-153, 378.

[38] Marić 1963; cf. Zaninović 2001.

[39] Popović 1996; Majnarić-Pandžić 1996a, see also Perić 1995 on La Tène pottery south of the Sava River.

[40] It is possible that archaeologically invisible ways of burial were used, such as exposure of the corpse or dispersion of ashes if cremation was used.

[41] Paškvalin 2008 (Kamenjača), cf. Čović 1987, 484-490, 496-502 and Perić 2002, 187-193 on these burials. It is worthy to note that the cremations in Kamenjača reveal different treatments of ashes, from burial in stone urns to simple in-ground deposition. On Illyrian helmets as status-symbols in indigenous contexts see Blečić 2007.

[42] Čović 1987, 480.

[43] Suda fr. 122 (= Polyb. fr. 64). There might have been other sources from the pre-Roman era. The 33rd book of Pompeius Trogus gave a fuller account on the origins of the *Pannonii*, as its abbreviated epitome by Justin reports.

BC: the expedition of Aurelius Cotta and unnamed Metellus in 119 BC against the Segestani. This was ultimately without lasting results, so no other attempts were made to conquer these communities until the very end of Republican period.[44]

What the Hellenistic perception of this space was, remains insufficiently known. Besides the fragment of Polybius, no other mention of the Pannonians has been preserved.[45] Hellenistic ethnography was certainly aware of them – Appian says that the Greeks consider as 'Illyrians' all those communities between Epirus (Chaonians and Tesprotians) and the Danube, marking the longitude of 'their country', and from Macedonians and Thracians to the Pannonians (*Paeones*), the Adriatic and foothills of the Alps, marking its latitude. This perception probably reflects a source earlier than the second century BC, but we cannot be absolutely certain whether Appian was thinking of the *Pannonii* or the Paeones from the north of ancient Macedonia (modern-day FRY Macedonia, and western Bulgaria).[46] There is a curious ongoing mix-up between the Paeones and the *Pannonii* and Pannonians in the Greek-language literature, which certainly derives from this period, or even earlier. Appian and Cassius Dio both stated that Paeones is the Greek name for Pannonians, whom the Romans call *Pannonii*. Dio added that the terms *Pannonii* and Paiones were self-designations of both groups in his time. The use of the term 'Paiones' for the *Pannonii* and Pannonians remained much more entrenched in the Greek-language literature until the late Empire, than was *Pannonioi*,[47] although Pannonia, not 'Paeonia', was the name for the province and the region. The reason for this mix-up was probably the cultural generalisations and stereotyping in the Greek world, perceiving communities northwest from Macedonia 'Paiones', as their stereotypical view of the western neighbors was 'Illyrioi'.

We can only guess the reasons why those communities were seen as a group by external observers. Dio mentions the etymology, apparently coming from the Pannonian word for a sleeved tunica stitched from pieces of old clothes. As is usual with ancient etymologies, it is rather difficult to believe. If he is indeed by any chance right, it might be used as evidence that the early Roman perception of the *Pannonii* was constructed from elements of shared cultural *habitus*, rather than a self-expressed sense of shared identity.[48] Whether the *Pannonii* shared specific language/dialect or not, will probably never be clear. Tacitus hints that the groups from Roman Pannonia, the Cotini and Osi, are not regarded to be Germans, because the Cotini speak Gallic and Osi speak the Pannonian language. However, we will probably never be absolutely certain what it was that Tacitus described with the words *Pannonica lingua*.[49]

Not much can be learnt about Pannonian identities from the sources dated before the Roman conquest. The communities, which were soon to be seen as *Pannonii*, consisted of different Iron Age communities which negotiated their identity between La Tène templates and existing local traditions. We might see them as typical later Iron Age communities in continental Europe, regionally interconnected heterarchical communities, where power was negotiated vertically and horizontally between different groups, rather than being hierarchically stratified societies.[50] Closely-knit regional networks of kinship and clientship, characteristic of the elites of heterarchic societies, probably influenced outsiders to see them as 'ethnic', or an ethnically-based group.[51] Also, the appearance of the Pannonian name in the fragment of Polybius and text of Appian, shows us that the ancients had a certain perception of the common culture of the population, probably developed from contacts with the most western of them – the Segestani.

4. The Conquest

A cluster of knowledge about the indigenous population started to form in the literary discourse of the early Principate. The most important reason for this was the strong political interaction of the wider region with Rome, which resulted in the conquest of the region between the Adriatic and Danube. The *Pannonii* were present in all three major conflicts which resulted in the Roman conquest: Octavian's expedition 35-33 BC, where, amongst others, he and his legates defeated the Pannonian Segestani, and also their allies who are known only as the unnamed, generic *Pannonii*, and probably the Daesitiates from the Dinaric hinterland, hidden under the term *Daisioi* used by Appian. The next event was the Pannonian war of Tiberius 12-9 BC, which strengthened Roman control in the region south of the Sava and subjugated the political alliance known as the Breuci, which controlled the region between the Sava and Drava at that time. Finally, the conflicts ended with the indigenous uprising, known as the *bellum Batonianum* AD 6-9, led by the Daesitiates and Breuci, but also involving the groups we know as the Maezaei, Pirustae, Andizetes and

44 Radman-Livaja 2004, 15-16; Šašel Kos 2005, 383-392; Dzino 2010a, 72-73, earlier Morgan 1971, 1974.

45 The earliest Mediterranean perceptions of the Danubian regions are visible through Greek mythology, Domić Kunić 2005. See also Domić Kunić 2006, 67-68 for the perceptions of this region in Hellenistic ethnography and geography.

46 App. 1.1; Šašel Kos 2005, 108-110. Šašel Kos argues that Appian's source was implying the Paeones, but this matter is inconclusive, taking into account that Appian writes "*from Macedonians and Thracians **to** Paeones, Alps and Adriatic*" (cf. App. 1.1), placing his Paeones on the opposite side from the Macedonians and Thracians, where we could expect to find the *Pannonii*.

47 App. 6.14, 14.40; Dio 49.36.6 see Grassl 1990; Šašel Kos 2005, 379-381.

48 Dio 39.36.5.

49 Tac. *Germ.* 43, Mócsy 1967. Anreiter (2001, 9-21) assumed the existence of the Pannonian language/dialect, which might be closer to the languages of the Iapodes, Delmatae and southern Illyrians, than that of the La Tène inhabitants of the south-eastern Alps. This view is disputed for a good reason, Adamik 2003. Pannonian population used significant amount of words from the Celtic language (Colombo 2010, 193-202), relying mostly on the evidence from imperial period

50 Heterarchy is the theoretical concept based on the ideas of Crumley, see e.g. Crumley 2007. The new opinions and approaches in research of the European Iron Age based on the concept of heterarchy are surveyed in Thurston 2009, 360-367.

51 Cf. Šašel Kos 2005, 377-378.

Delmatae.[52] The written evidence from this period derives from the autobiography of Octavian/Augustus, preserved in Appian, and parts of Strabo's *Geography*, which were probably based on similar autobiographical and formal reports relating to these conflicts.[53] There are also the eyewitness reports from the *bellum Batonianum* by Velleius Paterculus, and Cassius Dio's processing of the sources in his historical narrative, with the addition of Florus' epitomes of Livy. We also have contemporary stereotyping of the Pannonians as generic 'Barbarian' people in the poetry of Ovid and in the *Panegyricus Messallae*.

Two different perceptions of the indigenous population appear in the written sources from this period. The first shows a perception of the Pannonian communities stretching from the Drava deep into the Dalmatian mountains, and can be most clearly seen in Appian and Strabo. Appian, who admits to using the *Memoirs* of Octavian/Augustus, says that the *Pannonii* live all the way from the Iapodes in modern-day Lika to the Dardani in modern-day Kosovo, comprising almost all of the Adriatic hinterland.[54] In many ways, his general description corresponds with Strabo's which determines the *Pannonii* by spatial points represented by Segestica, the Danube, Dalmatia, the Scordisci and Ardiaei.[55] Strabo views Pannonian *ethne* mostly as political units: "*those ethne whose hegemon is/was Bato*" (i.e. Bato the leader of the Batonian war), naming them individually: the Breuci, Andizetes, Ditiones, Pirustae, Maezaei, and the Daesitiates.[56] A similar perception is also shown in the *Res Gestae*, where Augustus states that the *gentes* of the *Pannonii* were conquered by his (adopted) son Tiberius.[57] To these perceptions can be added views of the *Pannonii* in Roman poetry, namely the *Panegyricus Messallae* and by Ovid, as the inhabitants of the mountain peaks – hardly a characteristic of the plains between the Sava and Drava.[58] The mention of the mountains in the poetry shows that in Augustan times, Roman imagination sometimes projected a vision of the *Pannonii* as the inhabitants of the Dinaric mountains.[59]

In later Augustan and Tiberian times a different picture of indigenous Pannonian identity developed in the Roman perception. The earliest attestations are the writings of Velleius Paterculus, Tiberius' staff officer in the *bellum Batonianum*, who distinguished the Pannonians and Dalmatians as different groups of the enemy in the *bellum Batonianum*. His criterion for distinguishing indigenous communities was their location in regard to the provinces of Dalmatia or Pannonia, formed from the previous province of Illyricum, probably in the late Augustan or early Tiberian times. The division line between the provinces was drawn slightly south of the Sava River and accordingly, Velleius saw all communities south of that line as Dalmatian, and all those north as Pannonians. He does mention by name only the Daesitiates, Pirustae and Breuci.[60] Dio, who wrote in the late second and early third centuries, also used these criteria to distinguish indigenous participants of the *bellum Batonianum*, whether as clear anachronism and the projection of provincial identities from his own times, or because his sources used this terminology and Dio was simply following the source.[61]

The question of whether the northern part of Illyricum was initially called lower Illyricum (Illyricum *inferius*) after the division is still open to debate. The argument for lower Illyricum is bleak, as it is based on a now non-existent part of a damaged inscription from Epidaur (Cavtat near Dubrovnik), allegedly mentioning the '*civitates* of upper Illyricum',[62] while Pannonia was called Illyricum as late as the early 60s as evidenced by military diplomas.[63] The fact that Velleius calls the indigenous population the Dalmatians and Pannonians, and even calls the war *Pannonicum et Delmaticum bellum*, shows beyond any doubt that the terms Pannonia and Dalmatia were already being used either as formal or informal terminology in the 20s, when he was writing his history.[64] It is very likely that the use of the term Pannonia by Velleius is a consequence of his informal terminology, while in fact these provinces were officially called either: Dalmatia and Illyricum, or less likely, Illyricum *superius* and *inferius*.

More recent scholarship agrees that in the period of the great territorial expansion in the late Republic and early Principate, the Romans arbitrarily drew lines between certain regions in continental Europe, combining Hellenistic perceptions of indigenous populations with

---

52 Recently Dzino 2010a, 99-155; Domić Kunić 2006, 91-115. See also Šašel Kos 2005, 393-472, and Šašel Kos 2011 for Octavian's expedition.

53 Augustus in Strabo: Dzino 2008, 180. There were more autobiographies reporting on Octavian's expedition, such as Valerius Messalla Corvinus (Welch 2009, 207), or even Agrippa, Roddaz 1984, 568-571.

54 App. 14.40; Šašel Kos 2005, 375-379.

55 Strab. 7.5.2, 7.5.3, 7.5.10. Under Dalmatia, Strabo probably sees the land controlled by the Delmatian alliance, and views the territory of the Hellenistic Illyrian kingdom as the Ardiaei, see Dzino 2008, 182-183.

56 The more common Latin spelling of these groups is used. It has been suggested that Strabo took a description of the defeated groups from Tiberius' triumph, Colombo 2007, 7.

57 Aug. *RG* 30.1; see commentary in Ridley 2003, 154-157.

58 Tib. *Paneg.* 3.108-109; Ov. *Cons.* 385-390. Colombo 2007, 5 objects, preferring to view these mountains as Mt. Papuk (*Mons Claudius*). While we will never know what Ovid and Tibullus were thinking, and whether they wanted to be accurate or not – to their audience this would not matter.

59 However, that was really a fluid perception, as Ovid himself shows: *Adde triumphatos modo Paeonas, adde quietis subdita montanae brachia Dalmatiae*, Ov. *Ep.* 2.2.75-76.

60 The actual date of division is disputed and is dated from AD 8 to early Tiberian times. See: Novak 1966; Nagy 1970; Fitz 1988; 1993/95, 1.32-41; 2000; Kovács 2008; Dzino 2010a, 160; Šašel Kos 2010.

61 Noticed as early as Vulić 1933, 84-86, or Syme 1971 [1937], 19-21.

62 *CIL* III 1741 = *ILS* 938. The missing part was described by the Ragusian public notary, Marcus Sylvius, in 1547 and it was already missing when Theodore Mommsen saw the inscription in 1873. The arguments in favor of the historicity of the missing part were put forward by Novak 1965; Bojanovski 1988; Glavičić 2008, 45-48 and Šašel Kos 2010, esp. 125-126, but there are also good reasons to dispute this opinion: Fitz 1993/95, 1.33-34; Wilkes 1996, 565 n. 47.

63 Diplomas: *CIL* XVI 2; XVI 4; Dušanić 1998. The earliest epigraphic use of the term Pannonia from Pisidian Prostanna is dated to the time of Nero, see Kovács 2007; 2008. Whether it was the formal or informal use of the term remains disputed and inconclusive. See the recent discussion of Šašel Kos 2010, who argued that this inscription used informal terminology.

64 Pannonia: Vell. Pat. 2.104.4; 2.109.3; 2.110.2; 2.114.4; Pannonians: 2.110.5; 2.121.2; *pannonico et delmatico bello* 2.117.1, and the triumph *ex Delmatis et Pannoniis* 2.121.2.

the needs of their political geography and reorganization of conquered communities. One reason for this was the lack of established indigenous political institutions with developed systems of taxation and regional administration, comparable to the Eastern Mediterranean. Another was the attempt of the late Republican conquistadores, such as Julius Caesar, to explain and justify their conquests to public opinion in Rome. This landscaping of conquered regions with political power helped the Romans to construct new regions, which would become foundation-stones for imperial provincial organisations such as Gaul, Germany, Britain or Illyricum.[65] The creation of these geo-political constructs directly affected later Roman perceptions of heterogeneous indigenous communities from those regions as distinctive identities such as the 'Gauls' or 'Illyrians', although they originally did not share a sense of common identity.

The two discussed perceptions of the indigenous communities such as the *Pannonii* most certainly belong to the same context, showing two successive stages of Roman contact. The first one is the ethnographic generalisation of the indigenous groups. It probably originated in Hellenistic ethnographic perceptions, but the argument that Octavian/Augustus used it within the narrative of his *Memoirs* also seems convincing.[66] As the narrative of the *Memoirs* is now lost, it is difficult to argue why Augustus used it and what its function was within the narrative. Knowing the propagandistic purpose of this work, it was probably used as a narrative strategy for the justification of his unprovoked attack on the Segestani, and also for presenting his conquests as important to his audience.[67] By presenting the Segestani as part of a much larger Pannonian ethnic community, stretching from the south-eastern Alps to the Dalmatian hinterland, Augustus might have shown his conquest to be much more significant, while the presentation of the *Pannonii* as an amorphous mass of insufficiently civilised folk dehumanizes them and subtly justifies his actions in civilizing them - in other words conquering them militarily.[68] The other perception is somewhat easier to explain. It plainly shows new geographic and ethnographic discourse developed by a Roman colonial power, which submerges space with force, and divides it in order for easier rule and administration. Classifying indigenous communities into the predetermined categories of Dalmatians and Pannonians recreates their identities within an imperial and colonial context, accomplishing the conquest.

From available evidence there is nothing to suggest the existence of a common Pannonian identity before the Roman conquest, especially not as a shared identity-denomination of those communities. For a short while, the articulation of a common cultural *habitus* might have occurred during the Batonian war, but only as a result of shared historical experiences and opposition to the Romans, which did not last for long. The perception of the *Pannonii* as a loose group living south and north of the Sava valley probably arose in Hellenistic ethnography, and was accepted by the Romans who incorporated it into their own view of the indigenous population. The Romans first used the term to depict the population of the Dinaric Alps living north of the Delmatae and those inhabiting the valley between the Sava and Drava. Later, this perception was used to provide the name for the new region of Pannonia which was constructed as a distinct provincial space in the context of the wider colonial landscaping of conquered spaces in Continental Europe. The Romans divided communities previously perceived as the *Pannonii*, most of which joined to challenge Roman supremacy in the *bellum Batonianum* AD 6-9, into the two newly constructed geo-political provincial spaces of Pannonia and Dalmatia.[69]

5. Becoming Roman

The indigenous societies in Pannonia can be characterized as societies under stress after the Roman conquest. Significant transformations brought about by the establishment of Roman rule affected all aspects of daily life, including the construction of identities inside the Roman political and ideological frameworks in the period after the conquest – so-called 'Romanization'.[70] There are two basic views on the process of transformation in Pannonia defined in the earlier scholarship as either a top-down acceptance of Roman identity and the indigenous assimilation into this general Roman identity, or as resistance-discourse as continuity of indigenous culture and identity under the surface, with minimal influence of 'Romanization'.[71] Neither of these views can today provide us with sufficient insight into the multiple narratives of change occurring in any Roman provincial society, including Pannonia. Redefining culture as an inventive and continuing process rather than a system based on determined values and symbols allows us to see the development of provincial societies within the Roman Empire as the appearance of new and distinctive cultural practices. Provincials used symbols and artifacts related to what we view today as either indigenous or Roman culture, but combined them in forms which represented neither the continuance of indigenous, nor the assimilation

65 E.g. Woolf 1998, 242; Riggsby 2006, 28-32, 47-71; Krebs 2006; Osgood 2009 (Gaul); O'Gorman 1993 (Germany); Stewart 1995; Rutledge 2000; Clarke 2001 (Britain); Purcell 1990 (Cisalpine Gaul); Dzino 2010a, 80-84 (Illyricum). Nevertheless, it was not an invention of the period – the same attitude towards the landscaping of conquered spaces occurs as early as early 3rd century BC in Italy, Dench 2005, 162-165.

66 App. 22, taken from the *Memoirs*, fits nicely into the ethnographic prelude to Octavian's attack on the Segestica after he took Metulum – a chief stronghold of the Transalpine Iapodes. The same statement used earlier in *Illyrike* 14 looks out of context.

67 Dio, 49.36.1 clearly said it was unprovoked, while Augustus in App. 22.62 said that they were attacked for *"their arrogance"*. Segestica was considered to be an important point of attack for the Dacians, which Octavian might have considered, or pretended to consider App. 22; Strab. 7.5.3; Šašel Kos 2005, 397-398, 438-439; Dzino 2010a, 106.

68 *"These Paeones did not live in cities, but rather according to clans throughout the countryside and in villages. They did not gather for consultation, nor did they have collective leader"*, App. 22.63 (transl. Šašel Kos).

69 Dzino 2010a, 161 – see also below.

70 See basic literature listed in n. 4.

71 Mócsy 1974, 250-263; cf. Wilkes 1992, 254-259 (assimilation); Rendić-Miočević 1989 (surface Romanization).

into 'Roman' culture, but rather, new cultural practices and identities.[72]

The ways in which imperial provincials were 'becoming Roman' are hotly debated in recent scholarship. The idea that 'Roman' imperial culture was the exclusive result of elite-consensus and elite-negotiation[73] has been successfully challenged as too narrow, focusing only on a limited section of society and excluding the majority of provincials who did not have elite status. The view which slowly emerges is that the creation of 'Roman' imperial culture encompassed different and parallel processes which resulted in the appearance of a number of distinctive subcultures and regional cultures within provincial societies.[74] The evidence from Pannonia confirms these views. The process of interaction after Roman conquest caused the rapid transformation of existing identities, resulting in the construction of new ones by the descendants of the immigrants, whether they arrived from Italy, other parts of the Empire, or *Barbaricum* outside imperial borders.

The formation of provincial society in Pannonia began with the division of Illyricum into the provinces of Dalmatia and Pannonia.[75] As said before, the actual creation of the province of Pannonia and its administrative reshaping by the Roman authorities - first into peregrine *civitates*, and later into municipal territories - were active and deliberate interventions by the colonial power in an attempt to control the conquered space and population more efficiently. However, these measures are crucially important for our inquiry, as they directly affected the creation of future identities within the province. The division and reshaping of the space introduced and defined new spatial and political units, which were used as a key reference point for identity-perception by outside observers and as building blocks for identity-construction by the local population.

There are many questions which remain unanswered, such as the relationship between the indigenous and immigrant populations, which was certainly strained for at least an entire generation after the conquest, if not longer. From the evidence we cannot see if the friction and initially separate settlement places of these societies created opposed 'Roman' and 'Indigenous', or conflicting identity-narratives. On the contrary, in the long run these two societies worked in accordance with one another to bridge the differences and create new identity-narratives, affected not by origins, but rather by social contexts; joint experiences of living in the Pannonian provincial space and within the Empire. Elements we might today regard as 'Roman' or 'indigenous' were incorporated into forms and meanings recognizable to both societies in order to bridge their initial differences. The construction of imperial identities in Pannonia was dependant on social contexts, and sometimes even individual choices. There are too many variables that impact ethnicity and identity during the Roman period such as status, wealth, location, form of administration, the degree to which Rome asserted her presence, religion, gender that it becomes futile to pursue comprehensive approaches to this problem.[76] There are three identity-templates which can be distinguished from the available evidence: 'Roman' (i.e. global), provincial (Pannonian), and regional (*civitas* and municipal) identities. Each of them was fluent and depended significantly on social contexts and historical circumstances, which are in general much less emphasized in the scholarly analysis of provincial identities.[77]

6. Romanness/globalness

It is perhaps misleading to observe Romanness as a distinct and fixed set of values and symbols as such a thing did not exist: Romanness was a sum of all provincial cultures, united by imperial ideology developed in Augustan times.[78] It is more useful to redefine 'Romanness' here as a set of global practices and symbols established in the ancient Mediterranean world, some of which were articulated in Roman imperial ideology.[79]

The view that 'Romanization' was a top-down process of cultural assimilation, a gradual cultural 'progress' from 'native' to 'Roman', meant that the narratives of elite Romanness from Pannonia (especially those observable in the material record), were to be the prime focus of earlier scholarship.[80] Pannonia did not exist in a vacuum, isolated from the global Mediterranean world. The region experienced cultural and economic interactions with Italy and the Mediterranean world before its actual political inclusion and administrative re-arrangement. Nevertheless, that interaction significantly increased only in the period just before the conquest.[81] Velleius Paterculus mentions, whether over-exaggerating or not, the strong impact of the Latin language and Roman military training (*disciplina*) on the Pannonian and Dalmatian auxiliaries at the beginning of the *bellum Batonianum*.[82] The fast rate of acquiring the Latin language amongst lower class provincials is not necessarily evidence for a wish to participate in imperial culture, as it might be amongst the elite, but rather a practical matter, which enables opportunities due to

[72] The literature which goes beyond the dichotomy 'Romanization'/ resistance has grown over the last decades see e.g. Barrett 1997; Laurence 2001a; Woolf 1997; 2002; Van Dommelen 2007; Jiménez 2008; Hales 2010, to mention just a few.

[73] E.g. Terrenato 1998; 2001; Woolf 1998.

[74] Best defined in Hingley 2005, 91-116, see also the criticism of elite-centric approaches in Scott 1998.

[75] See above.

[76] Mattingly 2004, 10-11.

[77] As Oltean 2009, 93 pointed out: the issue of why a particular identity was expressed received much less attention compared to the issue of how a particular identity was expressed.

[78] Hales 2010, 242, cf. Mattingly 1997b, 17; Hingley 1996, 41-47.

[79] See Wallace-Hadrill 2008 on imperial culture and ideology, cf. Hingley 2005, 50-72, or Woolf 2001.

[80] Mócsy 1974 or the *ARP*. See the more recent and more insightful perspectives in Burns 2003, 194-247.

[81] See e.g. Tassaux 2004, 170-174 (trade); Dušanić 2008 (Roman business ventures before the conquest), and the evidence for trade with Aquileia in Strabo 5.1.8. For numismatic evidence on late Republican trade see: Popović 1987, 125-126, 140-147 (Pannonia); Kos and Šemrov 2003; Miškec 2004 (Eastern Alps).

[82] Vell. 2.110.5, see Mócsy 1983.

extended possibilities for communication.[83] It is very likely that through compulsory hostage-giving[84] and auxiliary service in the Roman army, the indigenous population - and in particular the indigenous elite - experience the earliest exposure to imperial culture. It does not seem adequate to regard the indigenous population of the conquest period and immediately after as a compact anti-Roman block. The initial Roman political interaction with this region resulted in the development of many personal and formal relationships between the Roman and indigenous elites, but also with the creation of a bi-lingual and bi-cultural people vertically throughout all levels of society who acted as cultural mediators.[85] Continually increasing numbers of military diplomas from the first century indicates a higher rate of homecoming for surviving indigenous auxiliaries in Pannonia (especially in the *civitas Breucorum*), so they should be regarded as important cultural mediators, similar to the returning Batavian auxiliaries.[86]

The placement of permanent camps for three Roman legions and some auxiliary units in southern and western Pannonia in the Iulio-Claudian era without doubt played the most significant role in managing Roman-indigenous relationships in the early period after the conquest.[87] In addition to the legions, civilians and military veterans settled in significant numbers during the post-conquest period.[88] The foundation of colonies, such as Poetovio, Siscia or Sirmium in the mid-later first century propelled the early immigrants into these ideological 'islands' of Romanness, which also gave a significant impetus to the development of the urban culture in Pannonia in the later period.[89] The cities were certainly part of a deliberate policy to impose a tighter Roman hold on this region. The 'ideal' Roman town, that colonies frequently represented, was an excellent template for establishing a new, universal Roman (i.e. global imperial) identity through newly conceived imperial images of public spaces.[90] The projection of ideological global Romanness functioned as necessary social glue, holding together cosmopolitan urban societies which formed in Pannonian cities.[91] Nevertheless, it is important to bear in mind that Roman Pannonia had a significant watershed in its history during the Marcommanic wars (166-180), when the province suffered a degree of depopulation, 'Barbarian' settlement and a subsequent new wave of immigration from Thrace, North Africa and Syria in the third century.[92]

Material evidence from scattered individual finds indicates the development of a cultural bricolage in the years after the conquest, with the use and recombination of Mediterranean and 'Roman' artifacts in indigenous contexts of southern Pannonia. It is interesting to note, for example, the recent finds of Mediterranean foods, such as grape seeds or figs besides local pottery in indigenous graves, soon after the conquest. In the same context we can observe individual examples of the indiscriminate use of La Tène and Roman-type weapons and the evidence for local production, which appears to be a cross-over between the two.[93] However, the use of Roman or Mediterranean objects is not a sign of cultural change *per se*. The use of these objects varies in different contexts and we are unable to understand the meanings and contexts; they were possibly statements of unity/disunity with the Roman invaders, or the Roman artefacts and symbols were used in order to articulate and justify local claims to power.[94]

The Roman soldiers, settlers and administrators could not make the Pannonian indigenous population 'Roman' by their presence alone or by force of Roman arms. Although they were in a dominant position over the indigenous societies in terms of power, they were lacking in numbers, and soldiers in particular formed as separate and distinct social groups which projected their own narrative of Romanness.[95] Changes in identity-construction occurred through a process of cultural mediation. We can see the developments of certain cultural interfaces[96] and cultural practices, which probably helped in crossing boundaries and mediating between the descendants of immigrants and the indigenous population: the worship of Silvanus, one

[83] For example, the rise of Latin proficiency and literacy amongst the Batavians: Derks and Roymans 2002; Hingley 2010, 64-69.
[84] *CIL* III 3224; Šašel Kos 2005, 455-458.
[85] See also Dzino 2009, 35-39 for members of the regional indigenous elite finding themselves between two worlds in the period of conquest, and Woolf 2009; 2011, 8-31 on individuals in the early Empire who acted as cultural mediators, operating in both worlds – indigenous and Roman.
[86] See Derks and Roymans 2006 for returning Batavians. The discovery of military diplomas from the first century is most frequent around Sirmium (*civitas Breucorum*), see Dušanić 1978; 1998; Dorn 1984; Miškiv 1998; Koledin 2000, and also *CIL* XVI 2; XVI 17; cf. *CIL* XVI 31 (*civitas Iasorum*).
[87] See Dzino 2010a, 167-168 with bibliography.
[88] See Fitz 1993/95, 1.119-124; Ferjančić 2002, 21-55 for military settlement and colonisation.
[89] On the cities in Pannonia see the comprehensive *Pannonia* I, II.
[90] See Whittaker 1997, 144-148; Zanker 2000; Hingley 2005, 77-87 on the role of cities and Carroll 2003; Romans 2004, 196-205 on Roman city-planting as a deliberate policy of control in the western frontier provinces.
[91] See the evidence for population structure in early imperial Siscia, from the names on lead *tesserae* (commodity tags) Radman-Livaja 2010, 541-547.
[92] See Barkóczi 1964; Fitz 1980, 151-154. It is important to note that the present state of evidence is insufficient to provide any definitive conclusions on the degree of depopulation the Marcommanic wars might have caused in the cities of southern Pannonia.
[93] Dizdar *et al.* 2003 (grapes, figs and north Italian products in graves from Ilok); for the wider context of imports to the Danubian provinces see Tassaux 2004, 174 f. and Dizdar and Radman-Livaja 2004 (weapons). It is difficult to agree that the use of local pottery or 'Celtic' swords reflect a particular identity (e.g. 'Celtic' soldiers in the Roman army), and that it can be used for the ethnic identification of individuals, as concluded in Dizdar *et al.* 2003 and Dizdar and Radman-Livaja 2004.
[94] Laurence 2001a, 95-98, see also the more comprehensive exploration of links between Roman objects and culture in Hingley 2005, 72-90.
[95] See e.g. Haynes 1999; James 1999b; 2001; Mattingly 2006, 166-224 on the identity of the Roman soldiers as a separate social group with its own identity-narratives.
[96] Similar to the concept of *Colonial Middle Ground* developed in the research of American frontier society – accomodation and mutual incomprehension, a new means of discourse developing in a neutral context, showing the inability of both sides to gain the upper hand through force, White 1991. See Malkin 2004, who used it in the context of Greek colonial encounters, Woolf 2009, esp. 209-210 in the early imperial West, and Ulf 2009 for comparative perspectives.

of the major divinities in Roman Pannonia for example.[97] The other is the 'epigraphic habit', and in particular the introduction of figured and Latin-inscribed funerary *stellae*, previously unknown in this region. This typically Italic custom and medium of expression was used by the migrants and indigenous population as another cultural interface.

However, the accuracy of the evidence coming from inscriptions is questionable. A significant corpus of personal names found on a lead *tesserae* in Siscia shows differences in the proportion of indigenous (La Tène or 'Illyro-Pannonian') personal names, when compared with the evidence from public and funerary inscriptions. The portion of Roman names found on inscriptions implies that the population of Siscia presented a strong sense of Roman identity in this particular context, as the proportion of indigenous personal names appears to be negligible.[98] The evidence from the lead *tesserae* is dated mostly to the first and second century. It still shows the dominance of Roman and Italian names (53%) but at the same reveals a much more significant proportion of indigenous names (28%) than shown in epigraphic evidence.[99]

7. Pannonianess

The construction of Pannonia as a single and distinctive space in the Roman administrative scheme had a far reaching consequence for the development of identity-discourses, especially on the formation of outside perceptions of Pannonians and Pannonianess. As already said, the Augustan period was characterized by the reshaping of conquered spaces, but it also had a decisive influence on the perception of those regions in the future, regardless of later administrative changes.[100] This is best seen in the written sources. Tacitus, for example, has no doubt in perceiving the space as Pannonia and its population as Pannonians.[101] Other sources usually follow the same perspective, although the perception of the wider region as Illyricum must be noted. This was based on the earlier province of Illyricum and the central Danubian tax-zone, *portorium Illyrici,* sees writers of the 2nd and 3rd modernsometimes inconsistently applying the terms 'Illyricum' and 'Illyrians' to the whole region, including Pannonia and its inhabitants.[102] The outside perception of the Pannonians and Pannonianess evolved slightly from what it had been during the conquest period. They were still excluded from imperial identity-narratives for being peripheral and culturally inferior. Now, however, they were not complete outsiders but internal Barbarians of the Empire, customarily 'thick' and slow-witted, but brave-hearted and warlike.[103]

The formation of a Pannonian province also had an impact on its population. The pre-conquest Pannonian space, which Roman 'ethnography' perceived as a compact 'ethnic' space and populated in their perception with the *Pannonii*, was now divided into Dalmatia and Pannonia, so that the descendents of the Daesitiates, Maezaei and Pirustae started to form their identities in a different way, within the mindset of the Dalmatian province. Also, the formation of Pannonia brought together the communities between the Drava and Sava valley (which belonged to the fringe of the La Tène world), with the communities in western Pannonia, such as the Boii or Eravisci, which were much more firmly entrenched within the cultural practices of the La Tène world. The difference in cultural traditions might have affected the internal regionalisation of Pannonian space, especially between its south and north/west regions in the early years of the province, resulting in different ways of renegotiating indigenous identities in imperial and provincial contexts, which is discussed below.

Pannonian identity appears clearly expressed in some diasporic contexts, through tombstone epitaphs or votive dedications. The sailors in the Misene and Ravennate fleet developed specific identity-narratives of Pannonianess in the later first and second century, visible in the frequent self-statement of identity as *natione Pannonius,* advertised on their tombstones in Italy. This identity-discourse is comparable to the identity-statement *natione Delmata,* shown on the tombstones of their fellow sailors from the Dalmatian province in the same period. We cannot see them as an original affiliation but rather as a reinvented identity of indigenous sailors who accepted an outside identity-label and reshaped it to form their own social and sub-cultural identity in this very particular context.[104] However, this was not the only way of expressing Pannonianess. Some individuals preferred to express their identity and origins in accordance with the provincial structure, such as 'from' or 'born in' *provincia Pannonia inferiore/superiore* after the division of Pannonia in Trajan's time, while others stated Pannonian identity besides municipal identity, or combined provincial with municipal identity.[105] Pannonian identity was also expressed in accordance with the stereotypes about the Pannonians; we have a good example of one Murranus, who referred to himself as the 'Barbarian from Pannonia'.[106] Pannonianess in relation to female identities is pretty rare.

[97] See the more advanced approaches to the problem in Perinić-Muratović and Vulić 2009 although failing to notice the hybrid aspects of Silvanus in Pannonia, and its role as a cultural interface. There were other cults in the wider region, which can be seen in the same context, such as Vulcanus in Dalmatia (Sanader 1996) or Hercules in the central Balkans (Gavrilović 2007).

[98] See Mócsy 1959, 24-26, 211-212; Barkóczi 1964, 259-261, 329-331.

[99] Radman-Livaja 2010, 541-548, and 549-555 (dating).

[100] Talbert 2004. There was a general approach taken by Roman authors to project post-conquest boundaries of the regions into the past, Isayev 2010.

[101] Tac. *Ann.* 15.10; Tac. *Hist.* 2.14; 2.17; 3.12; Tac. *Germ.* 1.

[102] Gračanin 2005.

[103] Stat. *Silv.* 1.4.79-80 (warlike); Front. *Princ. Hist.* 2.13 (stupid); Dio 49.36 (brave, bloodthirsty and primitive); Hdn. *Hist.* 2.11 (skillful in battle but slow-witted).

[104] See Domić Kunić 1996a; 1996b for the sailors from Dalmatia and Pannonia in Ravennate and the Misene fleet; Dzino 2010b for the discussion on Dalmatian identity amongst these sailors. Certainly the identity *natione Pannonius* was not limited to the navy, and appears in different military contexts: e.g. *CIL* VI 2488, 3184, 3241, 3239, 3285, 15011, etc. or even in civilian contexts *CIL* XIII 7247.

[105] For example *CIL* VI 3297: *Ulpius Cocceius ex Pannonia Superiore natus ad Aquas Balizas pago Iovista vico Cocconetibus.* See Noy 2000, 217-218 (the Pannonians in the city of Rome), and also below, n.121.

[106] Mancini 1933. Burns 2003, 36 sees Murranus as a descendent of 'Barbarian' immigrants or "*bypassed people from interior*".

It follows male models: *natione Pannonia*, or *ex provincia Pannonia*.[107] It also appears as a cognomen-nickname, such as two women called Iulia Pannonia, who both died in Mauritania.[108]

Pannonianess develops as a specific set of identities from early Imperial times, evolving in different social and geographical contexts. Pannonian identity appears as a result of the common living and shared historical experiences of both the indigenous population and immigrants in the peripheral province, which were excluded from participation in wider imperial power-networks. It was a product of the creation of Pannonia as a type of a *Colonial Middle Ground* zone - a way to negotiate the common identity of immigrants and the indigene population, and to express it in foreign surroundings, especially in the specific social context of the Roman army. On the other hand, the development of this identity was also affected by a cultural change associated with the creation of a wider Pannonian space as a part of the Empire. The significance of the Empire as a unified political and economic space significantly increased the importance of mobility, which in turn created the need to specify personal identity, especially in a diasporic context, and present it in ways other than just a place of birth.[109] Nevertheless, Pannonian identity was only one level of identity-construction, fragmented into a number of different regional narratives. These regional identity-narratives made Pannonian identity within the province very complex and probably sense of Pannonian identity held less importance for those who did not leave the province.

8. Regional identities

Peregrine administrative *civitates* in Pannonia were probably formed at the same time as the organisation of the province. Thanks to Pliny the Elder, the early Pannonian administrative structure of the mid-first century is known. It consisted only of peregrine *civitates*, unlike Dalmatia, where *conventus* was placed above *civitas* in the administrative structure.[110] It is important to say that we are unable to establish how accurately these *civitates* reflect the pre-Roman political situation or shared identities of the indigenous population. Comparative research on other indigenous communities in the Roman west reveals that peregrine *civitates* were not necessarily the exact picture of pre-Roman identities, either political or ethnic.[111] We can assume that there were some interventions of imperial authorities. It seems very possible that the important Breucian alliance was broken into smaller parts, while the name of the *civitas* – Colapiani – which appears in Pliny's list of Pannonian *civitates*, suggests that interventions occurred in the case of the polity dominated by the inhabitants of Segestica – the Segestani – who capitulated to Octavian in 35 BC.[112] Some, or most of these *civitates* in the beginning were administered by the *praefecti civitatis* (Roman military officers), but control passed entirely to the civilian indigenous elite sometime during the Flavian period.[113] The Pannonians at first served in 'ethnic' auxiliary units which were locally recruited, such as *VIII cohortes Breucorum*, *cohors Varcianorum et Latobicorum*, or were more general such as *cohortes Pannoniorum* (sometimes combined as *coh. Gallorum et Pannoniorum or Dalmatarum et Pannoniorum*). Thus, it is not surprising that the indigenous auxiliaries from Pannonia identified primarily with their *civitas* identity in this early period in diasporic contexts.[114]

A more liberal attitude to the enfranchisement of the peregrines, which began in the Flavian and accelerated into the Antonine period, resulted in the development of a municipal structure in Pannonia. The *municipia*, such as Mursa, Siscia, Andautonia, Sirmium, Savaria and Sopianae, developed in existing indigenous settlements, or former military bases.[115] The epigraphic evidence for the identity of the inhabitants of Pannonia from this diasporic context, shows that a municipal identity takes over from the earlier *civitas* identity, especially in the 2nd century, whether as a full identity expression or combined with a provincial or Pannonian identity.[116] Allegiances towards old *civitates* were not important for the new generations. However, they still remained faithful to their local identities – but in a different ways. It is interesting to note that the inhabitants of Pannonia, similar to those from Thracia and Moesia, strongly emphasised their local origins. In inscriptions they often denoted the *vicus*, *pagus* or *pedes* of their birth, depicting the importance that they attached to those local identities.[117] The civic cults, established within the province, contributed to the further binding of the local population to joint civic identities through the experience of common practices and the use of common images and symbols.[118]

Local identities are also reflected by those Pannonians who chose to keep their indigenous names in the epigraphic record, whether 'Pannonian' or La Tène. The choice of

[107] *CIL* VI 13336: *Aurelia Iustina nat(ione) Pannonia*, cf. 6.2708; *CIL* VI 2501: *Aurelia Crescentina civis Pannoniae*; or VI 37271: *Iul(ia) Carnuntilla ex pr(ovincia) P(annonia) Super(iore)*. Carnuntilla's cognomen clearly shows her local identity - Carnuntum.

[108] *CIL* VIII 3799, the same person featured on the mother's tombstone *CIL* VIII 3588. There is another Iulia Pannonia in Mauritania *CIL* VIII 4277.

[109] Laurence 2001b, 90-91 using Roman Britain as an example. See also Woolf 1995 on the formation of distinct and discrete provincial cultures within the Empire.

[110] Plin. *HN* 3.148.

[111] See comparative examples in Britain in Jones 1997, 29-36 for example.

[112] Mócsy 1957; 1974, 53-55; Fitz 1993/95, 1.30-31, 1.116-118; Domić Kunić 2006, 76-81.

[113] We have evidence of the praefects for *civitas* of the Colapiani (*CIL* III 14387) and the Boii and Azali in the north of the province (*CIL* IX 5363). That is certainly not to say that all *civitates* were administered by Roman military personal in the Iulio-Claudian period. The abolition of military control might have coincided with the foundation of colonies such as Siscia in the early Flavian era, Mócsy 1974, 134-137.

[114] Domić Kunić 1988, see the earlier Holder 1980, 132 and Birley 1988 for Pannonians in Britain.

[115] Mócsy 1974, 112-115, 134-147; Fitz 1993/95, 2.414-427.

[116] Also, in comparative perspective we can assume that those units started to loose its 'ethnic' homogeneity, like for example the Batavians, see Van Rossum 2004.

[117] Noy 2000, 219-220 (the evidence from the city of Rome). The Thracians and Moesians mostly stated only *vicus*, but not *pagus* or *pedes*.

[118] See Scherrer 2004 on civic cult and identities in Pannonia.

names amongst the soldiers seemed to be a very personal decision, reflecting a personal identity-narrative, rather than the existing bipolarity of Roman vs. indigenous identity. An example to support this view is the epitaph of praetorian M. Aurelius Dasius. Dasius kept his indigenous, non-Roman cognomen, while his brother M. Aurelius Candidus (who was also a soldier, but serving amongst the *equites singulares*), used a Roman cognomen.[119] The use of indigenous names declines over time, and from the third century they are completely replaced with Roman names, after Caracalla's famous citizenship edict. There is sufficient evidence for the popularity of certain Roman names in Pannonia: their frequency in the epigraphic record. Even within Pannonia some names are much more frequent in certain municipalities than in the others.[120]

Apart from these strongly localised identities, there are also specific identity-patterns which distinguished communities of western and northern Pannonia from those in the southern parts of the province. We can clearly see the existence of a distinctive Norican-Pannonian cuisine,[121] or the so-called Pannonian-Norican dress, recognizable through characteristic elite female head-covers known as Pannonian-Norican 'turbans'.[122] These female head-covers are known from Roman-era tombstones and we do not know if they were a recent or traditional fashion. It is much more likely that they represented a recent innovation which manipulated traditional elements into new forms.[123] Also, we are unable to see more clearly the way these 'turbans' functioned as gendered identity-indicators for the society which used them. It might be that they articulated elements of existing regional cultural *habitus* to express a specific regional identity of the La Tène communities from western Pannonia and Noricum, in particular advertising the social status of individuals.[124] It is also possible that they functioned as identity-demarcation signifiers between new provincial/regional identities in western Pannonia and Noricum, for there are subtle differences between the representations of Pannonian and Norican 'turbans' (and clothes in general), as well as regional differences between the provinces and inter-provincial similarities.[125]

More evidence for this distinction between the western/northern and southern parts of Pannonia can be seen in the visual representations of individuals on funerary *stellae*. The funerary *stellae*, most of which are located in west, and north of Pannonia allowed the flexible expression of personal identities through the visual representations of the deceased, using Roman and/or indigenous names and the Latin language. These representations were often strikingly un-Roman in the choice of dress or in the presentation of social values. The funerary *stellae* reflect what we can describe as a frontier society – a complex contact zone between the 'inner core' of the Empire, the indigenous communities, and the population outside the Empire. These specific circumstances created internal division within Pannonian provincial society, making the direct frontier experience among the northern and western communities much more embedded within their identity-discourses when compared with those who lived in the south of the province. It established it on more egalitarian grounds with less emphasis on social status than in the 'inner core' of the Empire, such as Rome or Italy.[126] Roman elite culture was international, but in essence it was extremely socially exclusive,[127] so the egalitarian nature of (northern) Pannonian society could be a result of their exclusion (immigrants and indigenous population equally) from the main power-channels of the Empire in the first few generations, creating a more intense cultural mediation and the amalgamation of their identities.

We can see that the Roman period brings with it different identity-narratives which developed in different and often opposed contexts, such as domestic/diasporic, military/civilian, provincial/local, northern/southern. Inside the province, identities seem to develop through cultural bricolage, where different cultural symbols were recombined into new cultural practices. The evidence from the diasporic contexts may suggest that the identification with the *civitas* of birth in the early Principate was replaced with the identification with the *municipium*, but also in some cases more locally with the village of someone's birth. Outside observers - the members of the imperial elite - perceived the identity of provincials as Pannonians in accordance with a world-view which developed in the early Principate, abandoning the earlier Hellenistic ethnographic perception of the *Pannonii* from the time of conquest. Outside of the province, the provincials (mainly soldiers or sailors) identified themselves in different ways in the early Empire. The auxiliaries used to identify themselves with the *civitas* of their birth, while sailors preferred affiliation with a Pannonian identity, which was in essence the reinvented military identity of peripheral peoples, similar to the construction of a military identity for the Batavians or Dacians.[128] From the second century onwards, soldiers from Pannonia combined different elements in the expression of their identity such as a general Pannonianess and/or statements of affiliation with the province, *municipium* or

[119] *CIL* VI 32680.
[120] Mócsy 1985.
[121] Swan 2009, 15-19.
[122] Garbsch 1965, 13-22; 1985, 559-562.
[123] Hales 2010, 234.
[124] In this context, social status seems to be understood in very fluid terms, more as sign of wealth, than of status, or origins. A stunning recent find of a *stela* from Donji Čehi near Zagreb, depicts the rich slave family of Valens and Melania from the mid-2nd century. Both of the females, mother and daughter wear type H3 of the Pannonian-Norican 'turbans', Migotti 2008a, esp. 459-460.
[125] It is important to note that significant similarities in dress from the imperial period exist throughout the Alpine and Danubian provinces, but also in Dalmatia and Moesia, Čremošnik 1964; Garbsch 1985, 547, 576; Martin-Kilcher 1993. For the differences between Norican and Pannonian 'turbans': Garbsch 1985, 560-561.
[126] Boatwright 2005, cf. Burns 2003, 224-225 on Pannonian *stellae*. The evidence is based mostly on the northern, north-western and western parts of the province, Mócsy 1974, 147-153 and the discribution of the tombstones on Fig. 26. On the frontier society developing around Roman frontiers see: Whittaker 1994, 98-131; Elton 1996; Mattern 1999, 109-114.
[127] Dench 2005, 34-35.
[128] Batavians: Van Driel-Murray 2002; Roymans 2004; Derks 2009, Dacians: Oltean 2009.

vicus – sometimes a more local and more general identity was created, or even more levels of their identity.[129] These identities did not exclude each other: one could belong to the local community or social class, his/her *civitas/ municipium*, and also feel Pannonian and Roman at the same time.

9. Later Antiquity

The donation of Roman citizenship by Caracalla in 212 AD removed the legal distinction between citizens and non-citizens inside the Empire, but also created an important legal demarcation between citizens and 'Barbarians' who were outside the empire. As Burns argues, a consequence of Caracalla's edict, the attraction of Roman citizenship and identity diminished and the late antique population became freer to construct their identities in a more diverse way than in the earlier period. The immigrants, who were regularly arriving in Pannonia from across the border, took different approaches towards the construction and presentation of their identities whether by maintaining Barbarian otherness, taking on a Roman identity or merging both. While this was underway, the population of the Danubian provinces, in which Pannonia belonged, began to use more indigenous elements in the construction of their identities.[130] Pannonian identity-narratives were expressed in even more fluent ways in later antiquity, connected only through the common sense of belonging to the Empire, regional sense of *patria Pannonia*, and participation in a new religious identity which spread quickly throughout the ancient world – Christianity.

The rise of a home-grown army (the so-called Danubization of Roman legions which began to fill the legions with local recruits in the second century), resulted in the establishment of a specific military identity within those legions: the *Illyriciani* (*exercitus Illyricianus*). These were the product of general trends within the Roman army, which resulted in its regional diversification, and the integration of the legions into the distinctive provincial cultures in which they were planted.[131] The initial heterogeneity of indigenous recruits in the Danubian legions (Pannonians, Dalmatians, Moesians, Thracians)[132] was bridged by the wider regional cultural *habitus* they shared and which acted as 'glue' in the construction of the *Illyriciani* identity. The construction of this identity was also articulated through ideological catch-phrases such as *virtus Illyrici*, or the symbols such as *genius exercitus Illyriciani*, advertised on the coinage of Trajan Decius and Aurelian. This identity-discourse originated from a wider ideological discourse represented by symbols, such as the *genius Hercules Illyricus*, the *terra Hillyrica* or the *genius portorius Illyricus*, which did not have military origins, but related to the entire Roman colonial construct of Illyricum.[133] We can also see expression of this military identity (and marker of status identity within this sub-cultural group), in material record, through appearance of the Cross-bow brooches with the portraits (Type 5), which are most frequently found in the Danubian provinces.[134] The *Illyriciani* were not unlike the Batavians: a constructed identity recruitment of 'peripheral' peoples as ethnic soldiery, excluded from the centres of political power and encouraging existing military traditions. The *Illyriciani* claimed a position for themselves within the imperial structures of power from which they had previously been excluded. Hence, the negative image of the Pannonians and *Illyriciani* in Roman literature was a reaction of the elites from the inner, 'cultured' core of the Empire to their rise in prominence.

The time of the Severan emperors represents the period after which the sources attest to the rising political power of the *Illyriciani*. The rise of military emperors resulted in a number of Roman emperors being chosen from amongst the ranks of the *Illyriciani*. They were the 'protectors' of Rome and Roman values and representatives of the new military elites that emerged in the Roman world of later antiquity; 'the best thing' for the Empire, as they liked to be called.[135] It seems that the power of the *Illyriciani* was crushed, or at least tamed and kept under control, by Constantine the Great and his successors. The election of Jovian as emperor, however, after Julian's death in Persia (363), and the election of Valentinian I after Jovian's death (early 364), returned power to the hands of the Danubian legions: more precisely, to the Pannonian emperors.[136]

A specific context of external and internal perceptions of the Pannonian identity-narrative is clearly visible during the time of the Pannonian emperors: Jovian, Valentinian I, Valens and Gratian, from 363 to 383. They were the offspring of a specific military culture of the *Illyriciani*, which we mentioned earlier. Fighting for Rome and protecting her frontiers, they developed their own version of Romanness, viewing themselves as the guardians of ancient Roman virtues.[137] In many ways this collided with the narrative of Romanness maintained by the sophisticated elites of Rome and the inner Mediterranean;

[129] E.g. Aurelis Verus from *CIL* VI 37213: *natione Pannonius pede Sirmese pago Martio vico Budalia*. Otherwise the minor *Vicus Budalia* (near Sirmium), became a known place when its native Trajan Decius became Roman emperor.

[130] Burns 2010.

[131] E.g. James 1999b; Mattingly 2006, 247-52; Gardner 2007.

[132] See Wilkes 1999.

[133] GENIUS EXERCITUS ILLVRICIANI: RIC IV/3, 16c; 17b (Trajan Decius); VIRTVS ILLVRICI: RIC V/1, 378 (Aurelian); GENIUS ILLVRICI: RIC IV/3, 15b, 116, 117 (Decius), RIC V/1, 110, 223 (Aurelian) and the coin minted by Herenius Etruscus AD 250-251 (Rendić-Miočević 1990, pl. V). *Hercules Illyricus*: *AE* 1948, 86.

[134] On Cross-bow brooches with portraits (Type 5) see Migotti 2008b, esp. 61-68 (status-symbol), and also Coulston 2004, for dress as a tool for expression of military identity.

[135] ... *optimi rei publicae fuere*, Aur. Vict. 39.26; Matthews 1975, 32-55; Mócsy 1977; Lenski 2002, 56-61; Brizzi 2004. *Illyricianus* replaced the simpler terms *Illyrius* and *Illyricus* and was used almost exclusively in a military context, with only a few exceptions, Kuntić-Makvić 1996.

[136] Persuasively argued by Caldwell 2007, 37-74. The rise of the Pannonian faction after Jovian's death might be a consequence of a power-struggle in Constantinople, rather than their political influence, Olariu 2005.

[137] The idea that people from the peripheral areas of the Empire, or even from outside the Empire were, in the words of Clarke, 'a depository of old Roman values' was as old as Tacitus, see O'Gorman 1993, 147-149; Clarke 2001, 106-109; 2003, 47-51.

they despised these 'new Romans' for their lack of manners and rusticity.[138] The literary perception of the Pannonians in later Antiquity changed slightly, allowing for the first time, the appearance of positive stereotypes, while the existing negative stereotypes were further elaborated. The positive stereotype emphasized the manliness, bravery and military valor of the Pannonians. This was probably developed and maintained by the *Illyriciani* themselves as a self-perception. The maintained negative stereotype saw them as a crude, uncultured and cruel bunch: the internal 'others' of the Empire.[139]

After the crisis of the later second century (with the invasion of the Marcomanni and Quadi), and the tumultuous third century, the fourth century evidence shows a blossoming of urban life in the southern Pannonian cities (Siscia, Mursa, Sirmium) and the rise of fortified *villae rusticae* in their countryside which represented the backbone of Pannonian economy.[140] The development of urban life cultivated fertile soil for the early development of Christianity in the region, making the picture of identities even more complex with the creation of dichotomies Christian-pagan, and when Pannonians took sides in frequent inter-Christian disputes.[141] St. Jerome is a good example of a personal identity-narrative from the region. He was born in Stridon, an unidentified town somewhere on the frontier between Pannonia and Dalmatia, to a rich landowning family.[142] St. Jerome's personal identity was complex. It was interwoven around his sense of Christianity, Romanness, and also his social class of provincial elite – the reason for his lamentations on the rusticity of his fellow-countrymen, used as a literary technique to reinforce his elite-status.[143]

Both the rebellion of the Gothic *foederates* in 378 and the gradual detachment of Pannonia from the empire represent the point where we end our enquiry. Despite significant emigration from the province,[144] after the *limes* on the Danube ceased to exist, the domestic population did not disappear– there is more than enough evidence to the contrary.[145] Nevertheless, attraction to the Roman identity in Pannonia gradually disappears, as the distinction between 'non-Romans' and 'Romans', imposed by the existence of the frontier, becomes meaningless. Militarised society, formed in an earlier period on both sides of the *limes*, offered other, more attractive ways of self-definition over the rigidly stratified late antique society.[146] The Pannonians are again perceived in some written sources as foreigners to the Mediterranean world. They are the 'Barbarians' and 'Others' in a poem of Martin of Braga, dedicated to St. Martin of Tours, who was also a native of Pannonia.[147] The notion of Pannonian *patria* remained throughout Late Antiquity and it was used by Charlemagne's administration in the ninth century as an anachronistic projection of antiquity in a different historical context.[148]

10. Conclusion

This discussion of Pannonian identities throughout antiquity leads to some interesting observations and conclusions. The most important is, perhaps, an emerging view that internal and external perceptions of Pannonianess remained fluid and contextual. A careful reading of the sources shows how, under the label 'Pannonians', we can recognize a plurality of different identities hidden beneath the surface. These identities were constructed in different chronological and social contexts, externally by foreigners or internally as specific sub-cultural group designations.

The original perception of the indigenous Pannonians (*Pannonii*) in Hellenistic ethnography was formed through the outside recognition of existing cultural similarities within the wider region, rather than being a self-perception shared by those Iron Age groups. This outside perception, which can be clearly seen in the *Memoirs* of Augustus, preserved in Appian, and in Strabo, imposes a certain order and logic for external observers. It projects a single ethnographic stereotype onto political and kin-based groups, which participated in maintaining a joint wider cultural *habitus*. However, this cultural *habitus* of the *Pannonii* was visibly fragmented by a complex set of horizontal and vertical social networks established between the groups, which were defined through different mechanisms of inter-group negotiations, appearing as either social inclusions or exclusions.

The Roman conquest and imperial reshaping of the geographical space brought important changes, including a major discontinuity from pre-conquest identities. Roman administration divided ethnographic Pannonian space, creating a new region – Pannonia. This included only the northern communities of ethnographic *Pannonii*, and joined them with the La Tène population who lived closer to the Danube. In this arbitrarily established mixture of indigenous societies, military camps were planted. Settlements for the veterans and colonists, which grew into important cosmopolitan cities, projected the ideological discourse of the Roman Empire into provincial society as elsewhere in the Empire. The position of Pannonia as a frontier province significantly impacted upon the creation of specific identity-discourses within it. Contact with societies

[138] Alföldi 1952; Dzino 2005.

[139] Lenski 2002, 48-49, 86-87. Manliness and bravery: *Pan. Lat.* 10 [2] 2.2; 11[3] 3.8-9; Auson. *Epigr.* 3.3-4; 4.4; Amm. Marc. 21.12.22; 29.6.13. Crudeness, cruelty and a lack of culture: Aur. Vict. 39.17, 26, 40.12–13, 17; 41.26; *Epit. de Caes.* 20.10; Julian *Mis.* 348 c–d (the Moesians); see also den Boeft *et al.* 2008, 19 for perceptions of the Pannonians in Ammianus.

[140] Mócsy 1974, 297-338, and Lenski 2002, 37-44 for a more recent outlook – in addition see Thomas 1980, esp. 312-317 and more recently Mulvin 2002 on late Roman *villae*.

[141] Bratož 1990; 1996b; Kovács 2003 and for material evidence for Christianity see: Migotti 1997; Gáspár 2002.

[142] Hieron. *Vir. Ill.* 135. Stridon was not successfully located, despite many attempts, Šanjek 2005.

[143] Rusticity: Hieron. *Ep.* 7.5, see Dzino 2010c, 71.

[144] On emigration from Pannonia see Wilkes 1972; Bratož 2007, Periša 2009, 239-242.

[145] Bratož 1986, 377-78; Gračanin 2008, 71-77 (ecclesiastic infrastructure); Vida 2007, 31-38; Stadler 2007, 59-70 (archaeology).

[146] Amory 1997, 277 313.

[147] *In Basil.* l. 13.

[148] Anon. Raven. 4.15-22, cf. the mention of the Pannonians within the Avar khanate in Paul *HL* 4.37.

outside the Empire, especially in the region north of the Drava, influenced the formation of specific frontier zones on both sides of the Danube. It was at the same time a contact-zone with the outside societies, whether through exchange, immigration or conflict, but also a front line of imperial defense, leading provincials to develop specific narratives of Romanness, which differed from Romanness as constructed in the internal core of the Empire. On the other hand, the region of southern Pannonia was much less exposed to the direct experience of the frontier, and its inhabitants developed different identity-discourses.

Shared historical experiences of living in the same region, and different socio-economic circumstances in which those experiences were had, created identity-narratives of Pannonianess, which might be detected in the sources from the early Imperial period, before the third century AD. Imperial Pannonianess developed as a kind of *Colonial Middle Ground*; a result of complex trans-cultural negotiations which brought together different societies, the descendants of migrants settling from Italy and other parts of the Mediterranean and the indigenous population. In a way, neither group was able to fully dominate. The indigenous inhabitants were controlled by the power of the Roman army, while the immigrants and army were excluded from the imperial power-structures, because Pannonia was positioned as a peripheral region in relation to the distribution of imperial power in the early Principate. Being Pannonian in the early Empire represented an affiliation with a set of local traditions, established as a response to the global, imperial culture. In that context, inhabitants of Pannonia were using 'Pannonian' as a joint denomination to position and self-identify themselves within the Empire in different contexts, at home, or in diasporic communities, mostly amongst the soldiers serving in other provinces.

The changes in the mechanisms of imperial power-distribution in the third century enabled some Pannonians to actively renegotiate their position within the Empire. The construction of the identity of the *Illyriciani* was a product of a military subculture of the Danubian legions, which now used local recruits. It enabled provincials to take a position in imperial-structures as 'defenders' of Roman values, and important power-brokers who now participated in the election of the emperors – often chosen within their own subculture. The rest of the population continued to identify with their local and social identities, but also through other mechanisms of social inclusion and exclusion created by the spread of Christianity. When the relationship with Roman identity lost its significance, especially after the *limes* on the Danube ceased to play any role in the division of societies, the Pannonians shifted the ways they expressed their identity to that of a frontier society, which developed around the *limes*. Pannonian identity was formed as a local response to the global 'Roman' world in order to articulate the claims of its inhabitants to participate in that global culture. The fragmentation and regionalization of this global world, epitomized within the Roman empire, made the active maintenance of Pannonianess obsolete and useless – its inhabitants in Late Antiquity and early medieval times found it much more beneficial to associate with other identities, while the term 'Pannonian' remained for some time a regional designation.

Bibliography

Adamik 2003 Béla Adamik, Review of Anreiter 2001, *AAntHung* 43, 262-268.

Alföldi 1952 Andreas Alföldi, *A Conflict of Ideas in the Late Roman Empire: The Clash Between the Senate and Valentinian I*, Oxford.

Alföldy 2005 Géza Alföldy, Romanisation – Grundbegriff oder Fehlgriff? Überlegungen zum gegenwärtigen Stand der Erforschung von Integrationsprozessen im römischen Weltreich, in: Z. Visy (ed.), *Limes 19. Proceedings of the XIXth International Congress of Roman Frontier Studies held in Pécs, Hungary*, September 2003, Pécs, 25-56.

Amory 1997 Patric Amory, *People and Identity in Ostrogothic Italy, 489-554*, Cambridge.

Anreiter 2001 Peter Anreiter, *Die Vörromischer Namen Pannoniens*, *Archaeolingua, Series Minor* 16, Budapest.

Antonaccio 2001 Carla M. Antonaccio, Ethnicity and Colonization, in: I. Malkin (ed.), *Ancient Perspectives of Greek Ethnicity*, Washington DC–Cambridge MA, 113-157.

Antonaccio 2010 Carla M. Antonaccio, (Re)defining ethnicity, in: Hales and Hodos 2010, 32-53.

Appadurai 1996 Arjun Appadurai, *Modernity at Large, Cultural Dimensions of Globalization*, Minneapolis.

Balen-Letunić 2004 Dubravka Balen-Letunić (ed.), *Warriors at the Crossroads of East and West*, Zagreb.

Barkóczi 1964 László Barkóczi, The population of Pannonia from Marcus Aurelius to Diocletian, *AArchHung* 16, 257-356.

Barrett 1997 John C. Barrett, Romanization: a critical comment, in: Mattingly 1997a, 51-66.

Barth 1969 Fredrik Barth, Introduction, in: F. Barth (ed.), *Ethnic Groups and Boundaries*, Bergen, 9-38.

Barth 1989 Fredrik Barth, The analysis of culture in complex societies, *Ethnos* 54, London, 120-142.

Bauman 1999 Zygmunt Bauman, *Culture as Praxis*, London.

Bayart 1996 Jean-François Bayart, *L'illusion identitaire*, Paris.

Benac 1987a Alojz Benac (ed.), *PJZ* 5: *Željezno doba*, Sarajevo.

Benac 1987b Alojz Benac, O etničkim zajednicama starijeg željeznog doba u Jugoslaviji, in: Benac 1987a, 737-802.

Bentley 1987 G. Carter Bentley, Ethnicity and practice, *Comparative Studies in Society and History* 29/10, 25-55.

Bevan 1999 Bill Bevan, *Northern Exposure: Interpretative Devolution and the Iron Age in Britain*, *Leicester University Archaeology Monograph* 4, Leicester.

Birley 1988 Eric Birley, Pannonians in Roman Britain, *ZPE* 73, 151-155.

Blečić 2007 Martina Blečić, Status, symbols, sacrifices,

offerings. The diverse meanings of Illyrian helmets, *VAMZ* 40, 73-116.

Blečić Kavur and Kavur 2010 Martina Blečić Kavur and Boris Kavur, Grob 22 iz beogradske nekropole Karaburma: Retrospektiva i perspektiva, *Starinar* 60, 57-84.

Boatwright 2005 Mary T. Boatwright, Children and parents on the tombstones of Pannonia, in: M. George (ed.), *The Roman Family in the Empire: Rome, Italy and Beyond*, Oxford, 287-318.

den Boeft *et al.* 2008 Jan den Boeft, Jan Willem Drijvers, Daniël den Hengst, and Hans C. Teitler, *Philological and Historical Commentary on Ammianus Marcellinus* XXVI, Leiden-Boston.

Bojanovski 1988 Ivo Bojanovski, Ad CIL III 1741, Obod pokraj Cavtata (Epidaurum), in: Ž. Rapanić (ed.), *Arheološka istraživanja u Dubrovniku i njegovoj okolici*, *IzdHAD* 12, 101-110.

Bourdieu 1977 Pierre Bourdieu, *Outline of a Theory of Practice*, Cambridge.

Božič 1981 Dragan Božič, Relativna kronologija mlajše železne dobe v jugoslovanskem Podonavju, *AV* 32, 315-347.

Božič 1999 Dragan Božič, Die Erfoschung der Latènezeit in Slowenien seit jahr 1964, *AV* 50, 189-213.

Brather 2004 Sebastian Brather, *Etnische Interpretationen in der frühgeschichtlicher Archäologie: Geschichte, Grundlagen und Alternativen*, Berlin-New York.

Bratož 1986 Rajko Bratož, Razvoj organizacije zgodnjekršćanske cerkve na ozemlju Jugoslavije od 3. do 6. stoletja, *ZČ* 40/4, 363-395.

Bratož 1990 Rajko Bratož, Die Geschichte des frühen Christentums im Gebiet zwischen Sirmium und Aquileia im Licht der neueren Forschungen, *Klio* 72/2, 508-550.

Bratož 1996a Rajko Bratož (ed.), *Westillyricum und Nordostitalien in der Spätromischen Zeit*, *Situla* 34, 185-192

Bratož 1996b Rajko Bratož, Christianisierung des Nordadria- und Westbalkanraumes im 4. Jahrhundert, in: Bratož 1996a, 299-366.

Brizzi 2004 Giovanni Brizzi, Ancora su Illyriciani e “Soldatenkaiser”: qualche ulteriore proposta per una messa a fuoco del problema, in: Urso 2004, 319-342.

Burns 2003 Thomas S. Burns, *Rome and the Barbarians, 100 BC – AD 400*, Baltimore-London.

Burns 2010 Thomas S. Burns, Negotiating a serviceable identity and a pathway to power in Late Antiquity, in B. Sidwell and D. Dzino (eds.), *Power and Emotions in the Roman World and Late Antiquity*, Piscataway NJ, 167-200.

Caldwell 2007 Craig H. Caldwell, *Contesting late Roman Illyricum: Invasions and transformations in the Danubian-Balkan province*, Unpublished PhD thesis, Princeton University.

Carroll 2003 Maureen Carroll, The genesis of Roman towns on the lower Rhine, in: P. Wilson (ed.), *The Archaeology of Roman Towns*, Oxford, 22-30.

Cecconi 2006 Giovanni A. Cecconi, Romanizzazione, diversità culturale, politicamente corretto, *Mélanges de l'Ecole française de Rome, Antiquité* 118/1, Rome, 81-94.

Chapman 1992 Malcolm Chapman, *The Celts: the Construction of a Myth*, New York.

Clarke 1999 Katherine Clarke, *Between Geography and History: Hellenistic Constructions of the Roman World*, Oxford.

Clarke 2001 Katherine Clarke, An island nation: re-reading Tacitus' *Agricola*, *JRS* 91, 94-112.

Clarke 2003 Katherine Clarke, Ever-increasing Circles: Constructing the Roman Empire, in: T. Minamikawa (ed.), *Material Culture, Mentality and Historical Identity in the Ancient World: Understanding the Celts, Greeks, Romans and Modern Europeans*, Kyoto, 43-54.

Clifford 1988 James Clifford, *The Predicament of Culture: Twentieth Century Ethnography, Literature and Art*, Cambridge.

Collis 2003 John Collis, *The Celts: origins, myths & inventions*, Stroud.

Colombo 2007 Maurizio Colombo, Le tribù dei Pannoni in Strabone, *Tyche* 22, 1-8.

Colombo 2010 Maurizio Colombo, Pannonica, *AAntHung* 50, 171-202.

Coulston 2004 Jon C. N. Coulston, Military identity and personal self-identity in the Roman army, in: de Ligt *et al.* 2004, 133-152.

Crumley 2007 Carole L. Crumley, Historical ecology: Integrated thinking at multiple temporal and spatial scales, in: A. Hornborg and C. L. Crumley (eds.), *The World System and the Earth System: Global Socioenvironmental Change and Sustainability Since the Neolithic*, Walnut Creek CA, 15-28.

Curta 2007a Florin Curta, Some remarks on ethnicity in medieval archaeology, *Early Medieval Europe* 15/2, 159-185.

Curta 2007b Florin Curta (ed.), *The Other Europe in the Middle Ages: Avars, Bulgars, Khazars and Cumans. East Central and Eastern Europe in the Middle Ages (450-1450)* 2, Leiden-Boston.

Čović 1976 Borivoj Čović, *Od Butmira do Ilira*, Sarajevo.

Čović 1987a Borivoj Čović, Srednjobosanska grupa, in: Benac 1987a, 481-530.

Čremošnik 1964 Irma Čremošnik, Die einhemische Tracht Norcums, Pannoniens und Illyricums und ihre Vorbilder, *Latomus* 23, Bruxelles, 760-773.

Dench 2005 Emma Dench, *Romulus' Asylum: Roman Identities from the Age of Alexander to the Age of Hadrian*, Oxford.

Dauge 1981 Yves A. Dauge, *Le Barbare: Recherches sur la conception romaine de la barbarie et de la civilization*, *Collection Latomus* 176, Bruxelles.

Derks 2009 Ton Derks, Ethnic identity in the Roman frontier. The epigraphy of Batavi and other Lower Rhine tribes, in: Derks and Roymans 2009, 239-282.

Derks and Roymans 2002 Ton Derks and Nico Roymans, Seal-boxes and the spread of Latin literacy in the Rhine delta, in: A. Cooley (ed.), *Becoming Roman, Writing Latin? Literacy and Epigraphy in the Roman West*, *JRA* 48, 87-134.

Derks and Roymans 2006 Ton Derks and Nico Roymans, Returning auxiliary veterans: some methodological considerations, *JRA* 19, 121-135.
Derks and Roymans 2009 Ton Derks and Niko Roymans (eds.), *Ethnic Constructs in Antiquity: the Role of Power and Tradition*, *Amsterdam Archaeological Studies* 13, Amsterdam.
Dizdar 2003 Marko Dizdar, Prilog poznavanju kasnoga latena u istočnoj Slavoniji, *OpArch* 27, 337-349.
Dizdar 2004 Marko Dizdar, Grob LT 11 iz Zvonimirova – primjer dvojnog pokopa latenske kulture, *OpArch* 28, 41-90.
Dizdar and Potrebica 2002 Marko Dizdar and Hrvoje Potrebica, Latenska kultura na prostoru požeške kotline, *OpArch* 26, 111-131.
Dizdar and Potrebica 2005 Marko Dizdar and Hrvoje Potrebica, The late La Tène culture in central Slavonia (Croatia), in: H. Dobrzańska, W. Megaw, and P. Poleska (eds.), *Celts on the Margin: Studies in European Cultural Interaction VII c. BC – I c. AD: Essays in Honor of Zenon Woźniak*, Kraków, 57-66.
Dizdar and Radman-Livaja 2004 Marko Dizdar and Ivan Radman-Livaja, Warrior equipment from Vrtna street in Vinkovci as a contribution to understanding the process of the early romanization of eastern Slavonia, *PIAZ* 21, 37-53.
Dizdar *et al.* 2003 Marko Dizdar, Rajna Šoštarić, and Kristina Jelinčić, An early Roman grave from Ilok as a contribution to understanding a process of the Romanization of western Syrmia, *PIAZ* 20, 57-77.
Dobesch 1994 Gerard Dobesch, Zur Chronologie des Dakerkönigs Burebista, in: R. Göbl (ed.), *Die Hexadrachmenprägung der Gross-Boier. Ablauf, Chronologie und historische Relevanz für Noricum und Nachbargebiete*, Vienna, 51-68.
Domić Kunić 1988 Alka Domić Kunić, Augzilijari ilirskoga i panonskoga porijekla u natpisima i diplomama (od Augusta do Karakale), *ARR* 11, 83-114.
Domić Kunić 1996a Alka Domić Kunić, *Classis praetoria Misenatium* s posebnim obzirom na mornare podrijetlom iz Dalmacije i Panonije, *VAMZ* 28-29 (1995-1996), 39-72.
Domić Kunić 1996b Alka Domić Kunić, *Classis Praetoria Ravennatium* with special reflection on sailors that origin from Dalmatia and Pannonia, *Žant* 46/1-2, 95-110.
Domić Kunić 1996c Alka Domić Kunić, Rimske provincijalne flotile (s posebnim obzirom na udjel Panonaca), *ARR* 12, 83-100.
Domić Kunić 2003 Alka Domić Kunić, *Mons Claudius* – Pitanje identifikacije i etničkog određenja, *ObHAD* 35/2, 14.
Domić Kunić 2005 Alka Domić Kunić, The Danube region in legend – prehistoric trade across Pannonia, in: Šegvić and Mirnik 2005, 219-228.
Domić Kunić 2006 Alka Domić Kunić, *Bellum Pannonicum* (12. - 11. pr. Kr.). Posljednja faza osvajanja južne Panonije, *VAMZ* 39, 59-164.
Dorn 1984 Antun Dorn, Rimska vojnička diploma iz Negoslavaca, in: N. Majnarić-Pandžić (ed.), *Arheološka istraživanja u istočnoj Slavoniji i Baranji*, *IzdHAD* 9, 165-174.
Dušanić 1978 Slobodan Dušanić, Military diploma of A.D. 65, *Germania* 56/2, Mainz am Rhein, 461-475.
Dušanić 1998 Slobodan Dušanić, An early diploma militare, *Starinar* 49, 51-62.
Dušanić 2008 Slobodan Dušanić, The Valle Ponti lead ingots: notes on Roman notables' commercial activities in free Illyricum at the beginning of the Principate, *Starinar* 58, 107-118.
Dzino 2005 Danijel Dzino, *Sabaiarius*: beer, wine and Ammianus Marcellinus, in: W. Mayer and S. Trzcionka (eds.), *Feast, Fast or Famine: Food and Drink in Byzantium*, Brisbane, 57-68.
Dzino 2007 Danijel Dzino, The Celts in Illyricum – whoever they may be: the hybridization and construction of identities in southeastern Europe in the fourth and third centuries BC, *OpArch* 31, 93-112.
Dzino 2008 Danijel Dzino, Strabo 7,5 and imaginary Illyricum, *Athenaeum* 96/1, 173-192.
Dzino 2009 Danijel Dzino, The *Bellum Batonianum* in contemporary historiographical narratives/In a search for the post-modern Bato the Daesitiate, *ARR* 16, 29-45.
Dzino 2010a Danijel Dzino, *Illyricum and Roman Politics 229 BC – AD 68*, Cambridge.
Dzino 2010b Danijel Dzino, Aspects of identity-construction and cultural mimicry among the Dalmatian sailors in the Roman navy, *Antichthon: Journal of Australasian Society for Classical Studies* 44, Sydney, 96-110.
Dzino 2010c Danijel Dzino, *Becoming Slav, Becoming Croat: Identity-transformations in Post-Roman and Early Medieval Dalmatia*, *East Central and Eastern Europe in the Middle Ages (450–1450)* 12, Leiden-Boston.
Dzino 2011 Danijel Dzino, Indigene zajednice zapadnog i središnjeg Balkanskog poluotoka i 21. stoljeće: metodološki problemi, *GodCBI* 40(38), 197-206.
Elton 1996 Hugh Elton, *Frontiers of the Roman Empire*, London.
Emilov 2005 Julij Emilov, Changing paradigms: Modern interpretations of Celtic raids in Thrace reconsidered, in: Dobrzańska *et al.* 2005, 103-108.
Emilov 2007 Julij Emilov, La Tène finds and the indigenous communities in Thrace. Interrelations during the Hellenistic period, *Studia Hercynia* 11, Prague, 57-75.
Emilov 2010 Julij Emilov, 'Ancient texts on the Galatian Royal Residence of Tylis and the Context of La Tène Finds in Southern Thrace. A Reappraisal', in L. F. Vagalinski (ed.), *In Search of Celtic Tylis in Thrace (III C BC)*, Sofia, 67-87.
Ferjančić 2002 Svetlana Ferjančić, *Naseljavanje legijskih veterana u balkanskim provincijama I-III vek n.e*, Srpska akademija nauka i umetnosti – Balkanološki institut 79, Belgrade.
Fitz 1980 Jenö Fitz, Population, in: *ARP*, 141-160.
Fitz 1988 Jenö Fitz, La division de l'Illyricum, *Latomus* 47/1, Bruxelles, 13-25.

Fitz 1993/95 Jenö Fitz, *Die Verwaltung Pannoniens in der Römerzeit* (4 Vols.), Budapest.

Forcey *et al.* 1998 Colin Forcey, John Hawthorn, and Robert Witcher (eds.), *TRAC 97: Proceedings of the Seventh Annual Theoretical Roman Archaeology Conference*, Nottingham 1997, Oxford.

Garbsch 1965 Jochen Garbsch, *Die norisch-pannonische Frauentracht im 1. und 2. Jahrhundert*, Munich.

Garbsch 1985 Jochen Garbsch, Die norisch-pannonische Tracht, *ANRW* 2.12.3, 546-577.

Gardner 2007 Andrew Gardner, *An Archaeology of Identity: Soldiers & Society in Late Roman Britain*, Walnut Creek, CA.

Gáspár 2002 Dorottya Gáspár, *Christianity in Roman Pannonia. An Evaluation of Early Christian Finds and Sites from Hungary. BAR Intern. Ser.* 1010.

Gavrilović 2007 Nađa Gavrilović, The cult of Hercules in the central Balkans, in: R. Haeussler, A. C. King (eds.), *Continuity and Innovation in Religion in the Roman West* 1, *JRA Suppl.* 67, 143-152.

Glavičić 2008 Miroslav Glavičić, Epigrafska baština rimskodobnog Epidaura, *ArchAdr* 2/2, 43-62.

Goffart 1988 Walter Goffart, *The Narrators of Barbarian History (AD 550-800): Jordanes, Gregory of Tours, Bede and Paul the Deacon*, Princeton.

Goffart 2002 Walter Goffart, Does the distant past impinge on invasion age Germans?, in: A. Gillett (ed.), *On Barbarian Identity: Critical Approaches to Ethnicity in the Early Middle Ages*, Turnhout, 21-37.

Goffart 2006 Walter Goffart, *Barbarian Tides: The Migration Age and the Later Roman Empire*, Philadelphia.

Goldsworthy and Haynes 1999 Adrian Goldsworthy and Ian Haynes (eds.), *The Roman Army as a Community*, *JRA Suppl.* 34.

Gračanin 2005 Hrvoje Gračanin, Illyricum of the 2nd and 3rd centuries AD in the works of Latin and Greek historians, in: Šegvić and Mirnik 2005, 287-298.

Gračanin 2008 Hrvoje Gračanin, Kršćanstvo i crkva u kontinentalnoj Hrvatskoj u ranom srednjem vijeku (VI. – XI. st.), *Crkvena kulturna dobra* 6, Zagreb, 70-84.

Grassl 1990 Herbert Grassl, Pannonii – Pannonioi – Paiones: Zur Frage der Identifikation antiker Völkernamen, in: H. Vetters and M. Kandler (eds.), *Akten des 14. Internationalen Limeskongresses 1986 in Carnuntum* 2, Vienna, 539-544.

Graßl 2000 Herbert Graßl, Die Taurisker. Beiträge zur Geschichte und Lokalisierung eines antiken Ethnonyms, *Orbis Terrarum* 6, 127-138.

Graßl 2001 Herbert Graßl, Das Gebiet Drau-Mur-Raab im Lichte der antiken Überlieferung, in: H. Taubert (ed.), *Die Drau-, Mur- und Raab-Region im 1. Vorchristlichen Jahrtausend, Akten des internat. Interdisziplinären Symposiums* (Bad Radkersburg 2000), *Universitätsforschungen z. Prähist. Archäologie* 78, Bonn, 329-333.

Graves-Brown *et al.* 1996 Paul Graves-Brown, Siân Jones, and Clive Gamble (eds.), *Cultural Identity and Archaeology: The construction of European communities*, London-New York.

Guštin 1996 Mitja Guštin, Taurisci – Verknüpfung der historischen und archäologischen Interpretation, in: E. Jerem, A. Krenn Lebb, J.-W. Neugebauer, and O. H. Urban (eds.), *Die Kelten in den Alpen und an der Donau: Akten des International Symposions* (St. Pölten1992), Budapest-Vienna, 433-440.

Hales and Hodos 2010 Shelley Hales and Tamar Hodos (eds.), *Material Culture and Social Identitiets in the Ancient World*, Cambridge.

Hales 2010 Shelley Hales, Tricks with mirrors: remebering the dead of Noricum, in: Hales and Hodos 2010, 227-251.

Hammond 2000 Nicholas G. L. Hammond, The *ethne* in Epirus and upper Macedonia, *Annual of the British School at Athens* 95, 345-352.

Hartog 1980 François Hartog, *Le miroir d'Hérodote. Essai sur la représentation de l'autre*, Paris.

Haselgrove and Moore 2007 Colin Haselgrove and Tom Moore, New narratives of the later Iron Age, in: C. Haselgrove and T. Moore (eds.), *The Later Iron Age in Britain and Beyond*, Oxford, 1-15.

Haynes 1999 Ian Haynes, Introduction: Roman Army as a Community, in: Goldsworthy and Haynes 1999, 7-14.

Hill 1995 James D. Hill, The pre Roman Iron Age in Britain and Ireland: an overview, *Journal of World Prehistory* 9, 47-98.

Hill 2006 James D. Hill, Are we any closer to understanding how later Iron Age societies worked (or did not work)?, in: C. Haselgrove (ed.), *Les Mutations de la fin de l'age du fer; Celts et Gaulois IV*, Bibracte 12/4, Glux-en-Glenne, 169-180.

Hingley 1996 Richard Hingley, The "Legacy" of Rome: the rise, decline and fall of the theory of Romanization, in: J. Webster and N. J. Cooper (eds.), *Roman Imperialism: Post-colonial Perspectives*, *Leicester Archaeological Monographs* 3, Leicester, 35-48.

Hingley 2005 Richard Hingley, *Globalizing Roman Culture: Unity, Diversity and Empire*, London-New York.

Hingley 2010 Richard Hingley, Cultural diversity and unity: empire and Rome, in: Hales and Hodos 2010, 54-75.

Hodos 2010 Tamar Hodos, Local and global perspectives in the study of social and cultural identities, in: Hales and Hodos 2010, 3-31.

Holder 1980 Paul A. Holder, *The Auxilia from Augustus to Trajan*, *BAR Intern. Ser.* 70.

Hope 1997 Valery M. Hope, Constructing Roman identity: funerary monuments and social structure in the Roman world, *Mortality* 2, 103-121.

Hope 2001 Valery M. Hope, *Constructing Identity: The Roman Funerary Monuments of Aquileia, Mainz and Nimes*, *BAR Intern. Ser.* 960, Cambridge.

Hope 2003 Valery M. Hope, Remembering Rome: memory, funerary monuments and Roman soldier, in: H. Williams (ed.), *Archaeologies of Remembrance: Death and Memory in Past Societies*, New York-London, 125-140.

Isayev 2010 Elena Isayev, Unintentionally being Lucanian:

dynamics beyond hybridity, in: Hales and Hodos 2010, 201-226.
James 1999a Simon James, *Atlantic Celts: Ancient People or Modern Invention?*, London.
James 1999b Simon James, The community of the soldiers: a major identity and centre of power in the Roman Empire, in: P. Baker, C. Forcey, S. Jundi, and R. Witcher (eds), *TRAC 98: Proceedings of the Eighth Annual Theoretical Roman Archaeology Conference*, Oxford, 14-25.
James 2001 Simon James, Soldiers and civilians: identity and interaction in Roman Britain, in: S. James and M. Millett (eds.), *Britons and Romans: Advancing and Archaeological Agenda*, York, 77-89.
Jenkins 1997 Richard Jenkins, *Rethinking Ethnicity: Arguments and Explorations*, London.
Jiménez 2008 Alicia Jiménez, A critical approach to the concept of resistance: new 'traditional' rituals and objects in funerary contexts, in: C. Fenwick, M. Wiggins, and D. Wythe (eds.), *TRAC 2007: Proceedings of the 17th Annual Theoretical Roman Archaeology Conference*, Oxford, 15-30.
Jones 1996 Siân Jones, Discourses of identity in the interpretations of the past, in: Graves-Brown *et al.* 1996, 62-80.
Jones 1997 Siân Jones, *The Archaeology of Ethnicity: Constructing Identities in the Past and Present*, London.
Katičić 1963 Radoslav Katičić, Das mittledalmatische Namengebiet, *ŽAnt* 12/2, 255-292.
Katičić 1965 Radoslav Katičić, Zur Frage der keltischen und pannonischen Namengebieten im römischen Dalmatien, *GodCBI* 3/1, 53-76.
Kearney 1995 Michael Kearney, The local and the global: the anthropology of globalization and transnationalism, *Annual Review of Anthropology* 24, 547-565.
Keay and Terrenato 2001 Simon Keay and Nicola Terrenato (eds.), *Italy and the West: Comparative Issues in Romanization*, Oxford.
Koledin 2000 Jovan Koledin, Rimska vojnička diploma iz Vukovara, *Glasnik Srpskoga arheološkog društva* 15-16 (1999-2000), Beograd, 231-239.
Kos and Šemrov 2003 Peter Kos and Andrej Šemrov, A hoard of Celtic and Roman coins from the Ljubljanica river. A contribution to the chronology of the coinage of the Taurisci, *AV* 54, 381-395.
Kovács 2003 Petér Kovács, Christianity and the Greek language in Pannonia, *AAntHung* 43, 113-124.
Kovács 2007 Petér Kovács, A Pisidian Veteran and the First Mention of Pannonia, *Tyche* 22, Vienna, 99-107.
Kovács 2008 Petér Kovács, Some Notes on the Division of Illyricum, in: I. Piso (ed.), *Die Römischen Provinzen. Begriff und Gründung*, Cluj-Napoca, 237-248.
Krebs 2006 Cristopher B. Krebs, Imaginary geography in Caesar's *Bellum Gallicum*, *American Journal of Philology* 127, 111-136.
Kremer 2004 Gabrielle Kremer, Die norisch-pannonischen Grabbauten als Ausdruck kultureller Identität?, in: Schmidt-Colinet 2004, 147-159.
Kuntić-Makvić 1996 Bruna Kuntić-Makvić, Illyricianus: l'histoire du mot et l'histoire de l'Illyrique, in: Bratož 1996a, 185-192.
Laurence 2001a Ray Laurence, Roman narratives. The writing of archaeological discourse - a view from Britain?, *Archaeological Dialogues* 8, 90-101.
Laurence 2001b Ray Laurence, The creation of geography: an interpretation of Roman Britain, in: C. Adams and R. Laurence (eds.), *Travel and Geography in the Roman Empire*, London-New York, 67-94.
Lenski 2002 Noel Lenski, *Failure of Empire: Valens and the Roman State in the Fourth Century A. D*, Berkeley-Los Angeles.
Lica 2000 Vasile Lica, *The Coming of Rome in the Dacian World*, *Xenia* 44, Konstanz.
de Ligt *et al.* 2004 Luuk de Ligt, Emily A. Hemelrijk, and H. W. Singor (eds.), *Roman Rule and Civic Life: Local and Regional Perspectives*, Amsterdam.
Lippert 2004 Andreas Lippert, Das archäologische Bild der frühen Illyrer, in: Urso 2004, 11-21.
Lolić 2003 Tatjana Lolić, *Colonia Flavia Siscia*, in: *Pannonia* I, 131-152.
Lucy 2005 Sam Lucy, Ethnic and cultural identities, in: M. Díaz-Andreu, S. Lucy, S. Babić, and D. N. Edwards (eds.), *The Archaeology of Identity: Approaches to Gender, Age, Status, Ethnicity and Religion*, London, 86-109.
Lund 1988 Allan A. Lund, *Die Germania des Tacitus*, Heidelberg.
Lund 1991 Allan A. Lund, Versuch einer Gesamtinterpretation der 'Germania' des Tacitus; Kritischer Forschungsbericht zur 'Germania' des Tacitus, *ANRW* 2.33.3, 1858-1988; 1989-2222.
Majnarić-Pandžić 1996a Nives Majnarić-Pandžić, Nekoliko napomena o uvođenju ranolatenskog stila u sjevernu Hrvatsku i Bosnu, *ARR* 12, 31-53.
Majnarić-Pandžić 1996b Nives Majnarić-Pandžić, Einige Beispiele der spätlatènezeitlichen Siedlungen in Nordkroatien und ihre Beziehung zu den Zentren der frühen Romanisation, *AV* 74, 257-265.
Majnarić-Pandžić 2009 Nives Majnarić-Pandžić, On the South Pannonian Population in the Late Iron Age, in: G. Tiefengraber, B. Kavur, and A. Gaspari (eds.), *Keltske študije 2. Studies in Celtic Archaeology. Papers in honour of Mitja Guštin, Collection Protohistoire européenne* 11, Montagnac, 235-246.
Malkin 2002 Irad Malkin, A colonial middle ground: Greek, Etruscan and local elites in the Bay of Naples, in: C. L. Lyons and J. K. Papadopulos (eds.), *The Archaeology of Colonialism*, Los Angeles, 151-181.
Mancini 1933 Giuseppe Mancini, Iscrizione sepolcrale di Anversa, in: N. de Arcangelis (ed.), *Atti del Convegno Storico abruzzese-molisano (25-29 marzo 1931)*, Casalbordino, 449-452.
Marić 1963 Zdravko Marić, Keltski elementi u mlađem željeznom dobu Bosne i Hercegovine, *GZMS* 18, 63-83.
Marić 1964a Zdravko Marić, Problemes des limites septentrionales du territoire illyrien, in: A. Benac (ed.), *Symposium sur la delimitation Territoriale et*

chronologique des Illyriens a l'epoque Prehistorique, Publications spéciales ANUBiH/CBI 4/1, 177-214.

Marić 1964b Zdravko Marić, Donja Dolina, *GZMS* 19, 5-128.

Marincola 1997 John M. Marincola, *Authority and Tradition in Ancient Historiography*, Cambridge.

Martin-Kilcher 1993 Stefanie Martin-Kilcher, Römische Grabfunde als Quelle zur Trachtgeschichte im zirkumalpinen Raum, in: M. Struck (ed.), *Römerzeitliche Gräber als Quellen zu Religion, Bevölkerungsstruktur und Sozialgeschichte, Archäologische Schriften des Instituts für Vor- und Frühgeschichte der Johannes Gutenberg-Universität Mainz* 3, Mainz, 181-203.

Matijašić 2009 Robert Matijašić, *Povijest hrvatskih zemalja u antici do cara Dioklecijana*, Zagreb.

Mattern 1999 Sussan P. Mattern, *Rome and the Enemy. Imperial Strategy in the Principate*, Berkeley-Los Angeles.

Matthews 1975 John Matthews, *Western Aristocracies and Imperial Court AD 364-425*, Oxford.

Mattingly 1997a David J. Mattingly (ed.), *Dialogues in Roman Imperialism: Power, Discourse and Discrepant Experience in the Roman Empire, JRA Suppl.* 23.

Mattingly 1997b David J. Mattingly, Dialogues of power and experience in the Roman Empire, in: Mattingly 1997a, 7-26.

Mattingly 2004 David J. Mattingly, Being Roman: expressing identity in a provincial setting, *JRA* 17, 5-25.

Mattingly 2006 David J. Mattingly, *An Imperial Possession: Britain in the Roman Empire*, London.

Mellor 1999 Ronald Mellor, *The Roman Historians*, London, New York.

Migotti 1997 Branka Migotti, *Evidence for Christianity in Roman Southern Pannonia (Northern Croatia). A Catalogue of Finds and Sites*, *BAR Intern. Ser.* 684, Oxford.

Migotti 2008a Branka Migotti, Nadgrobni spomenik robovske obitelji iz Donjih Čeha kod Zagreba, *ArchAdr* 11, 453-465.

Migotti 2008b Branka Migotti, *The Crossbow Brooches with Portraits in the Roman Empire*, Zagreb.

Milin 2003 Milena Milin, Pitanje ilirske komponente stanovništva jugoistočnog dela Donje Panonije u savremenim istraživanjima, *Balcanica* 32-33, Beograd, 49-60.

Millett 1990 Martin Millett, *The Romanization of Britain: an Essay in Archaeological Interpretation,* Cambridge.

Miškec 2004 Alenka Miškec, The early romanization of the southeastern Alpine region in the light of numismatic finds, *AV* 54, 369-379.

Miškiv 1998 Jesenka Miškiv, Rimska vojnička diploma iz Slavonskog Broda, *VAMZ* 30-31 (1997-1998), 83-101.

Mócsy 1957 András Mócsy, Zur Geschichte der peregrinen Gemeinden in Pannonien, *Historia* 6/4, 488-498.

Mócsy 1967 András Mócsy, Die *Lingua Pannonica*, in: A. Benac (ed.), *Symposium sur les Illyriens a l'epoque antique*, Special editions ANUBiH-CBI 5/2, 195-200.

Mócsy 1974 András Mócsy, *Pannonia and Upper Moesia*, London–Boston.

Mócsy 1977 András Mócsy, Pannonien und die Soldatenkaiser, *ANRW* 2.6, 557-571.

Mócsy 1983 András Mócsy, The civilized Pannonians of Velleius, in: B. Hartley and J. Watcher (eds.), *Rome and her Northern Provinces*, Oxford, 169-178.

Mócsy 1985 András Mócsy, *Beiträge zur Namenstatistik. DissPann* III/3.

Morgan 1971 M. Gwyn Morgan, Lucius Cotta and Metellus: Roman campaigns in Illyria during the late second century, *Athenaeum* 49/3-4, 271-301.

Morgan 1974 M. Gwyn Morgan, Cornelius and the Pannonians, *Historia*, 23/2, 183-216.

Mulvin 2002 Lynda Mulvin, *Late Roman Villas in the Danube-Balkan Region, BAR Intern. Ser.* 1064.

Murphy 2004 Trevor M. Murphy, *Pliny the Elder's Natural History: the Empire in the Encyclopedia*, Oxford.

Nagy 1970 Tibor Nagy, Der Aufstand der pannonisch-dalmatinischen Völker und die Frage der Zweiteilung Illyricums, in: V. Mirosavljević, D. Rendić-Miočević, and M. Suić (eds.), *Adriatica Praehistorica et Antiqua: miscellanea Gregorio Novak dicata*, Zagreb, 459-466.

Novak 1966 Grga Novak, La province Illyricum était-elle au temps d'Octavien Auguste et de Tibère divisée en Superior provincia Illyricum et Inferior provincial Illyricum?, in: R. Chevallier (ed.), *Mélanges d'archéologie et d'histoire offerts à André Piganiol* 3, Paris, 1359-1366.

Noy 2000 David Noy, *Foreigners at Rome: Citizens and Strangers*, London.

O'Gorman 1993 Ellen O'Gorman, No place like Rome: identity and difference in the *Germania* of Tacitus, *Ramus: Critical studies in Greek and Roman Literature* 22/2, Brisbane,135-154.

Olariu 2005 Christian Olariu, Datianus, Valentinian and the rise of the Pannonian faction, *Historia* 54/3, 351-354.

Oltean 2007 Ioana A. Oltean, *Dacia. Landscape, Colonisation, Romanisation*, Routledge.

Oltean 2009 Ioana A. Oltean, Dacian ethnic identity and the Roman army, in: W. S. Hanson (ed.), *The Army and Frontiers of Rome: Papers Offered to David J. Breeze, JRA Suppl.* 74, 91-101.

Olujić 2004 Boris Olujić, Putovi i raskršća, prioriteti i perspektive u istraživanju etničkih skupina sjeverozapadnoga Ilirika, in: M. Matijević (ed.), *Spomenica Filipa Potrebice*, Zagreb, 87-96.

Osgood 2009 Josiah Osgood, The pen and the sword: writing and conquest in Caesar's Gaul, *Classical Antiquity* 28(2), Berkeley CA 328-358.

Paškvalin 2008 Veljko Paškvalin, Kamenjača, Breza kod Sarajeva – mlađeželjeznodobna i rimska nekropola, *GodCBI* 37/35, 101-179.

Pavic 2010 Anto Pavic, Zwischen den Welten – Antike Kulturlandschaften im Illyricum, Frankfurter elektronische Rundschau zur Altertumskunde 11, 1-21. http://s145739614.online.de/fera/ausgabe11/Pavic2.pdf

Perić 1995 Slaviša Perić, Celtic pottery in settlements of the Central Bosnian Culture Group, *Starinar* 45-46 (1994-1995), 113-131.

Perić 2002 Slaviša Perić, O problemu sahranjivanja u

gornjem toku rijeke Bosne tokom bronzanog i gvozdenog doba, *GodCBI* 32/30, 179-197.

Perinić Muratović and Vulić 2009 Ljubica Perinić Muratović and Hrvoje Vulić, Observations on the Cult of Silvanus in Pannonia in the Light of the Find of an Altar in Vinkovci, *PIAZ* 26, 165-180.

Periša 2009 Darko Periša, Rimski Delminij kao sjedište starokršćanske biskupije, *ARR* 16, 225-242.

Pitts 2007 Martin Pitts, The emperor's new clothes? The utility of identity in Roman archaeology, *AJA* 111/4, 693-713.

Pochmarski 2004 Erwin Pochmarski, Das sogenannte norische Mädchen. Ein Beispiel für den Ausdruck lokaler Identität in der provinzialrömischen Plastik, in: Schmidt-Colinet 2004, 161-174.

Popović 1987 Petar Popović, *Novac Skordiska / Le monnayage des Scordisques*, Belgrade, Novi Sad.

Popović 1993 Petar Popović, The territories of the Scordisci, *Starinar* 43-44 (1992), 13-21.

Popović 1996 Petar Popović, Early La Tène between Pannonia and the Balkans, *Starinar* 47, 105-125.

Potter 1999 David S. Potter, *Literary Text and the Roman Historian*, London.

Potrebica 2003 Hrvoje Potrebica, Požeška kotlina i Donja Dolina u komunikacijskoj mreži starijeg željeznog doba, *OpArch* 27, 217-242.

Purcell 1990 Nicholas Purcell, The Creation of a provincial landscape: the Roman impact on Cisalpine Gaul in: T. Blagg and M. Millett (eds.), *The Early Roman Empire in the West*, London, 7-29.

Radman-Livaja 2007 Ivan Radman-Livaja, *In Segestica*, *PIAZ* 24, 153-172.

Rendić-Miočević 1989 Duje Rendić-Miočević, Problemi romanizacije Ilira s osobitim obzirom na kultove i onomastiku, in: D. Rendić Miočević, *Iliri i antički svijet. Ilirološke studije. Povijest – arheologija – umjetnost – numizmatika – onomastika* (collected works), Split, 425-439.

Rendić-Miočević 1990 Duje Rendić-Miočević, 'Illyro-Pannonica' kao tema legendi u rimskoj numografiji, *VAMZ* 23, 75-96.

Ridley 2003 Ronald T. Ridley, *The Emperor's Retrospect: Augustus' Res Gestae in Epigraphy, Historiography and Commentary*, *Studia Hellenistica* 39, Leuven-Dudley.

Riggsby 2006 Andrew M. Riggsby, *Caesar in Gaul and Rome: War in Words*, Austin.

Roddaz 1984 Jean-Michel Roddaz, *Marcus Agrippa*, *Bibliothèque des écoles françaises d'Athènes et de Rome* 253, Rome.

Roymans 2004 Nico Roymans, *Ethnic Identity and Imperial Power: The Batavians in the Early Roman Empire*, Amsterdam Archaeological Studies 10, Amsterdam.

Rutledge 2000 Steven H. Rutledge, Tacitus in *Tartan*: textual colonization and expansionist discourse in the *Agricola*, *Helios* 27/1, Lubbock, 75-95.

Sanader 1996 Mirjana Sanader, Tragovi Vulkanova kulta u rimskoj provinciji Dalmaciji, *ARR* 12, 257-269.

Schmidt-Colinet 2004 Andreas Schmidt-Colinet (ed.), *Lokale Identitäten in Randgebieten des römischen Reiches*. Akten des internationalen symposiums (Wiener Neustadt 2003), Vienna.

Shennan 1989 Stephen J. Shennan, Introduction: archaeological approaches to cultural identity, in: S. J. Shennan (ed.), *Archaeological Approaches to Cultural Identity*, New York, 1-32.

Scherrer 2004 Peter Scherrer, Die Ausprägung lokaler Identität in den Städten in Noricum und Pannonien. Eine Fallstudieanhand der Civitas-Kulte, in: Schmidt-Colinet 2004, 175-188.

Scott 1998 Eleanor Scott, Tales from a Romanist: a personal view of archaeology and "equal opportunities", in: Forcey *et al.* 1998, 138-147.

Shuttleworth Kraus 1999 Christina Shuttleworth Kraus (ed.), *The Limits of Historiography*, *Mnemosyne Supplementum* 191, Leiden.

Stadler 2007 Peter Stadler, Avar chronology revisited, and the question of ethnicity in the Avar qaganate, in: Curta 2007b, 47-82.

Straub 2002 Johann Straub, Personal and collective identity: a conceptual analysis, in: H. Friese (ed.), *Identities. Time, Difference and Boundaries*, New York-Oxford, 56-76.

Stewart 1995 Peter C. N. Stewart, Inventing Britain: the Roman creation and adaptation of an image, *Britannia* 26, London, 1-10.

Swan 2009 Vivien Swan, *Ethnicity, Conquest and Recruitment: Two Case Studies from the Northern Military Provinces*, *JRA Suppl.* 72.

Syed 2005 Yasmin Syed, Romans and others, in: Harrison 2005, 360-371.

Syme 1971 [1937] Ronald Syme, Augustus and the south Slav lands, in: *Danubian Papers*, Bucharest, 13-25.

Šanjek 2005 Franjo Šanjek, À la recherche de Stridon, lieu de naissance de saint Jérôme, *Revue d'histoire ecclésiastique* 100/1, 146-151.

Šašel Kos 2005 Marjeta Šašel Kos, *Appian and Illyricum*, *Situla* 43.

Šašel Kos 2010 Marjeta Šašel Kos, Pannonia or Lower Illyricum?, *Tyche* 25, 123-130

Šašel Kos 2011 Marjeta Šašel Kos, The Roman conquest of Dalmatia and Pannonia under Augustus - some of the latest research results, in: G. Moosbauer and R. Wiegels (eds.), *Fines imperii - imperium sine fine?: römische Okkupations- und Grenzpolitik im frühen Prinzipat: Beiträge zum Kongress "Fines imperii - imperium sine fine?", Osnabrück vom 14. bis 18. September 2009, Osnabrücker Forschungen zu Altertum und Antike-Rezeption* 14, Rahden, 107-117.

Talbert 2004 Richard Talbert, Rome's provinces as framework for word-view, in: de Ligt *et al.* 2004, 21-38.

Tasić 1992 Nikola Tasić (ed.), *Scordisci and the Native Population in the Middle Danube Region*, *Srpska akademija nauka i umetnosti – Balkanološki institut* 48, Belgrade.

Tassaux 2004 Francis Tassaux, Les importations de l'Adriatique et de l'Italie du Nord vers les provinces dannubienes de César aux Sévères, in: Urso 2004, 95-140.

Terrenato 1998 Nicola Terrenato, The Romanization of Italy: global acculturation or cutural bricolage, in: Forcey *et al.* 1998, 20-27.

Terrenato 2001 Nicola Terrenato, Introduction, in: Keay and Terrenato 2001, 1-6.

Thomas 1980 Edi B. Thomas, Villa settlements, in: *ARP*, 275-322.

Thurston 2009 Tina L. Thurston, Unity and diversity in the European Iron Age: out of the mists, some clarity?, *Journal of Archaeological Research* 17, 347-423.

Todorović 1980 Jovan Todorović, Autohtona komponenta u keltskoj kulturi Podunavlja, in: T. Knez (ed.), *Zbornik posvečen Stanetu Gabrovcu ob Šesdesetletnici / Festschrift für Stane Gabrovec, Situla* 20-21, 383-387.

Ulf 2009 Cristoph Ulf, Rethinking cultural contacts, *Ancient West & East* 8, 81-132.

Urso 2004 Gianpaolo Urso (ed.), *Dall'Adriatico al Danubio. L'Illirico nell'età greca e romana, I convegni della fondazione Niccolò Canussio* 3, Pisa.

Van Dommelen 2007 Peter Van Dommelen, Beyond resistance: Roman power and local traditions in Punic Sardinia, in: P. Van Dommelen and N. Terrenato (eds.), *Articulating Local Cultures. Power and Identity under the Expanding Roman Republic, JRA Suppl.* 63, 55-70.

Van Driel-Murray 2002 Carol Van Driel-Murray, Ethnic soldiers: the experience of the lower Rhine tribes, in: T. Grünewald and S. Seibel (eds.), *Kontinuität und Diskontinuität: Germania inferior am Beginn und am Ende der römischen Herrschaft, Beiträge des deutsch-niederländuschen Kolloquiums in der Katholieke Universiteit Nijmegen* (2001), Berlin, 200-217.

Van Rossum 2004 J. A. Van Rossum, The end of the Batavian auxiliaries as 'national' units, in: de Ligt *et al.* 2004, 113-132.

Vasić 1987 Rastko Vasić, Daljska grupa; Bosutska grupa; Sremska grupa zapadnobalkanskog kompleksa, in: Benac 1987a, 533-558.

Vida 2007 Tivadar Vida, Conflict and coexistence: the local population of the Carpathian Basin under Avar rule (sixth to seventh century), in: Curta 2007b, 13-46.

Vinski-Gasparini 1987 Ksenija Vinski-Gasparini, Grupa Martijanec – Kaptol, in: Benac 1987a, 182-231.

Vulić 1933 Nikola Vulić, Nekoliko pitanja iz antičke prošlosti. *Glas Srpske Kraljevske Akademije*, 155/78, Beograd, 3-86.

Yelvington 1991 Kelvin A. Yelvington, Ethnicity as practice? A comment on Bentley, *Comparative Studies in Society and History* 31/1, 158-168.

Wallace Hadrill 2008 Andrew Wallace Hadrill, *Rome's Cultural Revolution*, Cambridge-New York.

Welch 2009 Kathryn Welch, Alternative memoirs: tales from the 'other side' of the civil war, in: C. Smith and A. Powell (eds.), *The Lost Memoirs of Augustus and the Development of Roman Autobiography*, Swansea, 195-223.

Wells 1999 Peter S. Wells, *The Barbarians Speak: How the Conquered Peoples Shaped Roman Europe*, Princeton NJ.

Wells 2001 Peter S. Wells, *Beyond Celts, Germans and Scythians. Archaeology and Identity in Iron Age Europe*, London.

White 1991 Richard White, *The Middle Ground: Indians, Empires and Republics in the Great Lakes Region, 1650-1815*, Cambridge.

Whittaker 1994 Charles R. Whittaker, *Frontiers of the Roman Empire*, Baltimore-London.

Whittaker 1997 Charles R. Whittaker, Imperialism and culture: the Roman initiative, in: Mattingly 1997a, 143-165.

Wiessner 1983 Polly Wiessner, Style and social information in Kalahari San projectile points, *American Antiquity* 48/2, Washington, 253-276.

Wiessner 1989 Polly Wiessner, Style and changing relations between the individual and society, in: I. Hodder (ed.), *The Meaning of Things: Material Culture and Symbolic Expression*, London-Boston-Sydney, 56-63.

Wiessner 1990 Polly Wiessner, Is there a unity to style? in: M. W. Conkey and C. A. Hastorf (eds.), *The Uses of Style in Archaeology*, Cambridge, 105-112.

Wilkes 1992 John J. Wilkes, *The Illyrians*, Cambridge, Oxford.

Wilkes 1996 John J. Wilkes, The Danubian and Balkan provinces, in: *CAH* 10, 2nd ed., 545-585.

Wilkes 1999 John J. Wilkes, The Roman army as a community in the Danube lands: the case of the Seventh Legion, in: Goldsworthy and Haynes 1999, 95-104.

Wilkes 2005 John J. Wilkes, The Roman Danube: an archaeological survey, *JRS* 95, 124-225.

Woolf 1992 Greg Woolf, The unity and diversity of Romanisation, *JRA* 5, 349-352.

Woolf 1995 Greg Woolf, The formation of Roman provincial cultures, in J. Metzler, M. Millett, N. Roymans and J. Slofstra (eds.), *Integration in the Early Roman West: the Role of Culture and Ideology*, Luxembourg, 9-18.

Woolf 1997 Greg Woolf, Beyond Romans and natives, *World Archaeology* 28(3), Routledge, 339-350.

Woolf 1998 Greg Woolf, *Becoming Roman: The Origins of Provincial Civilization in Gaul*, Cambridge.

Woolf 2001 Greg Woolf, Inventing empire in ancient Rome, in: S. Alcock, T. N. D'Altroy, K. D. Morrison, and C. M. Sinopoli (eds.), *Empires: Perspectives from Archaeology and History*, Cambridge, 311-322.

Woolf 2002 Greg Woolf, Generations of aristocracy: continuities and discontinuities in the societies of interior Gaul, *Archaeological Dialogues* 9, 2-15.

Woolf 2009 Greg Woolf, Cruptorix and his kind. Talking ethnicity on the middle ground, in: Derks and Roymans 2009, 207-218.

Zaninović 2001 Marin Zaninović, Jadranski Kelti, *OpArch* 25, 57-63.

Zanker 2000 Paul Zanker, The city as a symbol: Rome and the creation of an urban image, in: E. Fentress (ed.), *Romanization and the City: Creation, Transformation and Failures, JRA Suppl.* 38, 25-41.

The Archaeological Background to the Formation of Ethnic Identities

Marko Dizdar

1. Introduction

The knowledge of the political, social and economic development of proto-historical ethnic communities inhabiting the area of southern Pannonia is particularly important for understanding the processes that occurred in this territory during the Roman conquest and ensuing stabilisation in the late 1st century BC and 1st century AD. Information in Greek and Roman literary sources testifies to the complexity of a chain of processes that occurred during the period when centuries-long independent development of those communities was ended by the Roman conquest. At the same time, the material legacy available to the researcher shows that these ethnic groups were of different origin, having in common the loss of independence at the start of a new millennium, but reacting differently to the changes introduced with the establishment of a new political-administrative-economic order.

At present, the development of south-Pannonian protohistorical ethnic identities can be followed from the second half of the 4th century BC onwards, when the communities of presumed Celtic origin settled southern Pannonia in the course of the last prehistoric migration in the region, creating new identities in the contact with the autochthonous population – the Scordisci and Taurisci. Their appearance also designated the start of the Late Iron Age period, characterised by new dominant material heritage, the La Tène Culture, which also represented a noteworthy advancement in civilisation, since new technological achievements that would significantly improve the quality of everyday life in southern Pannonia (for instance, the use of the potter's wheel, the production of new types of iron objects, the appearance of coinage), unknown in the preceding period, were implemented together with new forms of weaponry, tools and jewellery. On the other hand, certain characteristics of the autochthonous material heritage influenced the creation of a particular material articulation of the newly created identities within the frame of the Central European La Tène Culture. At present, this fact is most easily recognisable in the forms and decoration of pottery, in certain items of dress accessories and jewellery, as well as in weaponry. An autochthonous origin is attributed to the largest number of south-Pannonian identities known from the Greek and Roman literary sources. Their material heritage is poorly known at present, with only a small number of archaeological finds testifying to the fact that these autochthonous communities also used the objects affiliated with the La Tène Culture. The process of their formation is still unknown, although certain finds from the later phase of the Early Iron Age (end of 7th – second half of 4th century BC), continuing into the Late Iron Age, suggest the possibility that it should be searched for within this chronological frame.

The development of all south-Pannonian identities (fig. 1) was finally terminated at the end of the 1st century BC, as they all became part of the Roman Empire. The name of the future province of Pannonia clearly suggests that autochthonous Pannonian identities had an active role in these processes, as well as in the formation of the new civilisational environment. This is also supported by the archaeological evidence from the earliest beginnings of Early Roman provincial contexts, showing that autochthonous south-Pannonian population was an important segment of society and that its heritage was incorporated into this new civilisational frame.

2. South-Pannonian identities in the pre-Roman period

Although Greek and Roman literary sources mention south-Pannonian identities, especially in relation to military conflicts during the Roman conquest, only the results of archaeological excavations, together with new studies of previously known material kept in museum collections and only partially published, will provide more information on the period of intensive transformations that occurred at the BC/AD transition. Greek and Roman sources provide a good insight into the definition of the spatial frame of south-Pannonian proto-historical identities, while the period of the Late Iron Age (from the second half of the 4th to the end of the 1st century BC/early 1st century AD) represents the chronological frame.

Greek and Roman historiographers reveal that the Scordisci and Taurisci, settled on the southeastern frontier of the Celtic World, were the most important identities in the territory of northern Croatia. The Scordisci were settled in the border area towards Macedonia and Greece, in the southeastern part of the Pannonian basin from the eastern slopes of the Slavonian mountainous range and the lower course of the Drava to the lower Morava basin. Their political and economic centre was in Syrmia and eastern Slavonia. Literary and anthropo-onomastical sources testify to their heterogeneous composition, in which the heritage

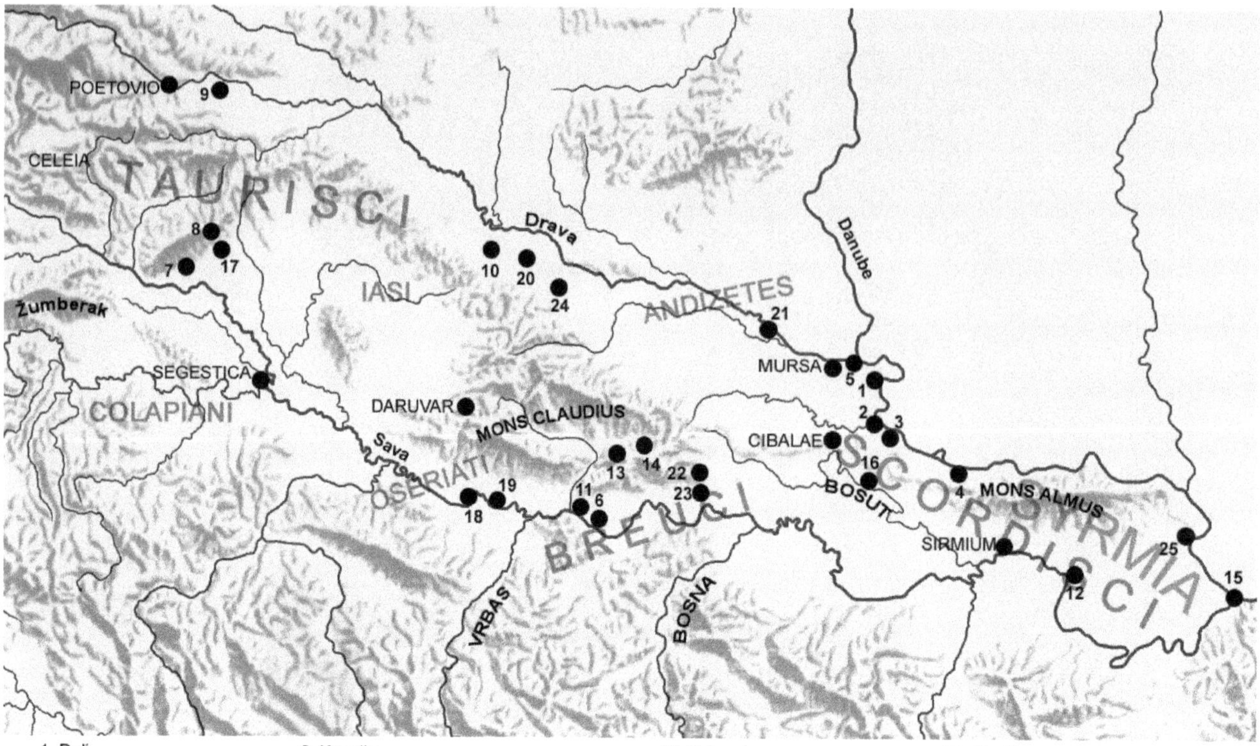

1. Dalj
2. Vukovar
3. Sotin
4. Ilok
5. Sarvaš
6. Donja Bebrina
7. Zagreb-Upper Town
8. Kuzelin
9. Varaždin
10. Virovitica, Kiškorija
11. Kupina near Slavonski Brod
12. Gomolava
13. Mali Bilač
14. Paka
15. Belgrade
16. Orolik
17. Sesvete
18. Stara Gradiška, Donja Varoš
19. Mačkovac
20. Orešac, Gaćište
21. Valpovo
22. Krndija
23. Stari Perkovci
24. Slatina, Trnovača
25. Novi Banovci

Fig. 1. Distribution of the protohistorical ethnic groups and sites at the time of the Roman conquest (made by M. Dizdar).

of the Central European La Tène Culture predominates.[1] Perhaps the most precise definition of the identity of the Scordisci has it that they represent a symbiosis of the newly arrived Celtic and various autochthonous ethnic communities, united by the military aristocracy. Moreover, with their inclusion into new society, the local elite took over the outer manifestations of their newly acquired position.[2] Celtic groups that would, together with local autochthonous population, form the tribe of the Taurisci, settled in southwestern Pannonia and the southeastern Alpine region.[3] The results of the excavations at the Carniola sites, such as Novo Mesto, are the best witnesses to the continuity of the strong autochthonous population that had incorporated part of its heritage into the new identity.[4] This situation resembles that of the Scordisci. An important confirmation for the distribution of individual ethnics in the area of study, besides the characteristic finds of material culture, are the finds of various coin types associated to the Scordisci and Taurisci. These, at the same time, indicate the intensity and direction of trade in the Late Iron Age. Coins were struck as imitations of Greek and Macedonian coinage, in the first place the coinage of Philip II, but certain issues characteristic for particular smaller regions can be discerned, such as the so-called eastern Slavonian type attributed to the western Scordisci,[5] which probably confirms the existence of differences in their complex ethnic composition (fig. 2).

Certain groups of autochthonous Pannonian origin are mentioned by Greek and Roman authors in the region between the Drava, Sava and Danube (such as the Iasi, Andizetes, Colapiani, Breuci). It is supposed that these were, as suggested by as yet inadequate archaeological research, exposed to military, political and economic domination and influences of the Scordisci and Taurisci, whose spheres of interest met at *Mons Claudius* (Plin. *NH* 3.25.148).[6] In the later period, during the Roman Empire, the boundary between Upper and Lower Pannonia passed through this area; the origin of this division is probably to be found in earlier boundaries (fig. 1). The Breuci, living in the region

[1] Papazoglu 1969, 212-213; Šašel Kos 2005, 136-141.

[2] Rustoiu 2005; Rustoiu 2008, 142-152 (on the example of the Late La Tène military aristocracy).

[3] Guštin 1984; Božič 1987; 1999.

[4] Križ 2001; 2005. More than 700 Late Iron Age graves were found at the site of Kapiteljska njiva, placed next to the Hallstatt burial mounds, supporting the notion of the continuity of burials and suggesting the probable continuity of the buried population; this is supported by certain metal and pottery finds from the Early and Late Iron Age graves.

[5] Popović 1987, 64. Coins of this type have been found in the important fortified settlements of the Scordisci, such as Osijek, Sarvaš, Stari Mikanovci, Privlaka, and Sotin.

[6] Šašel Kos 1986, 185; Dizdar and Potrebica 2002, 118-119; Šašel Kos 2005, 142-143, 416. It is supposed that it represents the central-*Slavonian mountainous* range. On the syntagm Slavonian Mountains, used in various chapters in this volume, see Migotti (in this volume), note 13.

Fig. 2. Coin of eastern Slavonian type from Stari Mikanovci (photo: I. Krajcar, AMZ).

on the mid-course of the Sava, the Andizetes (inhabiting the region on the lower Drava valley), and the Iasi (modern central Croatia from Varaždin to Daruvar) were the most important south-Pannonian ethnic groups mentioned by Strabo (Strab. 7.5.3). The material heritage of these groups, although rather unfamiliar, included a number of objects of the La Tène Culture, although local characteristics were still dominant. This fact distinguished them from the Scordisci or Taurisci, whose predominant traits came from the La Tène Culture. Therefore, the material legacy of the Andizetes and Iasi would probably display more obvious La Tène influences, that is, the greater number of finds pertaining to this culture, considering they were surrounded by groups of the presumed Celtic origin, than that of the Breuci, whose southern hinterland was protected by the groups of presumed Pannonian origins, inhabiting the region around the upper courses of the Rivers Bosna and Vrbas. The Pannonian War of Tiberius and the uprising of Bato, the latter threatening for a while the security of the Empire, testify to the fact that the ethnic identity of the Pannonian communities was not compromised by three centuries of exposure to the processes of *La Tèneisation* and *Celtisation*, taking place with various intensity in different regions in different periods. Nevertheless, even in the moments of greatest danger, they were not able to attain the level of firm political unity, which was one of the reasons for their losses of independence and the subordinate position within the Roman provincial setting, without any important centre such as, for instance, Sirmium, Mursa or Cibalae in their eastern areas, and Emona, Celeia and Poetovio in the western. The only exception is Siscia, founded in an originally Pannonian setting and owing its prominent position to its exceptional position at the intersection of important communications. Local elites, associated with the formation of fortified settlements, had an important role during the Roman conquest and the period of Romanisation of southern Pannonia and the southeastern Alpine region. They witness to important social and economic transformations that occurred in the first half of the 2nd century BC.

Our knowledge of Early and Middle La Tène settlements is very scarce, contrary to numerous finds from contemporary graves; at this moment, the existence of smaller-scale lowland settlements based on different economies is suggested.[7] The Scordisci have erected fortified settlements that became political, military, religious and economic centres of their respective areas, that is, smaller territorial districts. They were placed on important communication routes, on natural elevations, often on tel-type settlements formed during earlier prehistoric periods. In eastern Slavonia and western Syrmia, important settlements at Dalj, Vukovar, Sotin and Ilok were placed along the Danube, further at Osijek and Sarvaš along the Drava, and finally at Donja Bebrina on the Sava. The largest number of so-far documented fortified settlements was located along the river Bosut and its tributaries in the area of Vinkovci (fig. 3).[8] The characteristic of these fortified settlements was the existence of a fortification system consisting of an earthen rampart and a wide ditch. There were workshops in these settlements, producing various goods to satisfying local demands, but also for trade and exchange with the neighbouring communities. The lowland-type settlements and farms, most probably relying on agriculture as a primary production activity, gravitated towards fortified settlements, although smaller workshops for the production of certain goods, especially ceramic vessels, presumably existed in them. A similar settlement pattern was documented in the area settled by the Taurisci, although, considering the relief characteristics of the region, their fortified settlements were erected upon prominent elevations that

7 Dizdar 2001b (on the example of a Middle La Tène settlement at Ervenica in Vinkovci). The largest number of finds comes from Early and Middle La Tène graves of the Scordisci from Syrmia and eastern Slavonia: Majnarić-Pandžić 1970.

8 Majnarić-Pandžić 1984; Dizdar 2001b.

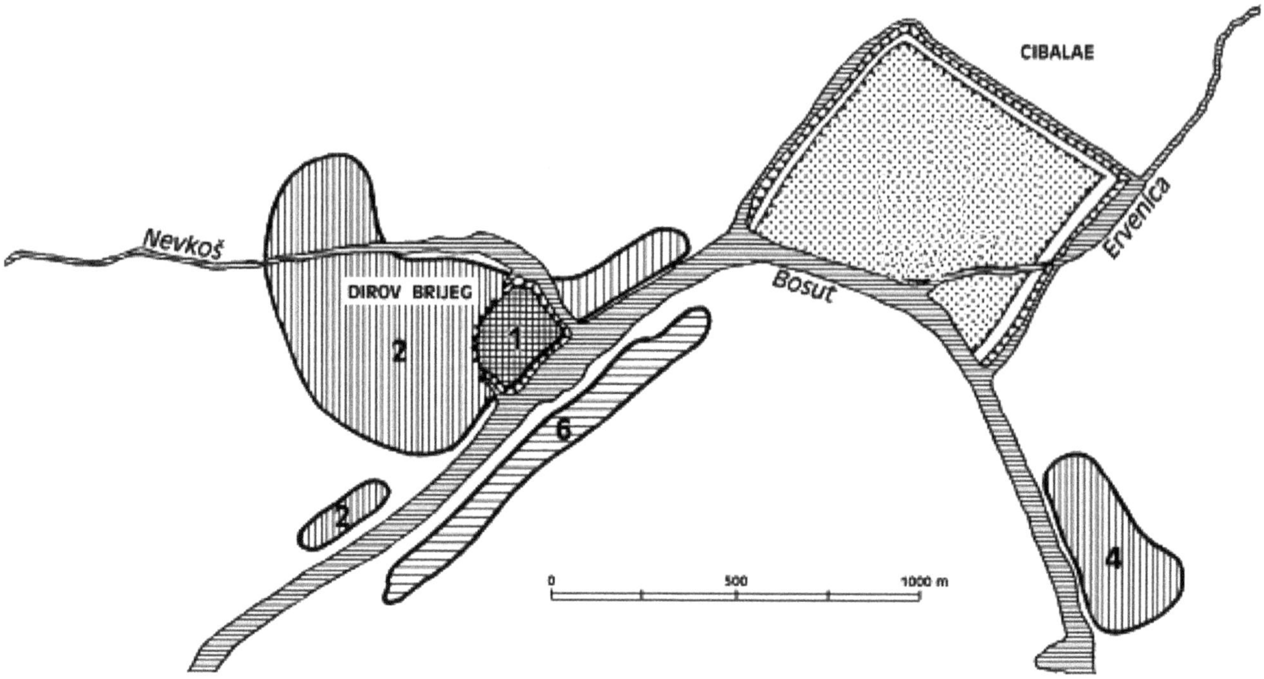

Fig. 3. Position of the fortified settlement at the site of Dirov brijeg in Vinkovci and its relation to the position of Cibalae (after Dimitrijević 1979, modified by M. Dizdar).

controlled important communication routes. This type of settlements was found at the sites of the Zagreb Upper Town and Kuzelin.[9] Contemporary lowland settlements, such as those excavated in the surroundings of Varaždin and Virovitica, were organized on family basis, comprising houses and other infrastructural facilities (fig. 4).[10] Similar settlements were documented also in the Sava and Drava valleys, as well as in their tributaries. On the other hand, the excavations of sites that can be connected to some of the Pannonian communities are extremely rare. Ceramic handmade vessels produced according to autochthonous traditions predominate at the site of Kupina near Slavonski Brod, while only an insignificant number of forms were produced on a potter's wheel in the La Tène tradition.[11] The settlement can probably be affiliated with the Breuci. Pannonian Segest/a/ica, located at the confluence of the Kupa and Sava, had a special role as a multi-regional centre attracting various goods and ideas even from distant regions, as is testified by numerous chance finds suggesting an urgent need for systematic excavations of this extraordinary site.[12]

Although numerous finds were discovered in the excavations of fortified settlements, such as that of the Scordisci at Gomolava[13] or of the Taurisci in Stična,[14] it is primarily burials that testify to a period of flourishing of the Late La Tène Culture in southern Pannonia and the southeastern Alpine region, because they offer a better insight into the social structure and economic relations of the ethnic communities of the area. The most prominent position had the warrior elite, with a prominent role of warrior-cavalryman; their burials are recognised by weaponry, horse gear or wagon elements, items of dress accessories, and rich offerings consisting of ceramic and bronze vessels. The importance of the cavalry aristocracy is further suggested by objects sacrificed to the war gods, most probably during initiation rituals or as a pledge for a success in battle. This group can be exemplified by the objects found in so-called fluvial contexts, especially those from the Ljubljanica[15] and the recently discovered Scordiscan sanctuary in Osijek on the Drava.[16]

Burials from the LT D1 phase (second half of the 2nd and early 1st centuries BC) are documented from the territory of the Scordisci, while burials from the phase preceding the loss of independence in the late 1st century BC are still lacking. This partially prevents the understanding of the processes developing immediately before the Roman conquest. The situation was probably similar to that of the Taurisci, where burials from the phase preceding the loss

9 Majnarić-Pandžić 1994; Sokol 1986; 1995.
10 Bekić 2006; Dizdar 2006b; 2007.
11 Minichreiter 1991. Similar conclusions can be drawn from rescue excavations in Slavonski Brod.
12 Majnarić-Pandžić 1970; Burkowsky 1999; Božič 2001.
13 Jovanović and Jovanović 1988.
14 Gabrovec 1994.
15 Gaspari 2009. The finds of the Novo Mesto type of helmet in the Sava at Sesvete (Sokol 2001) and Donja Varoš (Mihaljević and Dizdar 2007) probably have a similar meaning.
16 Filipović 2010. At the southern edge of the Late La Tène settlement in Osijek (the site of Vojarna) numerous finds of weaponry, horse gear, fragments of wagons and bronze vessels, as well as some coins, were discovered in 2008-2009, suggesting the existence of a sanctuary.

Fig. 4. Part of the infrastructure of the Late La Tène settlement Virovitica-Kiškorija (photo: M. Dizdar).

of independence have been documented. Even after the establishment of Roman power, burials in these cemeteries continued, suggesting continuity of the local ethnic communities regardless of the appearance of a new political-administrative government.[17] The offensive weaponry found in warrior graves of the Scordisci and Taurisci consisted of long iron swords carried in iron sheet scabbards and iron spears of various forms. Of the defensive weaponry metal fittings for the central part of the shield (*umbo*) survive, while only rarely iron and bronze helmets were found in the graves of prominent individuals in the southeastern Alpine area. In the second half of the 2nd century and the early 1st century BC (LT D1) the basic offensive weapon used by warriors was a two-edged sword with a uniformly wide blade, sometimes more than 1 m long (fig. 5). The blade ended in a curved top, indicating that these swords were used in combat only for slashing.[18] The only site in northern Croatia to yield this type of sword is Mali Bilač; it produced scabbard fragments provided with reinforcement in the form of two horizontal letters S on the upper part of the front side and a suspension loop on the back side, while the chape-end was in the form of a wide letter V with pointed clamps.[19] Swords in scabbards of the same type have been discovered both in the area of the Taurisci[20] and Scordisci.[21] The analogies with the finds from contemporary warrior graves of the Padea-Panagjurski kolonii group is especially important; this group is associated with warrior elite in the territory of southeastern Pannonia and neighbouring areas along the lower course of the Danube.[22] In the warriors' graves of the Scordisci long iron knives with mildly curved blade have also been found, such as the one from grave III from the site of Zmajevac in Sotin[23] or Paka.[24] The knives of the *sica* type are similar; they are often found in graves in Dacian territory, and their symbolic signification is presumed.[25] Spears of various forms are the most numerous finds in warrior graves. The type with a pronounced central rib and short sockets sometimes more than 50 cm long is exceptional; it was also used for close combat, while the short-leafed spears were probably used as throwing-spears (projectiles). Of the defensive weapons, central metal shield fittings have been found (*umbo*). In LT D1 shield bosses of circular form were common, such as that found in grave 3

[17] Windl 1975; Stare 1973; Dular 1991; Breščak 1989; Knez 1992; Božič 2008 (on the examples of Mihovo, Bela Cerkev, Verdun, and Novo Mesto).
[18] Dizdar 2009.

[19] Dizdar and Potrebica 2002, 113-114, 116, fig. 2.
[20] Božič 1987, 876, fig. 46.1, pl. LXXXVIII, 1; Božič 1999, 210.
[21] Todorović 1972; Božič 1981, 319, pl. 3.30.
[22] Rustoiu 2002, 15, fig. 2.2.
[23] Majnarić-Pandžić 1974, 58, pl. IV, 2.
[24] Dizdar and Potrebica 2005, 60, fig. 2.
[25] Rustoiu 2002, 57-61, figs. 38-42.

Fig. 5. Sword from Gaćište (photo: D. Doračić, AMZ).

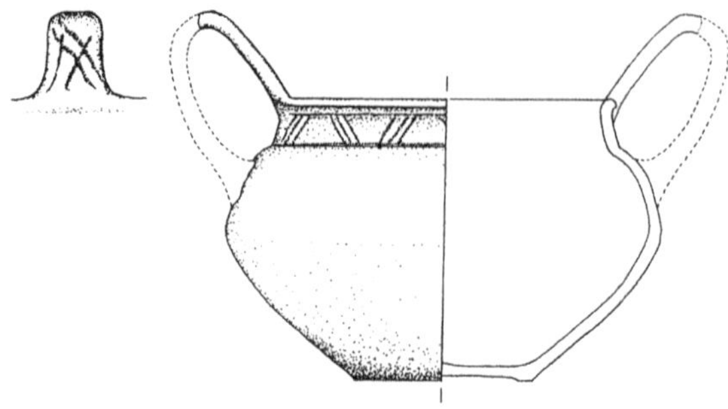

Fig. 6. Late La Tène kantharos from Vinkovci (after Dizdar 2001b).

in Sotin,[26] but an Arqua Petrarca-Mokronog type of shield boss was found in Mali Bilač[27]; it has parallels at the sites of the Taurisci and at Caput Adriae.[28] The find from Mali Bilač suggests the possibility of an exchange of military equipment or even a certain mobility of groups of warriors.

Items of horse gear and spurs were also found in warrior graves, occasionally also wagon parts, which portray them as warriors-horsemen,[29] thus emphasizing their prominent social status.

The Late La Tène warrior elite of the Scordisci and Taurisci, and probably also those of the Pannonian ethnic communities, especially the Breuci, further advertised their status by possessing prestigious goods, most often imported bronze vessels originating in north-Italian workshops. These imports anticipated the starting point of the conquest that would take place a century later, but also determined the social category of south-Pannonian ethnic communities that would take an active part in these events. Bronze vessels, as part of symposium sets, were found in graves of Scordisci together with ceramic vessels of local origin, in the first place the *kantharoi* (fig. 6), and were used for preparing and consuming wine, most probably during feasts. The finds in the graves of prominent warriors of the Scordisci are a proof to the connection of a social system with the circulation of goods; in other words, the possession of prestigious goods was a fulfilment of social needs in terms of displaying one's status and position within the community. The analysis of graves suggests that the number and type of weaponry corresponds to the number and types of imported bronze vessels.[30] Bronze vessels could have reach the local elite as items of trade but also as gifts to prominent members of the community in order to establish political and economic relations and the control of important communications used in trade and exchange of goods.[31]

The largest number of imported bronze vessels was documented in the rich graves of the Scordisci at the site of Karaburma in Belgrade,[32] but also at certain fortified settlements, such as Gomolava.[33] Bronze vessels were extremely rare in female burials, one stemming from the village of Jarak and a further one from grave 110 in Karaburma.[34] Bronze vessels were also found on some

26 Majnarić-Pandžić 1974, 66, pl. IV, 10, pl. V,3.
27 Potrebica and Dizdar 2005, 58, fig. 1.
28 Guštin 1991, 57-58, Abb. 30; Božič 1999, 198; Guštin 2001, 353; Guštin 2002, 15-16, fig. 2.

29 Todorović 1972; Božič 1984a; 1984b.
30 Egri and Rustoiu 2008 (on the examples from Karaburma).
31 Dizdar 2006a.
32 Todorović 1972; Božič 1981, 320; Guštin 1984, 327, 338.
33 Popović 1992.
34 Brunšmid, 1902a; Todorović 1972.

Fig. 7. Bronze vessels from Sotin (after Dizdar and Radman-Livaja 2004).

sites of the Scordisci in eastern Croatia: in graves from Sotin[35] (fig. 7) and Mali Bilač,[36] probably also in Dalj,[37] and in fortified settlements in Vinkovci and Orolik.[38] The most frequent forms are buckets, pans, ladles and sieves, while other forms appear less frequently. The buckets are most often of the type featuring two semicircular fittings on the neck supporting the handle attachments, with the handle ends slid through their openings. This type of bucket was found in grave I in Sotin and Mali Bilač, as well as a chance find in Karaburma.[39] Such buckets were distributed widely from Spain to France to Poland, and are dated to the late 2nd and first half of the 1st centuries BC.[40] The most frequent find in graves of the Scordisci are ladles of the Pescate type, considered to be a product of north-Italian workshops. Their cast handle is divided into two or three segments and ends in a small, stylised animal head, while its receptacle is spherical in form.[41] Bronze strainer-handles are also frequently found at eastern-Slavonian sites, since their bowls are rarely preserved due to fragility.[42] They are dated from the late 2nd century to Tiberius' reign and it is supposed that they were produced in Italic workshops, but their wide distribution suggests the possibility that they were also produced in Celtic settlements.[43] On account of the forms of bronze vessels, it is presumed that most of them were produced in north-Italian workshops, while some types could have been produced in local south-Pannonian workshops.[44] Bronze vessels reached the Scordiscan territory the trade route leading from Aquileia to Nauportus (Vrhnika), where the goods were loaded onto ships and transported by the Ljubljanica and Sava Rivers via Segestica all the way to the Danube region. This communication route is described by Strabo (Strab. 4.6.10) who also adds that wine and olive oil was exchanged for slaves, cattle and leather (Strab. 5.1.8). The same route can be discerned from the legend of the Argonauts, mentioned by Pliny (Plin. *NH* 3.148).[45] The trade was probably in the hands of rich Aquileian families, and was certainly very profitable. Velleius Paterculus (Vell. 2.110.4-5) probably refers to their presence in southern Pannonia when he, referring to a somewhat later period, mentions the merchants as victims at the start of the Pannonian-Delmataean rebellion. The Sava communication route gained in importance after Octavian's Illyrian War (35-33 BC), when the path towards the Danubian area was opened with the conquest of Segestica. This natural communication route between northern Italy and the southeastern Alpine region with the Danube basin is substantiated by coin finds of Greek cities of Apollonia and Dyrrachium, as well as by the finds of Republican coinage. The drachms of Apollonia and Dyrrachium probably first reached the Danube region at the beginning of the 1st century BC,[46] which is supported by the hoards from Dalj and Vukovar,[47] while Roman *denarii* started entering the region in larger numbers from the mid-1st century onwards.[48] This same route was followed by the Roman in conquering the area between the Sava, Drava and Danube in the late 1st century BC. In other words, economic relations preceded those military. The existence of regional trade network is further testified by the finds of the eastern Norican coin type (type Đurđevac), associated with the Taurisci, on the sites of the Scordisci,[49] while, on the other

35 Majnarić-Pandžić 1974.
36 Potrebica and Dizdar 2003.
37 Majnarić-Pandžić 1970.
38 Dizdar 2001b; Dizdar and Radman-Livaja 2004.
39 Potrebica and Dizdar 2005.
40 Bolla, Boube, and Guillaumet 1991, 11-13.
41 Todorović 1972; Popović 1992; Dizdar and Radman-Livaja 2004.
42 Dizdar and Radman-Livaja 2004.
43 Guillaumet 1977; Guillaumet 1991.
44 Majnarić-Pandžić 1974; 1996.

45 Domić Kunić 2004. The legend probably reflects the contacts in earlier periods that were taking place along the same communication route.
46 Popović 1987.
47 Brunšmid 1912; Dukat and Mirnik 1978; Mirnik 1981; Popović 1987.
48 Popović 1987. The suggestion that the appearance of the drachms of Apollonia and Dyrrachium should be dated not earlier than the mid-1st century BC is partly refuted by the finds from the site of Blato near Vinkovci. Some coins are found here in the LT D1 context finishing ca. 80/70 BC. I would like to thank to my colleague Hrvoje Vulić from the GMV for the information regarding the finds from Blato.
49 Dukat and Mirnik 1976, 190-192; Kos 1977, 46; Dizdar and Potrebica 2005, 63-64.

hand, the coinage of the Scordisci is found as far west as Donja Dolina in the mid-Sava region.[50]

In addition to prestigious objects intended for the elite, the existence of exchange and trade between northern Italy, the southeastern Alpine region and Central Europe on the one side, and southern Pannonia on the other, is documented by items of dress accessories and jewellery, that is, the objects owned by a wide array of social classes. A good example from the late 2nd and early 1st centuries BC is provided by the Vinkovci type of brooches with triangle- or leaf-shaped bow and unperforated rectangulae or trapezoidal foot, appearing on the sites of the Scordisci as a substitute for brooches of the Nauheim II type.[51] Since these brooches were especially numerous on eastern-Slavonian sites, it is supposed that they were produced in workshops situated in fortified settlements that were also craft and trade centres, and at the same time the source of the spread of the La Tène Culture. This is further supported by a number of variants of the type of brooches in question, some of which were distributed locally.[52]

Finds from the younger phase of the Late La Tène (LT D2), prevailing until the Roman conquest of the region between the Sava, Drava and Danube, are much less frequent in the territory of the Scordisci. For this reason, comparisons with sites of the Taurisci in the southeastern Alpine region are necessary, especially with sites in Carniola, where graves containing numerous items of weaponry and wear have been found. These are the graves of the warrior elite that would play an important role in Roman conquests of the southeastern Alpine and southwestern Pannonian region, but also in the subsequent period of stabilisation of the newly conquered areas. The existence of a higher social class is substantiated by Appian's account of Octavian's siege of Segestica, where he states that the defenders offered to surrender and lose independence, although lower classes were against it (App. 23).[53] Presumably, the local elite intended to keep their privileges, which is substantiated by the archaeological evidence from the southeastern Alpine region.[54] Although finds from the LT D2 period (80/70-15 BC) in northern Croatia are very rare, some of those suggest the presence of prominent individuals, similar to the previous phase of development. Thus, the neck guard of a bronze helmet was found near Sesvete,[55] while a completely preserved bimetal helmet was found in the Sava near Donja Varoš in the vicinity of Stara Gradiška. It is one of the most prominent European finds from the late phase of the development of the La Tène Culture (fig. 8).[56] In terms of their form and elements of construction, both helmets belong to the eastern Celtic type (Novo Mesto) of helmet. Their closest parallels come from the warrior graves in Carniola, dated from the mid and the second half of the

Fig. 8. Helmet of the Novo Mesto type from the Sava River near Donja Varoš (photo: D. Doračić, AMZ).

1st century BC.[57] The helmet found in the Sava River near Donja Varoš, along with the Đurđevac type of tetradrachms found in the nearby Donja Dolina and Mačkovac, testify to the importance of the communication route passing through the Sava valley. Battle knives with curved blade of the Pritoka-Bela Cerkev type are contemporary with the helmets (fig. 9).[58] Similar objects are expected to be found at Breucan sites. On the other hand, wagon parts and horse gear items identical to those from the Carpathian basin and the wider European contexts,[59] were discovered in Orešac (mid-Drava valley), testifying that the territory of northern Croatia was encompassed within the Central European La Tène Culture during the LT D2 period. The finds of brooches originating in northern Italy suggest that exchange and trade did not exclusively include prestigious goods. These finds prove the continuity of the established relations, as evidenced by imports of bronze vessels in the earlier periods. The number of brooches found in the southeastern Alpine and southwestern Pannonian area (especially in Siscia) is much larger in comparison with other territories, which is understandable, since these areas are much nearer to their place of origin. Nevertheless, an increasing number of LT D2 brooches have lately been discovered in the eastern section of the region between the Sava, Drava and Danube, such as those of Almgren 65 type.[60] Brooches of the Gorica[61] and Alesia[62] types are somewhat younger; it is supposed that the latter are associated with Octavian's campaign of 35-33 BC, that is, with the movements of the Roman army.[63] Brooches of the Idrija I type belong to the

[50] Popović 1987, 50.
[51] Demetz 1999, 78-81; Božič 2008, 50-65.
[52] Dizdar 2003; Potrebica and Dizdar 2005; Majnarić-Pandžić 2009.
[53] Šašel Kos 2005.
[54] Gaspari 2008, on the example of finds from the Posočje region.
[55] Sokol 2001.
[56] Mihaljević and Dizdar 2007.
[57] Schaaff 1986; 1988.
[58] Božič 1999, 211; Balen-Letunić 2006.
[59] Schönfelder 2002.
[60] Majnarić-Pandžić 1970, 21, pl. III, 7; Dizdar 2001a, pl. 4,12.
[61] Demetz 1999, 106-109.
[62] Feugère 1985, 299; Demetz 1999, 157-161.
[63] Guštin 1991.

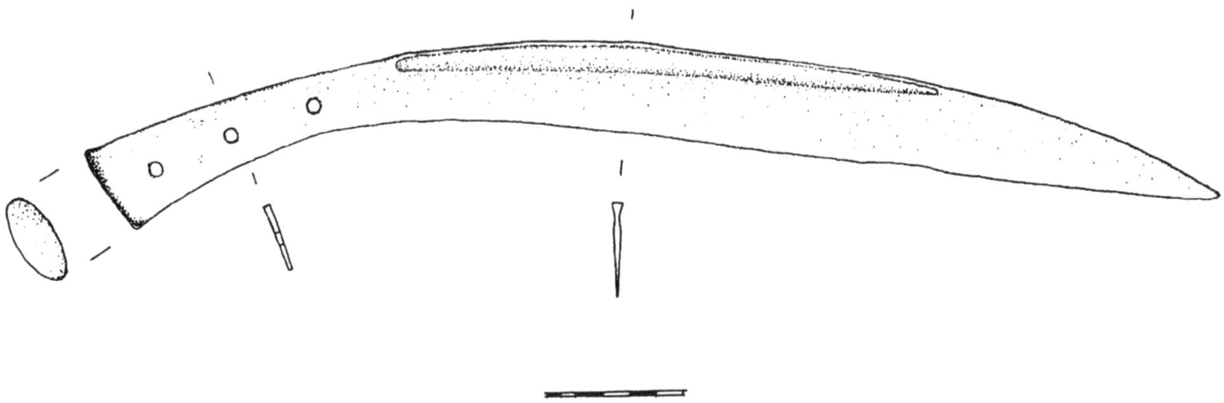

Fig. 9. Battle knife from Sisak (after Balen-Letunić 2006)

same phase of development,[64] together with very numerous Jezerine type examples.[65] Local imitations of these brooches produced in the workshops of the Scordisci have also been identified.[66] Another type of contemporary brooches is Almgren 18.[67]

Thus, the material evidence of south-Pannonian ethnic communities (especially the Scordisci) does not show any signs of decline during the 1st century BC. On the contrary, this is the period of their greatest rise, which partly contradicts hypotheses based on the analysis of literary sources, suggesting that after the defeat in 84 BC by the consul Lucius Cornelius Scipio Asiagenus, the Scordisci began to lose previously controlled territories. This was probably true only of the area of southern Serbia: an important communication route passed through this territory towards Macedonia and Greece, which the Scordisci controlled by having occupied important elevated fortified settlements, such as Kale at Krševica.[68] Also, their political dependence on the neighbouring Dacians[69] is still not archaeologically proven. It is more probable that the two peoples were in some kind of alliance, which is substantiated by a significant proportion of their common material heritage, and even a similar social organisation with the warrior elite as the leading group.

3. The archaeological evidence from the period of the Roman conquest

The Roman conquest encountered all Pannonian ethnic communities at the peak of their development, both in terms of material culture and social and economic organisation. Octavian's Illyrian War of 35-33 BC marks the beginning of the conquest of southern Pannonia. The Pannonian town of Segestica was conquered in the early phase of the war after a month's siege. It was positioned on an important strategic point that controlled the communication route along the Sava. Unfortunately, archaeological excavations have not yet proved any traces of these events; the same is true of the episodes that ensued during the following decades. In this period Siscia became an important Roman military stronghold, from which the conquest of the eastern part of the region started.[70] After Noricum and the adjacent territory of the Taurisci lost their independence in 15 BC,[71] the western portion of Pannonia became part of the Empire, contemporary with the successful conquest of the Alpine passes. The final conquest of the eastern part of the region between the Sava, Drava and Danube and the subjugation of the Pannonians was accomplished in the Pannonian War (12-11 BC), successfully concluded by Tiberius, who established the control of the communication route leading through the Sava basin, as the most direct route between northern Italy and the southeastern Alpine region and the Balkan provinces. The frontier of the Empire was established on the Drava and Danube. Another result of this war was the subjugation of the eastern Adriatic hinterland.[72] These military operations drew the attention of Greek and Roman historiographers, who now described certain previously unknown Pannonian ethnic communities that confronted the Roman armies. Tiberius used the Scordisci as allies in the Pannonian War (Dio 54.31.3); most probably in ca. 15 BC they recognised Roman supremacy and became Roman allies.[73] The only account of the loss of Scordiscan independence comes from Velleius Paterculus, who relates that Tiberius forced the final acknowledgment of the defeat from the Illyrians and Delmatae, and added to the Empire the provinces of Rhetia, Vindelicia, Noricum, and Pannonia, as well as the Scordisci (Vell. 2.29.3). The fact that the Pannonians were armed like the Scordisci is especially important, supporting the suggestion that the Pannonian ethnic communities accepted the accomplishments of the more advanced La Tène Culture, especially weaponry. Tiberius confiscated their weapons and, according to Florus (Flor. *Epit.* 2.24), threw them broken into a river.[74] The strong resistance of the Pannonian communities showed that

[64] Božič 2008, 99-100.
[65] Adam and Feugère 1982; Demetz 1999.
[66] Dizdar 2001a.
[67] Völling 1995, 187-188; Demetz 1999, 117-122.
[68] Popović 2009.
[69] Papazoglu 1969, 255-261; Šašel Kos 2005, 153.

[70] Šašel Kos 2005, 437-442.
[71] Šašel Kos 1986, 153; Šašel Kos 1997b, 32.
[72] Šašel Kos 1986 155-160; Šašel Kos 1997b, 31; Šašel Kos 1997a, 191; Šašel Kos 2005, 458-465; Domić Kunić 2006
[73] Šašel Kos 1986, 155-161; Šašel Kos 2005, 153.
[74] Šašel Kos 1986, 167-179, 183-189; Šašel Kos 2005, 506. A similar weapon-offering was earlier documented in the Ljubljanica: Gaspari 2009.

10a

10b

Fig. 10. Hoard of Roman coins from Valpovo/Osijek: a – averses, b - obverses (after Radman-Livaja and Dizdar 2010).

the three centuries of *Celtisation* did not jeopardise their ethnic identity, recognised as such by Greek and Roman authors, who clearly distinguish them from the Scordisci, described as of Celtic origin. The reason for their loss of independence could perhaps be found in the fact that the Pannonian peoples could not manage to attain a higher level of political unity even when faced with a gravest threat. Nevertheless, a later Breucan-Daesitiatean alliance did not manage to prevent the inevitable spread of Roman power to this strategically extremely important region, inhabited for centuries by the Pannonian ethnic communities.

In the period between Tiberius' conquest and the start of the Pannonian-Delmatean uprising in AD 6, a small number of auxiliary units were stationed in southeastern Pannonia; together with the Scordisci they controlled the basins of the Danube, Sava and Drava. This fact is substantiated by an account describing how Roman military units could not cope with the insurgents at the beginning of the uprising, ordering legions from other provinces to come to their assistance. Besides minor garrisons and legions stationed in the strategic positions (Siscia and Sirmium), the system of frontier security rested on political agreements with autochthonous communities governed by their elites, as was the case in other provinces. Perhaps the closest analogy can be found in the territories along the Rhine *limes*, where autochthonous local elites played an important role, at first in the conquest and later in the consolidation and defence of the frontier; besides weaponry, their graves yielded both local and imported prestigious objects.[75] There are almost no archaeological traces of the rebellion, a situation identical to the earlier Pannonian War, except for two coin hoards dated to this period and associated with the fights that took place in the region of Slavonia and Syrmia.[76] A hoard of Republican *denarii* (with two Augustan examples) was found in the late 19th century in the area between Valpovo and Osijek (fig. 10).[77] A similar hoard was found at the village of Krndija near Đakovo.[78] Some of the finds of Roman weaponry from Sisak can probably be associated with the uprising of AD 6.[79] Certain finds from the area of the future province of Dalmatia can also be associated with military operations, suggesting that the military engagement of the southeastern Alpine peoples, most probably the Taurisci, was not limited to the area of southern Pannonia. This applies to the hoard of the Đurđevac and Kapos coin types found at Balina glavica[80] and a grave from Tribanj near Starigrad Paklenica (fig. 11) in Liburnian territory, in which a bent *gladius* was found, with closest analogies in the graves in Mihovo and Verdun in Carniola.[81]

After the collapse of the Pannonian-Delmataean uprising and the initial consolidation, during the reign of Tiberius and especially that of Claudius, a firm Roman control

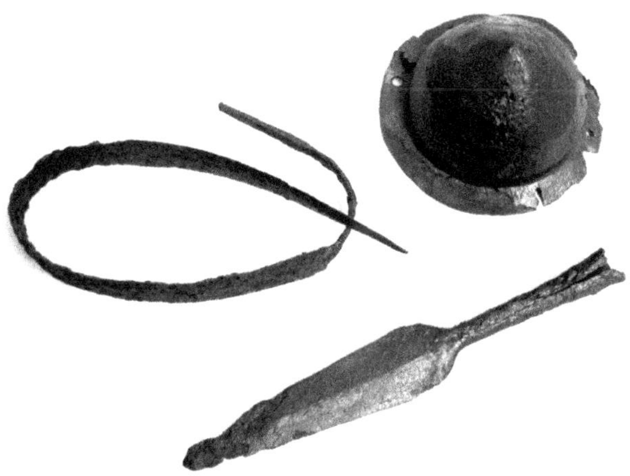

Fig. 11. Grave finds from Tribanj near Starigrad-Paklenica (photo: I. Krajcar, AMZ).

was established in southern Pannonia and the process of Romanisation spurred by stationing of the military and settling of new inhabitants, especially merchants and craftsmen that followed the armies, as well as by recruiting the autochthonous population in auxiliary units. The Scordisci were rewarded with the *civitas Scordiscorum* for their faithful alliance with Rome and their involvement in the subjugation of the insurgents, but this encompassed only the core of their previous territory in eastern Syrmia, that is, the area with their largest and most important sites predating the Roman conquest.[82] As a matter of fact, in this way the Scrodisci lost their former influence, and some autochthonous previously unknown Pannonian groups now appear in literary sources, such as the Andizetes or Amantini. The decline of the Scordisci is further recognisable in the separation of artificially created peregrine communities such as the Cornacati, Tricornense and Pincense, who all probably received their names after a main settlement or stronghold in their territory, and they represented parts of the former ethnic corpus of the Scordisci.[83] The importance of the Scordisci in the events that took place in southeastern Pannonia in the late 1st century BC and the early 1st century AD is further substantiated by the fact that the most important Roman centres (Sirmium, Mursa, Cibalae) or fortresses on the *limes* (Teutoburgium, Cornacum, Cuccium, etc.) were erected upon, or in the immediate vicinity of, their prominent settlements. The research of these settlements has revealed the material evidence, especially ceramic vessels, which originated in the earlier period, but also played an important role in the early Roman provincial culture. This means that the local population represented an important ethnic component in southern part of Roman Pannonia during the 1st century. In the newly founded centres, besides local forms, imported objects also appeared, such as thin-walled pottery, *terra sigillata*, and glass and bronze vessels. The earliest of those belong to the late Augustan and Tiberian periods and should

75 Roymans 1996; 2004.
76 Radman-Livaja and Dizdar 2010.
77 Brunšmid 1895.
78 Gardaš 1990.
79 Radman-Livaja and Dizdar 2010.
80 Kos 2002.
81 Windl 1975; Breščak 1989.

82 Dimitrijević 1961; Todorović 1974, 146-149; Šašel Kos 2005, 154, 513.
83 Šašel Kos 2005, 144.

be associated with the presence of the military.[84] Certainly, local elites should not be forgotten, who, by possessing prestigious goods, also wanted to advertise their status, as in earlier periods. These goods reached the region by the already known route leading through the Sava valley. The Siscia – Sirmium road was built along this course already in the Augustan and Tiberian periods.[85]

Archaeological excavations in the territory of study could be very important for the discussion of substantial transformations that were implemented with the establishment of a new political and administrative order, as well as for an understanding of a number of processes that unfolded during the period of the early Romanisation of southern Pannonia. At this stage, previous research and the finds at disposal only allow forming a temporary model that will have to be verified by systematic excavations and analysis of the material kept in the museums. Naturally, this research should be assisted by the study of similar contemporary processes from other frontier regions of the Empire, which disclose all complexities of Romanisation procedures. In the first place, by this term we normally understand the cultural levelling between different groups within the Empire. Nevertheless, some discrepancies conditioned by previously established different material, economic, social, and political characteristics of various regions remained, regardless of the spread of the Roman cultural forms in the language, ideas and values, as well as in the material culture.[86] Given all this, it is allowed to suppose that every individual protohistorical ethnicity was exposed to a different intensity of Romanisation that, in the end, resulted in the acceptance of the new, Roman form of living.

Certain finds dating from the early 1st century testify to the events that were taking place at the beginning of the new millennium and the establishment of different relations between the autochthonous groups and their new rulers. These processes can be indirectly associated to the Pannonian-Delmataean uprising in which the Scordisci and Taurisci fought on the side of the Romans, either as allies or recruits in auxiliary formations. At this stage, only a small number of finds from eastern Slavonia and Syrmia testifies to the involvement of the Scordisci in the already mentioned military operations as well as to their subsequent role in the protection of the frontier. Most often neither the circumstances of the discovery nor the original context of the finds are known. A good example is provided by the weaponry and bronze vessel fragments from the early 1st century, found at the site of Vrtna Street in Vinkovci, associated with the early Roman military presence in Cibalae. An iron sword with straight blade, rhombic cross section and gently thickening at the top, and two long spears with a narrow leaf, also rhombic in cross section, with a short socket, showing the Late La Tène typological characteristics, are prominent among the finds of weaponry. A bronze handle with a stamp ABVDVSF is especially important. The manufacturer's name, probably of Italic origin, has been identified in the inscription, while the letter F (for *fecit*) shows that it was produced in an Italian workshop or by an Italian craftsman active in southern Pannonia.[87] Bronze vessels were a standard part of Roman military equipment, but they were also found earlier in prominent Late La Tène Scordiscan warrior graves. These finds could represent the grave goods of a prominent local warrior in alliance with the Romans or a Roman soldier recruited in an auxiliary unit of Celtic origin, which participated in the conquest of the eastern part of the region.[88]

The exceptional finds from the Odescalchi Castle in Ilok testify to the Scordiscan involvement in the protection of the Danube frontier (fig. 12).[89] During the several years of excavation at the Upper Town in Ilok, remains of early Roman burials were found with grave architecture featuring traces of wooden caskets in large and deep rectangular grave pits. Grave inventories contained weaponry, imported pottery, glass and bronze vessels, items of dress accessories and jewellery, cosmetic sets, as well as ceramic vessels produced in the La Tène traditions. In grave 5 an iron sword (*gladius*) of the Mainz type was found, whose wooden sheaths were equipped with bronze fittings. A belt with decorated bronze fittings and strap ends (*cingulum*) was found along with the sword. A bronze *Aucissa* brooch from grave 3 also belongs to the military equipment. Numerous pottery, bronze and glass vessels and lamps were also deposited in graves (fig. 13). The largest proportion of ceramic vessels (Dressel 6B amphorae, *terra sigillata*, thin-walled pottery), together with bronze and glass vessels and lamps, arrived from Italy through trade both for the demands of the military and settlers and the higher classes of the pre-Roman population. These objects testify to the origin and routes of the ethnic, cultural and economic influences in the process of the early Romanisation of southeastern Pannonia. A certain number of the vessels were probably produced in local Syrmian early Roman workshops that copied imported originals. Nevertheless, the finds of ceramic vessels belonging by their form and ornamental techniques and motifs to the autochthonous heritage, are crucial for the establishment of the deceased's origin. These were the products of local potters that continued to produce, even after the conquest, the familiar forms of vessels for the needs of the autochthonous population. Foods and drinks were originally contained in these vessels, as is testified by the preserved seeds of grain and fruit, mostly grapevine. Remains of figs and olives suggest that, together with luxurious objects, Mediterranean products were also imported for the demands of soldiers and settled merchants and craftsmen.[90] Coins placed in graves as payment (*obolós*) to the ferryman of the dead together with other imported goods, show that

[84] Brukner 1981a, 144; 1981b, 48; 1987, 41.
[85] Bojanovski 1984, 248, 253-254.
[86] Roymans 2004.
[87] Dimitrijević 1979, 173; Majnarić-Pandžić 1996, 32-33; Dizdar 1999, 118-119; Dizdar and Radman-Livaja 2004.
[88] Dizdar and Radman-Livaja 2004.
[89] Tomičić *et al.* 2007. These finds were exhibited in the AMZ in 2010.
[90] Dizdar *et al.* 2003.

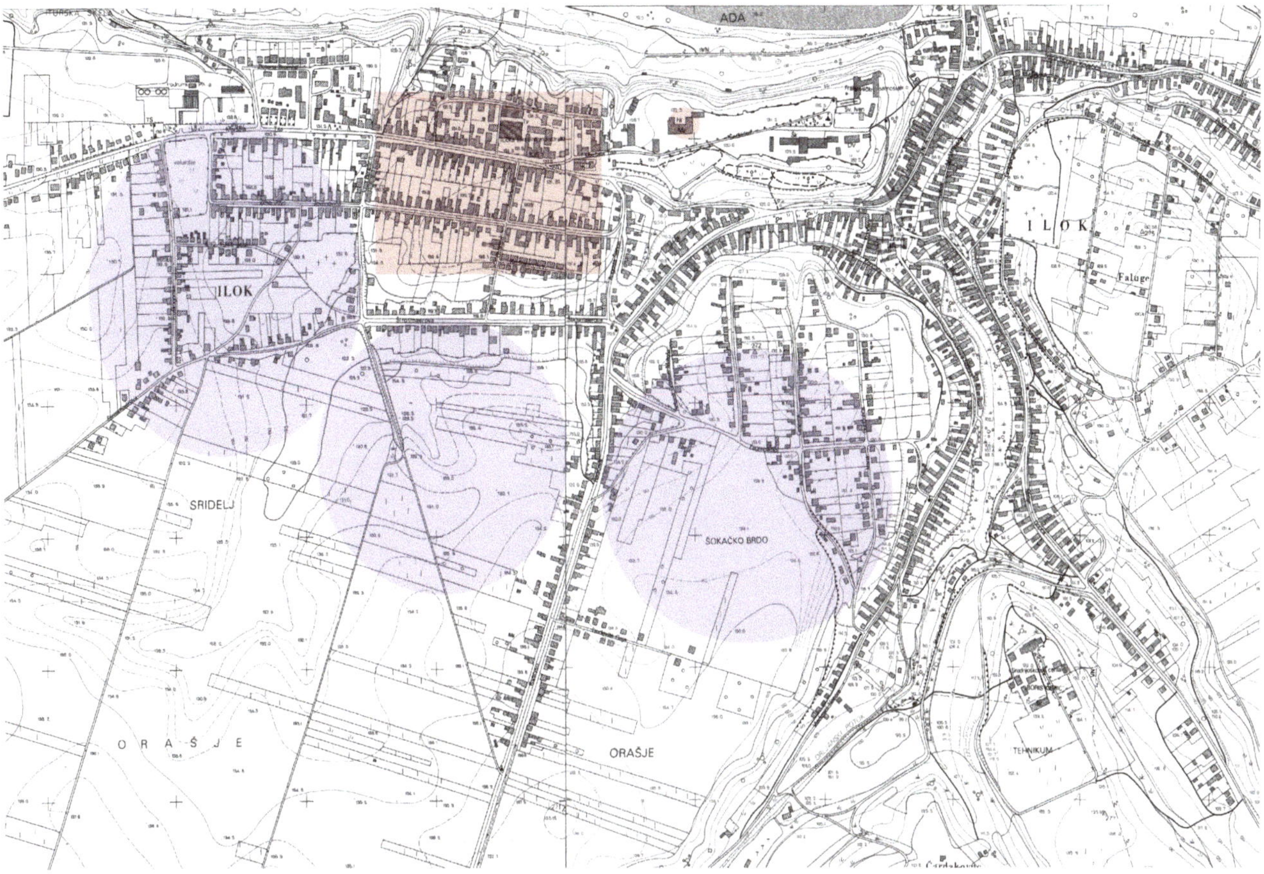

Fig. 12. Position of the Roman settlement Cuccium in Ilok with cemeteries: red dot marks the position of the early Roman cemetery (made by M. Dizdar).

Fig. 13. Ceramic finds from grave 5 in Ilok (photo: I. Krajcar, AMZ).

these are the graves from Tiberius' reign or, at the latest, from the beginning of Claudius'. Considering the finds of local pottery and offerings in the form of weaponry, it can be assumed that the deceased were members of the local Scordiscan aristocracy, who, as the allies of Rome, were entrusted with participation in the defence of the *limes*. They continued to be buried together with their personal weapons, now of Roman origin, in accordance with their ancient customs; these customs further included imported grave goods as well as certain pottery forms of local origin.

Graves of the veterans of Celtic origin and the locals that were the first to be Romanised were also found in the eastern cemetery of Sirmium, testifying to a complex ethnic and social composition of the population of Sirmium in the early 1st century. These burials are cremations in the form of a well. Besides weaponry, imported north-Italian products and ceramic vessels produced in the Late La Tène tradition were found in these graves. The earliest burials are dated in the late Augustan and Tiberian periods.[91] Several other sites from the 1st century with cremation graves are known in Syrmia; their autochthonous origins have been suggested on the basis of imported and local forms of pottery (Novi Banovci[92], Rakovac and Zemun[93], Surčin and Donji Tovarnik[94], Obrež[95], Zemun-polje[96]). These early Roman cremation burials with ceramic vessels in the forms produced in the Late La Tène tradition document a strong autochthonous heritage surviving until the end of the 1st century.[97] Grave offerings in the form of early Roman weaponry and imported bronze, ceramic and glass vessels and lamps, together with occasional ceramic forms of local origin, were also documented, similar as at Ilok and contemporary cemeteries in Carniola. Helmets of the Weisenau type, *gladii*, spears and shield bosses were found in graves from Mihovo and Verdun (some of the weapons were bended before deposition, which represents the continuation of burial traditions of the Mokronog group).[98] Similar finds were documented in graves in the Soča valley (Reka pri Cerknem, Idrija pri Bači)[99] and in other areas of the southern Alpine region, for instance in the Ornavasso and Giubiasco cemeteries.[100] Offerings of weapons prove that warriors equipped in Roman fashion were buried in these graves, most probably Roman allies or mercenaries engaged in conflicts that occurred during the Roman conquest of the southeastern Alpine and southwestern Pannonian regions in the late 1st century BC. Similar finds of early Roman graves with weapon offerings could be expected in northwestern Croatia, where the continuity of local population is at present suggested by the cremation cemeteries on the Žumberak and Samoborsko gorje (Samobor Mountains).[101]

The results of the excavations of settlements in the territory of eastern Slavonia and Syrmia suggest the existence of different models of relations between the newly founded Roman centres and earlier Scordiscan settlements, testifying, together with the abovementioned grave finds, to the process of Romanisation. Thus the most important Roman centres, such as Sirmium or Mursa, were founded on the sites or in the immediate vicinity of the fortified Scordiscan settlements. The same phenomenon is documented in relation to the settlements along the Danube, on which strongholds were erected along with civilian settlements that developed in their vicinity. Nevertheless, the greatest number of the fortified settlements of the Scordisci, representing the core of their human-resource and economic potentials, were abandoned during Tiberius' conquest in the late 1st century BC,[102] since they were probably considered as a threat to the newly founded Roman government. The abandonment of earlier centres is substantiated by the lack of finds of early Roman provincial forms in them; therefore, it is supposed that their inhabitants were resettled to the newly founded towns or other settlements (*vici*) where they continued to keep their autochthonous material culture, recognisable in the first place through forms of pottery.[103] The fate of lowland settlements, both those of the Scordisci and other Pannonian communities, of supposedly prevalent agricultural character, was completely different. Some of those continued to exist uninterruptedly in the early Roman period, preserving the prevalent autochthonous heritage, with only rare instances of imports of new provincial forms or prestigious goods. Some settlements were abandoned with the establishment of the Roman power, but the reasons behind this should probably be sought for in changed economic conditions.

One of the most important centres in southern Pannonia was Mursa, located in the territory of Donji grad (the Lower Town) in Osijek; rescue excavations in Osijek testify to the intensity of early Romanisation. At the moment of the outbreak of the Pannonian-Delmataean uprising, a Roman garrison was probably stationed in Mursa (*ala II Hispanorum Aravacorum*), controlling the important crossing of the Drava and the communication route towards the Danube.[104] Early Roman military presence is perhaps confirmed by the finds of vessels produced in the late Arretine and early Po valley workshops, dated to the late Augustan and early Tiberian periods,[105] especially the fragment of an ACO cup found at the site of Pristanište in one of the La Tène pits; the cup can be attributed to a soldier of a Roman military unit or an Italian merchant.[106] On the other hand, numerous forms of pottery produced in the Late La Tène traditions belong to the autochthonous heritage.[107] Some cremation burials from the northeastern cemetery, for example grave 8 with ceramic vessels

91 Milošević 1985; Milošević 1987, 14-17; Milošević 2001, 161-163.
92 Dimitrijević 1961, 93-94.
93 Brukner 1987, 40-41.
94 Srejović 1965, 58.
95 Dimitrijević 1961, 98.
96 Dimitrijević 1969, 86; Brukner 1987, 41.
97 Brukner 1987, 41.
98 Windl 1975; Breščak 1989; Božič 1999, 199-200.
99 Guštin 1991; Gaspari 2008.
100 Graue 1974; Pernet *et al.* 2006.
101 Gregl 2009.

102 Majnarić-Pandžić 1984, 27-28.
103 Dizdar 2001b, 104-105.
104 Klemenc 1961, 17-19; Klemenc 1963, 59-62; Pinterović 1968, 63-67, 73; Pinterović 1969, 61.
105 Gabler 1979, 206; Brukner 1981a, 143.
106 Tonc and Filipović 2010.
107 Bulat 1977, 48.

or the find of a coin of the Syrmian type C/D with an Augustan *as*, can also be associated with that tradition.[108] All of this shows that autochthonous population made an important part of Mursa's ethnic composition during the 1st century, since this was one of the most important centres of the Scordisci and probably their westernmost large stronghold. A similar situation was documented in the excavations of another large centre – Cibalae (Vinkovci) – in the hinterland of the Danubian *limes*, which spurred the economic development of the town. Numerous finds of pottery substantiate the important role of the autochthonous population in the earliest phase of the existence of Cibalae. The fortified settlement at the site of Dirov brijeg and the lowland settlement at the site of Ervenica offer analogies for those pottery finds; from these two settlements the population was most probably resettled to the newly erected centre located on the site of tel Tržnica next to the Bosut (fig. 3).[109] The continuity of the Late La Tène heritage, especially of ceramic vessels, is also recognised in other early Roman centres, such as Sirmium, where it appears along with imported objects,[110] but especially at Gomolava (phase VIc).[111] These data suggest the strength and the importance exercised by the autochthonous population in the first decades of early Roman provincial rule.

The continuity of autochthonous population, even as concerns settlement positions, is shown by the sites of the Scordisci located along the Danube. Military forts were now built upon them; the most important stronghold in eastern Slavonia was Teutoburgium in the territory of modern-day Dalj, probably erected not long after the conquest.[112] The settlement of the Scordisci in Sotin (Cornacum) was also overlain by a stronghold that, as the one in Dalj, protected the crossing of the Danube; a civilian settlement later developed around it.[113] A peregrine community of the Cornacati is associated to this stronghold. It is mentioned in ancient sources only after the establishment of Roman rule, and its existence was confirmed by the find of a military diploma in Srijemska Rača, issued before AD 54 to the veteran of a local Cornacatian origin.[114] Another important stronghold in eastern Croatia was located on the Upper Town in Ilok (Cuccium), with a civilian settlement also developing around it.[115] The hypotheses on the transformation into military fortifications of the Scordiscan settlements along the Danube *limes,* applies also to the remaining territory of Syrmia, as well as Transdanubia.[116]

A somewhat different picture has resulted from the excavations of lowland settlements, which remained outside of the initial Roman influence at the beginning of Romanisation. Because of that, only a small amount of either provincial or imported potteries was found in them, while the vessels produced on a potter's wheel in the Late La Tène tradition predominate.[117] It is especially important that certain Late La Tène settlements continued in the early Roman period, upkeeping their earlier infrastructural characteristics and, with those, the basics of the previous economic system. This type of settlement is very frequent in Syrmia, where it lasted as long as the 2nd century.[118] In eastern Croatia such settlements were found in the southwestern part of Vinkovci[119] and its vicinity. Excavated settlements near Slavonski Brod are especially important, such as Kupina[120] or Stari Perkovci near Đakovo.[121] These can be most probably affiliated to the Pannonian population (the Breuci), considering the dominant material heritage of autochthonous characteristics and a rather small proportion of La Tène objects, most often represented by tools and jewellery. The latest rescue excavations in the Drava valley near Slatina reveal a similar model. The Late La Tène settlement at Trnovača continued in the early Roman period, preserving the same types of buildings and therefore suggesting that this was a family-based settlement keeping its earlier agricultural character. Part of ceramic forms also shows continuity from the preceding period, while imports of thin-walled pottery and fine tableware appear in smaller numbers. All in all, the archaeological evidence suggests that the Trnovača settlement was an important centre of the Andizetes from the 2nd century BC to the 2nd century AD.

The analysis of small finds from the abovementioned sites is expected to testify to the important process of Romanisation of the autochthonous southern-Pannonian communities. A number of contemporary settlements of the Szakály type, showing continuous existence from the Late Iron Age to the early Roman period, were excavated along the course of the Kapos River, and they are attributed to the adjacent community of the Hercuniati.[122] This type of settlement is also frequent in the northern part of Transdanubia; it is supposed that these were important centres for the production of supplies used by locally stationed military units on the *limes*.[123] It is probable that the situation in eastern Slavonia and Syrmia was not much different. Here the autochthonous population, whether the Scordisci or the Pannonian groups such as the Breuci or Andizetes, continued with their way of living that was only enriched with some new forms, most often prestigious goods owned by the local elites. These were, in their turn, responsible for the establishment of the new administrative order, while at the same time retaining earlier privileges. At this stage, the local elite that managed to keep their previous status is most easily recognised in rich grave-mound burials in

108 Göricke-Lukić 2000, 100-101.
109 Dizdar 2001b, 93-96.
110 Brukner 1971; Brukner 1982; Brukner 1987.
111 Jovanović and Jovanović 1988; Brukner 1987, 34-35.
112 Klemenc 1961, 19-20; 1963, 62-63; Pinterović 1968, 67-69, 76-77.
113 Klemenc 1961, 20; Klemenc 1963, 63; Pinterović 1968, 70, 77.
114 Brunšmid 1897; Domić Kunić 1988, 90.
115 Dizdar, Šoštarić, and Jelinčić 2003.
116 Petres 1990, 7.
117 Brukner 1987, 37.
118 Dimitrijević 1961, 101-102; Dimitrijević 1969, 82; Brukner 1987, 37-38; Brukner 1988; Brukner 1995c, 17; Brukner 1995a; Brukner 1995b; Tapavički 2003.
119 Ožanić 2009.
120 Minichreiter 1991.
121 Haraša and Bekić 2007.
122 Gabler 1982a; Gabler 1982b; Gabler 1991b; Gabler 1991a; Gabler 1997.
123 Petres 1990; Gabler 2005; Ottományi 2005. Along with continuity of the material culture in the early Roman period, especially ceramic forms, a continuation of similar settlement infrastructure and economic base of the settlements is also presumed.

the neighbouring Transdanubian area,[124] identical to those in northwestern Croatia. All the excavations show that the autochthonous population, both urban and rural, preserved their ceramic forms with specific techniques of production and ornamental motifs. The S-profiled bowls, occasionally ornamented on the inside with wavy lines, are the most frequent type of ceramic form produced on a potter's wheel. Bowls of this type were discovered in Vinkovci,[125] Sremska Mitrovica,[126] Gomolava VIc[127] and in the so-called autochthonous settlements in Syrmia.[128] At the same time, S-profiled pots were also produced, often decorated with polished motifs on the vessel's shoulder. The find in grave 5 from Ilok (fig. 13) testifies to the continuity of production of this form of vessel. The *kantharoi* with high strapped handles, decorated with variously arranged polished motifs, are typical for the Scordisci, and their production continued in the 1st century AD. Rounded pots are the most frequent handmade type of vessel, occasionally decorated with a combed ornament. These are *situla*-like pots with a number of parallels in the Late La Tène settlements of the Scordisci in eastern Slavonia and Syrmia.[129] Two oval pots with abundant additions of quartz and chaff were found in grave 1 in Ilok.[130] Similar pots were found in early Roman layers in Vinkovci,[131] Sremska Mitrovica,[132] Gomolava VIc[133] and autochthonous settlements in Syrmia.[134] This pottery will persevere until the end of the 1st and the beginning of the 2nd century, with its autochthonous characteristics slowly diminishing with the increased Romanisation during the 2nd century; finally, it became replaced by provincial forms.[135]

4. Conclusion

The active role played by the Romans was always emphasised in previous studies of the process of Romanisation of southern Pannonia, while autochthonous communities were regarded as passive recipients of Roman culture. On the basis of available information and the results of the latest excavations, the important role in the organisation of the process played by the elites of autochthonous ethnic groups, especially the Taurisci and Scordisci, that crossed over to the side of the future conquerors during the Roman occupation, is being increasingly recognised, along with that of the military and the Roman settlers that followed them. Comparisons with contemporary events occurring in other frontier territories of the Empire, especially the Rhineland, where burials of local aristocracy equipped with weaponry and prestigious goods were found, thus testifying to their keeping of previously acquired status, are important for the understanding of the role played by the local elites in southern Pannonia. An identical model of Romanisation was thus applied from Gallia and Germania to the Danube basin, with some chronological variance corresponding to different dynamics of Roman conquests. Local elites were used as a base for recruiting of the allied or auxiliary units, which permitted the Romans to control the communities, but in the same time, to establish the protection of the new borders. This model is confirmed for the southeastern Alpine area through graves yielding offerings in the form of weaponry from Verdun and Mihovo, while the graves from Sremska Mitrovica and Ilok testify to the implementation of this model in the Scordiscan area. On balance, Romanisation was brought about through the ruling social class (the warrior elite) that was able to preserve some of their previously attained positions, as well as territories. On the other hand, the 'hostile' Pannonian communities experienced a different fate. The members of the highest social class were most probably sold into slavery (example of Bato the Breucan[136]) or were forced to give hostages (example of a boy of the Amantinian origin that drowned near Emona[137]), while men of military age were recruited into auxiliary units and sent to remote parts of the Empire. In the end, the Romans managed to carry out a forced conscript of men of military age, the propelling force of every society, which, according to the Greek and Roman historiographers, was the reason for the outbreak of the Pannonian-Delmataean uprising in AD 6. This is further suggested by the military diplomas and inscriptions on which the names of recruits of Pannonian origin appear already in Tiberius' reign.[138]

Finally, the Romanisation of southern Pannonia appears to have been a complex process including a variety of mutual relations between the autochthonous population with their own cultural expression and the newly arrived political-military-administrative order that ushered in a new civilisation. The continuity of material and spiritual autochthonous heritage thus had an important role in the process of Romanisation, and the autochthonous population was an important part of the ethnic corpus of the new province of Pannonia. Although the number of excavations is rather small at the moment, they nevertheless show that the autochthonous population of larger centres that emerged from the Late La Tène settlements was exposed to intensive Romanisation through the settling of Italic incomers and veterans, as well as through numerous imports; all this introduced the legacy of the Roman way of living. Rural autochthonous communities for a long time preserved characteristics of their own material culture, accepting only selected features of the newly established Roman cultural patterns. All these complex relations created a new heritage that will continue to be enriched by foreign ethnic, cultural, religious and economic components, yet the name of that specific identity, Pannonian, that offered the strongest

124 Petres 1990, 12-13; Gabler 1997.
125 Šaranović-Svetek 1981, 24.
126 Brukner 1987, pl. 25, 2-7.
127 Jovanović and Jovanović 1988.
128 Brukner 1995a, pl. 19: 192, pl. 22: 224.
129 Dizdar 2001b, fig. 12.
130 Dizdar, Šoštarić, and Jelinčić 2003, pl. 1: 1-2.
131 Šaranović-Svetek, 1981, pl. I: 13, pl. 7: 1.
132 Brukner 1971, pl. 5: 7; Brukner 1982, pl. 1: 7-10; Brukner 1987, pl. 33: 1-8.
133 Jovanović and Jovanović, 1988, 82-83.
134 Brukner 1988, fig. 8-9, pl. 1: 5-6; Brukner 1995a, pl. 21: 215, 217.
135 Brukner 1981b, 48; Šaranović-Svetek 1981, 24.
136 See Domić Kunić, in this volume, section 4.2.
137 See Domić Kunić, in this volume, section 3.2.
138 Domić Kunić 1988, 103-104.

resistance to the mightiest power of its time, will remain preserved until the present day.

Bibliography

Adam and Feugère 1982 Anne-Marie Adam and Michel Feugère, Un aspect de l'artisanat du bronze dans l'arc alpin oriental et en Dalmatie au Ier s. av. J.-C.: les fibules du type dit "de Jezerine", *AN* 53, 129-188.

Arheološka istraživanja Arheološka istraživanja duž auto puta kroz Srem (ed. Z. Vapa), Novi Sad 1995.

Balen-Letunić 2006 Dubravka Balen-Letunić, Kasnolatenski noževi tipa Pritoka-Bela Cerkev, *VAPD* 99, 63-70.

Bekić 2006 Luka Bekić, Zaštitna arheologija u okolici Varaždina, Arheološka istraživanja na autocesti Zagreb-Goričan i njezinim prilaznim cestama / Rescue Archaeology in the Varaždin Environs. Archaeological Research on the Zagreb Goričan Highway and its Access Roads, Zagreb.

Bojanovski 1984 Ivo Bojanovski, Prilozi za topografiju rimskih i predrimskih komunikacija i naselja u rimskoj provinciji Dalmaciji, *GodCBI* 22(20), 145-211.

Bolla, Boube, and Guillaumet 1991 Margherita Bolla, Christiane Boube, and Jean-Paul Guillaumet, Les situles, in: M. Feugère and C. Rolley (eds.), *La vaisselle tardo-républicaine en bronze*, Dijon, 7-22.

Božič 1981 Dragan Božič, Relativna kronologija mlajše železne dobi v jugoslovenskem Podonavju, *AV* 32, 315-336.

Božič 1984a Dragan Božič, Naoružanje ratnika mlađeg željeznog doba, in: D.

Božič and L. Bakarić (eds.), *KEATOI, Kelti i njihovi suvremenici na tlu Jugoslavije*, Ljubljana, 77-82.

Božič 1984b Dragan Božič, O starosti konjeniškega groba št. 16 z latenskega grobišča na beograjski Karaburmi, in: M. Guštin (ed.) *Keltski voz*, Brežice, 133-139

Božič 1987 Dragan Božič, Zapadna grupa, in: *PJZ* 5, 855-897.

Božič 1999 Dragan Božič, Die Erforschung der Latènezeit in Slowenien seit Jahre 1964, *AV* 50, 189-213.

Božič 2001 Dragan Božič, Ljudje ob Krki in Kolpi v latenski dobi, *AV* 52, 181- 198.

Božič 2008 Dragan Božič, *Poznolatensko-rimsko grobišče v Novem mestu. Ljubljanska cesta in Okrajno glavarstvo*, *KiM* 39.

Breščak 1989 Danilo Breščak, *Verdun pri Stopičah. Arheološke raziskave antičnega grobišča 1983-1988*, Novo mesto.

Brukner 1971 Olga Brukner, Osnovne forme i tehnike rimsko-provincijske keramike u Sirmijumu, *MatSADJ* 8, Zenica, 31-55.

Brukner 1981a Olga Brukner, Import of Roman Ceramic Ware in Slavonia, *AI* 20-21 (1980), 143-145.

Brukner 1981b Olga Brukner, *Rimska keramika u jugoslovenskom delu provincije Donje Panonije*, *DMBg* 24.

Brukner 1982 Olga Brukner, Prilog proučavanju urbanog razvoja Sirmijuma - prostorija 16 "carske palate" u Sremskoj Mitrovici, *GZPSKV* 11-12, 5-43.

Brukner 1987 Olga Brukner, Importovana i panonska keramička produkcija sa aspekta društveno-ekonomskih promena, in: *Počeci romanizacije*, 25-44.

Brukner 1988 Olga Brukner, Kontinuitet domorodačkih naselja u vreme rimske dominacije, *GZPSKV* 15, 95-102.

Brukner 1995a Olga Brukner, Domorodačka naselja, in: *Arheološka istraživanja*, 91-136.

Brukner 1995b Olga Brukner Rimska naselja i vile rustike, in: *Arheološka istraživanja* 137-174.

Brukner 1995c Olga Brukner, Rimsko doba, in: *Arheološka istraživanja*, 17-19.

Brunšmid 1895 Josip Brunšmid, Njekoliko našašća novaca na skupu u Hrvatskoj i Slavoniji II. Našašće rimskih obiteljskih denara između Valpova i Osijeka, *VHAD* 1, 108-114.

Brunšmid 1897 Josip Brunšmid, Rimski vojnički diplom iz Bijele Crkve kod Rače (Kotar Mitrovački), *VHAD* 2 (1896-1987), 1-6.

Brunšmid 1902 Josip Brunšmid, Colonia Aurelia Cibalae, *VHAD* 6, 117-166.

Brunšmid 1912 Josip Brunšmid, Nekoliko našašća novaca na skupu u Hrvatskoj i Slavoniji XXXIV. Nahođaj srebrnih ilirskih i rimskih novaca II i I stoljeća prije Kr. u Vukovaru, *VHAD* 12, 260-271.

Bulat 1977 Mirko Bulat, Nalazi s donjogradskog pristaništa u Osijeku, *OZ* 16, 11-77.

Burkowsky 1999 Zdenko Burkowsky, *Sisak u prapovijesti, antici i starohrvatskom dobu*, Sisak.

Demetz 1999 Stefan Demetz, *Fibeln der Spätlatène- und frühen römischen Kaiserzeit in den Alpenländern, Frühgeschichtliche und Provinzialrömische Archäologie, Materialien und Forschungen* 4, Rahden/Westf.

Dimitrijević 1961 Danica Dimitrijević, Nekoliko podataka o rimskom limesu u istočnom Sremu, in: *Limes*, 93-103.

Dimitrijević 1969 Danica Dimitrijević, Istraživanja rimskog limesa u istočnom Sremu s posebnim osvrtom na pitanje komunikacija, *OZ* 12, 81-120.

Dimitrijević 1979 Stojan Dimitrijević, Arheološka topgrafija i izbor arheoloških nalaza s vinkovačkog tla, in: *Corolla memoriae Iosepho Brunšmid dicata, IzdHAD* 4, Vinkovci, 201-269.

Dizdar 1999 Marko Dizdar, Željezno doba, in: *Vinkovci u svijetu arheologije*, Vinkovci, 39-48.

Dizdar 2001a Marko Dizdar, Nalazišta latenske kulture na vinkovačkom području, *PIAZ* 18, 103-134.

Dizdar 2001b Marko Dizdar, *Latenska naselja na vinkovačkom području*, *DMZg* 3.

Dizdar 2003 Marko Dizdar, Prilog poznavanju kasnoga latena u istočnoj Slavoniji, *OpArch* 27, 337-349.

Dizdar 2006a Marko Dizdar, Mlađe željezno doba, in: S. Mihelić (ed.), *Trgovina i razmjena u prapovijesti*, Zagreb, 63-70.

Dizdar 2006b Marko Dizdar, Nalazi staklenih narukvica latenske kulture u Podravini, *PIAZ* 23, 67-128.

Dizdar 2007 Marko Dizdar, Bikonični lonci sa stepeničasto raščlanjenim ramenom s groblja Zvonimirovo-Veliko polje, *PIAZ* 24, 121-144.

Dizdar 2009 Marko Dizdar, La Tène Sword from Gaćište near Virovitica, in: *Keltske študije*, 293-304.

Dizdar and Potrebica 2002 Marko Dizdar and Hrvoje Potrebica, Latenska kultura na prostoru Požeške kotline, *OpArch* 26, 111-131.

Dizdar and Potrebica 2005 Marko Dizdar and Hrvoje Potrebica, The Late La Tène culture in central Slavonia (Croatia), in: H. Dobrzańska, V. Megaw, and P. Poleska (eds.), *Celts on the Margin, Studies in European Cultural Interaction, 7th Century BC-1st Century AD, Dedicated to Zenon Woźniak*, Kraków, 57-66.

Dizdar, Šoštarić, and Jelinčić 2003 Marko Dizdar, Rajna Šoštarić, and Kristina Jelinčić, Ranorimski grob iz Iloka kao prilog poznavanju romanizacije zapadnog Srijema, *PIAZ* 20, 57-77.

Dizdar and Radman-Livaja 2004 Marko Dizdar and Ivan Radman-Livaja, Finds of Roman Bronze Ware on Celtic Sites in Eastern Slavonia, in: *Actes du XIVème Congrès UISPP, BAR Intern. Ser.* 1312, 49-53.

Domić Kunić 1988 Alka Domić Kunić, Augzilijari ilirskoga i panonskoga porijekla u natpisima i diplomama, *ARR* 11, 83-114.

Domić Kunić 2004 Alka Domić Kunić, Literarni izvori za iliričke provincije (Dalmaciju i osobito Panoniju) u *Naturalis Historia* Plinija Starijeg, *VAMZ* 37, 119-171.

Domić Kunić 2006 Alka Domić Kunić, Bellum Pannonicum (12.-11. pr. Kr.). Posljednja faza osvajanja južne Panonije, *VAMZ* 39, 59-164.

Dukat and Mirnik 1976 Zdenka Dukat and Ivan Mirnik, Pre-Roman Coinage on the territory of modern Yugoslavia, *Institute of Archaeology Bulletin* 13, London, 175-210.

Dukat and Mirnik 1978 Zdenka Dukat and Ivan Mirnik, Skupni nalazi novca u sjevernoj Hrvatskoj, *Arheološka istraživanja u sjeverozpadnoj Hrvatskoj, IzdHAD* 2, Zagreb, 197-208.

Dular 1991 Anja Dular, *Prazgodovinska grobišča v okolici Vinjega vrha nad Belo cerkvijo, KiM* 26.

Egri and Rustoiu 2008 Mariana Egri and Aurel Rustoiu, The Social Significance of Conviviality in the Scordiscian Environment, in: V. Sîrbu and S. L.Vaida (eds.), *Funerary Practices of the Bronze and Iron Ages in Central and South-Eastern Europe, Proceedings of the 9th International Colloquium of Funerary Archaeology*, Cluj-Napoca, 83-93.

Feugère 1985 Michel Feugère, *Les fibules en Gaule Méridionale de la conquête à la fin du Ve siècle après J.-C., Revue Archéologique de Narbonnaise,* Supplément 12, Paris.

Filipović 2010 Slavica Filipović, Lokalitet Osijek (Mursa) – Vojarna (Učiteljski fakultet), *HAG* 6 (2009), 37-42.

Gabler 1979 Dénes Gabler, Pannonia megszállásának néhény kérdése a terra sigillaták tükrében, *AÉ* 106/2, 199-217.

Gabler 1982a Dénes Gabler, Forschungen in der späteisenzeitlich-römerzeitlichen Siedlung von Szakály, *MAI* 10-11 (1980-1981), 71-99, 355-374.

Gabler 1982b Dénes Gabler, Aspects of the development of Late Iron Age Settlements in Transdanubia into Roman Period (Evidence based upon the excavations at Szakály in southern Hungary), in: *Studies in the Iron Age in Hungary, BAR Intern. Ser.* 144, 57-127.

Gabler 1991a Dénes Gabler, The shaping of the life of the Late La Tène settlements in the Roman period, *Antaeus* 19-20 (1990-1991), Budapest, 51-73.

Gabler 1991b Dénes Gabler, The survival of Late La Tène settlements in the Roman Period, in: V. A. Maxfield and M. J. Dobson (eds.), Exeter *Roman Frontier Studies 1989, Proceedings of the XVth International Congress of Roman Frontier Studies*, 424-431.

Gabler 1997 Dénes Gabler, Early Roman Occupation in the Pannonian Danube Bend, in: W. Groenman-van Waateringe, B. L. Van Beek, W. J. H. Willems, and S. L. Wynia (eds.), *Roman Frontier Studies 1995, Proceedings of the XVIth International Congress of Roman Frontier Studies, Oxbow Monographs* 91, Oxford, 85-92.

Gabler 2005 Dénes Gabler, Augusteische Sigillaten in Budaörs. Italischer Import in der vorrömischen Zeit im pannonischen Raum, in: *Die Grabungen in der spätlatènezeitlich-römischen Siedlung von Budaörs, AAntHung* 56/1-3, 133-175.

Gabrovec 1994 Stane Gabrovec, *Stična 1, Naselbinska izkopavanja, KiM* 28.

Gardaš 1990 Miro Gardaš, Skupni nalaz republikanskih denara pokraj sela Krndija u Đakovštini, *Obol* 42, Zagreb, 6-9.

Gaspari 2008 Andrej Gaspari, Posoški bojevniki med samostojnostjo in rimsko vojaško službo, in: K. Kofol (ed.), *Vojske, orožje in utrdbeni sistemi v Posočju*, Tolmin, 29-48

Gaspari 2009 Andrej Gaspari, Celtic warriors and the Ljubljanica, in: P. Turk, J. Istenič, T. Knific, and T. Nabergoj (eds.), *The Ljubljanica – a River and its Past*, Ljubljana, 72-78.

Göricke-Lukić 2000 Hermine Göricke-Lukić, *Sjeveroistočna nekropola rimske Murse*, Zagreb-Osijek.

Graue 1974 Jörn Graue, *Die Gräberfelder von Ornavasso, Hamburger Beiträge zur Archäologie*, Beiheft 1, Hamburg.

Gregl 2009 Zoran Gregl, *Žumberak i Latobici, Hommage jednoj planini*, Zagreb.

Guillaumet 1977 Jean-Paul Guillaumet, Les passoires de la fin de la La Tène ne Gaule et dans le monde celtique, *Gallia* 35/2, Paris, 239-248.

Guillaumet 1991 Jean-Paul Guillaumet, Les passoires, in: M. Feugère and C. Rolley (eds.), *La vaisselle tardo-républicaine en bronze*, Dijon, 89-95.

Guštin 1984 Mitja Guštin, Die Kelten in Jugoslawien, *JRGZM* 31, 305-363.

Guštin 1991 Mitja Guštin, *Posočje in jüngeren Eisenzeit, KiM* 27.

Guštin 2001 Mitja Guštin, I Celti. Argomenti per la definizione, in: *I Celti in Carnia e nell'Arco Alpino Centro Orientale, Tavola rotonda e dibattito*, Tolmezzo, 335-336, 353-354.

Guštin 2002 Mitja Guštin, I Celti dell'Adriatico. Carni tra fonti Storiche e Archeologia, in: *Gli Echi della Terra, Presenze celtiche in Friuli: dati materiali e momenti dell'immaginario*, Castello di Gorizia, 13-20.

Haraša and Bekić 2007 Ivana Haraša and Luka Bekić, Stari Perkovci-Sela, *HAG* 3(2006), 74-75.

Jovanović and Jovanović 1988 Borislav Jovanović and Marija Jovanović, *Gomolava, naselje mlađeg gvozdenog doba*, Novi Sad-Beograd.

Keltske študije G. Tiefengraber, B. Kavur and A. Gaspari (eds.), *Keltske študije II, Studies in Celtic Archaeology, Papers in honour of Mitja Guštin, Protohistoire Européenne* 11, Montagnac 2009. Klemenc 1961 Josip Klemenc, Limes u Donjoj Panoniji, in: *Limes*, 5-34.

Klemenc 1963 Josip Klemenc, Der pannonische Limes in Jugoslawien, *ARR* 3, 55-68.

Knez 1992 Tone Knez, *Novo mesto 2, Keltsko-rimsko grobišče Beletov vrt, Carniola archaelogica* 2, Novo mesto.

Kos 1977 Peter Kos, *Keltski novci Slovenije, Situla* 18.

Kos 2002 Peter Kos, A Find of Celtic Coins from the Area of Balina Glavica (*Sinotion*) in Dalmatia, *VAMZ* 35, 147-158.

Križ 2001 Borut Križ, *Kelti v Novem mestu*, Exhibition Catalogue, Novo mesto.

Križ 2005 Borut Križ, *Novo mesto VI, Kapiteljska njiva, Mlajšeželeznodobno grobišče, Carniola Archaeologica* 6, Novo mesto. *Limes Limes u Jugoslaviji* I, *Zbornik radova sa simposiuma o limesu 1960. godine*, Beograd.

Majnarić-Pandžić 1970 Nives Majnarić-Pandžić, *Keltsko-latenska kultura u Slavoniji i Srijemu*, Vinkovci.

Majnarić-Pandžić 1974 Nives Majnarić-Pandžić, Kasnolatenski keltski grobovi iz Sotina, *VAMZ* 6-7 (1972-1973), 55-74.

Majnarić-Pandžić 1984 Nives Majnarić-Pandžić, Prilog problematici kasnolatenskih naselja u Slavoniji, *OpArch* 9, 23-34.

Majnarić-Pandžić 1994 Nives Majnarić-Pandžić, Prapovijesna naselja na Gradecu, in: L.

Margetić and F. Šanjek (eds.), *Zagrebački Gradec 1242-1850*, Zagreb, 1-12.

Majnarić-Pandžić 1996 Nives Majnarić-Pandžić, *CORNACUM* (Sotin) and *CIBALAE* (Vinkovci) as Examples of the Early Romanization of La Tène Communities in Southern Pannonia, in: *Kontakte längs der Bernsteinstrasse (zwischen Caput Adriae und den Ostseegebiet) in der Zeit um Christi Geburt*, Kraków, 23-33.

Majnarić-Pandžić 2009 Nives Majnarić-Pandžić, On the South Pannonian Population in the Late Iron Age, in: *Keltske študije*, 235-245.

Mihaljević and Dizdar 2007 Marija Mihaljević and Marko Dizdar, Late La Tène bronze helmet from the river Sava near Stara Gradiška, *VAMZ* 40, 117-146.

Milošević 1985 Petar Milošević, Etnički i društveno-ekonomski aspekti kulta mrtvih na ranim nekropolama Sirmijuma, *MatSADJ* 22, 177-185.

Milošević 1987 Petar Milošević, Naoružanje i oprema rimskog ratnika u doba osvajanja i konsolidacije doline Save, in: *Počeci romanizacije*, 11-24.

Milošević 2001 Petar Milošević, *Arheologija i istorija Sirmijuma*, Novi Sad.

Minichreiter 1991 Kornelija Minichreiter, Arheološka istraživanja na dijelu auto-ceste Slavonski Brod-Lipovac, Godišnjak zaštite spomenika kulture Hrvatske 17, Zagreb, 179-192.

Mirnik 1981 Ivan Mirnik, *Coin Hoards in Yugoslavia*, *BAR Intern. Ser.* 95.

Ottományi 2005 Katalin Ottományi, Die spätlatènezeitlich-römische Siedlung von Budaörs. Die Grabungen in der spätlatènezeitlich-römischen Siedlung von Budaörs, *AAntHung* 56/1-3, 67-131.

Ožanić 2009 Ivana Ožanić, Rimski nalazi s lokaliteta Liskovac – Južna obilaznica Vinkovaca, *PIAZ* 26, 79-142.

Papazoglu 1969 Fanula Papazoglu, *Srednjobalkanska plemena u predrimsko doba*, *DjANUBiH* 30, *CBI* 1.

Pernet *et al.* 2006 Lionel Pernet, Eva Carlevaro, Luca Tori, Gianluca Vietti, Philippe Della Casa, and Biljana Schmid-Sikimić, *La necropoli do Giubiasco (Ti), vol. II, Les Tombes de La Tène Finale et d'Époque Romaine, Collectio Archaeologica* 4, Zürich.

Petres 1990 Éva F. Petres, The Problem of the Celtic Survival in Pannonia, *AR* 24, 7-15

Pinterović 1968 Danica Pinterović, Limesstudien in der Baranja und in Slawonien, *AI* 9, 55-82.

Pinterović 1969 Danica Pinterović, Problemi u istraživanju limesa na sektoru Batina Skela – Ilok, *OZ* 12, 53-69.

Počeci romanizacije Olga Brukner, Velika Dautova-Ruševljan, and Petar Milošević, *Počeci romanizacije u jugoistočnom delu provincije Panonije*, Novi Sad 1987.

Popović 1987 Petar Popović, *Novac Skordiska, Novac i novčani promet na centralnom Balkanu od IV. do I. veka pre n. e.*, Beograd-Novi Sad.

Popović 1992 Petar Popović, Italische Bronzegefässe im Skordiskergebiet, *Germania* 70/1, Mainz, 61-74.

Popović 2009 Petar Popović, Scordisci on the Fringes of the Hellenistic World, in: *Keltske študije*, 247-258.

Potrebica and Dizdar 2003 Hrvoje Potrebica and Marko Dizdar, Groblje latenske kulture u Malom Bilaču (Požega), *ObHAD* 35/2, 40-43.

Radman-Livaja and Dizdar 2010 Ivan Radman-Livaja and Marko Dizdar, Archaeological Traces of the Pannonian Revolt 6-9 AD: Evidence and Conjectures, in: R. Aβkamp and T. Esch (eds.), *Imperium – Varus und seine Zeit, Veröffentlichungen Altertumskommission für Westfalen*, 47-58.

Roymans 1996 Nico Roymans, The sword or the plough. Regional dynamics in the romanisation of Belgic Gaul and the Rheineland area, in: N. Roymans (ed.), *From the Sword to the Plough, Three Studies on the Earliest Romanisation of Northern Gaul, Amsterdam Archaeological Studies* 1, Amsterdam, 9-126.

Roymans 2004 Nico Roymans, *Ethnic Identity and Imperial Power, The Batavians in the Early Roman Empire, Amsterdam Archaeological Studies* 10, Amsterdam.

Rustoiu 2002 Aurel Rustoiu, *Războinici şi Artizani de Prestigiu în Dacia Preromană, Interferente Etnice şi Culturale în Milenile I a.Chr. – I p.Chr.*, Vol. III., Cluj-Napoca.

Rustoiu 2005 Aurel Rustoiu, Dacia şi Italia în sec. I. a. Chr. comerţul cu vase de bronz în perioada republicană târzie (Studiu preliminar), in: C. Cosma and A. Rustoiu (eds.), *Comerçt şi civilizaţie. Transilvania în contextul schimburilor comerciale şi culturale în antichitate*, Cluj-Napoca, 53-117.

Rustoiu 2008 Aurel Rustoiu, *Războinici şi societate în aria celtică Transilvăneană, Studii pe marginea mormântului cu coif de la Ciumeşti, Interferenţe etnice şi culturale în milenile I a. Chr. – I p. Chr.*, Vol . XIII, Cluj-Napoca.

Schaaff 1986 Ulrich Schaaff, Studien zur keltischen Bewaffnung, *JRGZM* 31, 622-625.

Schaaff 1988 Ulrich Schaaff, Keltische Helme, in: *Antike Helme, Monographien RGZ* 14, Mainz a. R., 293-317.

Schönfelder 2002 Martin Schönfelder, *Das Spätlatènezeit Wagengrab von Boé (dép. Lot-et-Garonne), Monographien RGZ* 54, Mainz a. R.

Sokol 1986 Vladimir Sokol, Donja Glavnica-Kuzelin, Sesvete, in: *40 godina arheoloških istraživanja u sjeverozapadnoj Hrvatskoj*, Koprivnica, 85-87.

Sokol 1995 Vladimir Sokol, Keltski novci na području Zagreba – sesvetskog Prigorja, *MV* 17, Čakovec, 19-25.

Sokol 2001 Vladimir Sokol, Neue Latènfunde aus der Umgebung von Zagreb (Croatia), *Instrumentum* 13, Montagnac, 16.

Srejović 1965 Dragoslav Srejović, Rimske nekropole ranog carstva u Jugoslaviji, *Starinar* 13-14 (1962-1963), Beograd, 49-88.

Stare 1973 Vida Stare, *Prazgodovina Šmarjete, KiM* 10.

Šaranović-Svetek 1981 Vesna Šaranović-Svetek, Ranocarska lončarska radionica u Cibalama, *Starinar* 31, 17-33.

Šašel Kos 1986 Marjeta Šašel Kos, *Zgodovinska podoba prostora med Akvilejo, Jadranem in Sirmijem pri Kasiju Dionu in Herodianu / A Historical Outline of the Region Between Aquileia, the Adriatic, and Sirmium in Casius Dio and Herodian*, Ljubljana.

Šašel Kos 1997a Marjeta Šašel Kos, Appian and Dio on the Illyrian wars of Octavian, *ŽAnt* 47, 187-198.

Šašel Kos 1997b Marjeta Šašel Kos, The End of the Norican Kingdom and the Formation of the Provinces of Noricum and Pannonia, in: *Akten des IV. internationalen Kolloquiums über Probleme des provinzialrömischen Kunstschaffens, Situla* 36, 21-42.

Šašel Kos 2005 Marjeta Šašel Kos, *Appian and Illyricum, Situla* 43.

Tapavički 2003 Milica Tapavički, Manche Überlegungen zur Romanisierung der Skordisker, *Balcanica* 34, Beograd, 1-9.

Todorović 1972 Jovan Todorović, *Praistorijska Karaburma I – nekropola mlađeg gvozdenog doba, Dissertationes et Monographiae Saveza arheoloških društava Jugoslavije* 13, Beograd.

Todorović 1974 Jovan Todorović, *Skordisci*, Novi Sad-Beograd.

Tomičić *et al.* 2007 Željko Tomičić, Marko Dizdar, Bartul Šiljeg, Hrvoje Kalafatić, Kristina Jelinčić, Kristina Turkalj, and Juraj Belaj, Rezultati zaštitnih arheoloških istraživanja Dvora knezova Iločkih 2006. godine, *AIA* 3, 7-16.

Tonc and Filipović 2010 Asja Tonc and Slavica Filipović, Novi osvrt na ACO pehar iz Osijeka, *VAMZ* 43, 503-518.

Völling 1995 Thomas Völling 1995, Studien zu Fibelformen der jüngeren voroömischen Eisenzeit und ältesten römischen Kaiserzeit, *Bericht der Römisch—Germanischen Kommission* 75 (1994), Frankfurt a. M., 149-282.

Windl 1975 Helmut J. Windl, Zur Chronologie des Gräberfeldes von Mihovo, Unterkrain (Dolenjsko), *Ann. Naturhist. Museum Wien* 80, Wien, 877-894.

A Review of South-Pannonian Indigenous Anthroponymy

Ivan Radman-Livaja

Hana Ivezić

A. Mócsy was the first to compile a detailed survey of the indigenous anthroponymy of southern Pannonia in the first two centuries AD in his seminal work, *Die Bevölkerung von Pannonien bis zu den Markomannenkriegen*.[1] A number of papers and several monographs by various authors complemented our knowledge on the subject in the intervening decades.[2] Even though these works undoubtedly laid the foundations for research into the local onomastics as well as, at least partly, into the ethnic picture of Roman Pannonia, contemporary researchers are still burdened with many unknowns and dilemmas. With this work we make no pretence at offering a new picture or a comprehensive review of these issues in southern Pannonia, but we shall nevertheless try, particularly through a review of certain recently published epigraphic finds, to ascertain whether and to what extent our understanding may have changed, at least when it comes to southern Pannonia, since the time when Mócsy and Barkóczi published their works (fig. 1).

Although ancient sources are all but mute when it comes to the ethnic composition of the population of southern Pannonia before the arrival of Rome or during the Roman conquest,[3] the picture of the people inhabiting the territory between the Drava and Sava is substantially complemented by the texts describing the situation in the 1st and 2nd centuries AD.[4] It is not insignificant that sources from the Imperial period mention tribes and ethnic groups that previously had not been referred to in any way whatsoever. This has prompted certain authors to conclude that following the conquest of southern Pannonia and the quelling of the local rebellions between 16 BC and AD 9, the Romans substantially influenced the transformation of the population structure of that area, not only by means of their drastic military measures but also through administrative action during the establishment and consolidation of their authority.[5] This hypothesis has recently been challenged by M. Colombo who argues that this theory is unfounded as he does not believe that the Romans originated the creation of new communities.[6] Indeed, it seems to us that the lack of information about some Pannonian tribes in the pre-conquest period should not be interpreted solely as a result of Roman involvement. It perhaps would not be far-fetched to hypothesise a displacement of tribes and/or the fractioning of their tribal territories into smaller communities after Rome quelled a large scale rebellion, like the one in 6-9 AD. However, we lack conclusive proof for that kind of policy in Roman-ruled Pannonia. It is however more than probable that the Romans dismantled the larger alliances led by the stronger tribes which likely existed before the conquest. Thus, smaller communities which lived in the shadow of their mightier neighbours could appear as *civitates* in the 1st century AD.

If we follow the sources, before the arrival of Rome the Taurisci lived in south-western Pannonia, the Breuci inhabited the area of the Sava district south and southeast of the Požega Valley, the Scordisci held south-eastern Pannonia, while the Andizetes resided in the eastern Podravina. Even though we can be fairly certain that these were the most powerful communities of the time, it is more than likely that several other tribes lived in the area prior to the Roman conquest. Certain communities, if not all of those mentioned by Pliny and Claudius Ptolemy, may have existed as organised groups even in the pre-Roman period, perhaps subordinate to the previously mentioned stronger tribes. It is precisely owing to Pliny (and far less to Claudius Ptolemy) that we have a reasonably clear picture of the ethnic-tribal map of southern Pannonia after the consolidation of Roman authority.[7] Although Pliny mentions the Taurisci and Scordisci among others, separated in his words by the Slavonian Mountains (*Mons Claudius*), it seems that with this sentence he merely made a reference to the situation before the conquest – probably also due to the depiction of Pannonia on Agrippa's map of the world (*orbis pictus*)[8] – when these two tribes apparently played a dominant role in the region. The territory directly or indirectly controlled by the Taurisci before the arrival of Rome, that is, through the best part of the 2nd and 1st centuries BC, covered approximately the upper and central Posavina and Podravina regions or, looking from the west eastwards, roughly the area from Nauportus to *Mons Claudius*, and from the north southwards, from the Drava River to the Kupa River. At the time, the Posavina

[1] Mócsy 1959; for the earlier literature dealing, among other things, with the issue of the indigenous anthroponymy in Pannonia, at least in part, cf. Krahe 1929; Gronovszky 1933; Mayer 1957.

[2] Barkóczi 1964; Lochner-Hüttenbach 1965; Katičič 1966; Katičić 1968; Fitz 1977; Fitz 1980; Domić Kunić 1988; Matasović 2003; Meid 2005.

[3] Polyb. *frg.* LXXVIII (122) (*FGrHist*); Strab. IV.6.10, VII.5.2-3, 10; App, 14, 22; Vell. II.96; Mócsy 1962, 527-535; Colombo 2010, 177-182.

[4] Plin. *NH* III.147-148; Ptol. II.14.2, II.15,2.

[5] Cf. Alföldy 1964, 254-255; Mócsy 1974, 53-56, 66; Fitz 1980, 141.

[6] Colombo 2010, 173-175, 184-185.

[7] Mócsy 1959, 15-16; Mócsy 1962, 604-610; Alföldy 1964, 253-255; Mócsy 1974, 53-55, 58-63, 66; Fitz 1980, 141-142; Meid 2005, 41-42; Domić Kunić 2006, 74-81; Colombo 2010, 172.

[8] Domić Kunić 2004, 130-131, 154-155, 164-166; Domić Kunić 2006, 74-75. On the syntagm *Slavonian Mountains* see Migotti in this volume, note 13.

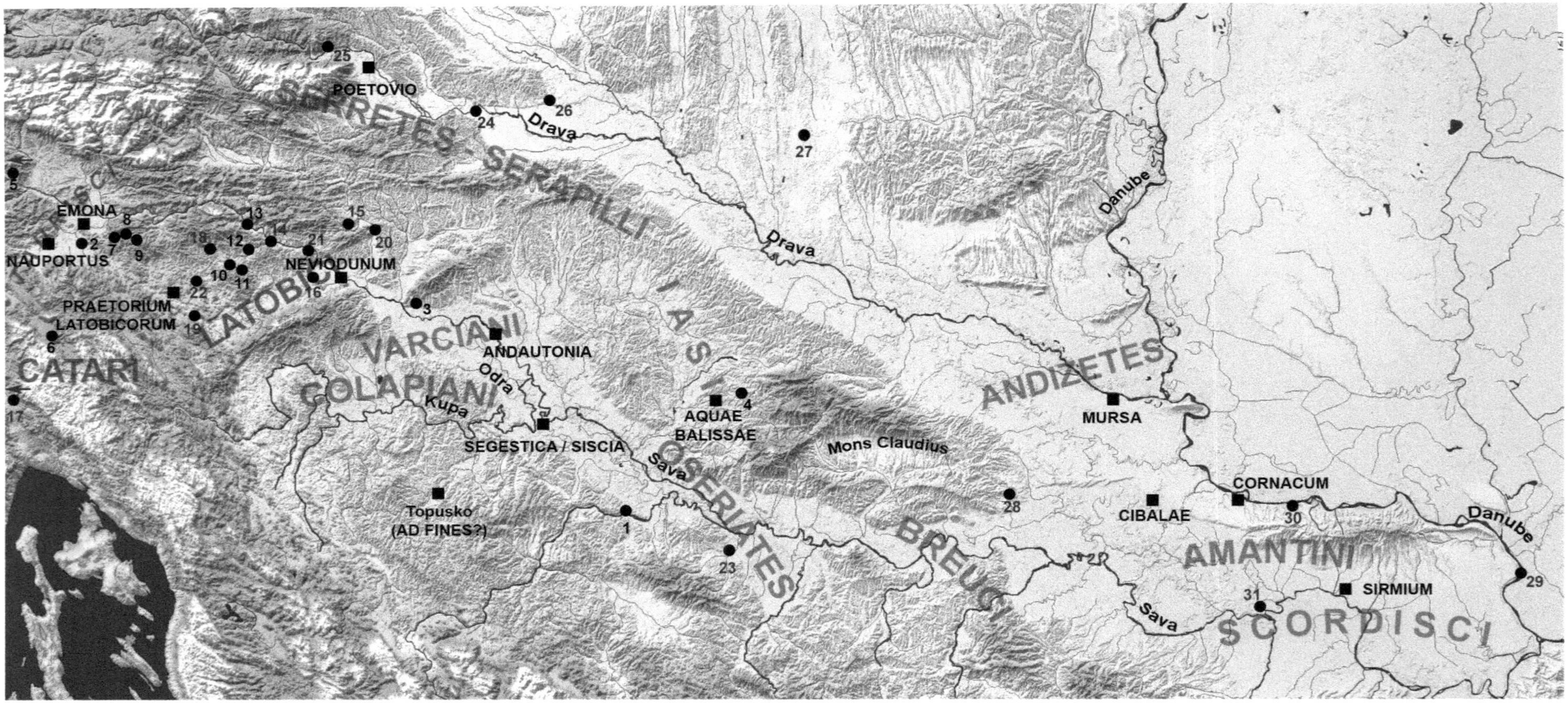

■ TOWNS AND LARGER SETTLEMENTS

● OTHER SITES

1. Dubica
2. Ig, Strahomer, Tomišlje, Sonneg Castle, Študenec, Golo, Iška Vas
3. Podsused
4. Brusnik
5. Radovljica
6. Šmarata
7. Gatina
8. Žaljna, Stična
9. Šentvid
10. Štatenberk
11. Mlada Gora
12. Teržišce
13. Radeče
14. Ševnica
15. Brezje
16. Križnica
17. Podgrad
18. Kamnje
19. Gorenji Globodol
20. Zgornja Pohanca
21. Brege
22. Straža
23. Podgraci Donji
24. Križovljan
25. Starše
26. Čakovec
27. Beleg
28. Grabarje
29. Belegiš
30. Susek
31. Bijela Crkva

Fig. 1. Map of the anthroponymic geography of southern Pannonia (after Google Maps, modified by T. Leleković).

region east of Segestica was most likely dominated by the Pannonians, whose precise tribal affiliation was not even discussed in the earlier sources, while the territory of Pannonia under Scordiscan rule probably covered a good part of present-day eastern Slavonia, as well as Syrmia. It should nevertheless be pointed out that due to the pressure of the Dacians and Pannonians the power and influence of the Taurisci and the Scordisci had already been considerably diminished several decades before the arrival of the Romans.[9] Considering that in the remainder of his text Pliny lists a number of Pannonian communities, mentioning also (with a higher or lesser degree of accuracy) where they lived, his words clearly point to the conclusion that in the 1st century AD the Taurisci and the Scordisci not only were no longer very important, but also that they were separated from *Mons Claudius* by a considerable distance as well as by a number of other communities. It is quite likely that the groups such as the Latobici, Varciani, Serapilli, Serretes, as well as perhaps the Colapiani, supposing that their identity was established at all in the pre-Roman period, in the past formed part of a wider tribal territory of the Taurisci, which the Romans later split into smaller communities.[10] The same applies to the Scordisci, whose territory in the Imperial period covered only the far southeast of Pannonia, that is, the area east of Sirmium, while the territory they had once controlled then passed into the hands of the Andizetes, Cornacates, and Amantini.[11]

Based on the mentioned ancient sources and the quoted literature, once Roman authority had been established and consolidated, we can be fairly certain, when it comes to southern Pannonia, about the geographic distribution of individual communities. Looking from the west eastwards, it was as follows:

- The Taurisci possibly continued to live only in the far southwest, in the neighbourhood of Emona (which, in all likelihood, can not be considered a Pannonian town[12]), but they would more rightly be considered as inhabitants of Noricum, where their name soon disappeared and they were never again mentioned after the 1st century AD.[13]
- The southern part of the territory of the Emona colony was inhabited by the Catari. Due to their relatively rich epigraphic record,[14] we have a fairly clear understanding of their anthroponymy, characterised by a blend of influences from the northern Adriatic, that is Venetian, and Celtic onomastics.
- The Celtic Latobici lived on the western border of Pannonia, in the area south of Poetovio and west of Andautonia, where during the latter half of the 1st century their *civitas* developed into *Municipium Flavium Latobicorum Neviodunum*.[15]
- Their eastern neighbours were the Varciani, another people of Celtic origin, or at least related to the Celts. Their territory bordered with the ager of Siscia. Their *civitas* apparently developed into *Municipium Andautonia* before the end of the 1st century.[16]
- The Colapiani lived south of the Varciani, in the valley of the Kupa River. They are considered an Illyrian tribe in the wider sense: their *civitas* continued into the second half of the 1st century (*CIL* III 14387 = *ILS* 9199), but after that they are no longer mentioned. In terms of ethnicity, they may have been related to the Segestani, that is, the original inhabitants of Segestica. It is possible that the latter were, at least partly, of Celtic origin – in the lack of concrete data, we can surmise that the inhabitants of pre-Roman Segestica were ethnically heterogeneous, which would be quite natural for a large urban centre, at the same time a commercial hub and traffic junction on the boundary of several tribal communities. Naturally, it is impossible to assess the proportion of the inhabitants of Illyrian-Pannonian and Celtic origin in the population of Segestica.[17] Nevertheless, it remains open to what degree the indigenous population formed the substrate of the inhabitants of the future Roman Siscia.[18] Presumably the Colapian *civitas* was fully incorporated into the *territorium* of the Flavian colony of Siscia,[19] although its western and southern borders have not been ascertained. The northern boundary of the Siscian territory stretched to the banks of the Sava River all the way to Andautonia, and in the east to Dubica (Servitium, *CIL* III 15200).[20]
- In the north, on the border with Noricum, lived the Serretes, while east of the *Serretes*, along the Drava River, lived the Serapilli, another tribe of Celtic origin.[21]
- The Iasi were the eastern neighbours of the Serapilli along the Drava River. They were probably an Illyrian-Pannonian tribe, but considering the presumed long-term subordination to the Taurisci, the Celtic cultural and linguistic influence was probably not insignificant. Their territory bordered in the east with Lower Pannonia, and their urban centre was in Aquae Balissae (Municipium Iasorum).[22]
- South of the Iasi and east of Siscia, the Posavina region was inhabited, probably all the way to the border with Lower Pannonia, by the Oseriates, another Illyrian-Pannonian tribe.[23]

---

[9] Cf. Mócsy 1962, 527-535; Mócsy 1974, 18-21; Domić Kunić 2006, 74-75.

[10] Guštin 2011, 120-122.

[11] Alföldy 1964, 255.

[12] Cf. Šašel Kos 2002, 373-382; Šašel Kos 2003, 11-19.

[13] Mócsy 1959, 17-18; Alföldy 1974, 25-27; Šašel Kos 1997, 25; Domić Kunić 2006, 76.

[14] Mócsy 1959, 16-17; Lochner-Hüttenbach 1965, 15-44; Katičić 1968, 61-118; Mócsy 1974, 59; Meid 2005, 27-29.

[15] Mócsy 1959, 21-24; Mócsy 1974, 135-136; Fitz 1980, 143; Lovenjak 2003, 93-95; Domić Kunić 2006, 76.

[16] Mócsy 1959, 22-24; Alföldy 1964, 254; Mócsy 1974, 136; Fitz 1980, 143; Nemeth-Ehrlich and Kušan Špalj 2003, 107-112; Šašel Kos 1997, 36; Domić Kunić 2006, 80-81.

[17] Cf. Šašel 1974, 730-731; Šašel Kos 1997, 34-35; Domić Kunić 2006, 80; Colombo 2010, 175.

[18] Radman-Livaja 2007, 160-161.

[19] Alföldy 1964, 254; Mócsy 1974, 114, 135; Fitz 1980, 143.

[20] Mócsy 1959, 24-26; Šašel 1974, 736-737; Lolić 2003, 132-133.

[21] Mócsy 1959, 28-30; Horvat *et al.* 2003, 158-161; Domić Kunić 2006, 76.

[22] Mócsy 1959, 26-28; Schejbal 2004, 99, 101-108; Domić Kunić 2006, 76.

[23] Mócsy 1959, 26; Alföldy 1964, 254-255; Domić Kunić 2006, 79-80; Colombo 2010, 173.

- The Andizetes inhabited the Podravina region south of the Iasi, that is from the border of Upper and Lower Pannonia, apparently all the way to Mursa. Based on Strabo's words (Strab. VII.5.3) it is likely that they were an Illyrian-Pannonian ethnic community, most likely exposed to Celtic influence.[24]
- During the Principate, the area of the Posavina region south of the Andizetes, east of the Oseriates and west of the Amantini, belonged to the Breuci, an Illyrian-Pannonian people whose separate identity and indubitable political importance was documented in texts from the time of the Roman conquests.[25]
- The Cornacates were a minor community, presumed to have been formed during the fragmentation of the territories of the major pre-Roman tribal communities in southern Pannonia. They were presumably ethnically related to the Breuci. Their main urban centre was, understandably, Cornacum.[26]
- The Amantini inhabited the south-eastern part of Pannonia, that is the area around Sirmium (*civitas Sirmiensium et Amantinorum*, Plin. *NH* III.148). It seems that they can also be considered a Pannonian-Illyrian population, possibly Celticised to a degree.[27]
- *Civitas Scordiscorum* occupied the far east of southern Pannonia. In spite of their original Celtic affiliation, until the Roman period the Scordisci were assimilated considerably with the indigenous Illyrian-Pannonian population. Consequently, in the scholarly literature they are sometimes described as an Illyrian-Celtic community.[28]

After this brief review of the presumed ethnic structure of southern Pannonia following the establishment of Roman authority, we return to the basic question of this paper: to what degree does the anthroponymy of the Imperial period as preserved in the epigraphic record reflect the ethnic structure of that part of Pannonia? The question is in fact more complex than it might appear at first sight. We should not equate the nomenclature, that is, the system of principles, rules and customs of naming, with a language or ethnicity. A number of languages in addition to the Latin and Greek were in use in the Roman Empire and it is needless to stress how variegated its ethnic composition was. Multilingualism, ethnic variability and multiculturalism were no less present in Pannonia. It is understandable, for instance, that a Pannonian whose parents spoke Celtic would often have a traditional Celtic name, or that a person born within an Illyrian-Pannonian community had a name typical for that ethnic group, but this was not necessarily a rule, especially once Romanisation got underway. In addition to the undeniable influence of tradition, ethnic background and the language spoken within the family, naming customs were influenced also by cultural, sociological and formal and legal aspects, which did not necessarily have any relation to the linguistic aspect.[29] Pannonian inscriptions in any case provide abundant example of families whose members had traditional names of Celtic and/or Illyrian origin, as well as Italic, that is, Latin names.[30]

Fig. 2. Funerary stone of C. Iulius Adietumarus (CIL III 10867, Photo Archive, AMZ).

One should, therefore, avoid simplified interpretations defining individuals or entire territories in ethnic terms based only on personal names documented in the

[24] Mócsy 1959, 74-75; Domić Kunić 2006, 77.
[25] Mócsy 1959, 75; Alföldy 1964, 254; Domić Kunić 2006, 78-79; Colombo (2010, 197-199) would rather consider them as a Celto-Illyrian population.
[26] Mócsy 1959, 76; Alföldy 1964, 254-255; Domić Kunić 2006, 79.
[27] Mócsy 1959, 76-77; Mócsy 1974, 114-115, 135; Domić Kunić 2006, 77.
[28] Mócsy 1959, 78; Domić Kunić 2006, 77-78.
[29] Cf. Dondin-Payre 2011, 17-23.
[30] Cf. Meid 2005, 311-318; Colombo 2010, 197.

inscriptions. There is no contention that names are an important indicator in the attempts to ascertain ethnic origin, but onomastics, just like any other historical source, should be used with caution, taking into consideration as many factors as possible, without limiting oneself only to the linguistic aspect, as important as it may be. It is sufficient to merely glimpse the annex of this paper to see that the available epigraphic material for southern Pannonia is relatively scant. Even though some areas yielded a relatively large number of inscriptions (that is, the number of epigraphic records of local anthroponymy reaching two-digit figures), the best part of southern Pannonia, particularly its central part, is a true epigraphic wasteland. In order to be relevant, statistics, as the science of collecting and organising data and of the methods for analysing the results, depends on the sample, that is, the quantity of data. In this particular case, basing the analyses on one, two or three finds is a fairly fruitless task. Nevertheless, even with a limited number of finds as in this case we can at least try to approximately compare the local anthroponymy with the presumed ethnic picture of southern Pannonia during the Principate, naturally, with due caution and without any ambition to put forward absolute claims.[31]

Although the area south and southwest of Emona lay almost certainly outside the borders of Pannonia, it has been included in this paper for at least two reasons. In addition to the fact that until recently this area was mentioned in the literature as a part of Pannonia, for instance by Mócsy, a relatively large number of inscriptions relevant for the topic from Ig and the surrounding places are a useful indicator of the ethnic picture of that border area with Pannonia. In terms of onomastics, this is indisputably an area of fusion of Celtic and northern Adriatic influences. Katičić's analysis leads to the conclusion about the indigenous character of northern Adriatic anthroponymy, but pointing also to a long-term Celtic influence. The south-western border of Pannonia might therefore represent also a linguistic, that is, ethnic border. All the relevant inscriptions from the territory of the Latobici contain generally indisputable (or, in a few cases, very probable) Celtic names. Twenty or so names do not make a particularly large sample, but it is nevertheless a solid indicator of the ethnic origin of the local population, if we accept the justified assumption that these are exclusively people of indigenous origin. The sample for the Varciani is significantly smaller (fig. 2). It is, however, necessary to stress that several inscriptions explicitly mention ethnic affiliation. In addition to the inscriptions from Podsused (*CIL* III 10867) and Ščitarjevo (*CIL* III 11463), which certainly point to Celtic anthroponymy (although we cannot be entirely sure that these were indigenous people, especially in the case of *Lusius Rucco*), there are also preserved military inscriptions confirming the Celtic character of the Varciani onomastics: *T. Flavius Bonio*, a native of Andautonia, has a cognomen of Celtic origin, *Vercaius Medi f(ilius)* and *Iantumarus Andedunis f(ilius)* are explicitly mentioned as Varciani, and it can

Fig. 3. Funerary stone of T. Flavius Ateboduus (after Rendić-Miočević in press, Photo Archive, AMZ).

also be surmised that *veteranus cohortis II Varcianorum equitatae*, *T. Flavius Ateboduus Aterigis fil(ius)* returned to his homeland once his military service was over (fig. 3). Still, caution is necessary in cases when a soldier's origin is not explicitly mentioned, because being a member of a unit does not necessarily imply ethnic affiliation, not even during the 1st century, let alone later. *Petitor Pirobori, miles cohortis II Varcianorum*, is a case in point.[32] His service in a Varciani cohort naturally does not necessarily mean that he himself was a Varcian. Since his name and patronymic appear to be of Daco-Thracian origin,[33] we can suppose that he was not a Pannonian or at least that he was not an

[31] Comments on the onomastics further in the text will refer to the inscriptions listed in the appendix with cited references.

[32] *CIL* XIII 8188; Dobó 1975, 462.

[33] Dana 2011, 44, 53.

inhabitant of south-western Pannonia. There are at least 4 preserved inscriptions of soldiers who define themselves as Colapiani, all of them, same as their fathers, with more or less typical Illyrian-Pannonian names (including some quite frequent ones such as *Bato* or *Liccaius*). Although there is no mention of the ethnic affiliation of *Scenus Scenobarvi*, the commander of a *turma* of the Colapian *Bato* in *ala Pannoniorum* (*CIL* III 4372), he was certainly a Pannonian, though not necessarily of the Colapiani tribe.

Although the epigraphic record of Siscia is not insignificant, at least when Pannonia is concerned, there are only few inscriptions containing names of possibly local character. Interestingly, they are entirely missing from the earlier inscriptions – with the possible exception of one graffito (*CIL* III 12041[529]) - and only starting with the end of the 2nd century and in the 3rd century do several examples appear that could possibly be attributed to the indigenous Pannonian anthroponymy. The *hapax Sipa* might be of Illyrian origin, but it is far from being certain (as a matter of fact, the local origin of the person whose name is incised on the vessel is also uncertain). The name *Arsaces* might be of Celtic origin, while the cognomen Ammianus may but need not be related to Celtic anthroponymy. Two praetorians, natives of Siscia, *M. Aurelius Dasius* and *M. Aurelius Tato*, have cognomina of Illyrian-Pannonian or Celtic origin. The sample is too small to fuel a serious discussion on the ethnic structure of the population of Siscia, all the more so as this is an insignificant number if compared with the total number of preserved inscriptions and personal names. Besides, Siscia is a large urban centre with a varying but stable influx of immigrants through the centuries, so we cannot claim, considering the dating of the mentioned monuments, that those were the descendents of the indigenous inhabitants of Pannonia, and that local tradition played any role in the choice of their personal names.

South and southeast of Siscia, Topusko and Podgraci yielded two examples of Celtic female names. Not only is this far too small a number to allow any interpretation, but based on the inscription from Podgraci we certainly cannot attribute Celtic anthroponymy to the Oseriates, even though this is the only example from their presumed territory. Poetovio with the surrounding area, that is the territory of the Serretes and the Serrapilli, likewise has yielded a limited number of inscriptions with recognisable indigenous anthroponymy, but these were exclusively Celtic names, which in principle corroborates the assumptions about the ethnic origin of the local population. The number of inscriptions relevant for this topic from the territory of the Iasi is even lower and their usefulness for interpretation is also questionable. Two relatively late inscriptions mention the cognomina *Maurus* and *Cocceius*, which might have a connection with Celtic anthroponymy, although we cannot claim with certainty that these were descendants of the indigenous population (this is perhaps somewhat more probable in the case of *Ulpius Cocceius*, while for *Marcus Flavius Maurus* one cannot exclude the possibility that his origin lies in the western part of the Empire, for instance in the Iberian peninsula or even in Africa, considering the high frequency of that name in these areas) and link their (possible) Celtic cognomens with the local onomastic tradition. In Brusnik near Daruvar, the patronymic of the mother of the family is indisputably Celtic, but since she was a spouse of a soldier, there is no indisputable proof that she was an indigenous member of the Iasi tribe. The most reliable evidence is the diploma issued to the soldier *Frontus*, who defined himself as belonging to the Iasi tribe, and whose father also had an Illyrian name.

When the tribes of the southern part of Lower Pannonia are concerned, the Breuci are by far the best represented in the epigraphic record, although merely a few epigraphic monuments were actually discovered in their presumed territory. Members of that tribe, however, were mentioned as soldiers in inscriptions and diplomas throughout the Empire. The available data leave no doubt whatsoever as to the Illyrian-Pannonian character of their anthroponymy, although few names might also belong to a Celticised, if not entirely Celtic onomastic tradition.[34] In any case, in the southern Pannonian context, Celtic influence on the onomastics of indigenous populations should not come as a surprise. The case of their neighbours in south-eastern Pannonia is nevertheless a bit more complex. Only one inscription can be securely attributed to the Andizetes, the now unavailable military diploma *RMD* 205, whose text mentions *Breucus Isticani f(ilius) princ(eps) Antizit(ium)*. The name Breucus is not only an ethnonym but also a fairly frequently documented anthroponym. It also seems that it was used as a personal name by other Pannonians,[35] leaving no room for wonder that a prominent member of the Andizetes should be named so. The inscriptions from Mursa do not contribute much to the knowledge about the anthroponymy of the Andizetes. The two known early inscriptions with Celtic names in all likelihood do not refer to indigenous persons, but to soldiers from the western part of the Empire.[36] The patron of the freedwoman *Urbana* and her son has an indisputably Celtic name (it is not entirely excluded that the remaining names from this inscription might be related with the Celtic onomastics), but nothing shows that Aterix was indigenous to that part of Pannonia. The decurion *C. Aelius Surinus* has a cognomen that could be associated with Celtic as well as Illyrian-Pannonian anthroponymy, although its ethnic origin cannot be defined with precision. Several names that appear on later inscriptions can perhaps be associated with Celtic (*Augenis, Messius*) or Illyrian (also *Messius*) anthroponymy, although there is no proof for reliable identification of these people as natives. When it comes to the soldiers that mention Mursa as their homeland, *Aurelius Dassius* has a cognomen of

[34] Colombo 2010, 199-200.

[35] Cf. Radman-Livaja 2010, 194-195.

[36] *Niger Bataronis f(ilius) Sve<l>trius, eques alae Aravacorum* (*CIL* III 3286 (10262); *ILJug* 294; Mócsy 1959, 217/3), Delamarre 2003, 68-69; Delamarre 2007, 38, s.v. Bataro, 174, s.v. *Suetrius*, 212; *Uelagenus Manti f(ilius) Ulattius, eques cohortis II Alpinorum* (*VHAD* 12, 1912, 2, no. 1; *AÉ* 1913, 135; Mócsy 1959, 217/4; *ILJug* 3104), the soldier was not of Pannonian, but of western Alpine origin, a member of the Ulatti tribe, cf. Meid 2005, 126; Delamarre 2007, 126, s.v. *Mantus, -ius, -ia,* 192-193, s.v. *Uelagenus*.

Fig. 4. Military diploma of Dases Dasmeni f. (CIL XVI 2, Photo Archive, AMZ).

Illyrian-Pannonian origin, while *P. Aelius Scantius* has a Celtic cognomen. The inscription *CIL* VI 32542 (2833) from AD 223 mentions a number of praetorians from Lower Pannonia, some of whom state Mursa and Cibalae as their homeland. Most have Italic cognomina, although those of Illyrian and Thraco-Dacian origin also appear. In view of the formulation *ex provincia Pannonia inferiore cives Cotini*, it is disputable whether any of them were in fact descended from the indigenous pre-Roman inhabitants of Pannonia.[37] The names of the Cibalitans mentioned on the curse tablet from Siscia, *Lucius Larcius* and *Secundus Vacarus*, can perhaps be associated with the Celtic anthroponymy, although it would be difficult to claim that they were indigenous to Pannonia. A Cornacatus whose name is mentioned on the diploma *CIL* XVI 2 has a typical Illyrian-Pannonian name, same as his father (fig. 4). It also seems probable that the female names on that diploma are of Illyrian-Pannonian origin (although *Turuna* might equally be a Celtic name). The inscriptions from present-day eastern Syrmia, that is, from the territory of the Ammantines and the Scordisci, mention Illyrian-Pannonian names, but also those closer to Celtic anthroponymy, in addition to a few names that could probably be associated with the Thraco-Dacian onomastic tradition.[38]

This survey allows us to conclude that it is possible to identify the dominant anthroponymic tradition for certain south-Pannonian communities (i.e. the Latobici, Varciani, Colapiani, Breuci), while in the case of others this remains an open issue. Taking into consideration that not in one case do we have a sample of any substantial size, we have to be satisfied with the available data. In optimal cases, the ethnic origin is clearly mentioned besides the names. When there are many such cases, and there are no discrepancies, this allows us to define the Varciani based on the onomastics

[37] Cf. Mócsy 1974, 57, 190, 192, 199.

[38] One votive monument from Sirmia, dated to AD 223 (Mirković 1971, cat. 11), mentions a signifer of the Second legion (*legio II Adiutricis*) *T. Gramon(ius) Lucilianus*. M. Mirković believes this to be a gentilicium from the Illyrian name Grabon/Gramon (Mirković 1971, 25). Even though this assumption is not implausible, the inscription was not considered in this discussion because nothing points with certainty to the possibility that this person was originally from Sirmium, or from that area, although this is not impossible.

as a Celtic ethnic group, and the Colapiani and the Breuci as Illyrian-Pannonian ethnic groups. When the origin of the people mentioned in the inscriptions is not explicitly stated, the linguistic criterion becomes exceptionally important in the attempts at defining the ethnic origin of the individuals, although it can never be considered decisive on its own. In the case of the Latobici or the area of Poetovio, it should be kept in mind that all the documented names of presumably indigenous character were of Celtic origin. In cases when they are not extremely rare, we can also be fairly certain that the local anthroponymy was based on a Celtic substratum. In the situations when the onomastic sample is markedly small, and in addition to this, linguistically varied, it becomes impossible to ascertain the ethnic character of the population under study. It is very difficult to ethnically define the Iasi based on onomastics, and even impossible in the case of the Oseriates, although we can in principle assume that these were both Illyrian-Pannonian communities. Although the hypothesis that the Andizetes and Ammantini are also Pannonian Illyrians is not contentious, the available epigraphic record allows us at best to assert that this is a population of a complex onomastic tradition, with elements of Celtic and Illyrian-Pannonian anthroponymy, though even this would be quite daring in view of the lack of concrete relevant epigraphic data. The same applies to the Scordisci, whom the sources in any case mention as a Celtic-Illyrian population. Still, all considered, in spite of the limited number of written sources, we can in principle accept the claim that the epigraphy in the main corroborates the existing assumptions about the ethnic and linguistic character of the communities inhabiting southern Pannonia.

At the end of this review, we feel it is necessary to highlight a few other details regarding the onomastic study of the Pannonian area, but also to draw attention to the fact that new discoveries can always significantly change the accepted picture, which we shall substantiate with a concrete example. When it comes to the indigenous anthroponymic tradition, the dating of inscriptions plays a major role. By far the most epigraphic monuments with names of local character are dated to the 1st century and the beginning of the 2nd century, which is quite understandable. Romanisation led to the gradual disappearance of local traditions, and in time local names became increasingly less present. This naturally does not mean that Celtic or Illyrian names would not occasionally appear on the inscriptions from the end of the 2nd or from the 3rd century, but then we cannot speak merely of the preservation of traditions, but also have to consider other factors. In the case of Pannonia, and Lower Pannonia in particular, we have to take into consideration the major population movements in the wake of the Marcomannic wars, which means that it is difficult to consider the inscriptions from the subsequent period as relevant evidence for assertions about the indigenous onomastic traditions of the given area. With the passage of time it becomes difficult to maintain indigenous anthroponymic traditions in large urban centres. In towns, especially in larger ones, the composition of the population was very much subject to change. Urban development went hand in hand with the influx of new inhabitants, both from the vicinity and from the more remote areas. As early as the 1st century the towns of southern Pannonia had a significant share of inhabitants from other parts of the Empire, from Italy to Gaul and Hispania, and even from further abroad. Long-distance immigration was particularly prominent in or nearby towns with military garrisons. Therefore, when major urban centres are considered, the appearance of Celtic or Illyrian names in the inscriptions but without mention of the origin of their bearers cannot necessarily be taken as evidence of the local anthroponymy. A person with an Illyrian name could, of course, be a Pannonian, but he could just the same be an immigrant from the neighbouring province of Dalmatia. This problem is even more prominent with Celtic names, because a person with a Celtic name in Poetovio, Siscia, Mursa, or Sirmium, could be an immigrant from northern Italy or from Noricum, from Gaul or even from Hispania, and not necessarily a local person. We have deliberately omitted an important element from this review, namely, the so-called assumed names and translated names (*Decknamen, noms d'assonance,* and *Übersetzungsnamen, noms de traduction*),[39] which, either as literal Latin translations of local names or as Latin names sounding similar to the traditional names, were often used among the Romanised indigenous population in Celtic and Illyrian areas. As useful as their detailed analysis might be, with this we would exceed the limits of this brief review. We shall therefore restrict ourselves to the observation that the frequency of names such as e.g. Avitus, Crescens, Exoratus, Ianuarius, Ingenuus, Secundus, Surus, Tertius or Verecundus are probably not only accidental, but represent evidence of Romanisation and adaptation of the local onomastics.

The issue of the number of monuments, that is, the available sample, has already been mentioned, although it would perhaps be useful to once again point out that we have at our disposal only a mere fraction of the once existing Roman inscriptions. The luck factor in their preservation certainly influenced our perception of the history of Roman Pannonia, including also the issue of the indigenous anthroponymy. We base some of our assumptions on only a handful of preserved inscriptions, sometimes even on merely one, but it is questionable whether these assumptions will still hold in the case of new finds. Not only are we faced with the fact that a large part of Roman inscribed monuments had disappeared before our time, but we also have to consider the probable fact that these monuments, even when observed comprehensively, could never have painted a realistic picture of the population of Roman Pannonia, because it was beyond reach of many people to commission inscribed stone monuments. Although it did not take a very wealthy person to commission such a monument, this was nevertheless available primarily to an affluent class of society. It is therefore small wonder that certain social categories are proportionally overrepresented in the inscriptions, and average visitors to the museum

[39] Cf. Dondin-Payre 2001, V-VII; Dondin-Payre 2011, 18-21.

collections of stone monuments could get the impression that most inhabitants of Roman Pannonia were either in the military or formed part of the urban elites. In any case, soldiers and urban elites, even when they are of local origin, were by and large partisans of Romanisation, including the onomastic aspect, so their monuments are not always of great use for the study of local anthroponymic traditions.

Nevertheless, there still appear finds that can shed an entirely new light on the questions that trouble us. A recent example is the corpus of lead tags from Siscia, recently analysed and awaiting publication in the near future.[40] These tags were used in the textile industry of Siscia, and their importance for our subject lies in the fact that the names of almost 900 women and men, the customers of the fullers and dyers of Siscia, have been preserved on them. Considering that this was not a luxury craft, it can be surmised that this corpus of inscriptions provides a certain cross-section of the names of Siscia's average citizens, of various financial standing, with a great probability that these include also those who could not afford carved stone monuments and who left no trace in the preserved epigraphic assemblage of Siscia. In addition to the size of the assemblage, representing a substantial statistical sample, it is very important that these inscriptions are chronologically fairly clearly defined: they are approximately dated from the Augustan period until the second half of the 2nd century, although most of them can be dated to a narrower chronological frame from the rule of Claudius to that of Hadrian.[41] It has already been mentioned that there are hardly any inscriptions among the monuments from Siscia that could be associated in any way with the local anthroponymy, whether of Celtic or Illyrian character. Based on the preserved inscriptions, the general impression used to be that during the 1st century Siscia received a considerable number of immigrants from northern Italy, but also from Dalmatia, Gaul and Hispania. Although nobody questioned the existence of inhabitants of local origin, they were all but invisible in the preserved inscriptions. The lead tags painted a considerably different picture: a total of 883 (possibly 884) individuals were documented, almost 40% of whom were women. These people use 692 different names (*praenomina, nomina, cognomina* and idionyms). These include 371 Latin or probably Latin names, 144 Celtic or probably Celtic names, 50 Illyrian-Pannonian names, 46 Greek names, 11 names of various origin (Punic, Semitic, African, Thracian etc.), as well as 70 names of indeterminate origin. A striking feature is the high proportion of Celtic and Illyrian names when compared with the known inscriptions. Although the Latin anthroponymy predominates (more than 53%), Celtic and Illyrian names make up almost 28% of the documented names and perhaps even more, considering that it is quite possible that some of the indeterminate names also formed part of the Celtic or Illyrian-Pannonian onomastic tradition. However, the statistical data presented in this way is deceptive to a degree, and it is necessary to explain how these names are distributed in the nomenclatures of those people. Out of 883 (or perhaps 884) individuals, 665 (or 666) have a Latin or partly Latin nomenclature. The latter group consists of citizens with a non-Latin cognomen or peregrines whose nomenclature contains a non-Latin name in addition to a Latin one. This would mean that the nomenclature of 76% of those people is fully Latin or at least partly Romanised. Although non-Latin names appear frequently on the Siscia tags, they are not the most popular ones: Celtic names are numerous and varied, but most appear only once, rarely on more than two occasions. The same is true of the Illyrian and Greek names, and if we look at the most common ones, 18 out of 20 most common names are definitely Latin.[42] Without going into too much detail, it suffices to underscore that this result substantially transforms the accustomed view of the population of Siscia as entirely Romanised or predominantly Italic. Although the comments about the mixed character of the population of major Pannonian towns are valid and there is in principle no doubt that the Latin anthroponymy was dominant, it is clear that the Celtic and Illyrian-Pannonian onomastics were far more strongly present in Siscia during the 1st century than is possible to assume based exclusively on the preserved inscriptions. The issue of the origin of the bearers of Celtic and Illyrian-Pannonian names remains open. Although it is possible that they are of local origin, it is not excluded that at least a part of these people may have arrived from other, not necessarily very remote areas. Given that the territory of Siscia is associated with the Colapiani, it is somewhat surprising that the Celtic element is more strongly present in the anthroponymy of the citizens of Siscia, but this only goes to show that each new find at the same time raises new questions. The example of the tags from Siscia proves to what degree our understanding is limited and subject to change, although we should not think that every new find could radically transform the accustomed view. The basic facts about the ethnic picture of southern Pannonia are probably valid, but numerous unknowns and dilemmas will not be resolved without new finds. We are left with the hope that future investigations will one day make it possible to discover new inscriptions, perhaps such as those on the lead tags from Siscia, that would shed new light on the population of major towns such as Mursa or Sirmium, as well as on the inhabitants of smaller south-Pannonian settlements.

[40] Cf. Radman-Livaja 2010.

[41] Radman-Livaja 2010, 549-555.

[42] Radman-Livaja 2010, 532-548.

Bibliography

Alföldy 1964 Géza Alföldy, Eine römische Strassenbauinschrift aus Salona, *AArchHung* 16, 247-256.

Alföldy 1969 Géza Alföldy, *Die Personnamen in der römischen Provinz Dalmatia*, Heidelberg.

Alföldy 1974 Géza Alföldy, *Noricum*, London.

Barkóczi 1964 László Barkóczi, The Population of Pannonia from Marcus Aurelius to Diocletian, *AArchHung* 16, 257-356.

Colombo 2010 Maurizio Colombo, *Pannonica*, *AAntHung* 50/2-3, 171-202.

Dana 2011 Dan Dana, L'impact de l'onomastique latine sur les onomastiques indigènes dans l'espace thrace, in: M. Dondin-Payre (ed.), Les noms de personne dans l'Empire Romain, Bordeaux, 37-87.

Degavre 1998 Jean Degavre, *Lexique gaulois, Recueil de mots attestés, transmis ou restitués et de leurs interprétations, Mémoires de la Société belge d'Études celtiques* 10, Bruxelles.

Delamarre 2003 Xavier Delamarre, *Dictionnaire de la langue gauloise*, 2ème éd., Paris.

Delamarre 2007 Xavier Delamarre, *Noms de personnes celtiques dans l'épigraphie classique*, Paris.

Dobó 1975 Arpad Dobó, *Inscriptiones extra fines Pannoniae Daciaeque repertae ad res earundem provinciarum pertinentes, editio quarta aucta et emendata*, Budapest.

Domić Kunić 1988 Alka Domić Kunić, Augzilijari ilirskog i panonskog porijekla u natpisima i diplomama, *ARR* 11, 83-114.

Domić Kunić 2004 Alka Domić Kunić, Literarni izvori za iliričke provincije (Dalmaciju i osobito Panoniju) u Naturalis historia Plinija Starijeg, *VAMZ* 37, 119-171.

Domić Kunić 2006 Alka Domić Kunić, Bellum Pannonicum (12.-11. pr. Kr.), posljednja faza osvajanja južne Panonije, *VAMZ* 39, 59-164.

Dondin-Payre 2001 Monique Dondin-Payre, L'onomastique dans l'Empire romain: questions, méthodes, enjeux, in: M. Dondin-Payre and M.-T. Raepsaet-Charlier (eds.), *Noms, identités culturelles et romanisation sous le Haut-Empire*, Bruxelles, I-XXIV.

Dondin-Payre 2011 Monique Dondin-Payre, *Introduction*, in: M. Dondin-Payre (ed.), *Les noms de personne dans l'Empire Romain*, Bordeaux, 13-36.

Evans 1967 D. Ellis Evans, *Gaulish Personal Names, A Study of some Continental Celtic Formations*, Oxford.

Fitz 1977 Jenő Fitz, Onomastique pannonienne: La population de la Pannonie sous l'aspect de l'onomastique et de l'archéologie, in: N. Duval (éd.), *L'onomastique latine*, Actes du Colloque International sur l'onomastique latine organisé à Paris du 13 au 15 octobre 1975, Paris, 395-402.

Fitz 1980 Jenő Fitz, *Population*, in: *ARP*, 141-158.

Forier 2001 Sarah Forier, Les anthroponymes formés sur des noms d'animaux en Gaule Narbonnaise et dans les provinces alpines, in: M. Dondin-Payre and M. Th. Raepsaet-Charlier (eds.), *Noms, identités culturelles et romanisation sous le Haut-Empire*, Bruxelles, 473-535.

Gronovszky 1933 Iván Gronovszky, *Nomina hominum Pannonica certis gentibus assignata, DissPann* I (2).

Guštin 2011 Mitja Guštin, On the Celtic tribe of Taurisci, in: M. Guštin and M. Jevtić (eds.), *The Eastern Celts, The communities between the Alps and the Black Sea, Annales Mediterranei*, Koper-Beograd, 119-128.

Holder 1896-1913 Alfred Holder, *Alt-Celtischer Sprachschatz*, 3 vols, Leipzig.

Horvat *et al.* 2003 Jana Horvat, Milan Lovenjak, Andreja Dolenc Vičič, Marija Lubšina-Tušek, Marijana Tomanič-Jevremov, and Zorka Šubic, Poetovio. Development and Topography, in: *Pannonia* I, 153-189.

Kajanto 1965 Iiro Kajanto, *The Latin Cognomina*, Roma.

Katičić 1962 Radoslav Katičić, Die illyrischen Personennamen in ihrem südöstlichem Verbreitungsgebiet, *ŽAnt* 12/1, 95-120.

Katičić 1963 Radoslav Katičić, Das mitteldalmatinische Namengebiet, *ŽAnt* 12/2, 255-292.

Katičić 1965 Radoslav Katičić, Zur Frage der keltischen und pannonischen Namengebiete im römischen Dalmatien, *GodCBI* 1, 53-76.

Katičić 1966 Radoslav Katičić, Keltska osobna imena u antičkoj Sloveniji, *AV* 17, 145-168.

Katičić 1968 Radoslav Katičić, Die einheimische Namengebung von Ig, *GodCBI* 6, 61-120.

Krahe 1929 Hans Krahe, *Lexikon altillyrischen Personennamen*, Heidelberg.

Krahe 1955 Hans Krahe, *Die Sprache der Illyrier*, Erster Teil: *Die Quellen*, Wiesbaden.

Križman 1991 Mate Križman, *Rimska imena u Istri, Osobna imena na istarskim natpisima iz rimskog doba*, Zagreb.

Lochner-Hüttenbach 1965 Fritz Lochner-Hüttenbach, Die Antiken Personennamen aus Ig bei Ljubljana, *Situla* 8, 15-45.

Lolić 2003 Tatjana Lolić, Colonia Flavia Siscia, in: *Pannonia* I, 131-152.

Lőrincz 1999 Barnabás Lőrincz, *Onomasticon Provinciarum Europae Latinarum*, Vol. II: CABALICIVS-IXVS, Budapest.

Lőrincz 2000 Barnabás Lőrincz, *Onomasticon Provinciarum Europae Latinarum*, Vol. III: LABAREVS-PYTHEA, Wien.

Lőrincz 2002 Barnabás Lőrincz, *Onomasticon Provinciarum Europae Latinarum*, Vol. IV: QVADRATIA–ZVRES, Wien.

Lőrincz and Redő 1994 Barnabás Lőrincz and Ferenc Redő, *Onomasticon Provinciarum Europae Latinarum*, Vol. I: ABA – BYSANVS, Budapest.

Lovenjak 2003 Milan Lovenjak, Municipium Flavium Latobicorum Neviodunum, in: *Pannonia* I, 93-105.

Nemeth-Ehrlich and Kušan Špalj 2003 Dorica Nemeth-Ehrlich and Dora Kušan Špalj, Municipium Andautonia, in: *Pannonia* I, 107-129.

Matasović 2003 Ranko Matasović, Jezični tragovi Kelta u Iliriku, *Latina et Graeca* 3, Zagreb, 5-23.

Mayer 1957 Anton Mayer, *Die Sprache der alten Illyrier*, Band I: *Einleitung, Wörterbuch der illyrischen Sprachreste, SchrBalk, Linguistische Abteilung* 15, Wien.

Meid 2005 Wolfgang Meid, *Keltische Personennamen in Pannonien*, Budapest.

Minkova 2000 Milena Minkova, *The Personal Names of the Latin Inscriptions in Bulgaria, Studien zur klassischen Philologie* 118, Frankfurt am Main.

Mirković 1971 Miroslava Mirković, *Sirmium, its history from the 1st century A. D. to 582 A. D., Sirmium* 1, Beograd, 5-94 (The Inscriptions from Sirmium and its territory, 60-90).

Mócsy 1959 András Mócsy, Die Bevölkerung von Pannonien bis zu den Markomannenkriegen, Budapest.

Mócsy 1962 András Mócsy, *s.v. Pannonia*, RE, Suppl. 9, 516-776.

Mócsy 1974 András Mócsy, *Pannonia and Upper Moesia*, London.

Mócsy 1983 András Mócsy, Nomenclator provinciarum Europae Latinarum et Gallia Cisalpinae, *DissPann* III/1.

Pape and Benseler 1870 Wilhelm Pape und Gustav Eduard Benseler, *Wörterbuch der Griechischen Eigennamen*, Braunschweig.

Radman-Livaja 2007 Ivan Radman-Livaja, *In Segestica...*, *PIAZ* 24, 153-172.

Radman-Livaja 2010 Ivan Radman-Livaja, *Les plombs inscrits de Siscia*, Unpublished PhD thesis, École Pratique des Hautes Études, Paris.

Raybould and Sims-Williams 2007 Marilynne E. Raybould and Patrick Sims-Williams, *A Corpus of Latin Inscriptions of the Roman Empire containing Celtic personal Names*, Aberystwyth.

Rendić-Miočević in press Ante Rendić-Miočević, The marble funerary stele of cohors II Varcianorum equitata's veteran, Titus Flavius Ateboduus, from the Odra village near Zagreb, in: N. Cambi and G. Koch (eds.), *Funerary Sculpture of the Illyricum and neighbouring Regions of the Roman Empire*, Split.

Rendić-Miočević 1971 Duje Rendić-Miočević, Ilirske onomastičke studije (III), Onomasticon Riditinum, *ŽAnt* 21 (Zbornik radova posvećen Mihajlu Petruševskom), 159-174 = Rendić-Miočević 1989, 785-800.

Rendić-Miočević 1981 Duje Rendić-Miočević, Dacico-illyrica, Neki aspekti ilirske kolonizacije Dacije u svjetlu natpisa na ceratama iz Alburnusa, *RJAZU* 20, 21-37 = Rendić-Miočević 1989, 751-767.

Rendić-Miočević 1989 Duje Rendić-Miočević, *Iliri i antički svijet*, Split.

Russu 1973 Ion I. Russu, *Dacia si Pannoia Inferior în lumina diplomei militare din anul 123*, Bucureşti.

Schejbal 2004 Berislav Schejbal, Municipium Iasorum (Aquae Balissae), in: *Pannonia* II, 99- 129.

Schmidt 1957 Karl-Horst Schmidt, Die Komposition in gallischen Personennamen, *Zeitschrift für celtische Philologie* 26, 33-301.

Simón and Rodá de Lanza 2008 Francisco Marco Simón and Isabel Rodá de Llanza, A Latin defixio (Sisak, Croatia) to the river god Savus mentioning L. Licinius Sura, Hispanus, *VAMZ* 41, 167-198. Solin 1996 Heikki Solin, *Die stadtrömischen Sklavennamen, Ein Namenbuch*, Stuttgart.

Solin 2003 Heikki Solin, *Die griechischen Personennamen in Rom, Ein Namenbuch, zweite, völlig neu bearbeitete Auflage*, Band 1-3, Berlin-New York.

Solin and Salomies 1994 Heikki Solin und Olli Salomies, *Repertorium nominum gentilium et cognominum Latinorum, Editio nova addendis corrigendisque augmentata*, Hildesheim-Zürich-New York.

Šašel 1955 Jaroslav Šašel, Nov antični nadgrobnik z Iga pri Ljubljani, *ŽAnt* 5/2, 373-382.

Šašel 1974 Jaroslav Šašel, *s.v.* Siscia, *RE*, Suppl. 14, 702-741.

Šašel Kos 1997 Marjeta Šašel Kos, *The end of the Norican Kingdom and the formation of the provinces of Noricum and Pannonia*, in: B. Djurić and I. Lazar (eds), *Akten des IV. Internationalen Kolloquiums über Probleme des provinzialrömischen Kunstschaffens*, *Situla* 36, 21-42.

Šašel Kos 2002 Marjeta Šašel Kos, The boundary stone between Aquileia and Emona, *AV* 53, 373-382.

Šašel Kos 2003 Marjeta Šašel Kos, Emona was in Italy, not in Pannonia, in: *Pannonia* I, 11- 19.

Zaninović 1966 Marin Zaninović, *Ilirsko pleme Delmati, I dio, Politička povijest i teritorijalna rasprostranjenost Delmata*, *GodANUBiH* 4, 27-92.

Annex: Indigenous anthroponymy in southern Pannonia

This catalogue contains only those inscriptions from southern Pannonia (and the border area around Emona) that mention names of indisputable Celtic or Illyrian-Pannonian origin, that is those that can be attributed with certainty to the indigenous inhabitants of southern Pannonia (this is why we did not exclude the names associated with the Venetian or Daco-Thracian anthroponymy). We did not include the so-called assumed names or translated Latin names, and we also did not include the hapaxes that could not be at least partly associated with the Celtic or Illyrian-Pannonian anthroponymy.

BORDER AREA (AGER EMONENSIS)

Emona

CIL III 3855; Mócsy 1959, 2/69 – *Buio*[43] *Vibi f(ilius), Lasciontia*[44] *Subloani*[45] *f(ilia), Fibio*[46]

[43] Holder I, 629; Krahe 1929, 25; Mócsy 1959, 167; Lochner-Hüttenbach 1965, 21-22; Katičić 1968, 72-73; Mócsy 1983, 55, *s.v. Buio*; Lőrincz and Redő 1994, 326, *s.v. Bvio*; Matasović 2003, 13; Delamarre 2007, 50, *s.v. Buio*.

[44] Holder II, 148; Krahe 1929, 63; Mócsy 1959, 178; Katičić 1968, 83-84; Mócsy 1983, 159, *s.v. Lasciontia*; Lőrincz 2000, 19, *s.v. Lasciontia*; Matasović 2003, 12; Meid 2005, 273-274.

[45] Holder II, 1651; Mócsy 1959, 191; Mócsy 1983, 275, *s.v. Subloanus*; Lőrincz 2002, 97, *s.v. Svbloanvs*; Matasović 2003, 12.

[46] Katičić 1968, 82; Mócsy 1983, 126, *s.v. Fibio*; Lőrincz 1999, 140, *s.v. Fibio*.

CIL III 3860; Mócsy 1959, 2/68; *AIJ* 185 – *Buiio*[47] *Senni*[48] *f(ilius), Eninna*[49] *Voltregis*[50] *f(ilia), Voltarons*[51] *Voltregis f(ilia)*

CIL III 3861 (10758); Mócsy 1959, 2/71 – *Enno*[52] *Secconis*[53] *f(ilius), Cetetiu*[54] *Rustici f(ilia), Valentio, Secu*[55]

CIL III 3866; Mócsy 1959, 2/73; *AIJ* 192 – *Oppalo*[56] *Firmi f(ilius), Rega*[57] *Buionis*[58] *f(ilia), Secundus, Severa Antoni(i) f(ilia)*

CIL III 3871; Mócsy 1959, 2/74; *AIJ* 195 – *Secco*[59] *Nammonis*[60] *f(ilius), Rega*[61] *Terti f(ilia), Rusticus, Quintus, Enignus,*[62] *Manu*[63] *avia*

CIL III 3872; Mócsy 1959, 2/65 – *C. Vedius Rufus, Secunda Epponis*[64] *f(ilia), Sabinus*

CIL III 3874; Mócsy 1959, 2/75; *AIJ* 197 – *Severus Sacciari*[65] *f(ilius), Pusilla*

CIL III 3877; Mócsy 1959, 2/78 – *Voltarons*[66] *Urbani f(ilius), Rusticus Secconis*[67] *(filius), Maxuma*

Radovljica

Mócsy 1959, 5/2; *AIJ* 217; *RINMS* 100 – *Masclus Ressimari*[68] *f(ilius), Bello*[69] *[Sec]undi (filia)*

Šmarata

CIL III 10722 – 10724; Mócsy 1959, 12/1; *AIJ* 123, 124 – *Feucon,*[70] *Planius*[71] *Sextilius, Sextilia Tatsoria,*[72] *Turoius*[73] *Pletor,*[74] *Pletor Poteius,*[75] *Planius Poteius, Volta*[76] *Lassonia,*[77] *Pletor Poteius*

CIL III 10725; Mócsy 1959, 12/2; *AIJ* 125 – *Turoius*[78] *Vot[. .],*[79] *Feva*[80]

---

[47] *Vide supra.*

[48] Holder II, 1479; Mócsy 1959, 190; Katičić 1966, 158; Mócsy 1983, 261, *s.v. Sennius, Sennus*; Lőrincz 2002, 67, *s.v. Sennivs, Sennvs*; Meid 2005, 205-206; Delamarre 2007, 165, *s.v. Sennius, Senius, -ia, Sennus.*

[49] Holder I, 1438; Krahe 1929, 46; Mócsy 1959, 173; Lochner-Hüttenbach 1965, 24; Katičić 1968, 77-78; Lőrincz 1999, 117, *s.v. Eninna*; Matasović 2003, 12; Meid 2005, 271; cf. Delamarre 2007, 95-96.

[50] Krahe 1929, 130; Mócsy 1959, 197; Lochner-Hüttenbach 1965, 38-39; Katičić 1968, 102-104; Mócsy 1983, 319. *s.v. Uoltrex*; cf. Križman 1991, 111-113, 133-135; Lőrincz 2002, 183, *s.v. Voltrex*; Meid 2005, 90-91.

[51] Krahe 1929, 129; Mócsy 1959, 197; Lochner-Hüttenbach 1965, 38-39;Katičić 1968, 102-104; Mócsy 1983, 318, *s.v. Uoltarons*; cf. Križman 1991, 111-113, 133-135; Lőrincz 2002, 182, *s.v. Voltarons.*

[52] Holder I, 1440; Krahe 1929, 47; Mócsy 1959, 173; Katičić 1968, 77-78; Mócsy 1983, 114, *s.v Enno*; Lőrincz 1999, 118, *s.v. Enno*; Meid 2005, 271.

[53] Holder II, 1424; Mócsy 1959, 189; Katičić 1966, 159; Mócsy 1983, 258, *s.v. Secco*; Lőrincz 2002, 56, *s.v Secco*; Matasović 2003, 13; Meid 2005, 285-286; Delamarre 2007, 163, *s.v. Secco.*

[54] Holder I, 1001; Mócsy 1959, 169; Katičić 1966, 158; Mócsy 1983, 76, *s.v. Ceteteiu*; Lőrincz 1999, 52, *s.v. Ceteteiv*; Matasović 2003, 13; Meid 2005, 263-264; Delmarre 2007, 64, *s.v. Ceteteiu.*

[55] Mócsy 1959, 189; Katičić 1966, 158; Mócsy 1983, 258, *s.v. Secu+*; Lőrincz 2002, 57, s.v. *Secv**; Matasović 2003, 13; Meid 2005, 285-286; cf. Delamarre 2007, 163, 231.

[56] Holder II, 862; Krahe 1929, 82; Mócsy 1959, 183; Lochner-Hüttenbach 1965, 30-31; Katičić 1968, 88-90; Mócsy 1983, 208, *s.v. Oppalo*; Lőrincz 2000, 114, *s.v. Oppalo*; cf. Delamarre 2003, 171; Matasović 2003, 12; Delamarre 2007, 145, *s.v. Oppalo.*

[57] Krahe 1929, 98; Mócsy 1959, 187; Lochner-Hüttenbach 1965, 42; Katičić 1968, 94-95; Mócsy 1983, 241, *s.v. Rega*; Lőrincz 2002, 24, *s.v. Rega*; Matasović 2003, 12; cf. Delamarre, 2007, 152.

[58] *Vide supra.*

[59] *Vide supra.*

[60] Holder II, 677; Schmidt 1957, 246-247; Mócsy 1959, 182; Lochner-Hüttenbach 1965, 29-30; Evans 1967, 234-236; Katičić 1968, 86-87; Mócsy 1983, 197, *s.v. Nammo*; Degavre 1998, 310; Lőrincz 2000, 94, *s.v. Nammo*; Matasović 2003, 13; Meid 2005, 278; Delamarre 2007, 138, *s.v. Nammo*; Radman-Livaja 2010, 364-365.

[61] *Vide supra.*

[62] Cf. Holder I, 1438; Mócsy 1959, 173; Lochner-Hüttenbach 1965, 24; Katičić 1968, 77-78; Mócsy 1983, 114, *s.v. Enignus*; Lőrincz 1999, 117, *s.v. Enignvs*; Matasović 2003, 13; Meid 2005, 127-28, 172-173; Delamarre 2007, 95, *s.v. Enignus, -ius.*

[63] Holder II, 412; Krahe 1929, 71; Mócsy 1959, 180; Katičić 1966, 159; Mócsy 1983, 177, *s.v. Manu*; Lőrincz 2000, 52, *s.v. Manv*; Matasović 2003, 13; Meid 2005, 275-276; Delamarre 2007, 126, *s.v Manu.*

[64] Holder I, 1455; Mócsy 1959, 173; Lochner-Hüttenbach 1965, 24-25; Katičić 1966, 158; Katičić 1968, 79-80; Mócsy 1983, 116, *s.v. Eppo*; Degavre 1998, 213; Lőrincz 1999, 121, *s.v. Eppo*; Delamarre 2003, 163; Matasović 2003, 12; Meid 2005, 195, 270; Delamarre 2007, 96, *s.v. Epo*, 97, *s.v. Eppo*, 221.

[65] Holder II, 1274; Mócsy 1959, 188; Katičić 1966, 159; Katičić 1968, 95-96; Mócsy 1983, 248, *s.v. Sacciarus*; Lőrincz 2002, 41, *s.v. Sacciarvs*; Matasović 2003, 13; cf. Delamarre 2007, 157.

[66] *Vide supra.*

[67] *Vide supra.*

[68] Holder II, 432-433, 1177; Schmidt 1957, 237-238, 258; Mócsy 1959, 187; Katičić 1966, 158; Evans 1967, 223-228, 249-251; Mócsy 1983, 242, *s.v. Ressimarus, Ressomarus*; Degavre 1998, 296, *s.v. maro-*, 352, *s.v. ress-*; Lőrincz 2002, 27, *s.v. Ressimarvs, Ressomarvs*; Matasović 2003, 13; Meid 2005, 114-116; Delamarre 2003, 218-219, *s.v. maros*, 256, *s.v. redsos, ressos*; Delamarre 2007, 153, *s.v. Ressimarus* ; Radman-Livaja 2010, 418.

[69] Holder I, 391, III, 841; Mócsy 1959, 166; Katičić 1966, 158; Mócsy 1983, 47, *s.v. Bello;* Lőrincz and Redő 1994, 286, *s.v. Bello*; Meid 2005, 220-222; Delamarre 2003, 72; cf. Delamarre 2007, 39-40, 213.

[70] Krahe 1929, 51; Mócsy 1959, 174; Katičić 1968, 105-106; Mócsy 1983, 126, *s.v. Feuco*; Lőrincz 1999, 140, *s.v. Fevco.*

[71] Holder II, 1012; Mócsy 1959, 184; Katičić 1968, 106; Mócsy 1983, 225, *s.v. Planius*; Lőrincz 2000, 145, *s.v. Planivs.*

[72] Holder II, 1753; Mócsy 1959, 192; Katičić 1968, 107; Mócsy 1983, 283, *s.v. Tatsorius*; Lőrincz 2002, 109, *s.v. Tatsorivs*; Matasović 2003, 12.

[73] Holder II, 2006; Krahe 1929, 120; Mócsy 1959, 193; Katičić 1968, 107-108; Mócsy 1983, 296, *s.v. Turoius*; Lőrincz 2002, 134, *s.v. Tvroivs,* Matasović 2003, 13; Delamarre 2007, 186, *s.v Turoiu.*

[74] Krahe 1929, 91-94; Mayer 1957, 272-274; Mócsy 1959, 184; Katičić 1968, 93-94, 106; Mócsy 1983, 225-226, *s.v. Plaetor, Pletor*; Križman 1991, 129; Lőrincz 2000, 145-146, *s.v. Plaetor, Pletor.*

[75] Mócsy 1959, 185; Katičić 1968, 107, Mócsy 1983, 230, *s.v. Poteius*; Lőrincz 2000, 156, *s.v. Poteivs.*

[76] Mócsy 1959, 197; Katičić 1968, 108; Mócsy 1983, 318, *s.v. Volta*; cf. Križman 1991, 111-113, 133-135; Lőrincz 2002, 182, *s.v. Volta;* Meid 2005, 90; cf. Delamarre 2007, 204.

[77] Mócsy 1959, 178; Katičić 1968, 106; Lőrincz 2000, 19, *s.v. Lasonia, Lassonia;* Matasović 2003, 13; Meid 2005, 273-274; cf. Delamarre 2007, 115.

[78] *Vide supra.*

[79] Katičić 1968, 108.

[80] Mócsy 1959, 174; Katičić 1968, 106; Mócsy 1983, 126, *s.v. Feua*; Lőrincz 1999, 140, *s.v. Feva.*

Strahomer

CIL III 3789 (10728); Mócsy 1959, 14/3; *AIJ* 128 – *Buco*[81] *Terti f(ilius), Tertia Sabini f(ilia)*

CIL III 3826; Mócsy 1959, 14/5; *AIJ* 144 – *Ursinus Buionis*[82] *(filius), Maxima Quinti f(ilia), Buio*

CIL III 10727 (3788); Mócsy 1959, 14/2; – *Bucio*[83] *Firmi f(ilius), Bucoa*[84] *Valentis, Maxima*

CIL III 10740; Mócsy 1959, 14/1; *AIJ* 131; Raybould and Sims-Williams 2007, PAN 058 – *P. Varisidius Hostius*,[85] *Epo*,[86] *Buquorsa*,[87] *Adnom[at]us*,[88] *Ingenuus*

Tomišlje

CIL III 3816 (10735); Mócsy 1959, 15/1; *AIJ* 142; *RINMS* 88 – *Tertius Epponis*[89] *Boleriani*[90] *f(ilius), Pusilla*

Sonneg Castle nearby Ig

AIJ 133; Mócsy 1959, 17/4 – *Nammo*,[91] *T(itus), Buquorsa*[92] *Ucci*[93] *f(ilia) ?*

CIL III 3791 (10729); Mócsy 1959, 17/3; *AIJ* 130; *RINMS* 83 – *Calvus Severi f(ilius), Voltia*[94] *Voltuparis*[95] *f(ilia), Maxima, Max(i)mila*

CIL III 3804 (10731); Mócsy 1959, 17/5; *AIJ* 134 – *Pletor*[96] *Laepius*,[97] *Moiota*,[98] *Rusticus*

CIL III 10745; Mócsy 1959, 17/7; *RINMS* 81 – *Voltaro*,[99] *[B]uia*[100] *Rustici f(ilia)*

Ig

CIL III 3787; Mócsy 1959, 18/4 – *Bucci*,[101] *Rega*,[102] *Bucci* . . .

CIL III 3790; *AIJ* 129; Mócsy 1959, 18/5; *RINMS* 82 – *Buio*[103] *Brocci*[104] *f(ilius), Maxima Epponis*[105] *f(ilia), Severus, O. Buciori*[106] *f., Quinctus Severi f(ilius), Lasonia*[107] *Voltani(s?)*[108] *[f.]*

CIL III 3793; Mócsy 1959, 18/6 – *Enignus*[109] *Plunconis*[110] *f(ilius), Enna*[111] *Oppalonis*[112] *f(ilia), Rega*,[113] *Plunco, Voltrex*[114]

CIL III 3794; Mócsy 1959, 18/7 – *Enni* . . .

[81] Holder, I, 625-626, III, 990; Šašel 1955, 374-379; Mócsy 1959, 167; Kajanto 1965, 63, 188, 225; Lochner-Hüttenbach 1965, 20-21; Katičić 1966, 158; Katičić 1968, 68-71; Mócsy 1983, 55, *s.v. Bucca*; Lőrincz and Redő 1994, 325, *s.v. Bvcca*; Solin and Salomies 1994, 304, *s.v. Buca, Bucca*; Degavre 1998, 98, *s.v. bocca*, 117, *s.v. bucco-/ bucca*; Matasović 2003, 13-14, 19; Delamarre 2003, 80, *s.v. bocca*; Meid 2005, 259-260; Delamarre 2007, 50, *s.v. Buccos, -us, -ius, -a, -ia*; Radman-Livaja 2010, 195-196.

[82] *Vide supra.*

[83] *Vide supra.*

[84] *Vide supra.*

[85] Cf. Holder II, 888; Mócsy 1959, 176; Lochner-Hüttenbach 1965, 26; Katičić 1968, 82-83; Mócsy 1983, 146, *s.v. Hostius*; Lőrincz 1999, 186, *s.v. Hostivs*; Matasović 2003, 12.

[86] *Vide supra.*

[87] Mócsy 1959, 167; Lochner-Hüttenbach 1965, 22; Katičić 1968, 73; Mócsy 1983, 56, *s.v. Buquorsa*; Lőrincz and Redő 1994, 327, s.v. Bvqvorsa; Meid 2005, 260; Delamarre 2007, 50, *s.v. Buquorsa.*

[88] Holder I, 43-44; Mócsy 1959, 162; Schmidt 1957, 114-115, 246-247; Lochner-Hüttenbach 1965, 15-16; Katičić 1966, 149, 155, 158; Evans 1967, 130, 235; Katičić 1968, 64-65; Mócsy 1983, 5, *s.v. Adnamatus*; Lőrincz and Redő 1994, 26, *s.v. Adnamatvs*; Delamarre 2003, 230; Matasović 2003, 13; Meid 2005, 159-160; Delamarre 2007, 12-13, *s.v. Adnamatus, -ius, Adnomatus.*

[89] *Vide supra.*

[90] Holder III, 902; Mócsy 1959, 167; Lochner-Hüttenbach 1965, 19; Katičić 1968, 67; Mócsy 1983, 52, *s.v. Bolerianus*; Lőrincz and Redő 1994, 308, *s.v. Bolerianvs*; Matasović 2003, 12; cf. Delamarre 2007, 45, 213.

[91] *Vide supra.*

[92] *Vide supra.*

[93] Holder III, 12; Mócsy 1959, 194; Katičić 1966, 159; Katičić 1968, 100; Mócsy 1983, 302, *s.v. Uccio, Uccius, Ucco, Uccus*; Lőrincz 2002, 150, *s.v. Vccio, Vccius, Vcco, Vccus*; Matasović 2003, 13; Meid 2005, 298-299; Delamarre 2007, 191, *s.v. Uccius, Ucco*; Radman-Livaja 2010, 488.

[94] *Vide supra.*

[95] Krahe 1929, 130; Mócsy 1959, 197; Lochner-Hüttenbach 1965, 38-39; Katičić 1968, 102-104; Mócsy 1983, 319, *s.v. Uoltupar*; cf. Križman 1991, 111-113, 133-135; Lőrincz 2002, 183, *s.v. Voltvpar*; Meid 2005, 90; Delamarre 2007, 237.

[96] *Vide supra.*

[97] Krahe 1929, 60; Mócsy 1959, 177; Lochner-Hüttenbach 1965, 27; Katičić 1968, 83; Mócsy 1983, 158, *s.v. Laepius*; Lőrincz 2000, 17, *s.v. Laepivs.*

[98] Krahe 1929, 77; Mócsy 1959, 182; Lochner-Hüttenbach 1965, 29; Katičić 1966, 159; Katičić 1968, 85-86; Mócsy 1983, 192, s.v. Moiota; Lőrincz 2000, 85, *s.v. Moiota*; Matasović 2003, 12.

[99] Krahe 1929, 129; Mócsy 1959, 197; Lochner-Hüttenbach 1965, 38-39;Katičić 1968, 102-104; Mócsy 1983, 318, *s.v. Uoltaro*; cf. Križman 1991, 111-113, 133-135; Lőrincz 2002, 182, *s.v.Voltaro.*

[100] *Vide supra.*

[101] *Vide supra.*

[102] *Vide supra.*

[103] *Vide supra.*

[104] Holder I, 617; Mócsy 1959, 167; Lochner-Hüttenbach 1965, 19-20; Katičić 1966, 158; Katičić 1968, 67; Mócsy 1983, 55, *s.v. Brocchus, Broccus*; Lőrincz and Redő 1994, 322, *s.v. Brocchvs, Broccvs*; Matasović 2003, 13; Meid 2005, 223; Delamarre 2007, 49, *s.v. Broccus.*

[105] *Vide supra.*

[106] Mócsy 1959, 167; Lochner-Hüttenbach 1965, 20-21; Katičić 1966, 158; Katičić 1968, 68-71; Mócsy 1983, 55, *s.v. Buccio*; Lőrincz and Redő 1994, 325, *s.v. Bvccio*; Matasović 2003, 13; Meid 2005, 259-260; Delamarre 2007, 50, *s.v. Buccio.*

[107] *Vide supra.*

[108] Mócsy 1959, 197; Lochner-Hüttenbach 1965, 38-39; Katičić 1968, 102-104; Mócsy 1983, 318, *s.v. Voltans*; cf. Križman 1991, 111-113, 133-135; Lőrincz 2002, 182, *s.v. Voltans*; cf. Meid 2005, 90.

[109] *Vide supra.*

[110] Holder II, 1020; Krahe 1929, 96; Mócsy 1959, 184; Lochner-Hüttenbach 1965, 32; Katičić 1966, 159; Katičić 1968, 94; Mócsy 1983, 227, *s.v. Plunco*; Lőrincz 2000, 147, *s.v. Plvnco*; Matasović 2003, 12.

[111] Holder I, 1439; Krahe 1929, 46; Mócsy 1959, 173; Katičić 1968, 77-78; Mócsy 1983, 114, *s.v. Enna*; Lőrincz 1999, 117, *s.v. Enna*; Matasović 2003, 12; Meid 2005, 271; Delamarre 2007, 95, *s.v. Enna.*

[112] *Vide supra.*

[113] *Vide supra.*

[114] *Vide supra.*

CIL III 3797; Mócsy 1959, 18/8 – *Firmus Provius, Venixema*[115] *Voltregis*[116] *f(ilia)*

CIL III 3796; Mócsy 1959, 18/9 – *Firmus Voltregis*[117] *f(ilius), Ovi*[118] *Ecconis*[119]

CIL III 3798; Mócsy 1959, 18/10; *RINMS* 84 – *Firmus Voltuparis*[120] *f(ilius), Tertia Secundi f(ilia)*

CIL III 3792; Mócsy 1959, 18/11 - ... *gius Coemo*[121] ... , *[La]sciontia*[122] ... *cumi f(ilia)*, ... *ius, Ronio*[123]

CIL III 3801; Mócsy 1959, 18/12 – *Maxima Buttonis*[124] *f(ilia)*

CIL III 3802; Mócsy 1959, 18/13 – *O. Decomonis*[125] *f(ilius), Enna*[126] *Voltanis*[127] *f(ilia), Voltarens*[128]

CIL III 3803; Mócsy 1959, 18/14 - ... *Oppaloni(s)*[129] *f(ilius), Vibius*

CIL III 3805; Mócsy 1959, 18/15 – *Plotius Maximi f(ilius), Quarta Voltregis*[130] *f(ilia), Sabinus, Geminus, Fronto, Voltrex,*[131] *Sextus*

CIL III 3810; Mócsy 1959, 18/17 – *Secco*[132] *Maxie (Maximus ?), Maxima Secu*[133]

CIL III 3811; Mócsy 1959, 18/18 – *Secundus Volturegis*[134] *Talsi*[135] *f(ilius), Voltia*[136] *Sex(ti) f(ilia)*

CIL III 3813; Mócsy 1959, 18/20 – *Sextus Rustici f(ilius), Oppa,*[137] *Quarta*

CIL III 3815; *AIJ* 141; Mócsy 1959, 18/21; *RINMS* 87 – *Surus Luci f(ilius), Quarta Firmi Galuni,*[138] *Sabinus*

CIL III 3817; Mócsy 1959, 18/23 – *Tertius Ottonis*[139] *f(ilius), Ennia*[140] *Maxumi f(ilia), Lasaiu*[141]

CIL III 3818; Mócsy 1959, 18/24 – *Tercio, Secundio Voltaronis*[142] *f(ilius)*

CIL III 3819 (10736); Mócsy 1959, 18/25; Raybould and Sims-Williams 2007, PAN 014 – *Veitro*[143] *Buttonis*[144] *f(ilius), Tetta*[145] *Adnomati*[146] *f(ilia), Butto*

CIL III 3820; Mócsy 1959, 18/26 – *Venix*[147] *Empetonis*[148] *f(ilius), Secunda Mossonis*[149] *f(ilia), Quartio, Romanus Terti f(ilius)*

CIL III 3821; Mócsy 1959, 18/27 – *Voltans*[150] *Oppalonis*[151] *f(ilius), Enna*[152] *Ruii*[153] *f(ilia), Surus*

CIL III 3822; Mócsy 1959, 18/28 – - ... *lvia, Voltia*[154]

---

115 Holder III, 171; Krahe 1929, 125; Schmidt 1957, 289-290; Mócsy 1959, 195; Katičić 1966, 159; Evans 1967, 277-279; Katičić 1968, 101-102; Mócsy 1983, 305, *s.v. Uenixema*; Lőrincz 2002, *s.v. Venixema*; Matasović 2003, 13; Meid 2005, 304-305; Delamarre 2007, 195, *s.v. Uenix(x)amus, -a.*

116 *Vide supra.*

117 *Vide supra.*

118 Krahe 1929, 83; Katičić 1968, 91; Mócsy 1983, 210, *s.v. Oui*; Lőrincz 2000, 118, *s.v. Ovi*; Matasović 2003, 13.

119 cf. Holder I, 1404; Mócsy 1959, 173; Katičić 1966, 158; Katičić 1968, 76; Mócsy 1983, 111, *s.v. Ecco*; Lőrincz 1999, 113, *s.v. Ecco*; Matasović 2003, 13; Meid 2005, 270; Delamarre 2007, 93, *s.v. Ecco.*

120 *Vide supra.*

121 Cf. Holder I, 1061, III, 1251; cf. Krahe 1929, 31; Mócsy 1959, 170; cf. Katičić 1968, 74-75; Mócsy 1983, 84, *s.v. Coemoius*; Lőrincz 1999, 68, *s.v. Coemoivs*; Matasović 2003, 13; Meid 2005, 191; Delamarre 2007, 69, *s.v. Coemo.*

122 *Vide supra.*

123 Holder II, 1228; Mócsy 1959, 188; Mócsy 1983, 245, *s.v. Ronio*; Lőrincz 2002, 32, *s.v. Ronio*; Matasović 2003, 13; Meid 2005, 279; Delamarre 2007, 155, *s.v. Ronio.*

124 Holder I, 646, III, 1011; Krahe 1929, 26; Mócsy 1959, 167; Katičić 1966, 158; Katičić 1968, 73-74; Mócsy 1983, 56, *s.v. Butto*; Lőrincz and Redő 1994, 330, *s.v. Bvtto*; Matasović 2003, 13; Meid 2005, 261; Delamarre 2007, 51, *s.v. Butto.*

125 Holder I, 1248-1249; Krahe 1929, 41; Mócsy 1959, 171; Katičić 1966, 158; Katičić 1968, 75; Mócsy 1983, 99, *s.v Decomo*; Lőrincz 1999, 95, *s.v. Decomo*; Matasović 2003, 13; Meid 2005, 268; cf. Delamarre 2007, 83, 219.

126 *Vide supra.*

127 *Vide supra.*

128 *Vide supra.*

129 *Vide supra.*

130 *Vide supra.*

131 *Vide supra.*

132 *Vide supra.*

133 *Vide supra.*

134 *Vide supra.*

135 Cf. Holder II, 1711; Krahe 1929, 110; Mócsy 1959, 192; Katičić 1966, 159; Katičić 1968, 99; Mócsy 1983, 281, *s.v. Talsus*; Lőrincz 2002, 106, *s.v. Talsvs*; Matasović 2003, 13; cf. Delamarre 2007, 177, 233.

136 Mócsy 1959, 197.

137 Krahe 1929, 82; Mócsy 1959, 183; Katičić 1968, 88-90; Mócsy 1983, 208, *s.v. Oppa*; Lőrincz 2000, 114, *s.v. Oppa*; Matasović 2003, 12.

138 Holder I, 1980; Mócsy 1959, 175; Lochner-Hüttenbach 1965, 25-26; Katičić 1968, 82; Mócsy 1983, 133, *s.v. Galunus*; Lőrincz 1999, 160, *s.v. Galvnvs*; Matasović 2003, 12; Meid 2005, 195-196; cf. Delamarre 2007, 102, *s.v. Gallunus.*

139 Holder II, 890; Mócsy 1959, 184; Katičić 1968, 90; Mócsy 1983, 210, s.v. Otto; Lőrincz 2000, 117, *s.v. Otto*; Matasović 2003, 12; cf. Delamarre 2007, 146.

140 Mócsy 1959, 173; Katičić 1968, 77-78; Mócsy 1983, 114, *s.v. Ennius*; Lőrincz 1999, 117, *s.v. Ennivs*; Matasović 2003, 12; Meid 2005, 271; Delamarre 2007, 95, *s.v. Enna.*

141 Holder II, 148; Krahe 1929, 63; Mócsy 1959, 177; Lochner-Hüttenbach 1965, 28; Katičić 1968, 83-84; Mócsy 1983, 159, *s.v. Lasaiu*; Lőrincz 2000, 19, *s.v. Lasaiv*; Meid 2005, 273-274.

142 *Vide supra.*

143 Holder III, 139; Krahe 1929, 124; Mócsy 1959, 195; Katičić 1968, 101; Mócsy 1983, 303, *s.v. Ueitro*; Lőrincz 2002, 151, *s.v. Veitro.*

144 *Vide supra.*

145 Holder II, 1801; Mócsy 1959, 193; Katičić 1966, 159; Katičić 1968, 99-100; Mócsy 1983, 286; Lőrincz 2002, 117, *s.v. Tetta*; Matasović 2003, 13; Meid 2005, 293; cf. Delamarre 2007, 180.

146 *Vide supra.*

147 Cf. *Venixema, Vide supra.*

148 Mócsy 1959, 173; Katičić 1968, 77.

149 Holder II, 645; Mócsy 1959, 182; Katičić 1966, 159; Katičić 1968, 86; Mócsy 1983, 193, *s.v. Mosso*; Lőrincz 2000, 88, *s.v. Mosso*; Matasović 2003, 13; cf. Delamarre 2007, 136.

150 *Vide supra.*

151 *Vide supra.*

152 *Vide supra.*

153 Krahe 1929, 99; Mócsy 1959, 188; Katičić 1968, 95; Matasović 2003, 12.

154 *Vide supra.*

CIL III 3823; *AIJ* 143; Mócsy 1959, 18/29; *RINMS* 89 – *Volterex*[155] *Buctoris*[156] *f(ilius), Ennina,*[157] *Buctor, Qu[ar]tio*

CIL III 3824; Mócsy 1959, 18/30 – *Voltrex*[158] *Lasonis*[159] *f(ilius), Quart[a?]*

CIL III 3825; Mócsy 1959, 18/31; *RINMS* 90 – *Voltrex*[160] *Plaetoris*[161] *f(ilius), Venix(e)ma*[162] *Plunconis*[163] *f(ilia)*

CIL III 3828; Mócsy 1959, 18/2 – *O(lus ?) C(laudius ?) Ampo,*[164] *....tia*

CIL III 10726 (3785); Mócsy 1959, 18/3 – *Amatu*[165] *Voltaronis*[166] *f(ilia), Opalo*[167] *Hostius, Moiota*[168]

CIL III 10748; Mócsy 1959, 18/22 – *Surus Voltieli,*[169] *Quarta*

29. *ŽAnt* 5, 373[170]; Mócsy 1959, 18/34; *ILJug* 297 – *Amatu*[171] *Buccioni[s]*[172]

Študenec

CIL III 10744; Mócsy 1959, 19/1 – *Quartus Voltaronis*[173] *f(ilius)*

CIL III 10746; *AIJ* 139; Mócsy 1959, 19/2 – *Saturna Neunti*[174] *lib(erta), Hostila, Veri(us)* (ili *Hostila Severi f(ilia)* ?)

ILSl 89 – *Sirus*[175] *Brogimari*[176] *f(ilius), Matugnata*[177] *Admatonis*[178]

Golo

CIL III 3784; Mócsy 1959, 20/1 – *Primus Enigni*[179]

Iška Vas

CIL III 3799 (10730); Mócsy 1959, 21/1 – *Maximus Buionis*[180] *f(ilius), Rustica, . . . anus*

CIL III 3806 (10732); *AIJ* 135; Mócsy 1959, 21/3 – *Quartio Ebonici f(ilius), Maxima Osti*[181] *f(ilia)*

CIL III 10741; *AIJ* 136; Mócsy 1959, 21/2 – *Q(uartio) Ebonici f(ilius), Amatu*[182] *Maximi f(ilia)*

CIL III 10734 (3814); Mócsy 1959, 21/5 – *Sextus Severi f(ilius), Tetiu*[183] *Buionis*[184] *f(ilia), Valens*

AIJ 140; Mócsy 1959, 21/4 – *Severus Buionis*[185] *f(ilius), Moiota*[186] *Eniconis*[187] *f(ilia)*

Gatina

CIL III 13402 (cf. p. 2328[27]); *AIJ* 221; Mócsy 1959, 22/1[188] – *M. Voltilius Firmus, Sabina Urbani f(ilia), M. Voltilius Maximianus, Voltilius Urbanus, Voltilii Voltognas*[189] *et Firmina heredes*

---

155 *Vide supra.*
156 Holder I, 627; Mócsy 1959, 167; Lochner-Hüttenbach 1965, 21; Katičić 1968, 71; Mócsy 1983, 55, *s.v. Buctor*; Lőrincz and Redő 1994, 326, *s.v. Bvctor*; Matasović 2003, 12
157 *Vide supra.*
158 *Vide supra.*
159 Holder II, 149; Krahe 1929, 63, Mócsy 1959, 178; Lochner-Hüttenbach 1965, 27-28; Katičić 1968, 83-84; Mócsy 1983, 159, *s.v. Laso*; Lőrincz 2000, 19, *s.v. Laso*; Matasović 2003, 12; Meid 2005, 273-274.
160 *Vide supra.*
161 *Vide supra.*
162 *Vide supra.*
163 *Vide supra.*
164 Holder I, 133; Mócsy 1959, 164; Katičić 1968, 66; Mócsy 1983, 17, *s.v. Ampo*; Lőrincz and Redő 1994, 104, *s.v. Ampo*; Matasović 2003, 12.
165 This female name is of latin origin (*Amatus, Amata*), but it has a celtic ending –u, cf. Katičić 1968, 66; Matasović 2003, 13; Meid 2005, 251.
166 *Vide supra.*
167 *Vide supra.*
168 *Vide supra.*
169 Mócsy 1959, 197; Katičić 1968, 102-104; Mócsy 1983, 318, *s.v. Uoltielus*; cf. Križman 1991, 111-113, 133-135; Lőrincz 2002, 182, *s.v. Voltielvs*; cf. Meid 2005, 90.
170 Šašel 1955, 373-381.
171 *Vide supra.*
172 *Vide supra.*
173 *Vide supra.*
174 Cf. Holder II, 739; Mócsy 1959, 183; Lochner-Hüttenbach 1965, 30; Katičić 1968, 88; Mócsy 1983, 200, *s.v. Neuntius*; Lőrincz 2000, 99, *s.v. Nevntivs*; Matasović 2003, 12.
175 Mócsy 1983, 268, *s.v. Sirus*; Lőrincz 2002, 85, *s.v. Sirvs*; Meid 2005, 242; Delamarre 2007, 169, *s.v. Sirus.*
176 Holder I, 620; Mócsy 1959, 167; Schmidt 1957, 157, 237-238; Evans 1967, 158-160, 223-228; Mócsy 1983, 55, *s.v. Brogimarus*; Lőrincz and Redő 1994, 323, *s.v. Brogimarvs*; Degavre 1998, 115, 296; Delamarre 2003, 90-91, 217; Meid 2005, 99-100; Delamarre 2007, 49, *s.v. Brogimarus.*
177 Schmidt 1957, 218, 239-240; Evans 1967, 207-209, 228-232; Degavre 1998, 240-241, 299; Lőrincz 2000, 67, *s.v. Matvgrata* (*sic*!); Delamarre 2003, 181-182, 221; Meid 2005, 129-130; Delamarre 2007, 129, *s.v. Matugnata.*
178 Schmidt 1957, 114; Delamarre 2003, 221; Meid 2005, 158; Delamarre 2007, 12, *s.v. Admato*; cf. Radman-Livaja 2010, 148-149.
179 *Vide supra.*
180 *Vide supra.*
181 *Vide supra, s.v. Hostivs.*
182 *Vide supra.*
183 Mócsy 1959, 193; Katičić 1968, 99; Mócsy 1983, 286, *s.v. Tetiunia*; Lőrincz 2002, *s.v. Tetivnia*; Matasović 2003, 13; cf. Delamarre 2007, 181, *s.v. Tetio.*
184 *Vide supra.*
185 *Vide supra.*
186 *Vide supra.*
187 Cf. Holder I, 1438; cf. Krahe 1929, 46; Mócsy 1959, 173; Lochner-Hüttenbach 1965, 23-24; Katičić 1968, 77-78; Mócsy 1983, 114, *s.v. Enico*; Lőrincz 1999, 117. *s.v. Enico*; Matasović 2003, 12; Delamarre 2007, 95, *s.v. Enico.*
188 These people are not necessarily natives
189 Mócsy 1959, 197; Katičić 1968, 102-104; Mócsy 1983, 319, *s.v. Uoltognas*; cf. Križman 1991, 111-113, 133-135; Lőrincz 2002, 183, *s.v. Voltognas.*

Žaljna

CIL III 13403; AIJ 222; Mócsy 1959, 23/1; Raybould and Sims-Williams 2007, PAN 060 – Avitus Severi f(ilius), Sabina Gai f(ilia), Exouna[190] *Optati f(ilia), Melesme liberta*

II. PANNONIA SUPERIOR

Stična

AÉ 1944, 133, Mócsy 1959, 24/2; *ILJug* 325; *ILSl* 69; Raybould and Sims-Williams 2007, PAN 072 – *Adnamo*[191] *C. Anonis f(ilius), Nemetomara*[192] *Germani f(ilia) / Adnamo Canonis*[193] *f(ilius), Sennemetomara*[194] *Germani f(ilia)* (after *IlSl* 69)

Šentvid

CIL III 10783; Mócsy 1959, 28/1; *RINMS* 143; *ILSl* 78; Raybould and Sims-Williams 2007, PAN 033 – *T. Flavius Rufinus, Claudia Coromara,*[195] *Flavia Rufilla*

Trebnje

CIL III 10786; RINMS 170; ILSl 115 – Mar(cius ?) Maroninus[196]

CIL III 3904; *RINMS* 162; *ILSl* 125 – *Aurelius Secundianus qui et Itriu(s)*[197]

Štatenberk

CIL III 10794; *AIJ* 236; Mócsy 1959, 31/1; *RINMS* 145; *ILSl* 85; Raybould and Sims-Williams 2007, PAN 076 – *Publius Nertomarius*[198] *Quartus, Meita*[199] *Optati f(ilia), Secundus, Iustus, Nigrinus*

Mlada Gora, Šmarjeta

AIJ 239; Mócsy 1959, 32/1; *RINMS* 146; ILSl 87 – *Secundus Victoris, Locita*[200]

Teržišce

CIL III 10795 (cf. p. 2328[28]); Mócsy 1959, 33/1; *ILSl* 93 – *Catio*[201] *Boudion(is)*[202] *f(ilius), Suadulla,*[203] *Vepo*[204] *Caioni(s)*[205]

Radeče

AIJ 256; Mócsy 1959, 34/1 – *Anta*[206] *Ressati*[207] *f(ilia), Castor*

Ševnica

CIL III 13406; Mócsy 1959, 35/1; Raybould and Sims-

[190] Cf. Holder I, 1489-1490; Mócsy 1959, 173; Katičić 1966, 158; Katičić 1968, 81; Mócsy 1983, 122, *s.v. Exounus*; Lőrincz 1999, 131, *s.v. Exovnvs*; Matasović 2003, 13; Meid 2005, 174; cf. Delamarre 2007, 100, 221.

[191] Holder I, 43-44; Mócsy 1959, 162; Katičić 1968, 64-65; Mócsy 1983, 5, *s.v. Adnamo*; Lőrincz and Redő 1994, 26, *s.v. Adnamo*; Matasović 2003, 14; Meid 2005, 159-160; Delamarre 2007, 12, *s.v. Adnamo.*

[192] Holder II, 433, 711; Schmidt 1957, 248; Mócsy 1959, 182; Mócsy 1983, 199, *s.v. Nemetomarus*; Degavre 1998, 296, 316; Lőrincz 2000, 97, *s.v. Nemetomarvs*; Matasović 2003, 14; Meid 2005, 112-113; Delamarre 2003, 232-233; Delamarre 2007, 140, *s.v. Nemetomarus, -a.*

[193] Schmidt 1957, 61; Mócsy 1983, 65, *s.v. Cano*; Degavre 1998, 133; Lőrincz 1999, 31, *s.v. Cano*; Delamarre 2003, 102; Meid 2005, 189-190; Delamarre 2007, 56, *s.v. Cano.*

[194] Schmidt 1957, 237-238, 248, 266-267; Evans 1967, 223-228; Delamarre 2003, 217-218, 232-233, 269-270; Meid 2005, 112-113; Delamarre 2007, 165, *s.v. Sennemetomara.*

[195] Holder I, 1055, III, 1246; Schmidt 1957, 183-184, 237-238; Mócsy 1959, 170; Katičić 1966, 157; Evans 1967, 223-228; Degavre 1998, 165, 296; Delamarre 2003, 126, 217-218; Matasović 2003, 14; Meid 2005, 102-103; Delamarre 2007, 74, *s.v. Coromara.*

[196] Schmidt 1957, 237-238; Evans 1967, 223-228; Lőrincz 2000, 59, *s.v Maroninvs*; Delamarre 2003, 217-218; cf. Delamarre 2007, 127, *s.v. Maronianus, Maronus, -ius*, 226.

[197] Holder II, 83; Katičić 1966, 157; Mócsy 1983, 154, *s.v Itrius*; Lőrincz, 1999, *s.v. Itrivs*; Meid 2005, 273.

[198] Holder II, 725; Schmidt 1957, 249; Mócsy 1959, 157, 183; Katičić 1966, 158; Evans 1967, 237; Mócsy 1983, 200, *s.v. Nertomarius, Nertomarus*; Degavre 1998, 296, 318; Lőrincz 2000, 99, *s.v. Nertomarivs, Nertomarvs*; Delamarre 2003, 217-218, 234; Matasović 2003, 14; Meid 2005, 113-114; Delamarre 2007, 140, *s.v. Nertomarus, -ius.*

[199] The origin of the name is far from being certain, Celtic or Pannonian?, cf. Holder II, 530; cf. Krahe 1929, 73; Mócsy 1959, 181; Katičić 1966, 157; Mócsy 1983, 184, *s.v. Meitima*; Lőrincz 2000, 73, *s.v. Meitima*; Matasović 2003, 14; Meid 2005, 113; Delamarre 2007, 131, *s.v. Meitima.*

[200] Mócsy 1959, 178; Katičić 1966, 157; Mócsy 1983, 165, *s.v. Locita*; Lőrincz 2000, 30, *s.v. Locita*; Matasović 2003, 14; Meid 2005, 274-275; Delamarre 2007, 225.

[201] Holder I, 841; Mócsy 1959, 168; Katičić 1966, 157; Mócsy 1983, 71, *s.v. Catio*; Lőrincz 1999, 43, *s.v. Catio*; Matasović 2003, 14; Meid 2005, 190-191; Delamarre 2007, 60-61, 215-216.

[202] Holder I, 497, III, 917; Schmidt 1957, 154; Mócsy 1959, 167; Katičić 1966, 157; Evans 1967, 156-158; Mócsy 1983, 53, *s.v. Boudio;* Lőrincz and Redő 1994, 314, *s.v.Bovdio;* Matasović 2003, 19; Meid 2005, 188; Delamarre 2007, 47, *s.v. Boudio*; cf. Radman-Livaja 2010, 194.

[203] Holder II, 1643; Mócsy 1959, 191; Katičić 1966, 158; Mócsy 1983, 275, *s.v. Suadullus*; Lőrincz 2002, 96, *s.v. Svadvllvs*; Matasović 2003, 14; Meid 2005, 206-207; Delamarre 2007, 172, *s.v. Suadulla.*

[204] Holder III, 177; Mócsy 1959, 195; Katičić 1966, 158; Mócsy 1983, 306, *s.v. Uepo*; Lőrincz 2002, 155, *s.v. Vepo*; Matasović 2003, 14; Meid 2005, 245-247; Delamarre 2007, 195, *s.v. Uepo.*

[205] Holder I, 684, III, 1040; Mócsy 1959, 167; Katičić 1966, 157; Mócsy 1983, 60, *s.v. Caio*; Lőrincz 1999, 21, *s.v. Caio*; Matasović 2003, 14; Meid 2005, 224; Delamarre 2007, 53, *s.v. Caio.*

[206] Holder III, 633; Mócsy 1959, 164; Katičić 1966, 156; cf. Mócsy 1983, 21, *s.v. Anta*; cf. Lőrincz and Redő 1994, 122, *s.v. Anta*; cf. Matasović 2003, 14; Delamarre 2007, 23, *s.v. Anta*; cf. Radman-Livaja 2010, 159.

[207] Holder II, 1177; Mócsy 1959, 187; Katičić 1966, 157; Mócsy 1983, 242, *s.v. Ressatus*; Lőrincz 2002, 27, *s.v. Ressatvs*; Matasović 2003, 14; Meid 2005, 203; Delamarre 2007, 153, *s.v. Ressatus.*

Williams 2007, PAN 040 – *Titus Ambisavi*[208] *f(ilius), Uxela*[209] *Griponi(s ?)*[210] *f(ilia)*

Brezje

CIL III 3930; Mócsy 1959, 37/1; Raybould and Sims-Williams 2007, PAN 015 – *Lucius Boniati*[211] *f(ilius), Iulia Prisca, Bussugnata*[212]

Križnica

CIL III 10815; Mócsy 1959, 38/1; *ILSl* 88 – *Severin(us) Maron(is),*[213] *Maximi(na ?), Ausca*[214]

Podgrad

ILSl 75 – *[Tib(erius) ? Cl]aud[ius . .]mat[. .], Claudia Ti(beri) f(ilia) Conerta*[215]

Kamnje kod Šentruperta

ILSl 71 – *Lucius Baibius Sempronius, Escinga*[216]

Gorenji Globodol

CIL III 10793; *ILSl* 82; Raybould and Sims-Williams 2007, PAN 034 – *Litugena*[217] *Cano[nis]*[218] *f(ilia), Mugia*[219]

Zgornja Pohanca

ILJug II 1121; *ILSl* 90 – *Su[r]us Sironis*[220] *f(ilius), Claudia Longinia, Sabina*

Brege

CIL III 10813; Barkóczi 1964, 18/1; *ILSl* 37 – *Mog(etius ?)*[221] *Ma[ter]nus, Suadr[a]*[222]

Straža kod Šentruperta

ILJug 334; *ILSl* 91; Raybould and Sims-Williams 2007, PAN 073 – *Togivepus*[223] *Optati f(ilius), Melissa*[224] *Cai, Speratus*

Podsused

CIL III 10867; Mócsy 1959, 49/1; Raybould and Sims-Williams 2007, PAN 035 – *C. Iulius Maximi lib(ertus) Adietumarus,*[225] *Sporilla*

Andautonia (Ščitarjevo)

CIL III 3679; Mócsy 1959, 52/3 – *T. Flavius Bonio,*[226] *Andautonia, eques alae Frontonianae*

---

[208] Holder I, 122, III, 590; Krahe 1929, 4; Schmidt 1957, 122-125; Mócsy 1959, 163; Katičić 1966, 157; Evans 1967, 134-136; Mócsy 1983, 15, *s.v. Ambisauus*; Lőrincz and Redő 1994, 93, *s.v. Ambisavvs*; Delamarre 2003, 41-42; Matasović 2003, 14; Meid 2005, 162; Delamarre 2007, 19, *s.v. Ambisauus.*

[209] Cf. Holder III, 61-67; Mócsy 1959, 194; Katičić 1966, 158; Mócsy 1983, 322, *s.v. Uxelus*; Lőrincz 2002, 189, *s.v. Vxela*; Delamarre 2003, 329; Matasović 2003, 14; Meid 2005, 208-209; Delamarre 2007, 207, *s.v. Uxela.*

[210] Mócsy 1959, 176; Katičić 1966, 157; Mócsy 1983, 139, *s.v. Gripo*; Lőrincz 1999, 171, *s.v. Gripo*; Matasović 2003, 14; Meid 2005, 315; Delamarre 2007, 106, *s.v. Griponus*; cf. Radman-Livaja 2010, 305.

[211] Mócsy 1959, 167; Katičić 1966, 156; Mócsy 1983, 52, *s.v. Boniatus*; Lőrincz and Redő 1994, 309, *s.v. Boniatvs*; Matasović 2003, 14; Meid 2005, 258-259; Delamarre 2007, *s.v. Boniatus, -a.*

[212] Holder I, 644-645, III, 1010; Schmidt 1957, 158, 218; Mócsy 1959, 167; Katičić 1966, 156; Evans 1967, 207-209; Mócsy 1983, 56, *s.v. Bussugnatus*; Lőrincz and Redő 1994, 329, *s.v. Bvssvgnatvs*; Degavre 1998, 120, 240-241; Delamarre 2003, 94-95, 180-181; Matasović 2003, 14; Meid 2005, 128-129; Delamarre 2007, 50, *s.v. Bussugnata.*

[213] Holder II, 432; Mócsy 1959, 167; Mócsy 1983, 179, *s.v. Maro*; Degavre 1998, 296; Lőrincz 2000, 59, *s.v. Maro*; Delamarre 2003, 217-218; Matasović 2003, 14; Meid 2005, 198; Delamarre 2007, 127, *s.v. Maro.*

[214] Holder III, 760; Mócsy 1959, 165; Katičić 1966, 156; Mócsy 1983, 40, 40, *s.v. Auscus*; Lőrincz and Redő 1994, 251, *s.v. Avscvs*; Matasović 2003, 14; Delamarre 2007, 35, *s.v. Ausco, -us.*

[215] Mócsy 1983, 86, *s.v. Conertus*; Lőrincz 1999, 72, *s.v. Conertvs*; Meid 2005, 169-170; Delamarre 2007, 72, *s.v. Conertus*; Radman-Livaja 2010, 230-231; cf. Holder I, 1054-1055; Katičić 1966, 148; Mócsy 1983, 91, *s.v. Counertus*; Lőrincz 1999, 81, *s.v. Covnertvs*; Delamarre 2003, 121-122, *s.v. com- con- co-*, 122-123, *s.v. comnertos*, 132, *s.v. cuno-*; Delamarre 2007, 76, *s.v. Counertus*, 217-218.

[216] Schmidt 1957, 171-172, 212-213; Evans 1967, 177-179, 202-203; Mócsy 1983, 121, *s.v. Excingus*; Lőrincz 1999, 130, *s.v. Excingvs*; Meid 2005, 174; Delamarre 2007, 99, *s.v. Excingius, -us.*

[217] Schmidt 1957, 216-217, 232; Evans 1967, 203-207; Mócsy 1983, 165, *s.v. Litugenus*; Degavre 1998, 234, 280; Lőrincz 2000, 29, *s.v. Litugenvvs*; Delamarre 2003, 176-177, 204; Meid 2005, 124-125; Delamarre 2007, 118, *s.v. Litugenus, -ius, -a.*

[218] Mócsy 1983, 65, *s.v. Cano*; Lőrincz 1999, 31, *s.v. Cano*; Meid 2005, 189-190; Delamarre 2007, 56, *s.v. Cano.*

[219] Lőrincz 2000, 89, *s.v. Mvgia*; Meid 2005, 237; Delamarre 2007, 137, *s.v. Mugia.*

[220] Mócsy 1983, 268, *s.v. Siro*; Lőrincz 2002, 85, *s.v. Siro*; Meid 2005, 242; Delamarre 2007, 169, *s.v. Siro, -nius.*

[221] Holder II, 608; Mócsy 1959, 182; Mócsy 1983, 191, *s.v. Mogetius*; Degavre 1998, 292; Lőrincz 2000, 85, *s.v. Mogetivs*; Delamarre 2003, 212; Meid 2005, 200; Delamarre 2007, 134, *s.v. Mogetius.*

[222] Holder II, 1642; Mócsy 1959, 191; Barkóczi 1964, 325; Katičić 1966, 156; Mócsy 1983, 275, *s.v. Suadru, Suadrus*; Lőrincz 2002, 96, *s.v. Svadru, Svadrvs*; Delamarre 2003, 284; Matasović 2003, 14; Meid 2005, 206-207; Delamarre 2007, 172, *s.v. Suadrus, -a, -o.*

[223] Schmidt 1957, 279, 290; Katičić 1966, 157; Mócsy 1983, 292, *s.v. Togiuepus*; Degavre 1998, 414, 440; Lőrincz 2002, 126, *s.v. Togivepvs*; Delamarre 2003, 298, 312-313; Matasović 2003, 14; Meid 2005, 155, 245; Delamarre 2007, 183, *s.v. Togiuepus.*

[224] Katičić 1966, 157; Mócsy 1983, 185, *s.v. Melissus*; Degavre 1998, 302; Lőrincz 2000, 74, *s.v. Melissvs*; Delamarre 2003, 223; Matasović 2003, 14; Delamarre 2007, 131, *s.v. Melissus, -a.*

[225] Holder I, 42, III, 507; Mócsy 1959, 162; Mócsy 1983, 5, *s.v. Adietumarus*; Lőrincz and Redő 1994, 22, *s.v. Adietvmarvs*; Degavre 1998, 27; Delamarre 2003, 32-33; Matasović 2003, 14; Meid 2005, 94; Delamarre 2007, 12, *s.v. Adietumarus.*

[226] Mócsy 1959, 167; Mócsy 1983, 52, *s.v. Bonio*; Lőrincz and Redő 1994, 310, *s.v. Bonio*; Matasović 2003, 14; Meid 2005, 258-259; Delamarre 2007, 45, *s.v. Bonio*, 213-214.

CIL III 11463; Mócsy 1959, 51/2 – *L. Lusius Rucco*[227]

The Varciani

CIL III 9796; Mócsy 1959, 52/2; Dobó 1975, 221 – *Vercaius*[228] *Me[di*[229] *f(ilius)], domo Varcianus*

CIL XVI 4; Mócsy 1959, 52/1; Raybould and Sims-Williams 2007, PAN 045 – *Iantumarus*[230] *Andedunis*[231] *f(ilius) Varcianus*

Odra

Zagreb before Zagreb, Exibition Catalogue, cat. 147; *OpArch* 20, 1996, p. 132; Rendić-Miočević in press – *T. Flavius Aterigis*[232] *fil(ius) Ateboduus,*[233] *veteranus cohortis II Varcianorum equitatae*

The Colapiani

CIL III 4372; Mócsy 1959, 158/6; *RIU* 255 – *Bato*[234] *Buli*[235] *f(ilius) Colapianus, ala Pannoniorum, turma Sceni*[236] *Scenobarvi*[237] *f(ilii)*

CIL III 4376; Mócsy 1959, 158/8; *RIU* 258 – *[_ _ _] Crali*[238] *f(ilius) Colapianus, Lirus*[239] *Plassari*[240] *f(ilius) heres*

CIL III 11227; Mócsy 1959, 155/9 - *[Li]ccaius*[241] *Colapianus*

RMD 250 - *M[. . .]o Batonis*[242] *f(ilio) Cabelo*[243] *Colap[ia] n(o)*

Siscia (Sisak)

CIL III 3994; Barkóczi 1964, 29/34 - *Arsaces*[244]

CIL III 9962; Barkóczi 1964, 29/67; Dobó 1975, 261- *M. Plusi[us?] Ammianus*[245]

CIL VI 32680; Barkóczi 1964, 29/68; Dobó 1975, 61- *M. Aur(elius) Dasius,*[246] *Pann(onius) colon(ia) Siscia*

CIL VI 32640; Barkóczi 1964, 29/75; Dobó 1975, 63/g – *Aurelius M. f(ilius) Tato*[247]

---

227 His cognomen is certainly of Celtic origin (cf. Holder II, 1239; Mócsy 1959, 188; Mócsy 1983, 245, *s.v. Rucco*; Degavre 1998, 359; Lőrincz 2002, 32-33, s.v. Rvcco; Delamarre 2003, 262; Matasović 2003, 14; Meid 2005, 280; Delamarre 2007, 156, *s.v. Rucco*), but we cannot be certain that *L. Lusius Rucco* was a native; according to Mócsy this man could have been from northern Italy.

228 Holder III, 182; Mócsy 1959, 195; Alföldy 1969, 324; Mócsy 1983, 306, *s.v. Vercaius*; Lőrincz 2002, 156-157, *s.v. Vercaivs*; Delamarre 2007, 195, *s.v. Vercaius*.

229 Cf. Mócsy 1983, 184-185; cf. Lőrincz 2000, 72-73; Delamarre 2003, 221-222; Meid 2005, 235; Delamarre 2007, 130-131, *s.v. Medius, Medo, Medos, Medus*, 226.

230 Holder II, 9; Schmidt 1957, 223-224; Mócsy 1959, 176; Mócsy 1983, 148, *s.v. Iantumarus*; Lőrincz 1999, 188-189, *s.v. Iantvmarvs*; Delamarre 2003, 32, 217-218; Matasović 2003, 14; Meid 2005, 109; Delamarre 2007, 107, *s.v. Iantumarus*, 223, 226.

231 Holder III, 614; Mócsy 1959, 164; Mócsy 1983, 18, *s.v. Andedu*; Lőrincz and Redő 1994, 109, *s.v. Andedv*; Delamarre 2007, 154-155; Matasović 2003, 14; Meid 2005, 164-165; Delamarre 2007, 21, *s.v. Andedunis*, 211, 220.

232 Schmidt 1957, 136-141, 260-261; Evans 1967, 142-145, 243-249; Mócsy 1983, 33, *s.v. Aterix*; Lőrincz and Redő 1994, 195, *s.v. Aterix;* Delamarre 2003, 57, 259-260; Meid 2005, 78-79; Delamarre 2007, 29, *s.v. Aterix*, 211-212, 230.

233 Schmidt 1957, 136-141, 152; Evans 1967, 142-145, 151; Mócsy 1983, 32, *s.v. Ateboduus*; Lőrincz and Redő 1994, 191, *s.v. Atebodvvs;* Delamarre 2003, 57, 81; Meid 2005, 165; Delamarre 2007, 28, *s.v. Ateboduus*, 211-213.

234 Krahe 1929, 17-20, *s.v. Bato*; Mayer 1957, 80-82, *s.v. Bato*; Mócsy 1959, 55, 59, 166; Katičić 1962, 110; Katičić 1963, 260; Barkóczi 1964, 307; Katičić 1965, 70; Alföldy 1969, 163-164, *s.v. Bato*; Rendić-Miočević 1971, 167 = Rendić-Miočević 1989, 789; Mócsy 1983, 45, *s.v. Bato*; Domić-Kunić 1988, 87, 89, 101; Lőrincz and Redő 1994, 278-279, *s.v. Bato*; Meid 2005, 29; Radman-Livaja 2010, 187-188.

235 Krahe 1929, 25, *s.v. Bul(i)us*; Mayer 1957, 100 ; Mócsy 1959, 167; Mócsy 1983, 56, *s.v. Bulus*; Lőrincz and Redő 1994, 327, *s.v. Bvlvs*; Radman-Livaja 2010, 196.

236 Krahe 1929, 101; Krahe 1955, 65; Mayer 1957, 313; Mócsy 1959, 189; Katičić 1963, 274-275; Alföldy 1969, 289, *s.v. Scenus*; Mócsy 1983, 256, *s.v. Scenus*; cf. Križman 130; Lőrincz 2002, 55, *s.v. Scenvs*; Delamarre 2007, 162, *s.v. Scenus* ; Radman-Livaja 2010, 440.

237 Krahe 1929, 101; Mayer 1957, 312; Mócsy 1959, 189; Alföldy 1969, 289; Mócsy 1983, 256, *s.v. Scenobarbus*; Lőrincz 2002, 54, *s.v. Scenobarbvs*.

238 The name is a *hapax*, presumably of Pannonian origin, Mócsy 1959, 170; Mócsy 1983, 91, *s.v. Cralus*; Lőrincz 1999, 82, *s.v. Cralvs*.

239 Krahe 1929, 68, *s.v. Lirus*; Mayer 1957, 211; Mócsy 1959, 178; Mócsy 1983, 165, *s.v. Lirus*; Lőrincz 2000, 28, *s.v. Lirvs*; Radman-Livaja 2010, 329-330. The name is hardly of Celtic origin, cf. Holder, II, 239.

240 Krahe 1929, 92; Mócsy 1959, 184; Mócsy 1983, 226, *s.v.Plassarus*; Lőrincz 2000, 145, *s.v. Plassarvs*; Radman-Livaja 2010, 395.

241 Krahe 1929, 66-67, *s.v. Licaius, Licaus, Liccaeus, Liccaius, Liccaus*; Krahe 1955, 58; Mayer 1957, 210-211, *s.v. Lic(c)a(v)ius, Lic(c)avus*; Mócsy 1959, 55, 178; Katičić 1965, 70-71; Alföldy 1969, 230, *s.v. Licaeus, Liccaius*; Rendić-Miočević 1981, 27 = Rendić-Miočević 1989, 755; Mócsy 1983, 164, *s.v. Liccaius*; Lőrincz 2000, 26, *s.v. Liccaivs* ; Meid 2005, 29, 284 ; Radman-Livaja 2010, 327; it is more likely not a celtic name, cf. Delamarre 2007, 117, *s.v. Liccaius, -aeus*.

242 *Vide supra.*

243 *Cabelus* is a *hapax*, cf. Krahe 1929, 27, *s.v. Cabaletus*; Lőrincz 1999, 15, *s.v. Cabaletvs, Cabalio, Cabalionivs, Caballivs, Caballvs*; Delamarre 2007, 51, *s.v. Cabalio, Cabalius, -llus, Caballos, Cabellio, Cabillus*.

244 A Celtic origin is uncertain, but it cannot be excluded, cf. Barkóczi 1964, 306; Delamarre 2007, 25, *s.v. Aresaces*, 26, *s.v. Arsaces*.

245 The name is not necessarily Celtic but in some cases it is quite plausible, cf. Alföldy 1969, 148; Mócsy 1983, 16, *s.v. Ammianus*; Lőrincz and Redő 1994, 98, *s.v. Ammianvs*; Delamarre 2003, 43; Meid 2005, 215-216; Delamarre 2007, 19, *s.v. Ammianus*, 210.

246 Krahe 1929, 37-38, *s.v. Das(s)ius*; Mayer 1957, 112-114, *s.v. Das(s) ius)*; Mócsy 1959, 55, 57, 171; Katičić 1963, 268-269; Katičić 1965, 70; Zaninović 1966 , 53-54 ; Alföldy 1969, 185-186, *s.v. Dassius, Dassius*; Rendić-Miočević 1981, 26 = Rendić-Miočević 1989, 755; Mócsy 1983, 98, *s.v. Dasius, Dassius*; Solin 1996, 613, *s.v. Dasius*; Lőrincz 1999, 93, 215, *s.v. Dasivs*; Meid 2005, 29; Radman-Livaja 2010, 252-253; this name is most likely not related to Celtic anthroponymy, *contra* Delamarre 2007, 82, *s.v. Das(s)ius*.

247 Holder II, 1751; Krahe 1929, 111; Mayer 1957, 330; Alföldy 1969, 305; Barkóczi 1964, 325; Mócsy 1983, 283, *s.v. Tato, Tatto*; Lőrincz 2002, 109, *s.v. Tato, Tatto*; Meid 2005, 291-292; Delamarre 2007, 179, *s.v. Tato, Tatto*.

CIL III 12014(529) (graffito on a ceramic vessel) - *Sipa*[248]

Topusko

ILJug 3112 - *Bradua*[249]

Podgraci Donji

AIJ 591; Mócsy 1959, 58/1 - *[O]ginia*[250]

Križovljan

CIL III 4110; Mócsy 1959, 63/1; - *Ulpia Vinda*[251]

Poetovio (Ptuj)

AÉ 1948, 239; Mócsy 1959, 64/38 – *C. Novellius Vitalis, Valeria Satonia*[252]

CIL III 4083; *AIJ* 406; Mócsy 1959, 64/46 – *M. Ulpius M. f. Cutio,*[253] *Medullia*[254] *Matiera*[255]

CIL III 4084; *AIJ* 407; Mócsy 1959, 64/52 – *M. Ulpius Tertullus, Ulpia Antonia, Ulpius Tutor,*[256] *Ulpius Tertius, Ulpius Tutorinus,*[257] *Ulpia Tutoriana*[258]

CIL III 10883; Mócsy 1959, 64/68; Raybould and Sims-Williams 2007, PAN 036 - *Deuso*[259] *Agisi*[260] *f(ilius), Adbugiouna*[261] *Atnamati*[262] *f(ilia), Primio*

CIL III 4033; Barkóczi 1964, 38/19 – *Aur(elius) Ceionus,*[263] *Speciatia Aisia*[264]

CIL III 13412; *AIJ* 394; Barkóczi 1964, 38/25 – *Aur(elius) Verinus, Cartoria Calandina,*[265] *Aurelia Vera, Aurelia Verecunda*

AIJ 2654; Barkóczi 1964, 38/46 – *M. Messius*[266] *Messor*[267]

Starše

CIL III 4103; AIJ 428; Mócsy 1959, 66/1 - . . . *ius Suc. f., . . . mus, Suadra*[268]

Čakovec

CIL III 4116; *AIJ* 457; Mócsy 1959, 68/1 – *Claudia T. f(ilia) Dagumena*[269]

---

248 Krahe 1929, 105; cf. Radman-Livaja 2010, 459.

249 Mócsy 1983, 53, *s.v. Bradua*; Lőrincz and Redő 1994, 317, *s.v. Bradva*; Solin and Salomies 1994, 303; Delamarre 2007, 47, *s.v. Bradua.*

250 Mócsy 1959, 157; Mócsy 1983, 207, *s.v. Oginius*; Lőrincz 2000, 111, *s.v. Oginivs*; Meid 2005, 238; Delamarre 2007, 144, *s.v. Oginius,-ia.*

251 Mócsy 1959, 28, 59, 197; Katičić 1966, 152, 156; Evans 1967, 386-387; Mócsy 1983, 313, *s.v. Vindus*; Degavre 1998, 450, s.v. *vindo-* ; Lőrincz 2002, 172, *s.v. Vindvs*; Delamarre 2003, 320-321, *s.v. uindos*; Matasović 2003, 14, 20; Meid 2005, 210-211; Delamarre 2007, 200, *s.v. Vinda, Vindus.*

252 We cannot be certain that *Valeria Satonia* was a native; Holder II, 1375; Mócsy 1959, 189; Mócsy 1983; Lőrincz 2002; Matasović 2003, 14; Meid 2005, 283-284; Delamarre 2007, 161, *s.v. Sattonius.*

253 Holder, I, 1208-1209; Mócsy 1959, 28, 171; Kajanto 1965, 164; Katičić 1966, 156; Mócsy 1983, 96, *s.v. Cutio*; Solin and Salomies 1994, 321, *s.v. Cutio*; Degavre 1998, 179, *s.v. cutios*; Lőrincz 1999, 90, *s.v. Cvtio*; Delamarre 2003, 133-134, *s.v. cutios*; Matasović 2003, 14, 19; Meid 2005, 267, *s.v. Cutio* ; Delamarre 2007, 80, *s.v. Cutius*, 218; Radman-Livaja 2010, 247.

254 Mócsy 1959, 156; Katičić 1966, 156; Mócsy 1983, 184, *s.v. Medullius*; Lőrincz 2000, 72, *s.v. Medvllivs*; Matasović 2003, 14; Meid 2005, 235-236; cf. Delamarre 2007, 131, *s.v. Medulli, Medulio, Meduliuia.*

255 Mócsy does not exclude the reading *Materna,* (Mócsy 1959, 181); cf. Meid 2005, 236; Delamarre 2007, 128

256 Mócsy 1959, 193; Katičić 156; Mócsy 1983, 297, *s.v. Tutor*; Lőrincz 2002, 135, *s.v. Tvtor*; Matasović 2003, 14; Meid 2005, 297-298.

257 *Vide supra, s.v. Tutor.*

258 *Vide supra, s.v. Tutor.*

259 Holder I, 1273; Mócsy 1959, 172; Katičić 1966, 156; Mócsy 1983, 101, *s.v. Deuso*; Lőrincz 1999, 98, *s.v. Devso*; Matasović 2003, 14; Meid 2005, 269; Delamarre 2007, 85, *s.v. Deusus, -a, -o.*

260 Holder I, 59; Mócsy 1959, 163; Katičić 1966, 155; Mócsy 1983, 9, *s.v. Agisus*; Lőrincz and Redő 1994, 56, *s.v. Agisvs*; Matasović 2003, 14; Meid 2005, 250; Delamarre 2007, 15, *s.v. Agisus.*

261 Holder I, 38, 44, III, 503; Schmidt 1957, 111; Mócsy 1959, 162; Katičić 1966, 155; Mócsy 1983, 4, *s.v. Adbugiounus*; Lőrincz and Redő 1994, 20, *s.v. Adbvgiovnvs*; Delamarre 2003, 94; Matasović 2003, 14; Meid 2005, 157-158; Delamarre 2007, 11, *s.v. Adbugiounus, -a.*

262 Holder I, 43-44; Schmidt 1957, 114-115; Mócsy 1959, 162; Katičić 1966, 155; Evans 1967, 128-131, 234-236; Mócsy 1983, 5, *s.v. Adnamatus*; Lőrincz and Redő 1994, 26, *s.v. Adnamatvs*; Matasović 2003, 14; Meid 2005, 231; Delamarre 2007, 12, *s.v. Adnamatus.*

263 Barkóczi 1964, 308-309; Katičić 1966, 156; Mócsy 1983, 73, *s.v. Ceionius*; cf. Lőrincz 1999, 46, *s.v. Ceionivs*; Matasović 2003, 14; Delamarre 2007, 62, *s.v. Ceionus.*

264 Barkóczi 1964, 304; Katičić 1966, 155; Mócsy 1983, 10, *s.v. Aisia*; Lőrincz and Redő 1994, 61, *s.v. Aisia*; Matasović 2003, 14; Delamarre 2007, 15, *s.v. Aisia.*

265 Barkóczi 1964, 308; Katičić 1966, 156; Matasović 2003, 14.

266 The name is related both to the Romanised Celtic and Illyrian anthroponymy, cf. Krahe 1929, 73-75; Mayer 1957, 226-229; Evans 1967, 367-368; Alföldy 1969, 247, *s.v. Messius, Mesius,Messus*; Rendić-Miočević 1981, 27 = Rendić-Miočević 1989, 755; Mócsy 1983, 188, *s.v. Messius, Messus*; Križman 1991, 129; Solin and Salomies 1994, 118, *s.v. Messius*; Degavre 1998, 304, s.v. *međđi-*; Lőrincz 2000, 78-79, *s.v. Messivs, Messvs*; Delamarre 2003, 223, *s.v. međđu-, međđi-, messi-*; Delamarre 2007, 133, *s.v. Mesus*, 227; Radman-Livaja 2010, 355-356.

267 Outside of Italy this Latin cognomen is often present both in the Romanised Celtic and Illyrian-Pannonian anthroponymy, cf. Krahe 1929, 74; Mayer 1957, 229; Schmidt 1957, 240-241; Barkóczi 1964, 318, *s.v. Messor*; Kajanto 1965, 82, 361; Zaninović 1966, 49; Evans 1967, 367-368; Alföldy 1969, 247, *s.v. Messor*; Mirković 1971, 25; Mócsy 1983, 188, *s.v. Messor*; Solin and Salomies 1994, 362, *s.v. Mesor*, 363, *s.v. Messor*; Degavre 1998, 304, *s.v. međđi-*; Lőrincz 2000, 79, 178, *s.v. Messor*; Delamarre 2003, 223, *s.v. međđu-, međđi-, messi-*; Delamarre 2007, 227; Radman-Livaja 2010, 355.

268 *Vide supra.*

269 Holder I, 1216; Schmidt 1957, 186-187; Mócsy 1959, 171; Evans 1967, 188-189; Mócsy 1983, 97, *s.v. Dagumenus*; Lőrincz 1999, 91, *s.v. Dagvmenvs*; Delamarre 2003, 134, 224, 226-227; Matasović 2003, 14; Meid 2005, 149; Delamarre 2007, 80, *s.v. Dacumena*, 218, 227.

Aquae Balissae (Daruvar)

CIL III 4002; Barkóczi 1964, 30/4 - *M. Fl(avius) Maurus*[270]

CIL VI 3297; Barkóczi 1964, 30/5; Dobó 1975, 115 - *Ulpius Cocceius,*[271] *eques singularis, natus ad Aquas Balizas, pago Iovista*

Brusnik

AIJ 590; Barkóczi 1964, 30a/1 – *Aurelius Naso, Prisca Tatonis*[272] *f(ilia), Aurelius Proclus, Proclianus, Provincialis, Maximianus*

Beleg

CIL XVI 31; Mócsy 1959, 71/1 – *Fronto Sceni*[273] *f(ilius) Iasus*

III. PANNONIA INFERIOR

The Breuci

CIL III p. 940 (*tabella cerata Dacica* n. VII); Mócsy 1959, 234/6; Dobó 1975, 267 –*Dassius*[274] *Breucus*

CIL III 4377; Mócsy 1959, 158/7 – *Scilus*[275] *Batonis*[276] *f(ilius) Breucus*

AÉ 1915, 69; Mócsy 1959, 234/5; Dobó 1975, 214 – *T. Flavius Breucus*[277], *veteranus alae I Pannoniorum*

CIL VIII 20150; Mócsy 1959, 234/5; Dobó 1975, 266 – *M. Flavius Breucus* (the son of *Titus Flavius Breucus* ?)

CIL VIII 21041[278]; Mócsy 1959, 238/9; Dobó 1975, 213, 268a, 367a – *Licco*[279] *Burnionis*[280] *f(ilius) Panno[nius eques], Breucus Landionis*[281]

CIL XII 7801-7802; Mócsy 1959, 234/2; Dobó 1975, 217, 458 – *Dasmenus*[282] *Dasi*[283] *f(ilius) Breucus*

CIL XIII 5378, Mócsy 1959, 234/7; Dobó 1975, 269 – *Breuci l(iberti) Noni*

CIL XIII 7510; Mócsy 1959, 234/1; Dobó 1975, 212, 356a - *[B]reucus Blaeda[ri*[284] *f(ilius)] natione Breucus*

CIL XIII 8313; Mócsy 1959, 234/4; Dobó 1975, 218, 455 – *Sassaius*[285] *Liccai*[286] *f(ilius)*

CIL XIII 8693; *ILS* 2559; Mócsy 1959, 234/3; Dobó 1975, 216, 457 – *Marcinus*[287] *Surconis*[288] *f(ilius) Breucus*

CIL XIII 12062; Dobó 1975, 456 - *Scen[obarbus]*[289] *Licc[onis*[290] *f(ilius) Bre]uc[us mil(es) ex coh(orte) VIII Breucorum _ _ _ _]*

AÉ 1945, 78[291]; Dobó 1975, 215, 451a – *Pradus*[292] *Sarni*[293] *f(ilius) miles ex cohorte VII Breucorum*

RMD 79 – *Liccaius*[294] *Liccai f(ilius) Breucus*

---

[270] Barkóczi 1964, 318; Kajanto 1965, 50, 206; Mócsy 1983, 183, *s.v. Maurus*; Solin and Salomies 1994, 361, *s.v. Maurus*; Lőrincz 2000, 68, *s.v. Mavrvs*; Delamarre 2007, 130, *s.v. Maurus.*

[271] Mócsy 1959 , 27, 170 ; Mócsy 1983, 83, *s.v. Cocceius*; Lőrincz 1999, 67, *s.v. Cocceivs*; Minkova 2000 , 139-140, *s.v. Cocceius* ; Meid 2005, 226-227; Delamarre 2007, 69, *s.v. Cocceius* ; Radman-Livaja 2010, 227.

[272] *Vide supra.*

[273] *Vide supra.*

[274] *Vide supra.*

[275] Krahe 1929, 102, *s.v. Scilus*; Mayer 1957, 314; Mócsy 1959, 189; Mócsy 1983, 256, *s.v. Scilus*; Lőrincz 2002, 55, *s.v. Scilvs*; Meid 2005, 29; Radman-Livaja 2010, 441.

[276] *Vide supra.*

[277] Krahe 1929, 24, *s.v. Breucus*; Mócsy 1959, 55, 75, 167; Barkóczi 1964, 307; Rendić-Miočević 1981, 26, 28-30 = Rendić-Miočević 1989, 754, 757; Mócsy 1983, 54, *s.v. Breucus*; Lőrincz and Redő 1994, 319, *s.v. Brevcvs*; Solin 1996, 39, *s.v. Breucus*; Radman-Livaja 2010, 194-195.

[278] Considering his name, it does not seem implausible that *Breucus Landionis f(ilius)* was himself a native of the *Breuci* tribe. *Licco Burnionis f(ilius)* was certainly a Pannonian; we can assume that he was from southern Pannonia but we cannot be certain that he was a *Breucus*, although this is not unlikely.

[279] Krahe 1929, 67, *s.v. Licco;* Mayer 1957, 211; Mócsy 1959, 178; Mócsy 1983, 164, *s.v. Licco;* Lőrincz 2000, 26, *s.v. Licco*; the celtic origin of the name does not seem very likely, *contra* Delamarre 2007, 117, *s.v. Licco*

[280] Holder I, 641, III, 1007; Krahe 1929, 26; Mayer 1957, 102; Mócsy 1959, 167; Delamarre 2007, 50, *s.v. Burnio.*

[281] Krahe 1929, 62; Mayer 1957, 204; Mócsy 1959, 177; cf. *RMD* 20; cf. Mócsy 1983, 159, *s.v. Landio*; cf. Lőrincz 2000, 18, *s.v. Landio.*

[282] Krahe 1929, 36; Mayer 1957, 114; Mócsy 1959, 55, 171; Katičić 1963, 268-269; Mócsy 1983, 98, *s.v. Dasmenus*; Lőrincz 1999, 93, *s. v. Dasmenvs*; Meid 2005, 29; Radman-Livaja 2010, 253; X. Delamarre does not exclude the possibility of a Celtic origin for this name, but this does not seem likely, cf. Delamarre 2007 , 82, *s.v. Dasmenus, ius.*

[283] *Vide supra.*

[284] Krahe 1929, 22, *s.v. Blaedarus;* Mócsy 1959, 166; Mócsy 1983, 50, *s.v. Blaedarus;* Lőrincz and Redő 1994, 300, *s.v. Blaedarvs*; the name might be of Celtic origin (cf. Delamarre 2003, 78-79; Delamarre 2007, 43, *s.v. Blaedarus*), but W. Meid does not consider it in his corpus of Celtic names in Pannonia.

[285] It would seem that this name could be related both to the Celtic and Illyrian-Pannonian anthroponymy, cf. Holder II, 1373; Krahe 1929, 100, *s.v. Sasaius*; Mócsy 1959, 189; Mócsy 1983, 254, *s.v. Sassaius*; Lőrincz 2002, 49, *s.v. Sassaivs;* Meid 2005, 284; Delamarre 2007, 161, *s.v. Sassaius.*

[286] *Vide supra.*

[287] This name could belong to the Celtic anthroponymy; Mócsy 1959, 180; Mócsy 1983, 178, *s.v. Marcinus*; Lőrincz 2000, 56, *s.v. Marcinvs*; Delamarre 2003, 216-217; Delamarre 2007, 126, *s.v. Marcinus.*

[288] Holder II, 1673; Krahe 1929, 107, *s.v. Surco;* Mayer 1957, 325; Mócsy 1959, 192; Mócsy 1983, 278, s.v. Surco; Lőrincz 2002, 101, *s. v. Svrco;* a celtic origin is not to be excluded, cf. Delamarre 2007, 175, *s.v. Surco.*

[289] *Vide supra.*

[290] *Vide supra.*

[291] *Pradus* might have been a *Breucus.*

[292] This name is a *hapax,* perhaps of Illyrian origin, but Krahe and Mayer do not mention it; Mócsy 1983, 231, *s.v. Pradus;* Lőrincz 2000, 157, *s. v. Pradvs.*

[293] Krahe 1929, 100, *s.v. Sarnus*; Mócsy 1983, 253, *s.v. Sarnus;* Lőrincz 2002, 49, *s. v. Sarnvs;* Delamarre believes that this name could be celtic, cf. Delamarre 2007, 161, *s.v. Sarnus.*

[294] *Vide supra.*

RMD 202 – *Dasius*[295] *Carmai*[296] *f(ilius) Breucus*

RMD 204 – *Liccaius*[297] *Birsi*[298] *f(ilius), Marsunnia*

RMD 205 – *Licco*[299] *Davi*[300] *f(ilius) princ(eps) Breucor(um)*

Mursa (Osijek)

CIL III 3289; Barkóczi 1964, 155/20 - *Augenis*[301]

CIL VI 3373; Barkóczi 1964, 155/28; Dobó 1975, 158 - *Aurelius Dassius*[302]

CIL VI 32628; Barkóczi 1964, 155/37; Dobó 1975, 29 - *P. Aelius Scantius*[303]

CIL VI 32542 (2833); Barkóczi 1964, 155/33; Dobó 1975, 22 - *M. Aur(elius) M. f. Cassius, M. Aur(elius) M. f. Genia(lis), M. Aur(elius) M. f. Fuscinus, M. Aur(elius) M. f. Marcianus, M. Aur(elius) M. f. Festianus, M. Aur(elius) M. f. Vitalis, M. Aur(elius) M. f. Potazis*[304]

VHAD 6 (1902), 99; Barkóczi 1964, 155/23; *ILJug* 3097 - *Messius*[305]

ILJug 292; Raybould and Sims-Williams 2007, PAN 071 - *Urbana Aterigis*[306] *l(iberta), Gaetulus*[307] *Urbanae f(ilius) Aterigis l(ibertus), [A]rcto*[308] *heres*

ILJug 3108 - *C. Aelius Surinus,*[309] *dec(urio) col(oniae) M(ursensium)*

The Andizetes

RMD 205[310] – *Breucus*[311] *Isticani*[312] *f(ilius) princ(eps) Antizit(ium)*

Grabarje

CIL XVI 17; Mócsy 1959, 220/1 - *[_ _ _] Lensi*[313] *f(ilius) Pannonius*

Cibalae (Vinkovci)

CIL VI 32542; Barkóczi 1964, 159/5; Dobó 1975, 22, 29/c - *M. Aur(elius) M. f(ilius) Dalutius,*[314] *M. Aur(elius) M. f(ilius) Deazius,*[315] *M. Aur(elius) M. f. Dassianus,*[316] *M. Aur(elius) M. f. Dolea*[317](praetorians, 223. AD)

AIJ 557; Mócsy 1959, 57/10[318] – *Lucius Larcius,*[319] *Secundus Vacarus*[320]

Bassianae (Petrovci)

CIL III 15134; Mócsy 1959, 230/2; *ILJug* 1035 – *M. Ulpius Dasius*[321]

CIL III 10212; Mócsy 1959, 230/4 - *Dasmenus*[322]

Belegiš

CIL III 15138[4]; Barkóczi 1964, 166/1 – *Aur(elia) Disala*[323]

---

295 *Vide supra.*
296 Illyrian-Pannonian or Celtic name?, cf. Lőrincz 1999, 37, *s. v. Carmaevs.*
297 *Vide supra.*
298 Alföldy 1969, 165 (*s.v. Bersa*); Mócsy 1983, 49, *s.v. Birsus*; Lőrincz and Redő 1994, 296, *s.v. Birsvs*; ; Delamarre 2007, 42, *s.v. Birsus.*
299 *Vide supra.*
300 This name could be of Celtic or Illyrian origin, Lőrincz 1999, 94, *s.v. Davivs*; Delamarre 2007, 83, *s.v. Dauius*; Radman-Livaja 2010, 254-255.
301 This name could be related to the Celtic anthroponymy, cf. Barkóczi 1964, 306.
302 *Vide supra.*
303 Cf. Mócsy 1983, 256, *s.v. Scantius*; cf. Lőrincz 2002, 54, *s. v. Scantivs*; Delamarre 2007, 162, *s.v. Scantius.*
304 A cognomen of Thracian origin?, cf. Barkóczi 1964, 320.
305 *Vide supra.*
306 *Vide supra.*
307 Unless his name was not inspired by one army unit stationed in Pannonia (*ala I Flavia Gaetulorum*) and if we exclude an African origin (cf. Kajanto 1965, 50, 206; Mócsy 1983, 132, *s.v. Gaetulus*; Solin and Salomies 1994, 336, *s.v. G(a)etulus*; Lőrincz 1999, 158, *s.v.Gaetvlvs*; Radman-Livaja 2010, 301-302), the possibility remains that this was a Celtic name, cf. Delamarre 2007, 104, *s.v. Getulus.*
308 This name is a *hapax* and it is therefore difficult to tell more about its origin but we can presume its Celtic affiliation, cf. Delamarre 2007, 24-25, 211.
309 This name is a variation (cf. Kajanto 1965, 113-114) of a very common name, i.e. the cognomen *Surus* (which can be related both to the Celtic and Illyrian anthroponymy (as well as Thracian and semitic), cf. Pape and Benseler 1870, 1463, s.v. Σύρος; Holder, II, 1678; Krahe 1929, 108-109; Mayer 1957, 326; Mócsy 1959, 33, 192; Barkóczi 1964, 325; Lochner-Hüttenbach 1965, 36; Evans 1967, 472-473; Alföldy 1969, 303, *s.v. Surus*; Mócsy 1983, 278, *s.v. Surus*, 279, *s.v. Syrvs*; Križman 1991, 131, 142; Solin 1996, 376, *s.v. Syrus*, 614, *s.v. Sura, Surus*; Degavre 1998, 396, *s.v. sur-/suri-*; Minkova 2000, 259-260, *s.v. Sura, Surus*; Lőrincz 2002, 102, 197, *s.v. Svrvs*, 104, *s.v. Syrvs*; Solin 2003, 668-669, *s.v. Syrus, Syra*; Meid 2005, 289; Delamarre 2007, 175, *s.v. Surius*; Radman-Livaja 2010, 468-470.
310 Only the transcript of this diploma is known, while the patronymic seems doubtful.
311 *Vide supra.*
312 A *hapax.*
313 Illyrian-Pannonian name?, cf. Mócsy 1959, 178; Mócsy 1983, 161, *s.v. Lensus*; Lőrincz 2000, 22, *s.v. Lensvs.*
314 A cognomen of thracian or dacian origin?, cf. Barkóczi 1964, 310.
315 Likely a name of Illyrian origin, cf. Barkóczi 1964, 310.
316 Krahe 1929, 35-38; Mayer 1957, 111-112; Mócsy 1983, 98, *s.v. Dasianus, Dassianus*; Lőrincz 1999, 93 *s.v. Dasianvs, Dassianvs.*
317 A cognomen of Thracian or Dacian origin?, cf. Barkóczi 1964, 310.
318 According to the latest reading of this *defixio* from Siscia, six men are mentioned by name, and their origin is also stated for some of them (*G. Domitius Secundus*, two men from Cibalae, *Lucius Larcius* and *Secundus Vacarus*, *P. Citronius Cicorellius* from Narbo, the Spaniard *Licinius Sura* and *Lucilius Valens*), cf. Simón and Rodá de Lanza 2008, 167-182.
319 Cf. Delamarre 2007, 115, *s.v. Lartius.*
320 Mócsy 1983, 299, *s.v. Vacarus*; Lőrincz 2002, 137, *s.v. Vacarvs*; Meid 2005, 302 (he does not exclude the possibility that this might be a Pannonian name); Delamarre 2007, 187, *s.v. Vacarus.*
321 *Vide supra.*
322 *Vide supra.*
323 Delamarre 2007, 86, *s.v. Disala*; or Thracian?, cf. Barkóczi 1964, 310.

Susek

CIL III 10249 (3685); Mócsy 1959, 224/1; Mirković 1971, cat. 89 - *T. Flavius T. f(ilius) Severus Gogaenus,*[324] *dec. col. Sirm.*

Sirmium (Sremska Mitrovica)

CIL III 14507; Barkóczi 1964, 169/48; Dobó 1975, 175 - *M. Ulp(ius) Messor*[325]

CIL VI 3184; Barkóczi 1964, 169/51 [*Siscia (sic, recte Sirmio)*]; Dobó 1975, 103 - *P. Aelius Surio*[326]

CIL VI 32536; Barkóczi 1964, 169/43; Dobó 1975, 34 – *C. Flaviu[s] Messianus*[327] *Sir(mium)*

CIL X 3375; Mócsy 1959, 233/10; Dobó 1975, 124 – *L. Valerius Dazantis*[328] *f(ilius) (H)ispanus*

Mirković 1971, cat. 32; *ILJug* 273 - *Aurelia Bona*[329]

Mirković 1971, cat. 35; *ILJug* 274 – *Verina*[330]

ILJug 1048 - *P. Aelius Dasius,*[331] *dec(urio) m(unicipi) B(assianarum)*

Mirković 1971, cat. 59; *ILJug* 3022 - *[Bo?]nonia*[332]

Mirković 1971, cat. 74 – *Gresa*[333] *Dasm[eni],*[334] *Masti*[335] *liberta*

Dobó 1975, 198a; *RMD* 21 – *Glavus*[336] *Navati*[337] *f(ilius)*

Bijela Crkva

CIL XVI 2; Mócsy 1959, 237/1 – *Dases*[338] *Dasmeni*[339] *f(ilius) Cornac(atus), Iora*[340] *Prososii*[341] *f(ilia), Turuna*[342]

The Scordisci

CIL III 3400; Mócsy 1959, 189/2 – *Terso*[343] *Precionis*[344] *f(ilius) Scordiscus*

---

[324] This name could be related to the Celtic anthroponymy, cf. Holder I, 2031; Mócsy 1959, 176; Mirković 1971, 25; Matasović 2003, 15.
[325] *Vide supra.*
[326] Mócsy 1983, 278, *s.v. Surio*; Lőrincz 2002, 101, *s.v. Svrio*; Delamarre 2007, 175, *s.v. Surio*; Radman-Livaja 2010, 467-468.
[327] This cognomen is Latin but it can be related both to Illyrian and Celtic anthroponymy, cf. Holder II, 575; Krahe 1929, 73; Mayer 1957, 227-228; Barkóczi 1964, 318, *s.v. Messianus*; Kajanto 1965, 150; Evans 1967, 367-368; Alföldy 1969, 246-247, *s.v. Messianus;* Mirković 1971, 25; Mócsy 1983, 187, *s.v. Messianus;* Solin and Salomies 1994, 362, *s.v. Messianus;* Degavre 1998, 304, *s.v. međđi-*; Lőrincz 2000, 78, *s.v. Messianvs*; Delamarre 2003, 223, *s.v. međđu-, međđi-, messi-*.
[328] Krahe 1929, 34-35; Mayer 1957, 109; Mócsy 1959, 171; Katičić 1963,268; Katičić 1965, 70; Zaninović 1966, 48, 53; Alföldy 1969, 185, *s.v. Dasas, Dazas*; Mirković 1971, 25; Rendić-Miočević 1971, 167 = Rendić-Miočević 1989, 789; Rendić-Miočević 1981, 26 = Rendić-Miočević 1989, 755; Lőrincz 1999, 93, *s.v. Dasas*; Meid 2005, 29; Radman-Livaja 2010, 251.
[329] Although it could be a Latin cognomen, this name is quite likely related to the Celtic anthroponymy, cf. Kajanto 1965, 274; Mócsy 1983, 52, *s.v. Bonnus, Bonus*; Lőrincz and Redő 1994, 310, *s. v. Bonnvs*, 311, *s.v. Bonvs*; Solin and Salomies 1994, 303, *s.v. Bonus*; Delamarre 2003, 82; Meid 2005, 258-259; Delamarre 2007, 45-46.
[330] Schmidt 1957, 290-291; Mócsy 1959, 196; Barkóczi 1964, 295, 327; Kajanto 1965, 254; Evans 1967, 279-280; Alföldy 1969, 324, *s.v. Verinus*; Mócsy 1983, 307, *s.v. Verinus*; Solin and Salomies 1994, 420, *s.v. Verinus*; Degavre 1998, 452, *s.v. viro-, vero-*; Minkova 2000, 275, *s.v. Verina*; Forier 2001, 481; Lőrincz 2002, 158-159, *s.v. Verinvs*; Delamarre 2003, 307, *s.v. uarina*; Meid 2005, 305; Delamarre 2007, 236; Radman-Livaja 2010, 504.
[331] *Vide supra.*
[332] It is perhaps not a personal name, cf. Holder, I, 487; Schmidt 1957, 154; Mócsy 1959, 152; Barkóczi 1964, 300; Kajanto 1965, 50, 198 ; Katičić 1966, 154; for similar names cf. 148, 150, 152, 154, 156; Lőrincz and Redő 1994, 310, *s.v. Bononivs*; Solin and Salomies 1994, 36, 303, *s.v. Bononius*; Delamarre 2003, 84; Meid 2005, 258-259; Delamarre 2007, 45, *s.v. Bononius, -ia*; Radman-Livaja 2010, 193.

---

[333] Krahe 1929, 55; Mayer 1957, 152; Mócsy 1983, 139, *s.v.Gresa*; Lőrincz 1999, 171, *s.v. Gresa.*
[334] *Vide supra.*
[335] Thracian or Illyrian? It is far from being certain that she was a native, cf. Mirković 1971, 25; Radman-Livaja 2010, 347.
[336] Delamarre 2007, 105, *s.v. Glauus*; Ion Russu (1973, 71-72) assumed it was an Illyrian name while *Glavus'* father had a Celtic or Illyrian name.
[337] Delamarre 2007, 139, *s.v. Nauatus.*
[338] Krahe 1929, 34-35; Mayer 1957, 109; Katičić 1963, 268; Katičić 1965, 70; Zaninović 1966, 48, 53; Alföldy 1969, 185, *s.v. Dasas, Dazas*; Rendić-Miočević 1971, 167 = Rendić-Miočević 1989, 789; Rendić-Miočević 1981, 26 = Rendić-Miočević 1989, 755; Lőrincz 1999, 93, *s.v. Dasas*; Meid 2005, 29; Radman-Livaja 2010, 251; the orthography *Dases* or *Dasses* seems to have been more common in Pannonia, cf. Krahe 1929, 35; Mayer 1957, 111; Mócsy 1959, 55, 57, 171; Lőrincz 1999, 93, *s.v. Dases.*
[339] *Vide supra.*
[340] This name might be Illyrian, cf. Mócsy 1959, 177; Lőrincz 1999, 196, *s.v. Iora.*
[341] Illyrian?, cf. Mócsy 1959, 186; Lőrincz 2000, 168, *s.v. Prosostus*; Colombo 2010, 196.
[342] Holder, II, 2020; Mócsy 1983, 297, *s.v. Turuna*; Lőrincz 2002, 134, *s.v.Tvrvna*; Delamarre 2007, 186, *s.v. Turuna*, 234; This name could be Celtic but since it is a *hapax*, it is hard to be certain about its origin. Although it is not mentioned by Krahe and Mayer, these scholars list many names with the same root. Thus, we cannot exclude the possibility that the name *Turuna* could belong to the Illyrian-Pannonian anthroponymy, cf. Krahe 1929, 120-121; Mayer 1957, 345-347; Matasović 2003, 15.
[343] Illyrian?, cf. Holder, II, 1798; Mócsy 1959, 192; Mócsy 1983, 285, *s.v. Terso*; Lőrincz 2002, 114, *s.v. Terso.*
[344] Illyrian?, cf. Holder, II, 1042; Mócsy 1959, 185; Mócsy 1983, 231, *s.v. Precio*; Lőrincz 2000, 157, *s.v. Precio.*

The Roman Army

Ivan Radman-Livaja

1. Introduction

The history of the Roman army in Pannonia is certainly not a topic that has been neglected in the literature, either in older publications or those more recent. In addition to studies devoted to certain military camps or Roman military units and archaeological finds that may be connected to the presence of the Roman army in the respective areas, there are also several more comprehensive reviews dealing with this topic in more detail. One may, therefore, rightly ask what this work could offer to the well-informed reader.[1] The fact is, however, that the existing literature places far more emphasis on the Roman military in Pannonia north of the Drava River (today's Hungary). This is understandable considering the degree to which that area has been archaeologically researched, but it is also due to the unquestionably greater interest of Hungarian archaeologists for this topic than was the case in Croatia. This is why I feel it would be useful to present a brief synthesis of the Roman military units that were present in the Pannonian region south of the Drava for over 4 centuries, as it is precisely this segment of Roman military history that is under-represented in the literature (fig. 1; for the distribution of ethnic communities in southern Pannonia see in this volume: Dizdar, fig. 1; Dzino and Domić Kunić, fig.1; Radman-Livaja, Anthroponymy, fig. 1).

2. Before the conquest

Unlike the coastal area, the northern part of today's territory of Croatia came into the sphere of interest of the Romans rather late. The Pannonian area could never be as attractive to Roman traders as, for example, Noricum with its rich mineral deposits. Consequently, only exceptionally did the Roman Republic pay any attention to the peoples inhabiting what is today central and northern Croatia, namely, only when they would cause disturbance. No real threat existed before the 2nd century BC, but after the establishment of Aquileia the north-western part of Italy became an attractive target for the looting raids of the Taurisci and the Iapodes. If to this we add the problems created by the inhabitants of the Julian Alps on the roads to Noricum, it becomes clear that the Roman Republic had to start showing some force and organise occasional military campaigns, or rather, punitive expeditions targeted at the Pannonian area.[2] After the conquest of Histria in 177 BC, the Iapodes, it seems, became the principal target of Roman military operations. It thus happened that in 171 BC, acting independently and without approval from the Senate, the consul Gaius Cassius Longinus ravaged the land of the Carni in the Julian Alps and the Iapodian areas.[3] A few decades later, the Romans reached Pannonia for the first time, presumably in 159 BC or 156 BC, when Segestica seems to have been attacked.[4] This assumption is based mostly on a statement by Polybius regarding the war between Rome and Pannonia and Appian's mention of a failed campaign against the Pannonians under the leadership of one Cornelius, as well as on Appian's statement that the Romans had attacked Segestica on two occasions before Octavian.[5] Because of Polybius, this definitely cannot be dated after 146 BC, but also not before 167 BC, as in that case the event would most probably have been mentioned by Livy. After comparing the data on the Roman consuls of the time with the *gentilicium* (*nomen*) Cornelius, Zippel concluded that the only ones that could have been the leaders of a campaign against the Pannonians were Gnaeus Cornelius Dolabella and Lucius Cornelius Lentulus Lupus, consuls in 159 BC and 156 BC respectively. If this campaign was merely a part of larger military operations against the Scordisci in the north and the Delmatae in the south, the year 156 would be the more probable date, but it remains questionable whether the political influence of the Scordisci or the Delmatae reached so far as to warrant a Roman intervention in the Sava valley.[6] The Iapodians came under attack again in 129 BC, when consul Gaius Sempronius Tuditanus led a campaign against them.[7] Some authors believe that in this campaign the Romans could have penetrated further north and reached Segestica, provided that Appian's previously mentioned two Roman attacks against Segestica before Octavian can be considered to include Tuditanus' campaign.[8] Personally, I do not find this hypothesis convincing, because the sources do not contain any information on this, and there is also no mention of the reasons that may have prompted Tuditanus to extend his combat actions to the north on this occasion. Bearing in mind the fragmentary nature of the historical sources, the possibility that Tuditanus actually did attack Segestica cannot be completely excluded – after all, we do not know anything about the relations and possible alliances

[1] Cf. Wagner 1938; Kraft 1951; Radnóti and Barkóczi 1951; Fitz 1962; Dušanić 1968; Pinterović 1968; Fitz 1980; Holder 1980; Visy 1988; Spaul 1994; Spaul 2000; Lőrincz 2001; Visy 2003a; Visy 2003b.

[2] Wilkes 1969, 32-33; Mócsy 1974, 31-32.

[3] Liv. 43.5; Mócsy 1974: 32; Wilkes 1992, 200; Dzino 2010, 58-59.

[4] Zippel 1877, 135; Mócsy 1962, 527-528; Mócsy 1974, 12, 22, 32; Šašel 1974, 731; Hoti 1992, 135; Radman-Livaja 2004a, 15-16; Domić Kunić 2006, 85-88; Dzino 2010, 73; Radman-Livaja 2010a, 179.

[5] Polyb. frg. 122 (ed. Hultsch); App. 14 and 22.

[6] Šašel Kos 2005, 302-303.

[7] App. 10; Liv. Per. 59; Wilkes 1969, 32-33; Zaninović 1986, 60; Hoti 1992, 135; Wilkes 1992, 200; Radman-Livaja 2004a, 16; Domić Kunić 2006, 89; Dzino 2010, 69-71.

[8] App. 22; Klemenc 1963, 55; Zaninović 1986, 60.

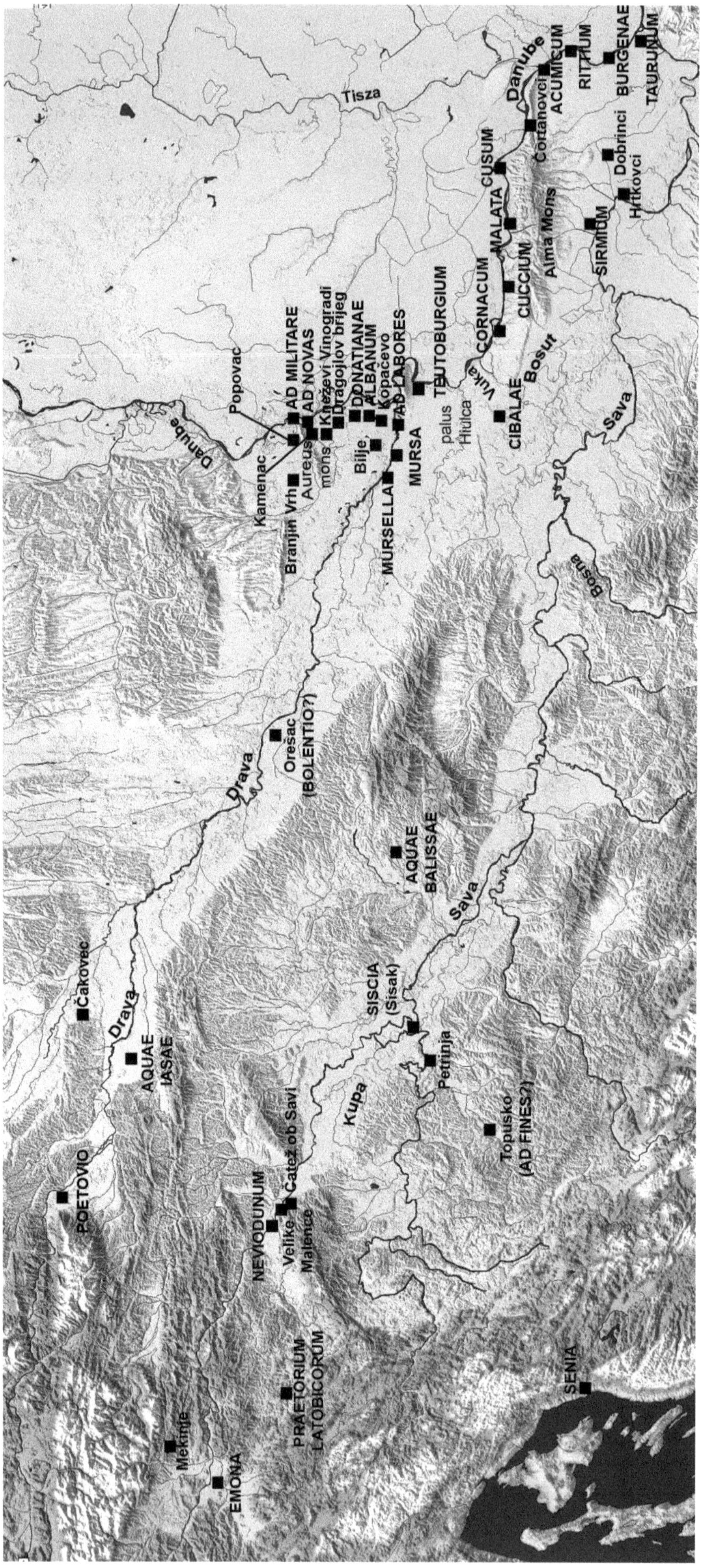

Fig. 1. Map of the military geography of southern Pannonia (after Google Maps, modified by T. Leleković).

between Segestica and the Iapodians – but this remains a hypothesis.

It is, however, likely that 10 years later, in 119, the Romans did in fact conduct military operations at Segestica during the Dalmatian War. The literature often states that both consuls, Lucius Aurelius Cotta and Lucius Caecilius Metellus, led the army that attacked Segestica, but this is a highly dubious assumption. Although there is no doubt that consul Lucius Aurelius Cotta was in command of the army, it is very unlikely that the campaign was deemed to be so important as to warrant the presence of both consuls. The other commander could have been Lucius Metellus Diadematus, as *propraetor* or legate. He was a cousin of the other consul, Lucius Caecilius Metellus, who very probably at the same time commanded the operations against the Delmatae.[9] Although Morgan's assumption is convincing, it is not impossible that this was some other Metellus, Diadematus' younger brother, for instance, Caius Metellus Caprarius, as suggested by D. Dzino. Appian left us only very scant information about this operation, but this was no doubt one of the two attacks that, according to Appian, preceded Octavian's siege of Segestica.[10] The assumption about a Roman attack against Segestica in 83 BC during the war against the Scordisci is not based on concrete arguments.[11] The Romans may well have defeated the Scordisci some time between 88 and 81 BC, but because the campaign progressed from the direction of Macedonia toward Moesia it is difficult to believe that the Pannonian area could have been the site of military operations and Klemenc's theory is, therefore, highly difficult to prove.[12] If we accept the possibility of a Roman military campaign in Liburnia in 84 BC or the assumption on military operations against the Iapodians perhaps around 78-76 BC, this still does not mean that the Romans advanced further into the hinterland, that is, all the way to Segestica.[13] Roman military operations in the area of today's central and northern Croatia before Octavian's campaign are generally very poorly documented and the available data is subject to various interpretations, creating many dilemmas. Still, it is quite certain that in this period the Roman army did not attempt to proceed deeper into Pannonian territory and it seems that it never got anywhere beyond Segestica. The documented Roman military campaigns in this area might not have been frequent, but were nevertheless significant. It seems that the consular army was engaged already for the first possible campaign, the one in 159 or 156 BC. This, at least theoretically, means at least 2 Roman legions and 2 *alae sociorum*. The same happened again in 119. On both occasions then, the Roman army reached Pannonia under the command of one of the consuls and theoretically cannot have counted less than some 19,000 men in this period, possibly even more, depending on whether the Romans used also mercenary troops (for instance, archers, slingers or cavalry) and possible local allies in addition to the legions.[14] In using so many troops, the Romans undoubtedly wanted to demonstrate the power of the Roman Republic, and with obvious success, as between 119 and Octavian's campaign there were probably no significant Roman military interventions so deeply in the hinterland of western Illyricum.

3. Octavian's campaign and the Augustan period

After long decades of peace, a military operation ensued that, unlike previous ones, was not merely a punitive expedition limited in area and duration, but was in fact the first step toward the final conquest of the area. As in previous campaigns, the goal was to take Segestica, but this was not to be the end of the operation. In 35 BC, the young Octavian set off with substantial military forces to conquer the hinterland of Illyricum. Although the written sources are not in complete agreement as to the reasons for his plan, there is no real doubt that Octavian embarked on this campaign as part of the preparations for the pending conflict with Mark Antony, for strategic as well as propaganda reasons. Julius Caesar's adopted son had to prove that he was worthy of his adoptive father, but also that he can measure up to Mark Antony, who by then had more than successfully proven and reaffirmed his military prowess. Therefore, a victory over the tribes of western Illyricum had the potential to strengthen Octavian's political power but also his popularity in Rome and Italy as a whole. From the strategic point of view, the control of Pannonia, that is, the Sava River basin, would mean that Octavian could block the best land communication Mark Antony had between the East and Italy, and this would significantly raise Octavian's odds for victory in case civil war broke out.[15] After the successful conquest of the Iapodes, when his warrior reputation was additionally boosted by his wounding at Metulum (evil tongues would add that it was a rather convenient wounding), he advanced on Segestica, demanding from its citizens not only to give hostages but also to accommodate a Roman garrison within the town walls. Although faced with Roman force, the town at first hesitated, but those in favour of putting up resistance eventually prevailed. The fact that it took Octavian 30 days to conquer the town despite all the forces at his disposal indicates that the town was well fortified and the resistance bitter.

The exact number of legions that took part in the campaign against Illyricum under Octavian's command remains

[9] Wilkes 1969, 33; Morgan 1971, 271-301; Mócsy 1974, 13, 22; Zaninović 1986, 59-60; Hoti 1992, 135; Radman-Livaja 2004a, 16; Šašel Kos 2005, 307-311; Domić Kunić 2006, 89-90; Dzino 2010, 72-73; Radman-Livaja 2010a, 180.

[10] App. 10, 22.

[11] Klemenc, 1963, 55.

[12] App. 5; Mócsy 1974, 15; Hoti 1992, 135; Domić Kunić 2006, 90-91.

[13] Wilkes 1969, 34-35; Dzino 2010, 73-74.

[14] Keppie 1984, 10-11, 19-20; Erdkamp 2007, 47-74; Hoyos 2007, 68-71.

[15] App. 16-21; Dio 49.35; Veith 1914, 17-49; Swoboda 1932, 3-29; Josifović 1956, 138-144; Wilkes 1969, 50-51; Mocsy 1974, 21-23; Barkóczi 1980, 87-88; Šašel Kos 1986, 134-139; Nagy 1991, 61-62; Hoti 1992, 136; Šašel Kos 1997a, 187-190; Šašel Kos 1999, 255-263; Radman-Livaja 2001, 132-135; Radman-Livaja 2004a, 16-17; Šašel Kos 2005, 393-437; Domić Kunić 2006, 91-93; Eck 2007, 35; Dzino 2010, 101-109; Radman-Livaja 2010a, 182.

unknown.[16] Nevertheless, the written sources furnish data on the overall strength of Octavian's forces before the campaign; it seems that after he took over the defeated Pompeian forces and Lepidus' men, he had at his disposal a total of around 200,000 legionaries, or 43 to 45 (depending on the sources and their interpretation) more or less fully manned legions, and around 25,000 cavalrymen and 40,000 light infantry, most probably auxiliaries of non-Italic origin.[17] He undoubtedly had an impressive military force at his disposal, although he certainly set off against Illyricum with only a part of his troops.[18] In addition, when advancing toward the hinterland, he would have left a part of the troops behind to secure the conquered areas and supply routes even before he reached Segestica. Therefore, it is quite certain that during the siege Octavian did not have at his disposal all the units he had left Italy with. Nevertheless, if we bear in mind that after conquering the town, he left a garrison of 25 cohorts there, which is equal to the strength of two and a half legions, it can be assumed that the army he commanded during the siege was much greater. In light of this, the assumption that Octavian could have around 5 legions at his disposal, as well as the accompanying auxiliary troops and river ship crews, is quite convincing.[19]

By leaving a strong garrison in the town under the command of Fufius Geminus, Octavian made it clear that the Romans intended to stay in Pannonia, but the conquest of Illyricum and the reinforcement of Roman power took many years. After more than a decade of some quiet, in 16 BC Pannonian and Norican raiders broke into Histria, and the following years were marked by frequent conflicts with the indigenous Pannonian population. There is mention, for example, of the Pannonian uprising in 14 BC, and the gravity of the situation in Pannonia is illustrated by the fact that in 13 BC Augustus decided to send Agrippa to the area. Following his sudden death, Tiberius took over command and the battles in the war that was to be called *Bellum Pannonicum* continued with varying intensity until 9 or 8 BC.[20] Despite inconclusive data from the sources, there is no doubt that permanent Roman military garrisons were set up in southern Pannonia, but we can only hypothesize about the identity of the Roman legions that stayed there during the rule of Augustus.[21] Although the Romans certainly used auxiliary units at the time, these probably had not yet acquired the character of permanent units and the lasting identity they would acquire during the rule of Augustus' successors; there is, generally, very little accurate knowledge about them in this period.[22] It is almost impossible to identify with certainty the auxiliary units that took part in the conquest and occupation of southern Pannonia during the rule of Augustus.[23] We should also not forget that the period in question is relatively long – over 4 decades, and during this time the security situation undoubtedly changed and the garrison strength varied accordingly. Thus the concentration of Roman troops may have been quite weak at certain times, if there was no serious unrest. At other times again, quite a substantial portion of the available Roman troops must have been in the area, for instance during the Pannonian War (*bellum Pannonicum*) from 13 or 12 to 9 or 8 BC, and, of course, during Bato's uprising (*bellum Batonianum*) from AD 6 to 9. It is not irrelevant to note that at the moment (somewhere between 32 and 27 BC) it became a province, Illyricum was established as a Senate province with only one legion (of an unknown garrison).[24] It is true that, in principle, the province would not be assigned to the Senate if stronger military forces had been concentrated there, but the possibility that for an extended period of time Illyricum had only one legion as a permanent garrison force and, of course, an unknown number of auxiliary units is rather doubtful.[25] This may have been possible for a shorter period immediately after the establishment of the province, but the military forces should have been significantly strengthened

16 We cannot know for certain which legions could have taken part in the campaign, but the literature mentions *legio XV Apollinaris* (cf. Ritterling 1925, 1747-1748; Šašel 1985, 549; Šašel-Kos 1995, 229; Wheeler 2000, 267). Although there is no confirmation that they took part in the campaign in 35 BC, because of their demonstrated loyalty and reliability in the conflicts before and after his campaign in Illyricum, Octavian may have also taken along the VII, VIII, IX and XI legions (cf. Ritterling 1925, 1664, 1690; Keppie 1984, 133, 207-209; Reddé 2000, 119-120; Strobel 2000, 526).

17 App. *B. C.* 5.127; Oros. *Hist.* 6.18; Vell. 2.80; Nagy 1991, 59.

18 G. Veith (1914, 108-109), with agreement from E. Swoboda (1932, 17), judges the strength of Octavian's army during the Illyricum campaign to have been app. 40-50.000 men, or 8 to 12 legions, which seems convincing.

19 App. 22-24; Dio 49.36; Veith 1914, 49-58; Swoboda 1932, 29; Mócsy 1962, 538-539; Wilkes 1969, 52-53; Mócsy 1974, 22; Šašel 1974, 732; Barkóczi 1980, 90; Šašel-Kos 1986, 139-142; Zaninović 1986, 62-63; Nenadić 1987, 73; Nagy 1991, 62-64; Hoti 1992, 137-138; Wilkes 1992, 206; Gruen 1996, 173; Wilkes 1996, 549-550; Šašel Kos 1997a, 190-196; Radman-Livaja 2004a, 17; Šašel Kos 2005, 437-442; Domić Kunić 2006, 92-100; Radman-Livaja 2007a, 161-162; Dzino 2010, 109-111; Radman-Livaja 2010a, 182-183.

20 Flor. *Epit.* 2.24; Vell. 2.96; Mócsy 1962, 539-541; Mócsy 1974, 34; Barkóczi 1980, 90-91; Šašel Kos 1986, 152-161; Nagy 1991, 64-84; Hoti 1992, 138-140; Gruen 1996, 174-175; Šašel-Kos 1997b, 31-33; Domić Kunić 2006, 100-118; Šašel-Kos 2009, 181-182; Colombo 2010, 171-193; Dzino 2010, 112-136.

21 Ritterling 1925, 1215-1238; Syme 1933, 20-23, 25-28.

22 Cheesman 1914, 12-20; Wagner 1938, 223-224; Holder 1980, 140-141; Saddington 1982, 27-82.

23 Alföldy 1962, 286. Based on an interpretation of the inscription on the tombstone of the veteran Tiberius Julius Rufus of Scarbantia, it is assumed in the older literature that his unit, *ala I Scubulorum*, may have been present in Pannonia already during the rule of Augustus, that is, that it may have taken part in Tiberius' campaigns in 12-10 BC (cf. *ILS* 9137; *AE* 1909, 198; Cichorius 1894, 1259; Schober 1923, 89-90, Cat. 191; Wagner 1938, 64-67; Kraft 1951, 158; Mócsy 1962, 620). Nevertheless, it seems that the presence of this ala in Pannonia can be traced with certainty only since the time of Claudius, after its transfer from Moesia (Beneš 1978, 12; Spaul 1994, 192-194; Lőrincz 2001, 23, 59; Lőrincz 2005, 53-55). This ala may have indeed been levied during Augustus, but there is no proof that it took part in the Pannonian wars and the crushing of the Pannonian rebellion, although this hypothesis is, generally speaking, not unconvincing. G. Alföldy assumed that during the late Augustan period, more precisely at the time of Bato's revolt, the following alae and cohorts may have been present in Illyricum, among the 10 alae and 70 cohorts not stated in Velleius' text: *ala I Parthorum, ala Pannoniorum, cohors III Alpinorum, cohors I Bracaraugustanorum, cohors I Campana, cohors II Cyrrhestarum, cohors XI Gallorum, cohors I Lucensium, cohors Montanorum, cohors VIII voluntariorum* (and probably a good part of the other *cohortes voluntariorum*, if not all of them); cf. Wagner 1938, 218-219; Alföldy 1969, 287; Knight 1991, 189-190.

24 Nagy 1991, 67; Gruen 1996, 174; Gayet 2006, 70; Dzino 2010, 119.

25 E. Ritterling therefore posited that in 27 BC the southern part of Illyricum was organised as a Senate province while the northern part, as a military frontier with significant forces (legions), was directly under Augustus, that is, his governor, who was supposed to be legatus pro praetore of consular rank (cf. Ritterling 1925, 1218-1219; Nagy 1991, 67-68, 79). It should be mentioned that there is no evidence indicating such an organisation in this period or the existence of these legates. (cf. Dzino 2010, 122-123).

at the latest in 16 BC, when there must have been more than one legion in Illyricum. After all, in 11 BC (or already in 13 BC?), Illyricum ceased being a Senate province and became an imperial province and this change was undoubtedly connected to the military situation.[26] It is not unlikely that the military presence in Illyricum was reduced to a minimum around 27 BC if at the moment of the establishment of the province Octavian did not consider Illyricum to be a military and security problem. The newly established province, however, could not have remained without a significant legionary garrison for long, that is, not more than roughly ten years. There is no consensus in the literature regarding the exact number or locations of the legions that stayed in Illyricum in the last century BC, that is, until Bato's rebellion broke out in AD 6, but there is no doubt that in this period several legions must have stayed there. Although it is difficult to say with any certainty how long and exactly where a certain legion may have stayed in Illyricum between 35 BC and AD 6, it is very probable that in this period at least 6 and perhaps even 7 legions spent a certain amount of time in the province (*VIII Augusta, IX Hispana, XI, XIII Gemina, XIV Gemina, XV Apollinaris,* and *XX*).[27] Although the VII legion is sometimes mentioned in this context, especially in the older literature, the possibility of their arrival to Illyricum before Bato's uprising can be rejected on safe grounds.[28] Despite serious doubts, it cannot be excluded that the VIII legion could have spent some time in Illyricum in the mentioned period, as there is no doubt that it took part in crushing Bato's uprising and it definitely stayed in Pannonia after AD 9. Nevertheless, there is no reliable evidence that it had been transferred to Illyricum before these events, although there are those who assume that it may have stayed in Liburnia, perhaps in Burnum.[29] The *IX legio Hispana*, seems to have been deployed from the Iberian peninsula to Illyricum between 19 BC and 15 BC, or by 13 BC at the latest, but we can only conjecture about its garrison before Bato's uprising. Nevertheless, the literature does mention Tilurium where it may have been stationed (depending on the interpretation of the inscription *CIL* III 13977) until the rebellion was quashed in AD 9.[30] Although it is quite probable that it stayed in Illyricum in this period, because of the uncertain reading of the mentioned inscription, the possibility cannot be excluded that the IX legion was stationed elsewhere in Illyricum, including the northern (Pannonian) part of the province or in Aquileia, which could have served as the base for possible interventions in Illyricum. The XI legion is assumed to have been stationed in the Balkans already after the battle of Actium, that is, immediately after Octavian reorganised the military, but there are no concrete data on its garrison in this period.[31] Ritterling assumed that prior to Bato's uprising, it could have, at some unidentified moment, been stationed in Pannonia (*Poetovio*?) or that it operated in that area.[32] Farnum assumes that between 27 and 16 BC it could have been in Tilurium, from where it was redeployed to Moesia (where it could have been already between 30 and 27 BC – *Naissus*?) and was stationed there until the end of Bato's uprising (*Viminacium* ?). As Farnum gives no arguments in favour of his assumptions, his opinion remains unreliable.[33] After Actium, *XIII Gemina* was most probably among the troops in Transpadania and Illyricum. The literature mentions Burnum and Emona as the possible garrisons, but there are no concrete arguments to support this; Emona can in any case probably be excluded as a possibility.[34] Based on the available data, in the period following Actium *XIV Gemina*, as well as the XIII legion, is placed in Illyricum or northern Italy. It may have stayed in Aquileia at a certain point in time, and Farnum also mentions *Poetovio* as a possibility. It probably left Illyricum (or northern Italy) by 13 BC at the latest, when it joined the legions on the Rhine and took part in Drusus' campaigns. It may have then returned to Illyricum, but it seems more probable that it did so only after AD 6, as one of the legions originally intended for the campaign against Maroboduus with which Tiberius had to crush Bato's uprising.[35] As mentioned earlier, *XV legio Apollinaris* is believed to have arrived to Illyricum already during Octavian's campaign and it seems to have operated there throughout the rule of Augustus, perhaps from its base in Aquileia or maybe from a camp in Illyricum itself (although it is sometimes mentioned as such, Emona is probably not a sound guess).[36] According to Velleius Paterculus, *XX Valeria Victrix* could also have been a part of the military in Illyricum before Bato's uprising, but the question remains as to when it arrived there. Before arriving to Illyricum, it seems to have stayed in Hispania, taking part in the Cantabrian Wars (*Bellum Cantabricum*), and it could have been redeployed after 19 BC, in which case it could have participated in the Alpine campaign in 16-15 BC and/ or Tiberius' Pannonian war in 13-9 BC. It was probably

[26] Mócsy 1962, 540-542; Mócsy 1974, 34; Nagy 1991, 68-79; Dzino 2010, 122-136.

[27] Syme 1933, 23, 25-26, 29-31; Keppie 1984, 208-211; Dzino 2010, 123-124. It should be mentioned that a temporary stay in this area, that is, participation in the military operations in this period, did not necessarily have to mean that a legion was more permanently stationed in Illyricum. The permanent garrisons of some legions were more probably located in northern Italy, from where they could intervene, if needed, in Illyricum. Possible garrisons are Aquileia (Ritterling 1925, 1645-1646, 1665, 1748; Mann 1983, 31-32; Wheeler 2000, 270), and perhaps also Emona (Saria 1938, 245-255; Mann 1983, 32; Visy 1988, 17), which, it seems, was never a part of Illyricum in administrative terms (cf. Šašel Kos 2002, 373-382; Šašel Kos 2003, 11-19). There are, however, well-founded opinions that seriously question the existence of a legion camp in Emona: Šašel 1968, 561-566; Mócy 1974, 43; Šašel-Kos 1995, 227-244; Vičič 2003, 22-24.

[28] Ritterling 1925, 1615-1617; Betz 1938, 8; Wilkes 1969, 94-95; Mitchell 1976, 301-303; Strobel 2000, 526-528; Sanader and Tončinić 2010, 45-46.

[29] Tac. *Ann.* 1.23, 30; Ritterling 1925, 1645-1647; Betz 1938, 5, 50-52; Wilkes 1969, 92-95, Oldenstein-Pferdehirt 1984, 397; Bojanovski 1990, 704; Reddé 2000, 120-121.

[30] Ritterling 1925, 1664-1665; Syme 1933, 23; Betz 1938, 52; Wilkes 1969, 92; Keppie 2000, 26; Farnum 2005, 21, 65.

[31] Betz 1938, 20.

[32] Ritterling 1925, 1691.

[33] Farnum 2005, 22, 64, 72-73.

[34] Ritterling 1925, 1711-1712; Syme 1933, 29; Wilkes 1969, 92-93; Farnum 2005, 22, 64-65; for Emona cf. note 27.

[35] Ritterling 1925, 1728; Wilkes 1969, 92-93; Franke 2000, 191-192; Farnum 2005, 23, 64-65.

[36] Ritterling 1925, 1747-1748 ; Šašel-Kos 1995, 236-237; Wheeler 2000, 261-268, 270-272; Farnum 2005, 23, 64-65; for Emona cf. note 27.

stationed in Burnum, or perhaps in Aquileia, that is its surroundings.[37]

Although there is at least some degree of agreement in the literature regarding the identification of the legions in Illyricum, despite the unresolved chronological issues, the question of their camps in the period before Bato's uprising remains open. It seems that for the time being Emona can be excluded from this story, but there are other possible legion garrisons, such as Burnum, Tilurium, Poetovio, Siscia and Sirmium, but also Aquileia, where the troops that operated as needed in Illyricum seem to have been stationed. Mostly the camps in Burnum (VIII, XIII, XX legion), Tilurium (IX, XI legion) and Poetovio (XI, XIV legion) are in the literature mentioned as the seats of certain legions in Illyricum before Bato's uprising, and it seems that in the same period some of the mentioned legions could have stayed in Aquileia or its vicinity (IX, XIII, XIV, XV, XX). While in the case of Aquileia, existing epigraphic material offers certain arguments in favour of this hypothesis, the assumptions on the existence of legion camps in this period at the other locations mentioned (with the exception of the somewhat dubious inscription *CIL* III 13977 in the case of Tilurium and some veteran inscriptions from the late Augustan period in the case of XX legion) are admittedly based solely on more or less convincing conjectures. I am inclined to believe that the legion camps in Burnum, Tilurium and Poetovio could have existed before Bato's rebellion broke out, but it is surprising that Siscia and Sirmium are less frequently mentioned in the literature in this context. Classical sources rightly stress the exceptional strategic significance of these two towns at the time when Bato's rebellion broke out, so it is difficult to believe that the Romans did not use them as military bases even before AD 6. Sirmium may not have had a Roman garrison before the Pannonian War (*Bellum Pannonicum*), but it would be sound to assume that it witnessed Roman military presence from that moment until the end of Augustus' rule.[38] As for Siscia, I am deeply convinced that in this period it was not left without a garrison for a single moment, although the garrison's strength certainly varied.[39] Although there is no concrete evidence to support this, the available data suggests that XV[40] and XX[41] legion may have spent some time in Siscia before Bato's uprising. Burnum[42] and Tilurium[43] could have become camps already in the second decade BC, if not even sooner.[44] The legion camp in Poetovio was presumably set up around 15 BC or a few years earlier.[45]

Considering the utter lack of reliable indications for the existence of legion camps in Illyricum during the rule of Augustus, that is, before AD 6, the issue of their location is better left in the realm of cautious conjecture. There is no doubt that they existed, but it is probably wrong to think of them as permanent military bases as was the case later on. It would perhaps be more correct to assume that during the first few decades of the occupation of Illyricum, the legions constantly circulated between certain locations and that no legion remained at any location permanently. At the time, Aquileia could have been a military base that was the starting point for military operations in the western Balkans, and certain legions and/or vexilations were deployed as the need arose within Illyricum, but also between Aquileia and Illyricum. The most important strategic locations undoubtedly very quickly acquired the character of more permanent Roman military strongholds, but their crews may have been subject to relatively frequent rotations of manpower and units, as seems to have been the common practice in the time of Augustus, when legions often changed garrisons.[46] It can soundly be assumed that in Pannonia these strategic locations were Siscia, Sirmium and Poetovio, and in Dalmatia Burnum and Tilurium, but all of these places cannot at the same time be claimed as legion camps. It is extremely difficult to establish when any of them could have become the seat of a legion, although it is quite possible that at least roughly between 13 BC (and even earlier in the case of Siscia) and AD 6 legionary garrisons could have stayed at any of the mentioned places. It is equally impossible to identify with certainty the crews of certain garrisons in this period, but we can probably bring the selection down to the mentioned 6 or 7 legions. In addition to the above-mentioned main strongholds of the military concentration, there were probably also smaller outposts with crews made up of auxiliaries and/or detached legionaries, and possibly also local allies, some of who perished when Bato's rebellion broke out.[47] If the question of the locations of legions in Illyricum during the rule of Augustus is complex and burdened with doubts and

37 Vell. 2,112; Dio 55.30 (does not expressly mention it); Ritterling 1925, 1769-1771; Betz 1938, 20, 56-58; Wilkes 1969, 93; Farnum 2005, 24, 65; Malone 2006, 28-31; Cambi *et al.* 2007, 13-16; Dzino 2010, 148; Miletić 2010, 120.

38 Mócsy 1959, 77; Klemenc 1961, 23; Klemenc 1963, 67; Mócsy 1971, 44-45; Mócsy 1974, 43. J. W. Eadie disagrees with this assumption (1977, 210). He doubts that any regular Roman units could have been stationed in Sirmium before AD 6. M. Mirković allows the possibility of Roman military presence in Sirmium at this time, but believes these were not permanent legion bases in the Pannonian area but only auxiliary units and legion vexilations. Cf. Saxer 1967, 5-6; Mirković 1971, 12; Mirković 1990, 639, note 42.

39 Mócsy 1974, 23, 43; Šašel 1974, 732-734; Radman-Livaja 2007a, 161-168; Radman-Livaja 2010a, 183-186.

40 As it is not associated with Burnum, Tilurium and Poetovio, it cannot be excluded that it operated between Aquileia and Siscia.

41 If Velleius Paterculus and Cassius Dio have been interpreted correctly (cf. Vell. 2.112; Dio 55.30), XX legion under the command of Valerius Messalla was sent by Tiberius as an advance contingent of his forces to Siscia as soon as he found out that a rebellion had broken out. We can only guess at his reasons, but it is not illogical to assume that XX legion was sent first because it had previously been stationed in Siscia and could get there fastest because it was familiar with the terrain. It is, of course, not excluded that Tiberius had other reasons and that XX legion was chosen, for instance, because of its combat efficiency and experience, or for another reason of a purely military and tactical nature.

42 Betz 1938, 57; Zaninović 1968, 121-122; Cambi et al. 2007, 13-16; Sanader 2008, 85-86; Miletić 2010, 120.

43 Zaninović 1984, 68-69; Sanader 2008, 87-88; Sanader and Tončinić 2010, 45-46.

44 The theory put forward by D. Periša (2008, 511-512) about 33 BC as the first possible date for Tilurium is not unconvincing, although the question remains as to whether the Romans remained permanently after the siege.

45 Saria 1951, 1170; Mócsy 1959, 28; Horvat *et al.* 2003, 156.

46 Syme 1933, 22.

47 Vell. 2.110.

insecurities, our knowledge about auxiliary units in this period is even more problematic.[48]

It seems that after Tiberius's victory in the Pannonian War (*bellum Pannonicum*) peace reigned in this area, at least on the surface, and the strong military presence no doubt helped keep it. The Romans obviously considered that in AD 6, after some 15 years free of any significant problems with the local population, the situation was stable enough for them to use the troops from the Illyricum garrisons for the war against the Marcomannic king Maroboduus, as well as to recruit the local population in significant numbers for this campaign. This turned out to be an entirely wrong assumption, because it seems that the mass uprising was brought on by the collection of a tribute and the recruitment of young men for auxiliary troops to be used in the war against the Marcomanni. Although the trigger that set off the rebellion is quite clear, discontent had been smouldering for years and only culminated in the great war from AD 6 to 9, known as Bato's uprising.[49] By all accounts, the main instigators of the uprising were the Daesidiates, headed by their ruler Bato, who found allies in the rulers of the Breuci, Bato and Pinnes. Although these rebellions did not realistically pose a serious threat to Italy itself, Augustus took them very seriously and mobilised significant human and material resources to quell the uprising. The gravity of the situation is confirmed by written sources, especially Suetonius and one of the participants in the war, Velleius Paterculus, the latter claiming that the rebels, whose number he judges to be several hundreds of thousands, were preparing to attack Italy. Although this number is certainly exaggerated, this must have been a widespread opinion among the Romans at the time.[50]

The news on the uprising in Illyricum reached Tiberius's headquarters in Carnuntum just before the beginning of the war against the Marcomanni in AD 6. The timing of the uprising was perfect because the greater part of the Roman troops stationed in Illyricum was temporarily deployed in the north under the command of Tiberius, so that they could participate in the mentioned campaign. The Daesidiates were joined by the Breuci, probably the strongest tribe in southern Pannonia, led by another Bato. Many smaller tribes also joined the rebels and they embarked on an all round massacre of the Roman citizens in the area. Velleius expressly mentions traders and military units who had the bad luck to be caught isolated in the midst of what had suddenly become an enemy territory. The rebels did not stop at killing all the Romans they could lay their hands on, but immediately moved on to larger settlements and even neighbouring provinces. Velleius make only a passing mention of the looting raids toward Macedonia and the preparations for the attack against Italy, but Cassius Dio explicitly mentions the attacks against Salona and Sirmium. It seems, therefore, that the rebels' plans were far more ambitious and far-reaching than merely killing and expelling the Romans from their tribal areas. Considering the circumstances, the Roman reaction was quick and quite effective. The Romans who happened to be in the settlements taken by the rebels could not, of course, be helped, and their fate certainly contributed to the panic that spread all the way to Italy. However, the larger settlements could still be saved and it was above all necessary to keep the rebels within the areas they already controlled and prevent them from advancing to other provinces, and especially Italy. Two key strategic points for controlling Pannonia were Siscia in the west and Sirmium in the east. As both sides were aware of this, the first year of the war was marked by the manoeuvring of forces between these two towns. The sources do not reveal whether Siscia was directly threatened when the uprising broke out, but it seems that as soon as he became aware of the situation, Tiberius urgently dispatched a legion to reinforce the town's defence. This was probably XX legion led by Valerius Messala, governor of Illyricum. A little later, Tiberius personally arrived to Siscia with the remainder of his large army. He set up camp there and the town remained under the firm hold of the Romans.[51] At the very beginning of the uprising, the rebels simultaneously marched on both Sirmium and Salona, but both operations failed. Sirmium was attacked by the Breuci, but the town was saved by Roman troops led by the governor of Moesia, Caecina Severus. Dio's description does not make it quite clear whether Caecina Severus broke the siege of Sirmium or whether he actually stopped the Breuci from approaching Sirmium, because the author explicitly states that Caecina Severus defeated the Breuci at the Drava River, which is quite a distance from Sirmium. It is possible that Caecina Severus broke the siege of the town and then drove the attackers to the Drava, where the Breuci decided to confront them in the battlefield, and were eventually defeated. Salona was besieged by the Daesidiates, but the town was well defended so the majority of the rebel army, headed by its wounded leader Bato, withdrew to the hinterland, although looters continued to raid the coastal area all the way to Apolonia. According to the written sources, the rebels soon embarked on another offensive, despite their defeat. After abandoning the siege of Salona, the Daesidiates tried to intercept the troops of Valerius Messala, but the Romans, although outnumbered, managed to defeat them. These were probably the troops that Tiberius had sent as reinforcements to Siscia, but it is not clear whether they were attacked on their way to Siscia or only after they had set up camp there and then embarked on an attack against the rebels. This defeat did not discourage the Daesidiates, as they headed east to meet

[48] Cf. note 21; Alföldy 1962, 286-287; Wilkes 1969, 139-144, 470-474; Lőrincz 2001, 57; Dzino 2010,124.

[49] Mócsy 1962, 541-544; Wilkes 1969, 65-69; Mócsy 1974, 34-37; Dzino 2010, 138-145.

[50] Vell. 2.110; Suet. *Tib.* 16; Crook 1996, 106-107; Gruen 1996, 176; Seager 2005, 33; Eck 2007, 133.

[51] In addition to the XX legion, we can assume that Tiberius also took IX, XIII, XIV and XV legions along to Siscia. These legions were supposed to participate in the campaign against Marobuduus and most of them were probably stationed in Illyricum also in the years leading up to Bato's rebellion, except for XIV legion which seems to have been transferred from the Rhine frontier because of Tiberius' campaign against the Marcomanni. Ritterling on the other hand thought that before the uprising, VIII, IX, XI, XV and XX legion were part of the Illyricum army; cf. Ritterling 1925, 1235; Wilkes 1969, 92-93; Franke 2000, 192; Keppie 2000, 26; Wheeler 2000, 271; Farnum 2005, 5.

up with the Breuci and, having joined forces once again, set off for Sirmium. They set up camp on *Alma Mons* (Fruška gora in Syrmia), where they began preparations for the attack against Sirmium. Caecina Severus, probably a very capable commander, sent the allied Thracian cavalry against them, led by king Rhoemetalces, in order to prevent them from attacking Sirmium. The Thracians succeeded and they drove off the rebels thus concluding the last big military operation in AD 6. At the year's end the Romans were in possession of the two most important towns: Siscia, defended by Tiberius' troops with reinforcements from Italy, and Sirmium, controlled by the army of Caecina Severus, as well as in control of the coastal towns. The rebels obviously controlled the entire territory in between, but the Romans definitely had the strategic advantage and they seemed to have believed that in the spring they would be able to crush the rebellion, owing to the numerous forces they had gathered. They started an offensive in AD 7, probably as soon as the weather conditions permitted. Three Moesian legions under Caecina Severus, two legions redeployed from Asia Minor led by Marcus Plautius Silvanus,[52] and the Thracian cavalry led by king Rhoemetalces, set off for Siscia from Sirmium, no doubt to secure the Sava River, the main communication route across Illyricum. The Romans, whose numbers gave them great confidence, probably were not too cautious and fell into an ambush in the marshy area known as *Hiulca Palus* (but also as *Volcae Paludes*), somewhere in today's south-eastern Slavonia. Despite their heavy losses they managed to break through and make it to Siscia, where the largest Roman military force since the time of the Civil Wars had just gathered.[53] As such a large concentration of troops probably created logistical problems, some of the troops were sent again to Moesia where their presence was, after all, needed because of the Dacian and Sarmatian looting raids. Although the Romans were not defeated, the result of the battles at the Volcian Marshes could hardly be described as a victory. This probably convinced Tiberius that the war could not be won easily on the battlefield and that a different strategy should be devised. It seems he decided on a „scorched earth" strategy, meaning the systematic destruction of crops and settlements with the aim of ultimately starving the population and crushing their will to fight. It seems that Augustus was not too happy with such an apparently passive strategy for crushing the rebellion, but Tiberius' plan proved to be quite adequate because it bore fruit the following year: on August 3 in AD 8, the starving Breuci led by their ruler Bato laid down their arms by the River *Bathinus* (present-day Bosna). Many of the Breuci loathed the idea of capitulation, and, understandably, their main allies the Daesidiates were also dismayed, so Bato of the Daesidiates launched an offensive towards Pannonia where he managed to overthrow and kill Bato of the Breuci. Nevertheless, this last rebellion of the exhausted Breuci was easily crushed that same year by Plautius Silvanus. In a counter-attack with troops from Sirmium he forced Bato of the Daesidiates to withdraw, thus finally crushing all resistance north of the Sava. Pannonia was thus pacified and the last resistance was put up by the rebels in Dalmatia where they were finally defeated the following year when Tiberius ended the war with a victory over Bato's surrounded forces in Andetrium.[54]

4. The 1st century AD: permanent garrisons, legionary camps and recruitment of natives

It was only after this last revolt was crushed with great difficulty that the real Romanisation of Pannonia actually began. It was in this period, in AD 8 or 10 or perhaps a year or two later (and even a decade or two later, according to some assumptions) that Pannonia became a separate province. This was the result of dividing Illyricum into two parts, whose names were at first *Illyricum Superius*, the future province of Dalmatia, and *Illyricum Inferius*, or Pannonia.[55] Gradually transferring its forces from the hinterland of the province to the border on the Danube, the Roman army would transform from an occupation force into a defence force and thus a guarantee of the peaceful development of Romanized Pannonia. This process took quite a long time and can be divided into several phases. Even after crushing Bato's revolt, the Romans did not immediately embark on building a series of fortifications and military stations on the border with the Barbarian world, so that during the reign of Augustus' immediate successors the Romans continued the practice of securing the borders of Pannonia by forging alliances with the neighbouring Barbarian peoples. This guaranteed peace and helped spare military resources, and the military remained mostly

[52] Wilkes assumed that VII, VIII Augusta and XI legion could have been under the command of Caecina Severus at that time, while *IV Scythica* and *V Macedonica* came from the East, but not all authors agree with this opinion. There are even doubts as to whether Caecina Severus took all three Moesian legions along on this campaign, and only two legions in Caecina's and Plautius's joint forces are assumed to have actually been Moesian. *IV Scythica* was probably one of Caecina's legions, because it seems to have been permanently stationed in Moesia during the rule of Augustus. *V Macedonica* and the VII legion were most probably transferred from Galatia under the command of Plautius Silvanus, but the question of the VIII and XI legion remains open. As for the VIII legion, depending on the author in question, it is assumed that it may have been stationed in Illyricum, Moesia, in the East, in northern Africa or Egypt, while the XI legion was most probably stationed in the Balkans at the time in question, but it is impossible to say exactly where. Plautius's third legion (if the assumption on three Eastern legions is correct in the first place) could perhaps be the VIII legion, but it could also have been one of the 3 Moesian legions, together with the IV and XI. The question also remains open as to the legion that Caecina Severus could have perhaps left behind in Moesia; cf. Ritterling 1925, 1234-1236, 1645, 1691; Wilkes 1969, 93; Oldenstein-Pferdehirt 1984, 397; Reddé 2000, 120-121; Speidel 2000, 327-328; Strobel 2000, 520, 525, 527; Farnum 2005, 4-5, 18-19, 20-22, 95.

[53] We can assume that these were Tiberius' IX, XIII, XIV, XV and XX, as well as *IV Scythica, V Macedonica*, VII, VIII and XI that arrived from Moesia.

[54] Dio 55.28-34; Vell. 2.110-116; Suet. *Aug.* 16, 25, *Tib.* 16, 20. For a more comprehensive review of the Pannonian-Dalmatian rebellion cf. Hirschfeld 1890, 351-362; Rau 1925, 313-346; Ritterling 1925, 1232-1236; Köstermann 1953, 345-378; Pavan 1955, 380; Pašalić 1956, 245-300; Mócsy 1962, 544-548; Wilkes 1969, 69-77; Mirković 1971, 12-13; Mócsy 1971, 43; Mócsy 1974, 37-39; Šašel 1974, 733-734; Barkóczi 1980, 88-89; Šašel-Kos 1986, 178-191; Zaninović 1986, 63; Gruen 1996, 176-178; Wilkes 1996, 553; Dizdar and Radman-Livaja 2004, 44-45; Sordi 2004, 221-228; Dzino 2005, 138-157; Seager 2005, 33-35; Šašel Kos 2009, 182-187; Dzino 2010, 137-155; Radman-Livaja and Dizdar 2010, 47-56.

[55] Klemenc 1961, 6; Mócsy 1962, 583; Mócsy 1974, 39; Barkóczi 1980, 89; Fitz 2003, 48-49; Dzino 2010, 159-162.

concentrated in the hinterland, away from the border.[56] There is no doubt, however, that already during the rule of Augustus smaller garrisons were stationed close to the Danube frontier as a sort of advance-guards, who would be helped in case of need by the troops stationed deeper in the hinterland. We can assume that in addition to the smaller legion detachments, these forward troops stationed close to the Danube border consisted mostly of auxiliary troops that probably arrived there from other parts of the Empire, such as Gaul or Hispania, but indigenous troops could have also been among them. Although this assumption is quite convincing, these units were probably not recruited directly by the Romans, but by the loyal local elites who placed their people in the service of Rome. At that time, such auxiliary troops responded to Roman requests as allies, subjects or mercenaries, mostly under their own commanders (the most important leaders of the revolt in AD 6, including Bato of the Daesidiates himself, could have previously been commanders of such units under Roman command).[57] Although it is more than likely that auxiliary troops were subsequently trained together with the Romans, in order to be able to function tactically and operationally as part of the Roman army, they definitely cannot be compared to the regular auxiliary units, such as those recruited since the 1st century AD that were fully incorporated into the Roman military system. Such local troops left no epigraphic traces and receive only passing mention in the sources, and we can only conjecture about their organisational structure and number.[58] Nevertheless, their existence can definitely be considered as the first attempt by the Roman authorities to use Pannonian human resources for military purposes.[59] These attempts, however, had very questionable results: while the Scordisci, for instance, if we are to judge by Cassius Dio, were evidently reliable allies and actively helped Tiberius, Bato's uprising was definitely largely caused by the unwillingness of the Pannonians to join the auxiliary troops that were supposed to participate in the war against Maroboduus. The fact that they operated at least partially under their own commanders and that they probably did not have to serve as long as the auxiliaries later on would, does not mean that they were very enthusiastic when mobilised by the Romans, especially if it was for a war that did not promise much in terms of booty and material gain. The question remains open, after all, as to how and under what conditions the peregrine communities filled the required quota of soldiers for manning such troops, as we may assume that there were not always enough volunteers. The bad experiences with Arminius in Germania and Bato in Illyricum undoubtedly led the Romans, probably already during the rule of Tiberius and in any case not much later than that, to change their approach in forming auxiliary troops and place them more firmly within their institutional military structures.

It seems that the large-scale recruitment of the Pannonian indigenous population into regular Roman auxiliary units in Pannonia and the raising of Pannonian auxiliary units ensued only several decades after Augustus. The only exception here would be *ala Pannoniorum,* a unit that is thought to have existed already at the beginning of Tiberius' rule, when it seems to have been transferred to Pannonia. It was probably raised already in Augustus' time, when it could have been stationed in southern Illyricum, that is, Dalmatia. We can assume that it was made up of members of Pannonian tribes, recruited probably before Bato's uprising, who remained loyal to the Romans in the war. It is quite possible that this ala was not formed from the ranks of loyal Pannonians after AD 9, but rather participated as part of the Roman army in crushing the rebellion. It seems likely, however, that already then, that is, toward the end of Augustus' rule and during the rule of Tiberius, it was manned with a number of Hispanians (and the odd German), which would indicate the real possibility that at the time there were simply not enough reliable Pannonian volunteers.[60] In any case, there is no concrete evidence that mass recruitment ensued immediately after the quelling of the Pannonian-Dalmatian revolt.[61] This is only understandable: the local male population suffered both in the battles and by being taken as slaves. The Romans were also not likely to start arming and training the local population that was hostile to say the least, directly after a great revolt was quelled in blood. Sending the surviving able-bodied men into slavery goes much further toward aiding a successful pacification.[62] Judging by the available data, it seems that at least one generation had to pass before more substantial recruitment took place.[63] Of course, the lack of epigraphic and other written sources does not constitute irrefutable proof that there was in fact no recruitment already at the end of Augustus' or at the beginning of Tiberius' rule, but I am more inclined to rely on what little evidence there is than to engage in conjecture. And the epigraphic finds tell us that the earliest documented

[56] Klemenc 1961, 7-8; Mócsy 1971, 41; Mócsy 1974, 39-41; Barkóczi 1980, 91-92.

[57] Dio 55.29; Vell. 2.110; Wilkes 1969, 69; Saddington 1982, 160.

[58] Dio 54.31; Vell. 2.110,5.

[59] Mócsy (1974, 39) considered that prior to AD 6 the Romans did not even attempt to recruit Pannonians, which does not really correspond to the picture presented by Velleius Paterculus and Cassius Dio. They probably approached recruitment selectively, supposedly seeking people from tribes more inclined to them, such as the Scordisci, or only volunteers, regardless of their tribal origins. Nevertheless, it is possible that there was no mass recruitment before AD 6, when mobilisation was conducted, probably unselective mass mobilisation, which was in contrast to the practice of previous decades and this may have triggered a wave of discontent.

[60] Cichorius 1894, 1255; Wagner 1938, 56-57; Kraft 1951, 25; Alföldy 1962, 262-263; Spaul 1994, 169; Spaul 1995, 63-73; Lőrincz 2001, 22.

[61] For an opposing view, cf. Mócsy 1959, 117, 120-121; Mócsy 1962, 645; Bogaers 1969, 29; Mócsy 1974, 39; Periša 2008, 514.

[62] Already in the previous war (*bellum Pannonicum*), after gaining victory, Tiberius ordered that the surviving young men of the defeated tribes be sold into slavery, so we can assume that after AD 9 he did not see fit to do otherwise, cf. Mócsy 1971, 44; Mócsy 1974, 34; Colombo 2010, 183.

[63] In her synthesis of the auxiliaries of Illyrian and Pannonian origin, A. Domić Kunić (1988, 86-92) lists all the known relevant inscriptions. However, with the exception of the diplomas, the dating of all the pre-Flavian inscriptions, mostly taken from the first publications, remains quite broad – she mostly speaks of the early Imperial period, the beginning of the Principate, the time of Claudius or Nero, the time until Vespasian, until Domitian etc, which is not a sound foundation for a hypothesis on mass recruitment after the quelling of Bato's rebellion. Saddington (1982, 160) also points out that the forming of Dalmatian and Pannonian auxiliary units cannot be definitely associated with specific historical events, beginning with the wars during Augustus up until AD 69.

auxiliaries of Pannonian and Dalmatian origin could have been recruited towards the very end of Tiberius' rule. For instance, Dases Dasmeni was recruited around AD 28-29, or perhaps a year or two earlier if we allow the possibility that he served for more than the regular 25 years, but even so this could not have been long before AD 25.[64] By all accounts, the Cornacatus Dases joined the Roman army some 20 years after the revolt was quelled, or maybe (a very strong maybe) somewhat earlier, but in any case this could not have been immediately after the end of the war, which Dases himself could have possibly experienced only as a small boy. Interestingly, he did not serve in a unit made up of his compatriots, but joined a unit set up on the Iberian peninsula and transferred to Pannonia evidently during the rule of Tiberius at the latest, if not even earlier (that is, in AD 28-29 at the latest, as this diploma is also the earliest mention of the II cohort of Hispanians).[65] The same is true of the Varcianus Iantumarus Andedunis, who joined the same unit some 5 to 6 years later,[66] as did Dasius Carmai, who joined his II ala of Hispanians and Aravaci around AD 36.[67] Dases, Iantumarus and Dasius were not recruited into any of the Pannonian, Breuci or Varciani units, probably because there was no massive recruitment of Pannonians and no raising of auxiliary units from that part of the Empire at the time they started serving voluntarily in the Roman military. In deciding for a military career, these young men understandably chose to serve in a unit that was stationed in Pannonia, that is, close to home, and which had obviously already had to fill its ranks by recruiting locals, as Hispanians were no longer easy to come by in sufficient numbers. Based on the available epigraphic material, it seems that the first Breuci, Pannonian and Varciani cohorts and Pannonian alae (with the exception of the mentioned *ala Pannoniourum*) were not raised before the very end of Caligula's and the beginning of Claudius' rule. Numerous authors believe that the cohorts of the Breuci and other Pannonian cohorts were raised immediately after the crushing of the rebellion in AD 9, in order to remove from Pannonia as many young and able-bodied men as possible.[68] As has been already mentioned, I cannot agree with this argument because there are no epigraphic and/or written sources to corroborate the assumption on such early raising of these Pannonian auxiliary units. If we compare the dates that diplomas were issued to the members of the Breuci cohorts, as well as the roughly dated epigraphic monuments mentioning these units, the earliest possible date for the raising of any Breuci cohort would be the one based on the diploma from Negoslavci.[69] It was issued on 17 June 65, to Liccaius, son of Liccaius (definitely a Pannonian in terms of onomastics and probably a member of the Breuci tribe), a soldier of the VII cohort of the Breuci, stationed at the time in Upper Germania; presumably Liccaius joined the army sometime around 39-40, if not a year or two earlier. This would mean that the VII cohort of the Breuci existed from this moment at the latest, if we assume that Liccaius was among the first recruits of the newly organised cohort, which I personally consider quite likely.[70] All the other known diplomas of members of the Breuci cohorts were issued at a later date so they do not provide us with any data on the possible Augustan or early Tiberian date of the foundation of certain Breuci cohorts.[71] It is indicative that the VII cohort of the Breuci existed at least from 39 or 40, which would mean that the cohorts numbered from I to VI were founded before it or at the same time. If we assume that the Breuci cohorts were raised in one wave, it is possible that they were formed for the purpose of the (unsuccessful) military operations of Caligula in Germania in 39, as well as for his frustrated invasion of Britain, because new legions as well as many new auxiliary units were also raised on that occasion.[72] The earliest epigraphic sources for the cavalry cohort of the Latobici and Varciani (*cohors Latobicorum et Varcianorum equitata*) prove that this unit undoubtedly existed at the beginning of Nero's rule, so it seems likely that it was formed already during Claudius, if not sooner.[73] The earliest proof of the existence of the second cavalry cohort of the Varciani (*cohors II Varcianorum equitata*) refers to Vespasian's rule, but its activities prior to AD 69 remain obscure.[74] The cohorts of the Pannonians (*cohors I Augusta Pannoniorum, cohors I Ulpia Pannoniorum, cohors I Pannoniorum equitata veterana, cohors I Pannoniorum, cohors I Pannoniorum equitata, cohors I Pannoniorum et Delmatarum equitata civium Romanorum, cohors II Pannoniorum* and *cohors II nova Pannoniorum*) also left no epigraphic traces that would unquestionably indicate their existence prior to Nero's rule, and some were definitely not formed before the 2nd century.[75] There is also no clear evidence of an early formation of the alae recruited in Pannonia, with the probable exception of the *ala Pannoniorum.*[76] *Ala I Pannoniorum Sabiniana* was perhaps founded during Tiberius, if we accept the rather dubious assumption that it received its name from the commander *Nymphidius Sabinus*;[77] inscriptions of a more definite dating, as well as diplomas, originate from the 2nd century.[78] *Ala II Pannoniorum veterana* existed definitely during Nero's rule; it was very probably founded even before his rule, but just how much earlier remains unknown.[79] *Ala Pannoniorum*

64 Brunšmid 1897, 1-6; *CIL* XVI 2; Saddington 1982, 64; Domić Kunić 1988, 90, cat. 20; Lőrincz 2001, 153, Cat. 1; Ilkić 2010, 267.

65 Wagner 1938, 152-153; Spaul 2000, 124-125; Lőrincz 2001, 36.

66 *CIL* XVI 4; Domić Kunić 1988, 89, cat. 18; Lőrincz 2001, 153, Cat. 2.

67 *RMD* 202; Dušanić 1998, 51-62; Koledin 2000, 231-239.

68 Mócsy 1959, 75, 120-121; Mócsy 1962, 645; Alföldy 1968, 48; Bogaers 1969, 36-46; Dušanić 1978, 467.

69 *RMD* 79; Dušanić 1978, 461-475; Dorn 1984, 165-174; Domić Kunić 1988, 90, cat. 19.

70 Dušanić on the other hand believes that Liccaius belonged to the second generation of the unit's recruits and that he was transferred from his homeland to Germania, as part of the reinforcements for that unit, which may have been the standard practice during the Julio-Claudian dynasty, cf. Kraft 1951, 35; Alföldy 1968, 99-100; Dušanić 1978, 467.

71 Kraft 1951, 23; Spaul 2000, 317, 320-325.

72 Dio 49.21-22; Ritterling 1925, 1244-1249; Holder 1980, 113; Knight 1991, 189; Wiedemann 1996, 226-228; Alston 1998, 54-55.

73 *CIL* XVI, 158; Spaul 2000, 328.

74 Spaul 2000, 329.

75 Spaul 2000, 330-338.

76 Cichorius 1894, 1255; Wagner 1938, 56-57; Kraft 1951, 25; Alföldy 1962, 262-263; Spaul 1994, 167-172; Spaul 1995, 63-73; Lőrincz 2001, 22.

77 *CIL* III 4269; the ala is not mentioned by name in this inscription.

78 Spaul 1994, 189-190.

79 Spaul 1994, 173-175; Ardevan 2007, 139-140.

Tampiana Victrix was founded perhaps already during Augustus, but as there is no confirmation for its existence before the Flavian period, perhaps the likelier assumption is that it was founded by the Pannonian legate *Tampius Flavianus* during the civil war in 69.[80] *Ala I nova Illyricorum* seems to have been formed in the 2nd century from the cavalry vexilations sent to Dacia from Illyricum, that is, probably from Dalmatia and Pannonia.[81] Like the inhabitants of the province of Dalmatia, the Pannonians also served in the navy, probably in considerable numbers, but here also the epigraphic sources do not supply indisputable evidence of the recruitment of seamen from Illyricum prior to the rule of Claudius.[82] This, of course, does not mean that they could not have served in the navy already during the rule of Augustus or Tiberius, but it does not speak in favour of the thesis that the indigenous inhabitants of Dalmatia were recruited in large numbers already after the crushing of Bato's uprising. It is not insignificant to note that, judging by the epigraphic material, we also cannot claim with certainty that the Delmatian cohorts were raised before the rule of Claudius, or in any case not much sooner than that; Caligula's recruitment of troops around 39 seems the earliest possible date of their formation.[83]

It seems that the second wave of auxiliary recruitment in Pannonia could have ensued during the rule of Claudius, around AD 50, and this is how it is put forward by Tacitus in his Annals.[84] This, of course, does not mean that some units could not have been raised much sooner, but here we enter the realm of pure conjecture with no solid arguments. Caution demands that we do not reject the possibility that some auxiliary troops in Pannonia were raised already during Augustus, which is actually quite probable at least for one ala, or during Tiberius. However, despite the common opinion that the Pannonian auxiliary units were levied soon after the crushing of Bato's uprising, the lack of concrete evidence makes me more inclined to believe that their formation, at least in more substantial numbers, did not occur before Caligula's rule. In any case, there can be no word of a more massive recruitment of Pannonians before the formation of real Pannonian auxiliary units. With regard to the period before Caligula or Claudius, we can only speak about individuals joining the Roman army. This, of course, does not mean that there were only small numbers of such individuals, but it is only the levying of auxiliary units from the area that can be seen to indicate a more substantial interest of the Roman army and state for the recruitment of the local population. In any case, the epigraphic material at this time does not suggest that the recruitment of the local population came into full swing already during Tiberius, but only gives us evidence that towards the end of Tiberius' rule some Pannonians definitely started serving in the Roman army. We must, of course, keep in mind that no diplomas were issued before the time of Claudius, that is before AD 52, so that they cannot serve as the only or key argument in an attempt to date the formation of certain auxiliary troops, at least not for the period before 26 or 27. Still, in the lack of other equally precise sources for dating, these diplomas provide very important information about the possible chronological frame for the existence of individual cohorts or alae.

The final crushing of Bato's uprising meant that there was no further need for a large concentration of troops in Illyricum, so that a great part of the Roman military forces was transferred to other provinces, especially the Rhine *limes*, where the situation had deteriorated significantly due to Varus' defeat and the loss of three legions. It seems that after AD 9 three legions remained in Pannonia: *VIII Augusta, IX Hispana* and *XV Apollinaris*.[85] Although there is no disagreement in the scholarly literature regarding their identification, the question of the garrison remains open.[86] According to Tacitus, at the moment of the military uprising in AD 14, all three legions shared a large summer camp, maybe near Poetovio or Siscia, but it is quite certain that they were otherwise not stationed together, at least not during the summer months.

The fewest dilemmas exist regarding the permanent camp (*hiberna*) of VIII legion, whose garrison in this period is placed with a great deal of certainty in Poetovio, as confirmed also by epigraphic evidence.[87] At the latest in AD 14 or 15, *XV Apollinaris* was stationed in the north, perhaps in Vindobona (in the older literature it was considered to have already been in Carnuntum at the time, but it seems that it arrived there during Claudius, around 39 or 40, or maybe as late as 49), but its location prior to AD 14 remains unknown.[88] It was previously believed that it could have been in Emona, but the opinion that prevails today is that Emona was not a legion camp.[89] Consequently, Siscia is suggested as the possible garrison. Siscia is most often connected to *IX Hispana*, which is believed to have been stationed in this town from the time of Tiberius until its departure for Britain in 43, that is, at the latest between 42 and 45, when Siscia no longer had a legion camp.[90] We also cannot exclude the possibility that Sirmium may have been

80 Spaul 1994, 215-216.
81 Spaul 1994, 150-151.
82 Alföldy 1965, 174-175; Domić Kunić 1996a, 52-65; Miškiv 1998, 92-98; Spaul 2002, 68, 73.
83 Kraft 1951, 22-24; Saddington 1982, 160; Knight 1991, 189, 191-192; Spaul 2000, 299-300, 302-314; for an opposing view cf. Alföldy 1965, 174.
84 Tac. *Ann.* 12.29; Mócsy 1959, 243-244, Cat. 158/ 2, 6-8; Mócsy 1971, 43, 45; Mócsy 1974, 51-52.

85 Tac. *Ann.* 1.16-30; Ritterling 1925, 1363; Mócsy 1959, 18, 25, 28, 77; Klemenc 1961, 7; Mócsy 1962, 612-613; Klemenc 1963, 57; Mócsy 1974, 42-43; Fitz 1980, 131; Šašel Kos 1995, 236-237; Wilkes 1996, 569-571; Ferjančić 2002, 29; Lolić 2003, 133-134; Farnum 2005, 60; Dzino 2010, 167-168.
86 The literature mentions Poetovio, Siscia, Sirmium, Savaria, Emon, and Carnuntum; for an overview of hypotheses cf. Wheeler 2000, 271.
87 Ritterling 1925, 1645-1646; Reddé 2000, 121; *CIL* III 10878 = *AIJ* 371, *CIL* III 10879 = *AIJ* 381, *CIL* III 4060 = 10869 = *AIJ* 260, *AIJ* 262.
88 Wheeler 2000, 271-274; Mosser 2003, 14, 16; Gugl 2003, 55-57; Harl 2003, 53.
89 Šašel Kos 1995, 227-244; Ferjančić 2002, 36-38.
90 It cannot be entirely excluded that the IX Hispana could have stayed also in Sirmium after Bato's rebellion, cf. Mócsy 1971, 46. the IX legion was temporarily (from 20 to 34 AD) transferred to Northern Africa, when it returned to Pannonia, probably to Siscia: Ritterling 1925, 1647, 1665-1666; Mócsy 1959, 25; Mócsy 1962, 613; Klemenc 1961, 8,10; Mócsy 1974, 43; Šašel 1974, 734-735; Fitz 1980, 131; Hoti 1992, 142; Keppie 2000, 26; Radman-Livaja 2010a, 190.

a legion camp after the crushing of the revolt and during the rule of Tiberius, at least for a short time, perhaps as the garrison of IX or XV legion.[91]

Roughly at the same time, between 43 and 45, *VIII Augusta* left Poetovio, where it was replaced by *XIII Gemina*.[92] Some time after that, in 62 or 63, *XV Apollinaris* headed east and was replaced in Carnuntum by *X Gemina*.[93] When the legion left Siscia, Poetovio remained the only legion camp in southern Pannonia, but the legion garrison (*XIII Gemina*) also left this town at the end of the 1st or the beginning of the 2nd century, some time between 97 and 106.[94] It is possible that in the late Flavian period Poetovio was perhaps not the only legion camp south of the Drava River, because, probably for a short period during Domitian's rule, a legion camp could have existed in Mursella near Mursa (today's Petrijevci). It is assumed that *XXI legio Rapax* was stationed there roughly from 89 or 90 to 92 or 93, when this legion was destroyed by the Sarmatians (Suet. *Dom.* 6). After this defeat, *XIV Gemina* may have taken its place for a short time, until it finally left the camp at the end of the 1st century. As there are no reliable archaeological traces, we cannot completely exclude the hypothesis that in this short period of time the legion garrison was stationed in Mursa. On the other hand, XXI legion might have never stayed in this place, or even in Pannonia.[95] Mursella and Mursa are not the only hypothetical legion camps in southern Pannonia at the end of the 1st century. It is not entirely impossible that Sirmium had a legion garrison in this short period.[96] The literature mentions the possibility that *legio II Adiutrix* could have stayed there.[97] It is also not impossible that *legio I Adiutrix* was stationed in Sirmium between 86 and 89.[98] As there is no definite evidence of the legions' stay in Sirmium itself,[99] it cannot be excluded that it may have been stationed in the vicinity. Archaeological investigations in Dobrinci in eastern Syrmia revealed traces of a large Roman camp, built probably in the 1st century, and it is quite possible that it served to accommodate a legion, perhaps *legio II Adiutrix*, during Domitian's rule.[100] Although after the IX legion left it, Siscia never again had the status of a permanent legion garrison, certain legion vexilations also could have stayed in the town occasionally in times

Fig. 2. Funerary stone of M. Mucius Hegetor, surgeon of the cohors XXXII voluntariorum (CIL III 10854, Photo Archive, AMZ).

[91] Mócsy 1971, 45-46; Mócsy 1974, 43; Šašel Kos 1995, 237.

[92] Ritterling 1925, 1647-1649, 1713; Klemenc 1961, 7; Mócsy 1962, 613; Mócsy 1974, 48; Fitz 1980, 131; Reddé 2000, 121; Wolff 2000, 203-204; S. Ferjančić (2002, 29-30) believes that this did not happen before 57; cf. also Sanader 2003b, 465; Sanader 2008, 97.

[93] Mócsy 1962, 613; Fitz 131; Wheeler 2000, 274-278.

[94] An earlier date, that is the period before the year 101, seems more probable: Ritterling 1925, 1714-1716; Syme 1928, 49-50; Alfföldy 1959, 126, 137, 140-141; Mócsy 1962, 614; Mócsy 1974, 88, 92; Ferjančić 2002, 30-31; Horvat *et al.* 2003, 156.

[95] There is also Klemenc's assumption on the presence of *X Gemina* in Mursa, perhaps during Galba or Trajan, before the Dacian campaign: Ritterling 1925, 1277-1279, 1736, 1788-1789; Klemenc 1928, 271-275; Syme 1928, 42-45, 49-51; Alföldy 1959, 124-126, 137-138, 140-141; Klemenc 1961, 9, 18-19; Mócsy 1962, 613-614; Klemenc 1963, 58, 60-62; Fitz 1980, 131; Lőrincz 1982, 285-286; Bérard 2000, 56-60; Sanader 2003b, 466; Perinić Muratović 2004, 98-99; Sanader 2008, 97, 99-101.

[96] It should be mentioned that in that same period, that is, from 86 to 106 at the latest, Sirmium, like the entire stretch of the territory along the frontier from the Drava estuary to the Sava estuary, was perhaps administratively attached to Upper Moesia, but this hypothesis has not been unanimously accepted: Alföldy 1959, 115; Dušanić 1967, 69; Mirković 1971, 28; Mócsy 1974, 92.

[97] Klemenc 1961, 23; Klemenc 1963, 67-68; Dušanić 1968, 87-90; Mirković 1971, 27-28, Eadie 1977, 210-212; Dušanić 1988, 85-86. Although Mommsen (*CIL* III, p. 420) assumed that it may have stayed in Acumincum, based on a probably mistaken Ptolomeus' statement, it seems perhaps more probable that for a shorter period during Domitian it stayed in Singidunum, if not in Sirmium. For an overview of these issues and the literature cf. Alföldy 1959, 119-121.

[98] Alföldy 1959, 121-123, 127-128; Klemenc 1961, 23; Klemenc 1963, 67; Mirković 1971, 28-30; Eadie 1977, 210-212.

[99] Mirković 1971, 30.

[100] Dušanić 1968, 89; Mirković 1971, 28; Petrović 1995, 222, 224.

of crisis; for example, a vexilation of legion *XIV Gemina* is thought to have stayed in Siscia during Domitian's wars on the Danube.[101]

While starting from the middle of the 1st century permanent legion garrisons were transferred from southern Pannonia, that is the area between the Sava and the Drava Rivers, the auxiliary units remained in the area, both in the hinterland and on the Danubian frontier. Although we know more about their deployment in the Pannonian region during the later Julio-Claudian period than during Augustus, many dilemmas and uncertainties still exist. As for southern Pannonia, it seems highly probable that in this period auxiliary units were stationed in Poetovio, Neviodunum, Siscia, Mursa and Sirmium, and we can assume that they stayed also at some other locations in the area between the Sava and the Drava, if not during Tiberius, then probably during Claudius at the latest. Because of the legion's presence, auxiliary units could be expected in Poetovio and Siscia,[102] but the epigraphic evidence is inconclusive. Siscia probably kept a garrison even after the legionaries had left, and it is assumed that in addition to the beneficiaries and seamen of the river fleet, an auxiliary garrison also remained at least for some time. It is possible that *cohors XXXII voluntariorum civium Romanorum* was stationed in the town throughout the best part of the 1st century. Presumably, it was in Siscia already during the stay of IX legion, and it seems that it remained in the town up until Vespasian's time, when it was transferred to Germania Superior.[103] This assumption is based on two funerary inscriptions (fig. 2). One stele comes from Siscia itself, and it belonged to Mucius Hegetor, the physician of the XXXII cohort,[104] and the other, erected for a veteran of the same unit, Nunnidius Successus, was found in Orešac.[105] The latter find cannot contribute significantly to the discussion on the garrison of XXXII cohort during its stay in Pannonia, because, notwithstanding the accepted dating, a veteran's funerary monument cannot be considered as reliable evidence that his former unit had been present in the vicinity of his last residence, all the more so in this case, as this vicinity is somewhat of a relative term considering that some 100 km lie between Orešac and Siscia. Nevertheless, if we discard the possibility that the physician Mucius Hegetor stayed in Siscia on a special assignment, separated from his home unit, his stele can be considered as quite reliable evidence that *cohors XXXII voluntariorum civium Romanorum* could have been stationed in Siscia at least in the mid 1st century, if not for longer. A further argument for its stay in Pannonia in the Claudian period is a votive inscription of the knight Gaius Antonius Rufus, who seems to have done each of his three turns of military service in Pannonia (*cohors XXXII voluntariorum, legio XIII, ala I Scubulorum*).[106] However, the hypothesis on the stay of this cohort in Siscia is not generally accepted: some consider that this unit was not founded before the rule of Commodus and the early dating of the two mentioned monuments is thus called into question.[107]

Although some auxiliary units were very probably stationed in Poetovio during the 1st century in addition to the legion crew, it is only in the Flavian period that the presence of the first Alpine infantry cohort (*cohors I Alpinorum peditata*) can be assumed with any degree of certainty. By all accounts, this unit also stayed in Lower Pannonia at the beginning of the 2nd century, probably in Lusonium, so it may be assumed that since the end of the 1st century there were no auxiliary units in Poetovio.[108] An exception to this is a brief period during the Marcomannic wars when the *cohors II Aurelia Dacorum milliaria equitata pia fidelis* seems to have stayed in the town.[109] A few early archaeological traces of the auxiliaries' presence were found in Mursa, namely of the second ala of Hispanians and Aravaci (*ala II Hispanorum Arvacorum*) and the second Alpine cavalry cohort (*cohors II Alpinorum equitata*). The stele of the cavalryman Niger Sve<l>trius is dated to the first half of the 1st century, but it seems that this ala was soon transferred from Mursa to Teutoburgium, as demonstrated by the tombstone of a decurion Tiberius Claudius Valerius, as well as a few fragmentary inscriptions and the countermarks *al(ae) (II) Ar(vacorum)* on coins, dated approximately from the late Tiberian to the early Claudian period.[110] The stele of a cavalryman Velagenus from the second Alpine cohort is dated to the Claudian period, that is, to the mid 1st century, so it can be assumed that this cohort replaced the second ala of Hispanians and Aravaci in Mursa, if it had not been in Mursa even earlier, during the stay of this ala.[111] Roman military presence can generally be assumed in Neviodunum, Mursa and Sirmium, as well as Acumincum and Burgenae, on the basis of early forms of *terra sigillata* finds.[112] He same is more or less true of the sites at which Augustan bronze money with the countermark AVG was found, but although such finds probably indicate the presence or at least the passage of soldiers, they cannot serve as a final proof that Roman military units

101 Dušanić 1983, 20-21; Hoti 1992, 143.

102 This could perhaps have been 3 alae and 6 cohorts: Vell. 2.117,1; Alföldy 1968, 137; Tóth and Vékony 1970, 155; Lőrincz 2001, 57.

103 Cichorius 1900, 356; Wagner 1938, 201; Kraft 1951, 199; Mócsy 1962, 624; Mócsy 1974, 81; Holder 1980, 156; Lőrincz 2001, 44.

104 It is dated to the mid 1st century; *CIL* III 10854; Zaninović 1981, 205; Lőrincz 2001, 297, Cat. 479.

105 It is dated to the Flavian period; *CIL* III 4006; Kraft 1951, 199, Nr. 3140; Lőrincz 2001, 297, Cat. 480.

106 The inscription was dated to Nero's time; *CIL* III 386 = *ILS* 2718; Lőrincz 2001, 208, Cat. 166.

107 Spaul 2000, 47-48.

108 Cichorius 1900, 238; Wagner 1938, 83-84; Radnóti and Barkóczi 1951, 200; Visy 1986, 491-492, 494-495; Spaul 2000, 262; Lőrincz 2001, 28, 70, 105.

109 Lőrincz 2001, 34.

110 *CIL* III 3286 and 10262 (Mursa), *CIL* III 3271 (Teutoburgium), *CIL* III 14039 (Pécs, allegedly comes from Dalj), *CIL* III 3273 = 10258 (Teutoburgium); Brunšmid 1909, 170-171, no. 368; Cichorius 1894, 1230; Wagner 1938, 47-48; Kraft 1951, 151; Mócsy 1962, 619; Pinterović 1968, 68-69; Tóth and Vékony 1970, 156; Eadie 1977, 214; Beneš 1978, 10; Holder 1980, 156; Saddington 1982, 67, 95; Kos 1984, 49-54; Visy 1986, 492-495; Lőrincz 2001, 21, 198-199, Cat. 133, 135, 136.

111 *ILJug* 3104; Cichorius 1900, 239; Wagner 1938, 84-85; Radnóti and Barkóczi 1951, 201; Mócsy 1962, 621; Lőrincz 2001, 27, 59, 229, Cat. 234.

112 Gabler 1979, 199-215; Lőrincz 2001, 57; Fitz 2003, 50; Gabler 2006, 87-88, 91-92, 95.

stayed there over an extended period of time.[113] There is a hypothesis, backed by very convincing arguments, that there was an auxiliary garrison in Andautonia during the rule of the Julio-Claudian dynasty, or at least during the first decades of the 1st century, but this has not been proved by archaeological evidence.[114] Although the lack of finds meant that until recently Cuccium was not mentioned in the literature in the context of early camps on the Danube *limes*, the find of a soldier's (or veteran's) grave from the 2nd quarter of the 1st century, as well as some other early Roman graves, indicate that there is a real possibility of Roman military presence at this location during the Julio-Claudian House, if not during the late Tiberian, then certainly during the Claudian period.[115] It can be convincingly presumed that *cohors II Hispanorum equitata* stayed in Pannonia during the 1st century, at least since the rule of Tiberius, if not before, all until the end of the century, when it was transferred to Upper Moesia, but its whereabouts during this period remain unknown.[116] Although there is no concrete evidence, the possibility that it was stationed in southern Pannonia cannot be excluded.

5. The development of the *limes*

This necessarily brief overview shows quite evidently how difficult it is to make far-reaching conclusions about the beginnings of the development of the Roman military system on the Pannonia's Danube frontier south of the Drava. Unfortunately, due to the lack of more comprehensive systematic investigations and the scartown of epigraphic material, it is no easier to trace its development after the Julio-Claudian period. While some sites can boast of at least some published material, others receive only summary mention in survey reports or archaeological rescue excavations reports. Therefore, any attempt at a synthesized chronological presentation of the development and establishment of this part of the *limes* from the Julio-Claudian period to the time of Hadrian's rule, and even later, is made substantially more difficult by the lack of sufficient reliable data. Still, some kind of global picture, even an incomplete one, can be gained from the individual sites, which can, with more or less certainty, be considered to have belonged to the Roman military fortification network on this part of the Danube *limes*. As an introduction to this subject, it should be stressed that although some fortifications undoubtedly existed already in the Claudian period, if not even sooner, it is quite probable that this stretch of the *limes* began acquiring its shape during the Flavian period, probably during the rule of Domitian, when the more substantial transfer of troops to the Danubian frontier began. It acquired, if not its final shape, then definitely a more lasting one in the Principate, during the rule of Trajan and Hadrian.[117] It was in fact during Trajan's rule, between 103 and 107 (probably in 106) that Pannonia was divided into two provinces, *Pannonia Superior* with the seat in Carnuntum, and *Pannonia Inferior* with the seat in Aquincum. This division was undoubtedly the result of the reorganisation of the Roman defence potential on the Danubian frontier, with Upper Pannonia confronting the Quadi and Lower Pannonia the Sarmatians.[118]

Epigraphic finds play a key role in the efforts to identify the military units that were stationed in certain camps on the Pannonian section of the Danubian *limes* at a certain point in time. Written sources are sometimes also helpful, but, with the exception of the *Notitia Dignitatum*, still do not provide substantial information. Unfortunately, at least in the case of that part of the *limes* discussed here, the epigraphic finds are scarce and contain relatively little information. Luckily, military diplomas from Lower Pannonia help in locating the camps of certain units at certain points in time. Two Hungarian archaeologists, A. Radnóti and L. Barkóczi, noticed a long time ago that the sequence of listing the units changed on the Pannonian diplomas around the middle of the 2nd century, that is, during the rule of Antoninus Pius. Until then, the units were listed by ordinal numbers within their groups (*alae* are mentioned first, followed by cohorts and then fleets). Although in the new scheme alae are listed before cohorts, the units' ordinal numbers are no longer necessarily connected to the sequence of the units on the diploma. Still, the repeated listing of the same or similar sequence on several Lower Pannonian diplomas issued at different points in time indicated that certain logic was respected in the process. The conclusion was that the sequence of the units' names corresponds to the geographic location of their fortifications, from the north southwards. Further analysis showed this „topographic" or „crypto-topographic"[119] sequence to be quite probable. Despite certain incongruities and exceptions,[120] and despite the fact that not all the units are mentioned on each diploma, this sequence is a very useful indication, although not reliable proof, in determining the garrisons of certain troops in Pannonia, beginning roughly at the time of Hadrian's rule all until the time of Marcus Aurelius.[121]

A few certain, probable or possible Roman fortifications situated between Altinum and Mursa were located north of the Drava and thus are not covered by the geographic frame of this paper. Nevertheless, they still deserve to be mentioned here, given that they belong to the Croatian part

113 Nauportus, Emona, Celeia, Poetovio, Siscia, Mursa, Teutoburgium, Cuccium, Acumincum, Burgenae, Taurunum, etc., cf. Kos 1984, 48.

114 Fitz 2003, 50.

115 Dizdar *et al.* 2003, 57-77; Tomičić *et al.* 2007, 14-15; Tomičić *et al.* 2008, 15-16; Dizdar 2010, 244-245.

116 Cichorius 1900, 299-300; Wagner 1938, 152-154; Kraft 1951, 178; Beneš 1978, 39-40; Saddington 1982, 64, 95; Visy 1986, 491-495; Spaul 2000, 124-125; Lőrincz 2001, 36, 57.

117 Mócsy 1974, 48-52, 80-103; Eadie 1977, 212-220; Barkóczi 1980, 93-95; Soproni 1980, 220-221; Petrović 1986, 91-93; Visy 1988, 26-27; Wilkes 2000, 580-581; Gabler 2003, 37-39; Visy 2003b, 109-110; Sanader 2010, 222.

118 Mócsy 1962, 586-587; Mócsy 1974, 92-94; Barkóczi 1980, 93-94; Wilkes 2000, 581.

119 A crypto-topographic sequence would be a sequence in which the units with the same ordinal number are ordered according to topography and not alphabetically, although the list itself is numerical, cf. Visy 1986, 484-485.

120 Cf. Roxan 1986, 775-777; Visy 1986, 483-490, 498-517; Lőrincz 2001, 84-85.

121 Radnóti and Barkóczi 1951, 191-226; Fitz 1959, 438-442; Dušanić 1968, 94; Roxan 1986, 768-777; Visy 1986, 483-517; Lőrincz 2001, 79-80, 84-85, 87-89; MacDonald and Pangerl 2003, 266.

of the Baranja. The northernmost place, close to today's Croatian-Hungarian border, is Batina (Batina Skela) or more precisely the site Gradac above Batina Skela. In Roman times it was called Ad Militare, and the existence of a fortification was confirmed also by some smaller-scale excavations at the beginning of the 1970-ies.[122] The camp is believed to have been built during Domitian's rule, around 85 or 86, but there are no reliable indications as to which unit may have been stationed there at the time. The first unit to have been stationed there beyond doubt is *cohors II Augusta Thracum equitata*, roughly from the beginning of Hadrian's rule until the beginning or the end of the Marcomannic wars.[123] *Cohors II Asturum et Callaecorum equitata* probably arrived there after the Marcomannic wars and stayed at least until the middle of the 3rd century, although it has sometimes been claimed that it was stationed in Ad Militare already during the first half of the 2nd century. This is probably not correct, but it cannot be entirely excluded that this cohort garrisoned the camp in the late Flavian period, as it was definitely present in Pannonia at the time.[124] Thanks to the *Notitia Dignitatum* we know that in the 4th century *equites Flavianenses* stayed at Ad Militare but it is impossible to identify a more precise chronological frame.[125] Ad Novas, today's Zmajevac, is located further to the south. Although there are assumptions that the camp was built in the 2nd century, investigations have so far discovered mostly traces of Late Antiquity, and there are also no epigraphic evidence which would indicate a military presence with any certainty earlier than the 4th century,[126] when a detachment of the *equites Dalmatae* and the *auxilia Novensia* stayed there.[127] Bansko brdo (Baranjska planina) or more precisely the village of Kamenac on Baranjska greda near Kneževi Vinogradi is the place where some authors place Aureus Mons, a location believed to have possibly been a military fortification or a station of a part of the *legio VI Herculia* in the 4th century. Despite the remains of Roman architecture and a 4th century cemetery, there is no reliable archaeological evidence for this assumption.[128] Dragojlov brijeg near the Jasenovac wasteland south of Kneževi Vinogradi is also believed to be a Roman site and a possible military fortification. Quite a while ago, a stone fragment with an inscription was discovered there, probably part of the base of a sculpture or an altar, which M. Bulat interpreted as follows: *Mar[ti et] | Victo[riae] | pro sal[ute] | Aug(usti) | n[ostri]*.[129] The Roman character of the site was confirmed during small-scale excavations in the 1980s. In addition to graves from Late Antiquity, traces of Roman architecture were also discovered, namely the foundations of walls of several rooms with a partially preserved floors and stone sewage. A brick with the stamp *c(ohortis) I Ulp(iae) Pan(noniorum)* was found in one of the foundations.[130] The site is most probably located on the

Fig. 3. Votive plaque of cohors I Alpinorum (after Radman-Livaja 2004b, Photo Archive, AMZ).

[122] Klemenc 1961, 17; Klemenc 1963, 59; Pinterović 1968, 59-60, 62, 73-74; Bulat 1969, 40-41; Visy 1988, 126; Minichreiter 1987, 88-90; Minichreiter 1989b, 182; Sanader 2003a, 137; Sanader 2008, 104; Sanader 2010, 225-227.

[123] Cichorius 1900, 338; Wagner 1938, 192-193; Radnóti and Barkóczi 1951, 201, 214-215; Nagy 1956, 68; Fitz 1962, 66; Mócsy 1962, 624; Spaul 2000, 367; Lőrincz 2001, 43, 89-90, 104.

[124] Cichorius 1900, 248; Wagner 1938, 91; Kraft 1951, 168; Radnóti and Barkóczi 1951, 201, 213; Nagy 1956, 68, Anm. 182; Fitz 1962, 64-65; Mócsy 1962, 621; Dušanić 1988, 87; Spaul 2000, 81; Lőrincz 2001, 29, 98, 100, 104.

[125] *Not. Dign. Occ.* XXXIII.45.

[126] *Not. Dign. Occ.* XXXII.9, 28, 40; Klemenc 1961, 17; Radnóti and Barkóczi 1951, 214-215; Fitz 1962, 66-67; Pinterović 1968, 58-61, 71, 73; Bulat 1969, 41; Minichreiter 1987, 90; Visy 1988, 126; Sanader 2003a, 138; Sanader 2008, 104-105; Filipović 2010, 10-27; Sanader 2010, 227-228. Although they are sometimes mentioned, there is no proof that *cohors I Montanorum, cohors III Alpinorum and cohors VII Breucorum* stayed there. *Cohors I Montanorum* was probably stationed in Cornacum from the time of Trajan until the Marcomannic wars, when it was transferred to Ad Latus (cf. Lőrincz 2001, 39), *cohors III Alpinorum* went from Dalmatia to Upper Pannonia in the 3rd century, more precisely to Quadrata (cf. Lőrincz 2001, 28), and *cohors VII Breucorum* arrived to Lugio in Pannonia from Moesia at the beginning of the rule of Antoninus Pius. However, because the bricks produced by *cohors VII Breucorum* can be found throughout Lower Pannonia, the sites where they were found are often wrongly recorded in publications as the cohort's camps. (cf. Dušanić 1988, 87; Lőrincz 2001, 31).

[127] The text mentions *auxilia Novensia, Arsaciana sive Novas*, but Arsaciana is probably a corruption of Antiana. Does this mean that this unit was deployed in both places? As Antiana is thought to have been in nearby Popovac, or, less likely, in Branjin vrh (Pinterović 1968, 58-59; Pinterović 1969, 57-58; Minichreiter 1989b, 182-183), it cannot be excluded that the same unit had troops in both locations. Or is this a pseudocomitatenses unit that appears in the list of units under the command of *Magister Peditum Praesentalis*, under the name of Antianenses?, cf. *Not. Dign. Occ.* V.112, 262.

[128] *Not. Dign. Occ.* XXXII.45; Pinterović 1968, 58-61, 75; Bulat 1969, 44; Dušanić 1968, 91; Pinterović 1969, 58 (giving also other, more or less convincing locations); Bulat 1977, 79; Minichreiter 1989bis, 182; Sanader 2003a, 138; Sanader 2008, 105; Sanader 2010, 228 (in the first two publications, M. Sanader gives the toponym Banovo brdo, probably a variant of the name Bansko brdo).

[129] Bulat 1960, 14-15; Pinterović 1960, 35; Pinterović 1968, 69; Bulat 1969, 41.

[130] Minichreiter 1987, 91, 120; Minichreiter 1989a, 102-103; Minichreiter 1989b, 182; Minichreiter 1990, 147-148. It should be mentioned that bricks with this cohort's stamp have been found at several locations in Lower Pannonia, and *cohors I Ulpia Pannoniorum* is thought to have been tasked with producing bricks for a certain period of time, probably during the Severan period. For the greater part of the 2nd century it was stationed in Solva (Esztergom), where it most probably stayed also during the 3rd century. This camp was originally located in Upper Pannonia, but after the frontier between Upper and Lower Pannonia was changed in 214, it ended up in the territory of Lower Pannonia: Lőrincz 2001, 40-41, 100, 116-117, Cat. 408.

route of a Roman road, which diverted westward on this section due to the alluvial areas near the Danube. This road connected Ad Novas (Zmajevac) with Mursa and with Ad Labores (Nemetin), as there was probably a junction near Kopačevo, with one road leading further toward the river crossing at Nemetin and the main road veering westward toward Bilje, and from there south towards Osijek.[131] All these finds and the position on the route of a Roman road along the *limes* were the reasons to assume the existence of a Roman military station on this location.[132] This hypothesis could have been finally confirmed by a recent chance discovery in the Baranja, but the circumstances of the find remain dubious (fig. 3). In 2002, the AMZ came into possession of a votive plaque dedicated to the battle insignia of the *cohors I Alpinorum peditata Antoniniana* by the unit's commander Allius Exsuperatus.[133] Because of the attribute *Antoniniana* it can be dated to the Severan period, probably to the rule of Caracalla or Elagabalus, and it is an unambiguous trace of the stay of this unit in this part of the *limes*.[134] Votive plaques dedicated to battle insignia are mostly found within military fortifications, where they originally stood, so it can reasonably be assumed that this plaque also stood in a shrine within the principia.[135] If this hypothesis is corroborated by future research, the inscription on the above-mentioned bronze plaque would shed completely new light on the issue of the location of the garrison of *cohors I Alpinorum peditata* during the Severan dynasty and would indicate the existence of at least one more Roman camp in the Baranja. There is no doubt that members of this unit were present in Mursa in connection to the visit of the imperial family at the beginning of the 3rd century,[136] but the question of the location of their permanent garrison at this time remains unresolved. Dragojlov brijeg lies at a mere 20 km as the crow flies from Mursa, and Branjin Vrh and Popovac are not much further away either. The inscription on the plaque in question indicates that at the time of the Severan dynasty *cohors I Alpinorum peditata* could have been stationed in this area. It should be emphasised that, despite the contradictory statements as regards the location of the finds, Dragojlov brijeg seems the likelier choice for a military fortification

Fig. 4. Funerary stone of a decurio alae II Aravacorum (CIL III 10258 = 3273, Photo Archive, AMZ)

than the area around Branjin Vrh, but only more thorough research of both locations can resolve this issue. The lack of research makes it impossible to say for how long this camp functioned, but at present we can conclude that, whether on Dragojlov brijeg (I personally find this the likelier variant) or in the vicinity of Branjin Vrh, a fortification existed where *cohors I Alpinorum peditata* stayed, at least for a brief period. The Roman name of this location also remains a great mystery. Our knowledge about the names and position of Roman settlements and fortifications in the area is based primarily on the written sources such as *Tabula Peutingeriana, Intinerarium Antonini*, the Ravenna Cosmography and the *Notitia Dignitatum*. The last to have dealt seriously with the issue was D. Pinterović, and although her assumptions on the location of Roman sites in the Baranja are for the most part acceptable, she was also not able to identify with any certainty the location of some settlements and fortifications.[137] Despite numerous hypotheses, the question of the location of the settlements Aureus Mons (*Itin. Ant., Not. Dign.*), Albanum (*Not. Dign.*), Antiana (*Itin. Ant., Tab. Peut.*, Anon. Raven., *Not. Dign.*?) and Donatianae (*Tab. Peut.*, Anon. Raven.) has still not been finally resolved. Interestingly, there has been no special mention of Dragojlov brijeg in the literature in connection to hypotheses about these issues. Bearing in mind the results of the research conducted in the 1980s and the existence of at least one epigraphic monument, or two if the above-mentioned plaque originates from this area, Dragojlov brijeg should not be left out of the discussions on the location of these Roman toponyms.

Albanum, perhaps today's Lug, is also a late Roman fortification where *equites Dalmatae* were stationed, and a certain number of archaeological finds confirms the Roman character of the site.[138] Lug is also presumed to have been

[131] Pinterović 1961, 44-45; Pinterović 1968, 56-61, Pinterović 1969, 56-59; Sanader 2003a, 135-136.
[132] Bulat 1969, 41; Pinterović 1961, 44; Pinterović 1968, 69; Pinterović 1969, 56, 61, 63; Minichreiter 1987, 91, 120.
[133] Unfortunately, the original finder has passed away in the meantime and before arriving to the Museum the plaque changed the hands of at least three owners. It is therefore very unlikely that the exact circumstances of the find will ever be disclosed. Although all its former owners agree that it comes from the Baranja, there are two different stories about the site where it was found. According to one version, it was found near Dragojlov brijeg, south of Kneževi Vinogradi, while according to the other it was found in a field not far from Branjin Vrh, that is, Popovac, cf. Radman-Livaja 2004b, 113-131; for *cohors I Alpinorum peditata* and *cohors I Alpinorum equitata* cf. Cichorius 1900, 237-239; Wagner 1938, 80-84; Radnóti and Barkóczi 1951, 199-200, 210-213; Nagy 1956, 63-65; Fitz 1962, 61-63, 65-66; Mócsy 1962, 620-621; Spaul 2000, 257, 259-263; Lőrincz 2001, 27-28; Gayet 2006, 68-69.
[134] Lőrincz 1985, 186-187; Lőrincz 2001, 146-147.
[135] Domaszewski 1895, 14, 19-20, 27; Radman-Livaja 2004b, 117-118, 127-128.
[136] *CIL* III 10269; Brunšmid 1900: 25; Pinterović 1960: 29-30; Pinterović 1968: 66; Lőrincz 2001: 233, Cat. 250; Radman-Livaja 2004b, 120, 129-130.
[137] Cf. cited bibliography in Pinterović 1968, 58-61; Pinterović 1969, 56-59.
[138] *Not. Dign. Occ.* XXXII.9, 29; Klemenc 1961, 17 (due to a printing error, it says Sug!); Pinterović 1968, 59-61; Bulat 1969, 41; Pinterović 1969, 59; Minichreiter 1987, 93; Sanader 2003a, 139; Sanader 2008, 105.

Table 1. Limes garrisons in southern Pannonia, 1st-4th centuries AD.

Fort	1st century AD	2nd century AD	3rd century AD	4th century AD
Ad Militare (Batina)	Fort built during Domitian's reign (85-86 AD) ? first garrison unknown	*cohors II Augusta Thracum* (118-180 AD ?) *cohors II Asturum et Callaecorum equitata* (180-250 AD ?)	*cohors II Asturum et Callaecorum equitata* (180-250 AD ?)	*equites Flavianenses*
Ad Novas (Zmajevac)	-	?	?	*equites Dalmatae auxilia Novensia*
Aureus Mons (Kamenac?)	-	-	-	*legio VI Herculia*
Dragojlov Brijeg	-	?	*Cohors I Alpinorum peditata* (Severan period)	?
Albanum (Lug?)	-	-	-	*equites Dalmatae*
Mursa (Osijek)	*ala II Hispanorum Arvacorum* (Augustan period ? – early Claudian period) *cohors II Alpinorum equitata* (Claudian period – 118 AD) *legio XXI Rapax* ? (89/90-92/93 AD ?) *legio XIV Gemina* ? (93/94-99 AD ?)	*cohors II Alpinorum equitata* (Claudian period – 118 AD)	?	*legio VI Herculia classis Histrica*
Mursella (Petrijevci?)	*legio XXI Rapax* ? (89/90-92/93 AD ?) *legio XIV Gemina* ? (93/94-99 AD ?)	-	-	-
Teutoburgium (Dalj)	*ala II Hispanorum Arvacorum* (Claudian period – late 1st century ?)	*ala I Praetoria civium Romanorum* (early 2nd century – 200 AD)	*ala I civium Romanorum* 200 AD –late 3rd century ?)	*equites promoti cuneus equitum Dalmatarum legio VI Herculia*
Cornacum (Sotin)	Fort built in the Flavian period at the latest; *ala I civium Romanorum* (70-101 AD ?)	*cohors I Montanorum* (106-180 AD ?) *cohors II Aurelia Dacorum milliaria equitata pia fidelis* (180 – 3rd century AD ?)	*cohors II Aurelia Dacorum milliaria equitata pia fidelis* (180 – 3rd century AD ?) *cohors I Lusitanorum ?* (early 3rd century ?)	*equites Dalmatae cuneus equitum scutariorum equites promoti*
Cuccium (Ilok)	?	*cohors I Lusitanorum* ? (180-early 3rd century ?)	*cohors I Lusitanorum* ? (180-early 3rd century ?)	*cuneus equitum promotorum equites sagittarii*
Malata/Bononia (Banoštor)	Fort built during Domitian's reign, first garrison uncertain; *cohors I Campanorum* ? (late 1st century ?)	*ala I Flavia Augusta Brittanica milliaria civium Romanorum bis torquata ob virtutem* (97/105-250 AD ?)	*ala I Flavia Augusta Brittanica milliaria civium Romanorum bis torquata ob virtutem* (97/105-250 AD ?)	*praefectus equites Dalmatae*
Cusum (Petrovaradin)	?	?	*ala Pannoniorum* ?	*equites Dalmatae*
Acumincum (Stari Slankamen)	Fort built in the Flavian period ? *cohors I Cantabrorum* ? (early Flavian period ?) *cohors I Britannica milliaria civium Romanorum equitata* ? (late 1st century ?)	*cohors I Campanorum voluntariorum civium Romanorum* (113/114-200 AD ?)	?	*cuneus equitum Constantium equites sagitarii*
Rittium (Surduk)	*cohors I Britannica milliaria civium Romanorum equitata* (70-89 AD ?)	*vexillationes equitum ex Syria* (106-113/114 AD) *ala I civium Romanorum* ? (113/114-118/119 AD ?) *ala I Augusta Ituraeorum sagittaria* (118/119-200 AD or later ?)	*ala I Augusta Ituraeorum sagittaria* ?	*equites Dalmatae*

Fort	*1st century AD*	*2nd century AD*	*3rd century AD*	*4th century AD*
Burgenae (Novi Banovci)	Fort built in the Claudian period; *cohors II Asturum et Callaecorum* ? (Flavian period ?)	*cohors V Gallorum* ? (106-118/119 AD ?) *ala I civium Romanorum* (118/119-138 AD) *cohors I Thracum civium Romanorum pia fidelis* (138-250 AD)	*cohors I Thracum civium Romanorum pia fidelis* (138-250 AD)	*legio V Iovia equites Dalmatae cuneus equitum Constantianorum*
Taurunum (Zemun)	-	*Classis Flavia Pannonica* ?	*Classis Flavia Pannonica cohors III Alpinorum equitata* ? (2nd half of the 3rd century ?)	*equites promoti auxilia ascarii*
Sirmium (Sremska Mitrovica)	*legio II Adiutrix ?* (late Domitian period ?) *legio I Adiutrix* ? (86-89 AD ?)	?	?	*praefectus militum Calcariensium ala Sirmensis praefectus classis primae Flaviae Augustae*

the location of Donatianae.[139] This Roman settlement was perhaps located in Vardarac and some believe that this was a late Roman fortification; although surveys have confirmed the existence of Roman remains, there is no concrete evidence of the existence of any sort of military facility.[140]

It has already been mentioned that Mursa had a garrison very early, possibly already in the Augustan period and very probably during Tiberius. *Ala II Hispanorum Arvacorum*, the first auxiliary unit whose presence in the town can be stated with great certainty, was transferred to Teutoburgium already during Claudius, and *cohors II Alpinorum equitata* was left in the town (see above). It seems that this cohort remained in Mursa until 118, when the town seems to have been left without a permanent garrison for an extended period of time.[141] It seems, however, that troops did occasionally stay in Mursa,[142] and in the 4th century, in addition to a detachment of *legio VI Herculia*, a river fleet base, that is *classis Histrica,* seems to have been stationed in the town.[143] Mostly owing to epigraphic finds, we have a more or less clear general idea of the units that stayed in Teutoburgium from the 1st century onwards.[144] As already stated, *ala II Hispanorum Arvacorum* arrived to Teutoburgium from Mursa during the rule of Claudius at the latest (fig. 4). It was perhaps the first unit to be stationed there and it seems to have remained in Teutoburgium until the end of the 1st century, when it was probably replaced by ala *I Praetoria civium Romanorum*, which remained there throughout the greater part of the 2nd century. It was followed by ala *I civium Romanorum*, which was stationed in Teutoburgium during the first half of the 3rd century, if not even longer. Thanks to the *Notitia Dignitatum,* we know that in the 4th century *equites promoti* as well as *cuneus equitum Dalmatarum* stayed there, as did detachments of the legion *VI Herculia.*[145] Archaeological evidence suggests that it was mostly cavalry units that were stationed in Teutoburgium.[146] In today's Sotin, a continuity of settlement dating back to the Neolithic can be traced archaeologically, and there is no doubt that Roman *Cornacum* was built somewhere around the settlement that developed already during the Iron Age.[147] Available data is not sufficient for establishing the date of the putting up of the fort, but there is not much doubt that a military garrison (presumably the *ala I civium Romanorum*) stayed there in the Flavian period at the latest.[148] This ala was a part of Trajan's army during the

[139] Minichreiter 1987, 93; Minichreiter 1989b, 182; Minichreiter 1990, 146-147. D. Pinterović (1969, 57) mentions various hypotheses on the location of the settlement Donatianae, concluding, however, that none of them is certain.

[140] Pinterović 1968, 58, 60-61; Bulat 1969, 41; Minichreiter 1989b, 183; Sanader 2003a, 139; Sanader 2008, 105.

[141] Cichorius 1900, 239; Wagner 1938, 84-85; Mirković 1971, 32; Visy 1986, 491, 494, 497, 507; Lőrincz 2001, 27, 51, 59, 63, 65, 68, 72, 75, 80, 83, 104; Perinić Muratović 2004, 97-99.

[142] For example, at the turn of the 2nd and 3rd centuries, cf. *CIL* III 10269; Brunšmid 1900, 25; Wagner 1938, 84; Pinterović 1960, 29-30; Pinterović 1968, 66; Lőrincz 2001, 233, Cat. 250; Radman-Livaja 2004b, 120.

[143] *Not. Dign. Occ.* XXXII.53; Visy 1988, 126-127.

[144] Cichorius 1894, 1230, 1237, 1258; Fluss 1934, 1171; Wagner 1938, 26-27, 47, 63; Kraft 1951, 144, 151; Radnóti and Barkóczi 1951, 198, 208-209; Nagy 1956, 59, 61-62; Klemenc 1961, 19-20; Fitz 1962, 67-69; Klemenc 1963, 62-63; Dušanić 1968, 99; Pinterović 1968, 67-69, 76-77; Tóth and Vékony 1970, 156-157; Mócsy 1974, 88; Eadie 1977, 214-216; Visy 1988: 127; Spaul 1994, 34-36, 85-86; 187-188; Lőrincz 2001, 18, 21, 23, 61, 63, 65, 68, 82, 89-90, 95, 98; Radman-Livaja 2004c, 62-71; Lőrincz 2005, 62-63; Radman-Livaja 2005, 941; Sanader 2010, 230.

[145] *Not. Dign. Occ.* XXXII.23, 30, 47; Dušanić 1968, 91.

[146] For the finds cf. Radman-Livaja 2005, 941-944. One monument, an altar dedicated to Hercules (*CIL* III 10255), today in the GMVu, is often claimed to have been found in Dalj, but in fact the precise location of its discovery is not certain. The altar, dated to the late 2nd or early 3rd century is sometimes stated as proof of the presence of one cohort in Teutoburgium. The altar was erected by T. Flavius Magianus, prefect of *cohors I Hispanorum equitata*, and then tribune of a *cohors milliaria equitata*, identified, depending on the interpretation, as *II Aurelia Dacorum* or as *II Aurelia Dardanorum*. Although the altar was undoubtedly erected by the commander of *cohors II Aurelia Dacorum*, there is no evidence that this cohort ever stayed in Teutoburgium. More probably, following the Marcomannic wars it was stationed, even if for a brief period, in nearby Cornacum (see below).

[147] Ilkić 2003, 1-16; Sanader 2003a, 141-142; Dizdar *et al.* 2009, 122-124; Sanader 2010, 228, 230.

[148] Cichorius 1894, 1237; Wagner 1938, 26-27; Kraft 1951, 144; Pinterović 1968, 67-68; Gabler 1971, 90, Abb. 3; Gabler 1979, 212, Abb. 4; Spaul 1994, 85-86; Lőrincz 2001, 18. An opposing view was put forward by Fitz (1962, 69) and Eadie (1977, 216-217); both authors doubt that a camp may have existed or that a military garrison may have been present before the 2nd half or the end of the 2nd century.

Fig. 5. Funerary stone of M. Ulpius Super, decurio alae praetoriae civium Romanorum (CIL III 10257 = 3272, Photo Archive, AMZ).

Dacian Wars, but there is no knowing which unit replaced it in Cornacum. If we are to judge by the so-called topographic or crypto-topographic sequence of the units listed on Lower Pannonian military diplomas, it seems that after the division of the province into Upper and Lower Pannonia, *cohors I Montanorum* probably took over the garrison in Cornacum around 106; it could have remained there at least until the end of the Marcomannic and Sarmatian wars, or roughly until 180.[149] With the end of the war, the garrison crew seems to have changed, as *cohors I Montanorum* went to Ad Latus and a unit formed during the rule of Marcus Aurelius and previously stationed in Poetovio, probably arrived to Cornacum: *cohors II Aurelia Dacorum milliaria equitata pia fidelis.*[150] Brick finds in Sotin bearing the inscription *C II AUR D ANT,* and especially the discovery of two lead seals with the abbreviation *COH II AUR D* are solid evidence of this unit's presence, even if for a shorter period.[151] The recent finds of two Lower Pannonian diplomas issued on 11 August 192 also confirm its stay in the province at the end of the 2nd century.[152] However, there is no knowing how long *cohors II Aurelia Dacorum milliaria equitata pia fidelis* (which during the Severan House, probably during the rule of Caracalla or perhaps Elagabalus, carried the epithet *Antoniniana*) remained in Cornacum. Based on a recent find from Ilok and an old but until now unknown epigraphic monument from Sotin, a hypothesis has cautiously been put forward on the possible, if even brief, stay of *cohors I Lusitanorum* in Cornacum, perhaps toward the end of the 2nd or during the 3rd century.[153] According to the *Notitia Dignitatum*, in the 4th century the garrison was made up of *equites Dalmatae, cuneus equitum scutariorum* and *equites promoti.*[154]

Until recently, *Cuccium* was mostly left out of the discussions on early military fortifications on the Danube *limes*, but the already mentioned recent research (see above) sheds completely new light on this site. Although the presence of some sort of military crew in this place during the Claudian period at the latest cannot be excluded, there are doubts about Roman military presence in Cuccium between the mid 1st and the 4th centuries, when, according to the *Notitia Dignitatum*, *cuneus equitum promotorum* and *equites sagittarii* were unquestionably stationed there.[155] There are occasional mentions of some sort of camp built there at the end of the 1st or the beginning of the 2nd century.[156] There is, however, no concrete evidence to back up this claim, although this is unquestionably a vantage point on the Danube River bank and an exceptionally favourable tactical position enabling control of the entire surrounding area. The lack of more comprehensive research makes it difficult to put forward a hypothesis on the existence of a military outpost in Cuccium during the 2nd and 3rd centuries. A settlement no doubt existed at that time, but this does not necessarily mean that a military crew was present. The Lower Pannonian military diplomas do not go toward confirming this hypothesis either: a military camp with a permanent auxiliary garrison in Cuccium does not quite fit the picture of the deployment of troops in this part of the *limes* as gleaned from 2nd century diplomas issued before the Marcomannic wars. This position may have had a military purpose in the early Principate until the Flavian period at the latest, when an auxiliary camp may have indeed been located there. However, the available archaeological data and the epigraphic and written sources so far have not indicated a realistic possibility that a

[149] Cichorius 1900, 316-317; Wagner 1938, 170-171; Kraft 1951, 181; Radnóti and Barkóczi 1951, 200-201, 214-215; Nagy 1956, 67-68; Alföldy 1962, 270; Fitz 1962, 66-67, 69; Mócsy 1962, 623; Eadie 1977, 216-217; Beneš 1978, 46-47; Šašel 1986, 782-786; Visy 1988, 127; Lőrincz 2001, 39; Ilkić 2005, 29-30; Gayet 2006, 70-71.

[150] Cichorius 1900, 279; Cheesman 1914, 154; Wagner 1938, 129-130; Radnóti and Barkóczi 1951, 215; Mócsy 1962, 622; Pinterović 1968, 69-70; Lőrincz 2001, 34; Ilkić 2005, 24-26; Lőrincz 2005, 63; Ilkić 2006, 64-66. In the older as well as in the more recent literature (Brunšmid 1901, 143-145; Wagner 1938, 131; Klemenc 1961, 20; Fitz 1962: 69, 95; Klemenc 1963, 63; Eadie 1977, 217; Spaul 1994: 109, 348, 350-351; Sanader 2003a, 142; Sanader 2008, 109) there are opinions that this was *cohors II Aurelia Dardanorum milliaria equitata* (the unit's full name and its ordinal number sometimes vary, especially in older authors, depending on the reading of the stamps on the bricks found in Sotin), but this is a highly questionable hypothesis. This cohort, probably formed during the Marcomannic wars, left relatively substantial epigraphic evidence in Upper Moesia, more precisely in Ravna (Timacum Minus), cf. Wagner 1938, 131-132; Wilkes 1969, 118; Beneš 1978, 30-31; Spaul 1994, 350-351. The only monument attributed to it in Lower Pannonia (sometimes Dalj is wrongly stated as its find spot, instead of Sotin, from where it was transferred to the Castle Eltz in Vukovar, cf. Ilkić 2006, 65) would be an altar dedicated to Hercules (*CIL* III 10255), but in all probability this is the result of a mistaken reading: it does not say *coh(ortis) Aur(eliae) D(ardanorum) Ant(oninianae) p(iae) f(idelis)* but *coh(ortis) Aur(eliae) Dac(orum) p(iae) f(idelis)*. Therefore, the hypothesis that the second cohort of Dardanians may have stayed in Lower Pannonia at the end of the 2nd century is not valid.

[151] Ilkić 2005, 24-26; Ilkić 2006, 64-66.

[152] *RMD* 446; *RMD* 447; Pferdehirt 2002, 247-260; MacDonald and Pangerl 2003, 259-271; ordinal number III on the diploma *RMD* 446 is almost certainly a mistake by the publisher of the diploma.

[153] Ilkić 2005, 30; Rendić-Miočević 2009, 92-98.

[154] *Not. Dign. Occ.* XXXII.3, 12, 22, 31; Visy 1988, 127.

[155] *Not. Dign. Occ.* XXXII.6, 13, 15, 32; Brunšmid 1901, 145-147; Klemenc 1961, 20; Visy 1988, 128.

[156] Sanader 2003a, 142; Sanader 2008, 109; Rendić-Miočević 2009, 100; Sanader 2010, 230-231.

permanent military garrison existed there after the 1st century and up to Late Antiquity, when the defence system on the Danubian *limes* was radically transformed and when the strategic position of Cuccium once again aroused the interest of the Roman army. Still, the recently discovered tombstone of a former centurion and veteran of *cohors I Lusitanorum* Marcus Aurelius Serenus, dated to the end of the 2nd and the beginning of the 3rd century, could perhaps indicate the existence of a garrison following the Marcomannic wars, that is the presence of this cohort in Cuccium,[157] if the possibility of its stay in Cornacum is rejected (see above).

The level of research on the Danubian forts and the Lower Pannonian *limes* in the territory of the present-day Republic of Serbia, that is, of Vojvodina, is far better than in the Croatian part of the *limes*; still, many questions remain open and many doubts remain as to the identification of the garrisons of individual camps. Unlike in the Croatian part of the *limes*, comprehensive investigations were conducted and they led not only to the discovery of auxiliary camps but also of the remains of minor military structures on the *limes*, that is, of watchtowers, observation posts and fortifications from Late Antiquity that did not leave any unambiguous traces in the written sources, as well as of the remains of Roman roads and minor civilian settlements.[158] The increased number of fortifications in Late Antiquity is particularly evident on the stretch from Cuccium to Taurunum, which is undoubtedly related to the exceptional importance of 4th century Sirmium, whose defence naturally could not be left to chance.[159] Banoštor had been the site of the Roman camp of Malata, which in Late Antiquity came to be known as Bononia. Even though the exact position of the fort itself has not been identified, traces of the settlement and the cemetery have been found.[160] In all likelihood, the camp was active from the Flavian period at the latest, more precisely from the rule of Domitian, though its first crew is unknown. It is possible that *cohors I Campanorum* stayed there for a short while, but there is no reliable proof of this. This unit probably arrived in Pannonia from Dalmatia, but not before the end of the 1st century, and it is believed to have been stationed in Aquincum from late Trajan's reign, meaning that its possible stay in Malata must have been relatively brief, i.e. definitely not exceeding fifteen years or so.[161] It appears that *ala I Flavia Augusta Brittanica milliaria civium Romanorum bis torquata ob virtutem* was stationed there from the beginning of the 2nd century. Although from that moment it took part in military campaigns outside Lower Pannonia on several occasions, it seems that it was permanently stationed in Malata, at least until the mid 3rd century.[162] In Late Antiquity this place was the seat of the *praefectus equites Dalmatae*, the fort was at the same time a river port, and in addition to this, the counter-fort *Onagrinum* (present-day Begeč)[163] was built on the opposite bank of the Danube. Rescue archaeological excavations at Petrovaradin, the probable site of Roman Cusum, yielded possible remains of a Roman fort, apparently unrelated to the fort from Late Antiquity.[164] For now it is difficult to be very specific when discussing the possible military camp and its crew earlier than the 3rd century, when the *ala Pannoniorum* was possibly stationed there.[165] In the 4th century the garrison consisted of *equites Dalmatae*.[166] Čortanovci yielded the remains of a fort from Late Antiquity, presumed to have been called Ad Herculem or Castra Herculis.[167] In Stari Slankamen (Acumincum), a Roman military camp was discovered in archaeological excavations on the bank of the Danube opposite the mouth of the Tisza River, at the position of Gradina (the excavations also revealed a fortified settlement from the La Tène period).[168] *Cohors I Cantabrorum* could have been garrisoned there in the early Flavian era.[169] There is also a hypothesis about the possible stay of *cohors I Britannica milliaria civium Romanorum equitata* towards the end of the 1st century,[170] but the theory about the arrival of *cohors I Campanorum voluntariorum civium Romanorum* towards the end of Trajan's reign, around AD 113 or 114, seems much more likely. This unit remained in Acumincum at least until the mid 2nd century, perhaps even until the beginning of the 3rd century.[171] Nothing is known about the garrison in the 3rd century, but in the 4th century *cuneus equitum Constantium* and *equites sagitarii* were stationed there.[172] Archaeological research proved the existence of a Roman military camp in Surduk, as well. This was undoubtedly the fort of Rittium, also situated on the bank

157 Lőrincz 2005, 63-64; Rendić-Miočević 2009, 80-101, cf. 101, note 23; for *cohors I Lusitanorum*, cf. Cichorius 1900, 312; Wagner 1938, 162; Kraft 1951, 180; Radnót and Barkóczi 1951, 199, 214; Fitz 1962, 65; Mócsy 1962, 623; Visy 1986, 491, 494-496, 507, 509-510, 512-514, 516; Spaul 2000, 61-62; Lőrincz 2001, 37-38.

158 For instance, Bač (smaller fort), Čortanovci (fort from Late Antiquity), Rakovac (sentry tower), Dumbovo (sentry tower), Susek (sentry tower), Dobrinci (large fort) etc., cf. Dimitrijević 1969, 82-117; Mócsy 1969, 71-75; Visy 1988, 129; Petrović 1995, 221-225; Vasić 2003, 143-146; Dautova Ruševljan and Vujović 2006, 60, 63.

159 Klemenc 1961, 20-23; Visy 1988, 128-130; Dautova Ruševljan and Vujović 2006, 58-60.

160 Brunšmid 1895, 183; Brunšmid 1901, 148; Klemenc 1961, 20-21; Visy 1988, 128; Vasić 2003, 144.

161 Eadie 1977, 217; Spaul 2000, 22-23; Lőrincz 2001, 33; Dautova Ruševljan and Vujović 2006, 64-65.

162 Cichorius 1894, 1235; Wagner 1938, 20-24; Holder 1980, 17; Saddington 1982, 157-159; Visy 1986, 510-516; Spaul 1994, 68-71; Lőrincz 2001, 16; Vasić 2003, 144.

163 *Not. Dign. Occ.* XXXII.14, 33, 41, 44; Klemenc 1961, 20-21; Dušanić 1968, 90-92; Visy 1988, 128; Dimitrijević 1996, 151-152; Vasić 2003, 143-144; Dautova Ruševljan and Vujović 2006, 64-66.

164 Vasić 2003, 145; Dautova Ruševljan and Vujović 2006, 63-64.

165 Dušanić 1968, 101-102; Spaul 1994, 167-172; Lőrincz 2001, 22, 98, 100.

166 *Not. Dign. Occ.* XXXII.15, 34; Dušanić 1968, 103.

167 *Not. Dign. Occ.* XXXII.20, 39, 42; Vasić 2003, 146; Dautova Ruševljan and Vujović 2006, 63.

168 Fröhlich 1892, 34-44; Brunšmid 1895, 181-182; Gorenc 1956, 150-154; Klemenc 1961, 21; Piletić 1986, 137-138; Visy 1988, 129; Dimitrijević 1996, 148-150, 153-154; Vasić 2003, 147; Dautova Ruševljan and Vujović 2006, 63.

169 Wagner 1938, 117-118; Klemenc 1961, 21; Spaul 2000, 99-100; cf. Dušanić 1988, 87 (points to the inaccurate reading of the stamp on the brick).

170 Even though the position of its camp has not been identified, it seems plausible that during the Flavian era it stayed in the southern part of the Pannonian limes, cf. Dušanić 1968, 95-96; Eadie 1977, 215.

171 Wagner 1938, 104-105, 114-116; Dušanić 1968, 97-99, 101; Mirković 1971, 32; Dušanić 1988, 87; Visy 1988, 129-130; Spaul 2000, 22-23, 193-194; Lőrincz 2001, 31-33, Vasić 2003, 147; 105; Dautova Ruševljan and Vujović 2006, 63

172 *Not. Dign. Occ.* XXXII.7, 16, 26, 36; Dušanić 1968, 103-106.

of the Danube.[173] Even though the finds of stamped bricks of *cohors VII Breucorum* and *cohors II Asturum et Callaecorum* were often considered in the older literature as proof that these cohorts were stationed in Rittium, these hypotheses should in all probability be abandoned, particularly in the case of the seventh cohort of the Breuci.[174] The already mentioned *cohors I Britannica milliaria civium Romanorum equitata* appears to be the most likely candidate for the first crew, at least during the first two decades of the Flavian era.[175] Nevertheless, there are no certain indications about the camp's crew before the time of Trajan, when for a short time, approximately between AD 106 and 113/114, a cavalry vexillation from Syria (*vexillationes equitum ex Syria*) was stationed there.[176] At that time, upon its return from Dacia, it was possibly replaced by an old Pannonian unit, *ala I civium Romanorum*, which stayed in Rittium only briefly, apparently until 118 or 119, when it moved to Burgenae.[177] Only then, apparently following a series of units that stayed only briefly at that place, did a unit arrive that would remain stationed there at least until the beginning of the 3rd century, if not a while longer: *ala I Augusta Ituraeorum sagittaria*.[178] Rittium kept its garrison in the 4th century, when *equites Dalmatae* were stationed there.[179] There is a rather extensive bibliography on the site of Burgenae, both regarding the archaeological finds and the site itself, as well as the Roman units that stayed there.[180] Despite many doubts, the basic course of historical events is known for the most part: the dominant view is that the fort was built during the Flavians at the latest, although the recently published finds of the remains of *lorica segmentata* from the AMZ indicate the realistic possibility that the camp was used at least from the Claudian era.[181] It is not certain which units may have been stationed there in the 1st century, but *cohors II Asturum et Callaecorum* may have been one of them.[182] The assumption that *cohors V Gallorum* resided there at the beginning of the 2nd century, under Trajan, is equally uncertain.[183] The stay of *ala I civium Romanorum* in Burgenae between 118 or 119 until 138 is somewhat more probable.[184] In any case, there is a unanimous view in the literature that *cohors I Thracum civium Romanorum pia fidelis* was stationed in Burgenae after 138, remaining there for a long time, probably until the second half of the 3rd century, when it was transferred to Bassianae, more precisely to Caput Bassianense.[185] The *Notitia Dignitatum* supplies information about the garrison of Burgenae in the 4th century: according to this source, a dettachment of *legio V Iovia* and two cavalry units, *equites Dalmatae* and *cuneus equitum Constantianorum* were stationed there.[186] Taurunum was situated in the area of present-day Zemun, near the confluence of the Sava and the Danube. Archaeological excavations have confirmed beyond doubt the remains of a Roman fort and cemetery. It seems that prior to Late Antiquity, that is, at least until the mid 3rd century, land forces had not been permanently present there, except for the occasional detachments. On the other hand, a river fleet base, in fact the headquarters of *Classis Flavia Pannonica*, which will be discussed later in the text, was probably present there at least from the first half of the 2nd century. There is a theory, based on the finds of bricks, that *cohors III Alpinorum equitata* was stationed there in the second half of the 3rd century or the first half of the 4th century, although this hypothesis remains inconclusive. In any case, the *Notitia* mentions two units stationed in Taurunum in the 4th century, *equites promoti* and *auxilia ascarii*, but interestingly, this place is no longer mentioned as a river naval base.[187] The possible military presence in Sirmium during the 1st century and the beginning of the 2nd century has already been discussed (see above), but it appears that troops occasionally stayed in the town in the later periods, i. e. in the second half of the 2nd century and in the 3rd century. Even though in Late Antiquity the town was surrounded by a network of military forts on the *limes* and in its hinterland, the *Notitia* notes that Sirmium itself also had a garrison (*militum Calcariensium* and *ala Sirmensis)*, a naval base (*praefectus classis primae Flaviae Augustae*), as well as military equipment manufactures (*Sirmiensis scutorum, scordiscorum et armorum*).[188]

Roman military presence in southern Pannonia was not

[173] Fröhlich 1892, 40; Brunšmid 1895, 180-181; Brunšmid 1901, 155-156; Dimitrijević 1961, 95; Klemenc 1961, 21-22; Klemenc 1963, 65-66; Visy 1988, 130; Dimitrijević 1996, 148; Vasić 2003, 149; Dautova Ruševljan and Vujović 2006, 62.
[174] Tóth and Vékony 1970, 156-157; Mirković 1971, 31-32; Lőrincz 2001, 29, 31.
[175] Dušanić 1968, 95-96; Spaul 2000, 193-194; Lőrincz 2001, 31-32.
[176] Dušanić 1968, 101; Lőrincz 2001, 45, 105.
[177] Spaul 1994, 85-86; Lőrincz 2001, 18, 88-89; Radman-Livaja 2004c, 67-68.
[178] Cichorius 1894, 1250; Wagner 1938, 52-54; Dušanić 1968, 99-101; Mirković 1971, 31; Eadie 1977, 218; Spaul 1994, 154-155; Lőrincz 2001, 21-22; Radman-Livaja 2004c, 69-71.
[179] *Not. Dign. Occ.* XXXII.17, 36; Klemenc 1961, 22; Klemenc 1963, 66; Dušanić 1968, 103-106; Visy 1988, 130; Vasić 2003, 149; Dautova Ruševljan and Vujović 2006, 62.
[180] Ljubić 1879, 99-100; Ljubić 1889, 62-63; Brunšmid 1895, 172-180; Fröhlich 1892, 40-44; Patsch 1899a, 1062; Brunšmid 1901, 156-168; Radnóti and Barkóczi 1951, 215-216; Dimitrijević 1961, 93-94; Klemenc 1961, 22; Piletić and Rašić 1961, 87; Klemenc 1963, 66; Piletić 1972, 7-14; Eadie 1977, 218; Piletić 1986, 138-140; Piletić 1989, 82-85; Dimitrijević 1996, 147-148; Lőrincz 2001, 61, 63, 66, 69, 72, 75, 80, 82; Vasić 2003, 149-150; Dautova Ruševljan and Vujović 2006, 61-62; Radman-Livaja 2008, 295-303; Radman-Livaja 2009, 1499-1506; Radman-Livaja 2010b, 189-192.
[181] Radman-Livaja 2010b, 190-192.
[182] Spaul 2000, 81; Lőrincz 2001, 29, 63. It has also been suggested that *cohors VII Breucorum* may have been stationed there at the end of the 1st century.
[183] Dušanić 1968, 101; Eadie 1977, 217; Spaul 2000, 168-169; Lőrincz 2001, 34-35, 80, 82; Gayet 2006, 88.
[184] Spaul 1994, 85-86; Lőrincz 2001, 18, 88-89; Radman-Livaja 2004c, 67-68.
[185] *Not. Dign. Occ.* XXXII.59; Radnóti and Barkóczi 1951, 215-216; Dimitrijević 1961, 94; Klemenc 1961, 22; Dušanić 1968, 89, 96-97, 101, 103; Eadie 1977, 218; Dušanić 1988, 87; Visy 1988, 130; Spaul 2000, 361-362; Lőrincz 2001, 42, 90, 99-101. There is also an assumption that after Marcus Aurelius it may have been replaced by *cohors III Alpinorum* , but this seems less probable, cf. Wagner 1938, 85-86; 189-190; Dimitrijević 1956-1957, 299-300; Klemenc 1963, 66; Dautova Ruševljan and Vujović 2006, 62.
[186] *Not. Dign. Occ.* XXXII.5=24, 18=37, 46; Dimitrijević 1961, 94; Klemenc 1961, 22; Dušanić 1968, 90-92, 103-106; Visy 1988, 130.
[187] Dušanić 1968, 103-106; Dušanić 1988, 86-87; Dimitrijević 1996, 144-147, 154-155; Spaul 2000, 266-268; Lőrincz 2001, 28; Vasić 2003, 150; Dautova Ruševljan and Vujović 2006, 61.
[188] *Not. Dign. Occ.* IX.18; XXXII.49, 50, 54; Mirković 1971, 30, 32, 44; Dautova Ruševljan and Vujović 2006, 67-68; for *fabricae* cf. James 1988, 257-294.

limited only to legionary and auxiliary camps – there was also a network of beneficiary stations, as well as ports, that is, river fleet bases. In many places in southern Pannonia, especially during peaceful times, such as the most part of the 2nd century, an average citizen was not likely to see any other soldiers apart from beneficiaries during his lifetime. Epigraphic traces of their presence were registered in Poetovio,[189] Čakovec,[190] Mekinje,[191] Varaždinske Toplice (Aquae Iasae),[192] Mihaljekov Jarak,[193] Trebnje (Praetorium Latobicorum),[194] Drnovo (Neviodunum),[195] Čatež ob Savi,[196] Krška Vas,[197] Velike Malence,[198] Mursa,[199] Dalj (Teutoburgium),[200] Siscia,[201] Daruvar (Aquae Balissae),[202] Topusko (Ad Fines),[203] Petrinja,[204] Sotin (Cornacum),[205] Hrtkovci[206] and Sirmium,[207] and their stations were also probably situated in or close to most of these places.[208] Roman armed forces in southern Pannonia did not consist only of land troops: the navy, deployed in several major or minor bases, commanded considerable forces in that area. It cannot be excluded that Siscia had been a major naval river stronghold already during the Augustan period, but its importance as a naval base was significantly diminished when the focus of operations moved to the Danube. The town perhaps remained the seat of a river navy unit, *Classis Pannonica* (or *Classis Flavia Pannonica* after Vespasian's victory and accession to power) even long after the Flavian period, possibly even until the 4th century, when it was mentioned in the *Notitia Dignitatum*, although no reliable proof exists that the naval base in Siscia was operational throughout four centuries. Even though the headquarters of the *Classis Flavia Pannonica* was in Lower Pannonia, its boats continued to patrol the Sava and Drava in Upper Pannonia. It is possible that the headquarters was located in Sirmium until the time of Hadrian, when it was transferred to Taurunum. This town is at the same time the only place that has been confirmed as a naval base during the Principate. Though it is also possible that the river fleet had a port in Burgenae, this hypothesis is somewhat disputable due to the small distance separating Burgenae and Taurunum. More information about the river fleet and its ports stem from the 4th century. The *Notitia*, for instance, mentions naval bases in Sirmium (*Classis prima Flavia Augusta*), a geographically unspecified place called Graium (*Classis secunda Flavia*, perhaps in the vicinity of Brčko), Mursa (*Classis Histrica*), Servitium (*Classis prima Pannonica*) and Siscia (*Classis Aegetensium, sive secunda Pannonica*). Although their existence in Late Antiquity is beyond dispute, it remains open to what extent the 4th century distribution of naval bases corresponds to those from the preceding centuries, as well as whether the data from te *Notitia* can be used at all for making assumptions about naval ports in the first three centuries AD, the more so as Taurunum is nowhere mentioned as a naval base in the 4th century.[209]

It should be pointed out that our knowledge about Roman military forces in southern Pannonia during the late Roman period relies mostly, if not exclusively, on the *Notitia Dignitatum*. As an official document listing officials and their staffs,[210] both civilian and military, as well as names of military units and their garrisons, the *Notitia* is undeniably a first rate historical source, but its data should be interpreted very cautiously. It is a compilation of chronologically disparate data, although the original text possibly comprised the whole area of the Empire, to be later divided into the western and eastern sections. Interestingly, although the text seems to have originated in the eastern part of the empire, the copy preserved nowadays is a western one, since the eastern portion of the text seemingly ceased being updated after 395 AD (or 401 at the latest).

While the original text quite certainly originated in the late 4th century, i.e. between 386 and 394 or 395 AD (the latter date being more plausible), its subsequent layers and later entries might have been added down to a period ranging from 401 to 408 AD, and quite possibly even few decades later. It would seem that this updating was done in a rather haphazard way, with many omissions, presumably because the preserved version derives from a copy likely used by the *officium* of the *magister peditum praesentalis* whose clerks only amended sections of interest for their *officium*. As a matter of fact, not only is the text not representative of the whole military forces of the Empire at one time but it is not even a complete record of all the garrisons and military units during that period. However, it is still the best source for the late Roman army and it is a reasonably reliable overview of the Roman military in the late 4th century, more so for the eastern part of the Empire, at least as far as the "paper strength" of the army is concerned.[211] As far as southern Pannonia is concerned, while we may presume that the *Notitia* describes more or less accurately the distribution of military forces in the region and the garrison

189 *CBI* 298-300.
190 *CBI* 274.
191 *CBI* 283.
192 *CBI* 357.
193 *CBI* 284-285.
194 *CBI* 338-355.
195 *CBI* 278-279.
196 *CBI* 275.
197 *CBI* 281-282.
198 *CBI* 358-361.
199 *CBI* 413-416.
200 *CBI* 392.
201 *CBI* 302-313; Zaninović 1981, 205.
202 *CBI* 277.
203 *CBI* 336-337.
204 *CBI* 287.
205 *CBI* 423; Brunšmid 1907, 111-112, KS 232; Mirković, 1971, 84-85, cat. 90.
206 *CBI* 407.
207 *CBI* 424; Mirković 1994a, 345-404; Mirković 1994b, 193-198; Nelis-Clément 2000, 141-148.
208 *CBI*, Karte 5, pp. 752-753; Bojanovski 1988, 360-361; Ott 1995, 215-216, Karte 5; Nelis-Clément 2000, 163-164, 183-186.
209 *Not. Dign. Occ.* XXXII.50-56; Klemenc 1961, 9; Mócsy 1962, 625; Dušanić 1968, 107-108; Šašel 1974, 734; Reddé 1986, 298-302; Dušanić 1988, 87-88; Hoti 1992, 143; Zaninović 1993, 56; Domić Kunić 1996b, 87-91; Dimitrijević 1996, 143-157.
210 Unless it was not a real administrative document but rather an antiquarian text written with an ideological and symbolic purpose, cf. Brennan 1997, 147-178; Kulikowski 2000, 358-359
211 Cf. Bury 1920, 131-154; Jones 1964, vol. III, Appendix II, 347-391; Varady 1969, 262-263; Mann 1976, 1-8; Mann 1991, 215-219; Southern and Dixon 1996, 1; Zuckermann 1998, 137-147; Richardot 2005, 79-80; Kulikowski 2000, 358-377; Le Bohec 2006, 11-12

network of the late 4th century, one can only wonder to what extent is the information supplied relevant for the earlier period. The situation must have changed a lot since the tetrarchy period, and considering the turmoil caused by the civil wars, first between Constatine and Licinius, and later on between Constantius II and Magnentius, and finally between Theodosius and Magnus Maximus, it would be wise not to apply indiscriminately the data presented in the *Notitia* to the earlier part of the 4th century in Pannonia.

6. The 2nd-4th centuries AD: campaigns and the military circumstances

It is clear from all of the above how many ambiguities and unknowns still remain regarding the Roman army in southern Pannonia, but also how poorly investigated this section of the *limes* is, particularly its Croatian part. Nevertheless, we do have a broad picture of the deployment of Roman troops in this territory over a period of several centuries. During this period these troops participated in a number of military operations, from the defence of the borders from Barbarian incursions, through wars of conquest to civil wars. The troops stationed in Pannonia did not fight exclusively in that territory, but were regularly sent to more or less remote battlefields, depending on the strategic and military requirements of the Roman Empire, that is, of its rulers, as well as the usurpers of the imperial throne.[212] As a detailed description of those historic events would go beyond the limits of this paper, I will conclude with only a summary review of the military operations in southern Pannonia from the 2nd to the end of the 4th century. Even though a large part of the Pannonian troops had a chance to participate in Trajan's campaigns against the Dacians, Pannonia itself was not jeopardized by that war, although it is quite likely that certain major Pannonian towns may have played a role as logistics bases in the hinterland, for instance Siscia or perhaps Sirmium.[213]

During most of the 2nd century, life in southern Pannonia proceeded peacefully until the beginning of the Marcomannic wars, which seriously threatened the stability and prosperity of both Upper and Lower Pannonia, literally wreaking havoc on certain parts of these provinces. Though most of the fighting took place in the northern part of Upper Pannonia, in 170 the Barbarians, having crossed Pannonia, undertook a looting raid all the way to Aquileia. As the Sarmatians simultaneously started their attack, Lower Pannonia, Upper Moesia and Dacia were not spared destruction either. The fighting took place with minor interruptions and changing war fortunes until the death of Marcus Aurelius in 180, when the situation briefly stabilized.[214]

Occasional battles with the Sarmatians (in 185 and 186) and with the Marcomanni and the Quadi (in 188 and 189)[215] continued during the reign of Commodus, but after his assassination in 193 the Roman Empire, after a long time, was once again threatened from inside. After Pertinax, too, was murdered, the main contenders for the throne immediately started a showdown. The weakest among them, Didius Julianus, who bribed the Praetorian Guard to secure the imperial title for himself, could count only on the garrison of the town of Rome. The remaining contenders had far superior military forces at their disposal, and were themselves experienced soldiers. Pescennius Niger had the eastern provinces on his side, while Septimius Severus was supported by the Danubian provinces, as well as a good part of the senators in Rome. Clodius Albinus, who later entered the struggle for power, could count with the forces from the western part of the Empire. Septimius Severus turned to be the most able and decisive among them. He first eliminated Didius Julianus, then in 196 he defeated Pescennius Niger in Asia Minor, and eventually also Clodius Albinus, thus finally consolidating his authority.[216] After the relative calm of the Severan period, disturbed by occasional clashes with the Barbarians on the Danubian *limes* and by campaigns outside Pannonia,[217] a period of crises ensued in which Pannonia again suffered considerably. Through most of the 3rd century Pannonia was exposed to Barbarian incursions as well as to civil wars that transformed the province into the scene of large-scale military operations on several occasions.[218] As soon as the great Barbarian incursion was repelled, civil war broke out between the usurper Ingenuus, the governor of Lower Pannonia, and the Emperor Gallienus. Ingenuus was defeated in the battle of Mursa in 261, but the fighting continued for a short while, because the troops in Upper Pannonia and Moesia proclaimed their candidate Regalianus as Emperor. He, too, was eventually defeated and eliminated.[219] In spite of intra-Roman conflicts, it seems that from Gallienus' time on the pressure on Pannonia eased somewhat, considering that until Diocletian there is little mention of stronger Barbarian attacks on that part of the Empire, with the exception of the successfully repelled Gothic incursion during Claudius II and Aurelian's victory over the allied Vandals, Suebi and Sarmatians.[220] From the mid 3rd century it becomes increasingly more difficult to follow the deployment of the Pannonian troops through the written and epigraphic sources. The situation was similar in most of the Empire, affected by the turmoil in which the army played the principal role. During that time Siscia again gained military significance, after Gallienus established a

212 Cf. Fitz 1962, 89-92, 96-98, 105-106, 111-112; Saxer 1967; Lőrincz 2001, 73-74, 81, 101.
213 Rostovtzeff 1957, 236; Mirković 1971, 30, 32; Hoti 1992, 143-144; Radman-Livaja 2004a, 19; Radman-Livaja 2010a, 194.
214 Fitz 1962, 32-36; Mócsy 1962, 555-562; Mócsy 1974, 183-194; Barkóczi 1980, 96-99; Birley 2000, 165-186; Wilkes 2000, 583-585; Lőrincz 2001, 92-95.
215 Fitz 1962, 83-89.
216 Birley 1988, 81-128; Southern 2001, 24-37; Campbell 2005, 1-6.
217 Fitz 1962, 83-112; Mócsy 1962, 563; Barkóczi 1980, 101-103; Šašel Kos 1986, 398-402; Hoti 1992, 146-147.
218 Starting with Maximinus' Sarmatian war in 236, a series of incursions and looting campaigns ensued, that of the Carpi between 242 and 247 standing out in terms of damage and human casualties, though the Sarmatian incursion of 260 was the most devastating one: Mócsy 1962, 565-566; Mócsy 1974, 202-206; Barkóczi 1980, 103-105; Šašel Kos 1986, 406-413.
219 Mócsy 1962, 568-569; Mócsy 1974, 206-208; Barkóczi 1980, 105; Drinkwater 2005, 43.
220 Mócsy 1974, 210-211; Barkóczi 1980, 105-106.

mint in the town, probably between 259 and 262.[221] This certainly imposed the need for stationing troops in Siscia, if they had not already been present in the town. Taking into consideration that the *Notitia* mentions *cohors III Alpinorum* as part of the town garrison,[222] there is no doubt that Siscia had a permanent crew in the 4th century, and the army was presumably stationed in the town permanently from the time the mint was founded, in order to safeguard this strategically exceptionally important institution.

After several decades of turmoil and war, the last in the series of military emperors firmly took hold of power and entered history as *C. Aurelius Valerius Diocletianus*. His reforms brought peace and stability to Pannonia for a certain period. In the new administrative division Pannonia was divided into four provinces. The territory of Upper Pannonia north of the river Drava became the province *Pannonia Prima*, while the southern territory became *Pannonia Savia*. The southern part of Lower Pannonia was named *Pannonia Secunda*, and the northern part *Valeria*. The *limes* was reconstructed and reinforced, and the Barbarian incursions were successfully repelled.[223] Nevertheless, Pannonia was not spared war clashes in the 4th century. The first devastations of war were not brought by the Barbarians, but by the Romans themselves, once again confronted in a new civil war: Constantine, with the troops from Gaul, clashed with Licinius, commanding the troops from Pannonia and Moesia.[224] It was long thought that the decisive battle between Constantine and Licinius took place near Cibalae on 8 October 314,[225] but some scholars, after re-examining both the written and numismatic sources, believe that Constantine's campaign to Pannonia should be placed in the year 316, and consequently the battle of Cibalae must have taken place on 8 October 316.[226] That same year Constantine took control of Siscia, while the defeated Licinius had to flee further east. He allied himself with the Goths and resumed war with Constantine, but was eventually defeated in 325. Having secured absolute power, Constantine again strengthened the defensive capabilities of the Empire, securing several decades of relative peace for Pannonia.[227] Conditions again deteriorated towards the mid-century, when Pannonia became involved in the civil war between Constantius II and the usurper Magnentius.[228] Constantius II positioned his troops near Cibalae,[229] waiting for an opportunity to attack Magnentius, but the latter had no intention to attack the well fortified imperial troops, directing his strategy instead at provoking Constantius to leave Cibalae and confronting him in a more convenient place. Magnentius' troops were laying waste to Pannonia, occupying Siscia[230] and unsuccessfully laying siege to Sirmium,[231] but Constantius was not leaving his well fortified camp. When autumn came, Magnentius started thinking about where his troops could spend the winter. Disheartened by the failed attempt to seize Sirmium, he eventually decided to capture Mursa. As the town was well fortified, Magnentius – perhaps less due to the lack of siege machines than the lack of time before the arrival of winter – could only hope to seize it by surprise before the defence was organized. In all likelihood, the surprise attack failed, the defenders were well prepared, and the attack on Mursa turned into a siege. Well aware of Magnentius' disadvantageous position, Constantius at last decided to move and headed for Mursa with an army allegedly as strong as 80,000 men. The forces commanded by Magnentius were half that number at best, and to top that, occupied a tactically extremely disadvantageous position between Constantius' troops and an enemy town. Magnentius' position was further aggravated by the fact that one of his officers, Silvanus, defected to Constantius immediately before the arrival of the imperial troops. In such a hopeless situation Magnentius was left with no choice but to accept battle on 28 September 351. Zosimus supplies a description of a bloody and long battle that ended with Constantius' victory and Magnentius' flight. The battle of Mursa was recorded as one of the most brutal battles of antiquity, with a huge number of casualties, allegedly 30,000 dead on Constantius' side and 24,000 on Magnentius'.[232] Such horrible losses no doubt contributed significantly to the diminution of the Empire's defensive potential. Soon after that, from 356 to 358, major Barbarian incursions took place, when the Sarmatians in alliance with the Quadi devastated Pannonia and Moesia.[233] New incursions ensued in 365 and 374.[234] A couple of years later Pannonia suffered the big incursion of the Goths and Alani, and possibly also the Huns, commanded by Alatheus and Saphrax.[235] It should be mentioned that the written sources for that period are very disordered and deficient, but it is nevertheless certain that the Barbarians took advantage of

221 Mócsy 1962, 566, 693; Šašel 1974, 720; Barkóczi 1980, 105; Póczy 1980, 268.

222 *Not. Dign. Occ.* XXXII.57; Šašel 1974, 734; Zaninović 1993, 54; Gayet 2006, 70.

223 Mócsy 1962, 570-571, 588; Mócsy 1974, 273-274; Barkóczi 1980, 109-110.

224 Zos. 2.18; Aur. Vict. 41; Eutr. *Brev.* 10.5; Oros. Hist. 7-28; *Anon.* Vales. 15, 16; Patsch 1899b, 2535; Brunšmid 1902, 126-127; Mócsy 1962, 571; Mócsy 1974, 277; Barkóczi 1980, 111; Iskra-Janošić 2001, 146; Cameron 2005, 93-96; Radman-Livaja 2007b, 114-115.

225 *Fasti Hydatiani, Volusiano II et Anniano consulibus*; Brunšmid 1902, 126-127; Seeck 1919, 163; Baynes 1939, 691-692; Korda 1961, 62; Mócsy 1962, 571; Jones 1964, 82; Mócsy 1974, 277; Barkóczi 1980, 111; Kienast 1996, 294, 299 (adducing the year 316 as a possible date of the battle).

226 Vogt 1960, 172-173; Bruun 1961, 10-22; Bruun 1966, 66, 76; Barnes 1973, 36-38; Barnes 1981, 62-68; Mirković 1997, 150-156. There is also a compromise opinion that the first civil war between Constantine and Licinius took place in two phases, in 314 and in 316, and that it was interrupted by a ceasefire and temporary conciliation in 315. According to this view, the battle of Cibalae still took place in 314, cf. Andreotti 1958-1959, 1002-1014.

227 Mócsy 1962, 571-572; Mócsy 1974, 277; Barkóczi 1980, 110-112.

228 Zos. 2.46-50; Aur. Vict. 42; Oros. *Hist.* 7.29; Eutr. *Brev.* 10.9-12; Brunšmid 1902, 127-129; Seeck 1919, 198; Mócsy 1962, 575; Šašel 1971, 205-216; Mócsy 1974, 286; Hoti 1992, 150-151; Hunt 1998, 20-21; Iskra-Janošić 2001, 146-147; Radman-Livaja 2007b, 115.

229 It seems that the stay of a large number of soldiers in Cibalae in 351 left traces, because among the coins from Late Antiquity discovered in Vinkovci the largest number are those of Constantius II, cf. Dimitrijević 1979, 183-185; Iskra-Janošić 2001, 147.

230 Zos. 2.49; Šašel 1974, 711, 737.

231 Mirković 1971, 38.

232 Zon. 13.8; Zos. 2.49-53; Eutr. *Brev.* 10.12; Aur. Vict. 43; Oros. *Hist.* 7.29; Brunšmid 1902, 128-129; Fluss 1933, 673; Jones 1964, 113; Richardot 2005, 68.

233 Barkóczi 1980, 112-113.

234 Mócsy 1962, 576; Barkóczi 1980, 115-116.

235 Zos. 4.34; Iord. *Get.* 140; Mócsy 1962, 578 ; Várady 1969, 19-33; Radman-Livaja 2007b, 116-117.

the disastrous Roman defeat at Hadrianopolis to pillage the poorly defended province. The attack perhaps took place only in 380, or we might be dealing with two successive incursions, one immediately following the battle of Hadrianopolis, and another one in 380. The Barbarians may have been temporarily defeated, but it seems that eventually they were allowed to settle in Pannonia, where, dissatisfied with the conditions, they took to pillaging.[236] It seems that this incursion devastated the settlements,[237] and one of the indicators of the general circumstances in Pannonia after 379 is the waning currency circulation.[238] These years undoubtedly mark the beginning of the collapse of towns in Pannonia. In time it was becoming increasingly more difficult to maintain normal urban life. It was not even necessary to conquer towns. In fact, there are but few examples of fortified Roman towns that the Germans and Huns, unaccustomed to siege warfare, managed to conquer without a prolonged siege and starving of the crew and the town population. It seems that their only successful attempts were sudden attacks on poorly defended towns and those without a garrison. Even when the sources do mention the Barbarian devastation of towns, they primarily refer to the suburbs outside the town walls.[239] Nevertheless, the general insecurity that took hold in Pannonia from that time, the interruption of trade currents and the collapse of agriculture and the economy in general undoubtedly influenced the dispersal of the domicile population and the gradual desertion of the towns that were incessantly targeted by the pillaging Barbarian groups, as well as by the mutually confronted Roman forces.[240] The civil war between Maximus and Theodosius, in which both sides hired Barbarians, finally brought an end to Roman Pannonia: armies pillaged Savia and Secunda, and in 388 Siscia was the scene of a big battle in which Theodosius succeeded in defeating Maximus' troops.[241] At the beginning of the 5th c. Pannonia was devastated, the Roman authority over the territory was mostly only nominal and the Roman military presence insignificant, not to say non-existent.[242] After more than four centuries, the Roman army ceased to play an indispensable role in the life of Pannonia and its history in this area came to an end.

Bibliography

Alföldi 1924 András Alföldi, *Der Untergang der Römerherrschaft in Pannonien*, 1. Band, Berlin-Leipzig.

Alföldy 1959 Geza Alföldy, Die Truppenverteilung der Donaulegionen am Ende des 1. Jahrhunderts, *AAntHung* 11/ 1-4, 113-141.

Alföldy 1962 Geza Alföldy, *Die Auxiliartruppen der Provinz Dalmatien*, *AAntHung* 14, 259-296.

Alföldy 1965 Géza Alföldy, *Bevölkerung und Gesellschaft der römischen Provinz Dalmatien*, Budapest.

Alföldy 1968 Géza Alföldy, *Die Hilfstruppen der römischen Provinz Germania Inferior*, *Epigraphische Studien* 6, Düsseldorf.

Alston 1998 Richard Alston, *Aspects of Roman History AD 14-117*, London.

Andreotti 1958-1959 Roberto Andreotti, *s. v.* Licinius (Valerius Licinianus), *Dizionario Epigrafico di Antichità Romane* IV, fasc. 31-33, Roma, 979-1041.

Ardevan 2007 Radu Ardevan, The Ala II Pannoniorum in Dacia, *Apulum* 44/1, Alba Iulia, 139-148.

Barkóczi 1980 László Barkóczi, History of Pannonia, in: *ARP*, 85-124.

Barnes 1973 Timothy D. Barnes, Lactantius and Constantine, *JRS* 63, 29-46.

Barnes 1981 Timothy D. Barnes, *Constantine and Eusebius*, Cambridge, Mass.

Baynes 1939 Norman Hepburn Baynes, Constantine, in: *CAH* 12, 678-699.

Beneš 1978 Jan Beneš, *Auxilia Romana in Moesia atque in Dacia*, Praha.

Bérard 2000 François Bérard, La légion XXIe Rapax, in: Y. Le Bohec and C. Wolff (eds.), *Les légions de Rome sous le Haut-Empire*, Lyon, 49-67.

Betz 1938 Artur Betz, *Untersuchungen zur Militärgeschichte der römischen Provinz Dalmatien*, *Abhandlungen des Archäologisch-epigraphischen Seminars der Universität Wien* 3, Wien.

Birley 1988 Anthony R. Birley, *Septimius Severus: The African Emperor*, London.

Birley 2000 Anthony R. Birley, Hadrian to the Antonines, in: A. K. Bowman, P. Garnsey, and D. Rathbone (eds.), *CAH* 11, *The High Empire, A.D. 70 – 192*, 132-194.

Bogaers 1969 Julianus Egidius Bogaers, Cohortes Breucorum, *Proceedings of the National Service for Archaeological Heritage in the Netherlands* 19, Amersfoort, 25-50.

Bojanovski 1988 Ivo Bojanovski, *Bosna i Hercegovina u antičko doba*, *DjANUBiH* 66.

Bojanovski 1990 Ivo Bojanovski, Legio VIII Augusta u Dalmaciji, *AV* 41, 699-712.

Bowman, Champlin, and Lintott 1996 Alan K. Bowman, Edward Champlin, and Andrew Lintott (eds.), *The Augustan Empire, 43 B.C.- A.D. 69*, *CAH* 10.

Brunšmid 1895 Josip Brunšmid, Arheološke bilješke iz Dalmacije i Panonije, *VHAD* 1, 148-183.

Brunšmid 1897 Josip Brunšmid, Rimski vojnički diplom iz Bijele Crkve kod Rače (kotar mitrovački), *VHAD* 2, 1-6.

Brunšmid 1900 Josip Brunšmid, Colonia Aelia Mursa, *VHAD* 4 (1899-1900), 21-42.

Brunšmid 1901 Josip Brunšmid, Arheološke bilješke iz Dalmacije i Panonije IV, *VHAD* 5, 87-168.

Brunšmid 1902 Josip Brunšmid, Colonia Aurelia Cibalae. Vinkovci u staro doba, *VHAD* 6, 117-166.

Brunšmid 1907 Josip Brunšmid, Kameni spomenici

[236] Seeck 1919, 250-260; Mócsy 1962, 577-578; Jones 1964, 156-157; Várady 1969, 33-41; Mócsy 1974, 339-342; Barkóczi 1980, 117; Wolfram 1988, 131-134; Heather 1991, 147-156.

[237] For instance, Mursa, cf. Fluss 1933, 674.

[238] Alföldi 1924, 16-20.

[239] Elton 1996, 168-171, 173-174; Richardot 2005, 220-223, 267-268.

[240] Mócsi 1974, 342-350; Barkóczi 1980, 117-118.

[241] Mócsy 1962, 578; Várady 1969, 58-61; Mócsy 1974, 342-343; Šašel 1974, 737-738; Barkóczi 1980, 117-118; Hoti 1992, 152-153; Curran 1998, 104-107.

[242] Mócsy 1962, 580-582; Mócsy 1974, 346-351; Barkóczi 1980, 118-120.

Hrvatskog narodnog muzeja u Zagrebu III, *VHAD* 9, 83-184.
Brunšmid 1909 Josip Brunšmid, Kameni spomenici Hrvatskog narodnog muzeja u Zagrebu IV, *VHAD* 10, 149-222.
Bruun 1961 Patrick Bruun, *Studies in Constantinian Chronology*, *Numismatic Notes and Monographs* 146, New York.
Bruun 1966 Patrick Bruun, *Roman Imperial Coinage VII*, London.
Bulat 1960 Mirko Bulat, Dva nova rimska natpisa u Muzeju Slavonije, *OZ* 7, 13-16.
Bulat 1969 Mirko Bulat, Topografska istraživanja limesa u Slavoniji i Baranji, *OZ* 12, 39-52.
Bulat 1977 Mirko Bulat, Stanje istraživanja antičkih naselja u Slavoniji, in: *Antički gradovi i naselja u južnoj Panoniji i graničnim područjima*, *MatSADJ* 13 Beograd, 63-87.
Cambi *et alii* 2007 Nenad Cambi, Miroslav Glavičić, Dražen Maršić, Željko Miletić, Joško Zaninović, and Allesandro Campedelli, *Rimska vojska u Burnumu / L'esercito romano a Burnum*, Drniš-Šibenik-Zadar.
Campbell 2005 Brian Campbell, The Severan dinasty, in: A. K. Bowman, P. Garnsey, and A. Cameron (eds.), *CAH* 12, *The Crisis of the Empire, A.D. 193 - 337*, 1-27.
Cheesman 1914 George L. Cheesman, *The Auxilia of the Roman Imperial Army*, Oxford.
Colombo 2010 Maurizio Colombo, Pannonica, *AAntHung* 50/2-3, 171-202.
Cichorius 1894 Conrad Cichorius, *s.v.* Ala, *RE* 1,1224-1270.
Cichorius 1900 Conrad Cichorius, *s.v.* Cohors, *RE* 4 231-356.
Crook 1996 John A. Crook, Political History, 30 B.C. to A.D. 14, in: Bowman, Champlin, and Lintott 1996, 70-112.
Dautova Ruševljan and Vujović 2006 Velika Dautova Ruševljan and Miroslav Vujović, *Rimska vojska u Sremu / Roman army in Srem*, Exhibition Catalogue, Novi Sad.
Dimitrijević 1957 Danica Dimitrijević, Rad zemunskog Narodnog muzeja na rimskom limesu u Sremu, *Starinar* 7-8 (1956-1957), 299-308.
Dimitrijević 1961 Danica Dimitrijević, Nekoliko podataka o rimskom limesu u Istočnom Sremu, *Limes u Jugoslaviji* I, Beograd, 93-107.
Dimitrijević 1996 Danica Dimitrijević, Die Häfen der Classis Flavia Pannonica des Donausektors von Bassianae (Taurunum – Cusum), in: P. Petrović (ed.), *Roman Limes on the Middle and Lower Danube, Đerdapske sveske*, Posebna izdanja 2, Beograd, 143-157.
Dimitrijević 1969 Danica Dimitrijević, Istraživanje rimskog limesa u istočnom Sremu s posebnim osvrtom na pitanje komunikacija, OZ 12, 81- 121.
Dimitrijević 1979 Stojan Dimitrijević, Arheološka topografija i izbor arheoloških nalaza s vinkovačkog tla, in: *Corolla Memoriae Iosepho Brunšmid dicata*, *IzdHAD* 4, Vinkovci, 133-200.
Dizdar *et al.* 2003 Marko Dizdar, Rajna Šoštarić, and Kristina Jelinčić, Ranorimski grob kao prilog poznavanju romanizacije zapadnog Srijema, *PIAZ* 20, 57–77.
Dizdar *et al.* 2009 Marko Dizdar, Daria Ložnjak Dizdar, Mato Ilkić, and Tino Leleković, Sotin, terenski pregled 2008., *AIA* 5, 122-124.
Dizdar and Radman-Livaja 2004 Marko Dizdar and Ivan Radman-Livaja, Nalaz naoružanja iz Vrtne ulice u Vinkovcima kao prilog poznavanju rane romanizacije istočne Slavonije / Warrior Equipment from Vrtna Street in Vinkovci as a Contribution to Understanding the Process of the Early Romanization of Eastern Slavonia, *PIAZ* 21, 37-53.
Dizdar 2010 Marko Dizdar, Inventar groba 5 / Inventory of grave 5, cat. 29, in: Radman-Livaja 2010c, 244-245.
Domaszewski, von 1895 Alfred von Domaszewski, *Die Religion des römischen Heeres*, Trier.
Domić Kunić 1988 Alka Domić Kunić, Augzilijari ilirskog i panonskog porijekla u natpisima i diplomama (od Augusta do Karakale), *ARR* 11, 59-164.
Domić Kunić 1996a Alka Domić Kunić,Classis Praetoria Misenatium s posebnim obzirom na mornare podrijetlom iz Dalmacije i Panonije, *VAMZ* 28-29 (1995-1996), 39-72. 83-114.
Domić Kunić 1996b Alka Domić Kunić, Rimske provincijalne flotile s posebnim obzirom na udjel Panonaca, *ARR* 12, 83-100.
Domić Kunić 2006 Alka Domić Kunić, Bellum Pannonicum (12.-11. pr. Kr.), posljednja faza osvajanja južne Panonije, *VAMZ* 39, 59-164.
Dorn 1984 Antun Dorn, Rimska vojnička diploma iz Negoslavaca, in: *Arheološka istraživanja u istočnoj Slavoniji i Baranji*, *IzdHAD* 9, Zagreb, 165-174.
Dušanić 1988 Milena Dušanić, Rimske opeke sa žigom sa panonskog limesa (odsek Cuccium – Taurunum), *Starinar* 39, 85-97.
Dušanić 1967 Slobodan Dušanić, Bassianae and its territory, *AI* 8, 67-81.
Dušanić 1968 Slobodan Dušanić, Rimska vojska u istočnom Sremu, *Zbornik Filosofskog fakulteta* 10/1, Beograd, 87-113.
Dušanić 1978 Slobodan Dušanić, A Military Diploma of A.D. 65, *Germania* 56/1, Mainz am Rhein, 461-475.
Dušanić 1983 Slobodan Dušanić, Moesia and Pannonia in Domitian's last war on Danube, *ŽAnt* 33, 13-21.
Dušanić 1998 Slobodan Dušanić, An early Diploma Militare, *Starinar* 49, 51-62.
Dzino 2005 Danijel Dzino, *Illyrian policy of Rome in the late Republic and early Principate*, Unpublished Ph. D. diss., University of Adelaide.
Dzino 2010 Danijel Dzino, *Illyricum in Roman Politics 229 BC – AD 68*, Cambridge, 2010.
Eadie 1977 John W. Eadie, The Development of the Pannonian Frontier South of the Drava, in: J. Fitz (ed.), *Akten des XI. Internationalen Limeskongresses*, Budapest, 209-222.
Eck 2007 Werner Eck, *The Age of Augustus*, Oxford.
Elton 1996 Hugh Elton, *Warfare in Roman Europe, AD 350-425*, Oxford.
Erdkamp 2007 Paul Erdkamp, Polybius and Livy on the

allies in the Roman army, in: L. de Blois and E. Lo Cascio (eds.), *The Impact of the Roman Army (200 BC – AD 476)*, Leiden-Boston, 47-74.
Farnum 2005 Jerome H. Farnum, *The Positioning of the Roman Imperial Legions*, *BAR Intern. Ser.* 1458.
Ferjančić 2002 Snežana Ferjančić, *Naseljavanje legijskih veterana u balkanskim provincijama, I – III vek n.e.*, Beograd.
Filipović 2010 Slavica Filipović, *Kasnoantička nekropola u Zmajevcu / A Late Antique Necropolis in Zmajevac*, Exhibition Catalogue, Osijek.
Fitz 1959 Jenő Fitz, Die Militärdiplome aus Pannonia Inferior in der zweiten Hälfte des 2. Jahrhunderts, *AAntHung* 7, 421-442.
Fitz 1962 Jenő Fitz, A Military history of Pannonia from the Marcomann Wars to the death of Alexander Severus (180-235), *AAntHung* 14, 25-112.
Fitz 1980 Jenő Fitz, Administration and army, in: *ARP*, 125-159.
Fluss 1933 Max Fluss, *s. v.* Mursa, *RE* 31, 670-676.
Fluss 1934 Max Fluss, *s.v.* Teutoburgion, *RE* 9, 1171-1172.
Franke 2000 Thomas Franke, Legio XIV Gemina, in: Y. Le Bohec and C. Wolff (eds.), *Les légions de Rome sous le Haut-Empire*, Lyon, 191-202.
Fröhlich 1892 Róbert Fröhlich, Emlékek és leletek. Acumincum vidéke és a régi Pannoniai Limes, *ArchÉrt* 12, 34-44.
Gabler 1971 Dénes Gabler, Die Eroberung Pannoniens im Spiegel der Sigillaten, *AAntHung* 23, 83-91.
Gabler 1979 Dénes Gabler, *Pannonia megszállásának néhány kérdése a terra* sigillaták tükrében, *ArchErt* 106/2, 199-217.
Gabler 2003 Dénes Gabler, *The Ripa Pannonica*, in: Visy 2003a, 37-42.
Gabler 2006 Dénes Gabler, Das Donautal in augusteischer Zeit, Zur Frage der frühen Militärstationen, *Carnuntum Jahrbuch 2006*, 81-101.
Gayet 2006 Frédéric Gayet, Les unités auxiliaires gauloises sous le Haut-Empire romain, *Historia* 55/1, 64-105.
Gorenc 1956 Marcel Gorenc, *"Gradina" u Starom Slankamenu*, *RVM* 5, 150-155.
Gruen 1996 Erich S. Gruen, The expansion of the Empire under Augustus, in: Bowman, Champlin, and Lintott 1996, 147-197.
Gugl 2003 Christian Gugl, Carnuntum castra legionis, in: Visy 2003a, 55-57.
Harl 2003 Ortolf Harl, *Vindobona castra legionis*, in: Visy 2003a, 53-55.
Heather 1991 Peter J. Heather, *Goths and Romans, 332-489*, Oxford.
Holder 1980 Paul A. Holder, *Studies in the Auxilia of the Roman Army from Augustus to Trajan, BAR Intern. Ser.* 70.
Horvat *et al.* 2003 J. Horvat, M.Lovenjak, A. Dolenc Vičič, M. Lubšina Tušek, M. Tomanič Jevremov, and Z. Šubic, Poetovio - Development and Topography, in: *Pannonia* I, 153-189.
Hoti 1992 Marina Hoti, Sisak u antičkim izvorima, *OpArch* 16, 133-163.
Hoyos 2007 Dexter Hoyos, The Age of Overseas Expansion (264-146 BC), in: P. Erdkamp (ed.), *A Companion to the Roman Army*, London, 63-79.
Hirschfeld 1890 Otto Hirschfeld, Zur geschichte der pannonisch-dalmatinischen Krieges, *Hermes* 25, Berlin, 351-362
Ilkić 2003 Mato Ilkić, *Cornacum, sotinski prostor i njegovo mjesto u organizaciji južnog dijela provincije Panonije*, Unpublished Ph. D. diss., University of Zadar.
Ilkić 2005 Mato Ilkić, Pečati na antičkim opekama i krovnim crepovima iz Sotina (Cornacum), *VAMZ* 38, 19-54
Ilkić 2006 Mato Ilkić, Antičke plombe iz Sotina (Cornacum), *RFFZd* 48, 57-80.
Ilkić 2009 Mato Ilkić, Rimska vojna diploma iz Sotina (Cornacum), *PIAZ* 26, 143- 164.
Ilkić 2010 Mato Ilkić, Rimske vojničke diplome s prostora Republike Hrvatske / Roman Military Diplomas from the territory of the Republic of Croatia, in: Radman-Livaja 2010c, 263-272.
Iskra-Janošić 2001 Ivana Iskra-Janošić, *Urbanizacija Cibala i razvoj keramičarskih središta*, Vinkovci.
James 1988 Simon James, The fabricae: state arms factories of the Later Roman Empire, in: *Proceedings of the Fourth Roman Military Equipment Conference, BAR Intern. Ser.* 394, 257-331.
Jones 1964 Arnold H. M. Jones, *The Later Roman Empire, 284-602*, London.
Josifović 1956 Stevan Josifović, Oktavijanovo ratovanje u Iliriku, *ŽA* 6/1, 138-162.
Keppie 1984 Lawrence Keppie, *The Making of the Roman Army, from Republic to Empire*, London.
Keppie 2000 Lawrence Keppie, Legiones II Augusta, VI Victrx, IX Hispana, XX Valeria Victrix, in: Y. Le Bohec and C. Wolff (eds.), *Les légions de Rome sous le Haut-Empire*, Lyon, 25-37.
Kienast 1996 Dietmar Kienast, *Römische Kaisertabelle, Grundzüge einer römischen Kaiserchronologie*, Darmstadt.
Klemenc 1928 Josip Klemenc, Nov rimski vojnički natpis iz Murse, *VHAD* 15, 271- 275.
Klemenc 1961 Josip Klemenc, Limes u Donjoj Panoniji, in: *Limes u Jugoslaviji* I, Beograd, 5-34.
Klemenc 1963 Josip Klemenc, Der pannonische Limes in Jugoslawien, in: *Fifth International Congress of Roman Frontier Studies / Quintus Congressus Internationalis Limitis Romani Studiosorum*, *ARR* 3, 55- 68.
Knight 1991 D. J. Knight, The movements of the Auxilia from Augustus to Hadrian, *ZPE* 85, 189-208.
Koledin 2000 Jovan Koledin, Rimska vojnička diploma iz Vukovara, *Glasnik Srpskog arheološkog društva* 15-16 (1999-2000), Beograd, 231-239.
Korda 1961 Josip Korda, Tragom limesa od Vukovara do Iloka s osobitim obzirom na Cibalae, *Limes u Jugoslaviji* I, Beograd, 59-65.
Kos 1984 Peter Kos, Ein pannonischer Gegenstempel der Ala II Arvacorum, *Germania* 62/1, Mainz am Rhein, 47-54.
Köstermann 1953 Erich Köstermann, Der pannonisch-

dalmatische Krieg 6-9 n. Chr, *Hermes* 81, Berlin, 345-378

Kraft 1951 Konrad Kraft, *Zur Rekrutierung der Alen und Kohorten an Rhein und Donau*, Bern.

Le Bohec and Wolf 2000 Yann Le Bohec and Catherine Wolff (eds.), *Les légions de Rome sous le Haut-Empire*, Lyon.

Lolić 2003 Tatjana Lolić, *Colonia Flavia Siscia*, in: *Pannonia* I, 131-152.

Lőrincz 1982 Barnabás Lőrincz, Some remarks on the history of the Pannonian legions in the late first and early second centuries A.D., *AR* 19, 285- 288.

Lőrincz 1985 Barnabás Lőrincz, Zu den Kaiserbeinahmen der römischen Truppen im 3. Jahrhundert, *AArchHung* 37, 177-189.

Lőrincz 2001 Barnabás Lőrincz, *Die römischen Hilfstruppen in Pannonien während der Prinzipatszeit*, Wien.

Lőrincz 2005 Barnabás Lőrincz, Zu den Besatzungen der Auxiliarkastelle in Ostpannonien, in: M. Mirković (ed.), *Römische Städte und Festungen an der Donau*, Beograd, 53-66.

Ljubić 1879 Šime Ljubić, Rittium (Surduk), *VHAD* 1/4, 97-101.

Ljubić 1889 Šime Ljubić, Rimske starine u Novih Banovcih, *VHAD* 11/2, 62-63.

MacDonald and Pangerl 2003 David MacDonald and Andreas Pangerl, A new diploma of Pannonia Inferior from 192 A.D., *AK* 33/2, 259-271.

Malone 2006 Stephen James Malone, *Legio XX Valeria Victrix, Prosopography, archaeology and history*, *BAR Intern. Ser.* 1491.

Mann 1983 John Cecil Mann, *Legionary recruitment and veteran settlement during the Principate*, London.

Miletić 2010 Željko Miletić, Burnum – Vojničko središte provincije Dalmacije / Burnum - a military centre in the province of Dalmatia, in: Radman-Livaja 2010c, 113-141.

Minichreiter 1987 Kornelija Minichreiter, Arheološko blago Baranje, *AnOs* 5 (1986-1987), 43-142.

Minichreiter 1989a Kornelija Minichreiter, Kneževi Vinogradi / Dragojlov brijeg, *AP* 28 (1987), 102-103.

Minichreiter 1989b Kornelija Minichreiter, Prilog poznavanju kasnoantičke arhitekture na području Slavonije i Baranje, *Lihnid* 7, 181-194.

Minichreiter 1990 Kornelija Minichreiter, Arheološka istraživanja u Slavoniji i Baranji od 1985. do 1988. godine, *AnOs* 7 (1989), 141-170.

Mirković 1971 Miroslava Mirković, Sirmium – its history from the 1st century AD to 582 AD, *Sirmium* 1, Beograd, 5-90.

Mirković 1990 Miroslava Mirković, Sirmium et l'armée romaine, *AV* 41, 631-641.

Mirković 1994a Miroslava Mirković, Beneficiarii consularis in Sirmium, *Chiron* 24, 345-404.

Mirković 1994b Miroslava Mirković, Beneficiarii consularis in Sirmium, in: *Die römische Weihebezirk von Osterburken* II, *Forschungen und Berichte zur Vor- und Frühgeschichte in Baden-Württemberg* 49, Stuttgart, 193-198.

Mirković 1997 Miroslava Mirković, Decennalia des Licinius und die Schlacht bei Cibalae, *ŽAnt* 47, 145-158.

Mitchell 1976 Stephen Mitchell, Legio VII and the garrison of Augustan Galatia, *The Classical Quarterly* 26/2, Cambridge, 298–308.

Miškiv 1998 Jesenka Miškiv, Rimska vojnička diploma iz Slavonskog Broda, *VAMZ* 30-31 (1997-1998), 83-101.

Mócsy 1959 András Mócsy, *Die Bevölkerung von Pannonien bis zu den Markomannenkriegen*, Budapest.

Mócsy 1962 András Mócsy, *s.v.* Pannonia, *RE*, Suppl. 9, 516-776.

Mócsy 1969 András Mócsy, Eine spätrömische Uferfestung in der Batschka?, *OZ* 12, 71-79.

Mócsy 1971 András Mócsy, Zur frühesten Besatzungsperiode in Pannonien, *AAntHung* 23, 41-46.

Mócsy 1974 András Mócsy, *Pannonia and Upper Moesia*, London-Boston.

Morgan 1971 M. Gwyn Morgan, Lucius Cotta and Metellus, Roman Campaigns in Illyria during the Late Second Century, *Athenaeum* 49, 271-301.

Mosser 2003 Martin Mosser, *Die Steindenkmäler der legio XV Apollinaris*, Wien.

Nagy 1956 Tibor Nagy, The Military Diploma of Albertfalva, *AAntHung* 7, 17-69.

Nagy 1991 Tibor Nagy, Die Okkupation Pannoniens durch die Römer in der Zeit des Augustus, *AAntHung* 43, 57-85.

Nelis-Clément 2000 Jocelyne Nelis-Clément, *Les beneficiarii: militaires et administrateurs au service de l'Empire (Ier s. a. C. – VIe s. p. C.)*, Bordeaux.

Nenadić 1987 Vesna Nenadić, Prilog proučavanju antičke Sisciae, *PIAZ* 3-4 (1986-1987), 71-102.

Oldenstein-Pferdehirt 1984 Barbara Oldenstein-Pferdehirt, Die Geschichte der Legio VIII Augusta, *JRGZM* 31, 397-433.

Ott 1995 Joachim Ott, *Die Beneficiarer, Untersuchungen zu ihrer Stellung innerhalb der Rangordnung des römischen Heeres und zu ihrer Funktion*, *Historia*, Einzelschriften 92.

Pašalić 1956 Esad Pašalić, Quaestiones de Bello Dalmatico Pannonicoque, *Godišnjak Istorijskog društva Bosne i Hercegovine* 8, Sarajevo, 245- 300.

Patsch 1899a Carl Patsch, *s.v.* Burgenae, *RE* 3, 1062.

Patsch 1899b Carl Patsch, *s. v.* Cibalae, *RE* 3, 2534-2535.

Pavan 1955 Massimiliano Pavan, *La provincia romana della Pannonia Superior*, *Atti della Academia Nazionale dei Lincei, Memorie, Classe di Scienze morali, storiche e filologiche*, ser. 8, vol. 6, fasc. 5, Roma.

Perinić Muratović 2004 Ljubica Perinić Muratović, Vojnički kultovi u Mursi, *VAMZ* 36, 97-112.

Periša 2008 Darko Periša, Je li delmatsko područje presjekao rimski limes?, *ArchAdr* 2/2, 507-517.

Petrović 1986 Petar Petrović, Odbrambeni sistemi u antici (jugoistočni sektor), *Odbrambeni sistemi u praistoriji i antici na tlu Jugoslavije*, *MatSADJ* 22, Novi Sad, 91-98.

Petrović 1995 Miodrag Petrović, Remains of Roman architectural ruins near Dobrinci – Caput Bassianense, in: D. Srejović (ed.), *The Age of Tetrarchs*, Belgrade, 219-229.

Pferdehirt 2002 Barbara Pferdehirt, Ein neues Militärdiplom für Pannonia Inferior vom 11.8.193. n. Chr., *AK* 32/2, 247-260.

Piletić and Rašić 1961 Dragoslav Piletić and Bogdan Rašić, Pregled radova Vojnog muzeja na Limesu od Novih Banovaca do Sremskih Karlovaca, in: *Limes u Jugoslaviji* I, Beograd, 87-92.

Piletić 1972 Dragoslav Piletić, Preliminarni rezultati istraživanja rimskog kastruma Burgenae u Novim Banovcima 1971. godine, *Vesnik Vojnog muzeja* 18, Begrad, 7-26.

Piletić 1986 Dragoslav Piletić, Acumincum i Burgenae - dva važna uporišta na donjopanonskom limesu, in: *Odbrambeni sistemi u praistoriji i antici na tlu Jugoslavije, MatSADJ* 22, Novi Sad, 137-144.

Piletić 1989 Dragoslav Piletić, Rimske fortifikacije na teritoriji SR Srbije, *Vesnik Vojnog muzeja* 33, Beograd, 79-104

Pinterović 1960 Danica Pinterović, Mursa za dinastije Severa, *OZ* 7, 17-42.

Pinterović 1961 Danica Pinterović, *O rekognosciranju baranjskog sektora limesa*, in: *Limes u Jugoslaviji* I, Beograd, 43-45.

Pinterović 1968 Danica Pinterović, Limesstudien in der Baranja und in Slawonien, *AI* 9, 55-82.

Pinterović 1969 Danica Pinterović, Problemi u istraživanju limesa na sektoru Batina Skela-Ilok, *OZ* 12, 53-69.

Rau 1925 R. Rau, *Zur Geschichte des pannonisch-dalmatischen Krieges der Jahre 6-9 n. Chr.*, *Klio* 19, 313-346.

Radnóti and Barkóczi 1951 Aladár Radnóti and László Barkóczi, The Distribution of Troops in Pannonia Inferior during the 2nd Century A.D., *AAntHung* 1/3-4, 191-230.

Radman-Livaja 2001 Ivan Radman-Livaja, Rimski projektili iz Arheološkog muzeja u Zagrebu, *VAMZ* 34, 123–152.

Radman-Livaja 2004a Ivan Radman-Livaja, *Militaria Sisciensia, Nalazi rimske vojne opreme iz Siska u fundusu Arheološkog muzeja u Zagrebu / Militaria Sisciensia. Finds of the Roman Military Equipment in Sisak in the Holdings of the Archaeological Museum in Zagreb*, Zagreb.

Radman-Livaja 2004b Ivan Radman-Livaja, Rimska brončana pločica iz Baranje, *VAMZ* 36, 113-133

Radman-Livaja 2004c Ivan Radman-Livaja, Rimske brončane sulice iz Dalja, *ARR* 14, 59-75.

Radman-Livaja 2005 Ivan Radman-Livaja, Finds of Roman Military Equipment from Teutoburgium, in: Zsolt Visy (ed.), *Limes XIX: Proceedings of the XIXth International Congress of Roman Frontier Studies,* Pécs, 941-953.

Radman-Livaja 2007a Ivan Radman-Livaja, *In Segestica...*, *PIAZ* 24,153-172.

Radman-Livaja 2007b Ivan Radman-Livaja, *Obrazina kasnoantičke kacige iz Vinkovaca*, *ARR* 15, 105-124.

Radman-Livaja 2008 Ivan Radman-Livaja, Roman belt-fittings from Burgenae, *Journal of Roman Military Equipment Studies* 16, 295-308.

Radman-Livaja 2009 Ivan Radman-Livaja, Roman horse harness fittings from Burgenae, in: Á. Morillo, N. Hanel, and E. Martín (eds.), *Limes XX, Estudios sobre la frontera romana / Roman Frontier Studies* (2006), *Anejos de Gladius* 13, 1499-1508.

Radman-Livaja 2010a Ivan Radman-Livaja, *Siscia kao rimsko vojno uporište / Siscia as a Roman military stronghold*, in: Radman-Livaja 2010c, 179-201.

Radman-Livaja 2010b Ivan Radman-Livaja, *Lorica segmentata fittings from Burgenae*, *Akten des 16. Internationalen Roman Military* Equipment Conference, *Xantener Berichte* 16, 189-194.

Radman-Livaja 2010c Ivan Radman-Livaja (ed.), *Nalazi rimske vojne opreme u Hrvatskoj / Finds of the Roman military equipment in Croatia*, Zagreb.

Radman-Livaja and Dizdar 2010 Ivan Radman-Livaja and Marko Dizdar, Archaeological Traces of the Pannonian Revolt 6–9 AD: Evidence and Conjectures, in: R. Aßkamp and T. Esch (eds.), *Imperium – Varus und seine Zeit, Veröffentlichungen der Altertumskomission für Westfalen Landschaftsverband Westfalen-Lippe* 18, Münster, 47-58.

Reddé 1986 Michel Reddé, *Mare Nostrum*, Rome.

Reddé 2000 Michel Reddé, Legio VIII Augusta, in: Y. Le Bohec and C. Wolff (eds.), *Les légions de Rome sous le Haut-Empire*, Lyon, 119-126.

Rendić-Miočević 2009 Ante Rendić-Miočević, Nadgrobna ploča Marka Aurelija Serena iz Iloka (Cuccium) / Gravestone of Marcus Aurelius Serenus from Ilok (Cuccium), *OpArch* 33, 79-103.

Richardot 2005 Philippe Richardot, *La fin de l'armée romaine, 284-476*, Paris.

Ritterling 1925 Emil Ritterling, *s.v.* Legio, *RE* 12(1-2) (1924-1925), 1186-1829.

Rostovtzeff 1957 Michael Rostovtzeff, *The Social and Economic History of the Roman Empire*, Oxford.

Roxan 1986 Margaret M. Roxan, Roman Military Diplomata and Topography, in: *Studien zu der Militärgrenzen Roms* III, *Vorträge des 13. Internationalen Limeskongresses in Aalen* (1983), Stuttgart, 768-778.

Saddington 1982 Denis Bain Saddington, *The development of the Roman auxiliary forces from Caesar to Vespasian (49 B.C. A.D. 79)*, Harare.

Sanader 2003a Mirjana Sanader, The Ripa Pannonica in Croatia, in: Zsolt Visy 2003a,135-142.

Sanader 2003b Mirjana Sanader, Rimske legije i njihovi logori u hrvatskom dijelu panonskog limesa, *OpArch* 27, 463-467.

Sanader 2008 Mirjana Sanader, *Imago Provinciarum*, Zagreb.

Sanader 2010 Mirjana Sanader, On the problem of topography of the Croatian part of the Danube limes based on recent archaeological excavations, in: Radman-Livaja 2010c, 221- 231.

Sanader and Tončinić 2010 Mirjana Sanader and Domagoj Tončinić, Gardun – the ancient Tilurium, in: Radman-Livaja 2010c, 33-53.

Saria 1938 Balduin Saria, *Emona als Standlager der Legio XV Apollinaris*, in: *Laureae Aquincenses* 1 (*DissPann* II, 10), 245-255.

Saria 1951 Balduin Saria, *s.v.* Poetovio, *RE* 21/1, 1167-1184

Saxer 1967 Robert Saxer, *Untersuchungen zu den Vexillationen des römischen Kaiserheeres von Augustus bis Diokletian*, *Epigraphische Studien* 1, Köln.

Schober 1923 Arnold Schober, *Die römischen Grabsteine von Noricum und Pannonien*, Wien.

Seager 2005 Robin Seager, *Tiberius*, Oxford.

Seeck 1919 Otto Seeck, *Regesten der Kaiser und Päpste für die Jahre 311 bis 476 n. Chr.*, Stuttgart.

Soproni 1980 Sándor Soproni, Limes, in: *ARP*, 219-238.

Sordi 2004 Marta Sordi, *La pacificazione del' Illirico e Tiberio*, in: G. Urso (ed.), *Dall'Adriatico al Danubio, Atti del convegno internazionale* (Cividale del Friuli, 25-27 settembre 2003), Pisa, 221-228.

Southern 2001 Pat Southern, *The Roman Empire from Severus to Constantine*, London.

Spaul 1994 John E. H. Spaul, *Ala*², Andover.

Spaul 1995 John E. H. Spaul, Ala I Pannoniorum, one or many?, *ZPE* 105, 63-73.

Spaul 2000 John E. H. Spaul, *Cohors*², *BAR Intern. Ser.* 841.

Spaul 2002 John E. H. Spaul, *Classes Imperii Romani*, Andover.

Speidel 2000 Michael Speidel, Legio IV Scythica, in: Le Bohec and Wolff 2000, 327-337.

Strobel 2000 Karl Strobel, Zur Geschichte der Legiones V (Macedonica) und VII (Claudia pia fidelis) under frühen Kaiserzeit und zurstellung der Provinz Gallatia in der augusteichen Heeregeschichte, in: Le Bohec and Wolff 2000, 515–28.

Syme 1928 Ronald Syme, Rhine and Danube Legions under Domitian, *JRS* 18, 41- 55.

Syme 1933 Ronald Syme, Some notes on the legions under Augustus, *JRS* 23, 14–33.

Swoboda 1932 Erich Swoboda, *Octavian und Illyricum*, Wien.

Šašel 1968 Jaroslav Šašel, *s.v.* Emona, *RE*, Suppl. 11, 540-578.

Šašel 1971 Jaroslav Šašel, The struggle between Magnentius and Constantius II for Italy and Illyricum, *ŽAnt* 21, 205-216.

Šašel 1974 Jaroslav Šašel, *s.v.* Siscia, *RE*, Suppl. 14, 702-741.

Šašel 1985 Jaroslav Šašel, Zur Frühgeschichte der XV. Legion und zur Nordostgrenze der Cisalpina zur Zeit Caesars, in: *Römische Geschichte, Altertumskunde und Epigraphik. Festschrift für Artur Betz zur Vollendung seines 80. Lebensjahres, Archäologisch-epigraphische Studien* 1, Wien, 547-555.

Šašel 1986 Jaroslav Šašel, Cohors I Montanorum, *Studien zu den Militargrenzen Roms III, Forschungen und Berichte zur Vor- und Frühgeschichte in* Baden-Würtemberg 20, Stuttgart, 782-786.

Šašel Kos 1986 Marjeta Šašel Kos, *Zgodovinska podoba prostora med Akvilejo, Jadranom in Sirmijem pri Kasiju Dionu in Herodijanu*, Ljubljana.

Šašel Kos 1995 Marjeta Šašel Kos, The 15th legion at Emona – some thoughts, *ZPE* 109, 227-244.

Šašel Kos 1997a Marjeta Šašel Kos, Appian and Dio on the Illyrian wars of Octavian, *ŽAnt* 47, 187-198.

Šašel Kos 1997b Marjeta Šašel Kos, The end of the Norican Kingdom and the formation of the provinces of Noricum and Pannonia, in: B. Djurić and I. Lazar (eds.), *Akten des IV. Internationalen Kolloquiums über Probleme des provinzialrömischen* Kunstschaffens, *Situla* 36, 21-42.

Šašel Kos 1999 Marjeta Šašel Kos, Octavian campaigns (35-33 BC) in Southern Illyricum, in: P. Cabanes (éd.), *L'Illyrie méridionale et l'Épire dans l'Antiquité* – III, Paris, 255-264.

Šašel Kos 2002 Marjeta Šašel Kos, The boundary stone between Aquileia and Emona, *AV* 53, 373-382.

Šašel Kos 2003 Marjeta Šašel Kos, Emona was in Italy, not in Pannonia, in: *Pannonia* I, 11-19.

Šašel Kos 2005 Marjeta Šašel Kos, *Appian and Illyricum*, *Situla* 43.

Šašel Kos 2009 Marjeta Šašel Kos, Mit geballter Macht - Die augusteischen Militäroffensiven im Illyricum, in: H. Kenzler *et al.* (eds.), *2000 Jahre Varusschlacht, Imperium,* Haltern am See-Stuttgart, 180-187.

Tomičić *et al.* 2007 Željko Tomičić, Marko Dizdar, Bartul Šiljeg, Hrvoje Kalafatić, Kristina Jelinčić, Kristina Turkalj, and Juraj Belaj, Rezultati zaštitnih arheoloških istraživanja dvora knezova Iločkih 2006.g., *AIA* 3, 7-16.

Tomičić *et al.* 2008 Željko Tomičić, Kornelija Minichreiter, Kristina Jelinčić, Kristina Turkalj, Gordana Mahović, Katarina Botić, Marko Dizdar, Hrvoje Kalafatić, Saša Kovačević, and Zorko Marković, Ilok - Dvor knezova iločkih, crkva Sv. Petra apostola, kula 8 i bedemi – rezultati zaštitnih arheoloških istraživanja 2007. Ilok, *AIA* 4, 7-22.

Tóth and Vékony 1970 Endre Tóth and Gábor Vékony, Beiträge zur Pannoniens Geschichte im Zeitalter des Vespasianus, *AAntHung* 22, 133- 161.

Vasić 2003 Miloje Vasić, The Ripa Pannonica in Serbia, in: Visy 2003a, 143-150.

Veith 1914 Georg Veith, *Die Feldzüge des C. Iulius Caesar Octavianus in Illyrien in den Jahren 35-33 v. Chr.*, *SchrBalk*, *Antiquarische Abteilung* 7, Wien.

Vičič 2003 Boris Vičič, Colonia Iulia Emona, in: *Pannonia* I, 21-45.

Visy 1986 Zsolt Visy, Die kryptotopographische Truppenaufzählung in den Auxiliardiplomen von Pannonien, in: W. Eck und H. Wolff (eds.), *Heer und Integrationspolitik, Die römische Militärdiplome als historische Quelle, Passauer historische Forschungen* 2, Passau, 482-517.

Visy 1988 Zsolt Visy, *Der pannonische Limes in Ungarn*, Stuttgart.

Visy 2003a Zsolt Visy (ed.), *The Roman Army in Pannonia, An Archaeological Guide of the Ripa Pannonica*, Pécs.

Visy 2003b Zsolt Visy, *The Ripa Pannonica in Hungary*, Budapest.

Vogt 1960 Joseph Vogt, *Constantin der Grosse und sein Jahrhundert*, München.

Wagner 1938 Walter Wagner, *Die Dislokation der*

römischen Auxiliarformationen in den Provinzen Noricum, Pannonien, Moesien und Dakien von Augustus bis Gallienus, Berlin.

Wheeler 2000 Everett L. Wheeler, Legio XV Apollinaris, From Carnuntum to Satala and beyond, in: Le Bohec and Wolff 2000, 259-308.

Wiedemann 1996 Thomas Ernst Josef Wiedemann, *Tiberius to Nero*, in: Bowman, Champlin, and Lintott 1996, 198-255.

Wilkes 1969 John Joseph Wilkes, *Dalmatia*, London.

Wilkes 1992 John Joseph Wilkes, *The Illyrians*, Oxford.

Wilkes 1996 John Joseph Wilkes, *The Danubian and Balkan Provinces*, in Bowman, Champlin, and Lintott 1996, 545-585.

Wilkes 2000 John Joseph Wilkes, The Danube Provinces, in: A. K. Bowman, P. Garnsey, and D. Rathbone (eds.), *CAH* 11, *The High Empire, A.D. 70 – 192,* 577-603

Wolff 2000 Catherine Wolff, *La legio XIII Gemina au 1er siècle*, in: Bohec and Wolff 2000, 203-204.

Wolfram 1988 Herwig Wolfram, *History of the Goths* (translated from the second German edition), Berkeley.

Zaninović 1968 Marin Zaninović, *Burnum, Castellum – municipium*, *Diadora* 4, Zadar, 119-129.

Zaninović 1981 Marin Zaninović, Siscia u svojim natpisima, in: *Arheološka istraživanja u Zagrebu i njegovoj okolici*, *IzdHAD* 6, Zagreb, 201-208.

Zaninović 1984 Marin Zaninović, *Vojni značaj Tilurija u antici*, in: *Cetinska krajina od prethistorije do dolaska Turaka*, *IzdHAD* 8, Split, 65-75.

Zaninović 1986 Marin Zaninović, Pojava antike u središnjoj Hrvatskoj, in: *Arheološka istraživanja na karlovačkom i sisačkom području*, *IzdHAD* 10, Zagreb, 59-67.

Zaninović 1993 Marin Zaninović, *Classis Flavia Pannonica*, in: *Arheološka istraživanja u Slavonskom Brodu i Brodskom Posavlju*, *IzdHAD* 16, Zagreb, 53-58

Zippel 1877 Georg Zippel, *Römische Herrschaft in Illyrien bis auf Augustus*, Leipzig.

URBANISM AND ARCHITECTURE

Tatjana Lolić
Zoran Wiewegh

1. Introduction

A systematic urbanisation, following the conquest, military disposition and re-arrangement of legions, is the most obvious component of the Romanisation of the Croatian part of the province of Pannonia. It was carried out in cycles, and was conditioned by a network of roads and waterways, forming an integral part of the landscape development. In the area of study Romanisation started in the late 1st century BC, following the Augustan Wars.[1] During the five centuries of Roman rule, in terms of geography, politics and administration, this area belonged to Illyricum, Pannonia, Lower and Upper Pannonia, Pannonia Savia and Pannonia Secunda.[2] Pannonia – or the area that would become Pannonia – was annexed, according to a widely accepted opinion, ca. 11 BC.[3] The Sava valley was always the southern border of this area in Roman territorial and political division,[4] while the eastern border was on the Danube. Although the later administrative division into four Pannonias (Prima, Secunda, Savia and Valeria) did follow the course of the latter, this border area functioned as a unique autonomous system throughout Roman rule, with its own urbanistic legislation differing from that of the civilian interior of the province. The Danube road route, as part of a huge military system (its Croatian section is still not studied in total, together with the settlements and military stations positioned along it), is a special case in the context of the construction of road infrastructure. The northernmost settlement on this route in Croatia was Batina (Ad Militare), while the southernmost was Ilok (Cuccium). The creation of Pannonia – that is, the division of Illyricum into Pannonia and Dalmatia – was directly connected to the gradual process of pacification and the early phase of Romanisation of the newly conquered territory. Trajan's and Diocletian's divisions of the province suggest the presence of an organised administration of this territory and reflect the well-advanced process of Romanisation and administrative, economical and military functioning of the region.[5] These factors were at the same time the base of the urban development of southern Pannonia, unfolding in several phases. The province frontier, the largest administrative unit outside of Italy, implied the existence of facilities controlling the flow of goods and the traffic system (customs, postal service, military). Customs offices were both located at the borders of Illyricum but also in the interior, while the customs services reached their peak in the reign of Hadrian.[6] The existence of customs services is, among other things, associated with an austere regulation of cross-border trade, and this fact reflects the positioning of *portoria*.[7] Although the character of the Sava valley differed from the integrated system of the Danube *limes*, it still retained a specifically frontier nature. Together with the characteristics pertaining to border areas, there was also a large amount of fluidity and economic reliance upon the nearby regions, notwithstanding the fact that some regions belonged to different administrative units. A circulation of goods and people existed at the level of the Empire, together with the growing of monocultures and production of goods at the level of provinces, depending on local climatic conditions, which were distributed throughout the Empire. Although the Roman world is often described as a complex of semi-independent units, the evidence shows that the interaction was strong; there were different regions within the Empire, which amounted to an urbanised system representing more than simply numerous individual administrative units.[8] The system functioned in the same time as regional and monocentric.[9] The dynamic and cycles of urbanisation were determined by economical and political situation in different parts of the Empire. A constant and long-term migration was the basis of urbanisation, together with a developed political network and the speed by which information arrived from Rome. The most important impulse for the first cycle of urbanisation of southern Pannonia (and other western provinces) was certainly the *Pax Romana* of the 1st and 2nd centuries, because Rome enabled the founding of the provinces and their functioning according to their own laws, on condition that they accept Roman taxes and military control.

Despite several administrative divisions, the province of Pannonia often functioned as a unity.[10] The level of urbanisation, the use of the landscape and its final appearance, were contemporary with and related to regional reorganisations. Administrative borders did not present a barrier for numerous services, or for economic administration of the province, which remained unified.[11]

1 Mócsy 1974, 33, 44; Barkóczi 1980, 100; Zaninović 1993, 53.
2 Mócsy 1974, 92, 273; Wilkes 2005, 41.
3 Mócsy 1974, 33.
4 Durman 1992, 118.
5 Mócsy 1974, 92, 273.
6 Mócsy 1974, 200; Hoti 1992.
7 Ørsted 1985, 257-258.
8 Morley 2005, 45.
9 Morley 2005, 42.
10 Fitz 1980, 126.
11 Fitz, 1980, 131-132.

The key event in the history of southern Pannonia was Octavian's capture of Segestica in 35 BC, which thus became an important strategic point for further expansion towards the east.[12] However, according to the archaeological evidence, the classical Roman scheme of urbanisation did not appear in the development of towns and other settlements of southern Pannonia during the first several decades of Roman occupation.[13] After the fall of Segestica a gradual process of pacification of the territory ensued, so that the process of founding of new towns, whose peak fell in the Flavian reign, was not to be expected at this early period.[14] The urbanisation of Pannonia is a unified concept closely associated with the process of Romanisation and the growth of autonomous towns and their areas of influence. This process, especially at the level of urban structures, was carried out in several longer cycles, and was directly associated with the activities of the Flavians, Antonines, and Severans.[15] During the 2nd half of the 1st century a number of autonomous towns were founded in Pannonia: Siscia (Sisak), Sirmium (Sremska Mitrovica), Andautonia (Ščitarjevo), Neviodunum (Drnovo), that is, more than twice in comparison to the period of the Julio-Claudian House. In the beginning the towns were founded along two vital communication routes, the Amber Road and the Sava valley. Systematic urbanisation of the interior of Pannonia and Dalmatia was conducted in the reign of Vespasian, resulting from a policy that tried to strengthen the Danube border by including this area in the regional economic structure through the integration of land traffic along the eastern edge of the Pannonian basin and the river traffic of the *Classis Flaviae Pannonicae*, into a unified system.[16] Vespasian granted the status of a colony to Siscia in AD 71 (a fact mentioned in a military diploma of a member of the Ravenna fleet)[17] and Sirmium (modern-day Sremska Mitrovica in Serbia), the latter confirmed by an ex-sailor's *deductio* (A. Mócsy believed this to be representing the concept of the foundation of the Sava fleet). The founding of Andautonia (a Flavian *municipium*), modern Ščitarjevo, was part of the same plan of stimulating the traffic along the Sava.[18] In Trajan's new division of Pannonia into two provinces (Superior and Inferior), implemented probably ca. 106, the more developed western part with numerous colonies was allocated to Pannonia Superior. Trajan's colonies were the last examples of such settlements created by a *deductio* in the history of the Empire. The only urban community in Pannonia Inferior at the time was the *colonia Flavia Sirmium*.[19] Hadrian's reign was characterised by a substantial programme of urbanisation, when the majority of previously existing settlements attained an urban level.[20] Although the rebuilding of Mursa (Osijek) had begun after the Dacian Wars, Hadrian stimulated its development through according it the status of colony (*colonia Aelia Mursa*).[21] While it is not clear whether it was created by a *deductio*, Mursa remains the last colony created in the Danube region.[22] The Italics, veterans and other Roman citizens, together with local population, organised themselves through the process of Romanisation into the *Municipium Iasorum* (regularly identified with the Roman town of Aquae Balissae on the place of modern-day Daruvar) and the *municipium Aelium Cibalae* (Vinkovci), both on the roads connecting Siscia and Mursa.[23] A decurion is known from each of the *municipia*, and government officials were elected from the ranks of the local aristocracy.[24] After a long standstill in the urban development of Pannonia, caused by the Marcomannic Wars, the period of the Severan emperors was characterised by new constructions added to the already existing buildings in Pannonan towns. Several towns were granted colonial status; such status was renewed for Siscia, which now became *colonia Septimia Siscia Augusta*.[25] The town of Cibalae, together with Carnuntum and Aquincum, was granted colonial status during Caracala's reign, becoming *colonia Aurelia Cibalae*.[26]

2. Urbanisation and road communications

According to the present state of research, in the final phase of urbanisation of the area of study, three urban centres had the status of colony (Siscia, Mursa, Cibalae) and two had the status of *municipium* (Andautonia and Aquae Balissae). Besides these autonomous towns, there were numerous other settlements (in terms of morphology), without their administrative status being confirmed either in the literary or epigraphic sources. The rest of the colonies and *municipia* known from the literary sources are at present unidentified. According to the latest archaeological evidence the autonomous town of Halicanum is located in modern-day Sv. Martin on the Mura in north-westernmost Croatia.[27] Marsonia was identified with modern-day Slavonski Brod on the Sava, but the (small-scale) excavations have so far not confirmed this hypothesis. Numerous attractive finds (a forum with temples and thermal complex, together with settlement remains) from Varaždinske Toplice, whose Roman name (Aquae Iasae) is epigraphically attested[28] and whose urban prerogatives are evident, suggest a higher level of autonomy, which, however, has not been confirmed either in the sources or in the material evidence. The urban layout and isolated public buildings found at *civitas* Iovia /Botivo (modern-day Ludbreg) belonged to a settlement whose name has not been ascertained epigraphically.[29] A road station of Varianae (Kutinska Lipa near Kutina) was

[12] Mócsy 1974, 22–23; Zaninović 1986, 63; Hoti 1993, 138.
[13] Cf. Gregl and Migotti 2004, 135-139; Lolić 2003, 138-143; 149 Nemeth-Ehrlich and Kušan 2003, 117-128.
[14] Zaninović 1996, 62.
[15] Poczy, 1980, 240.
[16] Zaninović 1993, 57; Zaninović 1996, 62.
[17] Brunšmid 1907, 106-108; Šašel 1974; Mócsy 1974, 112; Hoti 1993, 143; Zaninović 1996, 61.
[18] Mócsy 1974, 112.
[19] Mócsy 1974, 92-93.
[20] Póczy 1980, 241.
[21] *CIL* III 3279/10260; Pinterović1978, 53.
[22] Mócsy 1974, 119.
[23] Katančić 1824, 467-468; Kukuljević 1873, 92, 102; Pavić 1852, 340-341; Shejbal 2004, 100.
[24] Mócsy 1974, 143.
[25] *CIL* III, 4193; Šišić 1925, 130; Fitz 1980c, 152; Hoti 1993, 145.
[26] *CIL* III 14038; Brunšmid 1907, 136.
[27] Šašel 1955, 49; Tomičić 1999, 20.
[28] Vikić-Belančić, 1975, 165-170; Nemeth-Ehrlich and Kušan 1997, 8.
[29] Gregl and Migotti 2004, 131

placed on the Siscia – Mursa road; its urban character is suggested by the remains of an architectural complex, but its administrative status remains unknown.[30]

Generally, a town can be distinguished both administratively and morphologically, with certain economic and demographic prerogatives fulfilled. The Roman definition of a town emphasised the presence of autonomous administration warranting a firm social organisation. Thus Tacitus, in his description of Germania, claims it had no towns. In his words, an urban organisation, that is, a group of occupants of some space is a prerogative for an urban structure. Thus, the organisation of a settlement encompassing an area where individual families and clans remain isolated cannot be called a town.[31] The origin of the majority of Roman settlements and smaller towns was of a military character. However, an important role in the creation of towns was played by the continuity of prehistoric settlements and the presence of pre-Roman occupants. A large number of towns originated from specific political and socio-economical needs of the autochthonous population. Some of those were entirely new settlements fulfilling the demands of the military and serving as a sort of its logistics. A further development of these settlements depended upon specific natural resources, in the first place those with industrial potential.[32]

From the very beginning, the inhabitants of Pannonian towns formed urban and rural agglomerations composed of both Italics and autochthons, along with numerous Oriental immigrants arriving particularly during the reign of the Severans. The role of the autochthonous population is seen in a number of prefects of native pre-Roman stock, originating from a population that enjoyed a certain level of self-government and was exempted of military control. Siscia probably absorbed in itself a *civitas*; there is a notice of a certain Antonius Naso, a *praefectus* of Siscia in Nero's reign, of the Colapiani origin.[33] In the Flavian *municipium* of Andautonia, founded in the territory of the Variciani, north-Italian surnames (the Iuventii, Caesernii, Aconii, and Lusii) are epigraphically attested.[34] All the towns in area of study were sited on or along the main roads. The road network of the area was the product of the technical ingenuity of Roman engineers, but the routes were often inherited from prehistoric communications and were as a rule constructed upon prehistoric roads; this was evidently based on economic considerations.[35] The development and construction of roads and waterways in the territory of the future Roman province of Pannonia was initiated after the crushing of the great uprising of the two Batos (AD 6-9). Some of the roads were certainly constructed before that, especially when considering that after Octavian's campaign in Illyricum in 35 BC a military unit was stationed in Siscia, and a military and civilian harbour established there, as well as the headquarters for entire Illyricum.[36] Four main literary sources cover the topography and roads in the area between the Drava, Sava and Danube: 1. the Antonine Itinerary (*Itinerarium Antonini Augusti*), a list of road stations with distances between stations (3rd century); 2. the Bordeaux or Jerusalem Itinerary (*Itinerarium Burdigalense sive Hierosolymitanum*), a report of an anonymous Christian pilgrim from the 1st half of the 4th century on the voyage from Bordeaux to Jerusalem and back; 3. the Peutinger's table (*Tabula Peutingeriana*), a road map (an 11th century copy after the 4th-5th century original); 4. the Ravenna Cosmography (*Ravennatis Anonymi Cosmographia*), a list of settlements, dating from the 6th to 9th centuries, believed to be based upon earlier material. Epigraphic evidence is important in studying topography, especially the inscriptions on milestones and other stone monuments; yet, in this context archaeological excavations remain of the utmost importance, as they provide abundant and invaluable material evidences for the existence of settlements and roads, and have often produced the single confirmation of the literary data for the region between the Drava, Sava, and Danube. A review of traffic corridors and settlements sited along them is here given only summarily, classified by main communications and without any detailed analysis of the problems pertaining to the identification of individual settlements or stations (an abundant literature on these subject matter already exists[37]):

- the Ptuj (Poetovio) – Szombathely (Savaria) road;
- the Ptuj (Poetovio) – Osijek (Mursa) road;
- the Osijek (Mursa) – Sremska Mitrovica (Sirmium) road;
- the Ljubljana (Emona) – Sisak (Siscia) – Sremska Mitrovica (Sirmium) road;
- the Sisak (Siscia) – Ptuj (Poetovio) / Varaždinske Toplice (Aquae Iasae) – Ludbreg (Iovia) road;
- the Sisak (Siscia) – Daruvar (Aquae Balisae) – Virovitica / Daruvar – Varaždinske Toplice (Aque Iasae) road;
- the Sisak (Siscia) – Osijek (Mursa) / Vinkovci (Cibalae) road;
- the Danube *limes* road.

3. Pre-urban phase: the formation of settlements

The Celts started their conquest and colonisation of the major part of Europe, including the Middle Danube area, ca. 400 BC. The archaeological evidence from the Croatian part of the future Roman province of Pannonia suggests the existence of certain contacts between the autochthonous population and the Celts already in the mid-3rd century BC, and local cultures were almost vanquished by the turn of the 3rd and 2nd centuries, to be replaced by the more advanced

30 Bobovec 2005, 174-175.
31 Gros and Torelli 1988, 243.
32 Burnham and Wacher 1990, 7-14.
33 Mócsy 1974, 135.
34 *CIL* III 10866, 10826, 10824, 4009, 11463; Mócsy 1974, 136; Zaninović 1996, 63.
35 Soproni, 1980, 207; Egri 2007, 107. According to the archaeological evidence, the Roman roads between Andautonia and Siscia completely follow the prehistoric courses – cf. Bugar 2008.
36 Hoti 1992, 142.
37 Andrić 2002; Bojanovski 1974; Bojanovski 1988; Bojanovski 1993; Šimek 1997; Gračanin 2010.

La Tène civilisation.[38] The 2nd and 1st centuries BC in Pannonia are characterised by two types of settlements: of the open and of the fortified type. The latter, known as *oppida*, developed especially during the 1st century BC. With the Celts, the term *oppidum* designates certain fortified settlements that during the 1st century became the centres of social, political and military power, but also of important industrial activity.[39] Building materials for the Pannonian fortified settlements were sourced from the natural environment; thus the characteristic feature of *oppida*, the *murus gallicus* (a rampart made of stone and timber), does not appear and, therefore, the term itself is rarely used.[40] Fortified settlements in general occupied elevated positions and were surrounded by ramparts made of earthen banks with timber superstructures. Since they were in the majority of cases placed along watercourses, they were also surrounded by wide, water-filled ditches for additional protection. Archaeology has confirmed the existence of La Tène settlements under the later Roman towns in Vinkovci (sites of the *tel* Tržnica, Dirov brijeg and Ervenica),[41] Sisak (Pogorelec),[42] Osijek (the site of Donji grad / Lower Town),[43] and Ščitarjevo, Roman Andautonia, where a La Tène substratum appears in the entire territory occupied by the *municipium*.[44] Modern-day Daruvar (Aquae Balissae) was densely populated even before the arrival of the Celts, as is evidenced by the strong autochthonous linguistic survivals of pre-Celtic names on Roman monuments from the entire territory of the Iasi.[45] The layouts of prehistoric settlements are in general unknown because of the modern urban development. Settlement remains, if preserved, consist of scarce traces of timber constructions, accompanied by small finds, such as house pottery, coinage, jewellery, metal dress accessories and weaponry.[46] Prehistoric urbanism – the types, density and positions of Late Iron Age settlements – certainly influenced Roman landscape urbanism, furnishing it with already existing sites and a concomitant use of certain types of regional and local economy and raw materials familiar to prehistoric communities before their integration into the Roman Empire. The Iron Age in the Sava region was under the influence of the early invasion of Celtic groups, while the process of La Tèneisation was carried out even in areas without the actual presence of the Celts.[47] The Celts had already recognised and utilised the sites of earlier settlements, prominent in terms of economy, and the Romans took over the exploitation of communications and mines; such was the case with Trgovska gora 40 km to the south of Segestica/Siscia.[48] Mursa, Cibalae, and Andautonia were located in fertile agricultural areas, next to the navigable Rivers Sava, Drava and Danube, while Aquae Balissae exploited, besides its thermal springs, beds of quartz and iron ores, as well as limestone for the production of lime.[49] However, the analysis of numerous interconnections between prehistoric settlements and the newly-founded Roman towns in southern Pannonia has shown that, in terms of urbanism, the architectural substratum did not influence the classical Roman orthogonal street-grid, developed within the walls of the towns that mostly followed the lie of the land. This is clearly shown in the examples of Siscia, Andautonia, Cibalae, and Mursa. Prehistoric settlements had remained for a time within the urban periphery, until the town absorbed them with the facilities characteristic for a suburban area (cemeteries, workshops, and various services). This was due to the fact that, similar to other Roman provinces in Europe (Gallia, Noricum, Britannia), prehistoric construction techniques in Pannonia were on a significantly lower level in comparison with the entirely new and progressive civilisation.[50]

4. Urban geography of the landscape

Several important theoretical features of Roman urbanism remain unclear. Firstly, it is the scope of the influence of the administrative status on a town's urban profile, its size and spatial organization. Secondly, it is the question of the dependence of the public infrastructure upon colonial privileges, and accordingly, the possibility or otherwise that monumental archaeological remains suggest a town's administrative status.[51] Since not a single town or settlement in the area of study was excavated in its entirety, it remains to outline the morphology of the towns through a combination of the available evidence and the general rules of Roman urbanism, associated, among other things, with administrative status of a town. All Roman towns are now overlaid by densely populated modern towns; thus, future studies will not be any less difficult and the frontiers of our knowledge will not be easily moved towards a more competent synthesis of this subject.

On the basis of the archaeological evidence, four larger towns are at present known in the area of study, three *coloniae* and one *municipium*: Siscia (Sisak), Mursa (Osijek), Cibalae (Vinkovci), Andautonia (Ščitarjevo); the study of Aquae Balissae (Daruvar) at this stage does not allow any speculation on the character of its urban structure or its perimeter. The model for the setting up of all three colonies (Siscia, Mursa, Cibalae) was either directly or indirectly based on political considerations; all three were founded in order to pacify the area and control the newly acquired territories. The *municipia* of Andautonia and Aquae Balissae are of different character, as they were part of a general wave of urbanisation of Pannonia. The foundation of Andautonia was connected with the furtherance of the navigation on the Sava, while Aquae Balissae was founded as part of a general process of

38 Majnarić-Pandžić 1998, 329.
39 Jovanović 1987, 813.
40 Majnarić-Pandžić 1984, 24.
41 Dizdar 2001; Janošić 2001, 31.
42 Durman 1992, 120.
43 Šimić and Filipović 1997, 23.
44 Nemeth-Erlich and Kušan Špalj 2003, 113.
45 Shejbal 2003, 105.
46 Šimić and Filipović 1997, 23.
47 Jovanović 1987, 812.
48 Bojanovski 1988, 273; Durman 2002, 25.
49 Shejbal 2003, 119.
50 Cf. Durman 1992, 120.
51 Gros and Torelli 1988, 248.

strengthening civilian society.[52] As already mentioned, all three colonies were founded on the model of juxtaposition with an earlier settlement.[53] None of he colonies were founded independently of the existence of a prehistoric substratum, and all were sited next to a river (Kupa, Bosut, Drava), which was of a decisive influence on the formation of their urban landscapes. While Mursa and Cibalae were organised within a rectangular perimeter, the layout of Siscia is less orthodox, as it was determined by the course of the Kupa and the lie of the land.[54] The core of all three colonies was surrounded by town walls, with cemeteries situated immediately outside the walls, along the main roads.[55] All towns spread outside their perimeters, exerting their influence on the surrounding territory whose area could be only hypothetically determined due to the insufficient level of archaeological research and the dissolution of *ager* boundaries with the passage of time.

Since the influence of administrative status on urban spatial organisation in general remains undetermined, the excavations show that there was no essential difference between colonies and *municipia*. Moreover, there are examples from the western provinces that challenge this relation. One of the key reasons behind divergences from the standard pattern is the fact that Roman regulation of space was much more flexible in provincial colonies than in Italy. The canonical urban facilities, including the forum with its adjacencies, thermal complexes, dynastic temples, etc., were not always present in provincial towns. The reasons for this can be found in specific ethnic, topographical, financial, religious and economical circumstances.[56] In Pannonia, the relation between the layout of a colony and *municipium* was often inconsistent. The spatial organisation and quality of finds from the *municipium* of Andautonia are not different from parallel features and finds from the colonies in southern Pannonia. As archaeological evidence shows, southern Pannonia abounds in small towns and settlements of unknown administrative status; it is therefore impossible to juxtapose the category of status with the related architectural remains. However, in some cases there are enough arguments for guessing at the spatial organisation and the perimeter of some buildings. Although the information on spatial organisation is scarce, the internal morphology of smaller towns shows two patterns, one as a complex of individual buildings, and another as a more elaborate internal layout with buildings following the street-grid. There are at present some ten towns of this type, together with the above discussed colonies and *municipia* (Siscia, Mursa, Cibalae, Andautonia, Aquae Balissae, Aquae Iasae, Iovia/Botivo, Halicanum, Quadrata, and Varianae). The location of the four last-named from the list is not epigraphically attested, but a number of authors find them in Ludbreg, Sveti Martin na Muri, Topusko, and Kutinska Lipa, respectively.[57] Along with these, archaeology has further confirmed the existence of military stations on the Danube *limes* in Batina, Dalj, Sotin, and Ilok, as well as some 30 stations along the important roads, most of which are not securely identified with modern-day settlements, nor is their character (*mansio*, *mutatio*) known. Along with this network of towns, settlements, and road and military stations (fig. 1), large zones composed of exclusively rural places are also known, grouped in larger or smaller villages comprising from several to some 20 houses, accompanied by cemeteries and industrial facilities and clearly distinguished from smaller or larger urban settlements. This difference is even more conspicuous in Pannonia, owing to different construction techniques and spatial organisation: rural settlements were mostly built in timber, a local building material, while they were organised in an organic fashion determined by the proximity to water sources, woods, pastures, etc.[58]

At present, the names of Roman settlements in southern Pannonia are known, with some minor exceptions, almost exclusively from the road itineraries. Archaeological excavations have substantiated the fact that all urban settlements were placed on or along important roads. This is partly true of the rural settlements as well, but a number of them were sited away from the main traffic arteries, connecting to the urban centres by main and vicinal roads.

List of towns and other settlements

1. Sisak (colonia Flavia Siscia)
2. Osijek (colonia Aelia Mursa)
3. Vinkovci (colonia Aurelia Cibalae)
4. Ščitarjevo (municipium Andautonia)
5. Daruvar (Aquae Balissae)
6. Sv. Martin na Muri (Halicanum?)
7. Varaždinske Toplice (Aquae Iasae)
8. Ludbreg (civitas Iovia / Botivo)
9. Topusko (Quadrata?)
10. Kutina (Varianae?)
11. Slavonski Brod (Marsonia?)
12. Petrijanec (mansio Aqua Viva?)
13. Bartolovec (Popolis?)
14. Kunovec Breg (mutatio Sunista?)
15. Draganovac (mutatio Piretis?)
16. Virje (mansio Lentolis?)
17. Prugovac (mutatio Iovia? Carodunum?)
18. Turnašica (Iovia/Iovista?)
19. Špišić Bukovica (Cocconae?)
20. Virovitica – Taborište (Serota?)
21. Orešac (Bolentium?)
22. Donji Miholjac (Marinianae?)
23. Sveti Đurađ (Serena?)

[52] Cf. Mócsy 1794, 143.
[53] Lavendan and Hugeney 1966, 343.
[54] Marsigli 1726; Faber 1972-73, fig. 4; Lolić 2003, 138.
[55] Janošić 2001; Wiewegh 2003; Göricke-Lukić 2011.
[56] Gros and Torelli 1988, 253. Gros and Torelli adduce the examples of Arles and Nimes, with a significantly different municipal status, which is not reflected in material reality. For the province of Dalmatia they give the examples of municipia of Pelva and Salvum that never acquired the appearance of a town with public administrative buildings, while, on the other hand, some vici, such as Lausonna (present-day Laussanne), featured representative basilicas. There are also Gallo-Roman *vici* with a theatre, forum, basilica and porticoed streets.
[57] Gregl and Migotti 2004; Vidović 1989; Novinščak 2006; Šegvić 2006, 270; Bobovec 2005.
[58] Bugar 2009; Vujnović and Burmaz 2010; Leleković 2008.

24. Podgajci Podravski (Berebae?)
25. Valpovo (Iovalia?)
26. Petrijevci – Kraljevo brdo (Mursella?)
27. Bobota (Ad Labores Pontis Ulcae?)
28. Orolik (Cansilena?)
29. Tovarnik – Orašje (Ulmus?)
30. Komin (Pyrri?)
31. Dubovac (Romula?)
32. Mali Gradac (Ad Fines?)
33. Gornji Baćin (Ad Praetorium?)
34. Vragorilo (Ad Bassantem?)
35. Posavski Podgajci (Saldae?)
36. Baćin Dol (Menneianae, Maninianis, Mansianis?)
37. Tekić (Incerum?)
38. Gradac Našički (Stravianae?)
39. Ruševo (Picentium?)
40. Levanjska Varoš (Leuconum?)
41. Štrbinci (Certissia?)
42. Batina (Ad Militare)
43. Zmajevac (Ad Novas)
44. Dalj (Teutoburgium)
45. Sotin (Cornacum)
46. Ilok (Cuccium)

5. Town areas and population

Only some ten out of the total number of identified Roman settlements can be classified as urban in character; it is mostly *coloniae* and *municipia*, as well as several towns of an unknown administrative status, yet defined as such by archaeological remains (Halicanum, Aquae Iasae, Iovia/Botivo, Quadrata, Varianae). The determination of an average town area by categories is not a simple task. However, if we take the town perimeter with cemeteries and suburban zones as a factor, then the town surfaces in the area of study can be compared with other western provinces, whose towns were mostly founded *ex nihilo* and their Roman urban profile was not conditioned by an earlier architectural substratum. At this stage of research, the area of urban centres (including the cemeteries) in the final phases of their development encompassed from 40 ha (example of Andautonia, with the Kutelo cemetery included) to 200 ha (example of Siscia, taking 47 ha inside the walls and 200 ha with the cemeteries and suburban surroundings). Cibalae encompassed some 40 ha within the walls, Mursa 42 ha, while a smaller urban settlement such as Iovia/Botivo encompassed only 3.4 ha. This is comparable to the towns in the western provinces, for example Gallia, where the average town area during Late Antiquity encompassed from 50 to 80 ha. In Britania, the average town area measured from 40 to 96 ha.[59] In the Hungarian part of Pannonia, the town areas range from the minimum of 20-25 ha for Scarbantia (Sopron) and Sopianae (Pécs),[60] to almost 47 ha for Savaria (Szombathely),[61] to some 50 ha for Aquincum (Budapest) in Trajan's and Hadrian's reign.[62]

The size of the town population has never been an object of archaeological research, since the safe parameters for this calculation are lacking. It is not possible to associate the size of the population with the town area, as the town walls do not coincide with the population density. Sometimes residential areas are much smaller than the total area encompassed by the walls, and the suburban zone ultimately also belongs to the town. Thus the area encompassed by the walls represents only a tentative factor in calculating the population size, and one should take into account a more generous use of space.[63] Furthermore, the population size does not reflect the status of a settlement or represent a proof (or otherwise) of its administrative status. Thus neither Viminacium (modern-day Kostolac, Serbia) nor Carnuntum (Petronell and Bad Deutsch Altenburg, Austria) obtained a real urban outlook (neither of them has a forum or administrative adjacencies), yet both did have a network of orthogonal streets and were relatively densely populated; both served as logistical and economical support for the *limes* legions. On the contrary, the colonies such as Noviodunum (Nyon, France) and *Faventia Iulia Augusta Pia Barcino* (Barcelona, Spain) encompassed only ten ha each and had only a limited population, but they still played a crucial part in the urbanism of their respective regions, as demonstrated by a disproportion between a monumental centre and residential sections.[64] Several studies have tried to calculate the population of Siscia through the flow rates of the Roman water supply; the course of this aqueduct was documented in several villages in the vicinity of Sisak and in the town itself.[65] According to the calculation of the engineer Hlavinka,[66] the capacity of this aqueduct with a daily flow rate of 2,937 m^2 could have served some 20,000-25,000 inhabitants. However, this parameter is insecure because of the existence of other sources of drinking water, as well as the proximity of the Kupa and Sava and, finally, because of the system of uncontrolled drainage. If Chandler and Bairoch's model, according to which as much as 140,000 inhabitants could have lived on 1 ha of area in Roman times, is applied to the towns in the western provinces, then their average population varied from 5,000 to 20,000, with Trier's 50,000 and its special status being an exception to the rule.[67]

6. Components of urbanism and urban structure

The complexity of Roman society and institutions is reflected in the diversity of both private and public architecture. The uniformity of physical appearance of Roman architecture in the entire territory of the Empire makes it possible to determine the function of certain structures even though they are poorly preserved. The

[59] Gros and Torelli 1988, 250.
[60] Póczy 1980, 245; Visy 2003, 231.
[61] Póczy 1980, 239; Scherrer 2003, 64.
[62] Póczy 1980, 255.
[63] Gros and Torelli 1988, 250, 251.
[64] Gros and Torelli 1988, 247-249.
[65] Lapaine 1881, 60-61.
[66] Hlavinka, 1902, 31.
[67] Chandler 1987; Bairoch 1988.

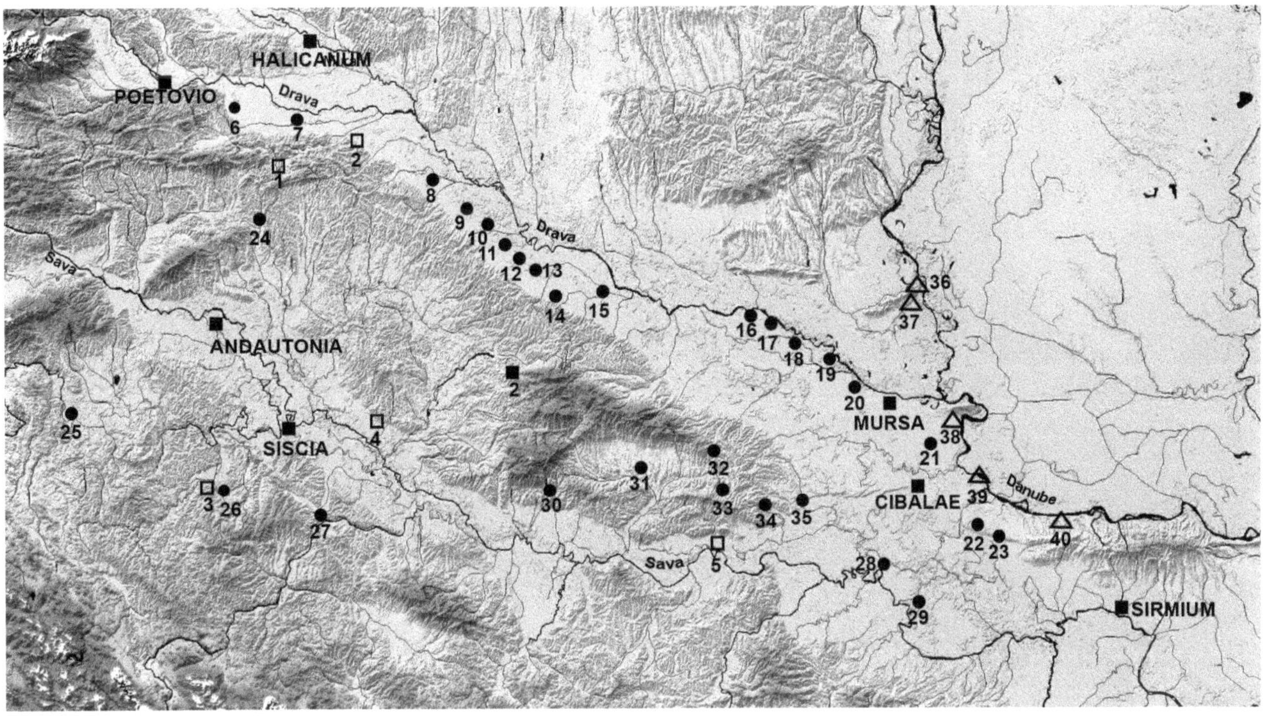

■ AUTONOMOUS TOWNS

□ SMALL TOWNS

1. Varaždinske Toplice (AQUAE IASAE)
2. Ludbreg (civitas IOVIA / BOTIVO)
3. Topusko (QUADRATA?)
4. Kutina (VARIANAE?)
5. Slavonski Brod (MARSONIA?)

● OTHER SITES

6. Petrijanec (mansio AQUA VIVA?)
7. Bartolovec (POPOLIS?)
8. Kunovec Breg (mutatio SUNISTA?)
9. Draganovac (mutatio PIRETIS?)
10. Virje (mansio LENTOLIS?)
11. Prugovac (mutatio IOVIA? CARODUNUM?)
12. Turnašica (IOVIA/IOVISTA?)
13. Špišić Bukovica (COCCONAE?)
14. Virovitica – Taborište (SEROTA?)
15. Orešac (BOLENTIO?)
16. Donji Miholjac (MARINIANAE?)
17. Sveti Đurađ (SERENA?)
18. Podgajci Podravski (BEREBAE?)
19. Valpovo (IOVALIA?)
20. Petrijevci – Kraljevo brdo (MURSELLA?)
21. Bobota (AD LABORES PONTIS ULCAE?)
22. Orolik (CANSILENA?)
23. Tovarnik – Orašje (ULMUS?)
24. Komin (PYRII?)
25. Dubovac (ROMULA?)
26. Mali Gradac (AD FINES?)
27. Gornji Baćin (AD PRAETORIUM?)
28. Vragorilo (AD BASSANTEM?)
29. Posavski Podgajci (SALDAE?)
30. Baćin Dol (MENNEIANAE?)
31. Tekić (INCERUM?)
32. Gradac Našički (STRAVIANAE?)
33. Ruševo (PICENTIUM?)
34. Levanjska Varoš (LEUCONUM?)
35. Štrbinci (CERTISSIA?)

△ SITES ON THE *LIMES*

36. Batina (AD MILITARE)
37. Zmajevac (AD NOVAS)
38. Dalj (TEUTOBURGIUM)
39. Sotin (CORNACUM)
40. Ilok (CUCCIUM)

Fig. 1. Distribution of Roman towns and settlements (after Google Maps, modified by T. Leleković).

same is generally true of the settlements in northern Croatia, where as a rule only the foundations and basic superstructures survive. This is especially true of the settlements that inherited the continuity of living from the ancient to the present day, with archaeological excavations mostly conditioned by the scope of modern-day city parcels, so that they only rarely result in complete research of a rounded and meaningful whole. Although these excavations regularly offer some information on the function of excavated structures, the complete urban layout of a settlement often remains unknown and poorly understood. Nevertheless, the century-long tradition of archaeological excavations of urban centres in northern Croatia, especially protective excavations carried out during the last 20 years, have enabled recognising of certain rules that were applied in the processes of urbanisation, as well as the typology of architecture in the area between the Rivers Sava, Drava, and Danube.

6.1. Urban matrix

The orthogonal plan, matched with the division into *insulae*, was consistently implemented and can be recognised in all urban centres. The dimensions of *insulae* can at this stage be determined by integrating the results of archaeological excavations into the known Roman standards applied in the western provinces. The ground plan of Siscia was organised about the *cardo maximus*, discovered at several points in modern Sisak.[68] The orientation of the urban complexes in the northern part of Siscia mostly followed the north-south axis, which was inclined from 8 to 30° towards the southern section (the greatest deviation is noted in the *insula* abutting on the southern town walls).[69] Certain sections of the *decumanus*[70] and *cardo*[71] have also been discovered and

[68] Lolić 2006, 12
[69] Lolić 2003, 149.
[70] Filipović and Katavić 2005, 8-9.
[71] Sunko, Katavić, and Jerončić 2007, 20-21.

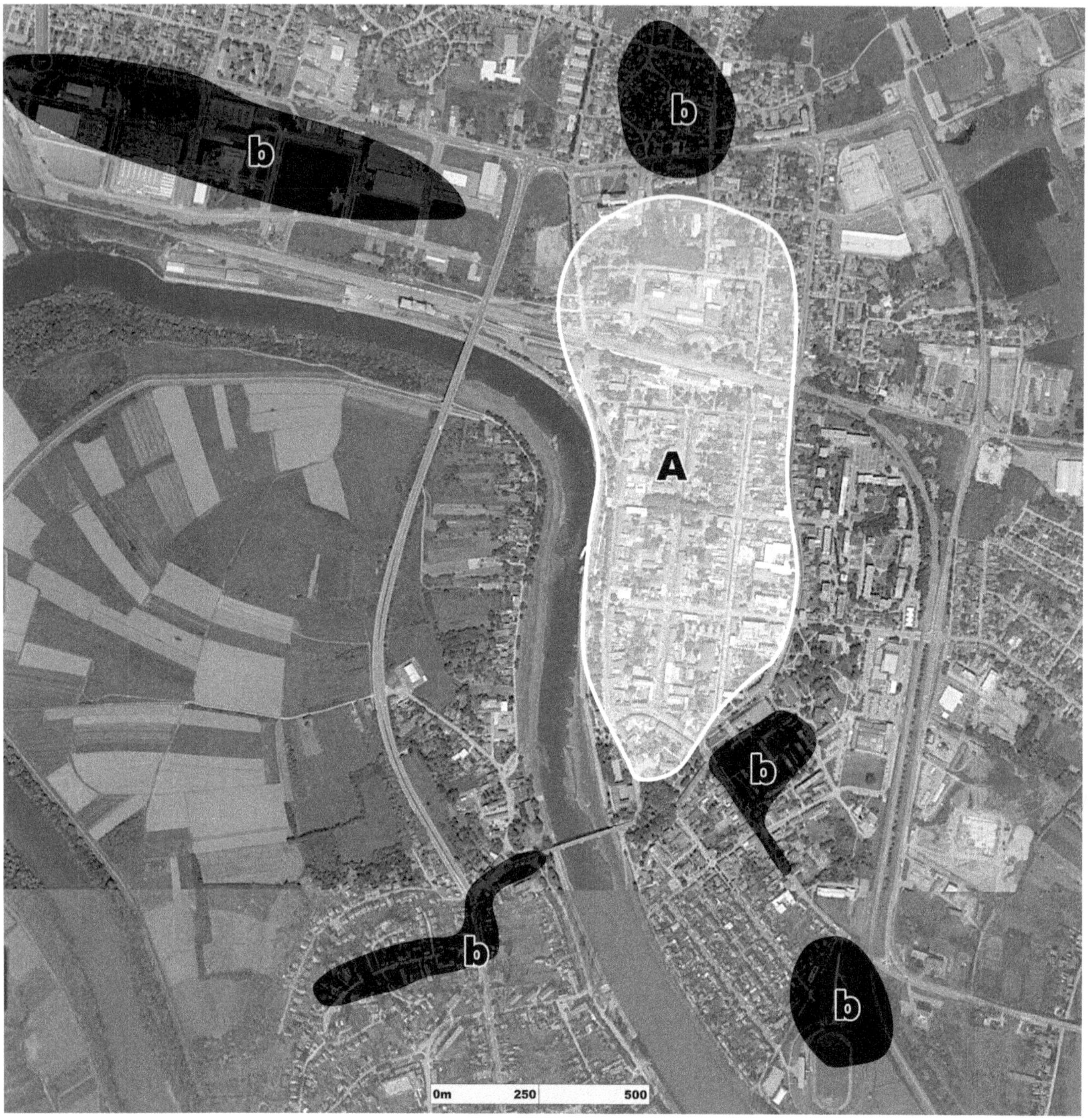

Fig. 2. Sisak : A – Siscia; b – cemetery (Lolić and Wiewegh).*

preserved in Mursa, and the complete rectangular ground plan of the town perimeter shows a mild inclination from the north-south axis.[72] In Andautonia a street network from different phases of construction can be discerned, while one systematically studied and presented *insula* shows an inclination of 12° from the north.[73] Town streets often formed a combined system with the drainage and water supply facilities in the form of vaulted canals. In Siscia and Andautonia the streets were paved with dressed limestone slabs and kerbstone blocks of various dimensions.[74] In Mursa the roads were based upon a timber grid structures, adapting to the evidently low-quality ground.[75] The streets discovered so far in Mursa and Siscia were 8- to 10-m wide, while the street in Andautonia near the thermal complex was 6.4 m wide. The dimensions of urban residential buildings lined by streets can be reconstructed from the data furnished in Mursa: at the site of Huttlerova Street, a larger part of an apartment building (*insula*) was discovered; it was bounded by an east-west road on the northern side and a north-south road on the western side, while in the south it abutted on the town walls (the eastern limit remains unexplored).[76] A dense urban structure was discovered in archaeological and geo-physical research at the St. Quirinus site in Sisak, as well as in the northern part of the town

72 Marsigli 1726; Pinterović 1978, K III; Filipović 2004, 161.
73 Nemeth-Ehrlich and Kušan 2003, 118-119; Nemeth-Ehrlich and Kušan 2007, 17.
74 Lolić 2006, 12; Nemeth-Ehrlich and Kušan 2007, 17.

75 Filipović and Katavić 2005, 8, 9.
76 Sunko, Katavić, and Jerončić 2007, 20-21.

Fig. 3. Sisak – southern town wall with a corner tower and part of a horreum (photo: M. Arbutina).

along the *cardo maximus*, which yielded a large surface with Roman apartment blocks.[77] The examples from Mursa and Siscia suggest that an urban matrix could have been divided into apartment blocks measuring 71 x 35.5 m (2 x 1 *actus*). A number of public buildings, both luxurious and more modest apartments, as well as baths, warehouses, and infrastructure facilities from all phases of urbanisation have been discovered in Siscia, Andautonia, Mursa, Cibalae, and Iovia/Botivo. The only site to yield a classic complex – a forum, temples and a basilica – is Varaždinske Toplice (Aquae Iasae), the important spa centre with hot water springs, where a forum with the Capitolium temples was discovered, dedicated to Jupiter (?), Juno, and Minerva.[78]

6.2. Public architecture

6.2.1. Town walls

Town walls are a certain indicator of the difference between urban and suburban sections of the inhabited zones of Roman towns.[79] Thus Siscia, Cibalae, Mursa, and Iovia/Botivo were all encompassed by walls that clearly separated them from their extra-urban peripheries. However, the towns such as Andautonia, with the urban core not surrounded by walls, could have had other type of marks separating them from their peripheries, such as those present in many towns in the western provinces.[80] Accordingly, the boundary of the town core of Andautonia is marked by its cemeteries and their alternating positions, that is, their subsequent expansion and overlapping with the urban structure.[81]

The perimeter of Siscia was determined by the discovery of longer sections of walls on several key locations – next to the northern and southern gates, at the bank of the Kupa

[77] Lolić 2010, 6.
[78] Nemeth-Ehrlich and Kušan 1997.

[79] Goodman 2007, 39.
[80] Goodman 2007, 86.
[81] Cf. Nemeth-Ehrlich and Kušan Špalj 2003, 117.

*Illustrations nos. 2, 5, 6, 7, and 11, marked by asterisks, were made by Lolić and Wiewegh on the basis of multiple referential literature for each individual town, adduced in the bibliography.

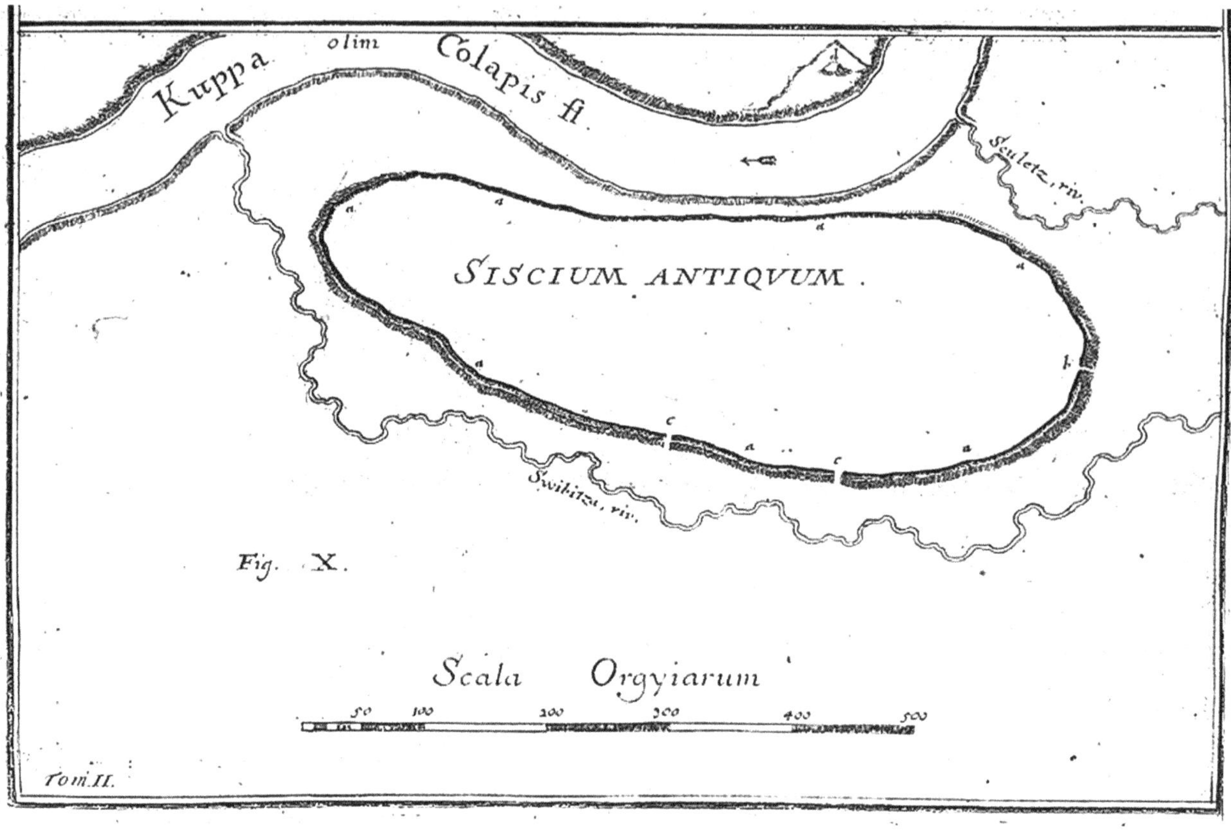

Fig. 4. Plan of Siscia (after Marsigli 1726).

on the western and eastern side of the town; the total area encompassed by the town inside the walls was 47 ha (fig. 2*). The excavated southern wall (the site of Bana Jelačića Square, where 30 m of the wall survives and is displayed) was built of brick (measuring 43 x 30 x 6.5 cm, some of them with the stamp *SISC*), with a smaller amount of mortared stone blocks, supporting the corner near the inner tower. The 2-m-wide wall was built at the turn of the 2nd and 3rd centuries, featuring additions and repairs from the 4th century. It was surrounded by ditches on the north, east and south sides (visible until the 19th century), with the Kupa River as its western "natural ditch". During a number of archaeological campaigns a system of inner rectangular towers, together with a single exterior tower, has been discovered on the west, east and south sides.[82] In 1726 the Austrian officer Luigi F. Marsigli (1726, fig. X) published the first illustration of the Siscian town walls with the position of gates, the Kupa and two streams that were surrounding the town on all sides; recent excavations have confirmed the accuracy of his drawing.

The walls of Andautonia (Ščitarjevo) (fig. 5*) have not yet been discovered. Still, an approximate perimeter of the town is known. In the first phase of development (1st to early 2nd centuries) the town extended both from north to south and from east to west for 250 m (the area occupied by the site of Gradišće with the parish church, and to the north of it). In the 2nd phase (2nd-3rd centuries) the town extended from north to south for 1 km (a significant expansion occurred towards the south), and 400 m from east to west. In the 3rd phase (late 3rd-4th centuries) the town area shrunk, as transpires from the graves found in places previously occupied by architecture.[83]

The spatial organisation of the fortified part of Iovia/Botivo (Ludbreg) (fig. 6*) suggests that it was a settlement of the *castellum* or *castrum* type.[84] It was fortified by masonry walls that went through several stages of construction and renovation, but the walls were throughout built of rubble stone and (rarely) half-dressed slabs, both bonded by lime mortar. The earliest walls were constructed at the turn of the 1st and 2nd centuries, and the later phase (when a new course of walls was added on the northern and western sides) is dated to the 4th century.[85] The area of the settlement encompassed by the walls was ca. 3.4 ha, measuring 170 (east-west) x 200 (north-south) m. Traces of double fortifications were found in the northern and western part of the town, with an empty space between the two stretches of walls; this mode of construction was not evidenced on the eastern and southern sides, which both followed the course of the Bednja River. The northern double wall was discovered at several places; at one of them it was 1.35 m thick. The western double wall was also found at several

[82] Vrbanović 1981; Lolić 2003; Wiewegh 2004; Burmaz 2007; Lolić 2007.

[83] Vikić-Belančić 1981; Nemeth-Ehrlich and Kušan Špalj 2007, 38-50.

[84] The form of the settlement can be indicative of its military origin, but cannot serve as the proof of it.

[85] Gorenc and Vikić 1984, 63, 64; Gregl and Migotti 2004, 133.

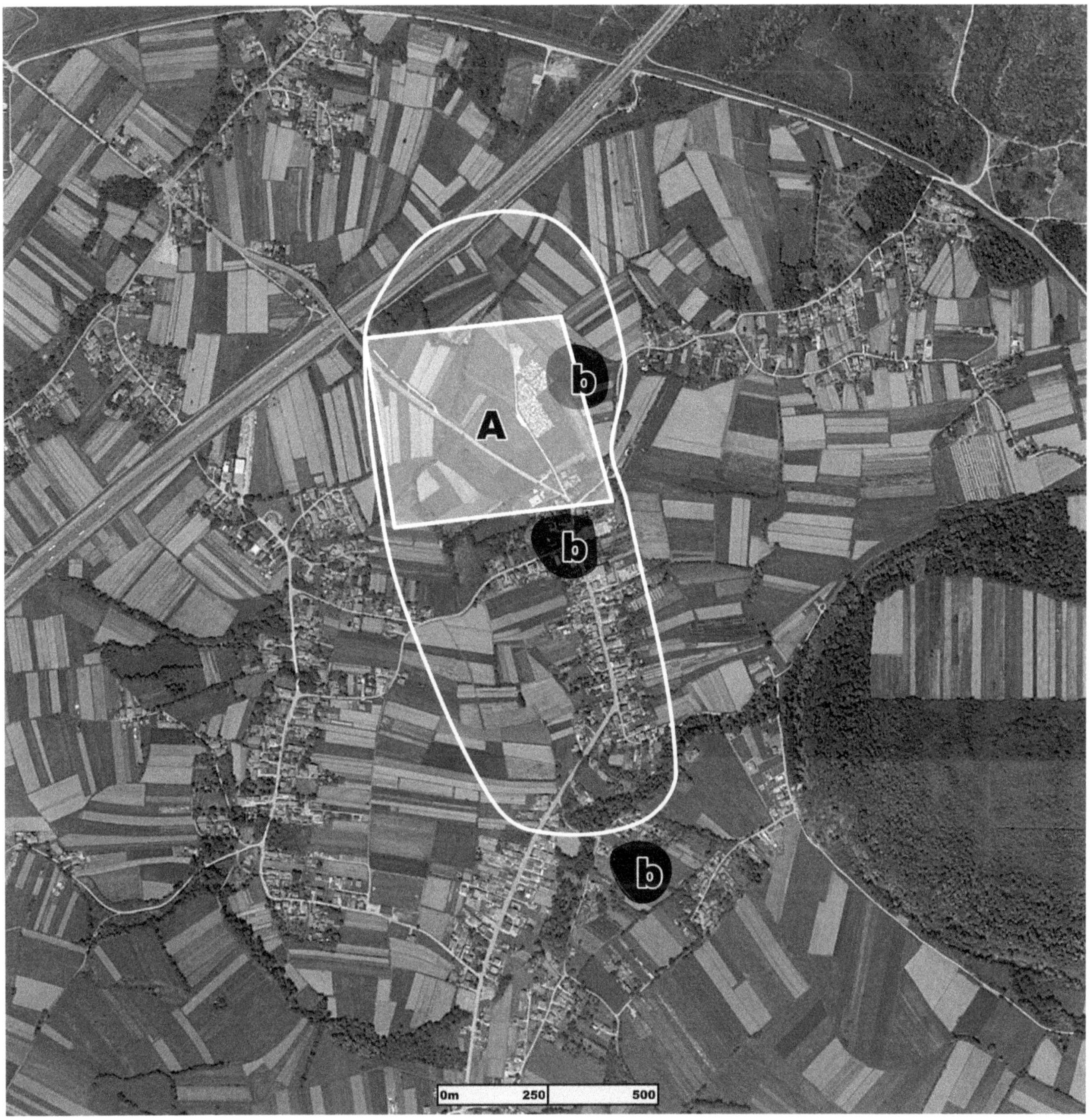

Fig. 5. Ščitarjevo: A – Andautonia; b – cemetery (Lolić and Wiewegh).*

points (1.1 and 1.3 m wide). Some 11 m of the southern wall was excavated, revealing its considerable width (4.6 m), while the eastern wall was 3.2 m wide.[86]

A single archaeologically confirmed segment of the Mursa (Osijek) walls is known at present (fig. 7*). It was recovered in 2007 during protective excavations in the southern stretch of Huttlerova Street (former Silos), at its junction with Frankopanska Street. The first phase of the site is roughly dated to 1st-2nd centuries, and the walls should certainly belong to the period before the Marcomannic Wars (169-180). A section of the trench belonging to the southern town wall and the accompanying defensive ditch were found there. Neither the wall nor the protective ditch were preserved in their original form due to the reuse of building materials and the modern constructions erected on the site. No masonry of the fortification system was found, apart from some broken bricks from the collapsed material that found its way to the ditch. Only two regular east-west parallel trenches were found. The northern (presumed) fortifications trench was ca. 1.5 m wide, while the southern one was 5 m wide in its upper section, tapering mildly towards the bottom. Besides its primary defensive function, this protective ditch probably also served as part of the town's drainage system.[87] It is possible that Mursa had double town walls,[88] but the above-mentioned

[86] Tomičić 1966; Tomičić 1997; Gorenc and Vikić 1984; Vikić-Belančić 1984; Vikić-Belančić 1998; Gregl and Migotti 2004, 136.

[87] Sunko, Katavić, and Jerončić 2008, 22, 23.

[88] Šimić and Filipović 1997, 54.

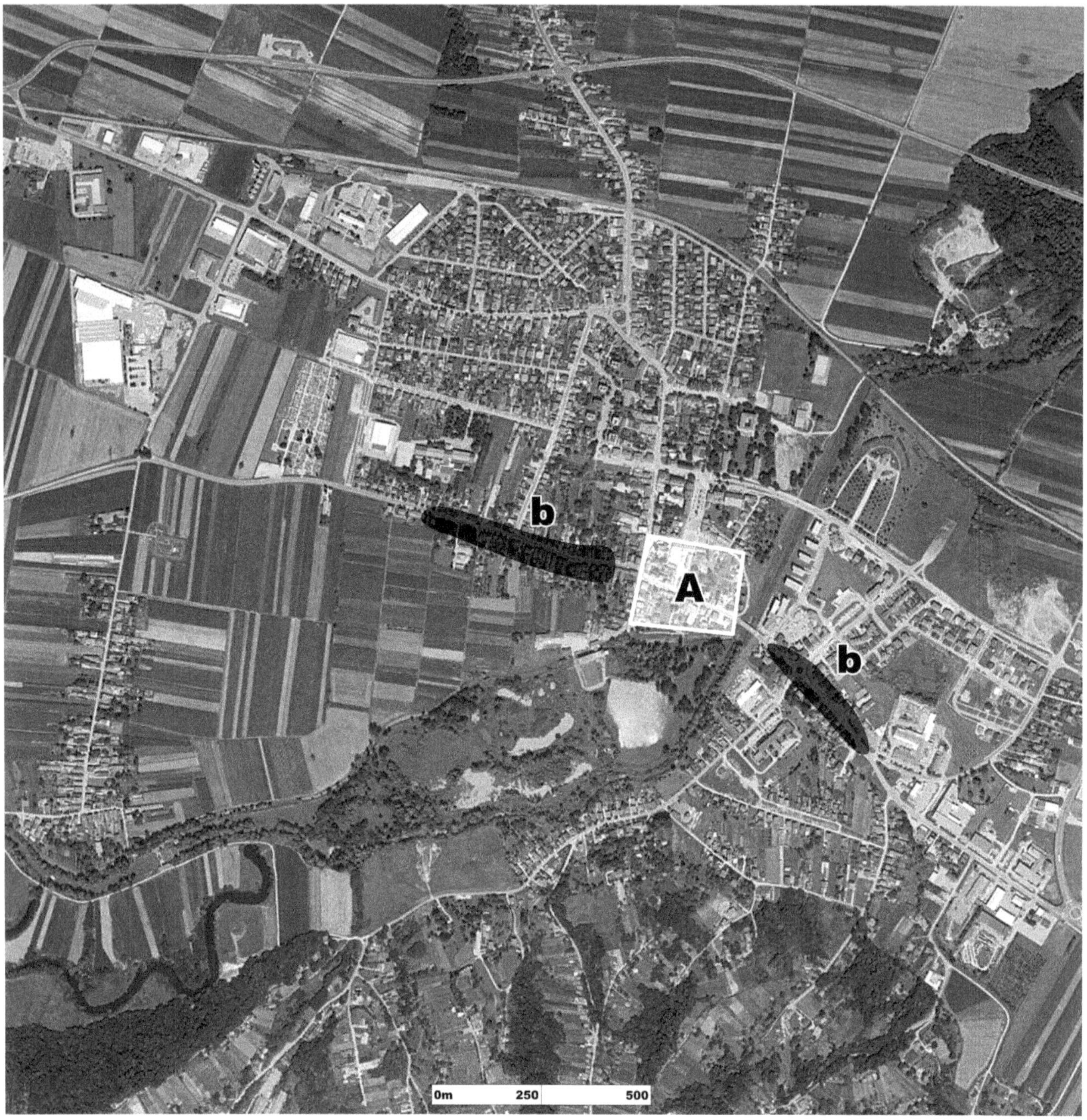

Fig. 6. Ludbreg: A – Iovia/Botivo; b – cemetery (Lolić and Wiewegh).*

excavations were unable to confirm this hypothesis since the area covered by the excavation trench was rather limited, especially in the section of the walls. The exact perimeter of Mursa is still unknown, and literary sources, together with maps and plans, have created space for different hypotheses. Accordingly, archaeological excavations so far have only been able to suggest the locations where the walls were *not* placed, rather than vice versa. Thus, for example, during the protective excavations on the future site of the market (Bana Jelačića Square), burials from two periods (1st to 2nd centuries and late 2nd to early 4th centuries) were found together with a stretch of a wide east-west road and some architecture (also featuring several phases) that was in use between the two periods represented by the graves. Arguably, this area was located outside the fortifications, which should be looked for somewhat more to the west than it is commonly supposed.[89] Somewhat to the north of this location, in the northern section of Zmaja J. Jovanovića Street, in front of the modern school building, an intact 4th-century sarcophagus was found, originally placed by a road leading from the bridge towards the northern or eastern wall.[90] Hermine Göricke-Lukić has discussed the position of the burials along the eastern wall, suggesting a possible position of fortifications in this part of the town, somewhat to the west in comparison with the earlier hypothesis.[91] A note (although rather imprecise) on the northern wall, found in the 1934 excavations in the area between the

89 Leleković 2009.
90 Pinterović 1978, 153.
91 Göricke-Lukić 2011.

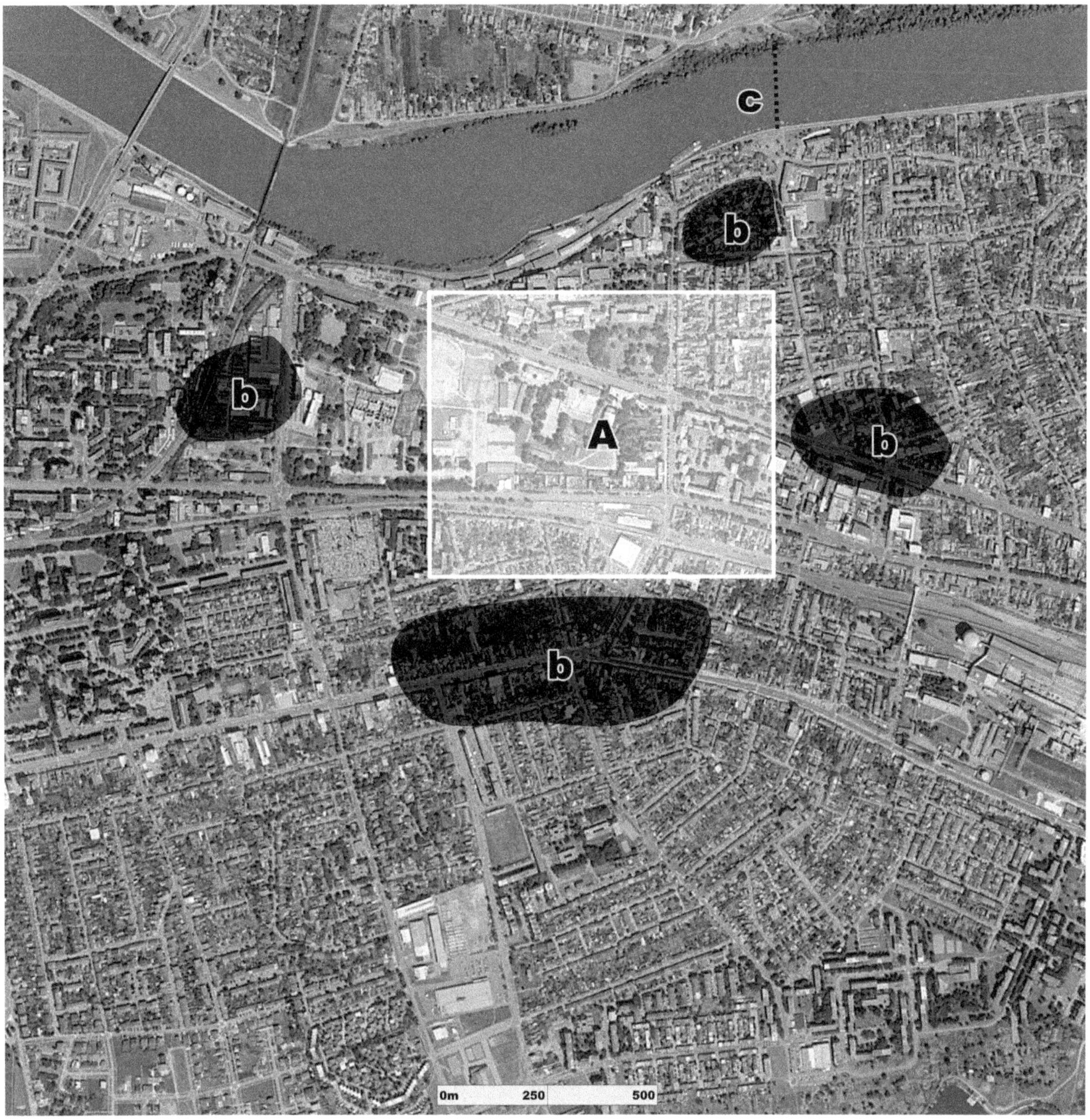

Fig. 7. Osijek :A – Mursa; b – cemetery; c – bridge (Lolić and Wiewegh*).

Fig. 8. Plan of Osijek with the remains of Mursa (after Marsigli 1726).

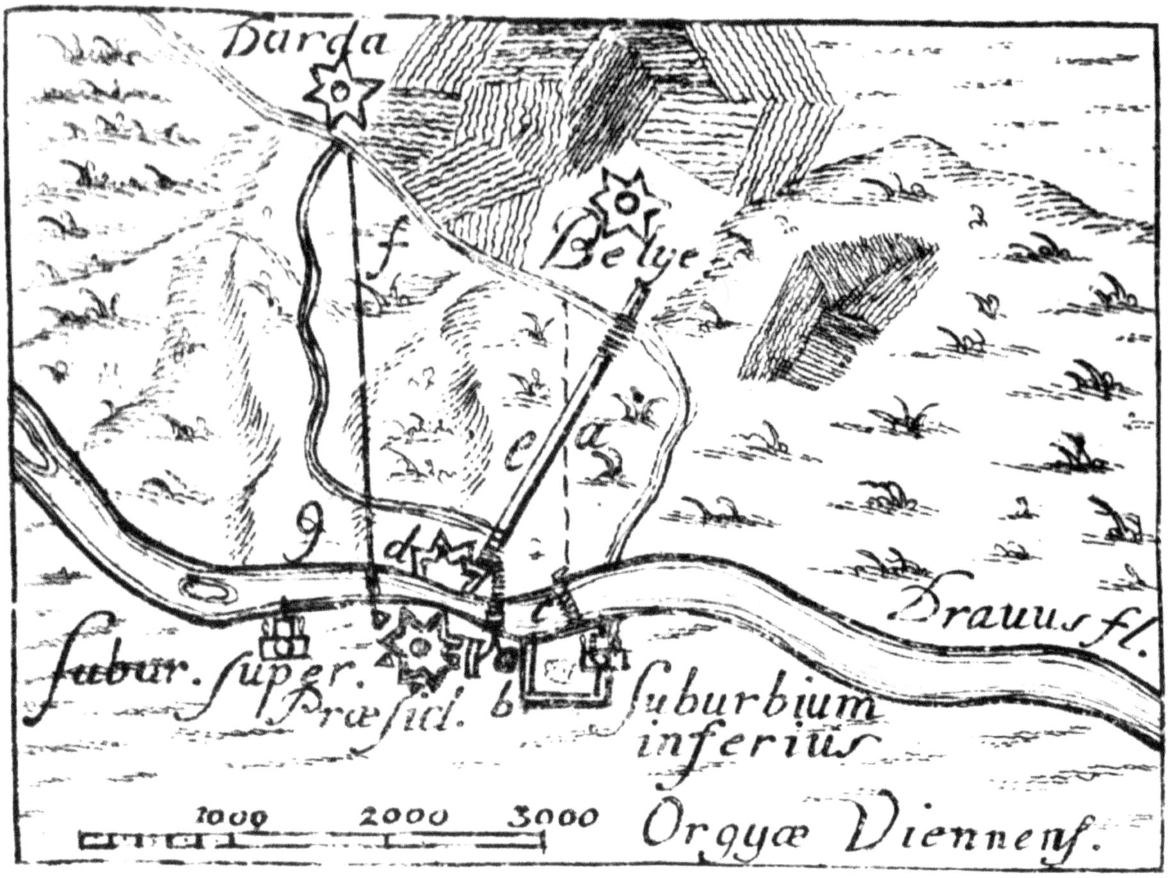

Fig. 9. Katančić's drawing of Osijek and its surroundings, 1782 (after Pinterović 1978).

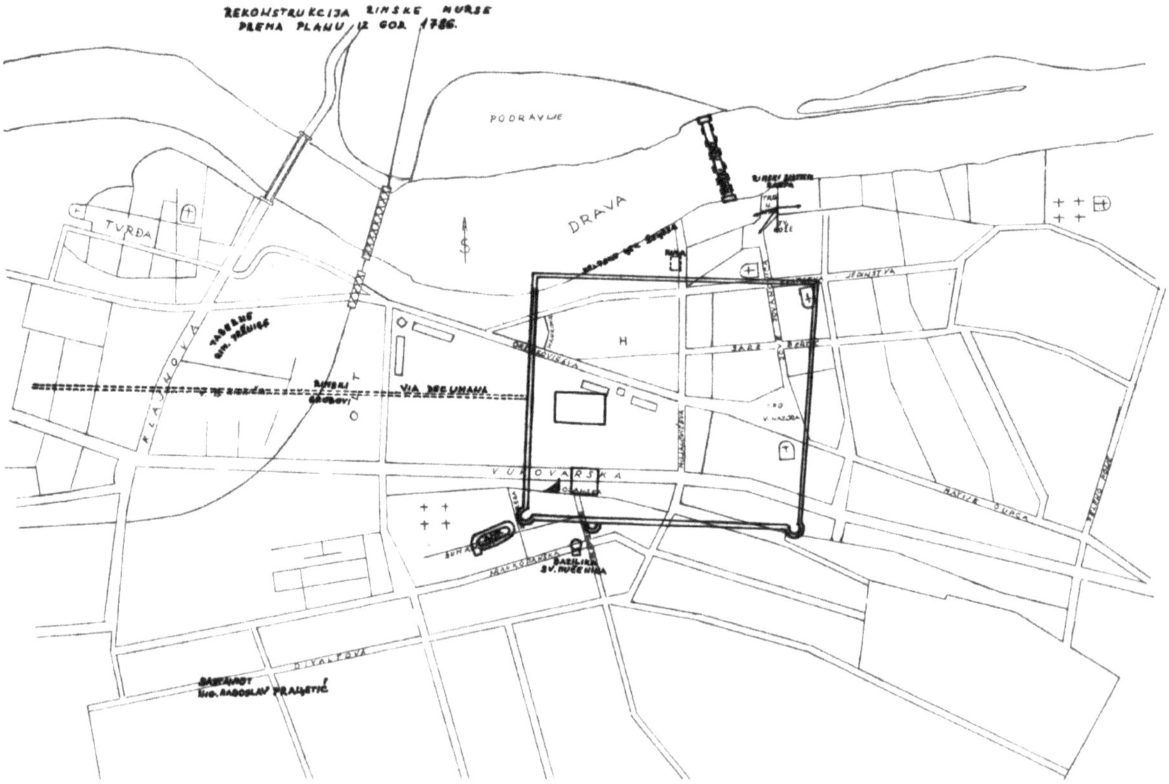

Fig. 10. Franjetić's reconstruction of Mursa (after Pinterović 1978).

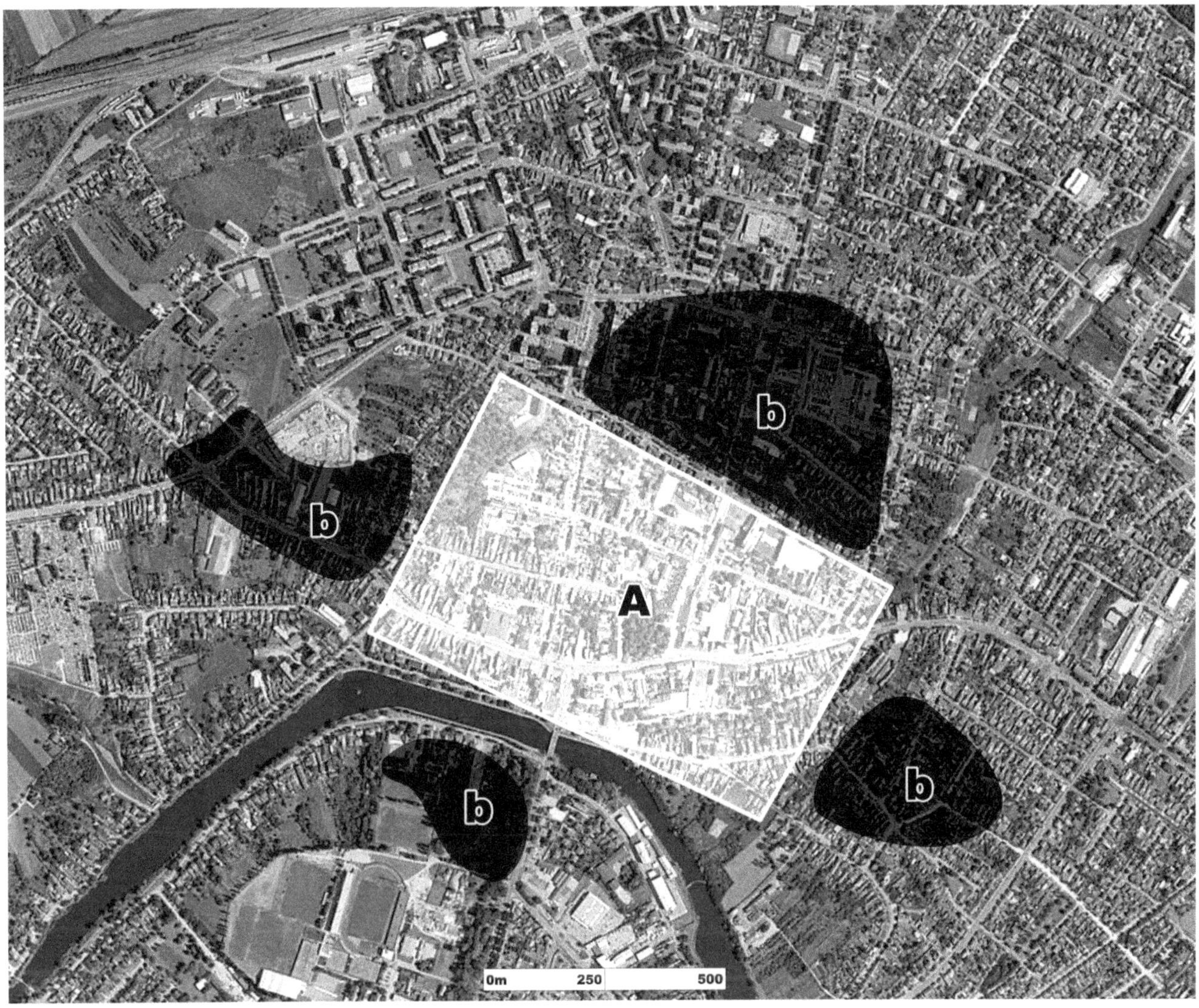

Fig. 11. Vinkovci: A – Cibalae; b – cemetery (Lolić and Wiewegh*).

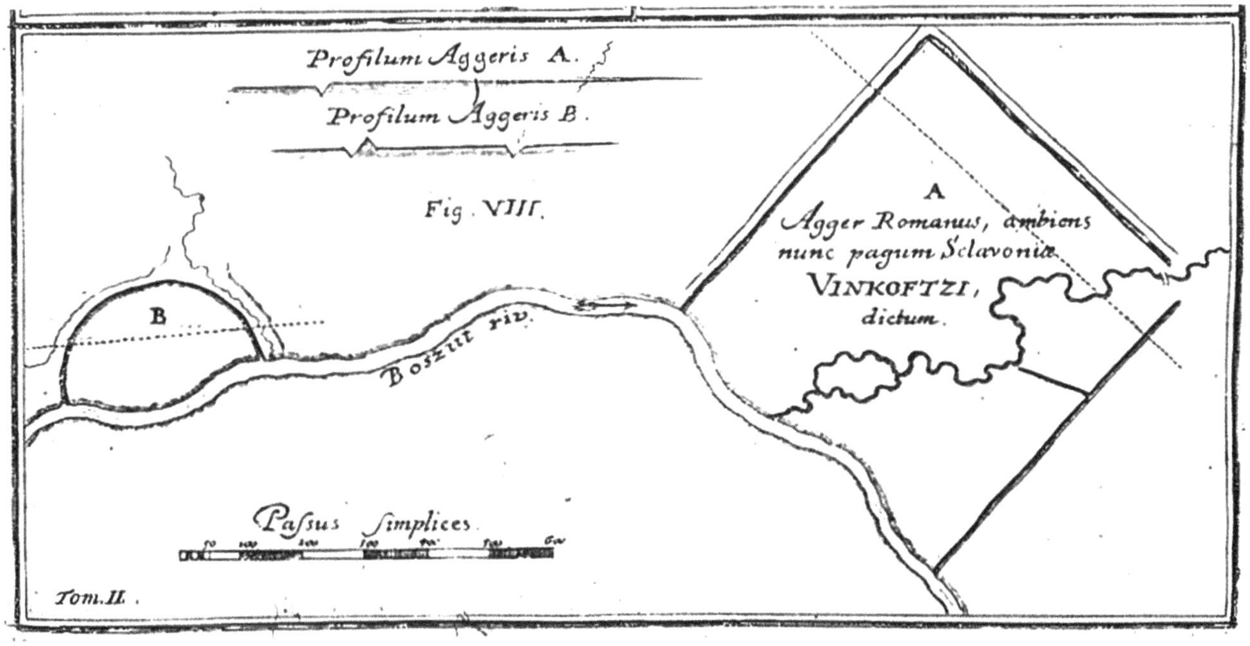

Fig. 12. Plan of Cibalae (after Marsigli 1726).

Fig. 13. Reconstruction of the fortification ditches of Cibalae (after Dimitrijević 1979).

customs office and the hospital facing the river bank, as well as a note on the western wall that extended towards the customs office, are nevertheless important.[92] Some essential information on the course of the town walls can be found in earlier studies; thus L. F. Marsigli (1726, fig. XI) published a graphic representation of the Turkish fort with five towers that he saw during his visit to Osijek in the late 17th century, and he also marked with a broken line the remains of the brick-built town walls of Mursa. In 1782, the Franciscan Matija Petar Katančić published a map of the wider Osijek area with a position of Roman Mursa and the bridge across the Drava between Mursa and the Baranja. Mursa is here represented as rectangular in form, surrounded by double walls on three sides with the Drava on the remaining northern side. The total circumference of the walls was 627.5 m; thus, the total area of the town inside the walls was little less than 40 ha.[93] In 1929 the engineer Radoslav Franjetić made a reconstruction of the layout of the town walls, based on the map of Osijek from 1786 and Katančić's drawing; it was published by Danica Pinterović.[94] According to Franjetić, Mursa was rectangular in form; the length of the northern and southern walls was each 680 m, while the length of the eastern and western walls was each 760 m, giving an area of 51.68 ha.[95] However, the analysis of the plans and data published by earlier researches, in combination with the results of more recent archaeological excavations, suggests that the total area of Mursa inside the walls encompassed 42 ha.

Archaeological excavations conducted in the Cibalae (Vinkovci) area (fig. 11*) have resulted in the discovery of fortifications at several points, but the exact perimeter of the town has not yet been accurately determined, due to the density of modern constructions. The town was subtrapezoidal in form, as the ideal rectangular form was somewhat transformed by the lie of the land.[96] The southern wall extended along the Bosut River, the western was discovered in I. Gundulića Street, while the northern was almost parallel with Jurja Dalmatinca Street. The position of the eastern wall remains somewhat perplexing; it was discovered at several points (Hrvatskih kraljeva, Bana Jelačića, and Istarska Streets), but its exact course towards the Bosut remains unknown.[97] Similarly as in Mursa, the size and course of the town walls of Roman Cibalae are variously plotted, with hypotheses based on the written sources, maps and plans, combined with archaeological excavations. F. Marsigli (1726, fig. VIII) published a drawing of ditches surrounding the town with clearly marked northern and western sides, the Bosut River representing their southern segment, while the eastern side is marked by a thin line, evidently representing only a hypothetical position of the ditch. The town was almost a regular rectangle, whose eastern side measured 610 m and the remaining three ca. 600 m; thus its area encompassed some 36.6 ha. Furthermore, according to this drawing, the town was traversed by the Ervenica-Barica Stream.[98] The site of Dirov brijeg with the position of the pre-Roman fortified settlement, is marked with the letter B and depicted to the east of the ditch. On the cross-section of this site as published by Marsigli, an earthen bank and a ditch connecting the Nevkoš Stream with the Bosut is

92 Pinterović 1978, 161, 165, 166. The position of the wall near the customs office building corresponds to a description of the church historian J. Koller (1804).
93 Pinterović 1978, 160, 161.
94 Pinterović 1978, K. III.
95 Pinterović 1979, 161; Filipović 2004, 160-161.
96 Dimitrijević 1979, 149.
97 On the walls see Dimitrijević 1979; Iskra-Janošić 2001; Iskra-Janošić, Krznarić Škrivanko, and Rapan Papeša 2006, 39; Krznarić Škrivanko 2007, 54, 55; Vulić, Krznarić Škrivanko, and Rapan Papeša 2008; Vulić, Krznarić Škrivanko, and Rapan Papeša 2010, 61-62.
98 Iskra-Janošić 2001, 33.

represented. Even so, a continuous occupation of the site after the arrival of the Romans has not yet been confirmed by excavations.[99] In the early 20th century Josip Brunšmid compared the analysis of Marsigli's plan and the 1789 town map with the actual situation in his time and tentatively outlined the town perimeter as covering an area of 56 ha. In 1979 Stojan Dimitrijević also hypothesised on the course of the town walls and the main gates. In the southeast part of the settlement a segment of the town was separated off by a canal, which was conditioned by the course of the Brook Ervenica. Dimitrijević further stated that the northern wall was 710 m long, the western 650 m, the southern (along the Bosut) 780 m, and the eastern 630 m, producing a total surface of 48.3 ha.[100] Ivana Iskra Janošić, after having studied the above noted hypotheses, and on the basis of the late 20th- and early 21st-century excavations, postulated a theory of two fortified areas of Cibalae, of which the smaller rectangular area conforms to Marsigli's ground plan. She further defined the length of the walls: 680 m each for the northern and southern sides, and 590 and 550 for the western and eastern sides, respectively, giving a total area of 38.1 ha. Iskra Janošić's larger area roughly confirms to the dimensions as given by Brunšmid and Dimitrijević: 800 m each on the northern and southern sides and 740 and 680 m on the western and eastern sides, respectively, giving a total area of 56.8 ha. The author further notes that the smaller fortified area was surrounded by double walls, varying in thickness from 13 to 14 m (outer stretches) and 10 to 14 m (inner stretches), while the ditch between them was 5 m wide (the eastern wall was not double since the Ervenica Stream offered sufficient protection); this phase of the construction of the walls was dated in Hadrian's reign.[101] The other fortified area, surrounded by single walls and dated to a later phase of construction, partially overlaid the earlier walls, but was partly built anew. Its width was 14 m on the northern side, and it was surmounted by a timber palisade 0.80 m wide.[102] This phase is dated to the period following the Marcomannic Wars (after 180 AD), when masonry fortifications were introduced on the *limes*.[103] According to the present state of archaeological research, the reconstruction of the town's layout as produced by Iskra Janošić can be accepted, with possible minor discrepancies allowed for.

Archaeological research of the fortifications of the larger towns in the area of study suggests that in the earliest phase of the Roman conquest these comprised of earthen defensive structures (banks and ditches) with palisades. Such structures have been discovered in archaeological excavations in Osijek and Vinkovci, while the earliest phase in Sisak is confirmed by literary sources claiming that after the capture of Segestica Tiberius had a large defensive ditch constructed.[104] Belts of (mainly) brick town walls, occasionally strengthened with stone, were erected during the 2nd century, as in other western provinces; they should be associated with imperial decrees, probably respecting regional political circumstances, either in terms of privileges or the needs of defence.[105]

6.2.2. Forums with accompanying adjacencies

No forum has been found in archaeological excavations in the autonomous towns of the area of study, mostly due to a dense urban structure developing gradually from the 18th to 20th centuries above the remains of Roman and mediaeval strata of modern towns.

A single public open-air complex has so far been excavated in Siscia (fig. 14). In the extreme northern section of the town an architectural complex was discovered, stretched along the *cardo maximus*, featuring dimensions and plan befitting a smaller square. Foundations of a temple of an unknown dedication were found on its western side (9.35 x 9.10 m), while an open space surrounded with temples on northern, western and southern sides, with the eastern side open towards a street, was discovered in front of it (square dimensions: 31 m along the east-west axis and 22 m along the north-south axis). In the western part of the square, at the temple entrance, remains of a base of a presumed sculpture-pedestal were discovered (the preserved segment of the base is 2 m long). Fragments of various types of marbles were found in the entire area of the square (*cipollino, verde antico* and *Carrara*) as well as moulded architectural stone fragments.[106] Remains of another building, perhaps a temple (see section 6.2.3.) sited in a public square, were found in Rimska Street next to the western town wall by the Kupa River, south of the remains of public baths. Fragments of an architrave inscription found near the remains of a supposed temple support the identification of this area as the forum – the following buildings are mentioned in various inscriptions related to this area: *basilica*, *macellum*, *porticus*, *curias*, all pointing to the buildings typical of a town centre.[107] However, none of these buildings have been proved archaeologically, due to the poor state of research of this part of Sisak.

The existence of a forum is archaeologically evidenced only in Aquae Iasae (fig. 15). Several phases of construction were established, of which the early 4th-century one, during Constantine's reign, is the best preserved. The entire architectural complex (ca. 6000 m^2 in area) is composed of a forum with the *Capitolium* and, south of the forum, a thermal complex with a basilica. The 1st-century remains are scarce, with the layout of the forum unknown and traces of a single temple found in its northern part. During the 2nd century the forum was structured as a central space framed by architectural constructions, that is, a porch on the eastern and western sides and with three juxtaposed temples on the northern side, somewhat more to the north in comparison to the temple erected in the first phase of construction. The

99 Dizdar and Radman-Livaja 2004, 46.
100 Dimitrijević 1979, 149.
101 Dizdar, Iskra-Janošić, and Krznarić Škrivanko 1999, 55.
102 Iskra-Janošić 2001, 39-42.
103 Iskra-Janošić 2001, 57.
104 Hoti 1992, 137.
105 Gros and Torelli 1988, 255.
106 Lolić 2010.
107 Faber 1973, 143-146, 155; Zaninović 1981, 206, 207.

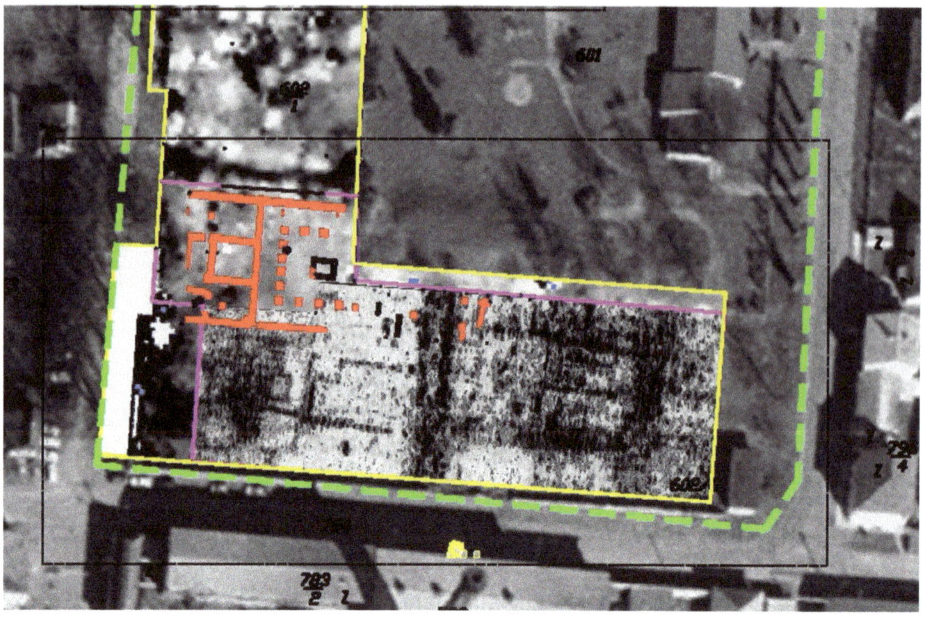

Fig. 14. Sisak, public open-air complex with a temple on the northern part of the town, excavated area supplemented with the results of geo-physical research (after Mušič, in Lolić 2010).

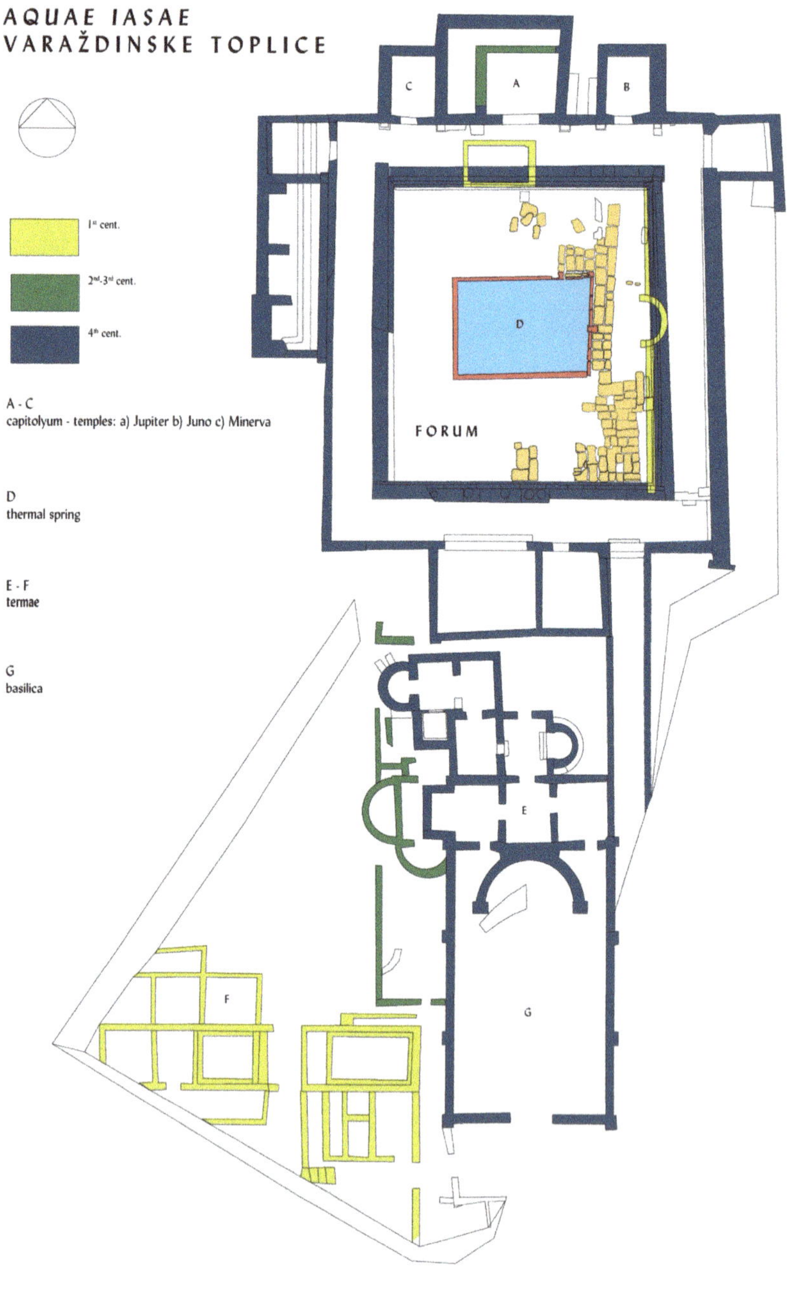

Fig. 15. Varaždinske Toplice, layout of the excavated area: forum with temple (north) and thermal complex (south) (after Nemeth-Ehrlich and Kušan Špalj 2011).

appearance of the construction enclosing a thermal-water spring is unknown at this time, but a *nymphaeum* probably existed in the forum, judging by the findings of *spolia* in the form of marble plates with reliefs depicting mythological scenes.[108] In the early 4th century, following Constantine's restoration of the entire thermal complex, the forum was surrounded by porches on all four sides. The ground plan of the three temples was modified, as now they shared the southern façade facing the forum and were covered by a common roof; at the farthest northern part of the eastern and western porches the entrances into two lateral rooms of unknown use were found. A rectangular stone-block enclosure (8 x 13.5 m) around the source of thermal water was constructed in the centre of the forum. The forum area measured 528 m² (22 x 24 m) and was paved with stone slabs, using also earlier monuments as building material in its construction; due to this, some inscriptions and reliefs originating in earlier phases of the complex were preserved. The 4th-century forum was surrounded by colonnades on three sides, with equally spaced columns, topped by horizontal beams. The columns of the northern colonnade were unequally spaced, which points to the fact that they were connected by arches accentuating the temple entrances. Two rooms of unknown function (*tabernae*?) were installed in the southern porch, as well as two lateral entrances to the forum, serving as a communication with the thermal part of the complex. Systems of drainage canals for the rainwater and the surplus of thermal water were discovered at several places throughout the forum and at a section of the western porch.[109]

The Roman forum of Iovia/Botivo has not been recovered in archaeological excavations, but judging from the urban layout in the form of a *castellum* or *castrum*, a forum of modest dimensions at the intersection of the *cardo* and *decumanus* (most probably north of Bana Jelačića Street) should be presumed; smaller sections of both *cardo* and *decumanus* have been attested archaeologically.[110] The forum in Mursa has also never been discovered; R. Franjetić hypothetically located it at the present-day Kraljice Kosače Square.[111] However, more probable location is the one somewhat further south, in the extreme part of the area bounded by Cara Hadrijana, Huttlerova, and Vukovarska Streets. An altar to the Capitoline Triad was found broken in two in two nearby locations – on the site of a former steam mill (Huttlerova Street) and in Vukovarska Street.[112] The altar might have originally been placed in the temple of the Capitoline Triad in the forum, the more so as this area was intersected by a wide east-west street (probably *decumanus maximus*), whose remains were found in the eastern[113] and western[114] parts of the town, and a north-south street (an extreme southern section of this street was discovered in archaeological excavations to the west of Huttlerova Street).[115] Recent excavations at the site of the future Academy of Education also suggest the location of the forum at this position: foundations of a large (probably public) building with porches, as well as sections of the town's main sewage system with connections, were found there.[116] An inscription mentioning one Gaius Aemilius Homulinus who, probably during Hadrian's reign, financed the construction of a porch with 50 *tabernae*, is an important argument in the discussion of the issue of the forum of Mursa.[117] Some authors doubt whether this was a single market building or a porch with *tabernae* in the forum. M. Suić argued for the *tabernae* placed along two longer sides of the forum; he further supposed, referring to the modules of forums and urban *insulae* from Trajan's and Hadrian's reign, that the forum's dimensions could have been 54 m (the sides with *tabernae*) x 44 m (the side with the basilica).[118] Judging from the so-far conducted smaller-scale trial trenches, the forum of Cibalae was located at the site that partially overlaps with the modern-day main town square (Bana Josipa Šokčevića Square); at its eastern side some structures were found that could be the remains of a porch and shops (*tabernae*). The fragments of *terra sigillata* pottery from this area date from the 2nd half of 2nd century.[119] The basilica has been located to the north of the forum, on the bases of late 18th-century finds, thus occupying the area between the forum and the hypothetical northern baths.[120]

6.2.3. Temples

Although there is no doubt about the existence of temples in all Roman towns in the area of study – a fact clearly attested in epigraphic finds, which are, however, often found devoid of any precise context; so far temples have been archaeologically evidenced only at Varaždinske Toplice and Sisak. Generally, temples yielding at least minimal structural remains to allow identification are scarce.

The foundations of a temple (9.35 x 9.10 m) were found on a small square in the extreme northern part of Siscia;

---

[108] Gorenc and Vikić, 1980, 10-12; Nemeth-Ehrlich and Kušan 1994, 10, 11. The original position of the forum nymphaeum was not determined; the ornamented plates could have been brought there later (in the 19th century) when this area was in function as a city park of Varaždinske Toplice, equipped with decorative and functional details (walking paths, gazeboes, benches, etc.).

[109] Nemeth-Ehrlich and Kušan 1997; Nemeth-Ehrlich and Kušan 2006a; Nemeth-Ehrlich and Kušan 2006b.

[110] The northern section of the cardo was discovered in the hotel yard, while another short segment was discovered in Bana Jelačića Street. Cf. Gorenc and Vikić 1984, 64; Vikić-Belančić 1998, 481.

[111] Filipović 2004, 162.

[112] Pinterović 1978, 76.

[113] Leleković 2009, 46. The road was ca. 10 m wide; two phases of construction were established (the first was dated by coins in Hadrian's reign, while the second dates from the late 2nd century)

[114] Filipović and Katavić 2006, 13. The road was 8.4 m wide between the kerbstones.

[115] Sunko, Katavić, and Jerončić 2008, 23.

[116] Filipović 2010, 37. The trench was excavated in the northeast section of an earlier military garrison, south of Cara Hadrijana Street and Kraljice Katarine Kosače Park.

[117] Pinterović 1978, 55, 56.

[118] Suić 1985, 58, 64, 65. He states that a temple and a basilica must have been erected in the area designated for the forum previous to 140, encompassing the shorter (frontal) side of the forum, which at the same time determined its width. He further believes that 50 Homulinus' tabernae were constructed somewhat later and that they were only distributed along the two lateral (longer) sides of the forum, with 25 of them to each side; according to the inscription the tabernae were binae (double), that is, some of them were interior and others exterior.

[119] Iskra-Janošić 2001, 67, 116.

[120] Iskra-Janošić 2001, 84.

however, data regarding its dedication is missing (see section 6.2.2.). Walls made of stone blocks more that 2 m in length, possibly the remains of a temple, were found next to the building known as Mali kaptol in Rimska Street, very close to the western town wall and south of the remains of public baths. In the 19th century an altar was also found there dedicated to Jupiter and Juno, further a base for the statue of Hadrian and fragments of an inscribed architrave, as well as numerous other moulded architectural fragments. This location was possibly occupied by a forum or some other public space.[121]

Several phases of temple-construction were distinguished at the northern edge of the forum in Aquae Iasae. The best-preserved architecture is the one from the early 4th century, from Constantine's reign, which at the same time represents the last phase of rebuilding at this site. Remains of a 1st-century temple were discovered at the spot where a staircase and a temple-atrium were later constructed (and in the 4th century also the northern porch). This temple was not in function during the 2nd century (its foundations were found beneath the pavement of the porch), and a new structure was erected at its rear side. It is believed that the central temple was built first, with the other two added subsequently at its sides. These three temples were at first separated from each other, and had their own roofs. The porches were most probably erected on the forum's eastern and western sides in this period.[122] In the early 4th century the layout of the temples was transformed. The central temple was expanded towards north and west and shared a wall with the western temple, while all three temples also shared the southern façade facing the forum, constructed as a single wall with three entrances.[123] Furthermore, all three buildings now had a common roof, and one entered the temples through a staircase in the northern porch. The columns of the northern porch were unevenly spaced, contrary to the situation at the remaining three porches, in order to emphasize the temple entrance. The purpose of the two lateral rectangular buildings (4 x 5 m) at the farthest northern part of the eastern and western porches is unknown; temples or rooms for the deposition of votive offerings have been conjectured. It is believed that the three temples were dedicated to the Capitoline Triad (Jupiter, Juno, and Minerva), as suggested by the archaeological evidence: a monumental statue representing Minerva, together with the associated pedestal, was found in the western temple, while a statue pedestal with a dedication to Juno and Fortuna was found in front of the eastern temple. Curiously, no such finds have been discovered in the central temple.[124]

No traces of temples have been discovered in Iovia/Botivo,

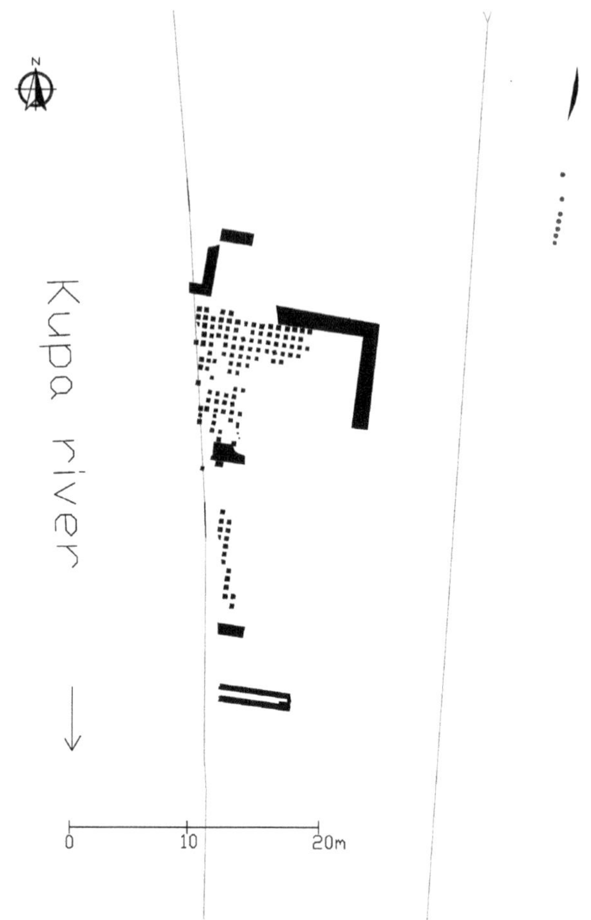

Fig. 16. Sisak, baths (after Faber 1973, adapted by T. Lolić).

but such building must have stood next to the forum, which was most probably located near the intersection of the *cardo* and *decumanus*.[125] Minor segments of these two streets have been recovered in archaeological excavations. The position of the Christian church is also undetermined. An adaptation of a part of the thermal complex into a church is the most frequently offered suggestion,[126] but the latest excavations seriously challenged this hypothesis.[127] A more plausible suggestion would be the site of the present-day parish church of the Holy Trinity in the far north-western corner of the town, which overlay a medieval church erected directly above the late Roman town walls. An additional

[121] Faber 1973, 143-146, 155; Zaninović 1981, 206, 207.

[122] Gorenc and Vikić 1980, 10-12; Nemeth-Ehrlich and Kušan 1997; Nemeth-Ehrlich and Kušan 2006a; Nemeth-Ehrlich and Kušan 2006b.

[123] The dimensions of the temples along the northern porch are the following: the central temple 9x7 m (prior to the Constantinian remodelling, it measured 6 by 5 m), the eastern and western temples 3.8 x 5 m each.

[124] Gorenc and Vikić 1980, 10, 12. The dedication to the Capitoline Triad has been questioned because no dedication of the central temple was found and, moreover, because the dedications to the two goddesses of the Triad were not formulated in the "Triad-scheme", but rather as dedications to individual goddesses, both defined by attributes that suggest their exceptionality and certain individuality; the monument dedicated to Minerva was erected by a decurio, while the one dedicated to Juno was erected by the community of Poetovio (Migotti 1998, 59). Lately, it has been presumed that the temples were not dedicated to the Capitoline Triad because Minerva should in that case have been placed in the temple on the eastern (right) side and not be represented as sitting; furthermore, the pedestal from the eastern temple is not solely dedicated to Juno; finally, not a single inscription from Varaždinske Toplice bears a dedication to Jupiter (Sinobad 2010, 252). Considering the number of exceptional finds (inscriptions, relief and statue fragments) recovered late in 2011 during the excavations of the northern part of the thermal-spring construction, the solution to this question might be forthcoming.

[125] Gorenc and Vikić 1984, 64.

[126] Tomičić 1997, 30; Vikić-Belančić 1984, 161; Vikić-Belančić 1998, 488; Migotti 1997, 52, 53.

[127] Pleše 2010.

Fig. 17. Ščitarjevo, aerial view of the remains of the thermal complex (left), street (middle) and remains of partly excavated architecture (right) (after Nemeth-Ehrlich and Kušan Špalj 2007).

argument in favour of such hypothesis is a *horreum*, whose remains were found in the hotel yard in the vicinity of the parish church, since such buildings often accompany early Christian complexes.[128]

The existence of a temple dedicated to the Capitoline Triad in Mursa has not been finally proved, but an inscription on an altar, possibly originating from a temple, does mention them[129] (see section 6.2.2.). The upper section of an altar with the names of Jupiter, Juno, and Minerva inscribed was taken out of the Drava at the area of the bridge in 1985;[130] it was probably built in as a *spolium* in one of the bridge reconstructions. Some votive altars dedicated to Jupiter by *beneficiarii*, found at the location of the modern-day school building at the southern side of the Bana Jelačića Square (next to Matije Gupca Street) points to the existence of a Jupiter's temple. The forum was presumably also located at this position,[131] but the site more probably held a *beneficiarii* station with a temple, and was situated outside the town walls, along the main road leading towards east. The possibility that Silvanus was worshipped in Musa is suggested by as many as five votive inscriptions to this god found in 2004 in the western part of the town, at the site of the University campus (the building of the Agricultural College).[132] This location is also sited outside the town walls, along the main road leading towards west. A damaged marble statue of Cybele, probably originating from a temple dedicated to this goddess, was found in 1966 in a trench at the intersection of Huttlerova and Frankopanska Streets.[133] Although neither the context of the find nor the exact position is known, later excavations on the same parcel have yielded sections of a road and of the town walls respectively.[134] Thus, the temple was clearly placed next to a road leading from the town towards the south, and in the vicinity of the town gate. The existence of a Jewish sanctuary is suggested by a late 2nd-early 3rd centuries inscription mentioning a renovation of the sanctuary (supposing that the incomplete word ... *seucham* is restored as *proseucha* – the word used for the Jewish community as well as the sanctuary), furthermore by the find of an altar inscribed as *Deo aeterno* (denoting possibly various divinities, though), and a brick with the Hebrew name of a month (modern-day September-October). The original position of these monuments remains unknown, but the sanctuary was probably placed outside the town walls, perhaps between the Drava and the northern wall.[135] The existence of a temple in Cibalae, perhaps dedicated to

[128] Vikić-Belančić 1984, 122-126; Vikić-Belančić 1998, 481; Migotti 1994, 53; Migotti 2004, 140.
[129] Pinterović 1978, 76, 129.
[130] Bulat 1988, 53.
[131] Pinterović 1978, 59, 60, 73.
[132] Filipović and Katavić 2005, 9.
[133] Pinterović 1978, 137.
[134] Sunko, Katavić, and Jerončić 2008.
[135] Pinterović 1978, 63-65; Migotti 1994, 43, 44.

Fig. 18. Ludbreg, excavated part of the late Roman thermal complex, 2010 (photo: T. Pleše).

the Capitoline Triad, has been hypothesised on the basis of thin evidence recovered in archaeological excavations on the site of the then PIK building, south of Jurja Dalmatinca Street; the temple was presumably destroyed in the erection of a Christian building at the same site. These constructions were located to the south of a 4th century *horreum* northeast of the forum, near the northern town gate.[136]

6.2.4. Public baths

A thermal complex in Siscia was discovered in Rimska Street, next to the Kupa, probably within the town walls, whose existence at this place, however, still remains unconfirmed. Only a small segment of the baths was excavated (a trench of 340 m^2), yielding remains of a hypocaust system, semicircular pool-apse, and a number of partitioning walls and drainage canals. The architectural characteristics of the structure point to the late 2nd-early 3rd centuries.[137] Other small segments of the baths were discovered in later protective excavations. A perimeter extending for 60 m north-south was thus determined, yielding a total surface of ca. 1800 m^2; however, these dimensions are not final.[138]

The bath complex was in Andautonia (fig. 17) constructed during the 1st half of the 2nd century in the place of an earlier, southern cemetery, damaged in a flood; previous to the erection of the baths the terrain was levelled for safety reasons. The baths were placed next to the north-south street, and were connected to it by a porch; they were in use until the late 3rd century. The building's layout was orthogonal, with elaborate inner articulation and with a semicircular pool on the northern side; it was made of mortared stone and was furnished with a system of floor heating and drainage canals.[139] The baths in Aquae Iasae (fig. 15) were placed to the south of the forum as its continuation on the north-south axis. Several phases of construction were archaeologically established, starting from the 1st century, but the best preserved is the one from the early 4th century, following the Constantinian renovation. The 1st-century baths are located in the southwestern section of the excavated area; they consisted of rectangular rooms and a pool, aligned along an east-west axis, with no heating system yet. During the 2nd century the baths were renovated and a floor heating system was added, whose remains were mostly found beneath the architectural complex following the Constantinian renovation. The foundations of two semicircular pools and some walls on the western side of the complex, outside the 4th century building, belong to this phase. The basilica was connected to the rest of the complex by a central passage, which was later closed. The 4th-century

[136] Iskra-Janošić 2001, 94-96; Perinić 2000, 421, 422. It is supposed that the altar found at the site of the PIK building was dedicated to Minerva. Presumably, it originally stood in the temple dedicated to the Capitoline Triad and was re-used in the 3rd or early 4th centuries.

[137] Faber 1973, 139-143.

[138] Documentation of the Conservation Department in Zagreb, 2002.

[139] Vikić-Belančić 1981; Nemeth-Ehrlich and Kušan Špalj 2003; Nemeth-Ehrlich and Kušan Špalj 2007.

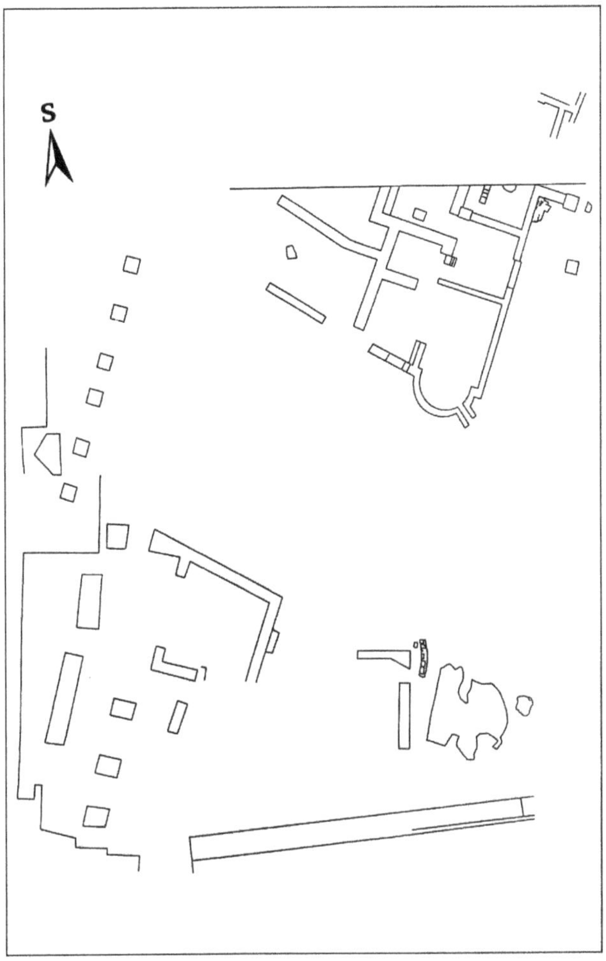

Fig. 19. Vinkovci – baths at the site of the Cibalae bank (after Iskra-Janošić 2001)

renovation partially took over the concept of the earlier complex with the basilica, but still changed it completely in some aspects. The rooms north of the basilica were used as changing rooms (*apodyterium*), a warm-air room (*tepidarium*), a cold-bath room (*frigidarium*), a hot-bath room (*caldarium*), and a sweating room (*sudatorium*). The rectangular longitudinal basilica measured 21.6 by 12.45 m, with the walls approximately 0.8 m wide in their upper sections. Eight pilasters measuring 0.98 m in width were regularly spaced along the walls: one to each corner and two each to the eastern and western walls. The building's height has been estimated at approximately 10 m, and it probably had a flat ceiling and gabled roof. The apse on the northern side was added later, when the central passage towards the baths was closed and two passages opened in lateral sides instead. Since a fragment of a wall painting depicting a saint's head with an aureole was found inside it, the basilica was probably at some point converted into a Christian church. The entire complex, the basilica and the baths, was lined on the eastern side by a porch, communicating with the southern porch of the forum by a door.[140] The layout of the late Roman baths complex (ca. 350 m^2), placed on an east-west axis, was discovered in recent excavations in Iovia/Botivo (fig. 18).[141] The complex was formed around two rectangular areas, exemplifying typologically the so-called block-type of baths.[142] They were placed to the east of the main street in the vicinity of the northern town gate and the suburbia, and were dated to the 2nd half of the 2nd century.[143] It is supposed that the baths were turned into a Christian church in Constantine's reign,[144] but this hypothesis was challenged by the results of the latest excavations (see section 6.2.3.). M. P. Katančić noted the remains of a *tepidarium* in Mursa, next to the Drava and the western town wall, suggesting the existence of baths.[145] Another public baths were probably located in the area next to Huttlerova Street (in the direction of Vukovarska Street). During the protective excavations conducted at 20-22 Huttlerova Street, remains of a multiple-room structure with a hypocaust system and the accompanying canals were found.[146] It spreads eastwards under the modern road and probably stands in some relation to the Roman architecture excavated in 1982-1983 on the eastern side of Huttlerova Street.[147] Unfortunately, apart from the information on the numerous remains of several structures, some of them very large, the results of these excavations remained unpublished.[148] This complex was probably located at the easternmost part of the area delineated by Cara Hadrijana, J. Huttlera, and Vukovarska Streets, thus near the presumed location of the forum. Remains of baths were found in Cibalae during the protective excavations in 1977-1978 on the site of the modern-day building of Cibalae bank, between Duga Street and the Bosut. These remains (hypocaust remains, *praefurnium*, *frigidarium*, bottom of a *piscina*, part of the yard porch) comprise only a small segment of the whole complex. The coins recovered during these excavations date from the 2nd half of the 2nd to the 2nd half of the 4th centuries.[149] Several protective excavations, as well as watching briefs, were conducted in the period 1956-1989 in the area occupied by residential buildings south of M. A. Reljkovića Street (between V. Nazora and Kralja Zvonimira Streets). These works have confirmed the possibility of the existence of another thermal complex in this area, to the north of the forum.[150]

6.2.5. Horrea

Next to the southern wall and the town gate of Siscia a monumental 4th-century building has been discovered. The southern part of a five-nave building with outer walls 1.2 m

140 Vikić-Belančić and Gorenc 1961; Gorenc and Vikić-Belančić 1980; Nemeth-Ehrlich and Kušan 1997; Nemeth-Ehrlich and Kušan 2006a; Nemeth-Ehrlich and Kušan 2006b.

141 Pleše 2010.

142 Ward-Perkins 1981, 72-73, passim; Yegül 1995, 48-91.

143 Vikić-Belančić 1998, 481.

144 Vikić-Belančić 1998, 488.

145 Pinterović 1978, 160; Šimić and Filipović 1997, 55; Katančić 1782, (translated by S. Srša 1991, 57).

146 Sunko and Katavić 2007.

147 Bojčić and Bulat 1985. Ca. 4000 m^2 were excavated (out of the planned 6000 m^2), but the published documentation offers only insufficient information on the exact position of the excavated structures.

148 Bulat 2007a, 372, 375-387. The author gives a summary review of the excavations.

149 Iskra-Janošić 2001, 75.

150 Vikić-Belančić 1970; Iskra-Janošić 2001, 75-82.

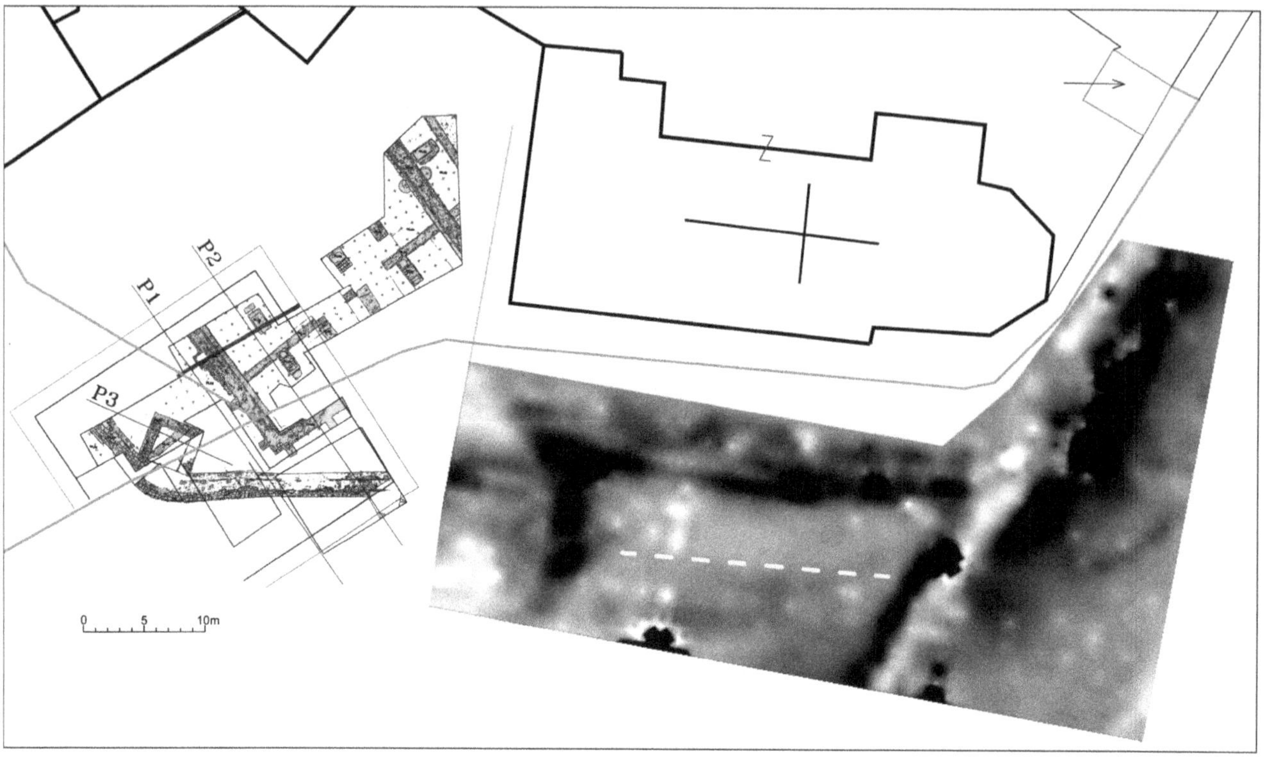

Fig. 20. Sisak – horreum, excavated area supplemented with the results of a geo-physical survey (after Mušič, in Lolić 2010).

thick, supported by a system of buttresses, was excavated there. The shorter side of the building is 23.5 m wide; it was impossible to determine the total extent of the longer side because it is now under the parish office (33 m of the wall were discovered in a geophysical survey). The inner space was divided into five naves, which were in their turn partitioned by four rows of massive pedestals 3.5 m apart, carrying the upper floor. The outer walls were made of half-dressed stone arranged in rows, supported on the outside by brick and ashlar buttresses. The interior was renovated in the late 4th century, and only the pedestals abutting on the walls were preserved in their original height, while others were levelled to the walking surface. After this renovation the building had no roof but only inner porches.[151]

Architectural remains of a massive wall with a pilaster, suggesting a *horreum*, were found in a hotel yard in Iovia/Botivo. It was built in the 2nd half of the 2nd-1st half of the 3rd centuries and remained in function until Late Antiquity.[152] No traces of a *horreum* have been found in Mursa, but such building was probably located in the peripheral area of the town, near the Drava, next to the as yet undiscovered northern town gate. The excavations conducted on a parcel at the eastern part of J. Dalmatinca Street in Cibalae (the site of the former PIK building) have yielded a segment of the town walls and a *horreum* (4th century); of the latter building ten column-base foundations (1.75 x 1.50 m) arranged in three rows were found.[153] The building was situated next to the northern town gate, while some 300 m to the east of it, at an extension of the Ervenica Stream' bed, the river port of Cibalae has been presumably located.

6.2.6. Amphitheatres, theatres, stadiums

Remains of these buildings have so far never been recovered in archaeological excavations. The presence of actors, thus probably also of a theatre in Siscia, is suggested by a funerary inscription (now kept in the MNM in Budapest).[154] It is supposed that there was also a stadium in Mursa; describing the events that took place in the early 2nd half of the 4th century in connection with the emperor Constantius II, the 5th-century Greek historian Zosimus mentions a stadium outside the town. Although this information has not so far been proved by archaeological excavations, R. Franjetić supposed that the stadium was built of timber and erected near the southeastern part of the town, outside the town walls, where the 19th century maps depict a large depression.[155] Today, this area is occupied by a cemetery and residential buildings (Vijenac I. Česmičkoga Street). The existence of either a theatre or amphitheatre in Cibalae has not been confirmed so far. If we discard the possibility that they never existed, one or both of these buildings could have been made of timber and the possibility of their discovery is very small due to the densely built structures of the modern town above them.

6.2.7. Bridges

The remains of a Roman bridge over the Kupa in Siscia

[151] Lolić 2003, 142.
[152] Vikić-Belančić 1984, 122-126, 161; Vikić-Belančić 1998, 481.
[153] Iskra-Janošić 2001, 93.
[154] Zaninović 1981, 207.
[155] Pinterović 1978, 72; Pinterović 1980, 63; Filipović 2004, 162.

leading towards the Roman settlement and thermal baths in Topusko have been found in the river bed at the southern part of the town, and the remains of some of the timber piles can even now be seen at low water.[156] A tower at the town gate was placed on the same course; it was discovered in 2002 at the southern section of the western town wall, along the Kupa.[157] A bridge must have existed in the northwestern part over the narrow bed of the Odra River, near to its confluence with the Kupa, which is substantiated by the course of the northwestern cemetery, stretching along the road towards Andautonia.[158] The Sava has in some places changed its course over time, which is documented on various maps and in geological studies. Such changes have been presumed for Andautonia as well, and the exact position of the riverbed in relation to the Roman town remains unknown.[159] Since an important road Siscia – Poetovio passed through Andautonia, a bridge over the Sava must have existed there. This has been substantiated by occasional finds of timber piles along the Sava, but no systematic research of them has even been conducted. The latest archaeological research of the river Drava in the Mursa area has resulted in the find of four stone pillars still standing in their original place in the river, ca. 20 m apart from each other, suggesting a monumental size of the bridge; it stretched up to the modern N. Tesle Square (often called Ribarski Square in the literature) on the right bank of the Drava.[160] The 18th-and 19th-century sources mention six pillars visible at low water; some of those were removed in the 20th century to enable a more secure navigation.[161] The date of the erection of the bridge has not been determined precisely; 2nd century, that is, Trajan's or Hadrian's reign, has been suggested, while a renovation in the early 3rd century, probably during Caracala's reign, is mentioned in an inscription.[162] A bridge over the Ervenica Brook in the eastern part of Cibalae was probably located near the findspot of the dedicatory inscription of Septimius Severus and Septimius Geta, perhaps originally standing at the town gate in M. Gupca Street;[163] traces of another bridge have been evidenced in Bana Jelačića Street.[164] A bridge was presumably erected in the southern part of the town, near the town walls, where the Bosut made a natural barrier for the roads entering the town. The Roman bridge probably stood near the modern one; the position of the southern cemetery (3rd-4th centuries) on the opposite side of the river suggests the existence of this communication.[165] It is possible that bridges were also erected at the western and northern side of the town, which was, according to both literary sources and some archaeological evidence, completely surrounded by ditches. The materials used in the construction of the bridges remain unknown; it was presumably timber.

6.2.8. Harbours

Remains of more than 200 oaken piles excavated in 1985 on the site of "Kovnica" in Siscia (on the right bank of the Kupa opposite the fortified part of the town) suggest the existence of a river dock.[166] The site is dated to the 2nd half of the 1st to the 4th centuries, on account of abundant ceramic finds; furthermore, a large quantity of brick, dressed stone, glass, and metal objects were also found (4th-century coins predominate among the coinage). The existence of a massive superstructure in the dry section of the port could be presumed, due to the density of piles and remains of building materials. Ceramic analysis has revealed a significant decrease in the number of forms starting from the 2nd half of the 3rd century; the structure was at that time possibly converted into a working facility.[167] Remains of trial imprints of unfinished dies were found there, which were perhaps produced exactly at the spot.[168] The important role of the Kupa (the confluence of the Kupa and Sava, a tributary of the Danube, is precisely at Siscia) was recognised from the start, as witnessed by the *deductio* of the sailors from the Misenum and Ravenna fleets during Vespasian's reign. Their arrival to Siscia was intended to increase the river traffic between two key harbours, Siscia and Sirmium (Sremska Mitrovica, Serbia).[169] The position of the harbour in Mursa has not been determined, but late 19th-century information on the find of a Roman shipwreck during the foundation works on the railroad bridge suggests that in Roman times the Drava was navigable at Osijek.[170] The shipwreck was positioned, according to R. Franjetić, some 325 m from the western edge of Mursa; Franjetić presumed that the harbour was located there, and he dated the shipwreck to the 2nd-3rd centuries by coinage. There is no information whether this was a mercantile or military ship. If we consider that the command of the Roman Danube fleet was in the 4th century located in Mursa[171] and that Mursa was only some 15 km distant from the Danube (approachable by the Drava), the existence of a harbour in the town should not be doubted. Cibalae was located on the Bosut River, and the mouth of the Stream Ervenica flowing into the Bosut was within the area of the town; the stream was connected by a canal with the Vuka, a river that flows into the Danube. The Bosut was connected to the Sava southeast from Cibalae (east of Županja) through a canal (ca. 2 km in length) mentioned by Zosimus, while it flows into the Sava 20 km to the east of Sirmium. All these waterways were navigable only in certain parts of the year, due to oscillating water level. Also, the riverbeds

156 Vrbanović 1981, 193;
157 Wiewegh 2004.
158 Wiewegh 2003, 9.
159 Nemeth-Ehrlich and Kušan Špalj 2003, 114; Nemeth-Ehrlich and Kušan Špalj 2007, 31.
160 Zubčić 2010.
161 Pinterović 1978, 159; Bulat 1988.
162 Filipović 2004, 160.
163 Dimitrijević 1979, 153.
164 Dimitrijević 1979, 153; Iskra-Janošić 2001, 49.
165 Dimitrijević 1979, 158.
166 Šarić 1986. The excavations were conducted by Marija Šmalcelj (Faculty of Philosophy, Zagreb University).
167 Wiewegh 2001.
168 Šipuš 1983. A mint was active in Siscia from the 2nd half of the 3rd century to the late 380s (according to some it was reopened during Honorius' reign, to last until the 420s), with several interruptions.
169 Zaninović 1993. The units patrolling the Danube, Sava and Drava in Vespasian's reign were called *Classis Flavia Pannonica*.
170 Pinterović 1978, 116.
171 Bulat 2007b, 324.

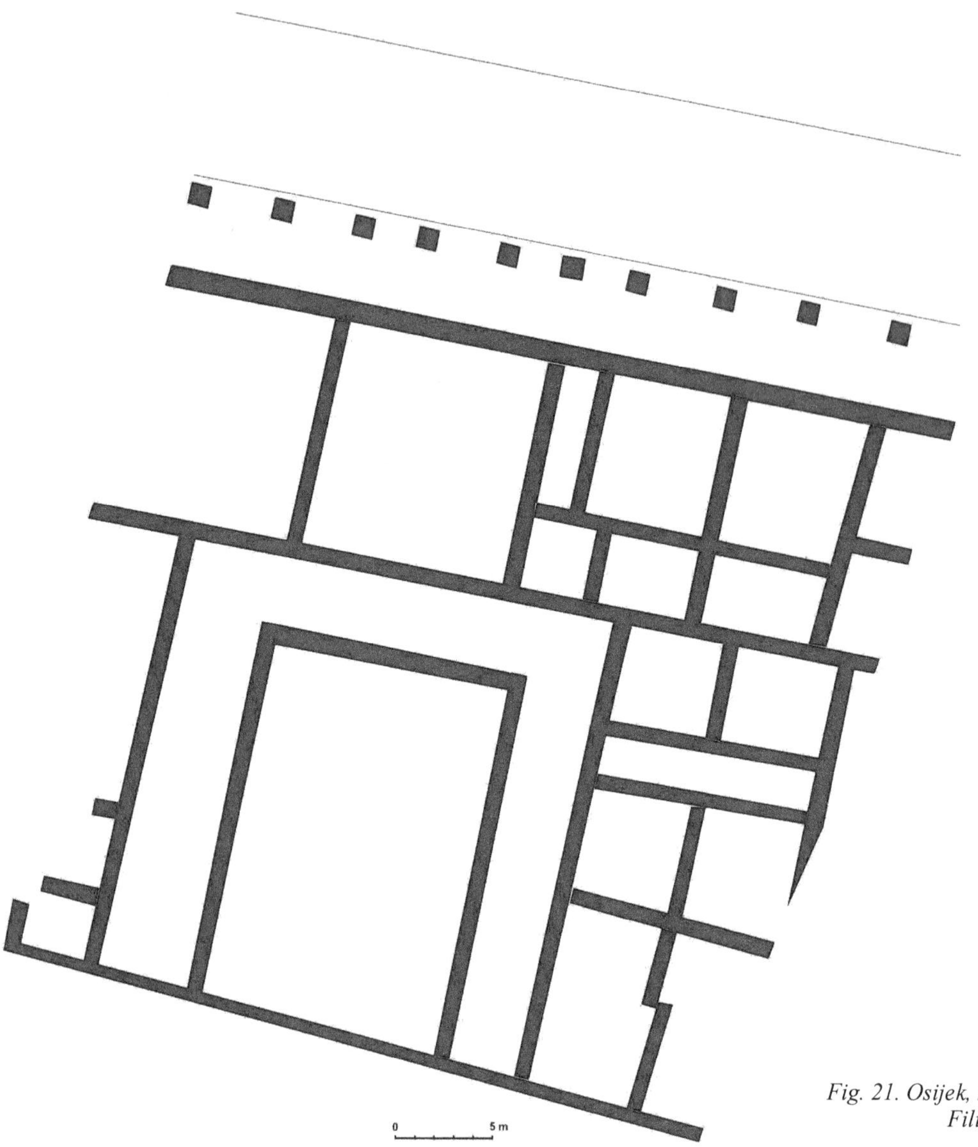

Fig. 21. Osijek, remains of a domus (after Filipović 2007).

were shallow and thus only allowed the navigation of small vessels. The port of Cibalae was probably located at the widened bed of the Ervenica, which also served as an outer ditch of a section of the eastern town wall (modern Barica). The bricks bearing the stamp of the river fleet C(*lassis*) F(*lavia*) P(*annonica*) are an important indication for the existence of a harbour.[172]

6.3. Residential architecture

Residential architecture has been archaeologically evidenced in all urban centres, but nowhere did the preservation of the structures or the size of the trenches result in a completely excavated whole. The information on the layout and function of individual rooms is mostly lacking in the published materials, which further hinders the establishing of the nature of the recovered architecture: a simple house, *domus*, *villa urbana*, *insula*, or palace. Residential architecture has mostly been deduced on the basis of wall-dimensions and basic layouts of the recovered sections, as well as small finds (domestic pottery and other utensils, as well as interior furnishings). Further, due to the poor preservation of the solid structures, it has never been possible to establish whether a building was multi-floored. All that is ever preserved is foundations and a low section of above-ground structures; with its domestic architecture preserved at places to the height of 1 m, Sisak makes an exception to the rule. In terms of building materials brick and tiles predominate, with frequent use of stone (mostly for the foundations) and a sporadic use of timber, typical mostly of the earliest phase of occupation. Nevertheless, on the present evidence all forms typical for the urban residential architecture appear to have been present.

At several sites in Siscia remains were found that display various levels of the quality of residential architecture, from modest to quite luxurious. The latter is, for example, exemplified by the remains of a 2nd-half of the 2nd-century house (*domus*) with a floor heating system, floor mosaic and frescoes, recovered near the eastern town walls.[173] Sections of two urban *insulae* separated from each other by a brick pavement were discovered at no. 10 A. and

[172] Iskra-Janošić 2001, 41, 46-49; Iskra-Janošić 2004, 182, 183.

[173] Baćani 2007.

S. Radića Street. The walls were built of rubble stone combined with brick, while one room was built of brick in the *opus spicatum* technique.[174] The function of the rooms remains undetermined, but the pottery dating from the 2nd half of the 1st century was found in every part of the complex. Represented were imitations of Arrentine *terra sigillata*[175] and a large quantity of thin-walled pottery, mostly produced in Italian workshops.[176] This construction phase should be associated with the period of prosperity and intensive Romanisation during Vespasian's reign. Next to the southern town wall of Mursa, in Huttlerova Street (the site of Silos), remains of a 2nd century *villa urbana* with an atrium were found (the entire structure was overlaid with later constructions).[177] In the western part of the town, at the site of the University campus (the building of the Civil engineering college), formerly occupied by a military base, a section of the *decumanus maximus* was found in 2007, with buildings of various functions on its both sides. A building was excavated to the south of the street, whose architectural features suggest that it was a residential object of *domus* type, owned by members of a superior social status. The residential complex 1,400 m^2 in area has a typical ground plan characteristic for a rich Roman house with an atrium and a peristyle, with 15 rooms, the largest of which were ca. 25 m^2 in area. A porch featuring ten pillars was discovered at the northern side of the house, facing the street.[178] The building can roughly be dated to the 2nd-3rd centuries and it was presumably placed immediately outside the western town walls.

7. The architecture of other urban centres

The state of research in other Roman settlements, situated under modern towns and villages, does not allow either a study of Roman urbanism or the layouts of individual buildings, while available information only allows a general overview of their archaeological potential: Sv. Martin na Muri (*Halicanum*?, possibly a *municipium*), Topusko (*Quadrata* or *Ad Fines*), Daruvar (*Aquae Balissae*) and Slavonski Brod (*Marsonia*?).

Sv. Martin na Muri is located in the vicinity of the Mura River and the borders with Slovenia and Hungary, respectively, and its Roman name was determined on the basis of its mention in the Antonine Itinerary. Small-scale excavations were conducted there in 1970s and 1980s and early in this millennium (unpublished). On the basis of the research so far, it transpires that the Roman settlement extended on a rectangular elevated plateau in the central part of the modern-day village, encompassing ca. 3 ha and featuring some traces of orthogonally arranged communications. The excavations have established architectural remains in several locations, and the settlement is dated from the 1st to the 4th centuries on the basis of pottery. An altar dedicated to Jupiter in the 2nd century by a consular *beneficiarius*, was found in 1977 in a trench next to the main village road.[179] A Roman settlement and a spa centre, situated on an important road leading towards Siscia, on the very border between Pannonia and Dalmatia, existed on the site of modern Topusko. Earlier finds, generally poorly documented, but also some more recent ones, although still modest, testify to a Roman urban settlement, focused on the facilities of a health resort. The exact positions of the town communications are unknown, and the same is true of public facilities (baths). However, the general position of the Roman settlement is known: it encompassed an area about the intersection of several streets (Spiegelbad, Školska, and Bana Jelačića Square). Finds of numerous altars dedicated to Jupiter (4), Silvanus (11), Liber and Libera, as well as the local divine couple of Vidasus and Thana (3), and other divinities, suggest the existence of temples and, in the case of local deities, open-air sanctuaries; the small finds date from the 1st to the 4th centuries. Late Antiquity is possibly represented by pottery fragments and remains of presumably late Roman walls identified on the hill of Nikolino brdo in the very centre of Topusko, immediately above the area with Roman architecture on the site of a prehistoric hill fort.[180] Daruvar (Aquae Balissae) was in pre-Roman times an important centre of the Pannonian-Celtic tribe of Iasi. It was situated next to the curative geo-thermal springs. With the arrival of the Romans in the 1st century the settlement started developing urban characteristics and became an important traffic centre and a healing resort, and was finally accorded municipal status (probably during Hadrian's reign). The urban topography of this Roman settlement can be reconstructed on the basis of old information on chance finds and occasional more recent small-scale protective excavations, while almost the entire surface of the Aquae Balissae is today overlaid with modern constructions. Overall, the urban structure of this settlement was only tentatively reconstructed, sadly lacking a sound support by archaeological evidence. The River Toplica crosses the area yielding Roman finds and presumably comprising urban structures. The forum was possibly located along right bank of the river, while the larger part of the settlement (the baths and residential area) was on its left bank. Remains of the presumed baths were found next to the river, while the residential section was further east, on a mild slope (the section of the city called Stari Slavik), which is the probable location occupied by the military camp in the earliest phase of Roman occupation. Remains of an earthen defensive system in the form of rampart-slopes and ditches can be seen at the site of Stari Slavik on the eastern side of the settlement; a segment of 2-m-thick masonry wall, belonging to the later phase of the fortification in the late 2nd century, was also noted in the literature. The position of the forum on the modern main town square (Kralja Tomislava Square) was identified on the basis of 19th-century finds, yielding remains of walls, four altars dedicated to Jupiter and several fragments of a large bronze horseman statue. Several

174 Šarić 1990.
175 Makjanić 1996, 101.
176 Lolić 2003, 142, 146.
177 Sunko, Katavić, and Jerončić 2008.
178 Filipović 2008; Filipović 2009; Lolić, Petrinec, and Wiewegh 2010.
179 Tomičić 1979; Tomičić 1984, 62, 63; Vidović 1989; Novinščak 2006.
180 Šegvić 1986; Šegvić 1991; Šegvić 2006, 270, 271; Čučković 2009, 15-20;

Roman walls and capital fragments, together with a small Jupiter statue and altars dedicated to Silvanus, suggesting the existence of a temple, were found in the area occupied by the 19th-century thermal baths, a probable location of the Roman thermal complex as well.[181] Slavonski Brod (Marsonia?) was an urban centre in the Roman period; its name is known from the Tabula Peutingeriana and a 1st-century military diploma found in 1997 (where it is mentioned as Marsunnia). The perimeter and the layout of its defensive system remain unknown. The so-far excavated remains of Roman architecture were found in the area occupied by the central town square (I. Brlić Mažuranić Square), opened with its southern side towards the Sava, and also to the east and west of it. In the area in front of the modern main town square, Roman cremation graves, predating the period of the urbanisation of the area, were found below the layer of Roman architecture, consisting of the remains of demolished buildings and building materials (stone, brick, tile, and mortar). The northern edge of the settlement was presumably positioned between M. Mesića and A. Starčevića Streets, the western at the confluence of the Mrsunja Stream and the Sava River, and the eastern between I. Zajca and T. Skalice Streets. The southern edge, although destroyed by the course of the Sava that forms a large bend there, probably extended to the Ada islet, where, as well as in the river itself, a large quantity of Roman finds was recovered. The small finds from the area of Slavonski Brod are dated from the 1st to the 4th centuries.[182]

8. The architecture of the *limes*

The region on the right bank of the Danube in the Croatian part of the province of Pannonia was part of the huge frontier system of the Roman Empire following the course of the river; its Croatian section was ca. 190 km long. The road extending along the border was at a segment of its northern section removed from the Danube, due to the marshy land. The road was in the first place constructed as a military supply route, and, indirectly, for commercial purposes. At present, neither every segment of this route nor every settlement and military outpost along it are known. The incursion towards the Danube started with Augustus' campaigns, and the Pannonian *limes* on the Danube was developed during Domitian's reign (late 1st century); it was completely finished during Trajan's reign, encompassing a developed system of fortifications (*castra* – legionary camps; *castellum* – garrison for auxiliary troops, *burgi* – lesser forts, *turres* – sentinels). In the area of study a 1st-century legionary camp was probably situated in Mursa or Mursela (Petrijevci), 15 km northwest of Mursa.[183] Later, military garrisons of the *castellum*, *burgus* and *turres* type grew along the Danube. The small-scale excavations conducted so far have rendered it impossible to determine the precise nature of individual structures. Form north to south, along the course of the Danube (with some fortresses removed from the Danube due to marshy land), the following sites are situated: Batina (Ad Militare), Zmajevac (Ad Novas), Kamenac (Mons Aureus?), Dragojlov brijeg, Lug (Albanum?), Vardarac (Donatianae?), Kopačevo (Ad Labores?), Bilje, Osijek (Mursa) and Petrijevci (Mursella?), Nemetin, Sarvaš, Aljmaš, Erdut, Dalj (Teutoburgium), Borovo, Vukovar, Sotin (Cornacum), Šarengrad, and Ilok (Cuccium).[184] Regrettably, excavations were few, using only small trenches at that; accordingly, the surfaces taken by some of the sites can be only tentatively estimated by the local land configuration.

Batina (*Ad Militare*) was a military garrison measuring 220 x 200 m, with remains of an adjacent civil settlement.[185] According to old data, Zmajevac (*Ad Novas*) was a rectangular fort measuring 150 x 120 m.[186] Dalj (*Teutoburgium*) was probably an auxiliary camp. Its remains were mostly destroyed by the local brickyard situated in the northern part of the modern-day village, next to the Danube. Although this was a large fort, the exact surface could not be determined due to a sustained damage. Remains of scattered building materials from the civil settlement adjacent to the camp were noted in the surroundings.[187] Sotin (*Cornacum*) was a Roman auxiliary camp. It was situated on an elevated rectangular plateau above the right bank of the Danube, on the site of Popino brdo next to the parish church, encompassing ca. 2 ha. The plateau is protected on three sides by natural slopes and deep trenches, allowing access only from the south. At first it was probably erected as a timber-earthen fortification, while the first important phase of urbanisation characterised by the use of solid materials started after the wars against the Marcomanni and Sarmatians.[188] Numerous Roman finds were discovered on this site, while the area next to the fort also yielded some remains, suggestive of the existence of a civilian settlement.

9. Phases of urban development

Several significant construction phases can be discerned in Pannonia from the period of the foundation of the towns in the 1st century to the late 4th century. Urban centres started developing urban outlook in the 2nd half of the 1st century. Building activity of the Flavians has been evidenced in all parts of Pannonia. In Siscia the road network could have been built even prior to this period, but it was in general constructed in the mid and late 1st century. The earliest urban building so far established in Siscia is represented by timber houses with economic sections, as well as a part of an industrial quarter from the first few decades of the

181 Bulat 1973; Pinterović 1975; Schejbal 2003; Schejbal 2004.

182 Miškiv 1998; Miškiv 2006; Bojanovski 1993; Miškiv, Miklik-Lozuk, and Lozuk 2006, 6-9; Miškiv 2007.

183 Sanader 2003.

184 Klemenc 1961; Bulat 1969; Pinterović 1969; Sanader 2003; Petrijevci, Osijek, and Nemetin are situated on the Drava, which flows into the Danube (Osijek is some 15 km from the confluence).

185 Bojčić *et al.* 2009, 9. Small-scale trial trenches were made in 1970.

186 Bulat 1969, 41; Pinterović 1969, 58, 59. These authors give different estimates for the area (thus Bulat gives 200 x 90 m2).

187 Bulat 1969, 43.

188 Ilkić 2009, 162. Numerous surface finds (construction materials and ceramics, the 2nd-century military diploma, altars) have on this location never been matched by systematic excavations.

1st century.[189] The river port was established in this first phase. The construction utilising solid materials became dominant in the 2nd half of the 1st century, although it was sporadically present in the earlier period. The basic matrix of the town was constructed during the reign of the Flavians (the frequency of pottery, both locally produced objects and a large quantity of fine ware of Italian provenance, was very high in this period).[190] The core of the settlement and the area initially encompassed by the town were created in Andautonia in the 1st and early 2nd centuries. The use of stone and other solid materials was begun during the reign of the Flavians. This construction phase can be traced through the remains of streets and residential architecture.[191] The archaeological evidence shows a fine progress in terms of construction in the early 2nd century, probably resulting from the accordance of the municipal status, and carrying with it the status of an administrative, commercial, and industrial centre of the wider surroundings. The network of paved streets, public buildings with wall frescoes, mosaic floors etc., all point to a high standard of living in the town. A 1st-century layer is also present in Iovia/Botivo, displaying architecture constructed of solid materials, overlaid by the early 2nd-century thermal complex.[192] The first phase of the baths in Aquae Iasae was also constructed in this period.[193] The rural areas in the zone of the strong influence of Siscia and Andautonia, especially along the main road leading through the Sava valley, were also involved in the process of Romanisation already in the 1st century, as is witnessed by a relatively high population density in the Turopolje and Pokuplje regions, indicated by a large number of Roman settlements and industrial country complexes (see Leleković and Rendić-Miočević in this volume). In Cibalae, the construction in solid materials was initiated in the late 1st and early 2nd centuries, the period of the town's particular economic growth; this is evidenced by workshop complexes along with residential areas and market places that were in function in the northwestern and southwestern part of Cibalae from the 1st to the 4th centuries. Cibalae was an important industrial centre for the production of pottery, brick and lime, and so far more than 50 Roman kilns were documented in the town territory.[194] The area of Mursa was probably the location of a legionary camp constructed in order to control the Drava Bridge even before the foundation of the civilian settlement.[195] The increase in building activity in Mursa was initiated after the Dacian wars and during Trajan's and Hadrian's rule.[196]

The second phase of construction began with Hadrian's rule. It is evidenced in Andautonia by the expansion of the town and the (excavated) town quarter with public buildings along the main north-south street (one of the buildings were the public baths). The expansion of the town core had caused the transfer of the cemetery to the extreme southern location of Kutelo, used from the 2nd to the 4th centuries.[197] In this phase the town expanded to reach the surface of ca. 40 ha, together with the cemeteries included; it extended for 1 km from north to south and ca. 400 m from east to west. The urbanisation of Cibalae was intensified in Hadrian's reign, which is probably connected to the town's newly acquired municipal status. The town was fortified by double earthen walls. In the late 2nd and early 3rd centuries Cibalae expanded and reached its final growth; along with public buildings single fortification walls with a palisade was erected in this period.[198]

The next phase falls in the reign of the Severans and is characterised by a rehabilitation of the province as a whole following the Marcomannic Wars. In this period public baths were constructed in Siscia, together with the town walls erected or rebuilt in the late 2nd and early 3rd centuries. A small town square with a temple was erected in one of the northern quarters along the *cardo maximus*, as well as an architecturally imposing northern gate. Structures erected during this phase were occasionally built upon the ruins of the buildings from the Flavian period, following their orientation and positioning. Thus it can be concluded that the original town matrix was respected in the following periods, with adaptations carried out within the already existing plan.[199] This phase of urbanisation is clearly evident in Andautonia, while a complex consisting of a forum with the Capitolium temples was erected in Aquae Iasae; the latter is the sole excavated complex of this type in the Croatian part of Pannonia.

The final phase of the development of south-Pannonian towns falls in the 4th century and has been evidenced in all urban centres. The military and administrative reforms of Diocletian, in the context of which Siscia became the capital of Pannonia Savia, initiated an overall revitalisation of south-Pannonian towns. The late Roman construction phase of the post-Constantinian period is most clearly seen in the granary (*horreum*) in Siscia and in the sections of the town walls strengthened with stone and repaired in typical late antique masonry fashion.[200] Several larger towns expanded during the 4th century, thus increasing and concentrating the population density.[201] The perimeter of Siscia in its final phase of Roman urbanism, encompassed by the town walls, measured 47 ha; with five cemeteries in the suburban area included, the total surface amounts to 200 ha. Remains of the final phase in Andautonia date from the 4th century and are documented in the entire town area. However, according to the archaeological evidence, the town's area shrunk, and after the 4th century there are no traces of either new occupation or construction.[202] During the 4th century a certain decadence is discernible

189 Lolić 2003, 149
190 Makjanić 1996, 101; Lolić 2003, 146
191 Nemeth-Ehrlich and Kušan Špalj 2007, 42.
192 Vikić-Belančić 1998.
193 Nemeth-Ehrlich and Kušan 2006b.
194 Iskra-Janošić 2001, 110.
195 Filipović 2004, 157.
196 Cf. Pinterović 1978, 168.
197 Nemeth-Ehrlich and Kušan Špalj 2007, 42.
198 Iskra-Janošić 2001, 67-75.
199 Cf. Faber 1973, 138, 156; Vrbanović 1981, 196. See also Lolić 2003, 149.
200 Lolić 2003, 143.
201 Mocsy 1974, 297; Poulter 1996, 101, 106.
202 Nemeth-Ehrlich and Kušan Špalj 2007, 49.

in Mursa; the town's population and surface decreased, and the urban planning was changed, which is evidenced by the altered orientation of the town architecture.[203] In Cibalae, the 4th-century phase is represented by the *horreum* near the northern town gate, and the monumental public architecture excavated during the several campaigns of protective archaeological excavations from 1956 to 1989 in M. Reljkovića Street.[204]

After the 4th century, there are no new architectural interventions, although the continuity of living can be discerned in several cities until the 6th century

10. Concluding remarks

In general, the processes of urbanisation and town development in the Croatian part of Pannonia correspond to the development of the entire province of Pannonia. Urban components, urban-matrix patterns and architecture fit in with the general typology to be found in the western provinces of the Empire. Nevertheless, regional differences do exist, conditioned by local resources and political, administrative, and economical function of the towns, with pre-Roman population possibly acting as a background to Romanisation. The basic economic pursuits in southern Pannonia were the trade in commodities and transfer and supply of the troops. Such background is reflected in the location of towns, and the existence of particular types of buildings and services meeting the needs of the military. The towns were mostly fortified and furnished with harbours and harbour installations, as well as huge granaries and storehouses. Responding to the basic activities of the region, the urban outlines and architectural structures reflect engagement in practical day-to-day matters, rather than leisure and luxury. On the other hand, the orientation towards the army did not decrease the quality of urban life in the whole area of study, particularly not in the section of Pannonia Superior. A somewhat superior position of the latter in terms of the quality of material culture and living conditions can be explained by the difference in terms of political geography. The eastern part of the province (later Pannonia Inferior) was namely conquered and urbanised later than the western, and moreover, remained a strongly militarised border area throughout. Accordingly, a superior position of the western part is evident in the quality of the equipment of public and private buildings, and in the field of the arts and crafts in the towns like Siscia, Andautonia, and Aquae Iasae, whose public buildings display a variety of marbles and carved stones. Their interiors were decorated in wall paintings and mosaics, preserved today in the local museums or *in situ*, with Andautonia and Aquae Iasae as the most outstanding examples. In Late Antiquity Siscia became the capital of Pannonia Savia, but the town gained a special status from the very beginning of the Romanisation process in Pannonia. Starting as a military outpost, it very early grew into a dominant commercial, spiritual and industrial centre, evidenced as such by monumental proportions of the urban structures and abundance in varieties of everyday utensils, not to mention the existence of the imperial mint and bishop's see.

The choice of construction materials was based on local resources: earth (early fortifications), wood and stone, with brick as the dominant material, whose production has been archaeologically proved in Siscia, Mursa, and Cibalae. All urban centres demonstrate a clear border between urban and suburban zones that were separated by walls and cemeteries, or, in the lack of the walls, just by cemeteries. At this stage it is not possible to define the patterns of the functional use of urban space. Nevertheless, judging by the example of Siscia, whose edge-zones have been rather intensively researched due to a lesser modern development, it transpires that public places and buildings, such as *horrea*, baths and squares, were preferably situated in the vicinity of the town walls, while residential buildings were dispersed throughout the town. Cibalae is another example of the location of public buildings (a *horreum* and one of the baths) next to the town walls. The town matrix comprises the orthogonal street grid, accompanied by sewages and water supply installations. The façades of the buildings facing the main streets often featured porches for the pedestrians, as is evidenced in Mursa, Siscia, Cibalae, Andautonia, Iovia/Botivo, and Aquae Iasae. The height of the preserved architecture rarely exceeds the foundations and the immediate aboveground layer; more often than not only foundation "negatives" survive. The majority of the sustained damage suffered by Roman architecture occurred during the architectural and infrastructural development of the towns from the 18th to the 20th centuries. In other words, the majority of the Roman urban structures are now in belowground archaeological layers, overlaid by modern buildings. An exception to this rule is Aquae Iasae, whose most attractive architecture of the forum and the baths survives to a considerable height and is preserved within the archaeological park of Varaždinske Toplice.

Urban archaeology in Croatia has a long tradition, resulting in the evaluation and preservation of archaeological remains during the whole of the 20th century. The majority of the archaeological research was instigated by the archaeological and conservational protection of historic structures, threatened by modern development. Archaeological heritage has in the last 30 years been included in the legislation on space planning, which means that excavations and preservation of archaeological sites have been imposed by law. In some cases (Andautonia, Siscia, and Aquae Iasae) it implies establishing of archaeological parks, and in others a presentation of finds at the level of the ground floors of modern buildings (Siscia, Mursa, and Cibalae).

Acknowledgements

We would like to express our gratitude to the following colleagues who improved this paper by sharing with us many useful information: Maja Krznarić Škrivanko, Anita Rapan Papeša, and Hrvoje Vulić of the GMV, Tea Tomaš Barišić and Iskra Baćani of the GMS, Dora Kušan Špalj and

[203] Filipović 2004.
[204] Iskra-Janošić 2001, 93.

Dorica Nemeth-Ehrlich of the AMZ, and Vedran Katavić of the Kaukal Archaeological Company (Split).

Bibliography

Andrić 2002 Stanko Andrić, Južna Panonija u doba velike seobe naroda, *SSlav* 2, 117-167.

Baćani 2007 Iskra Baćani, Sisak – igralište OŠ 22. Lipnja, *HAG* 3 (2006), 198-201.

Baćani, Gospodinović, Škrgulja, and Tomaš Barišić 2012 Iskra Baćani, Sanja Gospodinović, Rosana Škrgulja, and Tea Tomaš Barišić, Zaštitna arheološka istraživanja Gradskog muzeja Sisak od 2000. do 2010 / *Archaeological Conservation Rasearch of the City Museum Sisak since 2000. until 2010.*, Exibition Catalogue, Sisak.

Bairoch 1988 Paul Bairoch, *Cities and Economic Development from the Dawn of History to the Present*, Chicago.

Bobovec 2005 Ana Bobovec, Rimsko arheološko nalazište Kutinska lipa u Kutini, *ObHAD* 37/1, 98-103.

Bojanovski 1974 Ivo Bojanovski, *Dolabelin sistem cesta u rimskoj provinciji Dalmaciji*, Sarajevo.

Bojanovski 1988 Ivo Bojanovski, *Bosna i Hercegovina u antičko dobra*, Sarajevo.

Bojanovski 1993 Ivo Bojanovski, Neki problemi prometne infrastrukture Brodskog Posavlja i Slavonije u antici, in: *Arheološka istraživanja u Slavonskom Brodu i Brodskom Posavlju*, *IzdHAD* 16, Zagreb, 59-70.

Bojčić *et al.* 2009 Zvonko Bojčić, Marko Dizdar, Tomislav Hršak, and Tino Leleković, Bansko brdo, *HAG* 5 (2008), 7-11.

Bojčić and Bulat 1985 Zvonko Bojčić and Mirko Bulat, Mursa, Donji grad Osijeka – rimska kolonija, *AP* 24, 87-90.

Bugar 2008 Aleksandra Bugar, Naselje ranog srednjega vijeka Velika Gorica – Šepkovčica, in: M. Guštin (ed.), *Srednji vek / Mittellalter*, Ljubljana, 179-193.

Bugar 2009 Aleksandra Bugar, Šepkovčica, *HAG* 5 (2008), 269-273.

Bulat 1969 Mirko Bulat, Topografska istraživanja limesa u Slavoniji i Baranji, *OZ* 12, 39-52.

Bulat 1973 Mirko Bulat, Novi antički nalazi u Daruvaru, *GSM* 8, 24-26.

Bulat 1988 Mirko Bulat, Vađenje ostataka stupova rimskog mosta u Dravi kod Osijeka 1985, *GSM* 54-55, 51-54.

Bulat 1989 Mirko Bulat, *Mursa – Osijek u Rimsko doba*, Exhibition Catalogue, Osijek.

Bulat 2007a Mirko Bulat, Novija istraživanja antičke Murse, *Zbornik radova* (ed. M. Radić), Osijek, 370-387.

Bulat 2007b Mirko Bulat, Rimski brod na Dravi, *Zbornik radova* (ed. M. Radić), Osijek, 324-325.

Burmaz 2007 Josip Burmaz, Sisak – Euroherc, *HAG* 3 (2006), 196-198.

Burnham and Wacher 1990 Barry C. Burnham and John Wacher, *The Small Towns of Roman Britain*, Berkeley-Los Angeles.

Chandler 1987 Tertius Chandler, *Four Thousend Years of Urban Growth, An Historical Census,* Lampeter.

Čučković 2009 Lazo Čučković, Arheološka baština Topuskog, in: L Čučković, *Topusko,* Sisak-Topusko, 8-25.

De Laet 1949 Sigfried J. De Leat, *Portorium, étude sur l'organisation douanière chez les romains, surtout à l'époque du haut-empire*, Brugge.

Dimitrijević 1979 Stojan Dimitrijević, Arheološka topografija i izbor arheoloških nalaza s vinkovačkog tla, in: *Corolla memoriae Iosepho Brunšmid dicata, IzdHAD* 4, Vinkovci, 133-282.

Dizdar, Iskra-Janošić, and Krznarić Škrivanko 1999 Marko Dizdar, Ivana Iskra-Janošić, and Maja Krznarić Škrivanko, *Vinkovci u svijetu arheologije*, Vinkovci.

Dizdar 2001 Marko Dizdar, *Latenska naselja na Vinkovačkom području, DMZg* 3.

Dizdar and Radman-Livaja 2004 Marko Dizdar and Ivan Radman-Livaja, Nalaz naoružanja iz Vrtne ulice u Vinkovcima kao prilog poznavanju rane romanizacije istočne Slavonije, *PIAZ* 21, 37-53.

Durman 1992 Aleksandar Durman, O geostrateškom položaju Siscije, *OpArch* 16, 117-131.

Durman 2002 Aleksandar Durman, Iron resources and production for the Roman frontier in Pannonia, *Historical Metallurgy* 36/1 24-32.

Durman 2006 Saša Durman (ed.), *Stotinu hrvatskih arheoloških nalazišta*, Zagreb.

Egri 2007 Mariana Egri, The role of local elites in economic exploitation of the Danube provinces, in: Lyudmil F. Vagalinski (ed.), *The Lower Danube in Antiquity (VI C BC – VI C AD),* Sofia, 103-112.

Faber 1973 Aleksandra Faber, Građa za topografiju antičkog Siska, *VAMZ* 6-7 (1972-1973), 133-162.

Filipović 2004 Slavica Filipović, Colonia Aelia Mursa, in: *Pannonia* II, 157-168.

Filipović and Katavić 2005 Slavica Filipović nad Vedran Katavić, Mursa – Vojarna, *HAG* 1 (2004), 8-11.

Filipović 2008 Slavica Filipović, Osijek – Vojarna (Građevinski fakultet), *HAG* 4 (2007), 29-32.

Filipović 2009 Slavica Filipović, Osijek – Vojarna (Građevinski fakultet), *HAG* 5 (2008), 51-52.

Filipović 2010 Slavica Filipović, Osijek – Vojarna (Učiteljski fakultet), *HAG* 6 (2009), 37-42.

Fitz 1980 Jenő Fitz, Administration and Army, in: *ARP*, 125-140.

Gardaš 2001 Miro Gardaš, Neki novi podaci o rimskom mostu pokraj Murse, *AnOs* 17, Osijek, 7-26.

Goodman 2007 Penelope Goodman, *The Roman City and its Periphery From Rome to Gaul*, London-New York.

Gorenc and Vikić 1984 Marcel Gorenc and Branka Vikić, Antičko nasljeđe ludbreškog kraja, in: *Ludbreg* (ed. V. Mađarić), Ludbreg, 59-71.

Gorenc and Vikić-Belančić 1980 Marcel Gorenc and Branka Vikić-Belančić, *Varaždinske Toplice – Aquae Iasae u antičko doba*, Varaždinske Toplice.

Göricke-Lukić 2000 Hermine Göricke-Lukić, *Sjeveroistočna nekropola rimske Murse*, Zagreb-Osijek.

Göricke Lukić 2011 Hermine Göricke Lukić, *Nekropole rimskodobne Murske*, Osijek.

Gračanin 2010 Hrvoje Gračanin, Rimske prometnice i komunikacije u kasnoantičkoj južnoj Panoniji, *SSlav* 10, 9-69.

Gregl and Migotti 2004 Zoran Gregl and Branka Migoti, Civitas Iovia (Botivo), in: *Pannonia* II, 131-143.

Gros 2006 Pierre Gros, *L'architecture romaine du début du IIIe siècle av. J.-C. à la fin du Haut-Empire, 2. Maisons, palais, villas et tombeaux*, Paris.

Gros and Torelli 1988 Pierre Gros and Mario Torelli, *Storia dell'urbanistica, Il mondo romano*, Roma-Bari.

Hlavinka 1902 Vinko Hlavinka, Stari rimski vodovod preko Kupe u Sisku, *Viesti družtva inžinira i arhitekta u Hrvatskoj i Slavoniji* 3, Zagreb, 27-31.

Hoti 1992 Marina Hoti, Sisak u antičkim izvorima, *OpArch* 16, 133-163.

Ilkić 2009 Mato Ilkić, Rimska vojna diploma iz Sotina (Cornacum), *PIAZ* 26, 134-164.

Iskra-Janošić 2001 Ivana Iskra-Janošić, *Urbanizacija Cibala i razvoj keramičarskih središta*, Zagreb-Vinkovci.

Iskra-Janošić 2004 Ivana Iskra-Janošić, Colonia Aurelia Cibalae, Entwicklung der Stadt, in: *Pannonia* II, 169 – 195.

Iskra-Janošić, Krznarić Škrivanko and Rapan Papeša 2006 Ivana Iskra-Janošić, Maja Krznarić Škrivanko, and Anita Rapan Papeša, Vinkovci – zaštićena arheološka zona grada, *HAG* 2 (2005), 38-40.

Jovanović 1987 Borislav Jovanović, Keltska kultura u Jugoslaviji, Uvod, in: *PJZ* 5, 805-814.

Katančić 1782 (1987, 1989, and 1991) Matija Petar Katančić, *Dissertatio de columna milliaria ad Eszekum reperta / Rasprava o rimskom miljokazu pronađenom kod Osijeka* (Croatian translation: S. Srša), *OZ* 18-19 (1987), 349-374; 20 (1989), 299-318; 21 (1991), 51-76.

Klemenc 1961 Josip Klemenc, Limes u Donjoj Panoniji, in: *Limes u Jugoslaviji*, I, Beograd, 5-34.

Krznarić Škrivanko 2007 Maja Krznarić Škrivanko, Vinkovci – Ulica hrvatskih kraljeva 32, *HAG* 3 (2006), 54-55.

Lavendan and Hugueney 1966 Pierre Lavendan and Jeanne Hugueney, Histoire de l'Urbanisme, Antiquite, Paris.

Leleković 2008 Tino Leleković, Ivandvor, *HAG* 4 (2007), 12-15.

Lapaine 1881 Valentin Lapaine, O Sisačkom rimskom vodovodu, *VHAD* 3, 60-61.

Leleković 2009 Tino Leleković, Osijek – Trg bana Jelačića, *HAG* 5 (2008), 45-51.

Lolić 2003 Tatjana Lolić, Colonia Flavia Siscia, in: *Pannonia* I, 131-152.

Lolić 2006 Tatjana Lolić, *Roman City of Siscia, Sisak, Croatia, Preliminary Technical Assasment, Integrated Rehabilitation Project Plan/Survey of the Architectural and Archaeological Heritage* (IRPP/SAAH), Zagreb.

Lolić 2007 Tatjana Lolić, Siscija – Sv. Kvirin, *HAG* 3 (2006), 203-206.

Lolić 2010 Tatjana Lolić, *Roman City of Siscia, Sisak, Croatia – Business Plan, Integrated Rehabilitation Project Plan/Survey of the Architectural and Archaeological Heritage* (IRPP/SAAH), Zagreb.

Lolić, Petrinec, and Wiewegh 2010 Tatjana Lolić, Tomislav Petrinec, and Zoran Wiewegh, *Projekt konzervacije i prezentacije arheološkog lokaliteta Osijek – građevinski fakultet*, Zagreb.

Majnarić-Pandžić 1984 Nives Majanrić-Pandžić, Prilog problematici kasnolatenskih naselja u Slavoniji, *OpArch* 9, 23-34.

Majnarić-Pandžić 1998 Nives Majanrić-Pandžić, Keltsko-latenska kultura mlađeg željeznog doba u sjevernoj Hrvatskoj (od 4. st. pr. Kr. do romanizacije u 1. st. poslije Kr.), in: S. Dimitrijević, T. Težak-Gregl, and N. Majnarić-Pandžić, *Prapovijest*, Zagreb, 327-348.

Makjanić 1996 Rajka Makjanić, Sisak – terra sigillata, iskopavanja 1990., *OpArch* 20, 91-118.

Marsigli 1726 Luigi Ferdinando Marsigli, *Danubius Pannonico-mysicus* observationibus geographicis, astronomicis, hydrographicis, historicis, physicis perlustrarus et in sex tomos digestus, Hagae-Amstelodami.

Marta 1990 Roberto Marta, *Architettura Romana: tecniche construttive e forme architettoniche del mondo romano*, Roma.

Migotti 1994 Branka Migotti, Arheološka građa iz ranokršćanskog razdoblja u kontinentalnoj Hrvatskoj / The Archaeological Material of the Early Christian Period in Continental Croatia, in: Ž. Demo (ed.), *Od Nepobjedivog sunca do Sunca pravde, Rano kršćanstvo u kontinentalnoj Hrvatskoj / From the Invincible Sun to the Sun of Justice, Early Christianity in Continental Croatia*, Exhibition Catalogue, Zagreb, 41-67.

Migotti 1998 Branka Migotti, Od kulta Sola do kršćanstva u Varaždinskim Toplicama (*Aquae Iasae*), Radovi, *RFFZd* 37 (24), 51-88.

Miškiv 1998 Jesenka Miškiv, Rimska vojnička diploma iz Slavonskog Broda, *VAMZ* 30-31 (1997-1998), 83-101.

Miškiv 2006 Jesenka Miškiv, Slavonski Brod – antika, in: Durman 2006, 238-239.

Miškiv 2007 Jesenka Miškiv, Slavonski Brod – Šetalište braće Radić 16, *HAG* 3 (2006), 81-82.

Miškiv, Miklik-Lozuk and Lozuk 2006 Jesenka Miškiv, Lidija Miklik-Lozuk, and Josip Lozuk, *Zaštitna arheološka istraživanja „Trg Ivane Brlić-Mažuranić" Slavonski Brod 2001. i 2003.*, Slavonski Brod.

Mócsy 1974 András Mócsy, *Pannonia and Upper Moesia, A History of the Middle Danube Provinces of the Roman Empire*, London-Boston.

Morley 2005 Neville Morley, Cities in context: urban systems in Roman Italy, in: Helen M. Parkins (ed.), *Roman Urbanism, Beyond the Consumer City*, London-New York, 41-56.

Nemeth-Ehrlich and Kušan 1997 Dorica Nemeth-Ehrlich and Dora Kušan, *Aquae Iasae –Varaždinske Toplice, Vizualizacija rimske arhitekture / Visualizing Roman Architecture*, Exhibition Catalogue, Zagreb.

Nemeth-Ehrlich and Kušan 2006a Dorica Nemeth-Ehrlich and Dora Kušan, Gradski park (Varaždinske Toplice), *HAG* (2005), 135-137.

Nemeth-Ehrlich and Kušan 2006b Dorica Nemeth-Ehrlich and Dora Kušan, Konzervatorsko-restauratorski radovi i zaštitna arheološka istraživanja na antičkom kompleksu u Varaždinskim Toplicama od godine 2002. do 2005., *ObHAD* 38/1, 101-112.

Nemeth-Ehrlich and Kušan Špalj 2003 Dorica Nemeth-Ehrlich and Dora Kušan Špalj, *Municipium Andautonia*, in: *Pannonia* I, 107-129.

Nemeth-Ehrlich and Kušan Špalj 2007 Dorica Nemeth-Ehrlich and Dora Kušan Špalj, *Dvije tisuće godina Andautonije: od rimskog grada do arheološkog parka* / 2000 Years of Andautonia: From the Roman Town to the *Archaeological Park*, Zagreb.

Nemeth-Ehrlich and Kušan Špalj 2011 Dorica Nemeth-Ehrlich and Dora Kušan Špalj, The results of the latest excavations at the Roman forum in Aquae Iasae – Varaždinske Toplice, in: *Religion in Public and Private Sphere, Acta of the 4th International Colloquium The Autonomous Towns of Noricum and Pannonia, Annales Mediterranei,* Koper, 211-232.

Novinščak 2006 Petra Novinščak, Sveti Martin na Muri, in: Durman 2006, 262-263.

Ørsted 1985 Peter Ørsted, *Roman Imperial Economy and Romanization, A study in Roman imperial administration and the public lease system in the Danubian provinces from the first to the third century A.D.,* Copenhagen.

Perinić 2000 Ljubica Perinić, O zavjetnom natpisu Minervi, *OpArch* 23-24 (1999-2000), 417-424.

Pinterović 1969 Danica Pinterović, Problemi u istraživanju limesa na sektoru Batina Skela – Ilok, *OZ* 12, 53-69.

Pinterović 1975 Danica Pinterović, Nepoznata Slavonija, *OZ* 14-15, 123-166.

Pinterović 1978 Danica Pinterović, *Mursa i njeno područje u antičko doba*, Osijek.

Pinterović 1980 Danica Pinterović, „Basilica martyrum" u Osijeku, in: *Gunjačin zbornik*, (ed. I. Erceg *et al.*), Zagreb, 59-66.

Pleše 2010 Tajana Pleše, Ludbreg – Vrt Somođi, *HAG* 6 (2009), 204-207.

Póczy 1980 Klara Póczy, Pannonian cities, in: *ARP*, 239-274.

Poulter 1996 Andrew Poulter, Urbanism in the Danubian provinces, in: J. Rich (ed.), *The City in Late Antiquity,* London-New York.

Sanader 2003 Mirjana Sanader, Rimske legije i njihovi logori u hrvatskom dijelu panonskog limesa, *OpArch* 27, 463-467.

Schejbal 2003 Berislav Schejbal, Nova razmatranja o Aquae Balissae i narodu Jaza, *OpArch* 27, 393-416.

Schejbal 2004 Berislav Schejbal, Municipium Iasorum (Aquae Balissae), in: *Pannonia* II, 99-129.

Scherrer 2003 Peter Scherrer, Savaria, in: *Pannonia* I, 53-80.

Sinobad 2010 Marko Sinobad, Jupiter i njegovi štovatelji u svijetlu epigrafskih izvora na području Hrvatske, *OpArch* 34, 145-228.

Suić 1985 Mate Suić, Mursensia quaedam (ad *CIL* III 3288; *CIL* III 3289, 10260), *VAMZ* 18, 57-70.

Sunko and Katavić 2007 Ana Sunko and Vedran Katavić, Osijek – Huttlerova 20-22, *HAG* 3 (2006), 21-25.

Sunko, Katavić and Jerončić 2008 Ana Sunko, Vedran Katavić, and Tomislav Jerončić, Osijek – Huttlerova (Silos), *HAG* 4 (2007), 20-26.

Stamper 2005 John W. Stamper, *The Architecture of Roman Temples: The Republic to the Middle Empire*, Cambridge.

Šarić 1986 Ivan Šarić, Zaštitno arheološko istraživanje lokaliteta „Kovnica" u Sisku 1985. godine, *ObHAD* 18/1, 28-29.

Šarić 1990 Ivan Šarić, *Arheološka dokumentacija s lokaliteta MT-10 u Sisku, ožujak 1990*, Regionalni zavod za zaštitu spomenika kulture / The Regional Conservation Agency for the Protection of the Cultural Monuments (danas Konzervatorski odjel u Zagrebu / today The Conservation Department in Zagreb), Zagreb.

Šašel 1974 Jaroslav Šašel, *s. v.* Siscia, *RE* Supplbd. 14, 702-741.

Šegvić 1986 Marina Šegvić, Antički kultovi u Sisku i Topuskome, in: *Arheološka istraživanja na karlovačkom i sisačkom području*, *IzdHAD* 10, Zagreb, 95-101.

Šegvić 1991 Marina Šegvić, Topusko 1990., *ObHAD* 23/2, 46-51.

Šegvić 2006 Marina Šegvić, Topusko, in: Durman 2006, 270-271.

Šimek 1997 Marina Šimek (ed.), *Registar arheoloških nalaza i nalazišta sjeverozapadne Hrvatske*, Bjelovar.

Šimić and Filipović 1997 Jasna Šimić and Slavica Filipović, *Kelti i Rimljani na području Osijeka*, Exhibition Catalogue, Osijek.

Šipuš 1983 Nikola Šipuš, Neka razmatranja o radu rimskih provincijalnih kovnica novca, *NV* 37, Zagreb, 21-29.

Tomičić 1966 Željko Tomičić, Ludbreg – rimsko naselje, *AP* 8, 116-120.

Tomičić 1979 Željko Tomičić, Zaštitna arheološka istraživanja u Martinu na Muri, *MV* 2, 40-43.

Tomičić 1984 Željko Tomičić, Sumarni osvrt na rezultate arheoloških istraživanja područja Međimurja u razdoblju od 1972-1982. godine (I), *MV* 7, 56-65.

Tomičić 1997 Željko Tomičić, Arheološka topografija i toponimija (Iovia – Botivo –Ludbreg), in: K. Horvat-Levaj and I. Reberski (eds.), *Ludbreg, Ludbreška Podravina, Umjetnička topografija Hrvatske*, Zagreb, 21-41.

Tomičić 1999 Željko Tomičić, Arheološka topografija i toponimija Ludbrega i okolice, in: Željko Tomičić, *Panonski periplus, arheološka topografija kontinentalne Hrvatske*, Zagreb, 133-166.

Vidović 1989 Josip Vidović, *Arheološki odjel Muzeja Međimurja Čakovec* (vodič stalnim postavom), Čakovec.

Vikić-Belančić 1967 Branka Vikić-Belančić, *Rimska keramička proizvodnja na području Save i Drave od I do IV st. s osobitim obzirom na proizvodnju glinenih svjetiljki*, Unpublished PhD thesis, Zagreb.

Vikić-Belančić 1970 Branka Vikić-Belančić, Istraživanja u Vinkovcima 1966. godine, *VAMZ* 4, 159-176.

Vikić-Belančić 1981 Branka Vikić-Belančić, Etape urbanog razvitka Andautonije i antičko nasljeđe Zagreba, in: *Arheološka istraživanja u Zagrebu i njegovoj okolici, IzdHAD* 6, Zagreb, 129-154.

Vikić-Belančić 1984 Branka Vikić-Belančić, Sustavna

istraživanja u Ludbregu od 1968-1979. god., *VAMZ* 16-17 (1983-1984), 119-166.
Vikić-Belančić 1998 Branka Vikić-Belančić, Antičko nasljeđe Ludbrega, *Radovi Zavoda za znanstveni rad HAZU Varaždin* 10-11, Varaždin, 477-494.
Vikić-Belančić and Gorenc 1961 Branka Vikić-Belančić and Marcel Gorenc, Istraživanja antiknog kupališta u Varaždinskim Toplicama od 1956. do 1959., *VAMZ* 3, 181-223.
Visy 2003 Zsolt Visy, Sopianae, in: Z. Visy (ed.), *Hungarian Archaeology at the Turn of the Millennium*, Budapest, 230-231.
Vujnović and Burmaz 2010 Nikša Vujnović and Josip Brumaz, Okuje 3, 3a i 3c, *HAG* 6 (2009), 221-224.
Vrbanović 1981 Stjepan Vrbanović, Prilog proučavanju Siscije, in: *Arheološka istraživanja u Zagrebu i njegovoj okolici, IzdHAD* 6, Zagreb 187-200.
Vulić, Krznarić Škrivanko and Rapan Papeša 2008 Hrvoje Vulić, Maja Krznarić Škrivanko, and Anita Rapan Papeša, Vinkovci – Glagoljaška ulica b.b., *HAG* 4 (2007), 70-71.
Vulić, Krznarić Škrivanko and Rapan Papeša 2010 Hrvoje Vulić, Maja Krznarić Škrivanko, and Anita Rapan Papeša, Vinkovci – Ulica I. Gundulića 37, *HAG* 6 (2009), 60-62.
Ward-Perkins 1981 John B. Ward-Perkins, *Roman Imperial Architecture*, New Haven-London.
Wiewegh 2003 Zoran Wiewegh, *Jugoistočna nekropola Siscije*, Sisak.
Wiewegh 2004 Zoran Wiewegh, Otkriće kule zapadnih zidina rimske Siscije, *ARR* 14, 121-140.
Wiewegh 2007 Zoran Wiewegh, Nalaz građevine s apsidom u Drugoj ulici u Sisku, *ARR* 15, 91-103.
Wilkes 2005 John J. Wilkes, Roman Danube: An Archaeological Survey, *JRA* 95, 124-225.
Yegül 1995 Fikret Yegül, *Baths and Bathing in Classical Antiquity*, New York.
Zaninović 1981 Marin Zaninović, Siscija u svojim natpisima, in: *Arheološka istraživanja u Zagrebu i njegovoj okolici, IzdHAD* 6, Zagreb, 1981, 201-208.
Zaninović 1993 Marin Zaninović, Classis Flavia Pannonica, in: *Arheološka istraživanja u Slavonskom Brodu i Brodskom Posavlju, IzdHAD* 16, Slavonski Brod, 53-58.
Zaninović 1996 Marin Zaninović, Andautonia i Siscia u odrazu flavijevske politike, in: *Arheološka istraživanja u Zagrebu i zagrebačkoj regiji. Arheologija i obnova, IzdHAD* 17, Zagreb, 59-63.
Zubčić 2010 Krunoslav Zubčić, Osijek – rimski most, *HAG* 6 (2009), 35-37.

Construction Materials and Techniques

Ivan Knezović †

1. The natural and historical background to the architecture of southern Pannonia

1.1. Geographical and geological background

The Roman province of Pannonia – as well as the provinces created later by its division – encompassed the western and southern part of the Pannonian basin. This term, as opposed to the term Pannonian plain, also encompasses, in addition to the true lowland territories along the river courses, marginal mountainous tracts and mountain ranges. The extreme southern part of the province in modern-day northern Serbia, Bosnia and the Croatian regions Banovina and Kordun, also included the peripheral hills and mountains of the Dinaric Mountains. The major part of the territory of southern Pannonia occupied the region between the Danube, Drava and Sava Rivers. But even the lowland part of Roman Pannonia was not completely flat; isolated foothills and low hills break up the monotony of the lowland landscape creating a hilly or mildly mountainous landscape. Fruška gora in Syrmia, the mountainous range surrounding the Požega valley (Papuk, Psunj, Krndija, Požeška gora, and Dilj) and Moslavačka gora are the most prominent mountains in the region. The slopes in the western part of the Drava and Sava valleys, such as Kalnik, Ivanščica, Strahinjčica, and Medvednica, together with the Samoborsko and Žumberačko Gorje (the Samobor and Žumberak Mountains) south of the Sava, already belong to the peripheral sub-Alpine region, which becomes more and more elevated and intricate in terms of relief towards the west, that is, the Slovenian regions of Dolenjska and Štajerska. The relief dynamics is also increased by hilly elevations in the regions of Bjelovar, Bilogora Mountain and Hrvatsko zagorje.

The geological structure of southern Pannonia is very diverse. Its lowland regions – a true Pannonian plain – are the youngest in terms of geology, mostly originating in the Quaternary period. The lowland areas along the river courses are mostly formed of alluvial (Holocene) sediments that consist of gravel, sand and sandy clay deposits, which had been deposited by the rivers during the ice melting after the ending of the last Ice Age, that is, the last glacial period – stadial. Older sediments from the Pleistocene, mostly of Aeolian origin, predominate on the higher terrace areas between the river courses. They were made by sedimentation of particles brought by winds during the dry and cold glacial (stadial) periods; deep water-resistant layers of loess and clay were formed during these periods. During the Late Pleistocene period the Aeolian layers alternate with fluvial gravel-sand layers deposited during powerful tidal waves created during warmer interglacial and interstadial periods. The substratum of the Pleistocene Aeolian deposits is mostly made of older deposits of Tertiary period, mostly of clay, sand, large-granulate sand and, occasionally, gravel. These are all fresh-water sediments of the fluvial-lake type. Sediments creating low hilly elevations and lower slopes of higher elevations (mountainous tracts and mountain ranges) are similar in age and composition. Here the streams descending from higher mountain ranges occasionally deposit gravel material that merges with earlier Tertiary sediments.[1]

The builders in prehistory and the Roman period used natural building materials obtained from the surroundings of the building sites. Rich Pannonian woods in the river basins and mountainous tracts and mountain ranges were a rich source of timber. Lowland forests of Pendunculate and Sessile oak are especially important because of the superior quality of oak-wood. Geologically, later Holocene and Pleistocene layers and deposits do not create a solid rock mass and, therefore, building stone cannot be extracted from them, while the opposite is true of clay, sand and gravel. Potential building stone deposits are conditioned by the geological structure of Pannonian mountainous tracts and mountain ranges, as well as the peripheral areas of the Dinaric and Alpine mountain ranges. Although mountainous tracts, mountain ranges and precipices encompass a rather small area in comparison with the plains, their geological composition is diverse and varies considerably between different sites.

1.2 Petrography of southern Pannonia

The southern peripheral area of the Roman province of Pannonia (Gorski Kotar, Kordun, Banovina, northern Bosnia, northern Serbia) was contiguous with the Dinaric mountain range. Geologically, this area belongs to the Dinaric carbonate platform, mostly made of carbonate rocks, limestone and dolomite, from the Mesozoic period. In the Banovina and Kordun, younger deposits of Tertiary (Palaeogene and Neogene) age are also present, together

[1] Šimunić and Basch 1975, 155-158; Fürst-Bjeliš 1996, 36-38.

with considerably older metamorphic Palaeozoic rocks.[2] Mountains in northern Bosnia and north-western Serbia, such as Prosara, Motajica, Cer and Boranja, are in general made of granites and metamorphic rocks from the Mesozoic (Cretaceous) and Tertiary periods.[3] The western part of Roman Pannonia bordered on the Alpine area, formed of mostly carbonate rocks (limestone, dolomites) as well. On the other hand, the structure of Pohorje (Slovenia), a mountain above Maribor, comprises mostly magmatic and metamorphic rocks (marbles).

Geological structure of other Pannonian mountains is significantly different from both the Dinaric and Alpine region. The Pannonian mountain ranges and mountainous tracks are quite different between themselves in terms of petrography, and significant differences also exist within a single precipice.[4] A good example of this is Medvednica, a mountain on the north edges of Zagreb. The base of Medvednica's geological structure is formed of metamorphic Palaeozoic and Triassic rocks: slate, metacarbonate, metalimestone and marble. Magmatic Triassic and Jurassic rocks are also present: basalt, diabase, and gabbro. Triassic and Cretaceous complex is further formed of dolomite, limestone, sandstone, and conglomerate rocks. A band of Tertiary (Neogene) deposits of clay, marl, clastite and limestone is spread about Medvednica, with lithothamnium limestone (litavac) predominating. Layers of lithothamnium limestone, other types of limestone and dolomite rocks, are especially prominent in the western part of Medvednica, while palaeozoic green slate, metacarbonates and marble are predominant in its central part.[5] The nearby Samoborsko gorje (the Samobor Mountains) are rather different in stone composition. Most of the layers and deposits are formed of dolomite and various limestone rocks, with some sandstone sediments also present. A smaller part of the range is formed of older Palaeozoic basalt, diabase and conglomerate rocks, while the younger complex is formed of Neogene marlstone, clay and brackish limestone.[6] The hills of the Zagorje, as well as Ivanščica, Macelj and Kalnik, are formed of mostly Mesozoic limestone and dolomite sediments, and, in a smaller measure, eruptive rocks (basalt, diabase). Palaeogene and Neogene deposits of clastite, limestone and marlstone are especially prominent on the lower peripheral mountains' slopes and surrounding hilly elevations.[7] Moslavačka gora and the Slavonian Mountains have a significantly different geological composition. They are mostly formed of magmatic and metamorphic rocks of crystalline structure created during the Palaeozoic and Precambrian eras (granite, gneiss, amphibolite, marble limestone), combined with eruptive rocks. The peripheral lower parts are covered with various Neogene deposits similar to those on Medvednica and in the Hrvatsko Zagorje in north-western Croatia.[8] The geological structure of Fruška gora is also complex, comprising rock formations dating from the Palaeozoic era to the Neogene period. Crystalline background, slate, serpentinite and amphibolite formed the earliest rock complexes. The Mesozoic complex (mostly dating from the Triassic and Cretaceous period, in a lesser measure from the Jurassic) is formed by sandstone, limestone, conglomerate rock, breccias, marlstone and marl limestone. Coal deposits (Vrdnik) and marl and clay sediments were also formed in the Triassic period. The stone composition further comprises eruptive rocks from various geological eras (from the Palaeozoic to the Tertiary: trachyte, volcano tuff, andesite, diabase, and gabbro).[9] The existence of a deposit of Neogene (Tertiary) limestone on the lower slopes of mountainous tracts and mountain ranges should be mentioned as a common characteristic of the southern peripheral area of Pannonia. This type of stone is usually deposited so as to enable separation of larger blocks suitable for the manufacturing of architectural elements and individual monuments; this type of stone has therefore been utilised in architecture throughout history, from the Roman period onwards.[10]

1.3. Materials and techniques of Pannonian prehistoric architecture

Local traditions in construction of various structures in the territory of future Roman Pannonia originated in prehistory. From the Neolithic onwards, clay, timber and other natural materials (hides, reeds, rushes) were used in construction works. The first documented habitations were dugouts and these were constructed by the Starčevo culture population. Similar structures were created by the Eneolithic Baden culture population.[11] Another Neolithic/Eneolithic type of residential structure is the aboveground construction. Wattle and reed wickerwork with clay or loam mass spread over it was attached to a basic construction of wooden posts and beams. The floors were made of beaten earth, often intentionally fired on the surface. These types of structures were made by the Vinča-, Kostolac-, and Vučedol-culture populations.[12] Stone was used only sporadically in order to fasten the poles or posts planted into the earth. The construction of completely timber structures appeared in the Late Neolithic period. This type of structure is documented in the Sopot culture settlements in eastern Slavonia and Syrmia. The walls were built of horizontal beams supported by a row of vertical posts.[13] Timber and clay remained the main materials used for the construction of various structures in the Pannonian basin until the Iron Age. The use of stone as construction material originated in the early Iron Age in the southern and western border areas of the Pannonian basin along the mountainous tracts and lower mountains with accessible rock deposits. In the first place, this was the area of the Slovenian Dolenjska region, where the use of unworked stone and rubble on several hill-fort sites in mixed constructions made of

2 Crnković and Jovičić 1993, 142-145.
3 Pamić 1986, 219-237.
4 Crnković and Jovičić 1993, 141.
5 Crnković 1997, 11; Tomljenović 2002, 3.
6 Tomljenović 2002, 4.
7 Crnković and Jovičić 1993, 141-142.
8 Földvary 1988, 386-388. On the term *Slavonian Mountains* see Migotti in this volume, note 13, fig. 2.

9 Petković 1976; Földvary 1988, 387-388.
10 Rižnar and Jovanović 2006, 139-140.
11 Težak-Gregl 1998, 64-65, 129.
12 Dimitrijević 1979b, 304; Težak-Gregl 1998, 81, 129, 136-137.
13 Dimitrijević 1979, 270-273; Težak-Gregl 1998, 88.

clay, timber and stone has been evidenced.[14] It is believed that the local stone architecture was connected with the Mediterranean tradition that was introduced from Istria and the Slovenian karst region.[15] The use of stone as building material is documented at a hill-fort site in Lower Styria.[16] Coarse rubble stone and river pebbles were used in a grave mound construction in the contemporary Martijanec-Kaptol cultural group in north-western Croatia.[17] Because of the inadequate state of research there is no information regarding the use of stone as a building material in the settlements of the Martijanec-Kaptol cultural group. In the Donja Dolina-Sanski Most group, however, encompassing north-western Bosnia and the Bosnian Sava region, as well as more southern hilly areas, rubble stone was used in fortification construction at hill forts in the southern mountainous area populated by this group.[18] During the late Iron Age (La Tène) stone was used as building material at the Dolenjska hill forts.[19] More to the north, in Styria and north-western Croatia, there is no evidence for the use of stone in building during this period; earth (clay) and timber were used instead.[20] This period saw the incursion of the Celts from the north and north-west into Pannonia; they constructed characteristic settlements – the *oppida*.[21] Their fortifications were made of a combination of earth, wood and stone.[22] Due to the lack of stone in southern Pannonia, local *oppida* from the late 2nd and 1st centuries BC had earthen ramparts strengthened with palisades.[23] The use of stone is documented at only two sites in south-eastern Pannonia (a hill fort next to Slankamen, and Turski šanac near Bačka Palanka). In both cases dry-stone walls were the cores of earthen ramparts.[24]

1.4. Roman architectural innovations

In the period when Pannonia was becoming part of the growing Roman state, Roman architecture achieved high technological standards. Roman architectural heritage, with its numerous innovations, spread through the newly formed provinces together with the army and new settlers. In Pannonia, these innovations comprised new materials and building techniques as well as some novel architectural forms. The most important innovation in terms of materials was mortar (plaster), an indispensable component of Roman architecture. It was primarily used as bonding fabric in stone and brick masonry. Mixed with small-sized rubble stone it formed a mass with which the space between brick- or stone- paved walls or arch facades was filled; this mass could be also poured into a formwork, precisely as the modern concrete. This mass, known as *opus caementicium*, was also used as the core for wall construction of many Roman monumental buildings.[25] Roman technology of stone dressing and building, developed on a substratum of rich Mediterranean heritage, superseded in every respect the modest native tradition and stone masonry achievements of the area of study. The use of stone took two forms: as rubble stone for construction and as large blocks in architecture. Since a rock mass of special quality was not necessary for the production of rubble stone, it could have been quarried locally from a suitable deposit near the building site.[26] Technical and aesthetical qualities of raw stone material were not especially important for the most important Roman building techniques that utilised rubble stone (*opus incertum*, *opus caementicium*). The quarries, that is, stone deposits, with such characteristics that enabled the extraction of blocks suitable for constructions of superior quality or for the construction of larger architectural elements (architraves, pillars, monuments) are much scarcer. Their position is conditioned by geological factors such as rock composition, layering and geological intactness of the deposit. The most important Roman masonry techniques are *opus quadratum, opus incertum*, *opus reticulatum*, *opus vittatum*, and *opus spicatum*. *Opus quadratum* is a technique utilising regular blocks with no bonding material (mortar) that was used in construction of monumental structures such as fortifications, temples and other public buildings.[27] *Opus incertum*[28] utilised irregular rubble stone, occasionally hewed flat on the side exposed at the facade of the wall. The blocks were arranged irregularly with much use of mortar, without any horizontal layering, forming a rather solid irregular structure of the façade. Because of the wide availability of rubble stone and solidity of so constructed walls, this technique was often used in the construction of both private and public buildings. *Opus reticulatum* is a technique employing smaller blocks in the form of a four-sided pyramid fastened at the top onto the core of the wall. Rectangular bases of pyramidal blocks created a diagonal reticular pattern on the façade, which gave the technique its name.[29] *Opus vittatum* is a masonry technique utilising regular blocks of identical dimensions so that the upper-course blocks always covered the joints of the lower-course blocks; it was often used in constructing public buildings and defensive walls.[30] *Opus spicatum* imitates the fishbone pattern: relatively coarsely hewed blocks were arranged at an angle of 45° with every course being inclined in the opposite direction from the preceding. Wall footings and solid bases for floors and roads were mostly constructed in this particular technique.[31] The paving in rectangular brick pieces arranged in the fishbone pattern is also comprised by this term: terraces and yards were often paved in this way, as well as cistern-, pool- and fishpond-bottoms.[32] In all masonry techniques the core of the wall between the two façades was formed by *opus caementicium*. The

14 Gabrovec 1987, 81; Dular and Tecco Hvala 2007, 70-131.
15 Gabrovec 1987, 83-84.
16 Vinski-Gasparini 1987, 211-212.
17 Majnarić-Pandžić 1998, 230- 232; Šimek 2006, 142-143; Potrebica 2009, 99-101.
18 Čović 1987, 268.
19 Božič 1987, 882.
20 Božič 1987, 882-884.
21 The Latin word *oppidum* meaning *fortification, fort, town.*
22 *Murus Gallicus* – the Gallic wall, constructed of timber framing filled with stone and earth.
23 Majnarić-Pandžić 1998, 334-335.
24 Jovanović 1987, 822-824, 840-842.
25 Adam 2001, 79-81.
26 Adam 2001, 20-21.
27 Adam 2001, 106.
28 Latin *incertum* – insecure.
29 Latin *reticulatus* – reticular.
30 Adam 2001, 135-139.
31 Adam 2001, 144.
32 Lugli 1957, 49; Suić 2003, 178, 82.

unbaked brick was an important building material during the Republican period, while baked brick came in use more intensively only during the transition from the 2nd to the 1st century BC.[33] Vitruvius' remarks in his *Ten books on the architecture* on the use of this material are interesting in this context. In a section discussing the production and quality of bricks he exclusively considers the unbaked bricks (II.3). This suggests that in his time (40-32 BC) baked brick still did not become an important building material in Rome and northern Italy. Baked brick became increasingly more important in the Imperial period, while its role in the regions of the Empire with scarce sources of stone achieved special importance.[34]

Building technique utilising tiles, *opus testaceum* (*latericium*), became widespread in Tiberius' reign.[35] The combination of different masonry techniques, using both stone and brick, is called *opus mixtum* or *compositum*, and was mostly employed by combining corners and doorposts made of brick and wall facades made of stone. Or else, courses of bricks (usually 2 or 3) alternate in walls with stone courses. The core of both brick walls and those of mixed composition was made in *opus caementicium*, which contributed both to their solidity and an economical approach to building costs.[36] Along with the construction of walls, the brick was favoured, because of its water-resistant and thermal characteristics, in construction of subterranean conduits (water-supply, drainage) and under-floor heating structures (hypocaust). The roofs of Roman-period buildings were mostly covered in flat *tegulae* and saddle-form *imbrices*, covering the *tegulae* joints.[37] Water-supply conduits were also made of baked clay. Pillars and architraves of monumental buildings were usually composed of one or more stone pieces. Pillars could often be made of brick and/or small stones, and additionally plastered.[38] Among the main characteristics of Roman architecture were arches and vaults. The construction in regular stone blocks and bricks and the use of "Roman concrete" (*opus caementicium*) enabled the development and expansion of the construction of arches and vaults in previously unattainable dimensions; the bridging of previously impossible spans as well as the construction of large vaulted buildings such as basilicas and colossal imperial baths became possible. Roman architecture employed some earlier building techniques in clay and timber that were similar or almost identical to those typical of local Pannonian prehistoric traditions. The walls were constructed of moist clay rammed in timber framing. Another technique is identical to the one used in prehistoric Pannonia: a timber frame of pillars, beams and wattle was constructed and clay was applied to it. Both techniques demanded the addition of straw, chaff, sand and, occasionally, ashes, into the clay in order to solidify it.[39] Vitruvius mentions that wattle walls were built swiftly and economically, but they were not permanent and were extremely susceptible to water and fire (Vitruv. 2.8.20). With the spread of Roman architecture, wood became utilised in new ways. Besides in wall construction and in building of complete structures in combination with clay, brick and stone, it was also used in the construction of vaults and roof-frames. When multi-storey buildings started appearing in Rome, it was used for the upper-storey floors and staircases, and was indispensable in the construction of scaffoldings necessary for the erection of multi-storey buildings.[40]

2. Materials and techniques in south-Pannonian Roman settlements (fig. 1.)

2.1. Initial Roman constructions: military structures, traffic infrastructure and settlements

Roman construction works are mentioned for the first time during Octavian's campaign aimed at Segestica. After conquering the town, the Romans enclosed part of it with ramparts in order to accomodate 25 cohorts.[41] Archaeological traces of earlier military constructions have not been identified so far and the precise position of the military camp is unknown.[42] Military camps dating from the time of Octavian's military campaigns aimed at Segestica were discovered in the vicinity of Brežice (Slovenia), near Sv. Urh, Čatež and Obrežje.[43] The camp at Segestica was probably similar to those. A more permanent military presence in Siscia can be presumed between 35 BC and the beginning of the Pannonian War of 12 BC. During this war Siscia was an important base and it is probable that the building activities of the army were additionally amplified. In the Pannonian-Delmataean uprising Siscia was the main Roman military base, which surely implies new military constructions reflecting the need for the stationing of a large number of soldiers. Due to the strategic importance of the Sava crossing near the modern-day Ščitarjevo (Andautonia), one can presume that the foundation of Andautonia was due exactly to the war in question, which is further supported by some early Roman archaeological finds.[44] The area of Andautonia and the Sava valley must have been secured in order for the Roman army to be able to operate freely against the insurgents in modern-day Slavonia and Bosnia. After the hasty abandonment of the war with the Marcomanni, Tiberius passed through the territory of Andautonia with his army coming from the direction of Carnuntum; presumably, he additionally secured this important strategic point. The beginning of Roman building in Andautonia can be dated to the first half of the 1st century AD.[45] The growth of Roman Neviodunum (Drnovo near

[33] Adam 2001, 58-61.
[34] Adam 2001, 61.
[35] Adam 2001, 145-150.
[36] Adam 2001, 139-144.
[37] Adam 2001, 158-190; 213-214.
[38] Adam 2001, 156-157.
[39] Adam 2001, 58-61; Weber 2000, 81-82.
[40] Adam 2001, 81-101; 122-124, 196-213.
[41] App. 4.23, 4.24; Zaninović 1986, 63; Burkowsky 1999, 29.
[42] Lolić 2003, 140.
[43] Mason 2003a; Mason 2003b.
[44] Durman 1992, 118-120; Nemeth-Ehrlich and Kušan Špalj 2003a, 111-112.
[45] Nemeth-Ehrlich and Kušan Špalj 2003a, 117.

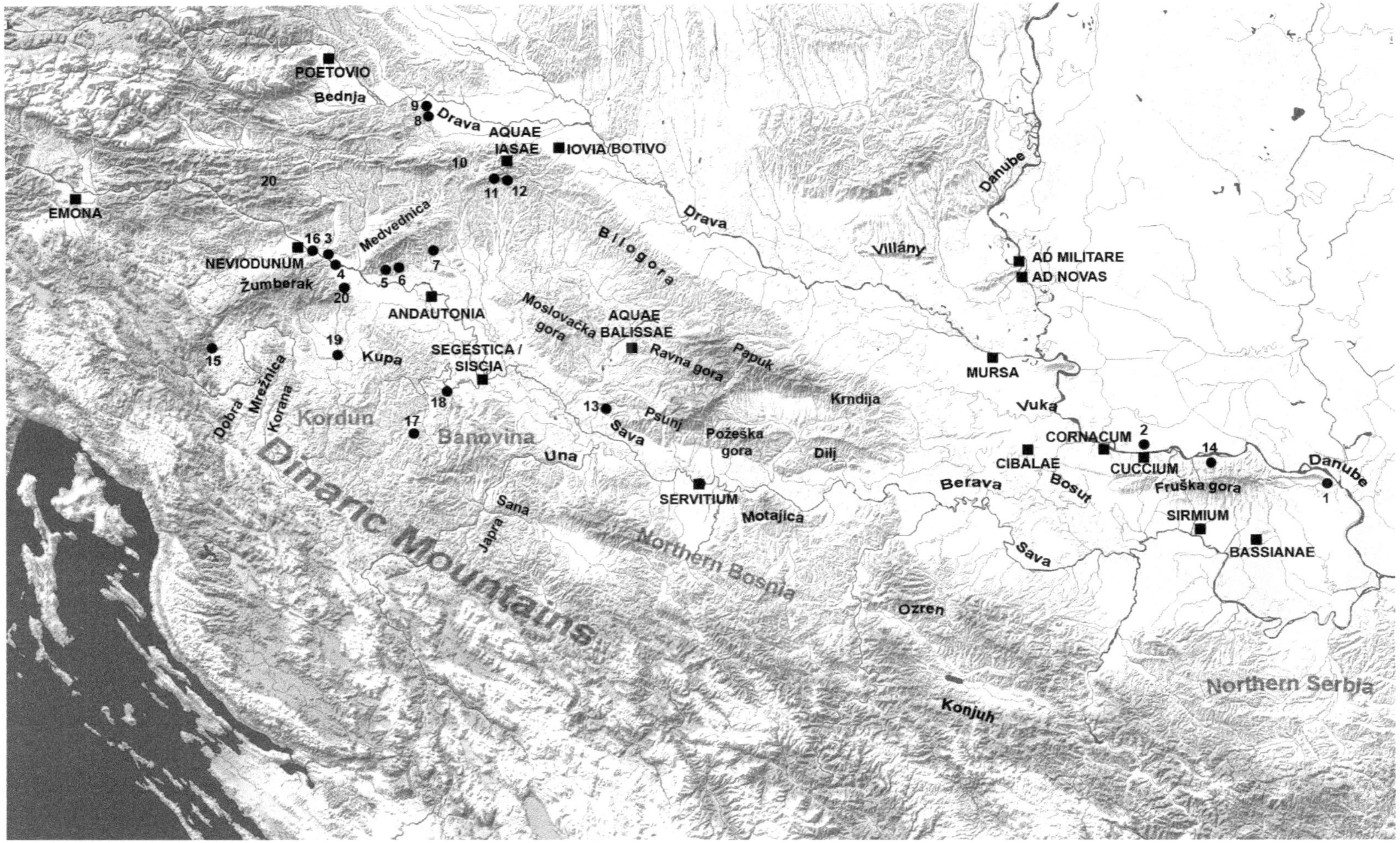

■ TOWNS AND LARGER SETTLEMENTS

● SMALLER SETTLEMENTS AND OTHER SITES

1. Slankamen
2. Turski šanac
3. Brežice
4. Čatež, Obrežje
5. Bizek
6. Gornje Vrapče (Vrapče potok)
7. Crna Voda
8. Vinica
9. Marčan
10. Novi Marof
11. Ljubešćica, Ljubelj Kalnički
12. Kamenik (Kamenjak)
13. Lipovljani
14. Carević, Beočin, Crveni Čot
15. Rožanec
16. Ribnica
17. Topusko, Gora, Hrastovica
18. Petrinja
19. Karlovac
20. Samobor

Fig. 1. Map of towns, settlements, and sites of natural resources for building materials in southern Pannonia and the neighbouring areas (after Google Maps, modified by T. Leleković).

Krško in Slovenia) was roughly contemporary, as the town was also placed on a site of some strategic importance for the control of land and river communications between Emona and Siscia.[46] In the eastern part of southern Pannonia between the Sava and Danube, the main Roman base during the Pannonian-Delmataean uprising was Sirmium, a former prehistoric settlement of the Celticised autochthonous population, conquered in the Pannonian War between 12 and 9 BC. Due to the importance and long history of occupation of the site of Sirmium, it can be presumed that the military immediately initiated construction works there, as a foundation of a Roman settlement. Besides these towns along the traffic route following the Sava valley, military centres and settlements were founded in the west of the province along the Amber Road during the 1st century AD. From south to north, these were Poetovio (Ptuj), Sala (Zalalövő), Savaria (Szombathely), Scarbantia (Soprony), Carnuntum (Bad Deutsch-Altenburg/Petronell near Vienna), and Vindobona (Vienna).[47] During the reign of the Flavians more intensive Romanisation started, with settlements developing into towns and receiving municipal and colonial status. Urbanisation was followed by an intensive construction of roads and the economical growth of the province. Neviodunum and Andautonia were the first to receive municipal status, while Siscia and Sirmium were the first to be granted colonial status.[48] Roman Iovia/Botivo (Ludbreg) was probably also founded in the 1st century. The foundation of this settlement is associated with the Roman military presence on the Drava, the temporary frontier of the Empire from the end of the Pannonian War in 9 BC until the start of the construction of the Danubian *limes* in Claudius' reign. It seems that Iovia obtained municipal status only in the 4th century.[49] In the reign of Augustus, a military camp was probably erected on the site of modern-day Daruvar in the territory of the Pannonian Iasi. After the relocation of the *limes* and the military on the Danube in the second half of the 1st century, a civil settlement was formed by the discharged Roman soldiers of local origin (the Iasi) and Roman settlers. The settlement was granted the status of a *municipium* during Hadrian's reign. The town is mentioned under the names of *Municipium Iasorum* and *Aquae Balissae*.[50] At the thermal springs in modern-day Varaždinske Toplice a thermal settlement of Aquae Iasae was developed during the 1st century AD, featuring thermal facilities from the earliest phase of Roman occupation. The settlement was of an urban character and encompassed the terraces and the foothill with the thermal spring.[51] It seems that Aquae Iasae never achieved the status of a *municipium* or colony but was part of the territory of Poetovio[52] or, perhaps, the much nearer Iovia.[53] In the area of Slavonia the processes of urbanisation and Romanisation progressed on a somewhat slower rate. The construction of the Danubian *limes* was initiated in the end of the first half of the 1st century AD, during Claudius' reign. This is also the time when the construction of the Slavonian-Syrmian part of the *limes* also began. From north to south, the auxiliary camps on the right bank of the Danube were sited at Batina, Dalj, Sotin, and Ilok.[54] The soldiers discharged from those units, often of local origin, after finishing their military service often settled in the rear of the frontier line, as a significant Romanising and urbanising factor in the area.[55] In modern-day Vinkovci there was an *oppidum*, a settlement of the local Celticised (La Tène-ised) population, most probably the Breuci, whose old settlement was probably destroyed in the Pannonian-Delmataean uprising. Because of its favourable position on the elevated left bank of the Bosut, the Roman settlement of Cibalae was founded there in the second half of the 1st century AD, contemporary with the initiation of more advanced processes of Romanisation during the Flavian emperors. Numerous archaeological finds suggest that a military unit, probably a local cohort composed of the Breuci recruits, was stationed in Cibalae in the early phase of Romanisation. This new settlement obtained the status of a *municipium* during Hadrian's reign, probably during his visit to Pannonia in 124; the town was elevated to the rank of a colony in Caracala's reign.[56] Roman Mursa also originated at the site of an earlier La Tène settlement; its origin should probably also be sought in an auxiliary camp. It is supposed that during the Dacian Wars in Domitian's and Hadrian's reign a legionary campexisted in Mursa. The construction of a civil settlement was begun after the end of the Dacian Wars in 106, and it obtained the status of a colony in Hadrian's reign.[57]According to A. Mócsy, whose classification is still widely accepted, the development and construction of Roman towns in Pannonia were carried out in four phases. The first phase was accomplished during the Flavian period. The second is connected to Hadrian's reign and the period ending with the Marcomannic Wars. The third is actually the restoration of the destroyed Pannonian towns in the Severan period, while the fourth is associated with the stabilisation during the reign of Diocletian, Constantine and his successors, after the turbulent period of the military emperors and Barbarian incursions into the province.[58] This classification is valid for the majority of urban centres, with minor corrections needed in individual cases due to certain geographical and historical specificities of their urban development.

Siscia

Modern-day Sisak was erected on the site of Roman Siscia, which is the main obstacle to archaeological excavations. Our knowledge of the Roman town therefore derives from numerous rescue excavations initiated by modern construction works. The position and perimeter of the walls, Roman bridges, baths, water supply and

46 Lovenjak 2003, 95-96.
47 Fitz 2003, 47-52.
48 Lengyel 1980, 479-480; Lovenjak 2003, 95; Nemeth-Ehrlich and Kušan Špalj 2003a, 112; Lolić 2003, 134; Mirković 2004, 147.
49 Gorenc and Vikić 1984, 63-65; Gregl and Migotti 2004, 132-133.
50 Schejbal 2004, 106-108.
51 Gorenc and Vikić 1980; Nemeth-Ehrlich 1993, 23-24.
52 Kukuljević 1873, 52-53.
53 Gregl and Migotti 2004, 135.
54 Sanader 2009, 106-108.
55 Iskra-Janošić 2004, 170.
56 Mócsy 1974, 143; Iskra-Janošić 2001a, 58-59; Iskra-Janošić 2004, 170-171.
57 Filipović 2004, 157.
58 Mócsy 1959, 16-96; Mócsy 1962, 598-600.

sewage installations, roads and granary (*horreum*), were all identified during rescue excavations. Along with monumental public buildings, remains of more modest private structures were also discovered in several places. The positions of Roman cemeteries outside the Siscia fortifications were also identified. Four Roman architectural phases can be generally distinguished in the remains of Roman buildings discovered in the Siscian territory. The first two show a temporal deviation from the accepted Mócsy's classification because the urbanisation of Siscia was begun earlier than that of the other Pannonian towns. The first phase is characterised by timber as the predominant construction material (1st half of the 1st century AD) and is associated with the presence of the Roman army. The remaining three phases were mainly evidenced by stone and brick structures. The second phase falls in the reign of the Flavians (later 1st-early 2nd centuries), while the third and fourth ran parallel to the standard Pannonian phases; the remains of the two last-mentioned are the best preserved ones.[59]

Traces of the earliest (timber-construction) phase in Siscia have been discovered at numerous positions in Sisak. Oaken posts 20-30 cm in diameter were planted fourth to seven metres deep into the marshy soil on the wider territory of the Roman town in order to solidify the soil and strengthen buildings' foundations. In this way the soil was solidified before the town walls were constructed.[60] A layer of soot probably suggests the existence of timber structures.[61] Remains of these structures are also preserved at places.[62] In some locations the timber phase, dated to the 1st half of the 1st century by pottery, is preserved under the later stone one.[63] During the low water level on the site of "Kovnica" (Mint) in the bed of the Kupa one can see oaken posts that carried the Roman harbour facilities. These posts were standing on an iron footing that eanbled their more effective fixing into the muddy riverbed. Dendrochronological analysis dated them from the 1st to the 4th centuries; they were replaced and supplemented every few years.[64] Besides some 200 piles, remains of a wooden boat were also found.[65] On the opposite bank of the Kupa a timber construction consisting of planted oaken posts and transversal beams has been discovered; it is believed that it served as a harbour facility or a bridge-support construction.[66] In several places timber framing has been recovered, showing a combination of brick and stone serving as a fill between wooden posts and beams; a special porous and light mortar was used in this type of construction.[67]

Solid constructions predominate in Siscia from the reign of Vespasian onwards, when the city attained the status of a colony, with brick and stone becoming the main construction materials. In the second construction phase at Siscia both materials were often used together in the *opus mixtum* technique.[68] Beneath the late Roman granary (*horreum*) remains of a structure were found, with a façade made of brick and semi-dressed stone and the core filled with rubble stone and mortar (*opus caementicium*), dated to the late 1st and early 2nd centuries.[69] The structures built in mixed technique could have had additional timber supports, such as the one discovered in the northern periphery of the town. Traces of vertical wooden beams are observable on its plastered walls, while holes for wooden posts were discovered both inside and outside of the wall; the structure was most probably part of the complex that also included the bridge at the northern town gate.[70] The foundations in this phase were made of rubble stone bonded with lime mortar, while monumental buildings' foundations were made of semi-dressed stone, stone plates and blocks.[71]

In the third phase, during the reign of the Severan dynasty, solid architecture continued to be built in Siscia. The most important structure from this period was certainly the town fortifications, which consisted of the walls and a double ditch running along them, as well as the free space within the fortifications that enabled free manoeuvring of the army during siege.[72] Segments of the town walls were discovered at various points along the entire perimeter.[73] The walls were of an average thickness between 1.85 and 2 m, while their foundations were placed on an average depth of 4.5 m. They were made of rubble stone and slanted bricks placed on their edges, and are somewhat reminiscent of *opus spicatum* – the so-called fishbone technique. The wall itself was made of rubble stone and bricks. On the outside, there is a brick structure ca. 1.40 m wide, abutting on the inside on a stone wall 0.4 m wide. The majority of the bricks are 43 x 43 x 6 cm in dimensions.[74] Some of them bear the mark *SISC*, confirming the existence of local brick production, which begun at the turn of the 2nd and 3rd centuries.[75] The fortifications were at places strengthened on the outside by means of box-form projections, whose walls are from 0.9 to 1.1 m thick, with the inside filled with construction debris and earth.[76] Rectangular keeps were added on the inside of the fortifications at the points where the direction of the course of the walls changed.[77] Structures associated with the fortifications' functions, probably of military and sentinel character, were added on the inside of the wall. Walls of these additions were made of both whole and rubble bricks bonded by mortar.[78] Remains of a bath complex in the present-day park on the left bank of the Kupa also date from this period. The structure was composed of two large halls

59 Lolić 2003, 141-143.
60 Burkowsky 2000, 10-11.
61 Baćani, Škrgulja, and Tomaš Barišić 2009, 69.
62 Vrbanović 1981, 190; Nenadić 1988, 83.
63 Burkowsky 2000, 42-44; Lolić 2003, 141.
64 Durman 1992, 120.
65 Lolić 2003, 141.
66 Burkowsky 2000, 15-16.
67 Faber 1973, 147-149; Adam 2001, 122 -124; Faber 1973.
68 Lolić 2003, 142.
69 Lolić 2001, 98.
70 Baćani, Škrgulja, and Tomaš Barišić 2009, 60-62.
71 Baćani and Tomaš 2008, 15-16.
72 Vuković 1994, 157.
73 Vrbanović 1981, 190-191; Burkowsky 2000, 10-14; Baćani, Škrgulja, and Tomaš Barišić 2009, 73-76.
74 Vuković 1994, 155; Lolić 2001, 95-99.
75 Vikić-Belančić 1967, 36; Lolić 2001, 95-99.
76 Burkowsky 2000, 12-13.
77 Lolić 2001, 95-99.
78 Burkowsky 2000, 12-13.

and several smaller rooms in which remains of a hypocaust, *caldarium* and a pool were discovered. The walls were made solely of bricks, which were extracted in the 19th century and reused as high quality building material. The hypocaust system was well preserved. It was founded upon a solid beaten mass consisting of stone and construction debris with fragments of wall plaster of an earlier structure, completely covered by this 1 m-thick levelling. This beaten layer was superimposed by a 7-cm-thick floor made of mortar with grinded brick, which, in its turn, was topped by a pavement consisting of bricks measuring 32 x 15 x 6 cm. The hypocaust pillars stand upon rectangular bricks (30 x 30 cm). They are made of ten circular bricks 28 cm in diameter and 0.98 m in height. Large bricks (64 x 64 x 8 cm) are placed upon them and they support the floor made of three layers of water-resistant mortar with grinded brick 31 cm thick. A *praefurnium* with large quantities of ashes and soot was discovered along with the hypocaust. The hall next to the one with the hypocaust most probably served as a *caldarium*. A pool could have been placed in one of the adjacent rooms because its walls were covered in water-resistant reddish plaster. A vaulted brick canal extended towards the Kupa (inner height: 98 cm, width: 45 cm), probably used as either air supply or exhaust, since it was not covered with water-resistant mortar.[79] Later excavations near the bath complex discovered remains of a cistern bottom with a multi-layer floor (bottom to top: rubble brick, stone and lime mortar base, lime mortar layer, water-resistant lime mortar layer with the addition of rubble brick, brick paving on top). The cistern was probably part of a thermal complex.[80] This building has yielded a large number of fragments of painted plaster imitating marble veneering, as well as dispersed black and white mosaic cubes.[81] Remains of several more structures belong to this phase, whose foundations were made of irregularly huddled slanted brick, placed on their edges, supporting a brick wall bonded by mortar.[82]

The latest building phase, from the Constantine and post-Constantine periods, is the best preserved one. In this phase earlier structures were demolished and new ones erected.[83] Remains of a granary (*horreum*) built in stone and brick, discovered next to the fortifications, belong to this period. The façades were made of rubble stone arranged in courses ca. 20 cm high, while the core in between them was filled with smaller rubble stones bonded by lime mortar (*opus caementicium*). The south-eastern wall has buttresses on both sides, while the north-western wall has them only on the outside. Remains of bases carrying pillars and a vault were discovered inside the granary. They were built in brick and dressed stone and supported a rectangular pillar in the middle and two circular lateral pillars. The granary was erected in two construction sub-phases, both dated in the 4th century. The granary was erected first, later to be turned into a building with porches and atrium, which probably functioned as a public building of uncertain use.[84]

Remains of structures built in permanent materials that cannot be precisely affiliated with one of Siscia's construction phases have been recovered on many locations. They belong both to public and private buildings, but their function and significance are not completely clear. Remains of monumental buildings with massive walls built of regular stone blocks or rubble stone have been found on several sites.[85] Massive walls were either built only in brick (*opus latericium*) or in *opus mixtum*, with alternating courses of brick and rubble stone.[86] A corner of one of these structures was built of regular hewn stone blocks bonded by mortar, with the addition of grinded bricks. A slot 29 cm in width and 5 cm in depth was carefully carved into the blocks 1.7 to 2.12 cm in length. Into these slots stone orthostats were inserted, perfectly fitting into the slot on the underlying block.[87] The bases of the pillars carrying porches were made of both complete and rubble brick.[88] Floors made of beaten earth and lime mortar, as well as those tile-paved, were found in these structures.[89] Numerous brick-constructed segments of drainage canals of the town sewage system have been found in Sisak. According to their dimensions, we can classify them into two basic types. Lesser, private drainage canals were rectangular or cubic in cross section, paved and coated in brick. Larger street canals were made of brick and vaulted by brick arches.[90] The bottom of some canals was made of rubble and grinded stone bonded by lime mortar.[91] The main (vaulted) drainage collector, running from north-east to south-west towards the Kupa, was made of brick bonded by mortar, with the floor paved in sesquipedal tiles; the opening measured 1.80 x 1.05 m.[92] Remains of Roman water-supply canals were discovered during digging the foundations for new buildings at several places inside the Roman fortifications. Judging from the positions of these finds, it is clear that the entire area of Roman Siscia was covered by the water-supply lines. In the town itself lead pipes were most frequently used, and they are often discovered during modern construction works. The water was brought to Siscia from the mountainous region of the Banovina south of the town. It was conducted into the Kupa mostly through subterranean canals built in rubble stone. The water bridged the river by an aqueduct, which was confirmed by two aqueduct pillar bases at the site of Zibel on the right bank of the river; they were made of bricks and rubble stone. There are two possible ways in which the aqueduct could have crossed the Kupa. Remains of lead pipes dredged out of the Kupa in 1902 speak in favour of a pressure conduit (siphon).[93] In 1988, besides lead pipes, parts of oaken pillars were also found, indicating

79 Faber 1973, 139-143.
80 Baćani, Škrgulja, and Tomaš Barišić 2009, 70.
81 Faber 1973, 139-143.
82 Baćani and Tomaš 2008, 15.
83 Lolić 2003, 142.
84 Lolić 2001, 95-99.
85 Faber 1973, 144-146; Baćani, Škrgulja, and Tomaš Barišić 2009, 69.
86 Burkowsky 2000, 8-9; Faber 1973, 135-138.
87 Faber 1973, 146.
88 Baćani and Tomaš 2008, 9-10.
89 Faber 1973, 135-138, 144-146.
90 Vuković 1994, 219-220.
91 Vrbanović 1981, 196.
92 Faber 1973, 150.
93 Vrbanović 1981, 194-195.

that the pipes could have been carried across the river on oaken supports.[94]

Remains of street sections have been discovered at various places in the territory of the Roman town. The substructure of the stone pavement and edge-stones was made of beaten coarse and sharp sand. A sewage collector piping was placed under the pavement surface following the course of the road axis. The water-supply conduit was also placed below the pavement surface, but nearer to the edge-stone.[95] The widest road section (0.2-9.6 m), belonging to the main town street (*cardo*), was discovered in the excavations on the northern periphery of the town, next to the St. Quirinus church; a vaulted drainage conduit was found below.[96] Since the town was surrounded by rivers, the main roads leading out of it inevitably ran across bridges. During the low water level of the Kupa, remains of two piles carrying a stone bridge were discovered on its right bank at the site of Pogorelec. Their foundations were made of rubble stone and water-resistant mortar. Today, they both lie outside the river-bed, suggesting that the course of the Kupa was 30 to 40 m nearer to its right bank in Roman times.[97]

Cibalae

Cibalae developed in four building phases. The first is characterised by building in timber and it is only sporadically documented through a burnt layer, while solid construction characterises the second and third phases. The fourth phase started somewhat later in Cibalae in comparison with the accepted classification of Mócsy. It is connected to the destruction of Cibalae in the late 4th and early 5th centuries and is documented only sporadically. Only some damaged structures in the centre of the town were rebuilt during this phase.[98] The Roman town lies under modern-day Vinkovci, making rescue excavations the only mean of obtaining information on its architecture.

Because of the characteristics of timber (susceptibility to fire, decomposition and decaying), the first phase is the least preserved and is documented mostly by layers of ashes and soot.[99] Clay (loam) was used alongside timber, which is a tradition inherited from prehistoric building techniques. Earth and timber were used in the construction of town fortifications, whose first phase date from Hadrian's reign when the town obtained the status of a *municipium*. Double earthen walls 10-14 m wide were separated by a 5-m-wide ditch. After the Marcomannic wars, the northern and eastern walls were relocated, resulting in the enlargement of the town's area (from 381 000 m^2 to 568 000 m^2). The new fortifications were also earthen, and were 14-15 m wide. As opposed to the earlier phase, these had a wooden palisade, similarly as in the military camps on the Danube *limes*. Brick was used only in peripheral sections and at the town gates. The water ditch was dug out only on the northern side of the town.[100]

In the second and third building phases of Cibalae brick highly predominated over stone.[101] The walls of buildings were most often built of standard sesquipedal bricks (1.5 x 1 ft, that is, 45 x 30 cm) in *opus testaceum*. Such bricks were also used for construction of drainage canals, wells, hypocausts, tombs, pillars and various pavements.[102] Besides the workshop stamps, various other ornamental motifs are appear on bricks, most frequently concentric circles and semicircles, wavy lines, tresses, spirals, imprinted circles and animal footprints, and in one case even a child's footprint.[103] Besides the sesquipedal bricks, other types were used, measuring 41 x 29 x 6 cm or 42 x 30 x 29 cm. Larger bricks (tiles) that were used for the construction of drainage canals and covers of walled tombs, were significantly thinner (3 cm) than those of smaller dimensions.[104] The bricks were of good quality, massive and heavy, from light to dark tones of red in colour. In order to produce a superior quality brick, clay should have been previously purified and silicate sand added to prevent crushing during abking. Bricks were frequently used since rather numerous clay deposits in the vicinity provided raw material easily accessible. The situation has not changed much until modern times, as even today large important brickyards exist near Vinkovci and in Slavonia in general.[105] The foundations of brick walls could have been made in two ways. In some cases, they were made of rubble brick arranged in slanted position, and were somewhat wider in comparison to the wall that they supported.[106] The foundations could also have been made of rubble stone, which was the most frequent use of stone in Cibalae.[107] Besides the foundations, lower wall-courses were also made of rubble stone, with a brick section extending above, producing *opus mixtum*.[108] In Late Antiquity both foundations and walls were occasionally built in rubble stone and brick.[109]

Besides its constructive purpose, stone had a decorative one, as suggested by the fragments of multicolour stone veneering. Some fragments are moulded at edges, while others feature remains of white and pink mortar with which they were bonded to the wall surface.[110] In wealthier houses stone was used for the construction of pillars, architraves, thresholds, and outer staircases.[111] In less assuming buildings the pillars were made of semicircular or quarter-circle bricks arranged in alternate courses, which gave them solidity and stability. The dimensions of semicircular bricks

94 Burkowsky 1999, 45.
95 Vuković 1994, 220; Baćani, Škrgulja, and Tomaš Barišić 2009, 60, 66.
96 Baćani and Tomaš 2008, 9-10.
97 Vrbanović 1981, 192-193.
98 Iskra-Janošić 2001a, 65-66; Iskra-Janošić 2005, 12-15.
99 Iskra-Janošić 1994, 59.
100 Iskra-Janošić 2001a, 33-43, 158; Iskra-Janošić 2005, 13-14.
101 Iskra-Janošić 1992, 208.
102 Iskra-Janošić 1992, 208.
103 Vikić-Belančić 1971, 162.
104 Iskra-Janošić 1992, 211.
105 Iskra-Janošić 1992, 208-212; Iskra-Janošić 2001a, 66-67.
106 Iskra-Janošić 1994, 58-60; Iskra-Janošić 2001a, 98; Iskra-Janošić 2006, 83-86; Jerončić and Katavić 2010, 65-67.
107 Brunšmid 1902, 137.
108 Vikić-Belančić 1971, 160.
109 Iskra-Janošić 2000, 60-63.
110 Vikić-Belančić 1971, 162.
111 Iskra-Janošić 1992, 208; Iskra-Janošić 2001a, 93.

were 37 x 18 x 8 cm, while those in the form of quarter-circles were actually a half of the former (18.5 x 18 x 8 cm).[112] Both stone and brick walls sat mainly on brick bases, as evidenced at several locations in Vinkovci.[113] Sesquipedal bricks were most often used for paving. Hexagonal (5 x 4.5 cm, 5 cm thick) and rectangular bricks (10.5 x 4 x 3.5 cm) arranged into fishbone pattern (*opus spicatum*) were used in more representative rooms.[114] Octagonal bricks (6.4 x 5.6) were also used.[115] Floors could have been made of beaten gravel coated in mortar.[116] Floors of better quality were composed of three different layers. The lowest layer was made of slanted rubble bricks, occasionally arranged in a fishbone pattern; a 35-cm-thick layer of brick was placed onto beaten loam, which was topped by a 20-cm-thick layer of rubble stone. The finishing layer was made of 4-cm-thick cover of gray-yellowish mortar, mixed with abundant lime and aggregates.[117] A hypocaust system was constructed under the floors of wealthier buildings. Bricks (30 x 30 x 6 cm or 30 x 25 x 6 cm) were placed on the lower floor level, which received pillars made of smaller bricks (ca. 20 x 18-20 x 6 cm), and, finally, a top layer of floor bricks.[118] Remains of similar heating systems have been found at several locations within the town perimeter.[119] Houses were covered in brick roof elements. Two *tegulae* (flat rimmed tiles) created a joint upon which a saddle-form *imbrex* was laid, thus sealing the joint. The dimensions of *tegulae* from Vinkovci were 55 x 41 cm, and that of *imbrices* 45 x 26 cm. Special smaller-size *tegulae* were made for corner segments. Antefixes were placed on the roof crown, sealing the rows of *imbrices*. A single specimen with the depiction of Hadrian's wife Sabina was found in Vinkovci, with analogies in Osijek (Mursa).[120] Subterranean infrastructure of Roman Cibalae was also mostly built of bricks. Brick in the form of circular segments was used in the construction of wells. The circumference of the outer arch of these bricks was 39 cm, and the circumference of the inner arch 30 cm; their width was 17.5 cm and thickness 7.5 cm. Dugout wells were constructed by arranging these bricks in a circular form; two such wells have been found in Cibalae.[121] Remains of a sewage piping built in stone and brick (*opus mixtum*) have also been found, whose bottom was paved in large tiles. Its total width was 2.3 m, while the width of the vaulted section of the canal was 0.65 m and its height 1.1 m; it was in use from the 2nd to the 5th centuries, as confirmed by coin finds.[122] Lesser canals built of both vertical and horizontal tiles with inner flow diameter of 0.5 m emptied into these large collectors. The houses were joined to the drainage system through canals of an even smaller inner diameter, built in bricks of three different dimensions: 27 x 27 x 6, 29 x 43 x 6, and 30 x 44 x 3 cm.[123]

Different materials were combined in street building. Remains of a Roman street built in several layers where found at the findspot of the abovementioned collector. The bottom layer was made of slanted rubble brick and crushed stone. It was superimposed by several layers: 1. gravel and yellow sand; 2. regularly arranged rubble brick; 3. grey beaten sand; 4. lime coating; 5. pavement formed of irregular flat stones as the top layer.[124] Segments of streets made of a single rubble-brick layer, without substratum, have also been discovered. Unassuming architecture mostly dates from the 4th century, when Cibalae was already hard damaged.[125]

Mursa

Building phases of Mursa are not precisely defined, as Roman architectural remains in the territory of modern-day Osijek are relatively poorly preserved. The accelerated development of the town in the 19th century mostly destroyed the surface and shallow underground traces of Roman Mursa. A large quantity of construction material (brick and stone) was during the 19th century built into the Drava embankments, streets of Osijek, and its yards and houses. Numerous remains of Roman architecture were discovered in many places in 19th- and 20th-century buildings; these were either destroyed or covered by later buildings.[126] Nevertheless, building phases as valid for Roman Pannonia in general can be discerned in Mursa as well, although slightly modified. The first phase encompasses chronologically the first two phases of Mócsy's classification, while Mursa's second and third phases correspond to the third and fourth Mócsy's phases.[127]

It has been conjectured that Mursa's predecessor was a military camp from the period of Domitian's and Trajan's Dacian wars.[128] It was probably built of timber and earth, according to the usual standard employed on the Danube *limes* in the 1st century.[129] Solid materials came into use in the time of Hadrian, when Mursa became a colony.[130] Numerous remains of buildings were discovered in small-scale rescue archaeological excavations after the Second World War,[131] but the excavated areas were of small size and dispersed over the entire town area, thus preventing a clearer assessment of Roman architectural wholes. An important advance in the knowledge on materials and techniques employed in the construction of Roman Mursa was brought forth by large scale protective excavations in the 21st century on the sites of Vojarna, Silos ("Sloboda" factory in Huttlerova Street) and Bana Josipa Jelačića

112 Iskra-Janošić 1992, 210.
113 Iskra-Janošić 1994, 58-60; Iskra-Janošić 2001 a, 72, 77, 89, 93.
114 Dimitrijević 1979a, 152; Iskra-Janošić 2001a, 66.
115 Brunšmid 1902, 134.
116 Iskra-Janošić 2001 a, 71-72.
117 Vikić-Belančić 1971, 160-161.
118 Iskra-Janošić 1992, 210.
119 Iskra-Janošić 2001a, 67, 72, 77; Vulić, Rapan Papeša, and Krznarić Škrivanko 2009, 98.
120 Iskra-Janošić 1992, 210-211.
121 Iskra-Janošić 1992, 211; Iskra-Janošić 2001a, 106.
122 Iskra-Janošić 1993, 65-67.
123 Iskra-Janošić 1992, 210.
124 Iskra-Janošić 1993, 65-67.
125 Iskra-Janošić 2001, 31-34; Iskra-Janošić 2002, 75-79.
126 Pinterović 1978, 162.
127 Filipović 2004, 157.
128 Klemenc 1961, 17.
129 Soproni 1980, 219-221.
130 Filipović 2004, 157; Filipović and Podrug 2007, 52.
131 Pinterović 1978, 164-166.

Square on the western, southern and eastern peripheries of Mursa.[132] The walls of the structures were mainly built of sesquipedal bricks, the most available building material in Mursa. Bricks of different dimensions were also used, for instance in the construction of hypocaust pillars (0.2 x 0,2 x 0.09 m and 0.28 x 0.28 x 0.05 m).[133] Walls' foundations made in brick were identical to those in Cibalae, and were constructed in two basic ways: on a footing made of either rubble stone or slanted rubble brick. Numerous structures and isolated walls with foundations of rubble stone and a brick superstructure were found in recent protective excavations.[134] This construction technique proved to be typical of the 2nd century.[135] The Severan phase is represented with the slanted-brick foundations and segments of roof tiling elements (*tegulae* and *imbrices*) made in *opus spicatum*. The last building phase (end of the 3rd to the early 4th century) is represented by foundation walls made of rubble-brick debris and stone reused from earlier buildings phases.[136] Foundation walls and (occasionally) the lowest brick courses are in general all that survives, which is true of all building phases in Mursa. Higher construction standards are visible in the remains of brick walls dating from the late 2nd and first half of the 3rd centuries; these were plastered and decorated with frescoes and stucco.[137] Sometimes this fact is only demonstrated in plaster fragments. Some walls were built in stone. Two massive 1.05-m-wide stone walls and two 0.3-m-wide brick walls were fund during some infrastructural works.[138] Two parallel walls, 2.8 and 3.7 m wide, respectively, discovered by the bank of the Drava, are considered to represent the remains of fortifications.[139] Sporadic finds of large stone blocks suggest the existence of wealthy and well-constructed public or private architecture.[140] Remains of timber structures on the eastern periphery of Mursa do not belong to the earliest building phase but to the mid-2nd century, when brick was used for construction in the town centre. These structures were residential and industrial, as transpires from clay pits and pottery kilns.[141] A number of pillar-base foundations, built in similar way to wall foundations, were discovered at the sites of Vojarna and Silos. The bases are mostly square in layout. Their footing was made of rubble stone and/or brick bonded by lime mortar, and the superstructure was made of regularly arranged tiles or well-dressed stone.[142] Upon these bases wooden or brick pillars were usually posted, but also more massive stone pillars. Massive bases and capitals of stone pillars, together with segments of an architrave, were found at the Vojarna site in a filled drainage canal under the Roman road.[143] Remains of structures with quality floors made of powdered brick and lime have been found at a number of sites within the Mursa perimeter. Their surface was smooth as an enamel and pale pink in hue (*terrazzo*).[144] Coarser floors, made of rubble stone and brick bonded by mortar, were discovered at several positions at the site of Silos.[145] A late Roman mosaic with floral motif stems from the same site.[146] Remains of mosaic floors were discovered in several other protective excavations in the area of Mursa.[147]Some of the buildings were equipped with floor heating (hypocaust); remains of such systems were found already in the 19th century when a modern road was being paved. The walking level of the floor, made of large tiles (mosaic cubes were also found), was placed upon a hypocaust *suspensura*, while frescoes were found on the walls, surviving up to 0.5 m high.[148] A hypocaust was also discovered at the Silos site; the bases of its *suspensura* were made of bricks measuring 0.20 x 0.29 m x 0.09 m, with terminal bricks of the same size. Two bases were made of stone and carried a stone pillar upon them, reaching a total height of 0.5 m.[149] A floor heating system was discovered at the Bana Jelačića Square site, consisting of canals made of brick (inner height 0.3 m), through which hot air was circulating. On the lateral walls of these canals, a 20- cm-thick floor was made of several layers of construction debris coated in mortar.[150] Roofs were covered in standard brick elements, *tegulae* and *imbrices*, as evidenced in many archaeological excavations. Decorative ceramic antefixes have also been found; twenty such pieces are kept in the MSO, mostly featuring a female head. The hairstyle on some of these heads is reminiscent of Hadrian's wife Sabina's portraits. All of them were manufactured in local workshops.[151]

Subterranean town infrastructure (water-supply and drainage) has been discovered at several sites within the town perimeter. Most frequent finds are smaller conduits empting waste waters from buildings to larger street canals.[152] They are square or rectangular in cross section and were made of horizontally and vertically arranged sesquipedal or other types of larger bricks bonded by water-resistant mortar. Lateral walls of smaller house canals could have been built of several courses of brick.[153] The main drainage conduits were built under the roads. A specific solution regarding roads and drainage canals was applied in Mursa, as attested in all three large-scale recent protective excavations (Vojarna,[154] Silos,[155] Bana Jelačića

[132] Filipović and Katavić 2005; Filipović and Katavić 2006a; Filipović and Katavić 2006b; Sunko and Katavić 2007; Filipović 2009; Jerončić and Katavić 2009; Leleković 2009.
[133] Pinterović 1978, 157; Sunko and Katavić 2007, 86, 88.
[134] Filipović and Katavić 2005, 111; Filipović and Katavić 2006b, 74, 77; Sunko and Katavić 2007, 82; Sunko, Katavić, and Jerončić 2008, 23-24.
[135] Jerončić and Katavić 2009, 57-58.
[136] Jerončić and Katavić 2009, 58.
[137] Leleković 2009, 47.
[138] Pinterović 1978, 164.
[139] Pinterović 1978, 166.
[140] Pinterović 1978, 157.
[141] Leleković 2009, 46-47.
[142] Filipović and Katavić 2005, 108, 111; Filipović and Katavić 2006a, 11; Filipović and Katavić 2006b, 78; Jerončić and Katavić 2009, 58.
[143] Filipović and Katavić 2005, 107-108; Filipović 2008, 29, 31.
[144] Pinterović 1978, 164.
[145] Sunko and Katavić 2007, 86.
[146] Jerončić and Katavić 2009, 58.
[147] Šimić 2008, 27-28.
[148] Pinterović 1978, 163-164.
[149] Sunko and Katavić 2007, 86.
[150] Leleković 2009, 47.
[151] Pinterović 1978, 158.
[152] Pinterović 1978, 165-166; Filipović and Katavić 2006, 12; Filipović and Katavić 2006b, 77; Jerončić and Katavić 2009, 57.
[153] Katavić and Sunko 2007, 84.
[154] Filipović and Katavić 2005b, 107-108; Filipović 2009, 51.
[155] Sunko and Katavić 2007, 86-88; Jerončić and Katavić 2009, 57.

Square).[156] The streets ranged in width from 8 to 10 m and were composed of a central causeway supported on wooden pillars and beams and canals placed on one or both edges, measuring ca. 2.5 m at the top and ca. 1.3 m deep. Wooden pillars carrying a timber construction on top were planted into the canals and alongside them; the embankments were finished by a surface of rubble brick. Under the street surface on the Silos site only one such canal was found. This kind of street was being constructed during the 2nd century, presumably in Hadrian's reign. All the excavated streets and canals pertaining to the later phases (2nd-4th centuries) were filled with construction debris.[157] The top layer of the street at the Vojarna site was made of rubble brick bonded with lime mortar.[158] A Roman road that led along the Danubian *limes* crossed the Drava at Osijek; remains of bridge stone pillars could be seen during the low water level. They presented an obstacle for the river traffic during the low water and therefore were demolished on several occasions in past times, while some fragments were taken to the MSO.[159] The survey conducted in 2008 once again confirmed the existence of the remains of pillars made of dressed and rubble stone, supported by timber piles.[160]

Andautonia

The remains of the Roman settlement are situated under the modern-day village of Ščitarjevo, which offers much more favourable conditions for research in comparison with Roman towns overlaid by modern cities. Given this, it was possible to conduct both trial trenches and systematic excavations, as well as a presentation of the architecture *in situ*.[161] Three basic construction phases are recognisable in archaeological remains in Andautonia, partly complementary with the four Mócsy's Pannonian phases of urbanisation.[162] The first phase started with the founding of the town in the first decades of the 1st century, probably soon after the Pannonian-Delmataean uprising, and lasted until the end of the first quarter of the 2nd century. Similar as in Siscia, the urbanisation of Andautonia began earlier in comparison with most other Pannonian towns, resulting in a longer first building phase. One can presume that even before this period, during the conquest of Pannonia, some military units were stationed there. Besides a favourable strategic position of Andautonia, this is also suggested by the finds of legionary *denarii* from the 4th decade BC, issued immediately after the conquest of Segestica, as well as by the finds of Augustan and Tiberian coinage. Traces of building works belonging to the first phase have been discovered on the Gradišće site, in the northern section of the Roman town. At that time Andautonia was granted the status of a *municipium* and became an important regional centre. The second phase corresponds to the abovementioned Mócsy's classification. The economic and architectural progress of entire Pannonia in Hadrian's reign was also felt in Andautonia. The town was enlarged and developed into an irregular oblong, ca. 1000 m long in N-S direction and ca. 300-400 m wide in E-W direction.[163] In its spreading towards the south and east, the town followed the natural elevation, thus assuming an elongated form. The third urban phase lasted from the later 3rd century until the abandonment of the town in the late 4th century; it encompassed the third and fourth postulated phases of urbanisation in Pannonia.

No architectural traces survive from the first half of the 1st century, yet sporadic small finds indicate their existence; the architecture of the first phase was presumably built in timber.[164] Timber constructions can in Andautonia only be glimpsed at in the remains of burnt layers discovered in archaeological excavations at several sites.[165] From the Flavian period onwards, solid materials, especially stone, were predominantly used in architecture.[166] The roofs were covered in standard brick elements – *tegulae* and *imbrices* – which were often found in building remains. An increase in building works is evident in the first half of the 2nd century. Remains of monumental stone structures discovered on the site of Gradišće belong to this period. The high standard of urban living and architecture is clearly seen in hypocaust remains, stucco decorations and wall paintings found in these structures. The main building material used was rubble stone, but occasionally partially dressed stone, and the vastly predominant building technique was *opus incertum*.[167] Earthen walls inside a stone structure also characterise this period; they were made of beaten clay in a timber frame and were additionally plastered and frescoed.[168] Stone wall-foundation trenches were dug into the loam soil and then partially filled with gravel (0.5-1 m), to receive a rubble stone and mortar footing.[169] The gravel underlay was meant to prevent the subterranean waters to penetrate into the walls and thus compromise their stability. A large-size walled tomb made of rubble stone discovered on the southern periphery of the city was constructed in the same fashion.[170] The corners of some structures were made of well-hewn stone blocks, which significantly increased their solidity and stability.[171] The techniques established during the first phase remained in use in the second. Rubble stone, used for the construction of walls in the *opus spicatum* technique, remained the most important building material; the foundations of these walls were also made in the same fashion as in the first phase. The largest number of structures pertaining to this phase was discovered in the parish yard, where the largest scale archaeological excavations in Andautonia were conducted and where

156 Leleković 2009, 46.
157 Sections of stone architraves and pillars (bases, capitals) were found in the canal fill on the Vojarna site.
158 Filipović and Katavić 2005, 107-108.
159 Pinterović 1978, 159; Bulat 1989, 22.
160 Zubčić 2009, 44-45.
161 Vikić-Belančić 1981; Vikić-Belančić and Gorenc 1984; Nemeth-Ehrlich and Vojvoda 1994; Nemeth-Ehrlich and Kušan-Špalj 2003a.
162 Vikić-Belančić 1981, 134; Nemeth-Ehrlich and Kušan-Špalj 2003a, 114-125.
163 Nemeth-Ehrlich and Kušan-Špalj 2003a, 114.
164 Nemeth-Ehrlich and Kušan-Špalj 2003a, 117, 123.
165 Vikić-Belančić and Gorenc 1984, trenches 2, 6, and 29.
166 Nemeth-Ehrlich and Kušan-Špalj 2003a, 123.
167 Nemeth-Ehrlich and Kušan-Špalj 2003a, 118-120.
168 Pintarić 2006, 6-7.
169 Vikić-Belančić 1981, 143.
170 Nemeth-Ehrlich and Kušan-Špalj 2003a, 116.
171 Vikić-Belančić 1981, 135; Nemeth-Ehrlich and Vojvoda 1994, 45.

an archaeological park was subsequently established. This architectural complex was erected during the town's expansion towards the south, covering the south cemetery from the first phase. A building recovered in the eastern section of the site was probably a thermal complex.[172] Remains of marble veneering, stucco decorations and wall paintings depict high living and architectural standards.[173] The town was now encompassed with a fortification system consisting of double walls ca. 1 m thick, built in larger-size and better-dressed rubble stone. Remains of the 2nd-century fortification system from the second building phase have so far been discovered only on the northern edge of the town.[174]

The third building phase represents a decrease in quality in comparison with both preceding phases. The walls were still made of stone, but they were of poorer quality. The foundations were laid more shallowly, on the remains of earlier demolished structures and a thick layer of gravel, representing, according to some hypotheses, the remains of a major flood.[175] It can, however, be interpreted as a heaped up material used for levelling of the terrain. The third-phase fortifications were discovered on the eastern side of the Gradišće site. They were built of large irregular stone blocks bonded by mortar with large-size gravel aggregate, with clay and timber once again utilised as building materials.[176] Foundation trenches with postholes were discovered in protective excavations, featuring burnt earthen walls with daub and wattle construction.[177] Foundation trenches for earthen walls were dug into a solid sterile layer of loam, used for levelling the preceding building layer. Pillar bases were constructed in a similar way, that is, of rubble stone and brick fragments coated in mortar and placed on a foundation made of gravel; this kind of footings was used in the construction of portico pillars lining the streets.[178] Fragments of stone pillars and capitals were discovered in trial trenches at several locations.[179] Floors were built in several techniques, with beaten clay often used as a solid base for floors of superior quality, due to its solidity and water-resistance.[180] Upper-floor layers could have been made of large-size gravel (pebbles) or rubble stone coated in mortar.[181] Pebbles coated in mortar could have served as the walking surface, but could have also been used as a substratum for other types of surfaces. Remains of stone-paved floors were found in several places. Stone slabs and irregular rubble stone, often in the form of plates, were used for paving.[182] Brick paving (most often with sesquipedal bricks) was also frequently employed.[183] Remains of hypocaust were also discovered in these structures. A partially reconstructed complete hypocaust section was recovered in the presumed public baths complex, featuring a brick *suspensura*, a large-size brick pavement, as well as hollow bricks for wall heating and hot air ventilation – *tubuli*.[184] Between two sections of the thermal complex, a drainage canal closed off by stone walls was discovered, as well as a rubble stone vault. Segments of two large sewage brick collectors from the first half of the 1st century AD were discovered at the site of Gradišće. One of them was completely preserved, featuring sesquipedal-brick walls and a vaulted roofing, as well as a tile-paved bottom. Twenty-five meters of the canal were excavated, revealing several preserved lateral rectangular openings for the attachment of smaller canals and the join of another large canal measuring 1.2 by 1 m.[185]

Streets were constructed in several different ways. Two bottom layers of lesser streets are identical to each other: the foundation layer consisted of a clay base mixed with gravel, while the upper layer was surfaced with small-sized rubble stone and pebbles. The finishing layer could have been made of large-sized gravel coated in mortar or it could have been stone-paved.[186] A section of the main street was excavated in the parish yard next to the supposed baths complex. Three of its layers survive, corresponding to three building phases of Andautonia. During the first phase (1st-2nd centuries) it served as a rural road surfaced with river gravel. During the second phase (2nd-3rd centuries) it was encompassed within the town, becoming the main street; yet it was still only gravelled. In the third phase (3rd-4th centuries) 30-cm-thick flat stone plates together with edge-stone blocks were placed upon the beaten gravel base.[187] The edge-stones and pavement with traces of cart ruts partially survive and can bee seen in the archaeological park.[188]

Other Pannonian urban centres

Iovia

Archaeological excavations so far conducted at the site of Roman Iovia in Ludbreg do not offer a complete picture of its building phases. The earliest Roman finds (1st century AD) suggest intensive building in the period from the late 1st to the first half of the 2nd centuries, when the town got the perimeter walls. According to the archaeological evidence, the continuity of building was interrupted in the second

[172] Vikić-Belančić 1981, 135, 139-140; Nemeth-Ehrlich and Kušan-Špalj 2003a, 116.
[173] *Zagreb prije Zagreba*, 125; Nemeth-Ehrlich and Kušan-Špalj 2003a, 119-120.
[174] Vikić-Belančić 1981, 141.
[175] Vikić-Belančić 1981, 141; Nemeth-Ehrlich and Kušan-Špalj 2003a, 121-122.
[176] Vikić-Belančić 1981, 143.
[177] The original field documentation from the 2003 research conducted by the Turopolje Museum.
[178] Nemeth-Ehrlich and Vojvoda 1994, 42; Pintarić 2006, 9.
[179] Vikić-Belančić and Gorenc 1984, trenches 10, 35, 36 and 37.
[180] Vikić-Belančić 1981, 141.
[181] Pintarić 2006, 9.
[182] Vikić-Belančić 1981, 143; Vikić-Belančić and Gorenc 1984, trenches 10, 30, 31, 32, 38, 40 and 41.
[183] Vikić-Belančić 1981, 135; Nemeth-Ehrlich and Kušan-Špalj 2003a, 115-116; Pintarić 2006, 9-10.
[184] Vikić-Belančić 1981, 135; Nemeth-Ehrlich and Vojvoda 1994, 42-45.
[185] Vikić-Belančić 1981, 135; Nemeth-Ehrlich and Kušan-Špalj 2003a, 118.
[186] Vikić-Belančić 1981, 141; Vikić-Belančić and Gorenc 1984, trenches 19 and 28.
[187] Nemeth-Ehrlich and Kušan-Špalj 2003a, 126.
[188] Nemeth-Ehrlich and Vojvoda 1994, 41-45.

half of the 2nd century, probably due to the Marcomannic Wars. During the middle and late Empire the town went through a phase of intensive rebuilding as well as building anew, suggested by the remains of monumental structures. The continuity of living in Iovia can be traced into the 5th century[189]

Almost all remains of Roman architecture in Ludbreg were made of stone. Residential (mostly late Roman) structures were built of inferiorly dressed rubble stone and large-size river pebbles, while wall fillings were made of smaller-size rubble stone (*opus incertum*). The walls of more massive and monumental buildings, probably public, were built of dressed or half-dressed stone, rubble stone and smaller-sized blocks arranged in relatively regular courses (*opus incertum*, partly *opus vittatum* – use of smaller-sized blocks). The filling between two better-executed wall façades consisted of smaller-sized rubble stone coated in abundant mortar.[190] Impressive remains of monumental Roman architecture at the Somođi garden site belong to a thermal complex,[191] as well as to a structure that probably served as a granary – *horreum*.[192] The fortifications of Iovia were also built in stone, using large and well-dressed pieces of rubble, as well as coarse flat blocks and (probably reused) fine-dressed stone blocks, with mortar containing larger-sized aggregate used for bonding. Since relatively small segments of fortifications were excavated, it is not possible to give a more precise description of its complete layout, constructed from the late 1st century to the late Imperial period. Therefore, it is not completely clear whether northern and western fortifications were constructed as double-wall system or whether the town perimeter was enlarged.[193] Brick was rarely used, examples being part of the pool and one apse of the thermal complex. In general, it was used for patching up and levelling the walls, as, for example, on a section of the western wall and in the construction of a late Roman residential building.[194] Roof brick elements (*tegulae*, *imbrices*) were found in several trenches, indicating a standard way of Roman-fashion roofing.[195] The bonding mortar was in early Imperial walls white-greyish in colour and filled with small-sized sand aggregate; in Late Antiquity it was yellowish-white, containing larger-sized sand and gravel aggregate, and was of lesser quality than before. Façade mortar was preserved on relatively small number of walls, usually containing two or three layers, each ca. 1 cm thick; it is yellowish-brown in colour and polished on the surface. Painted walls and marble wall veneering in some buildings testify to high living standards.[196]

Different types of Roman floors have been discovered in archaeological excavations, with beaten rubble stone, flat stone plates, and clay, most often used as an underlay for the upper layers. River pebbles and brick fragments were often fitted into the bottom layers. Upper-floor layers could have been made of different types of lime mortar with larger or smaller-sized aggregate or water-resistant mortar with brick fragments (for instance, in baths). Floors paved with brick and flat stone plates have also been found at several positions, as well as remains of floor heating system (hypocaust). The hypocaust remains in the baths complex is the best-studied example.[197] Under the floor level in some structures, obviously those more luxuriously built and equipped, remains of the sewage system were also found.[198] The excavated areas were small and failed to disclose remains of streets. Nevertheless, pavements made of pebbles, rubble stone and brick fragments, found between the outer and inner course of the northern wall, testify to the existence of public communications.[199] The area between the double walls probably had a defensive function and was arranged in a way to be easily accessible.

Aquae Balissae – Municipium Iasorum

Numerous Roman-period finds from the area of modern-day Daruvar speak in favour of a significant settlement. On the other hand, due to a nearly complete lack of archaeological excavations, the town topography can only be glimpsed at through scrappy surface indicators and chance finds of walls, mostly built of rubble stone. Chance finds of stone blocks further suggest that stone was probably the most important building material in this Roman *municipium*.[200] The hypothesis is further supported by numerous finds of stone monuments, pointing towards stone deposits near the town, from which construction stone could have been sourced.

Aquae Iasae

The remains of the Roman settlement of Aquae Iasae were discovered in Varaždinske Toplice around the thermal springs still in use. The settlement was named after the Iasi, a local prehistoric ethnic group, that probably also utilised the thermal-spring water. The Roman settlement existed from the 1st to the late 4th centuries AD, developing on terraced hill slopes. The central part of the settlement with the religious and baths complex was sited on the highest terrace around the thermal spring, while residential and industrial sections of the settlement developed on the lower terraces and foothill.[201] The settlement was ravaged in the Marcomannic Wars in the second half of the 2nd century and in Gothic incursions in the second half of the 3rd century, to be finally destroyed in the civil war between Theodosius I and his counter-emperor Maximus.[202] An inscription found in Varaždinske Toplice testifies to a rebuilding of the settlement in Constantine's reign. According to the

189 Vikić-Belančić 1984, 149-152, 159-162; Gregl and Migotti 2004, 139.
190 Vikić-Belančić 1984, 151-152.
191 Vikić-Belančić 1984, 132-135; Pleše 2009, 213-215.
192 Vikić-Belančić 1984, 122-126; Gregl and Migotti 2004, 138.
193 Vikić-Belančić 1984, 154-156; Gregl and Migotti 2004, 136-137, 139.
194 Vikić-Belančić 1984, 151.
195 Vikić-Belančić 1984, 122, 132, 137-139, 143.
196 Vikić-Belančić 1984, 151-152.
197 Vikić-Belančić 1984, 122, 132-133, 145, 151; Pleše 2009, 214.
198 Vikić-Belančić 1984, 151.
199 Vikić-Belančić 1984, 152-154.
200 Schejbal 2004, 115-116.
201 Gorenc and Vikić 1980, 3-4; Nemeth-Ehrlich 1993, 23.
202 Gorenc and Vikić 1980, 15, 20-22.

inscription, the emperor financed the rebuilding of the forum porticoes damaged in a fire. The inscription also indicates that the fair in Aquae Iasae was held regularly on Sundays.[203]

The central part of the settlement with its cultic, religious and thermal facilities is being systematically excavated from 1953 onwards, at the site of the city park (area of ca. 6000 m^2). The archaeological evidence testifies that the complex was constructed in three main phases: early Imperial (the 1st and the first half of the 2nd centuries), middle Imperial (the second half of the 2nd and the 3rd centuries) and late Imperial (4th century). The early Imperial period encompasses the first two Mócsy's phases. Two later building phases in Aquae Iasae correspond to Mócsy's third and fourth Pannonian building periods. The southern section of the central complex encompasses the baths with the thermal basilica that was from the start incorporated into its southern part. Rubble stone was the main construction material used for the building of the walls and foundations in all three phases, and the most frequent building technique utilised was *opus incertum*. Building in large well-dressed stone blocks has also been documented, and some of the walls are relatively well preserved, surviving to the height of 2 m. Bricks began to be used more intensively in the second phase, especially in the construction of the hypocaust heating system (vaulted *praefurnium*, *suspensurae*, *tubuli*), and also for the paving of floors. Water supply and drainage canals inside the baths were also built of bricks, arranged horizontally and vertically. The rooms were plastered, and water-resistant mortar was used in the pools. During Constantine's reign, in the third building phase, *opus mixtum* (stone and brick) was also utilised, which is well exemplified in the renovated basilica, now the most important building in the whole bathing complex. After Constantine, the basilica was separated off from the bathing complex and was converted into a Christian church, evidenced by the traces of wall paintings with Christian motifs. Wall paintings of superior quality were also discovered in several rooms in the baths complex.[204]

The temples of the Capitoline triad (Jupiter, Juno and Minerva) were discovered on the northern side of the excavated central complex. Three building phases were recognised in their construction, dating from the early 1st century AD to the reign of Constantine. The temples were erected, enlarged and rebuilt in rubble stone (*opus incertum*), just like the structures belonging to the baths complex; they survive to the height of 3 m. The interior of the temples was plastered and frescoed. Between the temple and bathing complex the forum was located (22 x 24 m), delimited with a large staircase leading towards the Capitolium and porticoes; remains of a stone wall – stylobate, stone pillars and stairs leading to the forum still survive. In Constantine's reign a porch in front of the Capitolium was erected on the northern side, functioning as the façade of the temple complex; a stone-paved forum was built in the same period. In 2006 a catchment pool of the thermal spring (ca. 13.5 x 9 m) was discovered there, as well ad drainage and overflow canals. The pool drainage canal was discovered in the south-eastern corner immediately below the forum pavement level. It was built of stone blocks with slots and covered with stone plates.[205]

Residential and industrial sections of the Roman settlements spread out on the terraces on the slopes of the hill. Because of the intensive later building, Roman architecture was only preserved and studied in fragments, which does not allow a reconstruction of the outlay of either the structures or the complete sections of the settlement.[206] Remains of several buildings situated in the mixed industrial-residential section of the settlement were excavated at the foothill. Two building phases could be discerned: the early, characterised by timber (1st and 2nd centuries), and the late Imperial, using stone (end of the 3rd and 4th centuries). In the early Imperial phase two sub-phases of timber construction are discernible: an earlier (end of the 1st century) and a later (2nd century). Timber architecture was proved by the remains of vertical and horizontal elements, oaken beams and boards, with an oaken dugout as one of the horizontal elements. The longevity of oak and the marshy soil were the key factors in preservation of these rare building remains. The building fashion of this structure reflects a strong local Iron-age tradition (basic frame, joining of edges and corners with cording and slotting), combined with Roman carpentry techniques. The floors were made of beaten clay, and the roofs probably of shake or straw. Hazel-wood cylinders found preserved inside the ruins could have carried such constructions. Not a single brick fragment was found in this timber construction phase; therefore, it is certain that roofs were not covered with *tegulae* and *imbrices*. Timber remains were covered by a thick burnt layer, which, together with numismatic finds, suggests the destruction of these structures in the Marcomannic Wars. The preserved foundations of timber structures were overlaid by a stone construction from the 3rd and 4th centuries, whose walls were built of coarse rubble stone in the *opus incertum* technique. The new buildings were roofed by standard Roman brick elements, recovered together with the remains of burnt roofing construction. The stone architecture was also damaged in wars, probably in the civil war between Theodosius and the usurper Maximus that was completed in 388 in Poetovio, after some heavy fighting taking place in this area.[207]

3. The origin, supplying and transportation of building materials

A review of building remains as described in the literature has shown that the most frequent and, at the same time, the most frequently documented building technique on southern

[203] *CIL* III 4121; *AIJ*, no. 469; Vikić and Gorenc 1966, 15.
[204] Vikić-Belančić and Gorenc 1958; Vikić-Belančić and Gorenc 1961; Vikić-Belančić and Gorenc 1971; Gorenc and Vikić 1980, 4-9; Nemeth-Ehrlich 1993; Nemeth-Ehrlich and Kušan Špalj 2003b; Nemeth-Ehrlich and Kušan Špalj 2006.
[205] Nemeth-Ehrlich and Kušan Špalj 2007, 150-153.
[206] Gorenc and Vikić 1980, 18.
[207] Vikić-Belančić 1973; Gorenc and Vikić 1980, 19-22.

Pannonian sites was building in rubble stone and brick with the use of mortar, most frequently attributed by the commentators to the reign of the emperor Hadrian and the Severan rulers. This is especially true of luxurious buildings made of stone blocks and architectural elements made of more expensive types of stone. The commentators suggest that the preceding period was characterised by building in timber, while the subsequent period was marked by a certain stagnation and decline, when the buildings were constructed by recycling the remains of earlier structures. Only in Siscia and Aquae Iasae the Constantinian building phase is well documented. All types of building materials utilised in the erection of southern Pannonian settlements were supplied and transported to the building sites in conform to the natural resources and transportation facilities of the region.

Clay, gravel, and sand

Clay, gravel and sand were mostly used, separately or in mixtures, for surfacing and levelling works, for beating of floors and roads, and as a bonding material (clay independently, sand and gravel as a component of mortar, along with lime and water). Clay was used as a building material already in prehistory, and it continued to be used in Roman times. Buildings' walls could have been made of beaten clay and in combination with timber frame and wattle. Straw, chaff, sand and ashes were added to clay in order to increase its solidity and compactness. This building technique was used in some areas even before the Roman occupation, and it is sporadically used even today.[208] These natural materials were relatively frequently found in the environment and, therefore, easily accessible; as a result, they were sourced not far from building sites. Due to the geological composition of southern Pannonia, clay deposits are often found along the river courses, at lower terraces and slopes of higher elevations. Gravel and sand can most often be found in riverbeds and in Holocene and Pleistocene deposits in river basins and valleys.[209] Therefore, these materials were relatively easily accessible to Roman builders in south-Pannonian towns. Due to a relative proximity of deposits of gravel, sand, and clay to building sites, their transport did not cause major logistic problems to Roman engineers.

Timber

Wood employed as a construction material in south-Pannonian Roman towns was sourced in the vicinity of these towns, since their surroundings were mostly wooded areas. Pendunculate oak-wood from the woods in marshy river plains was of special importance. Among others, Pliny described the exuberant Pannonian forests (Plin. *NH* 3.147). Beech, ash, willow and alder were also used in construction works in this area. The information on the use of each of these species of wood, as well as their purpose, was given by Vitruvius (Vitruv. 2.9-10). Due to the close proximity of rich wooded areas, the transport of building material was rather easy. Manpower or draught animals (mostly oxen) were employed for logging. Timber was transported on carts by paths and roads all the way to sawmills and carpentries. Rivers were used for long-range transport: logs were tied together into rafts and sent floating downstream or they were hauled upstream from the banks by draught animals. This kind of transport was used in subsequent periods wherever it could be employed until the appearance of railways.[210]

Stone

The stone required in the building of Roman towns and smaller settlements in Pannonia was supplied from quarries located on the nearby mountains. Traces of Roman quarrying have been discovered at several locations in southern Pannonia and bordering areas of neighbouring provinces. As the production of several Roman quarries outside Pannonia was oriented towards this province, they need to be included in the present discussion.

The main stone deposits exploited for the building of Andautonia were on Mount Medvednica surmounting Zagreb on the north, with Sljeme as its highest peak.[211] The positions of three Roman quarries have been were positively determined. The westernmost was located in the village of Bizek and the second in the village of Gornje Vrapče (Vrapče potok), where a segment of an unfinished pillar and several items of Roman mining tools were found.[212] The bulk of rock mass in both quarries is formed of bio-accumulated lithothamnium limestone, better known as *litavac*. Lime sandstone was also quarried at both locations.[213] These are the only types of stone in the wider surroundings of Andautonia suitable for extraction in large blocks, causing Medvednica stone to be used even outside Andautonian territory.[214] Large-sized edge-stones of the Roman road next to the Andautonia public baths were made of porous Bizek *litavac*.[215] The third Roman quarry on Medvednica is Crna Voda near Markuševac.[216] Two rock types predominate in this quarry: marbled limestone and marbled Palaeozoic slate, the latter being characterised by dark-grey and greenish colour (green slate).[217] The stone from the sub-Sljeme quarries was either conveyed or drawn on trodden paths and, wherever it was possible, river transport by the Sava was used. The Bizek and Vrapče quarries, situated reasonably close to the river, used this special advantage. Blocks of Medvednica *litavac* were probably transported further downstream from Andautonia, at least as far as Siscia,[218] and upstream to Neviodunum.

[208] Adam 2001, 58-61.

[209] Šimunić and Basch 1975, 155-158; Fürst-Bjeliš 1996, 36-38.

[210] This kind of timber transportation is even today employed in some countries (Adam 2001, 90).

[211] Vikić-Belančić 1981, 143; Gregl 1994, 36.

[212] Degmedžić 1957, 16, 21-22.

[213] Šikić 1995, 61-66; Crnković 1998, 32.

[214] Vuković 1994, 155; Burkowsky 1999, 65.

[215] The Bizek and Vrapče stone was used as a construction material in subsequent periods in the territory of Zagreb, for instance in the construction of the Zagreb cathedral and the Upper Town fortifications (Šikić 1995, 44-45; Crnković 1997, 11; Crnković 1998; 32).

[216] Degmedžić 1957, 14.

[217] Šikić 1995, 128-129.

[218] Burkowsky 1999, 65.

The transportation of building material by the Sava is confirmed by Roman bricks with Siscian stamps found in Neviodunum.[219] Archaeological finds from the mining and metallurgic region along the Sana and Japra Rivers (Bosnia and Herzegovina) testify to the use of lesser streams in transportation of heavier cargoes. Iron ingots (semi-finished products in the form suitable for transportation) and ore went downstream the Japra and Sana to the Una and Sava, finally reaching Sisak.[220] Together with other products, bricks bearing the Siscian stamp, such as those found in the Japra valley, were transported upstream by the same inland waterway.[221] Siscian bricks found in Neviodunum and the Japra valley testify to the large-distance upstream river transport of building materials.

On the eastern slopes of Ravna Gora west of Varaždin, near the villages of Vinica and Marčan, soft and porous limestone (*vinicit*) was quarried.[222] The Vinica quarries were ca. 5 km removed from the Drava and the stone could have easily been transported by that waterway. West of Ravna Gora stretches a hilly and mountainous landscape of Haloze (Slovenia), where sandstone used for the building of Poetovio was quarried in Roman times. White marble was quarried on Pohorje Mountain, west of Poetovio. Quarries are further documented on south and south-eastern slopes of the mountain in the vicinity of Šmartno.[223] The Pohorje marble quarries, together with the Norican quarries in the vicinity of Roman towns of Teurnia and Virnum (both in Austria), were, because of the rarity and superior quality of marble, also important for building in other provinces of the Empire. Since there are no marble deposits in Pannonia, Norican marble was imported into Roman towns in vast quantities, arriving by the Drava and Mura waterways, both flowing in the relative vicinity of the quarries. Further distribution of stone blocks and semi-finished monuments to the workshops in Pannonian towns was carried by the Danube and Sava waterways.[224]

Building stone could have been exploited at several locations in the vicinity of Varaždinske Toplice. At the site of Pisana pečina on the slopes of Mount Ivanščica, west of Novi Marof, several ancient quarries with traces of exploitation of soft and porous limestone have been identified. Similar limestone was extracted at the site of Ruopanec on a hill west of Varaždinske Toplice.[225] Both quarries could have been exploited in the Roman period and served as a stone-source for Aquae Iasae. On the northern slopes of Mount Kalnik, several kilometres to the south of Varaždinske Toplice, quarries of crushed stone in the vicinity of Ljubešćica and Ljubelj Kalnički are still active. Because of their proximity to Aquae Iasae, and their favourable position, rubble stone was perhaps quarried there in Roman times. The analysis of rubble stone from architectural remains of Roman Iovia (Ludbreg) has confirmed that local yellowish-white lithothamnium limestone (*litavac*) was extracted from the nearby quarry of Kamenik (Kamenjak) on the slopes of Kalnik. A very similar rock composition is found at other locations in the vicinity of Ludbreg, near the villages of Slanje and Leskovski Drenovac.[226] Because of their relative proximity, it is possible that both Roman settlements, Varaždinske Toplice and Ludbreg, sourced from the same quarries. Besides the land routes, the Bednja River could have been utilised for the transport of stone, since it runs through the valley from Varaždinske Toplice towards Ludbreg.

Geologists believe that Moslavačka gora (Moslavačka Mountain) could have been the source for stone blocks used in the construction of massive architectural elements and monuments in northern Croatia. The mediaeval Garić-grad Castle on Moslavačka gora was thus built of irregular granite, amphibolite and gneiss ashlars and larger blocks of lithothamnium limestone.[227] Although there is at the moment no real archaeological confirmation, several abandoned quarries, such as Pleterac, Mikleuška, Gornja Jelenska and Samarica, support the hypothesised stone quarrying at these sites in the Roman period. Granite blocks were quarried at Gornja Jelenska and Samarica.[228] The stone quarried on Moslavačka gora was probably transported to Siscia.[229] It is also probable that Roman quarries existed in the vicinity of Aquae Balissae. Rubble construction stone was quarried there, possibly also larger limestone and sandstone blocks, that is, the material for the production of various monuments and sarcophagi.[230] The quarries could have been located in the peripheral parts of Ravna gora (Ravna Mountain) near the small town of Sirač, and on Mountain Papuk.[231] The analysis of the materials used for the production of sarcophagi in the MSO has shown that limestone they were made from probably originated in the quarries at Sirač or Lipovljani near Novska, but possibly also from quarries situated in the wider Papuk Mountain area.[232] The analysis of pieces of pale yellowish stone veneering with grey granulates found in Cibalae has shown that they were made of gneiss-granite originating in the Slavonian Mountains. The same analysis has determined that whitish limestone with reddish veins and greyish marbled limestone, both used for veneering, originated somewhere in northern Croatia, perhaps also in the Slavonian Mountains (Papuk or Ravna gora).[233]

Quarries exploited by Roman settlement on the foothills

[219] Petru 1990, 90; Durman 1992, 127.

[220] 97 iron ingots of the average weight of 4.4 kg were found in the Una downstream of Dubica. This find supports the existence of a Roman waterway on this course (Durman 1992, 127; Durman 2002, 28-29).

[221] Basler 1977, tab. XVII/3; Durman 1992, 127.

[222] It was primarily used in the construction of Varaždin's Baroque buildings. It is possible that the stone was exploited in the Roman and the mediaeval period (Crnković 1993, 146; Crnković 1997, 11; Crnković 1998, 32).

[223] Djurić 1997, 76; Djurić 2004, 152-153, 157-158.

[224] Djurić 1997, 79; Djurić 2004, 162-165.

[225] Crnković and Jovičić 1993, 146.

[226] Vikić-Belančić 1984, 151.

[227] Crnković 1997, 11-12.

[228] Crnković and Jovičić 1993, 147.

[229] Burkowsky 1999, 65.

[230] Schejbal 2004, 119.

[231] Crnković and Jovičić 1993, 147; Jovičić, Oreški, and Kraljeta 1992, 127-138.

[232] Pinterović 1978, 154-155.

[233] Vikić-Belančić 1971, 162.

of Fruška gora (Fruška Mountain), above all Sirmium and Bassianae, but also other settlements from this section of the Danubian *limes*, certainly existed on this mountain. Sandstone used in the production of monuments was exploited from several sites.[234] Fruška gora stone was also used as a building material. The legend of five (or four) Roman stonemasons (*Passio sanctorum IV coronatorum*) working in imperial porphyry (granite) mines and their martyrdom in Diocletian's reign was often associated with the Fruška gora quarries.[235] Several presumed porphyry quarries on Fruška gora are mentioned in modern studies: the vicinity of the village of Carević, the neighbourhoods of Beočin and the locations around Crveni Čot mountain peak (Kipovno, Glavica and Gradac).[236] It seems that Roman quarries did exist on the sites of Glavica and Kipovo,[237] but porphyry was certainly not sourced from them. The geological and tectonic structure of Fruška gora excludes superior quality types of stone such as porphyry.[238] The complete lack of porphyry finds in the area of Fruška gora, Sirmium and the entire Pannonia also speaks against the theory of a local porphyry quarry. Rakovac, Srednje brdo, Srebro and Majdan near Stari Ledinci, Andrevlje – Debeli cer south of Beočin, and Beli Kamen near Bešenovo are the still active– or were active until recently – quarries on Fruška gora.[239] Limestone and trachyte in the form of crushed stone is (or was) exploited in those quarries, and rubble construction stone could also have been exploited there in antiquity.

The Sava and Kupa valleys slowly transform into borderline hilly and mountainous Dinaric areas in south-western and southern peripheral area of the province of Pannonia. Traces of Roman quarrying were discovered in the Slovenian provinces of Dolenjska and Bela Krajina. Near the village of Rožanec in the vicinity of Črnomelj in Bela Krajina a Roman limestone quarry was active, possibly exploiting large-sized blocks suitable for the construction and dressing of various monuments. A relief of Mithras was carved into the bedrock in the quarry, probably in the 2nd century.[240] The stone from a quarry in the north of Dolenjska, south of Čatež on the Sava, was probably exploited for the needs of the Roman settlement in the nearby Ribnica (probably Roman Romula) and nearby Neviodunum.[241] Traces of Roman quarrying were documented at several positions in the Croatian regions of Kordun and Banovina. In the area between the middle and lower courses of the Korana and Mrežnica on the Kordun, 37 Roman quarries have been documented; stone urns and sarcophagi in various stages of production were found in them. Besides the *in situ* quarries, ash-chests and sarcophagi were discovered as individual finds or as *spolia* in the surrounding area.[242] An unfinished sarcophagus, unattached from the rock, was found somewhat more to the west, beside the River Dobra in the village of Mračin near Duga Resa.[243] The karst plateau encompassing the area where these Roman limestone quarries were active during the 2nd and 3rd centuries belonged to the southern peripheral area of Roman Pannonia. Ash-chests and sarcophagi were produced for the inhabitants of smaller Roman nearby settlements, but were probably also aimed at larger settlements, such as Siscia or the Roman settlement in the area of modern-day Topusko; both land routes and river communications (the Dobra, Mrežnica, Korana, and Kupa) were used for transportation. Two sarcophagi discovered in the Korana and another one in the Kupa support the hypothesis of the river transportation of stone monuments.[244] Another proof of the Roman navigation on the Kupa is a boat's cargo found at Kobilić Pokupski, several kilometres downstream from Karlovac, containing bronze debris, probably intended for the Siscia melting-furnaces.[245] The Roman thermal centre at the site of modern-day Topusko sourced reddish limestone from a quarry nearby the settlement; the mediaeval Benedictine monastery was also built from this material. In the Roman period stone was quarried on the mountains of the Banovina, in the vicinity of Petrinja. Building stone exploited in these quarries was used for the foundations and masonry walls in Siscia, while the town's fortifications were built of this stone combined with bricks.[246] According to several studies, stone-masonry products in Siscian workshops were also produced from local limestone.[247] However, it is more probable that local limestone, due to its inferior quality, could have been used exclusively as rubble construction stone.[248] The Glina and Kupa Rivers could have been used for the transportation of stone from Topusko, Gora and Hrastovica, all three in the vicinity of Petrinja.

The analysis of the stone veneering fragments from Vinkovci suggest the possible locations of Roman quarries in the territory of northern Bosnia. It was determined that these were made of marble breccia, semi marble or slate. Yellowish stone with gray granulate additions is probably slate granite from Motajica, while reddish marble breccia probably originated from Ozren or Konjuh Mountains in north-eastern Bosnia.[249] If the stone from these locations was used in Roman Cibalae, it was certainly distributed to other Roman settlements in northern Bosnia and Slavonia. Motajica Mountain is in the vicinity the Sava, which facilitated the river stone transport towards Servitium, Marsonia and Sirmium. In Roman times, one could navigate to Cibalae on lesser streams (Berava, Bosut, Studva). There is a possibility that an inland waterway crossed from Cibalae through an artificial canal to the Vuka River towards modern-day Vukovar, thus connecting the Sava with the

[234] Milin 2004, 264.
[235] Tomović 1997, 229-230; Lalošević 2006, 60-64.
[236] Bulić 1908; Vulić 1931, Mijović 1966, 53-60; Lalošević 2006, 63-64.
[237] Vulić 1931, 368; Lalošević 2006, 64.
[238] Petković *et al.* 1976, 226-248; Tomović 1997, 231.
[239] http://www.eko.vojvodina.gov.rs/index.php?q=nacionalni_park_fruska_gora.
[240] Šušteršič 2002, 178-183.
[241] Ciglenečki 1983, 434; Klemenc 1938, 111-115.
[242] Šarić 1983, 45-52; Perkić 2002, 113, 117-120.
[243] Horvat 1964, 25-26.
[244] Perkić 2002, 113, 117-118.
[245] Šarić 1986b, 69-89.
[246] Vuković 1994, 155.
[247] Nenadić 1988, 97; Burkowsky 1999, 65.
[248] Crnković and Jovičić 1993, 142-145; Vuković 1994, 155.
[249] Vikić-Belančić 1971, 162.

Danube.[250] The largest excavated Roman quarries in Bosnia and Herzegovina are located near the village of Dardagani, 5 km north of Zvornik and ca. 4 km west of the confluence of the Rivers Sapna and Drina. In subterranean galleries at two sites (Sige and Bandera), granulate and porous limestone of Neogene (Tertiary) origin was exploited. Due to geological predispositions and superior quality of the rock at both sites, large blocks of stone suitable for dressing could have been exploited. The Dardagani quarries were situated in the border area between Pannonia and Dalmatia. It seems that, due to geographical conditions, their production was mainly oriented towards Pannonia and Sirmium, utilising the Drina and Sava as a waterway for stone transport. The analyses have confirmed that the stone from the vicinity of Dardagani was used in Sirmium. It seems that the stone of magmatic origin (andesite) from Rajići Mountain in eastern Bosnia (north-west of Srebrenica, Roman Domavia, in the province of Dalmatia) was brought to Sirmium by the same inland waterway.[251] Cer Mountain in north-western Serbia could also have held Roman quarries. The analysis of stone veneering found in Vinkovci has suggested that some of the slate granite fragments were extracted from Cer Mountain.[252]

Roman quarries in Transdanubia (the area between the Drava and Danube) could also have been used, due to their close proximity, as sources for building stone in southern Pannonian cities. In the Geresd mountainous tract, west of the auxiliary camp in Lugium on the Danube (Hungarian part of the Baranja, next to the modern-day settlement of Mórágy), red granite was exploited on several locations.[253] The stone coming from this area could have been utilised in the Lugium camp and the town of Sopianae (Pécs). From Lugium the stone could have been transported both upstream and downstream by the Danube towards other settlements on the limes along the river. Due to its close proximity (ca. 10 km) to the Drava and the lowland approach to the river, the stone quarried in the Villány Mountains south of Pécs could have been relatively easily transported to Mursa and other more distant Roman settlements and stations on the Danube.

Brick

The brick as a brand new building material appeared in southern Pannonia with the advance of the Roman army. It was used for many purposes, such as construction of walls and pillars, paving, drainage canals, etc. It was produced in brickyards from the end of the 1st and early 2nd centuries, at first in military workshops founded by the Roman military units stationed in the area. The military continued to produce brick until the late 4th century. The leading military brick-producing centres were Sirmium, Cibalae, Mursa and some smaller camps and military settlements along the Danube. Bricks produced by the army were stamped with the markings of the units owning the workshop, such as legions, cohorts, the Pannonian Flavian fleet (CL F P) and the Lower Pannonian army (EXER PAN INF). The largest number of this kind of bricks was found along the Danube limes, but bricks bearing military stamps were used in construction of civilian structures in new municipia and colonies as well.[254] Besides the military brickyards in larger Pannonian centres, such as Scarbantia, Vindobona, Carnuntum, and Sopianae, imperial brickyards were also being opened; one such yard was certainly operating in Mursa; bricks produced there were used in the area of Mursa, Cibalae, and Sirmium.[255] Mursan bricks are marked with the stamp M, the first letter of the town's name. The stamp DEC(VRIONVM) stems from Vinkovci, standing for the town council. Siscian bricks bear the stamp SISC.[256] Siscian workshops seem to have been especially thriving, since these bricks were found along the Sava from Neviodunum (near the modern-day Krško, Slovenia) all the way to Sirmium.[257] They were further found in the ancient Roman mining area in the Japra valley (Bosnia and Herzegovina).[258] Besides the military, imperial and town brickyards, private facilities were also operating. Their owners used their gentilicium initials as stamp marks.[259] The majority of bricks produced in private brickyards lack stamps and were supposedly used mostly in local surroundings.[260] Brick kilns were found on several sites in southern Pannonia. In modern Slovenia they were discovered in Ptuj, Donja Hajdina and Drnovo, and in northern Croatia in Sisak and Hrastovsko near Ludbreg, while pottery kilns are known from Vinkovci and Osijek, indicating brick production in these towns as well. In modern-day Serbia brick kilns were discovered in Sremska Mitrovica, Hrtkovci and Progar.[261] Findings of Siscian bricks from Sirmium and Neviodunum testify to the importance and significance of this brick-producing centre. They also confirm that this building material could have been transported over longer distances and that river transportation was convenient and profitable even for weighty cargoes.

4. Concluding remarks on the development, technology and economy of the Roman architecture in southern Pannonia

With the incorporation of southern Pannonia into the Roman Empire some important transformations took place in the domain of architecture that can be labelled as a true, if small-scale, revolution in the ways of building and, consequently, the culture of living. The preceding standards of prehistoric Iron-age architecture managed to survive alongside the novel building techniques, partially adapting to the new situation. The Roman town is, however, a completely new phenomenon in terms of communal life,

250 Iskra-Janošić 2004, 182-183.
251 Rižnar and Jovanović 2006, 139-151.
252 Vikić-Belančić 1971, 162.
253 Mócsy 1962, 674; Fitz 1980b, 325; http://www.mindat.org/loc-67874.html.
254 Šaranović-Svetek 1990, 48-63.
255 Bulat 1965, 9-10; Šaranović-Svetek 1990, 54-55, 63-64.
256 Šaranović-Svetek 1990, 55, 64; Iskra-Janošić 2001a, 66.
257 Šaranović-Svetek 1990, 55; Petru 1990, 90; Durman 1992, 127.
258 Basler 1977, pl. 17: 3; Durman 1992, 127.
259 Šaranović-Svetek 1990, 55-56, 64-65.
260 Brunšmid 1902, 134.
261 Šaranović-Svetek 1990, 43.

and it appeared almost immediately after the consolidation of Roman power, introducing new spatial layouts and building experiences. Timber and clay, known from earlier prehistoric periods, preserved their role in Roman architecture because they could be used in a fast and simple, that is, economical way. Traces of initial building phase, materialised in timber, in the towns discussed above are rather humble, which can be explained by the characteristics of the materials used, susceptibility to water and fire and their general perishable nature. This is the reason why the remains of this initial building phase in Andautonia, Cibalae, and Mursa are only evidenced as traces of soot and ashes. The remains of fortifications in Cibalae are the only significant traces of earthen and timber architecture. There are so far no substantiated traces of a "timber" phase in Iovia. In Siscia, small segments of timber structures and harbour facilities were preserved, while a complete timber phase of residential and industrial building was discovered and studied in Varaždinske Toplice. In Cibalae and Mursa, the early phase of earthen-timber architecture can be associated with military activities at the *limes* and with local tradition of building. It is questionable whether there was really a timber-building phase in Andautonia and Iovia, because solid materials predominate already during the Flavian reign, stone being the most important building material in both towns. At the same time, the building in stone of temple and thermal complex in Aquae Iasae started, while in Siscia there was a transition to solid construction through a combination of stone and brick. Building in brick began to predominate in Mursa and Cibalae in Hadrian's reign. In this period Mursa became a colony and Cibalae a *municipium*, which gave a new impetus to the development of these towns.

Urban layouts of south-Pannonian towns that will remain more or less unchanged until the end of Antiquity, were rounded in the mid 2nd century. Andautonia and Iovia were built in stone, most frequently rubble and most commonly in the *opus incertum* technique. Brick was used scarcely, only for the construction of special structures, such as sewage collectors, hypocaust systems and certain building segments. Combined stone and brick were used in Siscia's fortifications, but structures built exclusively in stone or in brick, respectively, were also discovered in this town. *Opus mixtum* was more frequently used in Siscia in the later Empire. Mursa and Cibalae were built in brick, with a small proportion of stone buildings.

Due to a rapid development of towns and the increased demand for every type of building material, supplying became a sophisticated and well-organised business. Sand and gravel used for the making of mortar, surfacing and levelling of terrain, as well as for the construction of floors and roads, were being extracted in the settlements' surroundings. Clay used for the making of earthen walls and beaten floor-foundations was also of local origin. Due to short distances, supply and transportation of these materials were not particularly demanding. Timber used for the construction of walls, floors and roof frames, as well as for the construction of scaffoldings and framings, was also available in the immediate vicinity of towns covered by thick forests, especially those of Pendunculate oak in river valleys. Timber was easily transported by rivers or other waterways.

Mortar is an indispensable component of Roman building; it was used for bonding in constructing foundations, wall-masonry and floors. Standard lime mortar was composed of an aggregate (sand or gravel), water and lime obtained by burning limestone. Lime-burning kilns and lime-hydrating pools were discovered in towns, suggesting that the process was carried out at building sites.[262] Along with the standard lime mortar, a water-resistant mortar with an addition of grinded brick was also in use in the construction of water supply and sewage canals as well as thermal complexes. Stone and brick coated in mortar were a complete novelty in Pannonian architecture. Building rubble stone and blocks of dimension stone were exploited from the newly opened quarries. Stone, an almost unknown building material in Pannonian prehistory, was now being exploited and used in this dominantly lowland province. Although it is relatively rarely found as a geological resource in the region, it became the most important building material, while brick, notwithstanding the deposits of superior clay exploited during entire prehistory, remained secondary.[263]

The quarries in southern Pannonia and peripheral areas of neighbouring provinces can be classified into two groups according to the stones extracted. The quarries providing rubble stone used in the construction of foundations and walls (the latter were in most cases merely covered with plaster) are much more numerous. Rubble limestone was also used in making of lime through burning. Rubble stone quarries could have been opened also on the sites where rock layers were of inferior quality. Building rubble stone was exploited in the immediate vicinity of settlements, usually at the nearest accessible deposit in the local mountainous tracts and mountain ranges, such as the Medvednica quarries supplying Andautonia[264] or the quarries in the mountainous tract at Gora and Hrastovica near Petrinja, which supplied Siscia.[265] The reason behind this was economical, that is, the profitability of the transport of this type of building material, which was not of particular quality but was relatively easily available even in Pannonia. Of another type were quarries where blocks of stone, that is, dimension stone used for load bearing and decorative architectural elements as well as various monuments, was exploited. Deposits of dimension stone were much scarcer. Depending on the quality and decorative value of the stone exploited in them, the dimension-stone quarries could have had much wider importance than meeting the needs of local building works and stone-masonry production. The finest example of this is Norican marbles from Pohorje and Styria that were used throughout Pannonia and other neighbouring

262 Iskra-Janošić 2001a, 72; Pintarić 2006, 5.
263 Mócsy 1962, 702; Fitz 1980b, 325-326; Gregl and Migotti 2004, 139.
264 Degmedžić 1957, 16, 21-22; Šikić 1995, 61-66; Crnković 1998, 32.
265 Vuković 1994, 155.

provinces. Lithothamnium limestone from Medvednica was also exploited beyond the local Andautonian needs and was used in Siscia and, probably, Neviodunum. Limestone blocks from Dardagani near Zvornik were in the first place used in building in the rather distant town of Sirmium. However, a superior quality and rarity of these stone types had their cost. Given this, architectural-stone quarries could have been located far from the settlements and building sites, since the transportation was profitable even over longer distances.

Brick production probably appeared in southern Pannonia at the turn of the 1st and 2nd centuries. Bricks and tiles were made in various forms and dimensions, depending on the purpose; they were used as building material in the construction of walls, vaults, subterranean canals, floors, street pavements and tombs. Because of the dominant stone building in Andautonia, Iovia and Aquae Iasae, brick production was rather insignificant in these places. Smaller local workshops were probably active there, and brick could also have been imported from larger brick-producing centres, such as Siscia, Mursa, and Cibalae, whose production outgrew the demands of their respective local markets, prompting exportation. Examples of this are findings of Siscian bricks in Neviodunum and Sirmium, as well as in the mining region in the Japra valley in western Bosnia.

The transport of stone and brick in southern Pannonia was one of the key components of the development business. The transport of building stone over relatively short distances could have been performed by land or water routes. The latter were favoured when dealing with longer distances because it was much more practical and efficient and, in the end, more profitable. The river transport was much easier downstream, but upstream transport was also practiced, by hauling the rafts and boats from the shore by draught animals or manpower; this type of transport was especially developed on larger Pannonian rivers. Pohorje and Carinthia marbles were transported downstream by the Mura and Drava to the Danube, by which the traffic continued both downstream and upstream. Siscian brick was transported by the Sava downstream to Sirmium and upstream to Neviodunum and by the Una upstream towards the Japra. Since heavier cargoes were in Roman times transported much easier by water in comparison with land, even smaller streams were used for navigation, as confirmed by archaeological indicators and written sources.[266] Lesser streams, such as the Ljubljanica in Slovenia, the Dobra, Mrežnica, Korana, Glina and Bosut in Croatia, and the Sana and Japra in Bosnia, were used for the navigation by rafts and smaller boats.

Architectural culture of southern Roman Pannonia shows certain continuity with the prehistoric tradition but new materials and techniques reflect important political and social transformations. The fast spread of the use of mortar, stone and brick as building materials suggests major changes in the ethnic composition of the area, that is, the arrival of a significant number of new settlers (soldiers, veterans and Italic civilians) that were accustomed to using these materials and techniques, so they stuck to them in the new surroundings. The first military structures were made of timber and earth. With the stabilisation of the government, the military introduced new materials and techniques. Stonemasons and brick-production craftsmen followed the military and brought with them new building skills. Some of them would permanently stay as veterans in this territory, settling in the developing civil settlements. The arrival of civilian settlers from Italy further reinforced this process. It seems that their influence was rather strong, since the transformations in architecture were introduced relatively early and on a large scale. Besides them, the local population undergoing Romanisation modified the old ways and accepted the new construction and urban traditions. Owing to that, in the 2nd century building works in southern-Pannonian towns experienced expansion and sudden raise in terms of quality. Also, the towns were rebuilt during the period ending with the late 4th century, following a lesser or a major decline caused by warfare and devastation in the Marcomannic, civil and Gothic wars.

[266] Šarić 1986b, 69-89; Durman 1992, 127; Šašel Kos 1994, 106; Perkić 2002, 113, 117-118.

Bibliography

Adam 2001 Jean-Pierre Adam, *Roman Building. Materials and Techniques*, London.

Baćani and Tomaš 2008 Iskra Baćani and Tea Tomaš, Pregled arheoloških istraživanja Gradskog muzeja Sisak u 2007. godini (I.), *GGMS* 8, 7-16.

Baćani, Škrgulja, and Tomaš Barišić 2009 Iskra Baćani, Rosana Škrgulja, and Tea Tomaš Barišić, Pregled arheoloških istraživanja Gradskog muzeja Sisak u 2008. godini, *GGMS 9*, 59-79.

Barkóczi 1980 László Barkóczi, History of Pannonia, in: *ARP*, 85-124.

Basler 1977 Đuro Basler, Rimski metalurški pogon i naselje u dolini Japre, *GZMS*, 30-31, 121-216.

Božič 1987 Dragan Božič, Zapadna grupa, in: Alojz Benac (ed.), *PJZ* 5, 855-897.

Brunšmid 1901 Josip Brunšmid, Arheološke bilješke iz Dalmacije i Panonije, *VHAD* 5, 87-168.

Brunšmid 1902 Josip Brunšmid, Colonia Aurelia Cibalae, *VHAD* 6, 117-166.

Brunšmid 1905 Josip Brunšmid, Kameni spomenici Hrvatskoga narodnoga muzeja u Zagrebu: nastavak, *VHAD* 8, 35-106.

Brunšmid 1979 Josip Brunšmid, Colonia Aurelia Cibalae, in: *Corolla*, 55-104.

Bulat 1965 Mirko Bulat, Rimska opeke i crjepovi s pečatima iz Muzeja Slavonije, *OZ* 9-10, 7-24.

Bulat 1989 Mirko Bulat, *Mursa – Osijek u rimsko doba*, Exhibition Catalogue, Osijek.

Bulat 1993 Mirko Bulat, Antička naselja u Slavoniji, in: *Slavonski Brod*, 173-180.

Bulić 1908 Frane Bulić, Proviene qualche cosa nel Palazzo di Diocleziano a Spalato, per es. qualche colonna, qualche pezzo architettonico od ornamentale, qualche statua, dalle lapidicine di Sirmium (Fruška Gora nella Slavonia)?, *Bullettino di archeologia e storia Dalmata 31, Split, 111-127.*

Burkowsky 1998 Zdenko Burkowsky, *Zaštitna arheološka istraživanja u Sisku,* Exibition Catalogue, Sisak.

Burkowsky 1999 Zdenko Burkowsky, *Sisak u prapovijesti, antici i starohrvatskom dobu,* Exibition Catalogue, Sisak.

Burkowsky 2000 Zdenko Burkowsky, *Pregled zaštitnih arheoloških iskopavanja 1990. – 2000.*, Exhibition Catalogue, Sisak.

Čakarun *et al.* 1987 Ivan Čakarun, Vinko Mraz, Željko Babić, Rozalija Mutić, Ana Sokač, and Domagoj Franić, Geološke i hidrogeološke specifičnosti kvartarnog vodonosnog kompleksa prisavske ravnice na dionici granica SR Slovenije – Rugvica, *Geološki vjesnik* 40, Zagreb, 273-289.

Ciglenečki 1983 Slavko Ciglenečki, Čateški Grič – Linhartov "Neviodunum"? (Arheološko-topografska skica prazgodovinske in antične poselitve v Brežiških vratih), *AV* 34, 431-441.

Corolla Corolla memoriae Iosepho Brunšmid dicata, IzdHAD 4, 1979, Vinkovci.

Crnković 1997 Branko Crnković, Geological and petrographic composition of Croatia, *HAnt* 3, 11-19.

Crnković 1998 Branko Crnković, Kamen zagrebačke prvostolnice, *Mineral: stručni časopis za područje rudarstva i graditeljstva*, 4/6, Zagreb, 32-33.

Crnković and Jovičić 1993 Branko Crnković and Dragomir Jovičić, Dimension stone deposits in Croatia, *Rudarsko-geološko-naftni zbornik* 5, Zagreb, 139-163.

Crnković and Šarić 2003 Branko Crnković and Ljubo Šarić, *Građenje prirodnim kamenom*, Zagreb.

Čović 1987 Borivoj Čović, Donja Dolina – Sanski Most, in: A. Benac (ed.), *PJZ* 5, 232-286.

Degmedžić 1957 Ivica Degmedžić, Sadržaj antiknih kamenih spomenika nađenih u Zagrebu i okolici, *Iz starog i novog Zagreba* 1, Zagreb, 91-117.

Dimitrijević 1979a Stojan Dimitrijević, Arheološka topografija i izbor arheoloških nalaza s vinkovačkog tla, in: *Corolla*, 133-200.

Dimitrijević 1979b Stojan Dimitrijević, Sjeverna zona, in: Alojz Benac (ed.), *PJZ* 2, 229-360.

Djurić 1997 Bojan Djurić, Eastern Alpine marble and Pannonian trade, in: B. Djurić and I. Lazar (eds.): *Akten des IV. Internationalen Colloquiums über Probleme des provinzialrömischen Kunstschaffens*, *Situla* 36, Ljubljana, 73-86.

Djurić 2004 Bojan Djurić, V Saksanovem svetu – Rimsko pridobivanje belega marmorja s Pohorja / In Saxanus Welt – Römische Gewinnung des weißen Pohorje Marmors, in: I. Lazar (ed.): *Rimljani: steklo, glina, kamen*, katalog razstav / *Glas, Ton, Stein*, Exhibition Catalogue, Celje-Maribor-Ptuj, 147-167.

Dizdar 2009 Marko Dizdar, Željezno doba, in: Kusin and Šulc 2009, 85-96.

Domić Kunić 2006 Alka Domić Kunić, Bellum pannonicum (12.-11. pr. Kr.): posljednja faza osvajanja južne Panonije, *VAMZ* 39, 59-164.

Dular and Tecco Hvala 2007 Janez Dular and Sneža Tecco Hvala, *South-Eastern Slovenia in the Early Iron Age. Settlement – Economy – Society*, Ljubljana.

Durman 1992 Aleksandar Durman, O geostrateškom položaju Siscije, *OpArch* 16, 117-131.

Durman 2002 Aleksandar Durman, Iron resources and production for the Roman frontier in Pannonia, *Historical Metallurgy* 36/1, Portsmouth, 24-32.

Durman 2006 Aleksandar Durman (ed.), *Stotinu hrvatskih arheoloških nalazišta*, Zagreb.

Faber 1973 Aleksandra Faber, Građa za topografiju antičkog Siska, *VAMZ* 6-7 (1972-1973), 133-162.

Filipović 2004 Slavica Filipović, Colonia Aelia Mursa, in: *Pannonia* II, 157-168.

Filipović 2008 Slavica Filipović, Mursa – Vojarna (Građevinski fakultet), *HAG* 4 (2007), 29-32.

Filipović 2009 Slavica Filipović, Mursa – Vojarna (Građevinski fakultet), *HAG* (2008), 51-52.

Filipović and Katavić 2005 Slavica Filipović and Vedran Katavić, Osijek – Vojarna – Poljoprivredni fakultet, *ObHAD* 37/3, 106-112.

Filipović and Katavić 2006a Slavica Filipović and Vedran Katavić, Mursa – Vojarna (Poljoprivredni fakultet), *HAG* 2 (2005), 11-14.

Filipović and Katavić 2006b Slavica Filipović and Vedran Katavić, Osijek – Silos 2005, *ObHAD* 38/1, 74-79.

Filipović and Podrug 2007 Slavica Filipović and Emil Podrug, Neobjavljene rimske opeke i tegule s radioničkim pečatima iz Muzeja Slavonije, *OZ* 28, 35-54.

Fitz 1980a Jenő Fitz, The way of life, in: *ARP*, 161-175.

Fitz 1980b Jenő Fitz, Economic life, in: *ARP*, 323-335.

Fitz 2003 Jenő Fitz, Die Städte Pannoniens, in: *Pannonia* I, 47-52.

Fitz 2004 Jenő Fitz, Gorsium, in: *Pannonia* II, 197-207.

Földvary 1988 Gábor Zoltán Földvary, *Geology of the Carpathian Region*, Singapore.

Fürst-Bjeliš 1996 Borna Fürst-Bjeliš, *Historijsko-geografska analiza prostornog pojma tradicionalne regije Turopolja*, Unpublished PhD thesis, Zagreb.

Gabrovec 1987 Stane Gabrovec, Dolenjska grupa, in: Alojz Benac (ed.), *PJZ* 5, 29-119.

Gömöri 2003 János Gömöri, Scarbantia, in: *Pannonia* I, 81-92.

Gorenc and Vikić 1980 Marcel Gorenc and Branka Vikić, *Varaždinske Toplice – Aquae Iasae u antičko doba*, Varaždinske Toplice.

Gorenc and Vikić 1984 Marcel Gorenc and Branka Vikić, Antičko nasljeđe ludbreškog kraja, in: Mađarić 1984, 59-72.

Gregl 1994 Zoran Gregl, Antičko (rimsko) razdoblje, in: *Zagreb prije Zagreba*, 33-36.

Gregl and Migotti 2004 Zoran Gregl and Branka Migotti, Civitas Iovia (Botivo), in: *Pannonia* II, 131-143.

Herkov 1973 Zlatko Herkov, *Naše stare mjere i utezi*, Zagreb.

Horvat 1964 Anđela Horvat, Presjek razvoja umjetnosti u karlovačkom pokuplju, *Zbornik gradskog muzeja u Karlovcu* 1, Karlovac, 25-26.

Hoti 1992 Marina Hoti, Sisak u antičkim izvorima, *OpArch* 16, 133-163.

Ilić 1993 Milan Ilić, Riječna luka u Zagrebu, *Geografski glasnik HGD* 55, Zagreb, 203-211.

Iskra-Janošić 1992 Ivana Iskra-Janošić, Građevinski materijali i njihova upotreba u Cibalama, *OpArch* 16, 207-218.

Iskra-Janošić 1993 Ivana Iskra-Janošić, Zaštitna iskopavanja u Vinkovcima u godini 1993. g., *ObHAD* 25/3, 65-67.

Iskra-Janošić 1994 Ivana Iskra-Janošić, Vinkovci – zaštitno istraživanje u zgradi muzeja, *ObHAD* 26/2, 58-60.

Iskra-Janošić 2000 Ivana Iskra-Janošić, Zaštitno iskopavanje u Vinkovcima – Duga ulica 40, *ObHAD* 32/2, 60-63.

Iskra-Janošić 2001a Ivana Iskra-Janošić, *Urbanizacija Cibala i razvoj keramičarskih središta*, Vinkovci.

Iskra-Janošić 2001b Ivana Iskra-Janošić, Vinkovci – zaštitno iskopavanje u Dugoj ulici 20, *ObHAD* 33/1, 31-34.

Iskra-Janošić 2002 Ivana Iskra-Janošić, Vinkovci – zaštitna iskopavanja 2001 godine, *ObHAD* 34/1, 75-79.

Iskra-Janošić 2004 Ivana Iskra-Janošić, Colonia Aurelia Cibalae, in: *Pannonia* II, 169-195.

Iskra-Janošić 2005 Ivana Iskra-Janošić, *Vinkovci u antici i srednjem vijeku*, Vinkovci.

Iskra-Janošić 2006 Ivana Iskra-Janošić, Zaštitno iskopavanje u Vinkovcima. Ulica Vladimira Nazora 3, *ObHAD* 38/1, 80-87.

Jerončić and Katavić 2009 Tomislav Jerončić and Vedran Katavić, Zaštitna arheološka istraživanja na južnoj periferiji antičke Murse, *ObHAD* 41/1, 48-59.

Jerončić and Katavić 2010 Tomislav Jerončić and Vedran Katavić, Zaštitno arheološko istraživanje na lokalitetu Glagoljaška 16, Vinkovci, *ObHAD* 42/1, 63-70.

Jovanović 1987 Borislav Jovanović, Istočna grupa, in: Alojz Benac (ed.), *PJZ* 5, 815-854.

Jovičić, Oreški, and Kraljeta 1992 Dragomir Jovičić, Ernest Oreški, and Biserka Kraljeta, Ležište arhitektonskog kamena granita "Zebrato", *Rudarsko-geološko-naftni zbornik* 4, Zagreb, 127-138.

Kandler 2004 Manfred Kandler, Carnuntum, in: *Pannonia* II, 11-66.

Karlovačko i sisačko područje Arheološka istraživanja na karlovačkom i sisačkom području, IzdHAD 10, Zagreb 1986.

Klemenc 1938 Josip Klemenc, *Archaeologische Karte von Jugoslavien*, Beograd.

Klemenc 1961 Josip Klemenc, Limes u Donjoj Panoniji, in: *Limes u Jugoslaviji I. Zbornik radova sa simposiuma o limesu 1960. g.*, Beograd, 5-34.

Knezović 2010 Ivan Knezović, The worship of Savus and Nemsesis in Andautonia, *AV* 61, 187-202.

Kukuljević 1873 Ivan Kukuljević, *Panonija rimska, RJAZU* 23.

Kuntić-Makvić and Šegvić 1992 Bruna Kuntić-Makvić and Marina Šegvić, Katančićev opis Siscije, *OpArch* 16, 165-181.

Kuntić-Makvić 1993 Bruna Kuntić-Makvić, Sisciensia Kercselichiana, *OpArch* 17, 263-271.

Kusin and Šulc 2009 Vesna Kusin and Branka Šulc (eds.), *Slavonija, Baranja i Srijem: vrela europske civilizacije*, Exhibition Catalogue, Zagreb.

Kušan and Nemeth-Ehrlich 1994 Dora Kušan and Dorica Nemeth-Ehrlich, Arheološki park "Andautonia" u Šćitarjevu, *Informatica museologica* 25, Zagreb, 30-34.

Lalošević 2006 Vesna Lalošević, Problemi vezani za mučeničku grupu Svetih Četvorice Ovjenčanih, *Radovi Zavoda za hrvatsku povijest* 38, Zagreb, 59-71.

Leleković 2009 Tino Leleković, Osijek – Trg bana J. Jelačića, *HAG* 5 (2008), 45-51.

Lengyel 1980 Alfons Lengyel, Chronological table, in: *ARP*, 477-487.

Lolić 2001 Tatjana Lolić, Lokalitet Trg bana J. Jelačića u Sisku, *ObHAD* 33/3, 95-99.

Lolić 2003 Tatjana Lolić, Colonia Flavia Siscia, in: *Panonnia* I, 131-152.

Lovenjak 2003 Milan Lovenjak, Municipium Flavium Latobicorum Neviodunum, in: *Pannonia* I, 93-105.

Lugli 1957 Giuseppe Lugli, *La tecnica edilizia romana* I, II, Roma.

Mađarić 1984 Vlado Mađarić (ed.), *Ludbreg*, Ludbreg.

Majnarić-Pandžić 1998 Nives Majnarić-Pandžić, Bronačano i željezno doba, in: *Prapovijest*, 160-369.

Makjanić 1981 Rajka Makjanić, *Antički kamenolomi na području Hrvatske, Dometi* 15/5, Rijeka, 71-76.

Mason 2003a Phil Mason, Obrežje MMP, in: Prešeren 2003, 202-203.

Mason 2003b Phil Mason, Rimska vojaška utrdba, in: Prešeren 2003, 66-71.

Mijović 1966 Petar Mijović, Sirmijski skulptori i kamenoresci – Quattuor coronati, *Starinar* 27, 53-60.

Milin 2004 Milena Milin, Bassianae, in: *Pannonia* II, 253-268.

Mirković 2004 Miroslava Mirković, Sirmium, in: *Pannonia*, II 145-156.

Mócsy 1959 András Mócsy, *Die Bevölkerung von Pannonien bis zu den Markomannenkriegen*, Budapest.

Mócsy 1962 András Mócsy, Pannonia, *RE* s. v. Suppl. IX, 516-776.

Mócsy 1974 András Mócsy, *Pannonia and Upper Moesia: A History of the Middle Danube Provinces of the Roman Empire*, London-Boston.

Mülle and Vogel 1999 Werner Müller and Gunther Vogel, *Atlas arhitekture* 1, Zagreb.

Muravljov 1989 Mihailo Muravljov, *Građevinski materijali*, Beograd.

Nemeth-Ehrlich 1991 Dorica Nemeth-Ehrlich, Šćitarjevo: antička Andautonija, *ObHAD* 23/2, 52-54.

Nemeth-Ehrlich 1993 Dorica Nemeth-Ehrlich, Varaždinske Toplice – Radovi na zaštiti antičkog lokaliteta, *ObHAD* 25/1, 23-26.

Nemeth-Ehrlich and Kušan Špalj 1999 Dorica Nemeth-Ehrlich and Dora Kušan Špalj, Konzervatorsko-resturatorski radovi i zaštitna istraživanja na antičkom kompleksu u Varaždinskim Toplicama u godini 1997. i 1998., *ObHAD* 31/1, 68-74.

Nemeth-Ehrlich and Kušan Špalj 2003a Dorica Nemeth-Ehrlich and Dora Kušan Špalj, Municipium Andautonia, in: *Pannonia* I, 107-129.

Nemeth-Ehrlich and Kušan Špalj 2003b Dorica Nemeth-Ehrlich and Dora Kušan Špalj, Konzervatorsko-resturatorski radovi i zaštitna arheološka istraživanja na antičkom kompleksu u Varaždinskim Toplicama od 1999. do 2001., *ObHAD* 35/1, 91-101.

Nemeth-Ehrlich and Kušan Špalj 2006 Dorica Nemeth-Ehrlich and Dora Kušan Špalj, Konzervatorsko-resturatorski radovi i zaštitna arheološka istraživanja na antičkom kompleksu u Varaždinskim Toplicama od godine 2002. do 2005., *ObHAD* 38/1, 101-112.

Nemeth-Ehrlich and Kušan Špalj 2007 Dorica Nemeth-Ehrlich and Dora Kušan Špalj, Varaždinske Toplice – Gradski park, *HAG* 3 (2006), 150-153.

Nemeth-Ehrlich and Vojvoda 1994 Dorica Nemeth-Ehrlich and Pavo Vojvoda, Andautonija – rimsko urbano središte, in: *Zagreb prije Zagreba*, 39-45.

Nenadić 1988 Vesna Nenadić, Prilog proučavanju antičke Sisciae, PIAZ 3-4, 71-102.

Pamić 1986 Jakob Pamić, Kredno-tercijarne granitne i metamorfne stijene u dodirnom području sjevernih Dinarida i Panonskog strukturnog kompleksa, *Geologija* 28-29, Ljubljana, 219-237.

Pamić 1990 Jakob Pamić, *Alpinski granitoidi, magmatiti i metamorfiti Moslavačke gore i okolne podloge Panonskoga bazena (sjeverna Hrvatska, Jugoslavija)*, Zagreb.

Perkić 2002 Domagoj Perkić, *Grad mrtvih u Bubijevoj* jami kod Barilovića, *HAnt* 8, 103-130.

Petković *et al.* 1976 Kosta Vladoje Petković *et al.*, *Fruška Gora. Monografski prikaz geološke građe i tektonskog sklopa*, Monografije Fruške Gore 2, Novi Sad.

Petru 1990 Sonja Petru, Neviodunum – Drnovo, *Arheološka najdišča Dolenjske*, Novo Mesto.

Pintarić 1998 Tatjana Pintarić, *Nekropole i groblja na području Ščitarjeva*, Exhibition Catalogue, Velika Gorica.

Pintarić 2000 Tatjana Pintarić, Arheološko i geofizikalno istraživanje u Ščitarjevu, *ObHAD* 32/3, 64-67.

Pintarić 2003 Tatjana Pintarić, Arheološko istraživanje u Ščitarjevu: lokacija župna livada, *ObHAD* 35/1, 107-113.

Pintarić 2006 Tatjana Pintarić, *Pronađeno vrijeme*, Exhibition Catalogue, Velika Gorica.

Pintarić and Knezović 2005 Tatjana Pintarić and Ivan Knezović, *Arheologija, Katalog stalnog postava Muzeja Turopolja*, Velika Gorica.

Pinterović 1978 Danica Pinterović, *Mursa i njezino područje u antičko doba*, Osijek.

Pleše 2009 Tajana Pleše, Ludbreg – Vrt Somođi, *HAG* 5 (2008), 213-215.

Póczy 1980 Klára Póczy, Pannonian cities, in: *ARP*, 239-274.

Potrebica 2009 Hrvoje Potrebica, Kaptol i starije željezno doba središnje i zapadne Slavonije, in: Kusin and Šulc 2009, 97-104.

Prapovijest Stojan Dimitrijević, Tihomila Težak-Gregl and Nives Majnarić-Pandžić, *Prapovijest*, Zagreb 1998.

Prešeren 2003 Damjana Prešeren (ed.), *Zemlja pod vašimi nogami. Arheologija na avtocestah Slovenije, Vodnik po najdiščih*, Ljubljana.

Rendić-Miočević 1994 Ante Rendić-Miočević, Katalog / Catalogue nos. 200 and 276, in: *Zagreb prije Zagreba*, 117-118, 131-132.

Riđanović 1993 Josip Riđanović, *Hidrogeografija*, Zagreb.

Rižnar and Jovanović 2006 Igor Rižnar and Divna Jovanović, Stone material of regional provenance from Sirmium, *Starinar* 56, 139-152.

Roma sul Danubio M. Buora and W. Jobst (eds.), *Roma Sul Danubio: Da Aquileia a Carnuntum Lungo La Via Dell'Ambra*, Udine 2002.

Sanader 2009 Mirjana Sanader, Vojne građevine na dunavskom limesu, in: Kusin and Šulc 2009, 105-110.

Schejbal 2004 Berislav Schejbal, Municipium Iasorum (Aquae Balissae), in: *Pannonia* II, 99-129.

Slavonski Brod Arheološka istraživanja u Slavonskom Brodu i Brodskom Posavlju, *IzdHAD* 16, Zagreb 1993.

Soproni 1980 Sándor Soproni, Limes, in: *ARP*, 219-238.

Starac 2006 Alka Starac (ed.), *Tragovima kamenoklesarstva: Arheološki nalazi u ulici Porta Stovagnaga*, Monografije i katalozi: Arheološki muzej Istre, 17, Pula.

Suić 2003 Mate Suić, *Antički grad na istočnom Jadranu*, Zagreb.

Suić and Zaninović 1980 Mate Suić and Marin Zaninović (eds.), *Materijali, tehnike i strukture predantičkog i antičkog graditeljstva na istočnom jadranskom prostoru*, Zagreb.

Sunko and Katavić 2007 Ana Sunko and Vedran Katavić, Zaštitna arheološka istraživanja u Huttlerovoj ulici 20 i 22 u Osijeku, *ObHAD* 39/3, 80-89.

Sunko, Katavić, and Jerončić 2008 Ana Sunko, Vedran Katavić, and Tomislav Jerončić, Osijek – Huttlerova (Silos), *HAG* 4 (2007) 20-26.

SZ Hrvatska Arheološka istraživanja u sjeverozapadnoj Hrvatskoj, *IzdHAD* 2, Zagreb 1978.

Šaranović-Svetek 1990 Vesna Šaranović-Svetek, *Ciglarstvo kao značajna privredna* grana na području jugoslovenskog dela provincije Donje Panonije, *RVM* 32, 41-79.

Šarić 1983 Ivan Šarić, Kamenolomi gdje su se u rimsko vrijeme izrađivali sarkofazi, *VAMZ* 15, 43-58.

Šarić 1986a Ivan Šarić, Zaštitno arheološko istraživanje lokaliteta "Kovnica" u Sisku 1985. godine, *ObHAD* 18/1, 28-29.

Šarić 1986b Ivan Šarić, Rimski bronačni materijal izvađen iz rijeke Kupe kod Karlovca, *Karlovačko i sisačko područje*, 69-89.

Šašel 1970 Jaroslav Šašel, Andautonia, *RE* suppl. 12, 71-72.

Šašel Kos 1994 Marjeta Šašel Kos, Savus and Adsalluta, *AV* 45, 99-122.

Šikić 1995 Krešimir Šikić, *Geološki vodič Medvednice*, Zagreb.

Šimek 2006 Marina Šimek, Jalžabet, in: Durman 2006, 142-143.

Šimić 2008 Jasna Šimić, Osijek – Park kraljice K. Kosače, *HAG* 4 (2007), 27-28.

Šimunić and Basch 1975 Antun Šimunić and Oto Basch,

Stratigrafija kvartarnih sedimenata zagrebačkog posavlja, *Geološki vjesnik* 28, Zagreb, 153-164.

Šterc 1979 Stjepan Šterc, Osnovne fizičkogeografske karakteristike savskog porječja između Radeča i Siska, *Radovi Geografskog odjela PMF*-a 14, Zagreb, 47-59.

Šušteršič 2002 France Šušteršič, Mitrej pri Rožancu – antični kamnolom, *Rast – Revija za literaturo, kulturo in družbena vprašanja* 2 (80), Novo Mesto, 178-183.

Taylor 2003 Rabun Taylor, *Roman Builders: A study in architectural process*, New York.

Težak-Gregl 1998 Tihomila Težak-Gregl 1998, Neolitik i eneolitik, in: *Prapovijest*, 57-158.

Tomljenović 2002 Bruno Tomljenović, *Strukturne značajke Medvednice i Samoborskog gorja*, Unpublished PhD thesis, Zagreb.

Tomović 1997 Miodrag Tomović, The Passio Sanctorum IV Coronatorum and the Fruška Gora hypothesis in the light of archaeological evidence, in: B. Djurić, I. Lazar (eds.): *Akten des IV. Internationalen Colloquiums über Probleme des provinzialrömischen Kunstschaffens*, *Situla* 36, 229-239.

Vikić-Belančić 1967 Branka Vikić-Belančić, *Rimska keramička proizvodnja na području Save i Drave od I. do IV. st. s obzirom na proizvodnju glinenih svjetiljki*, Unpublished PhD thesis, Zagreb.

Vikić-Belančić 1971 Branka Vikić-Belančić, Istraživanja u Vinkovcima 1966. godine, *VAMZ* 4, 159-176.

Vikić-Belančić 1973 Branka Vikić-Belančić, Prilog istraživanju antičkog naseobinskog kompleksa u Varaždinskim Toplicama, *VAMZ* 6-7 (1972-1973), 75-127.

Vikić-Belančić 1978 Branka Vikić-Belančić, Tipovi naselja u SZ Hrvatskoj (u svjetlu novih istraživanja), in: *SZ Hrvatska*, 159-176.

Vikić-Belančić 1981 Branka Vikić-Belančić, Etape urbanog razvitka Anadutonije i antičko nasljeđe Zagreba, in: *Zagreb i njegova okolica*, 129-154.

Vikić-Belančić 1984 Branka Vikić-Belančić, Sustavna istraživanja u Ludbregu od 1968. – 1979. god., *VAMZ* 16-17 (1983-1984), 119-166.

Vikić and Gorenc 1966 Branka Vikić and Marcel Gorenc, Varaždinske Toplice – Aquae Iasae, *VMKH* 11/5, 7-15.

Vikić-Belančić and Gorenc 1958 Branka Vikić-Belančić and Marcel Gorenc, Arheološka istraživanja antiknog kupališta u Varaždinskim Toplicama od 1953. – 1955. godine, *VAMZ* **1, 75-127.**

Vikić-Belančić and Gorenc 1962 Branka Vikić-Belančić and Marcel Gorenc, Istraživanja antiknog kupališta u Varaždinskim Toplicama od 1956. do 1959. godine, VAMZ 2, 181-223.

Vikić-Belančić and Gorenc 1971 Branka Vikić-Belančić and Marcel Gorenc, Završna istraživanja antičkog kupališnog kompleksa u Varaždinskim Toplicama, VAMZ 4, 121-157.

Vikić-Belančić and Gorenc 1984 Branka Vikić-Belančić and Marcel Gorenc, Analiza i valorizacija prirodnog ambijenta naselja Andautonija – Ščitarjevo i prikaz kontinuiteta naseljenosti, in: Nada Benić-Hlebec *et al.*, *Plan uređenja manjeg naselja – Ščitarjevo*, 2, Zagreb, Regionalni zavod za zaštitu spomenika kulture.

Vinkovci S. Jozić (ed.), *Vinkovci u svijetu arheologije*, Exhibition Catalogue, Vinkovci 1999.

Vinski-Gasparini 1987 Ksenija Vinski-Gasparini, Grupa Martijanec – Kaptol, in: Alojz Benac (ed.), *PJZ* 5, 182-231.

Vitruvije 1999 *Marcus Vitruvius Pollio*, *De architectura libri decem – Deset knjiga o arhitekturi*, translated by: M. Lopac, Zagreb.

Vojvoda 1997 Pavo Vojvoda, *Arheologija – rimsko doba, Katalog stalnog postava*, Muzej Turopolja, Velika Gorica.

Vrbanović 1981 Stjepan Vrbanović, Prilog proučavanju topografije Siscije, in: *Zagreb i njegova okolica*, 187-199.

Vulić 1931 Nikola Vulić, Fruškogorski mučenici, *Glasnik Istorijskog društva u Novom Sadu* IV/3, Novi Sad, 359-373.

Vuković 1994 Domagoj Vuković, *Siscija – vizija rimskog grada u Panoniji*, Sisak.

Vulić, Rapan Papeša, and

Krznarić Škrivanko 2009 Hrvoje Vulić, Anita Rapan Papeša, and Maja Krznarić Škrivanko, Vinkovci – Korzo, *HAG* 5 (2008), 95-99.

Wamser 2000 Ludwig Wamser (ed.), *Die Römer zwischen Alpen und Nordmeer: Zivilisatorisches Erbe einer europäischen Militärmacht*, München.

Weber 2000 Gerhard Weber, Von Holz zu Stein. Zum Bauwesen in den Nordwestprovinzen, in: Wamser 2000, 81-87.

Zagreb 1996 *Arheološka istraživanja u Zagrebu i zagrebačkoj regiji. Arheologija i obnova*, *IzdHAD* 17, Zagreb 1992.

Zagreb i njegova okolica Arheološka istraživanja u Zagrebu i njegovoj okolici, *IzdHAD* 6, Zagreb 1981.

Zagreb prije Zagreba Ante Rendić-Miočević (ed.), *Zagreb prije Zagreba. Arheološka baština Zagreba od pretpovijesti do osnutka biskupije 1094. godine / Zagreb Before Zagreb. The Archaeological Heritage of Zagreb from Prehistory to the Foundation of the Bishopric in 1094*, Exhibition Catalogue, Zagreb 1994.

Zaninović 1986 Marin Zaninović, Pojava antike u središnjoj hrvatskoj, in: *Karlovačko i sisačko područje*, 59-67.

Zlatunić 2006 Romuald Zlatunić, Povijest i razvoj tehnologije vađenja kamena i kamenoklesarstva, in: Starac 2006, 185-210.

Zsidi 2004 Paula Zsidi, Aquincum, in: *Pannonia* II, 209-230.

Zubčić 2009 Krunoslav Zubčić, Osijek – rijeka Drava (rimski most), *HAG* 5 (2008), 44-45.

Building Munificences of Septimius Severus in the Cities of the Pannonian Provinces: Epigraphic Evidence

Zsolt Mráv

In memoriam Géza Alföldy

1. Introduction

Building patronage of Roman emperors represented one of, if not the most spectacular and expensive type of imperial largesse not only to Rome but also to the towns of Italy and the provinces. To be a prolific builder was inherent in the judgment of a good *princeps*, who cares about town communities. Besides identified and archaeologically attested building projects outside Rome, the epigraphic documents (e.g. building inscriptions, brick stamps, *fistulae*) recording such projects present a source group of outstanding importance that inform us about the complex relationship system between Roman emperors and town communities across the Empire.[1]

The relationship would be defined by the interests and policies of emperors on many occasions, which were well reflected by Septimius Severus' provincial construction projects that concerned predominantly North African and Middle Danubian, mainly Pannonian towns.[2] It is a long known and well-emphasized fact that the emperor dominantly judged, supported, or punished towns of the Empire on account of their loyalty (or the lack of it) towards him in civil wars.[3] Furthermore, he lavished benefactions on those provincial towns that were related emotionally to him by his origin, or whose army served as an instrument and support of his power. The Pannonian provinces were a privileged territory of Septimius Severus' municipal policy from more aspects. Nevertheless, direct literary sources and epigraphic documents referring to it are fairly meagre in number. Such deficiency appears to be true of the building projects of the emperor as well, as merely one (assumed) restoration of a temple at Tác (county Fejér, Hungary) has been known for long (fig. 1).[4] In this study I shall therefore attempt to identify further three building donations from Septimius Severus and his sons by means of a new analysis and reinterpretation of two long known, but now lost inscriptions, one each from Aquincum (Budapest, Hungary) and Brigetio (Komárom-Szőny, Hungary), as well as two building inscriptions from Cibalae (Vinkovci, Croatia). Significant widening of the group of sources renders it possible to define some characteristic features of the specific relationship between Septimius Severus and Pannonian towns.[5]

2. The building inscription of Septimius Severus, Caracalla, and Geta from Aquincum

The 15th-century humanist and antiquarian with epigraphic interests, whom Theodor Mommsen named 'Antiquus' (who can most probably be identified with Felice Feliciano[6]), recorded an inscription that he noticed on a Roman stone slab reused as a medieval tombstone in the Franciscan Saint John Church of the Buda Castle (1-3 Színház street, District I, Budapest). This inscribed stone went missing during the wars against the Turks, or the baroque reconstruction of the church at the latest. The inscription or its fragments were not discovered on the occasion of the 1971 archaeological excavations either, which otherwise yielded more fragments of Gothic and renaissance inscribed tombstones.[7] Without seeing it, Mommsen identified this monument as a milestone and classified it among the milestones of Pannonia Inferior in *CIL* III. Its inscription was published on the basis of the *Antiquiores* in the following restored interpretation (fig. 2)[8]:

IMPP · L · SEPT · SEVERVS · PIVS
PERTINAX · A[R]A[B]ICVS · ADIA-
BENICVS · MAXIMVS · ET
M · AVR · ANTONINVS · AVGG · ET
5 [[- - - - - - -]]
[[- - - - - - -]]
[[- - - - - - -]]
[[- - - - - - -]]
SPLENDIDISSIMAE

[1] Therefore municipal building activity of Roman emperors and mainly the inscriptions commemorating them are regarded as distinguished fields of study; see McMullen 1959; Boatwright 1989; Waldherr 1989; Ward-Perkins 1993; Winter 1996; Horster 1997; Boatwright 2000; Scheithauer 2000; Horster 2001; Boatwright 2002; Alföldy 2002a; Alföldy 2002b; Fraser 2006. For the towns of the Pannonian provinces see Mráv 2000; Mráv 2001; Mráv 2002a; Mráv 2003a; Mráv 2006; Mráv 2010. Ibid.: Lőrincz 2007b, 213-217.

[2] For the military building inscriptions of Septimius Severus see Kovács 1999, 521-531 – cf. P. Kovács in *FPA* IV, 60.

[3] Haywood 1940, 175-186; Hasebroek 1921, 64-66; Birley 1971, 177-180; Kovács 2001, 42-43; Kovács 2007, 145-146.

[4] *RIU* 6, 1537 – cf. Alföldy 1997, 225-241. The inscription possibly commemorated the renovation of the Temple of *Dii Magni* at Tác, thus the grant indirectly addressed the town itself as one of the most important settlements in Aquincum's territory.

[5] The study is part of a larger, more comprehensive study that deals with the relationship between Septimius Severus and Pannonian towns.

[6] Szalay 1989, 65-75 - see further Kovács 2010, 117.

[7] For the excavation of the Saint John Church see Altmann 1973, 82 and Altmann 1994, 147. Publication of the recovered medieval tombs with inscription: Altmann and Lővei 2004, 11-34.

[8] *CIL* III 3706.

Fig. 1. Tác (Fejér County), Hungary: the imperial building inscription commemorating the restoration works of the templu(m) D[eorum Magnorum] from 197-209 AD. (Photograph of the remaining fragment is by O. Harl and a drawn restoration redrawn by Zs. Mráv, after Alföldy 1997, 235, Abb. 3.)

10 CVRANTE · L · BAEBIO · CAE-
CILIANO · LEG · AVG · PR · [PR].

The Roman stone monument of the quondam Franciscan Church of the Buda Castle, however, cannot be a *milliarium* for several reasons. Mommsen based the determination of the inscription type on the contemporary milestone of Bátaszék (Tolna County, Hungary),[9] which is a hardly plausible hypothesis, given that on the former stone the emperors are mentioned by their entire nomenclature in the ablative of time, with their lineages brought as well – from Marcus Aurelius in the case of Septimius Severus, and from Nerva in the case of Caracalla – and complete titles; at the end of the text the following formula of distance is indicated: *[a]b Aqu(inco) [m(illia) p(assuum) - - -]*. In the inscription from Buda, only the names of Septimius Severus, Caracalla (and of Geta) could be observed, but without the names of their adopted ancestors, and in the nominative case at that. The data indicating the distance, which was regarded as substantial in the case of milestones, is missing from the text, but it could have as well been chipped off. The only common feature of both of the two inscriptions is the appearance of the phrase *curante* and the governor's name, L. Baebius Caecilianus. The superlative form of the adjective *splendidissimae* with the ending in the genitive or dative case is also a unique example in the wording of milestone inscriptions. The findspot and the proof of the inscribed stones' reuse finally exclude their potential interpretation as a milestone. A stone cylinder, even in the chopped form, is unsuitable for covering a medieval grave sunk into the floor of a church, especially not in a form that let the inscription – which had been engraved onto an arched surface – be fully readable. In medieval times, such stone slabs were in use to cover graves inside churches that did not obstruct the passage; the same could have happened in the case of the mentioned grave of the Saint John church as well.[10]

This stone monument could have carried an imperial building inscription instead, as its text meets all of the required criteria. The names of the emperors were written

VIAE PANNONIAE INFER.

3706 Budae in aede S. Io. in lapide qui est supra sepulcrum ANT. in templo beato Ioanni dedicato fratrum minorum IUST.

IMP·L·SEPT·SEVERVS·PIVS
PERTINAX · *ArAbICVS* · ADIA
BENICVS · MAXIMVS · ET
M·AVR·ANTONINVS·AVGG·ET
5 . . . .
. . . .
. . . .
. . . .
SPLENDIDISSIMAE
10 CVRANTE·L·BAEBIO·CAE p. C. 199
CILIANO·LEG·AVG·PR·*pr*

Antiquus apud Ferrarinum n. 2 et Rambertum n. 2 (inde Don. 3, 37 ex Ramb., Mur. 244, 3 e sched. Farnes., 456, 2 e schedis Ramb. apud Bert.); Iustinianus ms.

1 IMP Ramb. (Flor.). — SEPTIMIVS Ramb. — 2 ALANICVS Ferr. Ramb. (Flor.), ALCINICVS Ramb. (Pat.). — 4 AVRELIVS Ramb. (Flor.). — 'Deficiunt quattuor lineae' Ramb., 'deficiunt multae litterae' Ferr. — 9 SPENDIDISSIMAE Ramb. (Pat.), SPLENDIDISSIME Ferr. — 10 BEBIO Ramb. — 11 PROCVRATORE pro PR·PR Ant. — Iustinianus sic habet: D·M|IMP·L·SEP·SEVER|VS·PIVS·PERTENAX|AN·II·CVRANDVM|BENICVS·MAXIM|VS·ET·M·AVR|ANTONIVS·V·CO|ET.

Eiusdem tituli alterum exemplum Bataszekense (unde constat hunc quoque cippum stetisse olim ad viam eam de qua agimus) infra proponitur n. 3733; hoc qui descripserunt non exceperunt nisi maioribus litteris perscripta et lectu faciliora. V. 9 SPLENDIDISSIMAE ortum est opinor ex Getae titulo nobilissimi Caesaris.

Fig. 2. Inscription CIL III 3706 issued by Th. Mommsen among the milestones of Pannonia Inferior.

[9] *CIL* III 3733.

[10] Apart from one sarcophagus, only inscribed stone slabs were recovered during the excavation of the Franciscan church of the Buda Castle: Altmann and Lővei 2004, 14-19.

in the nominative case[11], which indicates clearly that the construction was accomplished on the imperial family's own expense. The governor was in charge of the construction works, which is reflected by the word *curante*.[12] The same expression figures on Septimius Severus's and Caracalla's temple-building inscription from Tác (?),[13] and on two monumental imperial inscriptions from Cibalae (present-day Vinkovci, Croatia); in both cases the mentioned word stands exactly in front of the name of L. Baebius Caecilianus (see below).[14] The adjective *splendidissimae* in line 9 is rather in the dative than in the genitive case, which should be associated with the title *nobilissimus Caesar* of Geta in Mommsen's *CIL* III 3706 commentary, and refers to a construction relating to the town of Aquincum, too. Mommsen's hypothesis is not convincing for two reasons. Firstly, the adjective *splendidissimae* was neither written in the nominative case nor is its gender in agreement with Geta's name. Secondly, these two words were divided by at least three consecutive lines. It is unnecessary to regard the superlative *splendidissima* as an incorrect mode of expression, since it was a widely applied formula in inscriptions of the Roman Imperial period. Apart from a few exceptions, it was a specifying adjective of the *ordo decurionum* and settlements with municipal status (*municipium, colonia, civitas*), as well as their communities (*res publica, populus*). It was, however, often associated with provinces[15] and private associations (*collegia, corpores*) as well.[16] In inscriptions of private initiations, such expressions were recorded as adjectives of the words *puella* and *familia* on some occasions only.[17] The adjective *splendidissimus* was not unfamiliar in Aquincum, where it was associated with the town, too. A fragment of an altar was recovered from the foundation wall of the mayor's residence in Kiskunlacháza (Pest County), which was dedicated to I. O. M. Teutanus in 223 AD by [- - -] Maturus and Aur. Epictetianus, *[IIviri] col(oniae) splendidis[simae] Aq(uincensium)*.[18] The adjective appears in the sacral building inscription of C. Cornelius Corinthus, which can be dated to the 2nd century AD. In order to show gratitude for having become an *augustalis* of the town, the dedicator had *splendidissimo* a sanctuary constructed.[19] The adjective pronoun can possibly denote the local corporation of *augustales* or more likely the *municipium* of Aquincum that coincidentally also served as the spot for the *munus*.[20] In my opinion, in his inscription Corinthus called the municipium *splendidissimus* just to emphasise his personal sentimental bonds to his town.

In the first three lines of the inscription, the names of two *Augusti* – Septimius Severus and Caracalla, can be read. Triumphal cognomina of Septimius Severus were also indicated, however not in a correct form (see below). The conjunction ET that remained behind the name of the *Augusti* at the end of the 4th line was followed by Geta's name – if we assume the same letter size – that is most likely formulated by 18 letters in the following form: P. SEPT. GETA NOB. CAESAR (20 letters could fit in lines 1 and 2, 16 in line 3, and 19 in line 4.) Expectedly, from the restored text the name of Geta was erased. The person who executed the *abolitio nominis* did not stop after finishing with Geta's name, but – through either illiteracy or ignorance – he erased another three lines from the text. It cannot be excluded that lines 6-8 disappeared as a consequence of the stone's medieval reuse. The erased lines probably contained the subject matter and verb of the construction works, which remain unknown.

Those who were granted donation, the town or its inhabitants, were often mentioned in the imperial building inscriptions.[21] The inscription discovered in Cibalae, which named L. Baebius Caecilianus as governor overseeing the work, also states that the building was granted *Cibalensibus* by the emperors. The superlative adjective *splendidissimae* in line 9 can stand in the dative case, because of the usual wording and context, and can refer solely to the community for whom the construction was made: the town of Aquincum, which had been granted colonial status just before that and to which the adjective was actually attached in some inscriptions (see above). The town name *Aquincum* was, however, not indicated in the inscription, since the complete name with the adjective *colonia splendidissima Aquincensium* cannot be accommodated in the space available. Consequently, simpler phrases like *[coloniae] splendidissimae* or solely *splendidissimae* remain as the potential expression, thence for the audience they referred unequivocally to their own town.

The emperors assigned L. Baebius Caecilianus, who was the actual governor of the province Pannonia Inferior, to supervise the construction works, for which he was referred to by the word *curante* at the end of the inscription. The imperial construction was in progress and accomplished during his tenure in office some time between 198/199 and 202 AD.[22] Since January 198 AD, Septimius Severus used the triumphal cognomen *Parthicus maximus*,[23] which, however, is missing from his nomenclature in the inscription. The striking lack of this cognomen cannot be explained with the alternatives assuming that either the inscription had been made before Septimius Severus was given the cognomen *Parthicus maximus* or that the information of the latter had not reached Aquincum yet. This might have happened in the last months of 198 AD, as at that time Tiberius Claudius Claudianus, his predecessor in the

[11] Horster 2001, 39-48.
[12] Winter 1996, 157-160; Horster 2001, 191-198.
[13] *RIU* 6, 1537 – cf. Alföldy 1997, 225-241.
[14] Perinić Muratović 2004, 77-101 (*AE* 2004, 1135) – cf. Mráv 2006, 211-228 and in this study (see below).
[15] *AE* 1911, 107.
[16] E.g.: concerning *collegium*: *AE* 1965, 194; and *corpus*: *AE* 1955, 165; *AE* 2001, 1186.
[17] *AE* 2003, 352; *IDR* III/5, 441 (in this case only the initial letter S remained, thus its completion is uncertain).
[18] *CIL* III 10481 = *AE* 1993, 1313 = *TRH* 219 = Beszédes, Mráv, and Tóth 2003, Kat. Nr. D.
[19] *CIL* III 3579 = *TitAq*, 290 – cf. Gabler 1966, 22, Nr. 42.
[20] Gabler 1966, 30-33.
[21] Horster 2001, 64-67.
[22] For the governor L. Baebius Caecilianus's career see *PIR*[2], B14; Birley 1971, 338; Fitz 1993, 541-542, Nr. 323; Lőrincz 2004, 38-39, Nr. 13; Lőrincz 2007a, 184.
[23] Kneißl 1969, 142-148; Kienast 19962, 158.

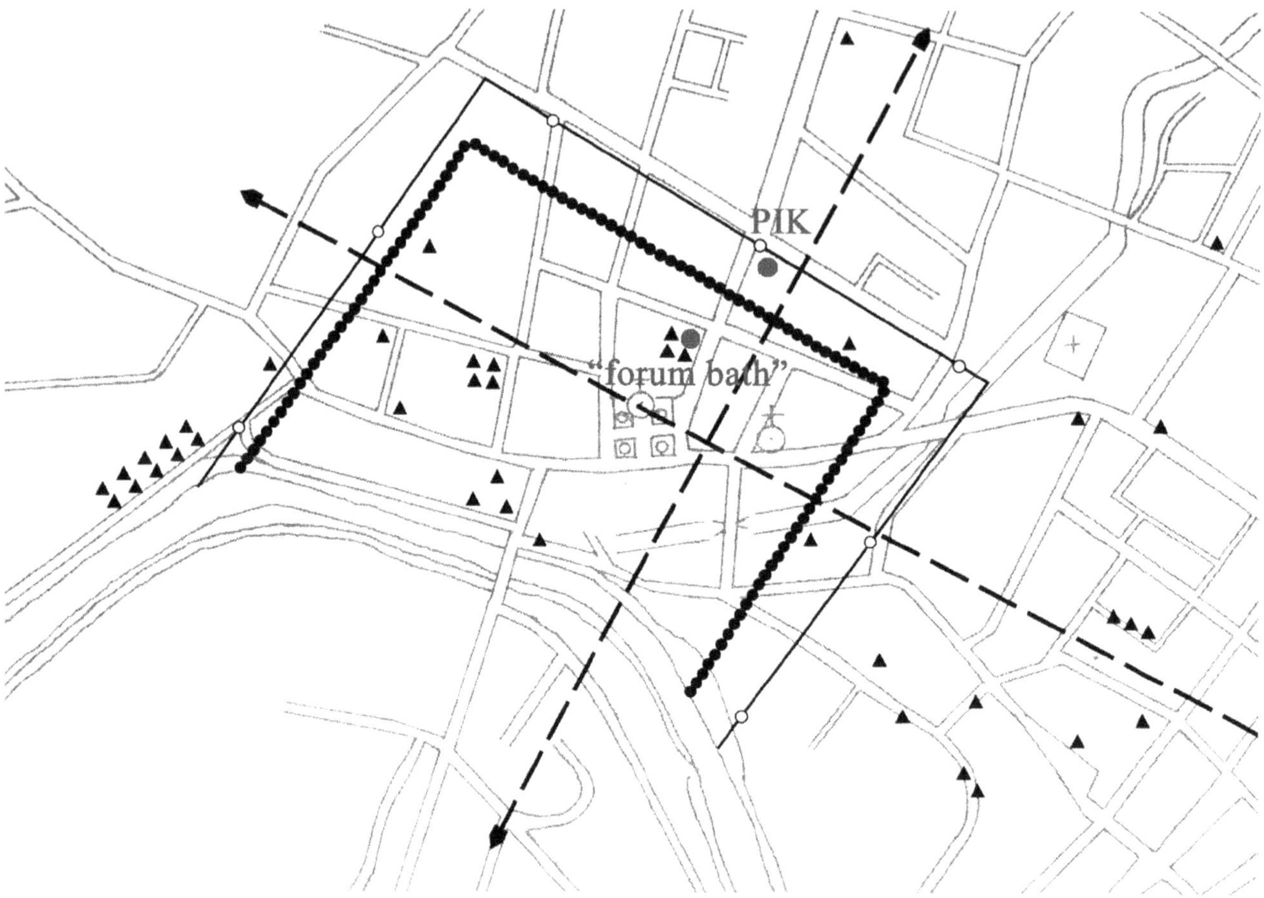

Fig. 3. The Roman town of Cibalae (Vinkovci), Croatia, and the PIK archaeological site (after Iskra-Janošić 2001).

office, was certainly acting governor of Pannonia Inferior.[24] We can arrive at the conclusion that the *scriptor* made the mistake by omitting the cognomen Parthicus, which is also verified by the presence of the superlative *maximus* that normally followed this triumphal cognomen. Septimius Severus began using this superlative in his nomenclature only later, together with the cognomen Parthicus (the Senate endowed him earlier with the cognomens 'Arabicus' and 'Adiabenicus', but not in superlative).[25] It appears likely, however, that the emperor had already decided about the construction before January 198 AD, and this was the reason why his name was inscribed on the slab in an incorrect form not long after that, perhaps in the second half of 198, or at the beginning of 199 AD. In this case, the governor could have received his assignment to supervise the constructions at the moment of his own appointment to governorship, thus this project might have been one of the first accomplished deeds in his province, Pannonia Inferior.

Taking everything into account, my restoration of the inscription is the following:

Im̂pp(eratores) L(ucius) Sept(imius) Severus Pius
Pertinax A[r]a[b]icus Adia-
benicus <Parthicus> maximus et
M(arcus) Aur(elius) Antoninus Augg(usti) et
[[P(ublius) Sept(imius) Geta nob(ilissimus) Caesar]]
[[- - - - - - -]]
[[- - - - - - -]]
[[coloniae?]]
splendidissimae
curante L(ucio) Baebio Cae-
ciliano leg(ato) Aug(ustorum) pr(o) [pr(aetore)].

The carrier of the inscription that had been reused in the quondam Saint John church of the Buda Castle could not have been a monumental, several-meter-long stone slab. The location and way of its reuse, and the fact that the lines of inscription remained undamaged and legible in full length, suggest that the original width of the inscription could not have exceeded 1 meter. The text's 11 lines, each consisting of ca. 20 letters, render the standing format of the slab probable. The titles of the co-emperors were not engraved on the slab, as was the case with the inscription

[24] *PIR*², C834; Fitz 1993, 509-513, 541 Nr. 300; Lőrincz 2004, 38 Nr. 12.
[25] Kneißl 1969, 126-138; Kienast 1996², 158.

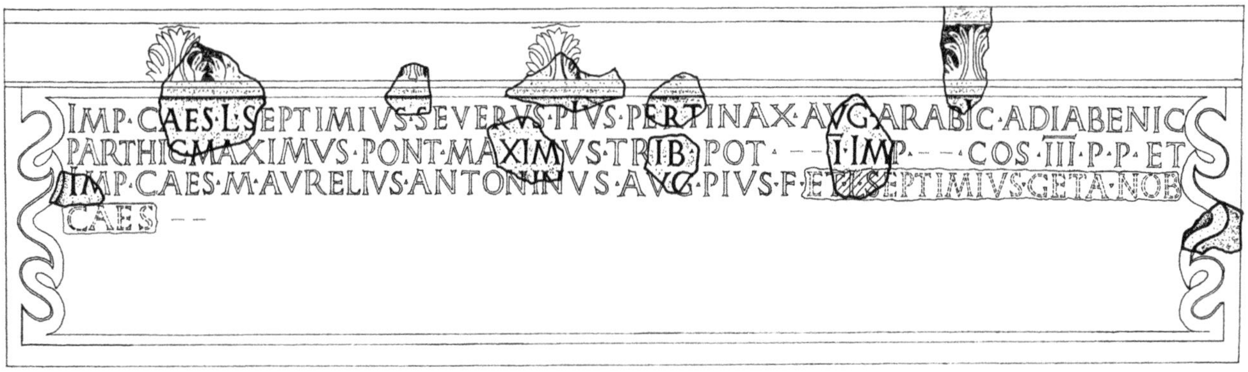

Fig. 4. Cibalae (Vinkovci), Croatia: the first and preliminary drawn restoration of the building inscription of Septimius Severus (after Mráv 2006, 225, fig. 3.).

from Tác, too; this can be explained by a limited width of the slab and a minor importance of the building. An inscription relating an imperial construction sponsored by Septimius Severus and his sons in Sarmizegetusa (Dacia) bears strong resemblance to this one – only its text is shorter.[26]

By means of the still unknown building – certainly not as the sole constructed donation – in Aquincum, Septimius Severus and his family contributed personally to the civil physiognomy of the town, which had been granted colonial status only a few years earlier. According to inscriptions and excavations, the town and the quondam *canabae*, which was attached to it, actually flourished during the reign of the Severi.[27]

3. The construction of Septimius Severus and his sons in Cibalae

Almost coincidentally with the building munificence of Septimius Severus and his sons to Aquincum, another construction was accomplished, which is recorded in at least two building inscriptions. In 1897, several stones, comprising marble statues of Neptunus and Hercules and 15 inscribed stone fragments, were recovered close to the town wall (14-18 Zvonimirova Street) in the northern periphery of Cibalae (present-day Vinkovci, Croatia). The inscription fragments, acquired by the AMZ, were first published by J. Brunšmid (here cat. no. I/2, 4-5, 7, 10, 12 and II/3).[28] Given the small number and a fragmentary state of the pieces, at the time hardly anything could be reasonably assumed on either the inscription they belonged to or the archaeological context. Fragment no. II/4 was certainly interpreted correctly by J. Brunšmid as relating to Septimius Severus and Caracalla: *[- - - Seve]rus [- - - | - - - m]axim[us - - - | - - - Anton]in[us - - -]*.[29] Near the place of their discovery, at the PIK archaeological site during the rescue excavations in 1977, remains of a *horreum* dated to the 4th century AD and a section of a wall and five pillars distributed in two rows were found (fig. 3).[30] Several 2nd- and 3rd-century stone reliefs and inscriptions were built into the foundations of the pillars as *spolia*, which were published by Ljubica Perinić Muratović in 2004 (cat. no. I/1, 3-4, 8, 9, 11 and II/1-3, 4-8). She observed that these fragments and those found in 1897 could originally be in connection with a building inscription of Septimius Severus.[31] Perinić Muratović published certain fragments but did not make an attempt to complete the inscription, assuming only that most of the fragments "belong to the same inscription which most probably figured at the front of a temple of the Capitoline triad". In 2006, I myself scrutinised the fragments of the inscription in a preliminary report, and made an effort to date and complete its text (fig. 4).[32] Since then, I was given the opportunity to examine the inscription fragments in Zagreb, in the Croatian Academy of Sciences and Arts[33] and the AMZ,[34] during which I not only managed to identify some new fragments, but also established that they belonged to two separate building inscriptions of nearly identical text, but had been made of different quality limestones and inscribed with letters of varying size. Owing to this fortunate circumstance and the fact that the fragments overlapped distinct sections of the two inscriptions, the remains of letters enabled mutual completing of each other's missing parts. This method, together with the drawing completions, permitted restoration of the entire text of the inscriptions. The two Severan building inscriptions read as follows:

I. Monumental imperial building inscription from hard, good quality limestone, of which 13 parts remained; some of them are composed of more fragments, thus the

[26] *CIL* III 1451 = *IDR* III 2, 21 = Horster 2001, 408 Kat. Nr. XXX 3, 2: *[Imp(erator) C]aes(ar) L(ucius) Sept(imius) Severus / [Pe]rt(inax) Aug(ustus) Arab(icus) Adiab(enicus) / Part(hicus) max(imus) et / Imp(erator) Caes(ar) M(arcus) Aurelius /5 Antoninus P(ius) Aug(ustus) et / [[P(ublius) Sept(imius) Geta]] nobiliss(imus) / Caesar fecerunt.*

[27] For this issue see Kovács 2001, 65-66 and Zsidi 2002.

[28] Brunšmid 1902, 80-81; Brunšmid 1912, 349-350.

[29] Brunšmid 1912, 350.

[30] Iskra-Janošić 2001, 92-94; Iskra-Janošić 2004, 180.

[31] Perinić Muratović 2004, 77-101 – cf. *AE* 2004, 1135.

[32] Mráv 2006, 211-228 = *AE* 2006, 1092.

[33] Fragments belonging to the collection of the GMVi were at the time temporarily stored in this institution.

[34] I would like to express my deep gratitude to Ljubica Perinić Muratović and Branka Migotti (The Croatian Academy of Sciences and Arts) and Ivan Radman-Livaja (AMZ) for their cordial assistance in the investigation of the inscription fragments.

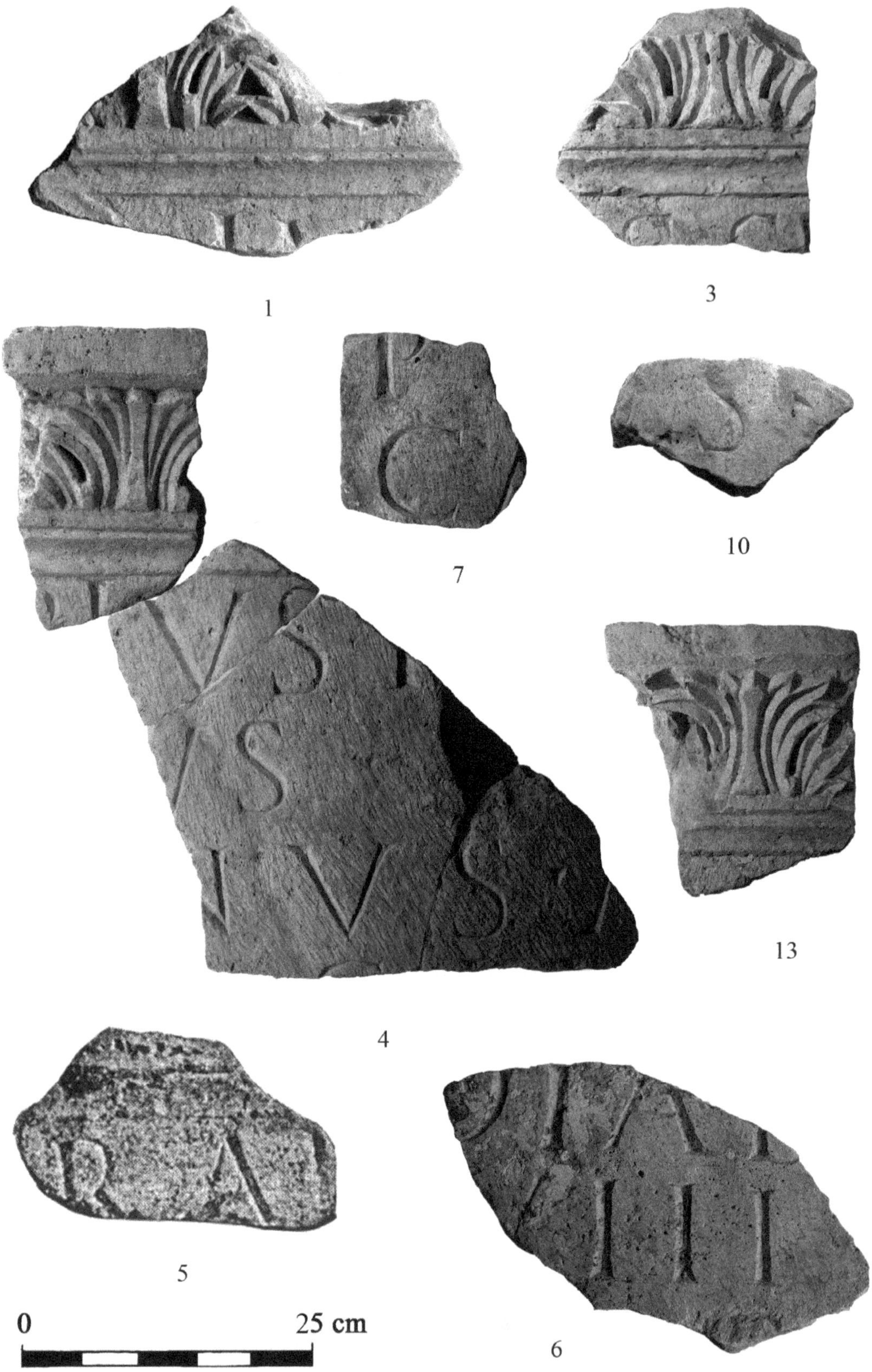

Fig. 5. Cibalae (Vinkovci), Croatia: photographs of the fragments belonging to the first Severan building inscription (I) (made by Zs. Mráv).

number of fragments attributed to the inscription field is 19 altogether.[35]

1. IM[- - -] | - - - - - - (fig. 5/1)

H.: 20 cm; w.: 37.5 cm; t. : 11.5-13 cm. GMVi.

A *cyma reversa*, framing the inscribed field, takes place on the top, above which a fragment of cyma with acanthus relief can be observed. On the left edge, the arched protrusion of the upper section of a Norico-Pannonian volute can be seen. Accordingly, the fragment carries the first two letters of the first line, which can hardly be anything else but the first two letters of the imperial *praenomen Im[p(erator) - - -]*.

Literature: Perinić Muratović 2004, 84, cat. no. 1, fig. 1.1.

2. [- - -]IM[- - - | - - -]NT[- - - | - - -]ẸS · +[- - - | - - -]+[- - -] | - - - - - - (fig. 6/2)

Upper fragment: h. 23.3 cm, w. 26.8 cm, t. 9 cm; lower fragment: h. 15.7 cm; w. 18.5 cm; t. 15.5 cm. AMZ.

Two fragments that fit together along the crack line with letters of the first three lines. The letters of the first line were cut more densely, while those in the second line are more loosely spaced.

Literature: Brunšmid 1902, 26, fig. 68.3 and 8; Brunšmid 1904-1912, 349 cat. no. 764.C.

3. [- - -]S [·] SẸ[- - -] | - - - - - - (fig. 5/3)

H.: 21.5 cm, w.: 22 cm; t.: 16 cm at moulding, 11.5 cm at inscribed field. GMVi.

A *cyma reversa* takes place above the inscribed field, above which a fragment of cyma enriched with acanthus leaves can be observed. It carries the letter fragments of the first line, whose S is the last one of Septimius Severus's *gentilicium* in the nominative case, whereas SE[- - -] are the first two letters of the cognomen Severus.

Literature: Perinić Muratović 2004, 86, cat. no. 16, fig. 3.12; Mráv 2006, 213, cat. no. 2.

4. [- - -] PIVS · P̣[- - - | - - -]VS · [- - - | - - -]NVS · A[- - - | - - -]C̣ỊḄ[- - -] | - - - - - - (fig. 5/4)

Upper fragment: h. 25.6 cm, w. 17 cm, t. 14.8 at the moulding, 5.5 cm at the inscribed field; middle fragment: h. 16.3 cm, w. 18.5 cm, t. 9 cm; lower fragment: h. 32.5 cm; w. 42 cm; t. 17 cm. AMZ and GMVi (middle fragment).

Three matching fragments with letters of the first three lines. A moulding ornamented with cyma and acanthus leaves runs in the upper section of the topmost fragment that carries the inscription [- - -]PI[- - -] with endings of the three name elements, which were in the nominative case. Only the topmost parts remained of the first three letters of line four, which have already been read as CIB by J. Brunšmid.

Litrature: Brunšmid 1902, 26, fig. 68.2 and 5; Brunšmid 1904-1912, 349-350, cat. no. 764. A and D.

5. [- - -]RAḄ[- - -] | - - - - - - (fig. 5/5)

J. Brunšmid only brought the letters' height (9.5 cm). Considering this height and a looser spacing of the letters, the fragment could only have belonged to inscription I (AMZ, lost).

A moulding with acanthus leaves running above the *cyma reversa* formed the upper frame of the inscribed field. Three letters of the triumphal cognomen Arabicus of Septimius Severus can be read on the fragment.

Literature: Brunšmid 1902, 26, fig. 68.6; Brunšmid 1904-1912, 349-350, cat. no. B.

6. [- - -]DIAB[- - - | - - -]VIII · [- - - | - - -] [[G̣]][- - -] | - - - - - - (fig. 5/6)

H.: 22.5 cm; w.: 36.5 cm; t.: 15.5 cm. GMVi.

Letters of the first line belonged to the triumphal cognomen Adiabenicus of Septimius Severus. Line 2 consists of iteration number VIII, which could refer to the eighth renewal of his *tribunicia potestas* on the basis of its position in the inscription. The horizontal stroke above the number was made with a single engraved line, which was certainly covered by stucco. Therefore, the stroke was presumably highlighted by painting. In the third line, the fragment of the upper arch of a C or G can be seen, which should certainly be identified as the first letter (G) of Geta's cognomen.

Unpublished.

7. - - - - - - | [- - -] P[- - - | - - -] CẠ[- - -] | - - - - - - (fig. 5/7)

H.: (15.8) cm; w.: (16) cm; t.: 8.9 cm. AMZ.

Literature: Brunšmid 1902, 26, fig. 68.4; Brunšmid 1904-1912, 349-350, cat. no. 764 G.

8. - - - - - - | [- - -]AN[- - - | - - -]NE[- - - | - - -]C[- - -] | - - - - - - (fig. 6/8)

H.: 18.5 cm; w.: 21.7 cm; t.: 11 cm. GMVi.

The letter in the third line can be G.

Literature: Perinić Muratović 2004, 88, cat. no. 33, fig. 6.29.

[35] Abbreviations used in the catalogue: h (height), t (thickness), w (width).

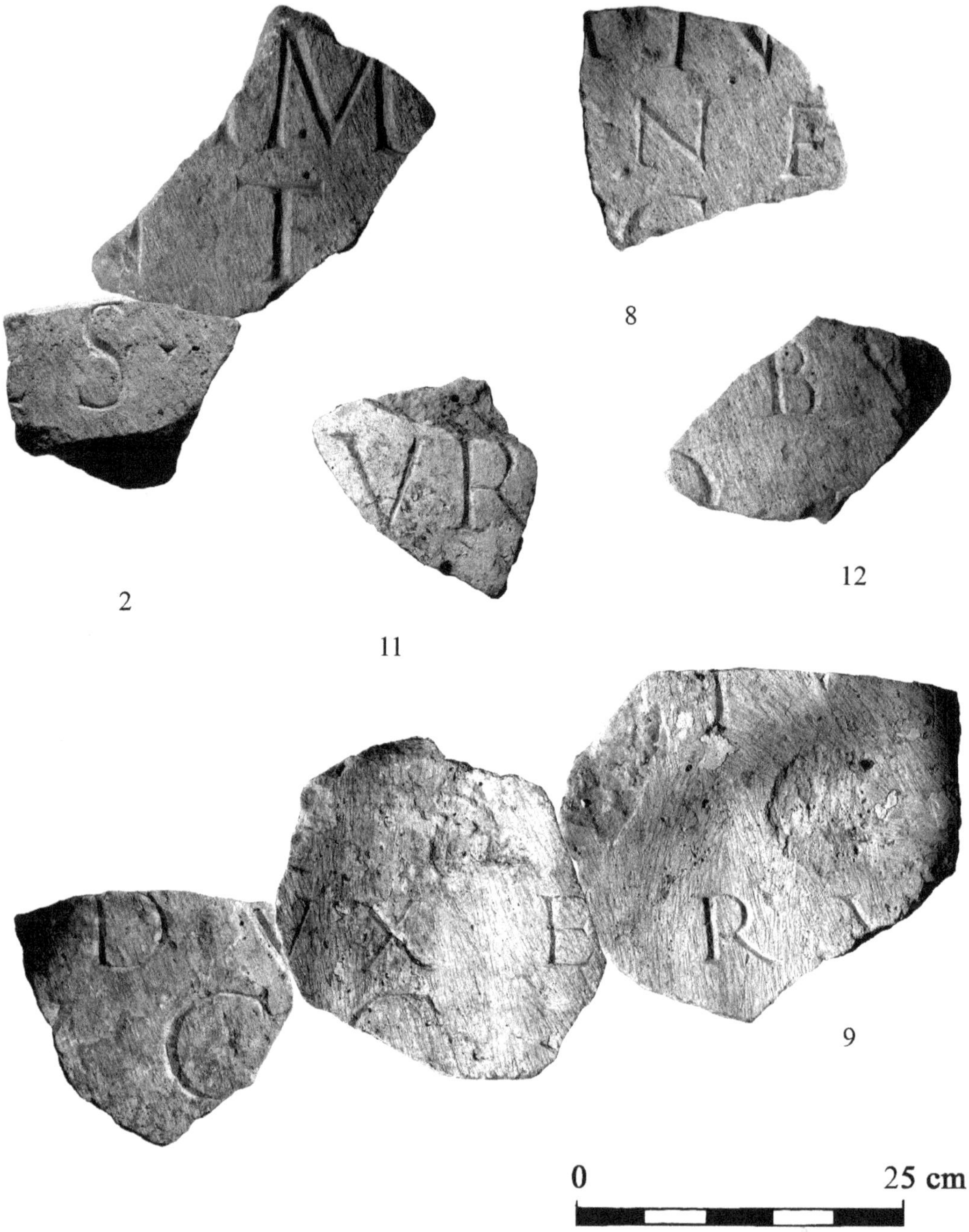

Fig. 6. Cibalae (Vinkovci), Croatia: photos of the fragments belonging to the first Severan building inscription (II) (made by Zs. Mráv).

9. - - - - - - | [- - -]+I · [- - - | - - -][[A · C]][- - - | - - -]
DVXERV[- - - | - - -]ṾGG̣ · [- - -] | - - - - - - (fig. 6/9)

The first fragment from the left: h. 19.8 cm, w. 20.7 cm, t. 15.5 cm; middle fragment: h. 26.5 cm, w. 25.5 cm, t. 16 cm; right fragment: h. 30 cm; w. 26 cm, t. 15 cm. GMVi.

Three matching fragments with letters of the last four lines. Number [- - -]I is an iteration number of Septimius Severus's titles, more exactly its last digit. Two erased letters of the following line remained from Geta's name, while beneath it the verb of construction is indicated with smaller letters. In the last line, letters of the governor's titles can be recognized.

Literature: the first two are unpublished; the third one was published by Perinić Muratović 2004, 87, cat. no. 28, fig. 5.28.

10. - - - - - - | [- - -]Ọ S · [- - -] | - - - - - - (fig. 5/10)

H.: 11.8 cm; w.: 21.3 cm; t.: 14.1 cm. AMZ.

Based on the *punctum* that was engraved after it, it might have belonged to a word with [- - -]OS ending, or an abbreviation, which was part of the titles in line 2 in accordance with the heights of the letters. Its completed version is most likely [- - - C]OS.

Literature: Brunšmid 1902, 26, fig. 68.10; Brunšmid 1904-1912, 349-350, cat. no. 764 F.

11. - - - - - - | [.]VR[- - -]. (fig. 6/11)

H.: 15 cm; w.: 13 cm; v.: 14 cm. GMVi.

It can be categorised as a fragment of inscription I on the basis of the quality of stone. Although the letters could belong to the 'Aurelius' *gentilicium* of Caracalla, given the larger empty space under it and its letter's height, it is likely to fit into the expression *curante* that is written at the beginning of the line (see below).

Literature: Perinić Muratović 2004, 86, cat. no. 20, fig. 4.20.

12. - - - - - - | [- - -]BV[- - - | - - -]O · [- - -] | - - - - - -? (fig. 6/12)

H.: 15.7 cm; w.: 25.4 cm; t.: 13.4 cm. AMZ.

Literature: Brunšmid 1902, 26, fig. 68.14; Brunšmid 1904-1912, 349-350, cat. no. 764 L.

13. [- - -] + [- - - | - - - - - - (fig. 5/13)

H.: 22.8 cm; w.: 21.8 cm; t.: 16 cm (at the moulding), 115 cm (at the inscribed field). GMVi.

A vertical fitting surface carved to flat can be seen on its right side. The fragment certainly belongs to the first line, because of the cyma with acanthus. An arched letter fragment remained (C, G, or S) with a larger empty space behind it, which indicates an ending of a word or abbreviation. Its location thus can be more varied.

Literature: Perinić Muratović 2004, 85, cat. no. 8, fig. 1.4.

On the top of fragments no. I/1 and I/3-4, above the *cyma reversa* of a 4.4 cm-wide upper frame, runs a vigorously figured moulding ornamented with rows of acanthus leaves and highlighted with the red background. A fragment of a Norico-Pannonian volute, which was partly painted in red, remained on the left edge of fragments no. I/1 and I/11. The inscribed field was thinly stuccoed. Lines 1, 2 and 5 fill in the available space completely, while lines 3 and 4 appear to be left-aligned. Apart from the evenly dense first line, beginnings of the rest of the lines are denser, whereas letters follow each other at more spaced intervals toward the line's ends. The *scriptor* obviously tended to widen the lettering through increasing intervals and spaces as the lines advanced. This unusual solution is one characteristic of the inscription's layout. The meticulously engraved, elegant letters were originally painted in red. Heights: 9.2-9.5 cm (line 1); 7.6-7.7 cm (line 2); 8-8.2 cm (line 3); 5.5 – 5.7 cm (line 4); ca. 7.8 cm (line 5). The letter *I* in the words *Imperator* (I/1) and *Pius* (I/4) are certainly *litterae longae*. The relatively small line spacing decreases slightly as going downwards: 2.2 cm (between lines 1 and 2), 2 cm (between lines 2 and 3) and 1.8 cm (between lines 3 and 4). Words and abbreviations are divided consistently with *puncta* represented by reversed triangles. The text of the inscription can be constructed as follows (fig. 7):

Ị̣ṃ[p(erator) · Caes(ar) · L(ucius) · Sept]iṃ[iu]ṣ · Ṣ[everus] · P̣ịus · P̣[ertinax · Aug(ustus) · A]raḅ[ic(us) · A]diaḅ[enic(us) · Parthic(us) | maximus ·] p̣[o]ṇt[ifex · maxim]us · [tribunicia · potest(ate) ·] ṾIII · [imp(erator)] · X̣I · [c]ọ(n)s(ul) · [II | p(ater) · p(atriae) · et · Imp(erator) ·] Cạẹs(ar) · [M(arcus) · Aur(elius) ·] An[toni]ṇus · Ạ[ug(ustus) · et [[· L(ucius) · Septimius ·] G̣[et]a · C̣[aes(ar)]] | extruxerunt · bal]ne[um? · et ·]C̣ịḅ[alensi]bu[s · aquam · per]duxeru[nt |5 [c]ur[ante · L(ucio) · Baebio ·] C[aecilian]o [· legato · A]ụgg(ustorum) · [pro · pr(aetore)].

The length of the inscribed field could have been 6.2 meters. Its height cannot be accurately ascertained, since we do not know whether more lines followed under the fifth line, or whether its surface was left empty; if there were more lines, they should have been arranged centrally. The fitting surface made on fragment no. I/13 indicates that the inscription was not written on a single limestone slab only, but was composed of more panels.

II. The second building inscription was made of poor quality, soft limestone, which contained a great deal of fossils, mostly shells. We know 9 inscribed sections of it, composed of twelve fragments. Below is the catalogue of fragments belonging to the inscription.

1. [- - -] C̣AES · L · S[- - - | - - -]C · MAX̣[- - - | - - - - - - (fig. 8/1)

H.: 30 cm; w.: 31.7 cm; t.: 17.5 cm. GMVi.

In the first line of this fragment admittedly L. Septimius Severus was mentioned. The letters of the second line – on the grounds of the fragment's position within the inscription field – can be explained only as the victory cognomen of the emperor: *[Parthi]c(us) ma[ximus - - -]*.

Literature: Perinić Muratović 2004, 85, cat. no. 15, fig. 3.15; Mráv 2006, 213, cat. no. 1.

2. [- - -]IM[- - - | - - -]Ṣ [·] P̣[- - - | - - - - - - (fig. 8/2)

H.: 8.2 cm; w.: 11.5 cm; h.: 8.2 cm. GMVi.

Unpublished.

3. [- - -]RVS [- - - | - - -]XIM[- - - | - - -]ṆIN[- - - | - - -] BAL[- - - | - - -]Ç[- - -]. (fig. 8/3)

Upper fragment: h. 15.8 cm, w. 20.7 cm, t. 12.5 cm. AMZ. Lower fragment: h. 14.3 cm, w. 18 cm, t. 16 cm. GMVi.

Relying upon the [- - -]RVS ending of the name, which can be read in the first line of this fragment, the name of Septimius Severus is given in the nominative case; the two S letters of fragment no. I/3 also prove this. Thus, the building munificence must have been carried out at the imperial family's own expense, that is, the stone can be interpreted as an imperial building inscription.

Literature: Brunšmid 1902, 26, fig. 68.16; Brunšmid 1904-1912, 349-350, cat. no. 765; Iskra-Janošić 2001, 59, 94; Perinić Muratović 2004, 86, cat. no. 21, fig. 4.21; Mráv 2006, 213, cat. no. 3.

4. [- - -]ERT[- - - | - - -]ṚIB · P̣[- - - | - - -]VG[- - -] | - - - - - - (fig. 8/4)

Upper fragment: h. 13.8 cm, w. 14.4 cm, t. 8.5 cm; lower fragment: h. 17.2 cm; w. 11.5 cm; t. 8 cm. GMVi.

The letters [- - -]ERT[- - -] in the first line cannot be interpreted as the title *consul tertium* of the emperor, as L. Perinić Muratović assumed; it can only be completed as the name *P]ert[inax*, which Septimius Severus took up on the occasion of his ascension, and used consistently later on.

Literature: Perinić Muratović 2004, 86, cat. no. 18, fig. 3.18 and cat. no. 1, fig. 3. 17; Mráv 2006, 213-214, cat. no. 5-6.

5. [- - -]VG · A[- - - | - - -]I · IM[- - - | - - -][[T · Ḷ · ṢẸ]] [- - -] | - - - - - - (fig. 8/5)

H.: 24 cm; w.: 17 cm; t.: 7.5 cm. GMVi.

Because of the fragment's small size, cyma part remained above line 1. The piece contains the letters of lines 1-3. [- - -]VG in the first line belongs to the cognomen 'Augustus' of Septimius Severus, while the letter A following it is the initial letter of his triumphal cognomen Arabicus. In line 2, iteration number [- - -]I can only be the number of the renewal of his *tribunicia potestas*, as transpires from the 'imperator' title behind it. In line 3, besides the erased initial letters of Geta's *praenomen* and *gentilicium*, the conjunction *et* can also be read.

Literature: Perinić Muratović 2004, 86, cat. no. 17, fig. 3.17; Mráv 2006, 214, cat. no. 6.

6. IM[- - -] | +[- - -] | - - - - - - (fig. 8/6)

H.: 11.5 cm; w.: 15.7 cm; t.: 18.3 cm. GMVi.

Fragment contains the beginning of a line. Based on the remains of the Norico-Pannonian volute, it cannot be fitted to the first, but only to the beginning of the third line. In this place, its letters can be identified with the first two letters of Caracalla's *praenomen* Imperator.

Literature: Perinić Muratović 2004, 87, cat. no. 24, fig. 4.24; Mráv 2006, 214, cat. no. 9.

7. C[- - -]. (fig. 8/7)

H.: 18.5 cm; w.: 26 cm; t.: 14.6 cm (at the volute) and 13.8 (at the inscribed field). GMVi.

It belonged to inscription II on the basis of the quality of stone. It is attributed to the beginning of line 5 according to the letter heights and the shape of the curve that remains of the Norico-Pannonian volute fragment. Its fitting in the last line is indicated by a large empty space beneath the letter.

Unpublished.

8. - - - - - - | [- - -][[VṢ]][- - - | - - -]VX+[- - - | - - -]G · P̣[- - -] (fig. 8/8)

H.: 26.3 cm; w.: 14.7 cm; t. 12 cm. GMVi.

In the erased line the last two letters of the *nomen gentile* Septimius of Geta remain, with two letters of the verb denoting construction beneath. And indeed, the letters of the last line can only be associated with the governor's title (see below).

Literature: Perinić Muratović 2004, 87 cat. no. 25, fig. 4.25.

9. - - - - - - | [- - -]NṬ | [- - -] PR. (Fig. 8/9)
Left side fragment: h. 24.5 cm, w. 15.8 cm, t. 14 cm; right side fragment: h. 24 cm; w. 22.5 cm; t. 14.5 cm. GMVi.

The usual PR abbreviation for the governor's title *praetore* is indicated before the Norico-Pannonian volute motif in the left section of the last line.

Literature: Perinić Muratović 2004, 87, cat. no. 26, fig. 5.26; 89, cat. no. 39, fig. 7.39.

The inscribed field is framed by a reversed cymatium. Above the upper cyma, the *cyma recta* moulding that corresponded completely to inscription I was ornamented with acanthus leaves. The acanthus leaves of the reliefs are highlighted by the red paint of the background. Smaller sections of Norico-Pannonian volutes were preserved on both II/6 and II/8 fragments. The entire surface of the slab was covered with stucco, which was applied to cover natural defects on the low quality limestone. The inscription is well composed, and is written in evenly, generally densely set letters, which fill the entire length of the line. The elegant

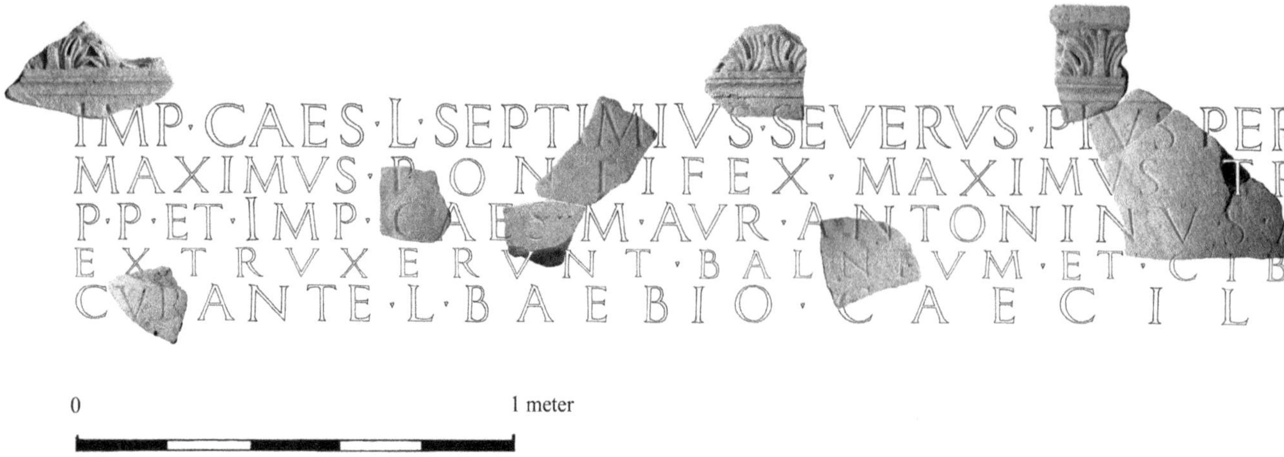

Fig. 7. Cibalae (Vinkovci), Croatia: drawing o

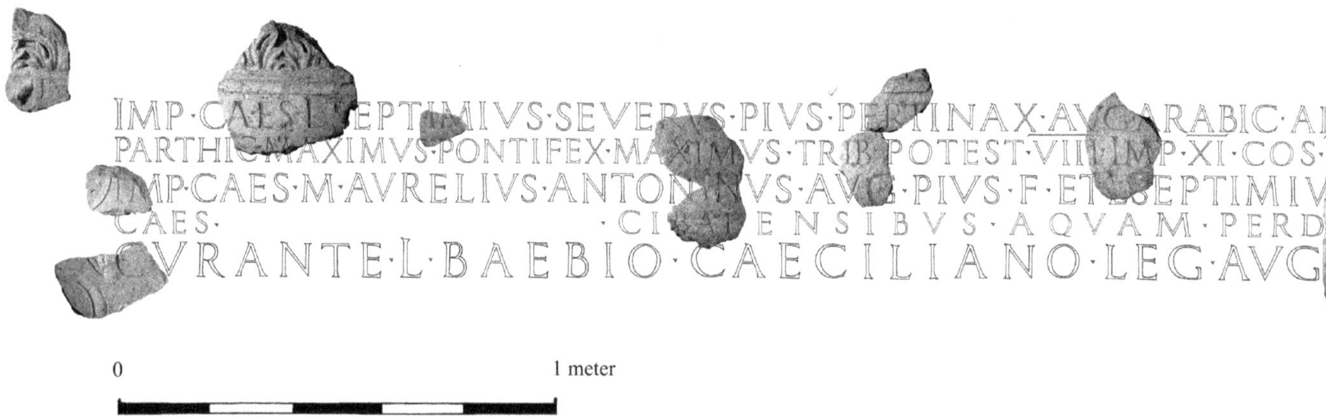

Fig. 9a. Cibalae (Vinkovci), Croatia: drawing of the restoration of building inscription II (made by Zs. Mráv).

Fig. 9b. Cibalae (Vinkovci), Croatia: drawing o

ation of building inscription I (made by Zs. Mráv).

ration of building inscription II (made by Zs. Mráv).

Fig. 8. Cibalae (Vinkovci), Croatia: photos of the fragments belonging to building inscription II (made by Zs. Mráv).

characters were carefully engraved and painted in red. Heights: 6 cm (line 1, however L-*longa* is 6.5 cm high), 5-5.1 cm (line 2), 6.5 cm (line 3), 5 cm (line 4) and 7.5 cm (line 5). The use of the *litterae longae* can be verified in case of the L of the *praenomen* Lucius (see fragments no. II/1 and 5) and of I in the word imperator (fragment no. II/5). The *scriptor* consistently applied dividing signs (*puncta*) in the shape of reversed triangles. By analogy with inscription I, the remaining fragments of inscription II can be restored as follows (fig. 9):

The original length of the inscribed field could have been 3.1 m, which is exactly half the size of inscription I. On the basis of the Norico-Pannonian volute of fragments II/6 and 7, its end point, which coincides with the bottom of the inscribed field, can be approximately established. Thence, the height of this space can be calculated at 60 cm, that is, about one-fifth the size of its length.

Comparing the two building inscriptions, some further differences can be observed apart from their size. The inscriptions are nearly identical, but their distinct dimensions influenced the line allotment of texts and word abbreviations. These were partly compensated with larger characters; nevertheless, the cutters did not manage to engrave inscriptions of perfectly identical layout. The first part of Septimius Severus's triumphal cognomen 'Parthicus maximus' could be fitted into the end of line 1 of the longer inscription I, but in inscription II it could only be placed at the beginning of line 2. The entire title of Septimius Severus could be fitted into line 2, where letters were set in a denser manner, whereas the *siglas* P. P., standing for *pater patriae*, could only be placed at the beginning of line 3. In inscription I the names of Caracalla and Geta filled in the line perfectly; whilst in inscription II the word 'Caesar' belonging to Geta's name slipped to the beginning of line 4, due to the insertion of the name 'Pius Felix' of Caracalla. The greatest challenge is, however, to complete the first section of line 4. This is not solely for the fact that only a few characters are available, but also because the

[Imp(erator)· C]aes(ar)· L(ucius)· S[ept]iṃ[ius·
Seve]ṛụṣ [· Pius· P]ẹrt[inax· A]ug(ustus)·
Ạ[rabic(us)· Adiabenic(us) | Parthi]c(us)·
mạ[ximu]ṣ · p̣[ont(ifex)· ma]xim[us·
tr]ib(unicia)· [pot(estate)· VIII]I· im[p(erator)·
XI· co(n)s(ul)· II· p(ater)· p(atriae)· et |
Iṃ[p(erator)· Caes(ar)· M(arcus)· Aurelius ·
Anton]ịṇ[us · A]ụg(ustus) [· Pius· F(elix)· e]ṭ [[·
L(ucius)· S[eptimiu]ṣ [· Geta | Caes(ar)·]] - - -·
Ci]bal[ensibus· aquam· per]dux[eru]ṇṭ |5
c[urante · L(ucio) · Baebio ·] C[aeciliano ·
leg(ato) · Aug]g(ustorum) · p[ro ·] pr(aetore).

same text cannot be accommodated, owing to the field's irregular shape. The sole distinction between the fifth lines, in which the governor is named, is that in inscription I the word *legatus* was not abbreviated, but was set in full length. Nevertheless, these distinctions are negligible, allowing both inscriptions to be ascribed to the same *scriptor*.

In spite of some differences, the designs and ornaments on the two inscriptions are practically homogeneous. We can observe the same type of cymatium enriched with a reliefed acanthus ornament above both inscriptions, which was carved from the same stone with the tablet.[36] The original position of a fragmentary cyma was certainly the left ending of inscription I or II (its size: h. 20.4 cm, w. 13.5 cm, t. 10 cm) (fig. 10). The background behind the acanthus leaves, analogously to the crown, was painted red in order to intensify the contrast and figurative effect. Although the length of inscription II is exactly half of that of inscription I, there is no difference between the heights and arrangement of the cymatia (the height of the surface carrying the cymatium with acanthus is 10 cm, and of the crown is 3.4 cm). On balance, both inscriptions were most probably installed on the same building.

Fragments I/1, as well as II/6-7 and 9, prove unequivocally that the cyma on both sides of the inscribed field ends in Norico-Pannonian volutes of identical form. This can be well restored in inscription II due to the great number of its fragments (Fig. 9). The volute, which is moulded and is axially symmetrical to the horizontal central line, and which forms a horn-shaped portion, is arched to opposite directions, and returns from the horn peaks in a doubled-volute form. Red paint can be observed on the internal stripe of the volute decoration of fragment I/11. The Norico-Pannonian type of volute decoration coincides with the inscription's dating, because similarly shaped volute motifs on both sides of the inscription field appear at the earliest in the Severan period not only on sarcophagi,[37] but on honorary[38] and building inscriptions as well.[39]

In 1977, a further relief fragment was discovered besides other Severan inscription fragments (h. 23.5 cm, w. 15.3 cm, t. 13.5 cm) (fig. 11), which proves that the inscribed field was on both sides flanked with reliefs representing Victoria in a niche. In the upper portion of the fragment the same cyma adorned with acanthus leaves can be seen as on the top of the inscriptions. This circumstance, as well as the same findspot, proves their common origin. The joint

[36] More fragments of this moulding remain (Perinić Muratović 2004, 95-96), but the position of the majority of them cannot be located accurately. Therefore, I included only those fragments into the restoration of the inscription, in whose lower portions the letter fragments defining their positions in the inscribed field remained.

[37] Pochmarski 2001, 204; Mráv and Ottományi 2005, 190-191.

[38] Mráv 2002a, 207-233.

[39] The building inscription of the Severan-period Iseum in Savaria, included in the marble frieze set above the two central columns of the *hexastylos* façade, commemorated the temple's reconstruction: Mráv 2005, 123-156. The building inscription of the *porticus* dedicated to Apollo and Hygieia in Brigetio was inscribed in 217 AD, and it also had a variant of this same volute motif on the two sides: *RIU* 2, 377, fig. 62; Barkóczi 1951, 226.

Table 1. Cibalae: comparative table of inscriptions I and II.

	I.	II.
Quality of the stone	High-quality hard limestone	Softer, poor quality limestone containing many fossils
Stucco covering	yes (thin)	yes (thick)
Heights of letters	larger	smaller
Spaces between letters and words	increasing toward ends	small
Lengths of inscribed fields	6.2 m	3.1 m
Ligatures	no	no
Puncta	reversed triangle	reversed triangle
Litterae longae	yes (in I certainly)	yes (L and I)
Overscoring	yes (with a thin line)	yes (with a thicker line)
Norico-Pannonian volute motifs on both ends of the inscribed fields	yes	yes
Cymatium enriched with acanthus relief above the tabula	yes	yes

Fig. 10. Cibalae (Vinkovci), Croatia: rim fragment of one of the inscriptions ornamented with acanthus leaves (photo: Zs. Mráv).

between the horizontal upper or lower cymatium and the Norico-Pannonian volute can be spotted in the fragment's lower right corner, left of which is a niche-like hollow with a fragmentary relief. The quality of the limestone suggests that the hollowed fragment could rather have belonged to inscription I, but it does not fit its fragment no. 1, since it contains the starting portion of the volute. Therefore, inscription II remains the only solution, as the relief fragment in the niche reminds us of the hand of Victoria figured in the left niche of an inscription from Székesfehérvár (Fejér County, Hungary); the goddess is leaning on the frame in a position as if holding a wreath (fig. 12). It cannot be a coincidence that this inscription was dedicated exactly to the salvation of Septimius Severus and his sons. In my opinion, the reliefs representing Victoria might have been positioned in the niches on both sides of the inscriptions from Cibalae, composed in the same way as those of the Victoria figures from Székesfehérvár (with palm-branches in their left hands and the right arms stretched upward). Although similar inscriptions can occasionally be found in other areas of the Empire, [40] the figure of the goddess Victoria applied for ornamental purpose is very typical of the Severan inscriptions from Pannonia (fig. 14).[41] A total of eight of them, including those from Cibalae, stem from this province (see table 2).

In the building inscriptions of the nymphaeum in Aquincum (1, table 2) and the Savarian Iseum (5, table 2), niches with reliefs of the goddess Victoria must have been combined with Norico-Pannonian volutes; therefore these are the closest analogies of the ornaments on the inscriptions from Cibalae. The features and the longer text of the inscription correspond to the imperial building inscriptions in the

[40] E.g. from Brougham (United Kingdom): *Roman Inscriptions of Britain*, 738. It should be emphasized that Victoria figures of the Severan period differ from representations of other periods. Therefore, the former constitute a closed and securely datable sequence.

[41] Mráv 1999, 306-310; Mráv 2005, 145-148. See also: Kovács 2001b, 234-235. The relief from Savaria should also be mentioned here, portraying Victoria in a niche holding a wreath and a palm branch: Steindenkmäler, 120, Kat. Nr. 187 (fig. 13). Because of its structure, it cannot be determined, however, whether it flanked an inscription, or whether it was part of the construction moulding.

Table 2. Building inscriptions from the Pannonian provinces flanked with decorative reliefs representing the goddess Victoria in a niche.

	Findspot	Dating	Description	Literature
1.	Ad Statuas (Ács-Vaspuszta, Hungary) (fig. 14/2)	during Caracalla	In the right niche, topped by a Norico-Pannonian volute. Victoria stands on a globe with a palm branch and a wreath in her hands.	*RIU* 3, 645
2.	Bátmonostor (Hungary), originally Aquincum (fig. 14/1)	during Caracalla	In a narrow, high niche framed by a Norico-Pannonian volute. Frontally positioned Victorias standing on globes, with wreaths in their raised right hands and palm branches in their left ones.	*CIL* III 6452 = 10496 = *ILS* 7124 = *TRH* 205 – cf. Kovács 2004, 79-85, fig. I.1.2
3.	Cibalae I. (Vinkovci, Croatia)	200 AD	It may be assumed on the basis of the niche fragments of inscription II.	-
4.	Cibalae II. (Vinkovci, Croatia) (fig. 11)	200 AD	Only the left hand of the Victoria relief, standing in the left niche, remained.	Published here.
5.	Savaria (Szombathely, Hungary), Iseum (figs. 13 and 14/6)	Severan period	In a niche framed by a Norico-Pannonian volute. Victoria is turned aside and holds a wreath in her raised right hand and a palm branch in her left. Her figure is highlighted by a red contour.	Steindenkmäler 95-96, Kat. Nr. 71; Mráv 2005, 140-150, 141, Abb. 9.
6.	Savaria (Szombathely, Hungary) (fig. 14/7)	Severan period	Only the lower right part of the right niche remained, with the clothes and the fragment of Victoria's left hand holding a palm branch. Due to their common findspot and size and material, it might have belonged to the Severan fragment *TRH* 3.	Mráv 2005, 148-149, fig. 13.
7.	Székesfehérvár (Fejér-megye, Hungary) (figs. 12 and 14/4)	197-209	Frontally depicted figures of Victorias in narrowly enclosed niches on both sides of the *tabula.* Their legs are crossed, and they imitate the posture of holding palm branches in one of their hands and wreaths in the others.	*RIU* 6, 1500
8.	Visegrád (Pest-megye, Hungary) (fig. 14/5)	197-209	Only a fragment of the left side niche remained, in which Victoria wearing a *chiton* was depicted in profile.	*RIU* 3, 815; Mráv 1999, 306-310; Kovács 2001b, 234-235.

Severan period, written in smaller letters and in more lines, unlike the earlier examples.[42] On the basis of completing the letters in the first line of fragment II/5 [- - -]ERT[- - -] to *tertius*, L. Perinić Muratović assumed that the inscription was set up in 202 AD, in the year of Septimius Severus' third consulship.[43] On the basis of this, she raised the possibility of an imperial visitation to Cibalae: On the grounds of the moulded upper frame, the fragment fitted in the first line is not an ordinal number written in letters, but can certainly be identified with the letters of the emperor's name Pertinax. Thus, there is no evidence for dating the inscription to 202 AD. Moreover, the emperor had his title *consul III* not only in 202 AD, but all the time until his death.[44] Based on the known fragments, in 2006 I endeavoured to establish the inscription's date as accurately as possible. All I could state with certainty was that it was inscribed when Caracalla was already 'Augustus' (after the autumn of 197 AD)[45] and Geta was still Caesar (between autumn 197 and September/ October 209 AD).[46] The latter was further supported by the fact that in fragment II/6 the conjunction *et* is directly followed by the *praenomen* L(ucius) of Geta and not by the names *Imp(erator) Caes(ar)*. Moreover, the letter C on recently identified fragment I/9 can only belong to the name 'Caesar' of Geta. I also pointed out that at the time when the inscription was made, the emperor had already worn

Fig. 11. Cibalae (Vinkovci), Croatia: fragment of the left hand of the goddess Victoria standing in a niche (photo: Zs. Mráv).

[42] Horster 2001, 15.
[43] Perinić Muratović 2004, 92.
[44] Kienast 1996², 157.
[45] Kienast 1996², 162.
[46] Kienast 1996², 166.

Fig. 12. Székesfehérvár (Fejér County), Hungary: building inscription enclosed by reliefs representing the goddess Victoria in a niche (made by Zs. Mráv).

the triumphal cognomina ‘Arabicus’, ‘Adiabenicus’ and ‘Parthicus maximus’, the last of which he bore officially from January 198 AD. Thus, the building inscriptions could generally be dated to the period between January 198 and the autumn of 209 AD. The possibilities were narrowed further by the last figure remaining in the tribune iteration number indicated in Septimius Severus’ titles in fragment II/5. At the time when the inscriptions were set up, this number ended in [- - -]I. From the possibilities that came into consideration, I could exclude the years 202 and 207 AD because at that time Septimius Severus had the tenth (*tribunicia potestate X*) and fifteenth (*tribunicia potestate XV*) tribune iteration.[47] The imperial construction and the inscription commemorating them cannot be associated with the year and events of the emperor’s and his family’s visitation to the Pannonian provinces in 202 AD. (For lack of data, we cannot at this stage argue in favour of the emperor’s personal visit to Cibalae either, which an imperial construction per se does not suppose anyway.) In accordance with its size and the quality of its stone, fragment I/6 that has just been ascribed to inscription I, resolves the issue of chronology. In the second line of the fragment, the figure VIII can be read beneath the triumphal cognomen ‘Adiabenicus’ of Septimius Severus (fig. 5/6). On the basis of the fragments’ positions inside the inscribed field, this numeral can only be the iteration number of the *tribunicia potestas*, which is in tune with the remaining digit [- - -]I of inscription II. Thus this building inscription was set up when Septimius Severus renewed his *tribunicia potestas* for the eighth time between December 10, 199 AD and December 9, 200 AD. Considering the construction season, inscriptions can be evidently dated to AD 200, more precisely, to the spring of this year at the earliest. All these assumptions are supported by the name of L. Baebius Caecilianus, the governor of Pannonia Inferior between the years 199 and 202 AD, who supervised the construction project (see below). Septimius Severus bore the titles

Fig. 13. Savaria (Szombathely), Hungary: relief of the goddess Victoria standing in a niche (photo: O. Harl).

[47] Kienast 1996², 157.

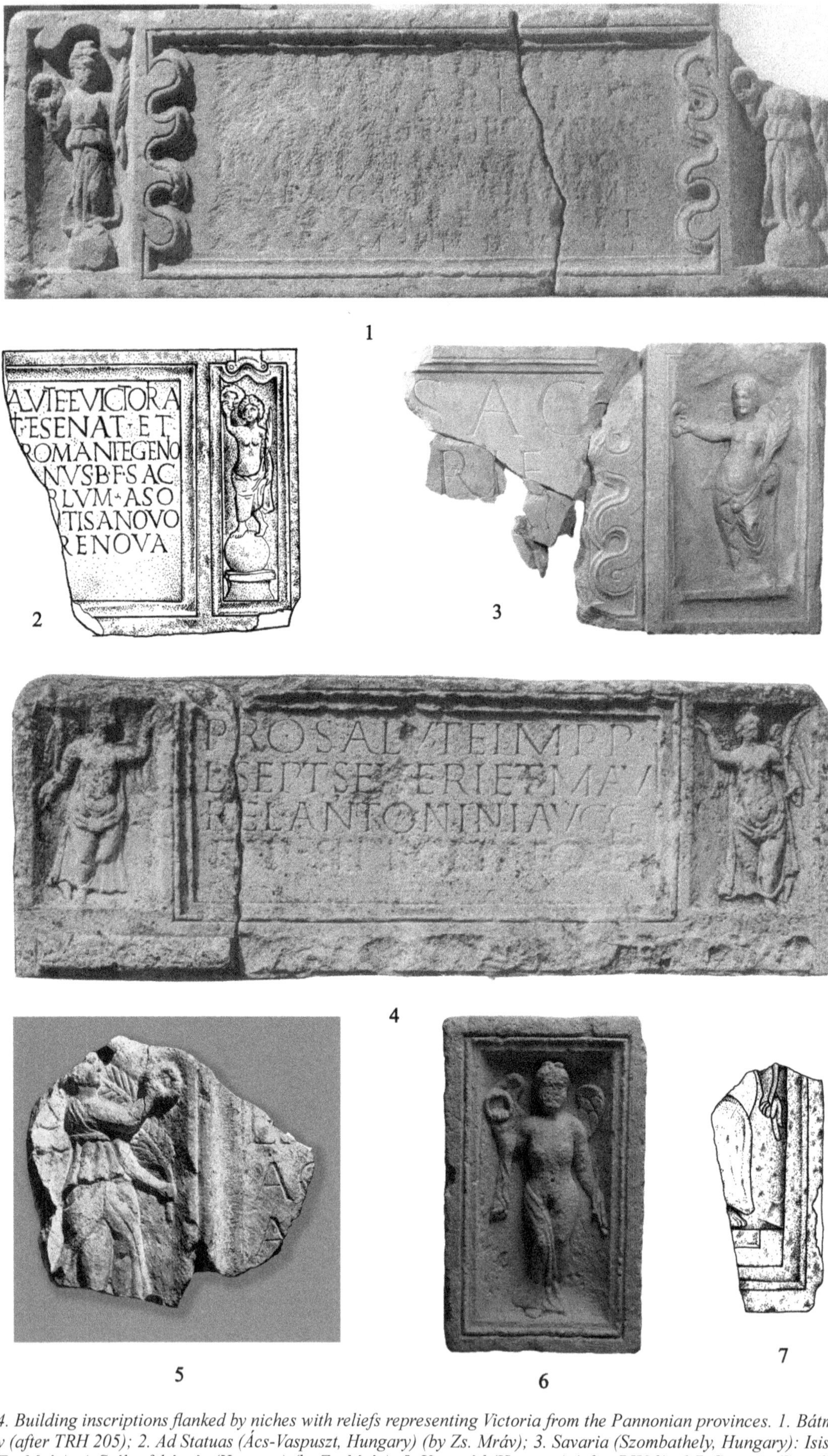

Fig. 14. Building inscriptions flanked by niches with reliefs representing Victoria from the Pannonian provinces. 1. Bátmonostor, Hungary (after TRH 205); 2. Ad Statuas (Ács-Vaspuszt, Hungary) (by Zs. Mráv); 3. Savaria (Szombathely, Hungary): Isis sanctuary (by Zs. Mráv); 4. Székesfehérvár (Hungary) (by Zs. Mráv); 5. Visegrád (Hungary) (after RIU 3); 6-7. Savaria (Szombathely, Hungary) (6: photo by O. Harl; 7: made by Zs. Mráv).

tribunicia potestate VIII, imperator XI, consul II, and *pater patriae* in 200 AD,[48] which provides the iteration numbers to complete the second lines of both inscriptions. These titles are also corroborated by the last numeral [- - -]I of the *acclamationes imperatoriae* on fragment I/9, as it can be completed in the form *[imp(erator) X]I*. The inscription's closest parallel is *CIL* VIII 20845 imperial inscription (Hr. Grimmidi, Maureatnia Caesariensis), in which Septimius Severus is mentioned with all his titles, but Caracalla and Geta have only their names inscribed.[49] The most significant distinction between the texts of the two inscriptions can be detected in the nomenclature of Caracalla. Due to the remaining letter A after the cognomen of 'Antoninus' on inscription I, and the position of fragment II/4 in inscription II, the name 'Augustus' followed. While in inscription I the name 'Augustus' was already followed by the conjunctive *et* and Geta's name, in inscription II a space adequate for 5-6 letters stayed blank. Into this space the names 'Pius Felix' of Caracalla can be inserted in *Pius F(elix)* form conveniently, which is exceptional but not unique way of abbreviation.[50] The distinction between the imperial nomenclatures on the two inscriptions can be traced in the date of the name adoptions. This event took place before April 200 AD,[51] the year when both inscriptions were accomplished, which verifies their unusual disposition and abbreviations. Thus inscription I might have been made before the news of the name adoption of Caracalla arrived in Pannonia Inferior. In inscription II Caracalla's name should be most probably restored by adding the names 'Pius Felix'. Therefore, inscription II is the later one.

The subject of the construction was known from indirect data only, that is, the size, shape and ornamentation of the presumably unique slab. Lj. Perinić Muratović associated the inscription with the town's Capitoline temple. According to her, this pagan sanctuary was destroyed by Christians in the 4th century and its parts built into the wall of a late Roman *horreum*. On the basis of the concordance in size with the building inscription of the Iseum in Savaria, which was (re?)built with a hexastylos façade, I also considered it possible that the inscription discussed here was inserted into the middle of the façade frieze of the sanctuary.[52] In the light of the new fragments, this hypothesis proved to be wrong, since the construction work was, in fact, of quite different nature. The name of the community that received the benefit, as well as the subject and verb of the construction, were engraved in the fourth lines of both inscriptions, which were written in the smallest letters.[53] In spite of the fact that no fragments remained of the word naming the subject of the building work, we may establish it with fair certainty by means of defining the verb at the end of the line. Of the verb in question, lines with the following letters [- - -] DVXERV[- - -] of fragment I/9 and [- - -]VX+[- - -]NT of fragment II/7 were preserved, which can only be completed as *perduxerunt* or *induxerunt*. This means that the water supply was procured as a sole benefit granted to the town, therefore the subject of the sentence would be the word *aquam*, or perhaps *aquas*. This solution can be verified not only by the fact that both words perfectly fit into the empty field before the verb, but also by the circumstance that the words *aquam* or *aquas* are in the accusative case and because of the verbs *in-* or *perducere* in the lettering of the inscriptions. To cite some examples, Augustus and Tiberius *aquas in coloniam perduxerunt*[54] in Brixia (Italia, reg. X), whereas Vespasianus and Titus *[aquam pe]rduxerunt*[55] in the *municipium* of Thysdrus (Africa Proconsularis). Consequently, Septimius Severus and his sons strove to improve the water supply of Cibalae by constructing a water conduit. We still do not know, however, whether it was an underground conduit or an aqueduct. At any rate, the existence and type of the building inscription renders an aboveground construction probable, which at the same time might have carried the inscription. Roman emperors often provided water-supply facilities (or had them renewed) for towns.[56] The reason for this was not only the fact that such undertakings required peculiar engineering expertise and high expenses; through the significance of such constructions for the public good, and through their outstanding engineering solutions, the emperor was able to advertise his piety and beneficence, too.

The letters [- - -] CIB [- c. 6 -]BV[- - -] remained in the middle of line 4 of fragments I/4 and 12 of inscription I, while the text fragment [- - -]BAL[- - -], which certainly refers to the same word, remained in fragment II/3 of inscription II. This word can hardly be understood otherwise than the name of the municipal community, *Cibalenses*, which was the beneficiary of the construction grant. Based on fragment I/12, the noun is in the dative case in both inscriptions: *Cibalensibus*, according to which the emperor and his sons installed water supply facility for the inhabitants of Cibalae. The mention of the town or its community that was honoured by the construction grant is not unusual in the wording of imperial building inscriptions. We can find several examples for such phrasing.[57] Moreover, the name of the town or the municipal community was most frequently mentioned in the dative case in imperial building inscriptions, when the subject matter of the construction

48 Septimius Severus bore his title *consul II* between 194 and 202 and the title *imperator XI* between the end of 197 and January 205: Kienast 1996², 157.

49 See also: *AE* 1985, 810; *AE* 1986, 683 = *AE* 1993, 1520; *CIL* III 14150 = *AE* 1896, 133 = *AE* 1897, p. 20 s. n. 52.

50 See e.g. *CIL* III 5720 – cf. Kienast 19962, 164; Mráv 2006, 218.

51 Mráv 2006, 219-220.

52 Inscriptions with this structure can be classified to Type 4 of texts by M. Horster (2001, Kat. Nr. X 3, 1.).
AE 1991, 1635 = Horster 2001, Kat. Nr. XXXIV, 4.

53 *CIL* V 4307 = *ILS* 114 = Horster 2001, Kat. Nr. X 3, 1.

54 *CIL* V 4307 = *ILS* 114 = Horster 2001, Kat. Nr. X 3, 1.

55 *AE* 1991, 1635 = Horster 2001, Kat. Nr. XXXIV, 4.

56 Horster 2001, 65-66.

57 For example: Augustus *Perusini[s]*: *AE* 1991, 666 = *AE* 1995, 499 = Horster 2001, Kat. Nr. VII, 7 (munic. Perusia, Italia, reg. VII); Hadrianus *Cingula[nis]*: *CIL* IX 5861 = Horster 2001 Kat. Nr. V, 3 (munic. Cingulum, Italia reg. V); Antoninus Pius *Agr[i]pp[i]n[ensibus]*: Horster 2001, Kat. Nr. XXIV, 1, 3 (*Colonia Claudia Ara Agrippinensium*, Germania Superior); Severus Alexander *Serteitanis*: *CIL* VIII 8828 = 20630 = *ILS* 6889 = *AE* 1973, 651 = Horster 2001, Kat. Nr. XXXVI, 4 (pagus Sertei, Mauretania Caes.); Valerianus *[F. Cl. Vallen]sibus*: *AE* 1977, 527 = *AE* 1982, 674 (Forum Claudii Vallensium/ Octodurum, Alp. Poeninae). For a summary see Horster 2001, 64-67.

project was the town wall or water supply.[58] The latter is applicable to the inscription of Cibalae, too. The text of a building inscription of Octodurum (Forum Claudii Vallensium, Alp. Poeninae) was similar; it related that Valerianus *[F. Cl. Vallen]sibus suis [aqua]m perduc[i - - -] ... iussit.*[59]

The text thus restored, though, only fills in the last three-fifths of line 4 of inscription I, whereas the restoration of the remaining part is still questionable. Nothing has remained of this part of the text in inscription II, while we only know the characters [- - -]NE[- - -] on fragment I/8 of inscription I. Assuming the correspondence to the content of the second half of the text, these two letters might have belonged to the word *bal(i)neum*. It seems logical, indeed, that the emperor's family had a public bath built coincidentally with the installation of water conduits; thus the water main could have provided partly the latter's supply, too. Scholarly terminology uses the term *bal(i)nea* for denoting such minor spas of both public and private use that did not include a *palaestra*, and only partly occupied its *insula* . On the other hand, the conception of a *thermae* implies a larger complex of spa possessed by the state or a town, which often took the entire *insula*, and also provided sport facilities.[60] Nevertheless, we do not know any proper Roman-period definition associated with these two words, and it appears that neither did their usage reflect firm consistency. Although there must have been a difference between meanings of the words *bal(i)nea* and *thermae* [61]– as there is between their modern definitions –, written sources often mistake and confuse them.[62] In any case, the term *bal(i)nea* as the subject of constructions or renovations is fairly frequently found in imperial inscriptions,[63] suggesting more significant buildings. Moreover, a *balineum* in Apamea did even hold an adjective revealing an imperial relation (*balineum Hadrianum*).[64] It appears likely therefore, that the small town of Cibalae in inner Pannonia was not granted a monumental spa complex, but rather a public bath, a larger size *balnea*, which was rather in accordance with the number of its inhabitants. Parts of a major public bath were unearthed around – or somewhat more southward of – the findspot of the fragmentary imperial inscriptions from Cibalae, which was named a forum-bath (fig. 15).[65] The location of this bath in the town, the dimensions of its halls – dominantly of its *tepidarium* – and its internal ornamentation permit its association with the above-mentioned *balneum*. According to the available space, the verb of the construction could have been *extruxerunt*.

The governor who supervised the work was mentioned in the last line of the inscription. The letters [- - -]VGG · [- - -] at the end of the line 5 of inscription I, and [- - -]G · P[..] PR at the same space of inscription II can only be restored as the general abbreviation of the governor's title LEG. AVGG. PRO PR. Based on the letters of fragments II/7 and I/11, the name of the governor was indicated after the phrase *curante*, as was the case in the inscription from Aquincum. (The letter C of fragment 7 of inscription II can be accommodated in line 5 only, whereas the letters VR on fragment I/11 can merely be assumed in the word *curante*.) Apart from the act of inserting the word, fragment I/12, whose letter O was the last one of a name standing in the ablative case, verifies the aforesaid in accordance with the postposition of *curante*. L. Baebius Caecilianus was the governor of Pannonia Inferior in 200 AD (see above). His name can be fitted perfectly in the last line of the inscription, with, however, only the initial C remaining of the cognomen 'Caecilianus' (in both inscriptions, see cat. nos. I/8 and II/3). It is transparent that the name of the governor was set in a remarkably larger format than those of the emperors,[66] notwithstanding that the attention was customarily paid to emphasise the emperors' name by means of the letters' size (see below).[67] This fact can be interpreted as a manifestation of a tighter relationship between Cibalae and the governor, or his outstanding role in the management of the construction project, or in gaining the emperor's support – in other words his mediation and initiative. On balance, Septimius Severus, Caracalla and Geta – through their legate L. Baebius Caecilianus – provided support for the construction of water-supply facility in Cibalae in 200 AD, and had also a bath built at the spot.

4. The building inscription of Septimius Severus from Brigetio, and the foundation of the town

E. Maionica and R. Schneider published a record of a monumental fragmentary inscription from Ószőny (present-day Szőny, district of Komárom, county Komárom-Esztergom, Hungary) in 1877,[68] and noted the following location: *in der Nähe der Kirche* (near the church). The mentioned church was certainly the Roman Catholic church of Szőny (today at 3 Széchenyi Street). However, it is not clear whether their report referred to the findspot of the inscription's reuse, or merely the contemporary place of its depository, where they found it in August 1877. In

58 Horster 2001, 64-66.
59 *AE* 1977, 527 = *AE* 1982, 674 – cf. Horster 2001, Kat. Nr. XVIII, 1.
60 For a summarized interpretation and problematic issues relating to the terms *balnea* and *thermae* see: Nielsen 1990, 3; Rebuffat 1991, 23-28; Yegül 1992, 43. The definition of a spa type might differ among authors.
61 Nielsen 1990, 3; Rebuffat 1991, 23-28; Yegül 1992, 43.
62 Nielsen 1990, 3; Yegül 1992, 43.
63 Trajan *balneum et plateas rep. mandavit*: *CIL* IX 5746 = *ILS* 5675 = Horster 2001, kat. Nr. V, 7,2 (munic. Ricina, Italia reg. V); Antoninus Pius *bal[in]eum ... [restituit]*: *CIL* XI 3363 = Horster 2001, Kat. Nr. 9 (munic. Tarquinii, Italia reg. VII); Marcus Aurelius és Lucius Verus *[te]mpla et baline[as] ... restit[uer(unt)]*: *CIL* XII 107 = *ILS* 5868 = Horster 2001, Kat. Nr. XVIII, 1 (Forum Claudii Ceutronum, Alp. Graiae); Caracalla *belneum ... ad pristin. faciem r[estituit]*: *AE* 1968, 157 = Horster 2001, Kat. Nr. IV, 3 (munic. Pinna, Italia reg. IV).
64 This (imperial) adjective does not denote an imperial building project however, but it only expressed loyalty of the town towards the Roman emperor. Cf. an example of Hadrian: Horster 2001, 104-105.
65 Iskra-Janošić 2001, 75-83; Iskra-Janošić 2004, 179-180.
66 Whereas letter-sizes in the name of Septimius Severus were 6 cm in line 1, and of those of Geta and Caracalla were 6.5 cm in line 3, the letters of L. Baebius Caecilianus' name were as high as 7.5 cm.
67 Alföldy 2004, 11.
68 Maionica and Schneider 1877, 154-155. The inscription was published on the basis of their description by Th. Mommsen in *Ephemeris Epigraphica* 4 (1881), n. 506, then by A. v. Domaszewski in *CIL* III Suppl. 11016.

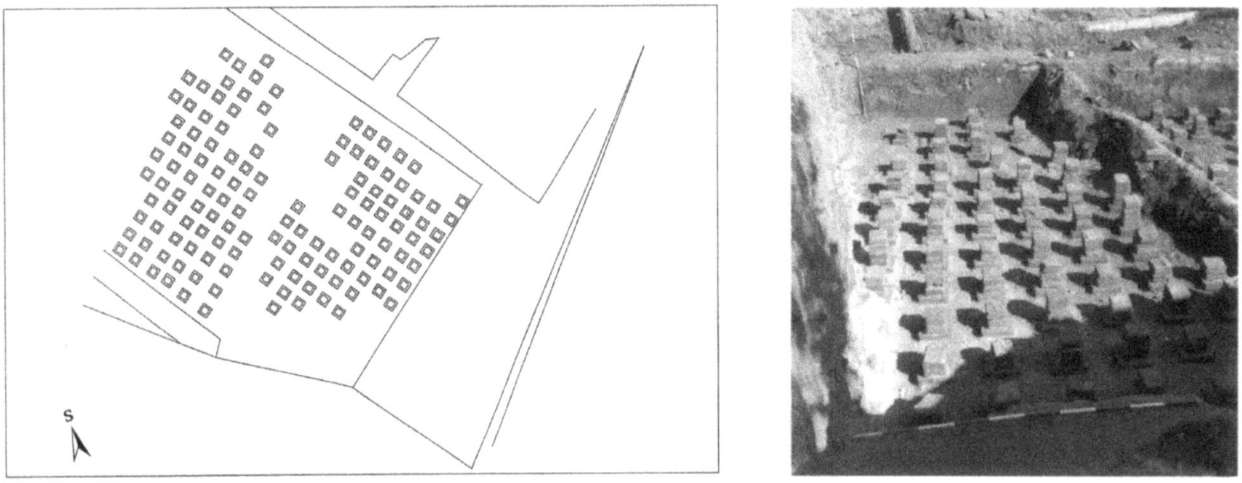

Fig. 15. Cibalae (Vinkovci), Croatia: excavated part of the so-called forum baths (after Iskra-Janošić 2001).

two lines of the fragment, which has gone missing in the meantime,[69] the authors identified the following letters and a figure (fig. 16):

\- - - - - -
[- - -]THIC[- - -]
[- - -]P VII · C[- - -]
\- - - - - -

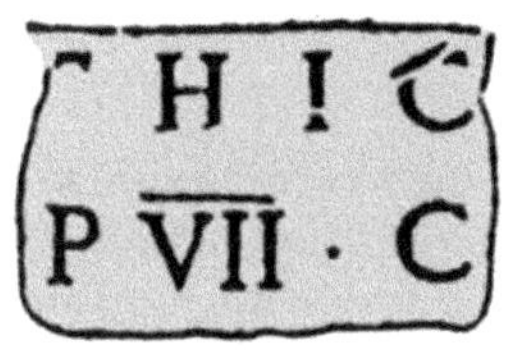

Fig. 16. Drawing of the fragmentary imperial inscription documented at the Catholic church at Ószőny (Komárom-Szőny presently) in August 1877 (after Maionica and Schneider 1877, 155).

E. Maionica and R. Schneider interpreted the first line of the inscription [- - -]THIC[- - -] as part of an emperor's triumphal cognomen 'Parthicus', and considered the iteration number VII between letters P and C in line 2 to be part of the title. They published no photo of the fragment, but a drawing (fig. 16) with the following dimensions: length 66 cm and height 49 cm.[70] In front of the letter C in line 2 a small dot was placed, representing a divider, but its accurate form was not indicated. Judging from the published drawing, the heights of the letters in both lines appear to be identical, measuring 10-12 cm. (Concerning the heights of letters only one local inscription can be compared with this one – an extremely long building inscription from Brigetio, featuring 10.2 cm high letters of the name of either Marcus Aurelius or Commodus, which could have commemorated the reconstruction of the gate of the Brigetio legionary fortress in the second half of the 2nd century AD.[71]) The nice engraving and unusual height of the letters were pointed out both by E. Maionica and R. Schneider: *mit schönen grossen Buchstaben* (with nice and large letters'),[72] as well as by Th. Mommsen and A. v. Domaszewski (*litteris magnis*).[73] It is not surprising at all that the name and imperial titles are in imperial inscriptions executed in oversized letters in relation to the local customs, as in Roman epigraphic traditions the strength of imperial power was represented by the size of letters in inscriptions related to emperors; the height of letters in inscriptions set up by private individuals was generally smaller.[74]

A suggestion to complete the inscription, attempted by E. Maionica and R. Schneider, was later adopted by L. Barkóczi in his commentary of *RIU* 2, 502, and reads as follows: - - - *Par]thic[o* - - - | - - - *trib(unicia)] p(otestate) VII c[o(n)s(ul)* - - - . E. Maionica and R. Schneider did not attempt to identify the emperor named in the inscription; they only narrowed down the group of possible persons to Lucius Verus, Septimius Severus, and Caracalla, on the basis of the cognomen 'Parthicus' and the letters, which

[69] Its discovery during the writing of the monograph of Brigetio (prior to 1944) was not acknowledged by L. Barkóczi, meaning that the fragment went missing some time between the years 1877 and 1940.

[70] Maionica and Schneider 1877, 154-155.

[71] *RIU* 2, 500. The original length of the fragment with the nomenclature of Commodus and his titles, which was discovered in the area of the legionary fortress in 1882, was certainly as long as 7 meters. Its alternative version, completed with the name of Marcus Aurelius, was somewhat shorter. The groundplan of the Brigetio legionary fortress' modernised type of gate represents the Antonine period characteristics of gate constructions (gate towers protruding from the wall plane by one fourth or one third of their lengths). Relying on dimensions of the inscription, and the almost same gatehouse type of the Brigetio legionary fortress, I suggest associating the *RIU* 2, 500 inscription with a gate construction of the legionary fortress: Mráv 2009, 263 69 and also Bechert 1971, 248-249. (In the first line the titles, preceding the COS III title of the mentioned emperor, should stand beside the name 'Augustus'.) A newly recovered inscription from Brigetio that contains imperial titles was written in 9 cm-high letters: Borhy 2006, 171, cat. no. 247.

[72] Maionica and Schneider 1877, 155.

[73] Th. Mommsen, in: *Ephemeris Epigraphica* 4 (1881), no. 506, and A. v. Domaszewski, in: *CIL* III Suppl. 11016.

[74] For more information about the secondary and symbolic meaning of letter dimensions see Alföldy 2004, 11.

were considered as relatively late.[75] The potential emperor, however, can be fairly securely determined on account of line 2 and the titles indicated in it. The greatest problem of the line's generally accepted restoration and interpretation is not only the fact that it presents the extremely shortened [- - - TRIB.] P. form for denoting the tribune power, but it also presumes the unusual sequence of the titles. It is clearly recognisable on the drawing of the fragment that the letters were written sparsely, at greater intervals. An adequate space was available for the text itself, therefore its *scriptor* did not need to squeeze the letters, or to omit certain elements of the titles, or use unusual, exaggerated abbreviations. Another problematic point is that the iteration number of consulship is not preceded by the date indicating the renewal of the tribune power; instead, the date of the imperatorial acclamations was used throughout. The same could apply to the inscription from Brigetio; therefore [- - -]P before the figure VII cannot be the last letter of TRIB. P., but that of the *imperator* title's abbreviation IMP. On the one hand, this rather meets the traditional requirements of the imperial title system, and furthermore, only by means of this alternative solution could the restoration drawing be produced (see below). On balance, what we know about the emperor mentioned in the inscription is that he used the triumphal cognomen 'Parthicus' in whatever form, and also that the iteration number of *acclamationes imperatoriae* was VII for a while. The only emperor who meets all these criteria is Septimius Severus, who used the title *imperator VII* between the summer and the end of 195 AD.[76] Now, it is true that Septimius Severus officially took the triumphal cognomen 'Parthicus maximus' in January of 198 AD.[77] Nevertheless, this title could appear in a Pannonian inscription erected exclusively in the second half of 195 AD, since for a little while in the summer of 195 AD (the year of the victorious accomplishment of the *expeditio Parthica*), the double triumphal cognomen 'Parthicus Arabicus' and 'Parthicus Adiabenicus' in the nomenclature of Septimius Severus appeared subsequently both on coins minted in Rome and in official inscriptions.[78] The Senate could have decreed to award the name in this form that could have been in use – predominantly in Rome, Italy and their proximal provinces – until it received the emperor's response with the disapproval of the name 'Parthicus' (*SHA*, *Sever*. 9, 10). As the circulation of information was fairly slow in that period, it could have taken months for this disapproval to be publicised. We can notice the same triumphal cognomen on the triumphal arch dedicated to Septimius Severus in the Forum Romanum in Rome.[79] Nevertheless, the arch itself was accomplished only in the year 203 AD; the cognomen displayed in its inscription indicates the name forms that were awarded in the Senate's decree of 195 AD.[80] Not long after its emergence was the double triumphal cognomen gradually substituted for the simpler names of 'Arabicus' and 'Adiabenicus', yet an appearance of the double triumphal cognomen involving the name 'Parthicus' is not only possible, but also more likely. Moreover, the restoration drawing can be completed solely with the entire name of the emperor, including his double cognomina. The remaining lettering of the fragment did not permit establishing whether the name and titles of Septimius Severus were written in the nominative or some other case. (Its restoration does not contribute to the solution of this problem at all, as both the nominative and dative cases could have been used equally in this inscription.) The unusual monumentality and the dating of the inscription speak for an imperial building inscription. Therefore, my restoration of the inscription's first three lines is the following (fig. 17a):

[Imp(erator) Caes(ar) L(ucius) Sept(imius) Severus Pius Pertinax Aug(ustus) | Parthicus Arabicus Par]thic[us Adiabenicus |³ pontif(ex) maximus trib(unicia) potest(ate) III im]p(erator) VII c[o(n)s(ul) II pater patriae | - - - - - -.

The length of the restored inscription – calculated on the basis of the preserved fragment – could have been approximately 6.3 meters, which classifies it as one of the longest inscriptions in Pannonia. The governor of the province – Pannonia Superior in this case – could have been in charge of the management of this construction, just as in other imperial construction works.[81] However, his identity in this period remains unknown.[82] Less probable is the formulation in the ablative case, which cannot be excluded, however, because of its disposition to fit in the gap (fig. 17b):

[Imp(eratori) Caes(are) L(ucio) Sept(imio) Severo Pio Pertinaci Aug(usto) | Parthico Arabico Par]thic[o Adiabenico |³ pontif(ici) maximo trib(unicia) potest(ate) III im]p(eratori) VII c[o(n)s(uli) II patri patriae | - - - - - -.

The basis of the inscription's restoration and thus the dating of building works rely on the triumphal cognomina 'Parthicus Arabicus' and 'Parthicus Adiabenicus' of Septimius Severus and his title *imperator VII*, related to the autumn of the year 195 AD. The execution of the building inscription from Brigetio could have taken place not long after the news of the seventh imperial acclamation reached Pannonia Superior, signalling the end of the emperor's first Parthian campaign. The governor could likewise have been informed of the cognomina 'Parthicus Arabicus' and 'Parthicus Adiabenicus', awarded to Septimius Severus earlier by the Senate, directly from Rome, as these triumphal cognomina had already appeared in this form associated with his title *imperator IIII*. In 195 AD the V-VII imperatorial acclamations of Septimius Severus were

75 Maionica and Schneider 1877, 155.

76 Kienast 19962, 157. We can observe that Trajan began to use the title *imperator VIII* from the autumn of 114 AD and the cognomen *Parthicus* only from February 116 AD. Neither Lucius Verus nor Caracalla did reach the 7th imperatorial acclamation: Kienast 1996², 144 (Lucius Verus); Kienast 1996², 163 (Caracalla).

77 Kneißl 1969, 142-148; Kienast 1996², 158.

78 Kneißl 1969, 126-138; Kienast 1996², 158.

79 *CIL* VI 1033.

80 Kneißl 1969, 131.

81 Horster 2001, 191-198.

82 Fitz 1993, 506; Lőrincz 2007a, 183.

IMP · CAES · L · SEPT · SEVERVS · PIVS · PERTINAX · AVG
PARTHICVS · ARABICVS · PARTHICVS · ADIABENICVS
PONTIF · MAXIMVS · TRIB·POTEST·III · IMP VII · COS·II · PATER · PATRIAE
- - - - - - -

a

IMP · CAES · L · SEPT · SEVERO · PIO · PERTINACI · AVG
PARTHICO · ARABICO · PARTHICO · ADIABENICO
PONT · MAXIMO · TRIB·POTEST·III · IMP VII · COS·II · PATRI · PATRIAE
- - - - - - -

b

Fig. 17. Brigetio (Komárom-Szőny), Hungary: restoration of the imperial building inscription with the name and titles of Septimius Severus in the nominative (a) and ablative (b) cases (made by Zs. Mráv).

decreed one after another quickly within a few months time, as tidings of his actual Parthian victory reached Rome.[83]

It remains to be answered whether the construction in Brigetio served military or civilian purposes, and accordingly, whether it took place in the legionary fortress or the town. We know that *legio I Adiutrix*, the garrison of Brigetio with its entire ranks, were involved in Severus' *expeditio Urbica*, in the military campaign against the usurper Pescennius Niger, and in his first campaign against the Parthian Empire,[84] whose successful conclusion happened in the summer of 195 AD. In the second half of that year, when the constructions in Brigetio were completed, the expedition army, together with the *legio I Adiutrix*, was still on their way marching from Asia Minor toward Europe. Hardly could have such a monumental construction been carried out by the military forces in the Brigetio legionary fortress in 195 AD, since the garrison was not in their permanent station at that time, and we do not have any information about their substitution either. Moreover, construction works of great importance concerning the legionary fortress, including the gatehouses as well, had been accomplished by that time. In addition to this, another episode concerning Brigetio in the Severan era suggests that the imperial construction in question was of civilian concern, and that it was associated with the civil town of Brigetio. The founder of *municipium Brigetionensium* has long been debated owing to the lack of evidence.[85] Only recently could the town foundation be dated prior to 205 AD, on the basis of indirect data, given that Q. Ulpius Felix, as an Augustalis of the town, already had a temple to Apollo and Hygieia built exactly in this year.[86] Yet this inscription does not denote more than the simple fact that Brigetio already possessed municipal status during the reign of Septimius Severus. However, another inscription that has been ignored in this respect renders it fairly probable that the town's foundations were laid in the Severan period. Th. Mommsen's corrected version of the inscription, which has gone missing by now, is the following (the town's name is still denoted as Breg. in W. Lazius's.): *Imp(eratori) Caesar(i) / L(ucio) Septimio / Severo Pio / Pertinaci /[5] Aug(usto) Arabic(o) / Adiabenic(o) / Parthico / maximo / Breg(etionenses) publice /[10] d(ecurionum) d(ecreto)*.[87] Its structure, the great number of short lines, and the lettering, suggest it could have been a statue base of Septimius Severus. The presence of the triumphal cognomen 'Parthicus maximus' – which could not have been inscribed later – and the absence of the name 'Britannicus maximus', date the base, which was originally erected in the civil town of Brigetio, to the period between January 198 AD and the turn of the years 209-210.[88] Based on A. Mócsy's opinion – viz. the inhabitants, collectively as *Brigetionenses*, had set up the statue before foundation of the town – the inscription was dismissed as a document verifying the municipal status of the settlement,[89] and Brigetio was long regarded as a town founded by Caracalla.[90] Corresponding to the practice of inscribing statue bases, which would be applied generally throughout

[83] Kienast 1996², 158.
[84] Lőrincz 2000, 157; Kovács 2007, 180-181.
[85] Mócsy 1962, 600.

[86] Galsterer-Kröll 1972, 82 Anm 213; *RIU* 2, p. 89; Mócsy 1990, 308, note 16; Kovács 2001, 43; Mráv 2006, 221; P. Kovács, in *FPA* IV, 79-80. My edited and modified version of the lettering is as follows: *[Apollini et H]ygiae san[ctis / pro sal(ute) Augg(ustorum duorum)] Q(uintus) Ulpius [Felix / August(alis) mun(icipii) B]reg(etionensium) temp[lum a / solo inpendis s]uis fecit [- - - / - - - Imp(eratore) Aur(elio) Anto]nino Aug(usto) II et [[Sept(imio) [Geta]] co(n)s(ulibus)].*
[87] *RIU* 2, 501 – cf. Mráv 2006, 221 38. j.; P. Kovács, in: *FPA* IV, 69.
[88] Mráv 2003, 351 Tab. 1a/Nr. 6.
[89] Mócsy 1962, 600: *Unter Septimius Severus haben seine bewohner noch als Brig(etionenses) dem Kaiser eine Basis errichtet.*
[90] Referring to inscription *RIU* 2, 450, in which the adjective *Antoninianum* can be read after the name of the town: Mócsy 1962, 600.

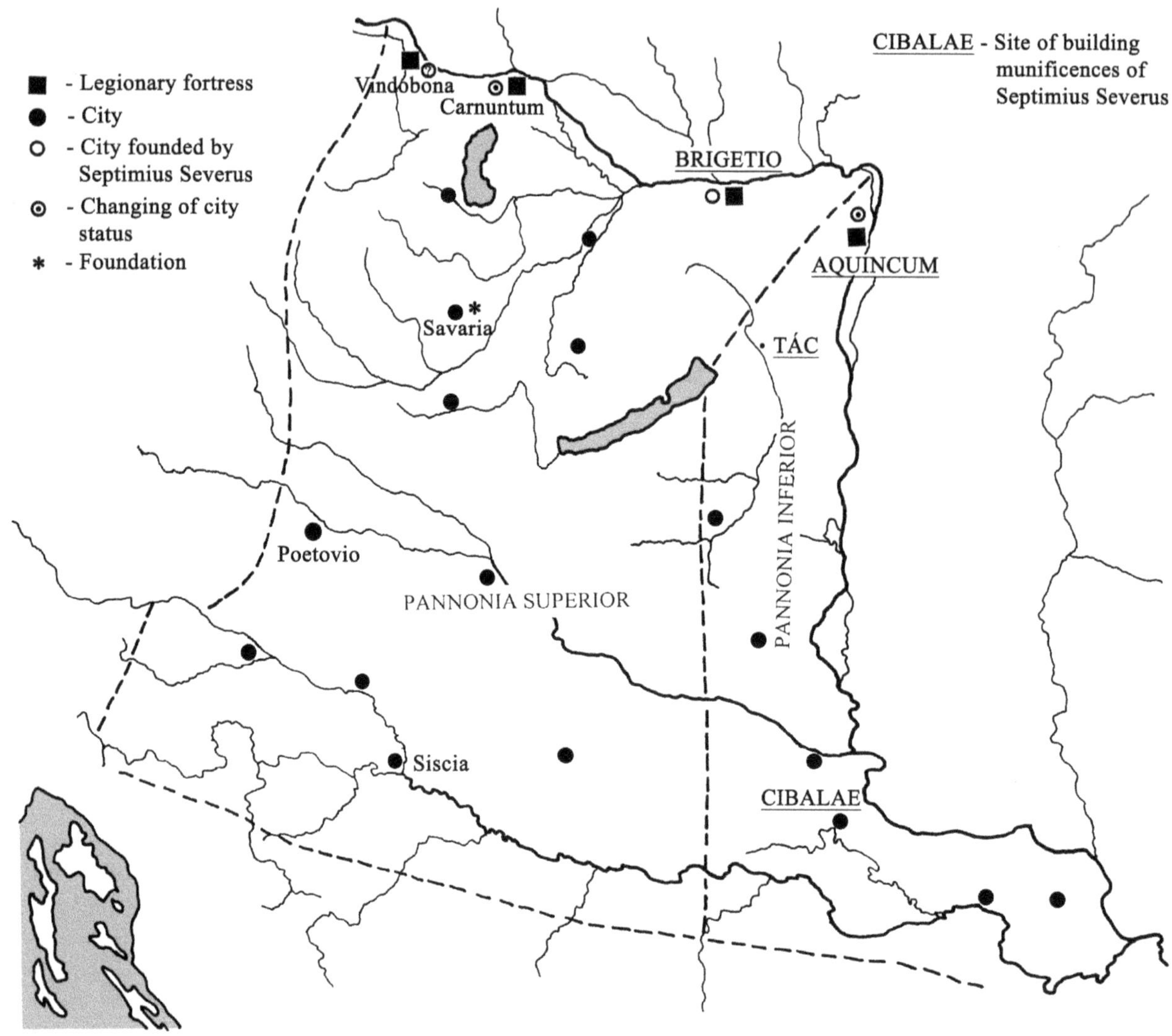

Fig. 18. Construction projects of Septimius Severus and his sons in the towns of the Pannonian provinces (made by Zs. Mráv).

the Empire, the collective noun *Brigetionenses* denoted undoubtedly the municipal community until the end of the 2nd century AD and still after that on some occasions.[91] Thus the statue base of Brigetio is a new proof of the settlement's municipal status granted by Septimius Severus. Its original spot might have been one of the public sections of the town, most probably its forum. But in what way does this statue base prove that the foundation of Brigetio occurred during the reign of Septimius Severus? Erecting of an emperor's statue by a municipal community was not associated with a major event of the imperial dynasty, such as imperial anniversaries, accessions to power, or triumphs. Towns merely fulfilled their fundamental political and religious obligations in their adoration of imperial statues, which was an official means of expressing loyalty to the ruler and the state.[92] The date of the accomplishment of the required task depended on the time the community needed to accumulate material resources to put up an imperial statue, which in any case burdened the town's treasury quite heavily. On the basis of accurately dated examples, the towns of Pannonia were generally able to fulfil the expectation of setting up statues within 1-2 years, or in the worst case, 3 years after the emperors' acceding to the throne (table 3.). This tendency was particularly true in the Severan era, when Pannonian towns could even get the wherewithal to erect statues representing female members of the imperial family.

Only one exception is known of an imperial statue erected by a Pannonian town remarkably later than the emperor's accession to the throne. The example in question is exactly the statue base of Septimius Severus from Brigetio, which was dedicated sometime between January 198 AD and the turn of the years 209 and 210 AD, at least five years after Severus acceded to the throne, or even later than that. This striking exception can only be explained by the extremely poor financial condition of town, or perhaps the town's foundation predating the emperor's accession and a slow development of communal life in the new place. The first

[91] Blanck 1983, 536; Alföldy 1984, 53-54; Mráv 2003, 339. To mention only Pannonian examples: The inhabitants of *colonia Aelia Mursa* erected sculpture to Divus Hadrianus simply as *Mursenses* (*CIL* III 3279 = 10262), and the dwellers of Poetovio colonia to Iulia Domna as *Poetovionenses* (*CIL* III 4054).

[92] Alföldy 1981, 206, 209; Alföldy 1984, 55.

Table 3. The duration of period between the imperial accessions and the time the towns of the Pannonian provinces could afford erecting statues dedicated to the emperors and the members of their families.

Within 1 year	- to Fulvia Plautilla in Siscia (*AIJ* 560) - to Fulvia Plautilla in Municipium Iasorum (Mráv 2007a) - to Iulia Aquilia Severa in Brigetio (*RIU* 3, 674) - to Traianus Decius and Herennia Etruscilla in Andautonia (*AIJ* 477-478)
Within 2 years	- to Traianus in Neviodunum (Lovenjak 2008, Nr. 4) - to Philippus Arabs in Mog(iones?) (Mráv 2003)
Within 3 years	- to Valerianus (*AE* 1965, 304) - to Sabinia Tranquillina in Municipium Iasorum (*AIJ* 587)
Over 5 years	- to Septimius Severus in Brigetio (*CIL* III 4309 = *RIU* 2, 501)

alternative can be excluded with great certainty; therefore the town must have been founded by Septimius Severus himself, not too long after his accession to power. The building inscription commemorating the emperor's own generosity, which refers to the fact that Septimius Severus supported the development of the town – which had been granted municipal status by him – with a monumental public construction, corresponds fairly well with the above arguments.[93] The fact that we may reckon with an accomplished imperial construction work in the town by the autumn of 195 AD, also proves that the civil town of Brigetio might have been created *ex novo* and awarded municipal status in 194 AD, at the same time when Aquincum and Carnuntum gained colonial status. (The civil town of Vindobona might have possibly been founded at the same time, as well.[94]) The suggestion seems to be logical, since there is no reason to assume that Septimius Severus would have discriminated the settlement near the fortress of the elite legion, the *I Adiutrix*, which contributed quite a lot towards his proclamation and in waging his wars; on account of the other two privileged Pannonian towns. It appears unlikely that Carnuntum and Aquincum were raised to the status of colony, while the settlement near Brigetio (and perhaps the *canabae* that had already been attached to it[95]) would have been denied municipal status.

5. The characteristics of the relationship between Septimius Severus and Pannonian towns as reflected in the new imperial building inscriptions

Besides establishing funds, new towns, and changing their municipal status,[96] Septimius Severus sponsored the town communities of the Pannonian provinces with constructions on his own expenses.[97] Through the publishing and/or reinterpretation of the four inscriptions discussed in this paper, the number of known imperial constructions has been raised to more than its double. With these four building works, a monumental array of constructions concerning Pannonian towns is delineated, comparable only with Severan buildings in the North African provinces.[98] The civil town of Carnuntum must have received outstanding support during the emperor's reign as well; however, we still do not possess any building inscription verifying imperial constructions. Carnuntum's distinguished significance is well attested by military and civilian building activity triggered by the prosperity of the Severan era, as well as by archaeological remains attesting to intensive construction works.[99] It is not at all surprising that the recently identified imperial inscriptions relating to Septimius Severus and his sons derive from Aquincum and Brigetio, two main spots that witnessed economic development in the Middle Danube region during the Severan era. The two towns were nearly immediately after Septimius Severus' accession to power raised to the status of colony or were founded *ex novo*. The first and most important patron of a town was always its founder, as the *Lex Coloniae Genitivae Iuliae* prescribed,[100] who effectively contributed to the promotion of civilian life, often with public constructions in favour of meeting social expectations. This was the case in Brigetio, too, where significant imperial buildings were begun contemporarily with its foundation, suggested by the extreme size of the imperial building inscription, and in Aquincum as well, which was raised to the status of colony at that time and also received constructions commissioned by Septimius Severus and his sons. The privileged roles of both towns were due to their situation in the vicinity of the fortresses of legions that supported the emperor's power and formed elite corpses in his warfare. (Suchlike privileged places can be traced in the Dacian provinces, e.g. beside the legionary fortresses of Potaissa and Apulum, as well as in Porolissum.) The abovementioned imperial benefits were not restricted exclusively to towns along the Danube, as Septimius Severus also paid attention to smaller towns inside Pannonia, such as Cibalae. These actions were probably aimed not only at the support of towns but also of their communities, which gave great numbers of enlistees. The town of Cibalae is an important example of the place

93 After a town's foundation, the *conditor* often contributed to its development with public construction works. This practice can be traced in the case of Pannonian towns, which were more in need of imperial support. Therefore, according to a building inscription, *colonia Ulpia Poetovionensium* was sponsored with a public building, most likely the forum of the town, between 103 and 106 AD by Trajan: Mráv 2000 (=*AE* 2000, 1189). The same occurred in Sarmizegetusa, also according to a new interpretation of the building inscription, which was placed on the entrance tetrapylon of the forum: Piso 2008, 151-161 (*AE* 2007, 1203).

94 On the debated date of the foundation of *municipium Vindobonense* see Piso 1991, 171-177; Kovács 2001, 43; Kovács 2007, 146.

95 Kovács 2001, 42-66.

96 Mócsy 1974, 219, 244; Mráv 2006; Kovács 2007, 145-146.

97 Mráv 2006.

98 Haywood 1940; Ward-Perkins 1993.

99 For a summary see Kandler 2000.

100 *Lex Coloniae Genitivae Iuliae*, XCVII.

Table 4. Known statue bases of Septimius Severus and his family erected by Pannonian towns.

	Sculptured or represented person	*Commemorator*	Dating	Findspot	Literature
1.	Divus Commodus	*res publica Iasoru(m)*	During the reign of Septimius Severus, after 195 AD	Municipium Iasorum / Aquae Balissae	*CIL* III 4000
2.	Divus Commodus	Possibly the *ordo* or inhabitants of Scarbantia	During the reign of Septimius Severus, after 195 AD	Scarbantia	*RIU* 1, 210 – cf . Szabó 2010, 213-222.
3.	Septimius Severus	*Brig(etionenses) publice d(ecurionum) d(ecreto)*	January 198 AD – end of 209 AD	Brigetio	*CIL* III 4309 = *RIU* 2, 501
4.	[[Fulvia Plautilla]]	*respubl(ica) Siscianorum*	202 AD	Siscia	*AIJ* 560
5.	[[Fulvia Plautilla]]	*respublica Ias(orum)*	202 AD	Municipium Iasorum / Aquae Balissae	Mráv 2007, 79-89 (*AE* 2007, 1144)
6.	Iulia Domna?	*Poetovionens(es)*	April 11, 195 A.D. – 211	Poetovio	*CIL* III 4054
7.	*[- - -] Imp(eratoris) L(ucii) Sep(timii) Severi Pii Pertinacis Aug(usti) et Imp(eratoris) Cae(saris) M(arci) Aur(elii) Antonini*	*[Aq]uinc(enses) publice*	197-209 A.D.	Aquincum	*CIL* III 3518= *RIU* 2, 397

where masses of recruits were enrolled, many of them serving in the imperial guard during the Severan era.[101]

Only three of the four presently known building munificences instigated by Septimius Severus and his family in Pannonian towns can be accurately or more or less accurately dated. The building inscription of Tác, which represents an exception, can be dated to a broader period of time between 197 and 209 AD, as the titles of the two 'Augusti' were not inscribed. Moreover, the name of the person who supervised the construction was chopped off; it must have been the governor in charge of the execution. We already know about the building works in Brigetio that were finished by the summer/autumn of 195 AD, while the building in Aquincum was accomplished some time during the governorship of L. Baebius Caecilianus in Pannonia Inferior (most probably in his early years in office) between 198/199 and 202 AD; the construction in Cibalae was also inaugurated in 200 AD. Therefore none of the buildings, which can be accurately dated, can be related to the emperor's visitation in 202 AD. Contrary to more of his predecessors, it was not the presence of Septimius Severus that played the most important role in his constructions.[102] On the contrary, each construction was in progress when the Pannonian legions were being engaged in some of Septimius Severus's wars away from their province. In 195 AD the soldiers of the elite *legio I Adiutrix* returned from the first Parthian campaign, at the time when the construction at Brigetio had already been accomplished.[103] The public buildings of Aquincum and Cibalae, which were supervised by L. Baebius Ceacilianus, had also long been accomplished by the time the Pannonian legions, summoned to the *expeditio secunda Parthica* at that time, arrived home in 202 AD.[104] It transpires that the emperor considered the welfare of Pannonian towns important and also lavished money on them, even during the course of costly civil wars and military campaigns.[105] The emperor, who was surrounded by Pannonian soldiers, might have certainly been more inclined to fulfil the claims of Pannonian towns, either through the governor or the towns' own delegates. Owing to a nearly complete absence of the armies in both Pannonian provinces, and the drop-off of the incomes, the emperor strove to compensate civil communities of Pannonia with benefits, mainly constructions of public interest, in order to maintain economic conjuncture. As a consequence of the fairly few literary sources, we learn only a few but informative elements of the tight relations system between Pannonian towns and the first Severi. If we regard inscriptions commemorating the constructions of the emperor and his sons as the most revealing documents that attest to Septimius Severus's specific attention to Pannonian towns, then we have to accept that civilian communities officially expressed their exemplary reverence to the Severan dynasty mostly by erecting statues of the emperors.

[101] Some soldiers from Cibalae certainly served among the praetorians: M. Aurelius Sabinus: *CIL* VI 32536,VS(28); Aurelius C(l)emens: *CIL* VI 32624,A(14). Praetorians of Cotinus origin recruited in Mursa and Cibalae even erected an inscription dedicated to the salvation of Severus Alexander. More of the mentioned praetorians might have certainly enlisted in the army during the reign of Septimius Severus: *CIL* 32542 (2833) = Dobó 1975, Nr. 22; *CIL* 32544 = Dobó 1975, Nr. 23; *CIL* VI 32557 = Dobó 1975, Nr. 24; *CIL* VI 32542 (frag. B, v. 9) = Dobó 1975, Nr. 29/c.

[102] Winter 1996, 108-118.

[103] Lőrincz 2000, 157; Kovács 2007, 180-181.

[104] *AE* 1976, 544 = *AE* 1978, 652 – cf. Németh 1976, 193-197.

[105] Imperial benefits – including erections of public buildings – were not being suspended even during large-scale and expensive external wars, if the emperor's policy held them important for some reasons.

Table 5. The forms of direct imperial patronage of the towns in the middle Danubian provinces during the reign of Septimius Severus.

	Foundation of towns	Change in attribute	Changes of town status	Imperial building patronage	Foundation	*ius Italicum*
Noricum	-	-	-	2 (Iuvavum, Celeia)	-	-
Dalmatia	-	-	-	-	-	-
Pannonian provinces	1 (Brigetio)	1 (Siscia)	2 (Carnuntum, Aquincum)	4 (Tác, Cibalae, Aquincum, Brigetio)	1 (Savaria)	-
Moesia superior	-	-	-	2 or 3? (Rudnik, Viminacium, Scupi?)	-	-
Dacian provinces	3 (Apulum, Potaissa, Porolissum)	-	1 (Drobeta)	1 (Sarmizegetusa)	-	5

Not only did the most important Pannonian towns set up statues of Septimius Severus and his sons, which was only expected at the time, but even towns of smaller importance inside Pannonia obviously 'surpassed' their obligations in this regard.

The most expressive example is the small town of Municipium Iasorum/Aquae Balissae (Daruvar, Croatia), where at least five statues can be associated with the imperial family during the reign of Septimius Severus. The statues of Septimius Severus, Caracalla and Geta can be regarded as obligatory dedications, so their erection was only expected; the lack of such findings so far must be an accident of archaeology. On the other hand, Municipium Iasorum yielded statue bases of Divus Commodus and Fulvia Plautilla. By means of setting up Fulvia Plautilla's statue, the towns responded instantly to the contemporary event concerning the imperial family: her sculpture was already set up *loco publico* of the town a few moths after her betrothal to Caracalla. Between 197 and 209 AD the *Aquincenses* even decided to erect a personified statue that can be associated with Septimius Severus and Caracalla, or one representing a god in a public place, which is the only known statue of this type in Pannonia.[106] The inscription on the statue base reads as follows: *[- - -] / Imp(eratoris) L(ucii) Sep(timii) Severi / Pii Pertinacis Aug(usti) / et Imp(eratoris) Caes(aris) M(arci) / Aur(elii) Antonini /*[5] *[Aq]uinc(enses) / publice*. Since the names of the emperors are in the genitive case, the base could not have carried the emperors' statues, because those did not fit on it either. Similar phrasings on statue bases are known from the examples representing a divine personification of the emperors' highly esteemed virtues (e.g. Concordia, Pietas, Providentia, Virtus, etc.[107]). The names of emperors can also take the genitive case if statues of deities related to the emperors were set up on the bases at the same time (e.g. I. O. M. Protector, Fortuna Redux, Victoria, etc.[108]). We should conceive of the broken off first line of the inscription of the base as a personification or a god's name, in contrast to the phrase *pro salute*, proposed by the *CIL*.[109]

6. Conclusion

The total number of Septimius Severus' known building munificences in the towns of the Pannonian provinces has been increased to four, as two new ones were added to those known before. This equals the number of imperial building inscriptions in North Africa, and therefore it may be assumed that Pannonian towns might have been bestowed donations of similar significance as the towns of North Africa, Septimius Severus' homeland. Among other available historical sources, imperial building inscriptions most clearly display a privileged status Pannonian towns were enjoying during the reign of Septimius Severus. The building works at Brigetio were accomplished in the autumn of 195 AD in the absence of the *legio I Adiutrix*, thus its location might be traced in the civil town. The reasons for the imperial grants could have been nothing else but to make the settlement a municipium, and then to develop its public life and its physiognomy as soon as possible. The date of the construction designates the date of Brigetio's gaining municipal status in 194 AD, the year in which Aquincum and Carnuntum were by Septimius

[106] The statue base cannot be dated to the time prior to the autumn of 197 AD (as it was done by P. Kovács in *FPA* IV, 63-64, and followed by B. Lőrincz in: *TitAq* 397), because on it Caracalla is certainly mentioned as the co-emperor of Septimius Severus, as transpires from the initial names 'Imperator Caesar' in his nomenclature. This is not contradicted either by the apparent absence of the name 'Augustus', since it was included in the same nomenclature, among others, on the statue base of Fulvia Plautilla in Municipium Iasorum in 202 AD. The town dedicated the statue to the Augusta as *sponsae Imp(eratoris) M(arci) Aurel(ii) Antonini*. It is even more plausible than Caracalla's being a Caesar if we presume that the base part carrying the name 'Augustus' had broken off the missing inscription, and remained only in the forms of paraphrases.

[107] See for example in the forum of Cuicul: Zimmer 1989, 54 C2 (*AE* 1916, 17), 57 C9 (*CIL* VIII 8300 + p. 1896), C10 (*CIL* VIII 20148), 67 C51 or in the territory of regio X Italiae: Alföldy 1984, 43. Nr. 119, 261, 117.

[108] Zimmer 1989, 55 C5 (*AE* 1916, 19), 56 C7 (*AE* 1916, 15).

[109] Mráv 2003, 339-340.

Severus raised to the rank of colony. The task of supervising the subsequent great construction projects was appointed to L. Baebius Caecilianus, the governor of Pannonia Inferior, again during the absence of Pannonian legions, between 198/199 and 202 AD. The governor supervised another unidentifiable building in Aquincum, whose inscription was preserved in medieval records only. Contrary to Th. Mommsen's opinion, this inscription could not have been inscribed on a milestone, as its text and the adjective *splendidissima* associated with the colony rather suggest that it is an imperial building inscription. Septimius Severus, together with his sons, appointed L. Baebius Caecilianus to execute a construction project in Cibalae, by which the town community was granted a water supply facility and a public bath. The buildings were opened to the public in 200 AD. The monumentality of the building inscriptions from Brigetio and Cibalae, both exceeding 5 meters, reflect the importance of Pannonian construction works. With his building projects in Pannonian towns, Septimius Severus strove to compensate all the losses the provinces suffered by the summoning of their entire armies with a view to making Pannonian soldiers, who provided the majority of the emperor's guard and the elite legions of his expedition army, be aware of the importance of their homeland.[110]

Bibliography

Alföldy 1981 Géza Alföldy, Bildprogramme in den römischen Städten des Conventus Tarraconensis - das Zeugnis der Statuenpostamente, in: *Homenaje García y Bellido IV*, *Revista de la Universidad Complutense de Madrid* 18 (1979), Madrid, 177-275.

Alföldy 1984 Géza Alföldy, *Römische Statuen in Venetia et Histria. Inscriptionische Quellen, Abh. der Heidelberger Akademie der Wissenschaften. Phil.-hist. Klasse*, Abh. 3.

Alföldy 1997 Géza Alföldy, Die Grossen Götter von Gorsium, *ZPE* 115, 225-244.

Alföldy 2001 Géza Alföldy, Pietas immobilis erga principem und ihr Lohn: Öffentliche Ehrenmonumente von Senatoren in Rom während der Frühen und Hohen Kaiserzeit, in: *Inschriftliche Denkmäler als Medien der Selbsdarstellung in der römischen Welt* (eds. G. Alföldy and S. Panciera), *Heidelberger Althistorische Beiträge und Epigraphische Studien* 36, Heidelberg, 10-46.

Alföldy 2002a Géza Alföldy, Zu kaiserlichen Bauinschriften aus Italien, *Epigraphica* 64, 113-145.

Alföldy 2002b Géza Alföldy, Roms Kaiser als Bauherren, *JRA* 15, 489-498.

Alföldy 2004 Géza Alföldy, Die Repräsentation der kaiserlichen Macht in den Inschriften Roms und des Imperium Romanum, in: L. De Blois, P. Erdkamp, O. Hekster, G. DE Kleijn, and S. Mols (eds.), *The Representation and Perception of Roman Imperial Power, Proceedings of the Third Workshop of the International Network Impact of Empire*, Amsterdam (2003), 3-19.

[110] The text was translated by Valér Bedő. My grateful thanks are due to Branka Migotti (The Croatian Academy of Sciences and Arts, Zagreb) for having corrected the English text.

Altmann 1973 Júlia Altmann, Előzetes jelentés a budavári ferences templom kutatásáról – Rapport préliminaire sur les fouilles executes dans l'église des franciscains de Budavár, *ArchÉrt* 100, 82-87.

Altmann 1994 Júlia Altman, Az óbudai és a budavári ferences templom és kolostor kutatásai, in: *Koldulórendi építészet a középkori Magyarországon. Tanulmányok. Művészettörténet – Műemlékvédelem* 7, Budapest, 137-152.

Altmann and Lővei 2004 Júlia Altman and Pál Lővei, Leletegyüttesek a budavári ferences templomból, *BudRég* 38, 11-34.

Ardevan 1998 Radu Ardevan, Viaţa municipală în Dacia romană, *Bibliotheca Historica et Archaeologica Banatica* 10, Timişoara.

Barkóczi 1951 László Barkóczi, *Brigetio, DissPann* II (22), Budapest.

Bechert 1971 Tilmann Bechert, Römische Lagertore und ihre Bauinschriften. Ein Beitrag zur Entwicklung und Datierung kaiserzeitlicher Lagertorgrundrisse von Claudius bis Severus Alexander, *BJb* 171, 201-287.

Beszédes, Mráv, and Tóth 2003 József Beszédes, Zsolt Mráv, and Endre Tóth, Die Steindenkmäler von Bölcske – Inschriften und Skulpturen – Katalog, in: *Bölcske. Römische Inschriften und Funde, Libelli Archaeologici* Ser. Nov. 2, Budapest, 103-218.

Birley 1971 Anthony Birley, *Septimius Severus, the African emperor*, London.

Blanck 1983 Horst Blanck, Rez. von J. P. Rollin, Untersuchungen zu Rechtsfragen römischer Bildnisse, *Gnomon* 55, München, 530-538.

Boatwright 1989 Mary T. Boatwright, Hadrian and the Italian Cities, *Chiron* 19, 235-271.

Boatwright 2000 Mary T. Boatwright, *Hadrian and the Cities of the Roman Empire*, Princeton.

Boatwright 2002 Mary T. Boatwright, Trajan outside Rome: construction and embellishment in Italy and the provinces, in: P. A. Stadter and L. Van der Stockt (eds.), *Sage und Emperor. Plutarch, Greek Intellectuals, and Roman Power in the Time of Trajan (98-117 A.D.)*, *Symbolae* Ser. A, 29, Leuven, 259-277.

Borhy 2006 László Borhy, Vezető Komárom város római kori kőemlékeihez, *Acta Archaeologica Brigetionensia*, Ser. I, 5, Komárom.

Brunšmid 1902 Josip Brunšmid, *Colonia Aurelia Cibalae. (Vinkovci u staro doba)*, Zagreb.

Brunšmid 1912 Josip Brunšmid, *Kameni spomenici Hrvatskoga narodnoga muzeja u Zagrebu* (1904-1912), Zagreb.

Buzov 2008 Marija Buzov, Plautilla, sudbina jedne princeze – Plautilla and the Fate of a Princess, *ArchAdr* 2, 473-488.

Dobó 1975 Árpád Dobó, *Inscriptiones extra fines Pannoniae Daciaeque repertae ad res earundem provinciarum pertinentes*, Budapest.

Fitz 1959 Jenő Fitz, Der Besuch des Septimius Severus in Pannonien im Jahre 202. u. Z., *AArchHung* 11, 237-263.

Fitz 1993 Jenő Fitz, *Die Verwaltung Pannoniens in der Römerzeit* II, Budapest.

FPA IV Péter Kovács (ed.), Fontes Pannoniae Antiquae in aetate Severorum, Budapest 2007.

Fraser 2006 Trudie E. Fraser, *Hadrian as Builder and Benefactor in the Western Provinces*, *BAR Intern. Ser.* 1484.

Gabler 1966 Dénes Gabler, Munera Pannonica. Magánszemélyek közcélú adományai Pannoniában, *ArchÉrt* 93, 20-35.

Galsterer-Kröll 1972 Brigitte Galsterer-Kröll, Untersuchungen zu den Beinamen der Städte des Imperium Romanum, in: *Inscriptionische Studien* Bd. 9, Bonn 1972, 44-145.

Halfmann 1986 Helmut Halfmann, *Itinera principum. Geschichte und Typologie der Kaiserreisen im Römischen Reich*, *HABES* 2, Stuttgart.

Haywood 1940 Richard M. Haywood, The African policy of Septimius Severus, *Translations of the American Philological Assosiation* 71, 175-186.

Horster 1997 Marietta Horster, Literarische Zeugnisse kaiserlicher Bautätigkeit. Eine Studie zu Baumaßnahmen in Städten des Römischen Reiches während des Prinzipats, *Beiträge zur Altertumskunde* 91, Stuttgart-Leipzig.

Horster 2001 Marietta Horster, *Bauinschriften römischer Kaiser. Untersuchungen zu Inschriftenpraxis und Bautätigkeit in Städten des westlichen Imperium Romanum in der Zeit des Prinzipats*, *Historia Einzelschriften* 157, Stuttgart.

IDR III/5 Ioan Piso (ed.), *Inscriptiones Daciae Romanae III/5*, Paris 2001.

ILSl 1 Milan Lovenjak (Hrsg.), *Inscriptiones Latinae Sloveniae* 1. *Neviodunum*, Ljubljana 1998.

Iskra-Janošić 2001 Ivana Iskra-Janošić, *Urbanizacija Cibala i razvoj keramičarskih središta – Die Urbanisation von Cibalae und die Entwicklung der Zentren für Keramikproduktion*, Zagreb-Vinkovci.

Iskra-Janošić 2004 Ivana Iskra-Janošić, Colonia Aurelia Cibalae. Entwicklung der Stadt, in: *Pannonia* II, 169-195.

Kienast 1996² Dietmar Kienast, *Römische Kaisertabelle. Grundzüge einer römischen Kaiserchronologie*, Darmstadt.

Kneißl 1969 Peter K. Kneißl, *Die Siegestitulatur der römischen Kaiser. Untersuchungen zu den Siegerbeinamen des ersten und zweiten Jahrhunderts, Hypomnemata* 23, Göttingen.

Kandler 2000 Manfred Kandler, Zur Bautätigkeit in Carnuntum unter den Severer, in: *Gentes, Reges und Rom. Auseinandersetzung - Anerkennung – Anpassung. Festschrift für Jaroslav Tejral*, Brno, 43-45.

Kovács 1999 Péter Kovács, Building inscriptions of Septimius Severus from Pannonia, in: *Atti del XI Congresso Internazionale di Epigrafia Greca e Latina*, Roma, 521-531.

Kovács 2001 Péter Kovács, The Pannonian canabae in the 3rd century, *Laverna* 12, 42-66.

Kovács 2007 Péter Kovács, Pannonia története a Severus-korban, in: *FPA* IV, 133-182.

Kovács 2010 Péter Kovács, Adatok a pannoniai feliratok gyűjtéséhez a 15-16. században [Datas to the collections of Pannonian inscriptions in the 15th and 16th century], *SEP* 2 (2009), 117-128.

Lőrincz 1990 Barnabás Lőrincz, A pannoniai csapatok részvétele a tartományon kívüli harcokban, in: J. Fitz and A. Mócsy (ed.), *Pannonia régészeti* kézikönyve, Budapest, 83-84.

Lőrincz 2000 Barnabás Lőrincz, Legio I Adiutrix, in: Y. Le Bohec (ed.), *Les légions de Rome sous le Haut-Empire*, Actes du Congrès de Lyon (17-19 septembre 1998), *Collection du Centre d'Études Romaines et Gallo-Romaines* 20, Lyon, 151-158.

Lőrincz 2004 Barnabás Lőrincz, Zur Statthalterliste der römischen Provinz Pannonia Inferior, in: C. Roman, Ch. Gazdac, *et al.* (eds.), *Orbis antiquus. Studia in honorem Ioannis Pisonis*, Cluj-Napoca, 35-40.

Lőrincz 2007a Barnabás Lőrincz, A pannoniai tartományok helytartói Kr.u.193-235 között, in: *FPA* IV, 183-186.

Lőrincz 2007b Barnabás Lőrincz, Császári téglavetők termékei Pannoniában, in: M. Bíró (ed.), *Orpheus búcsúzik. Tanulmányok Sarkady János emlékére*, Budapest, 213-217.

Maionica and Schneider 1877 Enrico Maionica and R. Schneider, Bericht über eine Reise im westlichen Ungarn, *Archäologisch-Epigraphische Mitteilungen aus Österreich-Ungarn* 1, Vienna, 145- 171.

Mastino 1981 Attilio Mastino *Le titolature de Caracalla e Geta attraverso de iscrizioni (indici), Studi di storia di antica* 5, Bologna.

McMullen 1959 Ramsey McMullen, Roman Imperial Building in the Provinces, *Harvard Studies in Classical Philology* 64, Cambridge MA, 207-235.

Mócsy 1962 András Mócsy, Pannonia, in: *RE* Suppl. IX, 515-776.

Mócsy 1974 András Mócsy, *Pannonia and Upper Moesia. A history of the Middle Danubian provinces of the Roman Empire*, London-Boston.

Mócsy 1990 András Mócsy, Municipiumok és coloniák, in: A. Mócsy and J. Fitz (eds.), *Pannonia Régészeti Kézikönyve*, Budapest, 61-64.

Mráv 1999 Zsolt Mráv, Die Bautafel eines Heiligtums aus Visegrád, *Antaeus* 24 (1997-1998), 306-310.

Mráv 2000 Zsolt Mráv, Building Inscription of Trajan from Poetovio, in: Gy. Németh and P. Forisek (eds.), *Epigraphica I. Studies on Epigraphy Hungarian Polis Studies* 6, Debrecen, 77-94.

Mráv 2001 Zsolt Mráv, Die Gründung Emonas und der Bau seiner Stadtmauer. (Zur Ergänzung der Inschrift *AIJ* 170B = *ILJug* 304), *AAntHung* 41, 81-98.

Mráv 2002a Zsolt Mráv, Eine *litterae aureae*-Inschrift aus Intercisa, *FA* 49-50 (2001-2002), 207-233.

Mráv 2002b Zsolt Mráv, Die Brückenbauinschrift Hadrians aus Poetovio, *CommArchHung* 2002, 15-57.

Mráv 2003a Zsolt Mráv, Kaiserliche Bautätigkeit zur Zeit Hadrians in den Städten Pannoniens, *AAntHung* 43, 125-137.

Mráv 2003b Zsolt Mráv, Die Statuenbasen des Philippus Arabs aus Környe, in: Á. Szabó and E. Tóth (eds.), *Pannonica. Provincialia et archaeologica. Studia sollemnia Eugenio Fitz octogenario dedicata. Libelli Archaeologici*, Ser. Nova 1, Budapest, 331-368.

Mráv 2006 Zsolt Mráv, An Imperial building inscription of Septimius Severus, Caracalla and Geta from Cibalae (Vinkovci, Croatia). Preliminary Report, in: Gy. Németh and P. Forisek (eds.), *Epigraphica III. Politai et Cives, Hungarian Polis Studies* 13, Debrecen, 211-228.

Mráv 2007a Zsolt Mráv, The statue base of Fulvia Plautilla from Municipium Iasorum, *ARR* 15, 79-89.

Mráv 2007b Zsolt Mráv, Adversus Parthos. Neue Angaben zur Teilnahme der pannonischen Legionen und Hilfstruppen an den Partherfeldzügen von Septimius Severus und Caracalla – Adversus Parthos. Új adatok a pannoniai legiók és segédcsapatok Septimius Severus és Caracalla parthus hadjárataiban való részvételéhez, *FA* 53, 119-154.

Mráv 2008 Zsolt Mráv, A pannoniai városok közösségi szoborállításai – Die Statuenaufstellungsbräuche pannonischer Städte, *Komárom-Esztergom Megyei Múzeumok Közleményei* 12, Tata, 107-158.

Mráv 2009 Zsolt Mráv, L. Cornelius Felix Plotianus, Pannonia inferior helytartója és az intercisai auxiliáris castellum építéstörténete. Két Commodus-kori monumentális katonai építési felirat Intercisából. In: *SEP* 1 (2008) 243-277 = *Id.*, L. Cornelius Felix Plotianus, der Statthalter von Pannonia Inferior und die Baugeschichte des Auxiliarkastells Intercisa (Dunaújváros, Komitat Fejér, Ungarn). Zwei commodischen Monumentalbauinschrift aus Intercisa, in: Sz. Bíró (ed.), *Ex Officina. Studia in honorem Dénes Gabler*, Győr 2009, 357-388.

Mráv 2010 Zsolt Mráv, Septimius Severus-kori császári építési feliratok Aquincumból és Brigetióból [Building inscriptions of Septimius Severus from Aquincum and Brigetio], *SEP* 2 (2009), 187-212.

Mráv and Ottományi 2005 Zsolt Mráv and Katalin Ottományi, *De{i}fu(n)c(tus) exp(editione) Germ(anica) Lauri(aci) mort(e) sua* - Sarkophag eines während der alamannischen Expedition Caracallas verstorbenen Soldaten aus Budaörs, *AArchHung* 56, 177-212.

Mráv in press Zsolt Mráv, Septimius Severus and the Norican City of Celeia (Celje, Slovenia). A „new" Severan imperial building inscription.

Németh 1976 Margit Németh, Septimius Severus 202. évi látogatásának újabb emléke Aquincumból – A Further Inscriptionic Monument from Aquincum of the Visit of Septimius Severus in the Year 202 A.D, *BudRég* 24, 193-197.

Nielsen 1990 Inge Nielsen, *Thermae et Balnea. The Architecture and Cultural History of Roman Public Baths*, Aarhus.

Perinić Muratović 2004 Ljubica Perinić Muratović, Odraz panonskog putovanja Septimija Severa u Cibalama. – The echo in Cibalae of the Pannonian journey (202 A.D.) of Septimius Severus, *ARR* 14, 77-101.

Piso 1991 Ioan Piso, Municipium Vindobonense, *Tyche* 6, 171-177.

Piso 2008 Ioan Piso, L'inscription monumentale de l'entrée dans le Forum Vetus de Sarmizegetusa, *Acta Musei Napocensis* 43-44/I (2006-2007), Cluj-Napoca, 151-161.

Rebuffat 1991 René Rebuffat, Vocabulaire thermal. Documents sur le bain romain, in: *Les thermes romains. Actes de la table ronde organisée par l'École française de Rome, 11-12 novembre 1988, Coll. de l'École Française de Rome* 142, Rome, 1-32.

RIU 1 László Barkóczi and András Mócsy, *Die römischen Inschriften Ungarns 1*, Budapest 1972.

RIU 2 László Barkóczi and András Mócsy, *Die römischen Inschriften Ungarns 2*, Budapest 1976.

RIU 3 László Barkóczi and Sándor Soproni, *Die römsichen Inschriften Ungarns 3*, Budapest-Bonn 1981.

RIU 6 Jenő Fitz, András Mócsy, and Sándor Soproni, *Die römsichen Inschriften Ungarns 6*, Budapest-Bonn 2001.

Szabó 2010 Ádám Szabó, Divus Commodus szobrának felirata Sopronból, in: *SEP* 2 (2009), 213-222 The inscription of the statue of Divus Commodus in Sopron, *ACD* 46 (2010), 83-90.

Szalay 1989 Ágnes Ritoókné Szalay, Der Kult der römischen Epigraphik in Ungarn zur Zeit der Renaissance, in: A. Buck, T. Klaniczay, and S. K. Németh (eds.), *Geschichtsbewusstsein und Geschichtschreibung in der Renaissance*, Budapest, 65-75.

TitAq 1 Péter Kovács and Ádám Szabó (eds.), *Tituli Aquincenses* I, 2009.

TRH Péter Kovács (ed.), *Tituli Romani in Hungaria reperti. Supplementum*, Budapest-Bonn 2005.

Waldherr 1989 Gerhardt Waldherr, *Kaiserliche Baupolitik in Nordafrika. Studien zu den Bauinscheriften der diokletianischen Zeit und ihre räumliche Verteilung in den römischen Provinzen Nordafrikas*, Frankfurt am Main.

Ward-Perkins 1993 John Bryan Ward-Perkins, *The Severan Buildings of Leptis Magna. An Architectural Survey, Society for Lybian Studies Monograph* 2, Manchester.

Winter 1996 Engelbert Winter, *Staatliche Baupolitik und Baufürsorge in den römischen Provinzen des kaiserzeitlichen Kleinasien*, Bonn.

Yegül 1992 Fikret Yegül, *Baths and Bathing in Classical Antiquity*, New York.

Zimmer 1989 Gerhard Zimmer, *Locus datus decreto decurionum. Zur Statuenaufstellung zweier Forumsanlagen im römischen Afrika. Mit inscriptionischen Beiträgen von G. Wesch-Klein, Bayerische Akademie der Wissenschaften. Phil.-hist. Klasse Abh. neue Folge* 102. München.

Zsidi 2002 Paula Zsidi, Transformation of the town structure in the Civil Town of Aquincum during the Severi (A. D. 193-235), *AArchHung* 53, 131-149.

Rural Settlements

Tino Leleković

Ante Rendić-Miočević

1. Introduction

Edith Thomas, a well-known expert in the field of the country architecture of Roman Pannonia, was the first author to make a list of rural settlements in the area of the Croatian part of Pannonia in her book on Roman villas in Pannonia. She also discussed them in the wider Pannonian context; at the time, out of a total of 152 villas only six were from Croatia.[1] These figures could serve as a starting point for the discussion of the settlement of rural areas, although not all of the structures Thomas discussed turned out to be villas, but rather some other form of settlements (see below). According to Dénes Gabler, until the year 2000 some 600 villas were recorded in Pannonia, as many as 400 in Hungary and only 21 in Croatia.[2] It transpires from this data that the number of villas in the whole of Pannonia has multiplied in the meantime. If one takes into account the disproportion in the number of villas in the area of study when compared with the rest of the province, it seems clear that northern Croatia significantly lags behind in the archaeological research of rural areas. The main reason for this is the fact that research outside the largest settlements was rarely conducted, and mostly on already known positions of larger rural villas. Yet this situation has changed following the year 2000, when a number of protective and rescue archaeological excavations and systematic surveys were conducted in advance of the construction of highways and other infrastructural projects. The number of rural sites in the Croatian part of Pannonia increased to ca. 250, 53 of which were actually excavated. Thus, it is clear that the nature, layout, and chronology of most of the settlements that had occupied these sites cannot be determined at this stage. Remains of larger rural settlements that could represent villages (*vici*) were found on nine of the excavated sites, as well as 21 rural villas, while the true nature of the settlement was impossible to determine on 23 sites. In view of this, this paper will discuss rural settlements classified into four categories: 1. *villae rusticae*; 2. villages (*vici*); 2. partly researched settlements of uncertain nature; 4. unexcavated and therefore unclassified settlements, with only the location and broad chronology known (fig. 1).

(T. Leleković)

[1] Thomas 1964, 336-352.

[2] Gabler 2003, 235.

2. *Villae rusticae*

2.1. *History of the research and examples of hypothetical villae rusticae*

The *villa rustica is* a relatively rare archaeological feature in northern Croatia, with a very small number of them properly excavated and presented, as opposed to the eastern Adriatic coastal and insular regions, where this type of architectural complexes was the main carrier of Roman agriculture and the Roman way of life outside towns.[3] It seems that this disproportion between the coastal and inland areas is the result of an insufficient level of archaeological research in the continental part of Croatia, as was noticed by E. Thomas.[4] Nonetheless, she claimed that "Villas situated between the Drava and Sava Rivers are eloquent witness of the important role this area played in the life of the late Roman Empire".[5] From this, one could draw the conclusion that country estates in southern Pannonia were more numerous than the actually researched and published sites demonstrate. E. Thomas substantiated her opinion on the then stage of the development of archaeological research of the area between the Drava and Sava with a number of examples. Among these was Batina Gornja, a village to the east of Belec, where sections of late Roman villa foundations of an unknown layout were discovered. The floor of one of the rooms was covered with small hexagonal tiles, while the floor of another room was covered with rectangular tiles. Her attention was next focused on the Sutla valley, more precisely, the part of the river's course from Bizeljsko and Klanjec, where significant (unexcavated) remains of Roman architecture – possibly a villa – were documented. She further discussed Gornji Stenjevec (the Lenarčić Castle), where clay fragments of water-supply system were found, while a small mediaeval church, probably erected upon a Roman villa, stands on the nearby hill of Crikvišće (Roman coins were found there). Remains of a "settlement with villas" from the 3rd-4th centuries, found in Nova Gradiška, were considered next. A mosaic discovered there in 1953 was described in detail on the basis of an oral report from Marcel Gorenc, a curator at the AMZ. Thomas also noted the site of Novigrad Podravski to the east of Koprivnica, which yielded remains

[3] Zaninović 1967, 358.

[4] Thomas 1964, 336-352. See also Vikić-Belančić 1968, 98 (note 42).

[5] Thomas 1980, 276.

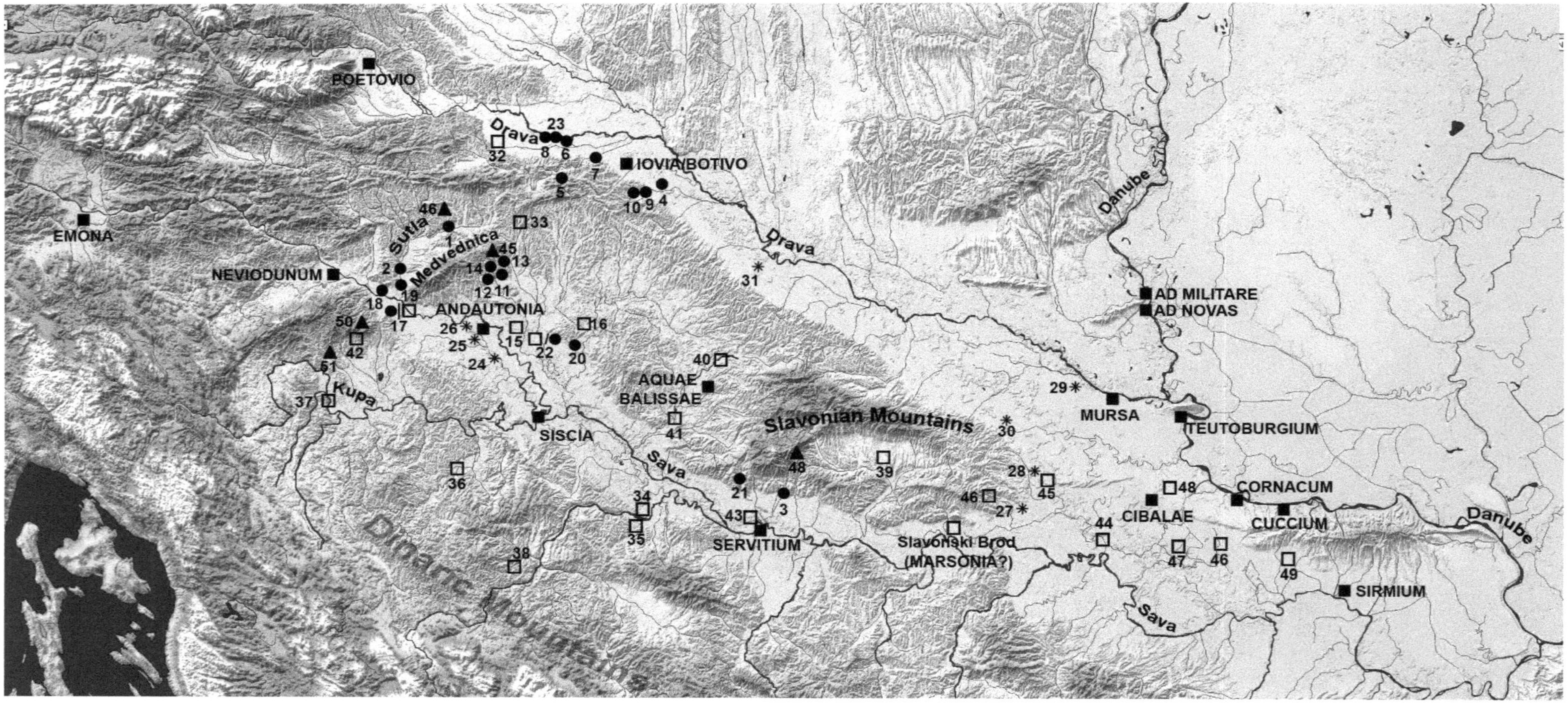

■ TOWNS AND LARGER SETTLEMENTS

● *VILLAE RUSTICAE*
1. Batina Gornja
2. Bizeljsko, Klanjec
3. Nova Gradiška
4. Novigrad Podravski
5. Tuhovec
6. Ferenčica
7. Jalžabet
8. Varaždin
9. Draganovac
10. Starigrad
11. Moravče-Draščica
12. Rošnica-Blaguša
13. Vurnovec-Crikvišče
14. Donja Glavnica-Frtić
15. Okešinec
16. Hercegovac
17. Drenje-Zaprešić
18. Laduč
19. Šibice
20. Kutinska lipa
21. Benkovac, Cage
22. Osekovo-Ciglenice
23. Kupinje

∗ *VICI*
24. Donji Vukojevac
25. Okuje
26. Šepkovčica
27. Sela
28. Ivandvor
29. Josipovac-Verušed
30. Našice
31. Kiškorija

□ UNCLASSIFIED RURAL SETTLEMENTS
32. Petrijanec
33. Komin
34. Hrvatska Dubica, Baćin
35. Suvaja
36. Topusko
37. Kamensko
38. Osječenica-Gorička
39. Tekić
40. Veliki Bastaji
41. Brusnik
42. Repišće
43. Pivare
44. Gudinci, Jaruga
45. Levanjska Varoš
46. Orolik
47. Privlaka
48. Cerić-Plandište
49. Tovarnik

▲ HILLTOP SETTLEMENTS
50. Popov dol
51. Ozalj
52. Kuzelin
53. Lobor
54. Rudina

Fig. 1. Distribution of rural settlements (after Google Maps, modified by T. Leleković).

of a Roman villa with thermal facilities, observing that she did not consult the ground plan of the structure; the same was mostly true of other Croatian sites in her book. Finally, she discussed the site of Tuhovec ca. 3 km from the well-known thermal centre of the Iasi, Varaždinske Toplice (Aquae Iasae). According to Thomas, remains of a richly ornamented Roman villa were found on this site, together with fragments of frescoes and tiles of the ruined floor mosaic.[6] However, it is clear that there is no reliable evidence regarding the character of some of these structures. It seems that Thomas' suggestions were based on indications and information unverified by archaeological research, so their credibility is occasionally doubtful. In this context it does not seem coincidental that Miloje Vasić, the author of a substantial study on Roman villas in the territory of the former Yugoslavia, was apparently not familiar with any country architecture from northern Croatia, although he was well acquainted with this type of structures from the rest of the country, as well as neighbouring areas.[7] However, E. Thomas was not alone in discussing the importance of country estates in the Croatian part of Pannonia. A number of Croatian authors who conducted excavations of Roman country estates in northwestern Croatia, especially in the wider Zagreb region, put forward the same opinion. This is evidenced by several surveys whose special concern is country estates, and which bring maps of the sites from the area in question. Some of these papers are rather early publications,[8] but recently some more ones have appeared.

The most complete presentation of the remains of Roman architecture in north-western Croatia was published in the catalogue of the exhibition *40 Years of the Archaeological Research in North-western Croatia* (Zagreb, 1986). Several authors contributed their papers on the current state of research in northern Croatia. Željko Demo, in a survey of the Roman-period research, stated that the number of excavated cemeteries in north-western Croatia was (at the time) significantly lower when compared with architectural remains. He also noted at least 30 structures in this area that were until then excavated, whether partly or systematically. Furthermore, he noticed that the principal type of the excavated sites was settlements, with larger urban centres much less frequently so, while smaller settlements were somewhat more frequently excavated, but were customarily labelled as *villae rusticae* both when elements for using this syntagm did exist and when they did not.[9] In the region along the Drava River Demo notes remains of a Roman structure in Ferenčica near Prelog, as well as a *villa rustica* between the villages of Jalžabet and Bartolovec. Partial remains of a villa were discovered on the latter site near Varaždin in 1962 and 1963; it was fortified already in its early phase of existence, due to the proximity of the Drava *limes*. The later and much better preserved phase conforms to the type characteristic for the 3rd and early 4th centuries.[10] Demo further discusses the remains of Roman structures in the villages of Draganovac and Starigrad in the Drava region, but also mentiones a number of other sites, presumed road stations or settlements, as well as several supposed *villae rusticae* in the area of Sesvete, especially on the slopes of Medvednica, the mountain on the northern periphery of Zagreb. He discusses a *villa rustica* on the site of Draščica in the village of Moravče, 17 km from Sesvete, where remains of an apsidal building with an elongated room were found, with lateral rooms belonging to production facilities.[11] Demo further discusss a partially excavated, well-preserved structure at the nearby site of Rošnica, next to the Blaguša brook.[12] In the catalogue of the abovementioned monograph, it is noted that the Moravče-Rošnica site is situated in the centre of the village and that some construction works were conducted there in 1980. The remains of a *villa* consisting of several rooms were excavated on this occasion.[13] Next in the line comes a Roman structure on the site of Crkvišče in the village of Vurnovec, 8 km north of Sesvete. The excavations on this site started in 1979, revealing some 200 m of walls, but the function and chronology of the building were not determined.[14] The most important rural building in the area is the *villa rustica* on the site of Frtić in the village of Donja Glavnica, 16 km from Sesvete, until then completely excavated (1976-1979). A source of mineral water is located approximately one km to the south, and the building dates from the 2nd-3rd centuries; presumably, it was abandoned in the second half of the 4th century. According to Demo, the villa was of a complex type, with residential and thermal parts, and possibly also a partially covered yard for economic facilities. In the thermal part of the complex the remains of three pools were found, ending in semicircular apses, together with a drainage canal, hypocaust, and other characteristic features.[15] In the next part of the introductory text the author discusses the remains of a settlement discovered during the excavations in Stenjevec and its vicinity, as well as the well-known fortified architectural complex on the Kuzelin hill above Donja Glavnica. In the region to the east of Zagreb Demo notes remains of Roman buildings in the villages of Okešinec and Hercegovac, both featuring parts of Roman settlements of undetermined function.[16] The author also mentions the site of Drenje west of Zaprešić, not far from the confluence of the Sutla and Sava, bringing a photograph and the ground plan. The excavations started on this site in 1980 and were resumed in 1984, resulting in the discovery of the remains of a large estate with a central building and adjacent facilities. It was suggested that the estate was in use from the 1st to the 4th centuries as a country house, as well as the centre of an agricultural estate.[17] Zoran Gregl claims that this place was

6 Thomas 1964, 336-352. This information was, among others, supplied to her by M. Gorenc.

7 Vasić 1970. Likewise, in discussing the typology of Roman villas in Pannonia M. Biró (1974) was not able to include any of the Croatian villas, as none had yielded any layout-elements by that time. Regrettably, nothing has changed in this respect so far.

8 Klemenc 1938; Vikić and Gorenc 1969.

9 Demo 1986, 29-32.

10 D. Nemeth-Ehrlich, in: Demo 1986, 116-117, no. 9.

11 V. Sokol, in: Demo 1986, 120-121, no. 15.

12 Demo 1986, 30.

13 V. Sokol, in: Demo 1986, 122, no. 16.

14 V. Sokol, in: Demo 1986, 126, no. 24.

15 Demo 1986, 30; V. Sokol, in: Demo 1986, 112, no. 1.

16 Demo 1986, 30-32.

17 Demo 1986, 30; Ž. Škoberne, in: Demo 1986, 115-116, no. 5.

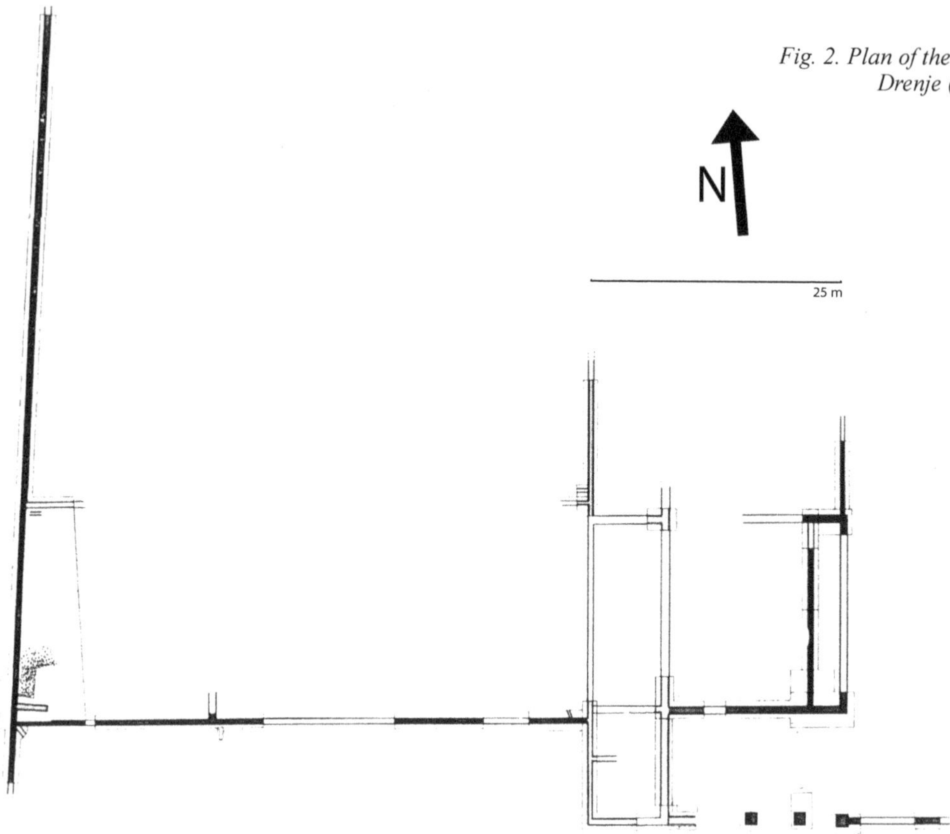

Fig. 2. Plan of the excavated part of the villa in Drenje (after Demo 1986).

the westernmost position in the wider Zagreb area that could have been used as a crossing over the Sava, and that it was located near the mouth of the Sutla, on the shortest route connecting to the Emona – Siscia road and passing along the right bank of the river. The author further notes villa remains in the nearby villages of Laduč and Šibice.[18] Several shorter papers by Gregl also cover the subject of country estates in the wider Zagreb area, stating that these were the centres of agricultural estates, featuring some characteristic villa structures. He points out that the convenient configuration of the land, together with some other elements favourable for the positioning of villas and similar architectural complexes, played a part in the distribution of Roman villas in any territory. Since the Zagreb area abounds in such favourable terrain, Gregl remarks that it is no wonder that a number of *villae rusticae* were discovered in this region.[19] They vary in size, number and arrangement of buildings and other structural elements, with residential sections built of solid materials (stone, brick) and characterised by painted walls, floor mosaics, heating systems, and accompanying economic facilities (stables, warehouses, etc.), often built in timber. Naturally, the proximity of water, particularly springs, was desirable, as well as pastures, woods, fields, and roads. According to Gregl's data, a total of 14 sites with architectural remains of presumed villas were discovered up to 1991 in the Zagreb area, including the regions of Turopolje, Sesvete Prigorje and the territory of Zaprešić; some of these sites were already mentioned above. They are as follows: Drenje, Laduč and Šibice (the Zaprešić area), Donja Glavnica-Frtić, Moravče-Draščica and Moravče-Rošnica (the Sesvete area), and Gornji Čehi, Mala Mlaka-Starče and Mala Mlaka-Ciglenica (the Turopolje). In the wider Zagreb area the sites in question are Gornji Stenjevec, Gračani, Gradišće near Remete and Gradišće near Vugrovec, and Resnik. In another paper Gregl brings the following sites as examples of rural architecture ("the so-called *villae rusticae* or countryside estates") in the wider Zagreb area: Gornji Stenjevac, Gračani, Remete, Gornji Čehi, Mala Mlaka, Glavnica, Moravče, etc.[20] Vladimir Sokol has on a number of occasions published the results of excavations conducted in the wider Sesvete area. It is important to note that the preserved architecture on these sites can with a high degree of certainty be interpreted as *villae rusticae*. In the Sesvete Prigorje area Sokol lists the *villae* in Donja Glavnica (with a thermal complex), Moravče, Vugrovec, Vurnovec, and Blaguša, labelling them as "the wealth of the Roman material culture of the 2nd to 4th centuries."[21] Given the overall state of research of the subject under discussion, it is important that relevant documentation (architectural drawings and photographs) for several of these sites exist. In his overview, Domagoj Tončinić catalogued the sites of rural character in the Zagreb area, as well as some recently excavated *villae* in the vicinity of Sesvete: Donja Glavnica-Frtić, Moravče-Rošnica and Moravče-Draščica, as well as Drenje near Zaprešić, to the west of Zagreb. He further contributes a discussion on the architectural remains lacking elements necessary for a reliable interpretation of their character, which is reflected in the question mark in the title

[18] Gregl 1991, 20-22.
[19] Gregl 1991, 20.
[20] Gregl 1994, 34.
[21] Sokol 2003, 201.

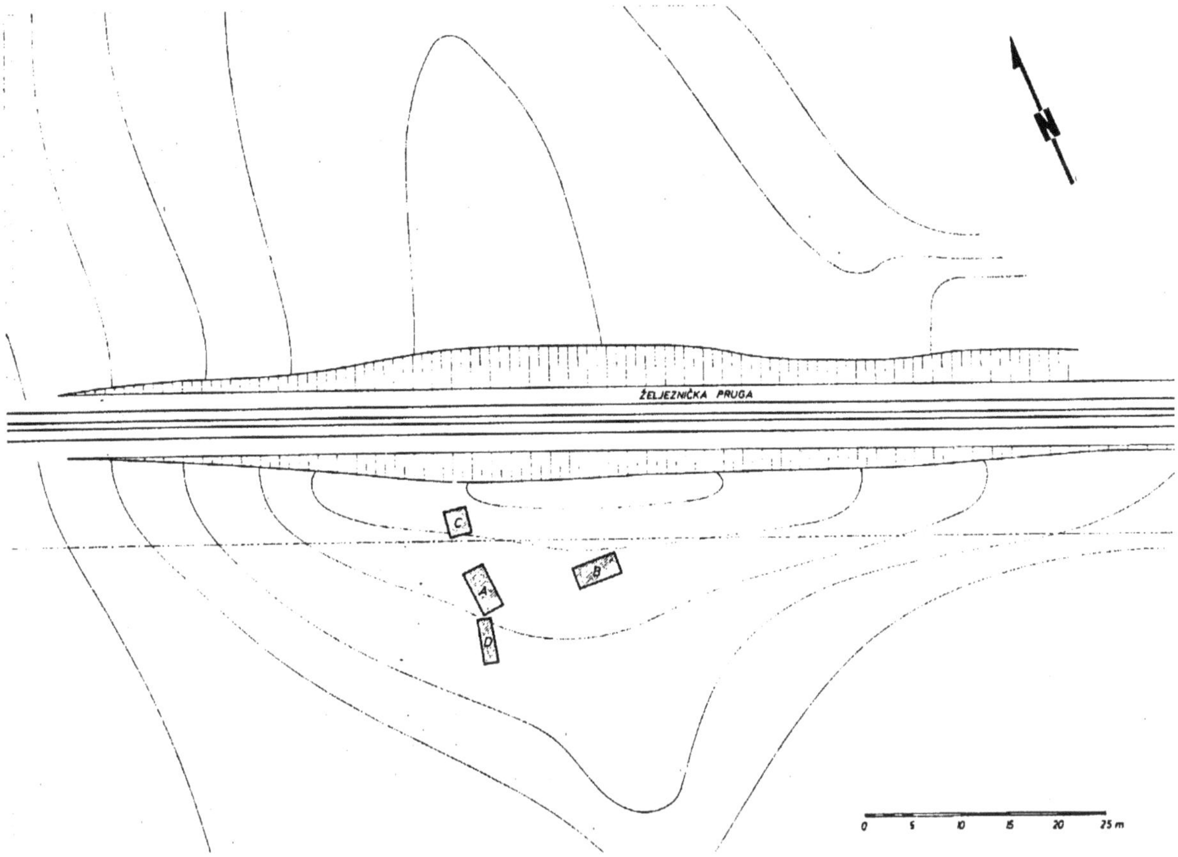

Fig. 3. Plan of the architecture at the site of Šibice (after Škoberne, Koščević, and Makjanić 1987).

of his article.[22] In this way he wanted to draw attention to the doubts accompanying the study of villa architecture in northern Croatia, which cannot be resolved without archaeological excavations. Twenty sites are labelled in Tončinić's paper as dubious in terms of the nature of their architecture: Gornji Grad / the Upper Town, Donji Grad / the Lower Town, Kaptol, Maksimir, Žitnjak, the intersection of the Savska and Vukovarska Streets, Zagorska Street, Stenjevec Donji, Stenjevec Gornji, Veternica, Štefanovec Donji, Markuševac, Trnava Markuševečka, Trnava Resnička, Gradište near Remete, Čučerje, Horvati, Gračani, Šimunčevec and Gradina near Šimunčevec.[23] When wall remains and fragments of unassuming furnishings are all that is preserved, the nature of the building in question cannot be precisely determined, contrary to the buildings which yielded mosaic floors and other typical features of villa architecture. The above-mentioned site with remains of Roman architecture in the village of Drenje counts among rather interesting examples, the more so as it belongs to a rare group of sites published in a monograph covering several excavation seasons. As a matter of fact, the site of Drenje yielded remains of a settlement complex of undetermined function: due to incomplete excavations it was impossible to precisely determine either the character or dimensions of several buildings that were excavated on this site, or their chronology. Nonetheless, three chronological phases were postulated: 1. Julio-Claudian; 2. the Flavians to the Marcomannic Wars ("the most intensive phase"); 3. the final phase, approximately up to the end of the 4th century.[24] Protective excavations in 2004 revealed foundation walls of an apsed building, dating from several phases of construction and suggesting that this countryside residential and agricultural structure was active from the 1st to the 4th centuries. (fig. 2).[25] It is certain that the position of this architectural complex was important due to the nearby crossing over the Sava River, as well as because of the proximity to one of the most important local communications: the Emona – Neviodunum (Drnovo near Krško) – Siscia road.[26] In the monograph on Drenje two further complexes of Roman architecture are discussed. The first is situated in the area of the village of Šibice and was excavated in 1984, yielding remains of walls, most probably of a *villa rustica* from the 1st and 2nd centuries, judging from small finds (fig. 3).[27] The other one (Drenje-Laduč) is situated in the northern part of the Drenje village, on the left side of the road connecting Laduč to Drenje; it was

[22] Sokol 2003, 201.

[23] Tončinić 2005, 270-274.

[24] Ž. Škoberne, in: Demo 1986, 114-116, no. 5; Škoberne, Koščević, and Makjanić 1987, 59-61.

[25] Koščević and Cvitković 2005, 107-108.

[26] Škoberne, Koščević, and Makjanić 1987, 59-60.

[27] Škoberne, Koščević, and Makjanić 1987, 10-13. The site is situated on the right bank of the Lužnica stream, and is partially traversed by the Zagreb – Ljubljana railroad.

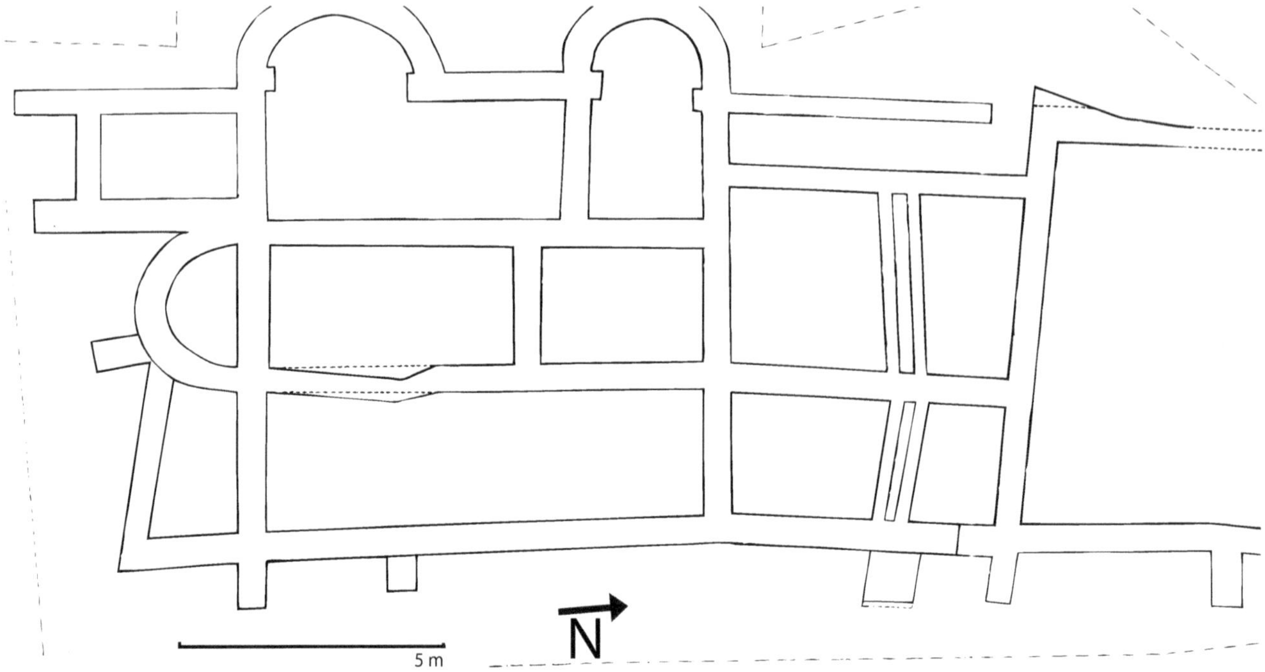

Fig. 4. Plan of the excavated part of the villa in Donja Glavnica (after Sokol 1981).

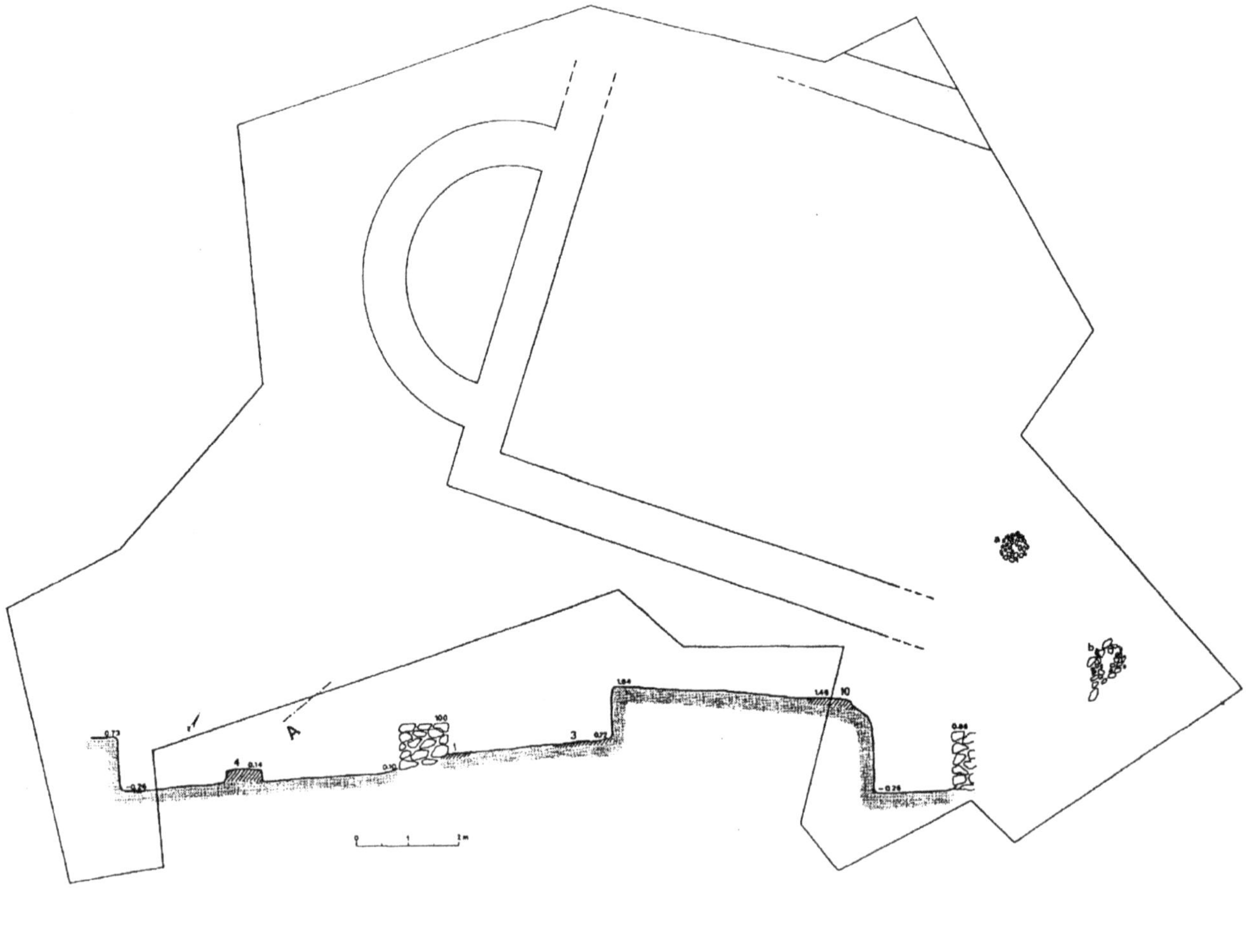

Fig. 5. Plan of the excavated part of the villa in Moravče-Draščica (after Sokol 1996).

Fig. 6. Remains of the villa in Kutinska Lipa (photo: Ante Rendić-Miočević).

excavated in 1986 with only trial trenches used, but they still seem to have recovered a noteworthy building. However, all these finds, that is, the arrangement of the structures in the wider Drenje area, perhaps suggest a rural settlement (a *vicus* or the like), rather than individual villas.[28]

On the bases of archaeological excavations it is possible to identify the function of some of the architectural complexes as agricultural country estates. This is mostly true of various architectural remains in the Sesvete area, such as the above-mentioned architectural complex in Donja Glavnica, which was intensively occupied in the 3rd and 4th centuries (fig. 4).[29] It seems, however, that the complex did not have an economic section, and it remains unclear whether this was a *villa rustica* at all, although this possibility should not be discarded. The site was excavated between 1976 and 1978, when an elaborate architectural complex with a yard was excavated in several trenches that enabled the determining of its perimeter. Baths with remains of a heating system and three vaulted pools with apses were discovered in the southern and western sections, as well as some residential buildings in the north-eastern part of the complex. Remains of economic facilities were not found; presumably, these are to be recognised in the structures dispersed around the villa up to several hundred meters from the central building.[30] The layout of this complex can be compared to the villa in Keszthely-Fenékpuszta (Hungary).[31] An architectural complex with an apsed building from the site of Moravče Draščica, excavated in 1980-1981, is significant in this context. Its function was not precisely determined, but it was at least partially of cultic nature (fig. 5).[32] In the Kutina area (the Moslavina in western Slavonia), on the site of Kutinska lipa (previously referred to as Crkvište or Fratrica, both toponyms indicating the mediaeval church or monastery located there) systematic excavations in 2006-2009 revealed the remains of a number of rectangular rooms. The results of these excavations are still not published in a complete form, at least not in the way that would allow reliable conclusions (fig. 6).[33] Therefore, the function of this complex cannot be precisely determined, but it seems that the rectangular rooms belonged to a minor settlement situated on the Roman road, although a *villa* cannot altogether be excluded.[34]

One among possible indications for determining the distribution of country estates in northern Croatia is the existence of a relatively dense network of Roman cemeteries in this area. These are especially numerous in the vicinity of larger towns, with the wider Zagreb area

28 Škoberne, Koščević, and Makjanić 1987, 13-14.
29 V. Sokol, in: Demo 1986, 178-184, nos. 15 and 16.

30 Sokol 1981.
31 Thomas 1980, 300, fig 44.
32 Sokol 1996, 34-35.
33 Bobovec 2008c.
34 Bobovec 2009, 26.

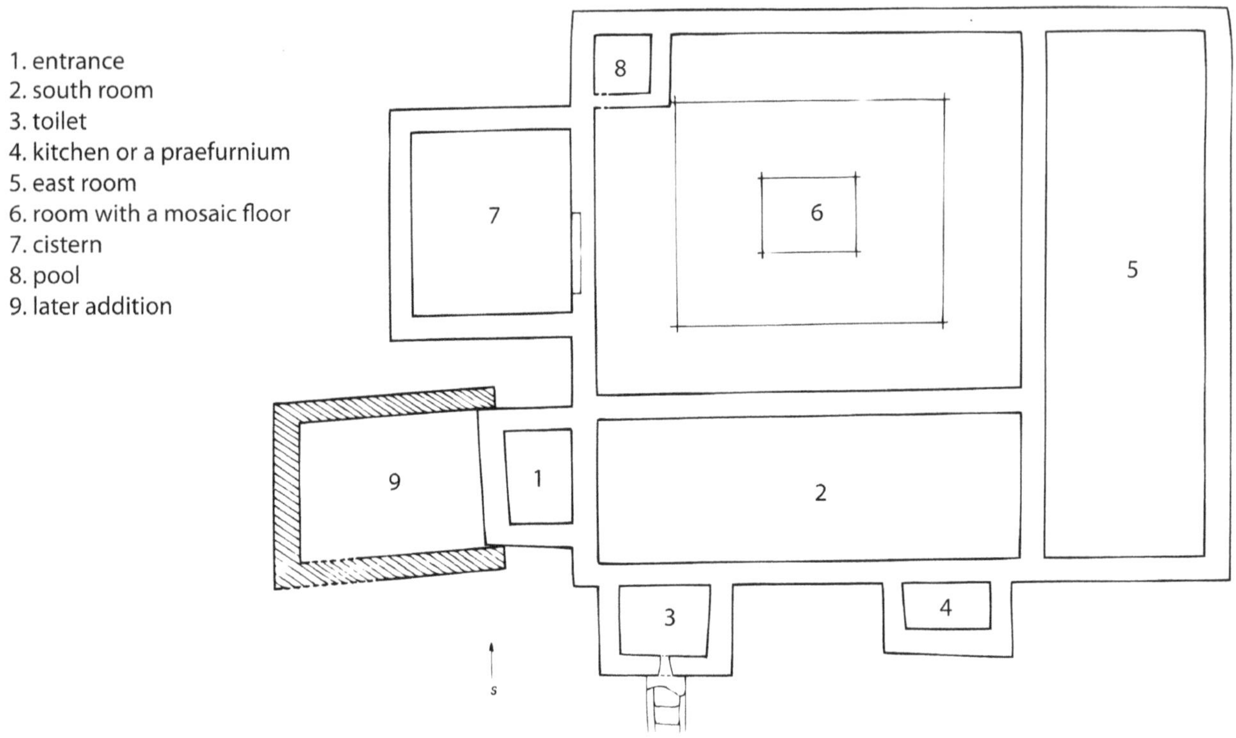

Fig. 7. Plan of the excavated part of the villa in Benkovac-Otrnci (after Gorenc and Damevski 1984).

as the most important in this regard. It was, probably not without good reason, fairly frequently in the focus of interest of archaeologists, which is probably the reason for the dense distribution of not only cemeteries, but various architectural remains. However, the determination of the nature of individual structures remains unsolved, because the discovery of a cemetery and some Roman walls can only indicate – but not prove – the existence of a villa.

On balance, the state of research in northern Croatia does not allow an easy comparison with other areas of Pannonia, regions with numerous and well-excavated, often also well-documented, country architectural complexes. This was noticed by András Mócsy, the author of an important and often cited book on Roman Pannonia. Besides some general remarks on early imperial villas – he had previously concluded that they were concentrated along the western borders of Pannonia and north of Lake Balaton (Pelso) – he also noted that excavations of Roman villas in the area between the Drava and the Sava were relatively scarce and, in any case, inadequate, pointing out that the dearth of discovered villas in this area did not actually mean they were non-existent or even present only in small numbers.[35] Thus, the majority of authors have postulated that the number of Roman villa settlements in northern Croatia reflects the poor state of research. Indeed, the state of research of country estates in the area of study is unsatisfactory and can hardly be compared with the results achieved in neighbouring countries that share the territory of the former Roman province of Pannonia with Croatia. Hungary is probably the leading state in this field of research, in the first place owing to many villas discovered and excavated near Lake Balaton, where the largest group of Roman country estates has been recorded.[36] The researches in this region have yielded precious insights into typological variations and character of country villas. At the same time, they made it possible, or at least easier, to study and compare similar architecture in other regions. It has been established that Pannonian villas were mostly constructed in several phases, commonly revealing adaptations of earlier buildings, dilapidated or destroyed in time.[37] In periods of great danger some villas were fortified with walls and towers, especially those positioned near the *limes*. When discussing the Croatian part of Pannonia, the fact should be borne in mind that some sites were excavated a relatively long time ago, often using techniques and methods not to modern standards, and almost always insufficiently funded. Research in general depended on a personal initiative, determination and enthusiasm of individuals, as well as on support and understanding of local authorities. The results of the excavations of individual sites were rarely and/or insufficiently published; only notes from surveys or unpublished excavation reports have survived from some of the excavations. Occasionally, such reports were supplied with simple and inadequate (in terms of quality and modern standards) documentation, that is, drawings and photographs. The information from yearly reports published in archaeological journals is rather summary, most often pertaining only to recently conducted excavations. All

[35] Mócsy 1974, 169.

[36] Thomas 1980, 275.

[37] Brukner 1982, 94.

Fig. 8. Villa in Benkovac: detail of the floor mosaic (Photo Archive, AMZ).

these facts have influenced, or rather, limited, some of the conclusions postulated above. Also, inadequate documentation is one of the main reasons for the lack of relevant papers and syntheses on country estates in northern Croatia. As already noted, in such circumstances it is often hard to determine whether certain buildings or complexes actually represent *villae rusticae*. In the majority of cases it seems that these structures had other functions and the arrangement of their architectural facilities varied. Due to the lack of features typical for country villas, it seems more appropriate, at least in some cases, to identify them as manors (occasionally called *latifundia*), estates, etc. Larger manors sometimes brought together a number of buildings, some of which could function as economic or residential structures. Some of the complexes should tentatively be interpreted as different types of road stations, often well equipped, distributed in large numbers in the vicinity of important roads. On balance, the problem of the identification of rural settlements in the area of study is not nearly resolved, although it was recognised a long time ago and is equally present in Pannonia in general. This was noticed by A. Mócsy, who postulated that the term *villa rustica* should refer to a single building or number of buildings with primarily agricultural function, whether supposed or proved. However, he further observed that buildings pertaining to road stations showed similar features since both types of buildings share Roman building techniques: stone walls, mortar, bricks etc. Finally, Mócsy observed that this type of architecture started to appear more frequently in the Pannonian countryside only in the 3rd and 4th centuries.[38]

2.2. Undisputed or probable *villae rusticae*

Only a small number of country buildings in northern Croatia can be with certainty interpreted as *villae rusticae* in the proper sense of the word. Thus, some uncertainty will be more or less always present when discussing the nature of a number of these structures. The architectural complex in Benkovac (north of Okučani, western Slavonia), excavated some time ago, is among the structures containing elements suggestive of a *villa rustica*. The remains of the villa, situated in the fertile valley surrounded by picturesque hills, were identified owing to the polychromous medallion forming a part of a Roman floor mosaic discovered in 1953 on the site of Otrnci on the right bank of the Sloboština River, near the bridge that connects the settlements Benkovac and Cage.[39] It is interesting to note that some remains of settlements and fortifications from prehistory to the late Middle Ages had previously been identified in this area. The road running parallel to the river was always an important traffic route connecting the modern-day region of the Bosanska Posavina with the areas north of it: central Slavonia and the Drava valley.[40] As far as the location of the villa at Otrnci-Benkovac is concerned, it is significant that the site of Selište in Cage is located on the opposite bank of the river, with another village of suggestive name somewhat to the south, Staro selo (Old village); all these data reveal the density of archaeological sites and suggestive toponyms of the area. South of the modern highway running from Zagreb towards east, archaeological excavations were conducted in several campaigns in the period between 1953 and 1983, revealing remains of an architectural complex.[41] The excavators believed that it functioned both as an economic and residential settlement from the end of

[38] Mócsy 1974, 169.

[39] Gorenc and Damevski 1984, 101; Mihaljević 2006, 202-203.

[40] Gorenc and Damevski 1984, 101.

[41] More detailed reports on the excavations can be found in excavation journals for the years 1953, 1955, 1960 and 1976-1984 kept in the AMZ archives, as well as in preliminary reports in *AP* 18 (1976), 79-81; *AP* 20 (1978), 80-81; *AP* 21 (1979), 98-100; *AP* 22 (1980), 80-82; *AP* 23 (1982), 84-86.

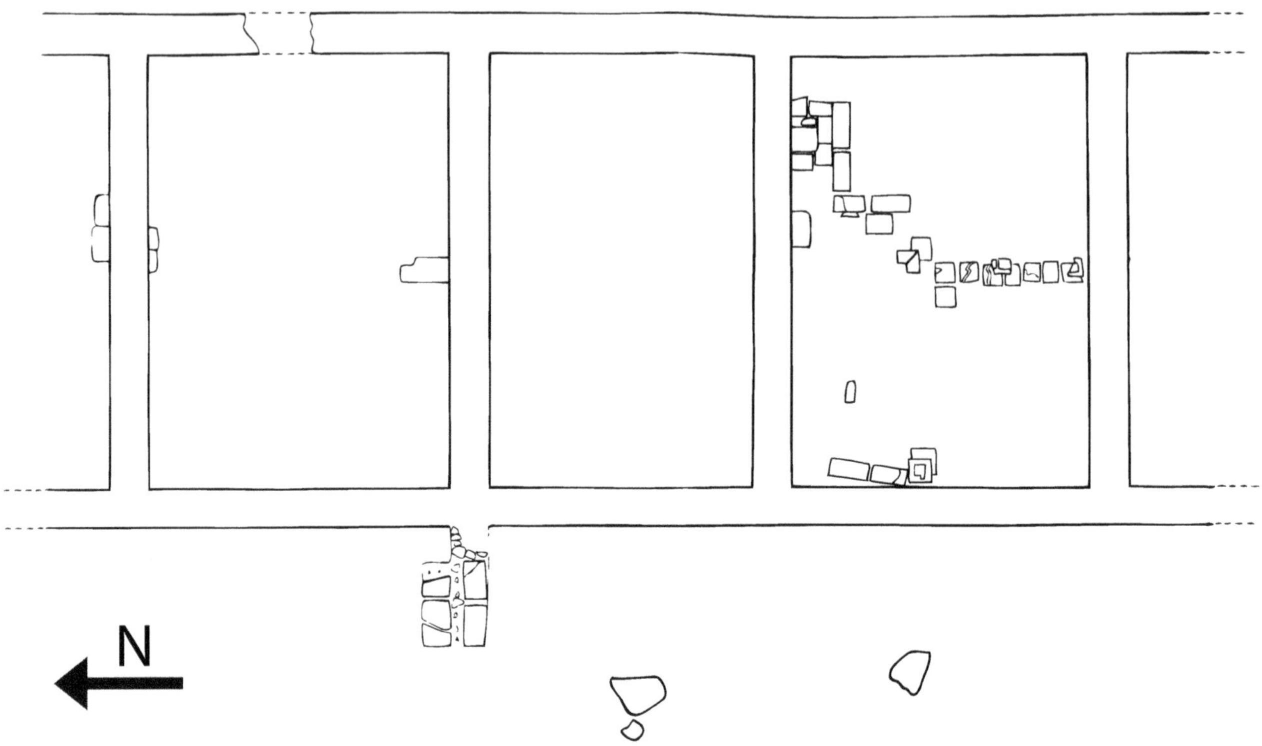

Fig. 9. Plan of the excavated part of the villa in Cage (Photo Archive, AMZ).

the 2nd to the second half of the 4th centuries.[42] Remains of brick walls on stone foundations and segments of drainage and water-supply systems were found, as well as a heating system.[43] The ground plan comprises several larger and smaller rectangular rooms, nine in total, arranged from west to east, with an entrance hall, sanitary facilities, and several rooms of various functions, as well as a large cistern and a small pool, and, finally, a late Roman annex in the western part, in front of the entrance section of the villa (fig. 7). The abovementioned mosaic with a geometrical-ornamental composition, stylised three-coloured meander and other decorative motifs, is among the most attractive Roman mosaics so far discovered in Slavonia. It was placed in the centre of the largest room in the northern section of the villa, whose walls, judging by the preserved fragments, were painted in polychrome frescoes (fig. 8). The excavators postulated that the structure was erected in ca. AD 200, but a later phase, lasting up to the mid-4th century, was represented by the abovementioned late Roman annex; the villa consisted of a residential area and accompanying economic facilities. It seems that it was erected inside a large yard of an economic character, since remains of a larger building of an unknown function and date were found nearby, probably constructed in the same period as the central hall.[44] The closest parallel for the Benkovac villa is the type of villa featuring a peristyle and portico, well-known from Hungary.[45]

On the left bank of the Sloboština River and opposite to the Benkovac villa is the site of Kućište (in the village of Cage near Nova Gradiška) with a water source. Further to the east is Staro selo (Old village), a site of a suggestive name with numerous Roman finds. It was partly excavated in 1981 to 1986, yielding remains of an architectural complex, described as a country estate with a central building and accompanying structures, connected to the Benkovac estate (*villa rustica*).[46] The building at Cage follows the route of the modern-day Okučani – Lipik – Daruvar road.[47] regrettably, it was only partially researched; disclosed were five rooms arranged in a row, of which the first and last were incompletely excavated. Nevertheless, it is almost certain that the complex was much larger (the total excavated area was 72 m^2) and that usual yard facilities were annexed to it.[48] The foundations of the building were of stone, while the brick walls were mortared and partially frescoed. Traces of a heating system were also discovered. Due to the insufficient state of research, the function of this building is hard to determine – it seems, however, that it was an economic country estate contemporary to the Benkovac villa (fig. 9). Small finds suggest that the building was at the same time a residential and economic estate, and its agricultural production presumably supplied the surrounding region; this, after all, was most frequently the main function of such estates.[49]

During the 1990s, surveys and excavations at the Osekovo-

42 No information on the total surface encompassed by the villa has been given in the reports.
43 Gorenc and Damevski 1984.
44 Gorenc and Damevski 1984, 112.
45 Thomas 1964, 358, figs. 172/1 and 173.
46 Excavation report 'Benkovac IV', Cage 1981 by V. Damevski and M. Gorenc.
47 Mihaljević 2006, 203.
48 Gorenc, Damevski, and Džaić 1985.
49 Gorenc, Damevski, and Džaić 1985.

Fig. 10. Aerial photograph of the site Osekovo-Ciglenice (after Bobovec 2008a).

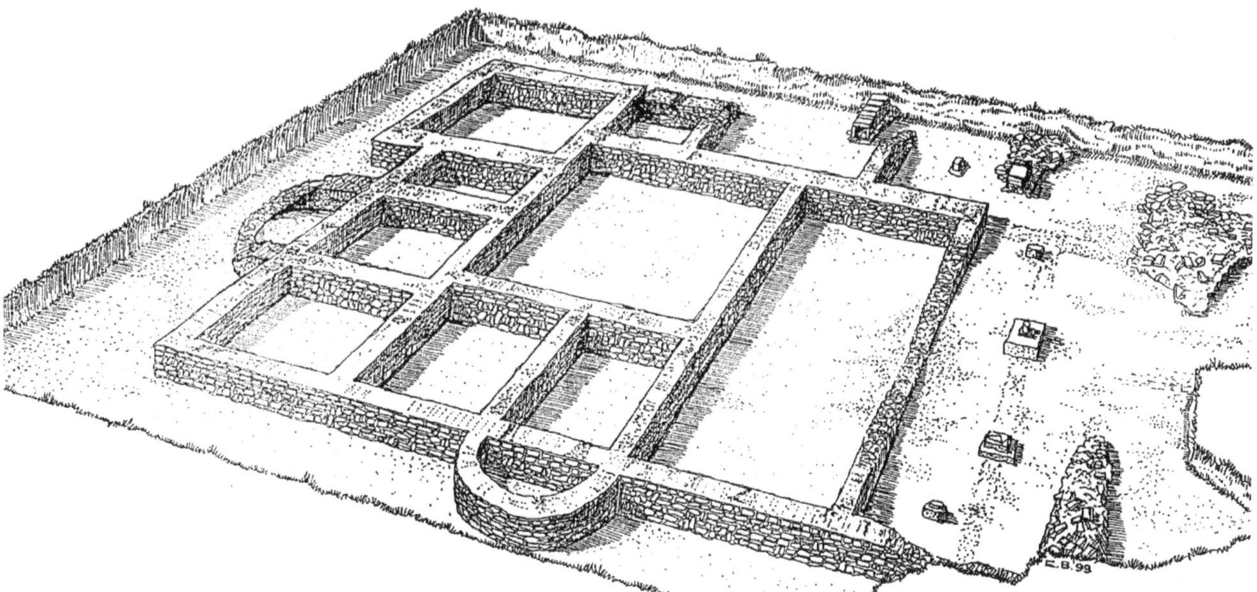

Fig. 11. Osekovo-Ciglenice: axonometric drawing of the villa (after Bobovec 2008a).

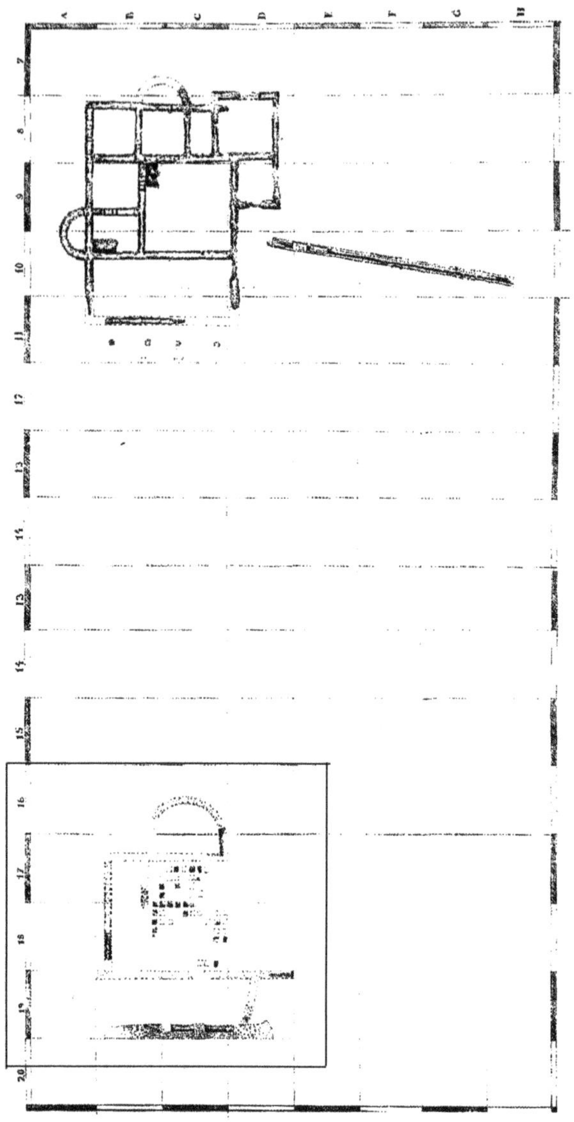

Fig. 12. Osekovo-Ciglenice: plan of the villa with the hypocausted room framed (after Bobovec 2008a, modified by T. Leleković,).

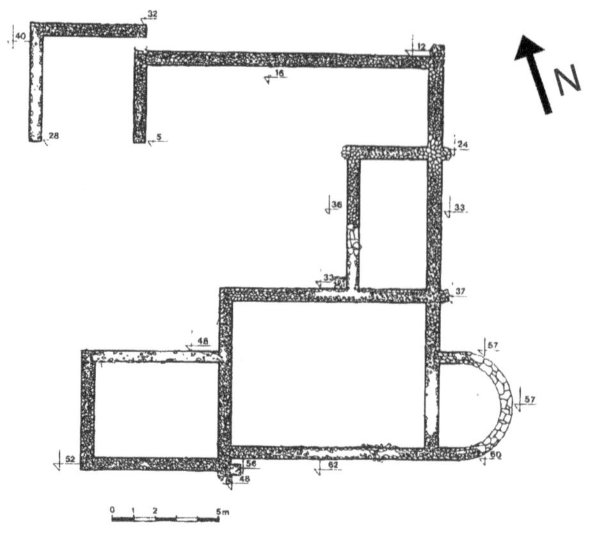

Fig. 13. Plan of the excavated part of the villa at Šemovec-Kupinje (after Šimek 1998).

Ciglenice site (the Moslavina region in western Slavonia), were conducted south of the modern-day Zagreb – Lipovac highway, in the area where "... the mild slopes of Moslavačka Gora meet the marshy lowland of the Lonjsko polje Nature resort", starting in 1998.[50] The total surface of the site was estimated at 3.1 ha. Remains of a residential building oriented in a southeast-northwest direction were found there, consisting of 11 rooms of various dimensions, two of them with apses.[51] The building had a porch, which is clear from the remains of pillar bases in the southeast section, and its sewage system is very well preserved. Roof- and floor-tiles were also found, together with coloured mortar, suggesting a high standard of living (figs. 10 and 11). Foundations of another structure with hypocaust were found in 2001 to the southeast, as well as numerous floor mosaic tiles used for paving the thermal section (fig 12).[52] The villa at Osekovo can be compared to the Pannonian porticoed villas.[53] It seems that the villa from Gyulafirátót-Pogánytelek (Hungary) makes a good analogy, especially in its second phase.[54]

During the preliminary works in advance of the construction of the Varaždin section of the Zagreb – Goričan highway in the mid-1990s, an architectural complex functioning as an economic facility or a farm-house estate, probably a *villa rustica*, was excavated. It was found at the site of Kupinje, near the village of Šemovec east of Varaždin, next to the River Plitvica, in the valley south of the Drava. Its perimeter is ca. 20 x 20 m; the residential section in the southern part of the complex consisted of three rectangular rooms of various dimensions and another room with a subsequently added apse, probably used as a pool. The latter is also the most representative room and the largest in size, with remains of a hypocaust system and *tubuli*.[55] It seems that a yard was in the centre, with another elongated room in the northern part and a further one in the northwestern part, all of them belonging to the economic section of the building. The ground plan reveals three functional complexes: a residential area, a yard, and an economic section (figs. 13, 14). The walls survive only in foundations, and the building seems to have lacked either luxurious details or fortification features, such as protruding corner stones (orthostats) that, for example, appear on a structure of similar function in Jalžabet (see below).[56] Judging by the small finds, the structure was in function from the early 2nd to the 4th century.

A further *villa rustica* was found on the site of Šarnica near Kelemen (Jalžabet), not far from Varaždin. Stone foundations of three rooms were found there in 1961 and

50 Minichreiter 2003; Bobovec 2006, 194-195.

51 Bobovec 2005, 113; cf. Bobovec 2008a; Bobovec 2008b.

52 Bobovec 2005, *ibidem*.

53 See Thomas 1964, 361, fig. 176: a 3rd-century villa.

54 Thomas 1980, 288, fig. 32 and 290, fig 34.

55 Bekić 2006, 289.

56 Šimek 1998, 63.

1962, probably pointing to a *villa rustica* or a fortified building, protecting the bridge over the Plitvica. The building was constructed in several phases, the latest falling to the 3rd-4th centuries, the period characterised by similar structures in the whole of Pannonia, but also in other provinces.[57] Archaeological excavations conducted in 1962 and 1963 revealed traces of a road, as well as walls attributed by the excavators to a *villa rustica*, dated to the 1st-4th centuries. It seems that the entire structure was burned down in the 2nd century and eventually renovated. During later Antiquity, probably due to its favourable position in the lowland near the Drava *limes*, the villa was fortified in the same way as similar complexes in Pannonia, Gallia, Germania, and elsewhere.[58] The first phase (1st and 2nd centuries) belongs to the period of the expansion of towns and a parallel appearance of a number of agricultural estates in fertile valleys in their surrounding; the second phase encompassed the 3rd and 4th centuries.[59] Judging by the remains discovered in the excavated southeast part, the Jalžabet building could typologically be defined as a fortified villa, featuring four corner towers in the shape of protruding stones (orthostats) (fig. 15).[60]

(A. Rendić-Miočević)

3. *Vici*

Until recently, *villae rusticae* were the only source for the study of the settlement structure of rural areas in the Croatian part of Pannonia. This situation has changed with large-scale protective excavations conducted during the last decade. Out of ten rural Roman sites recovered in these excavations, a *villa rustica* was found only at the site of Kupinje, while other sites yielded settlements of the type previously unknown in northern Croatia. Due to the lack of a better term, in this paper they are referred to as *vici*. Nevertheless, this term should be used cautiously, since it can be applied to a heterogeneous group of settlements, some of which were large industrial centres, while others even had a quasi-municipal organisation.[61] The existence of *vici* in southern Pannonia was until recently known only from epigraphic sources, which yielded information on their administrative organisation, but their locations and layouts remained unknown.[62] Owing to the recent constructions of public infrastructure, this has changed. Besides the fact that, for the first time in Croatia, a large number of archaeological excavations was conducted in the areas more or less distant from major Roman towns, these excavations are important for two reasons. Firstly, they encompassed exceptionally large areas, and secondly, owing to the modern methodology of excavation and fieldwork documentation, for the first time remains of timber architecture were clearly identified in rural settlements. In short, Roman rural settlements undoubtedly differing in two important elements from *villae rusticae*, were established in northern Croatia. The first difference is the size of these settlements. Clear information on the size of the (excavated) area of *villae* is usually lacking from the papers on *villae rusticae* in Pannonia, yet the graphic documentation in such studies clearly shows that the area of the majority of such structures does not surpass several hundred square meters, while only certain exceptions, such as Balaca and Keszthely-Fenékpuszta, encompassed areas measuring in hectares.[63] The sites identified in this paper as *vici* regularly encompass some 5 to 20 ha in surface, which is much more than an average Pannonian villa; thus, the large size of a settlement can indicate its nature in terms of a villa or *vicus*. The second difference is the use of timber as the main construction material, since the prevailing construction material in *villae rusticae* is stone and brick. Nevertheless, caution is necessary here for two reasons: *tegulae* are also found in *vici*, probably representing the remains of masonry; on the other hand, it is possible that timber constructions were simply overlooked in earlier excavations, due to inadequate excavation methodologies used. Taking into account the abovementioned various possible meanings of the term *vicus*, as well as the present level of archaeological research in northern Croatia, this term is here used to denote exclusively timber-built larger rural settlements with a clear agricultural function. A total of 9 *vici*, recognised so far, were excavated between 2005 and 2011 and not yet published in any detail. Therefore, it is at present possible to draw only a broad frame of the structures and functions of south-Pannonian *vici*.

Donji Vukojevac

The site is located on the Siscia – Emona road, at the location of the Ad Fines road station, 20 km to the northwest of Sisak. The excavations have only encompassed the peripheral section of the settlement, measuring 1 ha in area; the total area was not determined. It consisted exclusively of timber buildings, although some roof *tegulae* were found. Remains of a cremation cemetery were found in the vicinity (figs. 16 and 17).[64]

Okuje

The site is located on the Siscia – Emona road, 9 km to the southwest of Velika Gorica (Andautonia). The excavated surface (4.2 ha) revealed exclusively remains of timber structures. These were distributed on the entire surface, so that the excavations could not determine the total surface of the *vicus* or define its layout. The site is exceptional in that a La Tène settlement preceded the Roman one; a fenced area containing 13 cremations (1st century BC-1st century AD) was also found, probably a part of the La Tène cemetery.

[57] Vikić and Gorenc 1969, 8-10.
[58] Such structures were partially fortified, with protruding corner stones functioning as towers. Cf. Vikić-Belančić 1968, 96; Bekić 2006, 284.
[59] Vikić-Belančić, 1968, 96.
[60] Thomas 1964, 44, figs. 19 and 20, and 359, fig. 174. The fortified villa at Sümeg in Hungary (Thomas 1980, fig. 53) and the 3rd- to 4th-century villa from Gyulafirátót-Pogánytelek (Thomas 1980, 310, fig. 51 and 311, fig. 52) can be taken as parallels for this structure.
[61] Tarpin 1999, 5-6; Schulz 1999, 65.
[62] Pinterović 1975, 135-137; Schejbal 2004, 102.
[63] Thomas 1964; Thomas 1980; Gabler 2003.
[64] Dizdar, Tonc, and Ložnjak Dizdar 2011.

Fig. 14. Aerial view of the villa at Šemovec-Kupinje (Photo Archive, GMVa, photo: M. Šimek).

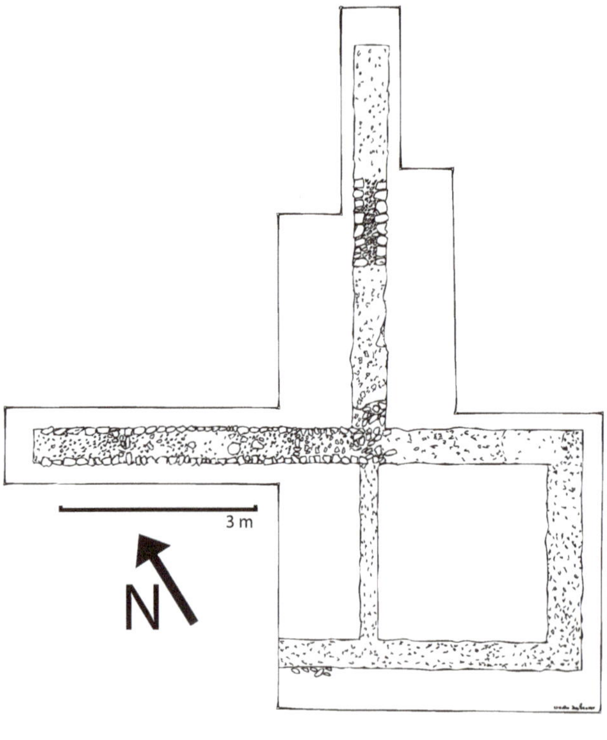

Fig. 15. Plan of the excavated part of the villa in Jalžabet (Photo Archive, AMZ).

Fourteen 1st-2nd-century cremations were found on another location, belonging to the Roman phase of the settlement.

Šepkovčica

The site was probably located on the Siscia – Emona road, 11 km to the southwest of Velika Gorica (Andautonia). Remains of a *vicus* 1.2 ha in surface were found on this site, with some additional structures outside of its perimeter. The settlement is dated from the 1st to the 4th centuries, but it was significantly enlarged in Late Antiquity. In the immediate vicinity two cemeteries numbering a total of 56 graves were found. One consisted of cremations from the 2nd half of the 1st and the 2nd centuries, buried into a destroyed tumulus, while the other consisted of both cremations and inhumations (1st-4th centuries) (fig. 18).[65]

Ivandvor

The site is located on the western outskirts of Đakovo, probably on the territory of Roman Certissia. Some 10 ha of the Roman settlement were excavated, while the total area encompassed was ca. 20 ha. Exclusively timber constructions were recovered with sporadic use of brick masonry, suggested by occasional finds of mortar, as well

[65] Bugar 2009, 271.

Fig. 16. Gornji Vukojevac: view of the remains of the Roman road (photo: Mario Vojtek).

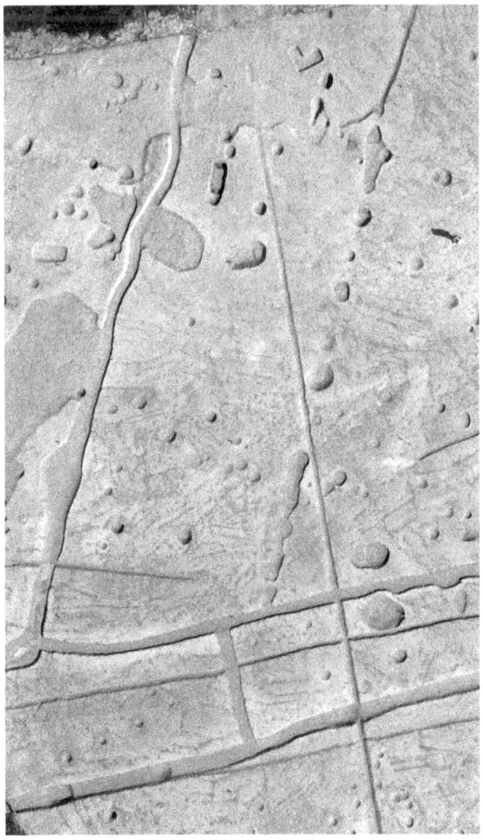

Fig. 17. Gornji Vukojevac: remains of the Roman settlement and road (photo: Cadcom d. o. o.).

as roof tiles and wall bricks. It is dated to the late 2nd to the mid 4th centuries, but the occupation was more intensive in the 4th century than in the 3rd (figs. 19 and 20).[66]

Sela

The site is located 11 km to the south of Đakovo, probably on the Siscia – Sirmium road (section from Marsonia to Certissia). Some 2 ha were excavated; the total area of the settlement remains unknown. The settlement consisted exclusively of timber constructions, and was dated to the 1st-4th centuries.[67]

Josipovac-Verušed

The site is located on the Poetovio – Mursa road (section from Mursa Minor to Mursa), 8 km to the northwest of Mursa. The settlement consisted exclusively of timber constructions, but the occasional finds of *tegulae* and bricks suggest the use of masonry. 17.1 ha of this site were excavated, yielding the remains of a prehistoric, Roman and mediaeval settlements. Neither the total area of the Roman settlement nor the area of the Roman settlement excavated in 2007 and 2008 has been mentioned in the published report.[68]

---

[66] Leleković 2008, 14.
[67] Haraša and Bekić 2007, 75.
[68] Filipec, Šiša-Vivek, and Roksandić 2009, 33.

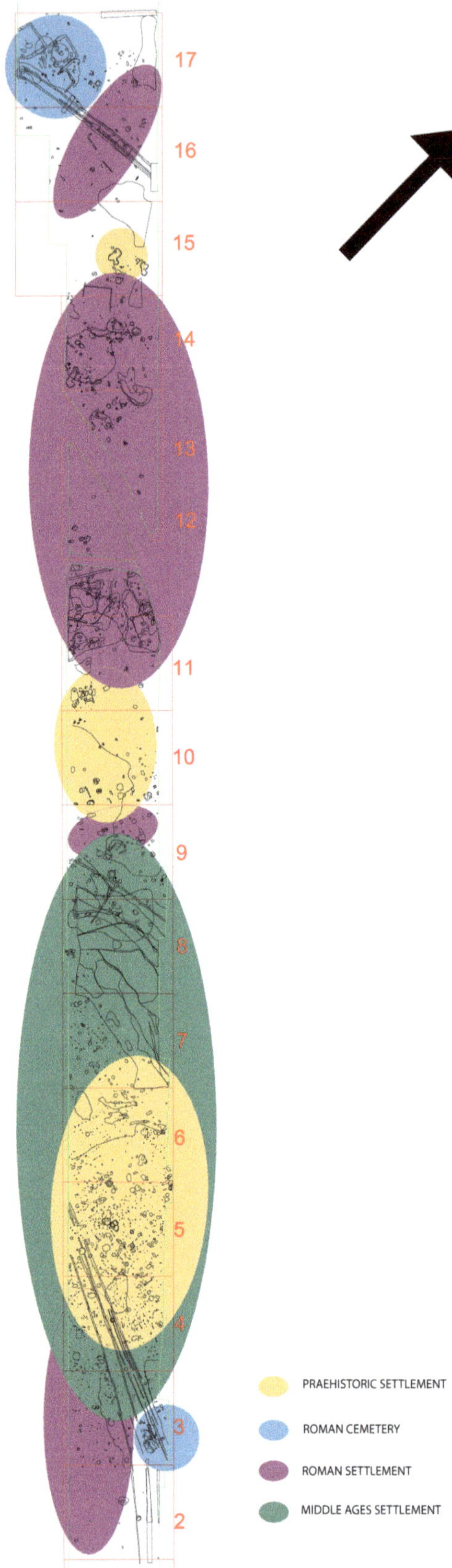

Fig. 18. Šepkovčica, with marked zones of: prehistoric settlement (yellow), Roman cemeteries (blue), Roman settlement (violet), medieval settlement (green) (after Bugar 2009).

Kiškorija

The site is located on the periphery of Virovitica, 30 km north of Daruvar (*Aquae Balissae*, *Municipium Iasorum*), on the Poetovio – Mursa road. A total of 1.7 ha of the *vicus* were excavated, yielding exclusively timber constructions (fig. 21).[69]

Našice

During the construction of the Našice bypass in 2009 two small-size rural settlements were excavated. The results of these excavations have not yet been published.[70]

As has already been mentioned, more detailed information on the listed settlements is not available. Even though, on the basis of preliminary publications several common features can be discerned that arguably represent some basic characteristics of the *vici* type of settlements in the area of study. When discussing these characteristics it must be borne in mind that all sites are located in areas of extensive cultivation. Therefore, walking surfaces and aboveground sections of buildings have been completely destroyed; only the layers below the surface and infrastructural features were better preserved. The traces of aboveground buildings consist of a number of randomly distributed postholes that do not allow the identification of the overall layout and function of the settlements. The determination of the layout of individual buildings is additionally hampered by the fact that the settlements developed in several construction phases during which the buildings were either replaced with new ones or renovated and enlarged. The dimensions and positions of the majority of the aboveground buildings could have been established only approximately, while the ground plans could have been evidenced only for the single-phase buildings. Nevertheless, these rare finds are at the moment the sole concrete remains of village settlements and can be considered as the base for a typology to be devised in future.

The most numerous group is represented by rectangular timber buildings made of parallel pairs of massive beams carrying a gabled roof. Their dimensions varied from 4 m^2 to as much as 300 m^2.The majority of the buildings were single-roomed (fig. 22, III-VI), while only few had partition walls (fig. 22, I, VII). Another group is represented by elongated buildings consisting of two or more rooms, made of densely packed beams that carried a gabled roof (fig. 22, VIII). An almost identical form of buildings was discovered in 4th-century Bajuwaric settlements in Germany.[71] Significantly, coarse stamp-decorated pottery was found at Ivandvor, suggesting a Germanic origin of its inhabitants.[72] Considering that in the 4th century Germanic tribes from the left bank of the Danube were settled on the territory of northeastern Croatia,[73] remains of rural

69 Jelinčić 2006, 63; Jelinčić 2007, 31.
70 Slavica Filipović, pers. com.
71 Fries-Knoblach 2006, 354-355.
72 Balen *et al.* 2009, 38.
73 Vingo 2010, 261-282.

Fig. 19. Plan of the vicus at Ivandvor (made by T. Leleković).

Fig. 20. View of the vicus at Ivandvor (photo: LDS d.o.o.).

Fig. 21. Kiškorija: aerial view (photo: Kristina Jelinčić).

architecture with Germanic features are no wonder. A notable characteristic of this type of buildings, witnessed at Ivandvor and Josipovac,[74] was an additional row of beams on the outside of the façade, creating a construction similar to a porch. The third group of structures consists of buildings and architectural complexes of a composite layout, 500 to 2000 m^2 in surface, testifying to the way of living not unlike that Roman. Judging by the structures with a discernible layout, these were building complexes grouped around a central yard (fig. 23). A somewhat chaotic distribution of postholes is a characteristic feature of the majority of this type of structures, suggesting renovations and enlargements, which obscured the layouts of the majority of them. It should be noted that brick and tile fragments, as well as traces of mortar, were found in the vicinity of these buildings, suggesting that some of them were mortared and tile-roofed, while certain portions, such as wall corners, doorsteps and pavements, were built in brick.[75] One such building from the site of Ivandvor was constructed by a double row of beams, probably a timber frame supporting unbaked-brick walls (fig. 22, IX). Fragments of roof tiles and pieces of façade plaster were found in its vicinity, which additionally challenges our perception of the structures typical of the *vici*. Therefore, these finds demonstrate that rural settlements did not necessarily provide living conditions inferior to those in the towns. In other words, they suggest that certain *vici* could have provided living conditions of the same level as those offered by the masonry buildings of *villae rusticae*. On balance, these finds open the possibility that some of the settlements designated here as *vici* were actually *villae rusticae*, differing from those described in the first part of this paper by the fact that they were completely timber-built.

Alongside the structures described above, a number of sunken features of different functions and dimensions were found within the *vici*. Most of them are pits of various depths, from 1 to 10 m^2 in surface, used as work areas. Among them the most notable are wells (some up to 7 m in diameter), due to the large quantity of finds recovered from them (figs. 24 and 25). Since the given dimensions are much larger in comparison with the standard dimensions of the Roman wells,[76] there is a possibility that these large circular pits were not proper wells, but rather a kind of cisterns for collecting subterranean waters and rainwater. Both pits and wells were regularly reused for the disposal and combustion of waste, which makes them the most important source of information on the way of living in *vici*. Two forms of sunken structures predominate, the first being evidenced at Ivandvor and Šepkovčica. At the former site two semi-sunken pits were found, containing a large-sized circular hearth and a bread oven (figs. 26 and 27). In both cases a well was found in the centre of the structure, so that the work area must have been located on a wooden platform above the well. Comparative examples that could enable determining the function of this type of structures are unknown to us, but small finds suggest their use as kitchens, apparently installed in detached self-standing buildings. A similar construction was found at Šepkovčica, which suggests that this type of facility was characteristic for rural settlements both in the eastern and the western part of the area of study. The second form of dugout structures were used as dwelling shelters, and they have been evidenced only recently in Croatia, one at the site of Kiškorija[77] and another in the vicinity of Našice.[78] In the Hungarian part of Pannonia similar residential objects are relatively frequent, and are characteristic of the settlements of the indigenous population during the 1st and 2nd centuries.[79]

Remains of agricultural buildings and facilities, as well as those used for animal husbandry, are rarely found. A network of 20-30 cm wide canals was found at Ivandvor, and within them smaller-size aboveground buildings of evident industrial function. The network was composed of 21 individual canals separating out parcels 1000 to 4000 m^2 in surface; in this way almost 4 ha were parcelled out

[74] Filipec *et al.* 2009, 50-53.
[75] Leleković 2008; Filipec *et al.* 2009, 49.
[76] Kovács 2003, 69; Szőnyi 2003, 145-146.
[77] Jelinčić 2006, 63; Jelinčić 2007, 31.
[78] The excavations were conducted in 2009 and 2010 by the MSO, but they have not been published.
[79] Gabler 2003, 241.

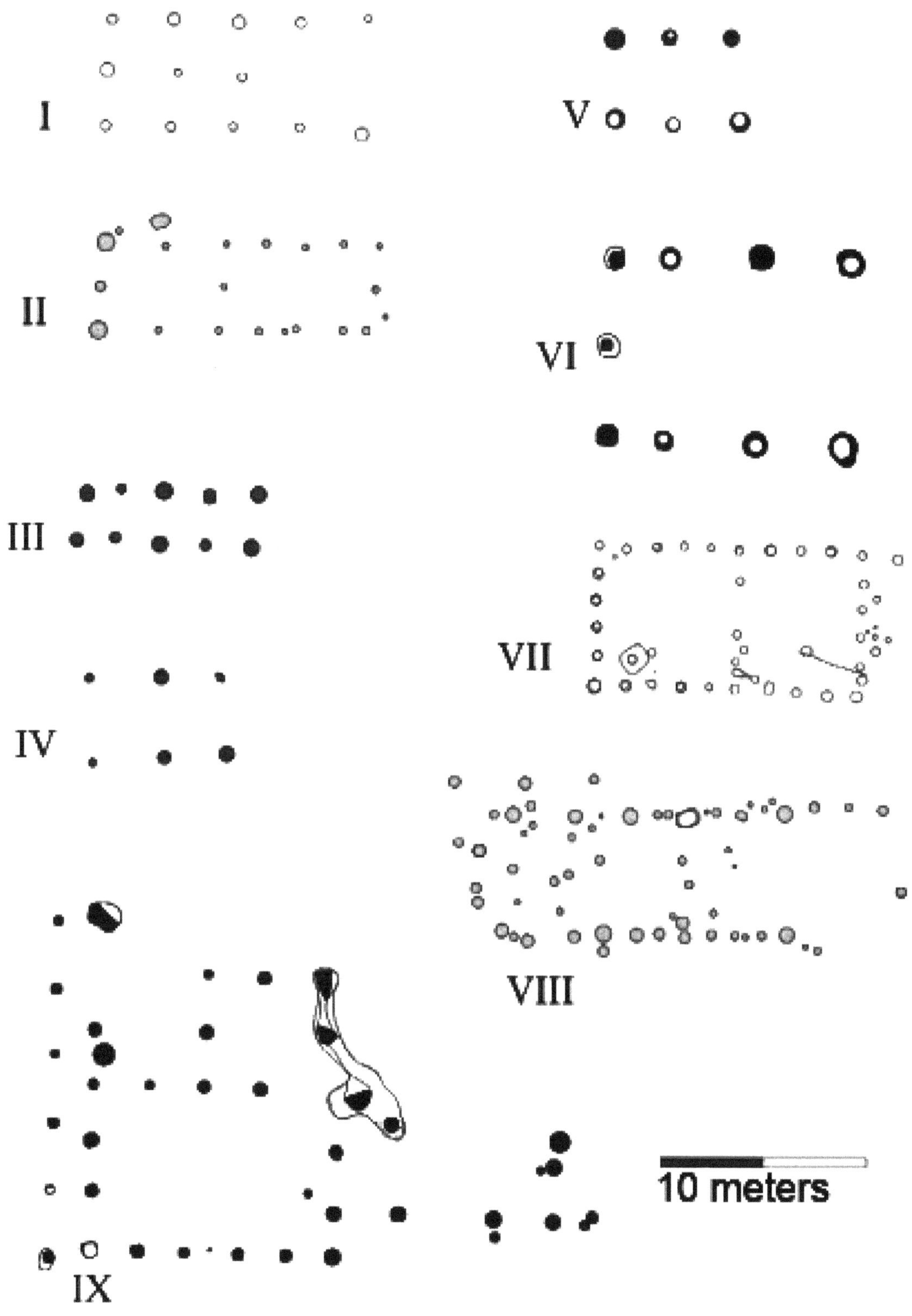

Fig. 22. Ivandvor: typological scheme of timber buildings (made by T. Leleković).

Fig. 23. Ivandvor: remains of aboveground buildings (photo: LDS d.o.o.).

Fig. 24. Šepkovčica: well no. 10 (photo: B. Rožanković).

Fig. 25. Ivandvor: pottery (photo: T. Leleković).

Fig. 26. Ivandvor: aerial view of kitchen A (photo: T. Leleković).

Fig. 27. Ivandvor: aerial view of kitchen B (photo: T. Leleković).

(fig. 28).[80] It was not possible to determine whether the canals were traces of horse pens or whether they were used as an irrigation system. The dilemma should be resolved by the analysis of animal bone gathered at the site: the horses are represented in an above-average proportion, suggesting that the fence system was used for keeping horses, that is, for commercial horse-breeding. This suggestion is further bolstered by the fact that Ivandvor is only 30 km from the Teutoburgium (Dalj) garrison, in which two *alae* were stationed in the 3rd century.[81] Foundations of buildings following the orientation of the canals were found within the canal network and immediately adjacent to it, suggesting that both these structures were in use contemporarily and in functional relationship. Since the fence system forms a single complex with the structures from the first building phase, it can be interpreted as a horse farm put up in the early Severan period for the supply of the nearby *limes* garrisons. Some unusual buildings consisting of pairs of massive beams were also recovered at Ivandvor (fig. 22, II), which could have served for hay drying or a similar agricultural activity; we are unfamiliar with any similar finds that could assist us in determining the function of these constructions.

80 Leleković 2008, 13.
81 Spaul 1994, 31-35.

Due to the large excavated surfaces, generally surpassing 5 ha, the recent excavations have enabled the study of the layouts of settlements and their inner structure, rather than individual architectural forms. The best example is offered by the excavations at Ivandvor, which resulted in a clear insight into the layout and arrangement of a Roman rural settlement of the *vicus* type, due to as much as 14 ha of excavated surface and the convenient "X" form of the dig (fig. 19). The settlement encompassed 10 ha, yet the structures were not found on the entire surface. The *vicus* consisted of three complexes, separated from each other by an empty space 30-100 m wide.[82] Similar uninhabited parts of the settlements were also evidenced at Josipovac[83] and Kiškorija[84] and, to a certain measure, at Šepkovčica. On the latter site uninhabited areas were taken by several pits, where some kind of industrial work took place. This shows that intermediate spaces could have been used for the activities undesirable within the residential area, such as tanning or waste combustion.[85] On the other hand, the excavations at Kiškorija have revealed that some sections of the settlement were left empty due to the lie of the land.

82 Leleković 2008, 13.
83 Filipec *et al.* 2008, 33.
84 Jelinčić 2007, 29.
85 Bugar 2009, 269.

Fig. 28. Ivandvor, remains of the fences (photo: T. Leleković).

The empty area in the settlement at Kiškorija was located at the lowest part of the site; it became obvious during the excavations that it was flooded for the most part of the year. Presumably, this part of the settlement remained empty because the terrain was unsuitable for residential purposes and not because of the intention to leave some empty space between certain inhabited sections.[86] Occasional flooding of bare sections of the settlement was also noticed at Ivandvor, but this site still suggests that such areas were perhaps intentionally used for separating out certain individual complexes within the settlement. The reason for this partition can be found in the definition of the *vicus* as a rural settlement in which a number of households lived together, sharing and/or communally cultivating the soil.[87] According to the results of the excavations presented here, it can reasonably be presumed that the main difference between the *villae* and *vici* lies in the division of the settlement in a number of separated complexes in *vici*, resulting from the fact that they were occupied by smaller village communities composed of several households that wanted to live separately and in their own premises. The site of Sela does not conform to this rule; a complex composed of timber-built structures less that 3 ha in size was discovered there. A division in several segments was not evidenced on this site, which therefore, according to the above parameters, should not represent a *vicus*. On the other hand, this complex can neither be classified as a *villa rustica*, as transpires from a simple timber architecture and scarcity of finds. A similar complex was found at the early stage of occupation at Ivandvor. A smaller settlement (2-4 ha) that functioned during the 2nd and 3rd centuries was found among the remains of a late Roman *vicus* – meaning that it had been occupied before the latter was built. The layout and organisation of the former settlement, that is, the first phase of the *vicus*, cannot as yet be established with any certainty, because the complex horizontal stratigraphy does not allow a clear distinction between the two settlement stages. Nevertheless, it was possible to establish that the settlement at Ivandvor was relatively small and compact in its first phase, consisting of central farm building with several utilitarian buildings. It has already been mentioned that the basic economic pursuit of the Ivandvor *vicus* in its first (3rd-century) phase was cattle or hoarse breeding.[88] It remains to be seen if the same was true of the nearby village at Sela (see above), since it also consisted of a compact complex of timber constructions. Therefore, when analysis of the animal bone from the site is once carried out, it should contribute towards establishing the basic economic occupation of the *vicus* at Sela.[89]

86 Kristina Jelinčić, pers. com.
87 Tarpin 1999, 5-6.
88 Leleković 2008, 14.
89 Haraša and Bekić 2007, 75.

4. Unclassified rural settlements

Besides the sites that were labelled – with more or less certainty – as *villae rusticae* or *vici*, there are those where remains of collapsed Roman walls and small finds, most frequently pottery, or Roman graves and/or stone monuments, were either accidentally found or recovered in small-scale excavations, lacking therefore arguments for a reliable classification into the category of either villas or *vici* (see also section 2.1 of this paper). The main reason for this uncertainty is the state of research, that is, too few excavations and their inadequate scope at that, which prevents conclusions on the layout, dimensions and nature of these structures. In most cases we are dealing with sporadic finds of remains (whether compact or scattered) of masonry-built constructions of larger dimensions, with some rather luxurious construction parts (stone pillars, pediments, mosaics, etc.). Remains of thermal baths or isolated masonry-built tombs are found occasionally, which cannot be classified either as *vici* (due to the level of luxury involved in their construction) or as *villae* (on account of their substantial dimensions). Large luxurious *villae rusticae* did exist in Pannonia, for example at Baláca, covering an excavated area of 8.79 ha.[90] However, such rural establishments are not particularly frequent in Pannonia. Therefore, when discussing partially excavated sites, it should be borne in mind that the settlements larger than 3 ha could be equally classified as a large villa or a small *vicus*, appearing in rural areas contemporarily with south-Pannonian towns. One of the obstacles to the interpretation of these sites is the lack of knowledge of settlement geography, that is, our ignorance of the exact positions of the settlements known from the itineraries. Therefore, the presumed locations often change in time (see below).

Remains of an architectural complex were found in the village of Petrijanec near Varaždin. It was interpreted as part of a luxurious *villa rustica* occupied in the 2nd and 3rd centuries.[91] Although a large part of the site area was excavated in 1969 and 1970, the results were never published extensively, leaving unknown even the exact excavated surface. The settlement or a road station of Aqua Viva, which was founded at the intersection of the roads Siscia and Poetovio – Mursa, is customarily located in Petrijanec.[92] This position conforms to the distances from the itineraries, and the suggestion is substantiated with abundant surface finds gathered on the site over the years.[93] Especially outstanding are two rich early 4th-century hoards consisting of coins, jewellery, and silver and gold dishes.[94] These hoards support the presumption that a major estate was located there, next to the road, whose traces were also found. On the other hand, the amount of finds from this site can be interpreted as belonging to a road station, functioning from the 1st to the 4th centuries on one of the most important road intersections in southern Pannonia.

A similar settlement was found near the village of Komin, on the road leading from Aqua Viva to Andautonia. No significant remains of residential architecture were evidenced, the main archaeological feature of the site being a cemetery composed of stone-fenced burial plots. On the basis of the latter, a major settlement was presumed on the site, most probably the road station Pyrri, known from the Roman itineraries.[95] Apart from Komin, in the region to the east of Zagreb remains of Roman architecture were found in the villages of Okešinec and Hercegovac, both situated along a section of the Poetovio – Siscia road. No settlements were mentioned on these locations in the itineraries, but remains of Roman architectural complexes of unknown functions could also be interpreted as the remains of smaller settlements developing along the roads.[96] A well preserved, but only partially excavated building on the nearby site of Rošnica, next to the Blaguša stream,[97] as well as the already discussed site of Drenje west of Zagreb (generally interpreted as a *villa rustica*), also belong to the group of sites of unknown function.[98] However, the results of recent excavations at Drenje have opened the possibility to interpret this architectural complex as a larger rural settlement. It occupied the site whose importance was enhanced by the vicinity of the nearby Sava crossing, as well as by its proximity to the Emona – Neviodunum – Siscia road; a smaller town could have easily grown there.[99]

Unclassified sites are also evidenced in the vicinity of Siscia. The settlement recorded in the itineraries as Praetorium, and the accompanying road station Ad Praetorium on the Siscia – Sirmium road, are good examples of this type of sites. Both are located in the wider area of Hrvatska Dubica southeast of Siscia. Ivo Bojanovski claimed that Ad Praetorium was situated on the left – today Croatian – bank of the Una River in the village of Baćin, next to the settlement named Praetorium, located in the village of Suvaja near Bosanska Dubica on the right bank of the Una.[100] An architectural complex measuring 200 x 200 m was found there, and was identified as a centre of a large – perhaps imperial – estate.[101] However, remains of a Roman cemetery at Hrvatska Dubica, only several km distant from Baćin and Suvaja, suggest another possible location for this settlement or a road station, opening a new perspective in interpreting the finds from this site.[102] In Topusko, 38 km to the southwest of Siscia, remains of modest baths were found in the vicinity of a thermal spring, in addition to stone altars and masonry tombs with luxurious grave goods. It had been previously believed that these represented the remains of the Roman settlement or

90 Palágyi 1995, 11.
91 Šarić 1977, 51-58.
92 Vikić and Gorenc 1969, 5-6; Šarić 1977, 51-58; Gračanin 2010, 18-20.
93 I. Šarić, in: Demo 1986, 124, no. 19.
94 Šiša-Vivek, Leleković, and Kalafatić 2006.

95 D. Nemeth-Ehrlich, in: Demo 1986, 118, no. 10.
96 Demo 1986, 31-32.
97 Demo 1986, 30.
98 Ž. Škoberne, in: Demo 1986, 114-116, no. 5; Škoberne, Koščević, and Makjanić 1987; Koščević and Cvitković 2005, 107-108.
99 Škoberne, Koščević, and Makjanić 1987, 59-61.
100 Bojanovski 1988, 336-337; Gračanin 2010, 17.
101 Busuladžić 2011, 120, 165.
102 Koščević and Makjanić 1986, 119-122.

Fig. 29. Aerial view of the site of Osekovo, with the excavated sections marked in red (after Google Maps, modified by T. Leleković).

a road station Ad Fines, but more recent finds and studies attribute the finds in Topusko to the road station Quadrata.[103] Remains of masonry, altars and numerous small finds were also found at Kamensko near Karlovac, not far from the confluence of the Mrežnica and Kupa Rivers. Near to this – as yet unexcavated site – a shipwreck was found, with a cargo of broken bronze statues and other massive bronze objects that were most probably being transported to Siscia for melting. This find suggests that a settlement and an important river port of as yet unknown name existed at this place, which was the farthest point where the River Kupa was still navigable.[104] Such sites urge for a re-examination of presently held opinions on the sites such as the already mentioned Osekovo east of Siscia. Although its architecture was interpreted as a luxurious *villa*, its layout, visible on aerial photographs (fig. 29), demonstrates that the excavated building was only a small segment of a complex similar to the one at Suvaja. Therefore, certain large architectural complexes, classified as large *villas* or centres of *latifundia*, could actually have been smaller settlements grown along the main roads. The site of Osječenica (Gorička), where a fragment of – possibly imperial – inscription was found, speaking for the importance of this location (settlement?), can be grouped together with Osekovo and Suvaja.[105] All these information suggest that the *ager* of Siscia, at least according to southern-Pannonian criteria, was densely packed with larger rural settlements.

The classification of several sites from the central part of northern Croatia is also doubtful. Remains of baths at the site of Tekić near Požega were ascribed to a larger settlement due to its dimensions and on account of the late Roman cemetery in the vicinity; such circumstances led the excavators to identify these finds with the town of Incerum, known from the itineraries.[106] At the same time, the unexcavated site of Veliki Bastaji 10 km to the north of Daruvar was classified as a large and luxurious *villa rustica*, although its dimensions and finds are on a par with those at Tekić.[107] The total area of the surface finds at Bastaji measures some 15 ha; a richly decorated marble sarcophagus is the most important of those finds.[108] A site in the village of Brusnik 17 km to the south of Aquae Balissae, where a *villa* or a smaller-sized settlement was located, can also be considered as unclassified.[109]

In addition to the settlements developing along or on the main roads, there were hilltop settlements in the western and central part of northern Croatia, erected on naturally protected positions, generally on locations away from the main roads and major towns. These smaller fortified settlements were otherwise characteristic of the Alpine and sub-Alpine area in Late Antiquity, and were in southwestern Pannonia probably more frequent than one would be led to believe on the basis of the present evidence.[110] One example of this type of settlements, a kind of a larger complex consisting of a number of smaller archaeological wholes, was found in the village of Repišće, on the slopes of Plješevica Mountain west of Zagreb. A Roman villa or *vicus* was found in the village itself, together with remains of a cremation cemetery composed of tumuli (1st-2nd centuries). On the site of Popov dol some 2.5 km from the *vicus*, a hilltop settlement was found, and a nearby 4th-century skeletal cemetery excavated.[111] So far, only two such hilltop settlements were systematically excavated. One of them is Kuzelin to the north of Zagreb, the best-known site of this type in northern Croatia. It features a layout in the form of an elongated ellipse, measuring some 200 x 100 m. It was occupied from the late 2nd to the early 5th century, but the majority of finds dates from the 2nd half of the 4th and the early 5th century. Besides several residential buildings and industrial facilities, remains of a massive defensive wall

[103] Čučković 1986, 14; Durman 1992, 127; Šegvić 1986, 97-98; Šegvić 2006, 270-271; Gračanin 2010, 14-16.
[104] Čučković 1986, 14-15; Šarić 198.
[105] Durman 1992, 126.
[106] Sokač-Štimac 2005; Sokač-Štimac 2006; Sokač-Štimac 2007; Sokač-Štimac 2008; Sokač-Štimac and Korać 2007.
[107] Migotti 1996, 150-151.
[108] Schejbal 2004, 119.
[109] Schejbal 2004, 121.
[110] Ciglenečki 1994; Migotti 1997, 28.
[111] Gregl 1992, 149-150.

encompassing the entire settlement were found.[112] The second excavated hilltop settlement is situated above the village of Lobor on Ivanščica Mountain north of Zagreb. Remains of a fortified plateau with residential and industrial buildings, and a temple to Diana or Magna Mater, as well as foundations of an early Christian 6th-century church, were found there. In addition to this, remains of a settlement at the foot of the hill were excavated below the fortified plateau; clearly, the majority of the population lived there, to retreat into the upper fortified part of in moments of severe danger.[113] The poorly excavated site of Ozalj near Karlovac can also be classified as a hilltop settlement, but it remains unknown whether it was a smaller settlement or a Roman fort.[114]

Interpretation of timber-built settlements faces similar doubts and insecurities as that related to poorly researched masonry remains. The abovementioned complex of timber buildings at the site of Sela could be classified as a small village, but also as a centre of a large rural estate. The case of Ivandvor, a site similar to Sela, can be of some help in solving the problem of the interpretation of the latter. The finds of bricks and tiles were suggestive of a large *villa rustica*, but the excavations have shown that neither villa nor any other form of solidly built masonry existed there. All the discussed examples lead to the conclusion that without larger-scale excavations and solid material evidence it is impossible to determine whether a complex represents a *villa rustica*, *vicus* or another kind of settlement. A similar uncertainty is attached to the interpretation of the results of rescue excavations in Slavonski Brod. In the centre of the modern-day town, next to the Sava River, a Roman settlement and station on the Siscia – Sirmium road was located, known from the literary and epigraphic sources as Marsonia/Marsunia.[115] The 2003 and 2004 excavations in Slavonski Brod revealed remains of brick- and timber architecture, a pottery kiln and a small number of cremation graves, attributed to Marsonia.[116] However, these finds are too insignificant to be taken as a final proof for the location of Marsonia. According to the published excavations in the area of Slavonski Brod, no substantial remains of masonry were found, but only wooden structures typical of rural settlements, especially *vici*. During the construction of a highway another rural settlement was excavated in the immediate vicinity of the town, but the results have not yet been published.[117]

Overall, the majority of rural settlements cannot be classified into any of the three categories discussed so far. These settlements were mostly registered during systematic surveys or were reported to the authorities as chance finds, while far less frequent are the cases of the remains of a settlement or a cemetery recovered in smaller-scale archaeological excavations or during construction works. Therefore, such structures lack the information on the dimensions, form and type, and can be studied solely through their position in terms of settlement geography. Accordingly, no interpretations of them could be provided in the following section of the paper; they will only be discussed in a summary fashion, within the analysis of the distribution of rural sites.

5. The distribution of rural sites

At present, the largest number of rural settlements is documented for the Zagreb area. The intensive development of the modern capital, situated in the territory of Roman Andautonia, has caused continuous discoveries of the sites in the vicinity of the Roman *municipium*. As much as 110 Roman sites from the area had been published previously,[118] while recent surveys, published in a preliminary form, have significantly enlarged that number.[119] The largest number of sites was documented along the Emona – Siscia road, which did not approach Andautonia itself, but passed ca. 15 km to the west and south of it.[120] Fourteen sites are located along its path: Drenje, Laduč, Šibice, Vugrovec, Gornji Čehi, Mala Mlaka I, Mala Mlaka II, Donji Vukojevac, Donja Lomnica, Zagreb-Stenjevec I, Zagreb-Stenjevec II, Gornji Čehi, as well as the abovementioned *vici* at Šepkovčica, Okuje and Donji Vukojevac. This shows that the majority of the settlements in this part of the territory of Andautonia were constructed along the abovementioned road. Another important road that passed through the territory of the town, but again, not through the town itself, is the one between Siscia and Poetovio. The number of Roman sites along this traffic route is significantly smaller: the six rural settlements discussed above (Donja Glavnica, Moravče I, Moravče II, Hecegovac, Zagreb-Šimunčevec I, Zagreb-Šimunčevec II), as well as the remains of two cemeteries of Norican-Pannonian tumuli on the sites of Dumovec and Đurđekovec.[121] The position of the site of Dumovec is especially significant, because of the presumed vicinity of the intersection of the main Poetovio – Siscia road with the local road connecting the former with the Emona – Siscia traffic artery. The importance of the mentioned local road is based on the fact that the town of Zagreb developed precisely in its corridor, so that 24 sites with various Roman finds suggesting some form of settlement were documented during the development of the town.[122] Thus, for example, single burials found at the sites of Borongaj, Maksimir, and Savska and Vukovarska Streets, suggest the existence of associated settlements.[123] On the other hand, excavations on the sites of Stenjevec and Držićeva Street have revealed the

[112] Sokol 1994, 199-209; Sokol 2006, 154-155.
[113] Filipec 2008, 58-60.
[114] Migotti 1997, 27-28.
[115] Miškiv 2006, 238-239.
[116] Miškiv, Miklik-Lozuk, and Lozuk 2006.
[117] Kornelija Minichreiter, pers. com.

[118] The list (Gregl 1991, 1-77) encompasses the wider area of Zagreb. Most of the sites were in the territory of Andautonia, but some, for example those on the Žumberak (west of Zagreb, on the border between Croatia and Slovenia), belonged to the territory of some of the neighbouring towns. It is not at present possible to affiliate the sites to their respective urban centres, because the precise territorial boundaries of the Roman towns in the area are unknown.
[119] Lolić 2003, 132; Burmaz and Lolić 2009; Ložnjak Dizdar 2008.
[120] Bojanovski 1976, 311-312; Lolić 2003, 132.
[121] Sokol 1981, 169-184.
[122] Gregl 1991; Gregl 1994; Tončinić 2005, 269-278; Gregl and Jelinčić 2010, 153-158.
[123] Vikić-Belančić 1981, 148.

existence of larger cemeteries: the one in Držićeva Street was continuously in use from the 1st to the 4th centuries, while that at Stenjevec was only in use in the first two centuries AD.[124] Almost nothing is known of the layout and the reasons behind the foundation of these settlements, but the fact that the stone from the quarries in the northern districts of Zagreb was evidenced in Andautonia, allows the assumption that a significant proportion of these settlements, as well as the vicinal roads constructed in this area, played a certain role in extracting and dressing stone for the needs of Andautonia.[125]

At present, 30 sites are known in the area gravitating towards Siscia. Their distribution is displayed on a map published in 2003, which demonstrates the grouping of rural settlements along the main roads leading from Siscia towards Incerum, Sirmium, Senia, and Emona/Poetovio.[126] In the future, large-scale protective excavations are planned on the highway corridor, but at this moment the main source of information are chance finds or field surveys conducted in the area so far. The already mentioned shipwreck near Karlovac suggests that the adjacent territory also gravitated towards Siscia.[127] This is an area with a rather small number of sites (such as an unclassified settlement of Sveti Petar Mrežnički[128] or the cave cemetery in Bubijeva jama near Barilović[129]). Even so, it can be presumed that this situation is a reflection of the present stage of archaeological research rather than the real state of things.

Rural cemeteries should also be associated to settlements, even if it is impossible to determine whether they belonged to villas or *vici*. Such finds testify the presence of settlements in an area, despite the fact that the settlements themselves remain undisclosed. Once again, this is mostly evident in the Zagreb area, where almost 80 rural cemeteries have been registered. It should be born in mind, however, that such density is perhaps caused by the state of research in the area, since this is a region that was often in the focus of archaeologists' interest. Therefore, a similar distribution of finds can be expected in other regions of southern Pannonia. Only a small number of tumuli cemeteries have been excavated; 35 were found on the Sesvete-Dumovec site, and more than 30 in the vicinity of the village of Đurđekovec.[130] The majority of tumuli at Sesvete-Dumovec (22) were arranged in two rows, while the remainder were scattered across a wider area. At Đurđekovec, on the other hand, the tumuli were arranged randomly, thus creating an irregular and inconsistent cemetery layout.[131] Between 1981 and 1983, six mounds were excavated on the site of Turopoljski lug in the territory of Andautonia, which is only a small section of the cemetery consisting of more than a hundred such mounds, dated from the 2nd half of the 1st to the early 3rd century.[132]

Cemeteries are especially important for the area north of Zagreb, as there they appear to be the only known remains of rural settlements. One of the first to have been excavated was the cemetery at Kunovec Breg near Koprivinica, which yielded 9 graves of various rites.[133] Significantly, two tumuli-cemeteries were found in this part of the Drava region: Gola, with 2 tumuli excavated, dated by the Westendorf *terra sigillata* and coins of Maximinus Thrax to the late 2nd and the 1st half of the 3rd centuries,[134] and another one near the village of Novačka, also with tumuli from the 2nd half of the 2nd and the early 3rd centuries.[135] These cemeteries suggest that the area was relatively densely populated in the second half of the 2nd century. On balance, it could be presumed that Roman buildings in Draganovec[136] and Jalžabet-Bartolovec[137] (see section 2.2. here) are only an indication for a possible reconstruction of the settlement pattern in this area during the Roman period. This is further substantiated by recent surveys, such as that of the area of Križevci, showing a different picture of a region previously believed to be very sparsely populated. On the contrary, it seems that small rural communities made the dominant form of settlement in this region; yet until recently such structures passed almost unnoticed.[138]

The central part of northern Croatia, delimited on the west by the Kutina – Bjelovar – Koprivnica line, on the east by the Donji Miholjac – Našice line, and on the north and south by the Drava and Sava Rivers, was less intensively studied in comparison to the eastern and western parts of the region; forty years ago Danica Pinterović referred to the central area as the 'unknown Slavonia'.[139] Nevertheless, this has partially changed in the meantime, especially as regards the Požega Valley. In the centre of the valley the site of Tekić is situated (probably Roman Incerum), while on the valley's margins, within a 10-km radius from Tekić, six lesser, unclassified settlements have been registered. These are probably only indicators of a much more substantial Roman heritage of this fertile valley.[140] The already mentioned site of Rudina is among those six sites; it might have been a hilltop settlement with a church – one of many settlements of this type presumed in the western part of the Slavonian Mountains.[141] A large stone stele stems from Novo Naselje, a village 16 km to the south of Aquae Balissae; it probably indicates a cemetery of a major villa or other kind of settlement.[142] The area to the north and south of the Slavonian Mountains, where the mountain

[124] Vikić-Belančić 1981, 148-149; Gregl 1994.
[125] Vikić-Belančić 1981, 148-149; Tončinić 2005, 269-278.
[126] Lolić 2003, 132.
[127] Čučković 1986, 14-15; Šarić 1986.
[128] *AIJ*, 218-219.
[129] Perkić 2006, 104.
[130] Sokol 1981, 169-178.
[131] Sokol 1981, 169-178.
[132] Koščević and Makjanić 1988, 35.
[133] Demo 1985, 120-123.
[134] Demo 1985, 119-120.
[135] Okroša-Rožić 2006, 186-187.
[136] Demo 1986, 114.
[137] D. Nemeth-Ehrlich, in: Demo 1986, 118, no. 10.
[138] Tkalčec *et al.* 2007, 8, 12, 16-17.
[139] Pinterović 1975, 123.
[140] Sokač-Štimac 2005, 2-4.
[141] Migotti 2009, 132. On the syntagm *Slavonian Mountains* see Migotti in this volume, note 13, fig. 2.
[142] Schejbal 2004, 121.

slopes descend to the banks of the Sava and Drava Rivers, could be described – according to the present state of research – as almost uninhabitable. The Poetovio – Mursa road passed along the right bank of the Drava. Besides the abovementioned site of Kiškorija,[143] the only notable site along this route is Orešac to the east of Virovitica. Although the position of this site can be associated to the road station Bolentio, while the excavations yielded archaeological material dating from the 1st to the 5th centuries, it is not possible to securely establish the type of this settlement.[144] Similarly, a small number of sites were documented on the southern slopes of the Slavonian Mountains, in the Croatian portion of the Sava valley. The Siscia – Sirmium road ran along the right bank of the Sava – now the territory of Bosnia and Herzegovina – so that such small number is not surprising. Together with the already described villas, four Roman sites have so far been evidenced in this region, among them Pivare near Nova Gradiška, located on the very bank of the Sava, opposite to the important river port of Servitium (Bosanska Gradiška, Bosnia and Herzegovina). The site remains unexcavated but it could turn out to be important as an argument for the existence of a road that connected Marsonia with Siscia passing along the left bank of the Sava, a traffic route not mentioned in the itineraries. Presumably, this road was connected by a ferry to the one passing along the right bank of the Sava, which was actually the main artery of this wider region. Some 60 km to the east, in the territory of Slavonski Brod, two rural settlements were found – the already discussed Marsonia[145] and an almost unknown rural settlement ca. 20 km from the site of Marsonia.[146] The field survey has revealed another settlement southeast of the Slavonian Mountains, probably on the Siscia – Sirmium road. It is located between the villages of Gudinci and Jaruga on the Berava brook, four km upstream from its confluence with the Sava, probably at the point where the Siscia – Sirmium road crossed the brook. An archaeological complex consisting of several smaller segments forming a large-sized rural settlement was found on this site. On the basis of the concentration of Roman brick and kitchen pottery, two separate complexes divided by the brook were presumed. Also, remains of a late Roman cemetery were identified, dated to the late 4th century on the basis of masonry tombs and grave offerings; it probably spread out along the road outside the settlement.[147]

When discussing the central part of the area of study, some progress in the state of research in the territory of Đakovo should be noted. Until recently we only possessed data for several sites, above all that of Štrbinci, probably Roman Certissia.[148] Field surveys of the area, together with ever increasing number of archaeological excavations, grew in number due to large-scale infrastructural projects, such as the Sredanci-Osijek highway, where the previously described *vici* were found. Still, the most important source for the study of the settlement pattern in the area is a recent paper on the archaeological topography of Roman sites in the territory of Đakovo, containing a gazetteer of 38 sites. Most of them were discovered through surface chance finds, which renders their classification impossible. Only at the sites of Budrovci, Krndija and Selci, Roman pottery was evidenced beyond doubt, but it still remains unclear whether we are dealing with the remnants of a villa or other types of rural settlements.[149] Levanjska Varoš is an especially important site, not yet excavated, possibly the site of Roman Leuconum, a station on the Certissia – Incerum road.[150] In contrast to the Đakovo region, the territory of Vinkovci (Cibalae) was studied more thoroughly. In 1979 a list of 21 sites was published in a paper on the archaeology of the region of Vinkovci, most probably belonging to the *ager* of the Roman colony of Cibalae. These sites were discovered owing to the reports on chance finds, but two of these should be further discussed.[151] The largest number of finds came from the site of Rajterovo brdo near Oriolik, to the east of Vinkovci. Surface finds were recovered from an area measuring almost 30 ha, suggesting a major settlement, most probably on the Siscia – Sirmium road. Southeast of Cibalae, near the village of Privlaka, another large-scale Roman site was found, perhaps a large villa or a *vicus*.[152] Remains of a *villa rustica* on the site of Cerić-Plandište northeast of Vinkovci should also be mentioned, probably located along the route that connected Cibalae with the Danube *limes* road. This site is known for a brick bearing a possible Christian inscription *Finctor vive dominis tvis*, but it was actually occupied from the 1st to the 4th century, as evidenced by small finds.[153] The 1979 list was supplemented during the last five years, owing to an increasing number of surveys of the corridor of the future Dunav – Sava canal,[154] as well as the corridors of local roads.[155] An unpublished site between the Cornacum (Sotin) fort and the site of Oriolik, probably a *villa rustica*, can also be added to the list.[156]

As the *ager* of Cibalae was located in the immediate hinterland of the Danube, the study of the settlement patterns in the area of Vinkovci is at the same time representative of such patterns at the *limes* periphery. The area of the *limes* was studied on a number of occasions in the past, but with the focus on the remains of military fortifications and adjacent civil settlements, while the territory in the immediate hinterland of the *limes* remained neglected.[157] Therefore, the recent excavations have provided a completely new picture of this region. A field survey of the area along the *limes* was conducted during the 1960s, which was one of the first studies of this type in Croatia, but it remained focused on the military infrastructure.[158] A new project of a systematic field survey of the Baranja

[143] See n. 69.
[144] Makjanić 1990; Salajić 2007; Jelinčić 2009a; Jelinčić 2009b.
[145] See n. 115.
[146] Kornelija Minichreiter, pers. com.
[147] Lozuk, Minichreiter, and Miškiv 1997, 40-47.
[148] Migotti *et al.* 1998.
[149] Knezović 2005, 63-83.
[150] Perinić 1998, 82; Knezović 2005, 71.
[151] Dimitrijević 1979, 185-187.
[152] Dimitrijević 1979, 186.
[153] Dimitrijević 1979, 186; Iskra-Janošić 2001, 46.
[154] Hutinec 2008; Krznarić Škrivanko 2007; Dizdar 2008.
[155] Dizdar and Ložnjak Dizdar 2009a; Dizdar and Ložnjak Dizdar 2009b.
[156] Mato Ilkić, pers. com.
[157] Pinterović 1969; Pinterović 1971; Minichreiter 1984; Filipović 2010.
[158] Pinterović 1970.

region is focused on the entire territory to the west of the *limes*, in order to acquire, for the first time, an insight into the organisation of the wider border area.[159] From the data gathered in field surveys, certain regularities and rules can be discerned even at this stage.[160] This is, above all, true of the *limes* itself. In the territory to the south-west of Ilok (Cuccium) 36 sites were recorded, 18 dating from the Roman period and distributed in three zones. The first group of sites is lined along the very bank of the Danube and these were interpreted as sentinels distributed between the major fortifications in this part of the *limes* (Cuccium and Cornacum). The second group is one to three km removed from the Danube and these were interpreted as smaller-sized civil settlements developed along the *limes* road on the points where it had to diverge from the river towards the interior, because of the lie of the land. The third group were civilian settlements constructed along the Cibalae – Sirmium road, which was only 10 km away from the *limes* in this region; the sites of Orašče west of Tovarnik (Ulmo?) and the so far unknown site near Ivanac (a large-sized villa or a *vicus*) are the largest and richest of those sites.[161]

In contrast with the area gravitating towards the Sava, that is, the territories of Certissia and Cibalae, the region along the Drava was considerably less studied. The Baranja region, for example, was surveyed several times, but never systematically, and the results of these surveys were only preliminary published.[162] The recently published survey of the area around the settlements/fortifications of Ad Novas (Zmajevac) and Ad Militare (Batina) has shown a similar regularity in the distribution of rural sites in the interior of the *limes*. Although the two-km wide belt of land, stretched along the edge of the plateau holding the fortification, was systematically surveyed, no smaller settlements were found. The only important site noted is Kotlina, 3 km distant from the limes.[163]

6. Conclusion

Some 250 rural sites have been included in this overview, showing however an uneven distribution across the area of study. Smaller or larger-scale excavations were conducted on only 53 sites, and they can be classified into three groups: 1. *villae* (21); 2. *vici* (9), properly excavated during large-scale construction works on the highway corridors; 3. unclassified sites. It should be noted that none of the *villae rusticae* were thoroughly excavated, so that at this stage there is not a single site that could give us a complete insight into the layout and functioning of this type of rural settlements. Accordingly, large architectural complexes (7 of these are known at present; see section 4) represent a special problem: given the insufficient surface and overall scope of the research, the true nature of these structures (quite large *villae rusticae* or smaller settlements?) remains unresolved.

Another difficulty in studying settlement patterns in the rural areas of southern Pannonia is the lack of evidence for determining, even approximately, the borders of the administrative units, as well as the precise routes of the main and local roads. The distribution of rural settlements, that is, a hypothetical attribution of these sites to settlements mentioned in the itineraries or other literary sources, is often one of the arguments in determining the settlement geography of the area. It should be borne in mind, though, that these attributions are often insufficiently substantiated. For example, the road station and important settlement of Marsonia is customarily and unquestionably located at the site of the modern-day town of Slavonski Brod, even though archaeological finds do not support this hypothesis. On the other hand, the site of Topusko, known for its relatively rich architectural remains and a cemetery, has been identified as two different settlements – the road stations Ad Fines and Quadrata, respectively. Finally, some of the (almost) definitely identified settlements, such as Certissia (Štrbinci near Đakovo) or Incerum (Tekić near Požega) still remain under a question mark because these identifications have not been finally proved in terms of archaeology. Arguably, future archaeological excavations could revise some generally accepted locations of itinerary settlements.

The unsatisfactory state of research notwithstanding, the rural area of the Croatian part of Pannonia can be divided into three regions. The first encompasses the area from the western slopes of the Slavonian Mountains to the border with Slovenia. Of a total of 160 rural sites, 132 are located in the vicinity of Zagreb and Sisak, that is, in the town territories of Siscia and Andautonia. A significant number of sites are located near Krapina and Varaždin (28), that is, in the area that most probably gravitated towards Poetovio. The majority of the sites are located near the most important roads (Emona – Siscia, Poetovio – Siscia, and Siscia – Sirmium), while their number significantly decreases in the areas removed from the main roads. Therefore, the number of Roman sites in the area of Bjelovar, Koprivnica, and Križevci is extremely small. In the territory stretching from the eastern slopes of the Slavonian Mountains to the Danube (the Croatian part of Pannonia Inferior), a significantly smaller number of sites are known (52). The majority of these remain unexcavated, but those along the highway corridor that were thoroughly researched, such as the *vici* Sela, Ivandvor and Josipovac, have yielded important data. All three settlements started in the late 2nd century and achieved their peaks in the 4th. This allows the conclusion, especially in the light of the comparison with the excavations in Mursa and Cibalae, that the Croatian part of Pannonia Inferior was in the 1st and 2nd centuries less densely populated than the western part (Pannonia Superior). This seems to have been so until the reign of the Severi, when the population density started to increase in the eastern region, so that in Late Antiquity the two areas became roughly equal in this respect. The same

[159] Ložnjak Dizdar, Dizdar, and Šiljeg 2004; Bojčić *et al.* 2009; Bojčić *et al.* 2010.
[160] Dizdar and Ložnjak Dizdar 2009.
[161] Dizdar and Ložnjak Dizdar 2009b, 78-79.
[162] Minichreiter 1984.
[163] Bojčić *et al.* 2010.

is true of the central area (the Slavonian Mountains), in the first place the Požega Valley, where the majority of the sites were most probably occupied in earlier periods, but only flourished in Late Antiquity. Two urban centres in this part of southern Pannonia are especially important: Municipium Iasorum and Incerum. The total number of sites (18) in this region is, however, rather small, which can be explained by the poor state of research, although one cannot discard the possibility that this number reflects a true situation. A feature particular to this region – again, if not resulting from the poor state of research – is the fact that so far only large *villae rusticae* were found in this area, and no other types of rural settlements.

The area of the Slavonian Mountains, together with the territory of Daruvar (Aquae Balissae) and the central parts of the Drava and Sava valleys, are the least studied parts of Croatian Pannonia. Recently, however, a certain progress has been achieved: a number of new sites were reported, if not yet excavated. The southern slopes of the Slavonian Mountains rapidly transform into a flooded terrace along the Sava. Although a road must have passed through this area, there is almost no information on the settlements along its presumed course. Besides the villas in Cage and Benkovac, there was a settlement in the area of Slavonski Brod, although its identification with Marsonia remains doubtful. It has been postulated that the centres of large estates in this area were placed near important river communications,[164] but the excavations in the regions along the Drava and Sava do not substantiate this suggestion. During the construction of the Virovitica bypass in 2005, a *vicus* was found at the site of Kiškorija north of the Slavonian Mountains.[165] In contrast to the above-discussed sites, this one is located in a poorly studied area, which gives it particular importance. The site is located on the main road that passed along the Drava, and since it is placed less than 30 km from the town of Municipium Iasorum (Aquae Balissae), it could have belonged to its territory. Therefore, it seems that the area around Municipium Iasorum was mostly composed of large estates with rich *villae rusticae* as their foci, while *vici* and lesser *villae* seems to be less frequent, compared to the rest of the province. Nevertheless, this hypothesis should not be taken at face value because it possibly reflects the state of research rather than the true situation.

On balance, the available information suggests that in the western part of southern Pannonia the majority of the rural population lived in the vicinity of the towns and generally near the main roads, while in the eastern part a significant portion of them lived in the interior of the *limes*. The rest of southern Pannonia seems to have been less densely populated, and the number of rural settlements decreases with the distance from large towns. However, this conclusion should not be taken as final, since some 40 sites discovered in the area of Siscia is three times less in comparison with the number of sites in the area of nearby Andautonia, which does not seem to be realistic; given the importance of Siscia, the number of sites in its area must have been larger. Moreover, the sites of Osekovo, Suvaja, and Dubica suggest that the most interesting examples of rural settlements can be expected exactly in the surroundings of Siscia. This assumption, however, as well as others suggested in this paper, cannot be taken at face value and should instead only stimulate future research of the subject of rural settlement in southern Pannonia, so far rather neglected in Croatian archaeology.

(T. Leleković)

Acknowledgements

We would like to express our gratitude to Aleksandra Bugar, Slavica Filipović, Mato Ilkić, Kristina Jelinčić, Branka Migotti, and Kornelija Minichreiter, for all their help and information shared with us.

Bibliography

Balen *et al.* 2009 Jacqueline Balen, Tomislav Bilić, Maja Bunčić, Ivan Drnić, and Ana Solter, Rezultati zaštitnih istraživanja na lokalitetu Ivandvor - šuma Gaj, *VAMZ* 42, 23 – 73.

Bekić 2006 Luka Bekić, Zaštitna arheologija u okolici Varaždina. *Arheološka istraživanja na autocesti Zagreb-Goričan i njezinim prilaznim cestama / Rescue Archaeology in the Varaždin Environs. Archaeological Research on the Zagreb Goričan Highway and its Access Roads*, Zagreb.

Biró 1974 Maria Biró, Roman Villas in Pannonia, *AArchHung* 26/1-2, 23-57.

Bobovec 2005 Ana Bobovec, Roman complex Ciglenice in Osekovo, in: *Illyrica antiqua. Ob honorem Duje Rendić-Miočević* (International conference on issues in Ancient Archaeology, Zagreb, November 6-8, 2003), Zagreb, 111-117.

Bobovec 2006 Ana Bobovec, Osekovo - Ciglenice, in: Durman 2006, 194-195.

Bobovec 2008a Ana Bobovec, *Rimski kompleks Ciglenice u Osekovu. Otkrivene tajne zlatnog doba Moslavine / Roman Complex Ciglenice in Osekovo. Revealed Secret of Golden Age of Moslavina*, Kutina.

Bobovec 2008b Ana Bobovec, Lokalitet: Ciglenice, *HAG* 4 (2007), 211-212.

Bobovec 2008c Ana Bobovec, Lokalitet: Kutina – Kutinska lipa, *HAG* 4 (2007), 215-216.

Bobovec 2009 Ana Bobovec, *Archaeological site Kutinska lipa, Kutina before Kutina*, Kutina.

Bojanovski 1974 Ivo Bojanovski, *Dolabelin sistem cesta u rimskoj provinciji Dalmaciji, DjANUBiH* 47.

Bojanovski 1988 Ivo Bojanovski, *Bosna i Hercegovina u antičko doba, Dj ANUBiH* 66.

Bojčić *et al.* 2009 Zvonko Bojčić, Marko Dizdar, Tomislav Hršak, and Tino Leleković, Terenski pregled područja Batine, *AIA* 5, 125-129.

Bojčić *et al.* 2010 Zvonko Bojčić, Marko Dizdar, Tomislav Hršak, Igor Vukmanić, Domagoj Dujmić, and Tino

[164] Mulvin 2002, 71-72.
[165] Jelinčić 2007, 27-34.

Leleković, Terenski pregled područja između Batine i Suze, *AIA* 6, 80-87.

Brukner 1982 Olga Brukner, Kudoš, Šašinci – villa rustica, *AP* 23, 93-95.

Bugar 2009 Aleksandra Bugar, Lokalitet: Šepkovčica, *HAG* 5 (2008), Zagreb, 269-273.

Burmaz and Lolić 2009 Josip Burmaz and Tatjana Lolić, Lokalitet: Autocesta Zagreb – Sisak, dionica Velika Gorica – Lekenik, *HAG* 5 (2008), Zagreb, 243-245.

Busuladžić 2011 Adnan Busuladžić, *Rimske vile u Bosni i Hercegovini*, Sarajevo.

Ciglenečki 1994 Slavko Ciglenečki, Höhenbefestigungen als Siedlungsgrundeinheit der Spätantike in Slowenien, *AV* 45, 239-266.

Čučković 1986 Lazo Čučković, Arheološka topografija karlovačke regije, in: *Arheološka istraživanja na karlovačkom i sisačkom području, IzdHAD* 10, Zagreb (1985), 9-18

Damevski 1980 Valerija Damevski, Istraživanja u Benkovcu kod Okučana 1978-1979, *VAMZ* 12-13 (1979-1980), 236-238.

Damevski 1981 Valerija Damevski, Istraživanja u Benkovcu kod Okučana 1980. god, *VAMZ* 14, 137-138.

Damevski 1982 Valerija Damevski, Terenska istraživanja 1981. godine, Benkovac, Cage, *VAMZ* 15, 267-268.

Damevski 1984 Valerija Damevski, Benkovac, istraživanje 1982/1983. god, *VAMZ* 16-17 (1983-1984), 301-303.

Damevski and Gorenc 1976 Valerija Damevski and Marcel Gorenc, Benkovac kod Okučana, Nova Gradiška – rimska ladanjska vila, *AP* 18, 79-81.

Damevski and Gorenc 1979 Valerija Damevski and Marcel Gorenc, Benkovac kod Okučana – rimska ladanjska vila, nastavak istraživačkih i zaštitnih radova, *AP* 20, 80-82.

Damevski and Gorenc 1980 Valerija Damevski and Marcel Gorenc, Benkovac kod Okučana – rimska ladanjska vila, *AP* 21, 98-101.

Damevski and Gorenc 1982 Valerija Damevski and Marcel Gorenc, Benkovac i Cage kod Okučana – rimska vila, *AP* 23, 84-87.

Demo 1985 Željko Demo, Grobni ritus u koprivničkoj Podravini tijekom 2. st. n. e, in: *Sahranjivanje pokojnika sa aspekta ekonomskih i društvenih kretanja u praistoriji i antici, MatSADJ* 20, Beograd, 111-126.

Demo 1986 Željko Demo, Antički period. Bilješke iz povijesti istraživanja u sjeverozapadnoj Hrvatskoj / The Roman period. Notes from the History of Archaeological Research in Northwestern Croatia, in: D. Balen-Letunić *et al.* (eds.), *40 godina arheoloških istraživanja u sjeverozapadnoj Hrvatskoj* / 40 Jahre archaölogischer Untersuchungen auf dem Gebiet des nordwestlichen Kroatien, Exhibition Catalogue, Koprivnica, 24-53.

Dimitrijević 1979 Stojan Dimitrijević, Archäologische Topographie und Auswahl archäologischer Funde vom Vinkovcer Boden, in: *Corolla memoriae Iosepho Brunšmid dicata, IzdHAD* 4, Vinkovci, 201-275.

Dizdar 2008 Marko Dizdar, Lokalitet: Kanal Dunav – Sava (op. Andrijaševci, Cerna, Babina Greda), *HAG* 4 (2007), 61-64.

Dizdar and Ložnjak Dizdar 2009a Marko Dizdar and Daria Ložnjak Dizdar, Terenski pregled trase zapadne zaobilaznice oko grada Belog Manastira, *AIA* 5, 98-101.

Dizdar and Ložnjak Dizdar 2009b Marko Dizdar and Daria Ložnjak Dizdar, Terenski pregledi područja općina Ilok, Lovas i Tovarnik u 2008. g., *AIA* 5, 117-121.

Dizdar, Tonc, and Ložnjak Dizdar 2011 Marko Dizdar, Asja Tonc, and Daria Ložnjak Dizdar, Zaštitna istraživanja nalazišta AN 6 Gornji Vukojevac na trasi auto-ceste Zagreb-Sisak, dionica Velika Gorica jug – Lekenik, *AIA* 7, 61-64.

Durman 1992 Aleksandar Durman, O geostrateškom pložaju Siscije, *OpArch* 16, 117-131.

Durman 2006 Aleksandar Durman (ed.), *Stotinu hrvatskih arheoloških nalazišta*, Zagreb.

Filipec 2008 Krešimir Filipec, *Arheološko-povijesni vodič po svetištu Majke Božje Gorske u Loboru*, Zagreb.

Filipec *et al.* 2009 Krešimir Filipec, Danijela Roksandić, Marija Šiša Vivek, and Morana Karneluti, *Arheološke slike iz Slavonije,* Exhibition Catalogue, Zagreb.

Filipec, Šiša Vivek, and Roksandić 2009 Krešimir Filipec, Marija Šiša Vivek, and Danijela Roksandić, Josipovac – Verušed (AN 15), *HAG* (2008), 30-33.

Filipović 2010 Slavica Filipović, *Kasnoantička nekropola u Zmajevcu – Mocsolás* / A Late Antique Necropolis in Zmajevac – *Mocsolás* , Osijek.

Fries-Knoblach 2006 Janine Fries-Knoblach, Hausbau und Siedlungen der Bajuwaren bis zur Urbanisierung, *Bayerische Vorgeschichtsblätter* 71, München, 338-430.

Gabler 2003 Dénes Gabler, Rural settlements in Pannonia, in: Visy 2003, 241-243.

Gorenc and Damevski 1981 Marcel Gorenc and Valerija Damevski, Benkovac kod Okučana – rimska ladanjska vila, *AP* 22, 80-83.

Gorenc and Damevski 1984 Marcel Gorenc and Valerija Damevski, *Rimska ladanjska vila u Benkovcu kod Okučana, in: Zbornik* radova, Četvrti znanstveni sabor Slavonije i Baranje, sv. 1, Osijek, 101-115.

Gorenc, Damevski, and Džajić 1985 Valerija Damevski, Marcel Gorenc, and Zorica Džajić, *Antičko ladanje i rimske vile u Benkovcu i Cagama – Antički arheološki lokaliteti na novogradiškom području* Exhibition Catalogue, Nova Gradiška.

Gračanin 2010 Hrvoje Gračanin, Rimske prometnice i komunikacije u kasnoantičkoj južnoj Panoniji, *SSlav* 10, 9-69.

Gregl 1991 Zoran Gregl, *Rimljani u Zagrebu*, Zagreb.

Gregl 1992 Zoran Gregl, Antičko nalazište Repišće kod Jastrebarskog, *VAMZ* 24-25 (1991-1192), 145-150.

Gregl 1994 Zoran Gregl, Antičko (rimsko) razdoblje / The Classical (Roman) Period, in: Rendić Miočević 1994, 33-36.

Gregl and Jelinčić 2010 Zoran Gregl and Kristina Jelinčić, O nekim manje poznatim antičkim lokalitetima u Zagrebu i okolici, *VAMZ* 43, 153-191.

Hanel and Suchany 1999 Norbert Hanel and Caty Suchany (eds.), *Colonia - municipium – vicus. Struktur und Entwicklung städtischer Siedlungen in Noricum, Rätien und Obergermanien, BAR Intern. Ser.* 783.

Haraša and Bekić 2007 Ivana Haraša and Luka Belić, Lokalitet: Sela, *HAG* 3 (2006), 74-75.
Hutinec 2008 Mirela Hutinec, Lokalitet: Kanal Dunav – Sava (op. Vukovar, Trpinja, Bogdanovci), *HAG* 4 (2007), 56-58.
Jelinčić 2006 Kristina Jelinčić, Lokalitet: Virovitica – Kiškorija jug, *HAG* 2, (2005), 83-87.
Jelinčić 2007 Kristina Jelinčić, Virovitica – Kiškorija jug, in: *Zapadna virovitička obilaznica, Zaštitna arheološka istraživanja 2005/2006*, Virovitica, 27-34.
Jelinčić 2009a Kristina Jelinčić, Orešac, arheološko istraživanje 2008. godine, *AIA* 5, 48-50.
Jelinčić 2009b Kristina Jelinčić, Lokalitet: Orešac – Dvorina, *HAG* 5 (2008), 149-150.
Klemenc 1938 Josip Klemenc, *Archaeologische Karte von Jugoslavien: Blatt Zagreb*, Beograd.
Knezović 2005 Ivan Knezović, Prilog antičkoj arheološkoj topografiji Đakovštine, *Zbornik Muzeja Đakovštine* 7, Đakovo, 61-90.
Koščević and Cvitković 2005 Remza Koščević and Ivan Cvitković, Lokalitet Drenje-Laduč, naselje Drenje, općina Brdovec, *HAG* 1 (2004), 107-109.
Koščević and Makjanić 1986 Remza Koščević and Rajka Makjanić, Neki noviji arheološki rezultati s područja antičke Siscije, in: *Arheološka istraživanja na karlovačkom i sisačkom području, IzdHAD* 10 (1985), Zagreb, 119 – 136.
Koščević and Makjanić 1988 Remza Koščević and Rajka Makjanić, Antički tumuli kod Velike Gorice i nova opažanja o panonskoj radionici glazirane keramike, *PIAZ* 3-4 (1986-1987), 25-70.
Kovács 2003 Péter Kovács, The roman well in the *principia* Matrica, *Antaeus* 26, 69-84.
Krznarić Škrivanko 2008 Maja Krznarić Škrivanko, Lokalitet: Kanal Dunav – Sava (op. Nuštar, Vinkovci), *HAG* 4 (2007), 58-61.
Leleković 2008 Tino Leleković, Lokalitet: Ivandvor, *HAG* 4 (2007), 12-15.
Lolić 2003 Tatjana Lolić, Colonia Flavia Siscia, in: *Pannonia* II, 131-152.
Lozuk, Minichreiter, and Miškiv 1997 Josip Lozuk, Kornelija Minichreiter, and Jesenka Miškiv, Rekognosciranje trase kanala Dunav-Sava (Dionica Šamac-Gundinci, Brodsko-posavska županija), *ObHAD* 29/2, 40-47.
Ložnjak Dizdar 2008 Daria Ložnjak Dizdar, Terenski pregled područja izgradnje HE Podsused, *AIA* 4, 109-112.
Ložnjak Dizdar, Dizdar, and Šiljeg 2004 Daria Ložnjak Dizdar, Marko Dizdar, and Bartul Šiljeg, Rezultati terenskog pregleda područja grada Iloka godine 2003., *ObHAD* 36/1, 45-50.
Makjanić 1990 Rajka Makajnić, Terra sigillata iz Orešca. Zbirka Vincek i Radijevac, *PIAZ* 7, 31-44.
Migotti 1996 Branka Migotti, Ranokršćanski grobni nalaz iz Velikih Bastaja kod Daruvara, *VAMZ* 28-29 (1995-96), 127-157.
Migotti 1997 Branka Migotti, *Evidence for Christianity in Roman Southern Pannonia (Northern Croatia). A catalogue of finds and sites, BAR Intern. Ser.* 684.
Migotti 2009 Branka Migotti, Kasna antika i rano kršćanstvo, in: V. Kusin and B. Šulc (eds.), *Slavonija, Baranja i Srijem. Vrela europske civilizacije*, Zagreb, 129-136.
Migotti *et al.* 1998 Branka Migotti, Ljubica Perinić, Mario Šlaus, and Zdenka Dukat, *Accede ad Certissiam. Antički i ranokršćanski horizont arheološkog nalazišta Štrbinci kod Đakova*, Zagreb.
Mihaljević 2006 Marija Mihaljević, Otrnci – Benkovac, in: Durman 2006, 202-203.
Minichreiter 1984 Kornelija Minichreiter, Rekognisciranja arheoloških lokaliteta Baranje, *ObHAD* 16/2, 32-36.
Minichreiter 2003 Kornelija Minichreiter, Osekovo – Ciglenice. Prva sustavna arheološka istraživanja antičkog lokaliteta u Moslavini, *Zbornik Moslavine* 5-6 (2002-2003), Kutina, 248-250.
Miškiv 2006 Jesenka Miškiv, Slavonski Brod – antika, in: Durman 2006, 238-239.
Miškiv, Miklik-Lozuk, and Lozuk 2006 Jesenka Miškiv, Lidija Miklik-Lozuk, and Josip Lozuk, *Zaštitna arheološka istraživanja, Trg Ivane Brlić-Mažuranić, Slavonski Brod 2002. i 2003.*, Slavonski Brod.
Mócsy 1974 András Mócsy, *Pannonia and Upper Moesia. A History of the Middle Danube Provinces of the Roman Empire*, London-Boston.
Mulvin 2002 Lynda Mulvin, *Late Roman Villas in the Danube-Balkan Region, BAR Intern. Ser.* 1064.
Okroša-Rožić 2006 Lana Okroša-Rožić, Novačka, in: Durman 2006, 186-187.
Ožanić Roguljić 2010 Ivana Ožanić Roguljić, Rimski nalazi s lokaliteta Liskovac – Južna obilaznica Vinkovaca, *PIAZ* 26, 79-142.
Palágyi 1995 Sylvia Palágyi, Schwerpunkte der neuen Ausgrabungen im Gutshof von Baláca, *BK* 3 (1994), 10-21.
Perinić 1998 Ljubica Perinić, Izvori o Certisiji, in: Migotti *et al.* 1998, 79-87.
Perkić 2006 Domagoj Perkić, Bubijeva jama kraj Barilovića, in: Durman 2006, 104-105.
Pinterović 1968 Danica Pinterović, Limesstudien in der Baranja und Slawonien, *AI* 9, 55-82.
Pinterović 1969 Danica Pinterović, Problemi na limesu - istraživanja na predjelu Batina Skela – Ilok, *OZ* 12, 53-69.
Pinterović 1971 Danica Pinterović, Batina - praistorijski i antički nalazi, *AP* 13, 55-58.
Pinterović 1975 Danica Pinterović, Nepoznata Slavonija, *OZ* 14-15 (1973-1975), 123-166.
Rendić-Miočević 1994 Ante Rendić-Miočević (ed.), *Zagreb prije Zagreba. Arheološka baština Zagreba od pretpovijesti do osnutka biskupije 1094. godine / Zagreb Before Zagreb. The Archaeological Heritage of Zagreb from Prehistory to the Foundation of the Bishopric in 1094*, Exhibition Catalogue, Zagreb.
Salajić 2007 Silvija Salajić, Lokalitet: Dvorina, *HAG* 3 (2006), 96- 97.

Schejbal 2004 Berislav Schejbal, Municipium Iasorum (Aquae Balissae), in: *Pannonia* II, 99-129.

Schulz 1999 Rüdger Schulz, Wirtschaftsbedingte Strukturen und Strukturwandel im Töpfervicus Rheinzabern, in: Hanel and Schucany 1999, 65-82.

Sokač-Štimac 2005 Dubravka Sokač-Štimac, Treštanovačka gradina, *HAG* 1 (2004), 48-49.

Sokač-Štimac 2006 Dubravka Sokač-Štimac, Treštanovačka gradina, *HAG* 2 (2005), 66-67.

Sokač-Štimac 2007 Dubravka Sokač-Štimac, Treštanovačka gradina, *HAG* 3 (2006), 92-93.

Sokač-Štimac 2008 Dubravka Sokač-Štimac, Treštanovačka gradina, *HAG* 4 (2007), 114-116.

Sokač-Štimac and Korać 2007 Dubravka Sokač-Štimac and Zrinka Korać, *Rimske kupke u Tekiću*, Požega.

Sokol 1981 Vladimir Sokol, Najnovija arheološka istraživanja u Prigorju, in: *Arheološka istraživanja u Zagrebu i njegovoj okolici, IzdHAD* 6, Zagreb, 169-185.

Sokol 1987 Vladimir Sokol, *Iz arheološke i starije prošlosti Zagreba – Sesvetsko Prigorje*, Sesvete.

Sokol 1994 Vladimir Sokol, Kasnoantičko i ranosrednjovjekovno razdoblje / The Late-Classical and the Early-Medieval Period, in: Rendić-Miočević 1994, 46-51.

Sokol 1996 Vladimir Sokol, *Nekropola kulture žarnih polja u Moravču kod Sesveta*, in: Arheološka istraživanja u Zagrebu i zagrebačkoj regiji i Arheologija i obnova, *IzdHAD* 17, Zagreb, 29-57.

Sokol 2006 Vladimir Sokol, Kuzelin, in: Durman 2006, 154-155.

Spaul 1994 John E. H. Spaul, *Ala: The Auxiliary Cavalry Units of the Pre-Diocletianic Imperial Roman Army*, Andover.

Szőnyi 2003 Eszter Szőnyi, Römische Brunnen in der Kleinen Tiefebene, *Antaeus* 26, 141-158.

Šarić 1977 Ivan Šarić, Antičko naselje u Petrijancu, in: *Antički gradovi i naselja u južnoj Panoniji i graničnim područjima, MatSADJ* 13, Varaždin, 49-63.

Šarić 1986 Rimski brončani materijal izvađen iz rijeke Kupe kod Karlovca, in: *Arheološka istraživanja na karlovačkom i sisačkom području, IzdHAD* 10 (1985), Zagreb, 69-90.

Šegvić 1986 Marina Šegvić, Antički kultovi u Sisku i Topuskom, in: *Arheološka istraživanja na karlovačkom i sisačkom području, IzdHAD* 10, Zagreb, *95-101*.

Šegvić 2006 Marina Šegvić, Topusko, in: Durman 2006, 270-271.

Šimek 1997 Marina Šimek (ed.), *Registar arheoloških nalaza i nalazišta sjeverozapadne Hrvatske*, Bjelovar.

Šimek 1998 Marina Šimek, Zaštitno istraživanje na trasi autoceste Zagreb-Goričan, dionica Varaždin, *ObHAD* 30/2, 59- 65.

Šiša-Vivek, Leleković, and Kalafatić 2006 Marija Šiša-Vivek, Tino Leleković, and Hrvoje Kalafatić, Ostava rimskog novca i srebrnog posuđa iz Petrijanca, *OpArch* 29, 231-245.

Škoberne, Koščević, and Makjanić 1987 Želimir Škoberne, Remza Koščević, and Rajka Makjanić, *Drenje, rezultati istraživanja 1980.-1985.*, Zagreb.

Tarpin 1999 Michel Tarpin, Colonia, municipium, vicus: Institutionen und Stadtformen, in: Hanel and Schucany 1999, 1-10.

Thomas 1964 Edit B. Thomas, *Römische Villen in Pannonien, Beiträge zur pannonischen Siedlungsgeschichte*, Budapest.

Thomas 1980 Edit B. Thomas, Villa settlements, in: *ARP*, 275-321.

Tkalčec *et al.* 2007 Tatjana Tkalčec, Snježana Karavanić, Bartul Šiljeg, and Kristina Jelinčić, Novootkrivena arheološka nalazišta uz rječicu Veliku kod mjesta Majur i Ladinec, *Cris: časopis Povijesnog društva Križevci* 9, Križevci, 5-25.

Tončinić 2005 Domagoj Tončinić, Das Stadtgebiet von Zagreb – eine Villenlandschaft in der Antike?, *BK* 9 (2004), Veszprém-Baláca, 20-23.

Valerija Damevski and Marcel Gorenc *Terenski dnevnici (izvješća) s arheoloških iskopavanja u Benkovcu i Cagama kod Okučana : 1953, 1955, 1960, 1976-1984. god.* (Arhiv AMZ / Archives of the AMZ).

Vasić 1970 Miloje Vasić, Römische Villen vom Typus der Villa rustica auf jugoslawischem Boden, *AI* 11, 52-81, pls. I-VIII.

Vikić and Gorenc 1969 Branka Vikić and Marcel Gorenc, *Prilog istraživanju antiknih naselja i puteva u sjeverozapadnoj Hrvatskoj*, Zagreb.

Vikić-Belančić 1968 Branka Vikić-Belančić, Istraživanja u Jalžabetu kao prilog upoznavanju života u zaleđu dravskog limesa, *VAMZ* 3, 75-101, pls. I-XI.

Vingo 2010 Paolo de Vingo, Shifting populations in Late Antiquity. Germanic populations, nomads and the transformation of the Pannonian limes, *AArchHung* 61/1, 261-282.

Zaninović 1967 Marin Zaninović, Neki primjeri smještaja antičkih gospodarskih zgrada u obalno-otočkom području Dalmacije, *ARR* 4-5, 357-371.

Cemeteries

Tino Leleković

1. Introduction

A comparison in the state of research of the Roman cemeteries in Pannonia as a whole reveals an obvious advantage of the northern (Hungarian) portion of the province. As far as the southern (Croatian) part is concerned, cremations of the 1st and 2nd centuries prevail in its western areas,[1] while 3rd- to 5th-century inhumation cemeteries are more typical of the eastern part.[2] Overall, although rural cemeteries are more numerous than urban, the cemeteries of the three colonies (Siscia, Mursa and Cibalae) have produced the majority of the so-far researched graves (fig. 1). Due to such imbalance in the state of research, it was not possible to apply here John Pearce's context-based classification of the cemeteries of Roman Britain, which comprises the following types: urban, of small towns, military, of villas, and rural. The evidence from northern Croatia so far allows only two basic categories of cemeteries to be identified: urban and rural. On the other hand, it was possible to apply the "British scheme" in terms of chronology and funerary ritual: early (from the later 1st century BC to the mid 2nd century AD, featuring almost exclusively cremations), transitional (from the mid 2nd century to the end of the 3rd century, with mixed cremations and inhumations) and late (4th and 5th centuries, with inhumations absolutely prevailing) cemeteries.[3]

2. Early cemeteries

The earliest evidenced cemeteries date from the beginning of the conquest of Pannonia and its subsequent integration, lasting to the mid 2nd century, when inhumations started to appear with greater frequencies. As this was also the period of intense Romanisation and urbanisation of the area under study, it should be pointed out that early cemeteries are an invaluable source for the study of these processes, the more so as early graves are often the only finds from the initial phase of Romanisation.

2.1. Funerary ritual

Funerary remains from the 1st half of the 1st century AD being exceptional, nearly all early cemeteries date from the later 1st and the 1st half of the 2nd centuries. Out of a total of 534 graves, around 200 stem from Siscia,[4] around 140 from Mursa,[5] and 129 from the rural cemetery at Stenjevec,[6] which means that these three cemeteries have yielded the majority of the finds (table 1). Recently, parts of the cemeteries of Mursa and Siscia have been investigated, with clear stratigraphies established, which has much improved our knowledge and understanding of the funerary archaeology not only of those two towns, but of the wider area. Two types of cremation ritual have been identified in the studied territory. One of them was cremation at the *ustrinum*, which implies washing the bones and isolating them from the pyre debris after the cremation, and finally depositing them in the grave called *ustrinatum*. The other ritual comprised pyre sites *in situ*, producing the *bustum* type of grave. In view of this, the first subject to be addressed is these two types of cremation rituals and the forms of graves resulting from them. Given that grave architecture and tumuli can house both *busta* and *ustrinata*, they will be tackled in separate sections.

2.1.1. Ustrinata

Such graves result from cremations in a specific place within the cemetery *(ustrinum)*, with the ashes isolated and put into an urn or wrapped in cloth, and deposited into the grave together with the grave goods.[7] *Ustrinata* are characterised by a variety of forms, due to the fact that the cremation ritual comprised of various components that changed over time, or could have been used only in certain phases of the funerary procedure.[8] Given that remains of only two pre-Flavian cemeteries have been identified so far, the funerary ritual and grave shapes of this period are difficult to assess in all their detail (fig. 2.1). Only five such graves dating from the 1st third of the 1st century have been preserved, stemming from Ilok (Cuccium), a site on the right bank of the Danube (fig. 6). The grave pits were rectangular, quite large (2 x 2 m) and regularly shaped, yielding an abundance of ceramic and glass vessels, clay-lamps and weaponry. Two of the graves feature prominently for yielding swords (*gladii*), which led the researchers to presume that the cemetery belonged to native soldiers serving in the Roman army as auxiliaries.[9] This hypothesis, however, remains inconclusive, as the poor knowledge of funerary rites of the autochthonous population in southern

1 Gregl 1989; Knez 1992; Gregl 1997; Istenič 1999; Wiewegh 2003; Gregl 2007; Božič 2008.
2 Milošević 2001,159-188.
3 Pearce 2008, 34; see also Philpott 1980, 58-59.
4 Wiewegh 2003; Leleković 2009b, 294-295.
5 Leleković 2009a, 46; Göricke-Lukić 2011,178.
6 Gregl 1989, 33.
7 Bechert 1980, 255-256; Jilek 1999, 27.
8 Abegg-Wigg 2008, 251.
9 Tomičić and Dizdar 2007, 40; Dizdar 2010.

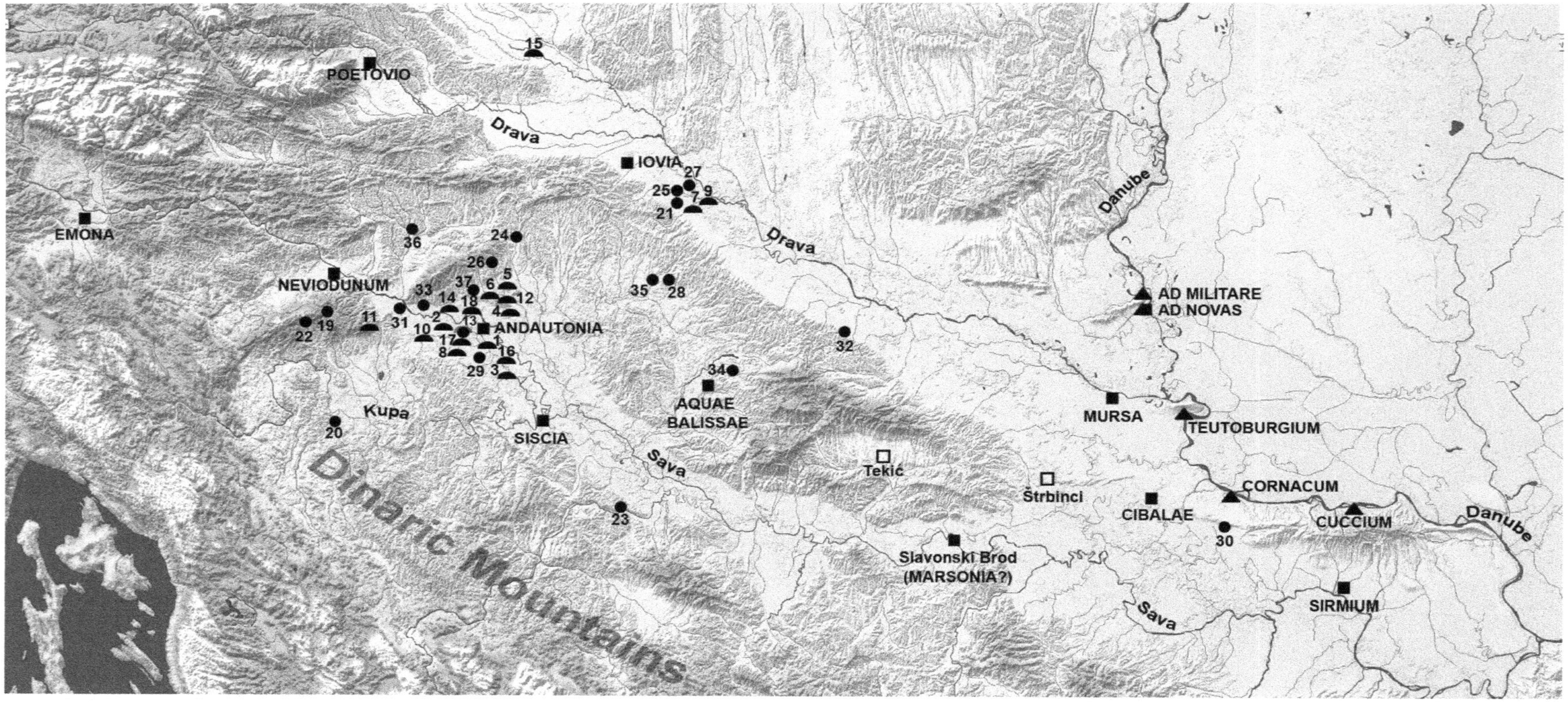

■ URBAN CEMETERIES
□ LATER ROMAN CEMETERIES WITH POSSIBLE URBAN CHARACTER
◖ TUMULI
1. Črnkovec
2. Donji Čehi
3. Donji Vukojevac
4. Dumovec-Dumovečki lug
5. Đurđekovec-Lušci
6. Glavničica
7. Gola
8. Mraclinska Dubrava
9. Novačka
10. Obrež-Brezje (Zagreb)
11. Repišće
12. Sesvete-Selčina
13. Šepkovčica
14. Trnava Resnička
15. Trnovčak
16. Turopljski Lug
17. Velika Gorica-Veliki brijeg
18. Zagreb-Borongaj

▲ CEMETERIES ON THE *LIMES*

● RURAL CEMETERIES
19. Bratelji
20. Bubijeva jama
21. Draganovec-Farkašić
22. Gornja Vas
23. Hrvatska Dubica
24. Komin
25. Kunovec Breg
26. Kuzelin
27. Mali Požari
28. Novoseljani-Ždralovi
29. Okuje
30. Rajterovo brdo
31. Novaki, Sveta Nedjelja
32. Sladojevci
33. Zagreb-Stenjevec
34. Veliki Bastaji
35. Veliko Korenovo
36. Vinagora
37. Zagreb-Držićeva, Zagreb-Maksimir

RURAL CEMETERIES WITH FLAT GRAVES AND TUMULI
43. Šepkovčica

Fig. 1. Distribution of Roman cemeteries (after Google Maps, modified by T. Leleković).

Table 1. Numbers of burials by sites.

	Burial mounds (tumuli)	Cremations	Inhumations	Total	Cremations of unknown date	Inhumations of unknown date	Early cemeteries	Transition cemeteries	Late Roman cemeteries	Total
MURSA										
Finds from unknown context		43	312	355	43	312				355
Bana Jelačića Square		98	308	406			92	314		406
120 Divaltova Street		12	70	82				79	3	82
18 Bana Jelačića Square				0						0
5 Huttlerova Street			20	20		20				20
Faculty of Medicine			6	6		6				6
19 Bana Jelačića Square			6	6		6				6
Cvjetkova Street			24	24		24				24
Hadrijanova Street / Obrtnička škola (Trade School)			4	4		4				4
18 Krstova Street			2	2		2				2
Vojarna (Barracks) / University campus			5	5		5				5
Total	0	153	757	910	43	379	92	393	3	910
SISAK										
Finds from unknown context, mostly stone chests				0						0
Sportski park (Sports Park)		166	2	168			166	2		168
Zeleni brijeg (Green Hill)			19	19		19				19
Kralja Tomislava Street			47	47					47	47
28 Gundulićeva Street (2007)		4	2	6			4	2		6
28 Gundulićeva Street (2008)		25	3	28			25	3		28
Zagrebačka Street bb (2003)				0						0
Zagrebačka Street bb (2009)		5		5			5			5
TOTAL	0	200	73	273	0	19	200	7	47	273
VINKOVCI										
Anina Street			12	12		12				12
Jurja Dalmatinca Street			20	20		20				20
Vladamira Nazora Street			27	27		27				27
Northern cemetery (unknown context)			8	8		8				8
Southern cemetery		1	72	73			1		72	73
49 Jurja Dalmatinca Street			3	3		3				3
Jurja Dalmatinca Street (2007)			6	6		6				6
32 I. G. Kovačića Street			9	9		9				9
Gundulićeva Street			5	5		5				5
Western cemetery (2009)		4	56	60	4	56				56
Kamenica			2	2				2		2
TOTAL	0	5	220	225	4	146	1	2	72	225

	Burial mounds (tumuli)	Cremations	Inhumations	Total	Cremations of unknown date	Inhumations of unknown date	Early cemeteries	Transition cemeteries	Late Roman cemeteries	Total
ŠČITARJEVO		8	3	11		3				0
DARUVAR			5	5				5±		0
TEKIĆ			129	129					129	129
ŠTRBINCI	0	0	163	163	0	0	0	0	163	163
LUDBREG				0						0
TUMULI										
ČRNKOVEC	1			1						
DONJI ČEHI	1			1						
DONJI VUKOJEVAC	1			1						
DUMOVEC-DUMOVEČKI LUG, Sesvete	35			35						
ĐURĐEKOVEC-LUŠCI	30			30						
GLAVNIČICA (near Sesvete)	4			4						
GOLA (near Koprivnice)	13			13						
MRACLINSKA DUBRAVA	24			24						
NOVAČKA (near Koprivnica)	35			35						
OBREŽ-BREZJE (Zagreb)	2			2						
REPIŠĆE	6			6						0
SESVETE-SELČINA	5			5						
ŠEPKOVČICA	1	11	36	48						
TRNAVA RESNIČKA	1			1						
TRNOVČAK (near Sv. Martin na Muri)	15			15						
TUROPLJSKI LUG	104			104						
VELIKA GORICA-VELIKI BRIJEG	0			0						
ZAGREB-BORONGAJ	1			1						
CEMETERIES ON THE LIMES										
ILOK		4	3	7			4	3		7
DALJ			1	1				1		1
ZMAJEVAC-MOCSOLÁS			175	175					175	175
RURAL CEMETERIES										
BRATELJI		4		4						
BUBIJEVA JAMA			29	29				29		29
DRAGANOVEC-FARKAŠIĆ (near Koprivnica)			1	1				1		1
GORNJA VAS	64									
HRVATSKA DUBICA			1	1				1		1
KOMIN				0						0
KUNOVEC BREG (near Koprivnica)		10		10			10			

	Burial mounds (tumuli)	Cremations	Inhumations	Total	Cremations of unknown date	Inhumations of unknown date	Early cemeteries	Transition cemeteries	Late Roman cemeteries	Total
KUZELIN			1	1				1		1
MALI POŽARI				0						0
NOVOSELJANI-ŽDRALOVI (near Bjelovar)			9	9				9		9
OKUJE										
RAJTEROVO BRDO			1	1				1		1
NOVAKI-SVETA NEDJELJA			1	1				1		1
SLADOJEVCI (near Slatina)			9	9					9	9
SLAVONSKI BROD		?		0			x			0
ZAGREB-STENJEVEC		129		129			128			128
ŠEPKOVĆICA		17		17			17			17
VELIKI BASTAJI			1	1				1		1
VELIKO KORENOVO			1	1				1		1
VINAGORA		1	1	2			1	1		2
ZAGREB-Držićeva Street		4		4			4			
ZAGREB-Maksimir			14	14						
TOTAL	343	534	1452	2265	47	544	453	448	423	2376

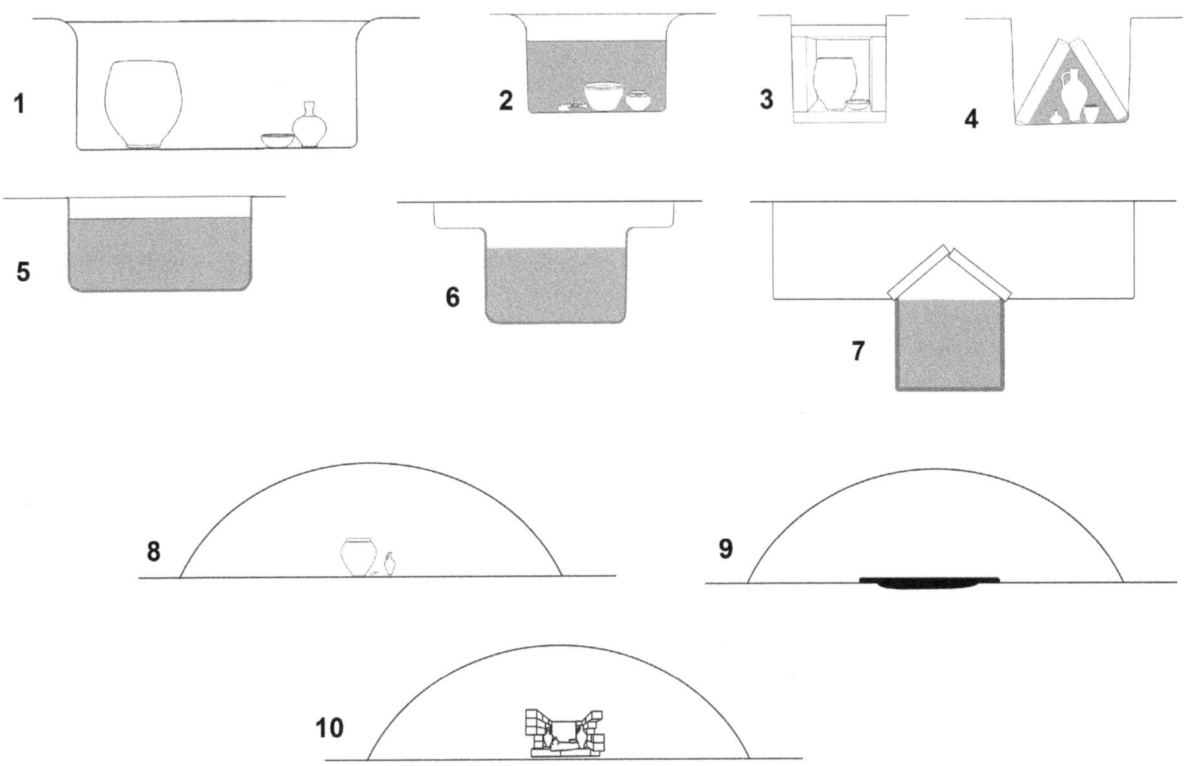

Fig. 2. Typological scheme of incineration graves (made by M. Mađerić).

Fig. 3. 28 Gundulićeva Street, Sisak, grave 6: ustrinatum with the pyre debris (photo: T. Leleković).

Pannonia prevents clear distinguishing between the native and Roman funerary habits.[10] A similar find occurred in Vinkovci (Cibalae), some 40 km west of Ilok. It contained 11 items of weaponry from the early 1st century, estimated to be local products made in imitation of Roman swords of the Augustan period. For a long time this find had been considered as a hoard of weaponry. However, in a recent reconsideration, its individual components were reinterpreted as swords stemming from destroyed cremation graves of an early Imperial cemetery.[11]

From the Flavian period on the number of graves grew, which gave rise to a variety of the forms of *ustrinata* in Pannonia as in all of the northwestern provinces.[12] Grave pits are of oval, round, or irregular form, with equally varied dimensions; predominant are fairly small pits up to 1 m long, although large and richly equipped graves more than 3-m long have been recovered, as in Andautonia.[13] An important argument for recognising grave pits is deposition of the pyre debris, known also from the majority of the northwestern provinces (fig. 2.2). This procedure has been evidenced in graves from the 1st to the 3rd centuries in the areas to the north and east of Italy, with regional and chronological variation. Accordingly, it is difficult to ascribe the habit of pyre deposition to a specific period or a population group.[14] In northwestern Croatia graves both with and without the deposition of pyre debris have been found, while the northeastern part has yielded only graves containing the pyre debris. The recent excavations of the southwestern cemetery of Siscia are quite important for a better understanding of this habit, as they have yielded two clearly separated layers from the early and mid 2nd

10 Cf. Dizdar in this volume, section 3.
11 Dizdar and Radman-Livaja 2004.
12 Jilek 1999, 27; Abegg-Wigg 2008, 251.
13 Nemeth-Ehrlich and Kušan Špalj 2007; Nemeth-Ehrlich and Kušan Špalj 2008, 199.
14 Bechert 1980; Jilek 1999, 27; Abegg-Wigg 2008, 251.

Fig. 4. 28 Gundulićeva Street, Sisak, grave 12: tile-lined ustrinatum before and after the removal of the grave goods (photo: T. Leleković).

century. The earlier layer produced graves with the pyre debris, which was absent from the later ones; this should be taken as the proof that graves containing the pyre debris were an early phenomenon, which died out by the mid 2nd century.[15] The excavations in Sisak have brought to light another feature: two forms of *ustrinata* constructed from tiles with no bondage, the first one resembling a box (fig. 2.3). The most frequent type of the box grave is rectangular in shape, measuring 0.60 x 0.60 x 0.30 m, its possible north-Italian origin being suggested by the frequency of comparable finds in north-Italian cemeteries. (fig. 4)[16] It should be noted, however, that this type of grave is exclusive to Pannonia Superior, while absent from Pannonia Inferior.[17] The latter area produced only elongated rectangular graves constructed from tiles with no bondage, which, however, only appeared in 3rd-century transitional cemeteries (see section 3.1.2.). Another type of early *ustrinata* is represented by box-shaped receptacles topped with a gable-wise arrangement of tiles. This type is otherwise frequent in the western provinces, but with some difference in comparison to Pannonia. While in Gallia and Germania grave pits were covered by a single tile or stone slab, graves composed of two tiles set gable-wise are typical of Pannonia and Noricum, and are often found in Aquincum, Savaria, Brigetio, and Carnuntum. (fig. 2.4)[18] Gabled *ustrinata* have been documented only in the southwestern cemetery of Siscia, with one grave standing out through its particular construction, in that its walls and floor remained bare of tiles, which, on the other hand, were used only for the gabled roof; the grave contained the pyre deposit.[19] Curiously, this type of grave is most frequently found in Carnuntum.[20] The remainder of the gabled boxes had their interiors lined with tiles, while lacking the pyre debris.

The cinerary urn, as a component of the funerary ritual, has not been frequently used in the Roman cemeteries of northern Croatia. In the majority of urned cremations the urn was placed in a pit; such graves both with and without the pyre debris have been evidenced. Apart from these, tile-constructed urned graves have been found, but they never contained the pyre debris. Such graves are otherwise typical of north-Italian cemeteries,[21] suggesting a funerary ritual influenced by Italian immigrants in Pannonia.[22] Urned graves in wooden boxes have until now been found only in the eastern cemetery of Mursa and the south-eastern cemetery of Siscia; these were pits with wood planks lining the walls and covering the opening; some of them produced iron nails. (fig. 5)[23] While the nails might have been part of the grave construction, this remains inconclusive as they were mostly found in the pyre debris, which means that they might equally have been parts of wooden coffins or biers. Stone urns are very rare finds in the region of study, and have so far been attested only in Siscia. (fig. 7) Since they were invariably made of low-quality local limestone, none has been completely preserved. As a rule, the stone urn contained both bones and grave goods: first, the larger fragments of (washed) bones were isolated and deposited at the bottom with the grave goods placed on top, to be finally closed with a lid. The urn of one such grave in the southwestern Siscian cemetery contained silver earrings, coins, a bronze-sheet coil, ceramic jugs and a lamp, and

[15] Leleković 2009b, 294-295; Leleković 2011.
[16] Gregl 1989, 13; Leleković 2009b, 295; see also Negrelli 2006, 70.
[17] Jovanović 1984, 39.
[18] Jilek 1999, 26-27.
[19] Leleković 2009b, 294-295.
[20] Jilek 1999, 26-27.
[21] For instance, at Verginese in the vicinity of Ferrara (Negrelli 2006, 93).
[22] Jovanović 1981,147; Gregl 1989, 13.
[23] Leleković 2009b.

Fig. 5. Bana Jelačića Square, Osijek, grave 103: enclosed ustrinatum (photo: A. Štefan).

Fig. 6. Ilok, ustrinatum, 1st half of the 1st c. (Photo Archive of the Institute of Archaeology in Zagreb).

a glass *arybalos*. The urn was placed in a rectangular pit filled with earth, and next to it was a ceramic censer. In the pits holding stone urns the pyre debris has never been evidenced.[24]

It is *ustrinata* that most extensively illustrate the scope and variety of grave goods, especially as concerns ceramic artefacts, as they usually survive undamaged. In spite of the lack of any uniformity in grave inventories, some basic characteristics can still be established, the most conspicuous of them being a gradual disappearance of vessels for serving solid foods. In early cemeteries vessels for liquids and solid foods are equally represented, while in the transitional phase drinking vessels predominate, to remain practically the only type of vessel as a grave good in late antique graves.[25]

2.1.2. Busta

Such graves are characterised by the fact that both the cremation and deposition were executed on the same place, which enables easy differentiation from the *ustrina*; *busta* feature burnt pit walls and their fill is composed of the pyre debris mixed with the human and animal bone, as well as the burnt and melted grave goods. Only very exceptionally have undamaged grave goods been found in such graves. Pits are usually of an elongated rectangular form, although some were sub-square, oval or totally irregular; the walls were mostly straight and perpendicular to the bottom, and only rarely were they slanted. As a rule, traces of burning are unequally distributed, being visible mostly on the bottom and in the upper portions of the walls. The rest of the walls usually feature only slight traces of burning, and are only exceptionally burnt on the whole surface (fig. 2.5-6).[26]

24 Leleković 2009b, 294-295.
25 See Istenič 1999, 83-166; Wiewegh 2003, 41-52; Božič 2008.
26 Gaitzsch and Werner 1993, 57-58.

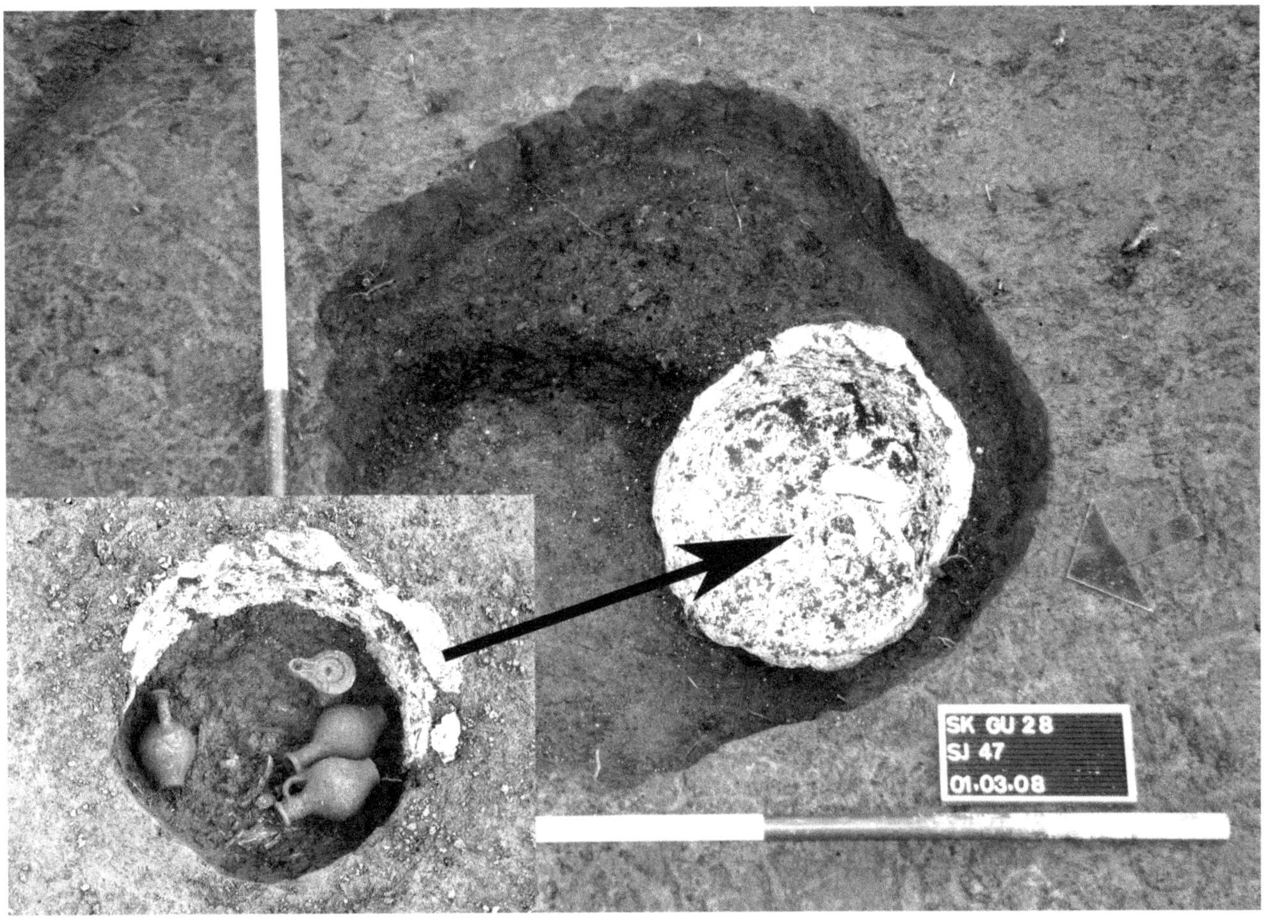

Fig. 7. 28 Gundulićeva Street, Sisak, grave 8: stone-urned ustrinatum (photo: T. Leleković).

The origin and interpretation of the *bustum* type of grave have not yet been satisfactorily explained. One of the theories postulates that such graves originated in north Italy and that during the 1st century AD legionaries spread them to the provinces, especially those of the Rhineland and the Danube/Balkan regions. On the other hand, given a strong concentration of the troupes from the East along the Rhine and Danube, it is possible to perceive of the *bustum* as a funerary feature imported to the West by the army and oriental immigrants.[27] According to Aleksandar Jovanović, the share of *busta* in the cemeteries of Moesia and southeastern Pannonia (the territory of Sirmium in this case) is from 60% to 90%, which distinguishes these areas from the rest of the Empire. On account of that, Jovanović believed that *busta* originated among the native Balkan (south-east Pannonia included) ethnic communities, to be subsequently spread to the rest of the Danube area. This theory has been further substantiated by the presumption that it was exactly the troupes stationed in the 1st and 2nd centuries in the Middle and Lower Danube that brought the *bustum* type of graves to the Rhineland and the rest of the Empire.[28] Jovanović does not give the chronology of the cemeteries featuring a large proportion of *busta*, which should be important in the light of the fact that the *bustum* became the predominant type of grave in the transitional phase (see section 3.1.2). This means that the claim for the autochthonous nature of this funerary ritual should be reconsidered by establishing the exact number of *busta* dating from the early and transitional phases, respectively. A possible predominance of *busta* in the transitional phase would seriously challenge the claim for their Balkan-Danube origin. Whichever the case, none of the *busta* from northern Croatia can be dated to the 1st century, suggesting that they were imported to the region at the end of the 1st century at the earliest.[29]

So far *busta* of the end of the 1st and the 2nd centuries have been found only in the eastern part of the territory of study, at the sites of Slavonski Brod[30], Osijek[31], Sotin,[32] and Batina.[33] Altogether four types of *busta* can be isolated, two classified by the grave shape and one each by the

27 Struck 1993b, 90.
28 Jovanović 2000, 205-206; Pirling 2002, 524-526.

29 In 2004 and 2005, 66 graves from the end of the 1st century were excavated in Győr (Hungary), the majority of which were busta. The results have not yet been published extensively, but it transpires from the preliminary report (Bíró 2007, 40-54) that this cemetery shares many common characteristics with the earliest phase of the east cemetery of Mursa, one of them being the fact that both contained burials of men and women.
30 Miškiv 2006, 6-9.
31 Göricke-Lukić 2000, 17; Leleković 2009, 46; Göricke-Lukić 2010, 26-27.
32 Hutinec 2010, 29-33.
33 Bojčić *et al.* 2010, 80-87.

Fig. 8. Bana Jelačića Square, Osijek, grave 31: bustum, 1st half of the 2nd c. (photo: T. Leleković).

Fig. 9. Bana Jelačića Square, Osijek, grave 392: bustum with the enlarged upper section, 1st half of the 2nd c. (photo: T. Leleković).

ritual procedure and grave goods. The most numerous are «classic» *busta* in oval or rectangular pits (fig. 2.5; fig. 8), as well as those whose pit narrows towards the bottom through steps cut in the upper walls of the two longer sides, or all four sides (fig. 2.6; fig. 9). The third type is represented by the *busta* in which larger bones were separated from the cremation remains and placed either on the cleared part of the bottom of the pit or on the burnt fill (fig. 2.7). In some examples the grave goods underwent the same procedure, possibly suggesting that some pyres were equipped with an iron (?) mobile construction that should have stabilised the pyre and prevent the larger bones and goods from falling into the pit. The fourth type is distinguished by the *busta* yielding whole vessels, mostly a jug or a cup, as grave goods (fig.6). Most often typical 3rd-century two-handled cups are found in such graves, indicating their origin in the transitional period.[34] This further complicates the issue of the chronology of this type of *bustum*, especially in the Croatian part of Pannonia. The majority of the Rhineland *bust*a date from the mid 1st century, and are mostly associated with military sites on the *limes*, while some finds, like those from the *tumuli* cemetery at Hunstrück-Eife, have been dated to the mid 2nd century (see section 3.1.2).[35] In the Danube provinces *busta* are considered as an autochthonous

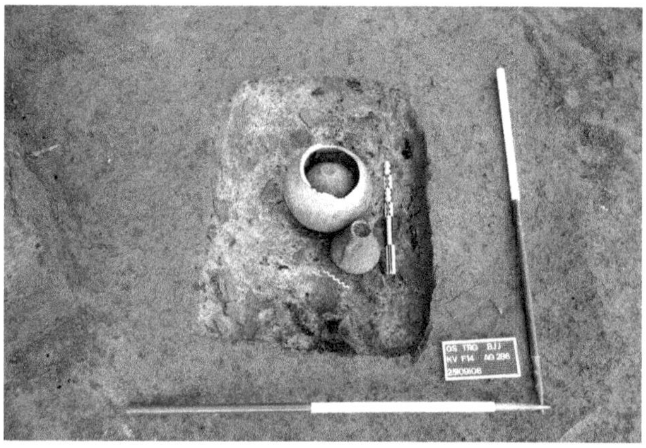

Fig. 10. Bana Jelačića Square, Osijek, grave 286: bustum with ceramic vessels, turn of the 1st and 2nd c. (photo: V. Mesarić).

[34] Brukner 1981, 41.
[35] Wigg 1993, 15-17, 107.

Fig. 11. 28 Gundulićeva Street, Sisak: grave enclosures in the southwest cemetery – aerial view (photo: T. Leleković).

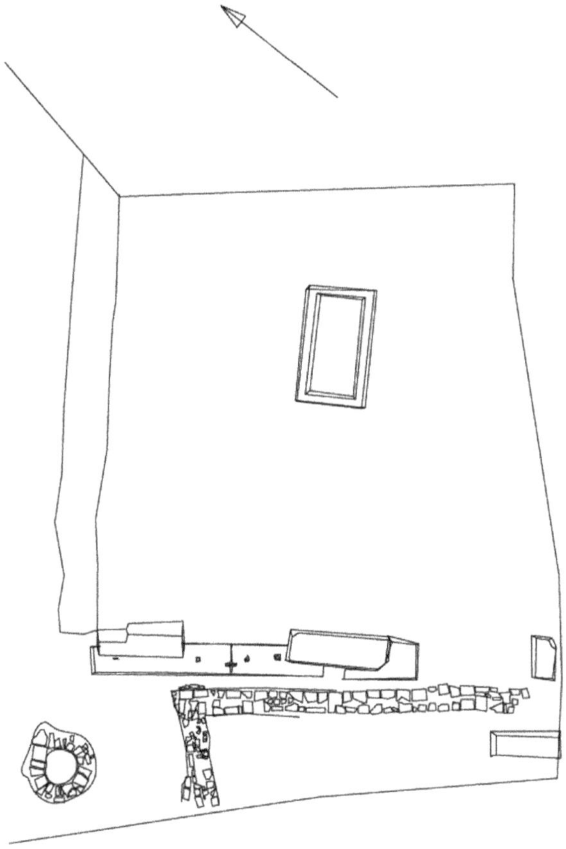

Fig. 12. 28 Gundulićeva Street, Sisak: grave enclosures in the southwest cemetery (drawing: T. Leleković).

feature dating from the pre-Roman period.[36] The earliest examples from northern Croatia come from Mursa and are dated to the turn of the 1st and 2nd centuries. Given that *busta* dating from the 2nd half of the 3rd century have also been evidenced in Mursa, on present evidence the chronological frame for this type of grave appears to be from the later 1st to the later 3rd centuries.[37]

2.1.3. Grave architecture

Two types of grave architecture are representative of the early phase: grave enclosures and individual tombs. Two enclosures were recovered in the southwestern Siscian cemetery (figs.11 and 12). One of them was enclosed by a brick-wall, with the mortared tile-constructed tomb measuring 2.20 x 2.16 m and preserved to the height of 1.40 m, stood in its middle. A rectangular stone urn was found in the tomb, holding burnt human bone and lavish grave goods: two *balsamaria*, two glass jugs, amber and gold jewellery, and four stamped clay lamps. The tomb was furnished with a niche containing burnt bone and remains of the pyre debris, testifying to a multi-burial construction. Finally, the skeleton of a dog was found on the floor, but its possible role in terms of funerary rite remains unclear.[38] A hypothetical reconstruction of this tomb on the basis of

36 Jovanović 2000, 205-206.
37 Leleković 2009a; Leleković 2010; Göricke-Lukić 2011, 178-181.
38 Pavleković and Škrgulja 2008, 17-18.

Fig. 13. Plan of Siscia (after Google Earth, modified by T. Leleković).

comparable Norican and north-Italian finds[39] suggests that the former was not an entirely belowground structure; the belowground grave chamber could have been surmounted by an aboveground architectural construction such as an aedicule or a similar feature holding portraits of the deceased and an epitaph. The other enclosure was encircled by a wall made of large stone blocks, with a stone chest in its middle[40]; its contents, as well as the lid, were found missing, and the chest bore no carvings or inscription. Behind the two enclosures, smaller tombs were found. Given that the enclosures were quite at a distance from the town gate, and in spite of the fact that the extent of the western Siscian suburb is generally unknown, it is possible to assume that the 600-m long section of the southwestern Siscian cemetery stretched along the road, holding several tens or even a hundred such monumental tombs (fig. 13). Individual simple tombs of the 1st and 2nd centuries have been found only in the southwestern cemetery of Siscia, located on an elevated ground behind the two enclosures. They were of a simple rectangular ground plan measuring approximately 1.30 by 1 m; one of the walls had an opening that was closed with a tile or a stone slab. The tombs were constructed in carefully cut rectangular pits with flat bottoms and with no foundations. The bottom was paved with tiles, whose function was both to cover the bottom and to support the tile walls. Apart from the paving tiles, large pieces were found lying on the bottom and by the walls, probably belonging to the construction of the cover/roof; nevertheless, they do not furnish enough elements for the reconstruction of the roofing. The last type of grave architecture is represented by *tumuli* - round masonry tombs, featuring 3 sub-types: 1. with no opening; 2. with the opening (doors); 3. with the doors and an entering passage (*dromos*). In northern Croatia such graves are dated from the 1st to the mid 2nd centuries, and are considered as a funerary feature of the autochthonous population of the area.[41]

2.1.4. Grave mounds (tumuli)

The tumulus burial is a funerary feature established in only two relatively small areas of northern Croatia. Of a total of 17 cemeteries, yielding altogether 279 tumuli, no fewer than 12 (with 216 tumuli) have been discovered in the area

[39] Luxurious examples of such tombs are known, among other places, in Aquileia in Italy (Tirelli 1998, 147) and Šempeter in Noricum (Kremer 2001, 13-54.)

[40] See notes 106 and 107.

[41] Gregl 2000, 166-167; Gregl 2007, 226-227.

of Zagreb, that is, along the River Sava and on the slopes of the nearby mountains of Medvednica and Vukomeričko gorje. In the Podravina region, next to the border with Hungary, three cemeteries containing altogether 63 tumuli have been found (see table 1).[42] All these finds are part of a wider phenomenon typical of the eponymous provinces of Noricum and Pannonia (*Norico-Pannonian tumuli*). The tumulus burial most probably originated in the 1st century in the modern-day Austrian regions of Styria and the south of Burgenland; from the later 1st till the mid 3rd centuries[43] this phenomenon spread further, covering the wider areas of eastern Austria, eastern Slovenia, southwestern Hungary, and northwestern Croatia.[44] Arguably, the Croatian grave mounds represent the southeastern peripheral branch of the Norico-Pannonian tumuli.

So far seven north-Croatian tumuli cemeteries have been investigated, yielding altogether 55 tumuli. Of a total of 13 tumuli found at the village of Gola to the east of Koprivnica, five have been actually excavated, with only two producing finds, while the remainder had been previously destroyed by ploughing. In one of the investigated tumuli a rectangular gabled stone urn was found, containing burnt human bone and grave goods. Another tumulus was composed of a flat *bustum* containing the pyre debris, remains of the cremated bodies and grave goods, covered with earth heaped up above. Both tumuli were dated to the late 2nd and the 1st half of the 3rd centuries by Westerndorf *sigillata* and the coins of the emperor Maximinus Thrax.[45] At the site of Novačka in the vicinity of Gola another cemetery of 35 tumuli was found, of which 21 were excavated. The tumuli varied in size from six to 15 m in diameter, while the preserved height was from 0.50 to 1.70 m. All of them were constructed by heaping up earth over the earmarked surface holding the cremated remains of the bodies and grave goods. However, the manner in which the remains were deposited exemplifies two main types of tumuli. The first one is characterised by an urn holding the cremated bone, placed on the burial spot together with the grave goods and the pyre debris (fig. 2.8). The other type is represented by the tumuli holding a mixture of the cremated bone, grave goods and pyre debris, all deposited directly on the burial spot (fig. 2.9). In only two examples of the second type were shallow pits found, containing the pyre debris mixed with the cremated bone and grave goods, possibly interpretable as *busta.* This type of tumuli dates from the later 2nd century to the 2nd half of the 3rd century (two of them were dated by coins of Caracalla and Gordian III).[46] Two cemeteries have been investigated in the eastern districts of Zagreb; the first one at the site of Sesvete-Dumovec produced 35 tumuli, four of which have been excavated. They were of equivalent dimensions (10 m in diameter and 1.5 m of the preserved height), with one slightly smaller, but significant for yielding a grave chamber. It was looted by the local inhabitants in the 1950s and the chamber was damaged. Nevertheless, enough of the construction survived to reveal the measurements (length: 2.40 m; width: 1.70 m; height: 0.70 m) and plastering of the walls; bronze fittings of a wooden chest and a stone urn were found in the chamber. The remaining three tumuli lacked chambers, and were constructed as those from Novačka, that is, by heaping up earth over the burial spot, which held the cremated remains of the bodies and grave goods. Two of the three tumuli each housed one central grave, while the third had two graves.[47] In the forest resort of Turopoljski lug, a site numbering 104 tumuli, six mounds have been investigated, ranging in measurements from five to 18 m in diameter and from 0.30 to 1.50 m in height. Two larger tumuli held poorly preserved tile-constructed and masonry grave chambers, while four smaller tumuli were constructed over the remains of the pyre debris. On the basis of small finds, the tumuli can be dated from the mid 1st to the early 3rd centuries.[48] Smaller clusters of tumuli recently found at the villages of Gornji Vukojevac[49] and Šepkovčica[50] south of Zagreb can be added to the above sites. Both were rural cemeteries holding cremated graves; there is no knowing whether they actually featured tumuli that could have been levelled by ploughing. If this is so, these tumuli can be defined as chamber-less mounds constructed over pit graves containing the pyre debris and featuring a certain peculiarity in the area. At the site of Donji Čehi to the south of Zagreb, three tumuli have been excavated as probably a part of a larger cemetery, whose importance is threefold. Firstly, a stone stele was found there during building works, testifying that tombstones made part of such constructions (fig. 2.10). Secondly, in tumulus 1 a coin of Tiberius was found, rendering it the earliest tumulus so far in northern Croatia. Thirdly, tumulus 2 testified to a *dromos* construction, as an exceptional example in northern Croatia.[51]

With few exceptions, the results of the excavations of Norico-Pannonian tumuli in northern Croatia have not been fully published, which hinders their detailed analysis. Even though, three main types of tumuli can be established with certainty: 1. with grave chamber housing the ashes and grave goods; 2. with deposited ashes and grave goods with or without (stone) urns; 3. with flat *busta* holding a mixture of the pure remains and un-selected remains of the cremated bodies. The differential chronologies of the tumuli cannot at this stage be established with any certainty; possibly,

[42] Gregl 1997,14-18.

[43] Similar type of burial has in the same period been noted in other parts of the Empire. For instance, in the valley of the River Moselle in the rear of the Rhine limes, of a total of 233 tumuli 55 date from the 2nd half of the 1st century and 44 from the 2nd to the mid 3rd centuries, while the remaining 99 tumuli have not been researched; arguably, the highest frequency of this funerary custom falls in the 1st century (Wigg 1993, 15-17, 107).

[44] Mikl-Curk 1997, 32-36; Palágyi 2003, 257-261; Palágyi and Nagy 2003, 50-51,154.

[45] Demo 1985, 119-120; Šarić 1986, 113, 116.

[46] Šarić 1979; Šarić 1990, 111-113; Okroša-Rožić 2006, 186-187.

[47] Sokol 1981, 169-178.

[48] Koščević and Makjanić 1988, 35. According to the authors, the tumuli lacking grave architecture do not enable differentiation between the pyre site and the place of deposition. Namely, it is possible that the red-hot pyre debris, mixed with fragments of the grave goods and burnt bone of the deceased, were redeposited from the cremation to the deposition spot, to be finally covered with earth in the form of tumulus.

[49] Dizdar, Tonc, and Ložnjak Dizdar 2011.

[50] Bugar 2009, 269, 281.

[51] Gregl 2000, 166-170; On the stele see Migotti 2008.

types 1 and 2 date from the 1st and 2nd centuries, while type 3 might be somewhat later (2nd half of the 2nd and 1st half of the 3rd centuries). Equally questionable is the possibility to classify the cemeteries in clearly defined typological groups, as the available evidence and documentation show that the tumuli cemeteries in the region of study do not share unified forms or organisational patterns. On the contrary, they are quite heterogeneous, sharing this characteristic with the rest of the province of Pannonia.[52] Some of the cemeteries form compact clusters with all or some of the tumuli lined along a road and arranged in one or more regular rows. One such example is the cemetery at Sesvete-Dumovec, whose 22 tumuli of a total of 35 are arranged in two rows, with the remainder unevenly scattered in the surroundings. Some of the cemeteries reveal irregular and inconsistent layouts, such as the one at Đurđekovec, whose 30 tumuli are dispersed irregularly across the burial area.[53] Sites with a lager number of tumuli, exemplified by Turopoljski lug and Novačka, feature smaller clusters of mounds dispersed across a broader space. The distance between some of the groups sometimes amounts to as much as 500 m, rendering it difficult to judge whether they formed one or more cemeteries. An example of this is the site of Turopoljski lug, whose 104 tumuli are arranged in ten clusters, with individual groups numbering from two to 24 tumuli, and the distance between them measuring several hundred metres. The available documentation reveals that some clusters of tumuli were arranged in rows, while others were scattered with no clear elements of organisation.[54] The same is true of the cemetery at Novačka, whose 35 tumuli were arranged in five clusters. Overall, it is impossible to establish the precise spatial relationships between the tumuli cemeteries on the one hand and the adjacent roads and settlements on the other. Nevertheless, the circumstances in the rest of the province, with tumuli mostly lining the roads at a distance of approximately 500 m from a related settlement or villa,[55] can be presumed for the north-Croatian cemeteries as well.

The origin of the tumulus burial remains a controversial issue; given the limited amount of the published material, the scrappy evidence from northern Croatia cannot contribute significantly towards clearing this controversy. Two confronted theories of the origin of the tumulus burial exist among the Croatian Roman archaeologists, as elsewhere. The first perceives the tumulus as a local custom, while the other posits that the tumulus burial resulted from the Romanisation of the local elites.[56] In view of this, it is significant that the majority of the north-Croatian tumuli stem from the presumed territory of the *municipium* Andautonia. At the same time, no such cemetery has been discovered in the presumed Siscian *ager*, hinting at the possibility of setting the territorial boundary between these two towns exactly on the basis of the distribution of types of burial.[57] On balance, it can be hypothesised that the north-Croatian Norico-Pannonian tumuli are characteristic of the cemeteries of rural settlements of the autochthonous origin, which explains their absence from the areas holding lands assigned to colonists and veterans. If to this we add the fact that the majority of the tumuli are dated to the 2nd and 3rd centuries, the phenomenon of the Roman tumulus in northern Croatia indeed seems to be connected with the Romanised local elites.[58]

2.2. *Urban cemeteries*

The study of the early phase of the urban cemeteries is much impacted by the fact that 60% of the finds come from Siscia and Mursa. This is not surprising, as these two towns were the only early *coloniae* and at the same time the only major towns in the territory of study. In the 1st and 2nd centuries the status of *municipium* was bestowed on Andautonia and Aquae Balissae, and probably also on Cibalae, but these towns produced only an insignificant amount of early burials. Therefore, the discussion of the layout and organisation of the early cemeteries, as well as their relation to the perimeters of their respective towns, has been sourced mainly from Mursa and Siscia.

Given that during Vespasian's rule Siscia became a colony and one of the leading urban centres and a starting point for the Romanisation of southern Pannonia, it should also bee a key site for the study of the early Roman cemeteries of northern Croatia.[59] Yet, the Siscia of the 1st century is very poorly known in terms of its dimensions and urban features, as well as the population number.[60] On the other hand, it is exactly its early cemeteries that have been researched in a measure allowing at least some study of the town in the 1st century. Cremation graves have been found on a fairly vast area comprising some 500 ha on the both banks of the River Kupa. Three foci of the concentration of cremation graves and tombstones have been established: north and southeast of the town walls, as well as southwest from the town, on the right bank of the Kupa. Even though, the exact extent of the area taken by the Siscian cemeteries has not been defined (fig. 13). The core of the northern cemetery was probably at the site of Zeleni brijeg, with further stretching towards north and northwest, but with northwestern limits unknown; it was bounded by the Rivers Kupa and Odra in the south and west, respectively. The southeastern cemetery was bounded by the rivers Kupa and Sava, while the cemeterial territory in the southwestern extraurban area, south of the Kupa, is more difficult to establish,

52 Palágyi and Nagy 2003, 53.
53 Sokol 1981, 169-170.
54 Koščević and Makjanić 1988, 26-27; Gregl 1997, 19.
55 Palágyi and Nagy 2003, 51-54.
56 Koščević and Makjanić 1988, 35; Gregl 1997, 23.
57 The administrative boundaries of Roman towns mostly remain in the realm of more or less plausible hypotheses. In northern Croatia this is particularly true of the boundary between Andautonia and Siscia, since the natural features of their respective terrains give no clues; on various opinions regarding the territorial boundaries between these two towns see Kušan Špalj and Nemeth-Ehrlich 2003, 107-110; Lolić 2003, 131-133.
58 On the settlement and the administrative organisation of this area see Mocsy 1974, 71-73.
59 Mócsy 1974, 112-114, 135.
60 Lolić 2003, 138-140.

with the Kupa still figuring as a broad boundary. [61]These three areas are consistent with the main town cemeteries, which, in their turn, correspond to the presumed routes of the main roads leading towards Siscia. Moreover, it is exactly owing to the established cemetery areas that the sections of the roads in the suburbs can be hypothesised. The northern cemetery grew along the road towards the station Ad Fines (the village of Buševec?) to the northwest of Siscia, near the place where the roads towards Emona and Neviodunum branched off from the road leading to Andautonia and Poetovio.[62] Along a still hypothesised route towards Ad Fines three separate clusters of cremation graves have been excavated, most probably sections of the same (northern) Siscian cemetery.[63] Regrettably, the results of the excavations have not been published. To the southwest of the town, on the right bank of the Kupa, grew a cemetery known in the literature as the "cemetery in Novi Sisak". It spread along the southern road heading towards Senia on the Adriatic coast, with an eastern branch leading towards Sirmium.[64] This cemetery had since long been outstanding for yielding a wealth of chance finds, among them stone sarcophagi.[65] Its importance has recently been further enhanced by excavations in 2007 and 2008, as it was for the first time that larger sections of a Siscian cemetery were disclosed and chronological relationships between various types of cremations established. These excavations are important for the study of early cemeteries, because they proved the existence of two chronological layers of burial, both belonging to the early phase (fig. 17). The first layer, dating from the early 2nd century, is represented by cremation earth pits, filled with pyres debris. The cemetery area was enclosed by a wooden fence, which followed the route of the Roman road at a distance of some 30-35 m. However, the limited scope of the excavations did not allow establishing whether the fence encircled the whole of the cemetery or just a larger funerary enclosure measuring 35 by 30 m (135 by 100 Roman ft.). The second layer, starting in the mid 2nd century, reshaped thoroughly the funerary area. Monumental enclosures dominating the cemetery grew next to the road, while in their rear smaller tombs of two types were put up: mortared tile-built tombs and chests constructed from rectangular tiles measuring 0.60 x 0.60 x 0.40 m. Unlike the earlier ones, the graves of the second layer contained no traces of the pyre debris. Among the finds from the southwestern cemetery an *ustrinum* stands out as a rare funerary feature documented in north-Croatian territory. It was a rectangular building measuring 1.80 x 2.10 m, and surviving only to the height of 0.50 m, which leaves its reconstruction in the realm of hypothesis. The fill was composed of a large amount of pottery fragments and crushed tiles, revealing traces of intense burning. Such traces were present also on the surface around the building, while to the south of it a thick deposit of ashes was evidenced.[66]

The southeastern cemetery grew along the road crossing the River Sava and leading towards Varianae (Kutina?) and further east. It was this cemetery that in the 1954 rescue excavations yielded the largest number (166) of early cremation graves so far in the Croatian portion of Pannonia. Although no phases could have been confirmed in terms of stratigraphy, they can be presumed on the basis of grave shapes. Overall, the remains of the three cemeteries discussed above can be taken as an indication of a gradual progress of the town, the influx of Italic immigrants and a profound Romanisation of its native citizens. The area to the east of the town walls remains completely unexplored, but it has nevertheless yielded an unknown number of cremation graves. Apparently, this cemetery did not grow along the main road, but a vicinal one that possibly connected the town with an as yet undiscovered settlement or a harbour on the Sava.[67]

Unlike Siscia, where all so far evidenced graves date to the period of the colony, nearly all cremations from Mursa predate the town's elevation to the rank of colony. The existence of a 1st-century fort or fortress has been presumed since long, but has never been archaeologically confirmed.[68] Yet, at the site of Bana Jelačića Square in the eastern section of Mursa, 100 cremation graves from the 2nd half of the 1st century and the 1st half of the 2nd century have been excavated so far, yielding as much as 60% of *busta*.[69] With this percentage Mursa figures among sites that can rightly claim the *bustum* as a local feature in terms of A. Jovanović's theory. If, on the other hand, the theory of the *bustum* as a military funerary custom is correct,[70] the abovementioned cemetery can be claimed as proof that Mursa was a military post, with the cremations of women suggesting an accompanying civilian settlement.[71] Among the most important of the graves are the *ustrinata* that yielded bronze wine strainers, otherwise typical of the

[61] The area on the right bank of the Kupa was in the 18th and 19th century known as Vojni (Military) Sisak, and from the mid 19th century on as Novi (New) Sisak. In New Sisak several toponyms exist that refer to particular archaeological sites, but are in the literature often used imprecisely and misleadingly. I would therefore like to use the opportunity to give precise locations and usage of these names, in order to avoid possible misunderstandings. The northern section of New Sisak, a plateau to the south of the right bank of the Kupa, is called Pogorelac. Before the arrival of the Romans, it was the site of the Celtic settlement Segesta, while in the Roman period it probably held an industrial suburb of the same name (see Radman Livaja 2007, 169-170). To the south of Pogorelec is a hilly area called Zibel, with roads running through the valleys, of which the most important is Gundulićeva Street, as it was the site of the southwest cemetery in the Roman period. In the literature, the sites of Pogorelac and Zibel are often equated and used imprecisely, with the most often mistake being the use of the name Pogorelac to cover Zibel as well. However, only the correct use of these two toponyms enables a meaningful study of the Siscian archaeological topography.

[62] Lolić 2004, 132; Gračanin 2010, 16.

[63] Burkowsky 1993, 78.

[64] Wiewegh 2003.The road network as brought in the Barrington atlas (Talbert 2000, 20-21) has been used. On the roads see also Durman 1992, 122-124; Lolić 2003, 132-133.

[65] Migotti 2007b; Vukelić 2007; Vukelić 2009.

[66] Leleković 2009b, 295-296.

[67] Nenadić 1987, 84-93; Burkowsky 1993, 74; Wiewegh 2003, 31.

[68] Klemenc 1928, 272; Spaul 2000, 265; Lőrincz 2001, 21; Pinterović 1977, 103-104, 112-113; Filipović 2004, 157; Dizdar and Radman-Livaja 2004, 45.

[69] Göricke-Lukić 2000, 152-154; Leleković 2009, 46; Göricke-Lukić 2011, 26, 229.

[70] See note 28.

[71] Leleković 2009, 46.

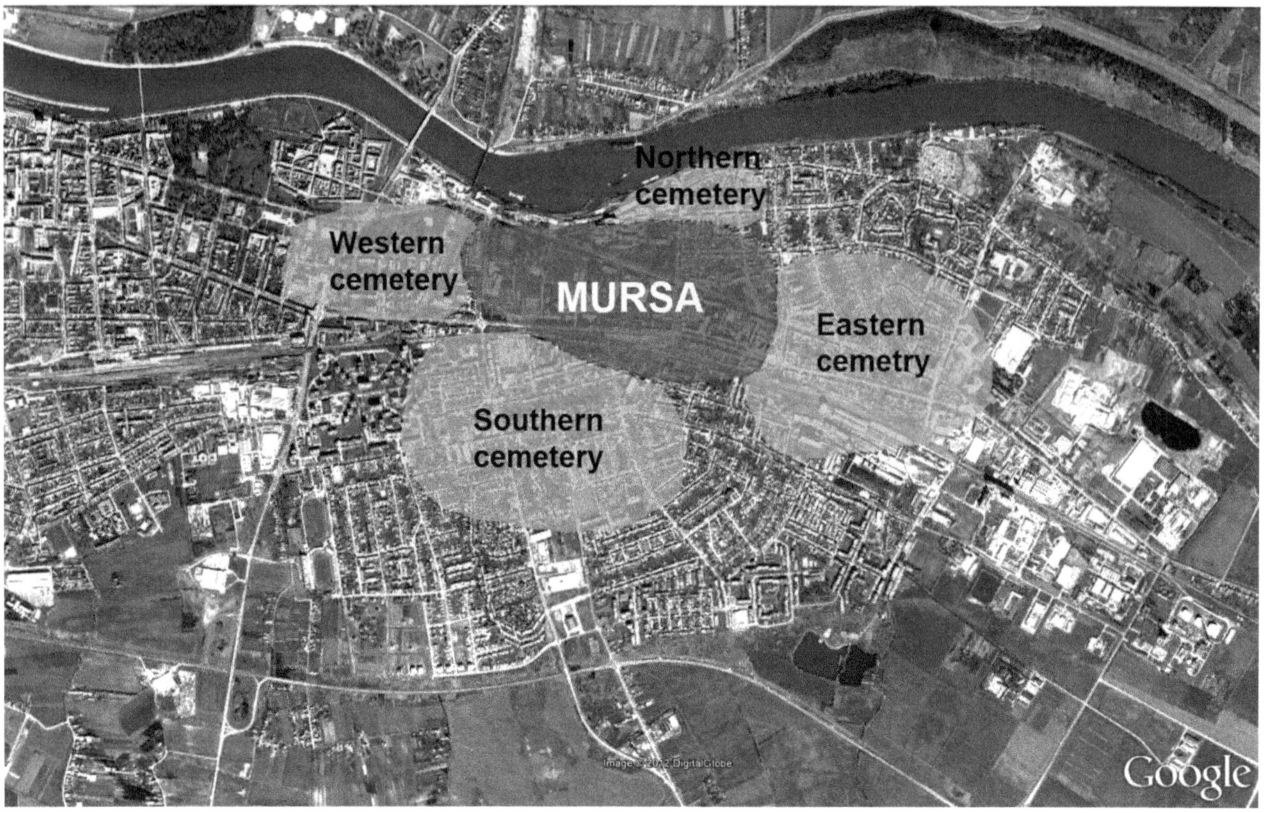

Fig. 14. Plan of Mursa in the Severan period (after Google Earth, modified by T. Leleković).

material culture of the Scordisci and testifying to the burials of natives. [72]

It seems that the *colonia* Mursa was planned so that the *cardo* and *decumanus* connected to the three main roads approaching the town: Poetovio – Mursa on the west, Mursa – Cibalae on the south, and the northern and eastern continuations of the *cardo* and *decumanus,* respectively, which linked to the *limes* road that approached Mursa from the north and proceeded eastwards towards Teutoburgium (Dalj). Accordingly, four main cemeteries grew around the town (fig. 14). At the moment, the northern cemetery seems to have been the smallest, due to the lack of free space between the town and the River Drava, the more so as the area north of the Drava was too marshy to accommodate a cemetery. The remaining three cemeteries spread freely along the roads up to one kilometre from the town. However, the sorest topographical issue of Mursa is the outlining of the outer borders of the cemeteries, as well as establishing the dividing lines between the cemeteries and urban districts. For a long time the course of the town walls of Mursa had been considered to be known, until it has been seriously questioned by the excavations in 2008 and 2009.[73] With the elevation to the rank of colony in Hadrian's time, the eastern border of the settlement was moved farther east, so that the eastern cemetery was overlaid by the inhabited area.[74] Such horizontal stratigraphy was further confirmed by the find of the fortification ditch in the same area, possibly also dating from Hadrian's time.[75] However, neither in this nor in any other of the town's cemeteries were graves dating form the time of Hadrian or the Antonines ever found. Consequently, there is no knowing which cemeteries were in use at that time or what kind of funerary ritual was practiced. This, in turn, brought to the conclusion that the border between the town and the eastern cemetery was not fixed, and that it was relocated according to the needs of the town. Overall, given that the earliest graves in the southern cemetery date from the 3rd century, it can be assumed that after the establishment of the *colonia* in the 2nd century only the eastern and western cemeteries existed; they were probably intensively used also during the Severan period.

No substantial remains of the early settlement phase of Cibalae have so far been discovered, yielding a nearly complete lack of early graves only expected. Apart from the already mentioned soldier graves,[76] Cibalae has so far

[72] Egri and Rustoiu 2009, 83-86.

[73] By using archival and field documentation produced by the Croatian historian and archaeologist Matija Petar Katančić (1750-1825), who noted the remains of Mursa prior to their usage as building materials in later constructions, the engineer and an amateur archaeologist Radoslav Franjetić made a reconstruction of the town walls at the beginning of the 20th century. The plan of Mursa as given by Franjetić, which had been used by subsequent authors (Pinterović 1978; Göricke-Lukić 2000; Filipović 2004), has only recently been more or less refuted by the most recent archaeological excavations. However, no new plan based on the results of the recent excavations has been made so far (see section 4.2.).

[74] Leleković 2009a, 46.

[75] Leleković 2009a, 46; Filipović 2007, 18.

[76] See section 2.1., and Dizdar and Radman-Livaja 2004.

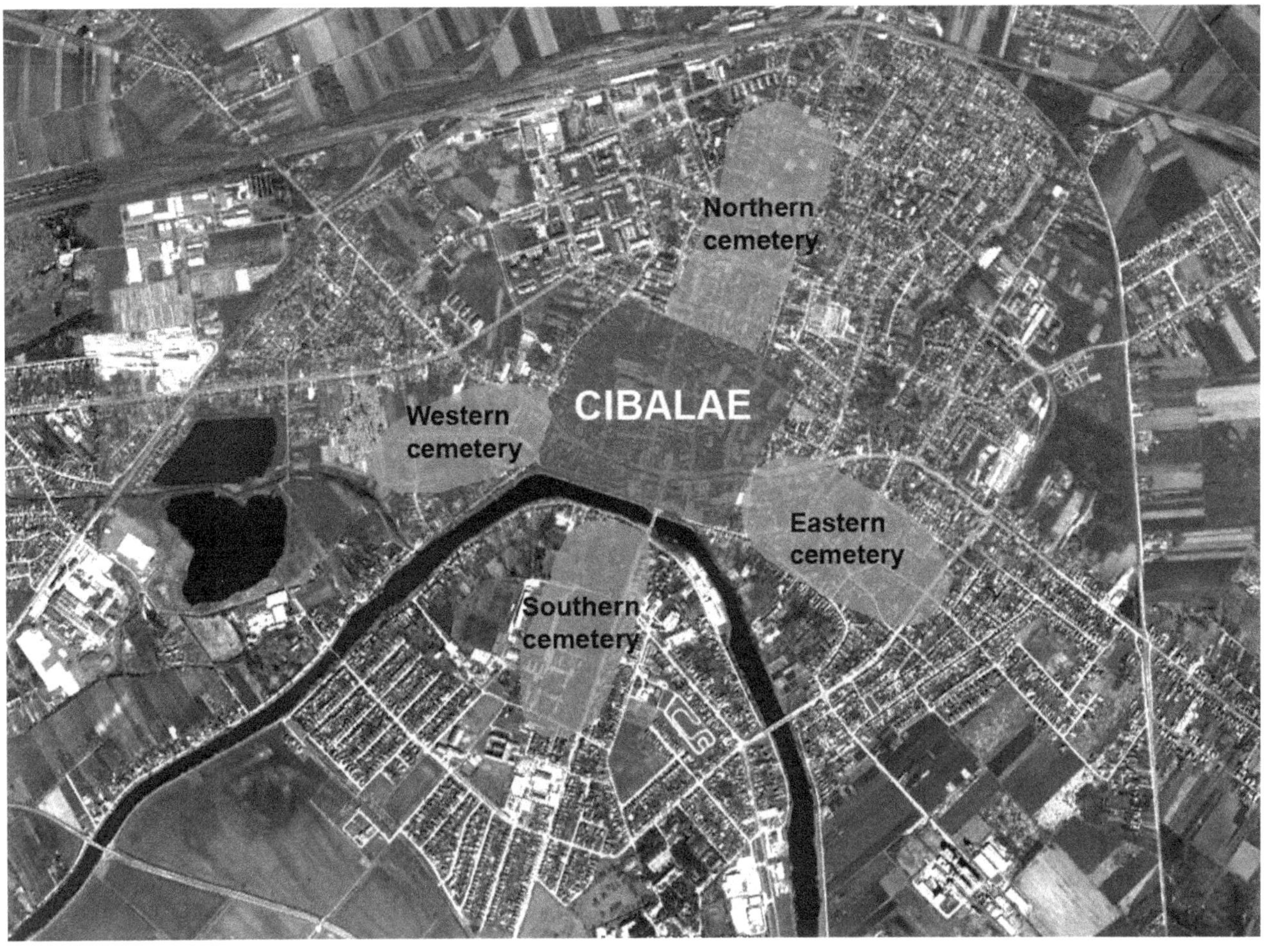

Fig. 15. Plan of Cibalae (after Google Earth, modified by T. Leleković).

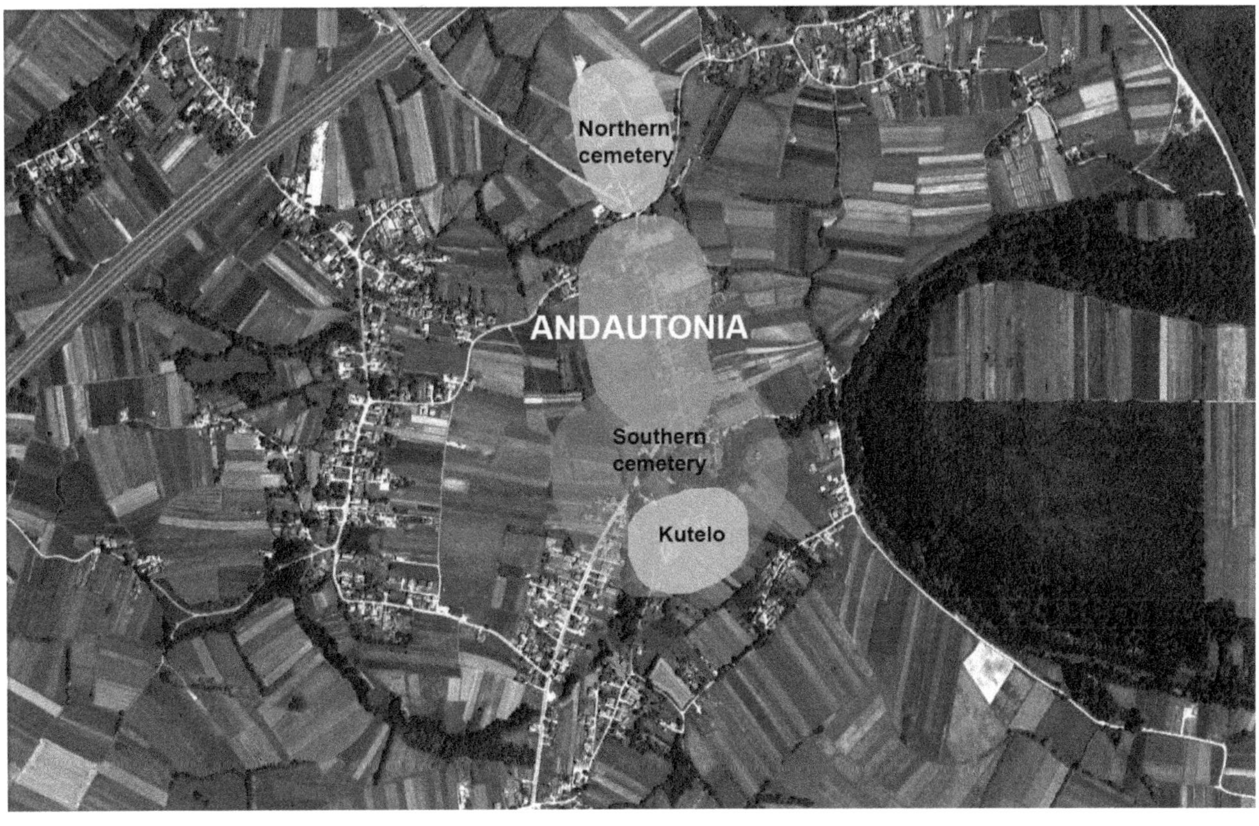

Fig. 16. Plan of Andautonia (after Google Earth, modified by T. Leleković).

produced only two early cremations, one in the southern cemetery and the other in the north (fig. 15). The latter was quite lavishly furnished and is considered to have belonged to a wealthy craftsman.[77] These two finds testify that in the 2nd century Cibalae was a settlement with at least two cemeteries, and that the dearth of early graves is probably due to the fact that they were overlaid or destroyed by later buildings, therefore escaping discovery. Andautonia (Ščitarjevo) became a *municipium* already during the Flavians, which renders it one of the earliest Roman towns in northern Croatia (fig. 16). In the last several decades a number of excavations have disclosed urban parts of the settlement with luxurious architecture, which suggests that the standard of living in smaller towns did not fall behind the major urban centres. Nevertheless, the cemeteries remained a poorer researched part of the town.[78] Only sections at the site of Kutelo (the southern cemetery) and the northern cemetery, recovered in the 2008 rescue excavations, are known so far; both were sited at the town's edges, along the major road Poetovio – Siscia running through the town. [79] The excavations have revealed that in the 120s the town extended over the southern cemetery, dislocating it some 500 m further south, to the site of Kutelo. Owing to this, it is possible to distinguish the cremations of the period predating the expansion of the town during Hadrian's rule from later burials. Excavations of the southern cemetery revealed early graves at two sites: in the vicinity of the parish church (dating from the 1st and 2nd centuries) and at Kutelo (dating from the 2nd to 4th centuries). Besides cremation graves, the latter site produced a masonry tomb.[80] Given that the results of the excavations have not yet been fully published, it remains unknown what were the shapes and the chronology of graves from the two sites, possibly representing two different sub-phases of the cemetery. According to the available literature, both sub-phases yielded cremation pits filled with the pyre debris, comparable to those found in the rural cemeteries in the surroundings of Andautonia.[81] A good example is a grave from the 2nd half of the 1st century, featuring a rectangular wooden construction and standing out for its measurements (1.6 x 3.2 m). It was fairly lavishly furnished, yielding several ceramic jugs and oil lamps, as well as glass beakers. In the 2009 excavations five cremations from the 2nd half of the 1st and the 1st half of the 2nd centuries were recovered, varying in size from 2.80 x 3.20 m to 2.90 x 4.50 m. Compared by the number of grave goods and dimensions of the pits, these five graves belong to the same layer as those found in 1996, both being basic for the study of the earliest Andautonian cemeteries.[82]

Besides the above mentioned towns, early cremations should be expected in several other urban settlements of the area of study, but such finds have not appeared there so far. The first of them is Aquae Balissae (Daruvar), which already in the Flavian period became a *Municipium Iasorum*. All graves discovered there so far date from the period after the mid 2nd century.[83] The same is true of lesser towns whose administrative status is unknown, such as (probably) Iovia/Botivo (Ludbreg). At some point in the 3rd century this settlement became a *civitas* in the late antique meaning of this term, and in the 4th or 5th century possibly even a bishopric. However, Iovia got the town walls already in the 2nd century, which provided it with an urban aspect in material terms, in spite of the lack of the legal municipal status. The cemeteries (or individual graves) of Iovia have not been studied, but the position of three funerary areas (northern, eastern, and western) have been located on the basis of the course of the town walls and scattered chance finds.[84] The situation in Aquae Iasae (Varaždinske Toplice) is even less encouraging in this respect, in that not a single funerary remain has ever been found there, in spite of the rich finds of the thermal baths and a forum.[85] Another thermal complex has been discovered at Topusko southwest of Siscia, resulting in the hypothesis of a small town at this site.[86] Masonry tombs have been found there, with only one (lavishly furnished) grave extensively published; it was dated to the turn of the 2nd and 3rd centuries by grave goods: ceramic vessels and clay lamps, as well as glass vessels and amber and gold jewellery.[87] However, such chronology should be questioned on the basis of comparison with similar finds from the southwestern cemetery of Siscia, dated to the mid 2nd century, as well as one equally dated amber figurine from a grave in the southeastern Siscian cemetery.[88] In a word, given the proximity of Siscia and Topusko, the abovementioned grave from the latter place should possibly be dated to the mid 2nd century, rather than its end, which would classify it among the early cemeteries, and not among those transitional. Another site that has yielded remains of a settlement and an affiliated cemetery is Slavonski Brod, a possible location of Roman Marsonia. Marsonia was in the Poetinger Table and the Antonine itinerary marked as a station on the road Siscia – Sirmium, and is considered to have been the centre of the *civitas Breucorum*; at this stage, however, it remains uncertain whether the settlement established at Slavonski Brod was a road station or a small town. Archaeological excavations there brought to light ten *busta* and two urned graves. Since the results of these excavations were published only cursorily, at this stage only two broad presumptions can be put forward. The first one concerns the chronology of the finds, probably dating from the 2nd and 3rd centuries, judging from the pottery. Secondly, the finds most probably stem from the part of the cemetery bordering on the inhabited area. Such position is suggested

77 Dimitrijević 1979, 159-160; Iskra-Janošić 2004,

78 Vikić-Belančić 1981, 129-154; Kušan Špalj and Nemeth-Ehrlich 2003, 116.

79 Pintarić 1998, 1-3; Kušan Špalj and Nemeth-Ehrlich 2003, 116. See also Lolić and Wiewegh in this volume.

80 Kušan Špalj and Nemeth-Ehrlich 2003, 116; Nemeth-Ehrlich and Kušan Špalj 2007, 44, 62, 65.

81 Gregl 1997, 24-25.

82 Nemeth-Ehrlich and Kušan Špalj 2007; Nemeth-Ehrlich and Kušan-Špalj 2008, 199-200.

83 See Schejbal 2003, 111-118.

84 Gregl and Migotti 2003, 13-43, 135-136.

85 Nemeth-Ehrlich 1997.

86 It has not been established with certainty whether on this spot Roman Ad Fines or Quadrata should be located: Šegvić 2006, 271; Gračanin 2010, 16, 39. See also Lolić and Wiewegh in this volume, section 7.

87 Šarić 1980; Šarić-Šegvić 1984.

88 Wiewegh 2003, 62; Leleković 2011.

by the finding of remains of a pottery workshop in later layers, indicating that at one point the town expanded and spread over the former cemetery area. The most characteristic feature of this cemetery is the *bustum*, which is otherwise more typical of the eastern part of the studied area. On the other hand, the pottery forms correspond to those from the Siscian cemeteries and are not found in the cemeteries of northeastern Croatia, above all Mursa. On balance, the cemetery at Slavonski Brod can be estimated as standing in the border area of two differing Pannonian traditions, southwestern and southeastern, and sharing the characteristics of both.[89]

2.3. Rural cemeteries

The study and understanding of rural cemeteries is above all affected by a generally poor and also uneven state of research of the rural areas of northern Croatia. This imbalance is most clearly reflected in the proportion of the rural cemeteries west of the Slavonian Mountains[90] and the rest of area of study. In the central and western northern Croatia 80 rural cemeteries have been catalogued so far, while only a few have been identified in the remaining area.[91] The imbalance in the state of research is evident also at the level of narrower zones, as is best illustrated by comparison between the areas of Zagreb (*ager* of Andautonia) and Sisak (*ager* of Siscia).[92] Despite the fact that these two towns were founded and integrated in the Empire at approximately the same time, there is a conspicuous difference in the level of rural settlement and the number of rural cemeteries between the territories of Siscia and Andautonia. The Andautonian area boasts of 25 early rural cemeteries, while they are completely absent from the Siscian.[93] Also, the former is characterised by a concentration of Norico-Pannonian tumuli strong enough to suggest that the southern limit of the distribution of tumuli cemeteries is at the same time the borderline between the Andautonian and Siscian territories.[94] Besides tumuli cemeteries, the Andautonian area has yielded 219 pit cremations, whose number is equal to that of tumuli, while some difference in their location has also been noted: tumuli are as a rule situated by the main roads Siscia – Emona and Siscia – Poetovio running through Andautonian territory, while pit graves were identified both along the roads and at some distance.[95] Although some differences have been observed between the variously located cemeteries, they cannot be clearly defined due to the inadequate scope of research. As an illustration of this, two cemeteries from the Zagreb area will be brought, both dated to the 1st and 2nd centuries. One of them was investigated in 1873 at the site of Stenjevec in the western district of Zagreb; it produced 123 incinerations of differing typology, with a predominance of tile-constructed graves and pit graves filled with ashes. The small finds from the graves count among the standard grave goods of the period.[96] Another cemetery has been discovered at the village of Gornja Vas in the mountainous Žumberak region to the west of Zagreb, inhabited by the community of the Latobici. It was composed of rectangular and circular dry-masonry tombs constructed of unworked stones and stone slabs, measuring approximately 1 by 1 m; a total of 63 tombs have been recovered on an area of 1100 square metres.[97] A peculiar feature of the Gornja Vas cemetery is the urns in the shape of a house (*house urns*); it has been presumed that they were brought by the Latobici from the area of Salla (Hungary) in the 1st century.[98] If now we compare the grave shapes of the cemeteries at Stenjevec and Gornja Vas, it transpires that they belonged to two completely different communities living at a small distance from each other. The former (Stenjevec) exemplifies the cemetery of a settlement which grew on the main road and whose inhabitants were Italian incomers and/or Romanised natives, while the latter (Gornja Vas) is an example of the cemetery of an autochthonous community who, in spite of Romanisation, continued practising traditional funerary rituals or developed new ones based on old customs. A significant example in this regard is a masonry tomb from Zagreb (Držićeva Street), resembling those found at the Žumberak villages of Gornja Vas and Bratelji. Namely, the cemetery at Držićeva Street was as distant from the main roads as were the cemeteries in the Žumberak, which points to the possibility that the native inhabitants of the Andautonian *ager* cherished identical or similar funerary customs.[99] This subject deserves to be further researched and the previous data reinterpreted, in order to enable distinguishing between the cemeteries with the prevailing immigrants from those mostly inhabited by the autochthonous population. Most important in this regard would be a further investigation of Norico-Pannonian tumuli, whose concentration is especially high in the surrounding of Andautonia.[100] At the moment, very important for the study of early cemeteries of the area are recent finds of burials, recovered during protective excavations of a section of the highway that partly follows the route of the main road Siscia – Emona. Along this corridor, three sites yielding Roman settlements and the related cemeteries have been excavated (Šepkovčica, Okuje, and Gornji Vukojevac). Since the results have not yet been fully published, the details of the number and shape of the graves are lacking. At the moment, the graves can be determined as smaller 2nd-century *ustrinata* filled with the pyre debris.[101] In broad terms, both the grave shapes and grave goods correspond to those found in Andautonia and Siscia, therefore supporting the presumption advanced above, that namely the inhabitants of the rural settlements along the main roads shared the same funerary rituals with the citizens of Andautonia and Siscia. Such conclusion

89 Miškiv 2006, 7-9.
90 On the syntagm *Slavonian Mountains* see Migotti in this volume, note 13, fig. 2.
91 Demo 1985; Gregl 1997; Gregl and Jelinčić 2010.
92 See notes 59 and 60.
93 Gregl 1997, 14-18; Gregl and Jelinčić 2010.
94 See note 58 and 59.
95 Gregl 1997, 22.
96 Gregl 1989.
97 Gregl 2007, 227-228.
98 Mócsy 1974, 17.
99 Vikić-Belančić 1981, 149-150.
100 Mócsy 1974, 136.
101 Bugar 2009, 271.

opens the question of the share of natives in the population of these two towns; this issue, however, remains unresolved at this stage.

In the rest of the western part of northern Croatia only a few randomly scattered early cemeteries have been discovered. Besides the already discussed tumuli (see section 2.1.4.) cemeteries in the Drava district (Gola, Novačka, Draganovec/Farkašić), the village of Kunovec Breg northwest of Koprivnica appears as an important site of the only so-far excavated cremation cemetery in that part of northern Croatia. A total of nine *ustrinata* were investigated, featuring rectangular pits filled with the pyre debris; a grave whose bottom was paved with fragments of amphorae and ceramic mortars stands out as a unique find.[102]

3. Transitional (intermediate) cemeteries

While in the core of the Empire inhumation became commonplace as early as the time of Hadrian's rule, as witnessed by the mass production of sarcophagi and the introduction of new forms of funerary architecture, at the same time incineration was still the most customary way of burial in Pannonia. Inhumation became more frequent there only from the 3rd century, although occasional incinerations occur as late as the 4th century. The 3rd-century cemeteries are named as transitional, on account of gradual abandonment of incineration and a concomitant introduction of inhumation. The fact that during a longer period two rituals existed side by side caused the transitional cemeteries to be characterised by a balanced mixture of incinerations and inhumations (fig. 23). It should be borne in mind that this process did not develop at the same pace or the same time in various regions of the area of study, as is evident from the fact that the investigated cemeteries show varied proportions of incinerations to inhumations, as well as some difference in ritual components.[103]

Until recently transitional cemeteries of northern Croatia have been a complete enigma, which comes as a surprise given that in the 3rd century the province of Pannonia was experiencing its age of flourishing, evidenced by a population increase and the emergence of new settlements or further growth of those already existing.[104] The only evidence of this period in northern Croatia in terms of funerary archaeology had until recently been stone sarcophagi, whose number increased exactly in the 3rd century. A problem attached to this category of monument in the present context is the uncertainty of use of individual specimens, which, on the base of their outer measurements, could have equally held the body of the deceased and ashes, pinpointing ritual duality of the period.[105] On the other hand, these sarcophagi/chests point to the fact that the change in funerary ritual was not based solely on the substitution of incineration by inhumation; this change introduced some new ideas and trends, whose most conspicuous appearance was funerary stone receptacles.[106]

3.1. Funerary ritual

The main characteristic of the transitional funerary ritual is a concurrent employment of incineration and inhumation. The lack of information on this phenomenon in northern Croatia has only recently been much improved by two protective excavations in Osijek (the sites of Bana Jelačića Square and 120 Divaltova Street), in which sections of Mursa's eastern and southern cemeteries, respectively, from the late 2nd and the mid-3rd centuries were investigated. These excavations revealed, for the first time, sections of transitional cemeteries of Mursa substantial enough to yield information on the arrangement of graves and a variety of their shapes, rituals employed, and the relationship between incinerations and inhumations. Overall, certain flexibility in the organisation of the cemeteries and the types of burials employed can be discerned. A lack of order is reflected in an uneven spacing within individual rows of graves, as well as in intersecting burials. Both these phenomena are typical of transitional cemeteries, whose possible initial tendency towards spatial order was subsequently disturbed by a mixture of incinerations and inhumations, due to the differences in both shapes and measurements between grave structures related to these two rituals.[107] Nevertheless, in the two investigated Mursan cemeteries inhumations prevailed, with only some 5% to 7% of incinerations. Funerary rites employed there can be divided in two basic categories: incineration- and inhumation-related. Besides Mursa, sections of transitional cemeteries have been investigated in Sotin (Cornacum) and Batina (Ad Militare) enabling, despite the limited scope, some insight into the transitional burial phase in the Danube *limes* area. In comparison with Mursa, those two *limes* sites help achieve a better understanding of the transitional phase in the Danube area and the differences between various sub-regions and sites (see section 2.4.). In the rest of north-Croatian territory finds of transitional cemeteries are extremely rare. Until recently, only stone sarcophagi testified to the existence of transitional burials.[108] It was only in the 2007 and 2008 excavations of the southwestern cemetery of Siscia that a cluster of graves presumably from the transitional phase was recovered (fig. 23). Regrettably, the graves had been robbed prior to discovery, with the skeletons missing. It therefore remains inconclusive whether they date from the transitional phase or Late Antiquity.[109] Beside the finds of individual rural graves and some Norico-Pannonian tumuli, both from the 2nd to 3rd centuries, an important discovery occurred in the cave Bubijeva jama in the vicinity of Karlovac, which yielded the first substantial cluster of transitional graves in northwestern Croatia.[110] Still, the

102 Demo 1985, 120-123.
103 Andreae 1963, 163-164; Heinzelmann 2001, 26; Pearce 2008, 34.
104 Hoti 1992, 145-146.
105 Some elements for distinguishing between the sarcophagi and ash-chests do exist, but are not always conclusive. Cf. Migotti 2005, 370.
106 On the stone sarcophagi and ash-chests of the area under study see Migotti 2007b; Göricke-Lukić 2011, 70-83.
107 Philpott 1993, 413-414;
108 Jovanović 1984, 45.
109 Cf. Göricke-Lukić 2011.
110 Perkić 2002.

Fig. 17. 28 Gundulićeva Street, Sisak: example of two separate burial layers in the southwestern cemetery (photo: T. Leleković).

overall dearth of evidence prevents a better understanding of the transitional phase of burial in the whole of the area under study, as well as the systematisation of the data provided by chance finds. Therefore, the so-far gathered knowledge on the transitional burial in northern Croatia will be outlined on the example of the cemeteries of Mursa, as the only resourceful evidence. Those rare other sites that have been investigated to a certain measure cannot serve but to delineate the differences and similarities, as well as a variety of 3rd-century funerary rituals in the towns and sub-regions of northern Croatia.

3.1.1. Inhumations

It seems that firm roots of incineration in Pannonia and other northwestern provinces caused a certain unwillingness to embrace inhumation, which was only begun there in the 2nd half of the 2nd century.[111] Therefore, inhumations from the early cemeteries that do exist in southern Pannonia should by no means be taken as the heralds of incipient transition, but rather as isolated examples of the graves of immigrants from the core of the Empire or its oriental regions.[112] The earliest instance of the transitional phase is represented by a section of the southern cemetery of Mursa, dating from the turn of the 2nd and 3rd centuries, with the ratio of 5% to 95 % between the incinerations and inhumations. This points to a high rate of the acceptance of inhumation among the inhabitants from the mid 2nd century, while this rite completely prevailed by the end of the 2nd century. Nevertheless, other settlements, such as Cornacum and Ad Militare, have produced different ratios, testifying that the transition was neither uniform nor simultaneous, while certain differences between the incineration and inhumation rituals at local levels also existed. Although a meagre sample does not allow plotting a systematic typology of differences, it is possible to isolate three main differentiating characteristics: the way the body was deposited, the choice of grave goods and the number of items, and finally, the grave form. Supine inhumations prevail (218 graves or 59%), with the legs stretched or slightly bent, and with the head straight up. Although the hands are usually on the pelvis, apparently there exists no single canonical position. Nevertheless, the position of the arms was certainly part of the ritual, as suggested by several recurrent variations: both arms along the sides; one of the hands on the pelvis and the other along the side; hands folded on the chest; hands to the shoulders. In the slightly less than 10% of the cases, the skeletons were placed on side with legs semi-flexed, associating sleep. Some of the skeletons were in foetal position, which is otherwise typical of the Germanic peoples from across the Danube.[113]

111 Nock 1932, 325-326; Jovanović 1984, 45; Faber and Fasold 2007, 11.
112 Jovanović 1984, 45; Leleković 2009, 46.

113 See Filipović 2010, 21.

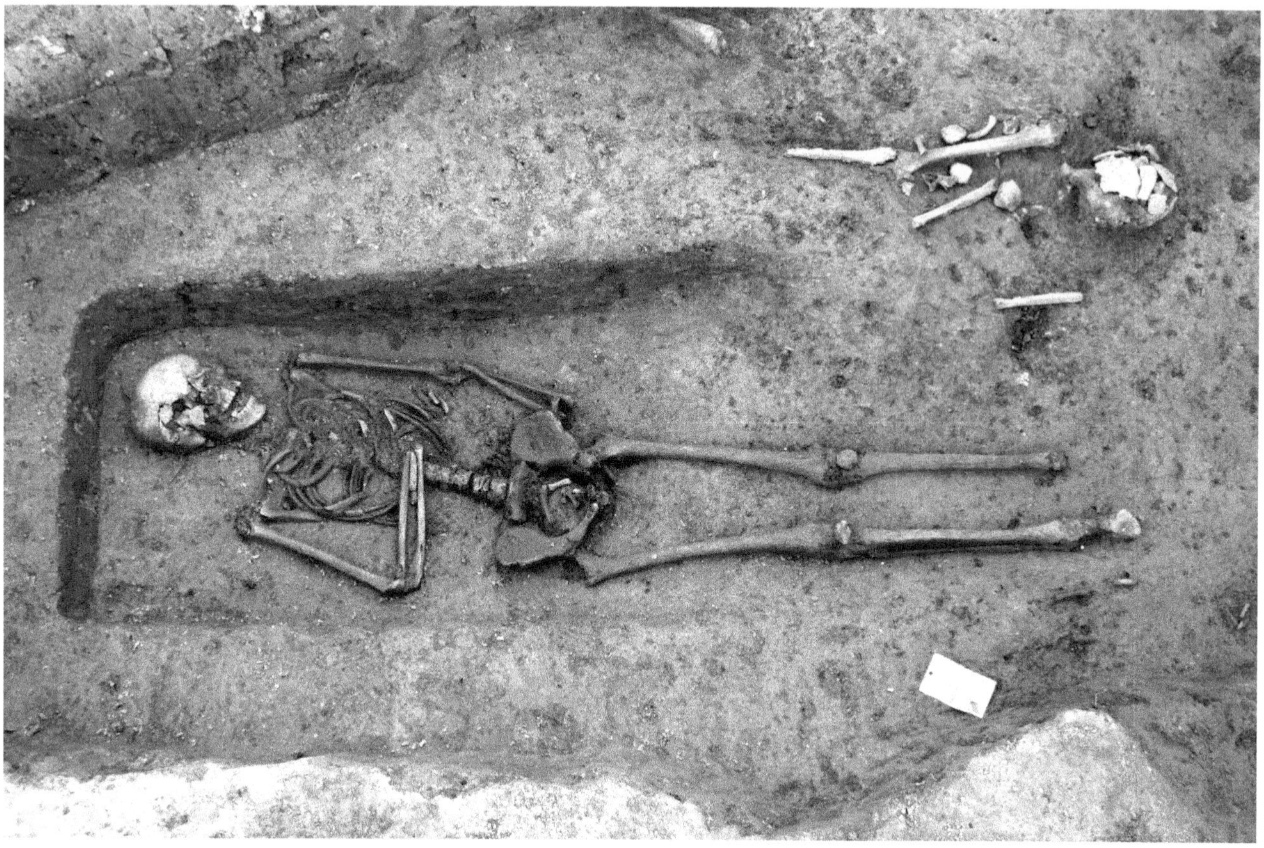

Fig. 18. Bana Jelačića Square, Osijek, graves 97 and 99: two inhumations of opposite orientation, 250-260 AD (photo: A. Štefan).

The same variety can be observed in the orientation of the skeletons and graves (fig. 18). The majority were oriented N-S (31%), W-E (30%), and NW-SE (11%), meaning that in nearly two fourths of all 3rd-century graves the deceased were buried with the head to the east, north, or south, while in 30% the head was to the NE-SW (5.4%), W-E (5.4%), SW-NE (4.4%), and SE-NW (4.1%). 16 graves held more than one body, of which 14 each held two, while two graves each held three bodies. A separate category of graves are those holding bodies buried with no indications of ritual, and placed on face; eight examples were found, adding up to 2% of all skeletons. In all eight of them the bodies were evidently dumped in the pits that can hardly be considered as proper graves. This is suggested by the fact that the pits' bottoms were unevenly cut and the bones found in non-anatomical position, displaying a complete lack of ritual observance. Two social categories can tentatively be connected with the burials disrespectful of the community's ritual norms: outcasts and the poor.

The transition from incineration to inhumation had its impact on the choice and number of grave goods. In the transitional phase burials lacking grave goods started to occur more frequently, marking a clear difference from the earlier graves, which customarily contained at least some goods. In the transitional cemeteries of Mursa nearly 60% of the burials lacked grave goods. The remainder did include associated artefacts, characterised by a conspicuous variety of forms and quantities. The excavations of a section of the southern Mursan cemetery at 120 Divaltova Street in Osijek yielded curious results: practically none of the 82 recovered graves produced an artefact that can be estimated as an item of dress accessory. The only similar items are the remains of shoes; in none of the cases were the shoes worn, but were always deposited next to the deceased, as a grave good in the basic meaning of this syntagm (figs. 20 and 21).[114] Presumably, it was the custom not to dress the deceased in the every-day clothes, except possibly a tunic.[115] If indeed this was the funerary custom of the Severan period, it was changed by the 2nd half of the 3rd century, as the later transitional graves yielded jewellery, shoes and dress accessories (belt buckles and brooches) in the positions as worn.[116] Among the vessels, ceramic forms suitable for drinking or serving drinks come most frequently. Vessels possibly intended for the solid food are rare, suggesting that the funerary ritual was altered in this respect as well.[117] The only grave good that deserves to be specifically mentioned is ceramic oil lamps, which were still commonplace in inhumations of the Severan period, to last deep in the 3rd century. In the 2nd half of the 3rd century, however, none of the inhumations contained such objects.

114 On the difference between grave goods and personal possessions cf. Migotti and Perinić 2001, 123-124.
115 Vukmanić and Hršak 2010, 34; Leleković 2010.
116 Leleković 2011.
117 Göricke-Lukić 2011, 162-167; see also note 26.

Fig. 19. Bana Jelačića Square, Osijek, double grave 17 of a man and a woman (photo: T. Leleković).

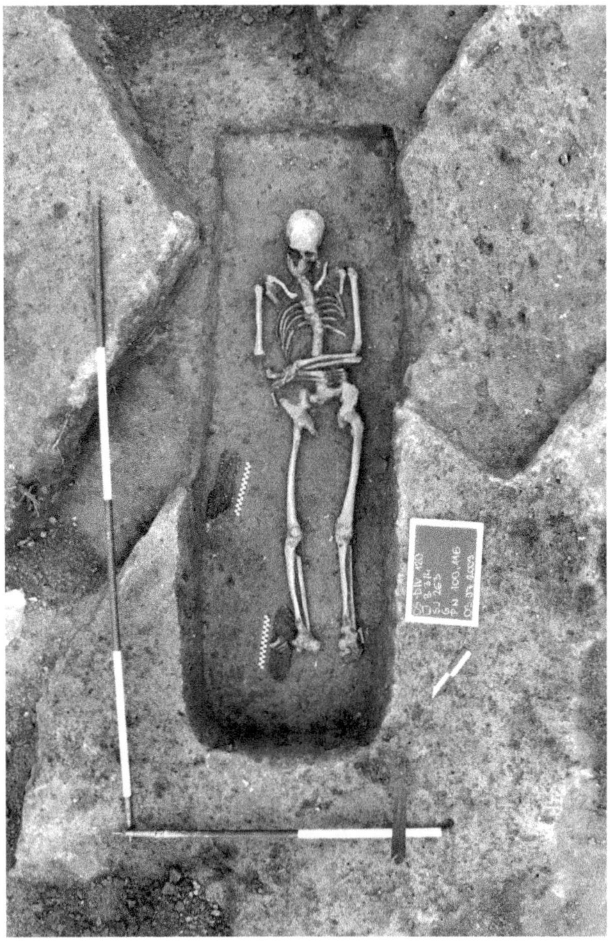

Fig. 20. 120 Divaltova Street, Osijek: inhumation grave 52 with the remains of the deposited shoes, 1st half of the 3rd c. (photo: V. Mesarić).

Fig. 21. 120 Divaltova Street, Osijek: detail of grave 41 (fig. 20) showing the manner of depositing the footwear (photo: V. Mesarić).

Fig. 22. 28 Gundulićeva Street, Sisak: southwestern cemetery, tomb 26 (photo: T. Leleković).

Curiously, during this time ceramic oil lamps come only with incinerations, which will be discussed below.

The study of the transitional cemeteries of Mursa has shown that pit graves (fig. 26.1) strongly predominate over tile tombs (figs. 26.3-5), with pits either rectangular or oval in form and with relatively rare burials featuring remains of wooden coffins. Of a total of 393 graves only 19 produced nails and remains of coffin fittings (fig. 26.2), of which 15 were grave pits and four were tombs. Tombs of this phase have been found in Siscia and Mursa, mostly constructed of tiles with no mortar, which does not qualify them as tombs in the narrow meaning. Three shapes of such constructions have been evidenced. The simplest variant was constructed of gable-wise arranged tiles, with the bottom sometimes paved with tiles, so that the perpendicularly set tiles lining the walls were either placed on the pavement tiles or directly on the earth (fig. 26.3). Another type is the grave in the form of chest or coffin, constructed of tiles lining both the bottom and the walls, as well as covering the receptacle (fig. 26.4). The third type of grave is also chest-shaped, but with gable-wise set tiles; tiles lined its longer sides both on the inside and outside, giving additional solidity to the roof (fig. 26.5). The short sides were closed with gable tiles, and the interior was never plastered.[118] Examples of "true" tombs were very poorly preserved, hardly ever featuring more than the bottom and foundations, suggesting therefore completely aboveground structures. The examples from Osijek and Sisak testify to substantial buildings measuring 2 by 3 m, with tiled and mortared walls, and even stone doorframes, as in the case of a tomb from the southwestern Siscian cemetery (fig. 22). The Siscian tombs produced stone chests, whose use in terms of funerary ritual remains inconclusive.[119]

3.1.2. Cremations

All cremations from the transitional phase appear to be *busta*, as they feature burnt walls and contain the pyre debris mixed with the ashes and burnt grave goods. Some of the graves lack traces of burning on their walls, but this should not deprive them of the definition as *busta* (see section 2.1.2.). One of the possible explanations is that the upper, presumably burnt portions of their walls, do not survive. The other possibility is that such graves were not *busta* in the narrow meaning of the word, as cremating

[118] Leleković 2011; Göricke-Lukić 2011.
[119] Leleković 2009b, 294-295; see also note 105.

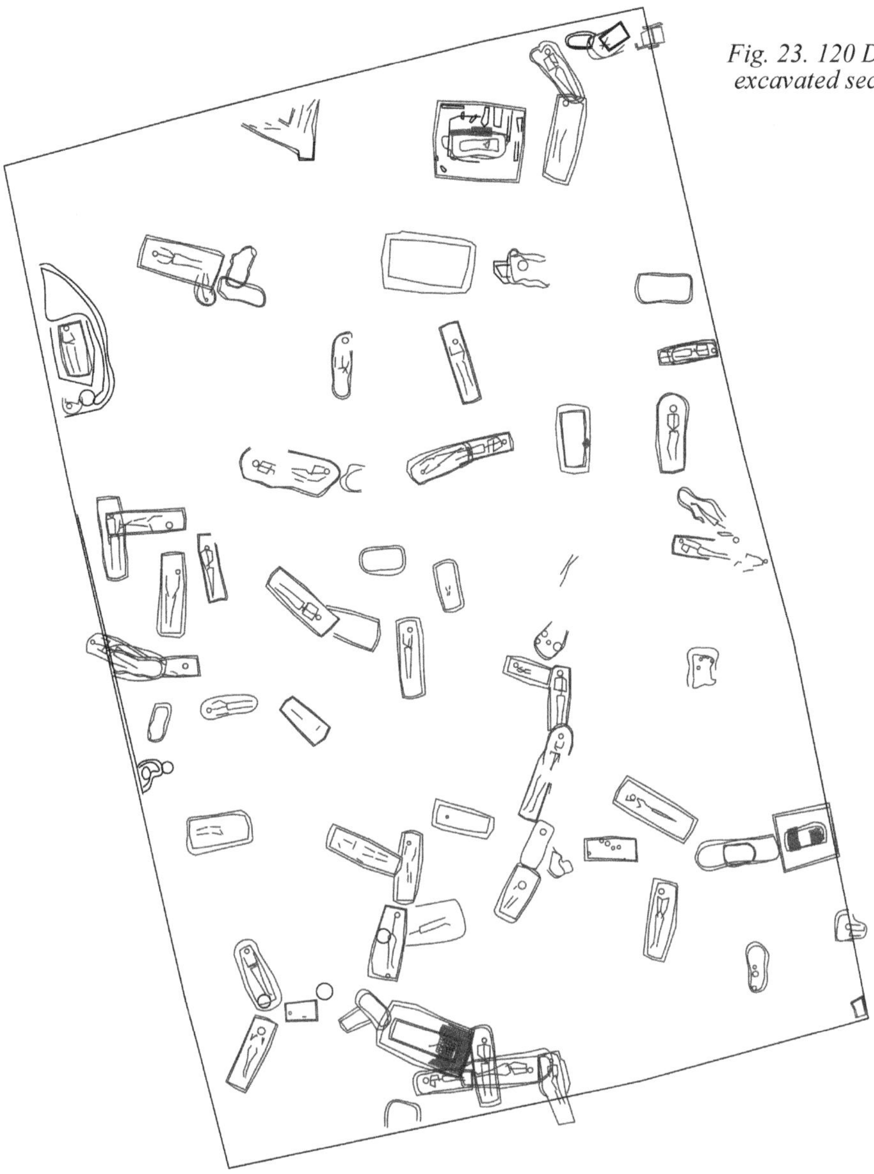

Fig. 23. 120 Divaltova Street, Osijek: plan of the excavated section of the south cemetery (photo: V. Mesarić).

Fig. 24. 120 Divaltova Street, Osijek, grave 31: stepped-side incineration grave, 3rd c. (photo: V. Mesarić).

Fig. 25. 120 Divaltova Street, Osijek, grave 38: stepped-side incineration grave, 3rd c. (photo: V. Mesarić).

might have occurred by the pit, which received the remains only subsequently. In general, the evidence testifies that the custom of cremating remained deep into the 3rd century, but underwent some changes, one of them being the shape of the *bustum*. While in the earlier period *busta* took shapes of oval or rectangular pits, characterised by intense burning of the walls and reduced grave inventories, *busta* of the transitional phase are of two types: 1. irregular in shape, with unevenly burnt walls; 2. featuring stepped sides (graves *en étage*). Besides these two types of pit-*busta*, another form is typical of the 3rd century: flat *busta* under tumuli (see section 2.1.4.).

So far irregular *busta* have been recovered only in recent excavations of the southern cemetery of Mursa; the graves were of the type with cremation by the pit (see above). As more evidence of such graves is still missing, they cannot be discussed at any length. On the other hand, the graves with stepped sides (*en étage*) and chambers on two levels, known also as Mala Kopašnica-Sase type (figs. 24 and 25), have been well studied in the Danube and Balkan provinces.[120] The upper chamber is usually preserved to the depth of 0.40 m, always taking rectangular shape and holding a smaller oval or rectangular pit (the lower chamber) in its midst. Accordingly, the cross section is stepped, giving the name to this type of grave. Three such graves were recovered in the 2008 and 2009 excavations in Osijek - one in the eastern Mursan cemetery (Bana Jelačića Square) and two in the southern (120 Divaltova Street). The eastern cemetery (early 2nd century) produced a stepped grave (sub-type A) whose upper chamber, measuring 2.72 by 1.15 m, was only slightly more spacious than the lower, which measured 2.54 by 0.74 m. The graves (sub-type B) from the southern cemetery (early 3rd century) differ from the type A in that the upper chamber was significantly more spacious then the lower, and that the latter was covered with tiles. The most conspicuous difference, however, is in the abundance of grave goods in graves of the type B. The chronological gap of nearly one hundred years between the discussed types of graves suggests that the sub-types A and B represent two chronological stages of development of stepped cremations. The sub-type B is presumably later, and is characterised by the development of wall extensions for the deposition of grave goods. Stepped graves of the sub-type B are widespread in southeast Pannonia, Upper Moesia, and in parts of the neighbouring provinces. A. Jovanović points out that the share of *busta*, especially the stepped type, is in these provinces disproportionately high in comparison with the rest of the Empire, with the percentage of 60% to 90% in the cemeteries of Moesia and Sirmium, the latter standing in the present context for southeast Pannonia (see section 2.1.2.).[121] Significantly, the percentage of *busta* in Mursa is also at about 60%, its first layer of burials in the eastern cemetery corresponding to the distribution as presented by Jovanović. On the other side, the share of *busta* in the transitional cemeteries of Mursa amounts to 100%.[122] On balance, the share of *busta* in north-Croatian Roman cemeteries cannot be studied in isolation, but within their chronological unfolding. A. Jovanović failed to separate *busta* of the early and transitional phases, rendering questionable his theory of the origin and distribution of *busta* in the Balkan-Danube area, at least as far as southern Pannonia is concerned. On the account of the concentration of *busta* in Moesia, especially the stepped type, Jovanović considered this burial custom as originating in the Balkan native communities, those in southern Pannonia included. Nevertheless, the way and manner of its importation to the rest of Pannonia remain unexplained, since *busta* appear also in northern and western Pannonia. Two stepped *busta* of the sub-type B have been evidenced in Intercisa; in one of them both chambers held the pyre debris, and in another the ashes were deposited in the lower chamber covered by a wooden plank, while the upper chamber held the grave goods covered by the pyre debris free of bone.[123] The suggestion has been put forward that it was the soldiers stationed in Pannonia in the 1st and 2nd centuries who disseminated *busta* to other regions of the Empire.[124]

120 Jovanović 2000, 205.

121 Jovanović 2000, 205-206.

122 Leleković 2010, 49-50; Leleković 2011.

123 Barkóczi, Erdélyi and Ferenczy 1954, 43, 50.

124 Pirling 2002, 524-526.

In the context of the above discussion, a masonry tomb from the western cemetery of Mursa, recovered in 1966, deserves attention. It was constructed of stone and tiles, covered by massive stone slabs, and plastered on the inside, measuring 2 (length) x 1.42 (width) x 0.60 (depth) m. Underneath was a lower chamber measuring 0.82 x 1.40 x 0.40 m, lined and covered with tiles and holding the pyre debris. Altogether, this tomb strongly resembles stepped *busta* of the sub-type B. The grave inventory (two bronze pitchers and a tray, inkwell cover, collapsible iron stool and pottery vessels) is unusually lavish for Mursa's standards, but the precise findspot of the grave goods within the tomb has not been recorded. The grave has tentatively been ascribed to a high military official from the turn of the 1st and 2nd centuries.[125] However, given its similarity to transitional-phase *busta*, its date should be reconsidered; the end of the 2nd or the 3rd centuries seems to be more realistic, although the grave goods are not diagnostic enough in terms of chronology. As for the occupation of the deceased, the inkwell and stool possibly suggest administrative function, be it within the military or civil administration. The tomb was part of a grave enclosure encircled by a masonry wall of unworked stone.[126] Given its placement in the vicinity of the suburban continuation of the *decumanus maximus*, this structure seems to be representative of monumental tombs lining the entrance to the town and probably featuring a stele or another monumental aboveground mark.

The prevailing theory in Serbian Roman archaeology postulates that stepped incineration graves were *ustrinata* with ritually burnt walls, with the pyre remains subsequently relocated from the *ustrinum* to the lower chamber.[127] The still prevailing A. Jovanović's opinion claims that graves of the type Mala Kopašnica-Sase II (stepped *busta*), are not *busta* at all, but graves with the walls ritually burnt for cleansing in preparation for receiving the burial. Although the author of this theory failed to explain his reasoning, it seems that the problem of recognising the Mala Kopašnica-Sase II type of graves as *busta* results from meagre remains of the pyre debris in those graves, as well as less intensive traces of burning on their walls.[128] This, however, can be questioned on the basis of W. Gaitzsch and A. Werner's experiment, which demonstrated that the amount of ashes and intensity of burning on the walls correspond to the findings from Mursa. In other words, the rejection of stepped *busta* in favour of *ustrinata* cannot be taken as conclusive.[129]

In the eastern Danube and Balkan provinces *busta* cannot be dated earlier than the beginning of the 2nd century. They come relatively frequently in the cemeteries from the 2nd and early 3rd centuries in the European part of the Empire, with some towns (Viminacium, Intercisa) cherishing this type of burial until the close of the 3rd century.[130] The end of the transitional phase in northern Croatia has been dated to the end of the 3rd century on the basis of a T-shaped brooch of hinged type (*Armbrustfibel*; Riha 6.4) from the latest *bustum* found in the eastern cemetery of Mursa.[131] Since these latest cremations produced pottery shapes untypical for the late 3rd century, such as stamped oil lamps, the deceased were possibly either older people or individuals who for some personal reasons insisted on tradition, preferring the "old-fashioned" type of burial, or having it chosen for them by somebody else.[132]

3.2. Urban cemeteries

The change in burial habit in south-Pannonian towns mostly occurred during the Antonine and Severan periods, that is, in times that witnessed these towns' greatest prosperity and territorial extent. Until recently, the national system for managing and protecting archaeological heritage established the protected archaeological zones of Roman towns in Croatia mostly in correspondence with their intramural areas. With such policy, the suburbs and cemeteries mostly remained outside the protected zone, and were only rarely excavated. Owing to this, the majority of graves of the period are chance finds that occurred in cultivation or building works.

The protection policy as mentioned above is particularly well reflected in the case of Siscia; although the protection of the majority of the archaeological zone was anticipated, it was mostly the area inside the Severan town walls that received protection. Consequently, the course of the walls and ditches is very well known, unlike the suburb and cemetery sections. This is especially true of the site of Pogorelac across the Kupa and the walled town area, which accommodated a production section of the town and most probably also the imperial mint (fig. 13). Given that the 3rd century was the period of the town's maximal prosperity, its spread across an area wider than that encircled by the fortifications can be reasonably presumed, while in Late Antiquity it probably shrank within the walled area.[133] The only indicator for the cemetery district of the transitional period is 22 stone chests and sarcophagi, recovered or identified as chance finds across a relatively wide area occupied by the southeastern and southwestern Siscian cemeteries. Although all the data on the findspots cannot be taken at face value, the broad distribution of the finds suggests that the area occupied by the cemeteries was at its largest exactly in the transitional period. The majority of the finds stem from the southwestern cemetery, pointing either to its considerable territorial scope or its social importance.[134] This suggestion seems to be strengthened by the results of the 2007 and 2008 excavations, which

[125] Bulat 1977, 79-83, Göricke-Lukić 2011, 105-111, 216.
[126] Bulat 1977, 79-83.
[127] Jovanović 2000, 205; Pop-Lazić 2002, 42.
[128] Jovanović 1984, 101-104.
[129] Gaitzsch and Werner 1993, 57-58.
[130] Struck 1993b, 93.
[131] Riha 1979, 167.
[132] Although stamped oil lamps were in use till the 4th century (Alram-Stern 1989, 43-44; Braidotti 2009, 107), their concentrated use in the east Mursan cemetery speaks in favour of the above hypothesis, especially as they were combined with by then an outdated type of burial, that is, cremations.
[133] On the course of the town walls and the extent of the urbanised area see Vrbanović 1981, 193-194; Nenadić 1987, 82; Vuković 2010.
[134] Migotti 2007b, 10-23.

revealed several 3rd-century masonry tombs. Regrettably, these structures were too badly damaged to assist in clearing the issue of the transitional burial rite in Siscia in terms of the ratio of inhumations and incinerations. If this is true of the more sumptuous constructions, attributable to the elite, it is even truer of common burials of (presumably) lower social classes, whose graves are mostly lacking from the evidence. The above finds testify that in the 3rd century the southeastern cemetery achieved its largest dimensions, which applies to all Siscian burial areas (see section 2.1.3.).[135] On the other hand, we should remain mindful of the fact that at this stage it is only possible to recognise graves of the transitional phase on the basis of their distance from the town walls.

During the transitional phase in the 2nd half of the 2nd and throughout the 3rd centuries, Mursa went through substantial modifications in both its extent and urban profile (fig. 14). The town was at the peak of its prosperity in the Antonine and Severan periods, as evidenced by the remains of buildings of the period. However, the protected archaeological zone encompassed only the urbanised area, leaving out the cemeteries of the 2nd and 3rd centuries. Due to this, until recently only those Severan graves that entered the protected area had been recovered. In the mid 3rd century, probably during emperor Gallienus' rule, the town seems to have shrunk, occupying the area within the town walls, which at that time were either renovated or built anew, enabling graves to enter deeper into the protected archaeological zone. This is especially well documented in the eastern cemetery, whose many burials date from this period, explaining at the same time the fact that the majority of the graves have been recovered from the stretch between the protected area and the town walls.[136] This, in its turn, demonstrates the stratigraphic and chronological division of Mursa's transitional cemeteries into two sub-phases: the Antonine-Severan and the Gallienic (see above). Nearly half of a total of 963 graves recovered in Osijek stem from the transitional phase, rendering Mursa one of the most suitable sites for the study of transitional cemeteries of the Croatian portion of Pannonia. It should be emphasised, though, that the majority of finds come from the only two excavated sites: Bana Jelačića Square (eastern cemetery) and 120 Divaltova Street (southern cemetery). Accordingly, the boundaries of the cemetery districts can only be outlined on the basis of chance finds, scattered across the area encircling the town and measuring some 300 ha. Of a total of 355 recovered graves, 312 can be classified as transitional inhumations. As, however, the majority lacked grave inventory, their affiliation can equally be transitional and late Roman. These graves were distributed in four large spatial clusters south, east and west of the town, corresponding to the presumed routes of the main roads leading towards Mursa and to the main cemeterial districts.[137] The only finds so far in the northern cemetery are six masonry tombs dated by coins to the Severan period. The northern cemetery developed in the limited area along the northern continuation of the *cardo* between the northern town edge and the Drava River bank. This area had previously been settled, as evidenced by Severan-period buildings, on a relatively large surface. The land on the opposite (left) bank of the Drava was extremely marshy and therefore unsuitable for burial. On balance, the northern cemetery was in the transitional period neither spacious nor important.

Archaeological excavations proved to be decisive in establishing the boundaries between cemeteries and inhabited areas, as well as in proving their shifting nature. This was most clearly evidenced in the case of the eastern cemetery, especially in the 2008 campaign at the site of Bana Jelačića Square, which revealed sections of both the occupied area and cemetery, the former represented by a spacious urban villa, while the area to the east of it was in the Severan period taken by the cemetery. In the 2nd half of the 3rd century the area of the former villa was abandoned and occupied by a westward extension of the eastern cemetery. There is no knowing at this stage whether the shrinking of the town area was accompanied by concomitant reducing of the cemetery and its withdrawal towards the town walls. The only Severan finds in the eastern cemetery were a small cluster of graves recovered in two minor excavations to the east of Bana Jelačića Square, delineating the then boundary between the town and the cemetery; no buildings have been found to the east of this site.[138] Further east only sporadic finds of graves and stone sarcophagi occurred, as well as 24 inhumations excavated in 2003. Given that the last-named finds have not been published, their chronoloy cannot be established with certainty; they may as well be late Roman.[139] As none of the inhumations found by chance in the eastern cemetery produced grave goods, they cannot be used in plotting the cemetery boundaries in the 3rd and the 4th centuries. It remains to presume that the most far-away inhumation graves, found on the site of Zeleno polje Street, date from the Antonine-Severan period, when the town reached its largest extent.[140] Similar circumstances have been recorded in the area of the western cemetery, whose only possibly Severan burial seems to be the abovementioned masonry tomb on the continuation of the main *decumanus*, probably located on the border with the urban district as it was set at the time (see section 3.1.2.).[141] Since this area is otherwise completely unexplored, it remains unknown whether this tomb within a burial enclosure was an isolated phenomenon or one among more such constructions. If the latter is true, it probably marked the beginning of a cemetery road comparable to that of the southeastern cemetery of

135 Leleković 2009b, 295-296.

136 On the town walls in short see Nenadić 1987, 75; Lolić 2003, 138-139; see also Vuković 2010.

137 The evidence of the chance finds and earlier excavations have been researched and expounded by H. Göricke-Lukić in her two monographs (2000 and 2011).

138 Göricke-Lukić 2000,167, 173; see also Pinterović 1978, 164-165.

139 Filipović 2004, 328.

140 Göricke-Lukić 2011, 18-19.

141 Bulat 1977, 79-90.

Siscia (see section 2.2.). To the west of this site, along the *decumanus* course, several tens of inhumation graves have been recovered, but the data on the remaining boundaries of the western cemetery is still lacking; it probably stretched for about 1 km towards west, resembling in its layout the eastern cemetery. The only significant number (several tens) of Severan burials was found in the 2009 excavations in the southern cemetery at 120 Divaltova Street.[142] Burials from the 3rd and 4th centuries were also found there, at some 500 m to the west of the continuation of the *cardo* and some 500 m to the south of the urban district. The large surface taken by this burial cluster possibly testifies to the vastness of the southern cemetery, which in this case would have occupied a 500-metre wide zone of the southern periphery of the town. Another possible explanation is that those graves were in effect one of the small cemeteries related to villages or villas that must have existed in the town's suburbs. Finally, a separate southwestern cemetery of Mursa could have been sited there, serving a town's community that for some reason preferred an isolated burial ground.

In addition to the burials of the eastern cemetery (Bana Jelačića Square), graves of the 2nd transitional phase have been identified also in the west part of the town. In the 2005 excavations, Severan architecture was discovered, and among it four inhumations. One of the graves was made of tiles, yielding among other grave goods a ceramic jug and an oil lamp. The find has not been fully published, leaving it inconclusive whether the grave goods correspond to those from the burials at the site of Bana Jelačića Square, which seems likely (see section 3.2.). If this is so, a shrinking of the town also in its western part can be presumed. This, in turn, would render it likely that the inhumations found on the site of Cara Hadrijana Street also date from this period, marking thereby the beginning of the western transitional cemetery in its second sub-phase. Some of the burials found at 120 Divaltova Street (southern cemetery) were dated to the 2nd half of the 3rd century by coins. Overall, all three cemeteries, southern, eastern and western, were in use during the whole of the transitional phase. In the northern cemetery only an insignificant number of scattered graves have been found, which can hardly testify to a substantial cemetery area to the north of the town at the time.

The transitional period most probably witnessed a considerable growth of Cibalae, which in the 2nd century became a colony, the third in line and the smallest in area of the three north-Croatian Roman colonies. As in Siscia, the urban area and suburban work districts have been much better researched than the cemeteries. Nevertheless, with new legislation on the protection of archaeological heritage, the circumstances have been changing for the better recently. Much in this respect has been achieved through the 2008 and 2009 excavations in the northern, western and southern cemeteries (fig. 15). Of all the finds, only 17 *busta* recovered in the 2009 rescue excavation in the southern cemetery date from the transitional phase.[143] Since the results have not yet been published, all that is known so far is the fact that the graves date from the turn of the 2nd and 3rd centuries, and that their lavish furnishings point to the town's elite.[144] Apart from the above *busta*, some inhumations recovered from the same site possibly date from the 3rd century, but they have not yet been studied and recognised as such. Small-scale excavations were also conducted in areas of the northern and eastern cemeteries, producing randomly scattered smaller clusters of inhumations whose chronological affiliation (transitional or late Roman) in most cases remains inconclusive.

Andautonia also counts among the towns that experienced prosperity in the 2nd and 3rd centuries, pushing its southern cemetery further south to the site of Kutelo (fig. 16). Some of the graves, lead coffins and sarcophagi possibly date from the transitional phase, but they have not yet been published. The same holds true for the cemeteries of *Aquae Balissae /Municipium Iasorum*, whose funerary stones and sarcophagi testify to a substantial transitional-period cemetery, or rather, two such cemeteries, as transpires from the findspots of the mentioned monuments along the roads north and south of the town. These cemeteries are, however, completely unexplored and their layouts and extent remain unknown.[145]

3.3. Rural cemeteries

Analysis of the transitional rural cemeteries in the area of study suggests an extremely varied picture of burial custom, as reflected both in burial ritual and grave shapes and inventories. At this stage of (insufficient) research, however, these facets can rather be indicated than fully grasped or defined, leaving obscure the specific characteristics of individual sites and regions. The above variety is best reflected in the case of *busta*, which are distributed throughout the region but in a variety of ways. In northwestern Croatia *busta* are always found under tumuli, appearing mostly as flat type (see section 2.1.4.). The only exceptions are two tumuli from Novačka, featuring *busta* in pits, which are otherwise typical of north-eastern Croatia. This is evidenced in Mursa, Cornacum, Ad Militare and Slavonski Brod (Marsonia?), all the sites having produced exclusively pit *busta*.[146] Nevertheless, the Danube *limes* sites in northeastern Croatia do not fit in completely in the above picture, as shown in transitional cemeteries that will be discussed below. In view of this, finds from Slavonski Brod are especially significant, given the town's placement at the border between the western and eastern parts of northern Croatia.[147]

142 Of a total of 82 burials only 12 can with certainty be established as Severan, while a further ten can be presumed as such by the context; the remainder can be dated to the mid 3rd and the 4th centuries.

143 Leleković 2010, 16.

144 Hrvoje Vulić, pers. com.

145 Schejbal 2004, 99-129, 112-114, 118.

146 Miškiv 2006, 6-9; Leleković 2009b, 294-295; Hutinec 2010, 29, 54; Vulić and Rapan Papeša 2010, 53.

147 Whichever opinion on the borderline between Pannonias Superior and Inferior in the area between the rivers Drava and Sava is accepted (see Migotti in this volume, note 9, fig. 2), Slavonski Brod remains in the border area. In view of this, the fact that this town is the westernmost

Transitional rural cemeteries only exceptionally produced inhumation graves, nevertheless testifying to the fact that the variety of burial custom in the region concerns both incinerations and inhumations. An example of this is the cave Bubijeva jama nearby Barilović, as one of the three caves in the area that were all probably used as burial places. So far, this cemetery has not been associated with any Roman settlement of a community living in the area. The skeletons (men, women and children) were found as deposited on flat ground, this being an atypical custom for Pannonia. Although the site is situated in the bordering area between the provinces of Pannonia and Dalmatia, it should be observed that neither the latter was home of such burial custom; possibly a prehistoric residue was at work in this case. The grave goods, Severan coins among them, do not vary from the usual, offering therefore no possibility of guessing at the ethnic or any other identity of the buried people.[148] Another example of miscellaneous funerary rituals of the period is the wagon burial from the village of Poljanec nearby Ludbreg, whose exceptional value comes from the fact that the ritual employed was typical of northern Pannonia.[149] So far, this grave remains the only such find south of the Drava. It was found by chance during ploughing in 1931, leaving behind no documentation either of the find itself or the context; a human skeleton was recovered in a wooden wagon provided with bronze and iron fittings and decorated with bronze figural sculpture. A tile-constructed grave was found in the vicinity, suggesting that the two graves were part of the cemetery of an unidentified settlement.[150]

3.4. Cemeteries on the limes

So far two *limes* settlements have produced remains of transitional cemeteries: Cornacum (Sotin) and Ad Militare (Batina). Both have been researched recently by means of targeted trial trenches outside the fort installations. The trenches in Sotin produced a Roman cemetery broadly dated to the 2nd and 3rd centuries, as well as a section of a cemetery featuring *busta*. Curiously, the layer predating the *busta* contained inhumations probably from the turn of the 1st and 2nd centuries. Since inhumation cemeteries of such early date are not typical of the area of study, the graves in question evidently belonged to Roman soldiers bringing with them a funerary custom that will became commonplace only a full century later than its introduction into the area.[151] It remains unclear whether these graves mark the beginning of the transitional phase or whether they exemplify isolated cases, as elsewhere in southern Pannonia.[152]

Archaeological excavations at the site of Ad Militare, 35 km north of Mursa, produced similar results. Two clusters of Roman graves were identified at two spots some 300 m away from each other, proving that the length of the cemetery that stretched along the road leading to the fort and *vicus* was several hundred metres.[153] Graves of the 2nd phase could be dated to the Severan or post-Severan periods by coins, and were laid out in regular rows, yielding both inhumations and incinerations; the latter were all of *bustum* type. Of the ten 3rd-century graves researched in 2010, two were incinerations and eight inhumations. Although the sample was relatively small, the finds so far suggest that all *busta* from Cornacum were of stepped type, and should be related to those from Mursa. The site of Ad Militare produced two smaller oval *busta* of the shape as evidenced in the early phase of the eastern cemetery of Mursa. All so far investigated *busta* in the *limes* area date from the Severan period; the sample is, however, too small for drawing conclusions on regional difference in funerary ritual.

4. Late antique cemeteries

In Late Antiquity incineration was finally abandoned in favour of inhumation as the exclusive burial ritual. Although the Roman history of Pannonia lasts till the end of the 6th century, the majority of late burials date from the 4th century, with rare examples of those from the 5th. The only site to yield considerable number of late antique burials is Cibalae, due to the investigation of larger sections of the town cemeteries on several occasions in the last decade or so. Nevertheless, their potential for study in the present context is much limited by the lack of published findings, which leaves the attribution of the graves to the transitional or late antique phases unresolved. Consequently, as the main source for the study and analysis of late antique burial remains the research of the cemeteries of three minor settlements at the sites of Tekić (Incerum?), Štrbinci (Certissia?), and Zmajevac (Ad Novas); so far Tekić has produced 124 graves, Štrbinci 161, and Zmajevac 175.

4.1. Ritual

The late antique graves of the area of study unexceptionally hold skeleton inhumations, producing a uniform picture of the funerary ritual and the appearance of the cemeteries. Some differences do exist in the choice and number of grave goods, as well as in a predilection for certain types of graves, but the cemeteries' layouts and the ways the graves were constructed are similar in all of them. The most conspicuous funerary procedures, such as a tendency towards aligning in burial rows and a uniformity of orientation of skeletons (mostly W-E, with head to the west), are commonplace in the whole of the late Roman Empire.[154] The most simple and at the same time the most

findspot of pit *busta*, opens the question of the correspondence between administrative borders and areas of specific funerary rites, possibly suggesting that the latter were drawn by taking into account ethnic territories.

148 Perkić 2002, 103-119; Perkić 2006, 104.

149 Mráv 2001; Zsidi 2003, 253-254.

150 Šeper 1962.

151 Hutinec 2010, 29-33.

152 Jovanović 1984, 158.

153 The excavations were carried out in 2010 in the organisation of the AMO and with the collaboration of Marko Dizdar of the Institute of Archaeology in Zagreb and Tino Leleković of the Archaeological Department of the Croatian Academy of Sciences and Arts (Zagreb).

154 Philpott 1993, 413-414; Pearce 2008. Still, some irregularities and deviation from the usual practise are commonplace everywhere, southern Pannonia included.

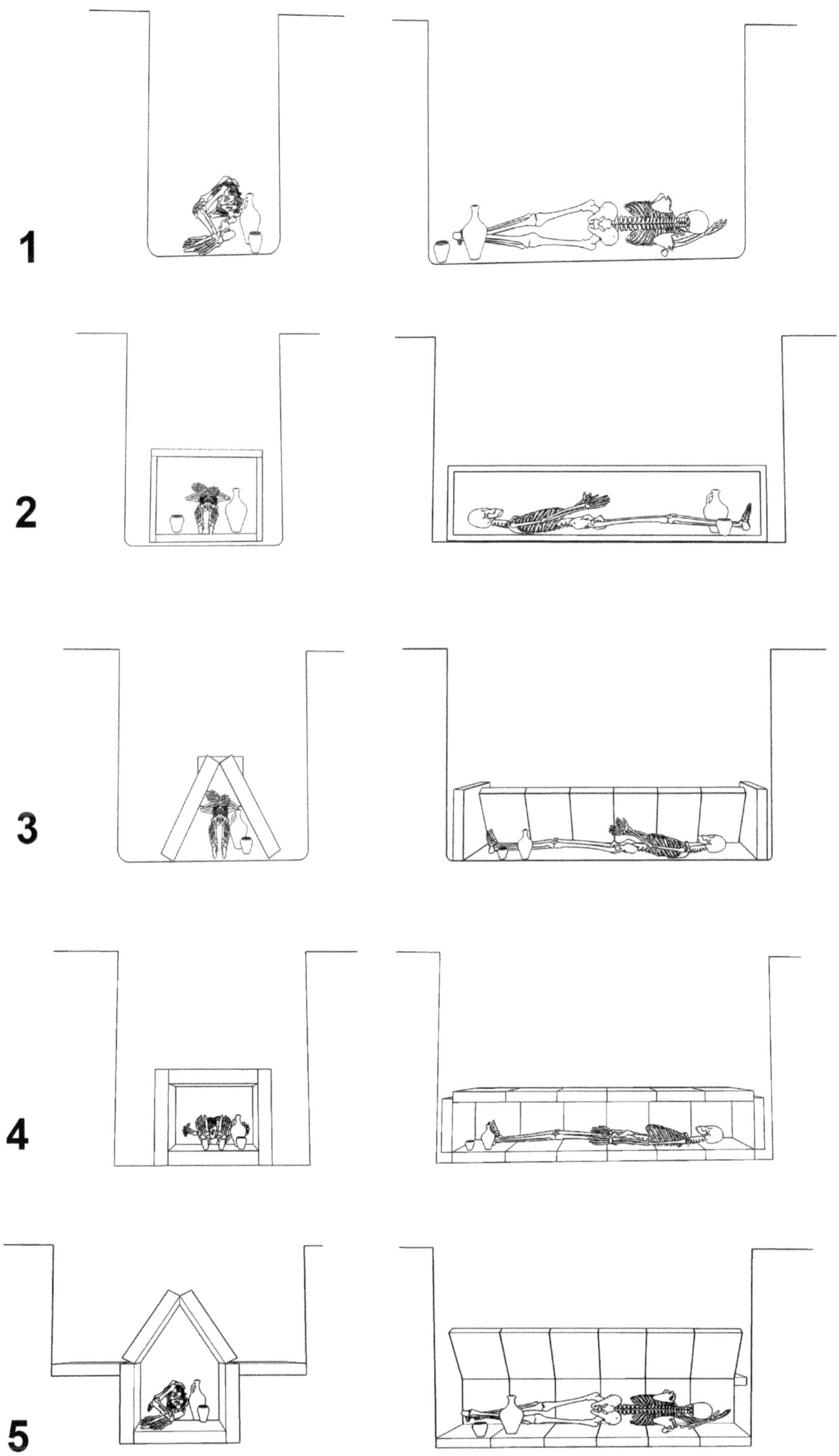

Fig. 26. Typological scheme of inhumation graves (made by M. Mađerić).

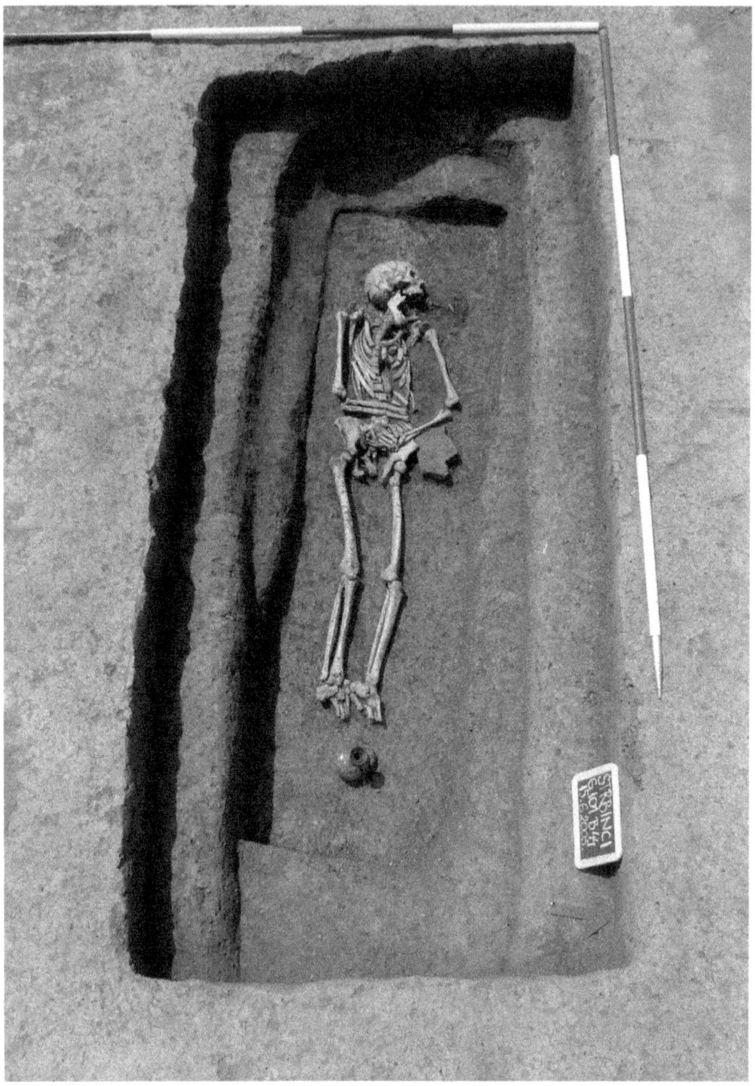

Fig. 27. Štrbinci, grave 101 featuring a wooden coffin (photo: T. Leleković).

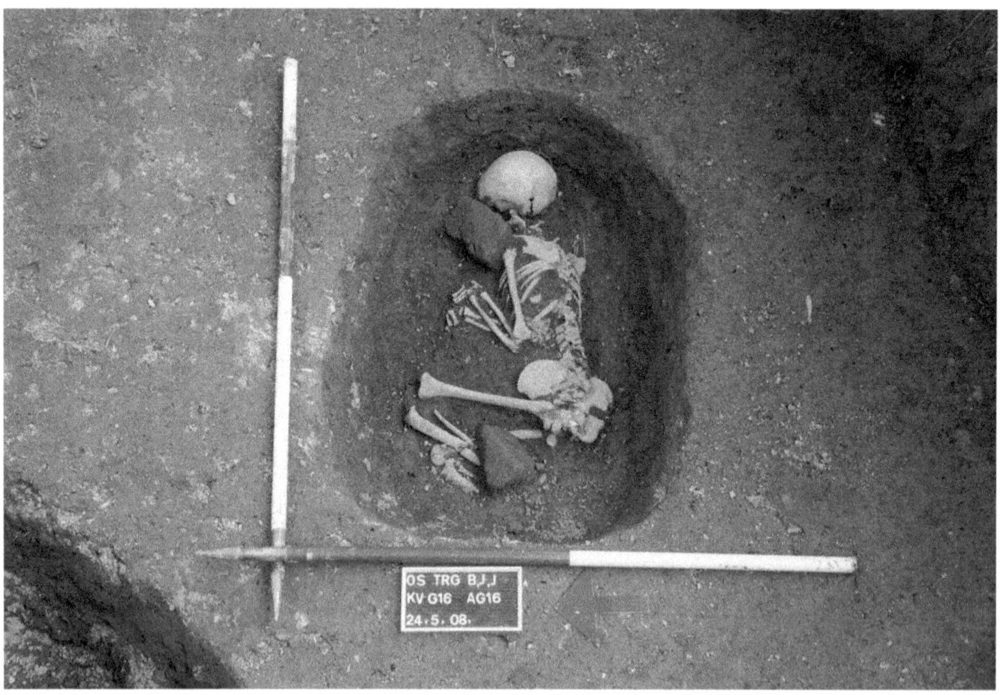

Fig. 28. Bana Jelačića Square, Osijek, grave 16: inhumation in foetal position, 2nd half of the 3rd c. (photo: T. Leleković).

Fig. 29. *Štrbinci, brick-masonry tomb 136, covered with horizontal tiles, 4th c.: before emptying of the fill (photo: T. Leleković).*

Fig. 30. *Štrbinci, brick-masonry tomb 136: inside (photo: T. Leleković).*

frequent shape of grave is a rectangular, oval, or triangular earth-cut pit. As transpires from the comparison between the late Roman cemeteries of Štrbinci and Zmajevac on the one hand and the transitional cemeteries of Mursa on the other, the employment of wooden coffins in late antique inhumations became more frequent (figs. 26.2 and 27). Coffins are usually evidenced by nails found in the corners of the pit, while in some cases they can be deduced from the earth colouring, revealing the coffins' dimensions.[155] The cemetery at Zmajevac is distinguished by a considerable pit-depth, measuring nearly 2 m, which is exceptional in comparison with other cemeteries featuring mostly shallow pits that rarely exceed 1 m in depth (Štrbinci, Tekić, Mursa). The Zmajevac cemetery is significant for some additional constructional features, unparalleled elsewhere in the region. Firstly, it is graves provided with holes to each long side, whose function was to facilitate the sinking of the coffin into the pit, further graves with niches in one of the longer sides to hold the bodies, and, finally, wooden roofs supported by posts at the four corners of the pit.[156] Another class of late antique graves in the area of study is represented by tile tombs constructed without mortar as a bondage material and with no inner plastering. In this group examples with gabled roofs (Siscia) are less frequent than those constructed in a rectangular pit by lining its sides with perpendicular tiles and covering the opening with horizontal tiles. Tiles could have been set on the rough bottom or in the grooves made along the edges, or on the tiled floor (fig. 26.4). When the roof tiles were set horizontally, the grave was of a box-like shape; examples from Mursa and Cibalae measure 2 x 0.40-0.60 m, while Siscia produced graves of a trapezoid cross section, resulting from slanted tiles of the sidewalls.[157] Some of the graves had gabled roofs constructed of a double row of tiles (Cibalae, Mursa, Zmajevac), measuring up to one metre in width. This type of grave has sub-variants, like that with lateral supports, evidenced in Mursa in both the transitional and late Roman periods. The support system could have contained one row of horizontal tiles, or more such superimposed rows. Another sub-variant of the gabled tomb features a horizontal tile to support those set gable-wise, producing a true triangular cross section (fig 26.3).[158] The Zmajevac cemetery has produced tombs with gabled roofs made of several rows of superimposed tiles, as well as those made of tiles to cut (or made in the shape?) to fit each other in the manner similar to dovetail wedging. In gabled-roofed

[155] For instance, grave 101 at Štrbinci, (2.10 x 0.60 m): Migotti 2009, 210.

[156] Filipović 2010, 19.

[157] Nenadić 1987, 86-87.

[158] Göricke-Lukić 2011, 27, 237-249.

Fig. 31. Štrbinci, brick-masonry tomb 74 covered with gable-wise set tiles, 4th c. (photo: T. Leleković).

Fig. 32. Štrbinci, brick-masonry tomb 70 covered with gable-wise set tiles (photo: T. Leleković).

Fig. 33. Štrbinci, detail of fig. 32: inside of tomb 70, 4th c. (photo: T. Leleković).

Fig. 34. Štrbinci, brick-masonry tomb 73 covered with gable-wise set tiles, 4th c. (photo: T. Leleković).

Fig. 35. Osijek (Obrtnička škola site): double tile-masonry tomb covered with gable-wise set tiles, 4th c. (after Filipović 2009).

tombs the bodies were put either directly on the earthen pit bottom or in wooden coffins, or, as evidenced in Siscia, in lead sarcophagi.[159] The third type is represented by tombs made of mortared tiles (fig. 26.6). The main part of such tombs is invariably a chest measuring 2-2.5 m in length by 0.60-2.50 m in width; their inner walls were mostly plastered (Štrbinci) and in some cases frescoed (Cibalae, Štrbinci), while the roofs were constructed either from horizontal (Štrbinci) (figs. 29 and 30) or gable-wise set tiles (Štrbinci, Zmajevac, Mursa, Cibalae, Siscia) (figs. 31-35). The site of Štrbinci produced tombs with the walls tapering towards the top to facilitate covering, producing the opening narrower than the bottom; the type of cover in most cases remained obscure due to the damage sustained by the upper parts of such tombs. Some of the tombs had stepped upper edges of the longer sides, evidently to support a gabled roof made of several superimposed rows of tiles (Mursa, Cibalae, Štrbinci) (figs. 31, 32, and 34).[160] A separate sub-type is represented by the tombs distinguished by relatively large dimensions. One such example from Štrbinci, preserved only in foundations, measuring 3.15 x 2.90 m. Its walls were made of whole and fragmented tiles and river pebbles bonded with mortar, with the floor made of mortared crushed tiles, and with the inner walls plastered and frescoed; its presumably gabled roof was constructed of tiles and *imbrices*. The eastern Mursan cemetery has produced a double-chambered tomb measuring roughly 3 x 4 m; the two chambers were connected along the longer sides, sharing the same gabled roof (fig. 35). This structure was encircled by a wall insufficiently preserved to allow safe identification: it might have been equally the enclosure wall and an aboveground building housing the tomb. No mention of possible inner decoration or grave goods of this tomb was made in the literature, leaving unclear its chronology in terms of burial phases.[161] Similar funeral structures have been evidenced in Cibalae, with one example featuring two connected graves whose dividing wall was cut in three places to provide communication.[162] Large tombs were invariably held more than one body, placed in individual compartments. Given their size, such tombs can be presumed as originally aboveground structures, either partly or completely. Large and sumptuously decorated *mausolea* such as those in Sopianae (Pécs, Hungary)[163] have not been found in northern Croatia. Foundations of a large building measuring 3.5 x 3 m, made of mortared stone and tiles, and with inner walls frescoed, have been recovered at Tekić, but its determination as a tomb or an object of funerary cult remains debatable.[164]

Irrespective of the type of grave, the way the body was placed in it was uniform: supine with the legs straight and the arms either at sides or with various positions of both hands or one of them, alternately. Hands could have been placed straight or folded on the pelvis or else folded on the chest; particular positions, such as hands to shoulders and the like, are exceptional. Exceptions from the usual supine skeletons were those in foetal position, deposited in oval

159 Nenadić 1987, 87, 85.
160 Filipović 2008, 17; Filipović 2010, 19; Göricke-Lukić 2011, 27, 220-223.
161 Filipović 2008, 17-18.
162 Šaranović-Svetek 1967; Dimitrijević 1979, 158-159.
163 Cf. Fülep 1977.
164 Sokač-Štimac 2008.

or triangular pits. Such graves have so far appeared only in Mursa (fig. 28) and Zmajevac; they held female deceased of possibly Germanic stock, as can be deduced by the grave goods.[165] Both graves with and without grave goods have been recovered, their ratio being approximately 40% to 60% in favour of the latter. Grave goods were prevalently represented by metal parts of dress accessories in men (brooches and/or belt buckles and strap ends), and items of jewellery in women. Both men and women, as well as children, were given vessels for drinking or serving drinks, most often a glass, dish, bottle or jug, with glass vessels prevailing. Children were often given apotropeic objects, mostly bullae and jewellery, and sometimes also brooches; jewellery is otherwise typical of female infants and sub-adults, as well as young women.[166] Some of the graves produced perfume bottles and *balsamaria*, continuing a tradition very typical of the transitional phase in Mursa. The same is true of utensils, although they now appear less frequently than before.[167] In general, grave goods were rather standardised in terms of categories and typology, bur with some exceptions in numbers and qualities of items.

4.2. Urban cemeteries

Archaeological research so far has not disclosed much about the urban profile and the extent of late antique Siscia, but has attested the existence of two cemeteries: northern and southeastern. In 1954 a cluster of 54 tombs were excavated in the latter (the site of Tomislavova Street), abutting on the town walls. A total of 47 tombs made of tiles without mortar were found; some of the skeletons were disposed in lead coffins. No data exists on the grave goods, except that one of the tombs contained an early Christian fingering, possibly dating the whole group to the 4th century (fig. 13). An additional argument for such date is a coin of Gordian III, recovered from the layer holding the tombs. Curiously, this layer consisted of a thick fill of the former fortification ditch, which came out of use in the 2nd half of the 3rd century to give place to a 4th-century cemetery. Some ten metres to the east of these tombs three sarcophagi were found, whose attribution to the late antique phase remains inconclusive.[168] Particular importance should be attached to tombs[169] and sarcophagi found in this area, several tens of metres to the west of the so far discussed finds, but within the town walls; they suggest the spread of the southeastern cemetery into the urban area, probably in relation to an early Christian site at the southern edge of the town.[170] A late antique section of the northern cemetery was discovered in 1943, when 19 tile tombs were excavated. Judging from the drawings, they were of the same shape as those in the southeastern cemetery. One of the tombs produced an early Christian clay lamp, probably dating the whole group to the 4th century.[171]

A bronze early Christian lamp in the form of a lamb holding a Chi-Rho on its head (4th to 5th centuries) stems from the area; it is a chance find from the 1970s, possibly originating from a destroyed grave.[172] No firm proof exists for a late antique phase of the southwestern Siscian cemetery on the site of Zibel, but it is suggested by the finds of masonry tombs holding stone chests; yet they could equally date from the transitional and late antique periods.[173] Apart from these finds, the site of Pogorelec, presumably holding industrial and harbour facilities, and from the 3rd century even the imperial mint, also produced quite a number of chance finds of inhumations. Unfortunately, these have not been dated, and those few archaeological excavations that have been conducted in the area have failed to yield any kind of burials. Arguably, the relationship between the town and its cemetery to the south of the right bank of the River Kupa, and a possible abandonment of that part of the town in favour of the cemetery, remain inconclusive.[174]

Two basic issues are the obstacles to a better understanding of the late antique phase of the cemeteries of Mursa. Firstly, it is the lack of larger clusters of graves securely dated to the 4th century; graves dated by coins are mostly individual finds lacking the archaeological context. Secondly, it is the fact that the outline of the town walls of Mursa still remains unknown. The location and layout of 4th-century Mursa has so far been based mostly on the town's mention in historical sources, one of them (Zos. 2.63-64) claiming that its strong walls resisted attacks during the battle between the emperors Constantius II and Magnentius in 351. Another set of data stem from the reconstruction made by R. Franjetić, which has been more or less refuted by the most recent archaeological excavations.[175] Until then, the above reconstruction was the only firm point for the chronology of burials, given that the majority of inhumations lacked grave goods. In other words, clusters of graves lacking finds would have been automatically attached to the nearest ones dated by coins. However, such procedure is highly questionable, especially if we bear in mind that the spread of the cemeteries at the expense of the settled area came about as early as the mid-3rd century. This means that inhumations recovered from abandoned buildings should not necessarily be dated to Late Antiquity and that at the same time the town's boundaries in the late phase should only be outlined on the basis of the distribution of graves securely dated to the 4th and 5th centuries. Such graves have been found in the entire Mursa's presumed area, that is, in all of its transitional cemeteries, attesting to their continuous use in Late Antiquity. The largest cluster of 4th-century burials have so far been found at the site of 5 Huttlerova Street in the northern cemetery, concentrated along the continuation of the main *cardo*. The western cemetery has yielded late graves dispersed across a relatively wide area, covering the places that in the Severan period held both a cemetery and a settlement, respectively. On the other hand,

165 Filipović 2010, 21.
166 Migotti 2007c.
167 Buzov 2002, 180, fig. 7.
168 Nenadić 1987, 85-90; Buzov 2002, 180-181; Lolić 2003, 141.
169 Lolić 2001, 96.
170 Migotti 1997 22-23; Artner 2003, 32-33; Vrbanović 1981, 189-199, Vukelić 2009, 117.
171 Nenadić 1987, 84; Buzov 2002, 179.
172 Migotti 1994, 92; Migotti 1994, 77-78; Migotti 1997, 77; Migotti 2011, 56.
173 Artner 2003, 33; Leleković 2009b.
174 See note 62.
175 See note 74.

the lack of a compact late antique cemetery at this site is possibly the result of inadequate research, conditioned by the vast damage to the area in the modern period. A similar dispersion of late graves can be discerned in the southern cemetery. However, excavations at 120 Divaltova Street proved that such circumstances did not result from a loss of archaeological material through damage, but probably revealed a realistic picture of the site. The eastern cemetery has so far yielded the scantiest traces of the late antique period. Some 4th-century graves have been found among burials of the later transitional phase, revealing horizontal stratigraphy resulting from the spread of the cemetery towards south in the 4th century. In the present context a historical source should be mentioned (Zon. 13. 8), claiming that the emperor Constantius II had 54,000 soldiers buried at the abovementioned battlefield nearby Mursa.[176] Neither such cemetery nor the stadium figuring as the site of battle has been archaeologically proved.

In terms of late Roman cemeteries Cibalae is the best researched town in northern Croatia and one that produced the greatest number of 4th-century burials. So far 220 inhumations have been recovered there, mostly grouped in clusters numbering from nine to 72 graves. Given that the majority of them lack data on the grave goods, it remains unknown how many of them can indeed be dated to the 4th century. Graves securely dated to the 4th century have been documented in all four cemeteries stretched along the roads leading to the town, but the relationship between the urban core and burial areas remains unresolved. Although the course of the town walls and ditches has been established, the chronology of the fortifications remains disputed in that it is not clear whether in the 4th century they were in function in their full perimeter, or whether they shrank.[177] In the northern part of the town inhumations have been evidenced within remains of architecture, possibly indicating shrinkage of the urban area in Late Antiquity and a concomitant spread of cemeteries over the town's peripheral sections. [178] The best researched so far is the northern cemetery, yielding 67 late graves distributed in three clusters along the northern town walls. These were probably situated in the vicinity of the main gates as parts of the cemetery that comprised several-hundred-metre broad zone stretched along the road leading towards Mursa and Cornacum; its largeness, as well as the number and quality of grave goods, renders this cemetery the most prestigious of all Cibalitan burial areas. Several among the finds recovered there should be pointed out: a frescoed masonry tomb of a child, the only burial in a lead coffin so far in Cibalae, several stone sarcophagi and tombstones, and several tile tombs measuring 2 by 2 m, holding some ten-odd deceased; the last-named were interpreted as mass graves of people who died in one of the epidemics or incursions that struck Cibalae in Late Antiquity. In recent years large-scale rescue excavations were carried out in the areas of the western and southern cemeteries, but the results have not yet been published, preventing a better insight into the finds. The western cemetery has yielded 52 inhumations, of which 34 in pits and 18 in tile-constructed tombs, revealing a sound concentration of graves in the centre of the area.[179] Prior to this, smaller clusters of inhumations placed along the road towards Siscia have been recovered, but more specific data is missing. Excavations in 2000 recovered 73 4th-century inhumations in the southern cemetery. The graves were arranged in sub-regular rows, with the majority of them oriented W-E (head to the west) and some placed along the N-S line.[180] The eastern cemetery has so far been the least explored, revealing randomly scattered clusters of inhumations and individual skeletal burials from the 3rd to 5th centuries. The core of the cemetery remains unknown, although it should have been concentrated along the road leading towards Sirmium.[181] An important site of Kamenica is placed 1,500 m to the east of the town walls. It has produced, mostly as chance finds and through one small-scale excavation, a considerable amount of stone and marble architectural fragments and sculpture, as well as inscriptions, supporting the hypothesis that it was the site of an early Christian church and a cemetery. On the basis of these finds, as well as the hagiographic quotation claiming that the local martyr Pollio was executed at a place one Roman mile (cca 1,500 m!) away from the town, a Christian martyrial complex has been hypothesised at Kamenica. It remains unclear whether the site of Kamenica was part of the eastern cemetery or a rural complex in the surroundings of Cibalae. [182]

The remainder of the north-Croatian Roman towns have yielded only very scanty remains of late antique cemeteries. Aquae Balissae/*Municipium Iasorum* has produced only two sarcophagi and three masonry tombs found by chance in 1819; the tombs were reported as having been decorated in mosaics. A sumptuous 4th-century cage cup (*vas diatretum*) was also probably recovered in one such tomb. The position of the above finds point to two main late Roman cemeteries of Aquae Balissae: the northern and the southern.[183] The same low rate of research holds true for Andautonia, with only few finds possibly, but not definitely, dating from Late Antiquity: remains of a masonry tomb (see section 3.2) and two sarcophagi, one each stone and lead, all found in the southern cemetery (Kutelo).[184] While no graves from the late phase could be determined as Roman, some of them were of Germanic affiliation, judging by the grave goods. One such 5th-century grave was found within the earlier urban area, probably testifying that in Late Antiquity the town shrank, giving place to burials along its fringes.[185] In Ludbreg (Iovia) stone sarcophagi have been recorded, although their chronology remains unknown. A bishopric

176 Caldwell 2007, 159.
177 Iskra-Janošić 2004, Abb. 10.
178 Iskra-Janošić 2004, 184.

179 Hrvoje Vulić, pers. com.
180 Iskra-Janošić 2004, 184; Vulić and Rapan Papeša 2009, 53-55; Vulić and Rapan Papeša 2009, 58-60.
181 Dimitrijević 1979, 158-160; Iskra-Janošić 2001, 185.
182 Migotti 1994, 48; Migotti 197, 22.
183 Schejbal 2004, 99-129, 112-114, 118.
184 Kušan Špalj and Nemeth-Ehrlich 2003, 122.
185 Vikić- Belančić 1981, 134-143; Nemeth-Ehrlich and Kušan Špalj 2007, 44.

has been hypothesised there, furnishing a broad context for a possible late date (4th and 5th centuries) of these finds.[186]

4.3. Cemeteries of small towns and rural settlements

In Pannonia, as elsewhere, Late Antiquity is the period of an increased growth of the settlements that cannot fully be designated as either rural or urban; the most appropriate term for them seems to be *small towns*. They mostly grew along the main roads, acquiring in time the importance and role of minor regional centres.[187] In the territory of study, such settlements, or presumable sites of small towns, have been tentatively identified on the basis of several elements. One of them is certainly a correspondence of a site with the data furnished by Roman itineraries and the Peutinger Table. Still, it mostly remains disputed whether so identified small towns were in effect late Roman *civitates* or settlements lacking administrative rank, or even just road stations (*mansio*, *mutatio*). Two such sites in northern Croatia, tentatively identified as itinerary settlements, have produced late antique cemeteries. One of them is Tekić (Incerum?) and another is Štrbinci (Certissia?). Unlike Štrbinci, with only the cemetery known, the site of Tekić has yielded architectural remains. Nevertheless, the dimensions of the cemeteries on both these sites prove that the related settlements were small towns rather than rural villas or villages.

The remains of the settlement and the cemetery at Tekić enable the study of the site as a potentially coherent whole, most probably a small town. Remains of thermal baths dating from the 1st to the 4th centuries prove an early start of the settlement, which, however, reached its full potential only in the 4th century, when it became the centre of an equally prosperous wider region.[188] The researched part of the cemetery encompassed an area of 1,600 square metres, yielding 125 graves, 25 of which were tombs of various forms, with all the types mentioned above represented and with some 60% of the graves holding grave goods. The most numerous were earth-cut graves, with 12 examples giving evidence for a wooden coffin; the remainder of the bodies was deposited directly in the pit. Components of the infrastructure have also been explored, above all the cemetery lane made of cobbles. Foundations of a larger building measuring 3 x 3.5 m have also been investigated, but its interpretation as a tomb or a funerary cult facility cannot be securely established. Regrettably, no layout of the cemetery exists in the literature, preventing the study of the spatial relationship between its various sections and components. All we know at this stage is that a 4th-century cemetery grew west of the town, probably along the road leading from Incerum towards Siscia.

At the site of Štrbinci near Đakovo late Roman cemeteries have been discovered at three various locations, with the distance among them amounting to 500 m. The site is a probable location of Certissia, a settlement figuring in the itineraries as a crossroad point on the road Emona – Sirmium, at which another road branched off, leading towards the Sava and further south into the province of Dalmatia. Remains of settlements have not been attested, but can be hypothesised on the basis of the cemeteries' locations and the lie of the land. The excavated section of the southeastern cemetery measures some 1,600 square metres, yielding 160 graves, 91 of which were earth-cut pits and 42 tile tombs, while in 10 of them remains of wooden coffins could be attested on the basis of nails and/or earth colouring. So far only the northern limit of this cemetery has been established, possibly corresponding to the course of the road leading from Certissia towards Cibalae. At the end of the 19th century a building with mosaic tiles, some of them gold-in-glass, was found by chance and destroyed, leaving no documentation apart from a mention in the literature. It should have been a sumptuous late Roman tomb, such as the one from Aquae Balissae,[189] or possibly even an early Christian church. In the vicinity, two tile tombs were excavated in the aftermath of the previous find, dated to the 4th century by grave inventories.[190] Some 300 m to the north of the southwestern cemetery, during a watching brief in 1966 two tile tombs were discovered and excavated. Both were dated by coins to after the year 320, and both featured lateral niches at each side, intended for the grave goods, giving them roughly the shape of the Latin cross.[191] Since the urban part of the settlement at Štrbinci has not been archaeologically confirmed, it cannot be finally established whether the burials in the southwest and northwest parts of the site were sections of one and the same (western) cemetery, or whether they made two separate burial areas. The latter suggestion seems more likely, so that we can expect altogether three cemeteries stretched along the roads leading towards Sirmium (southeastern), Incerum (northwestern) and Marsonia (southwestern).

In the vicinity of the village of Sladojevci, along the road Poetovio – Mursa, a section of a cemetery comparable to those just described has been discovered. It contained seven inhumations in earth-cut pits, oriented W-E and dated by coins to the 2nd half of the 4th century. While the grave goods were standard, such as bronze and glass bracelets and pottery vessels, this cemetery is distinguished by the fact that some of the graves held a considerable number of coins: 5, 12, 28, and even 35 to a burial.[192] Since these graves are the only trace of a settlement existing at the spot, there is no knowing whether they belonged to a village or villa, or perhaps a smaller settlement grown along the Poetovio – Mursa road. In the town of Hrvatska Dubica, the presumed location of the station Ad Praetorium on the road Siscia – Sirmium,[193] a 4th-century inhumation grave dated by coins was excavated in 1980. Apart from standard grave goods, such as two belt buckles and a brooch, it contained less common objects: a carpenter's axe and an iron clasp

186 Migotti 1997, 23-25.
187 Burnham and Wacher 1990, 1-7, 15-32.
188 Sokač-Štimac 2005, 9.

189 Schejbal 2003, 113-114.
190 Migotti 1998, 94.
191 Raunig 1980; Migotti 1997, 38.
192 Spajić 1967.
193 Gračanin 2010, 17.

knife. Besides this burial, a further six inhumation pit graves were excavated, as well as an unknown number of disturbed ones. If to these we add earlier finds of a lead sarcophagus and a frescoed tomb, of which only written records remain, a hypothesis of a small town's cemetery seems likely.[194] A plain, presumably 4th-century stone sarcophagus has been found in the village of Petrijanec (Aqua Viva?), suggesting a late antique cemetery of a settlement placed along the road Petovio – Mursa.[195]

In addition to the above sites, tens of others have been catalogued in all of the rural area of the Croatian part of Pannonia, producing either individual inhumations or small groups of them, or else finds interpretable as (late Roman) grave goods. It seems that sporadic late Roman chance finds, mostly tile-tomb inhumations, abound in the easternmost part of the area of study, especially around Cibalae, but also in the nearest hinterland of the *limes* and the settlements along the road Siscia – Cibale – Sirmium. Regrettably, their archaeological context mostly remains unknown. For example, late Roman graves found in the village of Stari Mikanovci were interpreted as indicating a *villa rustica*, while those in Oriolik are believed to suggest a road station.[196] Moreover, the lack of data on the grave goods prevents their classification into the transitional or late Roman phase. In spite of all that, chance finds of inhumations in the surrounding of Vinkovci testify to a 3rd-century population growth and probably also economic development in the territory of Cibalae, in comparison to the 1st and 2nd centuries. The same seems to be true of the Slavonian Mountains around the Požega valley, whose most important site in terms of late antique prosperity is Tekić, the presumed location of Roman Incerum. It has, among others, yielded quite a lot of chance finds of inhumations and late antique tombs, outnumbering those from the transitional phase. The hilltop settlement at the site of Rudina in the Slavonian Mountains can be added to that of Tekić, on account of a fragment of a sarcophagus found there.[197] On the other hand, the territories of Andautonia and Siscia seem to be underrepresented in terms of late Roman rural cemeteries. Apart from the cemeteries of the abovenamed small towns, data on inhumations in the rural areas is practically non-existent. The only exception is the surroundings of Zagreb, with three sites yielding late inhumations or probable grave goods of the period.[198] Featuring prominently among late Roman sites in the western part of northern Croatia are hilltop settlements, such as Popov dol and Kuzelin, both yielding sections of late cemeteries.[199] The rest of the area of study, such as the section along the Drava with the region of Baranja, are almost completely devoid of late Roman graves, which comes as a surprise given the importance of these regions in Late Antiquity. This, however, should be the result of the low rate of research, rather than the archaeological reality.

4.4. Cemeteries on the limes

Although numerous reports on chance finds of inhumations on a number of *limes* sites have been noted, only two of them reveal well defined late antique cemeteries, the best known being Zmajevac (Ad Novas). Next to the fort clearly outlined on the Várhegy hill, systematic field walking has revealed the sites of four late antique cemeteries. The best known of them is the cemetery at Mocsolás, a site situated on an elevated ground south of the road, some 300 m north of the fort. The so-far excavated area measures 1,700 square metres, yielding a total of 175 graves organised in neatly arranged rows running from NE to SW. In the southernmost fringe of the cemetery the orientation shows some slight deviation from the remainder of the cemetery. The most conspicuous features of this cemetery are extremely deep pits, some of which featured wooden constructions (see section 4.1), while there were also 17 tile-constructed tombs. In comparison with the cemeteries at the sites of Tekić and Štrbinci, this one seems to be richer in both the quantity and quality of grave goods, pointing to a high status the military enjoyed in the region. A Germanic component is discernible in some of the graves, especially those holding flexed female skeletons furnished with typical Germanic jewellery.[200] The remaining cemeteries have been found nearer to the *limes* road running along the foot of the fort, following the course of the gully cut into a loess terrace. Owing to a constant erosion of the terrace, landslides abound, often revealing graves and other features in the terrace's cross section. In this way two clusters of graves have been spotted, one on the southeast slope of the Várhegy hill, and the other to the north of the Mocsolás cemetery, possibly its integral part. Traces of a cemetery have been spotted as cropmarks in the opposite part of the valley to the northeast of the fort, on the account of their resemblance to tile tombs in shape and dimensions. Despite the lack of surface finds, a late antique cemetery can be reasonably presumed.[201] Some ten kilometres to the west of Zmajevac, nearby the village of Kamenac, another Roman fortified site has been located and tentatively identified with Mons Aureus (*Itin. Ant.* 243; *Not. Dign. Occ.* 32, 92, 45) Excavations carried out at the site during the Second World War brought to light remains of architecture and a related cemetery. Judging from available descriptions, the graves there are comparable to those from Mocsolás, evidently revealing appurtenance to the same chronological and cultural circle.[202] Some 2 km deeper into the *limes* hinterland, surface finds of a late antique cemetery have been traced at the site of Karanac, possible adjacent to a country villa or settlement.[203] Finally, several stone sarcophagi found in Ilok (Cuccium) possibly testify to a late

194 Mutavđić 1900, 226: Koščević and Makjanić 1985, 121-122.
195 Šimek 1997, 58.
196 Dimitrijević 1979, 185-187.
197 Migotti 1994, 61, 125; Migotti 1997, 27-28.
198 Gregl 1997, 18, 41 (Zagreb-Maksimir); Migotti 1997, 77-78 (Zagreb-Mirogoj).
199 Gregl 1992, 149-150; Sokol 1994, 199-209; Sokol 2006, 154-155.

200 Filipović 2010.
201 The field walking was carried out in 2008 and 2009: Bojčić *et. al.* 2010.
202 Bulat 1977, 63-86; Minichreiter 1989, 182.
203 Bulat 1969, 44; Minichreiter 1984, 34; Minichreiter 1989, 182.

Roman section of the cemetery stretched along the road of this important *limes* settlement.[204]

5. Conclusion

The development of funerary ritual in the researched area unfolded in several stages. During the 1st century it was fairly uniform, with the deceased cremated on *ustrinae* and mostly buried as *ustrinata*, that is, pits filled with the pyre debris. The Flavian period witnessed some change and variation in ritual, with *busta* starting to appear in the eastern and Norico-Pannonian tumuli in the western areas. The 2nd century is characterised by a considerable variety in burials, evident also in their territorial distribution. New types of graves appeared in towns, such as enclosures containing tile or masonry tombs, as well as sarcophagi, while both urban and rural 2nd-century cemeteries feature earth pits lined and closed with tiles. Changes are also evident in the equipment of rural tumuli, although it is not clear whether these changes started to occur in the 2nd or 3rd centuries. Also, due to the inadequate evidence, the period of the predomination of inhumations remains unsettled. Nevertheless, the example of Mursa suggests that inhumation came in use during the rule of Marcus Aurelius, to prevail over incineration in the Severan period. Incineration still remained in use deep into the 3rd century, in the form of *busta* (flat *busta* under tumuli in the western area and earth-pit *busta* in the eastern). The 4th century once again witnessed a homogenisation of the funerary ritual, with standardised inhumations in earth pits or tile tombs as the only types of grave.

The above survey should, however, be taken with some reserve, given the uneven state of research of various sections of the studied area. A good example is an unbalanced evidence for Siscia and Andautonia. Of a total of 51 cemeteries, as many as 25 come from the wider Zagreb area, that is, the Andautonian municipal territory, while the territory of Siscia has yielded only six cemeteries, four of which belonged to the town itself (Table 1). Another accidence of archaeology is well exemplified by the case of Siscia, given that in the 2nd half of the 1st century BC and the 1st half of the 2nd century AD its territory accommodated a legion camp and a civil settlement.[205] Even though, no cemeteries (presumably quite rich and long-lasting) of this period have been found so far, adding to a misleading impression of the earliest period of Roman Siscia in general, the cemeteries included. The lack of finds from the period of the Roman conquest in the whole of northern Croatia leads to the conclusion that it lacked Roman(ised) settlements of any importance in the pre-Flavian time. On balance, so far the Iulio-Claudian period remains the least explored part of the Roman history of southern Pannonia. Until recently, the same held true for the Severan and post-Severan periods as well, which would come as a surprise given the general prosperity of the province of Pannonia in the 3rd century. Until recently, the understanding of the 3rd-century funerary archaeology of northern Croatia was much hampered by the chronological frame outlined in the literature, which postulated that cremation graves should be dated to the first two centuries and inhumations to the 4th, with only rare examples of the latter that were dated to the 3rd century by coins or other grave goods. Within such frame, until 2000 only several tens of 3rd-century inhumation graves were known. Owing to the excavations during the last decade, this number grew to comprise 462 graves, to which an unknown number of a further 544 undated inhumations should be added. In a word, a decade of excavations has completely changed the state of research of this topic, as well as our understanding of the 3rd century in southern Pannonia, given that as many as 350 transitional graves date from the 250s to 260s; both through their sheer number but also their inventories, these graves testify to the fact that the economic growth, started with the Severans, continued in the 2nd half of the 3rd century. To a certain point, such conclusion is bolstered by the finds of 3rd- and 4th-century stone sarcophagi and tile and masonry tombs from Siscia.

The number of late antique burials approximates that from the transitional phase, featuring, however, inner disproportions at two levels. The most conspicuous imbalance is evident in the comparison between the western and eastern sections of the area of study. Of a total of 423 inhumations from the 4th and 5th centuries only 47 were excavated in Sisak (Pannonia Savia), while the rest mostly come from the cemeteries of smaller towns in Pannonia Secunda. Each of these sites has yielded more graves individually than has Siscia, which is evidently illogical and therefore certainly not accurate. In other words, the disproportion in the number of late burials both at the level of *urban centres versus small towns* and *the western versus the eastern portion* of the territory under study, can be considered as resulting from the state of research rather than reflecting the real state of matters.

On balance, the above survey of the state of research of the Roman-period funerary archaeology of northern Croatia should be taken as provisional and subject to a redefinition with new finds to come in future. At this stage, the evidence does not enable resolving of socio-historical issues, as an otherwise widely appreciated theoretical potential of funerary archaeology.[206]

Acknowledgements

I would like to express my gratitude to Branka Migotti, Željko Tomičić, and Hrvoje Vulić for the valuable information and suggestions that much improved my paper.

Bibliography

Abegg-Wigg 2008 Angelika Abegg-Wigg, Die Aschengruben im Kontext der provinzialrömischen

[204] In the Town Museum of Ilok sevearal fragments and two fully preserved sarcophagi are on display, while another nearly fully preserved piece is kept in the MSO; on the latter see Göricke-Lukić 2011, 71.

[205] Radman-Livaja 2010, 179-193.

[206] See Morris 1992.

Bestattungszeremonien. Problematik und Analysemöglichkeiten, in: Scheid 2008, 243-257.
Adler-Wölfl 2004 Kristina Adler-Wölfl, *Pannonische Glanztonware aus dem Auxiliarkastell von Carnuntum. Ausgrabungen der Jahre 1977-1988*, Wien.
Andreae 1963 Bernard Andreae, *Studien zur römischen Grabkunst*, Heidelberg.
von Artner 2003 Therese von Artner, Od Zagreba do Siska, *Riječi, Časopis za književnost, kulturu i znanost*, 1-2, Sisak, 31-38.
Barkóczi, Erdélyi and Ferenczy 1954 László Barkóczi, Gizella Erdélyi, and Endre Ferenczy, *Intercisa I (Dunapentele- Sztalinvaros): Geschichte der Stadt in der Römerzeit*, Budapest.
Bechert 1980 Thomas Bechert, Zur Terminologie provinzialrömischer Brandgräber, *AK* 10, 253-258.
Bíró 2007 Szylvia Bíró, Gräber und Funde. Römische Gräberfeld auf dem Gebiet der Waggonfabrik, in: Szylvia Bíró, Attila Molnár, and Andrea Nagy, *A vagongyár alatt - a vagongyár előtt: római temető és középkori település a győri Árkád területén: kiállitásvezető / Unter der Waggonfabrik - vor der Waggonfabrik: römisches Gräberfeld und mittelalterliche Siedlung auf dem Gebiet des Győrer Einkaufszentrums Árkád*, Exhibition Catalogue, Győr, 33-62.
Bojčić *et al.* 2010 Zvonko Bojčić, Marko Dizdar, Tomislav Hršak, Igor Vukmanić, Domagoj Dujmić, and Tino Leleković, Terenski pregled područja između Batine i Suze, *AIA* 6, 80-87.
Božič 2008 Dragan Božič, *Late La Tène-Roman cemetery in Novo Mesto, Ljubljanska cesta and Okrajno glavarstvo. Studies on fibulae and on the relative chronology of the Late La Tène period / Poznolatensko-rimsko grobišče v Novem Mestu: Ljubljanska cesta in Okrajno glavarstvo. Studije o fibulah in o relativni kronologiji pozne latenske dobe*, *KiM* 39.
Braidotti 2009 Elena Braidotti, Lucerne dall'area delle grandi terme di Aquileia. Scavi dell'univeristà di Udine, *AN* 80, 85-96.
Brukner 1981 Olga Brukner, *Olga, Rimska keramika u jugoslovenskom delu provincije Donje Panonije*, *DMBg* 24.
Bugar 2009 Aleksandra Bugar, Šepkovčica, *HAG* 5 (2008), 269-273.
Bulat 1969 Topografska istraživanja Limesa u Slavoniji i Baranji, *OZ* 12, 39-52.
Bulat 1977 Mirko Bulat, Stanje istraživanja antičkih naselja u Slavoniji, in: *Naseljavanje i naselja u antici*, *MatSADJ* 13, Beograd, 63-91.
Bulat 1977 Mirko Bulat, Rimski paljevinski grob iz Osijeka, *OZ* 14, 79-87.
Burkowsky 1993 Zdenko Burkowsky, Nekropole antičke Siscije, *PIAZ* 10, 69-80.
Burnham and Wacher 1990 Barry C. Burnham and John Wacher, *The Small Towns of Romain Britain*, Berkeley-Los Angeles.
Buzov 2002 Marija Buzov, Grad mrtvih uz grad živih – nekropole Siscije, *HAnt* 8, 175-192.
Caldwell 2007 Craig H. Caldwell III, *Contesting Late Roman Illyricum: Invasions and transformations in the Danubian-Balkan provinces*, Unpublished PhD thesis, Princeton University.
Demo 1985 Željko Demo, Grobni ritus u koprivničkoj Podravini tijekom 2. st. n. e., in: *Sahranjivanje pokojnika sa aspekta ekonomskih i društvenih kretanja u praistoriji i antici*, *MatSADJ* 20, Beograd, 111-126.
Dimitrijević 1979 Stojan Dimitrijević, Arheološka topografija i izbor arheoloških nalaza s vinkovačkog tla / Archäologische Topographie und Auswahl archäologischer Funde vom Vinkovcer Boden, in: *Corolla memoriae Iosepho Brunšmid dicata, IzdHAD 4*, Vinkovci, 201-268.
Dizdar 2010 Marko Dizdar, Inventar groba 5 / Inventory of grave 5, cat. 29, in: Radman-Livaja 2010a, 244-245.
Dizdar and Radman-Livaja 2004 Marko Dizdar and Ivan Radman-Livaja, Nalaz naoružanja iz Vrtne ulice u Vinkovcima kao prilog poznavanju rane romanizacije istočne Slavonije, *PIAZ* 21, 37-53.
Dizdar, Tonc and Ložnjak Dizdar 2011 Zaštitna istraživanja nalazišta AN 6 Gornji Vukojevac na trasi auto-ceste Zagreb-Sisak, dionica Velika Gorica jug – Lekenik, *AIA* 7, 61-64.
Durman 1992 Aleksandar Durman, O geostrateškom položaju Siscije, *OpArch* 16, 117-131.
Durman 2006 Aleksandar Durman (ed.), *Stotinu hrvatskih arheoloških nalazišta*, Zagreb.
Egri and Rustoiu 2009 Mariana Egri and Aurel Rustoiu, *The Social Significance of Conviviality in the Scordsican Enviroment. Funerary Practices of the Bronze and Iron Ages in Central and South-eastern Europe*, Cluj-Napoca, 83-93.
Ertel 1999 Christine Ertel, Die Notgrabungen im Gräberfeld südlich der Zivilstadt 1984 – 1986. Grabbbauten, in: Gassner *et al.* 1999, 32-34.
Faber and Fasold 2007 A. Faber and P. Fasold (eds.), *Körpergräber des 1.- 3. Jahrhunderts in der römischen Welt*, Internationales Kolloquium Frankfurt am Main, 19.-20.11.2004., Frankfurt am Main.
Filipović 2004 Slavica Filipović, Colonia Aelia Mursa, in: *Pannonia* II, 157-168.
Filipović 2008 Slavica Filipović, Osijek – Donji grad (Obrtnička škola), *HAG* 4 (2007), 17-18.
Filipović 2010 Slavica Filipović, *Kasnoantička nekropola u Zmajevcu-Mocsolás / A Late Antique Necropolis in Zmajevac-Mocsolás*, Exhibition Catalogue, Osijek.
Fülep 1977 Ferenc Fülep, *Roman Cemeteries on the Territory of Pécs, Sopianae*, Budapest.
Gaitzsch and Werner 1993 Wolfgang Gaitzsch and A. Werner, Rekonstruktion einer Brandbestattung vom Typ Bustum aufgrund archäologischer Befunde aus Siedlungen der Jülicher Lössbörde, in: Struck 1993a, 55-67.
Gassner *et al.* 1999 Verena Gassner, Christine Ertel, Sonja Jilek, and Herma Stiglitz (eds.), *Untersuchungen zu den Gräberfeldern in Carnuntum*, Wien.
Göricke-Lukić 2000 Hermine Göricke-Lukić, *Sjeveroistočna nekropola rimske Murse*, Zagreb-Osijek.

Göricke-Lukić 2011 Hermine Göricke-Lukić, *Nekropole rimskodobne Murse*, Osijek.

Gračanin 2010 Hrvoje Gračanin, Rimske prometnice i komunikacije u kasnoantičkoj južnoj Panoniji, *SSlav* 10, 9-69.

Gregl 1989 Zoran Gregl, *Rimskodobna nekropola Zagreb – Stenjevec*, Zagreb.

Gregl 1992 Zoran Gregl, Antičko nalazište Repišće kod Jastrebarskog, *VAMZ* 24-25 (1991-1992), 145-150.

Gregl 1997 Zoran Gregl, *Rimske nekropole sjeverne Hrvatske*, Zagreb.

Gregl 2000 Antičke kružne grobnice s dromosom u Hrvatskoj, *OpArch* 23-24, 165-169.

Gregl 2007 Zoran Gregl, Rimskodobna nekropola Gornja Vas na Žumberku, *VAMZ* 40, 2007, 221-331.

Gregl and Jelinčić 2010 Zoran Gregl and Kristina Jelinčić, O nekim manje poznatim antičkim lokalitetima u Zagrebu i okolici, *VAMZ* 43, 153-191.

Gregl and Migotti 2003 Zoran Gregl and Branka Migotti, Civitas Iovia (Botivo), in: *Pannonia* I, 131-144.

Heinzelmann 2001 Michael Heinzelmann, Einleitung, in: M. Heinzelmann, P. Fasold, J. Ortalli, and M. Witteyer (eds), *Römischer Bestattungsbrauch und Beigabensitten in Rom, Norditalien und den Nordwestprovinzen von der späten Republik bis in die Kaiserzeit, Palilia* 8, Wiesbaden, 11-30.

Hoti 1992 Marina Hoti, Sisak u antičkim izvorima, *OpArch* 16, 133- 163.

Hutinec 2010 Mirela Hutinec, Sotinska groblja u antici, in: *Arheološke spoznaje o Sotinu. Rezultati probnih istraživanja 2008- 2010*, Vukovar, 29-33, 50-62.

Iskra-Janošić 2001 Ivana Iskra-Janošić, *Urbanizacija Cibala i razvoj keramičarskih središta*, Vinkovci.

Iskra-Janošić 2004 Ivana Iskra-Janošić, Colonia Aurelia Cibalae, Entwicklung der Stadt, in: *Pannonia* II, 169-196.

Istenič 1999 Janka Istenič, *Poetovio, zahodna grobisca I: grobne celote iz Dezelnega muzeja Joanneuma v Gradcu / Poetovio, the western cemeteries I. Grave groups in the Landesmuseum Joanneum*, Graz, *KiM* 33.

Jilek 1999 Sonja Jilek, Die Notgrabungen im Gräberfeld südlich der Zivilstadt 1984 – 1986: Gräbertypologie, in: Gassner *et al.* 1999, 20-31.

Jovanović 1984 Aleksandar Jovanović, *Rimske nekropole na teritoriji Jugoslavije*, Beograd.

Jovanović 2000 Aleksandar Jovanović, Romanization and ethnic elements in burial practice in the southern part of Pannonia Inferior and Moesia Superior, in: J. Pearce, M. Millett, and M. Struck (eds), *Burial, Society and Context in the Roman World*, Oxford, 204-214.

Klemenc 1928 Josip Klemenc, Novi rimski vojnički natpis iz Osijeka, *VHAD* 15, 271–274.

Knez 1992 Tone Knez, Novo Mesto II: Keltisch-römisches Gräberfeld. Beletov vrt, Novo Mesto.

Koščević and Makjanić 1986 Remza Koščević and Rajka Makjanić, Neki noviji arheološki rezultati s područja antičke Siscije, in: *Arheološka istraživanja na karlovačkom i sisačkom području, IzdHAD* 10 (1985), Zagreb, 119-136.

Koščević and Makjanić 1988 Remza Koščević and Rajka Makjanić, Antički tumuli kod Velike Gorice i nova opažanja o panonskoj radionici glazirane keramike, *PIAZ* 3-4 (1986-1987), 25-70.

Kremer 2001 Gabrielle Kremer, *Antike Grabbauten in Noricum. Katalog und Auswertung von Werkstücken als Beitrag zur Rekonstruktion und Typologie*, Wien.

Lányi 1972 Vera Lányi, Die spätantiken Gräberfelder von Pannonien, *AArchHung* 24, 53-213.

Leleković 2009a Tino Leleković, Sisak – I. Gundulića 28, *HAG* 5 (2008), 293-296.

Leleković 2009b Tino Leleković, Osijek – Trg bana Jelačića, *HAG* 5 (2008), 45-51.

Leleković 2010 Tino Leleković, OS_DIV_120, *Rimska nekropola iz doba Severa*, Osijek.

Lolić 2001 Tatjana Lolić, Trb bana Jelačića u Sisku, *ObHAD* 23/3, Zagreb, 95-99.

Lolić 2003 Tatjana Lolić, Colonia Flavia Siscia, in: *Pannonia* II, 131- 152.

Lőrincz 2001 Barnabás Lőrincz, *Die römischen Hilfstruppen in Pannonien während der Prinzipatszeit, Teil I: Die Inschriften*, Wien.

Migotti 1994 Branka Migotti, Arheološka građa iz ranokršćanskog razdoblja u kontinentalnoj Hrvatskoj / The Archaeological Material of the Early Christian Period in Continental Croatia, in: Ž. Demo (ed.), *Od Nepobjedivog Sunca do Sunca Pravde*, Zagreb, 41-67.

Migotti 1996 Branka Migotti, Ranokršćanski nalazi iz Velikih Bastaja kod Daruvara, *VAMZ* 28-29 (1995-1996), 127-156.

Migotti 1997 Branka Migotti, *Evidence for Christianity in Roman Southern Pannonia (Northern Croatia). A catalogue of finds and sites,* BAR *Intern. Ser.* 684.

Migotti *et al.* 1998 Branka Migotti, Mario Šlaus, Zdenka Dukat, and Ljubica Perinić, *Accede ad Certissiam. Antički i ranokršćanski horizont arheološkog nalazišta Štrbinci kod Đakova*, Zagreb.

Migotti 2004 Branka Migotti, Kasnoantičko groblje u Štrbincima kod Đakova – iskopavanja 2001, *ARR* 14, 141-246.

Migotti 2005 Branka Migotti, The Ash-chest of Marcus Aurelius Glabrio from Siscia reconsidered, in: *Illyrica antiqua. Ob honorem Duje Rendić- Miočević*, Zagreb, 367-384.

Migotti 2007a Branka Migotti, Kasnoantičko groblje na Štrbincima kod Đakova-iskopavanja u 2002. i 2003, *ARR* 15, 141-246.

Migotti 2007b Branka Migotti, Rimski sarkofazi Siscije, *GGMS* 7, 7-30.

Migotti 2007c Branka Migotti, Rimska *bulla* u Panoniji, *VAMZ* 40, 187- 219.

Migotti 2008 Branka Migotti, Nadgrobni spomenik robovske obitelji iz Donjih Čeha kod Zagreba, *ArchAdr* 2/2, 453-465.

Migotti 2009 Branka Migotti, Kasnoantičko groblje u Štrbincima kod Đakova – iskopavanja 2004. i 2005, *ARR* 16, 107-224.

Migotti 2011 Branka Migotti, Prostor ranokršćanske biskupije Siscije i njegova arheološka ostavština, in: D.

Tepert and S. Jurić (eds.), *Sisačka biskupija - antiquam fidem*, Sisak, 35-69.
Migotti and Perinić 2001 Branka Migotti and Ljubica Perinić, Nekropola na Štrbincima kod Đakova u svjetlu kasnoantičkog horizonta Panonije, *ARR* 13, 103-204.
Mikl-Curk 1997 Iva Mikl-Curk, Rimljani ob Muri, in: I. Šavel *at al.*, *Pokrajinski muzej Murska Sobota – Katalog stalne razstave*, Murska Sobota, 23-36.
Milošević 2001 Petar Milošević, *Arheologija i historija Sirmijuma*, Novi Sad.
Minichreiter 1984 Kornelija Minichreiter, Rekognosciranja arheoloških lokaliteta Baranje, *ObHAD* 16 /2, 32-36.
Minichreiter 1989 Kornelija Minichreiter, Prilog poznavanju kasnoantičke arhitekture na području Slavonije i Baranje, *Lihnid* 7, 181-194.
Miškiv 2006 Jesenka Miškiv, Rimski sloj, in: Jesenka Miškiv, Lidija Miklik-Lozuk, and Josip Lozuk, *Zaštitna arheološka istraživanja, Trg Ivane Brlić-Mažuranić, Slavonski Brod 2002. i 2003.*, Slavonski Brod.
Mócsy 1974 András Mócsy, *Pannonia and Upper Moesia. A History of the Middle Danube Provinces of the Roman Empire*, London-Boston.
Morris 1992 Ian Morris, *Death-ritual and Social Structure in Classical Antiquity*, Cambridge.
Mráv 2001 Zsolt Mráv, Kaiserzeitliche Wagenbestattungen in Pannonia. In: F. Daim and Th. Kühtreiber (eds), *Sein und Sinn, Burg und Mensch*, St. Pölten, 123-141.
Mutavđić 1900 M. Mutavđić, Iskopavanje rimskih podataka iz Dubice, *VHAD* 4 (1899-1900), 225-228.
Negrelli 2006 Claudio Negrelli, Schede delle sepolture, in: Fede Berti (ed.), *Mors imatura, I Fadieni e il loro sepolcreto, Quaderni di Archeologia dell'Emilia Romagna* 16, Bologna, 69-102.
Nemeth-Ehrlich 1997 Dorica Nemeth-Ehrlich (ed.), *Aquae Iasae – Varaždinske Toplice. Vizualizacija rimske arhitekture / Aquae Iasae – Varaždinske Toplice.* Visualizing Roman Architecture, Zagreb.
Nemeth-Ehrlich and Kušan Špalj 2003 Dorica Nemeth-Ehrilch and Dora Kušan Špalj, Municipum Andautonia, in: *Pannonia I*, 107-130
Nemeth-Ehrlich and Kušan-Špalj 2007 Dorica Nemeth-Ehrlich and Dora Kušan-Špalj, *2000 godina Andautonije - Od rimskog grada do arheološkog parka / 2000 years of Andautonia. From the Roman Town to the Archaeological Park*, Zagreb.
Nemeth-Ehrlich and Kušan Špalj 2008 Dorica Nemeth-Ehrlich and Dora Kušan Špalj, Ščitarjevo – dvorište osnovne škole, *HAG* 4 (2007), 198-201.
Nenadić 1987 Vesna Nenadić, Prilog proučavanju antičke Sisciae, *PIAZ* 3-4 (1986-1987), 71-102.
Nock 1932 Arthur D. Nock, Cremation and Burial in the Roman Empire, *The Harvard Theological Review* 25/4, 321-359.
Okroša-Rožić 2006 Lana Okroša-Rožić, Novačka, in: Durman 2006, 186-187.
Palágyi 2003 Sylvia Palágyi, Roman period tumuli in Transpadania, in: Visy 2003, 257-261.
Palágyi and Nagy 2003 Sylvia K. Palágyi and Levente Nagy, *Römerzeitliche Hügelgräber in Transdanubien (Ungarn)*, Budapest.
Pavleković and Škrgulja 2008 Ida Pavleković and Rosana Škrgulja, Pregled arheoloških istraživanja Gradskog muzej Sisak u 2007. godini (II.), *GGMS* 8, 17-22.
Pearce 2008 John Pearce, Burial evidence from Roman Britain. The un- numbered dead, in: Scheid 2008, 29-42.
Perkić 2002 Domagoj Perkić, Grad mrtvih u Bubijevoj jami kod Barilovića, *HAnt* 8, 103-130.
Perkić 2006 Domagoj Perkić, Bubijeva jama kod Barilovića, in: Durman 2006, 104-105.
Philpott 1991 Robert A. Philpott, Burial Practices in Roman Britain. A survey of grave treatment and furnishing A.D. 43-410, *BAR Intern. Ser.* 219.
Philpott 1993 Robert A. Philpott, Personal ornaments and burial practices in Roman Britain, in: Struck 1993a, 167-180.
Pintarić 1998 Tanja Pintarić, *Nekropole i groblja na području Ščitarjeva*, Exibition Catalogue, Velika Gorica.
Pinterović 1978 Danica Pinterović, *Mursa i njeno područje u antičko doba*, Osijek.
Pirling 2002 Renate Pirling, *Busta* aus Krefeld-Gellep, *Germania* 80/2 Mainz, 491-527.
Pop-Lazić 2002 Stefan Pop-Lazić, Nekropole rimskog Singidunuma, *Singidunum* 3, Beograd, 7-100.
Radman-Livaja 2007 Ivan Radman-Livaja, In Segestica..., *PIAZ* 24, 153-172.
Radman-Livaja 2010a Ivan Radman-Livaja (ed.), *Nalazi rimske vojne opreme u Hrvatskoj / Finds of the Roman Military Equipment in Croatia*, Zagreb.
Radman-Livaja 2010b Ivan Radman-Livaja, *Siscia* kao rimsko vojno uporište / *Siscia* as a Roman Military Stronghold, in: Radman-Livaja 2010a, 179-201.
Raunig 1980 Branka Raunig, Dva kasnoantička groba iz okolice Đakova, *VAMZ* 12-13, (1979-1980), 151-168.
Rendić-Miočević 1994 Ante Rendić-Miočević (ed.), *Zagreb prije Zagreba. Arheološka baština Zagreba od pretpovijesti do osnutka biskupije 1094. godine / Zagreb Before Zagreb. The Archaeological Heritage of Zagreb from Prehistory to the Foundation of the Bishopric in 1094*, Exhibition Catalogue, Zagreb.
Riha 1979 Emilie Riha, *Die römischen Fibeln aus Augst und Kaiseraugst*, Augst.
Scheid 2008 John Scheid (ed.), *Pour une archéologie du rite: nouvelles perspectives de l'archéologie funéraire*, Rome 2008.
Schejbal 2004 Berislav Schejbal, Municipium Iasorum (Aquae Balissae), in: *Pannonia II*, 99-130.
Sokač-Štimac 2005 Dubravka Sokač-Štimac, *Tekić. Rimsko groblje i naselje*, Požega.
Sokač-Štimac 2008 Dubravka Sokač-Štimac, Treštanovačka gradina, *HAG* 4 (2007), 114-116.
Sokol 1981 Vladimir Sokol, Najnovija arheološka istraživanja u Prigorju, *Arheološka istraživanja u Zagrebu i njegovoj okolici, IzdHAD 6*, 169-186.
Sokol 1994 Vladimir Sokol, Kasnoantičko i ranosrednjovjekovno razdoblje / The Late-Classical and the Early-Medieval Period, in: Rendić-Miočević 1994, 46-51.

Sokol 2006 Vladimir Sokol, Kuzelin, in: Durman 2006, 154-155.

Spajić 1967 Emil Spajić, Kasnorimska nekropola u Sladojevcima, *OZ* 9, 101-132.

Spaul 2000 John Spaul, *Cohors²*, *BAR Intern. Ser.* 841.

Struck 1993a Manuela Struck (ed), *Römerzeitliche Gräber als Quellen zu Religion, Bevölkerungsstruktur und Sozialgeschichte*. Mainz.

Struck 1993b Manuela Struck, Busta in Britannien und ihre Verbindungen zum Kontinent. Allgemeine Überlegungen zur Herleitung der Bestattungssitte, in: Struck 1993a, 81-94.

Šaranović-Svetek 1967 Vesna Šaranović-Svetek, Vinkovci, Kamenica – antičko nalazište, *AP* 9,105-107.

Šarić 1980 Marina Šarić, Rimski grob u Topuskom, *VAMZ* 12-13 (1979-1980), 125-150.

Šarić 1986 Ivan Šarić, Gola, in: D. Balen-Letunić *et al.* (eds.), 40 godina arheoloških istraživanja u sjeverozapadnoj Hrvatskoj, Bjelovar-Koprivnica, 116.

Šarić 1990 Ivan Šarić, Rimskodobna nekropola zemljanih grobnih humaka u Novačkoj, in: *Arheološka istraživanja u Podravini i Kalničko-bilogorskoj regiji, IzdHAD* 14 (1989), Zagreb, 111-114.

Šarić-Šegvić 1984 Marina Šarić-Šegvić, Ponovo o rimskom grobu iz Topuskog, *VAMZ* 16-17 (1983-1984), 167-174.

Šegvić 2006 Marina Šegvić, Topusko, in: Durman 2006, 270-271.

Šeper 1962 Mirko Šeper, Rimska kola iz Poljanca kod Ludbrega, *ARR* 2, 335-428.

Šimek 1997 Marina Šimek (ed.), *Registar arheoloških nalaza i nalazišta sjeverozapadne Hrvatske*, Varaždin.

Talbert 2000 Richard J. A. Talbert, *Barrington Atlas of the Greek and Roman World,* Princeton.

Tirelli 1998 Margherita Tirelli, Horti cum aedificiis sepulturis adiuncti. I monumenti funerari delle necropoli di Altinum, *AN* 69, 137-204.

Tomičić and Dizdar 2007 Željko Tomičić and Marko Dizdar, Ilok - Dvor knezova Iločkih, *HAG* 3 (2006), 38-41.

Topàl 1981 Judit Topàl, *The Southern Cemetery of Matrica (Százhalombatta-Dunafüred)*, Budapest.

Topàl 1993 Judit Topál, *Roman Cemeteries of Aquincum, Pannonia. The Western Cemetery (Bécsi Road) I*, Budapest.

Vágó and Bóna 1976 Eszter Vágó and István Bóna, *Die Gräberfelder von Intercisa. Die Spätrömische Südfriedhof*, Budapest.

Vikić-Belančić 1981 Branka Vikić-Belanić, Etape urbanog razvitka Andautonije i antičko nasljeđe Zagreba, in: *Arheološka istraživanja u Zagrebu i njegovoj okolici, IzdHAD* 6, 129-154.

Visy 2003 Zsolt Visy (ed.), Hungarian Archaeology at the Turn of the Millenium, Budapest.

Vrbanović 1981 Stjepan Vrbanović, Prilog proučavanju Siscije, in: *Arheološka istraživanja u Zagrebu i njegovoj okolici, IzdHAD* 6, 187-199.

Vukelić 2007 Vlatka Vukelić, Prilog proučavanju razvoja sisačke arheologije u drugoj polovini 19. i prvoj polovini 20. stoljeća, *VAMZ* 40, 347-368.

Vukelić 2009 Vlatka Vukelić, Sarkofag Maksimijane Emilije iz Siska, *VAMZ* 42, 393–420.

Vukmanić and Hršak 2010 Igor Vukmanić and Tomislav Hršak, Osijek – Divaltova ulica 120,122, *HAG* 6 (2009), 32-35.

Vuković 2010 Domagoj Vuković, Sztariczkijeva karta Siska iz 1783. godine, *GGMS* 10, 121-144.

Vulić and Rapan Papeša 2010 Hrvoje Vulić and Anita Rapan Papeša, Vinkovci – Lapovačka ulica (Kaufland), *HAG* 6 (2009), 52-55.

Wiewegh 2003 Zoran Wiewegh, *Jugoistočna nekropola Siscije*, Sisak.

Wigg 1993 Angelika Wigg, *Die Grabhügel des 2. und 3. Jahrhunderts n. Chr. an Mittelrhein, Mosel und Saar*, Trier.

Zsidi 2003 Paula Zsidi, Burials, in: Visy 2003, 251-257.

Coin Circulation 3rd Century BC – AD 193

Tomislav Bilić

1. Introduction

Although it is estimated that only century 1% of coins from the Roman period was discovered so far[1] – one can presume that earlier periods do not differ much from this estimate – it still "illustrates well the general pattern of coin circulation in antiquity".[2] Both before and after the Roman occupation we can recognize three groups of findings: coin hoards, sporadic finds, and specimens coming from archaeological excavations.[3] Unfortunately, the third group of finds, the one most reliable, is rather poorly represented in the total amount of coins acquired from sites in southern Pannonia. There is not a single important site in this region with published coin finds; this paper will try to improve on this, starting with an analysis of both sporadic finds and systematically excavated coins in Sisak (*Siscia*), probably the most important of all southern Pannonian sites, apart from Sirmium, and also Osijek (*Mursa*), an important site in the hinterland the Danubian *limes*. Fortunately, the situation is somewhat different regarding the coins hoards: a seminal study by Mirnik (1981) offers rich data for analysis of coin circulation in the area. Two other works touch upon the problem of coin circulation in the region: Kos (1986), analyzing the south-eastern Alpine region but often touching upon Pannonia, and Bíró-Sey (1980), analyzing the data from the Hungarian part of Pannonia. Finally, Găzdac (2010) analyzed the circulation in Pannonia in the 2nd century in a chapter of his monograph on the monetary circulation in Dacia and surrounding Danubian provinces.[4]

Both Kos and Bíró-Sey group the finds from their respective areas of interest in similar periods (only the periods analyzed in the paper will be discussed, that is, ending with the year 193 AD):[5]

Bíró-Sey (1980):

1. Currency before the conquest;
2. Roman Republican and early Imperial *denarii* and their Barbaric imitations;
3. Early Empire to the Marcomannic wars.

Kos (1986):

1. Norican and Republican coinage;
2. Julio-Claudian period;
3. First and second centuries AD.

Thus we can see that both authors divide the period into three distinctive phases, but the border between 1 and 2 is rather uncertain and overlapping, as it should be. I will follow the two authors in this paper, since their divisions also reflect the present state of research in southern Pannonia, with occasional further subdivisions of the given phases.

2. Greek and Barbarian coinage in southern Pannonia

2.1. The coinage of Apollonia and Dyrrachium

The widespread circulation in southern Pannonia of the coinage produced by the Greek mints in Apollonia and Dyrrachium on the eastern Adriatic coast is a well-known phenomenon.[6] Although they were producing coins from the second half of the 5th century BC,[7] it is only from about 100 BC that they started to "export widely in the Danube basin".[8] Earlier, the mints were supposed to close down about the year 100 or somewhat later,[9] but this is no longer accepted. Thus, for example, Kos states that large quantities of coins were struck until the mid-1st century BC.[10] Therefore, it is probable that their circulation in the area started sometimes during the second half of the 2nd century BC. Apollonia and Dyrrachium produced several denominations of silver coinage (staters/tridrachms, drachms, later even *denarii*), but it was especially drachms (avg. weight 3,4 g), influenced by the Roman republican *denarius* (produced from the mid-3rd century onwards),[11] that circulated in the interior, thus also in southern Pannonia.[12] These were precisely the *illyrici argenti* of Gentius, captured by Romans in Scodra.[13]

1 Sutherland 1976, 85, cited in Kos 1986, 17.
2 Kos 1986, 17.
3 Bíró-Sey 1980, 337.
4 That is, Pannonia and Moesia: Găzdac 2010, 67-125 (ch. 3).
5 This corresponds to Fitz's (Fitz 1978) first period in his analysis of coin circulation in Pannonia (AD 14-193). Unfortunatelly, Fitz did not touch upon Southern Pannonia, which is the subject of present paper (see the review in Demo 1984).

6 The find of a pierced bronze Mamertine coin in the Zagreb Upper Town, coined after 288 BC, although interesting, does not reflect the circulation of such coins in Southern Pannonia (Dukat and Mašić 2005).
7 Dukat and Mirnik 1976, 181; Kos 1998, 33, 98.
8 Crawford 1985, 224.
9 Popović 1971, 151, 1987, 96; Dukat and Mirnik 1976, 181.
10 Kos 1986, 33, 99; cf. Popović 1971, 151, 1987, 97; Ceka 1972, 97, with a four-period division, from 325 to 37 BC.
11 Popović 1987, 96-97; Kos 1998, 33, 99.
12 Dukat and Mirnik 1976, 181, 196, 200; Crawford 1985, 224; Kos 1998, 33, 99.
13 Liv. XLV.43; Ceka 1972, 33 n. 27; Popović 1987, 97.

The coinage of Apollonia and Dyrrachium is present in several hoards from southern Pannonia: Dalj,[14] Dobra voda near Derventa,[15] Nova Pazova,[16] Pećinci,[17] Zaklopača,[18] and two hoards from Vukovar.[19] They were found together with Celtic coins – a rare example – in the hoard from Titel (69.6% Apollonia/Dyrrachium, 30.4% Barbaro-Celtic of the Syrmian type, but the total number of coins is very small (23))[20] and with Roman Republican coins in a hoard from Vukovar,[21] buried after 79 BC. It consisted of 136 drachms of Apollonia and Dyrrachium (97.1%) and mere four Republican coins (2.9%: RRC 299.1a, 301, 317.3, 383).[22] Popović separates out two areas with large quantities of the coinage of Apollonia and Dyrrachium, one of them being Syrmia with eastern Slavonia, associating this distribution with the area of the Scordisci during the 1st century BC.[23] This coinage is frequently found in the eastern part of southern Pannonia, but it is almost completely lacking in its western part (fig. 1).

It is not completely certain what was the reason for the introduction of coinage in the Pannonian basin, more specifically, the silver coinage of Apollonia and Dyrrachium; but it is not impossible that it arrived in the Danubian region because of the export of slaves and salt from the area, required by the then – after the year 167 BC – flourishing Roman-Epirote economy.[24] This would contradict Fitz's claim that the role of the coinage "was confined to local commerce" until the Roman conquest,[25] but also Brunšmid's belief that they were "circulating as money for trade".[26] The sheer number of Apollonia-Dyrrachium coinage circulating in the interior indeed suggests their dependence on Rome, reflecting Roman economical and political interests in the region.[27] The drachms of Apollonia and Dyrrachium were sometimes imitated by the Scordisci; they were made of poor quality silver, bronze or silvered bronze, with stylized representations and unintelligible inscriptions.[28]

2.2. Barbaro-Celtic coinage in the eastern part of southern Pannonia

Another (not exclusively, though) Pannonian phenomenon is the Barbaro-Celtic coinage. TheCeltic (Barbarian) tribes in the Pannonian basin used the tetradrachms of Philippus II of Macedon and, very rarely, golden staters[29] of his son and successor, Alexander,[30] until the year 168 BC, when the inflow was stopped.[31] It is possible that the posthumous coinage of Philippus II (struck until 294 BC) reflects precisely the economic needs of the "Balkan and Danubian peoples".[32] The shortage was remedied by imitations of *philippei* and staters, at first only by the Scordisci,[33] but later also by other tribes; the main period of minting of these Barbarian coins was between 150 and 50 BC.[34] On the other hand, it could be presumed that the first Barbarian imitations followed the stoppage in minting of posthumous *philippei* and this early coinage could be dated to the first half of the 3rd century[35] Barbaro-Celtic imitations of Alexander's tetradrachms were found in the hoard from Zemun,[36] imitations together with the coinage of Alexander and the Diadochi in the hoard from Krčedin,[37] while imitations of coins of Philippus II, Alexander and Larissa were found in another hoard from Zemun (or Jabukovac near Negotin?), together with various Hellenistic coinage (Athens, Alexander, Philippus III and the Diadochi).[38] These sites are situated in Syrmia, the area of the Scordisci, and the most common type of these imitations is called precisely 'Syrmian'.[39] From this area we know of a possible hoard from Sarvaš,[40] a hoard of Boiian gold staters from Obrovac Bački,[41] hoards of silver coinage from Surduk[42] and Neštin[43] and mixed silver and bronze coinage from Novi Banovci.[44] The easternmost find of a coin of the Syrmian type was

14 Brunšmid 1912, 264; Dukat and Mirnik 1976, fig. 3 on p. 194; Mirnik 1981, 42 no. 54; Crawford 1985, 325.

15 Ljubić 1888, 127; Dukat and Mirnik 1976, fig. 3 on p. 194; Mirnik 1981, 42 no. 55; Crawford 1985, 325.

16 Dukat and Mirnik 1976, fig. 3 on p. 194; Popović 1978, 21; Mirnik 1981, 46 no. 78; Crawford 1985, 325.

17 Popović 1978; Mirnik 1981, 46 no. 78; Crawford 1985, 325.

18 Dukat and Mirnik 1976, fig. 3 on p. 194; Mirnik 1981, 49 no. 98.

19 Dukat and Mirnik 1976, fig. 3 on p. 194; Mirnik 1981, 49 no. 96-97; Crawford 1985, 325.

20 Pink 1939, 147 no. 359; Dukat and Mirnik 1976, fig. 3 on p. 194; Mirnik 1981, 48 no. 89; Popović 1978, 21, 1987, 49, 80-81, 102-103.

21 Brunšmid 1912; *RRCH* 276; Dukat and Mirnik 1976, fig. 3 on p. 194; Popović 1978, 20-21, 1987, 102; Mirnik 1981, 49 no. 95.

22 The Vračevgaj hoard (Jankulov 1940, 250; Mirnik 1981, 48 no. 92; Crawford 1985, 325) consisted of some 400 drachms of Apollonia and Dyrrachium, together with "some Roman coins", probably Republican. Borić-Brešković and Popović 2006, 37 claim that as much as one third of the total number of coins in this hoard were Republican.

23 Popović 1978, 20; cf. the map on p. 19.

24 Crawford 1985, 225.

25 Fitz 1980, 332-333. For an argument in favor of the existence of the monetary economy in the southeast Alpine region, including – at least – southwestern Pannonia, in the mid-2nd century BC, see Kos and Šemrov 2003, 387.

26 Brunšmid 1912, 261.

27 Popović 1987, 98-99.

28 See for example the drachms from Osijek: Göricke-Lukić 2004, 39, 41-45, cat. no. 9-23.

29 Posid. *fr*. 240a.29-31, 37-40 Edelstein-Kidd (= *FGrHist* 87F48 = Ath. VI.25.234A-C), however, claimed that the Scordisci 'do not introduce gold into their country... they have abominated gold and do not introduce it into their country, since they have suffered many calamities on its account. Silver, on the other hand, they do use...' (χρυσὸν μὲν οὐκ εἰσάγουσιν εἰς τὴν αὐτῶν χώραν... ἀφωσιώκασι δὲ οὗτοι τὸν χρυσὸν καὶ οὐκ εἰσφέρουσιν εἰς τὰς πατρίδας, δι' ὃν πολλὰ καὶ δεινὰ ἔπαθον· ἀργύρῳ δὲ χρῶνται...). Compare also Popović 1987, 66.

30 Two such staters are kept in the AMZ: one from Ludbreg (inv. no. A4674) and another from Radoboj (inv. no. A1188).

31 Bíró-Sey 1980, 338; after that date Macedonian and Greek coinage is absent from hoards: Popović 1987, 82.

32 Popović 1987, 35.

33 From the mid-2nd century onwards: Kos 1998, 326.

34 Dukat and Mirnik 1976, 176; Bíró-Sey 1980, 338; Popović 1987, 77.

35 Popović 1987, 78.

36 Mirnik 1981, 36 no. 19.

37 Mirnik 1981, 36 no. 15a; Dautova-Ruševljan and Popović 1981; Popović 1983, 1987, 55, 79.

38 Saria 1927, 11; Preda 1973, 138; Dukat and Mirnik 1976, fig. 3 on p. 194; Mirnik 1981, 35 no. 15, 36 no. 20; Popović 1987, 41-42.

39 Dukat and Mirnik 1976, 192; Popović 1987, 45-53. See the distribution of sites in Popović 1987, fig. 10 on p. 47, fig. 12 on p. 49, fig. 13 on p. 51, fig. 14 on p. 53; cf. for the Krčedin type Popović 1987, 54-60 with fig. 21 on p. 58; for the distribution of sites, together with the distribution of sites with the Crişeni-Berchieş type, found as far west as Bjelovar, cf. Preda 1973, 101 fig. 8; for the distribution of sites, including Osijek and Syrmian sites, and for the "eastern Slavonian" type, contemporary with Syrmian phase B, see Popović 1987, 64-65 with fig. 22 on p. 65.

40 Dukat and Mirnik 1976, fig. 3 on p. 194; Mirnik 1981, 40 no. 41.

41 Mirnik 1981, 46 no. 80.

42 Mirnik 1981, 470 no. 88.

43 Popović 1987, 51, 81.

44 Dukat and Mirnik 1976, fig. 3 on p. 194; Mirnik 1981, 39 no. 34.

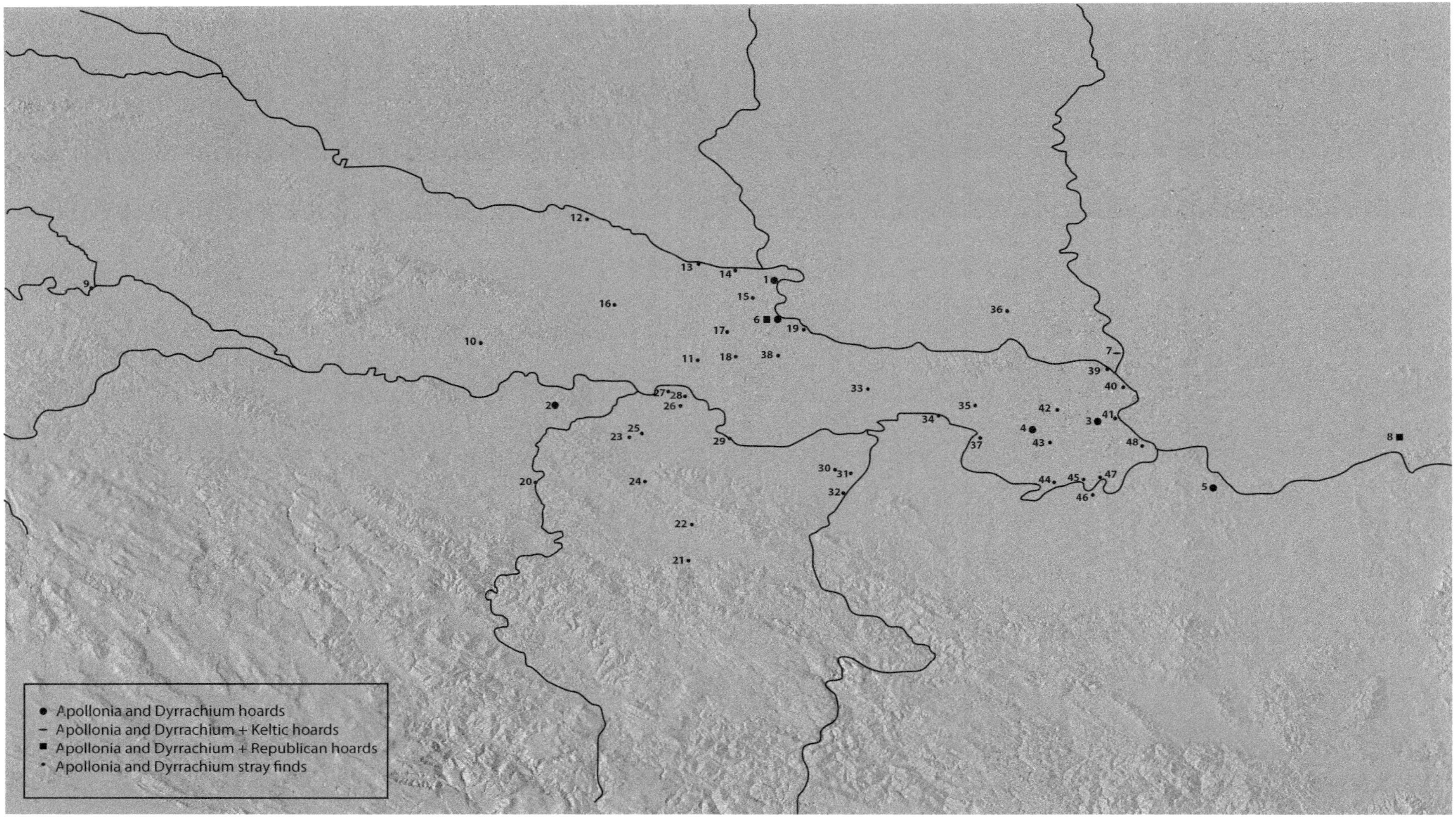

Fig. 1. Distribution of the coinage of Apollonia and Dyrrachium: (1) Dalj, (2) Dobra voda near Derventa, (3) N. Pazova, (4) Pećinci, (5) Zaklopača, (6) Vukovar, (7) Titel, (8) Vračevgaj (?), (9) Sisak, (10) Mihaljevići (Požega), (11) Cerna, (12) Podravski Podgajci, (13) Osijek, (14) Sarvaš, (15) Trpinja, (16) Gorjani, (17) Vinkovci, (18) Privlaka, (19) Sotin, (20)Usora, (21) Kovači (Tuzla), (22) Tuzla, (23) Gradačac, (24) Srebrnik, (25) Tramošnica, (26) Bok, (27) Domaljevac, (28) Tolisa, (29) Brčko, (30) Bijeljina, (31) Amajlije, (32) Janja, (33) Kukujevci, (34) S. Mitrovica, (35) Voganj, (36) Jarak, (37) Gomolava,(38) S. Karlovci, (39) St. Slankamen, (40) Surduk, (41) N. Banovci, (42) Golubinci, (43) Mihaljevci, (44) Kupinovo, (45) Progar, (46) Obrenovac (Šabac), (47) Boljevci, (48) Zemun (made by A. Solter).

found in Donja Dolina[45] – two unpublished specimens from Sicia (a drachm of phase C and a minimus of phase D,[46] both kept in the AMZ) can now be added – but their distribution is generally limited to the line Osijek-Đakovo in the west. Beside the already mentioned imitations of the drachms of Apollonia and Dyrrachium, the Scordisci also produced imitations of the coinage of Illyrian ruler Ballaios.[47]

2.3. Barbaro-Celtic coinage in the western part of southern Pannonia

In the western part of southern Pannonia the situation is somewhat different. The coins of the so-called Croatian group can be divided into three types: the Varaždin type, the Samobor type, and the Đurđevac type.[48] This division is today somewhat outdated,[49] and the typology of R. Göbl for the eastern Norican coinage should be used instead.[50] "The Croatian group" would belong to the so-called eastern Norican coinage,[51] according to earlier theories most probably produced by the Taurisci from century 70 BC to 21/20 or 16/15 BC,[52] or, according to latest analyses, from at least the mid-2nd century BC.[53] The hoard from Križovljan near Varaždin – giving the name of the type – consisted of 109 tetradrachms of the Varaždin A and B type, as well as of the Augentyp-Stamm type.[54] The Varaždin types are chronologically the earliest coinage of the Tauriscan group and were minted somewhere in the Varaždin area.[55] Coins of the Varaždin A type were also found in a hoard from Haimburg[56] and in Ogulin,[57] Sisak (*Siscia*),[58] and Kalnik,[59] while those of the Varaždin B type were found in Križevci,[60] Frauenberg,[61] in the Kärnten area,[62] Sopron[63] and Vács.[64] The hoard from Okić near Samobor – also giving the name of the type – consisted of century 1300 tetradrachms of the Samobor A type.[65] Coins of the Samobor A type were also found in Sisak,[66] Novi Banovci,[67] near Okić,[68] and on unspecified sites on the territory of Croatia,[69] while coins of the Samobor B type were found in Drnovo,[70] Križevci (Veliki Kalnik),[71] and Kuzelin near Zagreb.[72] The coins of the so-called Freie Samobor-Typen or Samobor C[73] were also found in various sites in Slovenia: Novo Mesto[74] and Drnovo[75] in Pannonia, Blatna Brezovica[76] and Bevke[77] not far from the Roman-period Emona,[78] Dobrna Retje,[79] Rifnik,[80] Celje[81] and Lemberg[82] in Noricum, Austria (Güttenbach, four pieces), Czech Republic (Stradonice),[83] and Hungary (Gógánfa, single piece),[84] with Croatian examples from Sisak,[85] Kuzelin near Zagreb,[86] and the Zagreb Upper Town – the Town Museum of Zagreb site.[87] Three more coins of the "Samobor" type were found at Kuzelin.[88] Dukat and Mirnik recognize a strong influence of the Samobor-type coins on certain eastern Norican/ Taurisci types (Augentyp-Stamm, Frontalgesichtstyp, Freie

45 Marić 1964, 49, pl. XXII.5, phase IIIc, 125 BC – early 1st century AD; Popović 1987, 50.
46 This coin is of a type exclusively associated with eastern Syrmia: Popović 1987, 53, 81.
47 Six examples from Osijek and its vicinity: Göricke-Lukić 2004, 39, 2004a, 158-159, 161, 165, cat. no. 498, 501-502, 505-506, 523. A Barbarian imitation of a Ballaios coin was also found in Carnuntum (Hahn 1976 = *FMRÖ* p. 171, no. 9784). The reign of Ballaios is dated differently, but Gorini's proposal (1984, 49, 1988, 16, 1991, 30) of 195-175 BC seems the most acceptable (Šašel Kos 2007, 127-131, 137); the imitations should chronologically follow the death of Ballaios (Gorini 1988, 19). The latest authoritative study of the subject, based on several thousand new coins found at Rhizon (Ciołek 2011, 73-76), dates the reign of Ballaios between ca. 260 and 230 BC (Ciołek 2011: 92).
48 Pink 1939, 113; Liščić 1978, 236; Dukat and Mirnik 1976, 190; Mirnik 1998, 485.
49 Cf. Kos 1977, 18.
50 *TKN* 30-38.
51 Rant (1987, 208) suggested the name "coins of the Sava type".
52 Kos 1977, 20, 33-35; Kos 1998, 249, 341. For different opinions on this question see Kos and Šemrov 2003, 384-385.
53 Kos and Šemrov 2003, 386-387; Kos and Mirnik 2011.
54 Pink 1939, 139 no. 154; Dukat and Mirnik 1976, 190; Mirnik 1981, 44 no. 65; Kos and Mirnik 2011. Yet the presence of the Augentyp-Stamm type coin in the hoard must be taken with caution (Kos and Mirnik 2011).
55 They were minted around the end of the first half of the second century BC (Kos and Mirnik 2011).
56 Kos 1977, 35; Ruske 2011, 64 (V8). Kos and Mirnik (2011) explicitly state that no Varaždin type coins were represented in the hoard, only the group with the Venetic inscription VES., separated out and somewhat earlier from the Varaždin A type. A smaller denomination – belonging to the East Norican small silver coinage – of this type was found in Celje (Kos 1977, 104, no. 739, cf. 1976, 220). Kos (1976, 220-221) discusses these East Norican small silver types, arguing for a mint in the area of Celje.
57 AMZ inv. no. 915; Kos and Mirnik 2011.
58 AMZ inv. no. 898; Liščić 1978, 250; Kos and Mirnik 2011.
59 Demo 1982, 85 nr. 1; Kos and Mirnik 2011. Cf. a possible Varaždin A coin from Zuglio and two more from the Friuli area (Kos and Mirnik 2011).
60 AMZ inv. no. 909; Demo 1982, 85 nr. 2; Kos and Mirnik 2011.
61 Kos and Mirnik 2011.
62 Dembski 1998, 90 no. 865; Kos and Mirnik 2011.
63 Kos and Mirnik 2011.
64 *TKN* 126, no. 22; Kos and Mirnik 2011. Cf. a possible coin of the Varaždin type from Manching (Kellner 1990, 60 no. 112; Kos and Šemrov 2003, 387 n. 39); a further specimen from the Teurnia hoard (Varaždin B) is mentioned by Ruske 2011, 70 (V25) (cf. Kos and Mirnik 2011); another possible coin of this type was found in Zollfeld (Kos and Mirnik 2011).
65 Pink 1939, 144 no. 285; Dukat and Mirnik 1976, 190; Mirnik 1981, 47 no. 84.
66 AMZ inv. no. A8866; Liščić 1978, 250.
67 Mirnik 1998, 486.
68 AMZ inv. no. G2146, Mirnik 1998, 486 n. 10.
69 AMZ inv. no. A8868-8869, 9582, Mirnik 1998, 486 n. 11.
70 Kos 1977, 120, no. 1.
71 AMZ inv. no. A917; Liščić 1978, 250, Đurđevac-type; Demo 1982, 85 nr. 3, Samobor B.
72 Sokol 1994a, 23 no. 3, 1994b, 99 no. 124, 2001, 10-11 no. 3; Mirnik 1998, 486. Smaller denominations of Samobor A/B or Đurđevac type were found in Celje (Kos 1977, 104, no. 741, 112, no. 991-992).
73 On the unsuitability of this name for this group see Kos and Šemrov 2003, 385.
74 Kos 1977, 136 no. 1-2.
75 Kos 1977, 120, 3-4.
76 Kos 1977, 84 no. 1, Kos and Šemrov 2003, 382-383 no. 4-56 (a hoard of 82 coins).
77 Kos 1983, 413 no. 35-38, Rant 1987, 210 no. 9-10 (a hoard of 48 coins).
78 Thus in Regio X of the province of Italia (Šašel Kos 2002, 378-379, 381-382; Šašel Kos 2003).
79 Kos 1977, 116-119, no. 28-33, a hoard of 553 coins, Pink 1939, 136 no. 56, Mirnik 1981, 42 no. 56; cf. Schachinger 2008, 77.
80 Kos 1977, 138, no. 1.
81 Kos 1977, 86-87, no. 13-14, 154-155, no. 1.
82 Kos 1977, 130-131, no. 57-58.
83 Dembski 1998, 93 no. 940.
84 See also the distribution map in Kos 1977, 43, Map 4.
85 AMZ inv. no. A6419; cf. Liščić 1978, 250; Mirnik 1998, 486 n. 9.
86 Sokol 1994a, 22-23 no. 4; Sokol 1994b, 99 no. 123; Mirnik 1998, 486-487; Sokol 2001, 11-12 no. 4.
87 Mirnik 1998, 487. A smaller denomination of the type were found in Celje (Kos 1977, 106, no. 772) and in a hoard from the Ljubljanica River near Blatna Brezovica (Kos and Šemrov 2003, 382-383 no. 57-59).
88 Sokol 2001, 12-14 no. 5-7.

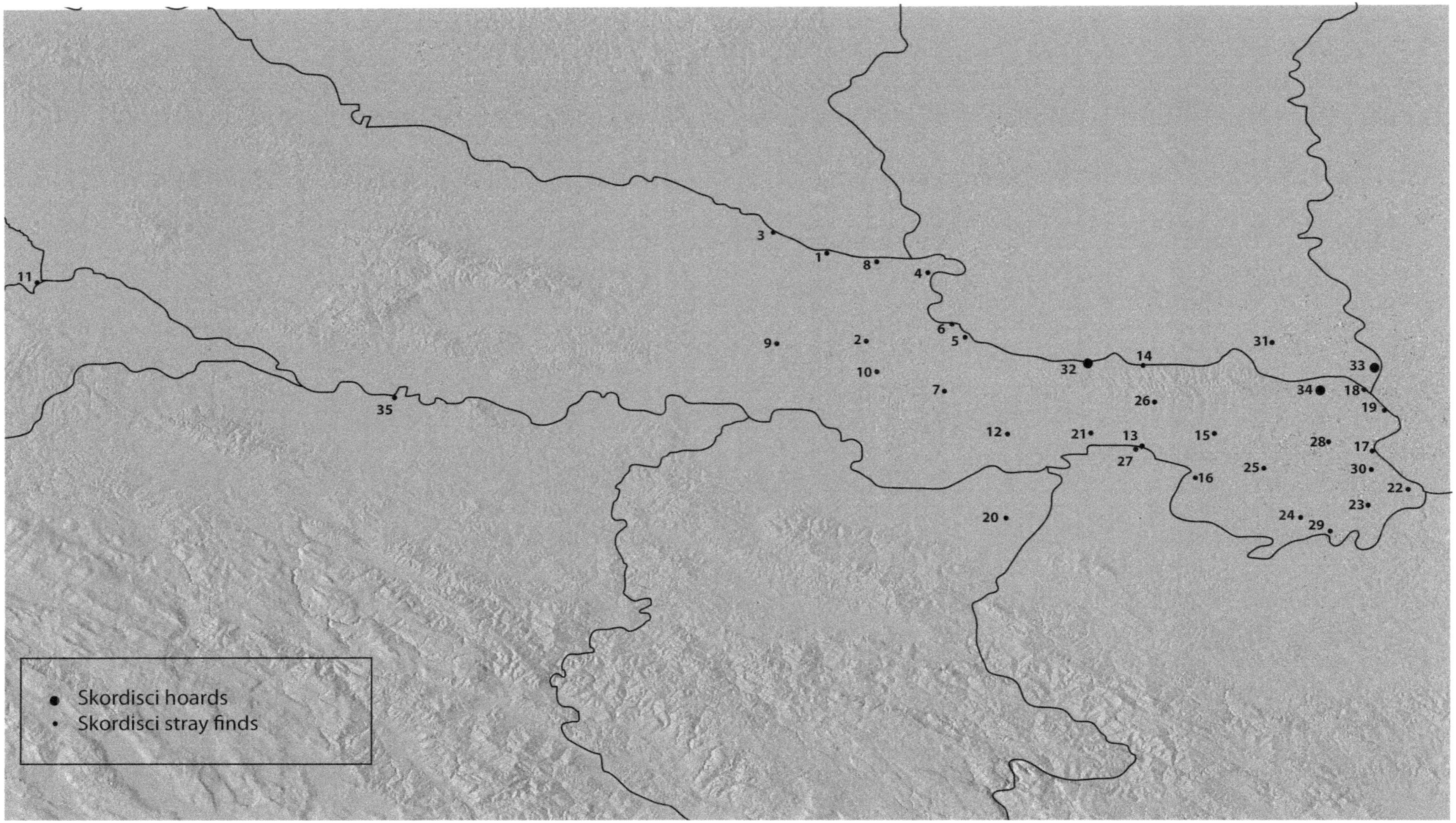

Fig. 2. Distribution of the coinage of the Scordisci: (1) Osijek, (2) Vinkovci, (3) Petrijevci, (4) Dalj, (5) Sotin, (6) Vučedol, (7) Nijemci, (8) Sarvaš, (9) St. Mikanovci, (10) Privlaka, (11) Sisak, (12) Morović,(13) S. Mitrovica, (14) Banoštor, (15) Ruma, (16) Gomolava, (17) N. Banovci, (18) St. Slankamen, (19) Surduk, (20) Bijeljina, (21) Martinci, (22) Zemun, (23) Surčin, (24) Ašanja, (25) Pećinci, (26) Grgurevci, (27) Mačvanska Mitrovica, (28) St. Pazova, (29) Progar, (30) Batajnica, (31) Budisava, (32) Neštin, (33) Titel, (34) Krčedin, (35) Donja Dolina (made by A. Solter).

Samobor-Typen).[89] The hoard from Šemovec near Đurđevac consisted of century 400 coins of the Đurđevac type.[90] The coins of this type were found in the wide area of Pannonia, Noricum and Dalmatia: from the hoard from Vrhnika in the west,[91] Velemszentvid (two pieces), Fenék, and Szávad[92] (Hungary),[93] as well as Oberleiserberg (Austria) in the north,[94] Gomolava (Serbia) in the east,[95] Donja Dolina[96] and Gračanica[97] in northern Bosnia,[98] to Dalmatia in the south: four pieces on the Balina Glavica near Drniš,[99] one piece in Dragišić, Šibenik hinterland,[100] with another on the island of Brač.[101] In north-western Croatia, occupying for the most part the future Pannonia Savia, this type was found in Varaždin,[102] Samobor,[103] Šćitarjevo,[104] and two pieces in Sisak,[105] with a hoard of seven coins of the type in Kozarevac near Koprivnica[106] and another hoard of twelve silver coins from Špišić Bukovica near Virovitica, where their presence has not been securely attested,[107] while in the Slovenian part of Pannonia it was found in Glavica pri Vrhpolju,[108] Škocjan pri Dobravi,[109] and Drnovo.[110] A hoard of eight silver tetradrachms found in Pokupsko south of Zagreb in 1991 consisted of two Đurđevac specimens, three specimens of the Samobor A and three specimens of the Samobor B type,[111] which is the only example of the two types found together, showing that they circulated contemporarily.[112] The existence of mints both in the area of Đurđevac and Samobor should probably be taken into account.[113] Several stray finds of later-phase large eastern Norican silvers are known from southern Pannonia, both from Croatia and Slovenia: Sisak,[114] Đurđevac,[115] Kuzelin near Zagreb,[116] and Mihovo,[117] as well as from nearby Norican sites of Kostrivnica[118] and Celje,[119] and from the Regio X of Italia (Ljubljana)[120] (Augentyp-Stamm), while the specimens of the Frontalgesichtstyp type were found in the Pannonian sites of Samobor, Tounj,[121] Vinji vrh,[122] and Stari Grad pri Kostanjevici.[123] They were also found in hoards at the Norican sites of Lemberg,[124] Malta[125] and Dobrna-Retje[126], as well as in the Pannonian site Varaždin.[127] A hoard of three coins of the Brezelohr type was found in Melje near Maribor (Noricum).[128] It is interesting that these later-phase large eastern Norican silver coins were found almost excluseively in the western part of southern Pannonia and the adjactent territories farther to the west (fig. 4), while the earlier types, those of the "Croatian group", circulated on a much wider area (fig. 3).

2.4. Barbaro-Celtic coinage – other phenomena

A rather peripheral phenomenon of Barbaro-Celtic coinage in southern Pannonia is represented by the hoard from Narta, with 36 coins of the Huşi-Vovrieşti type, a primarily Dacian occurrence.[129] Another interesting hoard is that from Ribnjačka,[130] consisting of century 112 imitations of tetradrachms of Philipp II.[131] One Turnierreiter specimen was found in Kobaš near Požega,[132] while two Dreieckhals

[89] Dukat and Mirnik 1976, 191.
[90] Pink 1939, 137 no. 89; Dukat and Mirnik 1976, 190; Mirnik 1981, 43 no. 60.
[91] *TKN* 129 no. 53; Kos 1977, 150-151, no. 2.
[92] Actually, Varasd, see Dembsky 1998, 91 no. 872.
[93] *TKN* 128-130, no. 20, 62a, 107; Kos 1977, 46 Map 6.
[94] Two AE tetradrachms (Karwowski and Militký 2011, 132 T. 1).
[95] Popović 1971, 149, 157 cat. no. 11; Kos 1977, 46 Map 6; Mirnik 1998, 487.
[96] 4 pieces, Truhelka 1904, 74 fig. 53-56; Marić 1964, 49, pl. XXII.4, phase IIIc, 125 BC – early 1st century AD; Mirnik 1998, 487.
[97] AMZ inv. no. A6758; Liščić 1978, 250; Mirnik 1998, 487 n. 20.
[98] Kos 1977, 46 Map 6.
[99] Kos 2002, 149-150.
[100] Kos 2002, 48 n. 2.
[101] Bonačić Mandinić 2003, 436, 438-439.
[102] 2 pieces: *TKN* 129-130, no. 56, 91.
[103] Dembsky 1998, 91 no. 870.
[104] Unpublished. The coin was found in a Roman-period grave near the Ščitarjevo elementary school in 2007 (SU 328, grave 2, quadrant a2, special find no. 493). The excavations were conducted by my colleagues from the Museum D. Nemeth-Ehrlich and D. Kušan-Špalj, to whom I am grateful for giving me the opportunity to study the coin. It will be published in a forthcoming study of the entire cemetery, prepared by Nemeth-Ehrlich and Kušan-Špalj.
[105] AMZ inv. no. A6757; Kos 1977, 46 Map 6; Mirnik 1998, 486 n. 8; AMZ main inventory of the Numismatic department no. 2195; Mirnik 1998, 487 n. 19.
[106] Pink 1939, 113, 139 no. 149; *TKN* 128-130, no. 12, 60, 68, 70, 84, 103, perhaps also 79; Mirnik 1981, 43-44 no. 64; in Mirnik 1998, 487 it is suggested that these coins perhaps belonged to the Đurđevac hoard.
[107] Mirnik 1998, 487.
[108] Kos 1977, 120-121, no. 1.
[109] Kos 1977, 146-147, no. 1.
[110] Kos 1977, 120-121, no. 2. A smaller denomination of the type was found in Celje (Kos 1977, 104, no. 740, cf. 1976, 220).
[111] Mirnik 1998, 488-489.
[112] Mirnik 1998, 485.
[113] *TKN* 56; Kos 1977, 35.
[114] AMZ inv. no. A7021; Mirnik 1998, 486 n. 9; cf. Liščić 1978, 250, Samobor-type. Perhaps from a hoard (see below).
[115] Dembsky 1998, 92 no. 910.
[116] Sokol 1994a, 22 no. 2; Sokol 1994b, 99 no. 125, 2001, 10 no. 2; Mirnik 1998, 487.
[117] Kos 1977, 134-135, no. 1.
[118] Kos 1977, 120-121, no. 2.
[119] Kos 1977, 86-87, no. 9.
[120] Kos 1977, 132-133, no. 2.
[121] AMZ inv. no. A8869, 918; Liščić 1978, 250-251.
[122] Kos 1977, 150-151, no. 1.
[123] Kos 1977, 144-145, no. 1.
[124] 9 pieces of Augentyp-Stamm (Kos 1977, 126-129, no. 27-35), 3 pieces of Frontalgesichtstyp (Kos 1977, 124-125, no. 8-10), 15 pieces of Brezelohr/A (Kos 1977, 124-127, no. 12-26), 20 pieces of Wuschelkopf-Gruppe (Kos 1977, 128-131, no. 36-55), 1 piece of Unscharfer Typ (Kos 1977, 130-131, no. 56).
[125] 2 pieces of Augentyp-Stamm (Ruske 2011, 67-68 / V16).
[126] Kos 1977, 114-115, no. 1-4 (Frontalgesichtstyp); Kos 1977, 114-115, no. 5-8 (Brezelohr/A and B); Kos 1977, 114-115, no. 9-11 (Verschwommener Typ); Kos 1977, 114-117, no. 12-22 (Wuschelkopf-Gruppe); Kos 1977, 116-117, no. 23-27 (Unscharfer Typ), together with six coins of the Samobor C type (see above). Cf. Schachinger 2008, 76-77 (5 pieces of Frontalgesichtstyp, 12-13 Wuschelkopf-Gruppe, 4 Brezelohr/A and B, 7 Unscharfer Typ, 5-6 Verschwommener Typ).
[127] Dembsky 1998, 92 no. 905; Kos and Mirnik 2011 no. 35 (Augentyp-Stamm). The presence of the Augentyp-Stamm type coin in the Varaždin hoard must be taken with some caution, though (Kos and Mirnik 2011).
[128] Pichler 1856, i.156; Mirnik 1981, 46 no. 77; Kos 1977, 134-135.
[129] Brunšmid 1895; Pink 1939, 35-36, 141 no. 205; Preda 1973, 122 no. 45; Dukat and Mirnik 1976, 192; Mirnik 1981, 39 no. 33.
[130] Lisičić 1957; Pink 1977; Dukat and Mirnik 1976, 192; Mirnik 1981, 39 no. 36; Kos and Mirnik 1999.
[131] Turnierreiter Typ: *OTA* 120, no. 1-44; Kos and Mirnik 1999; the type is associated by Pink (1977, 16-17) with the Scordisci, which is rejected by Kos and Mirnik (1999, 304), who opt for the Bjelovar region, while Popović (1987, 38) believes it should be associated with regions nearer to Macedonia in the southern Balkans; Kinnloser Typ: in Pink (1977, 13) called Übergangstyp zu Dreieckshals, in Kos and Mirnik Dreieckhals (Kos and Mirnik 1999, no. 45-64, *OTA* 318); Transitional Audoleon type, in Pink (1977, 13) called Wolljackentyp, in Kos and Mirnik 1999 Puppenreiter (Kos and Mirnik no. 65-71, *OTA* 417).
[132] *OTA* 120.3; Kos and Mirnik 1999, 305-306 with n. 39.

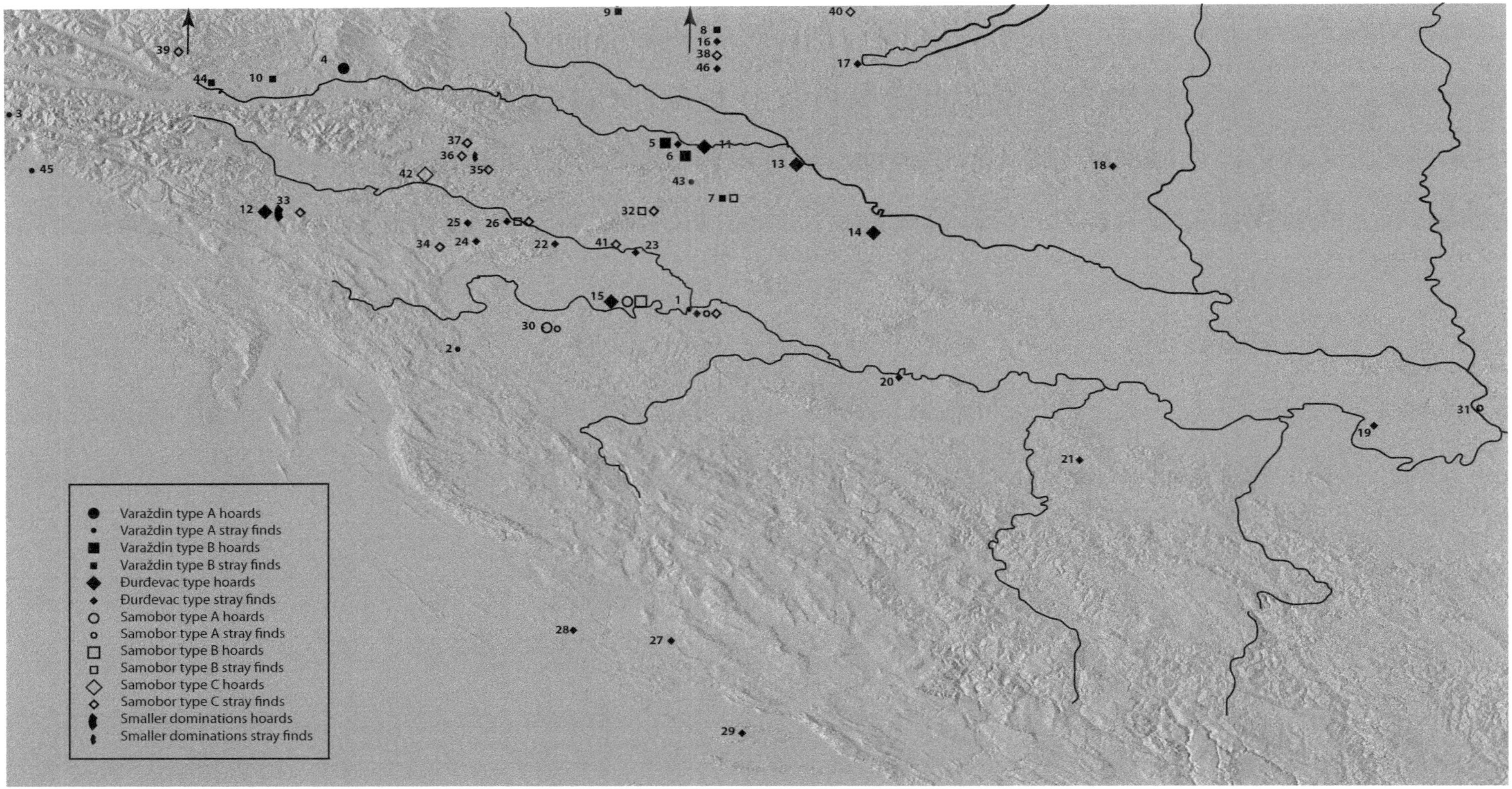

Fig. 3. Distribution of the coins of the Varaždin, Samobor and Đurđevac type: (1) Sisak, (2) Ogulin, (3) Zuglio, (4) Haimburg (?), (5) Varaždin, (6) Banjšćina (?), (7) Križevci, (8) Vács and Sopron, (9) Frauenberg, (10) Kärnten area, (11) Šemovec, (12) Vrhnika, (13) Kozarevac, (14) Špišić Bukovica, (15) Pokupsko, (16) Velemszentvid, (17) Fenék, (18) Szávad (actually, Varasd), (19) Gomolava, (20) Donja Dolina, (21) Gračanica, (22) Samobor, (23) Ščitarjevo, (24) Glavica pri Vrhpolju, (25) Škocjan pri Dobravi, (26) Drnovo, (27) Balina Glavica near Drniš, (28) Dragišić (Šibenik hinterland), (29) the island of Brač, (30) Okić, (31) Novi Banovci, (32) Kuzelin, (33) Blatna Brezovica, (34) Novo Mesto, (35) Rifnik, (36) Celje, (37) Lemberg, (38) Güttenbach, (39) Stradonice, (40) Gógánfa, (41) Zagreb – Gornji grad, (42) Dobrna Retje, (43) Kalnik, (44) St Peter in Holz, (45) Friuli area (?), (46) Oberleiserberg (made by Ana Solter).

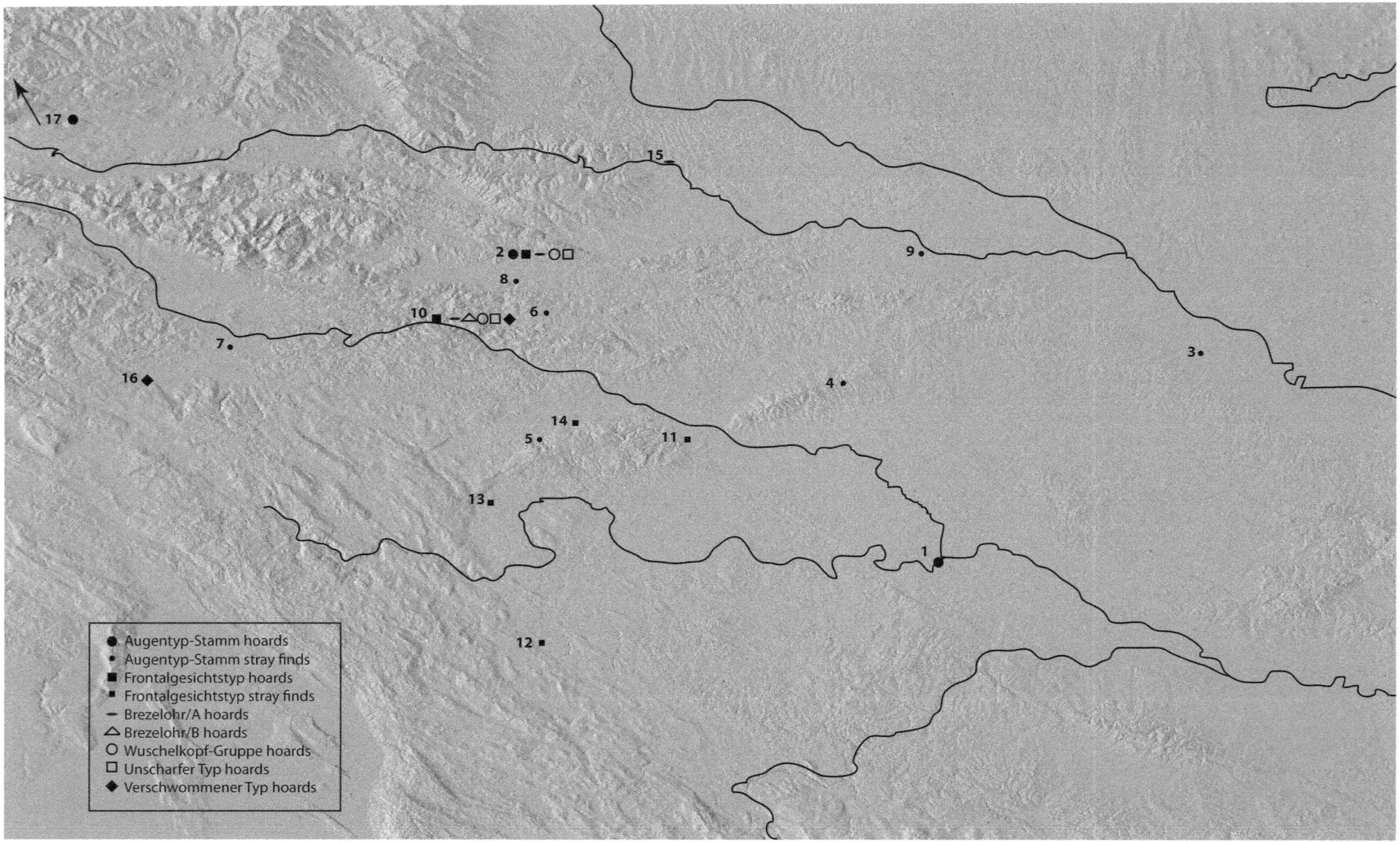

Fig. 4. Distribution of other Eastern Norican coin types: (1) Sisak, (2) Lemberg, (3) Đurđevac, (4) Kuzelin, (5) Mihovo, (6) Kostrivnica, (7) Ljubljana, (8) Celje, (9) Varaždin, (10) Dobrna-Retje, (11) Samobor, (12) Tounj, (13) Vinji vrh, (14) Stari Grad pri Kostanjevici, (15) Melje, (16) Vrhnika, (17) Malta (made by A. Solter).

coins were found at Munkács in Transcarpathian Ukraine.[133] These two hoards are thoroughly exceptional and show that the circulation of Barbaro-Celtic coinage was more complicated than usually presumed. On the one hand, coins stemming from Dacia could be found in large numbers – albeit sporadically – in western part of southern Pannonia, on the other a western Pannonian mint – if Kos and Mirnik are right – could produce "the most beautiful and skilfully executed barbarous coins in these parts of Europe",[134] yet with a relatively narrow area of circulation (although the specimen from Ukraine[135] speaks differently).[136] Another hoard of Eastern Celtic coins of the Kapostaler type was found in Sveti Petar Orehovec near Križevci.[137] Earlier Eastern Celtic coins are generally uncommon in southern Pannonia: a specimen similar to *OTA* 459-461 (Zickzackgruppe, Vogelreiter-Typ) was found at Kuzelin near Zagreb,[138] while another specimen of the Vogelreiter-Typ (OTA 441) was found in Sisak.[139]

2.5. African and Italic coinage

Another peripheral – at least for Pannonia – but extremely interesting phenomenon is the appearance of large quantities of Carthaginian, Numidian and Egyptian and other Greek and Italic/Roman currency (*aes rude, aes formatum, aes signatum, aes grave*) in the Balkans, found in large hoards of the so-called Mazin-type.[140] M. Crawford calls the context in which this coinage penetrated into the interior "almost wholly obscure"; he adds that it is to him "wholly baffling why this relatively small area suckled in bronze predominantly from Rome and North Africa from the middle of the third century onwards and then on the whole simply kept it in its original form for up to a century and a half…".[141] It is impossible to say whether this money entered the area through Dyrrachium; whether it entered the area in connection with the amber trade; and whether the hoards were concealed because of the insecurity caused by Roman military threat.[142] The list of coin hoards of this type shows that only a few of them are related to Pannonia, more precisely, with its border area on the middle course of the Una River: these are the two hoards from Bosanska Krupa,[143] Izačić,[144] and a pot burial from Vrankamen.[145] The first hoard from Krupa (Mirnik 1981 nr. 31) consisted only of Punic and Numidian coins; the hoard from Izačić of Punic, Numidian and Ptolomaeic currency; the second hoard from Krupa (Mirnik 1981 nr. 30) of Punic, Numidian and Celtic[146] coinage, together with some *aes formatum*, perhaps not of Italic origin after all, and *aes grave* and reduced Roman asses; and the hoard from Vrankamen consisted of Punic, Numidian, Ptolomaeic coinage, together with *aes rude, aes signatum, aes formatum, aes grave* and reduced Roman asses, one coin of Hiero II (Syracuse), and one coin of Hispanic origin (Emporiae). Thus the two latter hoards consisted of both African and Italic currency. A rather significant number of individual coin finds suggests that the coinage actually circulated over a large area, representing the very beginning of money circulation in the mountainous region of northern Dalmatia,[147] but not in Pannonia itself. Thus bronze coins of Hiero II were found at Donja Dolina and Osijek,[148] a coin of Micipsa in Glina, both of these coins in Kiringrad, three Ptolomaeic coins in Košaki, Muta,[149] and Spodnje Radvanje near Maribor and one in Surduk in Syrmia, two Carthaginian coins near Prijedor, *aes signatum* in Sisak and Trošmarija.[150] Five Neapolitan and two Syracusan bronzes were found together with one bronze of Apulian Teanum and Rome in Ptuj.[151]

3. Republican and early Imperial coinage

3.1. Circulation during the Republican era

We have seen that Roman Republican coinage was found together with that of Apollonia and Dyrrachium in the hoards from Vukovar (*RRCH* 276, Mirnik 1981, 95) and possibly Vračevgaj, although in rather small numbers (less than 3% in the former and a "small number" in the latter; but see n. 2). Two systematically excavated sites of the 2nd and 1st centuries BC have provided us with relevant, although scarce, information for this period. Thus several Republican coins were found in Donja Dolina on the middle Sava River, an anonymous *denarius*[152] and a *denarius* of M. Opeius,[153] belonging to the phase IIIc (125 BC – early 1st century AD), together with two Celtic coins mentioned earlier. In an earlier phase, IIIb (250-125 BC), a Syracusan coin (unpublished) was found together with an *as* of Opeimius.[154] Another excavated site, Gomolava in Syrmia, also produced some interesting finds. Three Apollonian coins were found,[155] two of them from the layer VIb,

[133] Pink 1939, 82, 141 no. 194; Kos and Mirnik 1999, 306.
[134] Dukat and Mirnik 1976, 192.
[135] "…a development of the 'Tournament Horseman' type" (Kos and Mirnik 1999, 306).
[136] A hoard of Celtic silver coinage from Banjšćina near Varaždin was described by Ivan Kukuljević Sakcinski in c. 1857, but the account of B. Zmajić (1967, 24) interprets the description as referring to both "Barbarian coinage of the Varaždin type" and as identical to the Ribnjačka coinage (Mirnik 1981, 40 no. 44, 1998, 486).
[137] Unpublished. Pink 1939, 103-105, no. 484-495; *OTA* 484-495; Dembski 1998, 113-114 no. 1413-1431. Two coins from the hoard are kept in the AMZ (inv. no. A12554-12555). Pink (1939, 104, 135 no. 48) mentions another coin of this type from Dalj.
[138] Sokol 1994a, 21 no. 1; Sokol 1994b, 98 no. 118; Mirnik 1998, 487; Sokol 2001, 9-10 no. 1.
[139] AMZ inv. no. A1176.
[140] Dukat and Mirnik 1976, 200; Mirnik 1987, 369.
[141] Crawford 1985, 222-223.
[142] Mirnik 1987, 373-374.
[143] Mirnik 1981, 38, no. 30 and 31; Crawford 1985, 222, 321; Mirnik 1987, 379.
[144] Mirnik 1987, 381.
[145] *RRCH* 146, p. 79-80; Mirnik 1981, 48-49, no. 93; Crawford 1985, 222, 321; Mirnik 1987, 382.
[146] Mirnik 1987, 379 with a question mark, cf. 371.
[147] Mirnik 1987, 371.
[148] AMZ inv. no. A6082; Göricke-Lukić 2004, 38, 40 cat. no. 2.
[149] Together with Brettian bronzes: Pichler 1865, i.210-212; Mirnik 1981, 46 no. 77a; Crawford 1985, 321.
[150] Mirnik 1987, 383-386 with fig. 1 on p. 379.
[151] Pichler 1865, i.212-214; Crawford 1985, 321.
[152] Marić 1964, 49, pl. XXII.3, *RRC* 222.1, 143 BC.
[153] Marić 1964, 49, pl. XXII.6, *RRC* 254.1, 131 BC.
[154] *RRC* 190.1, 169-158 BC; Marić 1964, 47, pl. XXII.2; Crawford 1985, 322 sees here "perhaps a post-semilibral sextans (*RRC* 41.9, 215-212 BC) overstruck on a semilibral uncial".
[155] Dautova-Ruševljan 1984, 48, 50-51, cat. no. 13-15.

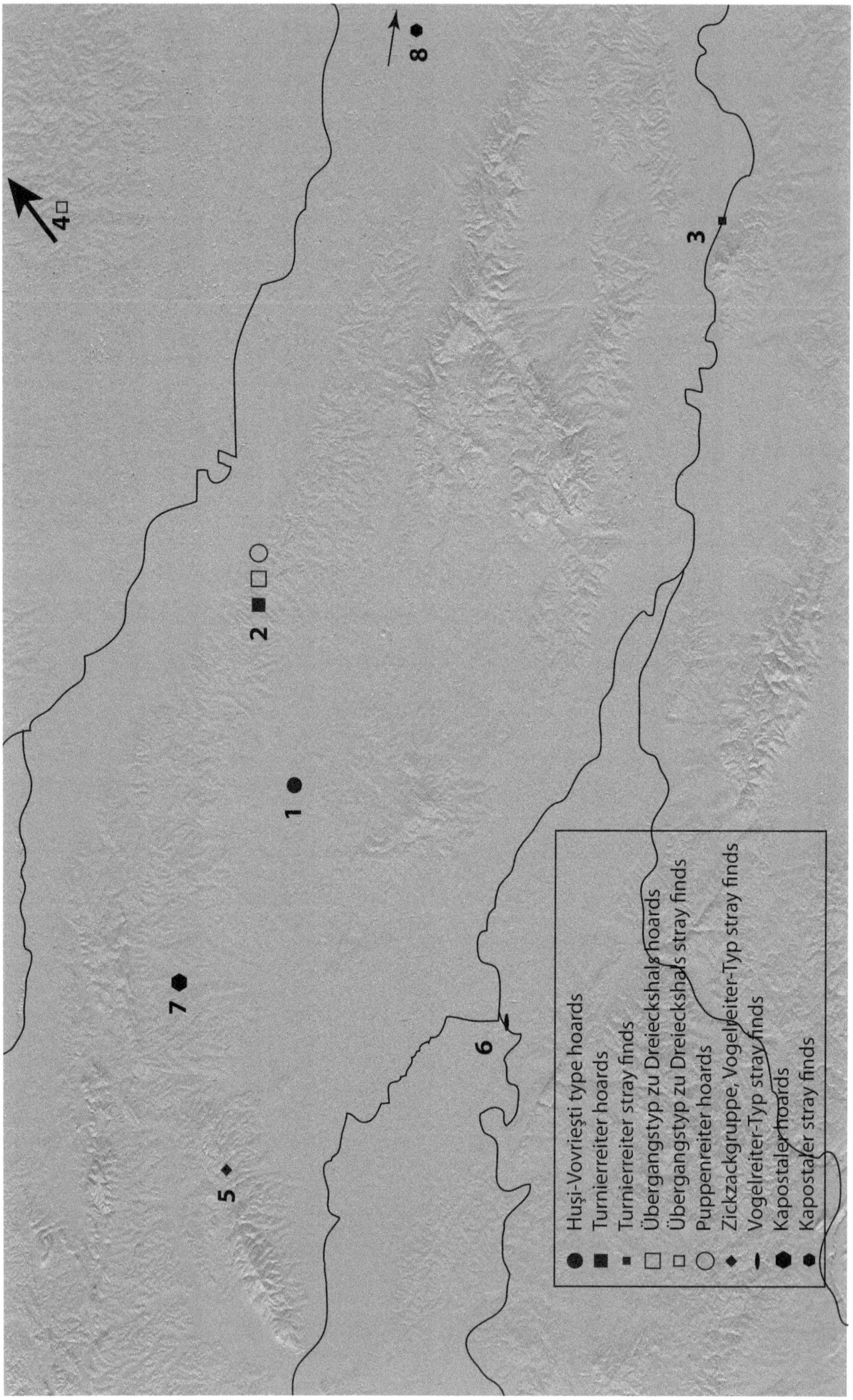

Fig. 5. Distribution of other Celtic coinage: (1) Narta, (2) Ribnjačka, (3) Kobaš, (4) Munkács, (5) Kuzelin, (6) Sisak, (7) Sv. Petar Orehovec, (8) Dalj (made by A. Solter).

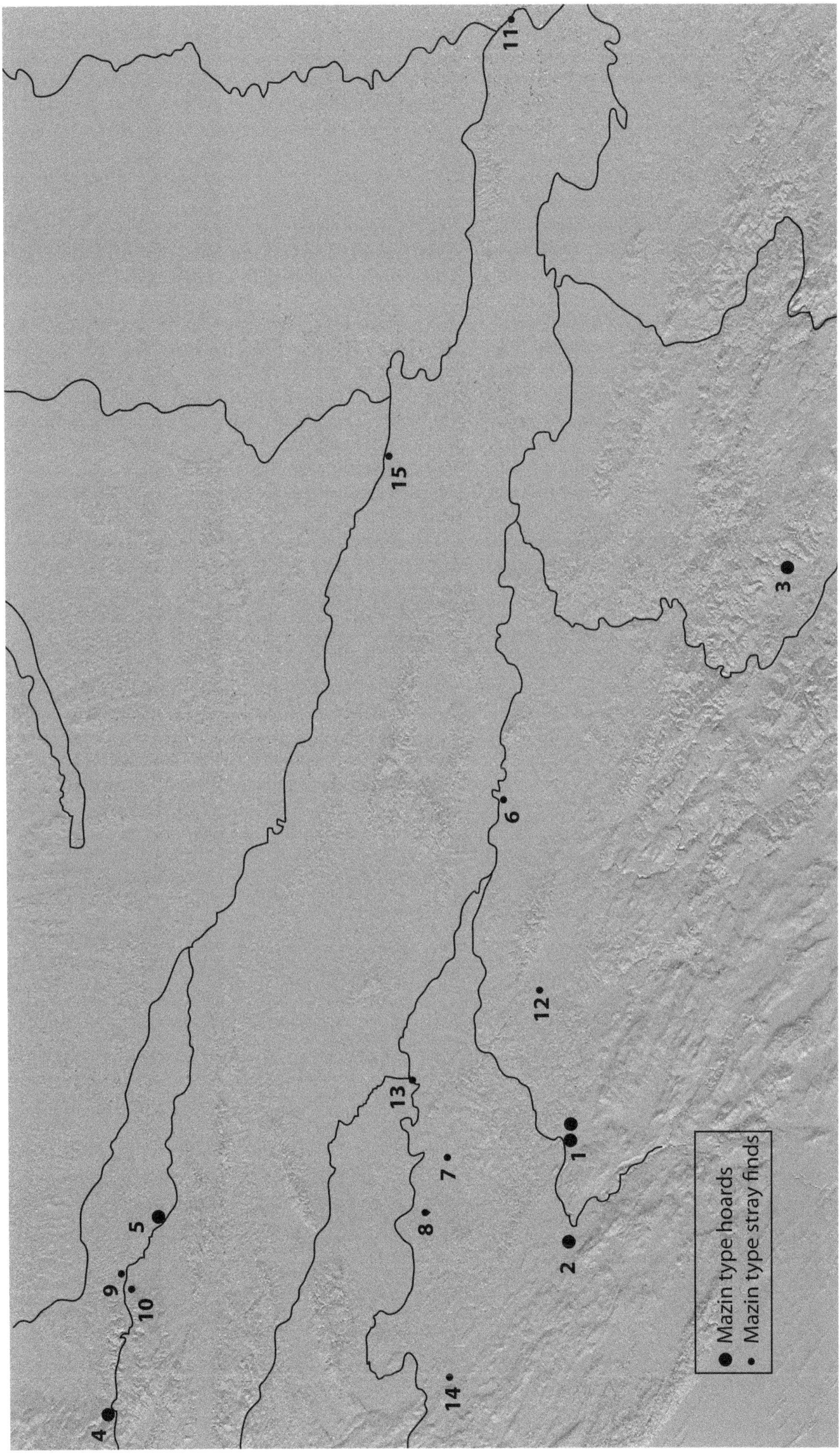

Fig. 6. Distribution of the types of coinage from the so-called Mazin type of hoards: (1) Bosanska Krupa, (2) Izačić, (3) Vrankamen, (4) Muta, (5) Ptuj, (6) Donja Dolina, (7) Glina, (8) Kiringrad, (9) Košaki, (10) Spodnje Radvanje, (11) Surduk, (12) vicinity of Prijedor, (13) Sisak, (14) Trošmarija, (15) Osijek (made by A. Solter).

dated to the second half of the 1st century BC,[156] several Republican *denarii*,[157] a *dupondius* of Nero Drusus,[158] three Barbaro-Celtic coins,[159] two of them[160] in the same layer with five Republican coins[161] and two Barbarian imitations of Republican coins[162] forming a small hoard, also in the layer VIb.[163] Popović associates the hoard with the destruction of the Celtic settlement at Gomolava in century 10 BC,[164] and believes that the Republican coinage arrived to Gomolava via Dacia.[165] Additional evidence for the introduction of Roman Republican coinage is provided by hoards. Beside the hoards from Vukovar, Vračevgaj, and Gomolava, we know of two other, both kept in the AMZ. During the dredging of the Kupa River at Sisak in 1912 a hoard of various coins circulating during the 2nd and 1st centuries BC in the southern Pannonia was found.[166] It consisted of 20 Roman Republican coins,[167] coined from 194-190 to 32-31 BC, together with a Barbarian imitation of a Republican coin,[168] an Augustan *quinarius*,[169] a large eastern Norican silver of the Augentyp-Stamm type, five Celtic coins of smaller denominations,[170] one Epirote[171] and one Thracian bronze.[172] Thus the coins in this hoard were produced between 233/232 BC and AD 12, and they were removed from the circulation probably during the fights in Pannonia in AD 6-9.[173] The second hoard is the one from Valpovo.[174] This hoard can also be associated with the great Illyrian-Pannonian uprising of AD 6-9 and was therefore interred during the early Imperial period. Yet the nature of this hoard reflects the earlier, pre-Augustan circulation, rather than the new Roman order of the 1st century AD. Furthermore, the Valpovo hoard "marks the beginning of the penetration of Roman coinage deep inland from the Dalmatian coast".[175] The hoard contained the following coins: RRC 199.1a, 256.1, 286.1, 289.1, 297.1b, 316.1, 317.3b, 340.1 (2 pieces), 341.2, 343.2, 344.3, 345.2 (a *quinarius*), 348.1, 350a.2 (2 pieces), 353.1a, 363.1b, 382.1a (2 pieces), 383.1, 387.1, 390.1, 393.1b, 407.2, 408.1b, 414.1 (2 pieces), 416.1a, 415.1, 426.1, 427.2, 433.1, 444.1a,[176] 458.1, 463.1a, 464.2 (2 pieces), 3a, 467.1a, 494.27, 543.1, 544.14, 18, 20, 28, 35, RIC i.45 no. 52a, i.52, no. 171a, i.54, no. 199, i.55 no. 207, i.65 no. 322. The earliest coin is, thus, of 155 BC, and the latest (*RIC* 207) from 2/1 BC.[177] This hoard should be related to a group of hoards found in the broader area, firmly placing the later Lower Pannonia, more precisely, Pannonia Secunda, into a region open to Roman mercantile and political interests in this period. We know of several hoards terminating with the *denarii* of C. L. Caesares from Illyricum (Dalmatia), Moesia, Noricum and north-eastern Transpadana, including the Regio X of Italia:[178] Lički Ribnik,[179] Kruševo near Obrovac,[180] Medvednica Hill near Ljubljana (Vrhnika, Bevke hoard),[181] Magdalensberg,[182] Vergnacco, Reana del Rojale, Udine,[183] Aquileia 1921,[184] and Viminacium.[185] P. Kos includes in this group the hoard from Sisak,[186] perhaps adding to the hoard an *aureus* from Sisak.[187] I see no reason to include it here, although it is possible that the concealment of the hoard could be associated with the Pannonian turmoil of the first decade of the 1st century AD. But a fraction of a hoard found in Treviso in 1913, kept in the AMZ as a part of the Benko Horvat collection, contained as much as eight pieces

[156] Popović 1971 cat. no. 9-10, 1978, 20. Brunšmid 1912, 265 (with n. 3), mentions three Apollonian drachms from Hrtkovci (Gomolava) in Zagreb, one kept in the Archaeological museum (inv. no. A7145), two in a private collection, together with one drachm of Dyrrachium (cf. Popović 1971, 158).

[157] *RRC* 335, 345, 443, 544.11, 96-32/31 BC; Popović 1971 cat. no. 8; Dautova-Ruševljan 1983, 51, cat. no. 4-5, 17-18. Again, Republican denarii from Hrtkovci (*RRC* 278.1, 442.1a) of 121 and 42 BC respectively, are kept in the AMZ (inv. no. B1040, 2161), probably also from Gomolava.

[158] Dautova-Ruševljan 1984, cat. no. 16.

[159] Dautova-Ruševljan 1984, 51, cat. no. 1-3.

[160] Popović 1971 cat. no. 11-12 – the former of the Đurđevac type, the latter similar to the Kapostaler type and the third of the Krčedin II type.

[161] Popović 1971 cat. no. 1-5; Dautova-Ruševljan 1984 cat. no. 6-10; *RRC* 306, 337.3, 436.16, 463.16 – 2 pieces, 485.1, from 108/107 to 46 BC.

[162] Popović 1971 cat. no. 6-7; Dautova-Ruševljan 1984 cat. no. 11-12 – the former is a hybrid of two coins, that of L. M. Censorinus of 82/81 BC and M. Volteius of 76 BC, the latter a *bigatus*.

[163] Popović 1971, 147-148; Mirnik 1981, 43 no. 61a; Crawford 1985, 236; Borić-Brešković and Popović 2006, 39.

[164] Popović 1971, 149, 158; cf. Dautova-Ruševljan 1984, 51; Borić-Brešković and Popović 2006, 39.

[165] It is occasionally mentioned that the Vrhnika hoard (Kos 1977, 150-153; Mirnik 1981, 49 no. 94) contained – along with Celtic coins of the Đurđevac (1), NEMET (1), *Verschwommener* (1), Eis (3), Magdalensberg (5), Karlstein (4), and ADNAMATI (1) type – a denarius of Marc Anthony of 32/31 BC, but this was probably not the case (Kos 1977, 54).

[166] Mirnik 1981, 47 no. 85; Crawford 1985, 236.

[167] *RRC* 132.1, 228.2, 233.1, 238.1, 262.1, 296.1h, 360.1b, 384.1, 414.1, 422.1b, 428.3, 441.1, 442, 443.1, 448.3, 496.1, 544.20 (2 pieces), 544.29, 544.

[168] Obv. *RRC* 465.5 or 494.38; rv. *RRC* 494.18, 19 or 25 (46-42 BC).

[169] *RIC* 276, 29-26 BC.

[170] Syrmian type phase C, phase D, and the Taurisci coinage of Magdalensberg IAg, IIh, and Karlstein IB2b types, coined during the 1st century BC. On the territory of the Croatian part of Pannonia two coins of the Magdalensberg type were found in Đurđevac (Kos 1977, 38) and three more in Sotin (Ilkić 2003, cat. no. 487-489).

[171] *SNG* De, München 12, 584, 233/232-168 BC.

[172] *SNG* Ita, Milano 6, 337-341; *SNG* XII, Hunt. 1, 794-795 (11 BC – AD 12). Six forgeries of 4th-century Constantinian coins found together with the former certainly do not belong to the same hoard, unless they are forgeries of, for example, Augustan coins from the third decade BC (Mirnik 2009, 307-308; another specimen of this type is in the private collection of R. Kavazović, Zagreb).

[173] Cf. Kos 1986, 30.

[174] Brunšmid 1886, 1895; *RRCH* 547; Mirnik 1981, 48 no. 91; Bilić 2009; Radman-Livaja and Dizdar 2010, 54-55.

[175] Crawford 1985, 236.

[176] Perhaps also 450.2 (48 BC), from the Brunšmid collection (inv. no. B846), not mentioned in Brunšmid (1886; 1895).

[177] The C. and L. Caesares issues are dated to this period in Fischer 1990, 352; Kehne 2000, 47-48 with n. 4; Wolters 2002, 298 with earlier literature in n. 4, 321 (Berger 1996, 13 opted for 2BC-AD1).

[178] Cf. Kos 1986, 30. I will discuss this group of hoards more thoroughly in a forthcoming paper.

[179] *RIC* 210; Klemenc 1935, 86, 103 no. 158, pl. I.24; *RRCH* 546; Mirnik 1981, 44-45, no. 68; Kos 1986, 30; Balen-Letunić, Dukat and Mirnik 1991/1993.

[180] *RIC* 207; *RRCH* 535; Mirnik 1981, 44, no. 66; Šeparović 2005, 67 no. 90, 70 no. 89-90, 70-71.

[181] *RIC* 207, 208 or 210; Müllner 1879, cxxxviii, c-d; *RRCH* 536; Mirnik 1981, 45, no. 73; Kos 1988, i.375, no. 204.3-4.

[182] *RIC* 207; Bruck 1961, 168-170, no. 22 in Schatzfund 1; *RRCH* 538; Bannert and Piccottini 1972, 54-55, no. 31 in Schatzfund II; Kos 1986, 30; Ruske 2011, 66 (V13).

[183] Gnecchi 1902, 551, no. 99; *RRCH* 548; Kos 1986, 30; Backendorf 1998, 126-127, 453-454. Backendorf (1998, 126 and n. 548) dates the closing of the hoard with "*RIC* Aug. 207…212"; *RIC* 211-212 were minted in AD 5 according to Wolters 2002, 310-311, 322.

[184] *RRCH* 522; Kos 1986, 30; Backendorf 1998, 39, 242-246, 14 pieces of *RIC* 207 and 14 pieces of *RIC* 210. Kos (1986, 30), following *RRCH* 529, adds here the hoard from Concordia, but it closes with *RIC* 220 (13-14 AD), according to Backendorf (1998, 63 with n. 246).

[185] *RIC* 207, 210; Jacanović 1988, 31, 38, no. 80, 81.

[186] See above, Mirnik no. 85.

[187] Kos 1986, 30. The *aureus* in question is an unpublished specimen from the AMZ (inv. no. C17, RIC 206).

of *RIC* 207 (unpublished).[188] P. Kos does not believe that the date of concealment of these hoards can be established, but they do indicate "the penetration of Republican coins from Cisalpina to Pannonia, surely a result of Augustus' campaigns in the Balkans".[189] Similarly, Crawford states that the hoards of Augustan *denarii*[190] were "following in the wake of the penetration of Roman armies".[191]

In terms of coin finds, a Roman military camp of the Augustan period at Obrežje, on the Sava River on the Slovenian-Croatian border, presents a somewhat specific case. Beside five coins found in a grave (three Augustan asses, 11-12 AD, a small Celtic silver, a *denarius* of Marc Antony, 32-31 BC), dating from after the abandonment of the camp, 33 more coins were found: 70% of those were moneyers' coins (27-6 BC, half of them halved), 13 % were Republican asses (all halved), 7% Republican *denarii*, 7% coins minted in Lugdunum, and 3% the so-called CA types, minted in Asia Minor.[192] Strategically positioned on the important Emona-Siscia road, the camp was probably in use during the AD 6-9 uprising.[193] In the future, the sites of Sotin (*Cornacum*) and Novi Banovci (*Burgenae*) on the Danube will provide us with a much clearer picture regarding the coin circulation on the Danube *limes*.[194]

Certainly, as we have seen, the monetary situation in Pannonia was far from clear at the moment of the introduction if the Republican coinage: Greek coinage of Apollonia and Dyrrachium dominated the local economy, at least in "international" relations, but the indigenous coinage of local (Celtic) tribes was also present.[195] M. Bíró-Sey believes that Roman Republican coinage entered Pannonia in the 80s or 70s BC – with an increase during the latter half of the 1st century BC.[196] However, the presence of many examples of issues of the late 2nd century BC and of the 80s in the hoards from Romania does not mean that these coins were introduced during these periods: "both periods were characterized by massive issues which remained in circulation in enormous quantities in the first century..."[197] M. Crawford argues that the Republican *denarii* could not have circulated for a long time in the Danube basin without being hoarded, and dates the "massive penetration of Republican *denarii*" with the closing date of the earliest hoards from the area.[198] Popović dates the Apollonia-Dyrrachium/Republican replacement to the mid-1st century BC, and believes that the Republican coinage only arrived in larger numbers into Pannonia in the beginning of the second half of the 1st century BC.[199] It is most probable that the Republican coinage arrived to Pannonia from the west, along the Aquileia-Nauportus-Siscia axis.[200] The large number of Roman Republican *denarii* found in numerous hoards in Dacia,[201] contrasted with a rather modest number of hoards in Panonia, suggests that southern Pannonia was a "transit area" for the large quantities of coinage circulating between Italy and Dacia,[202] which probably started – in large quantities, that is – only in the mid- or late 60s BC.[203] Certainly, prior to the conquests of 35-33 BC,[204] 14-8 BC,[205] and 6-9 AD,[206] only a small number of Republican *denarii* had circulated in Pannonia; therefore, a significant trade with Italy is out of the question.[207]

It is certain that the Romans made significant changes in the economy of pre-Roman Pannonia. Their "oppressive fiscal regime" was certainly among the main reasons for the great Illyrian revolt of the two Batos in AD 6-9.[208] This is implied, at least, in a dialogue between Tiberius and Bato of the Daesitiates as related by Cassius Dio (Dio LVI.16.3): after Tiberius had asked Bato why his nation decided to rebel against the Roman and to fight against them for such a long time, the leader of the rebels replied to him that it was the fault of the Romans themselves, since they have sent wolves, and not shepherds or dogs, to guard their flocks.[209] I believe that Crawford is right in recognizing in these "wolves" Roman tax collectors, who executed, in words of Šašel Kos, "the ruthless exacting of tribute",[210] but, naturally other factors contributed to the rebellion.[211] After all, it was Roman merchants who were massacred right at the outbreak of the rebellion in AD 6,[212] which shows certain resentment towards new social circumstances

[188] It is unclear whether some Domitian coins in the Benko Horvat collection came from the same hoard. If they did, then the hoard, naturally, does not belong in this discussion. The AMZ keeps 12 more specimens of stray finds of the C. and L. Caesares issues (*RIC* 207, 9 specimens; *RIC* 210, 2 specimens; one specimen is undeterminable), with one Barbarian imitation of RIC 207. One specimen (inv. no. C22, undeterminable) is from Sremska Mitrovica (*Sirmium*), another from Novi Banovci (*Burgenae*, inv. no. C10484, *RIC* 207, pierced three times), one from Suhopolje (inv. no. C25512, *RIC* 207), and one from Sotin (*Cornacum*, inv. no. C26249, *RIC* 207). The share of silver coins from the mint of Lugdunum, according to the data from Slovenia, was 68% in the total number of silver coins of the Augustan period (Miškec 2009, 286 with fig. 6 on p. 287). The numbers cannot be much different in other areas discussed here.

[189] Kos 1986, 30.

[190] Adding the hoards from Celje (*RRCH* 462, Mirnik 1981, 41 no. 49, Kos 1988 ii.73-75, 360/4, 32-31 BC) and Poetovio (*RRCH* 492, Mirnik 1981, 47 no. 83, Kos 1988 = Kos ii.301-303, 434/22, 16 BC). A hoard from Krndija near Đakovo, published by Gardaš (1990), terminating with *RIC* i2.178a, 12-11 BC, can be added to this group (*RRC* 231.1, 292.1, 319.1, 382.1b, 408.1b, 425.1, 443.1, 494.23, 497.2a, *RIC* i².178a (138-12/11 BC)).

[191] Crawford 1986, 236.

[192] Miškec 2009, 288-289.

[193] Miškec 2009, 289, 291.

[194] See preliminary reviews in Mirnik (1999) and Ilkić (2007, cf. 2003 for a comprehensive catalogue of the coins from his private collection).

[195] Bíró-Sey 1980, 339.

[196] Bíró-Sey 1980, 339.

[197] Crawford 1985, 231 n. 31.

[198] Crawford 1985, 231.

[199] Popović 1987, 104, 106.

[200] Popović 1971, 150, 158; Popović 1987, 105.

[201] "...one of the most remarkable phenomena within the pattern of monetary circulation in antiquity" (Crawford 1985, 226-227).

[202] Popović 1987, 106.

[203] Crawford 1985, 232.

[204] Šašel Kos 1986, 134-147; Dzino 2010, 99-116.

[205] Šašel Kos 1986, 158-162; Domić Kunić 2006; Dzino 2010, 129-136.

[206] Šašel Kos 1986, 178-191; Dzino 2010, 137-155.

[207] Bíró-Sey 1980, 339.

[208] Crawford 1986, 236.

[209] καὶ τέλος ἐρωτηθεὶς ὑπὸ τοῦ Τιβερίου "τί ὑμῖν ἔδοξε καὶ ἀποστῆναι καὶ ἐπὶ τοσοῦτον ἡμῖν χρόνον ἀντιπολεμῆσαι;" ἔφη ὅτι "ὑμεῖς τούτων αἴτιοί ἐστε· ἐπὶ γὰρ τὰς ἀγέλας ὑμῶν φύλακας οὐ κύνας οὐδὲ νομέας ἀλλὰ λύκους πέμπετε"

[210] Šašel Kos 1986, 182.

[211] Cf. Mócsy 1974, 37; Dzino 2006, 147, 2010, 142-144, where he agrees with Dio's reasoning.

[212] Vell. II.110.5-6; Mócsy 1974, 71; Šašel Kos 1986, 182; Radman-Livaja and Dizdar 2010, 55.

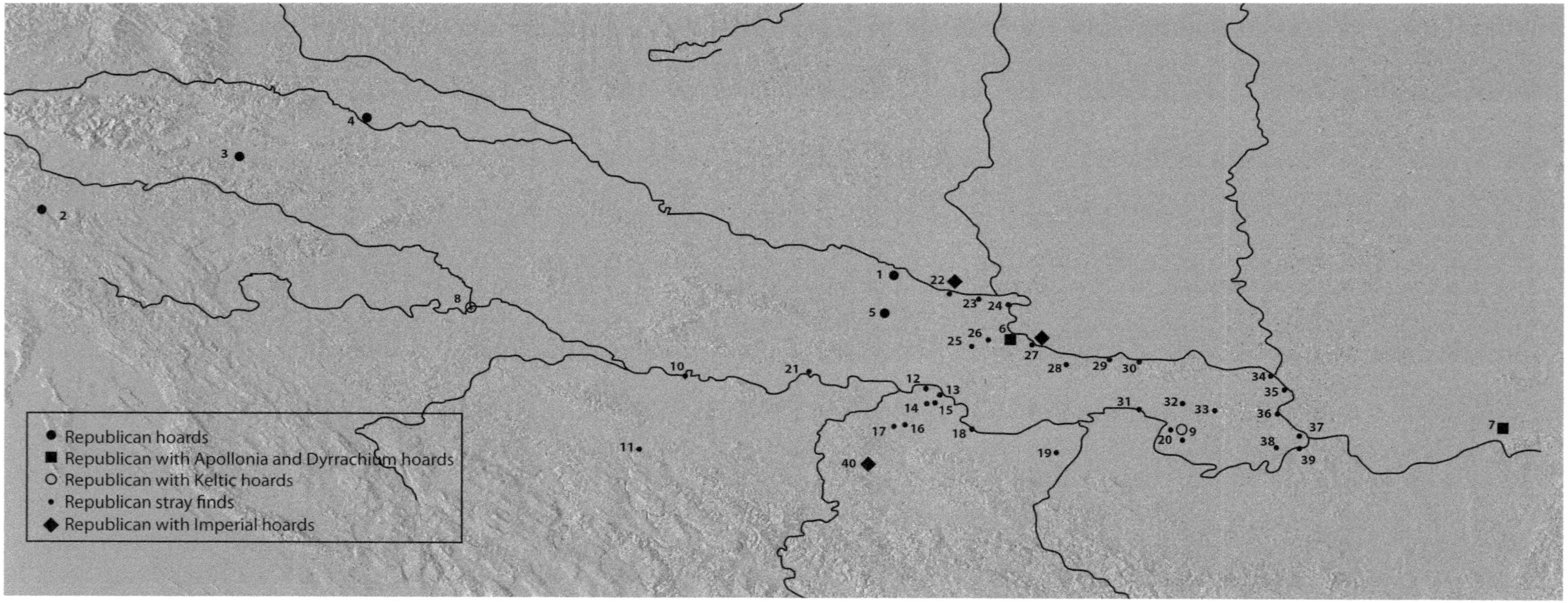

Fig. 7. Distribution of Roman Republican coinage: (1) Valpovo, (2) Vrhnika, (3) Celje, (4) Ptuj, (5) Krndija near Đakovo, (6) Vukovar, (7) Vračevgaj (?), (8) Sisak, (9) Gomolava (imitations, ?), (10) Donja Dolina , (11) Banja Luka, (12) Domaljevac, (13) Tolisa, (14) Luka, (15) Bok, (16) Tramošnica, (17) Gradačac, (18) Brčko, (19) Bijeljina, (20) Gomolava, (21) Sl. Brod, (22) Osijek, (23) Sarvaš, (24) Dalj, (25) Vinkovci, (26) Nuštar, (27) Sotin, (28) Bapska, (29) Ilok, (30) Banoštor, (31) S. Mitrovica, (32) Ruma, (33) Petrovci, (34) Stari Slankamen, (35) Surduk, (36) Novi Banovci, (37) Zemun, (38) Surčin, (39) Ada Ciganlija, (40) Gračanica (made by A. Solter).

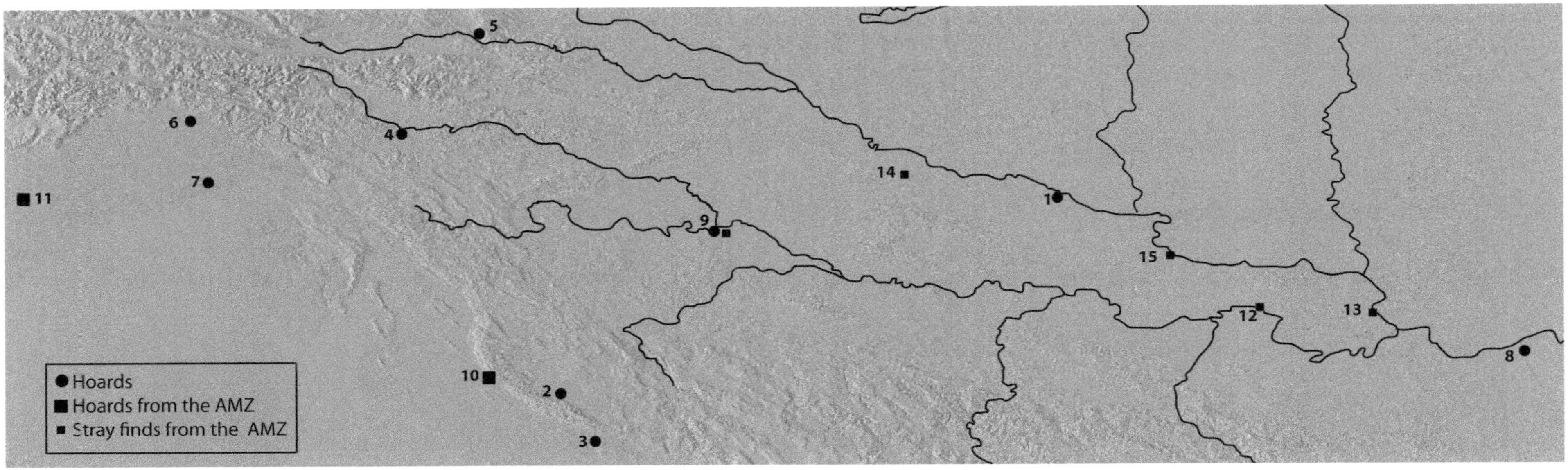

Fig. 8. The hoards terminating with the C.L. CAESARES issues, the C.L. CAESARES stray finds kept in the AMZ, and the C.L. CAESARES specimens from later hoards: (1) Valpovo, (2) Lički Ribnik, (3) Kruševo near Obrovac, (4) Medvednica Hill near Ljubljana, (5) Magdalensberg, (6) Vergnacco (Reana del Rojale, Udine), (7) Aquileia, (8) Viminacium, (9) Sisak (hoard?), (10) Cesarica, (11) Treviso, (12) S. Mitrovica, (13) N. Banovci, (14) Suhopolje, (15) Sotin (made by A. Solter).

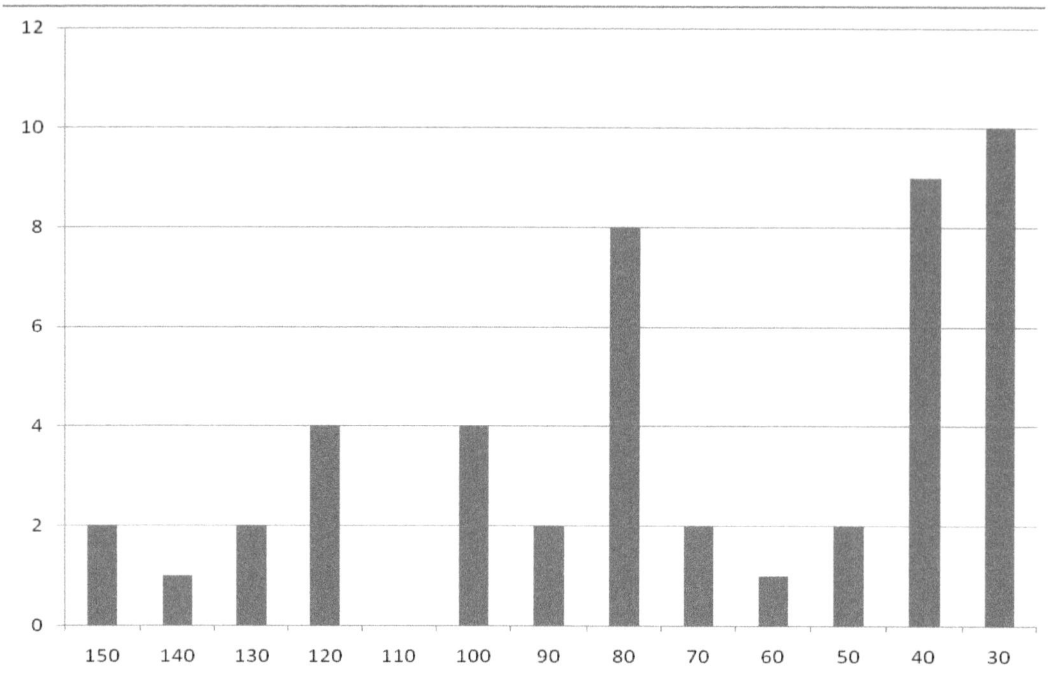

Chart 1. Roman Republican coinage from Sisak (Siscia) kept in the AMZ (total no.: 47).

introduced during the previous 40 years. Indeed, the first to arrive in the new founded province were precisely merchants, often preceding other groups, the army and administrators. [213] They were the ones responsible for the introduction of large quantities of money into Pannonia.[214]

For this paper I have analyzed the Republican coinage kept in the AMZ; due to the past political and geographical circumstances, the coinage found in Sisak and the rest of southern Pannonia, most notably the *limes* area, was regularly taken to the then only archaeological museum in this area. The coins were analyzed according to the *RRC*, and the dates used reflect this fact. Since Siscia was among the most important towsn of the Pannonian provinces, and probably among the first urban centres in contact with the Romans, it is interesting to see the data for the circulation of Republican coins in this town.

We can compare this diagram with the data for the circulation of Republican coins in southern Pannonia, consisting of coins from known southern Pannonian sites kept in the AMZ.

It transpires that both diagrams show two significant peaks: in the 80s, and 40s and 30s, the last characterised by a large number of the legionary *denarii* of Marcus Antonius. Chart 2 is somewhat different from the diagram brought by Popović,[215] probably because the latter included *all* the coins in the AMZ collection, but Popović's diagram also has the peaks in the 80s, 40s and 30s. The analysis of the 2nd century hoards from southern Pannonia shows a similar picture. The share of Republican *denarii* in the Gračanica hoard, concealed under Hadrian[216] is hard to determine, since the hoard was dispersed immediately following its finding in 1941, but the share of Republican *denarii* in the Osijek hoard, concealed under Marcus Aurelius[217] was 15.73% (439 out of total 2791), and in the Sotin hoard, also concealed under Marcus Aurelius,[218] the share was 12.65%.[219] Compared with the Emona hoard, concealed under Nero,[220] with 43.75% of Republican coins,[221] the share is, naturally, much smaller. This evaluation also stands if we further compare the Osijek and Sotin hoards with three hoards concealed during the reign of Domitian in the Danubian region of Serbia: Boljetin (61.76% of Republican coins, total no. 102), Bare (47.48% or 48.03%, total no. 278 or 279),[222] and Tekija (23.42%, total no. 111).[223] The hoards from the same region concealed during the reign of Marcus Aurelius and Commodus, though, show a different picture: the percentage of Republican coins, that is, the legionary *denarii* of Marcus Antonius, in the hoards from northern Serbia (Belgrade 1839, Bela Reka, Grocka,

[213] Suić 1966, 185-186; Rosada 2004, 191.
[214] Bíró-Sey 1980, 340.
[215] Popović 1987, 108, fig. 28.

[216] Pegan and Šporn 1960 for the Republican *denarii*; Mirnik 1981, 54 no. 122a; Kos 1986, 25; Popović 1987, 108 fig. 28.
[217] Mirnik 1981, 56 no. 134; Kos 1986, 25.
[218] Brunšmid 1911; Mirnik 1981, 57 no. 143; Kos 1986, 25.
[219] 237 out of total 1873, according to Mirnik (1981); Brunšmid (1911, 242-243) mentions 1938 coins, with 270 Republican, = 13.93%.
[220] Mirnik 1981, 51 no. 107; Kos 1986, 25; Kos 1988, i.207-209, no. 155/22.
[221] Kos (1986, 25) has 44.8%; 54.17% with legionary *denarii* of Marcus Antonius, 26 out of total 48.
[222] One *denarius* of M. Antonius is undocumented, and the percentage also depends on whether two *denarii* of Juba I are included within the total number of Republican coinage.
[223] Borić-Brešković 1994a, 108-110; Borić-Brešković and Popović 2006, 40, cf. 21 n. 52, 53, 56.

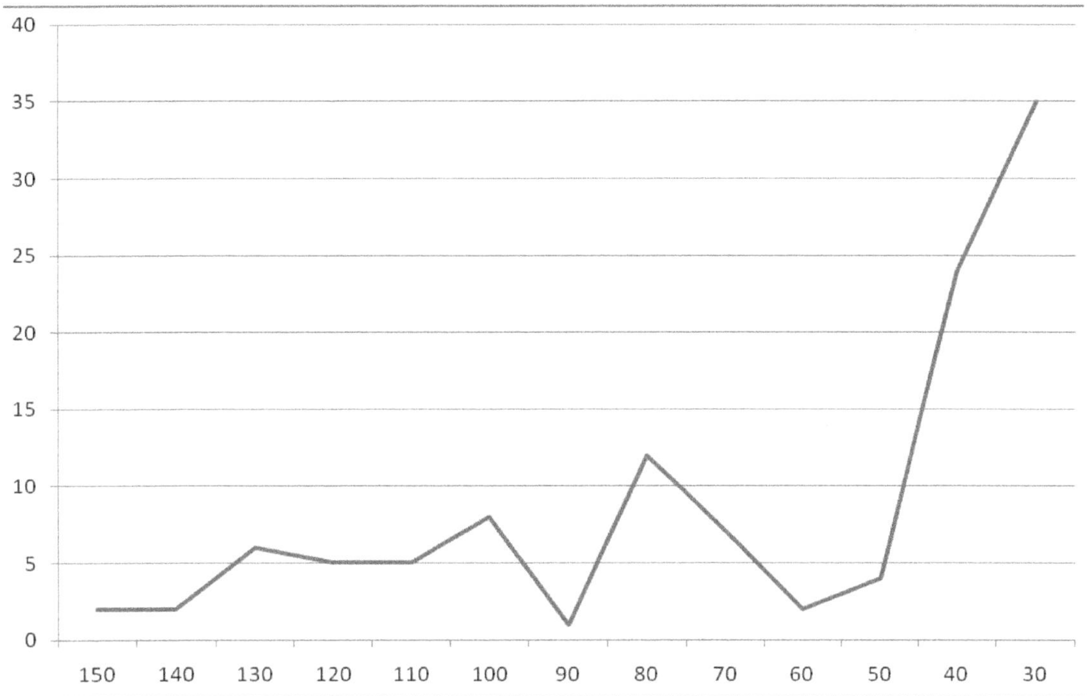

Chart 2. Roman Republican coinage from southern Pannonia kept in the AMZ (total no.: 113).

Nemenikuće, Mali Zvornik/Radalj) is between 0.37 and 0.84%.[224] But the evidence from the Croatian hoards clearly shows that Republican coins regularly circulated together with the Imperial coins of the first two centuries AD in the territory of southern Pannonia, at least until the reign of Marcus Aurelius. On the other hand, because of this they are „rather unreliable indicators for the establishment of monetary circulation in the region".[225] This is especially true for the coins of Marcus Antonius, circulating in „enormous numbers" under the Principate, sometimes surviving until the Severan age.[226]

4. Circulation during the first two centuries AD

It is during the Augustan period that we encounter an increased quantity of coinage – especially bronze – in the south-western Alpine and Pannonian area.[227] The *aes* coinage predominated in normal day-to-day monetary circulation during the Julio-Claudian period,[228] which is true for the south-eastern Alpine area,[229] as well as for southern Pannonia. According to P. Kos, the percentage of *denarii* in stray finds of Augustan coins in this region was 3-7%, *sestertii* 1-6%, *dupondii* 4-11%, *quadrantes* 3-15%, and asses from 61 to 89%, while the combined tally for the period from Augustus to Claudius shows an overall average of 2.8% (*denarii*), 4.1% (*sestertii*), 3.8% (*dupondii*), 5.4% (*quadrantes*), 83.7% (asses).[230] Compared with the Augustan coinage from Siscia, we see in this important Pannonian site a somewhat higher share of *dupondii* (17.28 ÷ max. 11%) and *denarii* (8.64 ÷ max. 7%), the presence of *quinarii* and *aurei*, not mentioned by Kos,[231] and a somewhat smaller share of asses (55.56 ÷ min. 61%).

The high percentage of *quadrantes*, related to a comparatively smaller percentage of halved Augustan asses, is a characteristic feature of the south-eastern Alpine region. Among the Augustan *aes*, the moneyers' asses of the mint of Rome have the greatest share, similar to neighbouring regions,[232] while the coinage of Gallic mints of Nemaesus and Lugdunum is, logically, rather scarce, and should not be associated with the transfer of Germanian and Raetian legions to Illyricum for the pacification of rebellious Pannonian tribes in AD 6-9.[233] The analysis of Augustan assesfrom Siscia (45 pieces, only five - 11.11% - halved),[234] compared to a much higher percentage on south-eastern Alpine sites (22.22-58%), does not support the conclusion of P. Kos, who claimed that the halved Augustan moneyers' asses „performed the major role in circulation of the Augustan *aes* in the area of the southeastern Alps and Illyricum", including the southern Pannonia, due to a

224 Borić-Brešković 1994b, 83; Borić-Brešković 2003, 19; Borić-Brešković and Arsenijević 2008, 98; Borić-Brešković and Crnobrnja 2005, 7-8, 10, 12-13; Borić-Brešković and Popović 2006, 12, 19 with n. 45, 21 with n. 51, 55, 57. The hoard from Ušće near Obrenovac, concealed during the reign of Marcus Aurelius (403 pieces in total), did not contain any Republican coins (Borić-Brešković and Crnobrnja 2008, 14-15). Neither did the hoard from Vršac-Ludoš, concealed in the last year of Commodus' reign (Bakić 2001, 14).

225 Kos 1986, 25.

226 Crawford 1985, 232.

227 Miškec 2009, 284.

228 Kos 1986, 32, citing Sutherland 1976, 81ff.

229 Kos 1986, 32.

230 Kos 1986, 34.

231 Kos 1986, 34.

232 Kos 1986, 36; Miškec 2009, 286-287 with fig. 3.

233 Kos 1986, 37.

234 Similar to Kos (1986, 38), who gives a value of 10.34%.

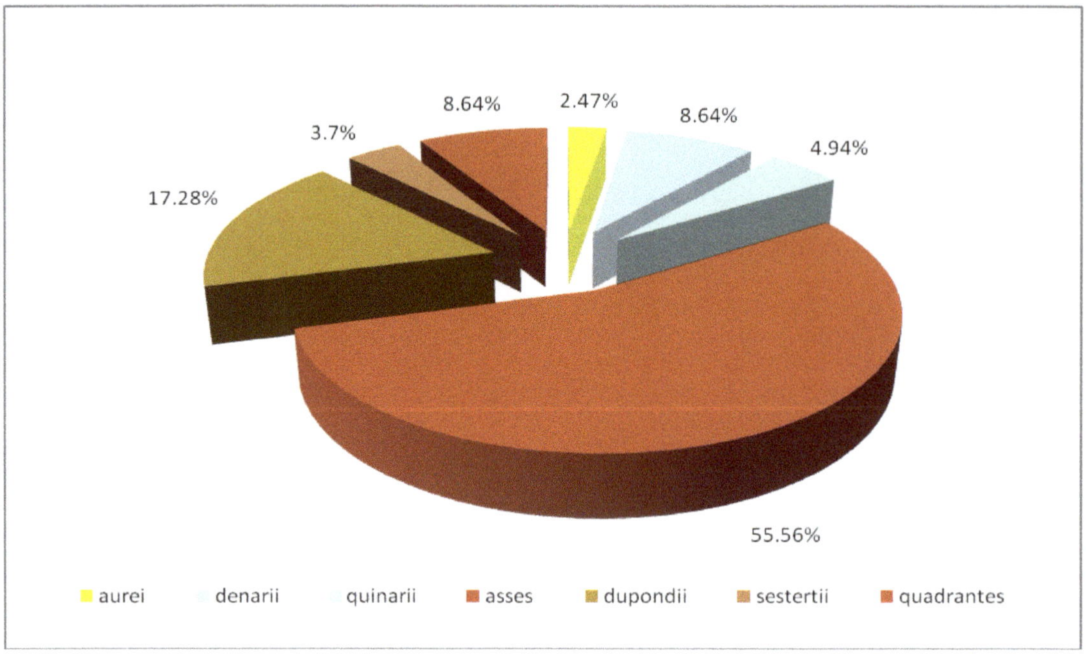

Chart 3. Denominations of Augustan coinage from Siscia (total no.: 81).[235]

shortage of low-value coins.[236] This discrepancy with the south-eastern Alpine data (covering the area of southern Noricum and the north-eastern part of the Regio X of Italia) could be a consequence of a different nature of the settlement in Siscia,[237] or of inadequate data available to the author of the present paper. On the other hand, the analysis of post-Augustan halved *aes* in Siscia – that is, their almost complete lack – concords well with the south-eastern Alpine data.[238]

The evidence from Sisak clearly supports the conclusion that the phenomenon of countermarking was not confined to the military sites along the Rhine, as argued by some commentators.[239] Eleven out of a total of 58 Augustan bronze coins were countermarked (18.96%),[240] which is a somewhat higher percentage compared with the corresponding values from the sites in the south-eastern Alpine area (4.16-14.87%).[241] The most frequent countermarks in Sisak were AVG (2 examples), with 28 more examples from other Pannonian sites, belonging to the third area (south-eastern Alpine-Balkan) of distribution of this countermark,[242] and CAE (4 examples, 1 coin is countermarked twice), with 7 more examples from other Pannonian sites.[243] Other countermarks appearing on the coinage from Pannonia are CAESAR, MP, DVA, MO, APOL (one from Sisak and another from Sotin; cf. Ilkić 2003, cat. no. 513), which shows that the phenomenon of countermarking was widespread in southern Pannonia.[244] An interesting countermark (MP VES), associated with Asia Minor, more precisely, Ephesus,[245] is found on a *denarius* of Marcus Antonius from the Sotin hoard.[246]

4.1. The first and second centuries – stray finds

The analysis of stray finds from Siscia shows trends basically similar to the monetary circulation in the south-eastern Alpine region as analyzed by Kos,[247] which, on the other hand, corresponds to "a very uniform monetary circulation throughout the Empire".[248] Thus one can detect a large increase of the intensity of monetary circulation during the reign of Vespasian (the increase is 333%, comparing the intensity in 54-69 with that of 69-81, which corresponds to that of Aventicum, 281%, Carnuntum, 302%, Vindobona, 479%, Poetovio, 303%, Neviodunum, 371%),[249] resulting from the imperial monetary policy, rather than some local

[235] Coins from the AMZ and several sites in Sisak (Burkowsky (2008); Koprivnjak (2011a, b)), as well as the (now lost) collection of the Town Museum of Sisak [I would like to thank I. Mirnik for allowing me to use his unpublished diploma thesis on this subject (Mirnik 1969)].
[236] Kos 1986, 38.
[237] Cf. Găzdac 2010, 118.
[238] Kos 1986, 39.
[239] Cited by Kos 1986, 40.
[240] Coins from the sites published in Koprivnjak (2011a, b) were not taken into account here.
[241] Kos 1986, 40.
[242] Kos 1986, 47-48.
[243] Kos 1986, 49.
[244] Kos 1986, 49-53.
[245] Mattingly 1923, xxviii, 1930, xvii, Howgego 1985, 293, no. 839, Martini 2003, 171, no. 100, AD 74-79.
[246] AMZ inv. no. B1104; cf. Brunšmid 1911, 245. Compare the same countermark on a *denarius* of Octavianus of 42 BC (*RRC* 497.3) from the Belgrade collection (Borić-Brešković and Popović 2006, 30, 294 cat. no. 1539, pl. LXXV), on a *denarius* of Augustus of 29-27 BC (*RIC* i².265a) from the Bare hoard (Borić-Brešković 1994b, 116, 154-155, 180 cat. no. 144, pl. VII), on a *denarius* of Caesar of 47-46 BC (*RRC* 458.1) from the Boljetin hoard (Borić-Brešković 1994b, 116, 180), and on a *denarius* of of 44BC from the Gračanica hoard (retrograde; Pegan and Šporn 1960, 125, no. 331; Borić-Brešković 1994b, 116, 180). For this countermark on legionary *denarii* of M. Antonius see Bahrfeldt 1876, 367-368, 371, 1887, 72-73.
[247] Kos 1986, 57-69.
[248] Kos 1986, 60.
[249] Kos 1986, 69. Cf. Bíró-Sey 1980, 340.

Table 1. The coins by number and period of issue and a corrected percentage comparison of the ten sites on an annual basis (total no.: 10863).

Period of issue/site*	*Aquileia no./%*	*Emona no./%*	*Celeia no./%*	*Neviodunum no./%*	*Poetovio no./%*	*Vindobona no./%*	*Carnuntum no./%*	*Brigetio no./%*	*Siscia no./%*	*Mursa no./%*
27 BC**-41 AD	716 0.955	233 0.311	157 0.209	21 0.028	125 0.167	13 0.017	177 0.236	8 0.011	149 0.199	20 0.027
41-54	279 1.834	76 0.5	33 0.217	21 0.138	98 0.644	7 0.046	106 0.697	4 0.026	92 0.605	3 0.02
54-69	134 0.782	20 0.115	7 0.04	8 0.046	28 0.161	7 0.04	54 0.311	6 0.034	60 0.345	3 0.017
69-96	611 2.009	87 0.286	44 0.145	57 0.187	155 0.51	72 0.237	301 0.99	43 0.141	366 1.203	7 0.023
96-117	412 1.724	63 0.264	63 0.264	32 0.134	173 0.724	69 0.289	276 1.155	66 0.276	267 1.117	17 0.071
117-138	408 1.707	49 0.205	55 0.23	42 0.176	150 0.628	91 0.381	275 1.151	47 0.197	239 1	47 0.197
138-161	557 2.136	66 0.253	61 0.24	54 0.207	176 0.675	86 0.33	443 1.7	75 0.288	120 0.46	24 0.092
161-180	607 2.794	51 0.235	17 0.078	47 0.216	135 0.621	61 0.281	363 1.671	74 0.341	103 0.474	48 0.221
180-193***	170 1.118	17 0.112	7 0.046	12 0.079	54 0.355	22 0.145	121 0.796	17 0.112	18 0.118	8 0.053

* The periods of issue are distributed according to Reece (1971).
** Kos (1986) has 23BC as the starting year.
*** The data supplied by Kos (1986), as well as the analysis of the coins from Sisak both end in AD 192.

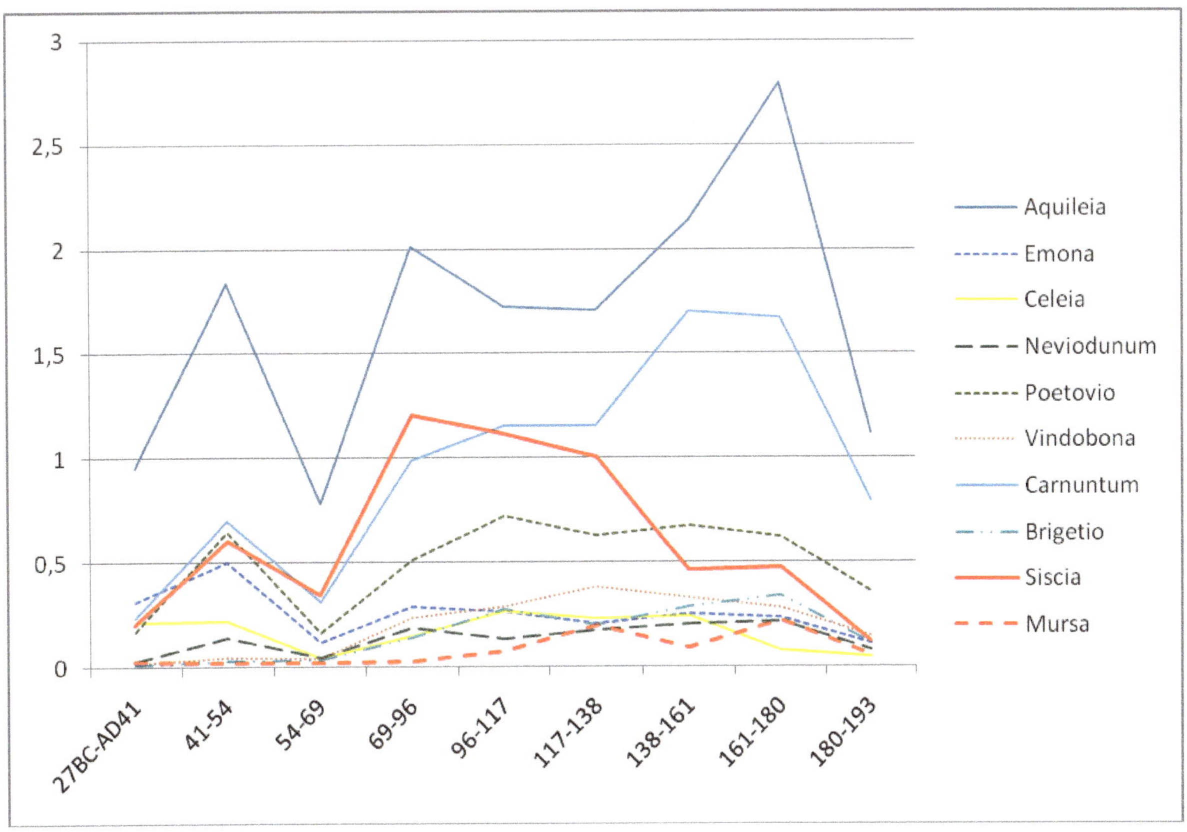

Chart 4. the data from Table 1 plotted on a chart.

Table 2. The coins from Siscia by the period of issue (total no.: 1414).

Period of rule	**Frequency no./%**	**Duration**	**Corrected frequency**	**Cumulative**
27 BC - AD 14	71 5.02	41	1.73	71
14-37	54 3.82	25	2.16	125
37-41	24 1.7	5	4.8	149
41-54	92 6.51	14	6.57	241
54-68	39 2.76	15	2.6	280
68-69	21 1.49	2	10.5	301
69-79	142 10.04	11	12.91	443
79-81	69 4.88	3	23.0	512
81-96	155 10.96	16	9.69	667
96-98	58 4.1	3	19.33	725
98-117	209 14.78	20	10.45	934
117-138	239 16.9	22	10.86	1173
138-161	120 8.49	14	8.57	1293
161-180	103 7.28	20	5.15	1396
180-192	18 1.27	13	1.38	1414

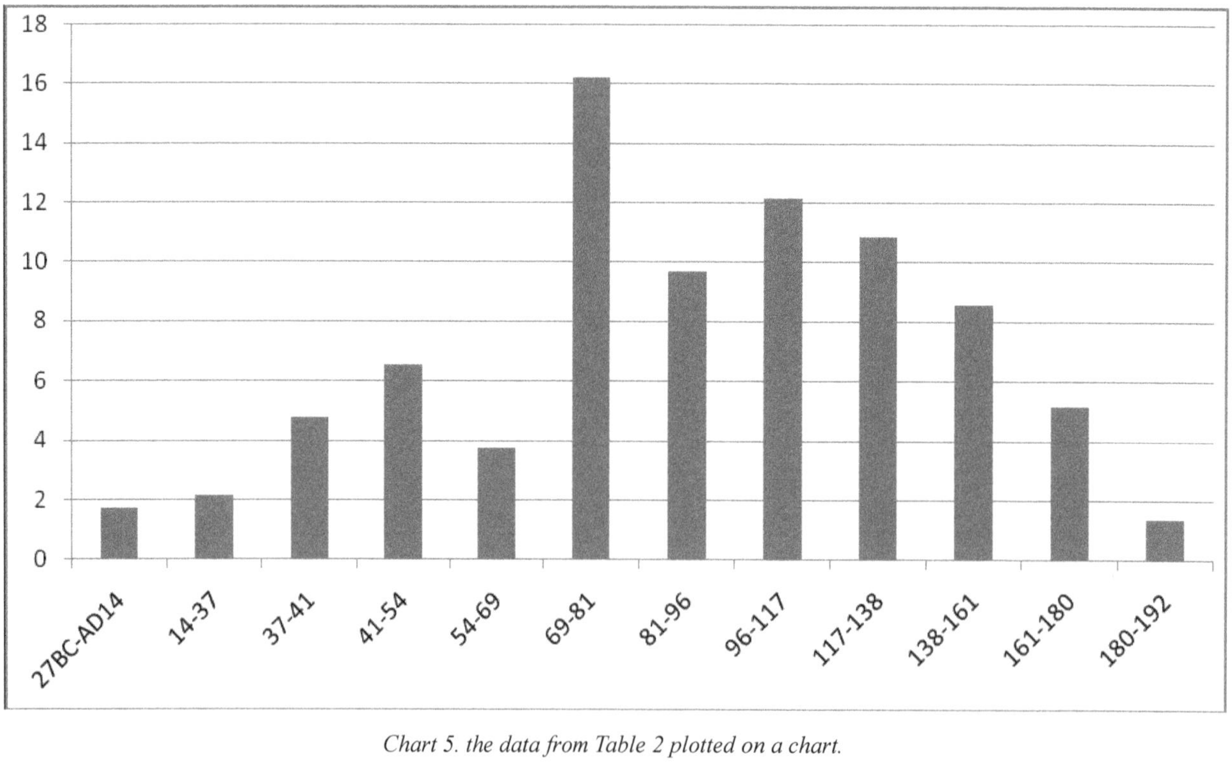

Chart 5. the data from Table 2 plotted on a chart.

*Table 3. The coins from Mursa by the period of issue (total no.: 177).**

Period of rule	*Frequency no./%*	*Duration*	*Corrected frequency*	*Cumulative*
27 BC - AD 14	7 3.95	41	0.17	7
14-37	11 6.21	25	0.44	18
37-41	2 1.1	5	0.4	20
41-54	3 1.7	14	0.21	23
54-68	1 0.56	15	0.07	24
68-69	2 1.1	2	1	26
69-79	3 1.7	11	0.27	29
79-81	0 0	3	0	29
81-96	4 2.3	16	0.25	33
96-98	5 2.8	3	1.67	38
98-117	12 6.78	20	0.6	50
117-138	47 26.55	22	2.14	97
138-161	24 13.56	14	1.71	121
161-180	48 27.12	20	2.4	169
180-192	8 4.5	13	0.62	177

* The data consists of the coins kept in the AMZ, as well as those published by Mažuran (1954) and Göricke-Lukić (1987, 1989, 1993, 1997).

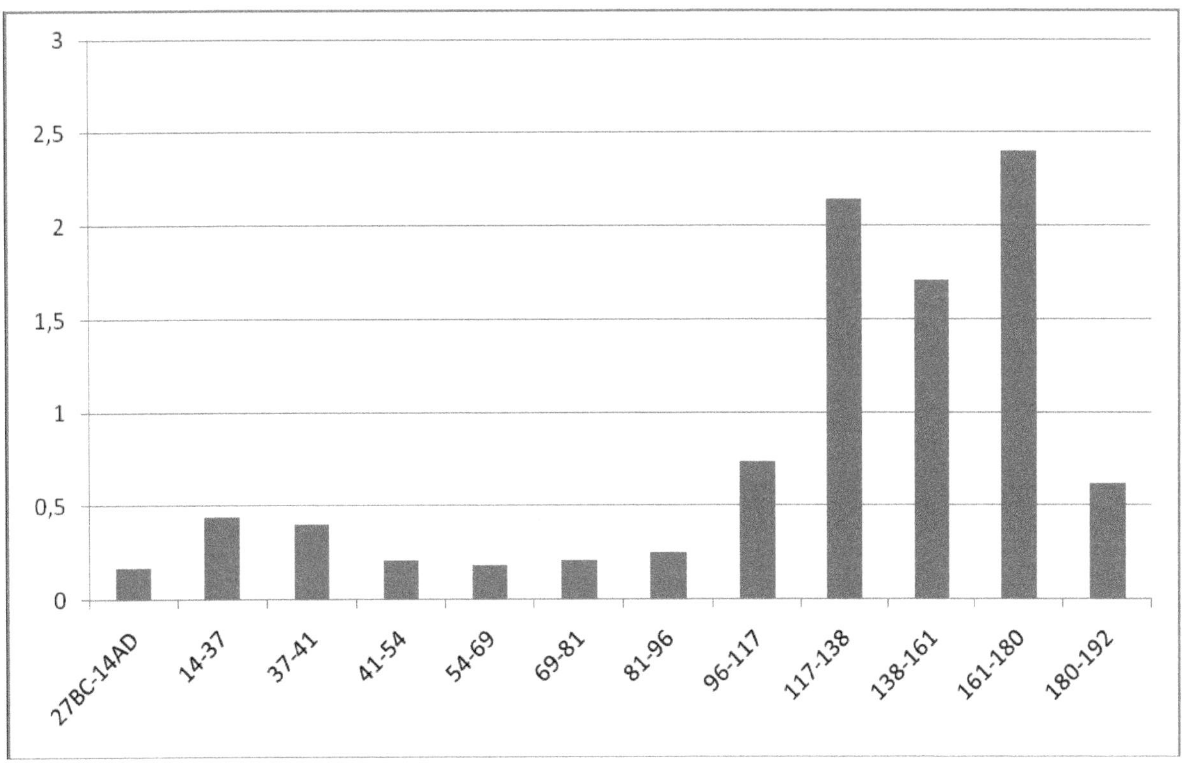

Chart 6. the data from Table 3 plotted on a chart.

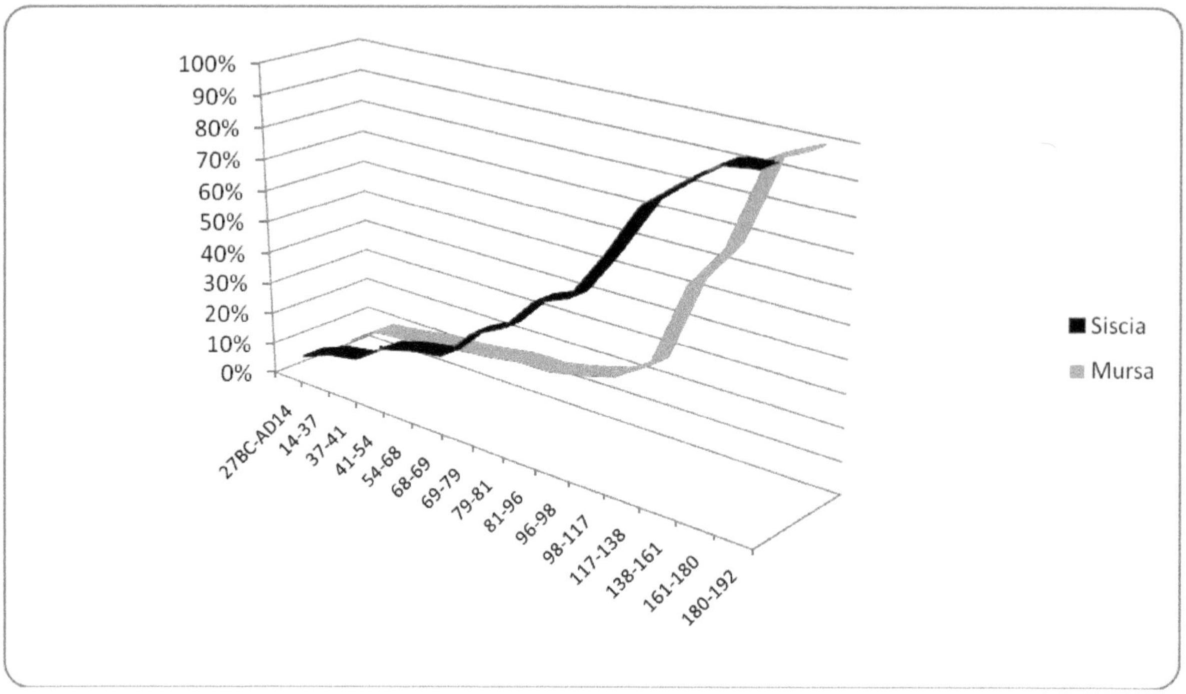

Chart 7. cumulated percentages of coin per period of rule for Siscia and Mursa.

circumstances.[250] When compared with the coin finds from other sites in this part of the Empire (Aquileia,[251] Emona, Celeia, Neviodunum, Poetovio, Vindobona, Carnuntum, and Brigetio),[252] the monetary circulation in Siscia reveals similar patterns of the circulation intensity: the increase during the reign of Claudius followed by a certain fall during the reign of Nero, the above-mentioned significant increase during the reign of Vespasian, maintained until the reigns of Antoninus Pius and Marcus Aurelius, when a rather large decline occurred (which is exact the opposite of the situation in Aquileia and Carnuntum), and, finally, another significant decline during the reign of Commodus, a feature common to all the sites discussed in this context. Thus the decrease of the intensity in Siscia was 72.3%, comparing the reigns of Marcus Aurelius and Commodus, compared with 47.2% in Emona, 34.5% in Celeia, 59.6% in Neviodunum, 36.7% in Poetovio, 68.4% in Savaria, 48.4% in Scarbantia, 42.9% in Vindobona, 47.3% in Carnuntum, 63.6% in Brigetio, 66% in Aquincum, and 51% in Intercisa.[253] Yet this trend, characteristic for the reign of Commodus, can be perceived throughout the Empire, and it should not be associated directly with either the Marcomannic wars or any local political or economic cause.[254]

The situation in Mursa is somewhat different. A large increase of the intensity of the monetary circulation during the reign of Vespasian is not evident in Mursa (an increase of mere 23% occurred between the intensity in 54-69 compared with that of 69-81). A slight increase occurred only with Nerva's and Trajan's reign, but the most significant increase occurred during the reign of Hadrian (254% comparing the reigns of Trajan and Hadrian), precisely in the period when the Colonia Aelia Mursa was founded.[255] The circulation remained stable during the reign of the next two emperors, while there is again a predictable decrease during Commodus' reign (73.6% comparing the reigns of Marcus Aurelius and Commodus).

Second-century hoards

There are a number of hoards concealed during the second century AD in the Croatian part of Pannonia. They are to be compared with contemporary hoards from Slovenia (yielding data for parts of Noricum and Pannonia, as well as the Regio X of Italy), most conveniently analyzed by P. Kos.[256] The following list shows their composition as well as the latest coin in each hoard, wherever the data is available:

A. Concealed during Trajan's reign

1. Zemun.[257] The region of eastern and central Syrmia is not a part of the modern-day Republic of Croatia, but it was formerly under the jurisdiction of the Triune Kingdom of Croatia, Slavonia and Dalmatia, and its territory was, therefore, under the jurisdiction of the today's AMZ. Thus, numismatic material was gathered, and is now kept in the

250 *Colonia Flavia Siscia*, for example, was founded early in Vespasian's reign (Mócsy 1974, 112, 130, 135).
251 Data from Reece 1971, 171.
252 Data from Kos 1986, 60, 63.
253 Kos 1986, 97.
254 Kos 1986, 98; cf. Găzdac 2010, 115, 119, 122.

255 Brunšmid 1900, 23-24; Pinterović 1967, 30, 43, 56; Mócsy 1974, 119, 222.
256 Kos 1986, 74-79, 83-85.
257 Mirnik 1981, 52 no. 112; Mirnik (2008) with earlier literature; Găzdac 2010, 67.

AMZ, and this is the reason why this territory has been included in the present analysis. The hoard, concealed in a clay pot, originally consisted of ca. 250 *aurei*, but the data is only available for 230 specimens. The latest dated coin in the hoard was issued in 98/99.[258]

The composition of the hoard:

Nero	6
Vespasianus	25
Titus	3
Domitianus	187
Nerva	3
Traianus	6

B. Concealed during Hadrian's reign

(?) 2. Deč near Stara Pazova.[259] For the reasons for the inclusion of this hoard see under (1).

Two *dupondii* of Hadrian is all that the AMZ managed to acquire. The original number of coins in the hoard cannot be ascertained.

3. Dalj 1924[260]

Two Lycian drachms is all that the AMZ managed to acquire. The original number of coins in the hoard cannot be ascertained.The latest dated coin in the hoard was issued in 98-99.

The composition of the hoard:

Domitianus	1 (1 drachm)
Traianus	1 (1 drachm)

4. Osijek 1982[261]

Five *sestertii* and one *dupondius* were found concealed in a pot. The latest dated coin in the hoard was issued in 119.[262]

The composition of the hoard:

Traianus 5 (1 dupondius)	
Hadrianus	1

5. Garešnica[263]

Some 157 *denarii* were found, but the exact data is available only for 37 coins that arrived to the AMZ.

The composition of the hoard:

Vespasianus	6
Domitianus	2
Nerva	1
Traianus	15
Hadrianus	13

6. Gračanica.[264] This hoard was found in the territory of the modern-day Republic of Bosnia and Herzegovina, formerly a part of the Roman province of Pannonia. Pegan and Šporn (1960) have published only the Republican *denarii* from the hoard (508 out of supposed total of ca. 1000);[265] together with them, a number of Imperial *denarii* were found, the latest having been minted during the reign of Hadrianus. The latest dated coin in the hoard was issued in 122/123.[266]

C. Concealed during Antoninus Pius' reign

7. Ladislav near Bjelovar[267]

Sestertii and *dupondii*, from Titus to Antoninus Pius. The original number of coins in the hoard cannot be ascertained.

8. Osatina near Đakovo[268]

Sestertii and *dupondii*, issued from the reign of Trajan (perhaps the Flavians) to Antoninus Pius.[269] Some data is available for only 16 specimens. The latest dated coin in the hoard was issued in 160.[270]

The composition of the hoard:

Traianus	2
Hadrianus	7
Antoninus Pius	7

(?) 9. Osijek 1882[271]

A hoard of silver and bronze (perhaps also gold) coins. There is some data only for two coins of Antoninus Pius.

10. Marija on the Mura near Čakovec[272]

A hoard of silver coins found in 1966. Only five *denarii* were acquired by the Varaždin Town Museum. The latest dated coin in the hoard was issued in 157.

The composition of the hoard:

Titus	1

[258] Mirnik 2008, 88, 95.
[259] Mirnik 1981, 54 no. 121.
[260] Unpublished, kept in the AMZ. I would like to thank I. Mirnik for drawing my attention to this hoard.
[261] Göricke-Lukić 2001.
[262] Göricke-Lukić 2001, 72.
[263] Mirnik 1981, 54 no. 122.

[264] Mirnik 1981, 54 no. 122a.
[265] Pegan and Šporn 1960, 111.
[266] Pegan and Šporn 1960, 111.
[267] Mirnik 1981, 55 no. 127.
[268] Brunšmid 1909; Mirnik 1981, 56 no. 133; Găzdac 2010, 70.
[269] Brunšmid 1909, 223.
[270] Brunšmid 1909, 223, 1911, 277. The concealment of this hoard was associated with the beginning of the Marcomannic wars, but it could belong to the group of hoards concealed during the reign of M. Aurelius (Găzdac 2010, 70).
[271] Mirnik 1981, 56 no. 135.
[272] Tomičić 1985. I would like to thank M. Nađ for drawing my attention to hoards no. 4 and 10.

Domitianus	1
Hadrianus	1
Antoninus Pius	2

D. Concealed during Marcus Aurelius' reign

11. Ivanovo Selo near Grubišno Polje[273]

Ca. 50 of *sestertii* and asses, but the exact data is available only for 25 coins.

The composition of the hoard:

Flavii	3
Traianus	7
Antoninus Pius	14
Marcus Aurelius	1

12. Novoselo near Bjelovar[274]

Ca. 300 *sestertii*, issued from the reign of Hadrian to that of Marcus Aurelius, concealed in a pot. Nothing more is known of this hoard._

13. Božjakovina near Dugo Selo[275]

A hoard of bronze coins (39 pieces were brought to the AMZ), issued from the reign of Vespasian to that of Marcus Aurelius. The original number of coins in the hoard cannot be ascertained.

14. Prugovac near Koprivnica[276]

Ca. 260 *sestertii* and *dupondii* were found in a pot in 1887. The exact data is available only for 77 coins. The latest dated coin in the hoard was issued in 164.[277]

The composition of the hoard:

Vespasianus	1
Domitianus	1
Traianus	13
Hadrianus	30
Antonius Pius	26
Marcus Aurelius	6

15. Osijek 1921[278]

The data is available for 439 Republican and 2352 Imperial coins. The latest dated coin in the hoard was issued in 165.[279]

The composition of the hoard (Imperial coinage only):

Nero	24 (1 drachm[280])
Galba	12
Otho	5
Vitellius	18
Vespasianus	465
Titus	78
Domitianus	174 (3 drachms)
Nerva	46
Traianus	420 (3 drachms)
Hadrianus	451
Antoninus Pius	505
Marcus Aurelius	154

16. Sotin[281]

More than 1937 silver coins, together with one as, were found in a bronze jar in 1909. Almost the complete hoard arrived to in the AMZ.[282] It consisted of 270 Republican coins (13.93%; see above on this figure) and 1667 Imperial *denarii* and one as of Hadrianus. The latest dated coin in the hoard was issued in 166.[283]

The composition of the hoard (Imperial coinage only):

Nero	19
Galba	5
Otho	11
Vitellius	8
Vespasianus	220
Titus	32
Domitianus	90
Nerva	27
Traianus	231 (5 drachms)
Hadrianus	217
Antoninus Pius	406
Marcus Aurelius	401[284]

Although it is hard to make any general conclusions on the basis of this material, with only one hoard probably reflecting its original composition (no. 16) – perhaps, less plausibly, also the hoards no. 6 and 15 – precisely these three hoards do show two important facts.[285] Firstly, the Republican coinage was in general and widespread use during the reign of Hadrian (no. 6), as well during the reign of Marcus Aurelius (no. 15 and 16). As we have noted earlier, the composition of these hoards clearly

[273] Mirnik 1981, 54 no. 124.
[274] Mirnik 1981, 55 no. 132.
[275] Mirnik 1981, 53 no. 118.
[276] Mirnik 1981, 57 no. 139.
[277] Brunšmid 1911, 277.
[278] Wruck 1937; Mirnik 1981, 56 no. 134.
[279] Wruck 1937, 315.

[280] Provincial issues – especially silver – were never a significant factor in the monetary circulation in southern Pannonia; the site finds discussed by Găzdac (2010, 103, Tab. E 2) show that they represented from 0 to 1.6% of the total coinage in the period AD 98-180, rising to 2.7-4.9% during the reign of Commodus (the data is for Pannoniae Superior and Inferior, respectively). The percentage of drachms in the hoards was from 0.7% to 1.7%, 12 out of total of 21 Pannonian drachms coming from the Osijek and Sotin hoards (Găzdac 2010, 108, Tab. A 2-3).
[281] Brunšmid 1911; Mirnik 1981, 57-58 no. 143.
[282] Brunšmid 1911, 242.
[283] Brunšmid 1911, 244.
[284] Vikić-Belančić 1974, 124 mentions a hoard of 54 coins found in Varaždinske Toplice terminating with coins of M. Aurelius. I was not able find any additional information on this hoard.
[285] The hoard no. 1 (the Zemun *aurei*), although almost complete, is truly exceptional and does not reveal much about everyday circulation of coinage in Southern Pannonia.

shows that Republican coins regularly circulated together with Imperial in the territory of southern Pannonia, at least until the reign of Marcus Aurelius. Secondly, no Augustan period – more precisely, no pre-Neronian reform[286] – coins were found in either of these hoards,[287] which means they were removed from circulation already during the reign of Hadrianus, perhaps even earlier.[288] Duncan-Jones believes that Republican coinage disappeared from the hoards during Hadrian's reign, although it does 'tend to re-emerge in later hoards in Britain and the Danube, although it had vanished from hoards in other parts of the Empire'.[289] The fact is that early Imperial *denarii* had a higher intrinsic value and were rapidly removed from circulation, much faster than the coinage of post-reform Nero and Vespasian, but also the legionary *denarii* of Marc Anthony – 'last of the major blocks of pre-Neronian *denarii* to disappear'[290] – and earlier Republican coinage, which had a lower intrinsic value (in the case of the latter perhaps only slightly lower) and thus circulated for a (sometimes much) longer period.[291] This is a textbook case of the so-called Gresham's or Copernicus' law, which states that "bad money drives out good". On the other hand, it is possible that other factors were at work, such as a deliberate removal of pre-Neronian reform Julio-Claudian coins occurring at the end of Domitian's reign and a removal of Republican coins during the reign of Trajanus.[292] The reminting of both Julio-Claudian and Republican coins was certainly profitable, which was the main reason for their withdrawal from circulation, while Marc Anthony's *denarii* were simply not worth melting down – at least not in Trajan's reign – and this is why they were allowed to remain in circulation for so long.[293] These two facts are also important for an analysis of coin finds from archaeological sites: since Republican coinage circulated in large quantities well into the 2nd century AD, it is impossible to date a site relative to the coins' issue date. Also the lack of early Imperial silver coinage does not mean the absence of Roman authority, since this coinage was simply removed from circulation shortly after being introduced into the area. Early Imperial bronze coinage, on the other hand, was constantly present in southern Pannonia, since its intrinsic value was more or less stable during the first two centuries of the Empire.[294]

We could only speculate on the reasons for the concealment of every particular hoard catalogued above, since none of them, as we have already stated, is complete.[295] But one notices that latest coins in hoards no. 14-16 were issued in 164, 165 and 166 respectively. P. Kos has concluded, based on the analysis of the Slovenian hoards possibly concealed during this period, that the incursion of Germanic tribes into the south-eastern Alpine region "had not caused any greater amount of concealment of the inhabitant's fortunes that would have taken place in politically more stable periods…".[296] He attributes this fact to the unexpectedness of the incursion. Yet he does associate the deposition of some hoards from the Carnuntum region to "different incursions across the *limes*" or "the strained political relations in that area".[297] As a conclusion of his discussion on the Marcomannic wars, Kos claims that Germanic incursion did not cause "an overall catastrophic devastation of the area or significant interruption of its cultural and economic development".[298] This is further corroborated by the comparison in the intensity of the coin supply to various settlements in the south-eastern Alpine region: during the period of reign of Marcus Aurelius, in comparison with the period of reign of Antoninus Pius, the intensity of the coin supply decreased by only 7.2% in Poetovio and by 6.2% in Emona, but by a significant 66.3% in Poetovio; on the other hand, the intensity of the coin supply actually increased at Neviodunum by 5%.[299] Siscia, with an increase of 3.9%, shows the greatest similarity to the nearby Neviodunum. Therefore, Kos' conclusion, valid for the south-eastern Alpine region, can be extended to include at least the western part of southern Pannonia with its largest centre.[300] Its eastern part, however, could have been affected more significantly with Iazygian incursions and the concealment and non-recovery of the hoards in this region could have been caused by these events.[301]

5. Conclusion

On the present evidence, the coin circulation in the western part of southern Panonnia starts with the introduction of the silver coinage of Apollonia and Dyrrachium, most probably during the second half of the 2nd century BC. It was especially the drachms, dependant on the Roman Republican *denarius*, that circulated in the Balkan interior, but the dependence did not stop with regards to the denomination: it is probable that this coinage actually reflects Roman economical and political interest in the region at this period. General reasons for the introduction of this coinage are therefore to be sought across the Adriatic, but it is much harder to search for a more specific reason. Another type of coinage was produced by certain Barbarian tribes in the interior, more specifically by the eastern Pannonian tribe of Scordisci and the western Pannonian Taurisci, as well as by the Norici farther to the west. These Barbarian tribes were at first using the coinage of Philippus II – his posthumous issues possibly reflect the economic needs of the Balkan interior – but when the inflow of this

286 The Neronian reform of 64 AD reduced the weight of aureus and denarius (*RIC* i².3-5, 136-137; Duncan-Jones 1994, 100, 218, 221, 237).
287 Brunšmid 1911, 242 (Sotin); Wruck 1937, 316 (Osijek); Pegan and Šporn 1960, 110 (Gračanica).
288 For the melting down of badly worn coins under Trajan see Dio LXVIII.15.3. Compare also Borić-Brešković 1994b, 84, 138; Găzdac 2010, 71.
289 Duncan-Jones 1994, 196.
290 Duncan-Jones 1994, 196.
291 Cf. Bíró-Sey 1980, 340; Găzdac 2010, 71.
292 Duncan-Jones 1994, 195-196; cf. Borić-Brešković 1994b, 84, 138.
293 Duncan-Jones 1994, 199-200.
294 Bíró-Sey 1980, 341.
295 Găzdac 2010, 72-73 believes that the large hoards (no. 15-16 in our case) belonged to wealthy landowners.

296 Kos 1986, 85.
297 Kos 1986, 85-86.
298 Kos 1986, 91; cf. Găzdac 2010, 93-94.
299 Kos 1986, 90.
300 Kos (1986, 91 n. 48) already suggested that his conclusion is also valid for Pannonia.
301 Găzdac 2010, 73, 94.

coinage stopped, they started minting their own currency. There is some controversy regarding the date of the beginning of minting of imitations of Macedonian coinage – the first half of the 3rd century or the mid-1st century – but the main period of minting seems to have started closer to the latter date. It is possible that the imitations of Phillip II's tetradrachms found in the Narta hoard were actually minted somewhere in the area of Bjelovar. Barbarian tribes in the western part of southern Pannonia started minting their coinage much later, during the first half of the 1st century, and the existence of mints in the area of Samobor and Đurđevac can be presumed. The coins of the latter type circulated across the wide area of Pannonia and Noricum, but eastern Norican tetradrachms were also in common use in western Pannonia. The appearance of Carthaginian, Numidian, Egyptian and other Greek and Italic/Roman currency is a phenomenon characteristic of the mountainous area south of Pannonia proper (that is, north-western Dalmatia), and only its peripheral influence on the southern-Pannonian coin circulation of the period can be observed. But the Republican coinage appears both in excavated sites of Donja Dolina and Gomolava and in a number of hoards, often with other coinage, Apollonian/Dyrrachian or Celtic. Several of the hoards interred during this period reflect Roman mercantile and political interests in this region, indicating the penetration of Republican coins that followed the armies, undoubtedly to be associated with Augustus' campaigns. But it is also possible that Roman merchants preceded both the armies and administration, and that they were responsible for the introduction of larger quantities of money into Pannonia. It is probable that Republican coinage only replaced that of Apollonia and Dyrrachium in the local economy during the mid-1st century, but it was undoubtedly present in the region somewhat – although not much – earlier. Another important consequence of the Roman conquest was a significant change in economy of native populations, which was certainly one of the reasons for the great uprising of AD 6-9. After the pacification, and continuing the process that had already started before the uprising, Augustan bronze coinage became predominant in everyday monetary circulation, and it remained so during the early Imperial period. The coin circulation in southern Pannonia in this period corresponds to a rather uniform overall circulation in the Empire. The most important trends observable through the analysis of stray finds from Sisak shows a large increase in the intensity of monetary circulation during the reign of Vespasian and a large decline during the reign of Commodus. Both of these trends reflect similar patterns of the circulation intensity when compared with neighbouring areas, but also when compared with the Empire as a whole. Other important facts regarding the circulation during the late 1st and the whole 2nd century AD are the general and widespread use of Republican coinage during this entire period and the absence of Julio-Claudian silver coinage predating the reform of Nero in AD 64. Finally, the analysis of stray finds from Sisak shows that the intensity of coin circulation actually somewhat increased during the reign of Marcus Aurelius, thus excluding the possibility of a significant interruption of the region's cultural or economic development due to the Marcomannic Wars.

Abbreviations

RIC = Sutherland and Carson 1984
RRC = Crawford 1974
RRCH = Crawford 1969
TKN = Göbl 1973

Bibliography

Backendorf 1998 Dirk Backendorf, *Römische Münzschätze des zweiten und ersten Jahrhunderts v. Chr. vom italienischen Festland*, *Studien zu Fundmünzen der Antike* 13, Berlin.

Bahrfeldt 1876 Max Bahrfeldt, Contremarken Vespasians auf Römischen Familiendenaren, *ZN* 3, 354-374.

Bahrfeldt 1887 Max Bahrfeldt, Contremarken Vespasians auf römischen Denaren, *ZN* 14, 67-74.

Bakić 2001 Ljiljana Bakić, Ostave Rimskih denara iz II. stoljeća n.e. pronađenih na lokalitetu Ludoš u Vršcu [The roman denarii hoard from the 2nd century A.D. from the locality of Vršac-Ludoš], in: J. Dobrinić (ed.), *Proceedings of the 3rd International Numismatic Congress in Croatia*, Pula, 13-31.

Balen-Letunić, Dukat, and Mirnik 1991/1993 Dubravka Balen-Letunić, Zdenka Dukat, and Ivan Mirnik, Tesoretto di Lički Ribnik, cat. no. 307, in: *Tesori nazionali della Croazia – Capolavori del musei di Zagabria: Arheološki muzej*, Exhibition Catalogue, Arezzo, 1991, 208-211 = *Arte e cultura in Croazia – Collezioni del Museo archeologico di Zagabria*, Exhibition Catalogue, Torino, 1993, 208-211.

Bannert and Piccottini 1972 Herbert Bannert and Gernot Piccottini, *Die Fundmünzen vom Magdalensberg*, FMRÖ Kärnten 1 (*Archäologische Forschungen zu den Grabungen auf dem Magdalensberg* 2, *Kärntner Museumsschriften* 52), Klagenfurt.

Berger 1996 Frank Berger, *Kalkreise 1. Die römischen Fundmünzen*, Mainz.

Bíró-Sey 1980 Katalin Bíró-Sey, Currency, in: *ARP*, 337-348.

Bonačić Mandinić 2003 Maja Bonačić Mandinić, Nekoliko primjeraka novca 2. i 1. stoljeća prije Krista s Brača - uz razmišljanje o keltskim nalazima u Dalmaciji [A collection of the 2nd and 1st century BC coins from the island of Brač (Reflecting on the Celtic finds in Dalmatia)], *OpArch* 27, 435-441.

Borić-Brešković 1994a Bojana Borić-Brešković, Rimski republikanski i carski denari u ostavi iz Bara / Roman republican and imperial denarii in the Bare hoard, in: Ivana Popović and Bojana Borić-Brešković, *Ostava iz Bara / The Bare hoard*, Beograd, 105-201.

Borić-Brešković 1994b Bojana Borić-Brešković, *Rimski denari u ostavi iz Bele Reke / The Roman denarii from Bela Reka hoard*, in: Ivana Popović and Bojana Borić-Brešković, *Ostava iz Bele Reke / The Bela Rekas hoard*, Beograd, 77-148.

Borić-Brešković 2003 Bojana Borić-Brešković, The Hoard of Roman Denarii from Grocka near Belgrade, *Numizmatičar* 22-23 (1999-2000), 11-58.

Borić-Brešković and Arsenijević 2008 Bojana Borić-Brešković and Mirjana Arsenijević, The Roman denarii hoard from Radalj near Drina, *Numizmatičar* 26-27 (2003-2004), 93-156.

Borić-Brešković and Crnobrnja 2005 Bojana Borić-Brešković and Nikola Crnobrnja, A find of Roman denarii of the emperors of the I – II century, from Nemenikuće on Mt. Kosmaj, *Numizmatičar* 24-25 (2001-2002), 7-94.

Borić-Brešković and Crnobrnja 2008 Bojana Borić-Brešković and Nikola Crnobrnja, The first and second century denarii hoard from Ušće near Obrenovac, *Numizmatičar* 26-27 (2003-2004), 9-82.

Borić-Brešković and Popović 2006 Bojana Borić-Brešković and Petar Popović, *Coins of the Roman Republic. Collections of the National Museum in Belgrade and Belgrade University*, Belgrade.

Bruck 1961 Guido Bruck, *Münzfunde*, in: Rudolf Egger, Die Ausgrabungen auf dem Magdalensberg 1958 und 1959, *Carinthia* 151/1, Klagenfurt,168-172.

Brunšmid 1886 Josip Brunšmid, Skrovište rimskih obiteljskih denara izmedju Valpova i Osieka [A hoard of Roman family denars found between Valpovo and Osijek], *VHAD* 8, 105-114.

Brunšmid 1895 Josip Brunšmid, Njekoliko našašća novaca na skupu u Hrvatskoj i Slavoniji. II. Našašće rimskih obiteljskih denara između Valpova i Osieka [A few finds of coin-hoards in Croatia and Slavonia. II. A find of Roman family denars between Valpovo and Osijek], *VHAD* 1, 108-114.

Brunšmid 1900 Josip Brunšmid, Colonia Aelia Mursa, *VHAD* 4, 21-42.

Brunšmid 1911 Josip Brunšmid, Nekoliko našašća novaca na skupu u Hrvatskoj i Slavoniji, XXXIII. Nahođaj rimskih denara I. i II. stoljeća u Sotinu [A few finds of coin-hoards in Croatia and Slavonia. XXXIII. A hoard of Roman *denarii* of the 1st and 2nd century from Sotin], *VHAD* 11, 241-277.

Brunšmid 1912 Josip Brunšmid, Nekoliko našašća novaca na skupu u Hrvatskoj i Slavoniji, XXXIV- XXXVI [A few finds of coin-hoards in Croatia and Slavonia. XXXIV-XXXVI], *VHAD* 12, 260-286.

Burkowsky 2008 Zdenko Burkowsky, *Numizmatički nalazi na novijim arheološkim istraživanjima u Sisku [Numismatic finds at recent archaeological excavations in Sisak*], Exhibition Catalogue), Sisak.

Ciołek 2011 Renata Ciołek, *Emisje króla Ballaiosa. Początki mennictwa w Ilirii*, Warsaw.

Ceka 1972 Hasan Ceka, *Questions de numismatique illyrienne*, Tirana.

Crawford 1969 Michael Crawford, *Roman Republican Coin Hoards*, London.

Crawford 1974 Michael Crawford, *Roman Republican Coinage*, London.

Crawford 1985 Michael Crawford, *Coinage & Money under the Roman Republic*, London.

Dautova-Ruševljan 1984 Velika Dautova-Ruševljan, Novac sa iskopavanja rimskog naselja na lokalitetu Gomolava (1953-1978) [Coins from the excavations of the Roman settlement on the Gomolava], *RVM* 28 (1982-1983), 47-79.

Dembski 1998 Günther Dembski, *Münzen der Kelten*, Wien.

Demo 1982 Željko Demo, Prilog topografiji Križevačke regije u antičko doba s osvrtom na numizmatičke nalaze [A contribution to the topography of the Križevci region with an emphasis on coin finds], *KZ* 2, 75-92.

Demo 1984 Željko Demo, Review of Fitz 1978, *AV* 34, 464-468.

Domić-Kunić 2006 Alka Domić Kunić, Bellum Pannonicum (12.-11. pr. Kr.): Posljednja faza osvajanja Južne Panonije [Bellum Pannonicum (12-11 BC): The final stage of the conquest of southern Pannonia], *VAMZ* 39, 59-164.

Dukat and Mašić 2005 Zdenka Dukat and Boris Mašić, Nalaz grčkog novca u Zagrebu, *VAMZ* 38, 133-137.

Dukat and Mirnik 1976 Zdenka Dukat and Ivan Mirnik, Pre-Roman coinage on the territory of modern Yugoslavia, *Bulletin of the Institute of Archaeology* 13, London, 175-210.

Duncan-Jones 1994 Richard Duncan-Jones, *Money and Government in the Roman Empire*, Cambridge.

Dzino 2006 Danijel Dzino, Velleius Paterculus and the Pannonii: Making up the numbers, *GodANUBiH* 35, 145-160.

Dzino 2010 Danijel Dzino, *Illyricum in Roman Politics 229 BC-AD 68*, Cambridge.

Fischer 1990 Thomas Fischer, Bemerkungen zur Gajus-Lucius-Prägung des Augustus (*RIC* I² 205- 12), *Geldgeschichtliche Nachrichten* 140, Frankfurt, 352.

Fitz 1978 Jenő Fitz, *Der Geldumlauf der römischen Provinzen im Donaugebiet Mitte des 3. Jahrhunderts*, Bonn-Budapest, 2 vols, 1978.

Fitz 1980 Jenő Fitz, *Economic life*, in: *ARP*, 323-335.

Gardaš 1990 Miro Gardaš, Skupni nalaz republikanskih denara pokraj sela Krndija u Đakovštini [A hoard of Republican denarii found in the vicinity of the Krndija village in the Đakovo area], *Obol* 28.42, 6-9.

Găzdac 2010 Cristian Găzdac, *Monetary circulation in Dacia and the provinces from the Middle and Lower Danube from Trajan to Constantine I (AD 106-337)*, Cluj-Napoca².

Gnecchi 1902 Francesco Gnecchi, Ripostiglio di Vergnacco, *Rivista Italiana di numismatica* 15, Milano, 548-551.

Göbl 1973 Robert Göbl, *Typologie und Chronologie der keltischen Münzprägung in Noricum*, Wien.

Göricke-Lukić 1987 Hermine Göricke-Lukić, Nalazi rimskog novca s donjogradskog pristaništa u Osijeku [Roman coins from the Lower-town pier in Osijek], *OZ* 18-19, 77-87.

Göricke-Lukić 1989 Hermine Göricke-Lukić, Nalazi rimskog novca u Donjem gradu u Osijeku, zaštitna heološka istraživanja (1982.-1986.) [Roman coins from the Lower town in Osijek, archaeological rescue excavations (1982-1986)], *OZ* 18-19, 77-87.

Göricke-Lukić 1993 Hermine Göricke-Lukić, Novac iz nekropole antičke Murse [Coins from the necropolis of Roman Mursa], *NV* 46, 8-18.

Göricke-Lukić 1997 Hermine Göricke-Lukić, Novac iz grobova rimske Murse [Coins from the graves of Roman Mursa], *NV* 50, 19-32.

Göricke-Lukić 2001 Hermine Göricke-Lukić, Skupni nalaz mjedenog novca Hadrijanovog vremena iz Murse [A hoard of brass coins from the period of Hadrian's reign from Mursa], *OZ* 24-25, 67-74.

Göricke-Lukić 2004 Hermine Göricke-Lukić, Novčani opticaj u Osijeku i okolici prije rimskog osvajanja [Coin circulation in Osijek and its vicinity before the Roman conquest], *OZ* 27, 35-48.

Gorini 1984 Giovanni Gorini, Re Ballaios: una proposta cronologica, in: *Il crinale d'Europa. L'area illirico-danubiana nei suoi rapporti con il mondo classico*, Roma, 43-49.

Gorini 1988 Giovanni Gorini, Bilješke o novcu kralja Baleja (Ilirija) [Notes on the coinage of King Ballaios (Illyria)], *Numizmatika* 1/7, 16-21.

Gorini 1991 Giovanni Gorini, The Ballaeus Hoard from Rizan in the Ashmolean Museum, Oxford, *Schweizer Münzblätter* 41.162, Zürich, 25-30.

Hahn 1976 Wolfgang Hahn, *Carnuntum, FMRÖ* III/1, Wien.

Howgego 1985 Christopher J. Howgego, *Greek Imperial Countermarks. Studies in the Provincial Coinage of the Roman Empire*, London.

Ilkić 2003 Mato Ilkić, *Cornacum. Sotinski prostor i njegovo mjesto u organizaciji južnog dijela provincije Panonije* [*Cornacum. The area of Sotin and its place in the organization of the southern part of the province of Pannonia*], Unpublished PhD dissertation, University of Zadar, 2003.

Ilkić 2008 Mato Ilkić, Rimski Cornacum u svijetlu numizmatičkih nalaza/Roman Cornacum in the Light of the Numismatic Evidence, *Proceedings of the 5th International Numismatic Congress in Croatia*, Rijeka, 51-75.

Jacanović 1988 Dragan Jacanović, Ostava republikanskih rimskih denara iz Viminacijuma [The hoard of Roman Republican denarii from Viminacium], *Viminacium* 2, Požarevac, 25-39.

Jankulov 1940 Borislav Jankulov, Halštatska i latenska nalazišta u Vojvodini [The Hallstatt and La Tène sites in Vojvodina], *Glasnik Istorijskog društva u Novom Sadu* 13/3-4, Novi Sad, 225-250.

Karwowski and Militký 2011 Maciej Karwowski and Jiři Militký, *The Oberleiserberg types in the context of Taurisci influences*, in: M. Guštin and M. Jevtić (eds.), *The Eastern Celts. The comminities between the Alps and the Black Sea*, Koper-Beograd, 131-136.

Kehne 2000 Peter Kehne, Zur Datierung der Fundmünzen aus Kalkriese und zur Verlegung *des Enddatums des Halterner Hauptlagers in die Zeit der Germanienkriege unter Tiberius und Germanicus (10-16 n. Chr.)*, in: R. Wiegels (ed.), *Die Fundmünzen von Kalkriese und die frühkaiserzeitliche Münzprägung*, Möhnesee, 47-79.

Kellner 1990 Hans-Jörg Kellner, *Die Münzfunde von Manching und die keltischen Fundmünzen aus Südbayern*, Stuttgart.

Klemenc 1935 Josip Klemenc, Ostava u Ličkom Ribniku [Coin hoard from Lički Ribnik], *VHAD* 16, 83-125 with pl. 1.

Koprivnjak 2011a Vedran Koprivnjak, Dio nalaza rimskog novca s lokaliteta Sisak – Lučka kapetanija [A selection of Roman coins from the Sisak – Lučka kapetanija site], *GGMS* 10, 63-94.

Koprivnjak 2011b Vedran Koprivnjak, Antički numizmazički nalazi s lokaliteta Sisak – Dunavski Lloyd [Roman coins from the Sisak – Dunavski Lloyd site], *GGMS* 10, 25-61.

Kos 1976 Peter Kos, Mali noriški srebrniki s konjički [Small Norican silver coins with a horse], *AV* 25, 213-224.

Kos 1977 Peter Kos, *Keltski novci Slovenije. Keltische Münzen Sloweniens, Situla* 18.

Kos 1983 Peter Kos, Die Schatzfund norischer Großsilbermünzen aus Bevke, *AV* 34, 411-417.

Kos 1986 Peter Kos, *The monetary circulation in the southeastern Alpine region ca. 300 BC – AD 1000, Situla* 24.

Kos 1988 Peter Kos, *Die Fundmünzen der Römische Zeit in Slowenien* I-II, Berlin.

Kos 1998 Peter Kos, *Leksikon antičke numizmatike* [*Lexicon of Greek and Roman Numismatics*], Zagreb.

Kos 2002 Peter Kos, A find of Celtic coins from the area of Balina Glavica *(Sinotion)* in Dalmatia, *VAMZ* 35,147-158.

Kos and Mirnik 1999 Peter Kos and Ivan Mirnik, The Ribnjačka Hoard (Bjelovar, Croatia), *NC* 159, 298-306 + pls. 28-33.

Kos and Mirnik 2011 Peter Kos and Ivan Mirnik, Coin hoards from Croatia XVII. The Križovljan (Varaždin) hoard of Celtic tetradrachms (1843), *VAMZ* 44, 77-130.

Kos and Šemrov 2003 Peter Kos and Andrej Šemrov, Skupna najdba keltskih in rimskih novcev v reki Ljubljanici: Doprinos h kronologiji novcev plemena Tavriskov (A hoard of Celtic and Roman coins from the Ljubljanica River. A contribution to the chronology of the coinage of the Taurisci), *AV* 54, 381-395.

Liščić 1978 Vladimir Liščić, Nalazi noričkih tetradrahmi u posjedu Arheološkog muzeja u Zagrebu ž [Norican tetradrachms in the Archaeological museum in Zagreb], *VAMZ* 10-11 (1977-1978), 235-259.

Ljubić 1888 Šime Ljubić, Razne viesti [Various news], *VAHD* 10, 127.

Marić 1964 Zdravko Marić, Donja Dolina, *GZMS* 19, 5-128.

Martini 2003 Rodolfo Martini, *Collezione Pangerl: Contromarche imperiali romane (Augustus – Vespasianus)*, Nomismata 6, Milano.

Mattingly 1923-1930 Harold Mattingly, *Coins of the Roman Empire in the British Museum*, Vol. I: *Augustus to Vitellius*. London, 1923. Vol. II: Vespasian to Domitian, London, 1930.

Mažuran 1954 Ive Mažuran, Rimski novci nađeni u Osijeku

(1949-1954) [Roman coins found in Osijek (1949-1954)], *OZ* 4, 32-40.

Mirnik 1969 Ivan Mirnik, *Katalog dijela numizmatičke zbirke Gradskog muzeja u Sisku* [*Catalogue of a part of the numismatic collection of the Town Museum Sisak*]. Unpublished diploma thesis, Faculty of Philosophy, Zagreb.

Mirnik 1981 Ivan Mirnik, *Coin hoards in Yugoslavia, BAR Intern. Ser.* 95.

Mirnik 1987 Ivan Mirnik, Circulation of North African etc. currency in Illyricum, *AV* 38, 369-392.

Mirnik 1998 Ivan Mirnik, Der Schatzfund von Pokupsko, in: U. Peter (ed.), *Stephanos nomismatikos: Edith Schönert-Geiss zum 65. Geburtstag*, Berlin, 485-489.

Mirnik 1999 Ivan Mirnik, Prolegomena to the numismatic topography of *Burgenae*, in: K. Bertók and M. Torbágy (eds.), *Emlékköny Bíró-Sej Katalin és Gedai István 65. születésnapjára. Festschrift für Katalin Bíró-Sej und István Gedai zum 65. Geburtstag*, Budapest, 221–235.

Mirnik 2008 Ivan Mirnik, Zemunski skupni nalaz zlatnog novca [The 1875 Zemun hoard of the 1st century A.D. gold coins - revision], VAMZ 41, 77-128.

Mirnik 2009 Ivan Mirnik, Patvorine rimskog carskog novca nađene tijekom jaružanja Kupe u Sisku 1912. god. [Forgeries of Roman coins found during the dredging of the Kupa river at Sisak in 1912], in: L. Bekić (ed.), *Jurišićev zbornik. Zbornik radova u znak sjećanja na Marija Jurišića*, Zagreb, 291-309.

Miškec 2009 Alemka Miškec, The Augustan conquest of southeastern Alpine and western Pannonian areas: coins and hoards, *AV* 60, 283-296

Mócsy 1974 András Mócsy, *Pannonia and Upper Moesia*, London-Boston.

Müllner 1879 Alfonz Müllner, Archäologische Excurse nach Süd-Steiermark. XI. Der Münzfund am Laibacher Moore, *Mittheilungen der K.K. Central-Commission zur Erforschung und Erhaltung der Kunst- und Historischen Denkmale*, N.F. 5, Wien, cxxxviii.

Pegan and Šporn 1960 Efrem Pegan and Janez Šporn, Najdba Gračanica. I. del: Novci rimske Republike [The Gračanica hoard. Part I: Roman Republican coins], *NVe* 4, 110-129

Pichler 1865 Friedrich Pichler, *Repertorium der steierischen Münzkunde I: Die keltischen und consularen Münzen der Steiermark*, Graz.

Pink 1939 Karl Pink, *Die Münzprägung der Ostkelten und ihrer Nachbarn*, *DissPann* II/15.

Pinterović 1967 Danica Pinterović, Mursa u svjetlu novih izvora i nove literature [Mursa in the light of newly discovered sources and latest discussions], *OZ* 11, 23-65.

Popović 1971 Petar Popović, Nalazi novca iz latenskog naselja na Gomolavi 1970. Godine [Coin finds from the La Tène settlement at Gomolava], *RVM* 20, 147-160.

Popović 1978 Petar Popović, Ostava drahmi Apolonije i Dirahiona iz Pećinaca [A hoard of Apollonian and Dyrrachian drachms from Pećinci], *Numizmatičar* 1, 9-22.

Popović 1987 Petar Popović, *Novac Skordiska* [*Coinage of the Scordisci*], Monografije Arheološkog instituta 19, Beograd-Novi Sad.

Preda 1973 Constantin Preda, *Monedele geto-dacilor*, Bucharest.

Radman-Livaja and Dizdar 2010 Ivan Radman-Livaja and Marko Dizdar, *Archaeological Traces of Pannonian Revold 6-9 AD: Evidence and Conjectures*, in R. Aßkamp and T. Esch (eds.), *Imperium – Varus und seine Zeit*, Beiträge zum internationalen Kolloquium des LWL – Römermuseums am 28. und 29. April 2008 in Münster, Aschendorff, 47-58.

Rant 1987 Andrej Rant, Zakladna najdba Bevke in njen pomen v proučevanju keltskega novčarstva na ozemlju Slovenije, *NVe* 15, 207-222.

Reece 1971 Richard Reece, Roman Coinage in Northern Italy, *NC* 11, 167-179.

Rosada 2004 Guido Rosada, Le vie di comunicazione e di commercio in epoca post classica: il caso della Postumia, *HAnt* 12, 191-204.

Ruske 2011 Alexander Ruske, Die Fundregesten der antiken Münzhorte aus Österreich, *NZ* 118, 53-181.

Saria 1927 Balduin Saria, Numismatischer Bericht aus Jugoslawien, *NZ* 60, 10-19.

Schachinger 2008 Ursula Schachinger, Richard Knabl und seine Schenkungen an die Grazer Universität – Nachträge zu den keltischen Schatzfunden von Dobrna-Retje und Lemberg, *NZ* 116-117, 73-86.

Sokol 1994a Vladimir Sokol, Keltski novci na području Zagreba – Sesvetskog prigorja [Celtic coins from the area of Zagreb – the Sesvete Prigorje], *MV* 17, 19-25.

Sokol 1994b Vladimir Sokol, in: A. Rendić-Miočević (ed.), *Zagreb prije Zagreba / Zagreb before Zagreb,* Exhibition Catalogue, Zagreb, 98-99, cat. no. 118, 123-125.

Sokol 2001 Vladimir Sokol, Keltski novac s Kuzelina [Celtic coins from Kuzelin], *NV* 54, 4-16.

Suić 1966 Mate Suić, Marginalije uz issejsko poslanstvo Cezaru [Marginal notes on the Issean delegation to Cesar], *VAHD* 68, 181-194.

Sutherland 1976 Carol H.V. Sutherland, *The emperor and the coinage: Julio-Claudian studies*, London.

Sutherland and Carson 1984 Carol H.V. Sutherland and Robert A.G. Carson, *Roman Imperial Coinage*, I (revised edition). *From 31 BC to AD 69*, London.

Šašel Kos 1986 Marjeta Šašel Kos, *Zgodovinska podobo prostora med Akvilejo, Jadranom in Sirmijem pri Kasiju Dionu in Herodijanu / A historical outline of the region between Aquileia, the Adriatic, and Sirmium in Cassius Dio and Herodian*, Ljubljana.

Šašel Kos 2002 Marjeta Šašel Kos, The boundary stone between Aquileia and Emona, *AV* 53, 373-382.

Šašel Kos 2003 Marjeta Šašel Kos, Emona was in Italy, not in Pannonia, in: *Pannonia* I, 11-19.

Šašel Kos 2007 Marjeta Šašel Kos, The Illyrian king Ballaeus – some historical aspects, in: D. Berranger (ed.), *Épire, Illyrie, Macédoine: mélanges offerts au professeur Pierre Cabanes*, Clemont-Ferrand, 125-138.

Šeparović 2005 Tomislav Šeparović, Revizija skupnog nalaza rimskog novca iz Kruševa kod Obrovca [Revision of a hoard of Roman coins from Kruševo near Obrovac], *VAPD* 98, 63-72.

Tomičić 1985 Željko Tomičić, Osvrt na jedan skupni nalaz antičkog novca iz Međimurja [A reflection on a Roman coin hoard from Međimurje], *Glasnik Gradskog muzeja Varaždin* 7, Varaždin, 87-96.

Truhelka 1904 Ćiro Truhelka, Der vorgeschichtliche Pfahlbau im Savebette bei Donja Dolina, *Wissenschaftliche Mitteilungen aus Bosnien und derHerzegowina* 9, Sarajevo, 3-156.

Vikić-Belančić 1974 Branka Vikić-Belančić, Prilog istraživanju antičkog naseobinskog kompleksa u Varaždinskim Toplicama, *VAMZ* 6-7 (1972-1973), 75-132.

Wolters 2002 Reinhard Wolters, Gaius und Lucius Caesares als designierte Konsuln und principes iuventutis. Die lex Valeria Cornelia und RIC I² 205 ff., *Chiron* 32, 297-323.

Wruck 1937 Waldemar Wruck, Der Denarfund von Osiek (Mursa), *Deutsche Münzblätter* 57, Berlin, 289-293, 311-317.

Zmajić 1967 Bartol Zmajić, Ivan Kukuljević-Sakcinski (1816-1889) i numizmatika u Hrvatskoj [Ivan Kukuljević-Sakcinski (1816-1889) and the numismatics in Croatia], *NV* 14-15, 22-26.

Coin Circulation AD 193-423

Miroslav Nađ

1. Introduction

This paper will be an attempt to give a short overview of numismatic material from some major archaeological sites as well as hoards from southern Pannonia; their composition will be studied and comparisons with some of the already published sites given. Chronologically, historically and archeologically specific periods of the 3rd and 4th centuries will be discussed, that is, the period from AD 193 (the beginning of the rule of the Severan House) to the year AD 423 (the end of the reign of Honorius).

Published surveys and papers on Roman monetary circulation in southern Pannonia are scarce, especially as concerns coin finds from major archaeological sites. A major work was contributed by Peter Kos (1986), who discussed the circulation of the southeastern Alpine region from c. 300 BC to AD 1000, including some of the sites from southern Pannonia (Neviodunum[1] and Poetovio[2]), Noricum (Celeia[3]) and Regio X of Italia (Emona [4]).[5] Jenő Fitz's "Der Geldumlauf der römischen Provinzen im Donaugebiet Mitte des 3. Jahrhunderts" (1979) analyses the currency from the Hungarian part of Pannonia from 14 AD to 268 AD; in spite of an attempt to cover southern Pannonia as well, the data is mostly lacking.[6] Alice Sz. Burger's survey (1981) furnishes data on the late Roman monetary circulation in southern Pannonia (AD 324-378), but covering only the Hungarian part of the province, with only two sites from Croatia and Bosnia and Herzegovina included: the hoards of Lupoglav and Vinkovci. Katalin Bíró-Sey (1980) discussed only the Hungarian part of Pannonia, but with some reference on the Croatian part and a broader chronological spectrum covered: from the pre-Roman period to the end of the 4th century. An important discussion was produced by Christian Găzdac (2010). In his "Monetary circulation in Dacia and the provinces from Middle and Lower Danube from Trajan to Constantine I (A.D. 106 – 337)" the author discusses the coin circulation in the provinces of Pannonia (Superior and Inferior) and Moesia (Superior and Inferior), using sites there as comparisons for the area/sites of Roman Dacia.[7] A specific type of circulation of coin imitations in southern Pannonia from the 1st to the 4th centuries was studied by Željko Demo (1982), covering stray finds as well as coins from hoards and collections. The situation is somewhat different concerning the hoards. A seminal study by Ivan Mirnik (1981), which lists all hoards and their composition in ex Yugoslavia up to 1979, provides a lot of information on hoards from the region of study. A studious paper about hoards from the time of Gallienus was contributed by Željko Demo (1983), giving the composition of some newly found hoards, reconsidering some published ones and interpreting them with the rest of the hoards form the same period, and, finally, putting them all into a historical context.[8] A summary discussion and a partial interpretation of the site finds (gold, silver and bronze[9]) and bronze hoards of the Danube and Balkan provinces from AD 237-578 were given by George L. Duncan (1993).[10]

2. Circulation

Site finds discussed in this paper mostly come from Siscia and Mursa, two large centres of southern Pannonia.[11] Of smaller sites, Andautonia (Ščitarjevo) has been included, as well as Cornacum (Sotin); the latter serving as an example of the circulation on the Croatian section of the Danubian *limes*. The sites have been chosen depending on the accessibility of the material for study and the published finds. Notwithstanding the estimation that only 1% of the Roman coins circulating in the Roman Empire have been found so far,[12] this still enables establishing a general pattern of currency for individual regions.[13] Although probably not all material from the area under discussion is covered, a general picture of the circulation in the future will not deviate considerably from that presented on tables and charts here. All analysed sites show a similar pattern of currency in comparison with the southeastern Alpine region, as analysed by Kos.[14] However, Siscia in the period 253-259 and Mursa in the period 249-259 are exceptions to this rule. On the other hand, the circulation in Siscia is consistent with the increased circulation in Poetovio in the same period.

1 Modern-day Drnovo, Republic of Slovenia.
2 Modern-day Ptuj, Republic of Slovenia.
3 Modern-day Celje, Republic of Slovenia.
4 Modern-day Ljubljana, Republic of Slovenia.
5 This work will be the main reference for comparison and its methodology will be applied in this paper.
6 Demo 1984, 465.
7 Croatian sites included in this paper were not discussed by Găzdac 2010.
8 These hoards will be discussed later in the text.
9 That is, bronze coins from Mursa, Sirmium, and Burgenae.
10 For the Croatian part of Pannonia and the rest of ex Yugoslavia Duncan has used only the inventory books of the AMZ. I would like to thank I. Mirnik for this information.
11 Finds from another two major south-Pannonian towns, Sirmium and Cibalae, have not been included; the coins from Cibalae are currently under study.
12 Sutherland 1976, 85, cited in Kos 1986, 17.
13 Kos 1986, 17.
14 Kos 1986, 92-110.

Table 1. The coins by numbers and periods of issue and a corrected percentage comparisons of the eight sites on an annual basis (total no. 13123).[*]

Period of issue**/site	***Siscia no./%***	***Mursa no./%***	***Andautonia no./%***	***Cornacum no./%***	***Poetovio no./%***	***Neviodunum no./%***	***Celeia no./%***	***Emona no./%***
193-211	47 0.018	15 0.005	1 0.003	21 0.008	105 0.042	21 0.008	31 0.012	7 0.002
211-218	17 0.018	5 0.005	1 0.001	12 0.013	13 0.014	10 0.010	5 0.022	7 0.007
218-222	14 0.026	10 0.010	- 0	12 0.022	36 0.068	12 0.003	12 0.022	12 0.022
222-235	31 0.018	20 0.011	1 0.001	21 0.012	88 0.051	41 0.024	12 0.007	26 0.015
235-238	6 0.015	1 0.002	- 0	7 0.017	19 0.048	5 0.012	2 0.005	3 0.007
238-244	26 0.032	149 0.189	- 0	5 0.006	81 0.102	16 0.020	6 0.007	21 0.026
244-249	19 0,028	270 0.411	2 0.003	7 0.010	49 0.074	15 0.022	5 0.007	9 0.013
249-251	17 0.064	72 0.274	- 0	5 0.019	28 0.106	3 0,011	2 0.007	2 0.007
251-253	17 0.064	50 0.190	- 0	8 0.030	23 0.087	4 0.015	5 0.019	2 0.007
253-259	23 0.029	5 0.006	1 0.001	3 0.007	30 0.038	42 0.053	8 0.010	10 0.012
259-268	288 0.243	43 0.152	1 0.001	26 0.019	318 0.269	144 0.121	68 0.057	76 0.064
268-270	194 0.739	45 0.171	5 0.019	25 0.095	192 0.731	64 0.243	42 0.160	48 0.182
270-275	204 0.310	53 0.080	2 0.003	47 0.071	252 0.384	156 0.237	32 0.048	70 0.106
275-282	122 0.132	45 0.048	3 0.003	39 0.042	89 0.096	65 0.070	20 0.021	18 0.019
282-285	29 0.073	3 0.007	3 0.007	13 0.032	22 0.055	11 0.027	1 0.002	5 0.012
285-313	136 0.036	57 0.015	4 0.001	59 0.016	64 0.017	20 0.005	82 0.022	168 0.045
313-337	688 0.218	156 0.049	21 0.006	130 0.041	475 0.150	306 0.097	86 0.026	208 0.066
337-364	935 0.263	206 0.058	11 0.003	103 0.028	642 0.181	499 0.140	160 0.045	486 0.137
364-383	556 0.222	100 0.040	4 0.001	83 0.033	447 0.179	307 0.123	94 0.037	495 0.198
383-402	214 0.085	7 0.002	-	27 0.010	42 0.016	9 0.003	17 0.006	294 0.117
402-423	12 0.004	1 0.001	-	-	-	-	10 0.003	71 0.025
TOTAL:	**3595**	**1313**	**60**	**652**	**3015**	**1750**	**700**	**2038**

[*] Including coins from Spajić 1975.
[**] The periods of issue are distributed according to Kos (1986) and they are merged when required.

Sisak (Siscia)

During the reign of Caracalla there was a decrease of 2% in Siscia in comparison with the intensity of the monetary circulation during the reign of Septimius Severus. In the period of Elagabalus' rule there was an increase of 44,62%, while during the reign of Alexander Severus the decrease amounted to 47.05%. At the beginning of the crisis of the Empire with the reign of Maximinus I Thrax (235-238) there was a decrease of 19 % in comparison with the preceding period, and during the period 238-244 there was an increase of 116.5%. During the reign of Philippus I a decrease of a mere 13.94 % is evident. During the reign of Trebonianus Gallus there is no visible increase or decrease in the intensity of the circulation in comparison with the increase of 123.36 % during Traianus Decius. The joined rule of Valerianus and Gallienus witnessed a decrease of 121.93 %, but during Gallienus' sole reign there was a drastically significant increase of 735.5 %, similar to that of 203.12% during the rule of Claudius II Gothicus; this can

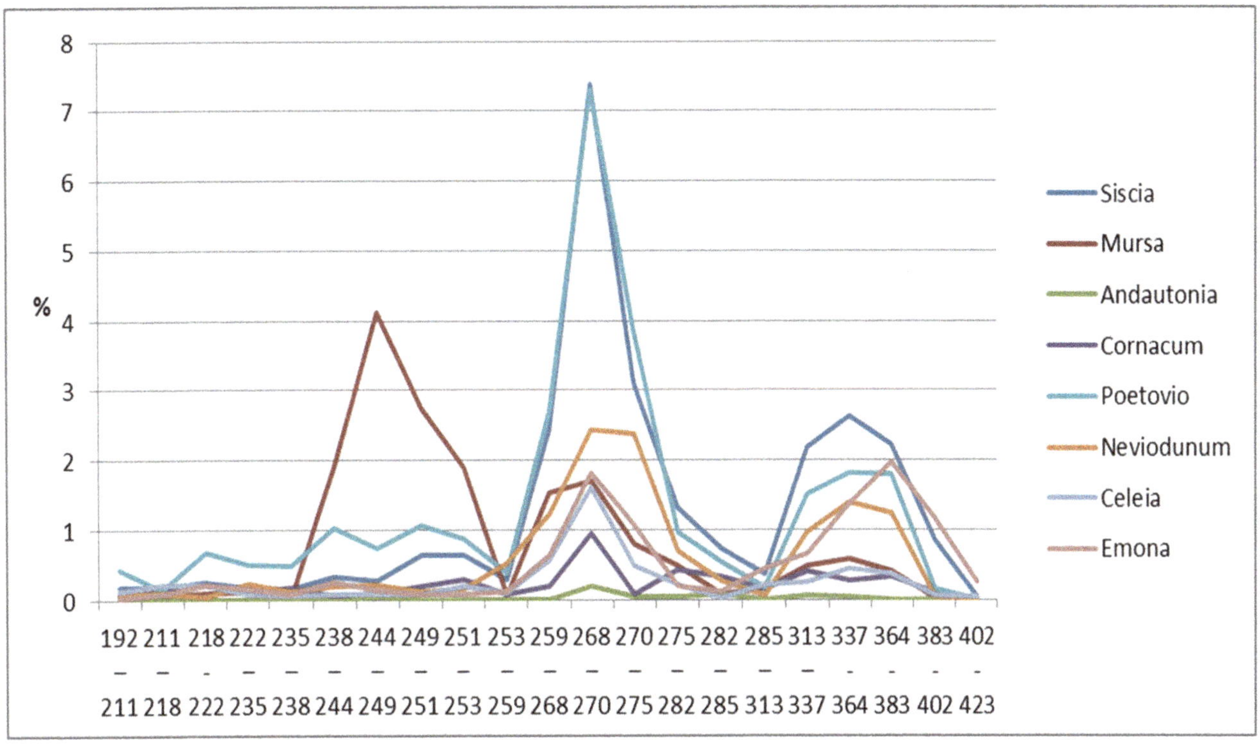

Chart 1. The data from Table 1 plotted on a chart.

*Table 2. The coins by numbers and periods of issue and a corrected percentage comparisons of the eight sites on an annual basis (total no. 12754).**

Period of issue /site	***Siscia no./%***	***Mursa no./%***	***Andautonia no./%***	***Cornacum no./%***	***Poetovio no./***	***Neviodunum no./%***	***Celeia no./%***	***Emona no./%***
193 – 211	47 0.019	15 0.006	1 0.001	21 0.008	105 0.043	21 0.008	31 0.012	7 0.002
211 – 218	17 0.019	5 0.005	1 0.001	12 0.013	13 0.014	10 0.011	5 0.005	7 0.007
218 - 222	14 0.027	10 0.019	0 0	12 0.023	36 0.070	12 0.023	12 0.023	12 0.023
222 – 235	31 0.018	20 0.012	1 0.006	21 0.012	88 0.053	41 0.024	12 0.007	26 0.015
235 – 238	6 0.015	1 0.002	0 0	7 0.018	19 0.049	5 0.013	2 0.005	3 0.007
238 – 244	26 0.033	50 0.065	0 0	5 0.006	81 0.105	16 0.020	6 0.007	21 0.027
244 – 249	19 0.029	64 0.100	2 0.003	7 0.010	49 0.076	15 0.023	5 0.007	9 0.014
249 – 251	17 0.066	17 0.066	0 0	5 0.019	28 0.109	3 0.011	2 0.007	2 0.007
251 – 253	17 0.066	44 0.172	0 0	8 0.031	23 0.090	4 0.015	5 0.019	2 0.007
253 – 259	23 0.030	4 0.005	1 0.001	3 0.003	30 0.039	42 0.054	8 0.010	10 0.013
259 – 268	288 0.376	41 0.035	1 0.008	26 0.022	318 0.277	144 0.125	68 0.059	76 0.066
268 – 270	194 0.760	45 0.176	5 0.019	25 0.098	192 0.752	64 0.250	42 0.164	48 0.188
270 – 275	204 0.319	53 0.083	2 0.003	47 0.073	252 0.395	156 0.244	32 0.050	70 0.109
275 – 282	122 0.316	45 0.050	3 0.003	39 0.043	89 0.099	65 0.072	20 0.022	18 0.020
282 – 285	29 0.075	3 0.007	3 0.007	13 0.033	22 0.057	11 0.028	1 0.002	5 0.013

Period of issue /site	*Siscia no./%*	*Mursa no./%*	*Andautonia no./%*	*Cornacum no./%*	*Poetovio no./*	*Neviodunum no./%*	*Celeia no./%*	*Emona no./%*
285 – 313	136 0.038	57 0.015	4 0.001	59 0.016	64 0.017	20 0.005	82 0.022	168 0.047
313 – 337	688 0.224	156 0.050	21 0.006	130 0.042	475 0.155	306 0.099	86 0.028	208 0.067
337 - 364	935 0.271	206 0.059	11 0.003	103 0.029	642 0.186	499 0.144	160 0.046	486 0.141
364 - 383	556 0.229	100 0.041	4 0.001	83 0.034	447 0.184	307 0.126	94 0.038	495 0.204
383 - 402	214 0.089	7 0.006	0 0	27 0.011	42 0.017	9 0.003	17 0.007	294 0.121
402 - 423	12 0.004	1 0.001	0 0	0 0	0 0	0 0	10 0.003	71 0.026
TOTAL:	**3595**	**944**	**60**	**652**	**3015**	**1750**	**700**	**2038**

* Coins from Spajić 1975 have been excluded.

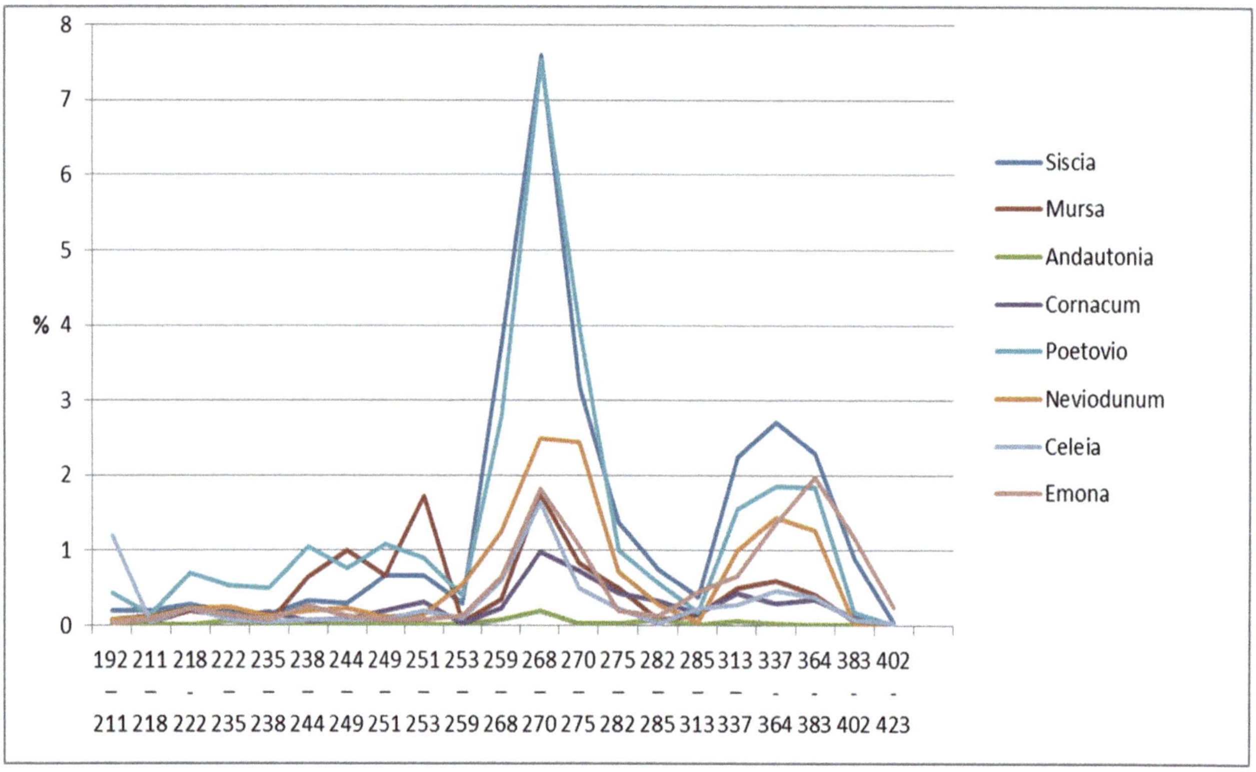

Chart 2. The data from Table 2 plotted on a chart. The coins from Spajić 1975 have been excluded.

be attributed to the opening of the Siscia mint or a rapid economic growth. During the short reign of Quintillus and the reign of Aurelianus there was a decrease of 137.74 %. A decrease of 134.21 % is evidenced in the period of 275 to 282, and the same is true of the reign of Carus, Carinus, and Numerianus (a decrease of 80.33 %.). The period of 285 to 313, as compared to the previous one, witnessed a decrease of 99.17 %. The period of 313 to 337 witnessed a strong increase of 490.92 % as does the period of 337 to 364, with an increase of 20.79 %. After that, a decrease was constant, that is, 18.31 % in the period of 364 to 383, 159.85 % in 383 to 402 and finally 1875.43 % in 402 to 423.

Osijek (Mursa)

There was a decrease of 9.85% during the reign of Caracalla, as compared to the intensity of the monetary circulation during the reign of Septimius Severus. In the reign of Elagabalus an increase was 100 % and in the reign of Alexander Severus it was 7.74 %. The reign of Maximinus I Thrax witnessed a large decrease of 363.63 % and during the period of 238-244 there was a drastic increase of 2424.24 %. Such increase must have been connected with the opening of the Viminacium mint[15]

[15] See Viminacium coins (P M S COL VIM).

Table 3. The coins from Siscia by periods of issue (total no. 3595).[*]

Period of rule	Frequency no./%	Duration	Corrected frequency	Cumulative
193-211	47 1.30	19	2.47	47
211-218	17 0.47	7	2.42	64
218-222	14 0.38	4	3.50	78
222-235	31 0.86	13	2.38	109
235-238	6 0.16	3	2.00	115
238-244	26 0.73	6	4.33	141
244-249	19 0.52	5	3.8	160
249-251	17 0.47	2	8.5	177
251-253	17 0.47	2	8.5	194
253-259	23 0.63	6	3.83	217
259-268	288 8.01	9	32	505
268-270	194 5.39	2	97	699
270-275	204 5.67	5	40.8	903
275-282	122 3.39	7	17.42	1025
282-285	29 0.80	3	9.66	1054
285-313	136 3.78	28	4.85	1190
313-337	688 19.13	24	28.66	1878
337-364	935 26.0	27	34.62	2813
364-383	556 15.46	19	29.26	3369
383-402	214 5.95	19	11.26	3583
402-423	12 0.33	21	0.57	3595

[*] The data has been provided by the coins kept in the AMZ, as well as those published by Mirnik 1969, Mirnik 1972, Burkowsky 2008; Burkowsky 2010; Koprivnjak 2011a; Koprivnjak 2011b.

and the presence of the military forces in the area, but the material still awaits a thorough analysis and better founded conclusions. An increase of 53.66% is evidenced in the period of 238 to 244. During the reign of Traianus Decius there was a decrease of 50.58% as compared with the previous period and again an increase of 158.82 % during Trebonianus Gallus' rule. The period of 253 to 259 witnessed a drastic decrease of 3,233.33%, probably due to the opening of the Viminacium mint, which had a great impact on the circulation in this period. Gallienus' sole reign witnessed a large increase of 589.39 %, similar to that of 394.5% during the rule of Claudius II Gothicus; this can be attributed to the opening of the Siscia mint. There was a decrease of 112.26 % in the period of 270 to 275. The next period (275-282) still witnessed a decrease of 65.1 % and the same tendency (542%) continued in the reign of Carus, Carinus, and Numerianus. In the period of 285 to 313, as compared to the previous one, there was still a decrease of 103%, almost the same as in Siscia. During the period 313-337 a large increase of 220.19 % occurred, while during the

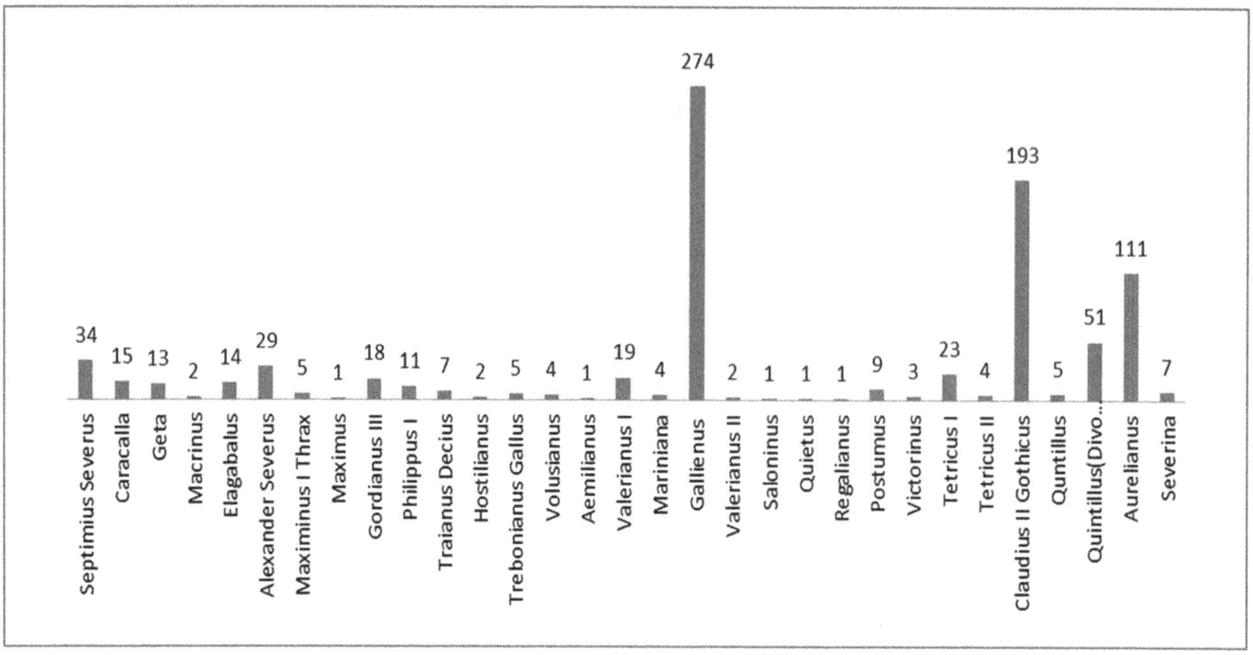

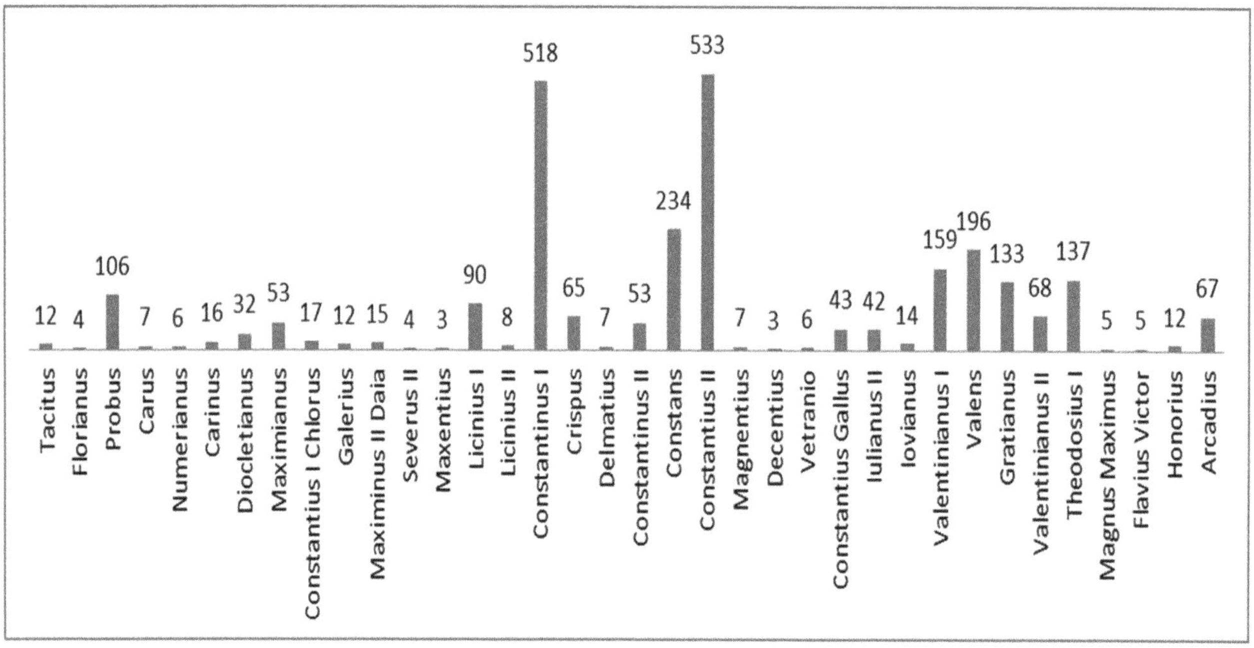

Charts 3-4. The coins from Siscia by emperors.

period 337-364 an increase was 17.23%. After the latter period a decrease was constant (as in Siscia), that is 44.86 % in 364-383, 1361.11 % in 383-402 and finally 800 % in 402-423.

Viminacium coins (P M S COL VIM)

As the chart above shows, the influx of the Viminacium coins during the period from Gordianus III to the reign of Valerianus was very intense. During that period the Viminacium coins made 90.4 %, or 25.95 % (compared to those of imperial mints) out of the total number of Mursa coins. [16] The coins of Viminacium from Spajić 1975 are not included although they are all found in Mursa, because "... they could have provided an erroneous perspective of the overall picture and resulted in invalid conclusion".[17] This is consistent with the opinion, or, rather, the fact that the Viminacium mint was the main supplier of the bronze currency in the Balkans during its running time and had a wide circulation in Pannoniae Superior and Inferior, Moesia Superior, and Dalmatia.[18] This can be exemplified

[16] That is, 944 coins without those from Spajić 1975.

[17] Găzdac 2008, 138. Table 1 and Chart 1 include coins from Spajić 1975, in order to present a general perspective of the site.

[18] Kos 1986, 109, 110.

*Table 4. The coins from Mursa by periods of issue (total no.1313/944).**

Period of rule	Frequency no./%	Duration	Corrected frequency	Cumulative
193-211	15 1.14/ 1.58	19	0.78	15
211-218	5 0.38/ 0.52	7	0.71	20
218-222	10 0.76/1.05	4	1.42	30
222- 235	20 1.52/2.11	13	1.53	50
235-238	1 0.07/ 0.10	3	0.33	51
238-244	149/ 50 11.34/ 5.29	6	24.83/8.33	200/ 101
244-249	270/ 64 20.56/ 6.77	5	54/12.8	470/ 165
249-251	72/ 17 5.48/ 1.8	2	36/8.5	542/ 182
251-253	50/ 44 3.80/ 4.66	2	25/22	592/ 226
253-259	5/ 4 0.38/ 0.42	6	0.83/0.66	597/ 230
259-268	43/ 41 3.27/4.34	9	4.77/4.55	640/ 271
268-270	45 3.42/ 4.76	2	22.5	685/ 316
270-275	53 4.03/ 5.61	5	10.6	738/ 369
275-282	45 3.42/ 4.76	7	6.42	783/ 414
282-285	3 0.22/ 0.31	3	1	786/ 417
285-313	57 4.34/ 6.03	28	2.03	843/ 474
313-337	156 11.88/ 16.52	24	6.5	999/ 630
337-364	206 15.68/ 21.82	27	7.62	1205/ 836
364-383	100 7.61/ 10.59	19	5.26	1305/ 936
383-402	7 0.53/ 0.74	19	0.36	1312/ 943
402-423	1 0.07/ 0.10	21	0.04	1313/ 944

* The data has been provided by the coins kept in the AMZ, as well as those published by Mažuran 1954, Spajić 1975, Göricke-Lukić 1993a, Göricke-Lukić 1993b, Göricke-Lukić 1997, and Göricke-Lukić 2011. The second column of Table 4 displays the percentages with and without the data from Spajić 1975. The fourth column displays the corrected frequency and the fifth the cumulative number of coins; the data from Spajić 1975 are included starting with the period 238-244.

by the case of Carnuntum, where during the period from Gordianus III to that of Trebonianus Gallus the Viminacium coinage made 51% of the total number of coins.[19] A complete analysis of the coins from Mursa would give a better and amore accurate picture of the Viminacium coins in comparison to that of the imperial coinage. The evidence presented here suggests that the bronze provincial coinage of Viminacium had a dominant role in the circulation in Mursa throughout its period of existence. The juxtaposition of only two sites from Pannonia and the argument of Fitz that the share of Viminacium coins made 38% in Pannonia Superior and 58% in Pannonia Inferior (for the period of 238-253)[20]clearly show that the circulation of the Viminacium currency was much higher.

Viminacium/imperial - Mursa[21]

Gordian III – 70.83 % / 29.16%; 12.59 %; 3.6 %

[19] Fitz 1980, 333.

[20] Fitz 1979, 74; cited in Kos 1986, 110.

[21] The following percentages are given: by the reign of emperors; of the Viminacium coins by emperors in the period Gordianus III to Valerianus; of the Viminacium coins in the total number of the Mursa currency.

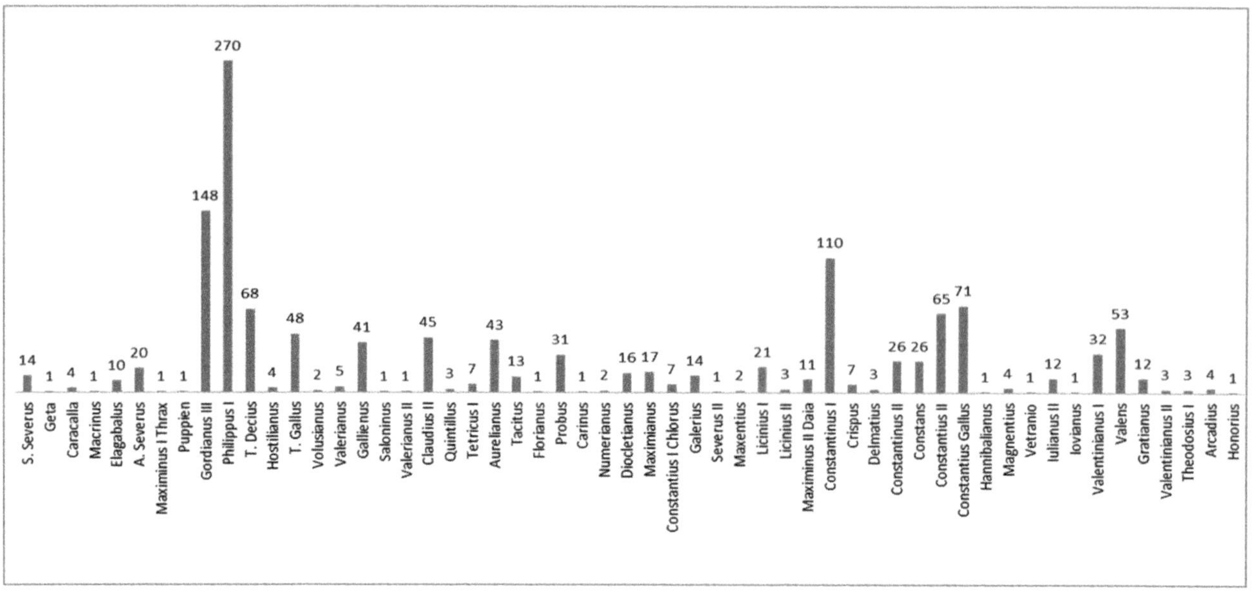

Chart 5. Coinage from Mursa by emperors.

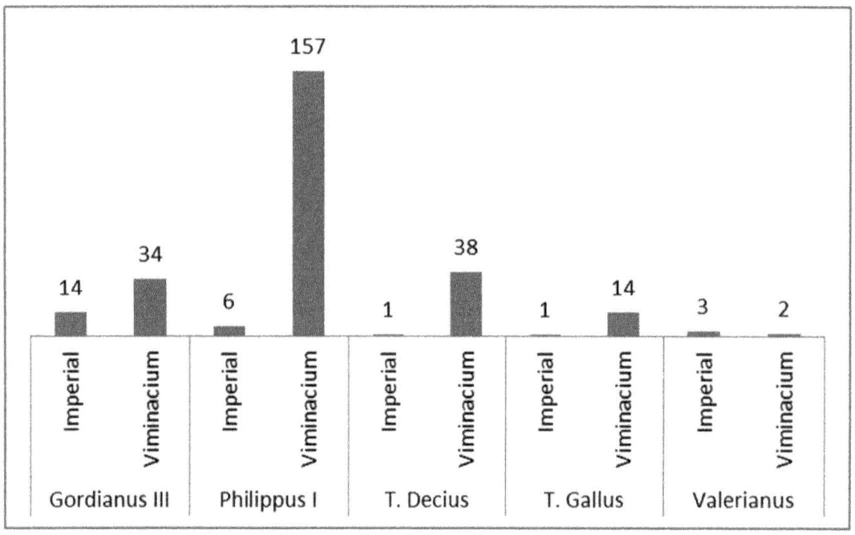

*Chart 6. Comparison of imperial and provincial mints from Mursa.**

*Coins from Spajić 1975 are not included.

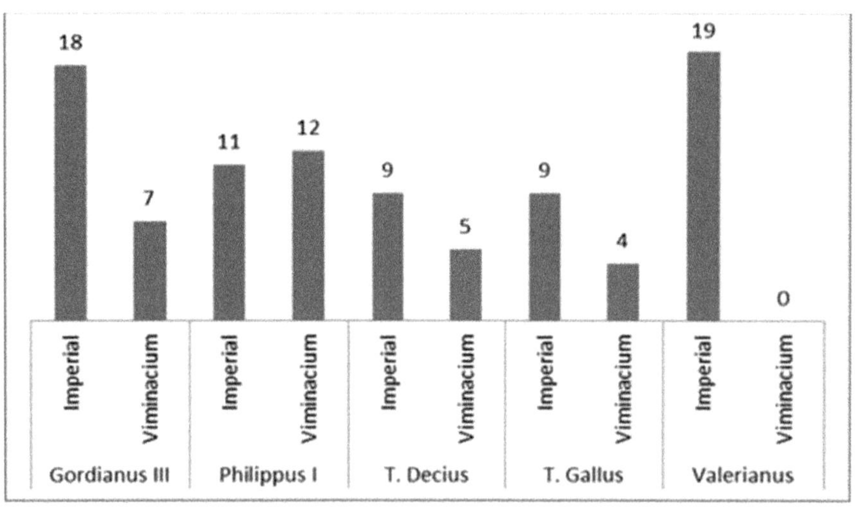

Chart 7. Comparison of imperial and provincial mints from Siscia.

*Table 5: The coins from Ščitarjevo (Andautonia) by periods of issue (total no. 60).**

Period of rule	*Frequency no./%*	*Duration*	*Corrected frequency*	*Cumulative*
193-211	1 1.66	19	0.05	1
211-218	1 1.66	7	0.14	2
218-222	-	4	0	2
222-235	1.66	13	0.07	3
235-238	-	3	0	3
238-244	-	6	0	3
244-249	2 3.33	5	0.4	5
249-251	-	2	0	5
251-253	-	2	0	5
253-259	1 1.66	6	0.16	6
259-268	1 1.66	9	0.11	7
268-270	5 8.33	2	2.5	12
270-275	2 3.33	5	0.4	14
275-282	3 5	7	0.42	17
282-285	3 5	3	1	20
285-313	4 6.66	28	0.14	24
313-337	21 35	24	0.87	45
337-364	11 18	27	0.40	56

Philippus I – 96.31 % / 3.68 %; 58.14 %; 16.63 %
T. Decius – 97.43 % / 2.56 %; 14.07 %; 4.02 %
T. Gallus – 93.33 % / 6.66 %; 5.18 %; 1.48 %
Valerianus – 40 % /60 %; 0.74 %; 0.21 %

As a comparison, in the same period the Siscian currency shows a decreased flow of the Viminacium coins compared to those imperial, but still intensive. The percentage is 29.78 % for the period and 0. 77% out of the total number of the Siscia coins. The consistency of the general pattern shows that the number of the Viminacium coinage is decreasing as the distance from the centre of its production increases. The coins of Viminacium analysed by Kos (1986), who discussed the sites in the southeastern Alpine region west of Sisak, show the same pattern.[22]

Viminacium/imperial - Siscia[23]

Gordianus III – 28.0 % / 72 %; 7.44%; 0.19 %
Philippus I – 52.17 % / 47.82 %; 12.76 %; 0.33 %
T. Decius – 35.71 % / 64.28 %; 5.31 %; 0.13 %
T. Gallus – 30.76 % / 69.23 %; 4.25 %; 0.11 %

Valerianus – 0 % / 100%; 0 %; 0 %

Sotin (Cornacum)

An increase of 55.45 % has been evidenced in Caracalla's reign at Cornacum, in comparison with the intensity of circulation in the reign of Septimius Severus. In the reign of Elagabalus an increase of 75.43 % occurred, while in the reign of Alexander Severus there was a decrease of 86.33%. The reign of Maximinus I Thrax witnessed an increase of 44.72 % and in the period of 238-244 a decrease was 180.72 %. Such circulation deviates from those in Siscia and Mursa in the same periods. There was an increase of 68.67 % in the period of 244-249, while during the reign of Traianus Decius the increase was 78.57 %. An increase of 60% has been suggested for the reign of Trebonianus Gallus. In the period of 253-259 there was a decrease of 300%, but the next period (259-268) witnessed an increase of 155 %. The same tendency occurred in the reign of Claudius II Gothicus; the increase was as much as 390.19 %. In the short reign of Quintillus and that of Aurelianus there was a decrease of 32.97 %. The same tendency was continued in the period of 275-282 (68.76%), as well as during the reign of Carus, Carinus and Numerianus (28.63 %). The period 285-313 witnessed a further decrease of 106.19 %. In the period of 313-337 there was a large increase of

[22] Kos 1986, 110.
[23] The following percentages are given: by the reign of emperors; of the Viminacium coins in the period from Gordianus III to Valerianus; of the Viminacium coins in the total number of the Siscia currency.

Period of rule	*Frequency no./%*	*Duration*	*Corrected frequency*	*Cumulative*
364-383	4 6.66	19	0.21	60
383-402	-	19	0	60
402-423	-	21	0	60

* The data has been furnished by the coins kept in the AMZ, as well as those published by Vikić-Belančić 1981, Dukat 1995, Knezović and Pintarić 2004, and Pintarić 2007. I would like to thank Dorica Nemeth-Ehrlich and Dora Kušan Špalj for the opportunity to examine unpublished coins from the 2007 excavations of Andautonia (the school yard site).

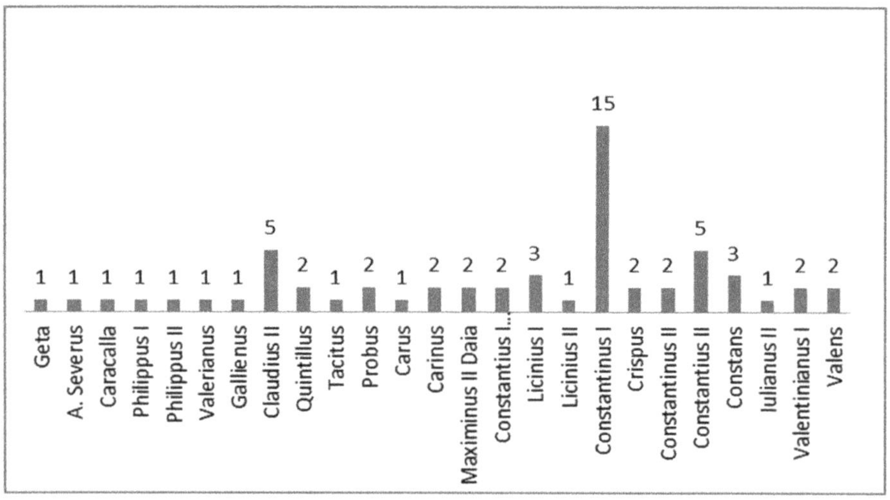

Chart 8. The coins from Ščitarjevo (Andautonia) by emperors.

157.61%, comparable to that at other analysed sites. On the other hand, in the following period (337-364) there was an increase of 43.50 %, which is contrary to the other analysed sites. Curiously, there was an increase of 15.64 % in the period of 364-383 and a decrease of 207.04 % in the period of 383-402. The circulation was completely terminated during the period of 402-423.

3. Hoards

The hoards of ancient coins are one of the most important sources for the study of ancient economic and political history of a particular area. The money is usually hidden in the times of an impending threat of war and the reasons for concealing are sought in unrest, hostile raids and invasions. A great caution, however, is required in interpretation of the hoards and the reasons for their concealment, because the concealment is not always caused by an immediate danger. The occurrence of the hoards can further be associated with monetary reforms. The introduction of the lower value coinage usually resulted in the Gresham law effect, that is, the money of better quality is withdrawn from the circulation and deposited because of its higher intrinsic value; only more or less complete hoards can give relevant information. The risk of analysing an incomplete hoard can result in a misinterpretation and wrong conclusions about the historical events and the reasons for the concealment of the hoard. This can happen if the earliest coins from the hoard are missing (often those better preserved and very interesting to the numismatic market) and the *terminus post quem* is indeterminable or false.[24] A hoard found in an archaeological context and fully documented can enable the exact determination of the time of the archaeological layer in which it was found; furthermore, it can clarify a number of questions connected with the flow of money and monetary history in general.[25] The conclusions derived from interpreting only one hoard are less reliable in comparison with the analysis of several hoards from the same period. A comparison of several hoards provides a possibility to determine the relative chronological sequence of minting of some coin emissions, what kinds of coins were in circulation at certain times, how fast the certain kinds of money were withdrawn from circulation, and what mint was dominant in circulation in a particular area in a particular period. And when that analysis is taken into account, it must be taken into consideration that the money that was buried

[24] Kos 1998, 328.
[25] Kos 1997, 114.

Table 6. The coins from Sotin (Cornacum) by the periods of issue (total no. 652).[*]

Period of rule	*Frequency no./%*	*Duration*	*Corrected frequency*	*Cumulative*
193-211	21 3.22	19	1.10	21
211-218	12 1.84	7	1.71	33
218-222	12 1.84	4	3	45
222-235	21 3.22	13	1.61	66
235-238	7 1.07	3	2.33	73
238-244	5 0.76	6	0.83	78
244-249	7 1.07	5	1.4	85
249-251	5 0.76	2	2.5	90
251-253	8 1.22	2	4	98
253-259	6 0.92	6	1	104
259-268	23 3.52	9	2.55	127
268-270	25 3.83	2	12.5	152
270- 275	47 7.20	5	9.4	199
275-282	39 5.98	7	5.57	238
282-285	13 1.99	3	4.33	251
285-313	59 9.04	28	2.10	310
313-337	130 19.93	24	5.41	440
337-364	102, 15.64	27	3.77	542
364-383	83 12.73	19	4.36	625
383-402	27 4.14	19	1.42	652
402-423	-	21	0	652

[*] The data has been provided by the coins kept in the AMZ, as well as those from Ilkić 2003 and Ilkić 2008.

in a particular area may have come from quite another part of the Roman Empire.[26]

Some new hoards have been found in southern Pannonia since the time when *CHY*[27] was published; so far this study was the main reference to the hoards of the area. The aim of this overview is to present a summary of the composition of both published and newly found hoards, as well as to provide new information on the already published hoards. The majority of the hoards stem from Croatian territory; those from other areas are specifically noted.

3.1. 3rd century hoards

1. Bakovci (Krog) [28]

This hoard, also known as the Krog hoard (Republic of Slovenia; Pannonia Superior), was found in 1901 as a pot burial containing 2290 *denarii* and *antoniniani*. The following emperors and members of the imperial family are represented: Commodus (1 *denarius*), Clodius Albinus (1 *denarius*), Septimius Severus (16 *denarii*), Iulia Domna (3 *denarii*), Caracalla (6 *denarii*), Plautilla (1 *denarius*),

26 Kos 1998, 329.
27 Mirnik 1981.

28 Mirnik 1981, 59, no. 148; Kos 1986, 116, no. 16; Kos 1988, 466.

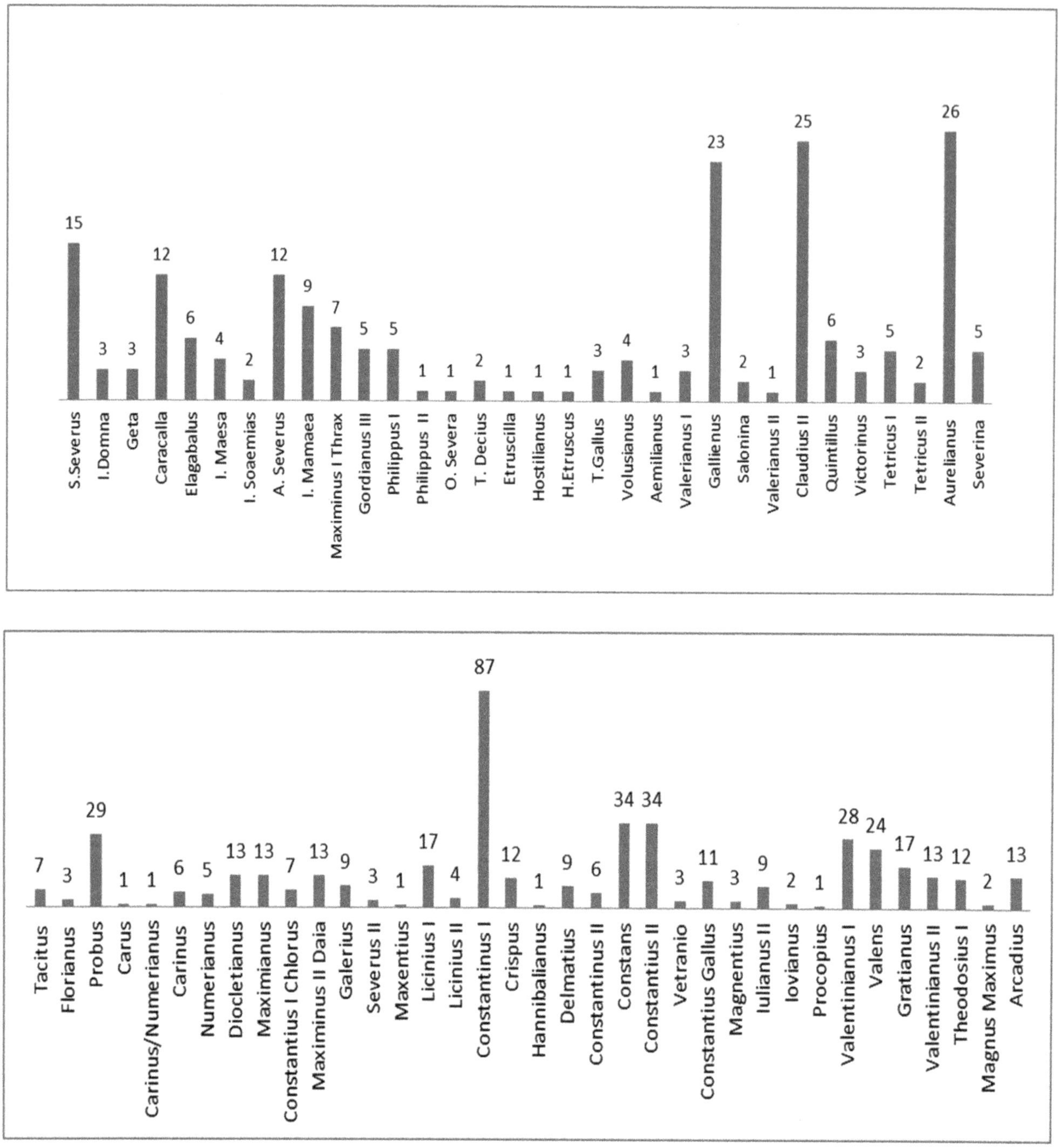

Chart 9-10. The coins from Sotin (Cornacum) by emperors.

Iulia Domna (1 *denarius*), Geta (1 *denarius*), Macrinus (1 *antoninianus*), Elagabalus (34 *denarii* and 9 *antoniniani*), Iulia Paula (3 *denarii*), Iulia Soaemias (3 *denarii*), Iulia Maesa (6 *denarii* and 2 *antoniniani*), Severus Alexander (36 *denarii*), Iulia Mammaea (11 *denarii*), Maximinus I Thrax (4 *denarii*), Balbinus (2 *antoniniani*), Pupiens (2 *antoniniani*), Gordianus III (8 *denarii* and 545 *antoniniani*), Philippus I (313 *antoniniani*), Otacilia Severa (73 *antoniniani*), Philippus II (53 *antoniniani*), Traianus Decius (120 *antoniniani*), Herennia Etruscilla (35 *antoniniani*), Herennius Etruscus (23 *antoniniani*), Hostilianus (10 *antoniniani*), Trebonianus Gallus (181 *antoniniani*), Volusianus (101 *antoniniani*), Aemilianus (16 *antoniniani*), Valerianus I (308 *antoniniani*), Mariniana (5 *antoniniani*), Valerian II (32 *antoniniani*), Gallienus (205 *antoniniani*), and Salonina (108 *antoniniani*). The *terminus post quem* is the year 259/260.[29]

2. Bartolovec[30]

The hoard was found before 1937, with Probus'AE *antoniniani* confirmed and one of them now in the AMZ.

[29] Kos 1986, 116, no. 16; Kos 1988, 466.
[30] Mirnik 1981, 60, no. 150.

The rest of the hoard has been dispersed, with no additional information available.[31]

3. Beltinci [32]

The hoard (the Republic of Slovenia; Pannonia Superior) contained 28 *antoniniani,* containing coins of Gordianus III (6), Philippus I (3), Hernnius Etruscus (1), Trebonianus Gallus (4), Volusianus (1), Aemilianus (1), Valerianus I (4), Mariniana (1), Gallienus (4), Saloninus (1) and inconclusive (2).[33]

4. Berak[34]

The hoard was found before 1902, with Maximianus' coins confirmed. They are now in the AMZ, while the rest of the hoard has been dispersed.[35]

5. Brđani[36]

More than 26 coins of Gallienus and Claudius II were found in a pot burial in 1938. All information about this hoard stems form the AMZ archive.[37]

6. Bušetina[38]

The hoard was found in 1873 as a pot burial containing more than 8 *aurei*, circa 200 silver coins and jewellery; it was part of the Prince of Schaumburg-Lippe collection. The following emperors and members of the imperial family are represented: Gordianus III, Tranquillina, Philippus I, Otacilia Severa, Iotapianus, Pacatianus, Traianus Decius, Herennia Etruscilla, Herennius Etruscus, Hostilianus, Trebonianus Gallus, Volusianus, Aemilianus, Cornelia Supera, Valerianus I, Mariniana, Gallienus, Salonina, Saloninus. All information about this hoard stems form the AMZ archive.[39] 30 coins are now in the AMZ: Domitianus (3 *denarii*), Traianus (2 *denarii*), Hadrianus (3 *denarii*), Antoninus Pius (1 *denarius*), Faustina I (1 *denarius*), Marcus Aurelius (1 *denarius*), Septimius Severus (1 *denarius*), Caracalla (2 *denarii*), Elagabalus (2 *denarii*), Severus Alexander (1 *denarius*), Maximinus I Thrax (2 *denarii*), Gordian III (2 *antoniniani*), Philippus I (3 *antoniniani*), and Gallienus (1 *aureus* and 5 *antoniniani*).[40] The *terminus post quem* is the year 258.[41]

7. Dolga Vas (Lendava) [42]

The hoard (the Republic of Slovenia; Pannonia Superior) was briefly mentioned in the literature as containing over 100 gold and silver coins from the period of the reign of M. Aurelius to Valerianus. Other data is missing.[43]

8. Dolnje Ponikve [44]

The hoard (the Republic of Slovenia; Pannonia Superior) was found as a clay pot of 30 *antoniniani* of Gordianus III (2), Philippus I (2), Otacilia Severa (1), Traianus Decius (1), Herennia Etruscilla (1), Volusianus (1), Aemilianus (1), and Gallienus (1). The *terminus post quem* is the years 253/254.[45]

9. Draganić (Karlovac)[46]

The hoard was found in 1897, containing more than 5 AE coins, of which 4 *sestertii* and 1 *aes* of Gordianus III (3 *sestertii* and 1 *aes*) and Valerianus I (1 *sestertius*). The hoard (5) is now in the AMZ.

10. Dvor[47]

The Dvor hoard (near Novo Mesto, the Republic of Slovenia; the province of Pannonnia Superior) was found in 1934 in a vineyard, as a pot burial containing ca. 400 *antoniniani*. Half of it was sold to the NMLj, 97 very badly preserved coins were sold to the AMZ,[48] and the rest has been dispersed.[49] The *terminus ante quem* is the opening of the Siscia mint (before 262 AD).[50] The composition is as follows: Etruscilla (1), Trebonianus Gallus (4), Volusianus (5), Valerianus I (45), Gallienus (19), Salonina (17), Saloninus (3), Valerianus II (2), and Quietus (1).[51] According to P. Kos revision of the hoard, its composition is as follows: Herennia Etruscilla (1), Trebonianus Gallus (4), Volusianus (5), Valerianus I (120), Gallienus (113), Salonina (72), Saloninus (3), Valerianus II (2), and Quietus (2).[52] The terminus post quem is 267/268.[53]

11. Garčin I[54]

The hoard was found in 1896 in two pots containing 252 *denarii* and *antoniniani*; its burial is dated to the year 261. The composition is as follows: Commodus (1), Septimius Severus (1), Elagabalus (2), Iulia Maesa (1) A. Severus (2), Balbinus (1), Gordianus III (15), Philippus I (8), Philippus II (3), Traianus Decius (3), Etruscilla (2), Herennius Etruscus (2), Trebonianus Gallus (17), Volusianus (10), Aemilianus (1), Valerianus (79), Mariniana (1), Gallienus (30), Salonina

31 *Ibid.*; Dukat and Mirnik 1978a, 205, 3rd c., no.1.
32 Kos 1986, 119, no. 18; Kos 1988, 459.
33 *Ibid.*
34 Mirnik 1981, 60, no. 154.
35 *Ibid.*
36 Dukat and Mirnik 1978b, 17, no. 14; Mirnik 1981, 60, no.155.
37 *Ibid.*
38 Mirnik 1981, 61, no. 159.
39 *Ibid.*
40 Mirnik 1981, 73, no. 246; Demo 1981b, 214, 215; Demo 1983b, 359-361, 403-405.
41 Demo 1981b, 214, 215; Demo 1983b, 359-361, 403-405.
42 Kos 1986, 119, no. 20; Kos 1988, 213.
43 *Ibid.*
44 Kos 1986, 115, no. 13; Kos 1988, 253.
45 *Ibid.*
46 This hoard was found in the B. Horvat collection during the revision of the Numismatic department of the AMZ. I would like to thank I. Mirnik for this information.
47 Mirnik 1981, 62, no. 167.
48 Klemenc 1936, 131.
49 Mirnik 1981, 62, no. 167.
50 Klemenc 1936, 131; Mirnik 1981, 62.
51 Demo 1983b,374, 375.
52 Kos 1986, 119-120, no. 20.
53 Kos 1986, 120.
54 Mirnik 1981, 63, no. 170

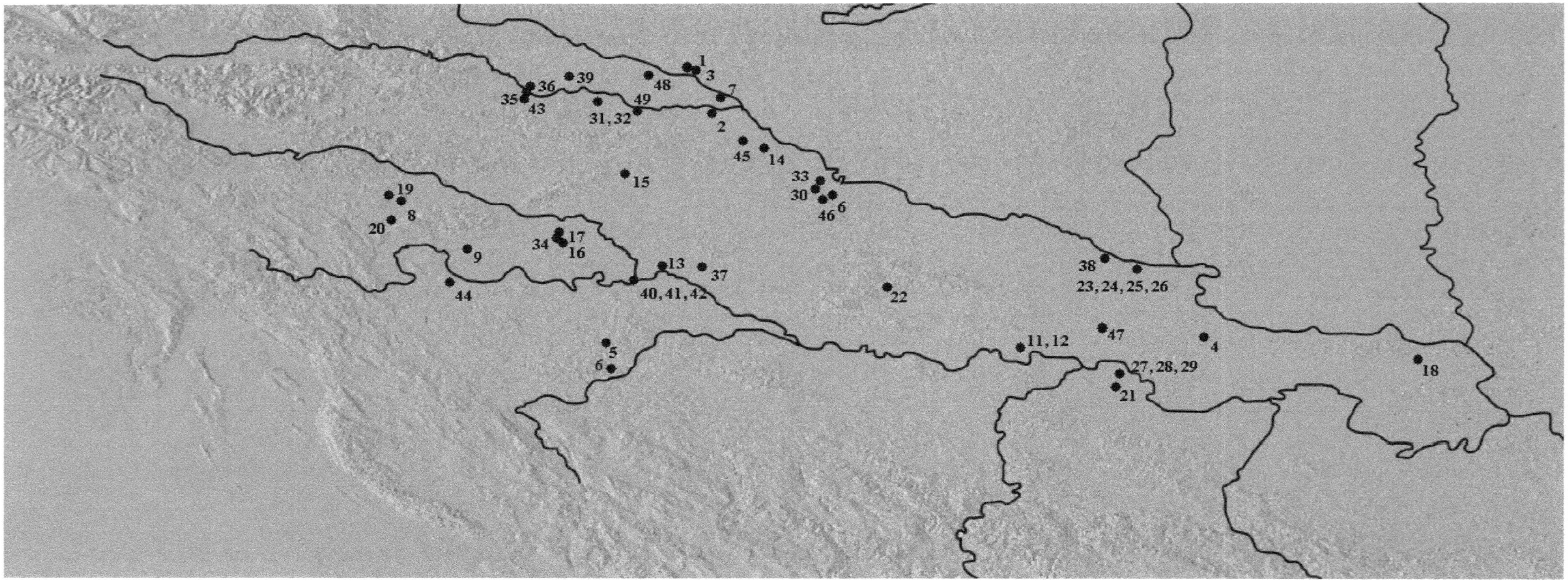

Fig. 1. Distribution of 3rd century hoards: (1) Bakovci (Krog), (2) Bartolovec, (3) Beltinci, (4) Berak, (5) Brđani, (6) Bušetina, (7) Dolga Vas, (8) Dolnje Ponikve, (9) Draganić, (10) Dvor, (11) Garčin I, (12) Garčin II, (13) Gušće, (14) Imbrijovec, (15) Komin, (16) Kravarsko, (17) Kurilovec, (18) Maradik, (19) Mokronog, (20) Novo Mesto, (21) Obudovac, (22) Orljavac, (23) Osijek, (24) Osijek, (25) Osijek, (26) Osijek, (27) Oštra Luka I, (28) Oštra Luka II, (29) Oštra Luka III, (30) Otrovanec, (31) Petrijanec, (32) Petrijanec, (33) Pitomača, (34) Podvornica, (35) Ptuj, (36) Ptuj (Rabelčja Vas), (37) Repušnica, (38) Satnica, (39) Sejanci, (40) Sisak, (41) Sisak, (42) Sisak, (43) Spodnja Hajdina, (44) Sveti Petar Mrežnički, (45) Virje, (46) Virovitica, (47) Vođinci, (48) Vratišinec, (49) Zbelava (made by M. Nađ).

(37), Saloninus (7), Valerianus II (3), Macrianus (1), and Quietus (1).[55]

12. Garčin II

The hoard was found in 1968, and its *terminus post quem* is the year 258; 51 *antoniniani* are kept in the MBP. The composition is as follows: Gordianus III (7), Philippus I (6), Otacilia Severa (2), Traianus Decius (4), Herennia Etruscilla (1), Herennius Etruscus (1), Trebonianus Gallus (3), Volusianus (3), Valerianus I (14), Gallienus (5), Salonina (2), and Valerianus II (3).[56]

13. Gušče[57]

This hoard was merged with the Repušnica hoard.[58]

14. Imbriovec[59]

The hoard was found in 1952 in a pot containing more than 420 *denarii* (1) and *antoniniani* (the rest). 321[60] or 315 coins are kept in the MGK, and the rest has been dispersed.[61] The composition is as follows: Elagabalus (1), I. Maesa (1), A. Severus (1), Maximinus I Thrax (1), Gordianus III (3), Philippus I (8), O. Severa (1), Philippus II (3), T. Decius (6), Herennia Etruscilla (1), Trebonianus Gallus (7), Volusianus (7), Valerianus I (57), Mariniana (2), Gallienus (115), Salonina (80), Valerianus II (12), Saloninus (10)[62], Quietus (2), Macrianus (1), Postumus (1).[63] According to Ž. Demo's revision of the hoard, its composition (274) is as follows: Caracalla (1 *denarius*.), Gordianus III (4), Philippus I (2), Otacilia Severa (1), Philippus II (3),Traianus Decius (4), Herennius Etruscus (1), Trebonianus Gallus (3), Volusianus (2), Valeianus I (55), Gallienus (102), Salonina (73), Valerianus II (12), Saloninus (8), Postumus (2), Macrianus (1), and Quietus (1); the *terminus post quem* is 265.[64]

15. Komin[65]

The hoard of Komin was found in1918 as a pot burial of ca. 30000 pieces (90 kg), mostly AE *antoniniani*. A large part of the find was acquired by the AMZ [66] (ca. 19755 initially, now amounting to 20197),[67] 401 coins were acquired by Z. Habuš,[68] and the rest has been dispersed. The *terminus post quem* is 276, and the composition is as follows: Marcus Aurelius (1), Caracalla (6), Gordianus I (1), Gordianus III (5), Philippus I (2), Otacilia Severa (1), Traiaus Decius (1), Herennia Etruscilla (1), Herennius Etruscus (1), Trebonianus Gallus (6), Volusianus (4), Aemilianus (2), Valerianus (223), Mariniana (14), Gallienus (1396), Salonina (241), Valerianus II (25), Saloninus (9), Postumus (12), Macrianus (2), Quietus (1), Claudius II (1656), Victorinus (3), Tetricus (4), Quintillus (228), Aurelianus (459), Severina (13), Vaballathus (5), and Tacitus (1).[69]

16. Kravarsko[70]

This hoard has been merged with the Podvornica hoard.[71]

17. Kurilovec[72]

The hoard was found as a pot burial in 1944 and its *terminus post quem* is the year 259. It contained more than 2000 mostly *antoniniani*, of which 980 (?) coins came to the AMZ, 83[73] plus 14 coins were acquired by P. Broz, and 202 coins make part of other (unknown) collections.[74] A partial composition of the hoard is as follows: Severus Alexander (1 *denarius*), Gordianus III (20 *antoniniani*), Philippus I (37 *antoniniani*), Otacilia Severa (4 *antoniniani*), Philippus II (6 *antoniniani*), Traianus Decius (12 *antoniniani*), Herennia Etruscilla (4 *antoniniani*), Herennius Etruscus (4 *antoniniani*), Aemilianus (5 *antoniniani*), Volusianus (22 *antoniniani*), Valerianus (75 *antoniniani*), Mariniana (7 *antoniniani*), Gallienus (66 *antoniniani*) Salonina (21 *antoniniani*), and Saloninus (12 *antoniniani*).[75]

18. Maradik

The hoard Maradik (the Republic of Serbia; the province of Pannonia Inferior) was found during 1968 and 1969 as a burial of more than 248 *antoniniani*; the *terminus post quem* is the year 259. The hoard is now divided between the MVi (230) and the Macković collection (18). Its composition is as follows: Gordian III (6), Filip I (10), Otacilia Severa (2), Filip II (2), Traianus Decius (3), Herennia Etruscilla (2), Hostilianus (1), Trebonianus Gallus (4), Volusianus (5), Aemilianus (3), Valerianus I (84), Mariniana (4), Gallienus (59), Salonina (49), Valerianus II (11), and Saloninus (2). [76]

---

[55] Brunšmid 1900, 89; Dukat and Mirnik 1978a, 206, no. 16; Mirnik 1981, 63, no.170; Demo 1983b, 377-381, 451-459, pls. 16 and 17.
[56] Demo 1983b, 362.
[57] Dukat and Mirnik 1978b, 19 no. 37; Mirnik 1981, 63, no. 173.
[58] Demo 1983b, 371.
[59] Dukat and Mirnik 1978b, 19 no. 38; Dukat and Mirnik 1978a, 206 no. 24; Demo 1979, 157-185; Mirnik 1981, 63, no. 175,
[60] Zmajić 1955,16.
[61] Mirnik 1981, 63 no. 175.
[62] Mirnik 1981, 63, no.175.
[63] Zmajić 1955, 16.
[64] Demo 1979, 157; Demo 1983b, 382.
[65] Dukat and Mirnik 1978a, 206 no. 30; Mirnik 1981, 64, no. 180; Demo 1981a, 12, no. 7.
[66] Klemenc and Saria 1936, 11; Klemenc 1936, 125, Mirnik 1981, 64.
[67] Mirnik 1981, 65. The revision of the Numismatic department of the AMZ, currently in progress, gave a new number of coins in the Komin hoard: 20197. I would like to thank Anja Bertol and Karmen Farac for their help in establishing this number.
[68] Habuš 1970, 19; Mirnik 1981, 65.
[69] Mirnik 1981, 65.
[70] Mirnik 1981, 65, no. 182.
[71] Demo 1983b, 363.
[72] Dukat and Mirnik 1978b, 21, no. 54; Dukat and Mirnik 1978a, 206, no. 33; Mirnik 1981, 65, no. 185.
[73] Broz 1953, 18; Habuš 1969, 19-20; Mirnik 1981, 65.
[74] Broz 1964, 12.
[75] Dukat and Mirnik, 1978, 21, no. 54; Mirnik 1981, 65, no. 185; Demo 1983b, 369.
[76] Dautova-Ruševljanin 1980, 21, 27; Demo 1983b, 271, 272, 276, 282, 283, 284, 303, 308, 313, 327, 328, 329, 330, 333, 334, 336, 339, 340, 374, 375, 376, 377, 498.

19. Mokronog[77]

This hoard was found in 1879 as a pot burial containing more than 4500 *antoniniani*, concealed after the year 284.[78] Only 13 coins from this hoard are kept in the AMZ, and the rest has been dispersed. A partial composition based on Brunšmid's description is as follows: Gallienus (332), Salonina (22), Saloninus (2), Postumus (2), Victorinus (3), Claudius II (390), Quintillus (8), Aurelianus (473), Severina (5), Tetricus I (14), Tetricus II (6), Tacitus, Florianus, Carus (7), Numerianus (6), and Carinus (3).[79]

20. Novo Mesto[80]

The hoard has been merged with the Dvor hoard.[81]

21. Obudovac[82]

The hoard (the Republic of Bosnia and Herzegovina; the province of Pannonia Inferior) was found before 1955. It contained more than 805 AR and AE *antoniniani*, now in the ZMS (805)[83] or (786).[84] Its composition is as follows: Gordianus III (2), Philippus I (1), Philippus II, Traianus Decius, Trebonianus Gallus (4), Volusianus (2), Aemilianus (1), Valerianus I (459), Mariniana (3), Gallienus (581), Salonina (119), Valerianus II (17), Saloninus (1), Postumus (1), Etruscilla (1)[85]. Ž. Demo revisited composition of the hoard: Gordianus III (2), Philippus I (1), Herennia Etruscilla (1), Trebonianus Gallus (4), Volusianus (2), Aemilianus (1), Valerianus I (69), Mariniana (3), Gallienus (39), Salonina (34), Valerianus II (9), Saloninus (8), Postumus (1), Gallienus (537), and Salonina (75).[86]

22. Orljavac[87]

According to the AMZ archive, 26 coins, presumably AE, came to the Museum. This was a part of the hoard found in 1895 in two pots, while the rest has been dispersed. In the meantime those 26 coins have mixed with the rest of the third-century coins in the collection.[88]

23. Osijek[89]

The hoard was found in 1897, and it comprised jewellery and gold and silver coins, issued from Elagabalus to Gordianus III.[90] All *aurei* were minted under Elagabalus and Alexander Severus, while silver coins were minted under Gordian III.[91] One *aureus* of Elagabalus is now in the AMZ, two of Elagabalus' *aurei* and one of Alexander Severus are in the MSO, while the rest has been dispersed.[92]

24. Osijek[93]

Information about this hoard is scarce.[94] It consisted of an unknown number of AE Roman imperial coins up to Gordianus III, as well as Roman provincial/Greek imperial coins.

25. Osijek [95]

As in the case of the previous hoard, information about this one is scarce.[96] It contained AE coins of Philippus I, Gordianus III, Traianus Decius, Herennia Etruscilla, Hostilianus, Trebonianus Gallus, Volusianus, Aemilianus, and Gallienus.

26. Osijek

The hoard was found in 1982 as a burial of several hundred AR *denarii* and *antoniniani* of Elagabalus (1), Gordianus III (13), Philippus I (13), Philippus II (1), Traianus Decius (4), Herennia Etruscilla (2), Herennius Etruscus (1), Trebonianus Gallus (9), Volusianus (9), Valerianus I (5), Gallienus (19), Salonina (8), and Valerianus II (3). A part of the hoard (98) is now in the MSO, while the rest has been dispersed.[97]

27. Oštra Luka I[98]

The hoard (the Republic of Bosnia and Herzegovina; the province of Pannonia Inferior) was found during ploughing in 1901; it contained more than 204 *antoniniani*, 229 of which are now in the ZMS. The composition of the hoard is as follows: Titus (1), Septimius Severus (1), Caracalla (1), Elagabalus (2), Gordianus III (43), Philippus I (63), Otacilia Severa (1), Philippus II (3), Traianus Decius (22), Herennia Etruscilla (4), Herennius Etruscus (2), Hostilianus (2), Trebonianus Gallus (32), Volusianus (10), Aemilianus (11), Valerianus (28), Mariniana (12), and Gallienus (9).[99]

28. Oštra Luka II

The hoard (the Republic of Bosnia and Herzegovina; Pannonia Inferior) was found in 1902 as a burial of probably more than 33 coins of Gordianus III (6), Philippus I (4), Otacilia Severa (1), Philippus II (2), Traianus Decius (4), Herennia Etruscilla (2), Trebonianus Gallus (2), Volusianus

77 Mirnik 1981, 66, no. 191; Demo 1981a, 12, no. 7; Kos 1986, 124, no. 33; Kos 1988, 223/2.
78 Brunšmid 1880, 31-32.
79 Brunšmid 1880, 31-32; Pegan 1967, 209; Mirnik 1981, 66; Kos 1986, 124, no. 33; Kos 1988, 223/2.
80 Mirnik 1981, 67 no. 196.
81 Mirnik 1981, 62, no.167.
82 Dukat and Mirnik 1978b, 23, no. 72; Mirnik 1981, 67, no. 197.
83 *Ibid.*; Petrović 1955, 181.
84 Demo 1983b, 384.
85 Petrović 1955, 181; Mirnik 1981, 67.
86 Demo 1983b, 384.
87 Mirnik 1981, 67, no. 198.
88 Mirnik 1979, 83.
89 Mirnik 1981, 67, no. 199.
90 Celestin 1899, 525, 526; Mirnik 1981, 67, no. 199.
91 Celestin 1897, 497.
92 Mirnik 2009b, 519.
93 Mirnik 1981, 67 no. 200
94 Celestin 1904, 21-29; Noe 1925, 91.
95 Mirnik 1981, 67, no. 201.
96 Celestin 1904, 16-21.
97 Lukić 1996, 93-102.
98 Mirnik 1981, 67, no. 202.
99 Patsch 1902, 425-426; Babić 1979, 279; Mirnik 1981, 67; Demo 1983b, 353-357, 398-402.

(3), Aemilianus (2), Mariniana (1), and inconclusive (6); 33 pieces are now in the ZMS.[100]

29. Oštra Luka III

The hoard (the Republic of Bosnia and Herzegovina; the province of Pannonia Inferior) was found in 1902 as a burial of 43 coins of Gordianus III (8), Philippus I (7), Traianus Decius (3), Herennia Etruscilla (2), Herennius Etruscus (1), Trebonianus Gallus (4), Volusianus (9), Aemilianus (2), Mariniana (2) and Gallienus (5). It is now in the ZMS.[101]

30. Otrovanec[102]

The hoard was found concealed in a pot containing ca. 500 *antoniniani*, issued during the reign of Elagabalus, Gordianus III, Philippus I, Otacilia Severa, Traianus Decius, Trebonianus Gallus, Volusianus, Valerianus I, Mariniana, Gallienus, and Salonina. 350 coins are now in the AMZ [103]

31. Petrijanec, 1805[104]

The hoard (formerly the Labaš collection) was found in 1805 as a pot burial of ca. 230 gold coins, medallions and gold jewellery. It has been mostly dispersed today; one *aureus* of Maximianus is in the AMZ, and one medallion of Carus and Carinus in the KHM.[105] The *terminus post quem* is the year 285[106] or 305/6.[107] A partial composition of the hoard is as follows: Hadrianus (1), Antoninus Pius (6), Caracalla (4), Marcus Aurelius (5), Lucius Verus, Faustina II, Iulia Domna, Tacitus (7), Gordianus III, Aurelianus, Probus (35), Carus, Carinus (29), Magnia Urbica (4), Numerianus (12), Iulianus I (1), Diocletianus (10),[108] and Maximianus (1).[109]

32. Petrijanec, 2005

The hoard was found in 2005 as a wooden "barrel" burial of 27,735 silvered *antoniniani* and *nummi*, as well as 3 silver plates; it is now in the GMVa (27,735). The *terminus post quem* is the year 294, and the composition is as follows: Gallienus (76), Salonina (6), Macrianus (1), Claudius II (52), Aurelianus (1382), Severina (351), Tacitus (870), Florianus (151), Probus (10567), Carus (829), Magnia Urbica (99), Numerianus (985), Carinus (1616), Iulianus I (5), Diocletianus (4279), Maximianus (2851), Constantius I Chlorus (48), Galerius (14), and inconclusive or corroded (3534).[110]

33. Pitomača[111]

The hoard was found in 1966 as a pot burial containing ca. 2000 *antoniniani*, with the following rulers represented: Trebonianus Gallus, Gallienus, Salonina, Postumus, Tetricus, Claudius II, Quintillus, and Aurelianus. The AMZ managed to acquire 1644 coins, while the rest has been dispersed. [112]

34. Podvornica[113]

This hoard has been merged with the Kravarsko hoard.[114] It was found in 1899 as a pot burial of more than 75 *antoniniani*. The *terminus post quem*, and the year of the concealment, is 259. 63 coins are now in the AMZ, while the rest has been dispersed. The composition, according to Brunšmid, is as follows: Gordianus III (2), Philippus I (4), Philippus II (1), Traianus Decius (1), Trebonianus Gallus (3), Volusianus (1), Valerianus I (32), Mariniana (8), Gallienus (15), Salonina (7), Saloninus (6) and Valerianus II (1).[115] Today composition of the hoard looks differently: Gordianus III (2), Philippus I (4), Philippus II (1), Traianus Decius (1), Trebonianus Gallus (3), Volusianus (3), Valerianus I (35), Mariniana (2), Gallienus (24), Salonina (10), Valerianus II (4), and Saloninus (2), altogether 92 *antoniniani*.[116] Hoard is buried at the end of 259.[117]

35. Ptuj[118]

The hoard has been merged with the Spodnja Hajdina hoard.[119]

36. Ptuj (Rabelčja Vas) [120]

The hoard of 13 *denarii* (the Republic of Slovenia; Pannonia Superior) was found during archaeological excavations with the following rulers represented: Vespasianus (1), Septimius Severus (5) and Caracalla (7). The latest coins were issued in the period 211-217.[121]

37. Repušnica (Tucilača)[122]

This hoard has been merged with the Gušće hoard.[123] According to the AMZ archive, it was found in 1941 as a pot burial of ca. 200 (ca. 0.45 kg [124]/1 kg[125]) silver

[100] Babić 1979, 280, 281; Demo 1983b, 357-358, 387.
[101] Babić 1979, 281; Demo 1983b, 268, 358-359.
[102] Dukat and Mirnik 1978b, 24, no. 78; Mirnik 1981, 67, no. 203.
[103] Dukat 1974, 134-135.
[104] Dukat and Mirnik 1978a, 206, no. 54; Mirnik 1981, 68, no. 205.
[105] Mirnik 2009a, 447.
[106] Ljubić 1880a, 111-113; Mirnik 1981, 68, no. 205.
[107] Mirnik 2009a, 449.
[108] Ljubić 1880a, 111-113; Mirnik 1981, 68, no. 205.
[109] Mirnik 2009a, 449.
[110] Šiša-Vivek 2006, 132 - 133; Šiša-Vivek, Leleković, and Kalafatić 2005, 231-233, 238-243.

[111] Dukat and Mirnik 1978b, 24, no. 83; Dukat and Mirnik 1978a, 206, no. 55; Mirnik 1981, 68, no. 206; Demo 1981a, 12, no. 3.
[112] Dukat 1967, 8; Pegan 1968, 86; Dukat 1975, 135; Mirnik 1981, 68.
[113] Dukat and Mirnik 1978a, 206, no. 56; Mirnik 1981, 68, no. 208.
[114] Mirnik 1981, 65, no. 182.
[115] *Ibid.*; Brunšmid 1901, 237; Klemenc 1938, 34.
[116] Mirnik 1981, 73, no. 246; Demo 1983b, 363-364.
[117] Demo 1983b, 366.
[118] Mirnik 1981, 69, no. 120a.
[119] Mirnik 1981, 70, no. 222.
[120] Kos 1986, 111, no. 1; Kos 1988, 434/18.
[121] Kos 1986, 112.
[122] Dukat and Mirnik 1978b, 25, no. 94; Mirnik 1981, 69, no. 211; Demo 1981a, 1, no.1.
[123] Mirnik 1981, 63, no. 173.
[124] Mirnik 1981, 69, no. 211.
[125] Demo 1983b, 372.

coins.[126] The *terminus post quem* is the year 260, and the composition of the hoard, containing 126 *antoniniani*, is as follows: Gordianus III (3), Philippus I (4), Otacilia Severa (1), Traianus Decius (3), Herennius Etruscus (1), Trebonianus Gallus (5), Volusianus (5), Valerianus (58), Mariniana (2), Gallienus (22), Salonina (17), Valerianus II (4), and Postumus (1).[127]

38. Satnica[128]

The hoard was found in 1906 during ploughing as a pot burial (ca. 6000 – 7000 coins) comprised of 3 *denarii* (Caracalla and Elagabalus) and *antoniniani* (the rest). The latest dated coin in the hoard was issued in 265, while the burial presumably occurred in the year 267 during Gallienus' campaigns against the Goths; the AMZ (formerly the NMZ) acquired 626 coins.[129] The composition of the hoard is as follows: Caracalla (2), Elagabalus (1), Gordianus III (7), Philippus I (6), Pacatianus (1), T. Decius (6), Etruscilla (3), Trebonianus Gallus (12), Volusianus (7), Aemilianus (1), Valerianus I (133), Mariniana (5), Gallienus (297), Salonina (112), Saloninus (30), Postumus (2), and contemporary imitations (?).[130] According to Ž. Demo's revision, the composition of the hoard is as follows: Composition of the hoard is: Caracalla (2), Elagabalus (1), Gordianus III (8), Philippus I (6),Philippus II (1), Pacatianus (1), Traianus Decius (6),Herennia Etruscilla (3), Trebonianus Gallus (12), Volusianus (7), Aemilianus (1), Valerianus I (118), Mariniana (5), Gallienus (55), Salonina (66), Valerianus II (25), Saloninus (7), Postumus (2), Gallienus (238), and Salonina (44).[131]

39. Sejanci[132]

The hoard (the Republic of Slovenia; Pannonia Superior) was found in 1892 or 189, containing 300-500 *denarii* and *antoniniani* issued during the period from Vespasianus to Severus Alexander; it was immediately dispersed.[133]

40. Sisak, Kačića Street[134]

The hoard, found in 1940, has been completely dispersed; (scarce) information has been furnished by the AMZ archive. The composition of the hoard is as follows: Iulia Mamaea, Pupienus, Claudius II, Aurelianus, Probus, and Carinus.[135]

41. Sisak, Socijalističkog saveza Street (today Rimska Street)[136]

The hoard was found in 1953 during the digging of the ditches for sewer pipes; it contained 2000 silver *argentei*, 3 silver plates and one silver vessel, all in a silver receptacle.[137] The *terminus post quem* is 295/6. The hoard is now distributed between the GMSi (421), NMB (8), NMLj (?), and unknown places (?). The following emperors are represented: Diocletianus (ca. 363), Constantius I (ca.133), Maximianus (ca.307), and Galerius (ca.263)[138]

42. Sisak, Maršala Tita Street (today Antuna and Stjepana Radića Street)

The hoard was found in 1990 as a pot burial (money box) of 41 *antoniniani*, containing coins of Gallienus (11), Claudius II (14), Quintillus (Divo Claudio) (6), Tetricus I (9), and one reminted coin. The hoard is now in the GMSi and its *terminus post quem* is the year 274.[139]

43. Spodnja Hajdina[140]

The hoard (the Republic of Slovenia; the province of Pannonia Superior) was found in 1900, containing 60 *denarii* and *antoniniani*. Its *terminus post quem* is 228/29 and it was buried between 230 and 235. Its composition is as follows: Vespasianus (1), Traianus (1), Faustina I (1), Commodus (2), Septimius Severus (8), Iulia Domna (2), Caracalla (2), Elagabalus (9), Iulia Paula (1), Iulia Soaemias (2), Iulia Maesa (2), Severus Alexander (17), and Iulia Mamaea (3.)[141]

44. Sveti Petar Mrežnički[142]

The hoard was found in or before 1896. It contained more than 8 *denarii*, now in the AMZ. Its composition is as follows: Marcus Aurelius (1), Septimius Severus (3), Iulia Domna (3), and Caracalla (1).[143]

45. Virje[144]

The hoard was found in 1878 as a pot burial of more than 1000 *antoniniani*. The *terminus post quem* of the hoard is the year 269, and its biggest part (978 pieces) is now in the AMZ, while rest has been dispersed. The composition is as follows: Philippus I (1), Traianus Decius (1), Trebonianus Gallus (1), Volusianus (1), Valerianus I (2), Gallienus (62), Salonina (3), and Claudius II (12)[145]

[126] Mirnik 1981, 69, no. 211.
[127] Demo 1983b, 373.
[128] Mirnik 1981, 69, no. 212; Demo 1981a, 11, no. 1.
[129] Brunšmid 1907, 210-211.
[130] Mirnik 1981, 69.
[131] Demo 1983b, 386, 470-489, pl. 20-22.
[132] Kos 1986, 112, no. 4; Kos 1988, 434/12.
[133] Kos 1986, 112.
[134] Dukat and Mirnik 1978b, 26, no. 103; Mirnik 1981, 70, no. 217.
[135] *Ibid.*
[136] Dukat and Mirnik 1978b, 26, no. 104; Dukat and Mirnik 1978a, 206, no. 63; Mirnik 1981, 70, no. 218.
[137] Broz 1955, 1-2; Jeločnik 1961.
[138] Mirnik 1981, 70.
[139] Dukat and Šarić 1996.
[140] Mirnik 1981, 70, no. 222.
[141] Skrabar 1906, 195-196; Klemenc and Saria 1936, 59; Mirnik 1981, 70.
[142] Mirnik 1981, 72, no. 233.
[143] Brunšmid 1900, 86; Mirnik 1981, 72.
[144] Dukat and Mirnik 1978a, 205, 3rd c., no. 13; Mirnik 1981, 73, no. 245; Demo 1981a, 12, no.1.
[145] *Ibid.*; Ljubić 1879, 90.

46. Virovitica[146]

This hoard has been merged with the Bušetina hoard.[147]

47. Vođinci[148]

The hoard was found in 1938 as a pot burial containing between 800 and 1000 *antoniniani*; 107 coins are now in the AMZ, while rest has been dispersed. The composition is as follows: Valerianus I (1), Gallienus (78), Salonina (9), Claudius II (17), Quintillus (2), and Aurelianus(?).[149]

48. Vratišinec[150]

The hoard was found in ca.1922 as a pot burial of one silver *antoninianus* and ca. 50 AE *antoniniani*; it is now in the AMZ. The emperors represented are from Gallienus to Claudius II.[151]

49. Zbelava

The hoard was found in 1994 as a burial of 2586 AE silvered *antoniniani* of Gallienus (24), Salonina (5), Claudius II (13), commemorative (5),[152] Aurelianus (1564), Severina (43), Tacitus (240), Florianus (122), Probus (532), and inconclusive (19). The majority of the hoard (2567) is now in the GMVa, while 19 coins have been held by the finder.[153]

3.2. 4th century hoards

50. Aljmaš

The hoard was found in 1903 as a pot burial of ca. 3800 AE coins of a „medium module" (AE2/AE3?). Of the former Oskar Friml collection containing some 3000 coins, 71 are now in the AMZ, while the rest has been dispersed. All information about this hoard stems from the AMZ archive. The hoard's composition is as follows: Magnentius (3), Vetranio (350), Constans (595), and Constantius II (1795).[154]

51. Bapska[155]

The hoard was found in 1952 as a pot burial of ca. 50 AE coins of a „medium and small module" (AE3/AE4 (?), composed of the coins of Constans and Constantius II; 11 pieces are now in the GMVu.[156]

52. Batina[157]

The hoard was found in ca.1910 as a burial of ca. 250 AE coins ranging from Septimius Severus to the 4th century.[158]

53. Bijeljina[159]

The hoard (the Republic of Bosnia and Herzegovina; the province of Pannonia Secunda) was found in 1931 as a burial of ca. 40 kg of AE coins of the emperors Valentinianus I, Valens, and Gratianus. The *terminus post quem* is the year 378, and the hoard is now in the ZMS.[160]

54. Bikić Dol[161]

The hoard (the Republic of Serbia; the province of Pannonia Secunda) was found in 1952 as a pot burial of 11,459 AE coins. Represented are the following emperors: Licinius I (595), Licinius II (274), Constantinus I (4996), Crispus (2901), Constantinus II (2588), as well as barbarous (13) and inconclusive (105) coins. The *terminus post quem* is the year 324, and a part of the hoard (11,353) is now in the MS.[162]

55. Bilići[163]

A reportedly large hoard (the Republic of Serbia; the province of Pannonia Secunda) was found in 1952 as a pot burial of AE coins. The emperors represented all belong to the Constantinian House. The majority of the hoard is now in the MS, while 148 pieces are in the GMVu, and the rest has been dispersed.[164]

56. Bjelovar

The hoard was found in 2006 as a pot burial of more[165] than 3970 AE coins (*nummi*), representing the following emperors: Maximinus II Daia, Licinius, I and Constantinus I; the *terminus post quem* is the Constantinian tetrarchy and the hoard is now in the GMB (3970).[166]

57. Bosanski Brod[167]

This hoard [the Republic of Bosnia and Herzegovina; the province of Pannonia Inferior (?)] was found in 1880 as a pot burial, consisting of coins of Constantius II, Constantius Gallus, and Iulianus II. The *terminus post quem* is the year 360. The hoard is now divided between the Prince of

146 Mirnik 1981, 73, no. 246.
147 Demo 1983b, 359.
148 Dukat and Mirnik 1978b, 29, no. 134; Dukat and Mirnik 1978a, 205, 3rd c., no. 14; Mirnik 1981, 74, no. 248; Demo 1981a, 12, no. 7.
149 *Ibid.*; the AMZ archive.
150 Mirnik 1981, 74, no. 250.
151 *Ibid.*; Žganec 1922.
152 *Divo Claudio*?
153 Težak 2010, 533-534.
154 Mirnik 1981, 76, no. 255.
155 Dukat and Mirnik 1978a, 206, no. 2; Mirnik 1981, 76, no. 256.
156 *Ibid.*
157 Mirnik 1981, 76 no. 257.
158 *Ibid.*; *Numizmatikai Közlöny* 9, 1910, Budapest, 135.
159 Mirnik 1981, 76, no. 261.
160 *Ibid.*; Sergejevski 1932, 23; Klemenc 1936, 130, no. 29.
161 Mirnik 1981, 76, no. 262.
162 *Ibid.*
163 Mirnik 1981, 76, no. 263.
164 *Ibid.*; Bauer 1953, 116.
165 I. Mirnik, pers. com.
166 Jakovljević 2006, 96-97; Jakovljević 2008, 57-58.
167 Mirnik 1981, 77, no. 264.

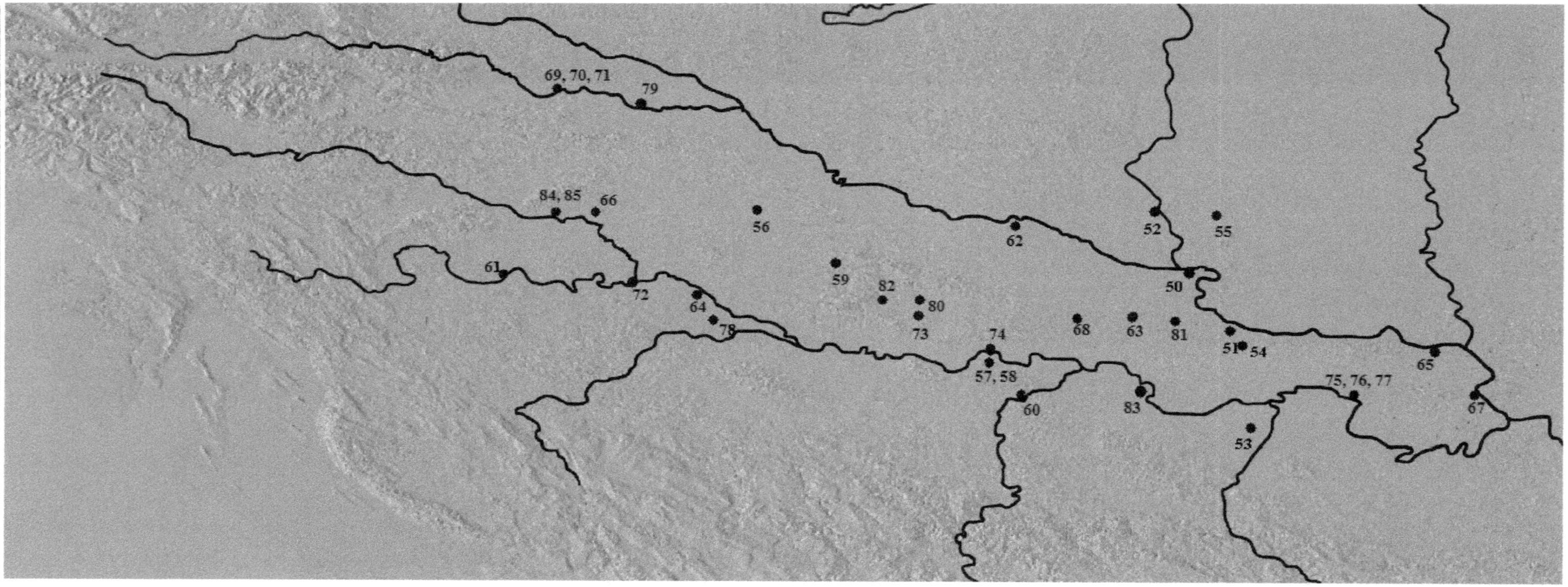

Fig. 2. Distribution of 4th century hoards: (50) Aljmaš, (51) Bapska, (52) Batina, (53) Bijeljina, (54) Bikić Dol, (55) Bilići, (56) Bjelovar, (57) Bosanski Brod, (58) Bosanski Brod, (59) Daruvar, (60) Dobra Voda, (61) Donja Kupčina, (62) Donji Miholjac, (63) Gaboš, (64) Graduša Posavaska, (65) Krčedin, (66) Lupoglav, (67) Novi Banovci, (68) Novi Mikanovci, (69) Ptuj, (70) Ptuj (Krčevina/ Vičava), (71) Ptuj (Panorama), (72) Sisak, (73) Požega, (74) Slavonski Brod, (75) Sremska Mitrovica, (76) Sremska Mitrovica, (77) Sremska Mitrovica (Jalija), (78) Sunja, (79) Šand(o)rovec, (80) Treštanovci, (81) Vinkovci, (82) Vranić, (83) Vidovice, (84) Zagreb, (85) Zagreb (made by M. Nađ).

Württemberg collection (?) and the AMZ (?), while the rest (?) has been dispersed.[168]

58. Bosanski Brod[169]

The hoard (see no. 51) was found in 1880 as a burial of a ca. 200 AE coins of Gratianus, Valentinianus II and Theodosius I. The AMZ acquired 4 coins (2 each of Valentinianus II and Theodosius I), and the rest has been dispersed.[170]

59. Daruvar[171]

The hoard was found in ca. 1905, containing ca. 100 AE coins of Constantnus I. It has been dispersed since, and the information was furnished by the AMZ archive.[172]

60. Dobra Voda[173]

This hoard [the Republic of Bosnia and Herzegovina, formerly the province of Pannonia Secunda (?)] was found in 1886 as a burial of ca. 50 AE coins of Constantinus I.[174]

61. Donja Kupčina[175]

The Hoard was found in 1900; the rulers represented are Tetricus I, Claudius II and Valentinianus II, and the *terminus post quem* is 375-392. The hoard was formerly part of the M. Turković collection (ca. 600 AE coins); it is now in the AMZ (?).[176]

62. Donji Miholjac[177]

The hoard of 16 AE coins was found in 1975, comprising 3 coins of Constans, 7 of Constantius II, and 6 inconclusive. It has since been dispersed.[178]

63. Gaboš[179]

The hoard was found in 1886 as a pot burial of more than 300 AE coins of Constans (3), Constantius II (90), Vetranio (6), Magnentius (4), Constantius Gallus (23) and inconclusive (10). The *terminus post quem* is the year 351, and the hoard has been dispersed.[180]

64. Graduša Posavaska[181]

The hoard was found before 1905; a single *solidus* was acquired by the AMZ, while the rest has been dispersed.[182]

65. Krčedin[183]

The hoard (the Republic of Serbia; the province of Pannonia Secunda) was found as a burial of coins of Constantinus II and Constantius II.[184]

66. Lupoglav[185]

The hoard was found in 1910 as a burial of more than 315 AE coins of Constantius II (11), Iulianus II (6), Iovianus (1), Valentinianus I (60), Valens (61), Gratianus (15) and inconclusive (158). The *terminus post quem* is the year 376.[186] A part of the hoard (?) is now in the AMZ, while the rest has been dispersed.[187]

67. Novi Banovci [188]

The hoard was found in 1918, containing AE coins of the Constantinian House. Information on the hoard, furnished by the AMZ archive, is scanty.[189]

68. Novi Mikanovci[190]

The hoard was found in 1952 as a pot burial of ca. 1600 AE coins of the Constantinian House; it is now in the GMVi.[191]

69. Ptuj[192]

The hoard (the Republic of Slovenia; the province of Noricum) was found in 1936 as a burial of more (?) than 24 AE coins of Constantinus I.[193]

70. Ptuj[194] (Krčevina/Vičava [195])

The hoard (the Republic of Slovenia; the province of Noricum) was found in 1933, containing 3 *solidi* of Valentinianus I (2) and Gratianus (1).[196]

168 *Ibid.*
169 Mirnik 1981, 77, no. 265.
170 *Ibid*; Ljubić 1880b, 124.
171 Mirnik 1981, 77, no. 269.
172 *Ibid.*
173 Mirnik 1981, 77, no. 271.
174 *Ibid*; Patsch 1902, 420-421.
175 Dukat and Mirnik 1978a, 206, no. 12; Mirnik 1981, 78, no. 274.
176 Mirnik 1981, 78, no. 274.
177 Dukat and Mirnik, 18, no. 28; Mirnik 1981, 78 no. 274.
178 Mirnik 1981, 78, no. 274.
179 Dukat and Mirnik 1978a, 206, no.14; Mirnik 1981, 78, no. 279.
180 *Ibid.*; Brunšmid 1886, 10-11; Brunšmid 1897, 82.

181 Mirnik 1981, 79, no. 283.
182 *Ibid.*
183 Mirnik 1981, 79, no. 287.
184 *Ibid.*; Vasić 1967, 75.
185 Dukat and Mirnik 1978a, 206, no. 37; Mirnik 1981, 80, no. 289.
186 *Ibid.*; Brunšmid 1912, 284-285; Klemenc 1938, 28.
187 Mirnik 1981, 80.
188 Mirnik 1981, 80, no. 296.
189 *Ibid.*
190 Dukat and Mirnik 1978b, 22, no. 67; Dukat and Mirnik 1978a, 206, no. 43; Mirnik 1981, 81, no. 297.
191 *Ibid.*; Bauer 1953, 116.
192 Mirnik1981, 81 no. 298a.
193 *Ibid.*; Klemenc and Saria, 1936, 59.
194 Mirnik 1981, 81, no. 299.
195 Kos 1986, 146, no.17; Kos 1988, 434/23.
196 Mirnik 1981, 81, no. 299; Kos 1986, 146; Kos 1988, 434/23.

71. Ptuj (Panorama)[197]

The hoard (the Republic of Slovenia; Noricum) contained 31 *antoniniani, nummi* and *centenionales* from Gallienus to Valens; its finding circumstances are unknown.[198]

72. Sisak, near the „Siscia mint"

The hoard was found in 1989 on the right bank of river Kupa, near the site of the supposed Siscia mint. It contained 235 AE coins of Constantinus II (1), Constans (7), Constantius II (158), Constantius Gallus (1), Iulianus II as a Caesar (57), and Iulianus II as emperor (9). The *terminus post quem* is the year 363, and the hoard is now in the Šipuš collection (235).[199]

73. Požega (earlier Slavonska Požega)[200]

The hoard was found before 1895 in the surroundings of Slavonska Požega. It comprised AE coins of Constantine's sons and a part of it is now in the AMZ.[201]

74. Slavonski Brod[202]

The hoard was found in 1883 as a pot burial of ca. 400 AE coins of Constantius II, Constantius Gallus, and Iulianus II. 172 coins are now in the AMZ and the rest has been dispersed.[203]

75. Sremska Mitrovica[204]

The hoard (the Republic of Serbia; the province of Pannonia Secunda) was found in 1894 as a burial of more than 857 AE coins of Helena (5), Theodora (2), Constantinus I (129), Constantinopolis (1), Urbs Roma (1), Hannibalianus (1), Constantinus II (3), Constans (501), and Constantius II (362). The *terminus post quem* is the year 350, and the hoard is now in the AMZ.[205]

76. Sremska Mitrovica[206]

The hoard (the Republic of Serbia; the province of Pannonia Secunda) was found in ca. 1800 as a burial of more than 8 AR coins of Constantinus I (3), Constantinus II (2), and Constantius II (3). It is now in the NMP, formerly in the Dr. A. Heinrich collection, Brno (5).[207]

77. Sremska Mitrovica (Jalija)[208]

The hoard (the Republic of Serbia; the province of Pannonia Secunda) was found in 1912, containing coins of Constantius II, Constans, Constantius Gallus, and Iulianus II. A part of it (?) is now in the AMZ.[209]

78. Sunja[210]

The hoard (probably equated with the hoard Graduša Posavska) was found before 1905, containing AV and AE coins of Theodosius I, Constantinopolis, and Urbs Roma.

79. Šand(o)rovec[211]

The hoard consisted of an unknown number of AR coins. Some of them are now in the AMZ: Iulia Domna (2), Philippus I (1), Valerianus (2), Gallienus (2), Salonina (1), Constantius II (1), and Valentinianus I (3); the rest has been dispersed.[212]

80. Treštanovci[213]

The Hoard was found in 1938 as a burial in six pots of ca. 6000 AE coins, three silver ones, one copper bracelet, a silver ring, a key, and a weight. The composition of the hoard is as follows: Constantine I (151), Constantinopolis (55), Urbs Roma (36), Constantine II (95), Constantius II (602), Constans (749), Crispus (1), Delmatius (22), Hannibalian (2), Helena (10), Magnentius (5),Vetranio (9), Decentius (1), Constantius Gallus (87), inconclusive (17), and imitations (4). The *terminus post quem* is 28 September 351, and a part (1846) of the hoard is now in the AMZ.[214]

81. Vinkovci, Krnjaš Street[215]

The hoard was found in 1909 as a burial of 1957 (ca. 3 kg) AE coins of Constantinus I (2), Constantinus II (2), Constans (8), Constantius II (195), Constantius Gallus (7), Iulianus II (49), Iovianus (2), Valentinianus I (579), Valens (900), Gratianus (188), Valentinianus II (2), and inconclusive (23). The *terminus post quem*, and the possible year of the burial, is 376; the hoard is now in the AMZ.[216]

82. Vranić[217]

The hoard was found in 1897 as a pot burial of ca. 20,000 AE coins of Theodora (1), Constantinus I (4), Constantinus II (5), Constans (3982), Constantius II (842), and Vetranio

197 Kos 1986, 146, no. 16; Kos 1988, 434/15.
198 *Ibid.*
199 Šipuš 1995, 46.
200 Mirnik 1981, 81, no. 301.
201 *Ibid.*
202 Mirnik 1981, 81, no. 302.
203 *Ibid.*; Ljubić 1883, 94-95.
204 Mirnik 1981, 81, no. 303.
205 *Ibid.*; Brunšmid 1897.
206 Mirnik 1981, 82, no. 303a.
207 *Ibid.*
208 Mirnik 1981, 82, no. 304.
209 *Ibid.*
210 Mirnik 1981, 82, no. 306.
211 Dukat and Mirnik 1978a, 205, 3rd c., no. 10; Mirnik 1981, 82, no. 307.
212 *Ibid.*
213 Dukat and Mirnik 1978b, 28, no.121; Mirnik 1981, 82, no. 309.
214 I would like to thank Mr. Davor Margetić for the information on the hoard of late Roman bronzes from Treštanovci, found in 1938, supplied from his unpublished manuscript.
215 Dukat and Mirnik 1978a, 206 no.79; Mirnik 1981, 83 no. 313.
216 *Ibid.*; Brunšmid 1912, 272-273.
217 Dukat and Mirnik 1978a, 206, no. 84; Mirnik 1981, 83 no. 317.

(51). The *terminus post quem* and most probably the year of the burial is 351. A part of the hoard is now in the AMZ (3765) and the MSO (1120), while the rest has been dispersed.[218]

83. Vidovice

The hoard was found in 1908 on the bank of river Sava as a pot burial of more than 2528 AE coins of Constantinus I, Constans, Iiulianus II, Valentinianus I, Valens, Constantius II, and Gratianus. The *terminus post quem* is the year 378; a part (2528) of the hoard is now in the ZMS, and the rest has been dispersed. [219]

84. Zagreb, Donji Štefanovec[220]

The hoard was found in 1925 as a burial of ca. 1000 AE coins of Constantius II (96), Constans (7), Constantius Gallus (8), and Iulianus II (48); 159 coins are now in the AMZ, while the rest has been dispersed[221]

85. Zagreb, St. Šimun (now Markuševec)[222]

The hoard of AE coins was found in 1893. A partial composition of the hoard, based on the coins acquired by the AMZ, is as follows: Gallienus (3?),[223] Constantius II (6), Valentinianus I (12), Valens (12), and Gratianus (6)[224]

The third century hoards are very diverse in their composition. The broader chronological spectrum characteristic for some hoards (Garčin-Commodus to Quietus, Petrijanec 1805-Hadrianus to Maximianus, Sisak, Kačića Street- I. Mamaea to Carinus) may be due to various reasons. One of the reasons might be the fact that chronologically earlier money was longer in circulation and hoarded together with the coins circulating during the time of the concealment of the hoard. Gold coins were kept out of circulation because of their intrinsic value, which explains the broad time range represented by the coins in the Petrijanec 1805 hoard. The same applies to high quality silver money that was withdrawn from circulation and hoarded over a longer period of time. The hoards with large quantities of currency (such as Komin–ca.30000 coins, Petrijanec 2005-27,735 coins, Satnica-ca.6000-7000 coins, etc.), it was argued, belonged to wealthy landowners,[225] but we can also speculate that these represent the savings of a group of people. On the other hand, the Vratišinec hoard is relatively small (ca.50 coins) and only two emperors are represented in its composition (Gallienus and Claudius II). It probably represents one's savings or currently held property hidden in haste. Another example of a hoard gathered in short period of time is the Sisak hoard (no. 41). All coins in the hoard are high quality *argentei*; only the first tetrarchy emperors are represented in the composition of the hoard. Together with 3 silver plates and 2 silver bowls, the coins suggest that the hoard was the property of a wealthy owner.

As shown above, the hoards of the first half of the fourth century have very compact composition. Again, in comparison with the southeastern Alpine region, we can see similarities in the number of *antoniniani* hoarded after the Diocletian's monetary reform. *Antoniniani* were tolerated in everyday monetary circulation but were only unintentionally hoarded during the first half of the fourth century.[226] One of the few examples is the Donja Kupčina hoard (no. 61). The hoards concealed during the second half of the fourth century still show a fairly consistent composition. In comparison with Ljubljana-Extra Muros[227] where *maiorinae* of Magnentius and Decentius predominate (62.9 %), the hoards of Gaboš (no.63.) and Aljmaš (no.50.) show only sporadic incursions of that type of currency. However, some hoards show presence of the currency uncharacteristic for the period such as the Zagreb, St. Šimun hoard (no. 85), where pre-294 coins were hoarded together with regular currency of Constantius II and the Valentinian House. This is consistent with the situation in the south-eastern Alpine region where, although sporadically, that type of currency was hoarded until the end of the fourth century.[228] We can only speculate on the exact reasons for the concealment of every particular hoard catalogued above, since only a few of them are found complete.

As indicated in the introduction, this is a general overview with an aim to present essential numismatic information concerning southern Pannonia. The data was obtained from the existing literature as well as the available numismatic material. The basic analysis extracted from this material gives a general picture of the circulation of currency through a period of a little more than two centuries of Roman rule in this area. The general picture of political conflict and turmoil during the 3rd century, the relative stabilization of the Empire at the end of the 3rd and during the 4th century and its disintegration at the end of the 4th and during the 5th century, is reflected on the presented numismatic material and is analyzed through tables and charts. The apparent increase in local provincial emissions of Viminacium coins as the dominant currency in the period of the mint's functioning is shown (Mursa 235-253 etc.), as well as certain local deviations in the proportion of Imperial coins on some sites (Siscia 259-270 etc.), but in general all sites presented show basically similar trends in monetary circulation. The presented hoards give an insight into the composition and distribution of the finds. Only basic information is provided for some hoards, with sometimes merely their existence documented, while others are analysed in detail. This is understandable if we consider

218 *Ibid.*; Brunšmid 1912, 170-171.
219 Babić 1979, 276-279.
220 Dukat and Mirnik 1978a, 206, no.73; Mirnik 1981, 84, no. 320.
221 Klemenc 1936, 128, no. 17; Klemenc 1938, 73.
222 Mirnik 1981, 84, no. 321.
223 *Ibid.*
224 *Ibid.*; Brunšmid 1912, 272; Klemenc (1938, 28-29) reports on a somewhat diffrent composition: Constantius II (2), Valentinianus I (6), Valens (9), and Gratianus (1).
225 Găzdac 2010, 72-73
226 Kos 1986, 160.
227 Kos 1986, 145, 160; Kos 1988 155/54.
228 Kos 1986, 160.

that some of the hoards were published 130 years ago. Such publications often do not provide detailed information for further analysis, and these hoards should be analysed more thoroughly and re-published in order to provide a better understanding of the relations between the hoards, their concealment (if possible) and monetary history.

This is just a modest beginning of a complex study of this subject. Further detailed and deeper analyses of the circulation will offer a more substantiated political and economic picture of the historical events and economic trends in southern Pannonia.

Bibliography

Babić 1979 Marko Babić, Nalaz rimskog novca u Vidovicama, *ŽAnt* 29, 275-285.

Bauer 1953 Antun Bauer, Neki skupni nalazi novca, *Numizmatika* 5, 116.

Bíró-Sey 1980 Katalin Bíró-Sey, Currency, in: *ARP*, 337-348.

Broz 1953 Petar Broz, Nalaz rimskog carskog novca iz Kurilovca kod Velike Gorice 1944.god., *Numizmatika* 5, 17-26.

Broz 1955 Petar Broz, Nalaz rimskog novca (argenteusa) I. tetrarhije u Sisku 6.V. 1953., *NV* 5, 1-6.

Broz 1964 Petar Broz, Nalaz rimskog carskog novca III. Stoljeća n.e. Kurilovec kraj Velike Gorice godine 1944, *NV* 21, 12-22.

Brunšmid 1880 Josip Brunšmid, O odkriću rimskih novaca kod Mokrenoga u Kranjskoj, *VHAD* 2, 31-32.

Brunšmid 1886 Josip Brunšmid, Skrovište rimskih novaca u Gabošu, *VHAD* 8, 10-14.

Brunšmid 1897 Josip Brunšmid, Nekoliko našašća novaca na skupu u Hrvatskoj i Slavoniji V-VII; VI. Našašće rimskih carskih novaca u Mitrovici, *VHAD* 2, 82-93.

Brunšmid 1900 Josip Brunšmid, Nekoliko našašća novaca na skupu u Hrvatskoj i Slavoniji V; VIII-XI. VIII. Našašće rimskih carskih denára kod Sv. Petra na Mrijeznici; IX. Našašće rimskih srebrnih novaca kod Garčina, *VHAD* 4, 86-155.

Brunšmid 1901 Josip Brunšmid, Nekoliko našašća novaca na skupu u Hrvatskoj i Slavoniji XII. Našašće rimskih antoninijana u Podvornici (kotar Velika Gorica), XIII. Našašće rimskih bakrenih novaca četvrtoga vijeka u Gornjem Hruševcu (kotar Velika Gorica), *VHAD* 5, 239-242.

Brunšmid 1902 Josip Brunšmid, Nekoliko našašća novaca na skupu u Hrvatskoj i Slavoniji V, XVI-XVII; XVI. Našašće rimskih bakrenih novaca iz sredine četvrtoga vijeka u Vraniću (kotar Požega), *VHAD* 6, 170-178.

Brunšmid 1907 Josip Brunšmid, Nekoliko Našašća novaca na okupu u Hrvatskoj i Slavoniji XXVI-XXVII, *VHAD* 9, 210-240.

Brunšmid 1912 Josip Brunšmid, Nekoliko našašća novaca na skupu u Hrvatskoj i Slavoniji. XXXIV-XXXVI; XXXV. Nahođaj rimskih bakrenih novaca IV. stoljeća posl. Kr. u Vinkovcima; XXXVI. Nahođaj rimskih bakrenih novaca IV. stoljeća posl. Kr. u Lupoglavu (kotar Dugoselo), *VAMZ* 12, 272-286.

Burger 1981 Alice Sz. Burger, *Late money circulation in South Pannonia, Régészeti Füzetek*, ser. II(22), Budapest.

Burkowsky 2008 Zdenko Burkowsky, *Numizmatički nalazi na novijim arheološkim istraživanjima u Sisku*, Exhibition Catalogue, Sisak.

Burkowsky 2010 Zdenko Burkowsky, Značajne prinove u Numizmatičkoj zbirci Gradskog muzeja Sisak, *GGMS* 10, 415-423.

Celestin 1897 Vjekoslav Celestin, Nalazak zlatnih novaca u Osijeku, *GZMS* 9, 497-499.

Celestin 1899 Vjekoslav Celestin, Goldmünzenfund in Essegg, *Wissenschaftliche Mitteilungen aus Bosnien und der Hercegowina* 6, Wien, 525-526.

Celestin 1904 Vjekoslav Celestin, Grčki i rimski kolonijalni novci nađeni u Osijeku, *VHAD* 7 (1903-1904), 15-29.

Dautova-Ruševljanin 1980 Velika Dautova-Ruševljanin, Ostava antoninijana iz Maradika, *Numizmatičar* 3, 21-45.

Demo 1979 Željko Demo, O skupnom nalazu antoninijana Galijenova vremena iz Imbriovca , *PZ* 79, 157-185.

Demo 1981a Željko Demo, Iz Numizmatičke zbirke Arheološkog muzeja u Zagrebu – Postum (260 – 268 g.), *NV* 35, 11-28.

Demo 1981b Željko Demo, Skupni nalaz novca i nakita 3. stoljeća iz Bušetine kraj Virovitice, *PZ* 81, 214-227.

Demo 1983a Željko Demo, Einige Beispiele für die Nachahmung antiker Münzen auf der Grundlage von Münzfunden im Slawonische-Syrmischen Raum, *VAMZ* 15, (1982), 211-250.

Demo 1983b Željko Demo, Münzfunde aus der zeit Gallienus in gebeit zwischen den flüssen Sava und Drava (Ein Beitrag zur militärpolitischen, ökonomischen und wirtschaftlich-monetären Geschichte des südlichen Pannoniens um die Mitte des 3. Jh. n. Ch.), *AV* 33 (1982), 258-498.

Demo 1984 Željko Demo, Rewiew of Fitz 1978, *AV* 34 (1983), 464-468.

Dukat 1967 Zdenka Dukat, Nalaz rimskog bakrenog novca u Pitomači, *Bilten Hrvatskog numizmatičkog društva* 11, Zagreb, 8.

Dukat 1975 Zdenka Dukat, Skupni nalaz antoninijana iz Otrovanca, *VAMZ* 8, 134-135.

Dukat and Mirnik 1978a Zdenka Dukat and Ivan Mirnik, Skupni nalazi novca u sjevernoj Hrvatskoj, in: *Arheološka istraživanja u sjeverozapadnoj Hrvatskoj, IzdHAD* 2, Zagreb, 197-206.

Dukat and Mirnik 1978b Zdenka Dukat and Ivan Mirnik, Skupni nalazi novca od 1936. godine do danas, *Numizmatika* 6, 15-33.

Dukat 1995 Zdenka Dukat, Nalazi rimskog novca, in: A. Rendić-Miočević (ed.), *Zagreb prije Zagreba, Arheološka baština Zagreba od pretpovijesti do osnutka biskupije 1094. godine / Zagreb Before Zagreb. The Archaeological Heritage of Zagreb from Prehistory to the Foundation of the Bishopric in 1094*, Exibition Catalogue, 37, 112-115, 138-144.

Dukat and Šarić 1996 Zdenka Dukat and Ivan Šarić, Rimska štedna kasica iz Siska, *VAMZ* 28-29, 97-110.

Duncan 1993 George L. Duncan, *Coin circulation in the Danubian and Balkan Provinces of the Roman Empire, A.D. 294-578*, London.

Fitz 1978 Jenő Fitz, *Der Geldumlauf der römischen Provinzen im Donaugebiet Mitte des 3. Jahrhunderts*, Bonn-Budapest.

Fitz 1980 Jenő Fitz, Money Circulation, in: *ARP*, 332-335.

Găzdac 2010 Cristian Găzdac, *Monetary circulation in Dacia and the provinces from the Middle and Lower Danube from Trajan to Constantine I (AD 106-337)*, Cluj-Napoca.

Găzdac and Alföldy Găzdac 2008 Christian Găzdac and Ágnes Alföldy Găzdac, The management of monetary crisis? The 'P M S COL VIM' and 'PROVINCIA DACIA' coinages in the Roman monetary policy of the 3rd century A.D., *NZ* 116- 117, 135-171.

Göricke-Lukić 1987 Hermine Göricke-Lukić, Nalazi rimskog novca sa donjogradskog pristaništa u Osijeku, *OZ* 18-19, 77-87.

Göricke-Lukić 1993a Hermine Göricke-Lukić, Rimski zlatni novac u Muzeju Slavonije, *OZ* 22-23, 45-70.

Göricke-Lukić 1993b Hermine Göricke-Lukić, Novac iz nekropole antičke Murse, *NV* 46, 8-18.

Göricke-Lukić 1996 Hermine Göricke-Lukić, Skupni nalaz antoninijana Galijenovog vremena iz Osijeka, *PIAZ* 10, 93-102.

Göricke-Lukić 1997 Hermine Göricke-Lukić, Novac iz grobova rimske Murse, *NV* 50, 19-32.

Göricke-Lukić 2011 Hermine Göricke-Lukić, *Nekropole rimskodobne Murse*, Osijek.

Habuš 1970 Zvonko Habuš, Prilog nalazu antoninijana u Kominu, *NV* 28, 18-21.

Jakovljević 2006 Goran Jakovljević, Slučajni zbirni nalaz rimskog novca iz Bjelovara, *ObHAD*, 38/3), 96-98.

Jakovljević 2008 Goran Jakovljević, *Bjelovar prije Bjelovara, Radovi Zavoda za znanstvenoistraživački i umjetnički rad u Bjelovaru* 1, Bjelovar, 55-70.

Habuš 1969 Zvonko Habuš, Prilog nalazu antoninijana iz prve polovine III. st. u Kurilovcu, *NV* 27, 19-20.

Ilkić 2003 Mato Ilkić, *Cornacum, sotinski prostor i njegovo mjesto u organizaciji južnog dijela provincije Panonije*, Unpublished PhD thesis, University of Zadar.

Ilkić 2008 Mato Ilkić, Rimski Cornacum u svijetlu numizmatičkih nalaza, *Proceedings of the 5th International Numismatic Congress in Croatia*, Rijeka, 51-75.

Jeločnik 1961 Alekasandar Jeločnik, *Najdba argenteusov zgodnje tetrarhije v Sisku*, *Situla* 3.

Klemenc 1936 Josip Klemenc, Nalazi novaca u Jugoslaviji 1910 – 1936, *Numizmatika* 2-4 (1934-1936), 124-133.

Klemenc 1938 Josip Klemenc, *Archeologische karte von Jugoslavien: Blatt Zagreb*, Beograd.

Klemenc and Saria 1936 Josip Klemenc and Balduin Saria, *Archeologische karte von Jugoslavien: Blatt Ptuj*, Beograd-Zagreb.

Knezović and Pintarić 2004 Ivan Knezović and Tatjana Pintarić, *Arheologija*, Exhibition Catalogue, Velika Gorica.

Koprivnjak 2011a Vedran Koprivnjak, Dio nalaza rimskog novca s lokaliteta Sisak – Lučka kapetanija, *GGMS* 10, 25-61.

Koprivnjak 2011b Vedran Koprivnjak, Antički numizmatički nalazi s lokaliteta Sisak – Dunavski Lloyd, *GGMS* 10, 63-94.

Kos 1986 Peter Kos, *The monetary circulation in the southeastern Alpine region ca. 300 BC – AD 1000*, *Situla* 24.

Kos 1988 Peter Kos, *Die Fundmünzen der römischen Zeit in Slowenien* I, II, Berlin.

Kos 1997 Peter Kos, Interpretacija (antičnih) novčanih najdb. Metodologija - njene možnosti in pasti, *AV* 48, 97-115.

Kos 1988 Peter Kos, *Die Fundmünzen der Römische Zeit in Slowenien* I-II, Berlin.

Kos 1998 Peter Kos, *Leksikon antičke numizmatike*, Zagreb.

Lukić 1989 Hermine Lukić, Nalazi rimskog novca u donjem gradu u Osijeku (Zaštitna arheološka istraživanja 1982-1986), *OZ* 20, 53- 82.

Ljubić 1879 Šime Ljubić, Razne viesti, *VHAD* 1, 24-29, 88-96.

Ljubić 1880a Šime Ljubić, Arkeologičke crtice iz mog putovanja po njekojih predjelih Podravine i Zagorja god. 1879., *VHAD* 2/4, 110-120.

Ljubić 1880b Šime Ljubić, Razne viesti - Odkriće rimskih novaca u Bosanskom Brodu, *VHAD* 2/4, 123-128.

Ljubić 1883 Šime Ljubić, Dopisi, razne viesti - Još jedno odkriće rimskih novaca i drugih predmeta kod Broda, VHAD 5, 94-95.

Mažuran 1954 Ivo Mažuran, Rimski novci nađeni u Osijeku (1949 -1954), OZ 4, 32-40.

Mirnik 1969 Ivan Mirnik, *Katalog dijela numizmatičke zbirke Gradskog muzeja u Sisku*, Unpublished diploma thesis, Faculty of Philosophy, Zagreb.

Mirnik 1972 Ivan Mirnik, O numizmatičkoj zbirci Gradskog muzeja u Sisku, VMKH 21/1, 28-31.

Mirnik 1979 Ivan Mirnik, Novčane ostave požeške kotline, *Vjesnik Muzeja Požeške kotline* 2-3, Slavonska Požega, 83-91.

Mirnik 1981 Ivan Antun Mirnik, *Coin Hoards in Yugoslavia*, *BAR Intern Ser.* 95.

Minik 2009a Ivan Minik, Maksimijanov aureus iz skupnog nalaza u Petrijancu 1805. god., *ArchAdr* 2/2, 433-452.

Mirnik 2009b Ivan Mirnik, Tragovi nalaza rimskog carskog novca iz Slavonije, *VAMZ* 42, 519-524.

Noe 1925 Sydney P. Noe, *A Bibliography of Greek Coin Hoards*, *Numismatic Notes and Monographs* 25, New York.

Patsch 1902 Karlo Patsch, Nahogjaji novaca, *GZMS* 14, 391-438.

Pegan 1967 Efrem Pegan, Najdbe novcev v Sloveniji, *AV* 18, 203-222.

Pegan 1968 Efrem Pegan, Najdbe novcev v Sloveniji, *AV* 20, 257-264.

Petrović 1955 Jozo Petrović, Rimski novac iz Obudovca, *GZMS* 10, 181-197.

Pintarić 2006 Tatjana Pintarić, *Pronađeno vrijeme* (Exhibition Catalogue), Velika Gorica.

Sergejevski 1930 Dimitrije Sergejevski, Nalazi rimskog novca kod Livna, *GZMS* 42/2, 119-132.

Skrabar 1906 Viktor Skrabar, Fund römischer Denare in Unter – haidin, *Mitteilungen der K.K. Central-Commision zur Erforschung und Erhaltung der Kunst- und Historischen Denkmale* 5/1-2, Wien, 195-196.

Spajić 1967 Emil Spajić, Kasnorimska nekropola u Sladojevcima, *OZ* 11, 101-132.

Spajić 1975 Emil Spajić, Novci kovnice Viminacijum nađeni u Osijeku *OZ* 14-15 (1973-1975), 173-220.

Sutherland 1976 Carol H.V. Sutherland, *The Emperor and the Coinage, Julio-Claudian Studies*, London.

Šipuš 1995 Nikola Šipuš, Skupni nalaz sitnih rimskih brončanih novaca iz Siska, *NV* 48, 46-51.

Šiša-Vivek 2006 Marija Šiša-Vivek, Lokalitet Petrijanec - Ulica V. Nazora 98, *HAG* 2 (2005), 132-133.

Šiša-Vivek, Leleković, and Kalafatić 2005 Marija Šiša-Vivek, Tino Leleković, and Hrvoje Kalafatić, Ostava rimskog novca i srebrnog posuđa iz Petrijanca, *OpArch* 29, 231-245.

Težak 2010 Spomenka Težak, Skupni nalaz antoninijana iz Zbelave, *VAMZ* 43, 531–544.

Vasić 1967 Miloje Vasić, Ostava iz Jablanice, *Starinar* 18, 63-82.

Vikić-Belančić 1981 Branka Vikić-Belančić, Etape urbanog razvitka Andautonije i antičko nasljeđe Zagreba, in: *Arheološka istraživanja u Zagrebu i njegovoj okolici, IzdHAD* 6, Zagreb, 129-154.

Zmajić 1955 Bartol Zmajić, Nalaz rimskih antoninijana u Imbriovcu kod Koprivnice, *NV* 5, 16-17.

Early Mediaeval Barbarian Elements in Late Antique Southern Pannonia

Anita Rapan Papeša

1. Introduction

The period characterised by the coexistence of Late Antiquity and the beginning of the Early Middle Ages, in its turn distinguished by migrations of Barbarian tribes, is one of the most intriguing epochs in the history of Europe. Cultural production and social organisation of the 5th and 6th centuries left their strong mark on this period, but also on Europe's modern outlook. The cohabitation of late antique Romanised population and the newly arrived Barbarian peoples is a cultural manifestation noted throughout the Carpathian basin. Excavations in northern Croatia so far have not revealed any specific elements characterising this period and the situation in neighbouring countries is only somewhat clearer. Since the context for most of the finds is unknown, the spatial and temporal distribution of particular Barbarian peoples in the territory of modern-day Croatia can be defined only in combination with historical literary sources. Nevertheless, even this poor state of research allows the interpretation of the data as suggesting a high assimilation rate of Barbarian peoples by the late Roman cultural frame.

In terms of geography the southern part of Pannonian basin belongs to the sub-Pannonian area. Highland-, foothill- and lowland-areas alternate from west to east. Foothills, indented with smaller fluvial basins that gradually descend into larger river valleys forming large plains in the extreme east, continue towards the east onto the relatively low hills in the west. In terms of geography the territory in question is divided into Central and Eastern Croatia.[1] Historically, during Late Antiquity this area was divided between the provinces of Pannonia Savia, with its centre in *Siscia* (Sisak), and Pannonia Secunda,[2] with its centre in *Sirmium* (Sremska Mitrovica). The former encompassed parts of modern-day Croatia, Bosnia and Herzegovina and Slovenia; besides its metropolis, more important towns in the Croatian part of the province were *Aquae Balissae* (Daruvar), *Iovia* (Ludbreg), *Andautonia* (Ščitarjevo) and *Aquae Iasae* (Varaždinske Toplice). The latter encompassed parts of modern-day Croatia, Bosnia and Herzegovina and Serbia; more important towns in the Croatian part of the province were *Mursa* (Osijek) and *Cibalae* (Vinkovci). Along with a number of lesser settlements[3] (*Marsonia* – Slavonski Brod, *Certissia* – Štrbinci), as well as frontier camps[4] (Batina – *Ad Militare*, Zmajevac – *Ad Novas*, Dalj – *Teutoburgium*, Sotin – *Cornacum* and Ilok – *Cuccium*), they formed the frame (fig. 1) encountered by the newly arrived peoples upon entering Pannonia. This paper will discuss the part of Pannonia Savia from the western border of modern-day Croatia to the western limit of Pannonia Secunda (along the line stretching from Donji Miholjac via the eastern foothills of Krndija and Dilj to the confluence of the Vrbas and Sava). With regards to Pannonia Secunda, this paper will discuss the area stretching from the eastern border of Pannonia Savia to the eastern border of modern-day Croatia.

The territory of these provinces was already in the Imperial period the scene of confrontations among various emperors and usurpers. From the late 4th century onwards a number of primarily Germanic peoples appeared in this area (fig. 2), partly arriving forcefully, through incursions and plundering, and partly peacefully, settling in the area as Roman allies. All this led to a demise of Roman power in 5th-century Pannonia, which became only symbolical, while the real control was exercised by the newly arrived peoples.[5] The chronology of this period is rather well defined through a combination of literary sources and archaeological evidence. In this paper the Migration period will be limited by AD 568 when the Lombards and other Germanic peoples left Pannonia for Italy,[6] given that the Germanic element no longer played any significant role in the following period in the Croatian part of Pannonia. The phase usually styled the Migration period by archaeologists and historians partially overlaps with Late Antiquity; two parallel processes were simultaneously in progress in this area: the demise of the old and the birth of the new cultural and historical landscape. The impulse for this process came from the East, when the Huns crossed to the right side of the frozen River Don in 375 AD. Their invasion gave the initial impulse that triggered migrations of various Barbarian tribes towards the south-west and the Roman Empire. The Western Goths (Visigoths) were the first to seek shelter

1 *VAH*, 225, 255.

2 The so-called Verona List (*Laterculus Veronensis* 6.2-3, 5-6, in: *Not. Dign. occ.*; 314-324 AD) enumerates the provinces *Saviensis*, *Valeria* and *Pannonia Inferior* (encompassing Pannoniae Secunda and Superior). The last two received their names in the mid-4th century since the *Short review of the history of Roman people* by Festus (ca. 370 AD) mentions the province of Pannonia Secunda. The name of *Panonnia Sirminesis* (Cassiod. *Var.* 3.23.2, 4.13.1) appeared during the second phase of the Ostrogothic rule, when Pannonia Savia was united with Dalmatia. I would like to thank Hrvoje Gračanin for clarifying this matter for me.

3 Vulić 2009, 121-122.

4 Sanader 2009, 107.

5 Andrić 2002, 131.

6 Paul *HL* 2.26.

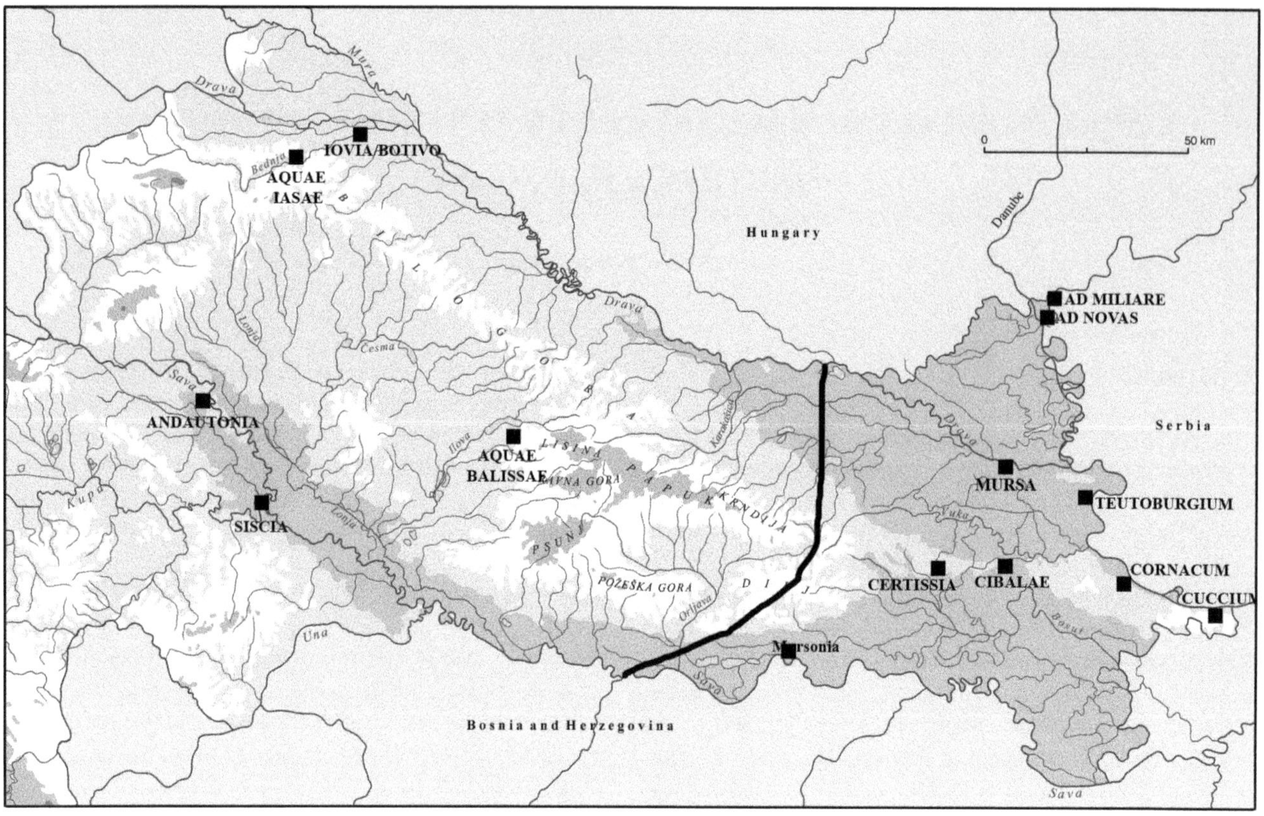

Fig. 1. Distribution of late antique provincial sites and finds (made by A. Rapan Papeša).

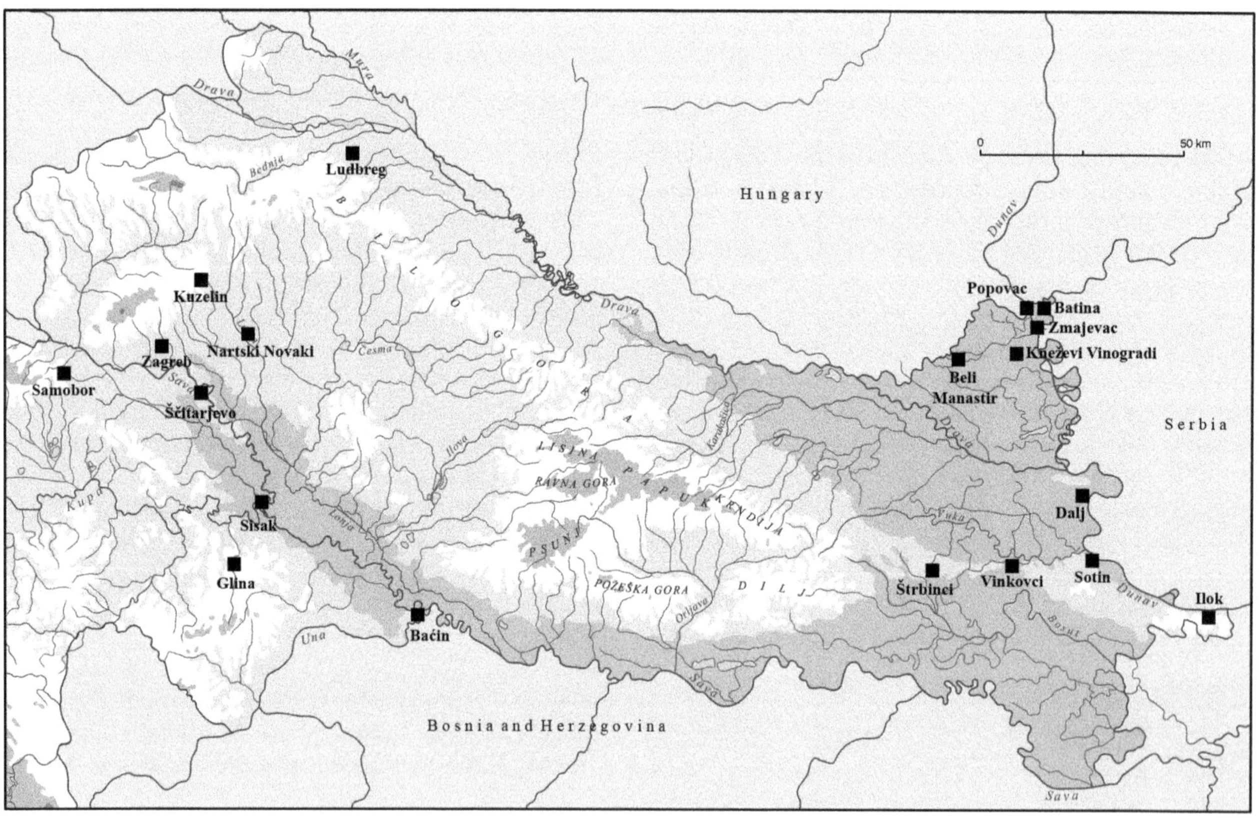

Fig. 2. Distribution of Barbarian finds and sites (made by A. Rapan Papeša).

from the Huns within the borders of the Eastern Roman Empire in 376 and they settled on Roman territory as allies (*foederati*), with permission from the emperor Valens, born in Cibalae. He paid his decision costly since he was killed during the battle with the insurgent Goths at Hadrianopolis in August of 378. The Visigoths had continued to roam and plunder Pannonia until the early 5th century, when they departed towards Italy.[7] The first literary testimony[8] of Germanic presence in Lower Pannonia describes the settling of a larger Barbarian group consisting of the Huns, Ostrogoths and Alani. This happened during the reign of the emperor Gratian (375-383), who settled them as allies north of the Drava in 380.[9] In the early 380, during a Gothic incursion, the destruction of Mursa and Stridon is explicitly mentioned.[10] In the vicinity of Sisak in Pannonia Savia a battle took place in the summer of 388 between the emperor Theodosius (379-395) and the usurper Magnus Maximus (383-388), involving *foederati*. On this occasion, or perhaps before the decisive battle near Poetovio (Ptuj, Slovenia), the fortress at Kuzelin and Aquae Iasae were also devastated.[11] Some years later the Western Roman Empire renounced all its rights on Eastern Illyricum, together with Pannonia, through the engagements and subsequent marriage of the emperor Valentinian III (424-455) with Eudoxia, the daughter of the Eastern Roman emperor Theodosius II, in 437.[12] The large and powerful Hun state under Atilla (406-453) most probably did not control either Pannonia Savia or the western part of Pannonia Secunda, but extended only to Syrmia (Srijem) towards the west.[13] Not long after Atilla's death, Germanic tribes, among which the Gepids and Ostrogoths were the most important, broke the power of the Hun state in the battle on the Nedao River in 454 and they further forced the Huns into retreat towards the Black Sea.[14] The Ostrogoths exploited their moment of power by occupying Pannonia,[15] more precisely, a smaller part of Pannonia Savia and the majority of the territory of Pannonia Secunda.[16] Between 454 and 473 they ruled this area until, driven by the lack of provisions, they split into two groups; one group departed towards Italy and the other towards the Eastern Empire.[17] The vacant territory of Pannonia Secunda was occupied by the Gepids (473-504), while Pannonia Savia was controlled by the Suevi, who were pressed from the north by the Ostrogoths.[18] A group of the Ostrogoths that had crossed the border of the Eastern Empire in 473, led by Theoderich, departed in the summer of 488 towards Italy, on the initiative of the emperor Zeno, in order to occupy the land.[19] They have spent the winter in the area of Sirmium and continued their movement in February of 489, confronting the Gepids, who controlled this territory, in the swamps of the Vuka River between Mursa and Cibalae. The whole-day battle ended in victory of the Goths, who even managed to seize the supplies wagons of the Gepids.[20] The Goths continued along the Drava road and arrived in northern Italy in August of 489. Theoderich assumed the complete power in Italy in 494.[21] The Ostrogoths returned to Pannonia in 504, after the situation in Italy had stabilised, and they formed the province of Pannonia Sirmiensis, a legal successor of Pannonia Secunda. A smaller part of the province was ceded to the Eastern Empire.[22] Pannonia Savia was somewhat earlier (ca. 497) united with Dalmatia into a province governed by a *comes* of Dalmatia and Pannonia Savia from the centre in Salona.[23] But with the crowning of the emperor Justinian and his well-known policy of the so-called reconquest, they were forced to withdraw all their available forces back to Italy.[24] Information on Pannonia Sirmiensis during Gothic rule are found in three decrees addressed in 509-510 by Theodeorics to the *comes* Colossus. Two of the decrees deal with military arrangements, that is, issues of defence and army supplies. The most interesting, however, is the decree addressing *all the Barbarians and Romans in Pannonia*, because it indirectly testifies to the survival of the autochthonous population.[25] Documents referring to Pannonia Savia are also preserved, in which Theoderics sends his emissaries Fridibad (507-511) and Severinus (525-526), primarily to solve the issues of property.[26] The Suevi remained in Pannonia Savia until the Lombards arrived in this territory in 547 with the allowance of the emperor.[27] The repeated withdrawal of the Ostrogoths was once more exploited by the Gepids, who returned to Pannonia Sirmiensis and expanded their rule in comparison with their initial phase of dominance in this area.[28] The renewal of the kingdom of the Gepids did not conform to the intentions of the Eastern Empire, hoping to recover the territories it had lost a long time ago, and it renewed in 547 the alliance with the Lombards, established in 526.[29] The period characterized by the alternation of war and peace between the Lombards and Gepids lasted from 549 to 567, when the last Gepid ruler Cunimund was slain in battle; the Eastern Empire interfered in these conflicts according to its needs. This is the reason why the Lombards turned to the Avars for assistance; the latter agreed to help them, demanding the territories controlled by the Gepids in return.[30] The Germanic period in the history of southern Pannonia ended with the departure of the Lombards and other Germanic peoples, but also of a part of autochthonous population, for Italy in April of 568.[31] With the establishment of the great Avar state – khanate – in the territories discussed in this paper, a direct Roman and Germanic influence subsided. Roman influence continued

7 Gračanin 2006, 83-89.
8 Jord. *Get.* 141-142; Zos. 4.34.2.
9 Gračanin 2005, 14-15; Gračanin 2006, 85.
10 Gračanin 2005, 13.
11 Gračanin 2005, 16-17; Gračanin 2006, 37-38.
12 Gračanin 2005, 28-31; Gračanin 2006, 54-58.
13 Gračanin 2005, 32, n. 61.
14 Andrić 2002, 133.
15 Jord. *Get.* 264.
16 Gračanin 2006, 95.
17 Andrić 2002, 140.
18 Gračanin 2006, 100.
19 Gračanin 2006, 104.
20 Andrić 2003, 297.
21 Gračanin 2006, 108.
22 Gračanin 2006, 108-112.
23 Andrić 2002, 145.
24 Gračanin 2007, 27.
25 Andrić 2003, 308.
26 Andrić 2002, 147; Gračanin 2006, 111.
27 Gračanin 2007, 37.
28 Gračanin 2007, 30.
29 Gračanin 2007, 37.
30 Gračanin 2007, 40-44.
31 Andrić 2002, 163.

to arrive from the Eastern Empire – Byzantium, while the Germani are recognizable in terms of archaeology through their jewellery and weapons, even though they were assimilated into the conglomerate of ethnic groups ruled by the Avars.[32]

2. A review of previous research

The majority of evidence of Barbarian provenance from the territory in question are chance finds, and, therefore, cannot be precisely culturally affiliated, since the differences, for instance, between the Gothic and Gepid material are truly minor. More often than not historical literary sources are used in the interpretation of archaeological finds in order to affiliate them to a particular Germanic people.[33] Recently, the question of ethnicity of the owners of archaeological finds is being put forward in their interpretation, because the tendency is, especially in scholarship outside Croatia, not to speak of the ethnicity of the owners but of the affiliation of the object itself to a particular cultural group.[34] Besides the objects that ended up in museums without any knowledge of the circumstances of their acquisition, there are other categories of finds: those acquired through survey/reconnaissance and those acquired in rescue excavations. Systematic archaeological excavations in Croatia, however, of the sites belonging to this period were never conducted, with the exception of Kuzelin[35] and Lobor,[36] where late Roman and early mediaeval layers were excavated.

The most substantial category of objects is that belonging to non-circumstantial finds kept in the mediaeval collection of the AMZ. These are mostly artefacts originating from Sisak and, in a lesser extent, eastern Slavonia and western Syrmia. The surveys carried out on sites from which individual objects of Germanic provenance originated were primarily conducted in order to clarify the circumstances of these finds. The best example of such practice is the site of Zmajevac in the Baranja: fittings and a vessel were found there in 1932 and brought to the AMZ, while the 1954 survey supplemented this find with a *solidus* of Theodosius II that, along with other similar evidence from other sites, helped with dating the original 1932 find.[37] Well documented archaeological excavations were conducted in Ludbreg (in 1952 and 1976),[38] Ozalj (in 1992),[39] Kuzelin (from 1981 onwards),[40] and Vinkovci (regularly from 1972 on).[41]

[32] As opposed to, for instance, some Hungarian cemeteries, the excavated cemeteries of the Avar period in Croatia do not show these characteristics.
[33] Cf. for instance Brunšmid 1905, 210.
[34] A critical review of the question can be found in Babić 2010, 137-145.
[35] Sokol 1998, 5-7.
[36] Early medieval Croatian cultural layer overlaps the Late Roman one, while no Germanic finds were discovered. Cf. Filipec 2008.
[37] Vinski 1957a, 36-38.
[38] Vikić 1973, 51; Vikić 1976, 83.
[39] Čučković 1992, 49-51.
[40] Sokol 2006, 155.
[41] Iskra-Janošić 1993.

3. Archaeological finds

3.1. Roman proincial culture (late 4th-end of 6th century)

Some of the finds kept in the medieval collection of the AMZ can be characterised as products of the late Roman provincial culture, with some of them bearing certain Byzantine traits. Since these objects arrived to the Museum as non-circumstantial finds, it is not easy to speculate on the ethnic background of their owners. The 5th- and 6th-century population was partly formed of Barbarian peoples and partly of the Romanised autochthonous population, while these finds can be affiliated with both groups. Therefore, they will be discussed as an introduction into the main subject. The following objects from the area of modern-day Sisak, kept in the AMZ, belong to this temporal and cultural frame: a cast bronze earring (fig. 3) with ends terminating in a polyhedron and a point respectively,[42] a gold earring (fig. 4) with a pendant in the form of a hollowed basket with a threaded pearl bead in the centre,[43] a thick silver wire torques (fig. 5),[44] three bronze pins (figs. 6, 7, and 8) with the polyhedral head,[45] two silver pins (figs. 9 and 10) with the polyhedral head,[46] two bronze pins (figs. 11 and 12) with the head in the form of a bird,[47] one bronze (fig. 13) (29 x 29 mm; 5th c.)[48] and one silver (fig. 14) brooch (42 x 21 mm; 5th c.)[49] in the form of a naturalistically rendered pigeon, two bronze brooches (figs. 15 and 16) in the form of a naturalistically rendered peacock (34 x 17 mm; 4th-5th c.),[50] a cast bronze brooch (fig. 17) in the form of a small horse (37 x 23 mm; 5th c.),[51] three cast bronze brooches (figs. 18, 19, and 20) with bent foot,[52] a cast bronze brooch (fig. 21) with equal arms,[53] two cast bronze brooches (figs. 22 and 23) in the form of a bird of prey (37x 25 mm; 6th c.),[54] six bronze buckle loops (figs. 24, 25, 26, 27, 28, and 29) of different forms,[55] one silver buckle loop (fig. 30),[56] two iron buckle loops (figs. 31 and 32) (30 x 15 mm; 5th c.),[57] an ornamental bronze belt mount (fig. 33),[58] seven bronze (figs. 34, 35, 36, 37, 38, 39, and 40)[59] and one iron buckle (fig. 41),[60] a bronze appliqué (fig. 42) in the form of a lioness,[61] four bronze bracelets, [62] and a lead pendant in the form of an openwork equal-armed cross

[42] Simoni 1989, 108 (1), 119 (1), pl. 1: 1.
[43] Simoni 1989, 108 (2), 119 (2), pl. 1: 2.
[44] Simoni 1989, 108 (3), 120 (3), pl 1: 3.
[45] Simoni 1989, 108 (6, 7), 120 (6, 7), pl. 1: 6, 7.
[46] Simoni 1989, 108(4, 5, 8 = 4, 5, 8), 120(4, 5, 8), pl. 1: 4, 5, 8.
[47] Simoni 1989, 109(10, 11), 120(10, 11), pl. 2: 2, 3.
[48] Simoni 1989, 109(15), 120(15), pl. 2: 7.
[49] Simoni 1989, 109(14), 120(14), pl. 2: 6.
[50] Simoni 1989, 110(16, 17), 120(16), 121(17), pl. 2: 8, 9.
[51] Vinski 1974, 17, pl. IX,2; Simoni 1989, 110(18), 121(18), pl. 2: 10.
[52] Simoni 1989, 110(20), 112(22,23), 121(20,22,23), pl. 3: 1, 3, 4.
[53] Simoni 1989, 111(21), 121(21), pl. 3: 2.
[54] Simoni 1989,114(33, 34), 122(33, 34), pl. 4: 5, 6.
[55] Simoni 1989:114(36,37,38,40), 115(41,42), 122(36,37,38), 123(40,41,42), pl. 5: 2, 3, 4, 6, 7, 8.
[56] Simoni 1989, 114(35), 122(35), Pl. 5.1.
[57] Simoni 1989, 114(39), 115(44), 123(39, 44), pl. 5: 5, 10.
[58] Simoni 1989, 115(46), 123(46), pl. 5: 12.
[59] Simoni 1989, 115(47), 116(48, 49, 51, 52, 53), 117(54), 124(52, 53, 54), pl. 6.1, 2, 3, 5, 6, 7, 8.
[60] Simoni 1989, 116(50), 123(50), pl. 6: 4.
[61] Simoni 1989, 117(59), 124(59), pl. 6: 13.
[62] Simoni 1989, 110(16, 17), 120(16), 121(17), pl. 2: 8, 9.

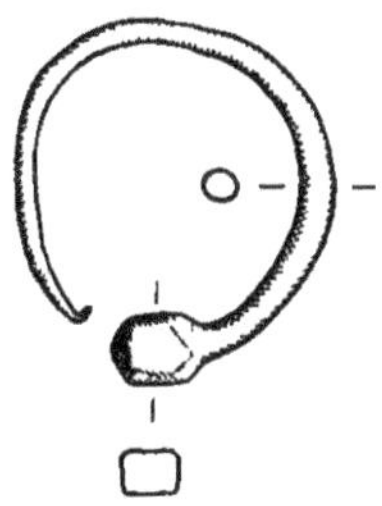

Fig. 3. Earring

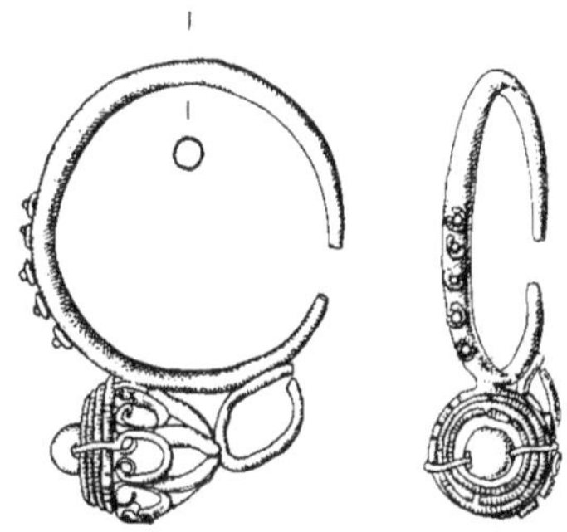

Fig. 4. Earring

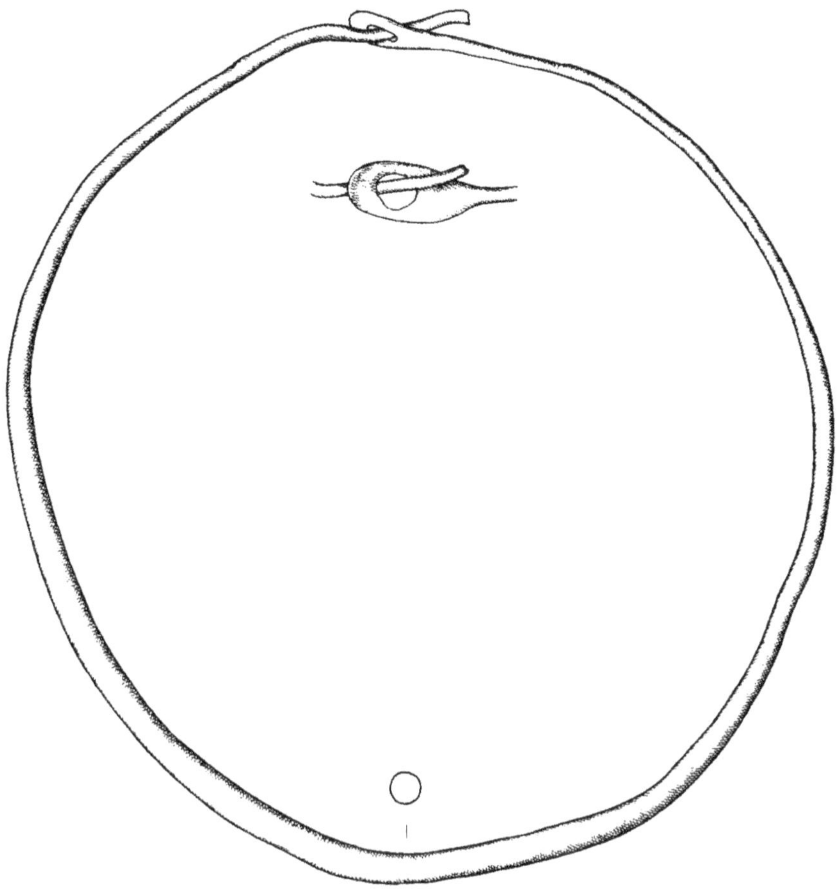

Fig. 5. Torques

Fig. 6. Hairpin

Fig. 7. Hairpin

Fig. 8. Hairpin

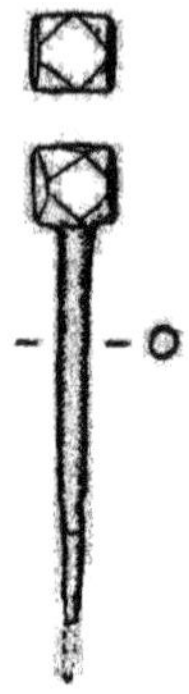

Fig. 9. Hairpin

Fig. 10. Hairpin

Fig. 11. Hairpin

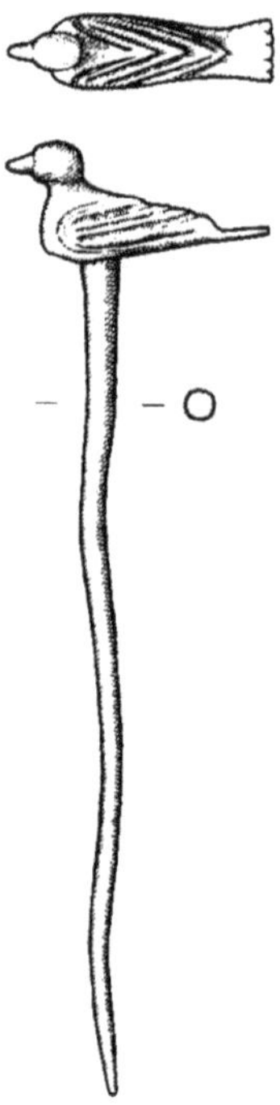

Fig. 12. Hairpin

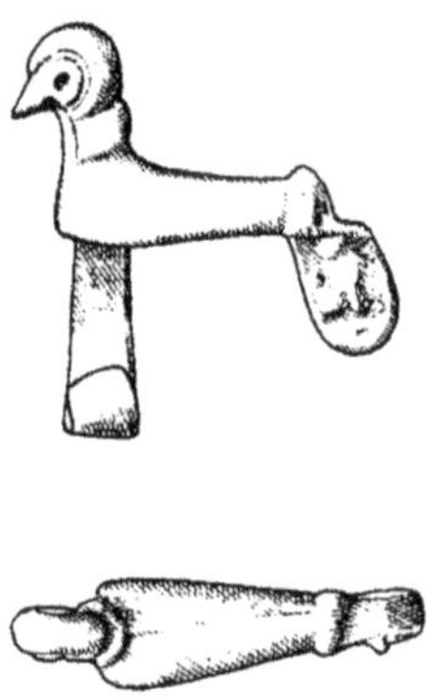

Fig. 13. Brooch

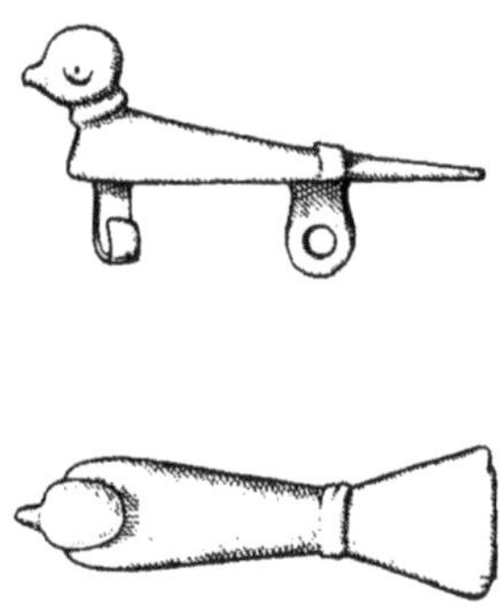

Fig. 14. Brooch

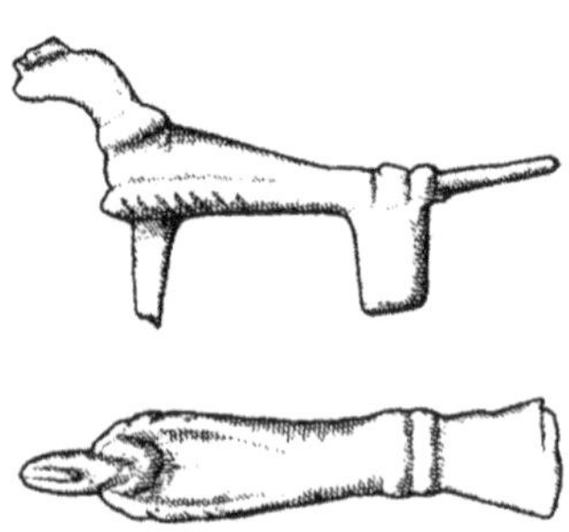

Fig. 15. Brooch

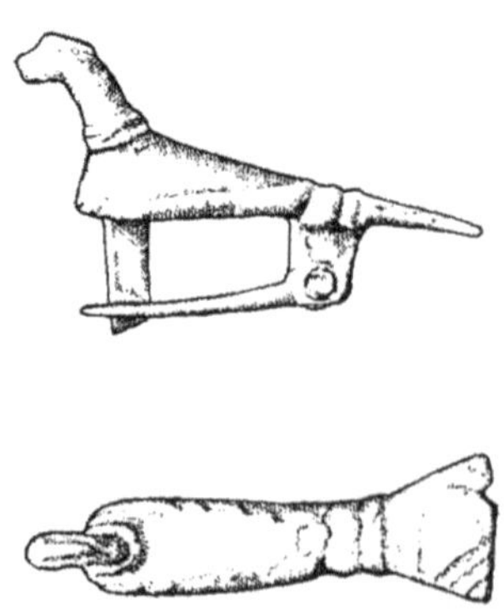

Fig. 16. Brooch

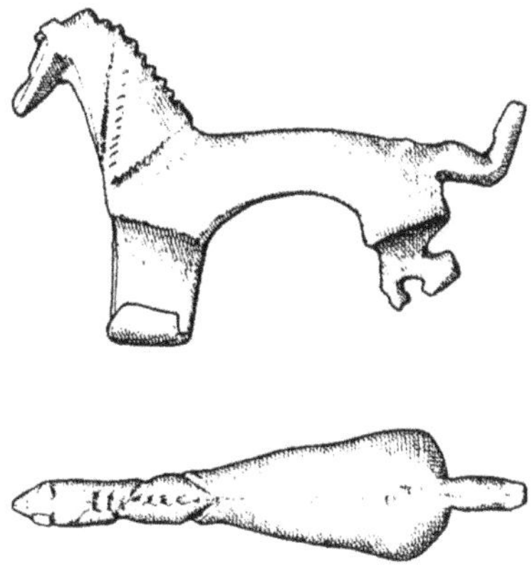

Fig. 17. Brooch

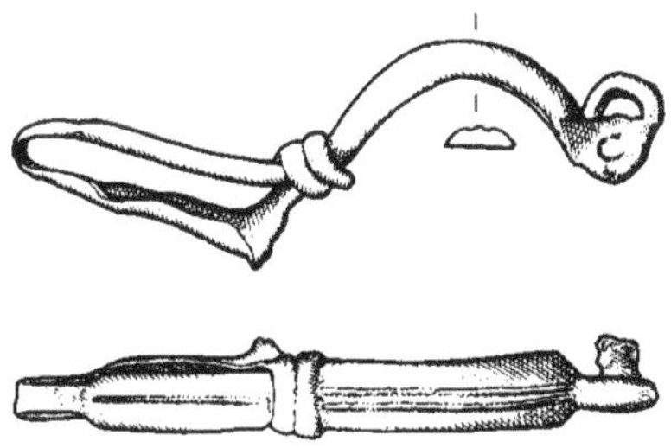

Fig. 18. Brooch

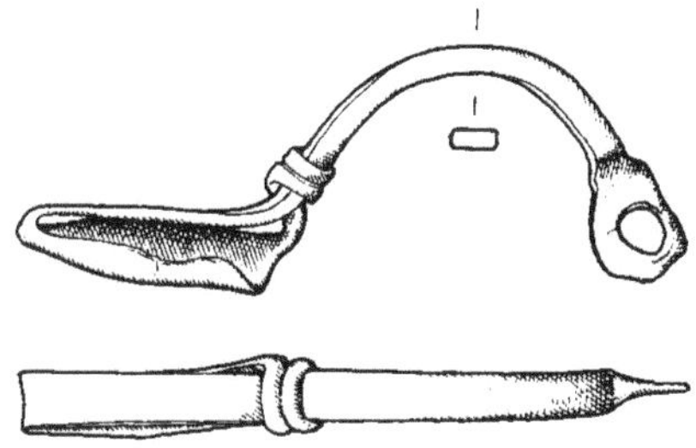

Fig. 19. Brooch

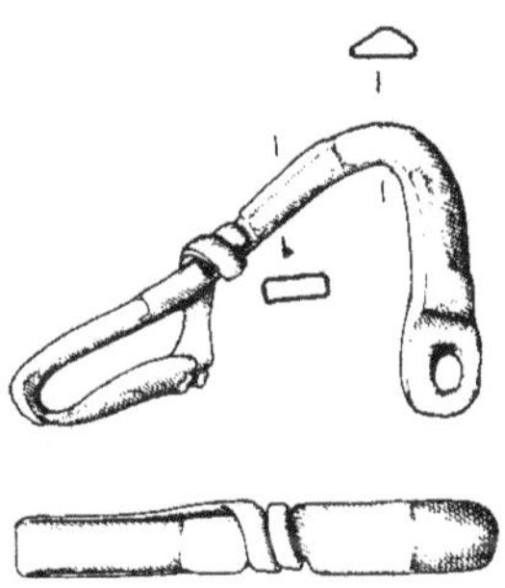

Fig. 20 Brooch

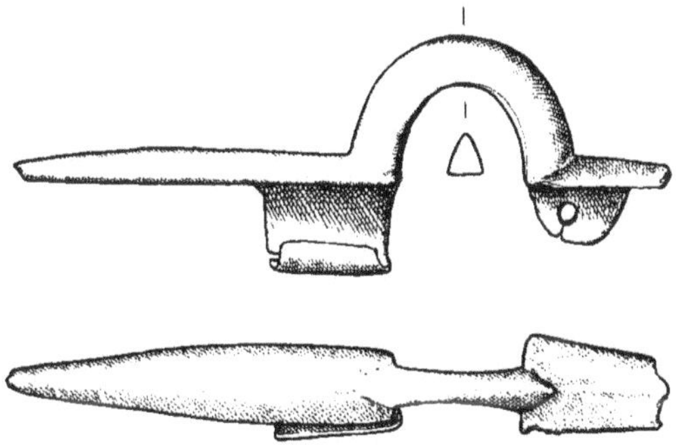

Fig. 21. Brooch

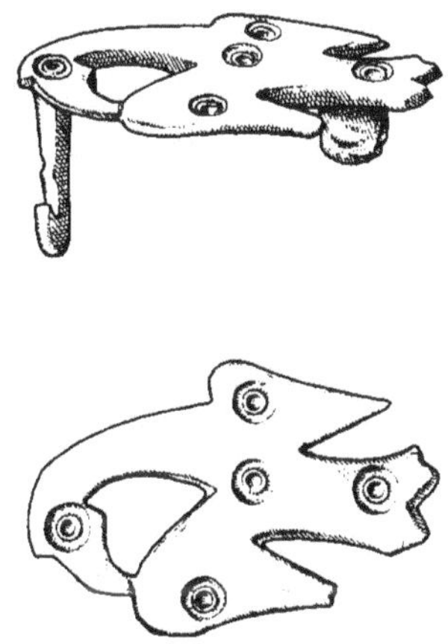

Fig. 22. Brooch

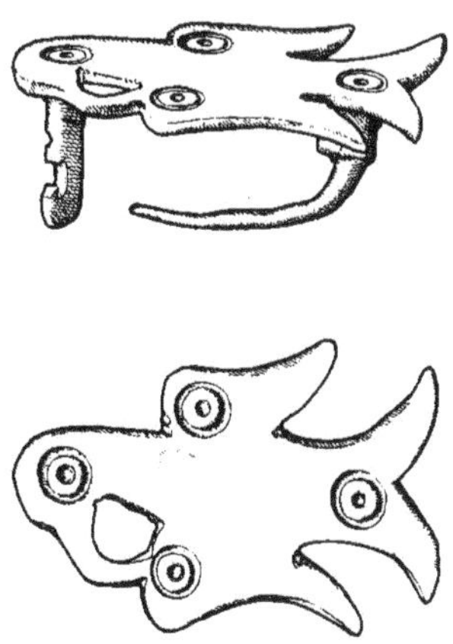

Fig. 23. Brooch

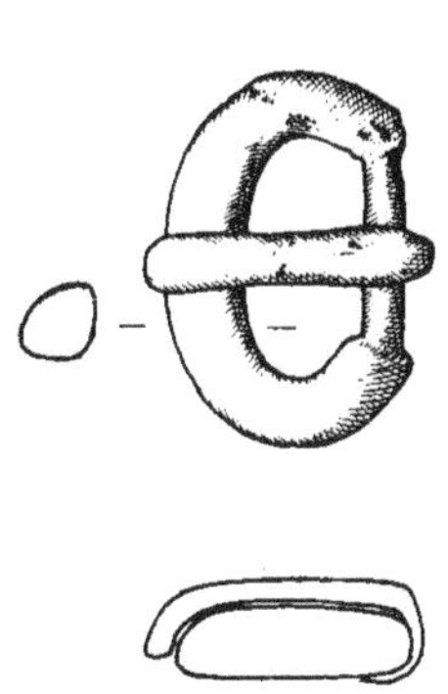

Fig. 24. Buckle loop

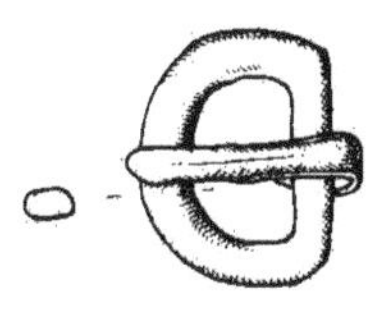

Fig. 25. Buckle loop

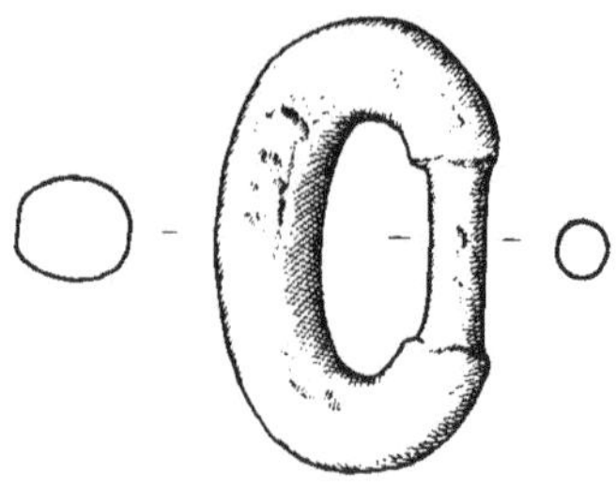

Fig. 26. Buckle loop

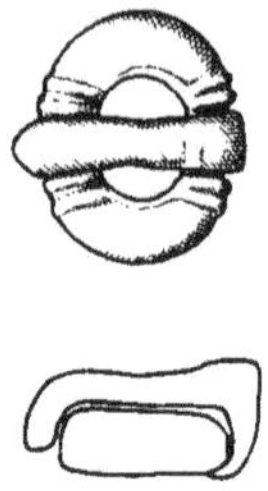

Fig. 27. Buckle loop

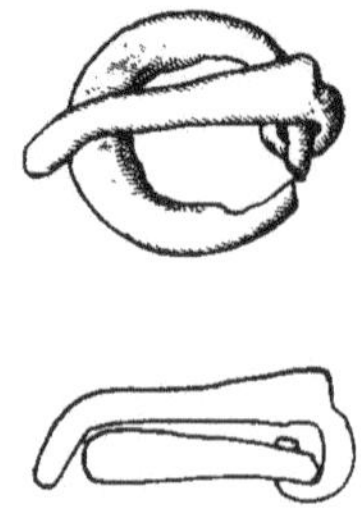

Fig. 28. Buckle loop

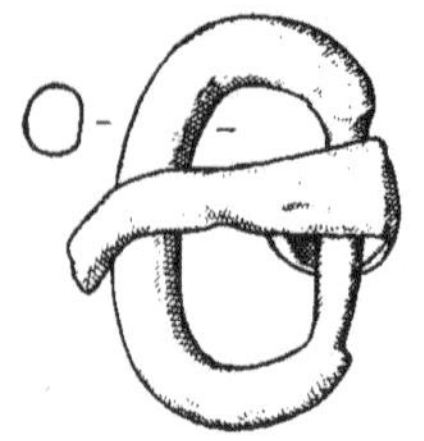

Fig. 29. Buckle loop

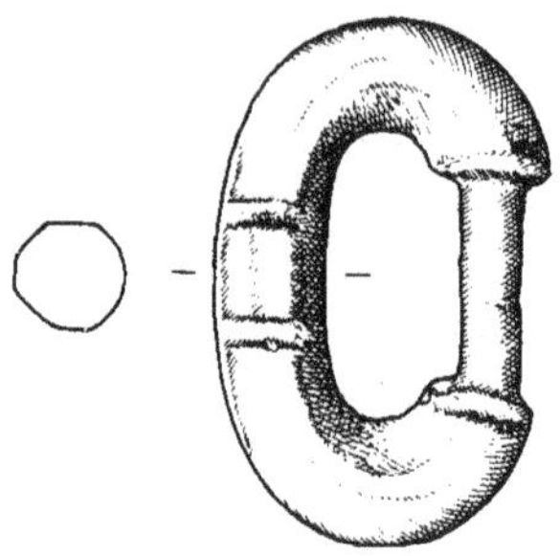

Fig. 30. Buckle loop

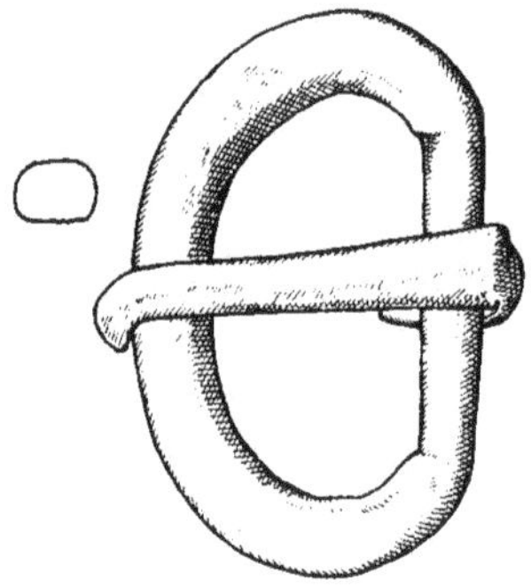

Fig. 31. Buckle loop

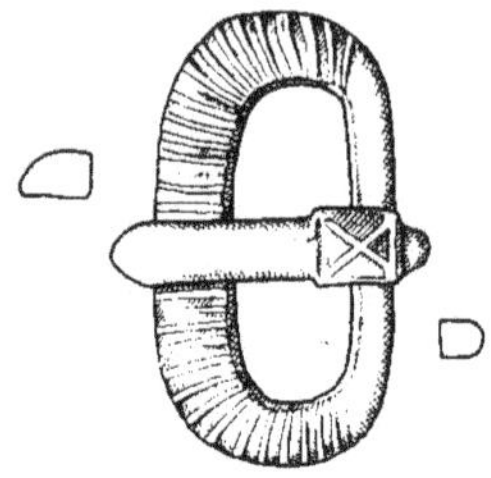

Fig. 32. Buckle loop

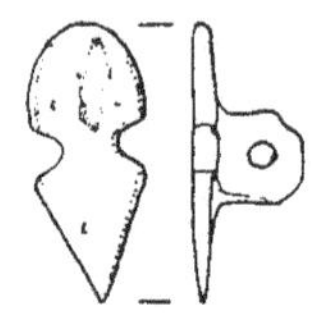

Fig. 33. Strap end

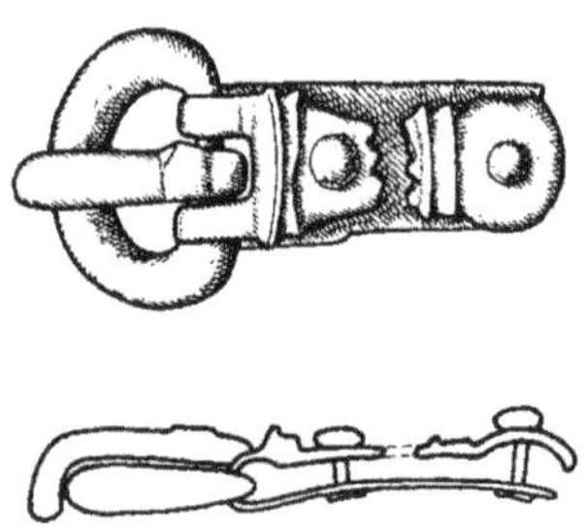

Fig. 34. Belt buckle

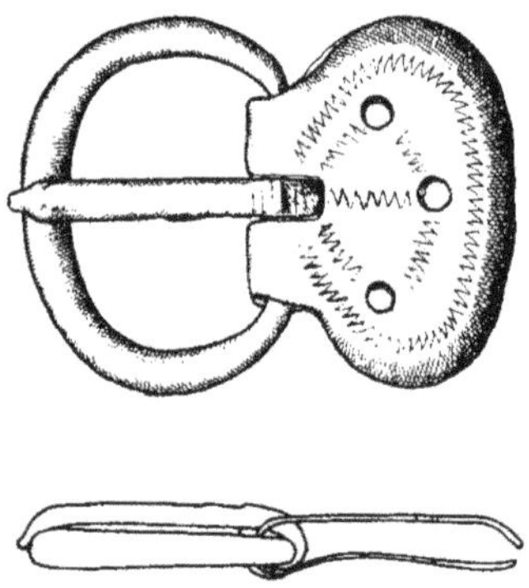

Fig. 35. Belt buckle

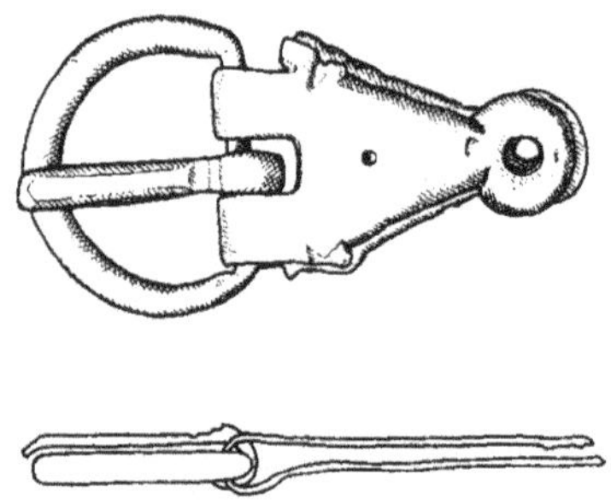

Fig. 36. Belt buckle

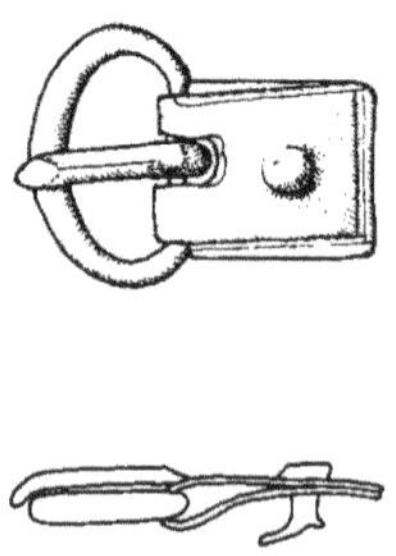

Fig. 37. Belt buckle

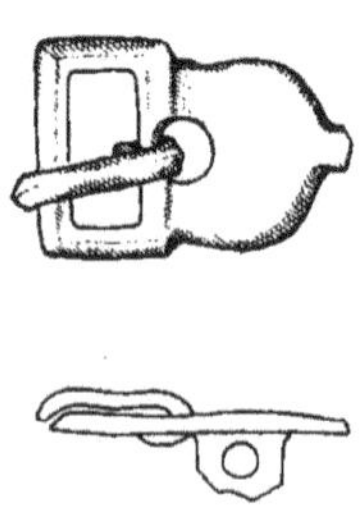

Fig. 38. Belt buckle

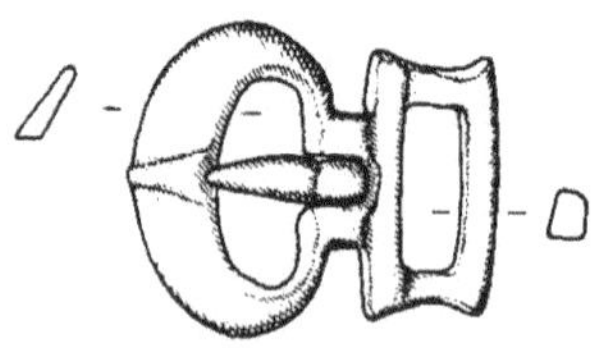

Fig. 39. Belt buckle

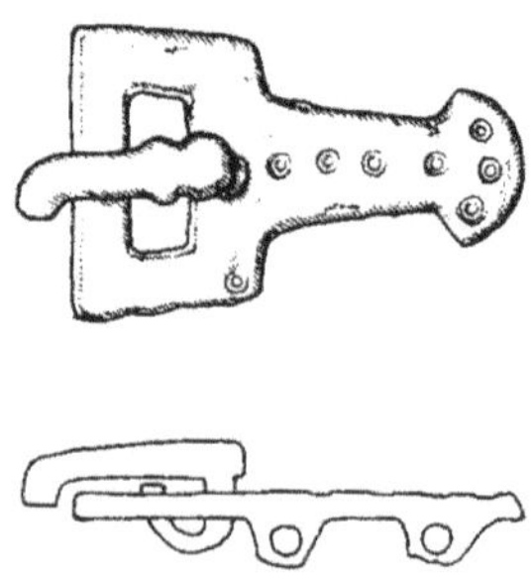

Fig. 40. Belt buckle

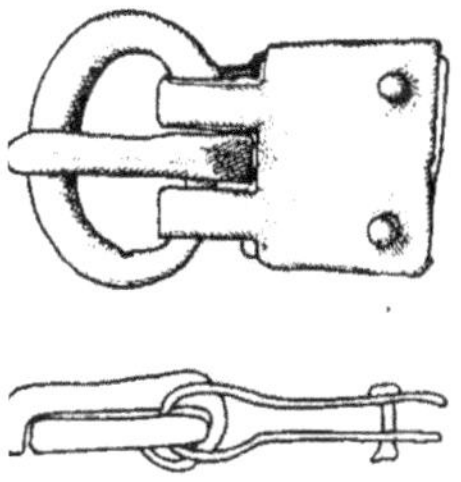

Fig. 40. Belt buckle

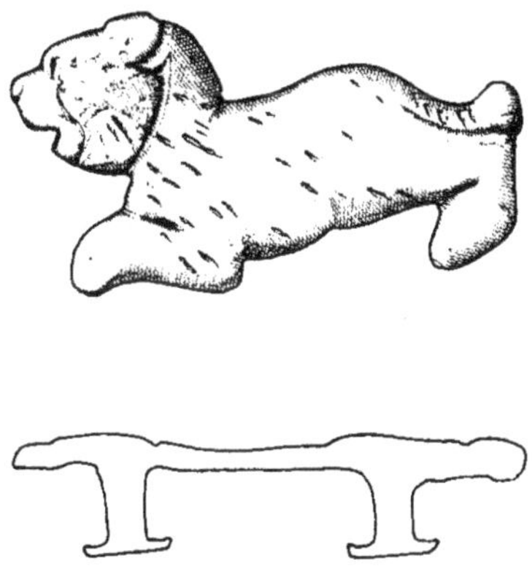

Fig. 42. Appliqué

Fig. 43. Shaft

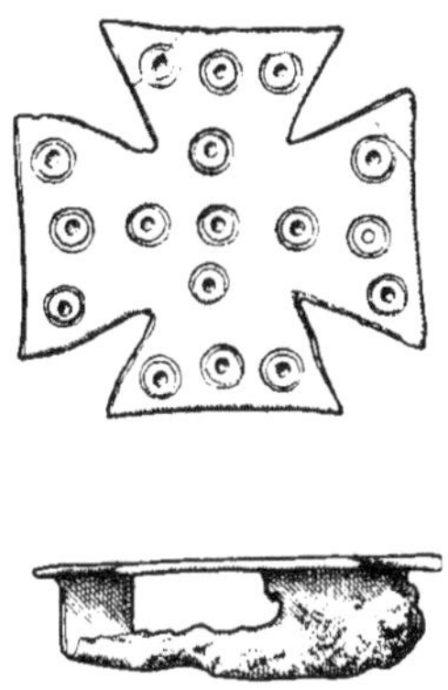

Fig. 44. Brooch

in a circular frame and a ribbed suspension loop (height: 28 mm, diameter: 22 mm; 4th-6th c.).[63] A bronze shaft with a collar-like moulding at the middle of its height, topped by a solid cast sitting bird and ending in a ring handle (fig. 43) also stems from Sisak (total length: 178 mm; 5th-6th c.; AMZ). It was interpreted as a Germanic cult object.[64] Arguably, it can be affiliated with the Roman provincial milieu, with analogies in numerous hairpins with pigeons' heads.[65] During rescue excavations in the old town of Ozalj, a silvered bronze brooch in the form of a Greek cross with expanded ends, decorated with dotted circles and provided with an iron pin, was found (6th c.; The Regional Museum Ozalj, Croatia).[66] It is a provincial Byzantine product. A number of pottery fragments were found at this site that can be similarly dated to the 6th century. Two chance finds of the same type stem from Sisak, both now in the AMZ: one bronze brooch[67]and a silver one (fig. 44) with a preserved iron pin with traces of fabric on the reverse side (30 x 30 mm; second half of the 6th c.).[68]

A rare example of a systematically excavated Germanic migration period site is the multi-layered site of Kuzelin near Zagreb. The fortified *castrum* functioned through four Roman phases, the third being dated to the end of the 4th and early 5th centuries and the fourth in the 6th century.[69] Several objects belong to this period and are all kept in the MP: rings (4 pieces), bracelets (3 pieces), arrows (13 pieces), a combat knife, a combat axe, 2 pins, a purse fitting, belt buckle parts – loops, mounts and tongues (9 pieces), as well as a number of construction elements (wedges). These finds were accompanied with coins struck from 365 to 423.[70] Also found at Kuzelin were a notched belt buckle, shallow ceramic vessels, and glasses that can be dated, by analogies, in the first quarter of the 5th century.[71] These objects belong to the Roman provincial material from the Germanic migration period.

3.2. Objects attributed to the Hun cultural circle (433-453)

Sporadic finds from Pannonia Savia that can be affiliated with material remains characteristic for the Hun period in Pannonia are mostly chance finds. Compared with literary sources, they arguably originated either in sporadic raids or through the settling of allied *foederati* groups in this territory. Since all this material comes from Sisak and its surroundings, one can argue for Siscia as a production centre for the jewellery made according to Barbarian

---

63 Sunce 1994, 89, cat. no. 68.
64 Simoni 1989, 119(71), 125(71), pl. 8: 1; Sunce 1994, 87, cat. no. 58.
65 Cf. for instance Brunšmid 1902, 144, sol. 69(1, 9).
66 Čučković 1992, 51; Sunce 1994, 114, cat. no. 147; Jurić 1998, 1093; Sokol 1998a, 1133, pl. 4:5; Čučković 2006, 204.
67 Vinski 1974, 22; Jurić 1998, 1092.
68 Simoni 1989, 113, 122, pl. 4: 4; Vinski 1974, 22, pl. XIII: 11; Vinski 1968, 107, pl. V: 8; Jurić 1998, 1093; Sunce 1994, 85, cat. no. 49.
69 Sokol 1998b, 8.
70 Sokol 1998b, 18-25.
71 Sokol 1998a, 1132, pl. 1: 1, 2, pl. 2: 1, 2.

patterns.[72] Considering the proximity of the centre of the Hun state, the objects from the period of Hun domination are somewhat more numerous in the territory of Pannonia Sirmiensis, when compared to Pannonia Savia. Almost all finds can be associated with the narrow belt of territory of the Danubian *limes*, and when further compared to literary sources, they reveal the association with the attacks on Roman fortifications along the *limes* or the sojourn of the groups of *foederati* in these fortifications.

Several chance finds of pieces of jewellery with the motif of cicada come from Sisak. This motif is characteristic for both the Hun and Ostrogothic periods, and can be properly affiliated only through its context. Zdenko Vinski believed these objects to be Ostrogothic, on the basis of other Ostrogothic finds from Sisak, but with the remark that some of them were made according to the late Roman tradition.[73] A cast silver hairpin (fig. 45) of a circular cross section with a head in the form of cicada with spread wings, ornamented with concentric circles, was found at an unknown site in the area of Sisak (total length: 93 mm; 5th c.; AMZ). The body of the cicada is triangular and separated from the head, on which the eyes are marked with concentric circles; the back side is smooth.[74] Another chance find from Sisak is a silver brooch (fig. 46) in the form of cicada (length: 22 mm; 5th c.; AMZ). The triangular body of the animal is separated from the head with two ribs, while the pin is missing.[75] A cast bronze ornamental fitting (fig. 47) also comes from Sisak. Its one end is made in the form of a cicada and another in the form of a semicircular numeral three; the back side is smooth (total length: 47 mm; 5th c.; AMZ).[76] A silver brooch (fig. 48) in the form of cicada with a mended bronze pin was found in the late Roman layer at Kuzelin; its head is separated off from the body and decorated with a triangle with a small incision (32 x 25 mm; 5th c.; MP).[77] A chance find of a cicada-formed appliqué (fig. 49) comes from the surroundings of Vinkovci (30 x 15 x 5 mm; first half of the 5th c.; The Ivan Vinkov Collection, Vinkovci, Croatia). The body of the cicada is marked with shallow incisions, its head is separated off from the body with a deeper incision, the wings are small and mildly expanded and the head is rendered in many details; a thrust is preserved on its lower side with which the appliqué was fastened to the base.[78] A bow brooch (fig. 50) made of struck silver wire (length: 83 mm; late 4th c.; MSO) is a chance find from Kneževi Vinogradi. Its degenerate crossbow on the head prolongs into an expanded and folded rhomboid foot, whose ending entwines around the bow; the foot has

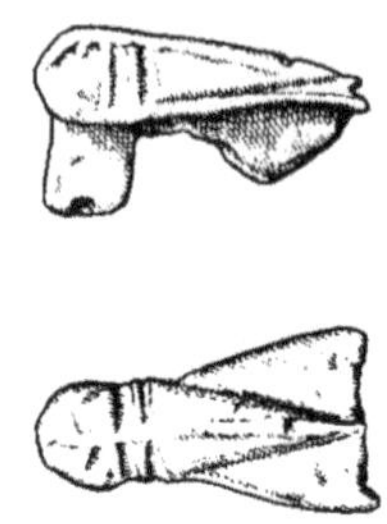

Fig. 46. Brooch

Fig. 45. Hairpin

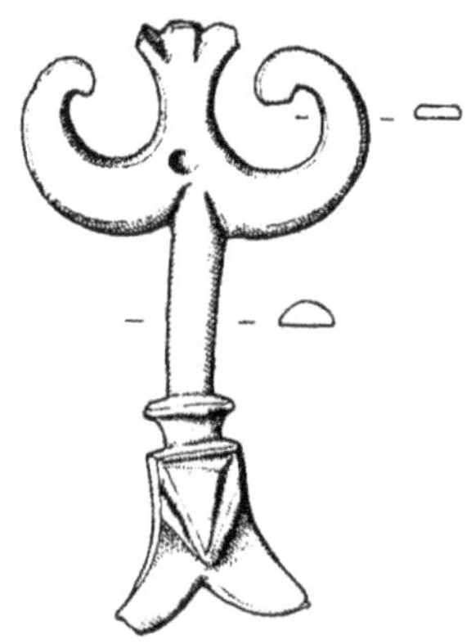

Fig. 47. Mount

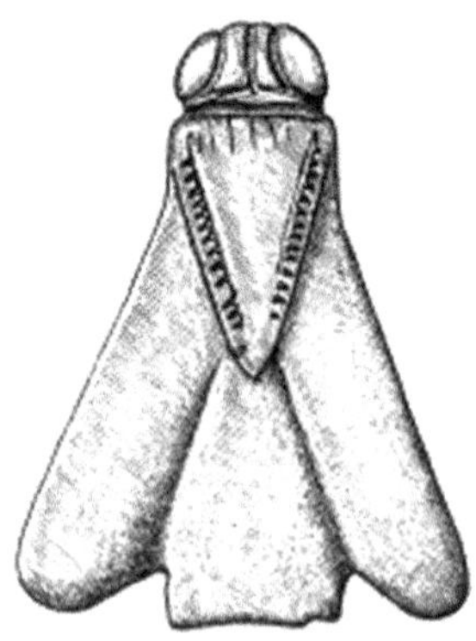

Fig. 48. Brooch

Fig. 49. Appliqué

[72] This information is based exclusively on the fact that necessary facilities were available in Siscia: workshops, artisans and technology that could have adjusted to demands and fashions of any period. In terms of archaeology, this suggestion remains unsubstantiated.

[73] Vinski 1957b, 156-157.

[74] Vinski 1957b, 136, Abb. 1; Simoni 1989, 109 (9), 120 (9), pl. 2: 1; Burkowsky 1999, 87.

[75] Vinski 1957b, 138, Abb. 52; Pülhorn 1987, 195, Abb. 4: 7.a; Simoni 1989, 109 (12), 120 (12), pl. 2: 4; Burkowsky 1999, 87.

[76] Vinski 1957b, 138, Abb. 2; Pülhorn 1987, 195, Abb. IV,7.b; Simoni 1989, 109 (13), 120 (13), pl. 2.5; Burkowsky 1999, 87.

[77] Sokol 1998b, 22, cat. no. 12.

[78] Rapan Papeša 2009, 136, cat. no. 433; Rapan Papeša 2010, 8, fig. 2.a, 2.b.

six symmetrical pellets.[79] A pair of brooches from Batina, made of hammered silver sheets, is similar to the above example: they have a degenerate crossbow on the head, irregular rhomboid feet rounded at the end and with a ridge along the middle; the pin fastener is missing (length: 70 mm each; 4th-5th c.; The Sümeg Museum, Hungary).[80] These brooches belong to the Sintana Mures-Tape phase, and their owners could have been the Visigoths of the pre-Hun period or the Gothic-Hun *foederati* settled in Pannonia in 380. The dredging of the Kupa at Sisak has always yielded numerous archaeological artefacts. Among others, six specimens of iron trilobite arrowheads (figs. 51, 52, 53, 54, 55, and 56) were recovered, conforming to the types of weapons used by both the Huns and Germans (5th c.; AMZ).[81] The best known archaeological finds from the period of the Hun rule in Pannonia are the fittings (fig. 57) from Zmajevac, made of thin fine gold (length: 49 mm; first half of the 5th c.; AMZ). They were discovered as a pair and have originally decorated a *spatha* scabbard as belt suspenders. Larger semicircular protuberances with six almandine inlays, inserted into cells and another one onto the loop (that is, 13 in total on one fitting), were soldered onto the somewhat narrower middle part. The almandine inlays are dark-red to light-purple in colour and smoothly whetted in the form of a triangle, rhomb and ovoid. The rims are decorated with pseudo-granulation as precisely hammered protuberances. The back side of the fitting is smooth and lined with a bronze plaque.[82] These objects are dated by an unused *solidus* (figs. 58a and 58b) of Theodosius II (408-450) struck in Constantinople in 443. The obverse bears the inscription DN THEODOSIUS PF AUG and the en face bust of the emperor, while the reverse bears the inscription XXXXII COS XVII PP CONOB and the representation of a seated helmeted figure holding a sceptre and a sword.[83] Together with the fittings and the coin, a handmade jug (fig. 59) with a handle (height: 85 mm), biconic body and kneaded rim, made of impure clay, was found.[84] The find as a whole is dated to ca. 400-453. Another ceramic biconic one-handled jug (figs. 60a and 60b) was discovered at Batina (height: 147 mm, rim diameter 73 mm; end of the 4th and first decades of the 5th c.; GMSo, Serbia).[85] Its rounded and spread rim prolongs into a concave skewed neck separated from the biconic receptacle. The handle is cylindrical, starting at the shoulder and ending at the mid-body, while the bottom is flat with an incised circular line. The jug is decorated with a polished ornament in the form of perpendicular and slanted lines on the neck and

Fig. 50. Brooch

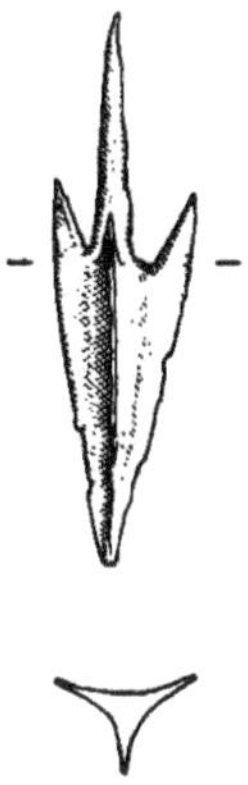

Fig. 51. Arrowhead

Fig. 52. Arrowhead

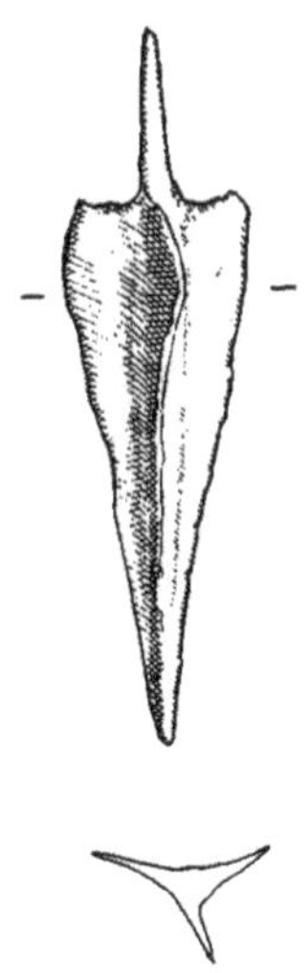

Fig. 53. Arrowhead

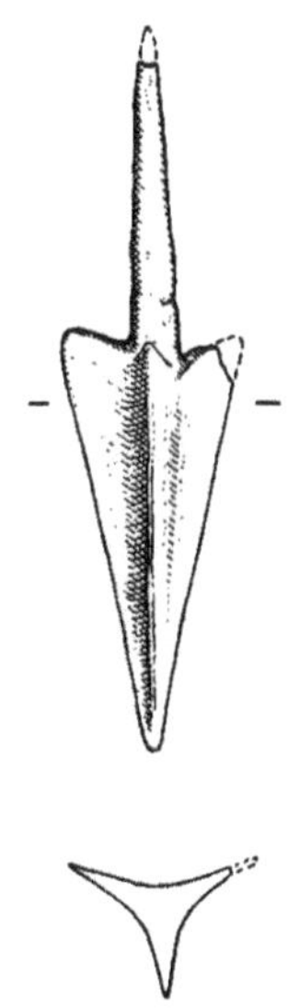

Fig. 54. Arrowhead

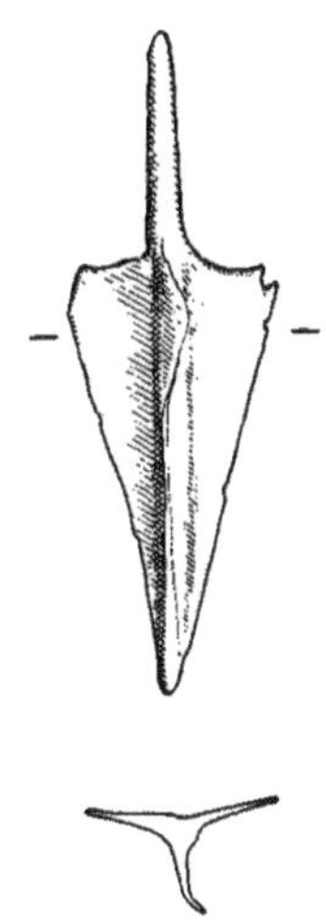

Fig. 55. Arrowhead

---

79 *Seoba*, 68; Mrkobrad 1980, 15, pl. IV,6; Bojčić 1984, 214; Radić 1997, 83, cat. no. 1; Radić 2009, 136, cat. no. 432.

80 Vinski 1957a, 32, pl. XXI, sl. 78, 79; Kovačević 1960, 41; *Seoba*, 67, fig. 1-2; Mrkobrad 1980, 15, pl. IV,5, 8; Bojčić 1984, 214.

81 Simoni 1989, 118(65-70), 125(65-70), pl. 7: 6-11.

82 Vinski 1957a, 38, pl. XXIV: 92a, 92b, 93a, 93b; *Seoba*, 69-70, pl. I: 1; Mrkobrad 1980, 18, pl. V: 11, 12; Bojčić 1984, 214; Simoni 1993, 179, cat. no. 254; Pülhorn 1987, 226, pl. 20: V/13.a; Demo 2009, 136, cat. no. 431.

83 Vinski 1957a, 37-38, pl. XXIV: 91a, 91b; *Seoba*, 70, pl. I: 1; Mrkobrad 1980, 20, pl. VIII: 7; Mirnik 1996, 192, cat. no. 225; Pülhorn 1987, 226, pl. 21: V/13.b.

84 Vinski 1957a, 37, pl. 24: 90; *Seoba*, 70, fig. 1; Pülhorn 1987, 226, fig. V: 13.c.

85 Mrkobrad 1980, 19, pl. 8: 2.

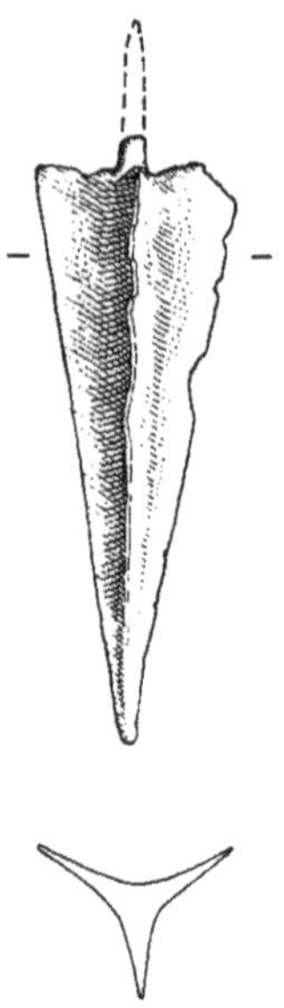

Fig. 56. Arrowhead

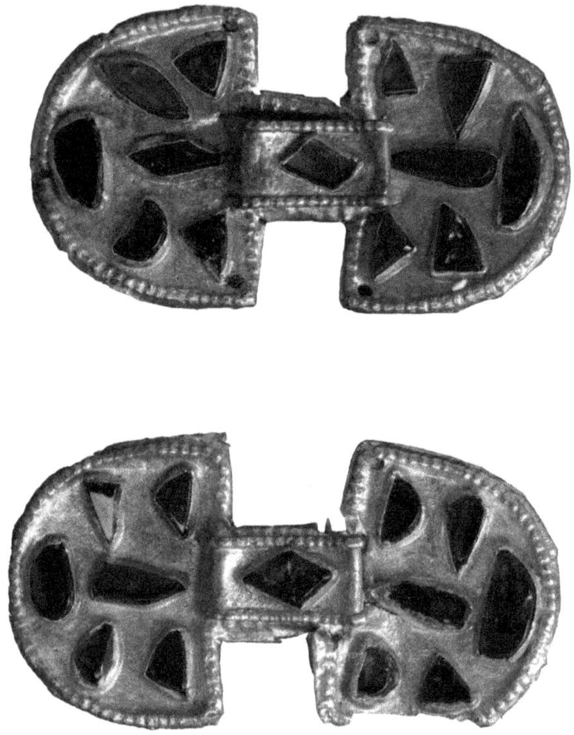

Fig. 57. Spatha fittings

Fig. 58 a, b. Solidus, obverse and reverse

Fig. 59. Pottery jug

perpendicular lines arranged into fields on the upper part of the receptacle. Below these is a polished wavy-line between two shallow flutes. The jug was made of well-purified clay on a potter's wheel, and it is grey in colour.[86] This type of vessel is characteristic for the Barbarian pottery of the territory along the Danubian *limes*.[87] A completely preserved mirror made of a silver alloy was also discovered in Batina (GMSo), certainly a facet of female cosmetic accessories.[88] Metal mirrors are characteristic of the Hun period and were often found as grave offerings in female graves.

3.3. Objects attributed to the Germanic cultural circle (5th and 6th centuries)

Similar to the Roman provincial archaeological evidence, there is a group of finds characteristic for all the Germanic tribes that passed through Pannonia. It is difficult to attribute the finds to a particular Germanic group precisely, unless we posses clearly presented historical sources or finds highly characteristic for a specific group. But even without these elements, this type of finds is very important for the

[86] I would like to thank for these information to Dragan Radojević, a senior curator at the GMSo.
[87] Parducz 1950, 167, pl. 109: 12.
[88] Mrkobrad 1980, 20, Pl. 8: 5. The inventory books of the GMSo do not contain any information pertaining to this object and, therefore, its very existence remains in doubt (pers. com. D. Radojević).

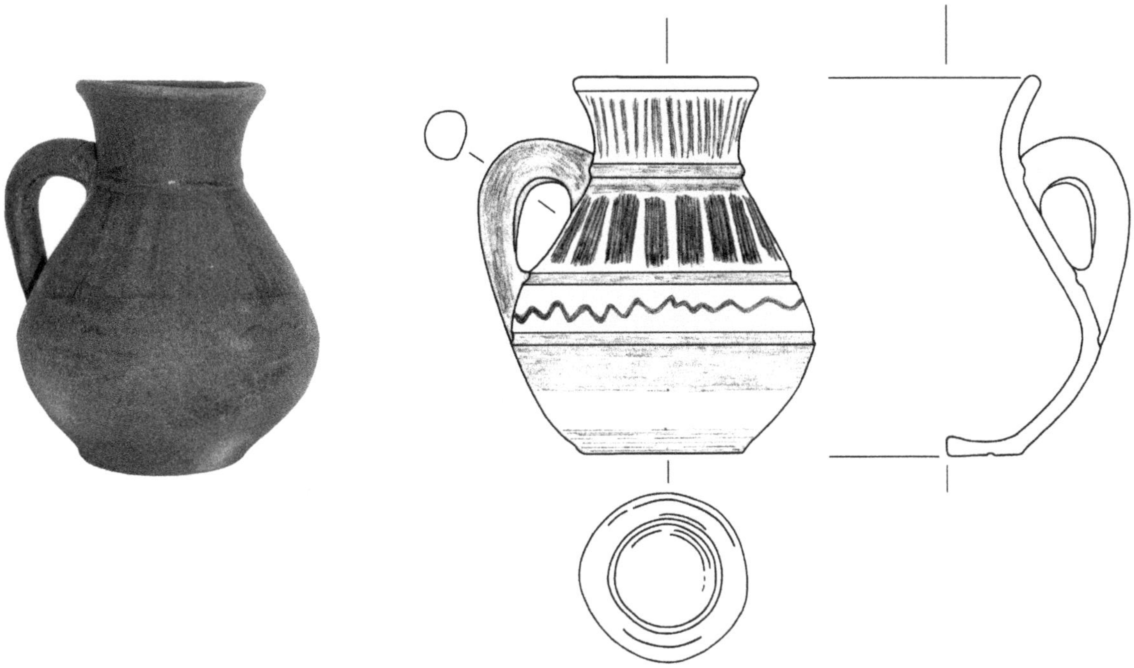

Fig. 60 a, b. Pottery jug

correct interpretation of the archaeological and historical context of Pannonia.

From Sisak stems a cast bronze bow brooch, smooth and undecorated, featuring a semicircular, almost triangle head with three protuberances (length: 49 mm; 5th-6th c.; AMZ). Its bow is short and elevated and the foot is oblong and rhombic with an end moulding. The find can be affiliated to the Visigoth circle.[89] Another bow brooch (fig. 61), made of silver with traces of gilding, was found in Sisak (first half of the 6th c.; AMZ). Its head is triangle in form and decorated with two notched tresses and three protuberances, while its smooth bow prolongs into a thin ribbed foot ending in a stylised animal head. A part of the hinge is also preserved, while the pin is missing.[90] Two skeletal burials situated within the ruins of the Roman town of Iovia/Botivo (modern-day Ludbreg), contained some grave goods with Germanic characteristics, now in the GMVa. The first (warrior's) grave, dated to the 4th-5th centuries, was discovered in 1952. The grave goods consisted of an iron double-edged sword – *spatha* (fig. 62) (length: 724 mm, blade width: 36 to 46 mm), a glass goblet (fig. 63) (height: 65 mm, mouth diameter: 83 mm) and a massive golden ring. The *spatha* is made of damasked steel with an exceptionally long tang for the grip socket. The goblet, typical for the late Roman period, is made of greenish glass and is bell-shaped with a moulding at the top of the mouth.[91] The gold ring was subsequently lost and there are no detailed information concerning it, but it is important because it was exactly such golden

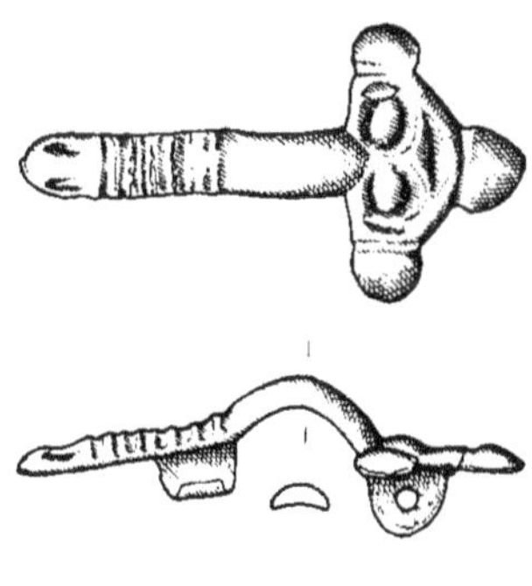

Fig. 61. Brooch

rings that Roman emperors bestowed onto distinguished individuals among the *foederati*.[92] Considering a mixture of late Roman and Germanic elements, it is presumed that in this particular grave a German *foederatus* in the service of the Roman army on the *limes* was buried. The second find is concerned with a female burial excavated in 1976. The grave goods (fig. 64) consisted of two small beads of greenish glass paste and a three-thread wire bracelet (5th c.; AMZ).[93] A warrior burial preliminary dated to the first third of the 6th century was found in Andautonia (modern-day Ščitarjevo). The deceased was placed in the grave together with his equipment: a shield, sword and a lance, and also belt buckles.[94] A battle axe (fig. 65) of the *francisca* type was found in the village of Nartski Novaki near Dugo Selo (length: 160 mm, blade width: 98 mm; second half of the 5th-early 6th c.; AMZ). This Frankish battle axe has a slender mildly curved S-shaped body and a broadened blade. Its

[89] Brunšmid 1905, 216, fig. 32,9; Simoni 1989, 112(25), 121(25), pl. 3: 6; Burkowsky 1999, 87.
[90] Simoni 1989, 113(28), 122(28), pl. 3: 9; Burkowsky 1999, 87.
[91] Simoni 1984, 73; Sunce 1994, 111, cat. no. 142; Tomičić 1997, 34; Šiša Wiewegh 2004, 297.
[92] Grane 2007, 179.
[93] Simoni 1984, 73, fig. 2, Vikić 1976, 83.
[94] Kušan Špalj and Nemeth-Ehrlich 2008, 200.

Fig. 62. Spatha

Fig. 63. Goblet

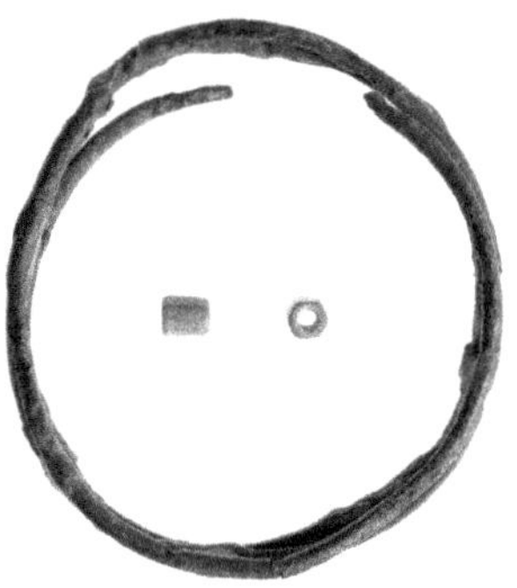

Fig. 64. Bracelet and beads

Fig. 65. Battle axe

socket hole is ovoid and the reverse side of the axe is placed at an angle in relation to the body. On the front side of the socket hole a small deltoid is incised, divided in half by a line.[95] This type of battle axe is characteristic for the graves of the Frankish Merovingian circle and it developed out of the axe used by the Romano-Germanic *foederati*. Since this particular axe belongs to the earliest type of *francisca*, it can be associated to the Germanic circle.

3.4. Objects attributed to the Ostrogothic cultural circle (454-473; 497/504-537)

Archaeological finds that can be affiliated by their style and typological characteristics with the Ostrogoths in the territory of Pannonia Savia almost exclusively come from Sisak, while they are somewhat more frequent in Pannonia Sirmiensis. Comparing the two periods of Ostrogothic rule in Pannonia, the finds belonging to the second are somewhat more frequent. The distribution of sites suggests that the Ostrogoths utilised the Roman settlement structure they inherited (settlements and stations on important river crossings and traffic junctions).

The following 3 finds all stem from Sisak and are now in the AMZ: a hammered silver sheet bow brooch (fig. 66) with a semicircular head featuring a mildly protruded rib in the centre and damaged foot (length: 82 mm; second half of the 5th c.); the bow is highly elevated and concave and the junction of the bow and foot is decorated with a threaded ribbed wire; on the back side of the head is a preserved part of a hinge, while the pin is missing,[96] a cast bow brooch (fig. 67) with a triangle head and a single protuberance, and with an elevated bow and rhombic notched foot (length: 64 mm; 5th-6th c.); on the back side a part of the hinge is preserved, while the pin is missing,[97] and a cast bow brooch (fig. 68) of the degenerative Karavukovo style, made of gilded silver with almandine inlays, decorated in notched tresses on the head and foot and framed with triangles (length: 55 mm; ca. 500); five moulded protuberances are on the head, the bow is short and the foot is rhombic.[98] Another chance find is a silver sheet bow brooch (fig. 69) from the site of Ciglana in Beli Manastir (length: 123 mm; second half of the 5th c.; MSO). It was produced by hammering with a palmette application at the junction of the bow and foot and lateral linear ornaments on brooch's head; the pin is missing, but large sections of transversal axes on the back side are preserved.[99] A pair of silver sheet massive bow brooches from Ilok was produced by hammering (lengths: 212 and 221 mm; mid-5th c.; AMZ). Lateral ornaments on semicircular heads are decorated in three lines, with another one in the middle of the head. The bow is short and elevated and the elongated thin foot has mildly pronounced rib in the middle. On the heads' back each there are two

[95] Simoni 1983, 252, 254, pl. 1.1; Simoni 1993, 182, cat. no. 264.

[96] Vinski 1957b, 143, fig. 48; Pülhorn 1987, 196, fig . 4: 7.c; Simoni 1989, 110(19), 121(19), fig. 1; Burkowsky 1999, 87.

[97] Brunšmid 1905, 219, no. 36; Vinski 1957b, 143, fig. 47, 156; Simoni 1989, 112(26), 121(26), pl. 3: 7; Burkowsky 1999, 87.

[98] Brunšmid 1905, 213, fig. 32,5, 216-217; Vinski 1962, 76, fig. 4; Vinski 1973, 195-196, pl. 5: 50; Pülhorn 1987, 196, fig. 4: 7.e; Simoni 1989, 112, 121-122, pl. 3: 8; Sunce 1994, 85, cat. no. 48.

[99] Vinski 1957a, 31-32, pl. 21: 77; Kovačević 1960, 20; *Seoba*, 68, fig. 1; Mrkobrad 1980, 27, pl. 10: 3; Bojčić 1984, 214; Radić 1997, 83,93, cat. no. 2; Radić 2009, 137, cat. no. 435.

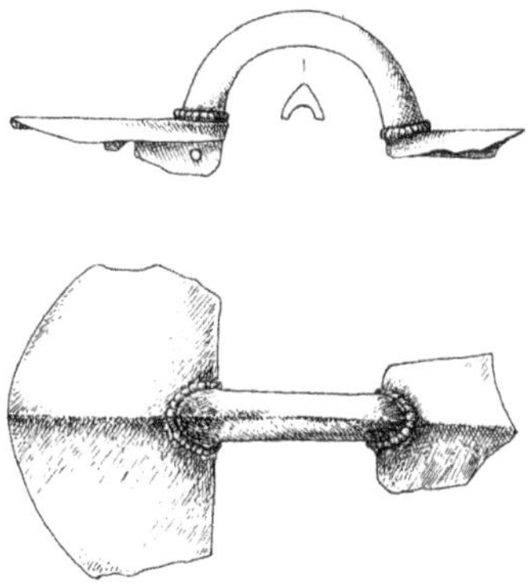

Fig. 66. Brooch

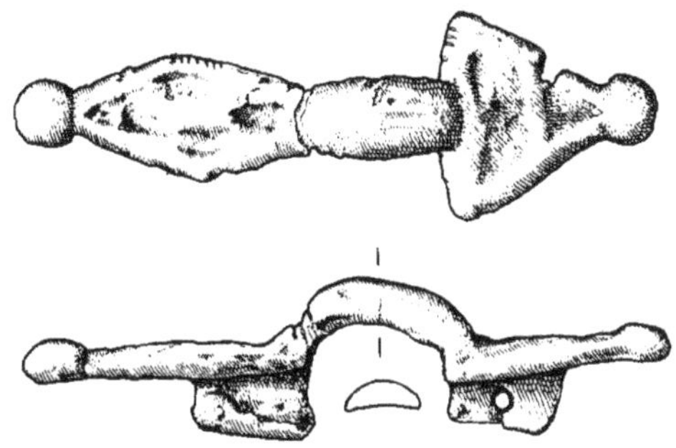

Fig. 67. Brooch

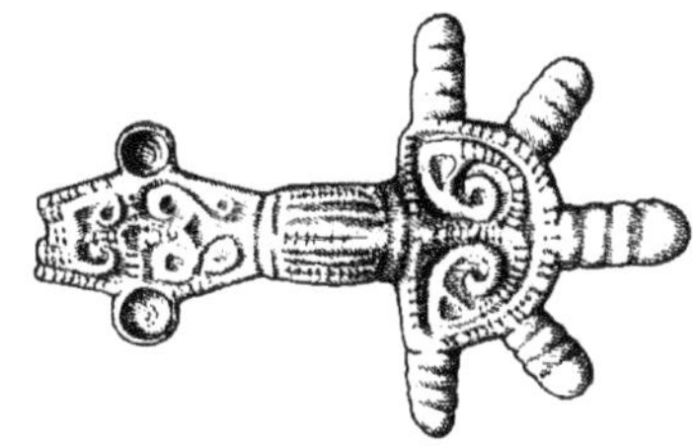

Fig. 68. Brooch

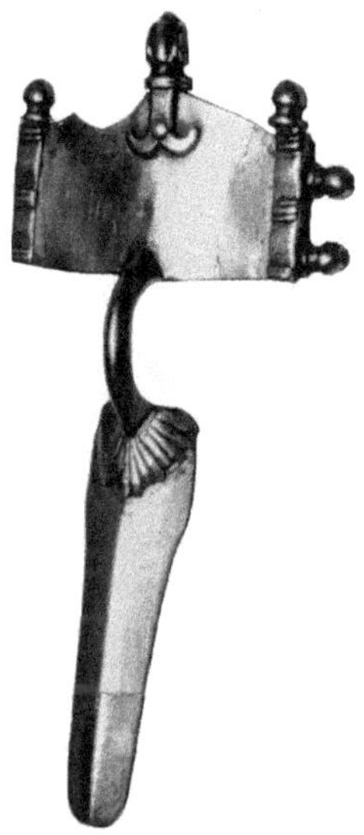

Fig. 69. Brooch

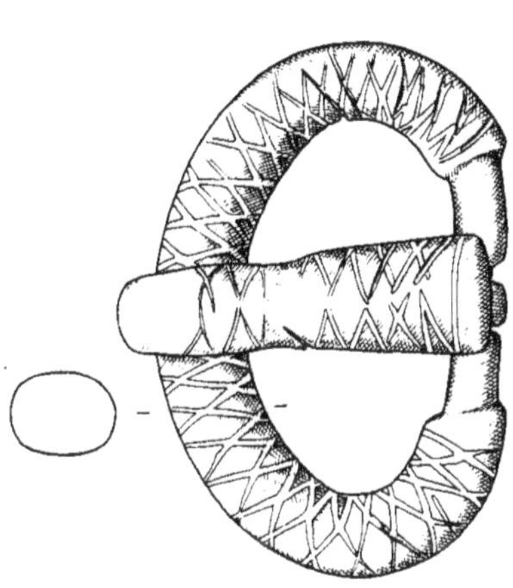

Fig. 70. Buckle loop

Fig. 71. Buckle mount

transversal axes covered by astragal scrolls. The pins are also of silver and the fasteners are in the form of spirally fluted tubes and.[100] A fragment of a small cast silver bow brooch was found in the area of Roman Cornacum, modern-day Sotin (17 x 10 x 4 mm; second half of the 5th c.; The Mato Ilkić Collection, Zadar, Croatia). A bow triangular in cross section and narrowed at the transition into a panel-shaped broadening is preserved, as well as a fragmented pin fastener on the underside of the bow.[101] This type of large silver sheet bow brooches are typical of the Ostrogothic female wear in the Danubian area, and they came into use after the mid-5th century. A massive bronze buckle loop (fig. 70) of ovoid form and cross section, with a damasked decoration in the form of the letter X, made of silver wire, stems from Sisak (5th c.; AMZ). The loop is wider at its front part and narrower at the tongue's base; the latter is massive, elongated, somewhat thinned in the middle.[102] A buckle mount (fig. 71) is a chance find from Dalj (51 x 57 mm; 504-536; MSO). It was made of gilded silver, ornamented with seven almandine inlays, two of which are missing. The decoration is notched; the motive is in the form of two rows of entwined and zigzag lines. Remains of a bronze plaque are preserved on the back side.[103] On the bases of some analogies, this object was probably made in an Italian workshop. A silver spoon (fig. 72) was found in Sisak (length: 229 mm; ca. 500-first half of the 6th c.; AMZ); its handle is long and massive and the ovoid shallow bowl is small and elongated. The bowl has five mildly protruding ribs on the reverse side, and there is a disk at the junction of the handle and bowl, ornamented in an incised Greek cross. A line of convex parallel ribs and concavities are found on the tapering handle. In view of these characteristics, the spoon has been classified as of the Ostrogothic Desana

[100] Brunšmid 1905, 210, fig. 31; Kovačević 1960, 33, fig. 68; *Seoba*, 76, pl. 2: 2; Mrkobrad 1980, 27, pl. 10: 2, 4; Bojčić 1984, 214; Pülhorn 1987, 224, fig. 5: 10; Simoni 1993, 180, cat. no. 258; Demo 2009, 137, cat. no. 434.

[101] Uglešić 1994, 146(1)-147, pl. I: 1, 1a, pl. II: 1, 1a; Ilkić 2007, 279, 282(4), pl. 1: 4.

[102] Simoni 1989, 115(45), 123(45), pl. 5: 11; Burkowsky 1999, 88.

[103] Bojčić 1984, 214-215, fig. 1; Radić 1997, 83, 93, cat. no. 3; Bojčić 2009, 32; Radić 2009, 137, cat. no. 436.

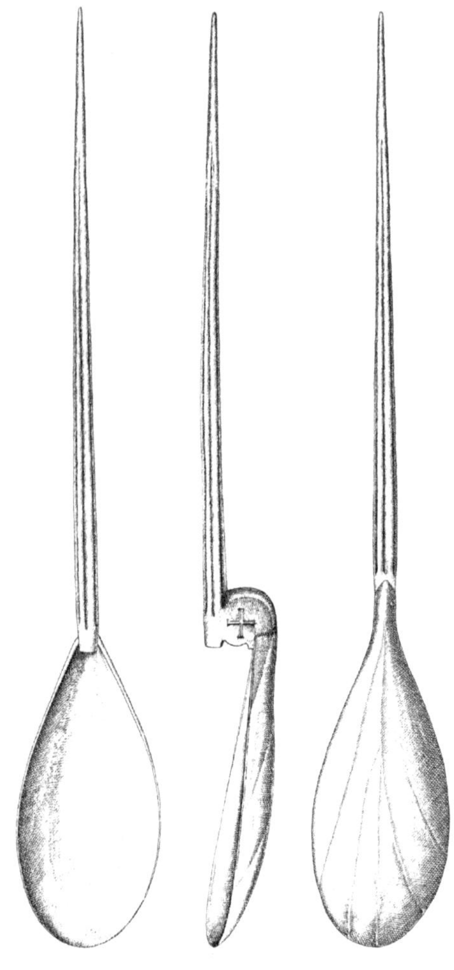

Fig. 72. Spoon

type.[104] Silver spoons developed out of Roman types; one such specimen was found in Cibalae.[105] In the period from the 5th to 7th centuries they were used in Eucharistic rituals or, more probably, had profane purposes, as a reflection of the wealth and social status of the owner.

Continuing the late Roman tradition, the Ostrogothic ruler Theoderic, among other things, regularly issued coins, most often on behalf of Byzantine emperors. The coinage of Theoderic was struck in Sirmium in the likeness of the coins issued from the Mediolanum mint; at this moment, only quarter-*siliqua*s issued on behalf of Anastasius I (491-518) and Iustin I (528-526) are known.[106] A *solidus* struck on behalf of the Byzantine emperor Justin I (518-527) by Theoderic (AMZ) was found in the village of Baćin near Dubica. The coin was struck in the Roman mint (the mintmark COMOB *) and bears the inscription DN IVSTI-NVS P F AVG on the obverse and VICTOR-IA AVGGG on the reverse.[107] The total number of Theoderic's coins found in Siscia is eleven (AMZ): 2 *tremisses*, 4 quarter-*siliqua*s and 5 *decanummi*. *Tremisses* and *siliqua*s were minted on behalf of Anastasius (491-518) with Theoderic's monogram on the reverse, while the *decanummi* bear the monogram of Ravenna.[108] A Theoderic's *tremis* minted in the Mediolanum mint on behalf of the emperor Anastasius was found in Glina.[109] Two silver quarter-*siliqua*s from unknown sites in Slavonia (MSO), minted by Theoderic on behalf of Anastasius, also belong to the Ostrogothic coinage.[110] A fragmented quarter-*siliqua* struck in the Sirmium mint (AMZ) was found in Dalj, but the only discernible inscription is Theoderic's monogram.[111] A silver quarter-*siliqua*, another chance find, was found at the site of Štrbinci in 1913 (AMZ); it was minted in Sirmium and has the depiction of the emperor Justin I (518-527) on the obverse and Theoderic's monogram on the reverse.[112]

3.5. Objects attributed to the Gepid cultural circle (473-504; 536-567)

Archaeological evidence that can be affiliated with the Gepids by their style and typological characteristics mostly comes from Cibalae (Vinkovci) and is kept in the GMVi. With regards to two periods of Gepid rule in Pannonia, the finds attributable to the second period are more numerous. Similar to the Ostrogoths, the Gepids usually occupied (destroyed) Roman settlements. In the early 20th century a damasked *spatha* with remains of a pommel was found in Ilok (length: 910 mm; 5th-6th c.; AMZ).[113] The *spatha* – a long double-edged sword – did not originate in the short Roman *gladius*, but rather in the late La Tène long sword, took over by the Goths on the Pontus from the Sarmatians and Alans. It is frequently found during the 5th and 6th centuries in graves of Germanic dignitaries, but if the circumstances of a find are not known, its ethnic affiliation remains obscure. Two silver quarter-*siliqua*s with the depiction of the emperor Justinian I (527-565) on the obverse and the monogram of the Gepid king Cunimundus (560-567) on the reverse come from Dalj or its vicinity. They were minted in Sirmium in ca. 560 or 565-567.[114] The remaining finds from the first phase all stem from Vinkovci and are kept in the GMVi. A cast bronze bow brooch (length: 55 mm; 6th c.) is a chance find; it is decorated with a simple notched ornament on the head plate and foot, with

104 Simoni 1988: 79 i d., Pl.1, pl.2; Simoni 1989: 119, 125, pl. 8: 1, 2, 2a, 2b; Simoni 1993, 181, cat. no. 261; Sunce 1994, 91, cat. no. 75; Burkowsky 1999, 88.
105 Brunšmid 1902, 158, fig. 88.
106 Demo 1994, 13.
107 Demo 1981, 455, 477(25), pl. 1: 25; Mirnik and Šemrov 1998, 205, cat. no. 819.
108 Demo 1981, 455, 477(18, 20, 21, 22), 478(33, 36, 37), 479(40, 41); Mrkobrad 1980, 41, pl. XXXIV: 1; Mirnik and Šemrov 1998, 203-205, cat. nos. 806, 810, 814, 815, 817, 823, 825, 826.
109 Mirnik and Šemrov 1998, 204, cat. no. 812.
110 Bojčić 2009, 21.
111 Demo 1981, 455, 477, cat. no. 24.
112 Kovačević 1960, 33; *Seoba*, 115; Mrkobrad 1980, 41, pl. XXXIV 34: 4; Demo 1981, 455, 478, cat. no. 29; Bojčić 1984, 214; Mirnik and Šemrov 1998, 206, cat. no. 832.
113 Vinski 1957a, 21, 34, pl. XXIII: 89; Csallany 1961, 241; *Seoba*, 77; Mrkobrad 1980, 51, pl. XLI: 5; Bojčić 1984, 214.
114 Mrkobrad 1980, 56, pl. XLVI: 10; Demo 1981, 464, 481, cat. nos. 67, 69; Bojčić 1984, 214; Mirnik and Šemrov 1998, 210, cat. no. 868.

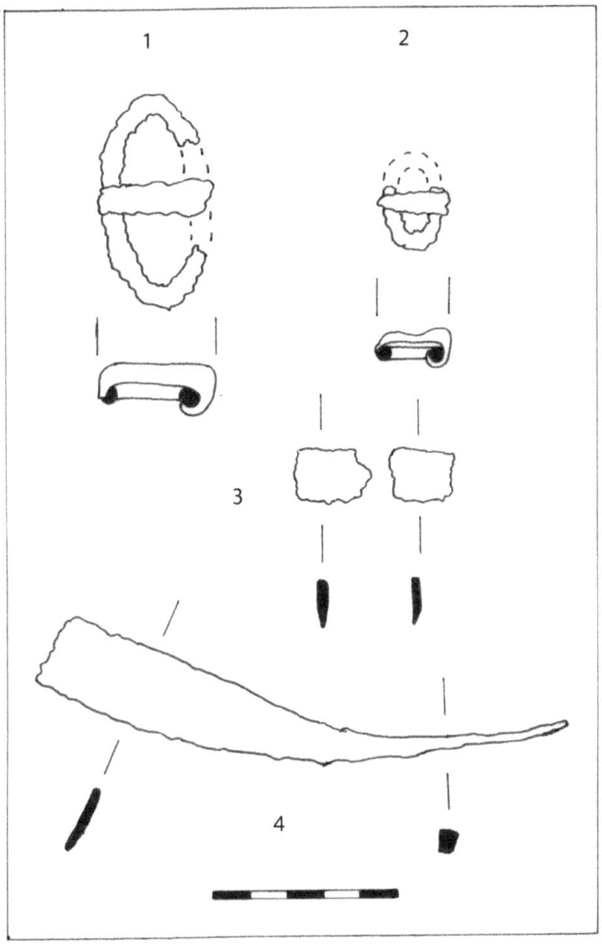

Fig. 73.1, 2. Buckle loops; 72.3, 4. Knives

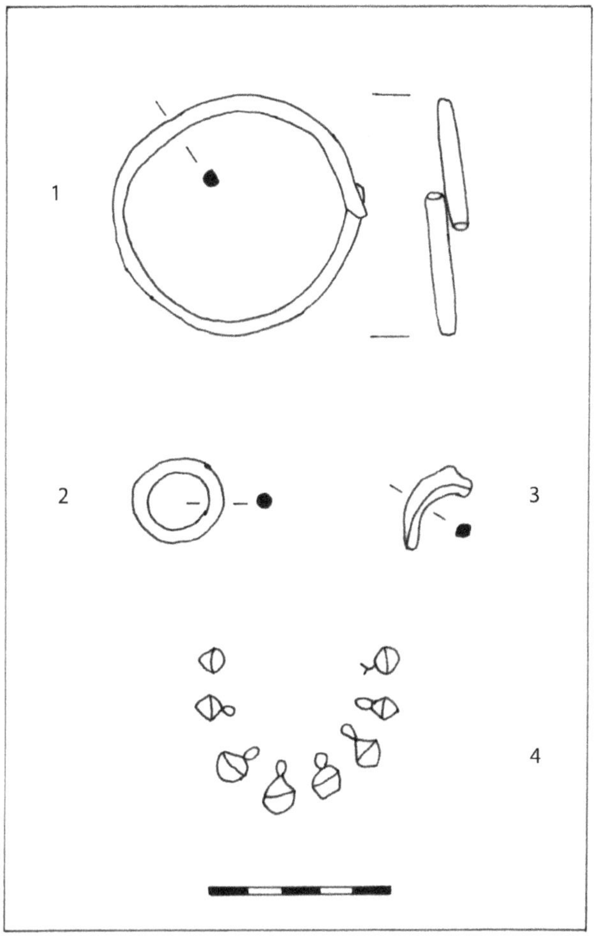

Fig. 74.1. Bracelet; 74.2,3. Rings; 74.4. Necklace beads

the head featuring three tongues.[115] Another chance find is a bronze single piece belt buckle of the Mediterranean type, decorated with engraved and chased ornament (length: 71 mm; 6th c.).[116] The remains pertaining to the second phase of Gepid domination in the territory of Croatia are also most comprehensively known from the area of Vinkovci. During rescue excavations in 1954-2007, 51 graves were recovered within the town perimeter of Roman Cibalae, without any obvious rules regulating the choice of a burial spot. Only 18 burials yielded grave goods: metal parts of dress accessories, jewellery, and utilitarian objects, all now in the GMVi.[117] Among dress accessories two iron ovoid belt buckles (figs. 73.1 and 73.2) were found (diameter: 15 and 57 mm; ca. mid 6th c.). Their tongue base is cut off and the tip of the tongue is bent over the loop.[118] Ovoid iron buckle loops have appeared as a late Roman heritage already in the 4th century and continued to be produced during the entire 5th, while in the 6th century they became a pan-Germanic property and are often found in graves of all Germanic peoples. Of the jewellery, two necklaces and one bracelet are known. The first of the necklaces (fig. 74.4) consists of nine bronze spheroid pendants with a suspension loop (diameter: 10 to 14 mm; 6th c.),[119] while the second (fig. 75.3) consists of 16 amber and seven glass paste beads (diameter: 4 to 12 mm; 6th c.).[120] The bronze bracelet (fig. 74.1) has mildly broadened and traversing endings (diameter: 69 mm; 6th c.).[121] All these types of jewellery are typical for early mediaeval Germanic burials. Bone combs are among the most numerous utilitarian objects; a total of six were found. Three of them (fig. 77) are of the two-row type,[122] which developed from late Roman bone combs, but were also used by Germanic tribes. The remaining three (fig. 76) are of a single-row type,[123] and are associated with Germanic peoples.[124] Bone combs as

[115] Vinski 1957a, 31, pl. XX: 76; Kovačević 1960, 33, fig. 69; *Seoba*, 97, fig. 1; Dimitrijević 1979, 190, pl. 27:1; Mrkobrad 1980, 48, pl. XXXV: 10; Bojčić 1984, 214; Vinkovci 1997, 151, cat. no. 317; Rapan Papeša 2009, 139, cat. no. 443.

[116] Dimitrijević 1979, 190, pl. 27: 8; Vinkovci 1999, 151, cat. no. 318.

[117] A more detailed analysis of these objects and their publishing is currently in progress.

[118] Vinkovci 1997, 151, 153, cat. nos. 319, 324; Rapan Papeša 2009, 139, cat. no. 444.

[119] Vinkovci 1997, 154, cat. no. 334; Rapan Papeša 2009, 139, cat. no. 441.

[120] Vinkovci 1997, 152, cat. no. 323; Rapan Papeša 2009, 138, cat. no. 440.

[121] Vinkovci 1997, 153, cat. no. 238; Rapan Papeša 2009, 138, cat. no. 439.

[122] Vinkovci 1997, 152, 153, 154, cat. nos. 321, 327, 332; Rapan Papeša 2009, 138, cat. no. 438.

[123] Vinkovci 1997, 152, 154, cat. nos. 322, 331, 333; Rapan Papeša 2009, 138, cat. no. 437.

[124] Dimitrijević 1960, 28.

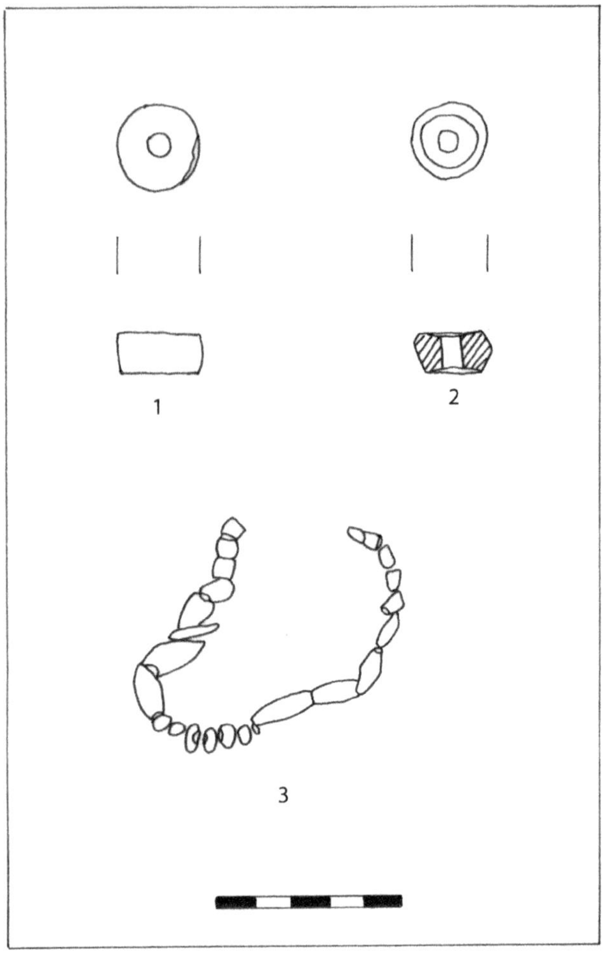

Fig. 75.1. Weight; 75.2. Bead; 75.3. Bead-necklace

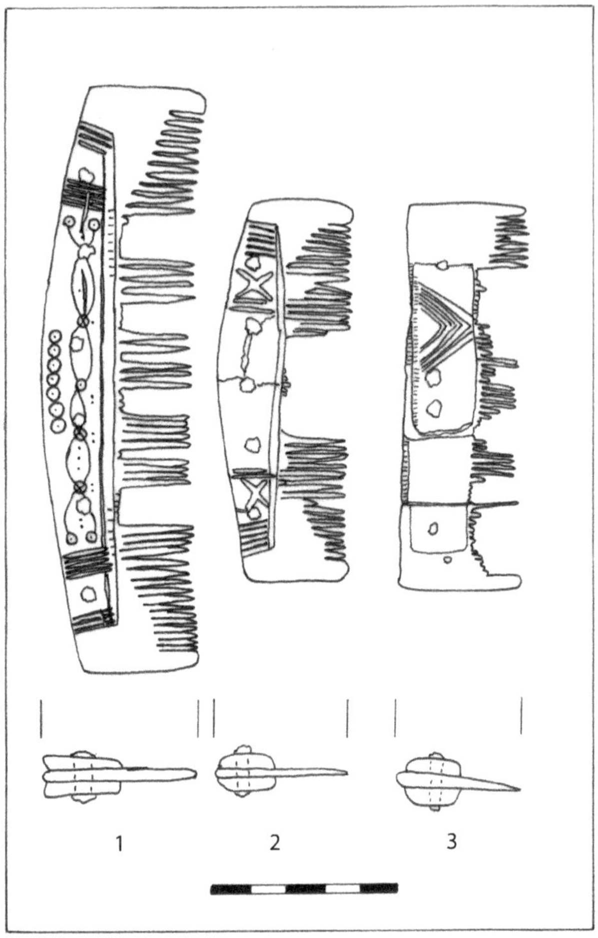

Fig. 76.1, 2, 3. Single-row combs

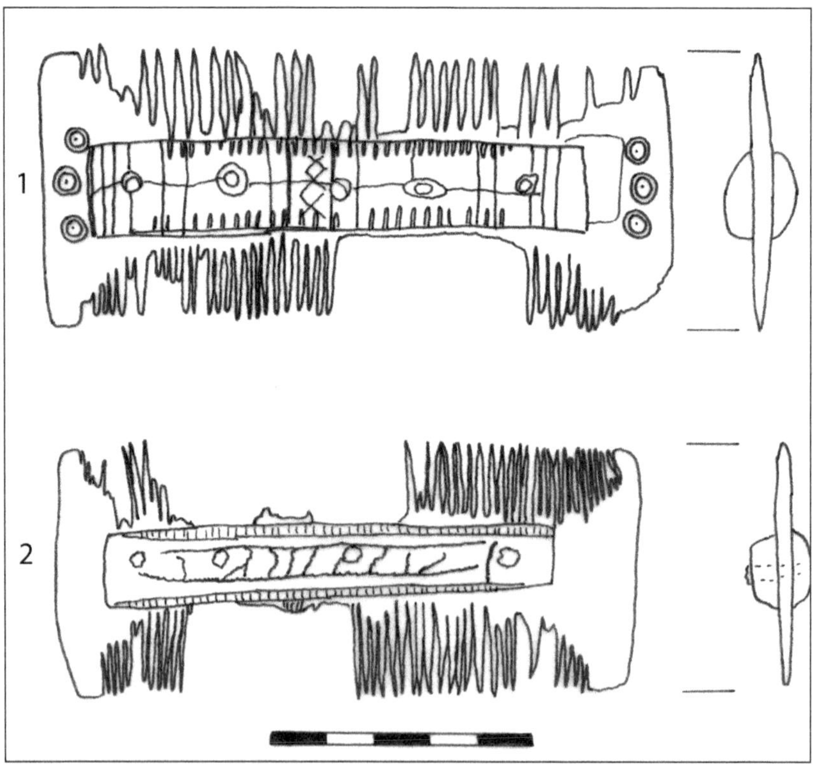

Fig. 77.1, 2. Two-row combs

Fig. 78. Pottery shards

grave goods are characteristic among the Gepids for all age groups, independent of the deceased's gender.[125] Several other items belong to the group of objects intended for everyday use: three iron knives (figs. 73.3 and 73.4) (length: 35 and 142 mm; 6th c.),[126] a ceramic spindle whorl (fig. 75.1) (diameter: 20 mm; 6th c.),[127] a chalk bead (fig. 75.2) used as a whorl (diameter: 23 mm; 6th c.),[128] as well as two bronze loops (fig. 74.2 and 74.3) (diameter 29 mm; 6th c.).[129] All these objects have direct analogies in Gepid graves from the territory of modern-day Serbia[130] and Hungary.[131] Besides the grave finds, more than 30 fragments of typical Gepid pottery (fig. 78) were found in Vinkovci, featuring stamp decorated patterns (some half of these have already been published),[132] but also fragments of Germanic pottery for everyday use.[133]

3.6. Objects attributed to the Lombard cultural circle

Archaeological objects that can be affiliated with the Lombards by their style and typological characteristics mostly come from Pannonia Savia, where the carriers of this cultural sphere lived for a longer period, while they appear only sporadically in Gepid-dominated Pannonia Sirmiensis. The distribution of sites suggests that the Lombards, although continuing the Roman tradition, mostly occupied the border areas of southern Pannonia, pointing towards their primary goal: defensive military fortification.

A massive bow brooch (fig. 79) of the Pannonian-Lombard type with a semicircular head and rhombic foot was found in Sisak (55 x 35 mm; mid-6th c.; AMZ). It has three moulded protuberances and notched ornament in the form of tresses on its head; the same motif is repeated on the short elevated bow. The foot is decorated with meanders and ends in a bulge, while its edges are decorated with small nielloed triangles.[134] A gold ring (fig. 80) with characteristic arcade crown is a chance find from Samobor (diameter: 24 mm; 6th c.; AMZ). The ring was hammered and granulated. The hoop is in the form of a low triangle in cross section and broadens towards the crown. The crown is in the form of a truncated cone and consists of a hexangular base upon which stand six small vertical arched columns, creating the appearance of an arcade carrying six small triangles made of three granulated beads. Every triangle carries upon itself another small hoop with a circular plate forming the upper cover of the crown. This is certainly a luxurious product made in a Mediterranean workshop, with analogies in Kranj

125 Csallány 1961, 251.
126 Vinkovci 1997, 151, 153, cat. nos. 320, 330. The third was discovered in 2007 and is currently being restored and conserved (Rapan Papeša and Vulić 2008, 72-73).
127 Vinkovci 1997, 153, cat. no. 326; Rapan Papeša 2009, 139, cat. no. 446.
128 Vinkovci 1997, 153, cat. no. 329; Rapan Papeša 2009, 139, cat. no. 442.
129 Vinkovci 1997, 152, cat. no. 325; Rapan Papeša 2009, 139, cat. no. 445.
130 Dimitrijević 1960, 44; Simoni 1978, 214; Zotović 1992-1993, 190.
131 Bóna and Nagy 2002; Bóna, Garam, and Vida 2005.
132 Dimitrijević 1966, pl. 17:9-17; Simoni 1978, 220-221, pl. V:1-12; Dimitrijević 1979, 190-191, pl. 27:2-5; Vinkovci 1999, 154, 155, cat. nos. 335-338.
133 These are currently being studied.

134 Brunšmid 1905, 213 , 216, fig. 32: 4; Simoni 1989, 113, 122, pl. 4: 1; Simoni 1993, 181, cat. no. 262; Sunce 1994, 85, cat. no. 48; Vinski 1972-73, 201, n.126, 210, pl. 12: 68.

(Slovenia), as well as Cividale and Castel Trosino (Italy).[135] A gilded cast silver S-brooch (fig. 81) decorated in notching was found in Vinkovci (length 26 mm; 540-560; AMZ).[136] It consists of two stylized antithetical interconnected bird's heads and is decorated in notched geometrical pattern. Although the pin is missing, there are remains of a hinge and a pin fastener on the back.[137] J. Werner named this type Varpalota 34/Vinkovci (Pannonian-Lombard type) and dated it to the first half of the 6th century. According to more recent and more precise studies, the brooch in question belongs to the North Danubian-Pannonian transitional phase; it must have been worn by the Lombards.[138] A gilded silver S-brooch (fig. 82) is a chance find from the site of Popovac – Šumarine (24 x 18 mm; 540-560; AMO).[139] It is decorated in notched triangles and two red almandine inlays are still preserved. It is an early predecessor of the Varpalota 34 type and also belongs to the above mentioned North Danubian-Pannonian transitional phase. A gold coin struck by Lombard rulers on behalf of Justinian I has been found in Osijek.[140] A decorative piece of horse gear (fig. 83) from Sisak has a form of an equal-armed cross, whose arms are decorated in deeply notched meander framed with small stamped rhombi (54 x 40 mm; 6th c.; AMZ). A rivet hole is seen on one of the arms, as well as two protuberances for strap fastening on the back side of the fitting, while its central part is made in the form of calotte, framed with engraved small bird's heads with strong beaks.[141] A bronze pommel (fig. 84) in the form of an elongated rhomb, decorated in vertical and diagonal plastic mouldings, has been found in Sisak (6th c.; AMZ). Its upper surface is in the form of an irregular quadrangle. A piece of an iron tongue is preserved in the middle, probably belonging to a Lombard *spatha*.[142] A handmade vessel (figs. 85 a and 85 b) of an irregular form made of greyish brown clay was found in the Kupa at Sisak (6th c.; AMZ). The rim is everted and the body is decorated with an ornament in the form of three circular stamps.[143] The coinage of the Lombard period in the territory of Pannonia Savia is represented with a *solidus* from Sisak and *tremisses* from Sisak and Petrinja respectively (AMZ); they were struck by unknown Lombard rulers on behalf of the emperor Justinian I (527-565).[144] A chance find of a bronze Byzantine *follis* from the second half of the 6th century, struck during the reign of the emperor Justin II and empress Sophia (565-578), was found at the site of Gradec in Zagreb. It was struck in Constantinople in 570-571, with depictions of Justin II and

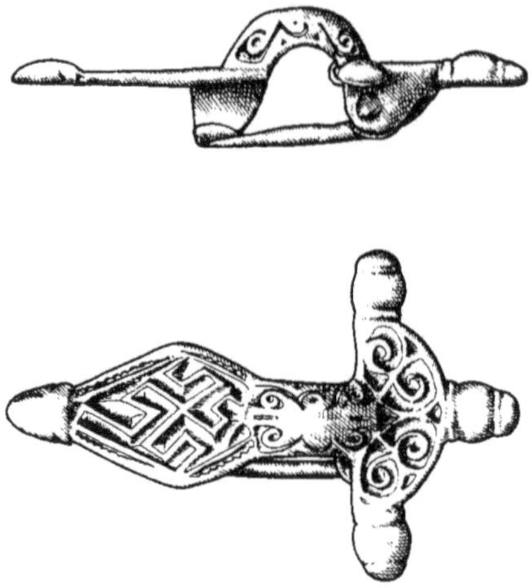

Fig. 79. Brooch

Fig. 80. Ring

Fig. 81. Brooch

[135] Vinski 1960, 50, pl. 11: 1; Simoni 1981, 156, fig. 1; Sunce 1994, 129, cat. no. 191; Pülhorn 1987, 196, pl. 16: IV/8.a; Sokol 1998a, 1133, pl. 4: 3.
[136] Brunšmid 1905, 219, fig. 37; Kovačević 1960, 33, fig. 69; *Seoba*, 97, pl. 10,3; Mrkobrad 1980, 63, pl. 50: 8; Bojčić 1984, 214; Pülhorn 1987, 230, fig. 5: 20; Demo 2009, 140, cat. no. 447.
[137] Simoni 1993, 182, cat. no. 266.
[138] Milavec 2007, 337.
[139] Bojčić 2007, 22, 32; Bojčić 2009, 16; Radić 2009, 140, cat. no. 448.
[140] Mirnik and Šemrov 1998, 208, cat. no. 855.
[141] Simoni 1989, 113(30), 122(30), pl. 4: 2; Simoni 1993, 181, cat. no. 263; Sunce 1994, 86, cat. no. 55; Burkowsky 1999, 88.
[142] Simoni 1989, 118, 124, pl. 7: 5; Burkowsky 1999, 88.
[143] Vinski 1954, 73-74, fig. D; Simoni 1989, 113(31), 122(31), pl. 4: 3; Burkowsky 1999, 88.
[144] Mirnik and Šemrov 1998, 208-209, cat. nos. 856, 857, 858.

Fig. 82. Brooch

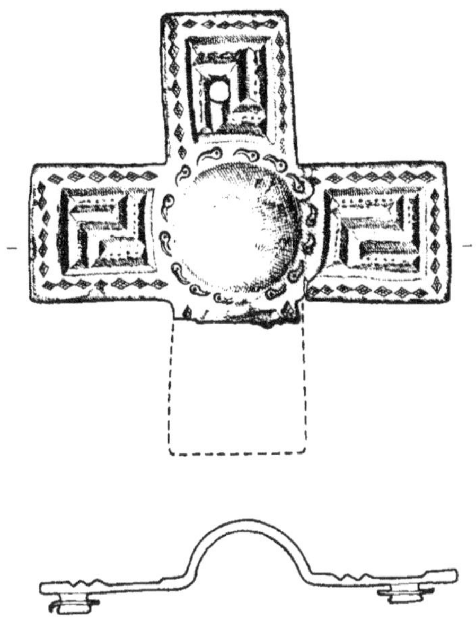

Fig. 83. Horse-gear piece

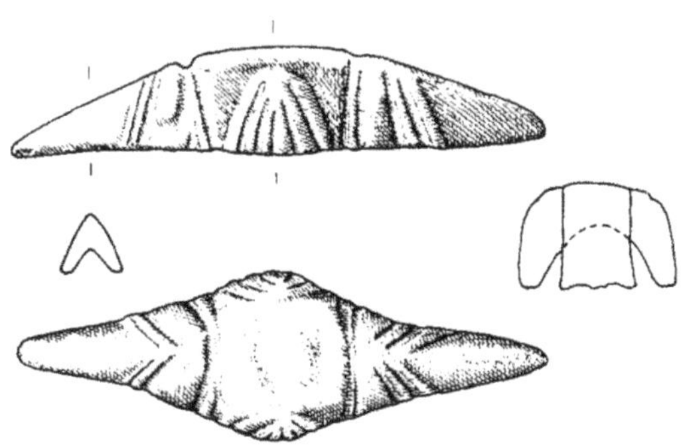

Fig. 84. Pommel

Fig. 85 a, b. Pot

Sophia on the obverse and the *officina* mark, Chi-Rho and the year of minting, on the reverse.[145]

4. Concluding remarks

The late Roman and early mediaeval period, that is, the period of Barbarian migrations, is a historical and cultural epoch characterized by a transition to the first steps towards the image of modern-day Europe. This period is exceptionally important in the area of the Carpathian basin. The cohabitation of the autochthonous Romanised population with the newly arrived Barbarian peoples was certainly not a painless process, and this development had encompassed a significantly wider area in comparison to the limited region of the present study. In terms of geography this region – the continental part of Croatia, that is, the Roman provinces of Pannoniae Savia and Secunda – stands at the very southern limit of the area encompassed by the process of Romano-Barbarian acculturation. After a meticulous analysis of archaeological material and a reinterpretation of literary sources, one can safely conclude that the theory of a complete collapse of life in the late 4th-century Pannonia is no longer valid. The present state of research cannot deny a significant reduction, compared to earlier centuries, of Roman finds from the mid-4th

[145] Simoni 1981, 155-156; Sunce 1994, 126, cat. no. 181; Dukat 1996, 69; Sokol 1998, 1133.

century onwards, but this is generally the characteristic of turbulent periods. The quantity of objects appearing in more recent excavations, as well as a more thorough analysis of some earlier finds, suggest that in the majority of late Roman urban centres – such as Siscia[146] and Cibalae[147] – life continued uninterruptedly until the first half of the 5th century, and occasionally even longer. This fact does not apply in its totality to rural areas, but even there some places[148] and settlements exist that continued life during this period.

The comparison of the distribution of archaeological finds and literary sources reveals that different Germanic groups that have roamed over this area in the majority of cases respected the pre-existing structures in the spheres of the settlement, economy and government. Their selection of settlement positions not only respected the topographically important Roman centres and other places, but demonstrated their wish to assimilate into the existing structures, to become a part of the Empire, and, finally, to reinstate it. Siscia and Cibalae are larger centres that have continued life under Germanic rule, while Andautonia, Iovia, Ad Novas, Cuccium and Cornacum, should be mentioned among lesser settlements of such nature.

Sporadic finds generally attributed to the Germani from the northern part of Pannonia Savia suggest that the Drava region was the main corridor for the movements of Germanic peoples through Pannonia, that is, that the road from Mursa to Poetovio was used as their path to Italy. The utilisation of this particular course, in comparison with the road Siscia-Sirmium along the Sava, was otherwise more opportune, considering the possibility of attack from the direction of Dalmatia. The rest of the finds, mostly stemming from the southern part of Pannonia Savia, testifies to a continuous life of Roman settlements in the period of the Germanic domination. However, because of the generally meagre evidence, and considering the questionable attribution of the finds, a more in-depth study is not possible at this stage.

Archaeological evidence attributable to the Huns consists mainly of chance finds and can be affiliated with either the groups of *foederati* known from literary sources or Hun forays. This is true of the finds from both Pannonian provinces – the finds from the area of Sisak and Kuzelin can be associated with the struggle between the emperor Theodosius (379-395) and the usurper Magnus Maximus (383-388), while the finds from the Danubian *limes* can be associated either with the *foederati* settled in frontier garrisons, or the Hun attacks on these garrisons. Taking literary sources into account, in this particular case, substantiated with archaeological finds, the westernmost point under direct Hun rule in the first half of the 5th century was Sirmium. But the proximity – or rather, source – of riches was in this case very near to them, and the area of southern Pannonia was certainly open to their forays. In their turn, these forays could have been responsible for the mixed evidence finds. Attila's army passed through this territory on its way to Gallia (451), as well as on its return from Italy (452), suggesting the possible date of some of those finds.

The finds attributed to two phases of the Ostrogothic occupation of Pannonian provinces are somewhat more frequent, although the majority of these can be associated with the time of Theoderic's rule (493-526). The brooches from Beli Manastir, Sotin and Sisak can be associated with the first phase of the Ostrogothic occupation of southern Pannonia, but the major part of the territory of Pannonia Savia was certainly not under direct Ostrogothic rule during this period. Pannonia was nearest to the image of Roman government during Theoderic' s rule, since in this period the provinces were restored, together with the government and economy. As part of this renaissance, archaeological finds suggest that the Ostrogoths took over the Roman logic of settlement patterns, that is, they used the sites of tested strategic importance: Siscia, as the metropolis of the province, as well as smaller settlements such as Dalj, Beli Manastir, Ilok and Sotin, along the Roman *limes* on the Danube, which made a frontier with the kingdom of the Gepids. Sporadic finds within these limits, such as Baćin, Glina, and Štrbinci, only testify to the movements along the corridors defined by the existing Roman roads.

The next in chronological development are the finds pertaining to two periods of Gepid supremacy in Eastern Pannonia, attested in literary sources. During the first period, the territory of today's northern Croatia probably only represented their sphere of interest, although an important battle with the Ostrogoths took place in the marshes of the Vuka. An archaeological find that could tentatively be associated with this period is the *spatha* from Ilok, in the light of possible "guerrilla" attacks of the Gepids on a detachment of the Ostrogothic army. The number of finds that can be associated with the second period of Gepid supremacy is much larger, and they were mostly found in Vinkovci. A renewal of Roman Cibalae, as evidenced by the distribution of the material, was rather chaotic and unplanned, and was probably a result of the strategic importance of this town as the westernmost limit of the Gepid kingdom, whose western frontier presumably followed the line Vinkovci-Dalj.

A review of Germanic finds ends with the Lombards. They followed the established strategic positioning in the territory of Pannonia Savia (Sisak), but also created some new sites (Samobor, Kuzelin), that enabled them to control the surrounding territory. In comparison with the Gepid archaeological evidence in Pannonia Sirmiensis, the Lombard finds are sporadically dispersed (the vicinity of Osijek, Vinkovci and Popovac).

[146] Cf. the Roman provincial objects from Siscia discussed in this study (Simoni 1989).

[147] For example the grave finds from the site of PIK, as well as a number of other movable finds from Cibalae (unpublished).

[148] For example, the cemeteries that suggest the existence of nearby settlements (Migotti 2009, 130-131).

It is necessary to touch here on the issue of ethnic changes that took place in the studied territory, although this phenomenon is not satisfactorily substantiated by archaeology, given that the majority of the evidence is chance finds. Recent studies have shown that the late Roman autochthonous population in Pannonia generally continued to live there unhindered until the mid-5th century. During generally turbulent periods, depopulation must have occurred, with the process of "Barbarisation" also taking place under the influence of mostly Germanic peoples. This is particularly obvious with respect to small finds, in the first place jewellery and dress accessories of the Roman provincial affiliation, which were increasingly acquiring the forms and techniques of decoration characteristic of Barbarian tastes. The same is shown in pottery, as well as in the general loss of urban character of major (Roman) towns. Yet precisely the archaeological material from the period of Germanic migrations, but also of later epochs, in the first place those dominated by the Slavs and Avars, suggest that the remaining Romanised autochthons, as well as the Barbarian population, were strong enough to convey to the early mediaeval period parts of their technology, customs and toponyms. Precisely for this reason – even though the first association when considering the transformation of the territory of northern Croatia from the Roman to the mediaeval period is discontinuity – it is possible to establish that in every respect (settlement positions, burial customs, traffic routes, economy, government, jewellery, weaponry and tool production techniques) there is a continuity of life in this area.

Sources of illustrations

Figs. 3-47, 51-56, 61, 66-68, 70, 72, 79, 83, 84, and 85 b (after Simoni 1989, M 1:1, except for no. 85 b = 2: 3); fig. 48 (after Sokol 1998b, M 1:1); figs. 73-77 (made by A. Rapan Papeša). The remainder were provided by courtesy of various museums: figs. 49, 78 and 81 (GMV; drawings: 1:1), figs. 50, 69, 71 and 82 (MSO), figs. 57-59, 62-65, 80 and 85 a (AMZ), and fig. 60 a, b (GMS, not to scale).

Bibliography

Andrić 2002 Stanko Andrić, Južna Panonija u doba Velike seobe naroda, *SSlav* 2, 117-167.

Andrić 2003 Stanko Andrić, Južna Panonija u doba ostrogotske prevlasti, *Kolo, časopis Matice hrvatske* 13, Zagreb, 291-311.

Babić 2010 Staša Babić, Arheologija i etnicitet, *Etnoantropološki problemi* 3/2, Beograd, 137-149.

Bojčić 1984 Zvonko Bojčić, Pregled istraživanja i rasprostranjenosti ranosrednjovjekovnih arheoloških nalaza u Istočnoj Slavoniji i Baranji, in: *Arheološka istraživanja u istočnoj Slavoniji i Baranji, IzdHAD* 9, 211-222.

Bojčić 2007 Zvonko Bojčić, *Izgradnja i otvorenje Arheološkog muzeja u Osijeku*, Osijek.

Bojčić 2009 Zvonko Bojčić, *Seoba naroda i srednji vijek*, in: Z. Bojčić (ed.), *Vodič kroz Stalni postav*, Osijek.

Bóna, Garam, and Vida 2005 István Bóna, Éva Garam, and Tivadar Vida (eds.), *Gepidische Gräberfelder im Theissgebiet II.*, *Monumenta Germanorum Archaeologica Hungariae* 2, Budapest 2005.

Bóna and Nagy 2002 István Bóna and Margit Nagy, *Gepidische Gräberfelder im Theissgebiet I.*, *Monumenta Germanorum Archaeologica Hungariae* 1, *Monumenta Gepidica*, Budapest.

Borovac 2002 Ivanka Borovac (ed.), *Veliki atlas Hrvatske*, Zagreb.

Brunšmid 1902 Josip Brunšmid, Colonia Aurelia Cibalae. Vinkovci u staro doba, *VHAD* 6, 117-166.

Brunšmid 1905

Josip Brunšmid, Starine ranijega srednjegijeka iz Hrvatske i Slavonije, *VHAD* 8, 208-220.

Burkowsky 1999 Zdenko Burkowsky, *Sisak u prapovijesti, antici i starohrvatskom dobu,* Exhibition Catalogue, Sisak.

Cambi and Marin 1998 Nenad Cambi and Emilio Marin (eds.), *L'Époque de Justinien et les problèmes de VIe et VIIe siècles*, *ACIAC* 13/1-3, Città del Vaticano – Split.

Csallány 1961 Dezsö Csallány, *Archäologische Denkmäler der Gepiden im Mitteldonaubecken (454-568 u. Z.)*, Budapest.

Čučković 1992 Lazo Čučković, Ozalj – zaštitna arheološka iskopavanja 1992. godine, *ObHAD* 24/3, 49-51.

Čučković 2006 Lazo Čučković, Ozalj, in: A. Durman (ed.), *Stotinu hrvatskih arheoloških lokaliteta*, Zagreb, 204-205.

Demo 1981 Željko Demo, Novac germanskih vladara druge pol. 5. st. do u drugu pol. 6. st. u Numizmatičkoj zbirci Arheološkog muzeja u Zagrebu, *AV* 32, 454-481.

Demo 1994 Željko Demo, *Ostrogothic Coinage from Collections in Croatia, Slovenia and Bosnia & Herzegovina, Situla* 32.

Demo 2009 Željko Demo, Nos. 431, 434, and 447, in: *Slavonija*, 136-137, 140.

Dimitrijević 1960 Danica Dimitrijević, Gepidska nekropola "Kormadin" kod Jakova, *RVM* 9, 5-50.

Dimitrijević 1979 Stojan Dimitrijević, Arheološka topografija i izbor arheoloških nalaza s vinkovačkog tla, in: *Corolla memoriae Iosepho Brunšmid dicata, IzdHAD* 4, Vinkovci, 133-200.

Dukat 1996 Zdenka Dukat, Nalazi novca na širem području grada Zagreba, *IzdHAD* 17, Zagreb, 65-71.

Gračanin 2005 Hrvoje Gračanin, Huni i Južna Panonija, *SSlav* 5, 9-47.

Gračanin 2006 Hrvoje Gračanin, Goti i Južna Panonija, *SSlav* 6, 83-126.

Gračanin 2007 Hrvoje Gračanin, Gepidi, Heruli, Langobardi i Južna Panonija, *SSlav* 7, 7-64.

Grane 2007 Thomas Grane, *The Roman Empire and Southern Scandinavia – a Northern Connection*, Copenhagen.

Ilkić 2007 Mato Ilkić, Nalazi seobe naroda i ranog srednjeg vijeka iz Sotina, *PIAZ* 24, 277-278.

Iskra-Janošić 1993 Ivana Iskra-Janošić, Povijesni pregled arheoloških istraživanja u Vinkovcima, *Godišnjak*

ogranka Matice Hrvatske Vinkovci 10 (1992), Vinkovci, 61-78.
Jurić 1998 Radomir Jurić, Le fibule a forma di croce in Croazia, in: Cambi and Marin 1998/2, 1091-1105.
Kovačević 1960 Jovan Kovačević: *Arheologija i historija varvarske kolonizacije južnoslovenskih oblasti od početka IV do kraja VII veka, Posebna izdanja Vojvođanskog muzeja* II, Novi Sad .
Kušan Špalj and Nemeth-Ehrlich 2008 Dora Kušan Špalj and Dorica Nemeth-Ehrlich, Šćitarjevo – dvorište osnovne škole, *HAG* 4 (2007), 198-201.
Migotti 2009 Branka Migotti, Kasna antika i rano kršćanstvo, in: *Slavonija*, 129-135.
Milavec 2007 Tina Milavec, Prispevek h kronologiji S-fibul v Sloveniji, *AV* 58, 333-355.
Mirnik 1996 Ivan Mirnik, Novac Istočnog Rimskog Carstva u numizmatičkoj zbirci Arheološkog muzeja u Zagrebu, *VAMZ* 28-29 (1995-1996), 159-228.
Mirnik and Šemrov 1998 Ivan Mirnik and Andrej Šemrov, Byzantine coins in the Zagreb Archaeological Museum Numismatic Collection. Anastasius I. (A. D. 497-518) – Anastasius II (A. D. 713-715), VAMZ 30-31 (1997-1998), 129-258.
Mrkobrad 1980 Dušan Mrkobrad, *Arheološki nalazi seobe naroda u Jugoslaviji, Fontes Archaeologiae Iugoslaviae* III, Monografije 6, Beograd.
Parducz 1950 Mihály Párducz, *A szarmatakor emlékei Magyarországon (Dänkmäler der Sarmatenzeit Ungarns)* III, Budapest.
Pülhorn 1987 Wolfgang Pülhorn (ed.), *Germanen, Hunnen und Awaren, Schatze der Völkerwanderungszeit*, Exhibition Catalogue, Nürnberg.
Radić 1997 Mladen Radić, Arheologija seobe naroda i srednjeg vijeka, in: M. Radić (ed.), *Blago Muzeja Sla*vonije, Exhibition Catalogue, Osijek, 81-93.
Radić 2009 Mladen Radić, nos. 432, 435, 436, and 448, in: *Slavonija*, 136-137, 140.
Rapan Papeša and Vulić 2008 Anita Rapan Papeša and Hrvoje Vulić, Vinkovci – Ulica Hrvatskih žrtava 11, *HAG* 4 (2007), 71-73.
Rapan Papeša 2009 Anita Rapan Papeša, nos. 433, 437-446, in: *Slavonija*, 136, 138-139.
Rapan Papeša 2010 Anita Rapan Papeša, Razmatranja uz jedan novi nalaz iz Vinkovaca, *SHP* 37, 7-15.
Sanader 2009 Mirjana Sanader, Vojne građevine na dunavskom limesu, in: *Slavonija*, 105-114.
Seoba Danica Dimitrijević, Jovan Kovačević, and Zdenko Vinski, *Seoba naroda – arheološki nalazi Jugoslavenskog Podunavlja*, Exhibition Catalogue, Zemun, 1962.
Simoni 1978 Katica Simoni, Dva priloga istraživanju germanskih nalaza seobe naroda u Jugoslaviji, *VAMZ* 10-11 (1977-1978), 209-233.
Simoni 1981 Katica Simoni, Zagreb i okolica u ranom srednjem vijeku, in: *Arheološka istraživanja u Zagrebu i njegovoj okolici, IzdHAD* 6, Zagreb, 155-168.
Simoni 1983 Katica Simoni, Skupni nalaz oruđa i oružja iz Nartskih Novaka, *VAMZ* 15, 251-266.
Simoni 1984 Katica Simoni, Ludbreg i okolica u ranom srednjem vijeku, in: V. Mađarić (ed.), *Ludbreg*, Ludbreg, 73-79.
Simoni 1988 Katica Simoni, Srebrna žlica iz Siska, *VAMZ* 21, 79-86.
Simoni 1989 Katica Simoni, Funde aus der Völkerwanderungszeit in den Sammlungen des Archäologischen Museums in Zagreb, *VAMZ* 22, 107-134.
Simoni 1993 Katica Simoni, no. 266: Fibula tipa Varpalota, in: Ante Rendić-Miočević (ed.), *Arte e cultura in Croazia – Collezioni del Museo Archeologico di Zagrabia*, Exhibition Catalogue, Zagreb.
Slavonija V. Kusin and B. Šulc (eds.), *Slavonija, Baranja i Srijem – vrela europske civilizacije*, Exhibition Catalogue, Zagreb 2009.
Sokol 1998a Vladimir Sokol, Panonija Savija u Justinijanovo doba, in: Cambi and Marin 1998/2, 1131-1143.
Sokol 1998b Vladimir Sokol: *Rimski metal s Kuzelina (iskapanja 1990.-1997.)*, Exhibition Catalogue, Muzej Prigorja Sesvete, Zagreb.
Sokol 2006 Vladimir Sokol, Kuzelin, in: A. Durman (ed.), *Stotinu hrvatskih arheoloških lokaliteta*, Zagreb, 154-155.
Sunce 1994 Ž. Demo (ed.), *Od nepobjedivog sunca do sunca pravde. Rano kršćanstvo u kontinentalnoj Hrvatskoj*, Exhibition Catalogue, Zagreb.
Šiša Wiewegh 2004 Marija Šiša Wiewegh, Iovia Botivo, *PZ* 30, 297-314.
Tomičić 1997 Željko Tomičić, Arheološka topografija i toponimija (Iovia – Botivo – Ludbreg), in: K. Horvat-Levaj and I. Reberski (eds.), *Ludbreg. Ludbreška Podravina*, Zagreb, 21-41.
Uglešić 1994 Ante Uglešić, Nalazi Seobe naroda iz Zbirke Mate Ilkića, *RFFZd* 33 (20), 145-152.
Vikić 1973 Branka Vikić, Jovia, Ludbreg – istraživanja u 1973. godini, *AP* 15, 50-51.
Vikić 1976 Branka Vikić, Ludbreg, kod Varaždinskih Toplica, *AP* 18, 81-83.
Vinkovci 1999 S. Jozić (ed.), *Vinkovci u svijetu arheologije*, Exhibition Catalogue, Vinkovci.
Vinski 1954 Zdenko Vinski, Gibt es frühslawishe Keramik aus der Zeit der südslawischen Landnahme?, *AI* 1, 74-75.
Vinski 1957a Zdenko Vinski, *Arheološki spomenici Velike seobe naroda u Srijemu, Situla* 2.
Vinski 1957b Zdenko Vinski, Zikadenschmuck aus Jugoslawien, JRGZM 4, 136-160.
Vinski 1960 Zdenko Vinski, Ranosrednjovjekovni arheološki nalazi u Zagrebu i njegovoj okolici, *Iz starog i novog Zagreba* 2, Zagreb, 47-65.
Vinski 1962 Zdenko Vinski, O značaju nalaza Seobe naroda iz Karavukova u Bačkoj, *VMKH* 11/3, 75-79.
Vinski 1968 Zdenko Vinski, Krstoliki nakit epohe seobe naroda u Jugoslaviji, *VAMZ* 3, 103-168.
Vinski 1973 Zdenko Vinski, O rovašenim fibulama Ostrogota i Tirinžana povodom rijetkog tirinškog nalaza u Saloni, VAMZ 6-7 (1972-73), 177-216.
Vinski 1974 Zdenko Vinski, Kasnoantički starosjedioci

i salonitanskoj regiji prema arheološkoj ostavštini predslavenskog supstrata, *VAHD* 69 (1967), 5-86.
Vulić 2009 Hrvoje Vulić, Rana antika, in: *Slavonija*, 119-125.
VAH I. Borovac (ed.), *Veliki atlas Hrvatske*, Zagreb, 2002.
Zotović 1994 Ljubica Zotović, Die Gepidische Nekropole bei Viminacium, *Starinar* 43-44 (1992-1993), 183-190.

www.ingramcontent.com/pod-product-compliance
Ingram Content Group UK Ltd.
Pitfield, Milton Keynes, MK11 3LW, UK
UKHW061843041026
15591UKWH00015B/890

9 781407 309859